WITHDRAWN

COLD SPRING HARBOR SYMPOSIA
ON QUANTITATIVE BIOLOGY

VOLUME LXXIV

http://symposium.cshlp.org

Institutions that have purchased the hardcover edition of this book are entitled to online access to the Symposium website. Please contact your institution's library to gain access to the website. The site contains the full-text articles from the 2009 Symposium and previous Symposia from 2004 onward.

If you received this book from Cold Spring Harbor Laboratory Press directly, the packing slip included with this book contains your activation instructions. If you received this book from your agent, please have your agent contact us for your activation instructions.

If you do not have an account number or are experiencing access problems, please contact Kathy Cirone, CSHL Press Subscription Manager, at 1-800-843-4388, extension 4044 (Continental U.S. and Canada), 516-422-4100 (all other locations), cironek@cshl.edu, or sqb-feedback@highwire.stanford.edu.

COLD SPRING HARBOR SYMPOSIA ON QUANTITATIVE BIOLOGY

VOLUME LXXIV

EVOLUTION
The Molecular Landscape

http://symposium.cshlp.org

Meeting Organized by Bruce Stillman, David Stewart, and Jan Witkowski

COLD SPRING HARBOR LABORATORY PRESS

2009

COLD SPRING HARBOR SYMPOSIA ON QUANTITATIVE BIOLOGY VOLUME LXXIV

©2009 by Cold Spring Harbor Laboratory Press
International Standard Book Number 978-0-87969-870-6 (cloth)
International Standard Book Number 978-0-87969-871-3 (paper)
International Standard Serial Number 0091-7451
Web International Standard Serial Number 1943-4456
Library of Congress Catalog Card Number 34-8174

Printed in the United States of America
All rights reserved
COLD SPRING HARBOR SYMPOSIA ON QUANTITATIVE BIOLOGY
Founded in 1933 by
REGINALD G. HARRIS
Director of the Biological Laboratory 1924 to 1936
Previous Symposia Volumes

I (1933) Surface Phenomena
II (1934) Aspects of Growth
III (1935) Photochemical Reactions
IV (1936) Excitation Phenomena
V (1937) Internal Secretions
VI (1938) Protein Chemistry
VII (1939) Biological Oxidations
VIII (1940) Permeability and the Nature of Cell Membranes
IX (1941) Genes and Chromosomes: Structure and Organization
X (1942) The Relation of Hormones to Development
XI (1946) Heredity and Variation in Microorganisms
XII (1947) Nucleic Acids and Nucleoproteins
XIII (1948) Biological Applications of Tracer Elements
XIV (1949) Amino Acids and Proteins
XV (1950) Origin and Evolution of Man
XVI (1951) Genes and Mutations
XVII (1952) The Neuron
XVIII (1953) Viruses
XIX (1954) The Mammalian Fetus: Physiological Aspects of Development
XX (1955) Population Genetics: The Nature and Causes of Genetic Variability in Population
XXI (1956) Genetic Mechanisms: Structure and Function
XXII (1957) Population Studies: Animal Ecology and Demography
XXIII (1958) Exchange of Genetic Material: Mechanism and Consequences
XXIV (1959) Genetics and Twentieth Century Darwinism
XXV (1960) Biological Clocks
XXVI (1961) Cellular Regulatory Mechanisms
XXVII (1962) Basic Mechanisms in Animal Virus Biology
XXVIII (1963) Synthesis and Structure of Macromolecules
XXIX (1964) Human Genetics
XXX (1965) Sensory Receptors
XXXI (1966) The Genetic Code
XXXII (1967) Antibodies
XXXIII (1968) Replication of DNA in Microorganisms
XXXIV (1969) The Mechanism of Protein Synthesis
XXXV (1970) Transcription of Genetic Material
XXXVI (1971) Structure and Function of Proteins at the Three-dimensional Level

XXXVII (1972) The Mechanism of Muscle Contraction
XXXVIII (1973) Chromosome Structure and Function
XXXIX (1974) Tumor Viruses
XL (1975) The Synapse
XLI (1976) Origins of Lymphocyte Diversity
XLII (1977) Chromatin
XLIII (1978) DNA: Replication and Recombination
XLIV (1979) Viral Oncogenes
XLV (1980) Movable Genetic Elements
XLVI (1981) Organization of the Cytoplasm
XLVII (1982) Structures of DNA
XLVIII (1983) Molecular Neurobiology
XLIX (1984) Recombination at the DNA Level
L (1985) Molecular Biology of Development
LI (1986) Molecular Biology of *Homo sapiens*
LII (1987) Evolution of Catalytic Function
LIII (1988) Molecular Biology of Signal Transduction
LIV (1989) Immunological Recognition
LV (1990) The Brain
LVI (1991) The Cell Cycle
LVII (1992) The Cell Surface
LVIII (1993) DNA and Chromosomes
LIX (1994) The Molecular Genetics of Cancer
LX (1995) Protein Kinesis: The Dynamics of Protein Trafficking and Stability
LXI (1996) Function & Dysfunction in the Nervous System
LXII (1997) Pattern Formation during Development
LXIII (1998) Mechanisms of Transcription
LXIV (1999) Signaling and Gene Expression in the Immune System
LXV (2000) Biological Responses to DNA Damage
LXVI (2001) The Ribosome
LXVII (2002) The Cardiovascular System
LXVIII (2003) The Genome of *Homo sapiens*
LXIX (2004) Epigenetics
LXX (2005) Molecular Approaches to Controlling Cancer
LXXI (2006) Regulatory RNAs
LXXII (2007) Clocks and Rhythms
LXXIII (2008) Control and Regulation of Stem Cells

Front Cover (*Paperback*): Illustration by Daniel W. Smith, popartportraitsplus.com.

Authorization to photocopy items for internal or personal use, or the internal or personal use of specific clients, is granted by Cold Spring Harbor Laboratory Press, provided that the appropriate fee is paid directly to the Copyright Clearance Center (CCC). Write or call CCC at 222 Rosewood Drive, Danvers, MA 01923 (508-750-8400) for information about fees and regulations. Prior to photocopying items for educational classroom use, contact CCC at the above address. Additional information on CCC can be obtained at CCC Online at http://www.copyright.com/

All Cold Spring Harbor Laboratory Press publications may be ordered directly from Cold Spring Harbor Laboratory Press, 500 Sunnyside Boulevard, Woodbury, New York 11797-2924. Phone: 1-800-843-4388 in Continental U.S. and Canada. All other locations: (516) 422-4100. FAX: (516) 422-4097. E-mail: cshpress@cshl.edu. For a complete catalog of all Cold Spring Harbor Laboratory Press publications, visit our World Wide Website http://www.cshlpress.com/

Website Access: Institutions that have purchased the hardcover edition of this book are entitled to online access to the companion website at http://symposium.cshlp.org/. For assistance with activation, please contact Kathy Cirone, CSHL Press Subscription Manager, at cironek@cshl.edu.

Symposium Participants

ALBÀ, MAR, Dept. of Health and Life Sciences, Fundació Institut Municipal d'Investigació Mèdica, Universitat Pompeu Fabra, Barcelona, Spain
ALBERT, FRANK, Dept. of Evolutionary Genetics, Max-Planck-Institute for Evolutionary Anthropology, Leipzig, Germany
ALBERTI, SIMON, Whitehead Institute for Biomedical Research, Cambridge, Massachusetts
ALFOLDI, JESSICA, Dept. of Genome Biology, Broad Institute of MIT and Harvard University, Cambridge, Massachusetts
ALFRED, JANE, *Development*, Company of Biologists, Cambridge, United Kingdom
AMARAL, PAULO, Insitute for Molecular Bioscience, University of Queensland, Brisbane, Australia
ANDAM, CHERYL, Dept. of Molecular and Cell Biology, University of Connecticut, Storrs
ANDERSSON, LEIF, Dept. of Medical Biochemistry and Microbiology, Uppsala University, Uppsala, Sweden
ANDREC, MICHAEL, BioMaPS Institute for Quantitative Biology, Rutgers University, Somerset, New Jersey
ANGELO, CLAUDIO, *Folha de S. Paulo*, Sao Paulo, Brazil
ARNOLD, FRANCES, Div. of Chemistry and Chemical Engineering, California Institute of Technology, Pasadena
ASANO, KATSURA, Dept. of Biology, Kansas State University, Manhattan
ATWAL, MICKEY, Cold Spring Harbor Laboratory, Cold Spring Harbor, New York
AVELAR, ANA TERESA, Dept. of Telomere and Genome Stability, Instituto Gulbenkian de Ciencia, Oeiras, Portugal
AYROLES, JULIEN, Dept. of Genetics, North Carolina State University, Raleigh
AYUB, QASIM, Dept. of Genome Research, Wellcome Trust, Sanger Instiute, Cambridge, United Kingdom
BAHCALL, ORLI, *Nature Genetics*, Nature Publishing Group, New York, New York
BAKER, CHRISTOPHER, Dept. of Microbiology and Immunology, University of California, San Francisco
BAKER, CLARE, Dept. of Physiology, Development, and Neuroscience, University of Cambridge, Cambridge, United Kingdom
BALDI, CHRISTOPHER, Dept. of Molecular Biology, School of Medicine, University of Medicine and Dentistry of New Jersey, Stratford
BALTRUS, DAVID, Dept. of Biology, University of North Carolina, Chapel Hill
BARTON, NICK, Institute of Science and Technology, Klosterneuburg, Austria
BARUA, BIPASHA, Dept. of Neuroscience and Cell Biology, Robert Wood Johnson Medical School, University of Medicine and Dentistry of New Jersey, Piscataway
BATISTA, PEDRO, Program in Molecular Medicine, School of Medicine, University of Massachusetts, Worcester
BAUCHET, MARC, Dept. of Evolutionary Anthropology, Max-Planck-Institute, Leipzig, Germany
BEHE, MIKE, Dept. of Biological Sciences, Lehigh University, Bethlehem, Pennsylvania
BEHRINGER, RICHARD, Dept. of Genetics, M.D. Anderson Cancer Center, University of Texas, Houston
BEJERANO, GILL, Dept. of Developmental Biology and Computer Science, Stanford University, Stanford, California
BELL, GRAHAM, Dept. of Biology, McGill University, Montréal, Québec, Canada
BELLOTT, DANIEL, Dept. of Biology, Massachusetts Institute of Technology, Cambridge
BENABENTOS, ROCIO, Dept. of Molecular and Human Genetics, Baylor College of Medicine, Houston, Texas
BENDESKY, ANDRES, Lab. of Neural Circuits and Behavior, The Rockefeller University, New York
BERNARDI, GIORGIO, Dept. of Molecular Evolution, Stazione Zoologica, Napoli, Italy
BERNHARDT, HAROLD, Dept. of Biochemistry, University of Otago, Dunedin, New Zealand
BERRY, ANDREW, Dept. of Organismic and Evolutionary Biology, Harvard University, Cambridge, Massachusetts
BERTRANPETIT, JAUME, Dept. de Ciències Experimentals i de la Salut, Institut de Biologia Evolutiva, Universitat Pompeu Fabra, Barcelona, Spain
BINA, MINOU, Dept. of Chemistry, Purdue University, West Lafayette, Indiana
BINGHAM, PAUL, Dept. of Biochemistry and Cell Biology, Stony Brook University, Stony Brook, New York
BLAKE, JUDITH, Dept. of Bioinformatics and Computational Biology, The Jackson Laboratory, Bar Harbor, Maine
BOOTH, LAUREN, Dept. of Microbiology and Immunology, University of California, San Francisco
BOSCH, ELENA, Dept. de Ciències Experimentals i de la Salut, Institut Biologia Evolutiva, Universitat Pompeu Fabra, Barcelona, Spain
BOWERS, EVELYN, Dept. of Anthropology, Ball State University, Muncie, Indiana
BRASLAU, ALAN, DSM-IRAMIS-SPEC, Commissariat à l'Energie Atomic, Gif-sur-Yvette, France
BRIDGHAM, JAMIE, Center for Ecology and Evolutionary Biology, University of Oregon, Eugene
BRIZUELA, LEONARDO, Dept. of Genomics, Agilent Technologies, Hamilton, Massachusetts
BROCCHIERI, LUCIANO, Dept. of Molecular Genetics and Microbiology, University of Florida, Gainesville
BROWN, SUSAN, Div. of Biology, Kansas State University, Manhattan

BROWNE, JANET, Dept. of the History of Science, Harvard University, Cambridge, Massachusetts
BULYK, MARTHA, Div. of Genetics, Dept. of Medicine, Brigham and Women's Hospital, Harvard University, Boston, Massachusetts
BURGESS, RALPH, Cold Spring Harbor Laboratory, Cold Spring Harbor, New York
BURT, DAVID, Dept. of Genetics and Genomics, Roslin Institute and Royal (Dick) School of Veterinary Studies, University of Edinburg, Roslin, United Kingdom
BUSTAMANTE, CARLOS, Dept. of Biological Statistics and Computational Biology, Cornell University, Ithaca, New York
CABUY, ERIK, Dept. of Single Cell Genomics, Friedrich Miescher Institute, Basel, Switzerland
CAPORALE, LYNN, Independent, *Darwin in the Genome*, New York, New York
CARLSON, ELOF, Dept. of Biochemistry and Cell Biology, Stony Brook University, Stony Brook, New York
CARPENTER, MEREDITH, Dept. of Molecular and Cell Biology, University of California, Berkeley
CARROLL, SEAN B., Lab. of Molecular Biology, Howard Hughes Medical Institute, University of Wisconsin, Madison
CARROLL, SEAN M., Center for Ecology and Evolutionary Biology, University of Oregon, Eugene
CECH, THOMAS, Dept. of Chemistry and Biochemistry, University of Colorado, Boulder
CHALKIA, DIMITRA, Dept. of Biological Chemistry, University of California, Irvine
CHANG, SHANG-LIN, Institute of Molecular Biology, Academia Sinica, Taipei, Taiwan
CHARLESWORTH, BRIAN, School of Biological Sciences, University of Edinburgh, Edinburgh, Scotland, United Kingdom
CHARLIER, CAROLE, Dept. of Animal Genomics, University of Liège, Liège, Belgium
CHEN, JACK, Dept. of Molecular Biology and Biochemistry, Simon Fraser University, Burnaby, British Columbia, Canada
CHEN, SIDI, Dept. of Ecology and Evolution, University of Chicago, Chicago, Illinois
CHICA, CLAUDIA, Structural and Computational Biology Unit, European Molecular Biology Laboratory, Heidelberg, Germany
CHODROFF, REBECCA, Genome Technology Branch, National Human Genome Research Institute, National Institutes of Health, Bethesda, Maryland
CHONG, SHAORONG, Dept. of Research, New England Biolabs, Ipswich, Massachusetts
CHOUARD, TANGUY, *Nature*, Nature Publishing Group, London, United Kingdom
CIBRIÁN, ANGELICA, Sackler Institute for Comparative Genomics, American Museum of Natural History, New York, New York
CLEMENT, YVES, Dept. of Computational Molecular Biology, Max-Planck-Institute for Molecular Genetics, Berlin, Germany
COLWELL, LUCY, Dept. of Applied Mathematics, Harvard University, Cambridge, Massachusetts
COMMINS, JENNIFER, Dept. of Genetics, Trinity College, Dublin, Ireland
CONACO, CECILIA, Neuroscience Research Institute, University of California, Santa Barbara
COOKE, BOB, Freelance journalist, Stowe, Vermont
CRESSY, MICHAEL, Cold Spring Harbor Laboratory, Cold Spring Harbor, New York
CRISCITIELLO, MICHAEL, Dept. of Veterinary Pathobiology, Texas A&M University, College Station, Texas
CRONK, QUENTIN, UBC Botanical Garden, University of British Columbia, Vancouver, British Columbia, Canada
DALAL, CHIRAJ, Dept. of Biology, California Institute of Technology, Pasadena
DARNELL, JAMES, Dept. of Molecular Cell Biology, The Rockefeller University, New York
DAVID, LIOR, Dept. of Animal Sciences, Hebrew University of Jerusalem, Rehovot, Israel
DAVIDSON, ERIC, Div. of Biology, California Institute of Technology, Pasadena
DAWSON, ELLIOTT, President, BioVentures, Murfreesboro, Tennessee
DAYAL, SANDEEP, Lab. of Molecular Biology, National Institute of Diabetes and Digestive and Kidney Diseases, National Institutes of Health, Bethesda, Maryland
DE GRAND, ALEC, Olympus America, Boston, Massachusetts
DEGNAN, BERNARD, School of Integrative Biology, University of Queensland, Brisbane, Australia
DEGNAN, SANDIE, School of Biological Sciences, University of Queensland, Brisbane, Australia
DEMOGINES, ANN, Dept. of Molecular Genetics and Microbiology, University of Texas, Austin
DENNETT, DANIEL, Center for Cognitive Studies, Tufts University, Medford, Massachusetts
DESAI, MICHAEL, Lewis-Sigler Institute for Integrative Genomics, Princeton University, Princeton, New Jersey
DIAZ-MEJIA, JUAN, Banting and Best Department of Medical Research, University of Toronto, Toronto, Ontario, Canada
DOEBLEY, JOHN, Dept. of Genetics, University of Wisconsin, Madison
DOOLITTLE, RUSSELL, Dept. of Biological Sciences, University of California, San Diego, La Jolla
DOOLITTLE, W. FORD, Dept. of Biochemistry and Molecular Biology, Dalhousie University, Halifax, Nova Scotia, Canada
DRAKE, JOHN, Lab. of Molecular Genetics, National Institute of Environmental Health Science, Research Triangle Park, North Carolina
DREWELL, ROBERT, Dept. of Biology, Harvey Mudd College, Claremont, California
DRUMMOND, DAVID, FAS Center for Systems Biology, Harvard University, Cambridge, Massachusetts
DUBNAU, JOSH, Cold Spring Harbor Laboratory, Cold Spring Harbor, New York
DUBOWITZ, VICTOR, Dept. of Child Health, University of London, London, United Kingdom
DUFOUR, HELOISE, Lab. of Genetics, Howard Hughes Medical Institute, University of Wisconsin, Madison
EAMES, MATT, Dept. of Biophysics, University of California, San Francisco
EBERT, JOAN, Cold Spring Harbor Laboratory Press, Woodbury, New York
EHRENREICH, IAN, Lewis-Sigler Institute for Integrative

Genomics, Princeton University, Princeton, New Jersey
EICHLER, EVAN, Dept. of Genome Sciences, Howard Hughes Medical Institute, University of Washington, Seattle
EICK, GEETA, Center for Ecology and Evolutionary Biology, University of Oregon, Eugene
ELENKO, MARK, Dept. of Genetics, Massachusetts General Hospital, Harvard University, Boston, Massachusetts
ELLAHI, AISHA, Dept. of Molecular Genetics and Microbiology, University of Texas, Austin
ELLIS, RONALD, Dept. of Molecular Biology, University of Medicine and Dentistry of New Jersey, Stratford
ENARD, DAVID, Dept. of Biology, Ecole Normale Supérieure Paris, Paris, France
ENGELKEN, JOHANNES, Dept. de Ciències Experimentals i de la Salut, Institut de Biologia Evolutiva, Universitat Pompeu Fabra, Barcelona, Spain
ESTEVEZ-TORRES, Andre, Dept. of Physics, Princeton University, Princeton, New Jersey
FALCO, MICHELE, Cellmates, New Haven, Connecticut
FARES, MARIO, Dept. of Genetics, University of Dublin, Trinity College, Dublin, Ireland
FAULK, CHRISTOPHER, Dept. of Biological Sciences, Louisiana State University, Baton Rouge
FELLMAN, LYNN, Fellman Studios, Minneapolis, Minnesota
FERGUSON, NIALL, Minda de Gunzburg Center for European Studies, Harvard University, Cambridge, Massachusetts
FERRERO, ENZA, Dept. of Genetics, Biology, and Biochemistry, University of Torino, Torino, Italy
FIEBIG, KLAUS, Research Programs, Ontario Genomics Institute, Toronto, Canada
FIETZ, SIMONE, Max-Planck-Institute for Molecular Cell Biology, Dresden, Germany
FISHER, DANIEL, Dept. of Applied Physics and Biology, Stanford University, Stanford, California
FLAVELL, MADLYN, Guilford, Connecticut
FLAVELL, RICHARD, Section of Immunobiology, School of Medicine, Yale University, New Haven, Connecticut
FLINSCH, ELISABETH, Dept. of Mathematics, Drexel University, Coatesville, Pennsylvania
FORREST, BARBARA, Dept. of History and Political Science, Southeastern Louisiana University, Hammond
FOSTER, KEVIN, Center for Systems Biology, Harvard University, Cambridge, Massachusetts
FOX, PAUL, Dept. of Cell Biology, Cleveland Clinic, Cleveland, Ohio
FRASER, HUNTER, Rosetta Inpharmatics, Seattle, Washington
FREUDENBERG, JAN, Dept. of Genomics and Genetics, Feinstein Institute for Medical Research, Manhasset, New York
FU, YONG-BI, Agriculture and Agri-Food Canada, Saskatoon Research Center, Saskatoon, Saskatchewan, Canada
GANN, ALEX, Cold Spring Harbor Laboratory Press, Woodbury, New York
GERBER, MAX S., Photography, Pasadena, California
GIGER, THOMAS, Dept. of Evolutionary Genetics, Max-Planck-Institute for Evolutionary Anthropology, Leipzig, Germany
GONÇALVES-SA, JOANA, Dept. of Molecular and Cell Biology, Harvard University, Cambridge, Massachusetts
GONZALEZ-GARAY, MANUEL, Human Genome Sequencing Center, Baylor College of Medicine, Houston, Texas
GOOD, MATTHEW, Dept. of Biochemistry and Biophysics, University of California, San Francisco
GORE, JEFF, Dept. of Physics, Massachusetts Institute of Technology, Cambridge
GOUGH, JULIAN, Dept. of Computer Science, University of Bristol, Bristol, United Kingdom
GRANT, SETH, Genes to Cognition Programme, Wellcome Trust Sanger Institute, Hinxton, United Kingdom
GRAY, JESSICA, Dept. of Systems Biology, Medical School, Harvard University, Boston, Massachusetts
GREENSPAN, RALPH, Dept. of Experimental Neurobiology, The Neurosciences Institute, San Diego, California
GRODZICKER, TERRI, Cold Spring Harbor Laboratory Press, Woodbury, New York
GUO, YIQING, Dept. of Molecular Biology, School of Medicine, University of Medicine and Dentistry of New Jersey, Stratford
GUTENKUNST, RYAN, Center for Nonlinear Studies, Los Alamos National Laboratory, Los Alamos, New Mexico
HANNON, GREGORY, Cold Spring Harbor Laboratory, Cold Spring Harbor, New York
HAQ, OMAR, BioMaPS Institute for Quanititave Biology, Rutgers University, Piscataway, New Jersey
HARMS, MICHAEL, Center for Ecology and Evolutionary Biology, University of Oregon, Eugene
HARPER, KRISTIN, Robert W. Johnson Health and Society Scholars Program, Columbia University, New York, New York
HARTENSTEIN, VOLKER, Dept. of Molecular, Cell, and Developmental Biology, University of California, Los Angeles
HAUSER, FRITZ, Project Segmenta, Bad Lippspringe, Germany
HAUSER, MARC, Depts. of Psychology, Human Evolutionary Biology, and Organismic and Evolutionary Biology, Harvard University, Cambridge, Massachusetts
HAUSSLER, DAVID, Center for Biomolecular Science and Engineering, University of California, Santa Cruz
HAYASHI, SHINICHIRO, Mycology Group, Institut National de la Santé et de la Recherche Médicale, Université Pierre et Marie Curie, Paris, France
HEBERT, PAUL, Biodiversity Insitute of Ontario, University of Guelph, Guelph, Ontario, Canada
HELLMANN, INES, Dept. of Integrative Biology, University of California, Berkeley
HENG, HENRY, Center for Molecular Medicine and Genetics, School of Medicine, Wayne State University, Detroit, Michigan
HENNELLY, SCOTT, Dept. of Theoretical Biology and Biophysics, Los Alamos National Laboratory, Los Alamos, New Mexico
HILLER, MICHAEL, Dept. of Developmental Biology, Stanford University, Stanford, California
HO, MARGARET, Dept. of Biology, Harvey Mudd College, Claremont, California
HOEKSTRA, HOPI, Museum of Comparative Zoology, Harvard University, Cambridge, Massachusetts
HORNER, JOHN ("JACK"), Dept. of Paleontology, Museum of the Rockies, Bozeman, Montana
HORNING, DAVID, Dept. of Molecular Biology and Chemistry, Scripps Research Institute, La Jolla, California

HOU, ZHUOCHENG, Perinatology Research Branch, Center for Molecular Medicine and Genetics, Wayne State University, Detroit, Michigan
HUGHES, TIMOTHY, Dept. of Molecular Genetics, Banting and Best Department of Medical Research, University of Toronto, Ontario, Canada
HUYNH, LYNN, Dept. of Human Genetics, Emory University, Atlanta, Georgia
IMUMORIN, IKHIDE, Dept of Biology, Spelman College, Atlanta, Georgia
INGLIS, JOHN, Cold Spring Harbor Laboratory Press, Woodbury, New York
JAHNEL, MARCUS, Max-Planck-Institute for Molecular Cell Biology and Genetics, Dresden, Germany
JAKIMO, ALAN, Dept. of Law, Hofstra University, Woodbury, New York
JANSSEN, KAAREN, Cold Spring Harbor Laboratory Press, Woodbury, New York
JAROSZ, DANIEL, Whitehead Institute for Biomedical Research, Cambridge, Massachusetts
JASTRZEBOWSKA, KAROLINA, Max-Planck-Institute for Molecular Cell Biology and Genetics, Dresden, Germany
JASZCZYSZYN, YAN, Institute of Neurosciences, Centre National de la Recherche Scientifique, Gif-sur-Yvette, France
JHA, AASHISH, Dept. of Medicine, University of California, San Francisco
JOYCE, GERALD, Dept. of Molecular Biology, Scripps Research Institute, La Jolla, California
JURKEVITCH, EDOUARD, Dept. of Plant Pathology and Microbiology, The Hebrew University of Jerusalem, Rehovot, Israel
KACZMAREK, LEONARD, Dept. of Pharmacology, Yale University, New Haven, Connecticut
KAMBEROV, YANA, Dept. of Genetics, Medical School, Harvard University, Boston, Massachusetts
KEENEY, DEVON, Dept. of Biological Sciences, Le Moyne College, Syracuse, New York
KELLER, ANDREAS, Dept. of Neurogenetics and Behavior, The Rockefeller University, New York
KENNEDY, WILLIAM, Dept. of Clinical Pharmacology, Merck, North Wales, Pennsylvania
KERÄNEN, SOILE, Dept. of Genome Sciences, Lawrence Berkeley National Laboratory, Berkeley, California
KHARE, ANUPAMA, Dept. of Molecular and Human Genetics, Baylor College of Medicine, Houston, Texas
KING, NICOLE, Dept. of Molecular and Cell Biology, University of California, Berkeley
KINGSLEY, DAVID, Dept. of Developmental Biology, Stanford University, Stanford, California
KISHORE, SANDEEP, Dept. of Microbiology and Immunology, Weill Cornell Medical College, New York, New York
KNOP, MICHAEL, Dept. of Molecular Cell Biology, European Molecular Biology Laboratory, Heidelberg, Germany
KOCH-NOLTE, FRIEDRICH, Institute of Immunology, University Hospital Hamburg-Eppendorf, Hamburg, Germany
KOLOKOTRONIS, SERGIOS-ORESTIS, Sackler Institute for Comparative Genomics, American Museum of Natural History, New York
KOONIN, EUGENE, National Center for Biotechnology Information, National Library of Medicine, National Institutes of Health, Bethesda, Maryland
KOONIN, TATIANA, Lab. of Viral Diseases, National Institute of Allergy and Infectious Diseases, National Institutes of Health, Bethesda, Maryland
KORTEMME, TANJA, Dept. of Bioengineering and Therapeutic Sciences, University of California, San Francisco
KOSCHWANEZ, JOHN, Dept. of Molecular and Cellular Biology, Harvard University, Cambridge, Massachusetts
KOSHIKAWA, SHIGEYUKI, Lab. of Molecular Biology, University of Wisconsin, Madison
KOZMIK, ZBYNEK, Dept. of Transcriptional Regulation, Institute of Molecular Genetics, Prague, Czech Republic
KRAUSE, JOHANNES, Dept. of Evolutionary Genetics, Max-Planck-Institute, Leipzig, Germany
KRAUSE, ROLAND, Dept. of Computational Molecular Biology, Max-Planck-Institute for Molecular Genetics, Berlin, Germany
KRISTENSEN, DAVID, Dept. of Bioinformatics, Stowers Institute for Medical Research, Kansas City, Missouri
KRUGLYAK, LEONID, Lewis-Sigler Institute for Intergrative Genomics, Princeton University, Princeton, New Jersey
KRYAZHIMSKIY, SERGEY, Dept. of Biology, University of Pennsylvania, Philadelphia
KUHL, JULIA, Cold Spring Harbor Laboratory, Cold Spring Harbor, New York
KUMAR, SANJAY, Dept. of Research, New England Biolabs, Ipswich, Massachusetts
KUNKEL, THOMAS, Lab. of Molecular Genetics, National Institute of Environmental Health Sciences, National Institutes of Health, Research Triangle Park, North Carolina
KUNTZ, ELEANOR, Dept. of Genetics, University of Georgia, Athens
KUNTZ, TACK, Dept. of Pharmaceutical Chemistry, University of California, San Francisco
KVITEK, DAN, Dept. of Genetics, Stanford University, Stanford, California
LABUDA, DAMIAN, Pavillon Charles Bruneau, CHU Sainte-Justine, Montréal, Québec, Canada
LACHANCE, JOSEPH, Dept. of Ecology and Evolution, Stony Brook University, Stony Brook, New York
LAMBERT, NELLE, Dept. of Developmental Neurobiology, Institute de Recherche Interdisciplinaire en Biologie Humaine et Moléculaire, Free University of Brussels, Brussels, Belgium
LANG, GREGORY, Lewis-Sigler Institute for Integrative Genomics, Princeton University, Princeton, New Jersey
LEGOURAS, IOANNIS, Dept. of Cell Biology and Biophysics, European Molecular Biology Laboratory, Heidelberg, Germany
LEMBERGER, THOMAS, *Molecular Systems Biology*, Heidelberg, Germany
LENSKI, RICHARD, Dept. of Microbiology and Molecular Genetics, Michigan State University, East Lansing
LESCH, BLUMA, Lab. of Neural Circuits and Behavior, The Rockefeller University, New York
LEVANDOWSKY, MICHAEL, Dept. of Biology, Pace University, New York
LEVINE, MICHAEL, Dept. of Molecular and Cell Biology, University of California, Berkeley

LEVY, RONALD, Dept. of Chemistry and Chemical Biology, BioMaPS Institute for Quantitative Biology, Rutgers University, Piscataway, New Jersey
LEVY, SASHA, Dept. of Biology, New York University, New York
LINDQUIST, SUSAN, Whitehead Institute for Biomedical Research, Cambridge, Massachusetts
LIPPMAN, ZACHARY, Cold Spring Harbor Laboratory, Cold Spring Harbor, New York
LITTLE, TOM, Institute of Evolutionary Biology, University of Edinburgh, Edinburgh, Scotland, United Kingdom
LIU, CHANG, Dept. of Chemistry, Scripps Research Institute, La Jolla, California
LONG, GUNHILD "GUNNI," Berger Health System, Circleville, Ohio
LOUIS, ED, Institute of Genetics, University of Nottingham, Nottingham, United Kingdom
LOWE, CRAIG, Dept. of Biomolecular Engineering, University of California, Santa Cruz
LUNDWALL, AKE, Dept. of Laboratory Medicine, Lund University, Malmo, Sweden
MA, JIAN, Center for Biomolecular Science and Engineering, University of California, Santa Cruz
MACPHEE, CAIT, School of Physics and Astronomy, University of Edinburgh, Edinburgh, Scotland, United Kingdom
MADERSPACHER, FLORIAN, *Current Biology*, Cell Press, London, United Kingdom
MAGENNIS, MARISA, School of Physics and Astronomy, University of Edinburgh, Edinburgh, Scotland, United Kingdom
MAJOVSKI, ROBERT, *Genome Research*, Cold Spring Harbor Laboratory Press, Woodbury, New York
MAKALOWSKI, WOJCIECH, Institute of Bioinformatics, University of Muenster, Muenster, Germany
MAKINO, TAKASHI, Smurfit Institute of Genetics, Trinity College, Dublin, Ireland
MALONE, COLIN, Cold Spring Harbor Laboratory, Cold Spring Harbor, New York
MANCERA, EUGENIO, Gene Expression Unit, European Molecular Biology Laboratory, Heidelberg, Germany
MARCUS, EMILIE, *Cell*, Cell Press, Cambridge, Massachusetts
MARQUES-BONET, TOMAS, Dept. of Genome Sciences, University of Washington, Seattle
MARTIENSSEN, ROBERT, Cold Spring Harbor Laboratory, Cold Spring Harbor, New York
MARTÍNEZ, MIGUEL, Lab. de Retrovirologia, Fundacio irsiCaixa, Universitat Autònoma de Barcelona, Badalona, Spain
MATHEWS, MICHAEL, Dept. of Biochemistry and Molecular Biology, School of Medicine, University of Medicine and Dentistry of New Jersey, Newark
MATTICK, JOHN, Institute for Molecular Bioscience, University of Queensland, St. Lucia, Australia
MAYROSE, ITAY, Dept. of Zoology, University of British Columbia, Vancouver, British Columbia, Canada
MCDONALD, MICHAEL, Institute of Molecular Biology, Academia Sinica, Taipei, Taiwan
MCGINTY, SUSAN, Dept. of Applied Science, London South Bank University, London, United Kingdom
MCKINNON, DAVID, Dept. of Neurobiology and Behavior, Stony Brook University, Stony Brook, New York
MEDINI, DUCCIO, Dept. of Systems Biology, Novartis Vaccines and Diagnostics, Siena, Italy
MELE, MARTA, Dept. Ciències Experimentals i de la Salut, Institut de Biologia Evolutiva, Universitat Pompeu Fabra, Barcelona, Spain
METTLING, CLEMENT, Institut de Génétique Humaine, Centre National de la Recherche Scientifique, Montpellier, France
MICKLOS, DAVID, Dolan DNA Learning Center, Cold Spring Harbor Laboratory, Cold Spring Harbor, New York
MILLER, JEFFREY, Dept. of Microbiology, Immunology, and Molecular Genetics, University of California, Los Angeles
MILLER, KENNETH, Dept. of Molecular Biology, Cell Biology, and Biochemistry, Brown University, Providence, Rhode Island
MINIS, ADI, Dept. of Biological Chemistry, Weizmann Institute of Science, Rehovot, Israel
MIRABEAU, OLIVIER, Alfred Fessard Institute for Neuroscience, Centre National de la Recherche Scientifique, Gif-sur-Yvette, France
MISHMAR, DAN, Dept. of Life Sciences, Ben-Gurion University of the Negev, Be'er-Sheva, Israel
MOCZEK, ARMIN, Dept. of Biology, Indiana University, Bloomington
MODE, CHARLES, Dept. of Mathematics, Drexel University, Coatesville, Pennsylvania
MOELLING, KARIN, Institute of Medical Virology, University of Zürich, Zürich, Switzerland
MOLARO, ANTOINE, Cold Spring Harbor Laboratory, Cold Spring Harbor, New York
MONACO, MARCELA, Cold Spring Harbor Laboratory, Cold Spring Harbor, New York
MORENO-HAGELSIEB, Gabriel, Dept. of Biology, Wilfrid Laurier University, Waterloo, Ontario, Canada
MOROZ, LEONID, Dept. of Neuroscience and Marine Bioscience, University of Florida, St. Augustine
MOROZOV, ALEXANDRE, Dept. of Physics and Astronomy, Rutgers University, Piscataway, New Jersey
MUELLER, JACOB, Dept. of Biology, Whitehead Institute for Biomedical Research, Cambridge, Massachusetts
MURALIDHARA, CHAITANYA, Dept. of Cellular and Molecular Biology, University of Texas, Austin
MURPHY, HELEN, Dept. of Biology, Wake Forest University, Winston-Salem, North Carolina
MUSTONEN, VILLE, Dept. of Theoretical Physics, University of Cologne, Cologne, Germany
NATHAN, MARCO, Dept. of Philosophy, Columbia University, New York, New York
NAVARRO, ARCADI, Dept. de Ciències Experimentals i de la Salut, Institut de Biologia Evolutiva, Universitat Pompeu Fabra, Barcelona, Spain
NEHER, RICHARD, Kavli Institute for Theoretical Physics, University of California, Santa Barbara
NIKOLAIDIS, NIKOLAS, Dept. of Biological Science, California State University, Fullerton
NOBREGA, MARCELO, Dept. of Human Genetics, University of Chicago, Chicago, Illinois
NOONAN, JAMES, Dept. of Genetics, Yale University, New Haven, Connecticut

NUGENT, PAUL, Old Saybrook, Connecticut
OGASAWARA, OSAMU, DNA Databank of Japan, National Institute of Genetics, Mishima, Japan
OHRSTROM, GERRY, Epicurus Fund, New York, New York
ORR, MARILYN, Berger Health System, Circleville, Ohio
ØSTMAN, BJØRN, Dept. of Computational Biology, Keck Graduate Institute, Claremont, California
OSTRANDER, ELAINE, Cancer Genetics Branch, National Human Genome Reseach Institute, National Institutes of Health, Bethesda, Maryland
OYIBO, HASSANA, Cold Spring Harbor Laboratory, Cold Spring Harbor, New York
PÄÄBO, SVANTE, Dept. of Evolutionary Genetics, Max-Planck-Institute for Evolutionary Anthropology, Leipzig, Germany
PADIAN, KEVIN, Dept. of Paleontology, University of California, Berkeley
PAGE, DAVID, Dept. of Biology, Howard Hughes Medical Institute, Massachusetts Institute of Technology, Cambridge, Massachusetts
PAIXAO, TIAGO, Dept. of Biology and Biochemistry, University of Houston, Houston, Texas
PALMER, JEFFREY, Dept. of Biology, Indiana University, Bloomington
PARKER, JOSEPH, Dept. of Genetics and Development, Columbia University, New York
PARROTT, ANDREW, Dept. of Biochemistry, School of Medicine, University of Medicine and Dentistry of New Jersey, Newark
PATEL, NIPAM, Dept. of Integrative Biology, University of California, Berkeley
PAYNE, SARAH, Dept. of Biology, California Institute of Technology, Pasadena
PE'ERY, TSAFI, Dept. of Medicine and Biochemistry and Molecular Biology, School of Medicine, University of Medicine and Dentistry of New Jersey, Newark
PEARSON, WILLIAM, Dept. of Biochemistry and Molecular Genetics, University of Virginia, Charlottesville
PEDERSON, THORU, Dept. of Biochemistry and Molecular Pharmacology, School of Medicine, University of Massachusetts, Worcester
PENG, WEIQUN, Dept. of Physics, George Washington University, Washington, D.C.
PENIGAULT, JEAN-BAPTISTE, Dept. of Development and Neurobiology, Institut Jacques Monod, Centre National de la Recherche Scientifique, Paris, France
PENN, OSNAT, Dept. of Cell Research and Immunology, Tel Aviv University, Tel Aviv, Israel
PENNISI, ELIZABETH, *Science*, American Association for the Advancement of Science, Washington, D.C.
PETRIE, KATHERINE, Dept. of Chemistry and Molecular Biology, Scripps Research Institute, La Jolla, California
PHILBRICK, WILLIAM, Dept. of Internal Medicine, School of Medicine, Yale University, New Haven, Connecticut
PINKER, STEVEN, Dept. of Psychology, Harvard University, Cambridge, Massachusetts
PITT, JASON, Dept. of Basic Sciences, Fred Hutchinson Cancer Research Center, Seattle, Washington
PLAVSKIN, EUGENE, Cold Spring Harbor Laboratory, Cold Spring Harbor, New York
PLOTKIN, JOSHUA, Dept. of Biology, University of Pennsylvania, Philadelphia
POLLOCK, MILA, Cold Spring Harbor Laboratory Library, Cold Spring Harbor, New York
POOL, JOHN, Dept. of Integrative Biology, University of California, Berkeley
PRICE, TOM, Cellmates, New Haven, Connecticut
PUURAND, TARMO, Dept. of Bioinformatics, University of Tartu, Tartu, Estonia
RAMAKRISHNAN, VENKI, Lab. of Molecular Biology, Medical Research Council, Cambridge, United Kingdom
RASSOULZADEGAN, MINOO, Institut National de la Santé et de la Recherche Médicale, University of Nice, Nice, France
RAUSCH, CATHERINE, Dépt. Systématique et Evolution, Muséum National d'Histoire Naturelle, Paris, France
RAY, PARTHO SAROTHI, Dept. of Biology, Indian Institute of Science Education and Research, Calcutta, India
REBEIZ, MARK, Dept. of Molecular Biology, University of Wisconsin, Madison
REYNOLDS, NOAH, Dept. of Microbiology, Ohio State University, Columbus
RIDDIHOUGH, GUY, *Science*, American Association for the Advancement of Science, Washington, D.C.
RIDLEY, MATTHEW, Blagdon Hall, Newcastle, United Kingdom
RIDLEY, SUSAN PORTER, Dept. of Molecular and Cellular Biosciences, National Science Foundation, Arlington, Virginia
RITTER, DEBORAH, Dept. of Bioinformatics, Boston College, Boston, Massachusetts
ROBINSON, GENE, Dept. of Entomology, University of Illinois, Urbana-Champaign
RODRIGUE, NICOLAS, Dept. of Biology, University of Ottawa, Ottawa, Ontario, Canada
ROLLINS, FRED, Cold Spring Harbor Laboratory, Cold Spring Harbor, New York
ROMERO, ROBERTO, Perinatology Research Branch, National Institute of Child Health and Human Development, National Institutes of Health, Detroit, Michigan
RUSSELL, PAMELA, Genome Sequencing and Analysis Program, Broad Institute of MIT and Harvard University, Cambridge, Massachusetts
RYTKÖNEN, KALLE, Dept. of Biology, University of Turku, Turku, Finland
SAAB-RINCÓN, GLORIA, Ingeniería Celular y Biocatálisis, Institute of Biotechnology, Universidad Nacional Autónoma de México, Cuernavaca, México
SACKLER, RICHARD, Dept. of Statistical Genetics, The Rockefeller University, New York
SAHA, NIL, Dept. of Molecular Genetics, Benaroya Research Institute, Seattle, Washington
SANBONMATSU, KEVIN, Dept. of Theoretical Biology and Biophysics, Los Alamos National Laboratory, Los Alamos, New Mexico
SATO, ATSUKO, Dept. of Zoology, University of Oxford, Oxford, United Kingdom
SCHAFFER, JAMES, Dept. of Biomedical Informatics, Philips Research, Briarcliff Manor, New York
SCHUTT, BARBARA, Nancy Lurie Marks Foundation, Wellesley, Massachusetts
SCHUTT, CLARENCE, Nancy Lurie Marks Foundation, Wellesley, Massachusetts
SCHWARTZ, JAMES, Independent scholar and writer,

Brookline, Massachusetts
SCOTT, EUGENIE, Executive Director, National Center for Science, Oakland, California
SECK, PIERRE, Faculty of Sciences, Technology and Communication, University of Luxembourg, Luxembourg
SEDEROFF, RON, Forest Biotechnology Group, North Carolina State University, Raleigh
SEMPLE, COLIN, Dept. of Bioinformatics, MRC Human Genetics Unit, Western General Hospital, Edinburgh, Scotland, United Kingdom
SHAPIRO, LUCY, Dept. of Developmental Biology, Stanford University, Stanford, California
SHIELDS, ROBERT, *PLoS Biology*, Public Library of Science, Cambridge, United Kingdom
SHOVAL, YISHAY, Dept. of Molecular Genetics, Weizmann Institute of Science, Rehovot, Israel
SIEGAL, MARK, Dept. of Biology, New York University, New York
SIKELA, JAMES, Dept. of Pharmacology, Anschutz Medical Campus, University of Colorado, Aurora
SIKORA, MARTIN, Institute of Evolutionary Biology, Universitat Pompeu Fabra, Barcelona, Spain
SMITH, GREGORY, Dept. of Applied Science, College of William and Mary, Williamsburg, Virginia
SNELL-ROOD, EMILIE, Dept. of Biology, Indiana University, Bloomington
SOLTIS, DOUGLAS, Dept. of Botany, University of Florida, Gainesville
SPIES, NOAH, Dept. of Biology, Whitehead Institute for Biomedical Research, Massachusetts Institute of Technology, Cambridge
STEIN, ARNOLD, Dept. of Biological Sciences, Purdue University, West Lafayette, Indiana
STERN, ADI, Dept. of Cell Research and Immunology, Tel Aviv University, Tel Aviv, Israel
STEWART, DAVID, Meetings and Courses Program, Cold Spring Harbor Laboratory, Cold Spring Harbor, New York
STEWART, FIONA, University of Western Ontario, London, Ontario, Canada
STILLMAN, BRUCE, President, Cold Spring Harbor Laboratory, Cold Spring Harbor, New York
STONE, JAMES, President's Council, Cold Spring Harbor Laboratory, Cold Spring Harbor, New York
SUNWOO, HONGJAE, Cold Spring Harbor Laboratory, Cold Spring Harbor, New York
SUSSMAN, HILLARY, *Genome Research*, Cold Spring Harbor Laboratory Press, Woodbury, New York
SUZUKI, IKUO, Lab. for DNA Data Analysis, National Institute of Genetics, Mishima, Japan
SWIGUT, TOMASZ, Dept. of Chemical and Systems Biology, Stanford University, Stanford, California
SZOSTAK, JACK, Dept. of Molecular Biology, Massachusetts General Hospital, Harvard University, Boston
TAIPALE, MIKKO, Whitehead Institute for Biomedical Research, Cambridge, Massachusetts
THOMSEN, GERALD, Dept. of Biochemistry and Cell Biology, Stony Brook University, Stony Brook, New York
THORNTON, JOSEPH, Dept. of Ecology and Evolutionary Biology, University of Oregon, Eugene
TISHKOFF, SARAH, Dept. of Genetics and Biology, University of Pennsylvania, Philadelphia

TOFT, CHRISTINA, Smurfit Institute of Genetics, Trinity College, Dublin, Ireland
TOLL-RIERA, MACARENA, Dept. of Health and Life Sciences, Universitat Pompeu Fabra, Barcelona, Spain
TOMILIN, NIKOLAI, Dept. of Chromosome Stability, Institute of Cytology, Russian Academy of Sciences, St. Petersburg, Russia
TOMOYASU, YOSHINORI, Dept. of Zoology, Miami University, Oxford, Ohio
TOTH, AMY, Dept. of Entomology, Pennsylvania State University, University Park
TOUCHON, MARIE, Dept. of Microbial Evolutionary Genomics, Institut Pasteur, Centre National de la Recherche Scientifique, Paris, France
TRAN, DIANA, Dept. of Biology, Harvey Mudd College, Claremont, California
TROTMAN, LLOYD, Cold Spring Harbor Laboratory, Cold Spring Harbor, New York
TURNER, PAUL, Dept. of Ecology and Evolutionary Biology, Yale University, New Haven, Connecticut
UMEN, JAMES, Lab. of Plant Biology, The Salk Institute, La Jolla, California
UNNEBERG, PER, Dept. of Biosciences and Nutrition, Karolinska Institutet, Huddenge, Sweden
VAISBUCH, EDI, Dept. of Obstetrics and Gynecology, Wayne State University, Perinatology Research Branch, National Institute of Child Health and Human Development, National Institutes of Health, Department of Health and Human Services, Detroit, Michigan
VALAS, RUBEN, Bioinformatics Program, University of California, San Diego, La Jolla
VAN DER SLUIS, ELI, Dept. of Chemistry and Biochemistry, University of Munich, Munich, Germany
VAN DITMARSCH, DAVE, Laboratoire de Biologie Chimique, Institut de Recherche de l'Université de Strasbourg, Strasbourg, France
VENTER, J. CRAIG, Dept. of Enviromental Genomics, J. Craig Venter Institute, Rockville, Maryland
VIEIRA-SILVA, SARA, Dept. of Microbial Evolutionary Genomics, Institut Pasteur, Paris, France
VINSON, CHARLES, Center for Cancer Research, National Cancer Institute, Bethesda, Maryland
VOLKER, CRAIG, Dept. of Computational Biology, GlaxoSmithKline, King of Prussia, Pennsylvania
WADE, NICHOLAS, Science Department, *The New York Times*, New York, New York
WAHRMUND, UTE, Dept. of Molecular Evolution, Institute for Cellular and Molecular Botany, University of Bonn, Bonn, Germany
WALLACE, DOUGLAS, Center for Molecular and Mitochondrial Medicine and Genetics, University of California, Irvine
WARE, DOREEN, Cold Spring Harbor Laboratory, Cold Spring Harbor, New York
WATANABE, HIDEMI, School of Information Science and Technology, Hokkaido University, Sapporo, Japan
WEINBERG, ERIC, Dept. of Biology, University of Pennsylvania, Philadelphia
WESSLER, SUSAN, Dept. of Plant Biology, University of Georgia, Athens
WHITE, TIM, Dept. of Integrative Biology, University of California, Berkeley

WILDENBERG, GREGG, Dept. of Molecular and Cellular Biology, Harvard University, Cambridge, Massachusetts
WILSON, EDWARD, Museum of Comparative Zoology, Harvard University, Cambridge, Massachusetts
WILSON, MICHAEL, Cambridge Research Institute, Cancer Research UK, Cambridge, United Kingdom
WITKOWSKI, JAN, Banbury Center, Cold Spring Harbor Laboratory, Cold Spring Harbor, New York
WONG, TERENCE, Dept. of Biology, Harvey Mudd College, Claremont, California
XUAN, ZHENYU, Cold Spring Harbor Laboratory, Cold Spring Harbor, New York
YAKLICHKIN, SERGEY, Dept. of Cellular and Developmental Biology, University of Pennsylvania, Philadelphia
YANCOPOULOS, SOPHIA, Feinstein Institute for Medical Research, Manhasset, New York
YONA, AVIHU, Dept. of Molecular Genetics, Weizmann Institute of Science, Rehovot, Israel
ZHAXYBAYEVA, OLGA, Dept. of Biochemistry and Molecular Biology, Dalhousie University, Halifax, Nova Scotia, Canada
ZILL, OLIVER, Dept. of Molecular and Cell Biology, University of California, Berkeley
ZNAMENSKIY, PETER, Cold Spring Harbor Laboratory, Cold Spring Harbor, New York
ZODY, MICHAEL, Dept. of Genome Sequencing and Analysis, Broad Institute of MIT and Harvard University, Cambridge, Massachusetts
ZOGRAFOS, LYSIMACHOS, School of Informatics, University of Edinburgh, Edinburgh, Scotland, United Kingdom

First row: J. Alfred, B. Stillman; P. Amaral; R. Behringer
Second row: J. Browne, V. Ramakrishnan; M. Bulyk; C. Bustamante; B. Charlesworth
Third row: M. Behe, A. Breslau, J. Dubnau; F. Doolittle; G. Bernardi, J. Bertranpetit
Fourth row: E. Dawson, L. Caporale; J. Williams and high school students; Q. Cronk

First row: Sean M. Carroll, Sean B. Carroll; J. Witkowski, E. Wilson;
Second row: Surviving the picnic; S. Pääbo
Third row: E. Ostrander; M. Levine; an attentive front row
Fourth row: N. Patel; L. Shapiro

Drawings of scientists by Julia Kuhl, 2009.

First row: E. Carlson; M. Gierszewska; G. Hannon; B. Forrest
Second row: G. Ohrstrom, J. Witkowski; K. Padian
Third row: C. Schutt; P. Hebert; P. Mitra; W. Pearson
Fourth row: D. Wallace; E. Davidson (seated); lunch at Blackford
Fifth row: M. Ridley, J. Horner; K. Miller; from fish to philosopher

First row: R. Martienssen, R. Doerge; V. Ramakrishnan, J. Szostak, A. Gann; N. King
Second row: Watching closed-circuit television; picnic
Third row: D. Dennett, E. Plavskin; E. Eichler, M. Mathews; S. Wessler
Fourth row: J. Watson, N. Wade; E. Koonin

First row: J. Thornton; D. Stewart; D. Kingsley; D. Page
Second row: J. Darnell; V. Hartenstein; D. Haussler
Third row: Evolution and the Public discussion: K. Miller, B. Forrest, E. Scott; M. Ridley, N. Ferguson
Fourth row: G. Joyce, F. Arnold; J. Schwartz, J. Stone; K. Sanbonmatsu, S. Hennelly
Fifth row: K. Padian, D. Stewart; S. Pinker, S. Grant; J. Drake, T. Kunkel

First row: J. Schaffer; Life Forms (ceramics by Christopher Adams); D. Wallace
Second row: P. Turner, G. Bell, H. Hoekstra; T. White, J. Horner
Third row: E. Scott, R. Lenski; R. Doolittle, R. Greenspan; R. Cooke
Fourth row: B. Charlesworth; questions following Ford Doolittle's talk

Drawings of scientists by Julia Kuhl, 2009.

Foreword

The 74th Cold Spring Harbor Laboratory Symposium on Quantitative Biology on *Evolution: The Molecular Landscape* was dedicated to Charles Darwin on the occasion of the bicentennial of his birth and the 150th anniversary of the publication of *On the Origin of Species*. The Laboratory celebrated the 100th anniversary in 1959 with its 24th Symposium on *Genetics and Twentieth Century Darwinism*. What was entirely absent from that Symposium and what dominated the Symposium 50 years later are the contributions molecular biology has made to our understanding of evolution. Even as the details of Darwin's ideas have been modified over the years, evidence from molecular studies has strengthened his fundamental thesis.

The 2009 Symposium set out to examine the current state of many of the ideas that Darwin developed in his four great books: *On the Origin of Species by Means of Natural Selection, The Variation of Animals and Plants Under Domestication, The Descent of Man and Selection in Relation to Sex,* and *The Expression of Emotions in Man and Animals*. Leading investigators were invited to present their latest research in a diversity of fields ranging from the origins of life (unicellular and multicellular) to speciation and domestication to the evolutionary basis of human attributes. An overarching theme of the meeting was the extent to which much of evolutionary biology can now be viewed in a molecular, and often genomic, framework and the extraordinary degree to which many of Darwin's insights remain profoundly relevant today.

The Symposium included two rather unusual sessions. Evolutionary concepts have had an impact far beyond the boundaries of science and there is hardly a field of human endeavor that has not been influenced by evolutionary thinking. To acknowledge this contribution of Darwin, there was a session on "Cultural Evolution" that included presentations on principles of natural selection applied to linguistics, ideas, and economics by, respectively, Daniel Dennett, Matt Ridley, and Niall Ferguson. In the second unusual session, "Evolution and the Public," Kevin Padian, Ken Miller, Barbara Forrest, and Eugenie Scott discussed so-called "intelligent design" and the threat such irrational and antiscientific attitudes pose to education in the United States and elsewhere.

In arranging this Symposium, the organizers were dependent on the guidance of a broad cadre of advisors including Drs. Nicholas Barton, Hans Ellegren, Claire Fraser-Liggett, David Haussler, Gerry Joyce, Susan McCouch, Sarah Otto, Svante Pääbo, Nipam Patel, Matt Ridley, James D. Watson, and Richard Wrangham. Opening night speakers included Janet Browne, Ed Wilson, David Kingsley, and Marc Hauser. Douglas Wallace presented the Reginald Harris Lecture on "Energetics in Eukaryotic and Human Origins." Kevin Padian enlightened a mixed audience of scientists and lay friends and neighbors with his Dorcas Cummings Lecture on "Darwin, Dover, & Intelligent Design," and Brian Charlesworth ended the meeting with a masterful and thought-provoking summary.

This Symposium was held May 27–June 1 and attended by 400 scientists from more than 25 countries. The program included 69 invited presentations and more than 175 poster presentations. For the first time, six Symposium fellows were selected from the submitted abstracts, and each of these young scientists gave a short research presentation.

We thank Val Pakaluk and Mary Smith in the Meetings and Courses Program Office for their assistance in organizing and running the meeting and John Inglis and his staff at Cold Spring Harbor Laboratory Press, particularly Joan Ebert, Rena Steuer, and Kathy Bubbeo, for publishing both the printed and online versions of the Symposium proceedings. Photographers Connie Brukin and Jan Witkowski captured candid snapshots throughout the Symposium, and artist Julia Kuhl sketched portraits of a number of the speakers.

Funds to support this meeting were obtained from the National Institutes of Health. Financial support from the corporate benefactors, sponsors, affiliates, and contributors of our meetings program is essential for these Symposia to remain a success and we are most grateful for their continued support.

Bruce Stillman
David Stewart
Jan Witkowski
March 2010

Sponsors

This meeting was funded in part by the **National Institutes of Health**.

Contributions from the following companies provide core support for the Cold Spring Harbor meetings program.

Corporate Benefactors

Amgen, Inc.
GlaxoSmithKline
Novartis Institutes for BioMedical Research

Corporate Sponsors

Agilent Technologies
Applied Biosystems
AstraZeneca
BioVentures, Inc.
Bristol-Myers Squibb Company
Genentech, Inc.
Hoffmann-La Roche, Inc.
IRX Therapeutics, Inc.
New England Biolabs, Inc.
OSI Pharmaceuticals, Inc.
Sanofi-Aventis
Schering-Plough Research Institute

Plant Corporate Associates

Monsanto Company
Pioneer Hi-Bred International, Inc.

Corporate Affiliates

Affymetrix

Contents

Symposium Participants ... v
Foreword ... xix

Introduction

Darwin the Scientist *J. Browne* ... 1
On the Future of Biology *E.O. Wilson* ... 9

RNA and Proteins

Evolution of Biological Catalysis: Ribozyme to RNP Enzyme *T.R. Cech* ... 11
Evolution in an RNA World *G.F. Joyce* ... 17
The Ribosome: Some Hard Facts about Its Structure and Hot Air about Its Evolution *V. Ramakrishnan* ... 25
Step-by-Step Evolution of Vertebrate Blood Coagulation *R.F. Doolittle* ... 35
How Proteins Adapt: Lessons from Directed Evolution *F.H. Arnold* ... 41

Cellular Evolution

Reconstructing the Emergence of Cellular Life through the Synthesis of Model Protocells
  S.S. Mansy and J.W. Szostak ... 47
Dynamic Chromosome Organization and Protein Localization Coordinate the Regulatory Circuitry
  that Drives the Bacterial Cell Cycle *E.D. Goley, E. Toro, H.H. McAdams, and L. Shapiro* ... 55
An Integrated View of Precambrian Eumetazoan Evolution *E.H. Davidson and D.H. Erwin* ... 65
The Dawn of Developmental Signaling in the Metazoa *G.S. Richards and B.M. Degnan* ... 81

Mutation

Evolving Views of DNA Replication (In)Fidelity *T.A. Kunkel* ... 91
Protein Folding Sculpting Evolutionary Change *S. Lindquist* ... 103
Predicting Virus Evolution: The Relationship between Genetic Robustness and Evolvability
  of Thermotolerance *C.B. Ogbunugafor, R.C. McBride, and P.E. Turner* ... 109
Genome-wide Mutational Diversity in an Evolving Population of *Escherichia coli*
  J.E. Barrick and R.E. Lenski ... 119
Selection, Gene Interaction, and Flexible Gene Networks *R.J. Greenspan* ... 131

Selection and Adaptation

The Oligogenic View of Adaptation *G. Bell* ... 139
Genetic Dissection of Complex Traits in Yeast: Insights from Studies of Gene Expression
  and Other Phenotypes in the BYxRM Cross *I.M. Ehrenreich, J.P. Gerke, and L. Kruglyak* ... 145
Measuring Natural Selection on Genotypes and Phenotypes in the Wild *C.R. Linnen and H.E. Hoekstra* ... 155
Exploring the Molecular Landscape of Host–Parasite Coevolution *D.E. Allen and T.J. Little* ... 169
Genetic Recombination and Molecular Evolution *B. Charlesworth, A.J. Betancourt, V.B. Kaiser,
  and I. Gordo* ... 177
Why Sex and Recombination? *N.H. Barton* ... 187

Diversity

Eradicating Typological Thinking in Prokaryotic Systematics and Evolution *W.F. Doolittle* — 197

The Phylogenetic Forest and the Quest for the Elusive Tree of Life *E.V. Koonin, Y.I. Wolf, and P. Puigbò* — 205

On the Origins of Species: Does Evolution Repeat Itself in Polyploid Populations of Independent Origin? *D.E. Soltis, R.J.A. Buggs, W.B. Barbazuk, P.S. Schnable, and P.S. Soltis* — 215

Evolution of Systems

Molecular Evolution of piRNA and Transposon Control Pathways in *Drosophila* *C.D. Malone and G.J. Hannon* — 225

Drosophila Brain Development: Closing the Gap between a Macroarchitectural and Microarchitectural Approach *A. Cardona, S. Saalfeld, P. Tomancak, and V. Hartenstein* — 235

A General Basis for Cognition in the Evolution of Synapse Signaling Complexes *S.G.N. Grant* — 249

Evolution of Development

Evolution in Reverse Gear: The Molecular Basis of Loss and Reversal *Q.C.B. Cronk* — 259

Darwin's "Abominable Mystery": The Role of RNA Interference in the Evolution of Flowering Plants *A. Cibrián-Jaramillo and R.A. Martienssen* — 267

Evolution of Insect Dorsoventral Patterning Mechanisms *M.W. Perry, J.D. Cande, A.N. Boettiger, and M. Levine* — 275

Lophotrochozoa Get into the Game: The Nodal Pathway and Left/Right Asymmetry in Bilateria *C. Grande and N.H. Patel* — 281

On the Origins of Novelty and Diversity in Development and Evolution: A Case Study on Beetle Horns *A.P. Moczek* — 289

Genetic Regulation of Mammalian Diversity *R.R. Behringer, J.J. Rasweiler IV, C.-H. Chen, and C.J. Cretekos* — 297

Domestication

Genomics, Domestication, and Evolution of Forest Trees *R. Sederoff, A. Myburg, and M. Kirst* — 303

Studying Phenotypic Evolution in Domestic Animals: A Walk in the Footsteps of Charles Darwin *L. Andersson* — 319

Fine Mapping a Locus Controlling Leg Morphology in the Domestic Dog *P. Quignon, J.J. Schoenebeck, K. Chase, H.G. Parker, D.S. Mosher, G.S. Johnson, K.G. Lark, and E.A. Ostrander* — 327

Human Evolution

Human Origins and Evolution: Cold Spring Harbor, Déjà Vu *T.D. White* — 335

Reconstructing the Evolution of Vertebrate Sex Chromosomes *D.W. Bellott and D.C. Page* — 345

The Evolution of Human Segmental Duplications and the Core Duplicon Hypothesis *T. Marques-Bonet and E.E. Eichler* — 355

snaR Genes: Recent Descendants of *Alu* Involved in the Evolution of Chorionic Gonadotropins *A.M. Parrott and M.B. Mathews* — 363

DUF1220 Domains, Cognitive Disease, and Human Brain Evolution *L. Dumas and J.M. Sikela* — 375

Mitochondria, Bioenergetics, and the Epigenome in Eukaryotic and Human Evolution *D.C. Wallace* — 383

Genetic Structure in African Populations: Implications for Human Demographic History *C.A. Lambert and S.A. Tishkoff* — 395

Social Interaction and Human Society

A Defense of Sociobiology *K.R. Foster* — 403

Evo-Devo and the Evolution of Social Behavior: Brain Gene Expression Analyses in Social Insects *A.L. Toth and G.E. Robinson* — 419

Social Interaction and Human Society (*continued*)

Cooking and the Human Commitment to a High-quality Diet *R.N. Carmody and R.W. Wrangham*	427
The Cultural Evolution of Words and Other Thinking Tools *D.C. Dennett*	435
When Ideas Have Sex: The Role of Exchange in Cultural Evolution *M.W. Ridley*	443
An Evolutionary Approach to Financial History *N. Ferguson*	449
The Religious Essence of Intelligent Design *B. Forrest*	455
Deconstructing Design: A Strategy for Defending Science *K.R. Miller*	463
Summary *B. Charlesworth*	469

Author Index — 475
Subject Index — 477

Darwin the Scientist

J. BROWNE
Department of the History of Science, Harvard University, Cambridge, Massachusetts 02138
Correspondence: jbrowne@fas.harvard.edu

> Charles Darwin's experimental investigations show him to have been a superb practical researcher. These skills are often underestimated today when assessing Darwin's achievement in the *Origin of Species* and his other books. Supported by a private income, he turned his house and gardens into a Victorian equivalent of a modern research station. Darwin participated actively in the exchange of scientific information via letters and much of his research was also carried out through correspondence. Although this research was relatively small scale in practice, it was large scale in intellectual scope. Darwin felt he had a strong desire to understand or explain whatever he observed.

During the 2009 Darwin commemorations, scientific researchers and scholars around the world eagerly seized the opportunity to reflect on Charles Darwin's many achievements. As a historical figure, Darwin can mean many things to many people. Most of us carry a particular affection for the young man who sailed in the *H.M.S. Beagle* and published his account of the voyage in 1839 as *Journal of Researches*, now a classic text in Victorian travel literature (Darwin 1989). In books, articles, and television documentaries, we have enjoyed catching glimpses of his developing scientific ideas, his incisive comments on the places he visited, and his clear, unaffected prose. Much later on in life, Darwin also wrote a charming autobiography (actually titled by him as *Recollections of the Development of My Mind and Character*) that has similarly caught the imagination of generations of readers, both inside and outside science. In those recollections, an elderly Darwin looked back over his life and times with generosity and much affection for his friends and family (Barlow 1958). In fact, Darwin left an exceptionally rich documentary record that somehow closes the gap between the centuries. We feel that he was a man we would like to know, someone uniquely special (Fig. 1).

Most of all, in 2009, Charles Darwin is justly celebrated for his magnificent book, *On the Origin of Species by Means of Natural Selection,* published in London in November 1859 to a storm of controversy (Darwin 1859, 1958). This book made Darwin one of the most prominent naturalists in the world, "first among the scientific men of England," as the socialist philosopher Edward Aveling put it. Even Alfred Russel Wallace, who independently formulated the same idea of evolution by natural selection, said that "Mr. Darwin has given the world a new science, and his name should, in my opinion, stand above that of every philosopher of ancient or modern times." As Wallace predicted, Darwin's ideas came to lie at the heart of fundamental shifts in opinion that swept through the 19th century and beyond (Browne 2002, 2006). The avenues of thought opened up by the theory of evolution by natural selection have guided and provoked research for 150 years. Today, greatly enhanced by the field of genetics and much diversified in intellectual scope, Darwinism stands securely at the cutting edge of scientific knowledge.

Nor do we need reminding that this book excited intense attention, not just for the powerful reach of the proposals, or Darwin's careful analysis of the problem of animal and plant origins, and his weighty accumulation of evidence, but also for the shocking absence—for some people, the *liberating* absence—of any allusion to the biblical story of creation. Darwin proposed an entirely natural process to explain the characteristics of the living beings that we see around us. This extraordinary combination of features was guaranteed to generate argument in Victorian England. Should God be banished as an explanation for the apparent design and harmony of the world? What should people think about their own origins? Was Darwin suggesting that humans emerged in a natural manner from ape ancestors? These were huge questions that continue to be asked today. *On the Origin of Species* must surely be regarded as a major publishing event in the 19th century that changed the way that people thought about themselves and helped to lay the foundations of the modern world.

What is less appreciated, even sometimes by biologists, is the range of Darwin's scientific interests—from geology and coral reefs through detailed taxonomic studies of barnacles, human expressions, domestication of plants and animals, to plant physiology. Darwin was a superb experimentalist who designed many effective techniques to probe the natural world. Carried out with simple tools, these experiments provided insights into processes as diverse as the role of earthworms in recycling the land and the movements of climbing plants in their search for support. The anniversary year provides an opportunity to look beyond the continuing uproar over evolution and celebrate Darwin, the practicing scientist.

Figure 1. Charles Darwin, in a portrait from life that has become iconic. Photogravure by Leopold Flameng, after the oil portrait by John Collier, 1881. The original hangs in the Linnean Society of London. (Reprinted, with permission, from the Wellcome Library, London.)

CELEBRITY

Darwin as a scientist often gets lost behind the heat and smoke of the controversy. During his own lifetime, to be sure, he became an icon, one of the first scientific celebrities. That celebrity razzamatazz has a pleasantly old-fashioned air—in those days, the media organization associated with the creation of fame was only just emerging. But there were the makings of a Darwin cult. In the years following publication of the *Origin of Species*, individuals with a few coins to spare could, if they wished, acquire a pottery statuette of a chimp contemplating a human skull. They could buy any number of photographic *cartes de visite* featuring Darwin's portrait. If their pockets ran slightly deeper, they could purchase one of a limited edition of portrait photographs of Darwin taken by the society photographer Julia Margaret Cameron and sold by the Bond Street gallery of Colnaghi's, each personally authenticated by the gallery's stamp and by the photographer. Or they might pay to gape at Julia Pastrana, the hairy lady who advertised herself as a "Missing Link," who toured Europe in 1862. British connoisseurs were able to commission an elegant piece of Wedgwood ware decorated with an evolutionary tree. They could sing a duet at the piano on the "Darwinian Theory." Even the agricultural firm of G.W. Merchant, of Lockport, near Rochester, New York, advertised its Gargling Oil with a Darwinian ape that sang:

> If I am Darwin's Grandpa,
> It follows, don't you see,
> That what is good for man & beast
> Is doubly good for me.

All these commercial products made the controversy about human origins fully tangible to Darwin's generation and the ones that followed (Browne 2003).

Darwin was also one of the few scientists to have been portrayed in an extraordinary variety of caricatures. Victorian cartoonists grabbed their chance. "Am I a Man and a Brother?" asked a gorilla in the May 1861 number of the British humorous magazine *Punch*, echoing the popular perception of Darwin's work (Clark 2009). Although Darwin did not mention human evolution or the likely ancestry of mankind in the *Origin of Species*, this was the subject that dominated debate after publication. The notorious confrontation between Bishop Samuel Wilberforce and Thomas Henry Huxley at the British Association for the Advancement of Science meeting in Oxford, in June 1860, made the point obvious. The question of whether we are descended from apes or angels quickly became the issue on which Darwin's and Wallace's theories were argued (Browne 2001).

> Am I satyr or man?
> Pray tell me who can,
> And settle my place in the scale.
> A man in ape's shape,
> An anthropoid ape,
> Or monkey deprived of his tail?

These cultural and social movements indicate strong contemporary interest in the man who was becoming known as the public "face" of evolution. Indeed, few other scientific theories at that time spread as far or as quickly as the theory of evolution by natural selection. Within 10 years of the publication of the *Origin of Species*, there were 16 different editions in English (including the British Isles and North America), and translations into German, French, Dutch, Italian, Russian, and Swedish, accompanied by important commentaries, criticisms, extracts, and supporting texts by other authors. There would be many more to come. To date, there have been 255 editions in English and translations into 29 other languages, including Turkish, Hindi, Ukrainian, and Yiddish, and one edition in Braille (Freeman 1977). Through these means, people all over the developed world increasingly encountered Darwin's work and were able, if they wished, to participate in what was to become one of the first truly international debates about science.

EXPERIMENTER

Where, however, is the experimental scientist in all this? Darwin was an active, and often inspired, experimentalist and observational naturalist. Although he worked at home, he turned his house and gardens into a Victorian version of a modern research station. His mind and hands never ceased working. In the years after the *Origin of Species* was published, Darwin continued at an astonishingly intensive rate. He consolidated the theory

Figure 2. Down House, Darwin's home in Kent for 40 years. (Reprinted, with permission, from the Wellcome Library, London.)

of sexual selection. He devised a theory of inheritance. He explored coadaptation between plants and insects for the purpose of fertilization, experimented on hybrid vigor in plants, documented the animal ancestry of mankind, and thought deeply about the adaptive purposes behind the evolution of the sexes. His major publications after the *Origin of Species* included an important study of variation under domestication (1868) that showed Darwin discussing heredity, variation, and transmission of what would come to be known as genetic information. He made lasting contributions to the sciences of mankind in his *Descent of Man* (1871) and the evolution of human expressions in *The Expression of Emotions in Man and Animals* (1872). His minor publications ranged very widely from seed dispersal to the transmission of ancestral characteristics such as stripes through many generations of horses (van Wyhe 2009).

This is the Darwin who said in his autobiography that he thought he was good at noticing those things that might usually escape attention, and in observing them carefully. He said of himself that he had to learn to be patient in getting results and that a good dose of ambition never did any harm. "From my early youth I have had the strongest desire to understand or explain whatever I observed—that is, to group all facts under some general laws" (Barlow 1958). He was characterizing himself as an experimentalist in an age when laboratories were hardly in existence.

In this regard, Darwin's house was without question his laboratory (Fig. 2). His study was its control center. This study was a scholarly space, but not directly comparable to an academic office. The room was completely domestic, full of an upper middle class Victorian household's furniture, with portraits of the people he most loved and respected above the fireplace, a comfortable chair or two, a red Turkish carpet, even a curtained lobby in a corner that marked off a small area that he used for washing and changing for bed (Fig. 3). There he kept his shaving things (in pre-beard days) and a chamber pot. Yet, in this study, he conducted indoor experiments on plants, worms, and seeds, worked at his microscope, dissected, planned projects, and wrote all his books. His filing shelves were kept beside the chair, his chemical implements and simple microscope close at hand. He did not use a desk. He wrote in a large leather-covered chair, with a board over his knees. Through these years, Darwin was also a family man, a loving father to 10 children, of whom seven reached full adulthood. His space was not entirely sacrosanct. Darwin's children felt able to run in and out of the study to ask for things that might be needed for their games. They sat on the wheeled stool, used by their father as his microscope seat, and punted themselves about the room with Darwin's walking stick. It was the room where Darwin spent so many years studying barnacles under the microscope that one of his sons, when a very little boy visiting another child to play, innocently asked where does *your* father do his barnacles (Darwin 1887 [1: 136]).

It was also the room from which Darwin ran his entire research and publishing ventures. Although Darwin did not actively engage in public debate, much preferring that quick-witted friends such as Thomas Henry Huxley should carry his theories into the eye of the storm, he nevertheless participated directly through correspondence. There are some 14,500 letters to and from Darwin still in existence, mostly housed in the University Library Cambridge, UK, and the American Philosophical Society in Philadelphia, that show the extent of Darwin's networks across the globe and the manner in which he used letters, on the one hand, to collect information and, on the other hand, as strategy to promote his views (Burkhardt et al. 1985). Current research into the history of Darwin's achievement, and the manner in which his views came generally to prevail, shows that he actively embedded himself in a web of correspondence that materially advanced the reception of his work.

Figure 3. Darwin's study. He used this room from 1842 to 1881. From a photograph taken after the opening of the house as a museum in 1929. (Reprinted, with permission, from the Wellcome Library, London.)

We may feel that letter writing was only to be expected in the 19th century. But it is useful to think of the world of Victorian correspondence as a major vehicle for scientific communication, in which experimental results were systematically circulated, proposals were assessed, conclusions were modified, and responses were gauged. Before the rise of the scientific periodical—the journal *Nature* was founded in 1869—it was customary to keep up to date through loosely structured correspondence networks. Of course, many scientific periodicals existed in the 19th century that served a diverse range of audiences (Cantor and Shuttleworth 2004). But recent studies suggest that research was, at that time, mostly published in books or in the transactions of learned societies. It is only since the early 20th century that scholarly journals have come to be seen as the primary place to publish. Hence, an exchange of letters was essential scientific procedure for Victorians like Darwin. At that time, too, there was not the same clear-cut dividing line between public and private, as there is now, and much "publication" took place in letters to various correspondents. One striking aspect of this pattern of 19th century communication remains today in the convention that journals such as *Nature* still publish important new results in a "letter" to the editor. Darwin keenly awaited letters every day, and he fixed a small mirror to the inside of his study window so that he could see the postman turning up the drive.

WORK AS DAILY LIFE

It is therefore possible to regard Darwin's working practices as an intensified form of daily life. He was fortunate to be privately wealthy and had no need to obtain paid employment. The research on which he based the *Origin of Species* reflected this world of personal wealth and privilege: a world of gentleman landowners who were sufficiently well financed and well educated to relish a few experiments in horticulture or land economy; a world in which plentiful correspondence was a vital link to contemporary opinion; a world of stability and prosperity where the advantages brought by rapid industrialization were subtly in evidence in the countryside (Desmond and Moore 1991; Browne 1995, 2002). Darwin's botanical and animal experiments were carried out in his own stable yard, greenhouse, or garden. He used ordinary organisms easily procured from the catalogs of nurserymen or through gentleman farmers. Any book research relied on his subscriptions to private libraries or his membership of the elite scientific societies of London. Occasionally, he yearned to acquire exotic species, and this yearning was a prime reason for Darwin's first exchange of letters with Alfred Russel Wallace in Malaysia in 1856 (Burkhardt et al. 1985 [vol. 6, letter to W.B. Tegetmeier, 26 November 1856]). Later, much of his botanical correspondence was dominated by the urge to obtain rare orchids or carnivorous plants for experimentation. Joseph Hooker at the Royal Botanic Gardens, Kew (in London) was a loyal friend, for example, and often supplied interesting materials from the government glasshouses. On one occasion, Darwin was distressed to realize that he had inadvertently destroyed a valuable Oxalis sent to him by Hooker for research into the "sleep" of plants. His was a highly domestic research environment. Investigations into the "fixity" of pigeon and poultry breeds turned into an admiration for the birds as pets. The family's dogs became the object of close observation when he began exploring animal expressions and emotions.

The point is worth emphasizing because it was partly this reliance on commonplace features of Victorian life—letters and small-scale experimental inquiries involving relatively accessible animals and plants—that generated the remarkable body of factual material on which the Origin of Species rests. Darwin took a certain pride in this ordinariness. Yet, we can retrospectively see something special. When compiling a chapter of recollections of his father for *The Life and Letters of Charles Darwin* (1887), his son Francis recorded that

> it was as though he [Darwin] were charged with theorising power ready to flow into any channel on the slightest disturbance, so that no fact, however small, could avoid releasing a stream of theory, and thus the fact became magnified into importance. In this way it naturally happened that many untenable theories occurred to him; but fortunately his richness of imagination was equalled by his power of judging and condemning the thoughts that occurred to him. He was just to his theories, and did not condemn them unheard; and so it happened that he was willing to test what would seem to most people not at all worth testing (Darwin 1887 [1: 149]).

He also favored ingenuity and frugality. When the French naturalist Alphonse de Candolle visited Darwin at home in the 1870s, he found the author of the *Origin of Species* pottering around his greenhouse working on carnivorous plants with almost no tools except his schoolboy chemical balance and some tin plant markers. All the experimental chemicals that Darwin used in order to discern the power of plant digestion came from around the house—ammonia, beer, urine, spittle, and nicotine. Experimental plants were bedded out in the kitchen garden and parts of the lawn were sectioned off for recording the number of species able to grow in a demarcated area. Such observations always ran hand in hand with hard thinking. Every day, Darwin would take a number of circuits around what he called his thinking path, the "Sandwalk," to ponder whichever question was uppermost in his mind at the time.

Darwin did not work alone. A prominent characteristic of these projects was the help he requested from friends, relatives, and even enemies. His children became assistants from an early age. Darwin usually spent the summer months researching insect pollination, for example, and might use the veil of his wife's hat to cover a particular plant. Or he could spend several hours a day closely watching ants or worms. At age five, his son Leonard ran to a garden flower and cried "I've got a fact to do"—a neat encapsulation of the way that Darwin's work was the focus of the household. The household staff was also accustomed to furthering their master's researches.

Indoors, Darwin's wife Emma often acted as his amanuensis, copying out sections of the *Origin of Species* before publication, as well as other works, and helping with his correspondence when he was unwell. Emma Darwin read the proof sheets of the *Origin of Species*, a sure sign that her religious beliefs were not holding Darwin back in any pragmatic sense. Later, when their daughter Henrietta was older, she acted as an editor for Darwin by going over his proof sheets for style. This female assistance is often ignored by historians (Harvey 2009). The proof sheets amended by Henrietta are now preserved in the Cambridge archive and we can see her comments in the margins: "this is a horrid sentence." Henrietta corrected the proof sheets of the *Descent of Man* for her father, who was so grateful that he sent her a substantial gift of money. All of the family were involved in his work in one way or another. Even the children's governess was prevailed upon to translate some difficult German biological tracts. In this manner, Darwin's researches became a family enterprise. He was one of the last gentlemen of science with sufficient private income to work at home, outside the developing academic institutions.

A TALENT TO EXPERIMENT

Francis Darwin was convinced that much of his father's talent lay in his urge to observe things for himself:

> There was one quality of mind which seemed to be of special and extreme advantage in leading him to make discoveries. It was the power of never letting exceptions pass unnoticed. Everybody notices a fact as an exception when it is striking or frequent, but he had a special instinct for arresting an exception. . . . Another quality which was shown in his experimental work, was his power of sticking to a subject; he used almost to apologise for his patience, saying that he could not bear to be beaten, as if this were rather a sign of weakness on his part. He often quoted the saying, "It's dogged as does it;" and I think doggedness expresses his frame of mind almost better than perseverance. Perseverance seems hardly to express his almost fierce desire to force the truth to reveal itself. (Darwin 1887 [1: 148,149])

The most important work Darwin did after the *Origin of Species* was undoubtedly that relating to human beings (Desmond and Moore 2009). During the *Beagle* voyage and beyond, he had made extensive notes on humankind, asking himself penetrating questions about physical and cultural anthropology, the mental and moral life of humans, metaphysics, history, and demography. During the 1860s and 1870s, he thoroughly reviewed these materials and expanded their scope dramatically. His preparatory research for the *Descent of Man* (1871) brought his concept of sexual selection to full development and required extensive documentation in the animal kingdom before he felt confident in applying it to explain the origin of human diversity (Darwin 2004). In that book, he drew on correspondence with anthropologists and travelers the world over to discuss the differences and similarities among humans. One objective was to show that the attributes of human beings could have derived from those of animals, and he corresponded widely about the origins of language, religious belief, and the moral sense in order to gather evidence that animals possessed similar traits to a lesser degree.

The following year, he brought out his book on the *Expression of the Emotions* (1872), a crucial follow-up to the *Descent of Man*. In this, he continued to argue for real links between humans and animals by documenting the mental life of human beings and suggesting that the facial musculature, and by implication the emotions underneath, could be connected with those of our presumed animal ancestors (Darwin 1988, 2009). This book required a huge international research project that called on artists and photographers as well as anthropologists. One element of research support came from the many female members of the extended Darwin family who made personal observations for him on their children. Darwin also observed his own children very carefully when they were babies, having no conceptual problem in comparing them to the baby orangutan that he saw in the London Zoo (Keynes 2001).

Time after time, he searched out appropriate people to help him. Oscar Reilander, the Swedish art photographer, lived in London and participated in Darwin's research. Reilander was a pioneer in composite photography and eager to experiment with different techniques for Darwin. He photographed himself mimicking the exaggerated emotional expressions used by actors on the London stage, several of which Darwin reproduced in the resulting book. In one composite image forwarded to Darwin, Reilander pictured himself laughing in one shot and expressing sorrow in the other, a piece of early technical wizardry to facilitate comparison of the musculature that was not pursued in detail until much later in the century (Prodger 2009).

Yet however ingenious, Darwin's experiments were relatively small scale. The point was made with a sting when Francis Darwin went in 1875 to Württemberg to work with the great plant physiologist Julius Sachs. Sachs had the most advanced laboratory in Europe, filled with expensive physiological apparatus that made Francis envious. His father was not convinced that they needed any of these instruments for their researches back at Down House, although he did allow Francis to buy a Zeiss microscope that is still in the Darwin museum. On his return from Sachs' laboratory, Francis pursued a new line of investigation into the movements of plant roots and shoots, and he showed his father how to improvise smoked paper on a rotating drum by fastening a small cylinder on an open clock face. When their joint book on the *Power of Movement in Plants* (1880) was published, Sachs savagely criticized the results by claiming that work done in a country house simply could not match the new experimental results emanating from a laboratory (de Chadarevian 1996). Darwin was irritated, to say the least. Nevertheless, some of his proposals about tropisms were proved to be correct in the 1920s and 1930s.

CONCLUSION

What can be made of all this activity? In a year of celebration and commemoration, it is apt to praise Darwin as a magnificent and perceptive theorist and as a gifted writer. We can come to see him as someone who lived and worked in an intellectually vigorous social network rather than as a solitary heroic individual. We can admire his modesty, succinctly reflected in a few words written to Huxley at the height of the debate about the *Origin of Species*: "I wish I could feel all was deserved by me."

But it is also good to remember his ability to connect observation with theory, his persistence, ambition, and good humor, all excellent qualities to foster in modern laboratory practice. He was a fine experimental scientist. Writing after a visit to Darwin's home in 1878, the British journalist Edmund Yates felt sure that he had been in the presence of a very great investigative mind:

> Without an atom of scientific jealousy, he is always ready to expound his views, to narrate the result of the delicate experiments on which he is perpetually occupied, and to assist other investigators from the stores of an experience that has ranged over the whole field of natural science, and the conclusions of a mind trained to reason closely on such facts as have been ascertained by actual observation. No naturalist of this or any other time has confined himself more strictly to well-ascertained facts, and devoted more labour to original investigation. The reason of this excessive care is to be found in the keystone of the Darwinian philosophy—*La vérité quand même*; the pursuit of truth through all difficulties, and without regard to consequences (Hodgson 1878 [p. 224]).

REFERENCES

Barlow N, Ed. 1958. *The autobiography of Charles Darwin 1809–1882* (with original omissions restored. Edited with appendix and notes by his granddaughter Nora Barlow). Collins, London.

Browne EJ. 1995. *Charles Darwin: Voyaging*, Vol 1 of a biography. Knopf, New York.

Browne EJ. 2001. Darwin in caricature: A study in the popularisation and dissemination of evolution. *Proc Am Philos Soc* **145:** 496–509.

Browne EJ. 2002. *Charles Darwin: The power of place*, Vol 2 of a biography. Knopf, New York.

Browne EJ. 2003. Charles Darwin as a celebrity. *Sci Context* **16:** 175–194.

Browne EJ. 2006. *Charles Darwin's Origin of Species: A biography*. Atlantic, London.

Burkhardt FH, Smith S, et al., Eds. 1985–. *The correspondence of Charles Darwin* (16 volumes). Cambridge University Press, Cambridge.

Cantor GN, Shuttleworth S, Eds. 2004. *Science serialized: Representation of the sciences in nineteenth-century periodicals*. MIT Press, Cambridge, Massachusetts.

Clark CA. 2009. "You are here": Missing links, chains of being, and the language of cartoons. *Isis* **100:** 571–589.

Darwin CR. 1859. *On the origin of species by means of natural selection, or the preservation of favoured races in the struggle for life*. Murray, London.

Darwin F, Ed. 1887. *The life and letters of Charles Darwin, including an autobiographical chapter*, 3 volumes. Murray, London.

Darwin CR. 1958. *On the origin of species: A facsimile of the first edition* (with an Introduction by Ernst Mayr). Harvard University Press, Cambridge, Massachusetts.

Darwin CR. 1988. *The expression of the emotions in man and animals* (3rd edition, with Introduction, Afterwords, and Commentaries by Paul Ekman). Harper Collins, New York.

Darwin CR. 1989. *Charles Darwin's journal of researches (1839)* (ed. with an Introduction by J Browne and M Neve). Penguin, London.

Darwin CR. 2004. *The descent of man, and selection in relation to sex* (edited and with an Introduction by James Moore and Adrian Desmond). Penguin, London.

Darwin CR. 2009. *Expression of emotions in man and animals* (ed. J Cain and S Messenger). Penguin, London.

de Chadarevian S. 1996. Laboratory science versus country-house experiments. The controversy between Julius Sachs and Charles Darwin. *Br J Hist Sci* **29:** 17–41.

Desmond AJ, Moore JR. 1991. *Darwin*. Michael Joseph, London.

Desmond AJ, Moore JR. 2009. *Darwin's sacred cause*. Allen Lane, London.

Freeman RB. 1977. *The works of Charles Darwin: An annotated bibliographical handlist*, 2nd ed. Dawson, Folkestone, United Kingdom.

Harvey J. 2009. Darwin's "angels": The women correspondents of Charles Darwin. *Intellect Hist Rev* **19:** 197–210.

Keynes R. 2001. *Annie's box: Charles Darwin, his daughter and human evolution*. Fourth Estate, London.

Prodger P. 2009. *Darwin's camera: Art and photography in the theory of evolution*. Oxford University Press, New York.

van Wyhe J. 2009. *Charles Darwin's shorter publications, 1829–1883*. Cambridge University Press, Cambridge.

Yates EH. 1878. Mr. Darwin at Down. Celebrities at home. *The World*, 223–230. www.darwin-online.org.uk.

On the Future of Biology

E.O. WILSON
Museum of Comparative Zoology, Harvard University, Cambridge, Massachusetts 02138
Correspondence: ewilson@oeb.harvard.edu

Biology is driven by two strategies of research. The first, using functional studies at the molecule-to-organism transitions, utilizes the principle that for every problem in biology, there is an organism ideal for its solution. The second, comparative and evolutionary in orientation, consists of thorough studies of every biological aspect of a given taxonomic group of species and follows the principle inverse to the first that for every organism, there is a problem to the solution for which it is ideally suited. The two approaches are complementary and of increased connectivity to each other. In passing from the traditional, intensely focused reductionist approach of 20th century biology to a more synthetic and authentically holistic phase, new emphasis will be placed on the major transitions between levels of biological organizations and onto processes of multilevel evolution by which they have been achieved.

The science of biology is coming together. Enough of its disciplines have been linked by cause-and-effect explanation to suggest how some of the framework of a unified biology can be visualized. At its foundation, biological knowledge conforms to two principles arguably firm enough to be called scientific laws, i.e., they are all-inclusive and with no proven exceptions, and their consequences can be followed wide and deep. The first law is that all of the entities and processes defining life are ultimately obedient to the laws of physics and chemistry. The second law, still in contention due to claims for counterexamples, is that all of the diagnostic entities and processes of life, and all of its diversity, have evolved by natural selection.

The conformity of life's processes to the physical sciences undergirds molecular and cellular biology. Its universal application has been challenged by the arguments for a supernatural intelligent design. Its champions may believe that scientists are in a conspiracy against intelligent design, but the opposite is true. Every researcher would like to discover an extrasomatic force: This would be one of the greatest achievements in the history of science. The problem is that not a trace of evidence has ever been found, nor has the need for such a theoretical place marker arisen from any database. The confidence of biological researchers in the complete physicality of life remains adamantine.

The same can be said for the exclusive role of natural selection in genetic evolution. The ruins of alternative theories, from neo-Lamarkian to punctuated equilibrium, litter the archives of past decades. The mysterious forces that they imply, undenoted and extraneous to the empirical evidence of genetics, do not aid evolutionary theory. Modern evolutionary theory, rooted in molecular genetics, begins with the distinction between the units of heredity and the targets of natural selection. The units are the genes, respectively, the base-pair sequences of their codes, their degree of duplication in the genome by tandem repeat sequencing, their location in the genome, and the interaction of their products during epigenesis. The targets of natural selection are the heritable phenotypes prescribed by the genes, including the plasticity of response in the final products composing the phenotypes.

Natural selection is multilevel: A phenotypic target can exist at any level of biological organization, from macromolecules to chromosomes to eukaryotic cells to multicellular organisms, and onto organized social groups, populations of organisms and groups, and finally, arguably, entire ecosystems.

Other agents have been found that can drive evolution for very short distances, but they are neither creative in shaping adaptation to the environment nor significant in magnitude to compare with natural selection. Different rates of mutation between opposing alleles can shift allelic frequencies, but the effect is easily overwhelmed by even barely detectable degrees of selection. Similarly, genetic drift, when permitted by low selection pressures and small population size, can shift frequencies, but it tends to lose alleles, narrowing the potential for future evolution.

Multiple selection is subject to reverberation in its impact. When selection favors one form of heritable trait at any given level of biological organization, it favors or disfavors traits prescribed by the same allele (or ensemble of alleles) at other levels. A familiar example is the inherited predisposition to certain kinds of cancer: An allele predisposing the unregulated growth of cells is favored at the level of the cell carrying it but disfavored by the cancer caused at the level of the organism. Organisms with cancer are hosts to multitudinous healthy offspring of the original mutant cell, yet they lose in competition with healthy organisms lacking the mutant allele.

Multilevel selection typically places selection at adjacent levels in opposition. To achieve a new level, for example, a multicellular organism or a complex society, requires some degree of altruism on the part of individuals at the lower level. It is promoted by group selection and opposed by individual-level selection. Such major transitions across levels of organization have very rarely occurred in evolution and taken long periods of geological time to be achieved even once. The origin of the eukaryotic organisms took more than half the known history of life to be attained, from the beginning more than 3.5 billion to 1.5 billion years ago. Multicellular animals were not added for more than another half-billion years. The origin of insect superorganisms, in particular the tightly knit colonies of ants and termites or their equivalent, was not attained during the first great insect adaptive radiation in the late Paleozoic Era, ~350–250 years before the present, and occurred only well into the late Jurassic to very early Cretaceous periods 200–140 million years ago.

It is in the major transitions of evolution that biology will more predominantly dwell. The Age of Reduction in biology has largely passed. Although enormous amounts of new information will continue to be yielded by the cleavage and analysis of systems, the big problems in each discipline are those that require an exactitude of synthesis or, more accurately, resynthesis of reduced systems. As more biological systems are understood, and with them the major transitions, common principles of emergent evolution should become apparent. There appears to be no a priori way to create such a unified theory of biology at the present time. It awaits far more empirical information than we now possess, requiring immense amounts of hard work with real organisms.

Modern biology has long been addressed by two very different strategies of research. The first strategy follows the dictum that *for every problem in biology, there exists an organism ideally suited to solve it.* The bacterium *Escherichia coli* has thus triumphed for molecular genetics, the nematode *Caenorhabditis elegans* for the neuronal basis of behavior, the honeybee for the instinct and self-organization of animal societies, and human beings for the conscious mind.

If those faithful to this dictum are called the problem solvers, the second tribe may be called the naturalists. The research strategy of the naturalists is the inverse of the strategy of the problem solvers. It is that *for every organism, there exists a problem for which the organism is ideally suited.* The procedure of the naturalist is to adopt a group of species, such as conifers, diatoms, and orb-weaving spiders, and learn as much as possible about the group across all levels of biological organization.

The overarching goal of the scientific naturalist is the complete mapping of biodiversity while acquiring knowledge of it at each level of biological organization. The daunting nature of this task is made clear when an effort is made to estimate even roughly the total number of species on Earth. When the hyperdiverse microorganisms, fungi, and small invertebrates are put in the roster, it turns out that we have discovered and named probably 10% or fewer of the species living on Earth. The variety of genes prescribing all of them is astronomical and, of course, even less known.

The third great goal, yet even more daunting, is the history of all life. Because of the paucity of information on the living species and their genomes, the phylogenies of only a few of the best-known groups, such as birds and flowering plants, can be drawn with any confidence. And to a record of all that has happened in past ages, the infant discipline of paleobiology faces a journey likely to take centuries, if not millennia, and is likely never to end.

Problem solvers, at work mostly in the laboratory, explain the proximate causes of biological phenomena, usually those visible only at a microscopic level—in other words, *how* the system works. Naturalists, working typically in the field but also experimentally in the laboratory, stress the adaptation of biological phenomena to the environment, the *why* in its attempted full explanation.

Although the two approaches have seemed at times to represent independent cultures, they are in fact complementary and can with more information be fitted nicely together. The boundaries are in fact being erased. Scientific naturalists, as organismic and evolutionary biologists, use the methods of molecular and cellular biology, whereas molecular and cellular biologists grow more prone to address patterns of diversity and evolution. With increasing frequency, the two tribes collaborate in research projects. At the risk of oversimplification, it can be said that the naturalists discover the problems in nature that the problem solvers solve.

The trajectory of a unified biology can be visualized to be T shaped. The horizontal arm is biodiversity at the level of species and genetic strains within each species. One vertical arm is the model species of the problem solver, reaching from the populations that constitute the species or strains down through the levels of biological organization, from the ecosystems of which it is a part to the genes that prescribe its diagnostic traits. At present, although fewer than 100 such model species are under such scrutiny, in a decade or two, there will be thousands and then tens of thousands. The process will accelerate as DNA sequencing grows ever faster and cheaper. Existing knowledge will be immediately available and its growth followed in real time, as the Encyclopedia of Life gears up to provide complete information through a single portal for all species known.

The more the two-dimensional array of knowledge is filled in, the more the gaps in knowledge and the most promising new directions of research will become apparent. At the same time, paleobiology and phylogenetic analyses will add the dimension of time. A unified biology is a goal worth thinking about, even if still far from within our grasp. Biologists are stretching causal explanations across wider segments of the levels of biological organization and in broader arrays of species. In so doing, they are melding explanations of function and adaptation, as well as molecules and cells with organisms and species. As a result, the emphasis in research is shifting from one or two levels of biological organization to all of the transitions between levels.

The major questions of biological theory are those of systems biology. They take the following form: How do the elements and processes that define one level emerge from the level below it? And why, and under what circumstance? Then, were the evolutionary routes the only ones, and if not, were they at least the optimal ones?

A unified biology will be the one that maps the pathways from molecules to ecosystems in unbroken transits of causal explanation. It will disclose the still unimagined commonalities, if such exist, among the evolutionary transitions. It will also provide insight into whatever different genetic codes and rules of transitions are possible, and even which of these might exist in other worlds.

Evolution of Biological Catalysis: Ribozyme to RNP Enzyme

T.R. CECH

*Howard Hughes Medical Institute, Department of Chemistry and Biochemistry,
University of Colorado, Boulder, Colorado 80309-0215*

Correspondence: thomas.cech@colorado.edu

The enzymes that perform biological catalysis in contemporary organisms are usually proteins, occasionally ribonucleoprotein (RNP) complexes, and in rare instances pure RNA (ribozymes). Because RNA can serve as both an informational molecule and a biocatalyst, it has been attractive to consider a primordial RNA World in which RNA enzymes catalyzed the replication of RNA genomes and an array of other metabolic steps, before the advent of protein enzymes and DNA genomes. By what pathways, then, did the RNA World evolve to the present state? Here, I describe plausible pathways for the evolution of biological catalysis, with special emphasis on the origin of RNP enzymes. Recent findings support the argument that RNP enzymes are not undergoing extinction, but instead, they are continuing to evolve and to elaborate new functions.

The organizers of this Cold Spring Harbor Symposium made the bold and exciting decision to cover the topic of evolution from a broad perspective, ranging from prebiotic chemistry through the evolution of molecules, cells, organisms, cognitive systems, ecosystems, and societies. To this end, they convened a broad array of scientists who normally attend more specialized conferences and therefore had little previous opportunity to interact. Now, faced with the prospect of contributing a chapter to the Symposium volume, I decided to take a cue from the organizers and write something with minimal technical jargon that aims to be broadly accessible to a wide range of scientists. This approach carries the danger of superficiality, so to counteract this limitation, I have included some key citations to the technical literature. In addition, entire volumes have been written on my topic of the evolution of RNA and RNPs, and the reader is encouraged to explore *The RNA World* (Gesteland et al. 2006) for detailed descriptions of the topics covered here.

My goal, then, is to consider plausible pathways for the origins of the biocatalysts present in all living beings on Earth—the enzymes that allow cells and organisms to grow, move, take in nutrients, perform metabolism, respond to their environment, and reproduce. I begin by reviewing what we know about contemporary protein enzymes and RNP enzymes, which provides some guidance regarding how they might have evolved. I then summarize some missing information, the lack of which limits our confidence about early events in the evolution of catalysis. Notwithstanding these limitations, I then propose reasonable scenarios for the evolution of an "RNP World," in which complexes of ribonucleic acid and protein catalyzed the biochemical reactions necessary for life. Finally, I speculate about the future of RNP enzymes.

WHAT WE KNOW ABOUT BIOLOGICAL CATALYSTS

The most detailed knowledge we have concerning biological catalysis of course comes from contemporary organisms. Furthermore, even though this current knowledge is incomplete, we can continue to explore and to experiment to fill gaps in our understanding of today's biocatalysts, approaches that are much more difficult to apply to prehistoric events and especially inaccessible for events surrounding the origins of life some four billion years ago. What we know is that in the bacterium *Escherichia coli*, catalysis is performed by ~2000 different protein enzymes and just a few known RNP enzymes. Baker's yeast has more than a dozen known RNP enzymes, yet protein enzymes still outnumber RNP enzymes by perhaps 100 to 1.

The numerical inferiority of RNP enzymes, however, belies their importance. For example, consider the four major, essential activities of genetic material (DNA and RNA):

- Replication (reproducing genomic DNA)
- Transcription (copying DNA information into RNA)
- RNA splicing (connecting informational stretches of RNA and deleting interrupting segments) and other posttranscriptional modifications
- Translation (using RNA information to synthesize proteins)

Of these four activities, the first two are performed by protein enzymes (DNA polymerases and RNA polymerases), whereas the latter two rely on RNP enzymes (the spliceosome and the ribosome). Thus, RNP enzymes are not peripheral but are central to contemporary biology.

This mention of the spliceosome and the ribosome provides a good opportunity to define "RNP enzyme" as a catalyst that contains both essential RNA and protein components. Whether a component is essential for function can be determined by a genetic knockout experiment. For example, in the case of the spliceosome, five small nuclear RNAs (U1, U2, U4, U5, and U6) are essential for splicing (Guthrie 1991), as are numerous protein subunits including U2AF (U2 RNA-associated factor) (Ruskin et al. 1988) and Prp8 (Pena et al. 2008). In the case of the ribosome, catalysis of peptide bond formation requires the large-subunit rRNA, and several large-subunit ribosomal proteins (notably L2, L15 and L16) are also essential, although none of these is itself a peptidyl transferase (Noller 1993). Note that this definition of the RNP enzyme does not distinguish which macromolecule contains the active site that positions the reacting groups and accelerates the chemical transformation. In the case of the ribosome, the active site for peptidyl transfer is thought to be composed entirely or at least mostly of RNA (Noller 1993; Nissen et al. 2000; Voorhees et al. 2009). The spliceosome is also thought to have an RNA active site (Toor et al. 2008; Valadkhan et al. 2009). The RNP enzyme telomerase, which catalyzes the extension of DNA at the ends of chromosomes, has an active site for the chemical step in the reaction composed of three aspartic acid residues in its TERT (telomerase reverse transcriptase) protein subunit (Lingner et al. 1997), yet catalysis of the overall reaction cycle requires the intimate collaboration between protein and RNA (Greider and Blackburn 1989; Tzfati et al. 2000; Miller and Collins 2002; Qiao and Cech 2008). Thus, although RNP enzymes require both their RNA and their protein constituents, there is a diverse array of ways in which the two different types of macromolecule collaborate to achieve catalysis.

QUESTIONS REGARDING EARLY EVENTS IN THE EVOLUTION OF BIOLOGICAL CATALYSIS

What was the environment? Any hypothesis about the origin of the first self-replicating system requires assumptions about the environmental conditions: What gases, dissolved molecules, and ions were present, what the temperature, pH, and oxidation-reduction potential might have been, and what sort of radiation flux (visible, UV, and ionizing radiation) was present. More specifically, if RNA is to be considered plausible as a prebiotic self-replicator, the environment must have avoided high pH, high temperature, and high concentrations of certain divalent cations that hydrolyze RNA. Furthermore, oxidative conditions can destroy the bases in RNA, low pH can promote depurination, and UV irradiation results in pyrimidine dimer formation. Unfortunately, there is little certainty about the environmental parameters. For example, the original Miller and Urey (1959) prebiotic simulation experiments started with methane, ammonia, water, and hydrogen and resulted in production of high concentrations of natural amino acids. However, these highly reducing conditions were later considered unrealistic. Recent experiments with nonreducing mixtures including nitrogen and carbon dioxide have shown production of amino acids, especially when oxidation is inhibited (Cleaves et al. 2008).

What were the first self-replicating molecules? The criteria for a primordial self-replicator include having monomer units that could reasonably be assembled under prebiotic conditions (Powner et al. 2009), ability to undergo both random and template-dependent polymerization (Inoue and Orgel 1983), and ability to be replicated ($A \rightarrow A + A$) with enough fidelity to maintain information content (Eigen 1971). Another key criterion is that there must be a plausible evolutionary pathway from the initial self-replicators to contemporary nucleic acids; i.e., even if one can show that mineral surfaces can self-replicate, it is not clear how they would be replaced by nucleic acids, whereas the principle of complementary base pairing makes it much easier to conceive of one nucleic acid system evolving into another (RNA-RNA $\rightarrow$ RNA-DNA $\rightarrow$ DNA-DNA). Thus, the concept of a primordial RNA World, with RNA providing the information and catalyzing its own replication, is attractive. Although it has been argued that RNA is too sophisticated to be a plausible candidate for the first self-replicating informational molecule (Joyce and Orgel 2006), the recent breakthrough in recapitulating pyrimidine nucleotide synthesis by Powner et al. (2009) has generated optimism that RNA might in fact have arisen spontaneously. Alternatively, an RNA World may have been preceded by a simpler self-replicating molecule. Some non-RNA nucleic acids that have been considered as plausible prebiotic predecessors include pyranosyl-RNA (Pitsch et al. 1995), peptide nucleic acid (Egholm et al. 1993), and phosphoramidate-linked glycerol nucleic acid (Chen et al. 2009).

How was chirality established? Any organic molecule containing a carbon atom bonded to four nonidentical chemical groups has a mirror-image form that is distinct from the original molecule, just as one's right hand is distinct from one's left hand. Two mirror-image molecules are called enantiomers. If a molecule contains more than one chiral center, there will be multiple stereoisomers called diastereomers. Undirected chemical reactions produce mixtures of these enantiomers or diastereomers, whereas biological reactions are stereochemically pure and need to be steriochemically pure in order to be useful. The requirement for chiral purity arises because if X needs to interact with Y to form X-Y, the enantiomer of X (X') is unlikely to have any specific interaction with Y because its interacting atoms will be in the wrong special arrangement. Even worse, in a polymeric nucleic acid, the presence of even an occasional monomer unit of the opposite chirality would greatly perturb its structure and function. The discovery of some organic chemical reactions that give rise (randomly) to a single chiral product when subjected to phase transitions may provide clues about the origin of chirality in biochemical molecules (Viedma et al. 2008; Blackmond 2009).

How early were self-replicators encapsulated? It is frequently pointed out that it is difficult to launch natural selection in a complex mixture of freely diffusing molecules; only after self-replicating systems are isolated from their neighbors will their own success accrue to them and give them a selective advantage for survival and reproduction. Modern cells achieve this separation by means of a cell membrane consisting in part of a phospholipid bilayer and in some cases by a cell wall. Primordial self-replicating systems could have been encapsulated as "protocells," surrounded by fatty acid membranes (Mansy et al. 2008). Other means of separation, such as being sequestered within a porous matrix or being suspended as aerosol particles (Dobson et al. 2000), have been considered.

PLAUSIBLE SCENARIOS FOR THE EVOLUTION OF AN RNP WORLD

Despite our substantial ignorance regarding the key issues described above, we can still describe plausible pathways for the evolution of an RNP World because we can rely on more recent evolutionary events to provide paradigms. First, it seems reasonable that primitive ribozymes would have encountered and formed complexes with existing small molecules, peptides, and (later) proteins and that natural selection would then occur at the level of the RNP rather than the RNA (Noller 2004; Cech 2009). Second, the ribosome is so similar in all organisms that it must have arisen before the last common ancestor of extant life, and once there was ribosome-catalyzed protein synthesis, there would have been selection for the synthesis of peptides and polypeptides that enhanced RNA function. Finally and most recently, protein enzymes continue to encounter an enormous variety of RNA molecules, and if they can form a complex, once again selection would occur at the level of the RNP. These steps are summarized in Figure 1.

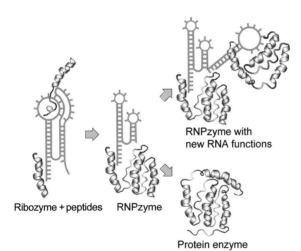

Figure 1. Model for the evolution of biological catalysis (Cech 2009). According to the RNA World hypothesis, early life-forms used RNA as both an informational molecule and a biocatalyst (ribozyme). However, given the likelihood that small molecules including peptides would be present in any environment in which ribonucleotides would be synthesized, the early RNA World may in fact have been a primitive RNP World (*left*). As described in the text, evolution led to a diverse array of RNP enzymes, some with their catalytic center still in the RNA and others with catalysis being performed by protein (*center*). RNP enzymes are seen as central to the evolution of biocatalysts; this is especially apparent because the ribosome, itself an RNP enzyme, synthesizes all protein enzymes in modern cells (*lower right*). However, RNA does not appear to be disappearing from biological catalysis, because there are numerous examples of old RNP enzymes gaining new RNA functions and new RNP enzymes arising (*upper right*).

The Early RNP World

Laboratory "test tube evolution" experiments are frequently performed with pure RNA, but it is difficult to conceive of a natural system in which RNA monomers and polymers would be synthesized without there also being an enormous collection of other organic molecules, including sugars, amino acids, and short peptides consisting of D and L amino acids (see discussion of chirality above) in branched chains as well as linear arrays. Many of these small molecules bind to RNA; natural examples include aminoglycoside antibiotics binding to rRNA, small metabolites binding to riboswitches, guanosine and arginine binding to group I introns, and arginine and short arginine-rich peptides binding to the human immunodeficiency virus *trans*-activating response region (HIV TAR) RNA (Yarus 1988; Puglisi et al. 1992).

Some protein enzymes bind cofactors (nicotinamide adenine dinucleotide, coenzyme A, pyridoxyl phosphate) and use them in catalysis. Many of these coenzymes are modified ribonucleotides, perhaps molecular fossils of the early RNA World (White 1976). Just as modern proteins expand their catalytic repertoire using these coenzymes, it seems obvious that RNA could have expanded its catalytic repertoire by using bound organic molecules. Indeed, because binding of small molecules by RNA seems chemically inevitable (see previous paragraph), natural selection must have occurred at the level of RNA–small-molecule complexes. If a prevalent small molecule inhibited an RNA molecule's function, it would have provided negative selection for that particular RNA, whereas any RNA that bound a small molecule that stabilized its active conformation or extended its functional repertoire would have undergone positive selection.

The Middle RNP World

Evolution of RNP enzymes was constrained when RNA had to make use of whatever amino acids and peptides were present in the environment. Thus, a key event occurred when RNA stumbled upon the ability to synthesize reproducibly those peptides that were useful to its functions. This step is sometimes envisioned as the advent of a very primitive version of today's ribosome: peptide bond formation being directed by a code provided by a separate mRNA. Although the invention of a primitive ribosome was clearly a breakthrough event, the earliest production of specific peptides may have involved RNAs that always bound and joined the same two activated amino acids. Indeed, the D-Ala-D-Ala ligase that contributes to synthesis of bacterial cell wall peptidoglycans is an example of non-mRNA-directed peptide synthesis. It provides a protein analogy to early RNA enzymes that may have catalyzed peptide ligation.

The Recent RNP World

RNP enzymes have been created more recently, since the advent of the common ancestor to extant life. Once created, they continue to evolve. I describe two examples of RNP evolution, one in which a ribozyme acquired a protein and another in which a protein enzyme acquired an RNA. Both stories are a bit like historical novels: The details are fictional but plausible, because they are embedded in a context of fact. In each case, the most parsimonious explanation for an observation is given preference over explanations that require multiple unsubstantiated steps.

The first story concerns the evolution of an RNP enzyme from an intron RNA that underwent self-splicing in the absence of protein. The ribozyme is considered the ancestral form, because all of the group I ribozymes share sequence, structure, and mechanistic features, whereas their associated proteins are diverse and largely unrelated. Consider a moderately active group I intron that was able to bind weakly to a preexisting protein (in this case, an aminoacyl-tRNA synthetase). Amino acid insertions in the synthetase protein that did not interfere with its synthetase activity but that enhanced its ability to interact productively with the RNA would be subject to positive selection. "Interact productively" could include increasing the rate or accuracy of splicing, allowing regulation of activity, or permitting activity under a wider variety of temperatures or other conditions. As protein took over part of the RNA-stabilizing function, portions of the RNA structure would become superfluous and would accumulate mutations and deletions (Paukstelis et al. 2008). The present-day intron RNA-CYT-18 protein RNP enzyme has been captured in a crystal structure (Fig. 2). This is presumably just today's snapshot of an ongoing evolutionary process.

The second story concerns the evolution of the telomerase RNP. I begin by noting that the TERT protein subunit is conserved among eukaryotes and that its reverse transcriptase domain is related to those of retroviral and retrotransposon reverse transcriptases (Nakamura and Cech 1998). Thus, the ancestral telomerase was presumably a reverse transcriptase protein that copied an RNA template that was only loosely bound. Insertions and mutations in both the protein and the RNA that enhanced their ability to form a stable productive complex would be advantageous. Thus, the reverse transcriptase acquired the RNA-binding domain seen in contemporary TERTs, and the RNA acquired TERT-binding sequences. Once the enzyme was acting as a stable RNP, any further additions to the RNA that allowed recruitment of accessory proteins were subject to positive selection (Zappulla and Cech 2004). Futhermore, any changes in the RNA that supplemented the protein's active site for DNA synthesis would also be subject to selection (Qiao and Cech 2008) and might even permit degeneration of portions of the protein. It is clear that telomerase evolution is a multibranched pathway, because different telomerases have very different-sized RNA subunits and very different protein to RNA ratios.

WHAT DOES THE FUTURE HOLD FOR RNPS?

Some would argue that RNPs are on the way out as biological catalysts and that protein enzymes will assume their functions. Indeed, there are some findings that can be interpreted as supporting such a proposition, although these putative "missing links" are few.

One case is the human mitochondrial ribosome, which has a ratio of rRNA to protein much less than other ribosomes (O'Brien 2002; Sharma et al. 2003, 2009). This has led some to wonder whether the rRNAs could disappear entirely and be replaced by protein. However, before we give too much weight to this fascinating observation, it must be pointed out that mitochondria are highly specialized and are under unusual selective pressures. For example, pressure to minimize the size of the mitochondrial genome, coupled with the difficulty of importing large RNAs through the mitochondrial membrane and the relative ease of importing proteins from the cytoplasm, would provide a strong driving force for a mitochondrial ribosome with a low RNA to protein ratio. Because mitochondrial ribosomes translate only a very small number of mRNAs, a simpler RNP machine may be quite adequate. In addition, before citing the mitochondrial ribosome as an example of "disappearing RNA," we must consider whether it could represent an ancient state rather than recent evolution. If the protein-rich mitochondrial ribosome were a fossil of a more ancient ribosome, frozen in time because of its limited functional requirements, it would represent a missing link in the acquisition of more RNA elements in the modern ribosome—exactly the opposite of the "disappearing RNA" hypothesis!

A dramatic example of disappearing RNA is the human mitochondrial RNase P. RNase P is an RNA-processing enzyme that makes a specific cut in the precursors to tRNAs, and bacterial RNase P enzymes consist of a catalytic RNA subunit supported by a small basic protein

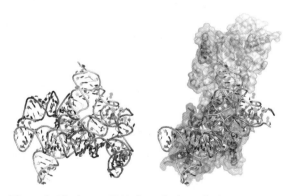

Figure 2. The intron RNA from the bacteriophage Twort can either self-splice or undergo splicing as a complex with the CYT-18 protein, suggesting one possible pathway for the evolution of RNP catalysis. (*Left*) Crystal structure of the Twort intron RNA (Golden et al. 2005). (Yellow) P4-P6 domain (i.e., RNA base-paired segment 4 through paired segment 6), (pink) P3-P8-P9 domain, (steel gray) P1-P2 domain, which contains the splice site that is cleaved during the reaction (*top left*), and the P7.1-P7.2-P9.1 domain, which buttresses the folded RNA structure (*bottom right*). (*Right*) Crystal structure of an active Twort intron–CYT18 protein complex (Paukstelis et al. 2008). Shown is the dimeric amino-terminal domain of the CYT-18 protein, with each monomer shown in a different shade of blue. (Green) Amino acid insertions relative to the conserved aminoacyl-tRNA synthetase domain. The orientation of the intron is the same in both panels, but the steel gray domains have been deleted from the right panel to facilitate visualization of the RNA–protein interaction.

that aids in RNA conformational stability. RNase P enzymes of Archaea and Eukarya are also RNPs, with the RNA subunits clearly related to those of Bacteria and therefore derived from a common ancestor (Evans et al. 2006). However, the human mitochondrial RNase P is composed entirely of three protein subunits, each unrelated to any known component of RNase P in other organisms (Holzmann et al. 2008). As in the case of the protein-rich mitochondrial ribosome described above, the unusual selective pressure experienced by the human mitochondrion may be driving evolutionary events that would be unlikely to be successful at the whole-cell level.

Balanced against these rare examples that could perhaps support the view of the demise of RNPs, abundant examples support the view that in the modern post-DNA world, many new catalytic RNPs have arisen. These are of two major types. The microRNAs are involved in translational repression, acting by binding to sequences in the 3′-untranslated regions of mRNAs and in some cases by directing cleavage of the mRNA at their binding site. The active entity is an RNP consisting of the small RNA bound to a protein of the Ago (Argonaute) class. The Ago protein has in fact been shown to have an active site for mRNA cleavage (Song et al. 2004), so it is a classical RNP enzyme.

The second major role for catalytic RNPs is seen in mammals, where long noncoding RNAs bind proteins including histone-modifying enzymes that promote transcriptional silencing. The most famous example is the Xist RNA that coats only one of the two X chromosomes in female mammals, binding proteins that promote chromosome condensation and transcriptional inactivation. The end result is gene dosage compensation—the same number of active X-chromosome genes in females as in males. A rather similar situation occurs much more locally along chromosomes. For example, a long noncoding RNA is transcribed upstream of and in the same direction as the mRNA from the human cyclin D1 promoter. It binds a protein TLS (translocated in liposarcoma), which in turn undergoes a conformational change, binds a histone acetyltransferase, and prevents it from activating the chromatin structure around the promoter of the cyclin D1 gene (Wang et al. 2008).

CONCLUSIONS

The RNP World is thus alive and well, and it is continuing to expand as evidenced by the new classes of RNPs seen only in Eukarya. If one could come back in a billion years and examine the state of biocatalysis, it seems likely that it would still be shared by protein enzymes and RNP enzymes.

ACKNOWLEDGMENTS

I thank Barb Golden (Purdue University) for preparation of Figure 2 and Jack Szostak (Massachusetts General Hospital) and Jamie Williamson (Scripps Institute) for comments on the manuscript.

REFERENCES

Blackmond DG. 2009. An examination of the role of autocatalytic cycles in the chemistry of proposed primordial reactions. *Angew Chem Int Ed Engl* **48:** 386–390.

Cech TR. 2009. Crawling out of the RNA world. *Cell* **136:** 599–602.

Chen JJ, Cai X, Szostak JW. 2009. N2′→p3′ phosphoramidate glycerol nucleic acid as a potential alternative genetic system. *J Am Chem Soc* **131:** 2119–2121.

Cleaves HJ, Chambers JH, Lazcarno A, Miller SL, Bada JL. 2008. A reassessment of prebiotic organic synthesis in neutral planetary atmospheres. *Orig Life Evol Biosph* **38:** 105–115.

Dobson CM, Ellison GB, Tuck AF, Vaida V. 2000. Atmospheric aerosols as prebiotic chemical reactors. *Proc Natl Acad Sci* **97:** 11864–11868.

Egholm M, Buchardt O, Christensen L, Behrens C, Freier SM, Driver DA, Berg RH, Kim SK, Norden B, Nielsen PE. 1993. PNA hybridizes to complementary oligonucleotides obeying the Watson-Crick hydrogen-bonding rules. *Nature* **365:** 566–568.

Eigen M. 1971. Self organization of matter and the evolution of biological macromolecules. *Naturwissenschaften* **58:** 465–523.

Evans D, Marquez SM, Pace NR. 2006. RNase P: Interface of the RNA and protein worlds. *Trends Biochem Sci* **31:** 333–341.

Gesteland RF, Cech TR, Atkins JF, Eds. 2006. *The RNA world*, 3rd ed. Cold Spring Harbor Laboratory Press, Cold Spring Harbor, NY.

Golden BL, Kim H, Chase E. 2005. Crystal structure of a phage Twort group I ribozyme-product complex. *Nat Struct Mol Biol* **12:** 82–89.

Greider CW, Blackburn EH. 1989. A telomeric sequence in the RNA of *Tetrahymena* telomerase required for telomere repeat synthesis. *Nature* **337:** 331–337.

Guthrie C. 1991. Messenger RNA splicing in yeast: Clues to why the spliceosome is a ribonucleoprotein. *Science* **253:** 157–163.

Holzmann J, Frank P, Löffler E, Bennett KL, Gerner C, Rossmanith W. 2008. RNase P without RNA: Identification and functional reconstitution of the human mitochondrial tRNA processing enzyme. *Cell* **135:** 462–474.

Inoue T, Orgel LE. 1983. A nonenzymatic RNA polymerase model. *Science* **219:** 859–862.

Joyce GF, Orgel LE. 2006. Progress toward understanding the origin of the RNA world. In *The RNA world* (ed. RF Gesteland et al.), pp. 23–57. Cold Spring Harbor Laboratory Press, Cold Spring Harbor, NY.

Lingner J, Hughes TR, Shevchenko A, Mann M, Lundblad V, Cech TR. 1997. Reverse transcriptase motifs in the catalytic subunit of telomerase. *Science* **276:** 561–567.

Mansy SS, Schrum JP, Krishnamurthy M, Tobé S, Treco DA, Szostak JW. 2008. Template-directed synthesis of a genetic polymer in a model protocell. *Nature* **454:** 122–125.

Miller MC, Collins K. 2002. Telomerase recognizes its template by using an adjacent RNA motif. *Proc Natl Acad Sci* **99:** 6585–6589.

Miller SL, Urey HC. 1959. Organic compound synthesis on the primitive earth. *Science* **130:** 245–251.

Nakamura TM, Cech TR. 1998. Reversing time: Origin of telomerase. *Cell* **92:** 587–590.

Nissen P, Hansen J, Ban N, Moore PB, Steitz TA. 2000. The structural basis of ribosome activity in peptide bond synthesis. *Science* **289:** 920–930.

Noller HF. 1993. Peptidyl transferase: Protein, ribonucleoprotein, or RNA? *J Bacteriol* **175:** 5297–5300.

Noller HF. 2004. The driving force for molecular evolution of translation. *RNA* **10:** 1833–1837.

O'Brien TW. 2002. Evolution of a protein-rich mitochondrial ribosome: Implications for human genetic disease. *Gene* **286:** 73–79.

Paukstelis PJ, Chen JH, Chase E, Lambowitz AM, Golden BL. 2008. Structure of a tyrosyl-tRNA synthetase splicing factor bound to a group I intron RNA. *Nature* **451:** 94–97.

Pena V, Rozov A, Fabrizio P, Lührmann R, Wahl MC. 2008. Structure and function of an RNase H domain at the heart of

the spliceosome. *EMBO J* **27:** 2929–2940.
Pitsch S, Krishnamurthy R, Bolli M, Wendeborn S, Holzner A, Minton M, Lesueur C, Schlonvogt I, Jaun B, Eschenmoser A. 1995. Pyranosyl-RNA ("p-RNA"): Base-pairing selectivity and potential to replicate. *Helv Chim Acta* **78:** 1621–1635.
Powner MW, Gerland B, Sutherland JD. 2009. Synthesis of activated pyrimidine ribonucleotides in prebiotically plausible conditions. *Nature* **459:** 239–242.
Puglisi JD, Tan R, Calnan BJ, Frankel AD, Williamson JR. 1992. Conformation of the TAR RNA-arginine complex by NMR spectroscopy. *Science* **257:** 67–80.
Qiao F, Cech TR. 2008. Triple-helix structure in telomerase RNA contributes to catalysis. *Nat Struct Mol Biol* **15:** 634–640.
Ruskin B, Zamore PD, Green MR. 1988. A factor, U2AF, is required for U2 snRNP binding and splicing complex assembly. *Cell* **52:** 207–219.
Sharma MR, Koc EC, Datta PP, Booth TM, Spremulli LL, Agrawal RK. 2003. Structure of the mammalian mitochondrial ribosome reveals an expanded functional role for its component proteins. *Cell* **115:** 97–108.
Sharma MR, Booth TM, Simpson L, Maslov DA, Agrawal RK. 2009. Structure of a mitochondrial ribosome with minimal RNA. *Proc Natl Acad Sci* **106:** 9637–9642.
Song JJ, Smith SK, Hannon GJ, Joshua-Tor L. 2004. Crystal structure of Argonaute and its implications for RISC slicer activity. *Science* **305:** 1434–1437.
Toor N, Keating KS, Taylor SD, Pyle AM. 2008. Crystal structure of a self-spliced group II intron. *Science* **320:** 77–82.
Tzfati Y, Fulton TB, Roy J, Blackburn EH. 2000. Template boundary in a yeast telomerase specified by RNA structure. *Science* **288:** 863–867.
Valadkhan S, Mohammadi A, Jaladat Y, Geisler S. 2009. Protein-free small nuclear RNAs catalyze a two-step splicing reaction. *Proc Natl Acad Sci* **106:** 11901–11906.
Viedma C, Oritz CE, de Torres T, Izumi T, Blackmond DG. 2008. Evolution of solid phase homochirality for a proteinogenic amino acid. *J Am Chem Soc* **130:** 15274–15275.
Voorhees RM, Weixlbaumer A, Loakes D, Kelley AC, Ramakrishnan V. 2009. Insights into substrate stabilization from snapshots of the peptidyl transferase center of the intact 70S ribosome. *Nat Struct Mol Biol* **16:** 528–533.
Wang X, Arai S, Song X, Reichart D, Du K, Pascual G, Tempst P, Rosenfeld MG, Glass CK, Kurokawa R. 2008. Induced ncRNAs allosterically modify RNA-binding proteins in *cis* to inhibit transcription. *Nature* **454:** 126–130.
Yarus M. 1988. A specific amino acid binding site composed of RNA. *Science* **240:** 1751–1758.
White HD. 1976. Coenzymes as fossils of an earlier metabolic state. *J Mol Evol* **7:** 101–104.
Zappulla DC, Cech TR. 2004. Yeast telomerase RNA: A flexible scaffold for protein subunits. *Proc Natl Acad Sci* **101:** 10024–10029.

Evolution in an RNA World

G.F. JOYCE

*Departments of Chemistry and Molecular Biology and the Skaggs Institute for Chemical Biology,
The Scripps Research Institute, La Jolla, California 92037*

Correspondence: gjoyce@scripps.edu

A long-standing research goal has been to develop a self-sustained chemical system that is capable of undergoing Darwinian evolution. The notion of primitive RNA-based life suggests that this goal might be achieved by constructing an RNA enzyme that catalyzes the replication of RNA molecules, including the RNA enzyme itself. This reaction was demonstrated recently in a cross-catalytic system involving two RNA enzymes that catalyze each other's synthesis from a total of four component substrates. The cross-replicating RNA enzymes undergo self-sustained exponential amplification at a constant temperature in the absence of proteins or other biological materials. Amplification occurs with a doubling time of ~1 hour and can be continued indefinitely. Small populations of cross-replicating RNA enzymes can be made to compete for limited resources within a common environment. The molecules reproduce with high fidelity but occasionally give rise to recombinants that also can replicate. Over the course of many "generations" of selective amplification, novel variants arise and grow to dominate the population based on their relative fitness under the chosen reaction conditions. This is the first example, outside of biology, of evolutionary adaptation in a molecular genetic system.

The last time the Cold Spring Harbor Symposium focused on evolution was in 1987, on the topic "The Evolution of Catalytic Function." I was happy to have attended that meeting. Being a postdoctoral fellow at that time, I felt obliged to write out my introductory remarks, which I have saved to this day. In my introduction I said, "I choose to interpret 'evolution of catalytic function' in the prospective sense, by which I mean the potential to evolve novel catalysts in the laboratory." I also said, "In the laboratory we focus on the problem of replication and on trying to copy genetic information without the aid of an external catalyst" (Joyce 1987).

I was hardly the first person to have had such thoughts. In that same meeting, Jeremy Knowles said, quoting from his paper in the 1987 Symposium volume, "We outline the first steps of an attempt to monitor the improvement in catalytic efficiency of an enzyme as its gene is mutagenized at random and more efficient catalysts are selected for" (Hermes et al. 1987). Knowles described what were some of the first directed evolution experiments, in which he randomly mutagenized the gene for triose phosphate isomerase and screened for variant enzymes with improved catalytic efficiency. Twenty years before that, Francis Crick discussed the possibility of replication of RNA genomes without the aid of an external catalyst. He said, "Possibly the first 'enzyme' was an RNA molecule with RNA replicase properties" (Crick 1968). In Crick's view, RNA was the Ur enzyme—the first enzyme to be capable of bringing about its own replication, thereby providing the basis for Darwinian evolution.

Since the time of the 1987 Symposium, the technology of directed evolution has advanced tremendously, for both proteins and RNA. My own laboratory has focused on the in vitro evolution of RNA enzymes, especially those relevant to the replication of genetic information. The technology itself has become so powerful, and yet so routine, that it can be practiced by any biochemist or molecular biologist. It is straightforward to amplify RNA molecules by a combination of reverse transcription, PCR (polymerase chain reaction) amplification, and forward transcription. One can impose selection constraints on the RNA molecules such that if they meet those constraints (e.g., binding a target ligand or performing some catalytic function), then they become eligible for amplification. And one can introduce random mutations, usually at the level of double-stranded DNA, through mutagenic or recombinogenic PCR procedures. Taken together, the ability to amplify, select, and mutate populations of RNA molecules gives one the opportunity to perform the Darwinian evolution of RNA-based catalytic function (Joyce 1989; Beaudry and Joyce 1992).

One of the first examples of the directed evolution of RNA enzymes concerned the same function that Francis Crick had talked about in 1968: the ability of RNA to catalyze the RNA-templated joining of RNA molecules (Bartel and Szostak 1993). This is fundamentally the same chemistry that is brought about by RNA-dependent RNA polymerase proteins. To discover RNA enzymes that catalyze this reaction, one can go searching in random sequence space. One can attach random sequence polynucleotides to an RNA template–substrate complex and install primer-binding sites at the 3′ end of the random sequence region and at the 5′ end of substrate (Fig. 1). Then, through selective reverse transcription–PCR (RT-PCR), one can amplify only those molecules that have catalyzed the joining of the substrate to themselves. The first application of this selection scheme, and the first case in which enzymatic function was derived starting from random sequence RNAs, was the work of David Bartel and Jack Szostak (1993), which resulted in the "class I" RNA

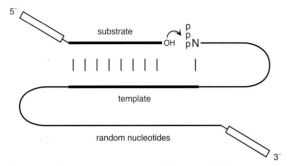

Figure 1. Scheme for selective amplification of RNA molecules that catalyze the RNA-templated joining of RNA. The putative catalytic domain consists of random sequence nucleotides that are attached to a template region that is complementary to the 3′ end of an oligonucleotide substrate and to the 5′ end of the population of RNAs. Any RNA molecule that catalyzes ligation of the substrate to itself (curved arrow) will contain two primer binding sites (boxed regions) that are necessary for reverse transcription and PCR amplification.

ligase enzyme. It is a robust enzyme, with a k_{cat} of 14 min^{-1} and K_m of 9 μM, obtained from a starting population of ~10^{15} random sequence 220 mers. This work demonstrates that Crick's notion of RNA-catalyzed RNA replication, together with Knowles' approach to the directed evolution of catalytic function, is experimentally viable.

CONTINUOUS IN VITRO EVOLUTION

More recently, but still more than 10 years ago, a new technology for the directed evolution of RNA was devised in our laboratory—what we termed "continuous in vitro evolution" (Wright and Joyce 1997). This method first was applied to the class I RNA ligase, which was challenged to attach an oligonucleotide substrate to the 5′ end of the RNA enzyme. The substrate had the sequence of the T7 RNA polymerase promoter, containing mostly deoxynucleotides but also a few ribonucleotides at its 3′ end. RNA enzymes that reacted with this substrate became reversed transcribed, in the same mixture, to yield double-stranded RNA-DNA molecules that contained a functional promoter element. The reaction mixture also included T7 RNA polymerase, which generated multiple copies of "progeny" RNA enzymes per reacted parental molecule. These progeny in turn could catalyze additional ligation reactions, and so on, resulting in the exponential amplification of functional RNAs. This cycle of events could be continued indefinitely, as long as one maintained a supply of the promoter-containing substrate and other reagents, usually accomplished through a serial transfer procedure.

The continuous in vitro evolution of RNA enzymes is analogous to the continuous culture of bacterial or eukaryotic cells, except that our culture medium is purely biochemical, containing two polymerase proteins (reverse transcriptase and T7 RNA polymerase), the four NTPs and dNTPs, salts, and buffer. This system enables longitudinal studies of the Darwinian evolution of RNA enzymes, analogous to the work of Richard Lenski and colleagues concerning the long-term experimental evolution of *Escherichia coli* (Elena et al. 1996; Blount et al. 2008).

One way to track the evolving population of RNA enzymes is to measure the concentration of RNA before and after each transfer throughout the course of a serial transfer experiment. This produces what we term "zigzag plots," reflecting repeated rounds of growth and dilution (Fig. 2). In the first of many such continuous in vitro evolution experiments, we performed 300 successive rounds of ~1000-fold growth and 1000-fold dilution, achieving an overall amplification of ~10^{300}-fold in 52 hours (Wright and Joyce 1997). The evolving population not only withstood this extreme dilution schedule, but also exhibited progressive improvement in its catalytic function. The most fit enzymes grew preferentially to dominate the population and had the opportunity to give rise to novel variants with even higher catalytic efficiency. The starting class I ligase enzyme exhibited a catalytic efficiency (k_{cat}/K_m) of 8×10^2 M^{-1} min^{-1}, whereas the evolved enzyme exhibited a catalytic efficiency of 1×10^7 M^{-1} min^{-1} (measured in the presence of 15 mM MgCl$_2$ at pH 8.5 and 37°C). This improvement of ~10^4-fold was attributable to 30 acquired mutations that improved both the k_{cat} and K_m of the ligase enzyme.

Continuous in vitro evolution, although a powerful method for witnessing the evolution of catalytic function in real time (Paegel and Joyce 2008), suffers from the fact that behind the curtain lurk two informational macromolecules: reverse transcriptase and T7 RNA polymerase, which themselves are not subject to evolution within the system. Reverse transcriptase, derived from a retrovirus, and T7 RNA polymerase, derived from a bacteriophage, are the products of biological evolution, and not what I had in mind at the 1987 Symposium when I discussed the imperative to "copy genetic information without the aid of an external catalyst" (Joyce 1987). Instead, what one wants is what Francis Crick talked about: an RNA enzyme that is "capable of bringing about its own replication" (Crick 1968). One wants a system in which the evolving RNA molecules adopt a structure that confers the ability to catalyze the amplifica-

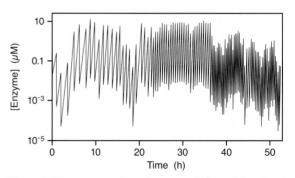

Figure 2. Time course of continuous evolution of the class I RNA ligase enzyme in a serial transfer experiment involving 100 successive rounds of ~1000-fold growth and 1000-fold dilution. The concentration of RNA enzymes was measured before and after each transfer (zigzag line). The time between transfers was decreased as tolerated, initially 1 hour and eventually 15 minutes. (Adapted from Wright and Joyce 1997.)

tion of RNA molecules, including the production of new copies of the enzymes themselves. Mutations will occur as a matter of course, and selection would be based on the differential replication rate of various RNA molecules in the population. In this way, the Darwinian evolution of RNA could be a self-sustaining process.

SELF-SUSTAINED REPLICATION OF RNA

In recent years, we have made substantial progress in developing RNA enzymes that catalyze their own replication. This work involves a different RNA ligase, the "R3C" RNA enzyme, which also was obtained by directed evolution starting from a large population of random sequence RNAs (Rogers and Joyce 2001). Like the class I ligase, the R3C ligase catalyzes the joining of two RNA substrates, one bearing a 3′-hydroxyl and the other bearing a 5′-triphosphate, forming a 3′,5′-phosphodiester and releasing inorganic pyrophosphate. The R3C ligase has a simple three-way junction architecture, consisting of three stem loops that are joined at a central location that contains the catalytic domain of the enzyme (Fig. 3A). Nucleotides within the catalytic domain are highly conserved in sequence, but those within the pendant stem loops are generic, as long as they form a stable duplex structure.

Two of the stem-loop regions within the R3C ligase are involved in binding the RNA substrates. Because these regions are generic in sequence, they can be designed to accommodate substrates whose sequences are identical to that of the enzyme itself. The two substrates (A and B) can be made to correspond to the 5′ and 3′ portions of the enzyme (E), so that when the substrates become ligated, they form another copy of the enzyme (Fig. 3B). In this way, at least in a formal sense, one can perform the self-replication of an RNA enzyme (Paul and Joyce 2002). The reaction does indeed proceed autocatalytically, but is not very efficient and does not reach a high maximum extent. For example, if one uses 1 μM starting concentration of ligase enzyme and 2 μM each of the two RNA substrates, there is an initial exponential burst that consumes ~5% of the substrates in 20 minutes, followed by a slow linear phase that proceeds at a rate of <0.01% min^{-1}. In the absence of any starting enzyme, there is no exponential burst, consistent with the autocatalytic nature of the system. However, even under optimal conditions, an incubation time of 17 hours is required to produce as many new enzyme molecules as the numbers that were present at the outset (Paul and Joyce 2002). Reaching this break-even point, and doing so many times over, is critical for achieving self-sustained replication of RNA.

Taking a lesson from the semiconservative nature of nucleic acid replication in biology, the next step was to devise two ligase enzymes: a plus-strand enzyme that directs the synthesis of a minus-strand enzyme, which in turn directs the synthesis of a new plus-strand enzyme (Kim and Joyce 2004). This approach causes replication to proceed in a cross-catalytic manner, with two enzymes (E and E′) catalyzing each other's synthesis from a total of four component substrates (A′ + B′ → E′ and A + B → E, respectively). Compared to self-replication, cross-

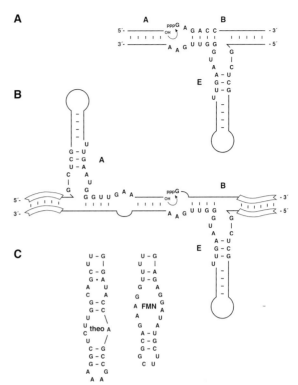

Figure 3. Sequence and secondary structure of various forms of the R3C ligase enzyme. (*A*) The enzyme (E) adopts a three-way junction structure upon binding two oligonucleotide substrates (A and B), which become ligated (curved arrow) to form the product. Conserved nucleotides that are essential for catalytic function are shown. (*B*) The self-replicating or cross-replicating enzyme ligates two substrates to yield a new copy of the enzyme or its cross-catalytic partner, respectively. (Open boxes) Regions of Watson-Crick pairing between enzyme and substrates that can have any complementary sequence. (*C*) The central stem loop of the enzyme can be replaced by an aptamer domain, configured such that binding of the corresponding ligand is required to stabilize the active structure of the enzyme. The aptamer domains for theophylline (theo) and flavin mononucleotide (FMN) are shown.

replication places fewer design constraints on the sequences of the replicating molecules. The self-replicating enzyme must be fully palindromic (in the molecular biology sense), whereas the cross-replicating enzymes need only have short regions of complementarity between the replicating partners. Furthermore, the extensive self-complementarity of the self-replicating enzyme is the chief reason for its limited extent of growth in the exponential phase of the reaction (Paul and Joyce 2002). This is because the two substrate molecules are complementary to each other (as well as to the parent) and therefore have a tendency to form a nonproductive substrate–substrate complex. The initial exponential phase consumes the readily available substrate molecules, and the subsequent linear phase reflects the slow dissociation of substrate molecules from the nonproductive complexes. Importantly, the step of product release is not rate-limiting, freeing the newly synthesized enzyme molecules to enter another round of replication.

Initial attempts to perform cross-catalytic replication were an improvement compared to self-replication but still disappointing with regard to the goal of reaching the break-even point. Using a starting concentration of 1 µM each of E and E′ and 2 µM each of the four RNA substrates, the exponential phase consumed ~25% of the substrates in 6 hours (Kim and Joyce 2004). Under optimized reaction conditions and using long incubation times, it would be possible to limp past the break-even mark, but this is hardly sufficient for sustained replication. One needs to think in terms of achieving 10–100-fold breakeven so that, like for protein-mediated continuous in vitro evolution, one can perform serial transfer experiments that allow replication to proceed indefinitely.

It thus became necessary to return to directed evolution methods to improve the rate and maximum extent of the cross-replicating RNA enzymes. This was done by evolving each enzyme separately, but seeking solutions that would apply to both members of the cross-replication pair. A quench-flow apparatus was used to select molecules that could react in times as short as 10 msec. The resulting E and E′ enzymes exhibited a 38-fold and 12-fold improvement in catalytic rate, respectively, and reacted to a maximum extent of ~90% in the initial fast phase. These optimized molecules were found to be capable of undergoing self-sustained replication, achieving 100-fold amplification in 5 hours at a constant temperature of 42°C (Lincoln and Joyce 2009).

A serial transfer experiment was performed using a starting concentration of 0.1 µM each of E and E′ and 5 µM each of the four RNA substrates, in the presence of 25 mM $MgCl_2$ and 50 mM EPPS (pH 8.5), but with no proteins or other biological molecules. Following 5 hours of incubation at 42°C, 4% of the reaction mixture was transferred to a new reaction vessel that contained a fresh supply of the substrates, but only those enzymes that were carried over in the transfer. This procedure was repeated for six rounds, resulting in an overall amplification of >10^8-fold in 30 hours (Lincoln and Joyce 2009). The corresponding zigzag plot was highly regular, with each round consisting of ~25-fold amplification of both E and E′ followed by 25-fold dilution (Fig. 4A). This process can indeed be continued indefinitely.

AN ARTIFICIAL GENETIC SYSTEM

Immortality can be rather dreary if it does not allow for the possibility of variation. What one wants is not a single replicating entity, but rather a heterogeneous population of replicators that can undergo mutation and selection. The cross-replicating RNA enzymes provide the opportunity to construct an artificial genetic system based on the transmission of sequence information from parent to progeny molecules. The replicating enzymes contain two "alleles," represented by the two regions of base pairing interactions between E and E′. Each allele encodes a corresponding trait, represented by the catalytic domain that is covalently linked to the allele.

In principle, the cross-replicating RNAs have the potential to transmit 30 bits of genetic information via the 15

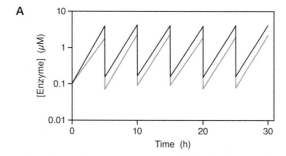

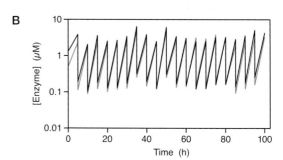

Figure 4. Self-sustained cross-replication of the R3C RNA ligase enzyme in a serial transfer experiment. The concentrations of E (black) and E′ (gray) were measured before and after each transfer. (*A*) A single cross-replicator was propagated for six successive rounds of ~25-fold growth and 25-fold dilution. (*B*) A starting population of 12 different cross-replicators were propagated for 20 successive rounds of ~20-fold growth and 20-fold dilution, with the opportunity for recombination throughout. (Adapted from Lincoln and Joyce 2009.)

base pairs (two bits per base pair) that comprise the two alleles. One of the alleles contains 7 base pairs (16,384 possible variants), and the other contains 8 base pairs (65,536 possible variants). However, not all sequences will be discriminated with high fidelity, especially at the extreme 5′ and 3′ ends of the molecule, thus reducing the information capacity of the system. Significantly, there is the opportunity for combinatorial diversity through recombination of the two alleles. This can occur due to occasional incorporation of a mismatched substrate, which results in a recombinant enzyme that also can cross-replicate. Recombinants can give rise to other recombinants, as well as revert back to nonrecombinants. During the course of many "generations" of selective amplification, novel replicators can arise through recombination and can grow to dominate the population, exhibiting Darwinian behavior in a nonbiological system.

As a test case, we constructed a model population of 12 different pairs of cross-replicating RNAs (Lincoln and Joyce 2009). Each pair had a different genetic sequence in the two allelic regions, which encoded different functional sequences in the corresponding catalytic domains of the E and E′ molecules. A coding relationship was established between a particular genetic allele and its associated phenotypic trait, implemented through the chemical synthesis of the various RNA molecules. Together, the 12 pairs of cross-replicators have the potential to give rise to 132 pairs of recombinants, which may be more or less fit than their progenitors. A serial transfer experiment was performed, starting with ~0.1 µM each of the 12 different E and E′ molecules and 5 µM each of the various A, B, A′, and B′ molecules. The population was subjected to 20 successive rounds of ~20-fold amplification and 20-fold dilution (~10^{26}-fold overall amplification) in 100 hours. In this case, the zigzag plot was not uniform, because novel variants arose and competed with existing members of the population, resulting in the preferential survival of the most efficient replicators (Fig. 4B).

After 20 rounds (86 doublings) of evolution, 100 individuals were cloned from the population and sequenced. The great majority of these (93%) were recombinants that were not present at the start of the experiment. Three such recombinants dominated the population, together accounting for one-third of all clones. These three recombinants all contained the A5 allele, together with the A2′, A3′, or A4′ allele. Overall, the A5 and A3′ alleles were the most enriched, whereas the A8, A11, and A11′ alleles were the most depleted among the evolved population of replicators (Lincoln and Joyce 2009).

What was the basis for the selective advantage of the dominant individuals? In the presence of their cognate substrates alone, the three dominant recombinants are less efficient replicators compared to the most efficient of the 12 starting replicators. The most efficient recombinant (A5-A3′) has an exponential growth rate of 0.68 h^{-1}, whereas the most efficient starting replicator (A1-A1′) has a growth rate of 0.75 h^{-1}. However, in the presence of the complete set of 48 substrates, the A5-A3′ recombinant amplifies more efficiently (0.33 h^{-1}) compared to the A1-A1′ starting replicator (0.10 h^{-1}). Furthermore, when the A5-A3′ recombinant is supplied with just the eight substrates that correspond to the enriched set of alleles, it has an exponential growth rate of 0.84 h^{-1}, the highest measured in the study.

It appears that the three dominant recombinants form a clique, not only replicating themselves efficiently, but also giving rise to one another through preferred mutational pathways. An analysis of predicted ΔG values for each combination of matched and mismatched substrates suggests that the most likely recombination events involve exchange of the A2′ and A3′ alleles and exchange of the A3′ and A4′ alleles, favoring the interconversion of the three dominant replicators (Lincoln and Joyce 2009).

REPLICATION CONTINGENT ON OTHER FUNCTIONS

Although replication efficiency is the ultimate measure of fitness, other traits may confer selective advantage to biological organisms through their indirect effect on fecundity. So too in an artificial genetic system it is possible to make reproductive fitness contingent on the execution of some other function. The cross-replicating RNA enzymes contain three generic stem loops, two that are committed to substrate binding and a third that can contain a functional domain (Fig. 3C). The functional domain might be an RNA aptamer that binds a specific ligand or a catalyst that has some function other than replication. The activity of this added functional domain must somehow relate to replication so that molecules that are better able to execute the secondary function will enjoy a replicative advantage.

It is straightforward to install an aptamer domain within the central stem loop of the replicating enzymes, configured so that the enzymes undergo exponential amplification in the presence, but not the absence, of the corresponding ligand. Such constructs are termed "aptazymes" and have been developed in the laboratory for simple RNA enzymes (Tang and Breaker 1997); they have also been discovered within naturally occurring "riboswitches" (Winkler et al. 2004). We installed aptamers that specifically recognize theophylline (Jenison et al. 1994) or FMN (Burgstaller and Famulok 1994) in either one or both members of a cross-replicating pair, causing exponential amplification to be dependent on the presence of one or both ligands (Lam and Joyce 2009). In the absence of the ligand, the aptamer is unstructured and cannot support the active structure of the enzyme, whereas in the presence of the ligand, the aptamer adopts a well-defined structure that stabilizes and therefore activates the adjacent catalytic domain.

Cross-replicating enzymes that contain the theophylline aptamer exhibited exponential growth in the presence of theophylline, with growth leveling off as the supply of substrates became depleted (Fig. 5A). In the absence of theophylline, or in the presence of the closely related molecule caffeine (which differs from theophylline by the presence of a single methyl group at the N7 position), no growth was detected (Lam and Joyce 2009). All-or-none, ligand-dependent, isothermal exponential amplification is highly unusual. The closest parallel is the isothermal exponential amplification of nucleic acids (Guatelli et al. 1990; Walker et al. 1992; Notomi et al. 2000), which can be highly specific for a particular target but only applies to nucleic acid targets.

The exponential growth rate of the cross-replicating aptazymes depends on the concentration of the ligand relative to the K_d of the aptamer domain (Fig. 5B). This provides a way to measure the concentration of ligand in an unknown sample, analogous to quantitative PCR, but for a broad range of ligands (Lam and Joyce 2009). It also provides a means for the replicating molecules to sense their local environment and to reflect this behavior in their reproductive fitness. Cross-replication can be made dependent on two different ligands by installing a different aptamer domain in the two members of a cross-replicating pair. This was done by installing the theophylline aptamer in E and the FMN aptamer in E′ (or vice versa). In the presence of just one ligand, linear growth was observed. This is because only one of the two enzymes was active but still able to operate with multiple turnover. In the presence of both ligands, however, both enzymes were active and exponential amplification occurred (Lam and Joyce 2009). In principle, multiple aptamer domains could be installed in series within E or E′, resulting in more complex ligand-dependent behavior. Such tandem aptazymes have been constructed previously in the laboratory (Jose et al. 2001), and tandem riboswitches have been found to occur in nature (Sudarsan et al. 2006).

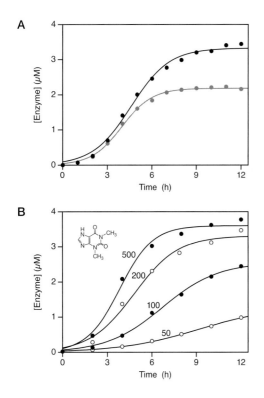

Figure 5. Ligand-dependent exponential amplification of cross-replicating RNA enzymes that contain the theophylline aptamer (see Fig. 3C). (*A*) Amplification of E (black) and E′ (gray) occurs in the presence of 5 mM theophylline, but not 5 mM caffeine. (*B*) The exponential growth rate depends on the concentration of theophylline, which was 50, 100, 200, or 500 µM. The chemical structure of theophylline is shown. (*A*, Adapted from Lam and Joyce 2009.)

IS IT ALIVE?

No. The artificial genetic system based on RNA enzymes that catalyze their own replication has many of the properties of a living system, but it lacks the ability to bring about inventive Darwinian evolution. The molecules can undergo self-sustained replication with exponential growth. "Self-sustained" in this context refers to their ability to operate without the aid of an external catalyst. All of the genetic information that is necessary for the system to replicate and evolve is part of the system that is undergoing replication and evolution. Genetic information within the system is represented by the two regions of base pairing interactions between the E and E′ enzymes, and that information is inherited through the process of cross-replication. The system is informational because many such genetic sequences can be represented, each of which can be maintained in a heritable fashion. Furthermore, this genetic information encodes complex phenotypic traits, reflected in the catalytic and ligand-recognition properties of the associated functional domain.

The opportunity exists for mutation through recombination within the artificial genetic system, and the resulting recombinants also are capable of propagating genetic information. However, the sequence space available to the system is meager, limited to the $n \times m$ combinations of the two genetic alleles. Sequence space in biology is far more generous due to the 4^n possible combinations for a nucleic acid genome of length n. In the artificial genetic system that we have demonstrated, n and m were chosen to be 12 and 12, resulting in 144 possible cross-replicating pairs (Lincoln and Joyce 2009). In principle, n and m each could be on the order of 10^4–10^5, giving 10^8–10^{10} possible combinations. However, not all of these potential genotypes would be discriminated with high fidelity. In addition, it would be difficult for any replicator to find its corresponding substrates among a mixture of tens of thousands of potential substrates. Complexities on the order of $10^3 \times 10^3$ are likely to be the maximum that can be achieved, unless one resorts to methods outside the system to reduce substrate diversity in a selective manner, for example, by using deconstructive PCR methods to convert the population of newly formed enzymes to a daughter population of substrates (Lincoln and Joyce 2009).

Even with a complexity of 12×12, it was possible to perform Darwinian evolution in the artificial genetic system, seen as the emergence of novel variants and survival of the fittest in response to a particular set of environmental conditions. Fitness can be made to reflect not just the replicative function, but also other functions that are linked to replication, such as ligand recognition. What the system cannot do, and the chief reason that it cannot be considered alive even in a molecular reductionist sense, is invent novel function within the system. There are evolved entities still lurking behind the curtain—not polymerase enzymes borrowed from biology, but the R3C catalytic motif and various aptamer motifs that were obtained by directed evolution conducted outside the system. Once placed within the synthetic genetic system, these preexisting motifs can be further evolved, but how could functional motifs be invented within the system?

A living system must not only be capable of undergoing Darwinian evolution in a self-sustained manner, but also have a broad inventive capability that enables the discovery of adaptive solutions to a variety of challenges imposed by the environment. The cross-replicating system based on the R3C ligase may indeed have the capacity for inventive Darwinian evolution, but this will depend on the degree of complexity that can be implemented through a simple $n \times m$ genetics. One can imagine many thousands of replicators, each with a particular genetic sequence encoding a different randomly chosen sequence within the corresponding functional domain. A diverse population of such replicating RNAs may provide the basis for the discovery of novel function, although the extent to which such inventive capability can lead to the emergence of complex and interesting behaviors remains to be seen.

ACKNOWLEDGMENTS

This work was supported by NASA grant NNX07AJ23G, National Institutes of Health grant R01GM065130, and National Science Foundation grant

MCB-0614614. I am grateful to Roslind Varghese for preparing a transcript of my lecture given at the 2009 Symposium, which was the basis for this manuscript.

REFERENCES

Bartel DP, Szostak JW. 1993. Isolation of new ribozymes from a large pool of random sequences. *Science* **261:** 1411–1418.

Beaudry AA, Joyce GF. 1992. Directed evolution of an RNA enzyme. *Science* **257:** 635–641.

Burgstaller P, Famulok M. 1994. Isolation of RNA aptamers for biological cofactors by *in vitro* selection. *Angew Chemie* **33:** 1084–1087.

Blount ZD, Borland CZ, Lenski RE. 2008. Historical contingency and the evolution of a key innovation in an experimental population of *Escherichia coli*. *Proc Natl Acad Sci* **105:** 7899–7906.

Crick FHC. 1968. The origin of the genetic code. *J Mol Biol* **38:** 367–379.

Elena SF, Cooper VS, Lenski RE. 1996. Punctuated evolution caused by selection of rare beneficial mutations. *Science* **272:** 1802–1804.

Guatelli JC, Whitfield KM, Kwoh DY, Barringer KJ, Richman DD, Gingeras TR. 1990. Isothermal, *in vitro* amplification of nucleic acids by a multienzyme reaction modeled after retroviral replication. *Proc Natl Acad Sci* **87:** 1874–1878.

Hermes JD, Blacklow SC, Knowles JR. 1987. The development of enzyme catalytic efficiency: An experimental approach. *Cold Spring Harbor Symp Quant Biol* **52:** 597–602.

Jenison RD, Gill SC, Pardi A, Polisky B. 1994. High-resolution molecular discrimination by RNA. *Science* **263:** 1425–1429.

Jose AM, Soukup GA, Breaker RR. 2001. Cooperative binding of effectors by an allosteric ribozyme. *Nucleic Acids Res* **7:** 1631–1637.

Joyce GF. 1987. Nonenzymatic template-directed synthesis of informational macromolecules. *Cold Spring Harbor Symp Quant Biol* **52:** 41–51.

Joyce GF. 1989. Amplification, mutation and selection of catalytic RNA. *Gene* **82:** 83–87.

Kim D-E, Joyce GF. 2004. Cross-catalytic replication of an RNA ligase ribozyme. *Chem Biol* **11:** 1505–1512.

Lam BJ, Joyce GF. 2009. Autocatalytic aptazymes enable ligand-dependent exponential amplification of RNA. *Nat Biotechnol* **27:** 288–292.

Lincoln TA, Joyce GF. 2009. Self-sustained replication of an RNA enzyme. *Science* **323:** 1229–1232.

Notomi T, Okayama H, Masubuchi H, Yonekawa T, Watanabe K, Amino N, Hase T. 2000. Loop-mediated isothermal amplification of DNA. *Nucleic Acids Res* **28:** e63.

Paegel BM, Joyce GF. 2008. Darwinian evolution on a chip. *PLoS Biol* **6:** 900–906.

Paul N, Joyce GF. 2002. A self-replicating ligase ribozyme. *Proc Natl Acad Sci* **99:** 12733–12740.

Rogers J, Joyce GF. 2001. The effect of cytidine on the structure and function of an RNA ligase ribozyme. *RNA* **7:** 395–404.

Sudarsan N, Hammond MC, Block KF, Weiz R, Barrick JE, Roth A, Breaker RR. 2006. Tandem riboswitch architectures exhibit complex gene control functions. *Science* **314:** 300–304.

Tang J, Breaker RR. 1997. Rational design of allosteric ribozymes. *Chem Biol* **4:** 453–459.

Walker GT, Fraiser MS, Schram JL, Little MC, Nadeau JG, Malinowski DP. 1992. Strand displacement amplification: An isothermal, *in vitro* DNA amplification technique. *Nucleic Acids Res* **20:** 1691–1693.

Winkler WC, Nahvi A, Roth A, Collins JA, Breaker RR. 2004. Control of gene expression by a natural metabolite-responsive ribozyme. *Nature* **428:** 281–286.

Wright MC, Joyce GF. 1997. Continuous *in vitro* evolution of catalytic function. *Science* **276:** 614–617.

The Ribosome: Some Hard Facts about Its Structure and Hot Air about Its Evolution*

V. RAMAKRISHNAN

MRC Laboratory of Molecular Biology, Cambridge CB2 0QH, United Kingdom
Correspondence: ramak@mrc-lmb.cam.ac.uk

By translating genetically encoded information to synthesize proteins, the ribosome has a central and fundamental role in the molecular biology of the cell. Virtually every molecule made in every cell was made either directly by the ribosome or by enzymes made by the ribosome. Although the ribosome was discovered half a century ago, progress in the field of translation has been revolutionized by the atomic structures of the ribosomal subunits determined in 2000. These structures paved the way not only for more sophisticated biochemical and genetic experiments, but also for the phasing and/or molecular interpretation of all subsequent structures of the ribosome by crystallography or cryoEM (cryo-electron microscopy). In addition to facilitating our understanding of ribosome function, these structures also shed light on the evolution of the ribosome.

Ribosomes from all species consist of approximately two-thirds RNA and one-third protein. Ribosomes from mammalian mitochondria are an exception, with the ratio of protein and RNA reversed (see Sharma et al. 2003). All ribosomes consist of two subunits, termed 50S and 30S in bacteria or 60S and 40S in eukaryotes. Together, they comprise the 70S ribosome in bacteria or the 80S ribosome in eukaryotes (Fig. 1A). The mRNA containing the genetic template binds in a cleft in the small subunit. The amino acids themselves are brought into the ribosome by aminoacylated tRNA substrates. The ribosome has three binding sites for tRNA: the A (aminoacyl) site that brings the new aminoacyl tRNA, the P (peptidyl) site that holds the nascent peptide chain, and the E (exit) site to which the deacylated P-site tRNA moves after peptide bond formation (Fig. 1B).

Translation in all species can be divided into three stages (Fig. 2) (for review, see Schmeing and Ramakrishnan 2009). During initiation, the small subunit of the ribosome binds mRNA at the start site of the coding sequence, in a precise manner that puts the start codon in the P site. This requires three initiation factors and a special initiator tRNA that binds to the P site.

Initiation is followed by the elongation cycle, which consists of three important steps: decoding, peptidyl transfer, and translocation. During decoding, the correct aminoacyl tRNA, which is delivered to the A site of the ribosome as a ternary complex with elongation factor Tu (EF-Tu) and GTP, is selected based on the codon on the mRNA in the A site. Selection of the tRNA leads to hydrolysis of GTP by EF-Tu and release of the factor from the ribosome. The aminoacyl end of the selected tRNA then swings into the peptidyl transferase center (PTC) in the 50S subunit of the ribosome, where peptide bond formation occurs rapidly and spontaneously.

Peptidyl transfer leaves the P-site tRNA deacylated, with the A-site tRNA now containing a nascent peptide chain that has been extended by one residue. The 3′ ends of the A- and P-site tRNAs then move first with respect to the 50S subunit to form an intermediate or hybrid state of the ribosome, followed by movement of the mRNA and tRNAs with respect to the 30S subunit, which requires the action of EF-G, another GTPase factor. This leaves the ribosome with an empty A site with a new mRNA codon ready to accept the next aminoacyl tRNA.

The elongation cycle continues until a stop codon is reached in the A site. The so-called class I release factors (RF1 or RF2 in bacteria, eRF1 in eukaryotes) recognize the stop codon and catalyze the cleavage of the polypeptide chain from the P-site tRNA. Finally, a factor known as ribosome recycling factor (RRF), with the help of EF-G, disassembles the ribosome so that a new round of protein synthesis can begin.

Most of these aspects of translation are common to all kingdoms of life. In eukaryotes, initiation is far more complex and involves a specifically modified mRNA with a 5′ cap and a poly(A) tail at the 3′ end, as well as almost a dozen factors, many of which are large multisubunit complexes themselves (Kapp and Lorsch 2004).

Recent structural and biochemical work has shed light on many aspects of translation. In particular, the high-resolution structures of the ribosomal subunits (Ban et al. 2000; Wimberly et al. 2000) were useful in the molecular interpretation and/or phasing of all subsequent structures, including a lower-resolution crystal structure of the 70S ribosome with mRNA and tRNA ligands at 5.5-Å resolution (Yusupov et al. 2001), more recent higher-resolution structures of the empty 70S ribosome from *Escherichia coli* (Schuwirth et al. 2005), and the 70S ribosome with mRNA and tRNAs from *Thermus thermophilus* (Selmer et al. 2006). These basic structures have been followed by high-resolution structures of the ribosome with protein factors, most notably with release factors (Laurberg et al.

---

*The title is a paraphrase of one by the late eminent crystallographer David M. Blow: "Hard facts on structure: Hot air about mobility" (*Nature* [1982] **297**: 454–455).

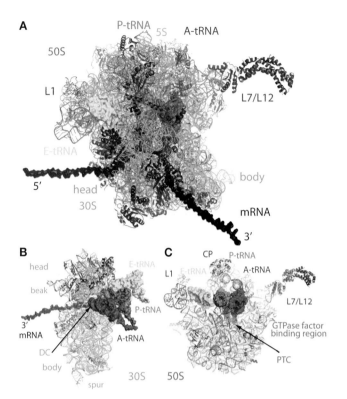

Figure 1. Structure of the ribosome. (*A*) Overview of the bacterial 70S ribosome with the 50S subunit on top and the 30S subunit on the bottom. The mRNA (dark gray) is shown wrapped round the neck of the 30S subunit. (Magenta) A-site, (green) P-site, (yellow) E-site tRNAs. (*B*) The 30S subunit showing the decoding center (DC) where codon–anticodon interactions are monitored during tRNA selection. (*C*) The 50S subunit showing the GTPase-factor-binding region and the PTC where peptide bond formation is catalyzed. (Reprinted, with permission, from Schmeing and Ramakrishnan 2009 [©Nature Publishing Group].)

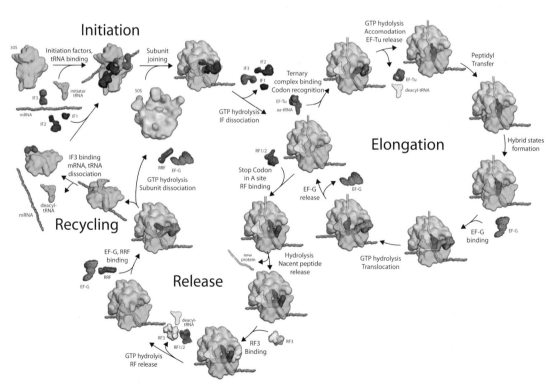

Figure 2. Overview of the translational pathway showing the phases of initiation, elongation, release, and recycling. (Reprinted, with permission, from Schmeing and Ramakrishnan 2009 [©Nature Publishing Group].)

2008; Weixlbaumer et al. 2008) and more recently with elongation factors EF-Tu and EF-G (Gao et al. 2009; Schmeing et al. 2009). In addition, many cryoEM structures of the ribosome represent different functional states at varying resolutions.

THE RIBOSOME AS AN RNA-BASED MACHINE

The ribosome itself is a large and complex assembly of RNA and more than 50 proteins. In addition, translation requires a host of protein factors and aminoacyl tRNA substrates. Thus, understanding the evolution of the ribosome poses a difficult challenge. To begin with, the system poses the standard "chicken or egg" question: If the ribosome consists of both RNA and protein, and is needed to make protein, how did it come about? The first attempt to address this was Crick, who presciently wrote, "It is tempting to wonder if the primitive ribosome could have been made *entirely* of RNA" (original italics) (Crick 1968). To my knowledge, this was the first idea that RNA could be both an information carrier and able to perform catalysis, and can be thought of as the origin of the "RNA world hypothesis," which postulates a primordial world consisting of replicating RNA molecules before the advent of proteins. However, in the absence of any known examples of catalysis by RNA, not even Crick could imagine that catalysis in the current ribosome would be RNA based.

It is clear that protein factors could have evolved later to make translation more efficient, because even today it is possible to get inefficient and limited translation without them. For instance, factor-free protein synthesis in vitro was demonstrated by Spirin and coworkers (Gavrilova and Spirin 1974; Gavrilova et al. 1976). But what about the ribosome itself? What are the relative roles of protein and RNA?

The earliest work on ribosome function focused on proteins for two reasons. Partly, proteins were thought to be the molecules responsible for catalytic function. Second, because the standard laboratory organism *E. coli* contains seven genes for rRNA, it was difficult to isolate RNA mutants, and many of the early mutations, such as those for antibiotic resistance, mapped to ribosomal proteins. However, there were hints from quite early on that RNA had a more important role than just providing a scaffolding for functional proteins. For instance, it was shown that chemical modification of rRNA but not proteins would abolish binding of tRNA to 30S subunits (Noller and Chaires 1972). In the absence of any prior evidence for the catalytic properties of RNA, the results were taken to suggest that tRNA-binding sites must therefore consist of both protein and RNA. Subsequent work on the ribosome, notably by Noller and coworkers, continued to provide evidence for the importance of rRNA, but in the absence of an intellectual framework in which RNA catalysis was a real possibility, it was hard to make definite progress.

This situation changed dramatically when catalysis by RNA was discovered in the context of the group I intron (Zaug et al. 1983) or RNase P (Guerrier-Takada et al. 1983). With evidence that RNA could in principle perform catalysis, the ribosome community was far readier to accept that rRNA might have crucial functions, and this prospect renewed interest in the field (Moore 1988). Subsequently, an experiment showed that 50S subunits from *Thermus aquaticus* treated extensively with proteinase K in the presence of SDS nevertheless preserved their peptidyl transferase activity (Noller et al. 1992). This experiment was a major step forward, but not conclusive proof for a variety of reasons. *E. coli* 50S subunits did not maintain activity with such protease treatment, nor did in-vitro-transcribed 23S RNA show activity. Moreover, in the *Thermus* 50S subunit, a large number of protein fragments remained bound after protease treatment, and indeed, subsequent work showed that the treatment left three proteins essentially intact (Khaitovich et al. 1999). A subsequent effort to provide conclusive proof of the role of RNA in peptidyl transfer using in vitro transcription of 23S RNA and its individual domains appeared to have narrowed down the activity to the RNA domain containing the peptidyl transferase center (Nitta et al. 1998), but this work was retracted a year later (Nitta et al. 1999). Thus, although work on the group I intron and RNase P showed that catalysis by RNA was certainly possible, conclusive evidence for a similar role in the ribosome proved to be difficult to obtain by purely biochemical means. It is striking that this limitation was recognized very early on by Crick, who said, "Without a detailed knowledge of the structure of present-day ribosomes it is difficult to make an informed guess" (Crick 1968).

High-resolution structures of ribosomal subunits and the whole ribosome have revealed in stunning detail the environment of the PTC-, tRNA-, and mRNA-binding sites, the intersubunit interface, and many other functionally important regions of the ribosome. These structures at long last allow us to provide conclusive insights into many aspects of ribosome function and, in particular, show unambiguously how widespread the role of RNA is in the contemporary ribosome.

DECODING BY tRNA

A crucial event in protein synthesis is the selection of the correct tRNA corresponding to the codon on mRNA. At a fundamental level, this involves base pairing between the codon and the anticodon on tRNA. However, the free energy of base pairing between codon and anticodon is not sufficient to explain the relatively low error rate of the ribosome (Ogle and Ramakrishnan 2005). Kinetic experiments show that binding of the correct tRNA leads to induced conformational changes in the ribosome that accelerate GTP hydrolysis by EF-Tu and tRNA selection (see Rodnina and Wintermeyer 2001).

Minor Groove Recognition by RNA

Discrimination against incorrect tRNA ultimately depends on recognizing mismatched base pairs. Because base pairing alone is insufficient to explain the accuracy of decoding, and the ribosome appears to have an active role, what is the ultimate nature of this discrimination? Experiments showed that the binding of a cognate tRNA induced three universally conserved bases in the decoding center of the 30S subunit to line the minor groove between codon and anticodon in a manner that would distinguish between canonical Watson–Crick and noncanonical base pairs at the first two positions but did not monitor the geometry of base pairing at the wobble position (Fig. 3) (Ogle et al. 2001). This finding at once explained the additional discrimination provided by the ribosome and the long-standing nature of the genetic code, in which mismatches were allowed at the wobble position but not at the first two positions (Crick 1966). The additional binding energy of cognate tRNA from the induced changes at the decoding center resulted in large-scale movements of the shoulder domain of the 30S subunit, whereas near-cognate tRNA failed to produce such movements (Ogle et al. 2002). This finding led to a model in which the conformational change from an open form to a closed form was required for tRNA selection; such a model could help to rationalize disparate biochemical genetic data on fidelity (Ogle et al. 2002). The rationale for why such a conformational change is needed for tRNA selection has recently been further clarified by the structure of the ribosome bound to elongation factor Tu and tRNA (Schmeing et al. 2009).

The Possible Role of Minor-groove Recognition in Evolution

DNA and RNA polymerases have exactly the same problem of discriminating against noncanonical base pairs. They are even more accurate than the ribosome. It is striking that both DNA and RNA polymerases use conserved amino acids to monitor the minor groove between the template and transcript strands. Indeed, it is possible for the polymerases to choose base pairs that form no hydrogen bonds between them, as long as they have the same shape as Watson–Crick base pairs at the minor groove (for review, see Kool 2000). Thus, minor-groove recognition is an important feature of ensuring proper base pairing complementarity.

The ribosome shows that such minor-groove recognition can be done by RNA alone. This is significant for the evolution of complexity from a primitive self-replicating RNA system. Such a system initially would have had a very high error rate, because base pairing alone would not have a sufficiently high free-energy difference to allow for substantial accuracy. However, if a primitive replicase also evolved to take advantage of minor-groove recognition of complementary base pairs, the error rate would be reduced by at least two orders of magnitude, allowing for the much more accurate replication required for complex systems to evolve.

Distortions in tRNA during Decoding

Structures of the ribosome complexed with EF-Tu and tRNA by cryoEM (Valle et al. 2002; Schuette et al. 2009; Villa et al. 2009) or more recently by crystallography (Schmeing et al. 2009) show that when tRNA is delivered to the ribosome by EF-Tu, it is distorted in the anticodon stem (Fig. 4). This suggests that the tRNA molecule has within itself considerable conformational variability. It is possible that this distortion could occur transiently in the absence of EF-Tu. Thus, tRNA in the absence of factors could have still performed decoding in a very similar way, with a transient bend during initial recognition followed by accommodation of the aminoacyl end into the PTC. However, GTP hydrolysis by EF-Tu, an important step in tRNA selection, would have been absent, so that the process would have been both slower and less accurate.

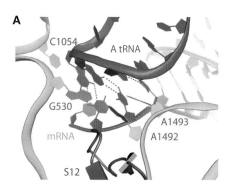
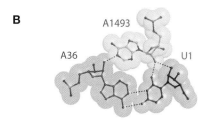
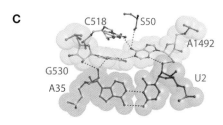
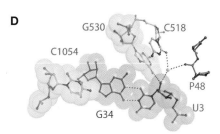

Figure 3. (*A*) Decoding center in the 30S subunit of the ribosome, showing how ribosomal bases interact with the minor groove of the codon–anticodon minihelix. (*B–D*) Interaction of ribosomal bases with the minor groove of the first, second, and third base pairs between the codon (pink) and anticodon (gray) (Ogle et al. 2001). (Reprinted, with permission, from Schmeing and Ramakrishnan 2009 [©Nature Publishing Group].)

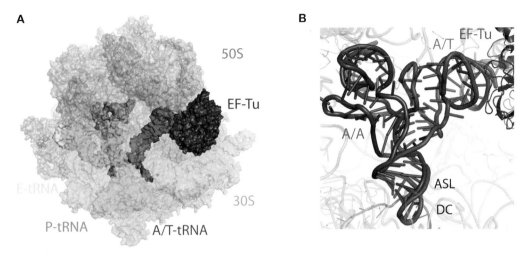

Figure 4. Structure of the complex of EF-Tu and aminoacyl tRNA bound to the ribosome. (*A*) Overview showing EF-Tu and aminoacyl tRNA (purple) bound to the ribosome. (*B*) Comparison of the conformation of the tRNA in the distorted state when it is bound to EF-Tu (purple) with the more canonical form after it has swung into the peptidyl transferase center (dark blue). (*A*, Reprinted, with permission from Schmeing and Ramakrishnan 2009 [©Nature Publishing Group]; *B*, reprinted, with permission from Schmeing et al. 2009 [©AAAS].)

PEPTIDYL TRANSFER

The central chemical event in translation is the formation of the peptide bond between the nascent polypeptide chain on P-site tRNA and the incoming amino acid on A-site tRNA. This catalysis occurs in the PTC of the ribosome that is located in the 50S subunit. As we have discussed above, the question of whether catalysis is RNA-based or whether proteins are involved could not be settled by purely biochemical experiments.

It was therefore striking that the structure of the 50S subunit in complex with a transition-state analog that defined the catalytic site showed that there were no proteins within 18 Å of the active site (Nissen et al. 2000). Although an acid-base catalytic mechanism involving specific ribosomal bases was proposed, this was disproved by subsequent biochemical work (for review, see Rodnina et al. 2006). Interestingly, none of the ribosomal bases appear to have a catalytic role in the chemical sense of contributing or accepting electrons or protons. Rather, the ribosome's contribution appears to be primarily entropic, by holding the substrates in the proper orientation (Sievers et al. 2004). The one moiety that appears to have a chemical role is the 2′-OH of the terminal adenosine of P-site tRNA itself, showing that the ribosome is an example of substrate-assisted catalysis (Weinger et al. 2004).

One interesting role for the ribosome is that of exposing the ester bond on the P-site tRNA to nucleophilic attack by an induced conformational change in the PTC on A-site tRNA binding (Schmeing et al. 2005b). This is an elegant way of protecting the nascent chain from hydrolysis by water except when the proper A-site substrate is bound, when certain conserved bases move to expose the ester bond that links the nascent chain to P-site tRNA. As in the case of decoding, the induced conformational changes all involve the RNA component of the ribosome.

Although no proteins were found near the active site of the archaeal PTC (Nissen et al. 2000), studies in bacteria have long implicated specific proteins in peptidyl transfer. In particular, two proteins, L16 (Moore et al. 1975) and L27 (Wower et al. 1998), were shown to aid peptidyl transfer. Of these, L27 was known to cross-link to the 3′ ends of both A- and P-site tRNAs, thus placing it right at the PTC (Wower et al. 1998), and just deletion of the first few amino-terminal residues reduced both the cross-linking yield and the rate of peptidyl transfer (Maguire et al. 2005).

The role of these proteins has recently been clarified by crystallography, in which it was shown that L16 becomes ordered and helps to stabilize A-site tRNA, whereas L27 has a long extension that places its amino terminus right at the PTC where it interacts with both A- and P-site tRNAs (Fig. 5) (Voorhees et al. 2009). Thus, at least in the bacterial ribosome, the PTC does have a protein component that has some role in facilitating peptidyl transfer by stabilizing tRNA substrates. In so doing, its role is not fundamentally different from that of rRNA. However, one should bear in mind that it is possible to delete L27 without affecting viability, but the deletion of many conserved RNA residues near the PTC is lethal. So the notion that the ribosome is fundamentally an RNA-based machine is unchanged, but clearly in viewing the contemporary ribosome, we are seeing a snapshot in evolution in which proteins are beginning to play a supporting role.

PEPTIDE RELEASE

The process of termination has analogies with both decoding and peptidyl transfer because the stop codon must be recognized and catalysis of the release of the peptide chain must take place at the PTC.

A significant advance has been made recently, owing to the crystal structures of both RF1 and RF2 bound to the

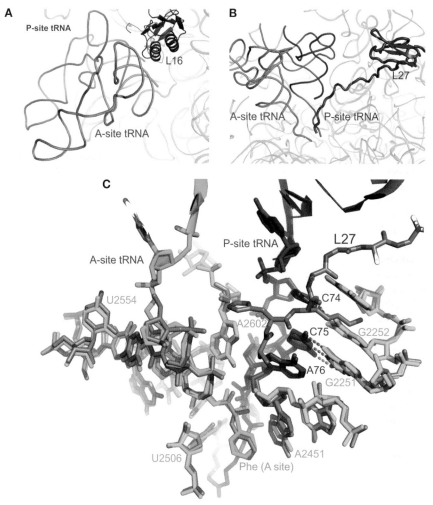

Figure 5. The peptidyl transferase center of the ribosome. (*A,B*) Location of two proteins that aid peptidyl transfer activity by stabilizing tRNA substrates. (*C*) Details of the bacterial PTC with A-site tRNA (green), P-site tRNA (red), and 23S RNA (cyan). A superposition with the archaeal peptidyl transferase center (1VQN, from Schmeing et al. 2005a) with the tRNAs in a slightly darker shade and the 23S RNA (orange) shows that bacterial and archaeal PTC are virtually identical. However, in bacteria, the amino-terminal tail of protein L27 is right at the PTC. (*A* and *B*, Reprinted, with permission, from Voorhees et al. 2009 [©Nature Publishing Group].)

ribosome (Korostelev et al. 2008; Laurberg et al. 2008; Weixlbaumer et al. 2008). These structures shed light on both the mechanism of codon recognition by these factors and the role of a conserved GGQ motif in catalysis. In particular, an induced fit of the same three nucleotides involved in decoding by tRNA is required for proper recognition of the stop codon by release factors. Moreover, a similar induced fit on the binding of the GGQ motif in the PTC is seen on release factor binding as was seen for tRNA binding, except in this case, instead of a nucleophilic attack by the amine on A-site tRNA, there is presumably an attack by a water molecule that leads to hydrolysis of the nascent chain.

It is striking that bacterial and eukaryotic release factors have no sequence or structural homology. This suggests that despite their common GGQ motif at the catalytic site, they evolved independently after the divergence of the three kingdoms. If this is true, it is likely that the role of termination was originally played by a tRNA.

Presumably, such a tRNA had anticodons complementary to a stop codon so that decoding could occur, but no synthetase was associated with them, so that they bound in the deacylated form by the more inefficient factor-free route, rather than as a complex associated with EF-Tu. They could then still induce a change in the PTC that would expose the ester bond to nucleophilic attack by water. This hypothesis is supported by the fact that even in the contemporary ribosome, deacylated tRNA promotes peptide release but not as efficiently as release factors (Zavialov et al. 2002). Presumably, release factors, in particular their properly positioned GGQ motif, more optimally coordinate a water for hydrolysis of the nascent peptide chain. They may also be more efficient at stop codon discrimination, because they have a very low error rate without the proofreading present in normal decoding (Freistroffer et al. 2000). Nevertheless, the structural and biochemical data clearly suggest how a protein factor has taken over a role once likely performed by tRNA.

TRANSLOCATION

The sequential nature of protein synthesis requires that the ribosome be able to move relative to mRNA and tRNA after each round of addition of an amino acid to the growing protein chain. This process, translocation, is highly complex and involves large-scale movements that must result in the precise movement by one codon to preserve the reading frame.

The idea that all ribosomes have two subunits because they need to move relative to one another was proposed a long time ago (Bretscher 1968; Spirin 1968). One of these proposed that the tRNAs move first relative to one subunit and only then with respect to the other to generate hybrid states (Bretscher 1968), an idea that was borne out almost two decades later in a landmark experiment using chemical footprinting of rRNA (Moazed and Noller 1989). More recent cryoEM experiments have shown that the ribosomal subunits "ratchet" or rotate relative to one another during translocation (Frank and Agrawal 2000; Valle et al. 2003), and the formation of hybrid states is indeed directly related to this ratcheting movement (Ermolenko et al. 2007).

Strikingly, the interface between the two subunits consists mainly of RNA. This suggests that the features required to ratchet as part of translocation may have existed even in a primordial protein-free ribosome. The finding that factor-free translation, however inefficient, can occur under certain conditions is in keeping with this idea (Gavrilova and Spirin 1974).

Energy Stored in tRNA

If GTP hydrolysis by EF-G is not strictly required for translocation, what determines the directionality of the movement? The progression of tRNA from A to P to E sites involves a progression of changes in its chemical state, from aminoacyl to peptidyl to deacylated. As has been pointed out previously (Spirin 1985; Noller 2005), the affinity of the various sites has evolved so that changes in the chemical state of tRNA would allow it to progress to the next site on thermodynamic grounds alone. Moreover, Noller has pointed out that the energy from peptide bond formation could be used to drive the process even in the absence of GTP hydrolysis by translational factors (Noller 2005).

Interestingly, it has been observed that the P-site tRNA is distorted relative to free tRNA in solution (Selmer et al. 2006). If tRNA is allowed to relax, e.g., after peptide bond formation when it becomes deacylated, the direction of relaxation would be such as to move it toward the E site. Therefore, some of the energy required may be stored in the distortion in P-site tRNA itself.

E Site: Conservation and Role

A particular role of the E site in this process could be to trap the intermediate state of translocation by binding the 3′ end of the tRNA that has moved from the P site of the 50S subunit, resulting in a hybrid P/E tRNA. By stably trapping this intermediate, the E site would facilitate translocation.

A comparison of the structures of the 3′ end of tRNA in a bacterial 50S subunit (Korostelev et al. 2006; Selmer et al. 2006) with that in an archaeal 50S subunit (Schmeing et al. 2003) reveals some interesting similarities and differences. The E site can only bind a deacylated tRNA, and the terminal adenine with its 2′-OH is required (Lill et al. 1989; Feinberg and Joseph 2001). This requirement is consistent with its role in trapping deacylated tRNA. Interestingly, both bacterial and archaeal E sites bind the terminal adenine in exactly the same way (Fig. 6). In both cases, the adenine base is intercalated between two conserved purines of 23S RNA and makes identical contacts with a conserved cytidine. This strongly suggests that the E site evolved even before the split among the three kingdoms. Outside the vicinity of the terminal adenine, E-site interactions are quite different. This divergence suggests that those interactions are less essential.

HOW DID THE RIBOSOME EVOLVE?

Recent structures of the ribosome have shown unambiguously that the essential functions of the ribosome such as decoding, peptidyl transfer, and translocation all appear to be mediated by RNA. The evolution of the ribosome has been much discussed (see, e.g., reviews by Moore 1993; Noller 2005), but there is little detailed understanding of how the process might have occurred. One interesting observation from the structure of the 50S subunit is that the PTC itself has a twofold symmetry that extends beyond the binding sites for A- and P-site tRNAs (Bashan et al. 2003). This suggests that an independently folded domain of RNA may have been duplicated to cre-

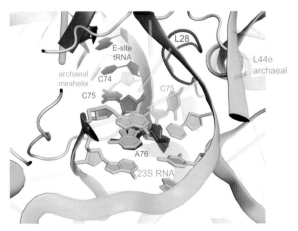

Figure 6. tRNA E site in the 50S subunit of the ribosome. A comparison of the bacterial E-site tRNA (reddish brown) (Selmer et al. 2006) with a minihelix representing the 3′ acceptor arm of tRNA in the archaeal ribosome (green) (Schmeing et al. 2003) shows that the terminal A76 is in an identical conformation making identical interactions with 23S RNA (cyan), suggesting that the E site evolved before the divergence of archaea and bacteria. However, the distinct conformations of C75 as well as the differences in the proteins (L28 in bacteria; L44e in archaea) suggest that other features of the E site have diverged significantly.

ate the precursor of the PTC. It has been noticed that independent modules that are duplicated and held by tertiary contacts involving precisely the same type of minor-groove interactions as found in decoding could be the basis for the evolution of RNA in the contemporary 50S subunit (Bokov and Steinberg 2009). The evolution of the 30S subunit and coded synthesis involving tRNA and mRNA is even less well understood despite decades of speculation. What is clear is that although the contemporary ribosome appears to be a highly complex assembly of RNA and protein, and additionally involves many different protein factors, the high-resolution structures of the ribosome provide strong support for the idea that the essential functions of the ribosome are mediated by RNA and that the ribosome evolved from a primordial RNA world. In so doing, it appears to have been a Trojan horse that accelerated the transformation of that world into the protein world that we know today.

ACKNOWLEDGMENTS

Work in the author's laboratory is supported by the Medical Research Council (U.K.), the Wellcome Trust, the Agouron Institute, and the Louis-Jeantet Foundation. I thank T. Martin Schmeing for making Figures 1–4 in connection with other publications.

REFERENCES

Ban N, Nissen P, Hansen J, Moore PB, Steitz TA. 2000. The complete atomic structure of the large ribosomal subunit at 2.4 Å resolution. *Science* **289:** 905–920.

Bashan A, Agmon I, Zarivach R, Schluenzen F, Harms J, Berisio R, Bartels H, Franceschi F, Auerbach T, Hansen HA, et al. 2003. Structural basis of the ribosomal machinery for peptide bond formation, translocation, and nascent chain progression. *Mol Cell* **11:** 91–102.

Bokov K, Steinberg SV. 2009. A hierarchical model for evolution of 23S ribosomal RNA. *Nature* **457:** 977–980.

Bretscher MS. 1968. Translocation in protein synthesis: A hybrid structure model. *Nature* **218:** 675–677.

Crick FHC. 1966. Codon-anticodon pairing: The wobble hypothesis. *J Mol Biol* **19:** 548–555.

Crick FHC. 1968. The origin of the genetic code. *J Mol Biol* **38:** 367–379.

Ermolenko DN, Spiegel PC, Majumdar ZK, Hickerson RP, Clegg RM, Noller HF. 2007. The antibiotic viomycin traps the ribosome in an intermediate state of translocation. *Nat Struct Mol Biol* **14:** 493–497.

Feinberg JS, Joseph S. 2001. Identification of molecular interactions between P-site tRNA and the ribosome essential for translocation. *Proc Natl Acad Sci* **98:** 11120–11125.

Frank J, Agrawal RK. 2000. A ratchet-like inter-subunit reorganization of the ribosome during translocation. *Nature* **406:** 318–322.

Freistroffer DV, Kwiatkowski M, Buckingham RH, Ehrenberg M. 2000. The accuracy of codon recognition by polypeptide release factors. *Proc Natl Acad Sci* **97:** 2046–2051.

Gao Y-G, Selmer M, Dunham CM, Weixlbaumer A, Kelley AC, Ramakrishnan V. 2009. The structure of the ribosome with elongation factor G trapped in the post-translocational state. *Science* **326:** 694–699.

Gavrilova LP, Spirin AS. 1974. "Nonenzymatic" translation. *Methods Enzymol* **30:** 452–462.

Gavrilova LP, Kostiashkina OE, Koteliansky VE, Rutkevitch NM, Spirin AS. 1976. Factor-free ("non-enzymic") and factor-dependent systems of translation of polyuridylic acid by *Escherichia coli* ribosomes. *J Mol Biol* **101:** 537–552.

Guerrier-Takada C, Gardiner K, Marsh T, Pace N, Altman S. 1983. The RNA moiety of ribonuclease P is the catalytic subunit of the enzyme. *Cell* **35:** 849–857.

Kapp LD, Lorsch JR. 2004. The molecular mechanics of eukaryotic translation. *Annu Rev Biochem* **73:** 657–704.

Khaitovich P, Mankin AS, Green R, Lancaster L, Noller HF. 1999. Characterization of functionally active subribosomal particles from *Thermus aquaticus*. *Proc Natl Acad Sci* **96:** 85–90.

Kool ET. 2000. Synthetically modified DNAs as substrates for polymerases. *Curr Opin Chem Biol* **4:** 602–608.

Korostelev A, Trakhanov S, Laurberg M, Noller HF. 2006. Crystal structure of a 70S ribosome-tRNA complex reveals functional interactions and rearrangements. *Cell* **126:** 1065–1077.

Korostelev A, Asahara H, Lancaster L, Laurberg M, Hirschi A, Zhu J, Trakhanov S, Scott WG, Noller HF. 2008. Crystal structure of a translation termination complex formed with release factor RF2. *Proc Natl Acad Sci* **105:** 19684–19689.

Laurberg M, Asahara H, Korostelev A, Zhu J, Trakhanov S, Noller HF. 2008. Structural basis for translation termination on the 70S ribosome. *Nature* **454:** 852–857.

Lill R, Robertson JM, Wintermeyer W. 1989. Binding of the 3′ terminus of tRNA to 23S rRNA in the ribosomal exit site actively promotes translocation. *EMBO J* **8:** 3933–3938.

Maguire BA, Beniaminov AD, Ramu H, Mankin AS, Zimmermann RA. 2005. A protein component at the heart of an RNA machine: The importance of protein L27 for the function of the bacterial ribosome. *Mol Cell* **20:** 427–435.

Moazed D, Noller HF. 1989. Intermediate states in the movement of transfer RNA in the ribosome. *Nature* **342:** 142–148.

Moore PB. 1988. The ribosome returns. *Nature* **331:** 223–227.

Moore PB. 1993. Ribosomes and the RNA world. In *The RNA world* (ed. RF Gesteland and JF Atkins), pp. 119–136. Cold Spring Harbor Laboratory Press, Cold Spring Harbor, NY.

Moore VG, Atchison RE, Thomas G, Moran M, Noller HF. 1975. Identification of a ribosomal protein essential for peptidyl transferase activity. *Proc Natl Acad Sci* **72:** 844–848.

Nissen P, Hansen J, Ban N, Moore PB, Steitz TA. 2000. The structural basis of ribosome activity in peptide bond synthesis. *Science* **289:** 920–930.

Nitta I, Ueda T, Watanabe K. 1998. Possible involvement of *Escherichia coli* 23S ribosomal RNA in peptide bond formation. *RNA* **4:** 257–267.

Nitta I, Kamada Y, Noda H, Ueda T, Watanabe K. 1999. Peptide bond formation: Retraction. *Science* **283:** 2019–2020.

Noller HF. 2005. Evolution of ribosomes and translation from an RNA world. In *The RNA world*, 3rd ed. (ed. RF Gesteland et al.), pp. 287–307. Cold Spring Harbor Laboratory Press, Cold Spring Harbor, NY.

Noller HF, Chaires JB. 1972. Functional modification of 16S ribosomal RNA by kethoxal. *Proc Natl Acad Sci* **69:** 3115–3118.

Noller HF, Hoffarth V, Zimniak L. 1992. Unusual resistance of peptidyl transferase to protein extraction procedures. *Science* **256:** 1416–1419.

Ogle JM, Ramakrishnan V. 2005. Structural insights into translational fidelity. *Annu Rev Biochem* **74:** 129–177.

Ogle JM, Brodersen DE, Clemons WM Jr, Tarry MJ, Carter AP, Ramakrishnan V. 2001. Recognition of cognate transfer RNA by the 30S ribosomal subunit. *Science* **292:** 897–902.

Ogle JM, Murphy FV, Tarry MJ, Ramakrishnan V. 2002. Selection of tRNA by the ribosome requires a transition from an open to a closed form. *Cell* **111:** 721–732.

Rodnina MV, Wintermeyer W. 2001. Fidelity of aminoacyl-tRNA selection on the ribosome: Kinetic and structural mechanisms. *Annu Rev Biochem* **70:** 415–435.

Rodnina MV, Beringer M, Wintermeyer W. 2006. Mechanism of peptide bond formation on the ribosome. *Q Rev Biophys* **39:** 203–225.

Schmeing TM, Ramakrishnan V. 2009. What recent ribosome structures have revealed about the mechanism of translation. *Nature* **461:** 1234–1242.

Schmeing TM, Moore PB, Steitz TA. 2003. Structures of deacylated tRNA mimics bound to the E site of the large ribosomal subunit. *RNA* **9:** 1345–1352.

Schmeing TM, Huang KS, Kitchen DE, Strobel SA, Steitz TA. 2005a. Structural insights into the roles of water and the 2′ hydroxyl of the P site tRNA in the peptidyl transferase reaction. *Mol Cell* **20:** 437–448.

Schmeing TM, Huang KS, Strobel SA, Steitz TA. 2005b. An induced-fit mechanism to promote peptide bond formation and exclude hydrolysis of peptidyl-tRNA. *Nature* **438:** 520–524.

Schmeing TM, Voorhees RM, Kelley AC, Gao Y-G, Murphy FV IV, Weir JR, Ramakrishnan V. 2009. The crystal structure of the ribosome bound to EF-Tu and aminoacyl-tRNA. *Science* **326:** 688–694.

Schuette JC, Murphy FV IV, Kelley AC, Weir JR, Giesebrecht J, Connell SR, Loerke J, Mielke T, Zhang W, Penczek PA, et al. 2009. GTPase activation of elongation factor EF-Tu by the ribosome during decoding. *EMBO J* **28:** 755–765.

Schuwirth BS, Borovinskaya MA, Hau CW, Zhang W, Vila-Sanjurjo A, Holton JM, Cate JH. 2005. Structures of the bacterial ribosome at 3.5 Å resolution. *Science* **310:** 827–834.

Selmer M, Dunham CM, Murphy FV IV, Weixlbaumer A, Petry S, Kelley AC, Weir JR, Ramakrishnan V. 2006. Structure of the 70S ribosome complexed with mRNA and tRNA. *Science* **313:** 1935–1942.

Sharma MR, Koc EC, Datta PP, Booth TM, Spremulli LL, Agrawal RK. 2003. Structure of the mammalian mitochondrial ribosome reveals an expanded functional role for its component proteins. *Cell* **115:** 97–108.

Sievers A, Beringer M, Rodnina MV, Wolfenden R. 2004. The ribosome as an entropy trap. *Proc Natl Acad Sci* **101:** 7897–7901.

Spirin AS. 1968. How does the ribosome work? A hypothesis based on the two subunit construction of the ribosome. *Curr Mod Biol* **2:** 115–127.

Spirin AS. 1985. Ribosomal translocation: Facts and models. *Prog Nucleic Acid Res Mol Biol* **32:** 75–114.

Valle M, Sengupta J, Swami NK, Grassucci RA, Burkhardt N, Nierhaus KH, Agrawal RK, Frank J. 2002. Cryo-EM reveals an active role for aminoacyl-tRNA in the accommodation process. *EMBO J* **21:** 3557–3567.

Valle M, Zavialov A, Sengupta J, Rawat U, Ehrenberg M, Frank J. 2003. Locking and unlocking of ribosomal motions. *Cell* **114:** 123–134.

Villa E, Sengupta J, Trabuco LG, LeBarron J, Baxter WT, Shaikh TR, Grassucci RA, Nissen P, Ehrenberg M, Schulten K, Frank J. 2009. Ribosome-induced changes in elongation factor Tu conformation control GTP hydrolysis. *Proc Natl Acad Sci* **106:** 1063–1068.

Voorhees RM, Weixlbaumer A, Loakes D, Kelley AC, Ramakrishnan V. 2009. Insights into substrate stabilization from snapshots of the peptidyl transferase center of the intact 70S ribosome. *Nat Struct Mol Biol* **16:** 528–533.

Weinger JS, Parnell KM, Dorner S, Green R, Strobel SA. 2004. Substrate-assisted catalysis of peptide bond formation by the ribosome. *Nat Struct Mol Biol* **11:** 1101–1106.

Weixlbaumer A, Jin H, Neubauer C, Voorhees RM, Petry S, Kelley AC, Ramakrishnan V. 2008. Insights into translational termination from the structure of RF2 bound to the ribosome. *Science* **322:** 953–956.

Wimberly BT, Brodersen DE, Clemons WM Jr, Morgan-Warren RJ, Carter AP, Vonrhein C, Hartsch T, Ramakrishnan V. 2000. Structure of the 30S ribosomal subunit. *Nature* **407:** 327–339.

Wower IK, Wower J, Zimmermann RA. 1998. Ribosomal protein L27 participates in both 50 S subunit assembly and the peptidyl transferase reaction. *J Biol Chem* **273:** 19847–19852.

Yusupov MM, Yusupova GZ, Baucom A, Lieberman K, Earnest TN, Cate JH, Noller H.F. 2001. Crystal structure of the ribosome at 5.5 Å resolution. *Science* **292:** 883–896.

Zaug AJ, Grabowski PJ, Cech TR. 1983. Autocatalytic cyclization of an excised intervening sequence RNA is a cleavage-ligation reaction. *Nature* **301:** 578–583.

Zavialov AV, Mora L, Buckingham RH, Ehrenberg M. 2002. Release of peptide promoted by the GGQ motif of class 1 release factors regulates the GTPase activity of RF3. *Mol Cell* **10:** 789–798.

Step-by-Step Evolution of Vertebrate Blood Coagulation

R.F. DOOLITTLE

Department of Chemistry and Biochemistry and Molecular Biology, University of California, San Diego, La Jolla, California 92093-0314
Correspondence: rdoolittle@ucsd.edu

The availability of whole-genome sequences for a variety of vertebrates is making it possible to reconstruct the step-by-step evolution of complex phenomena such as blood coagulation, an event that in mammals involves the interplay of more than two dozen genetically encoded factors. Gene inventories for different organisms are revealing when during vertebrate evolution certain factors first made their appearance in or, on occasion, disappeared from some lineages. The whole-genome sequence databases of two protochordates and seven nonmammalian vertebrates were examined in search of ~20 genes known to be associated with blood clotting in mammals. No genuine orthologs were found in the protochordate genomes (sea squirt and amphioxus). As for vertebrates, although the jawless fish have genes for generating the thrombin-catalyzed conversion of fibrinogen to fibrin, they lack several clotting factors, including two thought to be essential for the activation of thrombin in mammals. Fish in general lack genes for the "contact factor" proteases, the predecessor forms of which make their first appearance in tetrapods. The full complement of factors known to be operating in humans does not occur until pouched marsupials (opossum), at least one key factor still absent in egg-laying mammals such as platypus.

Blood coagulation in humans is a delicately balanced process involving more than two dozen extracellular proteins, many of which need to be converted from precursor forms during the process (Fig. 1). Almost half of the components are members of the serine protease family. Briefly put, the process acts as a biochemical amplifier in the forward direction, a small number of newly exposed tissue molecules acting as an input stimulus for an avalanche response converting a large number of fibrinogen molecules into a gelatinous clot. A delicate balance exists between the need for a fluid circulating state and the polymeric gel at a wound site; of necessity, the process is highly regulated, with various protease inhibitors and counteracting proteases involved. The question arises of how and when this complex process evolved.

Many of the proteins involved are clearly related to one another by gene duplications, and in the past, sequence-based phylogenies have offered insights into the relative order in which certain factors appeared (see, e.g., Doolittle and Feng 1987; Doolittle 1993; Hughes 2000). The relationship of paralogs can also be inferred by the kinds and arrangements of subsidiary domains associated with the catalytic domains (Fig. 2).

The subsidiary domains have very important roles in the blood-clotting process, serving as protein–protein interaction sites and localizing the clot at the site of injury. With regard to localization, GLA (γ-carboxyglutamic acid) domains and discoidin domains anchor certain of the factors to platelets at wound sites. Other subsidiary domains, including kringles, epidermal growth factor (EGF), and plasminogen-apple-nematode (PAN) domains, are well known to promote protein–protein interactions, leading in this case to the assemblage of factors needed for clot formation.

Which of the factors appeared first? As far as known, the thrombin-catalyzed conversion of fibrinogen to fibrin is unique to vertebrate animals (Doolittle 1961; Doolittle and Surgenor 1962), a group for which an accurately determined fossil record is available. As such, the times of appearance or disappearance of various genes can be gauged by examining genomes from various classes of extant vertebrates.

Although blood clotting is mostly an extracellular event centering around the thrombin-catalyzed conversion of fibrinogen into fibrin, blood platelets are also intimately evolved, being both thrombin sensitive, on the one hand, and having a strong affinity for fibrinogen and fibrin, on the other hand. Platelets also sequester some other coagulation factors as well. It must be kept in mind that in nonmammalian vertebrates, the equivalents of platelets are white cells commonly called thrombocytes.

SEARCHING WHOLE-GENOME SEQUENCE DATABASES

The whole-genome sequence (WGS) databases of sea squirt, amphioxus (lancelet), lamprey, puffer fish, frog, green anole lizard, chicken, platypus, and opossum were examined in search of ~20 genes known to be associated with blood clotting in mammals (Table 1).

The general strategy for identifying putative coagulation genes in WGS databases for nonmammalian organisms began with BLAST searching (Altschul et al. 1997) of the protein sequences of a queried factor as it occurs in mammals. The strongest "hits" were then "back-searched" against the standard NCBI nonrecombinant (nr) database. If the latter search returns the target protein as the highest scorer, the hit is presumed to be positive. If other proteins

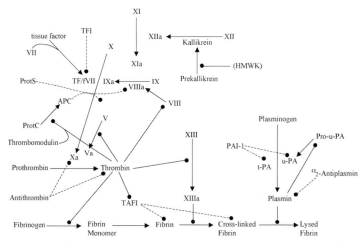

Figure 1. Schematic representation of blood coagulation pathway as it occurs in humans. (Dashed lines) Inhibitory actions. (TFI) Tissue factor inhibitor, (APC) activated protein C, (HMWK) high-molecular-weight kininogen, (TAFI) thrombin-activated fibrinolysis inhibitor, (t-PA) tissue plasminogen activator, (u-PA) urinary plasminogen activator.

have higher scores, it is generally thought that the hit is a paralog and not the targeted query.

The strategy is not foolproof. Changes in the rates of evolution after gene duplications can confound the process, and caution is the byword. The further back in time one tries to probe, the more challenging the process. Obviously, it is more difficult to prove the absence of a gene than its presence. Not all WGS databases are wholly complete or fully assembled. Genes can be missed.

The major difficulty in identifying genes in genome databases, however, is distinguishing orthologs from recently diverged paralogs. Several of the gene duplications that gave rise to new clotting factors took place very soon after the process got started in the narrow window between the appearance of protochordates and jawless vertebrates.

BASAL CHORDATES

Basal chordates, or protochordates, consist of a few groups of organisms that have a notocord at some stage in their development but do not have backbones. Two of these animals, the ascidian called sea squirt and the cephalochordate called amphioxus or lancelet, have had their genomes sequenced. Until recently, the ascidians were regarded as the earlier diverging species, compared with cephalochordates, but on the basis of their genomes, it is now argued that cephalochordates like amphioxus are the more primitive (Putnam et al. 2008).

Creatures in both groups have simple circulatory systems and primitive hearts, and it has long been thought that the clumping of circulating cells at wound sites is the only hemostatic mechanism needed, there being no evi-

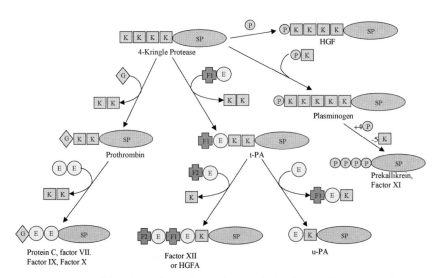

Figure 2. Step-by-step rearrangement of domains during early vertebrate evolution that could have given rise to assorted clotting factor proteases. Other pathways are possible. (K) Kringle, (G) GLA domain, (E) EGF domain, (SP) serine protease, (P) PAN domain, (F1) fibronectin type I, (F2) fibronectin type II. (Adapted from Jiang and Doolittle 2003.)

Table 1. Chordate Genome Sequences Used in This Study

Protochordates	*Ciona intestinales*	sea squirt
	Branchiostoma floridae	amphioxus
Jawless fish	*Petromyzon marinus*	lamprey
Bony fish	*Fugu rubripes*	puffer fish
Amphibian	*Xenopus laevis*	frog
Reptile	*Anolis carolinensis*	green anole lizard
Bird	*Gallus gallus*	chicken
Monotreme	*Ornithorrhychus anatinus*	platypus
Marsupial	*Monodelphis domestica*	opossum
Mammal	*Homo sapiens*	human

dence of a fibrin clot (Fry 1909). This view was borne out by a systematic survey of genes in the sea squirt (*Ciona intestinalis*), with no bona fide clotting factor genes being found (Jiang and Doolittle 2003).

The amphioxus genome is somewhat larger than that of the sea squirt (540 Mb vs. 330 Mb), the latter having lost many genes, and it was therefore important to search the amphioxus genome for putative clotting factor genes in case the reduced genome of sea squirts had lost them. In the end, no genes for authentic coagulation factors were found in amphioxus. There *are* numerous genes for fibrinogen-related domains (FREDs) in both genomes (sea squirt and amphioxus), some of which have been cloned (Fan et al. 2008), but none are clustered with other domains that are characteristic of fibrinogen molecules.

Similarly, there are genes that have sequences which resemble the serine protease domains of vertebrate prothrombins, but they do not have coding regions for the subsidiary domains that typify this factor, such as the GLA domain (Fig. 2).

Because fibrinogen, the mainstay of the clot, has never been found in any protochordate by any method, we must conclude that the invention of thrombin-catalyzed fibrin formation took place in the ~50–100-million-year window between the appearance of protochordates and the jawless fish.

JAWLESS FISH

Jawless fish are the earliest appearing vertebrates. Two genera are extant: hagfish and lamprey. Although there has been a long-standing debate as to whether these two are monophyletic, current opinion favors separate divergences, with hagfish being the more primitive (Janvier 1996). Genomic data for hagfish are sparse, however, and for the moment, our study of clotting factors in this group is mostly limited to the lamprey. Even so, the system in these creatures is decidedly simpler than in mammals and serves as an illustration of how such a system can become more complex.

As an example, mammals have two large, multidomain nonenzyme proteins—factors V and VIII—that have key roles in thrombin generation. These homologous proteins are descended from another blood plasma protein called ceruloplasmin. Sequence differences aside, factors V and VIII differ from ceruloplasmin and some other homologs in having two discoidin domains at their carboxy-terminal ends. The discoidin domains bind to platelet (thrombocyte) surfaces, whereas the main bodies of these proteins are situated normal to the surface and serve as "holders" for factor X in the case of factor V, or factor IX in the case of factor VIII, and bring about the proper orientation that allows these proteases to convert prothrombin to thrombin.

The situation in lampreys is simpler in that only one of these protease-holder duos appears to be present (Fig. 3). Neither factor IX (protease) nor factor VIII (holder protein) was found by searching the lamprey genomic databases (Doolittle et al. 2008), and efforts to clone these factors have been negative, even though other related clotting factors were found (Kimura et al. 2009).

The question may be asked, how can new factors be introduced into an existing pathway? It was long ago suggested that in the case of clotting pathways, new factors that are the products of gene duplications could easily be sandwiched into the middle of pathways where they initially were only performing the same operation as the original gene product. Only a few amino acid replacements were likely needed to broaden the proteolytic specificity to the point where the duplicon could itself activate the other surviving gene product (Doolittle 1961). In line with this thinking, all of the vitamin-K-dependent proteases (prothrombin, factors VII, IX, and X, and protein C) cleave after arginine residues in the same general regions of their homologous substrates. Comparison of thrombin generation in lampreys with that in other vertebrates illustrates such a scheme of events perfectly (Fig. 3).

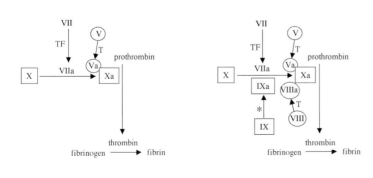

Figure 3. Comparison of thrombin generation in a jawless fish (lamprey) with that in other vertebrates. The lamprey system lacks factor IX (a protease) and factor VIII (a cofactor protein). (T) Thrombin, (TF) tissue factor. The asterisk (*) denoting the activator for converting factor IX to IXa would be factor VIIa in fish and a contact phase factor in tetrapods. (Adapted from Doolittle et al. 2008.)

In passing, we may also note that the lamprey also has some "extra" factors. The genome data make it clear that there are three factors VII, just as was found in the puffer fish (Davidson et al. 2003; Jiang and Doolittle 2003). Additionally, the lamprey has two factors X, the duplication in this case having occurred well after the divergence of lampreys from the rest of the vertebrate lineage.

THE EMERGENCE OF CONTACT FACTORS

Historically, thrombin generation has been regarded as a two-pronged process made up of an *extrinsic* pathway and an *intrinsic* pathway, although it is now recognized that considerable "cross-talk" exists between the two pathways. Classically, the extrinsic pathway begins with tissue factor combining with factor VII to activate factor X in conjunction with factor V. The intrinsic system involves platelet stimulation of factor IX in conjunction with factor VIII to activate factor X similarly. In mammals, factor IX is greatly activated by a subpathway called the "contact factor system," provoked by contact with "foreign" surfaces, as might occur in an injury circumstance. The contact system is composed of a high-molecular-weight kininogen (HMWK) and three proteases: factors XI and XII and prekallikrein (Fig. 4).

Although HMWK is found in fish (Zhou et al. 2008), the contact system proteases do not make their phyletic appearance until after the evolution of tetrapods (Ponczek et al. 2008). Moreover, the HMWK found in fish lacks the particular domain thought to be responsible for the contact activation, with that domain first appearing in HMWK at the level of tetrapods (Zhou et al. 2008).

The contact factor proteases obviously arose from gene duplications of other proteins. Factor XII has the same domainal arrangement as a protease known as hepatocyte growth factor activator (HGFA in Fig. 2). The amino acid sequences of prekallikrein and factor XI, which have strings of four PAN domains (Tordai et al. 1999), cluster with those of plasminogens and hepatocyte growth factor (HGF), both of which have PAN domains at their amino termini (Fig. 2).

Interestingly, reptiles have the gene for factor XII, but birds have lost it (Table 2). There has been a long-standing debate about whether or not factor XII in mammals is truly essential to effective blood clotting, a concern that intensified with the finding that it is also absent in some marine mammals (Robinson et al. 1969).

MORE REGULATION IN MAMMALS

Several additional regulatory devices for the clotting scheme have evolved among mammals during the past 100 million years. One of these was the appearance of a large protein called apolipoprotein(a), or apo(a), a recognizable paralog of plasminogen. But whereas plasminogen has five kringles, apo(a) is a polymorphic protein that can have as many as 50 kringles, almost all of which resemble a single kringle found in plasminogen (Lawn et al. 1997). It also has an inactive relic of a serine protease domain, having lost some key active site residues. The protein is intimately bound up with low-density lipopoteins (LDLs), a well-known threat to the vascular circulation. Somehow, apo(a) modulates the destruction of clots, interfering with the action of plasmin. Remarkably, it seems to have evolved independently on two occasions, once among insectivores such as the hedgehog, and again among higher primates (Lawn et al. 1997).

EVOLUTION OF THE FIBRINOGEN γ′ CHAIN

Another recently evolved device for regulating clot formation is the result of a new splicing site in the gene for the γ chain of fibrinogen. As it happens, the splicing at this site in humans is not wholly efficient, and 5%–10% of the time, a nonsplice leads to an alternative chain. The alternative chain, called γ′, has been shown to have two important functions; binding factor XIII in fibrinogen and binding thrombin once it has become a part of fibrin. Not all species have active γ′ sequences exhibiting both functions; some species such as the mouse do not bind thrombin at this site at all (Mosesson et al. 2009). WGS studies now reveal that the intron involved in this particular γ-chain splicing did not make its appearance until the evolution of reptiles and birds (Doolittle et al. 2009).

THE ORIGINS OF VERTEBRATE CLOTTING

The total absence of clotting factors in basal chordates and the existence of a relatively complex system in lampreys indicate that it is not likely we will find any extant creatures with truly rudimentary clotting systems, leaving us to speculate on how the system might have gotten started. With regard to the numerous serine proteases involved, sequence-based phylogenetic trees invariably show thrombin to be earliest appearing of the GLA-containing proteases. Moreover, in mammals, thrombin is known to have a relatively broad range of activities, targeting not only fibrinogen, but also factors V, VIII, and XIII and platelets. It seems unlikely that thrombin and fibrinogen would appear simultaneously; more reasonably, one already existed with an alternative function. For

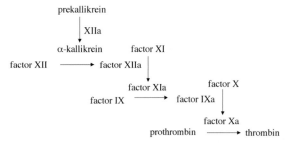

Figure 4. In mammals, contact factor proteases lead to activation of factor IX.

Table 2. Occurrence of Genes for Contact Factor Proteases and Some Paralogs

Organism	Factor XI	Prekallikrein	Factor XII	HGFA	HGF
Human	yes	yes	yes	yes	yes
Opossum	yes	yes	yes	yes	yes
Platypus	no	yes	yes	yes	yes
Chicken	no	yes	no	yes	yes
Green lizard	no	yes	yes	yes	yes
Frog	no	yes	yes	yes	yes
Zebra fish	no	no	no	?	yes
Puffer fish	no	no	no	yes	yes
Lamprey	no	no	no	yes	yes

(HGF) Hepatocyte growth factor; (HGFA) hepatocyte growth factor activator. HGFA and factor XII have the same domainal arrangement. Factor XI, prekallikrein, and HGF contain PAN modules. (Adapted and updated from Ponczek et al. 2008.)

example, fibrinogen may have had a role in cell–cell interactions, a property of many proteins with fibrinogen-related domains.

A more likely scenario, however, is that thrombin had an early role in agglutinating thrombocytes by proteolyzing cell surface proteins, something it is known to do today, attacking a set of G-protein-coupled receptors called PAR proteins (Vu et al. 1991; Coughlin 2005). According to this scenario, a tissue factor would become exposed during the course of injury, activating prothrombin that would then clump cells which were the ancient ancestors of mammalian platelets (Doolittle 1993). A GLA domain could have helped to keep thrombin localized on the surface of the thrombocytes.

The subsequent appearance of fibrinogen would allow thrombin to broaden its attack, generating a more durable clot composed of fibrin (Fig. 5). Duplications of the prothrombin gene would lead to the appearance of factors VII, X, and eventually IX.

The main problem with this simple scenario has to do with the kinds of subsidiary domains found in thrombin. Besides the GLA domain, thrombin has two kringle domains, usually thought to have an affinity for fibrin. The kinds of domains that interact with tissue factor, however, are the EGF domains found in factors VII, X, and IX. It may be that there was much domain shuffling in the early stages and that thrombin originally had EGF domains, or no peripheral domains at all. Kringle-containing serine proteases have been identified in both sea squirts (Jiang and Doolittle 2003) and amphioxus (Liu and Zhang 2009) and may have served as starting points for many of the proteases later to be involved in clotting (Fig. 2).

WHY DID THE CLOTTING SYSTEM BECOME MORE COMPLEX?

It is fair to ask why the clotting system is more complex in mammals than it is in earlier diverging vertebrates. Indeed, the quality and character of the fibrin clots generated in fish and mammals do not appear to be significantly different. The answers must have to do with regulating the response under a wider spectrum of environments. Higher blood pressure, more complicated cardiovascular systems, higher metabolic rates, and new organs such as lungs or the placenta all introduce more challenges for maintaining the balance between liquidity and gelation. Having more components in the amplification system may also make it possible to deliver a maximum response with a smaller input stimulus, and this in turn would necessitate having a faster turn-off or lytic response.

CONCLUSIONS

Vertebrate blood clotting evolved in parallel with the development of a pressurized and closed circulatory system with its all important red cell content. Its origin and evolution occurred within the 50–100-million-year window between the appearance of protochordates and vertebrates (Fig. 6). A census of component genes in the

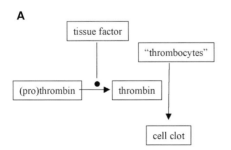
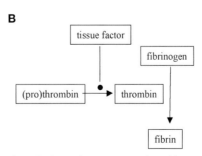

Figure 5. Two simple systems depicting how vertebrate clotting may have arisen. In *A*, a serine protease activated by exposure of a tissue factor provokes cell clumping at the site by attacking a cell surface protein. In *B*, the specificity of the protease is broadened to include an attack on fibrinogen that leads to its polymerization into fibrin.

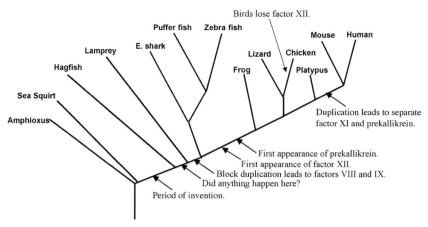

Figure 6. Time-line phylogeny for appearance (and disappearance) of various clotting factors during the course of vertebrate evolution.

lamprey genome has revealed a simpler, but still complex, system in jawed vertebrates. The system is expanded in tetrapods, more factors being added step by step until the time of marsupials. Even then, some additional regulatory devices have appeared along lineages leading to various mammalian groups, including the vulnerability of fibrin to fibrinolytic enzymes, a process that can be regulated by alternative splicing.

ACKNOWLEDGMENTS

I am grateful to students and colleagues who have helped with this study in recent years, including Sung Hong, Yong Jiang, Justin Nand, and Michal Ponczek.

REFERENCES

Altschul SF, Madden TL, Schaffer AA, Zhang J, Zhang Z, Miller W, Lipman DJ. 1997. BLAST and PSI-BLAST: A new generation of protein database search programs. *Nucleic Acids Res* **25:** 3389–3402.

Coughlin SR. 2005. Protease-activated receptors in hemostasis, thrombosis and vascular biology. *J Thromb Haemost* **3:** 1800–1814.

Davidson CJ, Tuddenham EG, McVey JH. 2003. 450 Million years of hemostasis. *J Thromb Haemost* **1:** 1487–1494.

Doolittle RF. 1961. "The comparative biochemistry of blood coagulation." PhD thesis, Harvard University, Cambridge.

Doolittle RF. 1993. The evolution of vertebrate blood coagulation: A case of Yin and Yang. *Thromb Haemost* **70:** 24–28.

Doolittle RF, Feng D-F. 1987. Reconstructing the evolution of vertebrate blood coagulation from a consideration of the amino acid sequences of clotting proteins. *Cold Spring Harbor Symp Quant Biol* **52:** 869–874.

Doolittle RF, Surgenor DM. 1962. Blood coagulation in fish. *Am J Physiol* **203:** 964–970.

Doolittle RF, Jiang Y, Nand J. 2008. Genomic evidence for a simpler clotting scheme in jawless vertebrates. *J Mol Evol* **66:** 185–196.

Doolittle RF, Hong S, Wilcox D. 2009. Evolution of the fibrinogen γ' chain: Implications for the binding of factor XIII, thrombin and platelets. *J Thromb Haemost* (in press).

Fan C, Zhang S, Li L, Chao Y. 2008. Fibrinogen-related protein from amphioxus *Branchiostoma belcheri* is a multivalent pattern recognition receptor with a bacteriolytic activity. *Mol Immunol* **45:** 3338–3346.

Fry HJB. 1909. Blood platelets and coagulation of the blood in marine chordates. *Folia Hematol* **8:** 467–503.

Hughes AL. 2000. Modes of evolution in the protease and kringle domains of the plasminogen-prothrombin family. *Mol Phylogenet Evol* **14:** 469–478.

Janvier P. 1996. The dawn of the vertebrates: Characters versus common ascent in current vertebrate phylogenies. *Paleontology* **39:** 259–287.

Jiang Y, Doolittle RF. 2003. The evolution of vertebrate blood coagulation as viewed from a compassion of puffer fish and sea squirt genomes. *Proc Natl Acad Sci* **100:** 7527–7532.

Kimura A, Ikeo K, Nonaka M. 2009. Evolutionary origin of the vertebrate blood complement and coagulation systems inferred from liver EST analysis of lamprey. *Dev Comp Immunol* **33:** 77–87.

Lawn RM, Schwartz K, Patthy L. 1997. Convergent evolution of apolipoprotein(a) in primates and hedgehog. *Proc Natl Acad Sci* **94:** 11992–11997.

Liu M, Zhang S. 2009. A kringle-containing protease with plasminogen-like activity in the basal chordate *Branchiostoma belcheri*. *Biosci Rep* (in press).

Mosesson MW, Cooley BC, Hernandez I, Diorio JP, Weiler H. 2009. Thrombosis risk modification in transgenic mice containing the human fibrinogen thrombin-binding γ' chain sequence. *J Thromb Haemost* **7:** 102–110.

Ponczek M, Gailani D, Doolittle RF. 2008. Evolution of the contact phase of vertebrate blood coagulation. *J Thromb Haemost* **6:** 1876–1883.

Putnam NH, Butts T, Ferrier DE, Furlong RF, Hellsten U, Kawashima T, Robinson-Rechavi M, Shoguchi E, Terry A, Yu JK, et al. 2008. The amphioxus genome and the evolution of the chordate karyotype. *Nature* **453:** 1064–1071.

Robinson AJ, Kropatkin M, Aggeler PM. 1969. Hageman factor (fXII) deficiency in marine mammals. *Science* **166:** 1420–1422.

Tordai H, Banyai L, Patthy L. 1999. The PAN module: The N-terminal domains of plasminogen and growth factor are homologous with the apple domains of the prekallikrein family and with a novel domain found in numerous nematode proteins. *FEBS Lett* **461:** 63–67.

Vu T-K, Hung DT, Wheaton VI, Coughlin SR. 1991. Molecular cloning of a functional thrombin receptor reveals a novel proteolytic mechanism of receptor activation. *Cell* **64:** 1057–1068.

Zhou L, Li-Ling J, Huang H, Fei M, Li Q. 2008. Phylogenetic analysis of vertebrate kininogen genes. *Genomics* **91:** 129–141.

How Proteins Adapt: Lessons from Directed Evolution

F.H. ARNOLD

Division of Chemistry and Chemical Engineering, California Institute of Technology, Pasadena, California 91125

Correspondence: frances@cheme.caltech.edu

Applying artificial selection to create new proteins has allowed us to explore fundamental processes of molecular evolution. These "directed evolution" experiments have shown that proteins can readily adapt to new functions or environments via simple adaptive walks involving small numbers of mutations. With the entire "fossil record" available for detailed study, these experiments have provided new insight into adaptive mechanisms and the effects of mutation and recombination. Directed evolution has also shown how mutations that are functionally neutral can set the stage for further adaptation. Watching adaptation in real time helps one to appreciate the power of the evolutionary design algorithm.

In making his case for the role of natural selection in evolution, Darwin started by pointing to the enormous phenotypic variation that could be achieved in just a few generations of artificial selection. In Darwin's day, the importance of good breeding practices in determining the productivity and quality of a farmer's stock or the crop size of a pigeon fancier's prize bird was clear to all, and Darwin's artificial selection arguments provided a powerful foundation for his idea that competition for limited resources could similarly tailor phenotypes, and ultimately create new species, by selecting for beneficial traits.

Today, we can use artificial selection to breed not just organisms, but also the protein products of individual genes. By subjecting them to repeated rounds of mutation and selection (a process usually referred to as "directed evolution"), we can enhance or alter specific traits and even force a protein to acquire traits not apparent in the parental molecule. And, just as in Darwin's day, these artificial selection experiments have the potential to teach us a great deal about evolution, only now at the molecular level. The remarkable ease with which proteins adapt in the face of defined selection pressures, from acquiring the ability to function in a nonnatural environment to degrading a new antibiotic, was largely unexpected when the first laboratory protein evolution experiments were performed two decades ago.

Directed evolution experiments can recapitulate different adaptive scenarios that may at least partially characterize natural protein evolution. But perhaps even more interesting is the opportunity to go where nature has not necessarily gone. Under artificial selection, a protein can evolve outside of its biological context. This allows us to explore the acquisition of novel features, including those that may not be useful in nature. In this way, we can distinguish properties or combinations of properties that are biologically relevant and found in the natural world from others that may be physically possible but are not relevant and not easily encoded, and therefore are not encountered in natural proteins.

Darwin's great insight long predated any understanding of the molecular mechanisms of inheritance and evolution. How DNA-coding changes alter protein function is new information that directed evolution experiments contribute to the evolution story. With access to the entire "fossil record" of an evolution experiment, we can determine precisely how gene sequences change during adaptation and can connect specific mutations to specific acquired traits. During the last 20 years, directed evolution experiments have revealed that useful properties such as catalytic activity or stability can frequently be enhanced by single-amino-acid substitutions and that significant functional adaptation can occur by accumulation of relatively few such beneficial mutations (changing as little as 1%–2% of the sequence). This contrasts with the large sequence distances—frequently 50% or more—that separate natural protein homologs, which have diverged and adapted to different functions or environments. Directed evolution can identify minimal sets of adaptive mutations, but the precise mechanisms by which adaptation occurs are still difficult to discern: The individual effects of beneficial mutations are usually quite small, and their locations and identities are often surprising (e.g., distant from active sites).

Directed evolution experiments have also elucidated a key feature of the fitness landscape for protein evolution. A common expectation has been that mutational pathways to new properties would be tortuous, reflecting a fitness landscape that is highly rugged. In fact, laboratory evolution experiments have demonstrated over and over again that smooth mutational pathways—simple uphill walks consisting of single beneficial mutations—exist and lead to higher fitness. Many interesting and useful properties can be manipulated by the accumulation of beneficial mutations one at a time in iterative rounds of mutagenesis and screening or selection.

The role of neutral mutations in protein evolution has also been explored. Directed evolution has demonstrated an important mechanism whereby mutations that are functionally neutral but stabilize the protein's three-dimensional structure can set the stage for further adaptation by providing the extra stability that allows functionally important but destabilizing mutations to be accepted. By this mechanism, stability contributed by functionally neutral mutations promotes evolvability. In addition, it has been demonstrated that accumulating mutations which are neutral for one function can lead to the appearance of others—a kind of functional "promiscuity"—that can serve as a handle for evolution of new functions such as the ability to bind a new ligand or to catalyze a reaction on a new substrate.

The molecular diversity on which artificial selection acts can be created in any number of ways in order to mimic natural mutagenesis mechanisms: Directed evolution experiments use random (point) mutagenesis of a whole gene or domain, insertions, and deletions, as well as other, more hypothesis-driven mutagenesis schemes. Another important natural mutation mechanism is recombination. We have explored how recombination can contribute to making new proteins, by looking at its effects on folding and structure as well as function. Recombination of homologous proteins is highly conservative compared to random mutation—a protein can acquire dozens of mutations by recombination and still fold and function, whereas similar levels of random mutation lead to loss of function. Although the mutations made by recombination are less disruptive of fold and function, they can nonetheless generate functional diversity. Experiments have shown that recombined, or "chimeric," proteins can acquire new properties, such as increased stability or the ability to accept new substrates, through novel combinations of the mostly neutral mutations that accumulated during natural divergence of the homologous parent proteins.

In the remainder of this chapter, I describe how directed evolution experiments performed in this laboratory on a model enzyme, a bacterial cytochrome P450, have provided support for these lessons. This is a personal account, and I apologize in advance for making no attempt to cover the large relevant literature and contributions from other laboratories.

DIRECTED EVOLUTION: A SIMPLE MOLECULAR OPTIMIZATION STRATEGY

Directed evolution starts with a functional protein and uses iterative rounds of mutation and selection to search for more "fit" proteins, where fitness is defined by the experimenter via an assay or some other test (e.g., a genetic selection). The parent gene is subjected to mutation, and the mutants are expressed as a library of protein variants. Variants with improved fitness are identified, and the process is repeated until the desired function is achieved (or not). Directed evolution usually involves the accumulation of beneficial mutations over multiple generations of mutagenesis and/or recombination, in a simple uphill walk on the protein fitness landscape (Romero and Arnold 2009).

Directed evolution relies on proteins' abilities to exhibit a wider array of functions and over a wider range of environments than might be required for their biological functions. This functional promiscuity, even if only at some minimal level, provides the jumping-off point for optimization toward that new goal. A good starting protein for directed evolution exhibits enough of the desired function that small improvements (expected from a single mutation) can be discerned reliably. If the desired behavior is beyond what a single mutation can confer, the problem can be broken down into a series of smaller ones, each of which can be solved by the accumulation of single mutations, for example, by gradually increasing the selection pressure or evolving against a series of intermediate challenges.

Epistatic interactions occur when the presence of one mutation affects the contribution of another. These nonadditive interactions lead to curves in the fitness landscape and constrain evolutionary searches. Mutations that are negative in one context but become beneficial in another are a ubiquitous feature of protein landscapes, where they create local optima that could frustrate evolutionary optimization. Directed evolution, however, does not find all paths to high fitness, only the most probable paths. These follow one of many smooth routes and bypass the more rugged, epistatic routes. Hundreds of directed evolution experiments have demonstrated that such smooth paths to higher fitness can be found for a wide array of protein fitness definitions, including stability, ability to function in nonnatural environments, ability to bind a new ligand, changes in substrate specificity or reactivity, and more (Bloom and Arnold 2009).

CYTOCHROME P450 BM3: A MODEL ENZYME FOR DIRECTED EVOLUTION

The cytochrome P450 enzyme superfamily provides a superb example of how nature can generate a whole spectrum of catalysts from a single shared structure and mechanism (Lewis and Arnold 2009). More than 10,000 P450 sequences have been identified from all kingdoms of life, where they catalyze the oxidation of a stunning array of organic compounds. These enzymes all recruit a cysteine-bound iron heme cofactor responsible for this activity. The widely varying substrate specificities of the P450s are determined by their protein sequences, which accumulated large numbers of amino acid substitutions as they diverged from their common ancestor. Despite differences in up to 90% of the amino acid sequences, the P450s all share a common fold.

P450 BM3 from *Bacillus megaterium* (BM3) is particularly attractive for laboratory evolution experiments. It is one of only a handful of known P450s in which the heme domain and the diflavin reductase domains (FMN and FAD) required for generation of the active oxidant are fused in a single polypeptide chain. Furthermore, it is soluble and readily overexpressed in *Escherichia coli*, an excellent host for directed evolution experiments. The substrates of BM3 are largely limited to long-chain fatty acids, which it hydroxylates at subterminal positions at high rates (thousands of turnovers per minute). During the past decade, we and others investigators have used directed evolution to alter the specificity of this well-behaved bacterial P450 family member so that it can mimic the activities of widely different P450s, including some of the human enzymes. These experiments have demonstrated that dramatic changes in substrate specificity can be achieved with just a few mutations in the catalytic (heme) domain (Landwehr et al. 2006; Rentmeister et al. 2008; Lewis and Arnold 2009; Lewis et al. 2009).

Is a P450 Propane Monooxygenase Physically Possible?

One of our early directed evolution goals was to generate a P450 that could hydroxylate small, gaseous alkanes such as propane and ethane. In nature, these are substrates of methane monooxygenases, enzymes that are mechanistically and evolutionarily unrelated to the cytochrome P450 enzymes. A P450 had never been reported to accept propane, ethane, or methane as a substrate. We were curious as to whether a P450 heme oxygenase was capable of binding and inserting oxygen into ethane or methane, whose C–H bond strengths are considerably higher than those of the usual natural P450 substrates.

P450 BM3 hydroxylates the alkyl chains of fatty acids containing 12–16 carbons and has no measurable activity on propane or smaller alkanes. We have never found any single mutation that confers this activity. To make a version of BM3 that hydroxylates propane, we therefore first targeted activity on a longer alkane (octane), a substrate that the wild-type enzyme does accept, albeit poorly (Glieder et al. 2002). We reasoned that variants of BM3 having enhanced activity on octane might eventually acquire measurable activity on shorter alkanes and thus that further mutagenesis and screening on progressively smaller substrates could ultimately generate enzymes with good activity on the gaseous alkanes. This reasoning assumed that the problem was mainly one of substrate recognition and that there is no inherent mechanistic limitation to hydroxylation of small alkanes at the heme iron.

Five generations of random mutagenesis of the heme domain, recombination of beneficial mutations, and screening for activity on an octane surrogate led to BM3 variant 139-3, which contains 11 amino acid substitutions and is much more active on octane (Glieder et al. 2002). The improved octane activity was in fact accompanied by measurable activity on smaller alkanes, including propane. Further rounds of mutation and recombination of beneficial mutations further enhanced activity on propane. Variant 35E11, with 17 mutations relative to BM3, was highly active on propane and even provided modest conversion of ethane to ethanol (Meinhold et al. 2005). Breaking down the more difficult problem of obtaining activity on very small substrates by first targeting octane and then propane lowered the bar for each generation and allowed the new activities to be acquired one mutation at a time.

This enzyme, however, was still not as efficient at hydroxylating the alkanes as is the wild-type enzyme with its preferred fatty acid substrates. Finely tuned conformational rearrangements within and among the heme and reductase domains mediate electron transfer and efficiently couple BM3-catalyzed hydroxylation to consumption of the NADPH (nicotinomide adenine dinucleotide phosphate) cofactor. When these processes are disrupted, either by mutations or by introduction of novel substrates, catalysis is no longer coupled to cofactor consumption: NADPH consumption instead produces reactive oxygen species that eventually cause the enzyme to self-destruct. To retune the whole system for oxidation of propane, we therefore also targeted the FMN and FAD domains of variant 35E11 for mutagenesis (individually, but in the context of the holoenzyme) and continued to screen for increased ability to convert propane to propanol. We then combined the optimized heme, FAD, and FMN domains to generate $P450_{PMO}$ (Fasan et al. 2007, 2008). This enzyme displayed activity on propane comparable to that of BM3 on fatty acids and 98% coupling of NADPH consumption to product hydroxylation. $P450_{PMO}$ thus became as good an enzyme on propane as the wild-type enzyme is on laurate with a total of 23 amino acid substitutions, amounting to changes in less than 2.3% of its (>1000 amino acid holoenzyme) sequence.

Creation of $P450_{PMO}$, a complex, multidomain enzyme finely tuned for activity on a substrate not accepted by the wild-type enzyme, demonstrates the remarkable ability of the cytochrome P450 to adapt to new challenges by accumulating single beneficial mutations over multiple generations.

The $P450_{PMO}$ Evolutionary Trajectory

Studying the evolutionary intermediates along the lineage of $P450_{PMO}$ revealed interesting features of adaptation to propane. Activity on propane first emerged in 139-3, a variant that is active on a wide range of substrates. But by the time the enzyme became highly active and fully coupled on propane, it had lost its activity on laurate—a more than 10^{10}-fold change in specificity, just from 139-3 to $P450_{PMO}$. Thus, becoming a good propane monooxygenase in $P450_{PMO}$ came at the cost of the native enzyme's activity on fatty acids, even though this property was not included in the artificial selection pressure. This is apparently the easiest route to high activity on propane.

Substrate specificity changes for selected variants along the lineage to $P450_{PMO}$ were also investigated on alkanes having one to 10 carbons. These activity profiles revealed that intermediate variants (e.g., 139-3, 35E11) acquired activity on a range of alkanes before ultimately respecifying for propane (Fig. 1) (Fasan et al. 2008). $P450_{PMO}$ is highly specific compared to its precursors: Its activity drops precipitously on alkanes having just one more or one less methylene group. Only positive selection (for high activity on propane) had been used to obtain $P450_{PMO}$; there was no selection against activity on any other substrate. One can conclude that it is easier to obtain very high activity on propane than it is to have high activity on a range of substrates; thus, a highly active "specialist" is easier to find than a highly active "generalist."

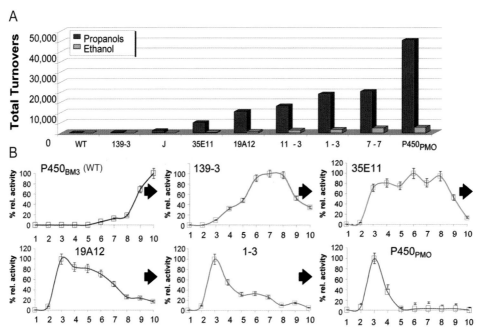

Figure 1. (*A*) Total turnovers catalyzed by selected variants along the P450$_{PMO}$ lineage on propane and ethane. (*B*) Relative activities on C$_n$ (n = 1–10) alkanes. (Reprinted, with permission, from Lewis and Arnold 2009 [© Swiss Chemical Society].)

Sequencing reveals the mutations acquired in each generation of directed evolution. The 21 amino acid substitutions in the heme domain of P450$_{PMO}$ (two of the 23 are in the reductase domain) are distributed over the entire protein (Fig. 2). Many are distant from the active site and influence specificity and catalytic activity through unknown mechanisms. The crystal structure of 139-3 (Fasan et al. 2008) revealed only small changes in the active site volume, consistent with its activity toward a wide range of substrates. Modeling studies, however, indicate much more dramatic reduction in the volume accessible to substrate in P450$_{PMO}$ (C Snow, unpubl.).

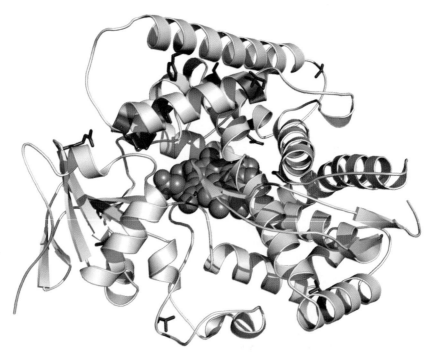

Figure 2. P450 BM3 heme domain backbone, showing locations of 21 of 23 mutations that convert P450 BM3 to a highly active, fully coupled propane monooxygenase (P450$_{PMO}$).

STABILITY PROMOTES EVOLVABILITY: A ROLE FOR NEUTRAL MUTATIONS IN ADAPTIVE EVOLUTION

It is useful to consider when this simple adaptive walk might fail. Of course, it will fail if the functional bar is set too high—this happens when the fitness improvements required to pass the screen or selection are not reached by single mutations. It also fails when the protein is not robust to mutation (Bloom et al. 2005, 2006). At one point during the evolution of $P450_{PMO}$, in fact at mutant 35E11, we could find no additional mutations that further enhanced the enzyme's activity on propane. Upon characterizing 35E11 and its precursors, the reason for this became clear: The enzyme had become so unstable that it simply could not tolerate any further destabilization and still function under the expression and assay conditions. Most mutations are destabilizing, and most activating mutations are also destabilizing, possibly more so than the average mutation. The process of enhancing P450's activity on propane had destabilized it so much that 35E11 simply could not accept any further destabilizing mutations. Once we incorporated mutations that stabilized the structure (but were neutral or nearly neutral with respect to activity), directed evolution of activity could continue as before, and significant additional improvements were achieved (Fasan et al. 2007). Stabilizing the structure made it robust to further mutation and opened up the ability to explore a whole spectrum of mutational paths that were previously inaccessible.

We demonstrated this key role of functionally neutral but stabilizing mutations in adaptive evolution with another experiment that directly compared the frequency with which a marginally stable and a highly stable cytochrome P450 enzyme could acquire activities on a set of new substrates upon random mutation (Bloom et al. 2006). A markedly higher fraction of mutants of the stable protein were found to exhibit the new activities. This increased evolvability could be traced directly to the enzyme's ability to tolerate catalytically beneficial but destabilizing mutations.

Directed evolution has thus shown the crucial role that stability-based epistasis can have in adaptive evolution. A protein that has been pushed to the margins of tolerable stability may lose access to functionally beneficial but destabilizing mutations. But this protein is still not stuck on a fitness peak, because it can regain its mutational robustness and evolvability via a neutral path, by accumulating stabilizing mutations that do not directly affect function. In natural evolution, such a process might require stabilizing mutations to spread by genetic drift (Bloom and Arnold 2009).

ADAPTIVE EVOLUTION RELIES ON FUNCTIONAL PROMISCUITY, WHICH CHANGES WITH NEUTRAL MUTATIONS

A well-recognized feature of proteins is their functional promiscuity. Enzymes, for example, often catalyze a much wider range of reactions, or reactions on a wider range of substrates, than are biologically relevant. Directed evolution experiments have shown that protein activities or functions present at a low level can often be improved via an adaptive pathway of sequential beneficial mutations. Protein functional promiscuity thus provides a stepping stone for generation and optimization of new functional molecules by adaptive evolution.

Directed evolution experiments with P450 BM3 have also demonstrated that promiscuous activities can emerge on mutations that are neutral with respect to a main (biological) function (Bloom et al. 2007). We performed a kind of neutral evolution by random mutagenesis and selection for retention of catalytic activity on a fatty-acid-like substrate. The variants containing these "neutral" mutations were then examined for activity on several other nontarget substrates. In many cases, the neutral mutations had led to changes in these promiscuous activities. Neutral mutations can also set the stage for adaptation by exploring a varied set of evolutionary starting points, at little or no cost to the current biological function.

I already discussed how neutral mutations can enhance a protein's stability, thereby increasing its tolerance for subsequent functionally beneficial but destabilizing mutations. Neutral mutations can also lead to changes in functions that are not currently under selective pressure but can subsequently become the starting points for the adaptive evolution of new functional proteins. A process that generates large numbers of mostly neutral mutations is recombination (of homologous proteins), which exploits the genetic drift that underlies the divergence of their sequences. As we discuss below, swapping these mutations in the laboratory can generate proteins different from the parent proteins, including those that are more stable or exhibit activities not present in the parents.

NOVEL PROTEINS BY RECOMBINATION

Recombination is an important mutation mechanism in natural protein evolution. We have studied the effects of mutations made by recombination of homologous proteins (that share a three-dimensional structure but may differ at hundreds of amino acid residues) by making and characterizing large sets of "chimeric" proteins. The probability that a protein retains its fold and function declines exponentially with the number of random mutations it acquires—random mutations are quite deleterious on average. By quantifying the retention of function with mutation level in chimeric β-lactamases made by swapping sequence elements between two homologous enzymes, we showed that the mutations made by recombination are much more conservative, presumably because they had already been selected for compatibility with the lactamase folded structure (Drummond et al. 2005). Recombination can generate proteins that have a high probability of folding and functioning despite having dozens of mutations compared to their parent sequences. Thus, recombination is conservative. But does it lead to new functions or traits?

We generated a large set of recombined P450 heme domains by swapping sequence elements among three natural P450 BM3 homologs sharing ~65% sequence identity

(Otey et al. 2006). A sampling of the functional P450s showed that they exhibited a range of activities, including activity on substrates not accepted by the parent enzymes (Landwehr et al. 2007). The chimeric P450s also exhibited a range of stabilities, with a significant fraction of them more stable than any of the parent enzymes from which they were constructed (Li et al. 2007). Like many proteins, P450s are only marginally stable, never having been selected for thermostability or long-term stability. Depending on the degree to which stability has already been maximized in the parent sequences, recombination can generate proteins that are less stable or more stable than the parent proteins.

Recombination shuffles large numbers of mutations that individually have little or no effect on function. Our experiments have shown that these mutations can generate proteins of widely varying stabilities and with a wide range of promiscuous activities, both of which can open new pathways for further functional evolution.

CONCLUSIONS

Directed evolution does not necessarily mimic natural evolution: Laboratory proteins evolve under artificial pressures and via mutation mechanisms that usually differ significantly from those encountered during natural evolution. These experiments nevertheless allow us to explore protein fitness landscapes, the nature of the evolutionary trajectories, as well as the functional features of individual protein sequences. Anything created in the laboratory by directed evolution is also probably easily discovered by natural evolution. Thus, knowing what functional features are accessible to evolution helps us to understand what biology cares about, i.e., what features are retained and encouraged by natural selection, and what biology tends to throw away. Laboratory evolution experiments beautifully demonstrate that biological systems, themselves the products of millions of years of evolution, readily evolve to meet new challenges.

ACKNOWLEDGMENTS

The author thanks all of her coworkers that have contributed to the work described here, and especially thanks Jesse D. Bloom, Phil Romero, Jared C. Lewis, and Rudi Fasan. Support is from the Jacobs Institute for Molecular Medicine, the Department of Energy, the U.S. Army, DARPA, and the National Institutes of Health.

REFERENCES

Bloom JD, Arnold FH. 2009. In the light of directed evolution: Pathways of adaptive protein evolution. *Proc Natl Acad Sci* **106:** 9995–10000.

Bloom JD, Silberg JJ, Wilke CO, Drummond DA, Adami C, Arnold FH. 2005. Thermodynamic prediction of protein neutrality. *Proc Natl Acad Sci* **102:** 606–611.

Bloom JD, Labthavikul ST, Otey CR, Arnold FH. 2006. Protein stability promotes evolvability. *Proc Natl Acad Sci* **103:** 5869–5874.

Bloom JD, Romero PA, Lu Z, Arnold FH. 2007. Neutral genetic drift can alter promiscuous protein functions, potentially aiding functional evolution. *Biol Direct* **2:** 17.

Drummond DA, Silberg JJ, Meyer MM, Wilke CO, Arnold FH. 2005. On the conservative nature of intragenic recombination. *Proc Natl Acad Sci* **102:** 5380–5385.

Fasan R, Chen MM, Crook NC, Arnold FH. 2007. Engineered alkane-hydroxylating cytochrome $P450_{BM3}$ exhibiting native-like catalytic properties. *Angew Chem Int Ed* **46:** 8414–8418.

Fasan R, Meharenna YT, Snow CD, Poulos TL, Arnold FH. 2008. Evolutionary history of a specialized P450 propane monooxygenase. *J Mol Biol* **383:** 1069–1080.

Glieder A, Farinas ET, Arnold FH. 2002. Laboratory evolution of a soluble, self-sufficient, highly active alkane hydroxylase. *Nat Biotechnol* **20:** 1135–1139.

Landwehr M, Hochrein L, Otey CR, Kasrayan A, Bäckvall J-E, Arnold FH. 2006. Enantioselective α-hydroxylation of 2-arylacetic acid derivatives and buspirone catalyzed by engineered cytochrome P450 BM-3. *J Am Chem Soc* **128:** 6058–6059.

Landwehr M, Carbone M, Otey CR, Li Y, Arnold FH. 2007. Diversification of catalytic function in a synthetic family of chimeric cytochrome P450s. *Chem Biol* **14:** 269–278.

Lewis JC, Arnold FH. 2009. Catalysts on demand: Selective oxidations by laboratory-evolved cytochrome P450 BM-3. *Chimia* **63:** 309–312.

Lewis JC, Bastian S, Bennett CS, Fu Y, Mitsuda Y, Chen MM, Greenberg WA, Wong C-H, Arnold FH. 2009. Chemoenzymatic elaboration of monosaccharides using engineered cytochrome P450 B_{M3} demethylases. *Proc Natl Acad Sci* **106:** 16550–16555.

Li Y, Drummond DA, Sawayama AM, Snow CD, Bloom JD, Arnold FH. 2007. A diverse family of thermostable cytochrome P450s created by recombination of stabilizing fragments. *Nat Biotechnol* **25:** 1051–1056.

Meinhold P, Peters MW, Chen MY, Takahashi K, Arnold FH. 2005. Direct conversion of ethane to ethanol by engineered cytochrome P450 BM3. *Chembiochem* **6:** 1765–1768.

Otey CR, Landwehr M, Endelman JB, Hiraga K, Bloom JD, Arnold FH. 2006. Structure-guided recombination creates an artificial family of cytochromes P450. *PLoS Biol* **4:** 0789–0798.

Rentmeister A, Arnold FH, Fasan R. 2008. Chemo-enzyme fluorination of unactivated organic compounds. *Nat Chem Biol* **5:** 26–28.

Romero PA, Arnold FH. 2009. Exploring protein fitness landscapes by directed evolution. *Nat Rev Mol Cell Biol* **10:** 866–876.

Reconstructing the Emergence of Cellular Life through the Synthesis of Model Protocells

S.S. Mansy[1] and J.W. Szostak[2]

[1]*Armenise-Harvard Laboratory of Synthetic and Reconstructive Biology, Centre for Integrative Biology, University of Trento, 38100 Mattarello (Trento), Italy;* [2]*Howard Hughes Medical Institute, and Department of Molecular Biology and Center for Computational and Integrative Biology, Massachusetts General Hospital, Simches Research Center, Boston, Massachusetts 02115*

Correspondence: szostak@molbio.mgh.harvard.edu

The complexity of modern biological life has long made it difficult to understand how life could emerge spontaneously from the chemistry of the early earth. The key to resolving this mystery lies in the simplicity of the earliest living cells, together with the ability of the appropriate molecular building blocks to spontaneously self-assemble into larger structures. In our view, the two key components of a primitive cell are not only self-assembling, but also self-replicating, structures: the nucleic acid genome and the cell membrane. Here, we summarize recent experimental progress toward the synthesis of efficient self-replicating nucleic acid and membrane vesicle systems and discuss some of the issues that arise during efforts to integrate these two subsystems into a coherent whole. We have shown that spontaneous nucleic-acid-copying chemistry can take place within membrane vesicles, using externally supplied activated nucleotides as substrates. Thus, membranes need not be a barrier to the uptake of environmentally supplied nutrients. We examine some of the remaining obstacles that must be overcome to enable the synthesis of a complete self-replicating protocell, and we discuss the implications of these experiments for our understanding of the emergence of Darwinian evolution and the origin and early evolution of cellular life.

The differences between chemical evolution and Darwinian evolution are clear-cut at the extremes, but they are less distinct at the interface where complex chemistry transforms into simple biology. A careful consideration of the transition from chemistry to biology reveals a series of stages, in which the inheritance of variation and the natural selection of advantageous phenotypes become progressively more powerful and open-ended. For example, the sequences of the first polynucleotides would have been biased by simple chemical considerations such as the concentration and intrinsic reactivities of the available monomers, and different sequences would have been degraded at rates dependent on chemical factors such as the intrinsic lability of different linkages, as well as the tendency of certain sequences to fold into more stable secondary structures. The sequence distribution of the population would thus change with time, due to differences in the rate of synthesis and degradation. At this level, we can see the emergence of genetically determined phenotypes, including differential survival, but the transmission of variation from generation to generation is still missing. This critical property comes into play as soon as some mechanism for template-directed replication arises. Because some sequences would undoubtedly be better templates than others, there would certainly be natural selection for sequences that were replicated more efficiently. This could include subtle competing effects such as selection for secondary structures that slow down reannealing with complementary strands while not being too difficult to replicate. Nevertheless, it seems unlikely that template copying in solution could lead to more complex outcomes, such as the emergence of catalysts of replication or metabolic reactions. To attain this highest and most open-ended level of natural selection, we have argued previously that spatial compartmentalization is required (Szostak et al. 2001).

Many potential modes of compartmentalization have been considered, e.g., replication on the surface of mineral particles or within porous rocks, but biology makes universal use of bilayer lipid membranes for cellular compartmentalization. Whether there was at some stage a jump to this mode of organization, or whether this form of spatial localization came first, is unclear. In any case, this was a momentous step in the emergence of life, because cell membranes keep genomic molecules and their products physically together so that the genomic molecules can reap the benefits of any useful functions that they encode. At this level, the amazing ability of natural selection to generate diverse and highly adapted new life-forms came fully into play. In summary, just as the history of prebiotic chemistry involved a series of stages of increasing molecular complexity, so we imagine the origin of cellular life as the culmination of a series of stages of increasingly complex organization at the molecular and supramolecular levels, with the final stage being a recognizably biological structure capable of open-ended Darwinian evolution (Fig. 1). In this chapter, we refer to a cell-like structure with the as yet unrealized potential to evolve useful functions as a protocell; once the evolution of sequences that encode useful functions has occurred, we refer to such a structure simply as a cell.

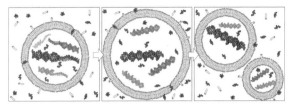

Figure 1. Schematic model of a protocell. A replicating vesicle enables spatial localization, and a replicating genome encodes heritable information. A complex environment provides nucleotides, lipids, and various sources of energy, including mechanical energy for division, chemical energy for nucleotide activation, and phase transfer and osmotic gradient energy for growth. (Reprinted, with permission, from Mansy et al. 2008 [Nature Publishing Group].)

PROTOCELL SYNTHESIS IN THE LABORATORY AND ON THE EARLY EARTH

We expect that efforts to synthesize model protocells in the laboratory will provide useful clues to the kinds of molecules and the nature of the physical environments that may have conspired to generate protocells on the early earth. Such efforts build on decades of pioneering work in prebiotic chemistry (Orgel 2004), the self-assembly and replication of membrane vesicles (Hanczyc and Szostak 2004), the nature of potential genetic polymers (Eschenmoser 1999), and the nonenzymatic template-directed copying of nucleic acid sequences (Joyce and Orgel 2006). Although ultimately interested in plausible scenarios for the origin of life, we have chosen to avoid, for now, the real or imagined constraints of prebiotic chemistry. Our immediate goal is simply to demonstrate the possibility of purely physicochemical replication schemes for model protocell membranes and genetic polymers and to show that these can be mutually compatible. Once such a system is devised, however artificial, we suspect that it will become much easier to construct related systems that are more realistic analogs of potential early earth systems.

NUCLEIC ACID REPLICATION

The initial polymerization of nucleotides into more or less random sequence polynucleotides has been shown to occur in at least two quite different ways: by local concentration effects arising from absorption to clay minerals (Ferris et al. 1996) or by freezing to eutectic conditions (Kanavarioti et al. 2001). The influence of minerals, such as montmorillonite, is particularly interesting because montmorillonite can catalyze both nucleic acid polymerization and membrane formation, thus bringing nucleic acid polymers and vesicle membranes together (Hanczyc et al. 2003). Although vesicles will encapsulate polynucleotide strands by spontaneous self-assembly around dissolved strands, the mineral-directed assembly and encapsulation of polynucleotides may significantly enhance the efficiency of this process. A recently described alternative is the simultaneous concentration of dilute fatty acids and nucleic acids by thermophoresis in thin channels such as those found in hydrothermal vents (Baaske et al. 2007), which leads to the assembly of vesicles containing encapsulated nucleic acids (Budin et al. 2009).

NONENZYMATIC TEMPLATE COPYING

Vesicle-encapsulated polynucleotides must replicate to be maintained during subsequent generations of vesicle growth and division; the ability of vesicles to acquire nucleotides from the environment provides a path for fueling template-dependent nucleic acid replication reactions inside vesicles. For replication reactions to occur without the aid of sophisticated catalysts such as protein enzymes or ribozymes, as may have occurred early in the origin of life, and as we wish to occur in our model protocells, it is necessary to consider modifications of standard nucleotide chemistry. For example, it has long been known that the substitution of imidazole for the pyrophosphate leaving group increases reactivity considerably, thereby allowing template-directed polymerization to occur spontaneously (Lohrmann and Orgel 1973, 1976). However, these more reactive nucleoside phosphorimidazolides are also subject to faster hydrolysis and cyclization, which compete with polymerization. Further increases in reactivity (without a corresponding increase in hydrolysis) can be obtained by substituting the hydroxyl nucleophile of the nucleotide with a better nucleophile, such as an amino group (Zielinski and Orgel 1985; Tohidi et al. 1987). The resulting phosphoramidate-linked polynucleotides are quite similar to the corresponding phosphodiester-linked polynucleotides. The template-directed polymerization of 3′-amino ribonucleoside phosphorimidazolides was initially examined by Orgel and colleagues (Lohrmann and Orgel 1976), but this line of inquiry was not pursued, apparently due to concerns that amino nucleotides were not prebiotically realistic. We have chosen to investigate a series of phosphoramidate polynucleotides (Fig. 2) with the goal of identifying a nucleic acid system capable of complete cycles of self-replication. Such a self-replicating nucleic acid would allow for the synthesis of interesting laboratory protocell models and might also provide clues to the nature of primitive chemically replicating genetic polymers on the early earth.

We have recently described the synthesis and base-pairing properties of the phosphoramidate version of glycerol nucleic acid (GNA) (Chen et al. 2009). Not surprisingly, this system is problematic due to the extremely rapid cyclization of its activated nucleotide monomers. However, we were able to circumvent this problem by synthesizing activated dinucleotides, which were in turn able to assemble into longer oligonucleotides by template-directed ligation. These experiments highlight the importance of a sterically constrained sugar moiety in a nucleotide monomer, so that the reactive nucleophile and leaving group are physically held apart from each other and cyclization cannot occur. One such system, which we have begun to study, is based on 2′-amino dideoxyribonu-

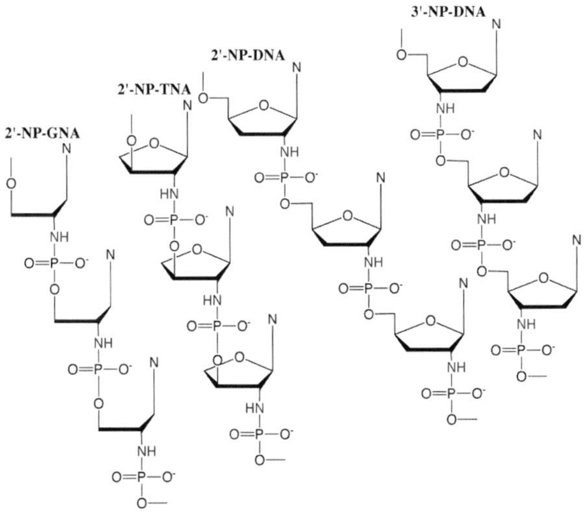

Figure 2. Phosphoramidate-linked nucleic acids. (*Left to right*) Phosphoramidate-linked glycerol nucleic acid (NP-GNA), threose nucleic acid (NP-TNA), 2′-5′-phosphoramidate DNA (2′-5′-NP-DNA), and 3′-5′-phosphoramidate DNA (3′-5′-NP-DNA).

cleotides, which cannot cyclize. Preliminary experiments showed that 2′-amino dideoxyguanosine, activated as the 5′-phosphorimidazolide, was able to rapidly polymerize on an oligo(dC) template (Mansy et al. 2008). We are continuing to explore the potential of this promising system as a sequence-general, chemically replicating genetic polymer.

NUTRIENT ACQUISITION BY PROTOCELLS

Modern cells are highly organized structures that interact extensively with their surroundings through the selective absorption and release of chemicals. In the absence of protein transport machinery, protocells would have had to rely on the selective permeability of their membranes for nutrient uptake and waste release. Because membrane permeability properties dictate which molecules are available for protometabolic processes, it is important to evaluate the similarities and differences between model protocell membranes, composed of simple, prebiotically plausible amphiphiles such as fatty acids and related molecules, and contemporary biological membranes, which are composed of more complex amphiphiles such as phospholipids and sterols. In general, it appears that permeabilities are similar for small, uncharged molecules, whereas large differences are observed for ionic solutes (Hargreaves and Deamer 1978; Chen et al. 2005; Sacerdote and Szostak 2005).

Permeability of Membranes to Sugars and Nucleotides

Fatty acid membranes are similar to phospholipid membranes in that they are both permeable to and can discriminate between small uncharged molecules (Hargreaves and Deamer 1978; Sacerdote and Szostak 2005). Differences in permeation rates among small neutral solutes depend strongly on hydrophobicity and molecular size (Gerasimov et al. 1996; Kleinzeller 1999). A particularly interesting example is the permeability of ribose in comparison with its diastereomers. Despite having equal molecular weight and number of hydroxyl groups, the permeability coefficient of ribose is fivefold larger than that of its diastereomers (Sacerdote and Szostak 2005). Recent molecular dynamics simulations suggest that the greater permeability of ribose reflects its internally satisfied H-bonding interactions and consequently less extensive water–solute interactions (Wei and Pohorille 2009). Membranes composed of lipids that flip quickly and slowly have similar permeabilities, but increased fluidity does correlate with increased rates of permeation without loss of selectivity for ribose (Sacerdote and Szostak 2005; Mansy et al. 2008). The selective permeability of model prebiotic membranes for ribose suggests an early kinetic advantage of ribose over other five-carbon sugars. The observed preference for ribose could have provided an advantage for a protocell that used ribose (as opposed to any of its diastereomers) to make nucleotides.

An important difference between fatty acid and phospholipid membranes is that phospholipid membranes are generally impermeable to charged solutes, whereas fatty acid membranes are permeable. Differences in solute permeability that depend on the characteristics of individual fatty acid molecules have been studied using phospholipid membranes that contain trace amounts of fatty acids (Kamp and Hamilton 1992). The presence of low concentrations of fatty acid renders phospholipid membranes permeable to protons and metal ions. The data are consistent with a carrier mechanism in which carboxylate head groups, neutralized by protonation or complexation with metal, flip from one leaflet to another and then release their solute (Kamp and Hamilton 1992). Subsequent studies of monovalent and divalent cation permeability of membranes composed entirely of fatty acids are mechanistically consistent with these earlier studies on mixed phospholipids–fatty acid systems. However, because more exploitable solute interaction sites exist and because lipid flipping dynamics are faster in pure fatty acid membranes, cation fluxes are orders of magnitude higher than that for mixed phospholipids–fatty acid membranes (Chen and Szostak 2004).

In addition to small cations, larger charged organic molecules, such as nucleotides, can traverse fatty acid, but not phospholipid, membranes (Chen et al. 2005; Mansy et al. 2008). At slightly alkaline pH, nucleotide monophosphates have a charge of 2–, which greatly decreases their ability to cross hydrophobic barriers. However, complexation with Mg^{++} significantly increases permeability. The importance of charge neutralization is exemplified by comparison between AMP and ADP permeabilities. Although ADP is more highly charged, it also has a higher affinity for Mg^{++} (Khalil 2000) and so it is more easily neutralized. Thus, in the presence of Mg^{++}, ADP crosses fatty acid membranes more quickly than AMP (Mansy et al. 2008). Although the mechanism of nucleotide permeation is not known, the difference in permeation through fatty acid membranes versus phospholipid membranes is consistent with a dependence on lipid dynamics, perhaps via a mechanism similar to the carrier mechanism described above for monovalent cations. It is striking that a charged and relatively large molecule, such as a nucleotide, can pass through fatty acid membranes, whereas larger polymers, such as proteins and nucleic acids, are retained. Large nonspecific pores must not be present in fatty acid membranes, because such pores would not be able to discriminate between mononucleotides and polymers.

Heterogeneous Membrane Composition

Primitive cell membranes must have been composed of mixtures of lipids, and the properties of such membranes can be quite different from those of membranes assembled from a single component (Mansy and Szostak 2008). Many of the permeability changes observed for mixed lipid membranes arise from decreased acyl-chain packing, and thus decreased van der Waals interactions. For example, permeability is increased by adding branched isoprenoid membrane components. Similarly, large bulky head groups result in a separation of acyl chains and thus decreased acyl-chain packing and increased solute permeability. The addition of 33 mol% of the glycerol ester of myristoleic acid (GMM) to a myristoleic acid (MA) membrane increases the permeability of the membrane to nucleotides by fivefold. However, it should be noted that head group influences may be much more complex than simple steric and shape effects. For example, charge repulsion between lipid head groups, solute–head group interactions, and the influences of the polarity of the head group on lipid flipping dynamics all affect the structure and permeability properties of membranes (Fig. 3). In addition, solute-lipid interactions could increase local solute concentrations thereby increasing permeation rates regardless of the mechanism invoked.

Heterotrophy

The permeability properties of primitive membranes are an important consideration for models of the nature of early cells. Highly impermeable membranes would have allowed for the retention of internally synthesized small molecules, but they would have prevented the absorption of nutrients from the environment as well as the release of chemical wastes. For such a system to survive, mechanisms for the catalysis of internal metabolic reactions, and for the replication of the catalysts, must have been present from the beginning. Thus far, little experimental progress has been made in creating laboratory models of such autotrophic protocells. Conversely, prebiotically plausible lipids self-assemble into semipermeable compartments capable of acquiring complex chemical nutrients without the aid of specific membrane transporters and without any danger of losing their genetic heritage. The resulting vesicle system lays the foundation for the construction of a laboratory model of a heterotrophic protocell capable of Darwinian evolution. However, for such a system to be realized, a full replication cycle including compartment growth and division, and nucleic acid replication, must be achieved.

TEMPLATE COPYING INSIDE VESICLES

The ability of fatty acid vesicles to passively absorb nucleotides coupled with the existence of an efficient nonenzymatic model of template-directed nucleotide polymerization allowed us to explore a simple protocell model involving nucleic acid replication inside a membrane vesicle. We were able to show that with only four components, excluding buffer and salts, a model protocell can be assembled that is capable of nutrient acquisition and internal nucleic acid copying (Mansy et al. 2008). The four components included two amphiphiles (the fatty acid myristoleic acid and its glycerol ester) and two oligonucleotides (a DNA primer and a DNA template). The fatty acids were used to prepare small unilamellar vesicles, which encapsulated the primer–template complex. When activated nucleotides were added to the outside of the vesicles, they diffused to the vesicle interior and polymerized by template-directed primer extension. The primer-extension reaction proceeded more slowly than analogous solution reactions, due to the time required for nucleotide permeation across the membrane but nevertheless progressed to completion in about 24 h.

Continued cycles of nucleic acid replication would result in increased internal osmotic pressure, because the polymer strands are trapped inside the vesicle, whereas the monomers can equilibrate between the internal compartment and external environment. This increased internal osmotic pressure can drive vesicle growth (Chen et al. 2004) at the expense of neighboring vesicles with less internal nucleic acid and thus a lower osmotic pressure. In effect, activated nucleotides can fuel both nucleic acid copying and the growth of fatty acid vesicles. Because this growth results from competition between vesicles for a limiting resource (fatty acids), any sequences that favored faster replication would come to dominate the population. This system is remarkable in that it suggests that it may be possible to create, from a relatively small number of components, a system that is capable of Darwinian evolution.

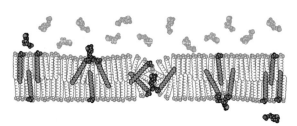

Figure 3. Model for spontaneous membrane permeation by charged solutes. Solute molecules form transient complexes with membrane amphiphiles through polar interactions with head groups and nonpolar interactions with acyl chains. Inversion across the membrane is followed by solute release. (Adapted, with permission, from Mansy et al. 2008 [Nature Publishing Group.])

STRAND SEPARATION

Complete nucleic acid replication requires cycles of template copying followed by strand separation. Assuming that some genetic polymer will be found that is capable of rapid and accurate strand copying, a mechanism for strand separation would then be required to complete the replication cycle. Strand separation can be achieved by a variety of means, including enzymatic, chemical, and thermal processes, but by far the simplest means of obtaining cycles of strand separation is through thermal fluctuations, as commonly used today in polymerase chain reaction (PCR). Indeed, the reconstitution of PCR with *Taq* polymerase (Oberholzer et al. 1995) inside impermeable phospholipid vesicles shows that strand separation can be thermally driven inside vesicles, as long as those vesicles are sufficiently stable and do not release their contents at high temperature.

Thermal Stability of Fatty Acid Vesicles

Fatty-acid-based vesicles are in general more delicate than phospholipid vesicles and are disrupted by conditions such as high ionic strength, the presence of divalent cations, and extremes of pH that do not affect phospholipid vesicles (Monnard et al. 2002). We were therefore surprised to observe that fatty acid vesicles can be quite thermally stable, depending on their composition (Mansy and Szostak 2008). Adding fatty alcohols or fatty acid glycerol esters to pure fatty acid vesicles can result in dramatic stabilization. For example, myristoleic acid membranes will retain encapsulated DNA for 1 h at a maximum temperature of ~50°C, whereas membranes containing monomyristolein (GMM) were stable for 1 h to 100°C. Less stable membrane compositions, such as decanoic acid vesicles, also retained encapsulated DNA at 100°C, albeit for much shorter periods of time. In general, decreasing acyl-chain length or disrupting acyl-chain packing through branching results in decreased thermal stability.

Duplex Melting Inside Fatty-acid-based Vesicles

The surprising thermal stability of fatty-acid-based vesicles suggested that the strands of duplex nucleic acids could be thermally separated inside such vesicles without the loss of genetic material to the external environment. We were able to monitor thermal DNA melting and annealing by fluorescence spectroscopy of encapsulated duplex DNA, of which a trace amount was labeled with a fluorescent dye on one oligonucleotide and a quencher on the other; in this state, the dye fluorescence was largely quenched (Fig. 4) (Mansy and Szostak 2008). However, after thermal denaturation and reannealing, the dye and quencher were randomly distributed between separate duplexes, leading to a greatly increased fluorescent signal. We were also able to show that no DNA leaked out of the vesicles during the high-temperature treatment that led to DNA melting and strand separation.

What is the most plausible source of temperature fluctuations on the early earth that could have driven strand separation? Although day–night cycles are frequently invoked, diurnal temperature fluctuations on the modern earth are generally small, suggesting that sufficient warming to cause strand separation would require an

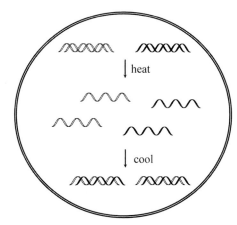

Figure 4. Strand separation inside vesicles. (Black lines) DNA strands labeled with donor and quencher dyes. When annealed to each other, fluorescence is low. (Open lines) Unlabeled DNA strands. Following strand separation and reannealing, the donor and quencher oligonucleotides are separated, resulting in a high fluorescence signal.

average ambient temperature close to the melting point of the nucleic acid duplex. Because template copying chemistry generally proceeds optimally at low temperatures, this scenario is problematic. An interesting alternative scenario involves geothermal heating near hydrothermal vents or hot springs. In a cold environment, ponds might be partially covered with ice while being locally heated by hot rocks, which would induce convection currents that would result in entrained particles being occasionally briefly heated and then rapidly cooled. Laboratory models of such convection currents are capable of mediating PCR by carrying nucleic acids through regions of high and low temperatures to melt and anneal nucleic acids, respectively (Krishnan et al. 2002; Braun et al. 2003). Similar geothermally driven temperature cycles on the early earth may have been responsible for duplex melting and strand separation inside of semipermeable vesicles.

Vesicle Permeability at High Temperatures

The permeability of fatty-acid-based membranes to nucleotides is dramatically enhanced at high temperatures. Nucleotide equilibration across fatty acid membranes is less than half complete after 24 h at 23°C for 100-nm-diameter unilamellar vesicles, whereas at 90°C, the reaction is complete in <10 min (Mansy and Szostak 2008). Permeability enhancement at elevated temperatures complements thermally driven cycles of nucleic acid replication because the acquisition of fresh nucleotides and the dissipation of waste material can occur concurrently with strand separation. Thus far, the myristoleic acid:GMM model membrane system is the system most amenable to the construction of a model protocell, because these vesicles are both highly permeable and highly thermostable while forming at a relatively low amphiphile concentration of ~1 mM. A more prebiotically reasonable membrane composition consisting of decanoic acid mixed with decanol and a decanoate-glycerol ester requires a higher amphiphile concentration to form vesicles (~20 mM) and can only withstand brief high-temperature excursions. Because the esterification of glycerol to a fatty acid can occur under model prebiotic conditions by dehydration condensation (Apel and Deamer 2005), the first step in constructing phospholipids from fatty acids would have been both chemically simple and advantageous.

COMPARTMENT DIVISION

An essential aspect of the design of a model protocell is the nature of the compartment division process. Until recently, no robust but simple cell division mechanism has been described. Modern cells rely on an intricate set of cytoskeletal proteins that coordinate membrane division with cell wall growth and nucleic acid partitioning (Margolin 2000). However, studies of pure phospholipid membranes demonstrate that budding and division can be induced with thermal or osmotic fluctuations without the aid of proteins (Hanczyc and Szostak 2004). The relevant physical forces arise from membrane asymmetries, such as differences in leaflet composition or area, the presence of phase-separated lipid domains, or changes in the ratios of surface area to volume. These processes are important in that they demonstrate the ability of vesicles to divide in the absence of protein machinery, but they cannot form the basis of a cell cycle because they rely on asymmetries that are lost upon division.

A crude but effective way of forcing vesicles to divide is by extrusion through small pores. Despite the very high shear stresses imposed by extrusion, only 30%–40% of encapsulated molecules are lost to the environment during division (Hanczyc et al. 2003). This is little more than the volume decrease that must occur upon division of a spherical vesicle into two daughter vesicles with the same total surface area. The retention of contents during extrusion suggests that division occurs by the pinching off of the daughter vesicles, as opposed to the disruption of the parental vesicle and the resealing of smaller membrane fragments. Because the exposure of acyl chains to aqueous solution is highly energetically unfavorable, the membrane resists tearing and any transiently induced pore rapidly seals. Although compartment division can be induced by extrusion, we have argued on hydrodynamic grounds that an analogous process is unlikely to have operated on the early earth (Zhu and Szostak 2009). However, the laboratory demonstration of vesicle extrusion as a division mechanism allowed us to demonstrate a cyclical process of growth via feeding with micelles followed by extrusion-induced division (Hanczyc et al. 2003). Such a cycle could be used as the basis of a laboratory model of a protocell, but a more efficient system that avoided the loss of a substantial fraction of the protocell contents during each division cycle would be highly desirable.

Most studies of the growth and division of fatty acid vesicles have focused on the behavior of small unilamellar vesicles, which are relatively homogeneous. However, spontaneous vesicle formation, either by rehydration of dried films or by the acidification of alkaline micelles, generates a heterogeneous mixture of vesicles of varying lamellarity and a wide size distribution. Recent fluorescence microscopy studies of large multilamellar vesicles revealed a surprising mode of vesicle growth and division (Fig. 5) (Zhu and Szostak 2009). Following the addition of excess micelles, multilamellar vesicles grow by forming a thin tail-like protrusion; during ~30 min, the thin protrusion elongates and thickens until the entire original vesicle is subsumed into the resulting long filamentous vesicle. This filamentous vesicle is quite fragile and divides into smaller daughter vesicles upon gentle agitation, with no loss of contents to the environment (Zhu and Szostak 2009). The daughter vesicles that form by fragmentation of the filamentous vesicle are still multilamellar, and they can grow again with the addition of more fatty acids in the form of micelles. The precise mechanism of division remains unclear, but highly elongated phospholipid vesicle structures are known to spontaneously divide into many daughter vesicles through the pearling instability (Bar-Ziv and Moses 1994), and the division of filamentous fatty acid vesicles may also proceed through the pearling instability. Growth and division through a filamentous intermediate state were observed for a wide range of membrane compositions, vesicle sizes, and environmental conditions. Two critical factors contribute to this growth mode: a transient imbalance between surface area and volume growth, and the multilamellar nature of the vesicles. In the presence of a highly permeable buffer such as ammonium acetate, volume growth is not limited by slow solute equilibration; as a result, the outer membrane of the initial vesicle swells up spherically, and no filamentous protrusion is formed. In contrast to multilamellar vesicles, large unilamellar vesicles grow more symmetrically into elongated structures but never form a thin initial protuberance. In addition, after growth, large unilamellar vesicles are disrupted by gentle agitation and lose much of their contents to the environment. Growth and division via a filamentous intermediate structure is appealing because it occurs with vesicles that form spontaneously, and therefore appears to be prebiotically plausible. In addition, growth and division are mechanistically coupled because it is the pattern of growth that predisposes the vesicle to division (Zhu

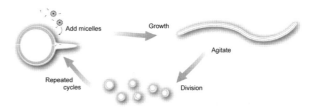

Figure 5. Cycles of vesicle growth and division. A spherical multilamellar vesicle grows after the addition of fatty acid micelles by the formation of a thin protuberance. This grows over time until the initially spherical vesicle transforms into filamentous vesicle. Gentle agitation leads to division into daughter vesicles, which in turn can grow and repeat the cycle. (Reprinted, with permission, from Zhu and Szostak 2009 [©American Chemical Society].)

and Szostak 2009). This vesicle growth–division cycle is very efficient at retaining encapsulated contents. The fact that multilamellar vesicles grow into filamentous vesicles that are predisposed to divide upon gentle agitation is a strong indication that essential life-like properties can emerge from simple chemical components in an appropriate physical environment.

CONCLUSIONS

Simple amphiphiles such as fatty acids and related molecules spontaneously assemble into membrane vesicles that have properties which are well-suited for models of primitive cells, in that they can grow by incorporating additional molecules, and they allow for the uptake of charged nutrients such as nucleotides from the environment. They are compatible with the chemistry of nucleic acid template copying, and we have been able to demonstrate such copying reactions using encapsulated templates. These primitive membranes are surprisingly thermostable, and they withstand the high temperatures required for strand separation of internal nucleic acid duplexes. Finally, multilamellar vesicles undergo a shape transformation, from spherical to filamentous, during growth, and the resulting filamentous vesicles are sufficiently fragile that they divide into daughter vesicles in response to gentle agitation.

Although this recent progress toward the synthesis of model protocells is quite encouraging, many challenges remain. Foremost among these is the design and synthesis of a genetic polymer capable of repeated cycles of chemical (i.e., nonenzymatic) replication; this is the central barrier to the synthesis of a complete protocell. Amino nucleotides, which polymerize into phosphoramidate-linked nucleic acids, are good candidates for the basis of such a chemical replication system, and we are currently exploring several such systems. A related and also potentially serious problem is strand reannealing, which competes with template copying. Because the reannealing of complementary strands is quite rapid, some means for slowing down reannealing to a timescale comparable to that of template copying must be identified. At this point, there do not appear to be major problems associated with the growth and division of model protocell membranes or with the integration of the protocell membrane with the internal nucleic-acid-copying chemistry. However, an important future goal is to identify simple means that will coordinate the replication of the nucleic acid genome with the replication of the membrane compartment.

Although our work has been directed toward the design and synthesis of a model protocell, our experiments do have implications for theories of the origin of life on the early earth. These implications stem primarily from the surprising physical properties of membranes composed of simple amphiphiles such as fatty acids and related molecules. These membranes seem to possess many of the properties that would be necessary for a primitive cellular system. However, as previously noted (Deamer 1997), fatty acid membranes are incompatible with the high salt and divalent cation concentrations of marine environments and can only exist in freshwater environments. The requirement for cycling between low and high temperatures for nucleic acid copying and strand separation strongly suggests that freshwater ponds or springs in a generally cold environment, locally heated by geothermal activity as in a volcanic region, would be an ideal incubator for the origin of life.

ACKNOWLEDGMENTS

We thank J. Schrum, T. Zhu, A. Ricardo, J. Chen, I. Budin, and the many other current and former members of the Szostak lab who have contributed to the work described in this chapter. We are also grateful to J. Iwasa for Figure 3. J.W.S. is an Investigator of the Howard Hughes Medical Institute. S.S.M. is supported by the Armenise-Harvard foundation. This work was supported by grants from the National Science Foundation (CHE 0434507) and the NASA Exobiology section (EXB02-0031-0018) and fellowships from the Harvard Origins Initiative.

REFERENCES

Apel CL, Deamer DW. 2005. The formation of glycerol monodecanoate by deydration/condensation reaction: Increasing the chemical complexity of amphiphiles on the early earth. *Orig Life Evol Biosph* **35:** 323–332.

Baaske P, Weinert FM, Duhr S, Lemke KH, Russel MJ, Braun D. 2007. Extreme accumulation of nucleotides in simulated hydrothermal pore systems. *Proc Natl Acad Sci* **104:** 9346–9351.

Bar-Ziv R, Moses E. 1994. Instability and "pearling" states produced in tubular membranes by competition of curvature and tension. *Phys Rev Lett* **73:** 1392–1395.

Braun D, Goddard NL, Libchaber A. 2003. Exponential DNA replication by laminar convection. *Phys Rev Lett* **91:** 158103.1–158103.4.

Budin I, Bruckner RJ, Szostak JW. 2009. Formation of protocell-like vesicles in a thermal diffusion column. *J Am Chem Soc* **131:** 9628–9629.

Chen IA, Szostak JW. 2004. Membrane growth can generate a transmembrane pH gradient in fatty acid vesicles. *Proc Natl Acad Sci* **101:** 7965–7970.

Chen IA, Roberts RW, Szostak JW. 2004. The emergence of competition between model protocells. *Science* **305:** 1474–1476.

Chen IA, Salehi-Ashtiani K, Szostak JW. 2005. RNA catalysis in model protocell vesicles. *J Am Chem Soc* **127:** 13213–13219.

Chen JJ, Cai X, Szostak JW. 2009. N2′ → P3′ phosphoramidate glycerol nucleic acid as a potential alternative genetic system. *J Am Chem Soc* **131:** 2119–2121.

Deamer DW. 1997. The first living systems: A bioenergetic perspective. *Microbiol Mol Biol Rev* **61:** 239–261.

Eschenmoser A. 1999. Chemical etiology of nucleic acid structure. *Science* **284:** 2118–2124.

Ferris JP, Hill AR, Liu R, Orgel LE. 1996. Synthesis of long prebiotic oligomers on mineral surfaces. *Nature* **381:** 59–61.

Gerasimov OV, Rui Y, Thompson TE. 1996. Triggered release from liposomes mediated by physically and chemically induced phase transitions. In *Vesicles* (ed. M Rosoff), pp. 679–746. Marcel Dekker, New York.

Hanczyc MM, Szostak JW. 2004. Replicating vesicles as models of primitive cell growth and division. *Curr Opin Chem Biol* **8:** 660–664.

Hanczyc MM, Fujikawa SM, Szostak JW. 2003. Experimental models of primitive cellular compartments: Encapsulation, growth, and division. *Science* **302:** 618–622.

Hargreaves WR, Deamer DW. 1978. Liposomes from ionic, single-chain amphiphiles. *Biochemistry* **17:** 3759–3768.

Joyce GF, Orgel LE. 2006. Progress toward understanding the origin of the RNA world. In *The RNA world* (ed. RF Gesteland et al.), pp. 23–56. Cold Spring Harbor Laboratory Press, Cold Spring Harbor, NY.

Kamp F, Hamilton JA. 1992. pH gradients across phospholipid membranes caused by fast flip-flop of un-ionized fatty acids. *Proc Natl Acad Sci* **89:** 11367–11370.

Kanavarioti A, Monnard PA, Deamer DW. 2001. Eutectic phases in ice facilitate nonenzymatic nucleic acid synthesis. *Astrobiology* **1:** 271–281.

Khalil MM. 2000. Complexation equilibria and determination of stability constants of binary and ternary complexes with ribonucleotides (AMP, ADP, and ATP) and salicylhydroxamic acid as ligands. *J Chem Eng Data* **45:** 70–74.

Kleinzeller A. 1999. Charles Ernest Overton's concept of a cell membrane. In *Membrane permeability: 100 years since Ernest Overton* (ed. DW Deamer et al.), pp. 1–18. Academic, San Diego.

Krishnan M, Ugaz VM, Burns MA. 2002. PCR in a Rayleigh-Bénard convection cell. *Science* **298:** 793.

Lohrmann R, Orgel LE. 1973. Prebiotic activation processes. *Nature* **244:** 418–420.

Lohrmann R, Orgel LE. 1976. Template-directed synthesis of high molecular weight polynucleotide analogues. *Nature* **261:** 342–344.

Mansy SS, Szostak JW. 2008. Thermostability of model protocell membranes. *Proc Natl Acad Sci* **105:** 13351–13355.

Mansy SS, Schrum JP, Krishnamurthy M, Tobe S, Treco DA, Szostak JW. 2008. Replication of a genetic polymer inside of a model protocell. *Nature* **454:** 122–125.

Margolin W. 2000. Themes and variations in prokaryotic cell division. *FEMS Microbiol Rev* **24:** 531–548.

Monnard PA, Apel CL, Kanavarioti A, Deamer DW. 2002. Influence of ionic inorganic solutes on self-assembly and polymerization processes related to early forms of life: Implications for a prebiotic aqueous medium. *Astrobiology* **2:** 139–152.

Oberholzer T, Albrizio M, Luisi PL. 1995. Polymerase chain reaction in liposomes. *Chem Biol* **2:** 677–682.

Orgel LE. 2004. Prebiotic chemistry and the origin of the RNA world. *Crit Rev Biochem Mol Biol* **39:** 99–123.

Sacerdote MG, Szostak JW. 2005. Semipermeable lipid bilayers exhibit diastereoselectivity favoring ribose. *Proc Natl Acad Sci* **102:** 6004–6008.

Szostak JW, Bartel DP, Luisi PL. 2001. Synthesizing life. *Nature* **409:** 387–390.

Tohidi M, Zielinski WS, Chen CH, Orgel LE. 1987. Oligomerization of 3′-amino-3′ deoxyguanosine-5′ phosphorimidazolidate on a d(CpCpCpCpC) template. *J Mol Evol* **25:** 97–99.

Wei C, Pohorille A. 2009. Permeation of membranes by ribose and its diastereomers. *J Am Chem Soc* **131:** 10237–10245.

Zhu TF, Szostak JW. 2009. Coupled growth and division of model protocell membranes. *J Am Chem Soc* **131:** 5705–5713

Zielinski WS, Orgel LE. 1985. Oligomerization of activated derivatives of 3′-amino-3′-deoxyguanosine on poly(C) and poly(dC) templates. *Nucleic Acids Res* **13:** 2469–2484.

Dynamic Chromosome Organization and Protein Localization Coordinate the Regulatory Circuitry that Drives the Bacterial Cell Cycle

E.D. GOLEY, E. TORO, H.H. MCADAMS, AND L. SHAPIRO

Department of Developmental Biology, Stanford University School of Medicine, Stanford, California 94305

Correspondence: shapiro@stanford.edu

The bacterial cell has less internal structure and genetic complexity than cells of eukaryotic organisms, yet it is a highly organized system that uses both temporal and spatial cues to drive its cell cycle. Key insights into bacterial regulatory programs that orchestrate cell cycle progression have come from studies of *Caulobacter crescentus*, a bacterium that divides asymmetrically. Three global regulatory proteins cycle out of phase with one another and drive cell cycle progression by directly controlling the expression of 200 cell-cycle-regulated genes. Exploration of this system provided insights into the evolution of regulatory circuits and the plasticity of circuit structure. The temporal expression of the modular subsystems that implement the cell cycle and asymmetric cell division is also coordinated by differential DNA methylation, regulated proteolysis, and phosphorylation signaling cascades. This control system structure has parallels to eukaryotic cell cycle control architecture. Remarkably, the transcriptional circuitry is dependent on three-dimensional dynamic deployment of key regulatory and signaling proteins. In addition, dynamically localized DNA-binding proteins ensure that DNA segregation is coupled to the timing and cellular position of the cytokinetic ring. Comparison to other organisms reveals conservation of cell cycle regulatory logic, even if regulatory proteins, themselves, are not conserved.

Cell duplication, whether in a mammal with complex organ systems or in a single-celled bacterium, must use rigorous cell cycle regulation to ensure that the cell is ready before proceeding from one step to the next. Premature entry into DNA synthesis (S) phase or exit from mitosis (M) could have drastic consequences, notably fatal damage to the genome. Once initiated, improper execution of physical aspects of the cell cycle such as chromosome segregation and cytokinesis is equally dangerous. Thus, the cell has robust mechanisms to assure fidelity of every step of the cell duplication process. The cell cycle of the aquatic α-proteobacterium *Caulobacter crescentus* (hereafter called *Caulobacter*) has been extensively studied at both the system and molecular levels.

Caulobacter has a life cycle characterized by precise developmental transitions and asymmetric cell division (Fig. 1A). Each division produces two cell types: a motile swarmer cell and a sessile stalked cell. After cytokinesis, a newborn swarmer cell is in the equivalent of G_1 phase of the cell cycle, unable to replicate its chromosome (Degnen and Newton 1972). The swarmer cell has a period of motility and then differentiates into a stalked cell identical to its stalked sibling. In this process, it sheds its polar flagellum and builds a stalk at the same site while simultaneously initiating replication of the chromosome from a single origin of replication (*Cori*). Replicated portions of the chromosome are then segregated to opposite ends of the cell as replication proceeds (Viollier et al. 2004). While this is occurring, a flagellum is constructed at the pole opposite the stalk. Constriction of the cell envelope at the incipient division site begins before completion of chromosome segregation (Jensen 2006). The cytoplasm is divided into two distinct compartments shortly after the duplicated chromosomes are fully segregated (Judd et al. 2003; Jensen 2006). The differentiation of the two daughter cells is triggered by this compartmentalization event, as the genetic programs in each of the compartments immediately diverge. About 20 minutes later, the daughter cells separate to yield a newborn swarmer cell that undergoes an obligate G_1 phase before becoming a stalked cell and the stalked cell that immediately reenters S phase.

Global regulatory paradigms governing the *Caulobacter* cell cycle have emerged from the analysis of specific cell-cycle-dependent events such as DNA replication initiation and cell division site selection that parallel those in other bacteria, as well as in eukaryotes. Here, we describe key events in the *Caulobacter* cell cycle and highlight conserved regulatory themes that reflect constraints guiding the evolution of cell cycle control mechanisms in diverse organisms.

ENTRY INTO S PHASE

Master Regulators Provide a Cell Cycle Timing Mechanism and Regulate Entry into S Phase

Whole-genome microarray studies on synchronized populations of *Caulobacter* revealed a transcriptional cascade governing cell cycle progression, wherein expression of functional gene modules (e.g., replication

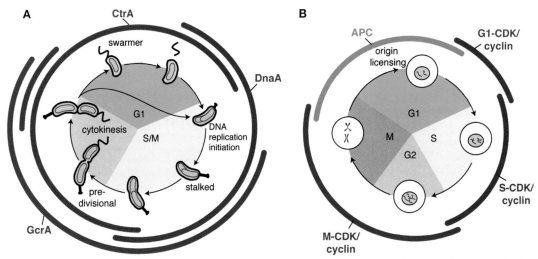

Figure 1. Replication initiation is limited to once per cell cycle by oscillating master regulators in both prokaryotes and eukaryotes. (*A*) *Caulobacter* cell cycle. The swarmer cell sheds its polar flagellum, and in the presence of low CtrA (red) and high DnaA (purple), DNA replication can initiate in the new stalked cell. (Colored arcs) Presence of the indicated master regulator protein, (curved ellipses) circular chromosome. (*B*) Generalized eukaryotic cell cycle. The APC is active from the metaphase-to-anaphase transition in late mitosis through most of G_1 phase, keeping CDK–cyclin activity low and allowing loading of pre-RC complexes and origin licensing. When S-phase CDK cyclins are activated at the onset of S phase, licensed origins fire simultaneously.

initiation, flagellum biogenesis, chemotaxis apparatus assembly, and cell division) is sequentially activated just in time to accomplish their cellular functions (Laub et al. 2000). Central to this genetic circuitry is a set of three master regulators (DnaA, CtrA, and GcrA) that together affect expression of ~200 cell-cycle-regulated genes (Laub et al. 2002; Holtzendorff et al. 2004; Hottes et al. 2005). The protein levels of these master regulators oscillate out of phase with one another (Fig. 1A) (Collier et al. 2006) in a manner similar to the oscillations of cell cycle regulatory cyclin-dependent kinase (CDK)–cyclin complexes and the anaphase promoting complex (APC) in eukaryotes (Fig. 1B). Each master regulator controls expression of the next. A fourth protein, the CcrM DNA methyltransferase, completes the closed cell cycle control circuit (Fig. 2) (Collier et al. 2007). The robustness of the circuit is increased by posttranscriptional regulation of the activities of the master regulators, including phosphorylation, proteolysis, and dynamic spatial positioning of the phosphosignaling proteins and proteases. This tightly integrated spatial and temporal choreography of regulatory factors advances the cell cycle while simultaneously driving *Caulobacter*'s asymmetric development.

The availability of a large number of bacterial genome sequences combined with systems-wide approaches to the identification of the genetic circuitry controlling the bacterial cell cycle has allowed analysis of the evolution of regulatory networks and the degree of plasticity of regulatory network structure (McAdams et al. 2004). An example of the plasticity of regulatory networks comes from the pathway controlled by the CtrA global regulator that directly activates 95 genes. Although the CtrA protein and multiple elements of the complex circuitry that controls the timing of CtrA expression and activation have been conserved through evolution of the α-class of proteobacteria, the portfolio of subsystem functions controlled by CtrA in each bacterial species differs to reflect their specific environmental niche (Bellefontaine et al. 2002; McAdams et al. 2004). Nevertheless, this analysis suggests that an intact genetic circuit can act as an evolutionary unit.

In addition to acting as transcription factors, DnaA and CtrA act coordinately to regulate entry into S phase. DnaA, an AAA^+ ATPase, is a broadly conserved replication initiator in bacteria that binds to the replication origin (*Cori*) and locally unwinds the DNA to allow loading of the replication machinery (Mott and Berger 2007). Conversely, CtrA acts as a repressor of replication initiation by binding to five sites in *Cori* (Quon et al. 1998). The cell can only enter S phase when active CtrA levels are low and active DnaA levels are high. These conditions exist only once per cell cycle in *Caulobacter*: in stalked cells, either at the swarmer-to-stalked cell transition or in the nascent stalked cell compartment of late-predivisional cells (Fig. 1A).

Multiple Modes of DnaA and CtrA Regulation: Phosphorylation, Localization, and Proteolysis

CtrA is only active in its phosphorylated form (CtrA~P) (Domian et al. 1997). Its phosphorylation state is controlled by a phosphorelay from the CckA histidine kinase through the ChpT phosphotransferase (Fig. 2) (Jacobs et al. 1999; Biondi et al. 2006). CtrA~P is present in swarmer cells and in predivisional cells, but it is redundantly inactivated by both dephosphorylation and proteolysis at the swarmer-to-stalked cell transition and in the stalked compartment of late-predivisional cells (Domian et al. 1997). Inactivation of CtrA is critical for licensing DNA replication initiation in newborn stalked cells;

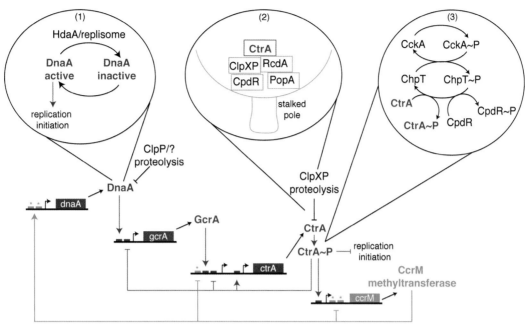

Figure 2. The *Caulobacter* core cell cycle transcriptional control circuit includes posttranslational regulation of master regulators. Each master regulator activates transcription of the next and, in the case of CtrA and CcrM, inhibits transcription of the previous master regulator in the cascade. Promoters and genes encoding the master regulators are depicted with regulatory and/or binding motifs (small boxes) indicated. The color of the regulatory motif correlates with the master regulator that governs it (i.e., small red boxes represent CtrA-binding sites, etc.). (Asterisks) CcrM methylation sites, (bubbles) posttranslational regulation of DnaA and CtrA. (*1*) DnaA is inactivated after HdaA joins the replisome upon initiation of replication. (*2*) CtrA is localized to the stalked pole, by the concerted action of RcdA, PopA, ClpXP, and CpdR, where it is subjected to ClpXP-mediated proteolysis. (*3*) CtrA is phosphorylated and activated by the same CckA-ChpT phosphorelay that phosphorylates and inactivates CpdR.

expression of a constitutively active, stable mutant of CtrA causes cell cycle arrest in G_1 (Domian et al. 1997).

To enter S phase, CtrA must be cleared from the cell. It is degraded by the essential ATP-dependent ClpXP protease complex (Jenal and Fuchs 1998), which proteolyzes CtrA in vitro in the absence of accessory factors (Chien et al. 2007). Remarkably, however, CtrA degradation in vivo requires not only that ClpXP be present, but that both ClpXP and its CtrA substrate be localized to the stalked pole (Ryan et al. 2002; McGrath et al. 2006). This colocalization is achieved through the action of the CpdR phosphoprotein, which localizes ClpXP to the pole (Iniesta et al. 2006), and the combined actions of the RcdA localization factor (McGrath et al. 2006) and the PopA cyclic di-GMP effector protein (Duerig et al. 2009) that together bring the CtrA substrate to polar ClpXP (Fig. 2). The response regulator CpdR only localizes to the stalked pole and recruits ClpXP in its unphosphorylated form. The timing of CtrA degradation is therefore controlled in part by the phosphorylation state of CpdR, which changes as the cell cycle progresses. CpdR is phosphorylated in swarmer and early-predivisional cells, but it is dephosphorylated at the swarmer-to-stalked transition and in the stalked compartment of late-predivisional cells, just in time to promote clearance of CtrA (Iniesta et al. 2006). Interestingly, CpdR is phosphorylated (thereby preventing its polar localization) by the same phosphorelay that activates CtrA, the CckA-ChpT pathway (Biondi et al. 2006). Thus, the pathway that activates CtrA by phosphorylation also prevents its degradation by inactivating the CpdR polar localization factor, providing robust control of entry into S phase.

Caulobacter DnaA is also posttranscriptionally regulated. At least two mechanisms are likely to be involved in DnaA inactivation to prevent early entry into S phase or overinitiation of replication (Fig. 2). First, the DnaA protein is relatively unstable, with a half-life of about one-third of the cell cycle (Gorbatyuk and Marczynski 2005). Its short half-life combined with cell-cycle-regulated transcription results in high DnaA levels in stalked cells and early-predivisional cells and low levels of DnaA in swarmer and late-predivisional cells (Gorbatyuk and Marczynski 2005; Collier et al. 2006). The second mode of inactivation of DnaA uses the HdaA protein, which binds to the replisome upon replication initiation and inactivates DnaA, rendering any remaining DnaA protein incapable of reinitiating replication (J Collier and L Shapiro, in press).

This multilayered and tightly regulated control of the initiation of DNA replication ensures that *Caulobacter* replicates its chromosome only once per cell cycle. This is in contrast to *Escherichia coli*, where overlapping rounds of replication can take place in each cell cycle when grown in rich media. These contrasting modes of regulation of DNA replication reflect the different niches that the two species occupy. *E. coli* is adapted to achieve very rapid growth under intermittent high nutrient conditions in the mammalian gut. *Caulobacter*, on the other hand, is adapted to survive under the low nutrient conditions found in lakes and streams.

Preventing Improper Replication Initiation in Other Bacteria

Replication of the genome is central to the cell cycle, and it is perhaps not surprising that a factor essential for initiation of replication, DnaA, is highly conserved in eubacteria. The activity of DnaA in *Caulobacter* (J Collier and L Shapiro, in press) and *E. coli* (Kato and Katayama 2001) is restricted after initiation by HdaA and Hda, respectively. Hda acts by promoting conversion of active ATP-DnaA to inactive ADP-DnaA in *E. coli*. Given the conservation of the AAA$^+$ ATPase activity of DnaA, it is likely that the control of replication via regulation of the ATPase activity of the initiator protein is broadly used. Indeed, the activities of eukaryotic origin recognition complex (ORC) AAA$^+$ ATPase subunits and the initiator protein Cdc6 are regulated by modulation of their nucleotide-bound states as well (Bell and Dutta 2002).

Unlike DnaA, CtrA is not found outside the α-proteobacteria, and even in closely related *Caulobacter* species, the number and position of CtrA-binding sites near the origin of replication differ considerably (Shaheen et al. 2009). However, restricting access to DnaA-binding sites is a commonly used mechanism for preventing improper initiation of replication. In *E. coli,* for example, reinitiation is restricted in part by binding of SeqA to hemimethylated sites on newly replicated origins and transiently preventing DnaA binding (Slater et al. 1995). Another inhibitor that is remarkably similar to CtrA is Spo0A from the spore-forming Gram-positive bacterium *Bacillus subtilis*. Like CtrA, Spo0A is a response regulator and transcription factor that acts as a master regulator for entry into sporulation only in its phosphorylated form (Piggot and Hilbert 2004). In addition to its transcription-related activities, Spo0A also binds to sites in the origin overlapping DnaA-binding sites and prevents replication initiation once sporulation has commenced (Castilla-Llorente et al. 2006). In the case of both CtrA and Spo0A, it appears that the cell has adapted an existing transcription factor to couple developmentally regulated changes in gene expression to duplication of the genome. This strategy might be readily evolved, because it requires only a series of base pair changes in the region around the origin of replication to introduce a new binding site for a pre-existing DNA-binding protein.

Replication Licensing in Eukaryotes

The basic requirement for high DnaA and low CtrA activities to allow replication initiation in *Caulobacter* is strikingly similar to the mode of origin licensing in eukaryotes. There, licensing occurs before entry into S phase, when APC levels are high and CDK–cyclin levels are low (Fig. 1B) (for review, see Diffley 2004; Arias and Walter 2007). Only under those conditions can prereplication complexes (pre-RCs) load at chromosomal origins. Pre-RCs consist of the ORC, the MCM2-7 presumptive helicase, and two additional factors, Cdc6 and Cdt1, each of which is essential for origin licensing. Upon inactivation of the APC and activation of S-phase CDKs, origins that have been preloaded with pre-RCs can fire simultaneously for a single round of replication. Reinitiation is prevented because CDK levels are high from S phase until late M phase, disallowing assembly of additional pre-RCs at origins until the next late M and G_1 phases. Unlike CtrA and DnaA in *Caulobacter*, however, the APC and CDKs exert their effects on replication licensing indirectly.

As described above, CtrA activity is regulated redundantly by phosphorylation–dephosphorylation and synthesis–proteolysis, which also relies on precise subcellular protein localization. Similar mechanisms regulate components of the pre-RC in yeasts and metazoans (for review, see Diffley 2004; Arias and Walter 2007). In the budding yeast *Saccharomyces cerevisiae*, for example, CDK phosphorylation of pre-RC components can target them for ubiquitin-mediated proteolysis, affect their subcellular localization, or regulate their enzymatic activities. In addition to CDK-mediated inactivation of pre-RC components, in metazoans the APC mediates destruction of the Cdt1 inhibitor geminin in M and G_1 phases, allowing pre-RC assembly precisely in that window of time. Variations on these regulatory events, with CDK-mediated phosphorylation inactivating components of the pre-RC and the APC inactivating CDKs, are found in all eukaryotes that have been studied.

The requirement in many species to limit replication initiation to once per cell cycle has led to the independent evolution of redundant, and remarkably similar, regulatory mechanisms in diverse organisms. The fundamental paradigm of oscillation between two mutually exclusive states as a component of control of replication initiation (e.g., high DnaA–low CtrA in *Caulobacter* and high APC–low CDK in eukaryotes) appears across distant domains of life as the driver behind origin licensing and replication initiation. The combination of this oscillation with the additional elements of spatial and temporal regulation via phosphorylation, proteolysis, and localization of key components produces a rigorous system with numerous fail-safes to restrict DNA replication initiation until all conditions necessary for successful completion are satisfied.

AFTER ENTRY INTO S PHASE: CHROMOSOME SEGREGATION AND ORGANIZATION

Timing of Replication and Segregation: All Things Are Not Equal

As we have noted, striking parallels exist in the logic, if not the specific proteins, that drives cell cycle progression in eukaryotes and bacteria. However, there are notable differences. One significant difference is the strict temporal separation of replication and segregation in eukaryotes (Philpott and Yew 2008); these processes are concurrent in *Caulobacter* and other bacteria (Viollier et al. 2004; Thanbichler and Shapiro 2006a).

The ability to synchronize *Caulobacter* cells enabled Viollier et al. (2004) to demonstrate concurrent replication and segregation directly by visualizing segregation in live cells. Using 10 separate strains, they showed that segregation of genomic loci follows a strict order that corresponds with the order of replication. Furthermore, comparing the timing of replication with that of segregation demonstrated that segregation does not wait for replication to finish. It is important to note that this distinction, with prokaryotes initiating segregation of loci immediately after they are replicated, is not absolute. An exception occurs in *E. coli*, where an origin-proximal portion of the genome undergoes "sister chromatid cohesion" for a short period of time and is segregated as a single unit (Bates and Kleckner 2005; Espéli et al. 2008). Once this origin macrodomain has segregated, however, the rest of the *E. coli* chromosome undergoes concurrent replication and segregation (Nielsen et al. 2006a).

In stark contrast to bacterial systems, eukaryotic entry into mitosis can only occur after all DNA replication is completed. In the presence of unreplicated DNA, the checkpoint inducer protein Chk1 initiates a phosphorylation signaling cascade that results in inactivation of mitotic CDK–cyclin and cell cycle arrest (Dasso and Newport 1990; Kumagai et al. 1998), thus assuring that eukaryotic DNA replication is completed before initiating mitosis.

Chromosome Segregation Is an Active Process

In addition to differences in the relative timing of segregation with respect to replication, a second major difference between the eukaryotic and prokaryotic cell cycles is commonly cited, namely, eukaryotes have a dedicated machinery that separates the chromosomes, whereas bacteria do not. However, in recent years, this view has been challenged.

In 1964, Jacob et al. published a possible mode of chromosome segregation in bacteria based on the hypothesis that the origin of replication is anchored to the cell envelope (Jacob et al. 1964). If, after replication, the cell wall grows in between two chromosome attachment sites, this model predicts that DNA separation could happen passively as a result of cell growth. This first and very influential model of DNA segregation was the accepted paradigm for a considerable time. With the advent of methods to visualize DNA loci in live bacterial cells using fluorescent fusions to sequence-specific DNA-binding proteins (Robinett et al. 1996; Straight et al. 1996; Webb et al. 1997; Nielsen et al. 2006b), however, several groups demonstrated that the movement of labeled DNA loci is too fast to be accounted for by cell growth (Mohl and Gober 1997; Webb et al. 1997; Viollier et al. 2004; Fogel and Waldor 2006), thus challenging the Jacob model. Furthermore, several species of bacteria have proteins that specifically affect the localization of segments of the chromosome (Ben-Yehuda et al. 2003; Yamaichi and Niki 2004; Gitai et al. 2005; Fogel and Waldor 2006; Bowman et al. 2008; Ebersbach et al. 2008), and in *Caulobacter*, the existence a centromere, i.e., a site of force exertion on the DNA during segregation, has been demonstrated (Toro et al. 2008). This centromere, the widely conserved *parS* sequence, is the only demonstrated exception to the rule of sequential segregation mentioned above. By necessity, the site of force exertion must move ahead of all other sites, and *parS* was shown to be invariantly the first locus segregated, even when the genomic distance, and therefore time of duplication, between the replication origin and *parS* was increased considerably. Interestingly, *Caulobacter parS* is situated very close to the origin of replication, and a comparative genomics study found that this arrangement is true of most bacteria (Livny et al. 2007). This suggests that concurrent replication and segregation may require the centromere to be placed near the origin of replication.

In short, at least some bacteria actively segregate their chromosomes and have a functional equivalent of a mitotic spindle. The detailed molecular mechanism that drives this movement is still unknown, but the ParA/ParB/*parS* partitioning system is likely to be involved (Fogel and Waldor 2006; Toro et al. 2008). In several bacterial species, however, the ParA/ParB/*parS* system is not essential, and some species (e.g., *E. coli*) do not have it at all, so bacteria must have evolved several independent DNA segregation systems.

The Chromosome Is Structured within the Nucleus and Nucleoid

An exciting recent development in cell biology has been the realization that chromosomal DNA is nonrandomly organized. In the case of bacteria, it was first discovered in *Caulobacter* and *B. subtilis* that the origin and terminus of replication held fixed positions within the cell (Mohl and Gober 1997; Webb et al. 1997). Later, a comprehensive study of 112 different loci in *Caulobacter* demonstrated that the cellular position of any given locus is linearly correlated with its position on the chromosome (Viollier et al. 2004), a finding that is likely to be true of other bacteria (Teleman et al. 1998; Niki et al. 2000). Interestingly, whereas *B. subtilis* and *Caulobacter* arrange their chromosomes so that the origin of replication is at one pole and the terminus at the other, *E. coli* arranges the left and right replichores along this axis, instead (Wang et al. 2006). The reason for this difference is unknown; an explanation will require an understanding of the mechanism that controls chromosome orientation within the cell. Toward this end, two factors have been identified that control chromosomal anchoring to the cell wall, one in *B. subtilis* and the other in *Caulobacter* (Ben-Yehuda et al. 2003; Bowman et al. 2008; Ebersbach et al. 2008; Toro et al. 2008). The relevant questions now are (1) how is chromosomal anchoring regulated and (2) are there other anchoring points in the cell?

In *Caulobacter*, the linear arrangement of genetic loci within the cell, and the simultaneous replication and segregation of these loci, is a factor in the temporal regulation of transcription of the *dnaA* and *ctrA* master regulator genes that orchestrate cell cycle progression (Reisenauer and Shapiro 2002; Collier et al. 2007). Replication initiates on a fully methylated chromosome. As the replication fork proceeds, the DNA copies become hemimethylated and remain in that state until the completion of DNA replication, because the CcrM DNA methyltransferase, which methylates the chromosome, is only present for a short period of time after completion of replication. The *dnaA* gene, which is adjacent to *Cori* on the chromosome, is preferentially transcribed from a fully methylated promoter. When the replication fork passes through *dnaA*, the two copies of the *dnaA* gene become hemimethylated, decreasing the transcription of *dnaA*. The *ctrA* gene, on the other hand, is farther from *Cori* and is preferentially transcribed from a hemimethylated promoter. This assures that *ctrA* transcription is activated later than *dnaA*, and the timing of activation of both transcripts is thus linked to the passage of the replication fork.

The realization that chromosomes are nonrandomly organized within the nucleus of eukaryotic cells originated from microirradiation experiments on synchronized hamster cells (Zorn et al. 1979). These authors showed that irradiation of a small section of the nucleus resulted in large amounts of damage to a small number of chromosomes, rather than a small amount of damage to all. This suggested that chromosomes occupy a small fraction of the nucleus, termed a "chromosome territory." Later experiments confirmed this interpretation (Cremer et al. 1993; Croft et al. 1999) and showed that different cell types arrange their chromosomes in different configurations. For example, mouse hepatocytes tend to keep chromosomes 5 and 6 together and chromosomes 12 and 15 apart, a situation that is reversed in lymphocytes (Parada et al. 2004). This observation is particularly interesting because mouse hepatomas frequently show translocations between chromosomes 5 and 6, whereas lymphomas tend to show translocations between 12 and 15. A second striking example is found in the organization of heterochromatin in rod photoreceptors of nocturnal mammals. The majority of eukaryotic cells examined thus far localize the bulk of the heterochromatin to the nuclear periphery. Rod photoreceptors in nocturnal mammals, however, are an exception to this rule, and computer simulations suggest that this inverted nuclear organization works as a lens that channels light more efficiently to the light-sensing segments of the rod cells (Solovei et al. 2009). This remarkable finding underscores the fact that DNA organization within the cell is under strong selection not only for regulatory efficiency, but also for the physical implications of its structure.

DNA Mobility within the Nucleus

Localization is not the only aspect of chromosome organization that is strictly controlled. In bacteria, the mobility of DNA segments is restricted to subcellular domains (Fiebig et al. 2006). The spatial organization of the chromosome thus appears to favor certain DNA–DNA interactions much more than others. In the case of *Caulobacter* and *B. subtilis*, interactions that are symmetric around the origin–terminus axis are favored, whereas *E. coli* favors interactions perpendicular to this axis. Comparative genomics as well as experimental manipulations have shown that there is strong selection for equal length along the two arms of the bacterial chromosome (Esnault et al. 2007). Hence, the orientation of the chromosome will determine the relative likelihood of neutral versus deleterious inversions. Taken together, these findings paint a picture in which the three-dimensional organization of the DNA within the nucleoid is highly regulated and has an important role in maintaining the structural and functional integrity of the genome.

DIRECTING THE DIVISION PLANE

MipZ Communicates Information from the Chromosomes and the Poles to the Cell Division Machinery in *Caulobacter*

The final stage of the cell cycle is the physical separation of the cell into two daughter cells, or cytokinesis. Selection of the site of division is tightly controlled to assure that each daughter receives a full, undamaged copy of the genome. Bacterial cell division is mediated by the concerted action of a molecular machine called the divisome (Harry et al. 2006). The first protein to localize to the incipient division plane is the tubulin homolog FtsZ, which polymerizes into a ring structure (the Z ring) that forms the basis for assembly of the remainder of the divisome. The subcellular position of FtsZ assembly thereby dictates the site of cell division. In all bacteria that have been characterized, positioning of the Z ring involves negative regulators of FtsZ assembly that prevent division everywhere except mid cell. In *Caulobacter*, the locations of the duplicated chromosomes are intimately linked to the placement of the Z ring through a Walker A cytoskeletal ATPase (WACA) called MipZ that is conserved in α-proteobacteria (Thanbichler and Shapiro 2006b).

MipZ interacts directly with FtsZ in vitro and stimulates its GTPase activity, thereby promoting depolymerization (Thanbichler and Shapiro 2006b). This activity is spatially controlled in vivo so that the highest concentration of MipZ activity is at the cell poles and the lowest is near mid cell. MipZ is localized to the cell poles by virtue of its interaction with the ParB partitioning protein (Thanbichler and Shapiro 2006b), which, in turn, binds to both the PopZ polar anchoring protein (Bowman et al. 2008; Ebersbach et al. 2008) and the *parS* centromere (Mohl and Gober 1997; Toro et al. 2008). Importantly, the site of MipZ localization therefore reflects both the location of the centromere(s) and the locations of the cell poles. These interactions, combined with the dynamics of the chromosome, lead to a defined spatial arrangement of the MipZ and FtsZ proteins over the course of the cell cycle (Fig. 3A) (Thanbichler and Shapiro 2006b). Before duplication and segregation of *parS*, MipZ (along with ParB/PopZ) is observed at the stalked pole and FtsZ is found in a focus at the opposite new pole. Upon segregation and bipolarization of the *parS*/ParB/MipZ/PopZ complex, FtsZ depolymerizes at the new pole and reassembles at the site of lowest MipZ concentration in the cell, roughly the middle. The accuracy of this system relies on two characteristics of the *Caulobacter* chromo-

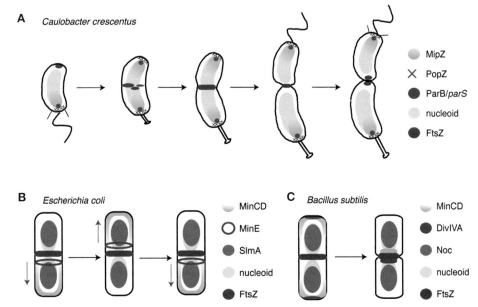

Figure 3. Placement of the division site integrates positional information from the poles and the chromosomes. (*A*) In *Caulobacter*, the MipZ complex is localized at the old pole and FtsZ is observed at the new pole. Upon segregation of *parS*, MipZ becomes bipolar, displacing FtsZ from the pole and targeting it to mid cell. (*B*) In *E. coli*, MinCD oscillates from pole to pole driven by MinE (which moves in the direction of the purple arrows). Additionally, SlmA on the segregated chromosomes inhibits FtsZ assembly over the bulk of the nucleoid. (*C*) *B. subtilis* also uses MinCD to inhibit polymerization of FtsZ next to newly formed poles but is localized by DivIVA. The activities of the Min system are complemented by Noc, which prevents Z-ring assembly over the chromosome(s).

some: (1) The *parS* centromeres are anchored at the cell poles after duplication and segregation and (2) the single chromosome is duplicated exactly once per cell cycle.

Other Bacteria Use Separate Systems to Sense the Cell Poles and the Replicating Chromosome

Mechanisms of division site selection in *E. coli* and *B. subtilis* also reflect the positions of the cell poles and the segregated chromosomes. However, they achieve this using two separate mechanisms, each of which is distinct from MipZ (Fig. 3B,C). Polar assembly of FtsZ is inhibited in these organisms by MinC, a protein that directly interacts with FtsZ and antagonizes lateral interactions between protofilaments that are thought to be important for formation of a stable divisome (Dajkovic et al. 2008; Scheffers 2008). MinC localizes by interacting with the membrane-associated WACA MinD. In *E. coli*, MinCD oscillates from pole to pole driven by the action of a third protein, MinE (Fig. 3B) (Hu and Lutkenhaus 1999, 2001; Raskin and de Boer 1999a,b; Fu et al. 2001), whereas in *B. subtilis*, MinCD localizes by binding to DivIVA, a protein unrelated to MinE that localizes to the septum and to the new pole (Fig. 3C) (Marston et al. 1998). Unlike MipZ, MinCD is not essential in *E. coli* or *B. subtilis* because a second system is in place to inhibit FtsZ assembly by a nucleoid occlusion mechanism. Nucleoid occlusion functions in *E. coli* and *B. subtilis* through unrelated effectors that act as negative regulators of Z-ring assembly: SlmA in the former and Noc in the latter (Wu and Errington 2004; Bernhardt and de Boer 2005). In both cases, the nucleoid occlusion protein associates asymmetrically with the chromosome(s) and inhibits assembly of the FtsZ ring on the membrane around the nucleoid (Fig. 3B,C). This limits initiation of cell division to areas of low DNA concentration, either close to the poles (which is inhibited by MinCD and therefore disallowed) or between the chromosomes after segregation.

Ancestors to the α-proteobacteria probably originally possessed MinC and/or MinD and lost them, because these proteins are present in diverse species including other classes of proteobacteria, Gram-positive organisms, and even some chloroplasts. MipZ may have arisen originally as a gene duplication of another WACA protein such as ParA. It could then adapt to direct FtsZ localization while the mechanisms allowing polar positioning of *parS* were evolved, eliminating the need for MinCD. It has not been determined if MinCD artificially expressed in *Caulobacter* can oscillate from pole to pole, but we speculate that the distinct biochemical identities of the poles in *Caulobacter* might affect the efficiency of such a system. Moreover, the MipZ division plane positioning mechanism allows FtsZ to assemble over the bulk of the nucleoid early in the cell cycle, where it can direct both elongation of the cell before invagination at the division site (Aaron et al. 2007) and the process of division. A nucleoid occlusion system would eliminate the ability of FtsZ to direct the early phase of growth at mid cell, and this activity may have supplied selective pressure for eliminating nucleoid occlusion in *Caulobacter* or for preventing its inception. On the other hand, MipZ alone would not function efficiently in *E. coli* or *B. subtilis* because these organisms can have multiple rounds of chromosome replication occurring per cell division. In *Caulobacter*, the cell division regulatory mechanism may in turn select against mutations that alter positioning of the centromere or that allow reinitiation of DNA replication before division in α-proteobacteria, leading to stable coupling of chromosome dynamics and division site selection.

CONCLUSIONS

We have described here the extraordinary lengths to which all cells go to preserve the integrity of their genomes. Intricate timing and synchronization mechanisms ensure accurate progression of the cell cycle into S phase only when necessary precursor processes are complete and nutrients are adequate. During and after replication, the chromosomes are highly organized and are actively shaped and moved during segregation. Finally, cell division occurs at a time and place in the cell that reflects constraints communicated from the replicated, segregated chromosomes. These general rules for preserving genome integrity are found throughout all domains of life, and the conservation of the regulatory logic and mechanisms underpinning cell cycle control in diverse organisms is truly remarkable.

In general, however, it is problematic that the great majority of our cell biological knowledge comes from a handful of organisms, particularly in the case of bacteria. The fact that *E. coli* and *B. subtilis* diverged about 1.5 billion years ago (Ochman and Wilson 1987) underscores the immense molecular variation that is contained in prokaryotes, but which is vastly underexploited. Exploration into other, divergent organisms will yield new insights into the conservation and evolution of the regulatory logic and genetic circuit design that drives cell cycle progression.

ACKNOWLEDGMENTS

Work in the L.S. and H.H.M. laboratories is supported in part by National Institutes of Health grants R01 GM51426 R24 and GM073011-04 (to L.S.) and Department of Energy grant DE-FG02-05ER64136 (to H.H.M. and L.S.). E.T. was funded by the Smith Stanford Graduate Fellowship. E.D.G. is a Helen Hay Whitney postdoctoral fellow.

REFERENCES

Aaron M, Charbon G, Lam H, Schwartz H, Vollmer W, Jacobs-Wagner C. 2007. The tubulin homologue FtsZ contributes to cell elongation by guiding cell wall precursor synthesis in *Caulobacter crescentus*. *Mol Microbiol* **64:** 938–952.

Arias EE, Walter JC. 2007. Strength in numbers: Preventing rereplication via multiple mechanisms in eukaryotic cells. *Genes Dev* **21:** 497–518.

Bates D, Kleckner N. 2005. Chromosome and replisome dynamics in *E. coli:* Loss of sister cohesion triggers global chromosome movement and mediates chromosome segregation. *Cell* **121:** 899–911.

Bell SP, Dutta A. 2002. DNA replication in eukaryotic cells. *Annu Rev Biochem* **71:** 333–74.

Bellefontaine AF, Pierreux CE, Mertens P, Vandenhaute J, Letesson JJ, De Bolle X. 2002. Plasticity of a transcriptional regulation network among α-proteobacteria is supported by the identification of CtrA targets in *Brucella abortus*. *Mol Microbiol* **43:** 945–960.

Ben-Yehuda S, Rudner DZ, Losick R. 2003. RacA, a bacterial protein that anchors chromosomes to the cell poles. *Science* **299:** 532–536.

Bernhardt TG, de Boer PA. 2005. SlmA, a nucleoid-associated, FtsZ binding protein required for blocking septal ring assembly over chromosomes in *E. coli*. *Mol Cell* **18:** 555–564.

Biondi EG, Reisinger SJ, Skerker JM, Arif M, Perchuk BS, Ryan KR, Laub MT. 2006. Regulation of the bacterial cell cycle by an integrated genetic circuit. *Nature* **444:** 899–904.

Bowman GR, Comolli LR, Zhu J, Eckart M, Koenig M, Downing KH, Moerner WE, Earnest T, Shapiro L. 2008. A polymeric protein anchors the chromosomal origin/ParB complex at a bacterial cell pole. *Cell* **134:** 945–955.

Castilla-Llorente V, Munoz-Espin D, Villar L, Salas M, Meijer WJ. 2006. Spo0A, the key transcriptional regulator for entrance into sporulation, is an inhibitor of DNA replication. *EMBO J* **25:** 3890–3899.

Chien P, Perchuk BS, Laub MT, Sauer RT, Baker TA. 2007. Direct and adaptor-mediated substrate recognition by an essential AAA$^+$ protease. *Proc Natl Acad Sci* **104:** 6590–6595.

Collier J, Shapiro L. 2009. Feedback control of DnaA-mediated replication initiation by replisome-associated HdaA protein in *Caulobacter*. *J Bacteriol* (in press).

Collier J, Murray SR, Shapiro L. 2006. DnaA couples DNA replication and the expression of two cell cycle master regulators. *EMBO J* **25:** 346–356.

Collier J, McAdams HH, Shapiro L. 2007. A DNA methylation ratchet governs progression through a bacterial cell cycle. *Proc Natl Acad Sci* **104:** 17111–17116.

Cremer T, Kurz A, Zirbel R, Dietzel S, Rinke B, Schröck E, Speicher MR, Mathieu U, Jauch A, Emmerich P, et al. 1993. Role of chromosome territories in the functional compartmentalization of the cell nucleus. *Cold Spring Harbor Symp Quant Biol* **58:** 777–792.

Croft JA, Bridger JM, Boyle S, Perry P, Teague P, Bickmore WA. 1999. Differences in the localization and morphology of chromosomes in the human nucleus. *J Cell Biol* **145:** 1119–1131.

Dajkovic A, Lan G, Sun SX, Wirtz D, Lutkenhaus J. 2008. MinC spatially controls bacterial cytokinesis by antagonizing the scaffolding function of FtsZ. *Curr Biol* **18:** 235–244.

Dasso M, Newport JW. 1990. Completion of DNA replication is monitored by a feedback system that controls the initiation of mitosis in vitro: Studies in *Xenopus*. *Cell* **61:** 811–823.

Degnen ST, Newton A. 1972. Chromosome replication during development in *Caulobacter crescentus*. *J Mol Biol* **64:** 671–680.

Diffley JF. 2004. Regulation of early events in chromosome replication. *Curr Biol* **14:** R778–R786.

Domian IJ, Quon KC, Shapiro L. 1997. Cell type-specific phosphorylation and proteolysis of a transcriptional regulator controls the G1-to-S transition in a bacterial cell cycle. *Cell* **90:** 415–424.

Duerig A, Abel S, Folcher M, Nicollier M, Schwede T, Amiot N, Giese B, Jenal U. 2009. Second messenger-mediated spatiotemporal control of protein degradation regulates bacterial cell cycle progression. *Genes Dev* **23:** 93–104.

Ebersbach G, Briegel A, Jensen GJ, Jacobs-Wagner C. 2008. A self-associating protein critical for chromosome attachment, division, and polar organization in *Caulobacter*. *Cell* **134:** 956–968.

Esnault E, Valens M, Espéli O, Boccard F. 2007. Chromosome structuring limits genome plasticity in *Escherichia coli*. *PLoS Genet* **3:** e226.

Espéli O, Mercier R, Boccard F. 2008. DNA dynamics vary according to macrodomain topography in the *E. coli* chromosome. *Mol Microbiol* **68:** 1418–1427.

Fiebig A, Keren K, Theriot JA. 2006. Fine-scale time-lapse analysis of the biphasic, dynamic behaviour of the two *Vibrio cholerae* chromosomes. *Mol Microbiol* **60:** 1164–1178.

Fogel MA, Waldor MK. 2006. A dynamic, mitotic-like mechanism for bacterial chromosome segregation. *Genes Dev* **20:** 3269–3282.

Fu X, Shih YL, Zhang Y, Rothfield LI. 2001. The MinE ring required for proper placement of the division site is a mobile structure that changes its cellular location during the *Escherichia coli* division cycle. *Proc Natl Acad Sci* **98:** 980–985.

Gitai Z, Dye NA, Reisenauer A, Wachi M, Shapiro L. 2005. MreB actin-mediated segregation of a specific region of a bacterial chromosome. *Cell* **120:** 329–341.

Gorbatyuk B, Marczynski GT. 2005. Regulated degradation of chromosome replication proteins DnaA and CtrA in *Caulobacter crescentus*. *Mol Microbiol* **55:** 1233–1245.

Harry E, Monahan L, Thompson L. 2006. Bacterial cell division: The mechanism and its precison. *Int Rev Cytol* **253:** 27–94.

Holtzendorff J, Hung D, Brende P, Reisenauer A, Viollier PH, McAdams HH, Shapiro L. 2004. Oscillating global regulators control the genetic circuit driving a bacterial cell cycle. *Science* **304:** 983–987.

Hottes AK, Shapiro L, McAdams HH. 2005. DnaA coordinates replication initiation and cell cycle transcription in *Caulobacter crescentus*. *Mol Microbiol* **58:** 1340–1353.

Hu Z, Lutkenhaus J. 1999. Topological regulation of cell division in *Escherichia coli* involves rapid pole to pole oscillation of the division inhibitor MinC under the control of MinD and MinE. *Mol Microbiol* **34:** 82–90.

Hu Z, Lutkenhaus J. 2001. Topological regulation of cell division in *E. coli*. Spatiotemporal oscillation of MinD requires stimulation of its ATPase by MinE and phospholipid. *Mol Cell* **7:** 1337–1343.

Iniesta AA, McGrath PT, Reisenauer A, McAdams HH, Shapiro L. 2006. A phospho-signaling pathway controls the localization and activity of a protease complex critical for bacterial cell cycle progression. *Proc Natl Acad Sci* **103:** 10935–10940.

Jacob F, Brenner S, Cuzin F. 1964. On the regulation of DNA replication in bacteria. *Cold Spring Harbor Symp Quant Biol* **28:** 329–348.

Jacobs C, Domian IJ, Maddock JR, Shapiro L. 1999. Cell cycle-dependent polar localization of an essential bacterial histidine kinase that controls DNA replication and cell division. *Cell* **97:** 111–120.

Jenal U, Fuchs T. 1998. An essential protease involved in bacterial cell-cycle control. *EMBO J* **17:** 5658–5669.

Jensen RB. 2006. Coordination between chromosome replication, segregation, and cell division in *Caulobacter crescentus*. *J Bacteriol* **188:** 2244–2253.

Judd EM, Ryan KR, Moerner WE, Shapiro L, McAdams HH. 2003. Fluorescence bleaching reveals asymmetric compartment formation prior to cell division in *Caulobacter*. *Proc Natl Acad Sci* **100:** 8235–8240.

Kato J, Katayama T. 2001. Hda, a novel DnaA-related protein, regulates the replication cycle in *Escherichia coli*. *EMBO J* **20:** 4253–4262.

Kumagai A, Guo Z, Emami KH, Wang SX, Dunphy WG. 1998. The *Xenopus* Chk1 protein kinase mediates a caffeine-sensitive pathway of checkpoint control in cell-free extracts. *J Cell Biol* **142:** 1559–1569.

Laub MT, McAdams HH, Feldblyum T, Fraser CM, Shapiro L. 2000. Global analysis of the genetic network controlling a bacterial cell cycle. *Science* **290:** 2144–2148.

Laub MT, Chen SL, Shapiro L, McAdams HH. 2002. Genes directly controlled by CtrA, a master regulator of the *Caulobacter* cell cycle. *Proc Natl Acad Sci* **99:** 4632–4637.

Livny J, Yamaichi Y, Waldor MK. 2007. Distribution of centromere-like parS sites in bacteria: Insights from comparative genomics. *J Bacteriol* **189:** 8693–8703.

Marston AL, Thomaides HB, Edwards DH, Sharpe ME, Errington J. 1998. Polar localization of the MinD protein of *Bacillus subtilis* and its role in selection of the mid-cell division site. *Genes Dev* **12:** 3419–3430.

McAdams HH, Srinivasan B, Arkin AP. 2004. The evolution of genetic regulatory systems in bacteria. *Nat Rev Genet* **5:** 169–178.

McGrath PT, Iniesta AA, Ryan KR, Shapiro L, McAdams HH. 2006. A dynamically localized protease complex and a polar specificity factor control a cell cycle master regulator. *Cell* **124:** 535–547.

Mohl DA, Gober JW. 1997. Cell cycle-dependent polar localization of chromosome partitioning proteins in *Caulobacter crescentus*. *Cell* **88:** 675–684.

Mott ML, Berger JM. 2007. DNA replication initiation: Mechanisms and regulation in bacteria. *Nat Rev Microbiol* **5:** 343–354.

Nielsen HJ, Li Y, Youngren B, Hansen FG, Austin S. 2006a. Progressive segregation of the *Escherichia coli* chromosome. *Mol Microbiol* **61:** 383–393.

Nielsen HJ, Ottesen JR, Youngren B, Austin SJ, Hansen FG. 2006b. The *Escherichia coli* chromosome is organized with the left and right chromosome arms in separate cell halves. *Mol Microbiol* **62:** 331–338.

Niki H, Yamaichi Y, Hiraga S. 2000. Dynamic organization of chromosomal DNA in *Escherichia coli*. *Genes Dev* **14:** 212–223.

Ochman H, Wilson AC. 1987. Evolution in bacteria: Evidence for a universal substitution rate in cellular genomes. *J Mol Evol* **26:** 74–86.

Parada LA, McQueen PG, Misteli T. 2004. Tissue-specific spatial organization of genomes. *Genome Biol* **5:** R44.

Philpott A, Yew PR. 2008. The *Xenopus* cell cycle: An overview. *Mol Biotechnol* **39:** 9–19.

Piggot PJ, Hilbert DW. 2004. Sporulation of *Bacillus subtilis*. *Curr Opin Microbiol* **7:** 579–586.

Quon KC, Yang B, Domian IJ, Shapiro L, Marczynski GT. 1998. Negative control of bacterial DNA replication by a cell cycle regulatory protein that binds at the chromosome origin. *Proc Natl Acad Sci* **95:** 120–125.

Raskin DM, de Boer PA. 1999a. MinDE-dependent pole-to-pole oscillation of division inhibitor MinC in *Escherichia coli*. *J Bacteriol* **181:** 6419–6424.

Raskin DM, de Boer PA. 1999b. Rapid pole-to-pole oscillation of a protein required for directing division to the middle of *Escherichia coli*. *Proc Natl Acad Sci* **96:** 4971–4976.

Reisenauer A, Shapiro L. 2002. DNA methylation affects the cell cycle transcription of the CtrA global regulator in *Caulobacter*. *EMBO J* **21:** 4969–4977.

Robinett CC, Straight A, Li G, Willhelm C, Sudlow G, Murray A, Belmont AS. 1996. In vivo localization of DNA sequences and visualization of large-scale chromatin organization using lac operator/repressor recognition. *J Cell Biol* **135:** 1685–1700.

Ryan KR, Judd EM, Shapiro L. 2002. The CtrA response regulator essential for *Caulobacter crescentus* cell-cycle progression requires a bipartite degradation signal for temporally controlled proteolysis. *J Mol Biol* **324:** 443–455.

Scheffers DJ. 2008. The effect of MinC on FtsZ polymerization is pH dependent and can be counteracted by ZapA. *FEBS Lett* **582:** 2601–2608.

Shaheen SM, Ouimet MC, Marczynski GT. 2009. Comparative analysis of *Caulobacter* chromosome replication origins. *Microbiology* **155:** 1215–1225.

Slater S, Wold S, Lu M, Boye E, Skarstad K, Kleckner N. 1995. *E. coli* SeqA protein binds oriC in two different methyl-modulated reactions appropriate to its roles in DNA replication initiation and origin sequestration. *Cell* **82:** 927–936.

Solovei I, Kreysing M, Lanctot C, Kosem S, Peichl L, Cremer T, Guck J, Joffe B. 2009. Nuclear architecture of rod photoreceptor cells adapts to vision in mammalian evolution. *Cell* **137:** 356–368.

Straight AF, Belmont AS, Robinett CC, Murray AW. 1996. GFP tagging of budding yeast chromosomes reveals that protein-protein interactions can mediate sister chromatid cohesion. *Curr Biol* **6:** 1599–1608.

Teleman AA, Graumann PL, Lin DC, Grossman AD, Losick R. 1998. Chromosome arrangement within a bacterium. *Curr Biol* **8:** 1102–1109.

Thanbichler M, Shapiro L. 2006a. Chromosome organization and segregation in bacteria. *J Struct Biol* **156:** 292–303.

Thanbichler M, Shapiro L. 2006b. MipZ, a spatial regulator

coordinating chromosome segregation with cell division in *Caulobacter*. *Cell* **126:** 147–162.

Toro E, Hong SH, McAdams HH, Shapiro L. 2008. *Caulobacter* requires a dedicated mechanism to initiate chromosome segregation. *Proc Natl Acad Sci* **105:** 15435–15440.

Viollier PH, Thanbichler M, McGrath PT, West L, Meewan M, McAdams HH, Shapiro L. 2004. Rapid and sequential movement of individual chromosomal loci to specific subcellular locations during bacterial DNA replication. *Proc Natl Acad Sci* **101:** 9257–9262.

Wang X, Liu X, Possoz C, Sherratt DJ. 2006. The two *Escherichia coli* chromosome arms locate to separate cell halves. *Genes Dev* **20:** 1727–1731.

Webb CD, Teleman A, Gordon S, Straight A, Belmont A, Lin DC, Grossman AD, Wright A, Losick R. 1997. Bipolar localization of the replication origin regions of chromosomes in vegetative and sporulating cells of *B. subtilis*. *Cell* **88:** 667–674.

Wu LJ, Errington J. 2004. Coordination of cell division and chromosome segregation by a nucleoid occlusion protein in *Bacillus subtilis*. *Cell* **117:** 915–925.

Yamaichi Y, Niki H. 2004. *migS*, a *cis*-acting site that affects bipolar positioning of *oriC* on the *Escherichia coli* chromosome. *EMBO J* **23:** 221–233.

Zorn C, Cremer C, Cremer T, Zimmer J. 1979. Unscheduled DNA synthesis after partial UV irradiation of the cell nucleus. Distribution in interphase and metaphase. *Exp Cell Res* **124:** 111–119.

An Integrated View of Precambrian Eumetazoan Evolution

E.H. Davidson[1] and D.H. Erwin[2]

[1]*Division of Biology 156-29, California Institute of Technology, Pasadena, California 91125;*
[2]*Department of Paleobiology, MRC-121, Smithsonian Institution, Washington, D.C. 20013-7012, and Santa Fe Institute, Santa Fe, New Mexico, 87501*
Correspondence: davidson@caltech.edu

The eumetazoan clade of modern animals includes cnidarians, acoels, deuterostomes, and protostomes. Stem group eumetazoans evolved in the late Neoproterozoic, possibly before the Marinoan glaciation, according to a variety of different kinds of evidence. Here, we combine this evidence, including paleontological observations, results from molecular and morphological phylogeny, and paleoecological considerations, with deductions from the organization of the gene regulatory networks that underlie development of the bilaterian body plan. Eumetazoan body parts are morphologically complex in detail, and modern knowledge of gene regulatory network structure shows that the control circuitry required for their development is hierarchical and multilayered. Among the consequences is that the kernels of the networks that control the early allocation of spatial developmental fate canalize the possibilities of downstream evolutionary change, a mechanism that can account for the appearance of distinct clades in early animal evolution. We reconstruct preeumetazoan network organization and consider the process by which the eumetazoan regulatory apparatus might have been assembled. A strong conclusion is that the evolutionary process generating the genomic programs responsible for developmental formulation of basic eumetazoan body plans was in many ways very different from the evolutionary changes that can be observed at the species level in modern animals.

The appearance of the Eumetazoa represents a significant evolutionary transition in the history of life. Most eumetazoans, or their ancestors, exhibit some form of bilaterial symmetry, an innovation often thought to be associated with the regionalization of sensory structures and the eventual elaboration of a central nervous system. In this chapter, we integrate data on the origin and early evolution of eumetazoans as observed in the fossil record and phylogenetic information on the relationships of major metazoan clades, with insights deriving from the gene regulatory networks (GRNs) that underlie the developmental process. Recent knowledge of GRN structure and function provides a primary focus on mechanisms of assembly and change of the genomic programs that control the developmental construction of eumetazoan body plans.

A robust phylogeny is crucial to our discussion, and as we discuss further below, the phylogenetic relationships between the principal clades of the cnidarians, acoels, and nematodermatids, and the protostomes plus deuterostomes, are now well established, and they provide this fundamental framework. However, the phylogenetic placement of several groups remains highly controversial, including the issue of whether sponges represent a single clade, or are paraphyletic, the affinities of the placazoan *Trichoplax* and the phylogenetic placement of the ctenophores (Fig. 1). The current level of controversy regarding the phylogenetic relationships of these basal metazoan groups indicates that it is premature to address developmental evolutionary processes at that level. Resolution of these phylogenetic issues is essential for determining characters present in the last common ancestor at various nodes along the more basal metazoan evolutionary tree, but they are not vital to eumetazoan evolution per se.

The fossil record provides important constraints on metazoan evolution in general. The earliest current record of metazoans comes from molecular fossils, biomarkers, preserved in rocks older than 635 million years ago (mya) in Oman. Early to Middle Ediacaran rocks of the Doushantuo Formation in southern China preserve a rich assemblage of algae and microfossils and an array of exquisitely preserved metazoan embryos. Several of these are most plausibly interpreted as representing bilaterian metazoan clades. The most obvious component of the fossil record of the Ediacaran Period (635–542 mya) are the assemblages of the soft-bodied Ediacaran macrofauna dating to 579–542 mya. Although many of these exhibit apparent bilaterial symmetry, only one of them, *Kimberella* (555 mya; White Sea, Russia), can reasonably be considered as representing a protostome; no deuterostomes are known. By the end of the Early Cambrian, ~530 mya, a diverse assemblage of bilaterian clades is well established in the fossil record, including virtually every major metazoan clade, from arthropods to vertebrates. This has given rise to the popular concept of the "Cambrian explosion."

A better understanding of the developmental evolutionary events that occurred during the late Neoproterozoic (~750–542 mya) is important not only for understanding the mechanisms associated with the construction of animal body plans, but also for improving interpretations of the fossil record. Developmental evidence suggests that the capacity to produce "bilaterian" morphological characters had evolved by the last common ancestor of cnidarians + other eumetazoans. Additional characters appeared with acoels and with protostomes + deuterostomes. Consequently, there is no unique association between the definitive suite of "bilaterian" characters and any single clade. Rather (and

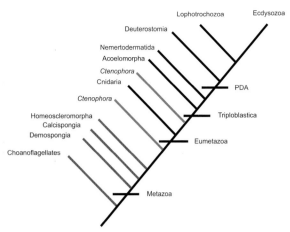

Figure 1. A consensus metazoan phylogeny, based on a variety of recent molecular studies as discussed in the text. The sponges are shown as three separate clades, consistent with several recent studies, and the Nemertodermatida are placed between the acoels and the protostome–deuterostome ancestor (PDA), as suggested by several recent studies discussed in text. The position of several groups remains uncertain. Two alternative placements of the ctenophores are shown; the phylogenetic position of the placazoans (*Trichoplax*) is probably between the sponges and cnidarians, but recent studies have given conflicting results so they are not shown on this tree.

indeed, unsurprisingly), "bilaterian" characters appeared progressively through eumetazoan evolution. This stands in contrast to a number of papers interpreting the nature of an "urbilaterian" (Kimmel 1996; Arendt and Wittbrodt 2001; Balavoine and Adoutte 2003; Hejnol and Martindale 2008). Recognition of the sequential evolution of bilaterian characters sheds new light on interpretation of fossil evidence for "bilaterians."

We begin with a discussion of the phylogenetic framework for Metazoa and then proceed to the precambrian fossil record of eumetazoan evolution, before integrating the fossil and developmental patterns within a phylogenetic context. In the later part of the paper, we turn to the issue of how the developmental GRNs associated with eumetazoans may have arisen and sketch out the evolutionary steps that are indicated by existing evidence. Several recent papers have proposed morphological scenarios for the early evolution of animals (Martindale et al. 2002; Finnerty 2005; Baguñà et al. 2008; Nielsen 2008; Martindale and Hejnol 2009). In contrast, we are interested here in the evolutionary changes in the structural character of developmental GRNs that may have underlayed eumetazoan evolution during the late Neoproterozoic and Cambrian, and the implications of these changes.

METAZOAN PHYLOGENY AND MOLECULAR CLOCKS

Metazoan phylogenies have used various combinations of morphological, developmental, and sequence data, the latter largely from 18S rRNA and conserved protein sequence. During the past 5 years, these analyses have become far more robust, sampling a greater variety of taxa, adding more gene sequences and other information including secondary structures and the presence or absence of specific markers such as microRNAs (miRNAs). More sophisticated phylogenetic algorithms have been introduced as well, including methods for statistical testing between alternative phylogenetic models. Consequently, metazoan phylogeny is increasingly robust, although some critical areas of controversy still remain.

Virtually all recent analyses agree on a topology with choanoflagellates as the closest living relatives of metazoans, followed by sponges, cnidarians, and acoel flatworms, and finally the last common ancestor of lophotrochozoans + ecdysozoans (protostomes) and deuterostomes (Fig. 1) (Douzery et al. 2004; Nielsen 2008; Peterson et al. 2008; Minelli 2009). Nielsen (2008) recently summarized the discordance among various recent molecular-based phylogenies of the most basal metazoan groups. As described briefly below, there is little consensus. Because our emphasis here is on the early evolution of eumetazoans, phylogenetic relationships among the various lophotrochozoan, ecdysozoan, and deuterostome lineages need not concern us.

Several analyses of morphologic and molecular data for sponges suggest that they are polyphyletic, having arisen multiple times at the base of the metazoa (Peterson and Butterfield 2005; Sperling et al. 2007; Nielsen 2008), although this is disputed (Dunn et al. 2008; Philippe et al. 2009; Schierwater et al. 2009). Differences in taxon sampling appear to have been a major reason for these divergent results (Sperling et al. 2009a). The most recent analysis (Sperling et al. 2009b) uses seven nuclear genes from 29 sponges and a variety of eumetazoan outgroups. The results strongly support sponge paraphyly, and most importantly, the phylogenetic tests applied are strongly inconsistent with sponges as a single clade, or with the existence of a clade of diploblastic organisms (Porifera + Cnidaria + placazoa). A significant result of this study is that the homoscleromorph sponges appear to be the sister group to the eumetazoans. Homoscleromorphs are a small group of unusual sponges that possess several characters not found in other sponges, including a basement membrane with collagen IV, in both adults and larvae, and other details of cell structure (Nielsen 2008). Placazoa (*Trichoplax*) are likely to be basal to Eumetazoa (Srivastava et al. 2008; Sperling et al. 2009a), despite a contrary claim that they are basal instead to a clade of diploblasts (Schierwater et al. 2009). The position of the ctenophores is even more contentious and remains unresolved. Some studies favor them as the sister group to Cnidarians, and others as the sister to acoels + remaining bilaterian clades (summarized in Nielsen 2008). One recent study suggests that they are the most basal metazoan clade (Dunn et al. 2007), although this receives no support from other recent analyses. There is little known of the developmental molecular biology of ctenophores, and their phylogenetic position is not relevant for our discussion here. Several interesting recent papers suggest that the nermatodermatids lie above acoels and below the PDA (Wallberg et al. 2007; Baguñà et al. 2008; Paps et al. 2009). As discussed further below, these groups may represent the surviving descendants of a once richer late Neoproterozoic group of early bilaterian clades.

Because the fossil record only provides minimal estimates of the appearances of distinctive and fossilizable morphologies, molecular clock estimates in principle offer critical additional information on the timing of the diversification of eumetazoan lineages (Aris-Brosou and Yang 2003; Douzery et al. 2004; Peterson and Butterfield 2005; Peterson et al. 2008). Recent molecular clock results indicate divergences that are largely congruent with the fossil record, in contrast to earlier results that suggested divergences occuring much earlier in the Proterozoic. For example, two recent studies, by Douzery et al. (2004) and Peterson et al. (2008), respectively, report dates for the origin of the Metazoa near 850/770 mya, eumetazoa near 695/680 mya, and the protostome–deuterostome divergence near 640 mya (all with uncertainties of tens of millions of years). Thus, sponges and cnidarians appeared during the Cryogenian, with the protostome–deuterostome ancestor close to the base of the Ediacaran, and bilaterian divergences during the Ediacaran (for geologic framework and time units, see Fig. 2).

THE FOSSIL RECORD OF BILATERIA

Environmental Effects

The origin of Metazoa and of eumetazoans occurred during an interval of considerable environmental change. Two extensive glaciations, affecting much of the globe, occurred beginning ~730 mya and ended at 635 mya (Kaufman et al. 1997; Condon et al. 2005). These are known to geologists as the Sturtian and Marinoan glaciations, respectively, after the regions in which they were first identified. Each may have had several pulses of glaciation, but more importantly, persuasive geologic evidence suggests that ice extended close to the equator and may have resulted in an essentially global "snowball Earth," although refugia evidently persisted, because the origins of animals, algae, and fungi lie much deeper than these events (Hoffman et al. 1998; Hoffman and Schrag 2002). The duration of these glacial intervals is not yet well constrained, but they appear to have lasted for millions of years, and durations of tens of millions of years have been claimed. A later, probably less-extensive, glaciation occurred near 580 mya (the Gaskiers glaciation; Thompson and Bowring 2000), and there was possibly an additional glacial event close to the Ediacaran–Cambrian boundary at 542 mya.

Aerobic respiration provides about an order of magnitude more energy for the same amount of food than does anaerobic metabolism. Consequently, oxygen is critical to the origin and diversification of metabolically active, complex organisms, and oxygen concentrations limit the maximum size of organisms. Many paleontologists have argued that rapid increases in oxygen levels were associated with the initial diversification of animals. However, as Butterfield has pointed out, it is difficult to unequivocally evaluate the required oxygen levels necessary for the evolution of metazoans, nor is the geochemical evidence easily interpreted (Catling et al. 2005; Butterfield 2009). Of course, attaining higher oxygen levels is no guarantee that complex animals will evolve, but it may function as a threshold. Oxygen is

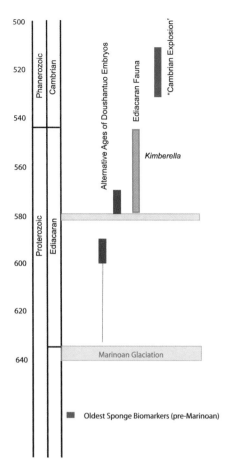

Figure 2. Timescale, geologic framework, and key fossil events associated with the evolution of Eumetazoa. The durations of the Sturtian and Marinoan glaciations are not well constrained, although the end of the Marinoan is well dated. Each may contain multiple events. The duration of the Gaskiers glaciation at 580 mya appears to be relatively short. Some organic walled remains from the Doushantuo that may be fossil metazoan embryos (*Tianzushania*) date to soon after the Marinoan glaciation, as shown by the gray line. The age of the most fossil-rich horizons not well resolved, however, and two alternatives are shown: 1, before the Gaskiers glaciation, or 2, following the Gaskiers. The Ediacaran fauna has been divided into three sequential but partly overlapping assemblages, which are not shown here. The position of the appearance of diverse skeletonized fossils and other metazoan assemblages in the Early Cambrian is shown as the "Cambrian Explosion."

also required for the biosynthesis of collagen (Towe 1970). Although simple animals could have evolved with relatively low oxygen levels, even the construction of large sponges requires considerable oxygen levels in their immediate environment for the synthesis of collagen.

Geochemical evidence records a remarkable change in the redox state of the oceans during the late Neoproterozoic, shifting the oceans from a largely anoxic, and probably iron- and sulfur-rich state, to the predominantly oxygenated state found through the past 550 million years. This evidence comes from measurements of shifts in isotopic ratios of carbon, sulfur, and more recently other stable isotope systems, as well as other measurements. Geologic measurements of the ratio of two isotopes of carbon, ^{12}C and ^{13}C, chronicle shifts between two large carbon reservoirs, one consisting of organic carbon, including living organisms and their buried remains such as coal, peat, and oil, and the other of inorganic or carbonate carbon. Shifts in the carbon isotope ratio reflect changes in the burial of organic carbon or the release of previously buried carbon, among other causes. The carbon ratio is measured relative to a standard and is reported as δ^{13}C. The late Neoproterozoic and Early Cambrian interval is characterized by generally positive δ^{13}C values, denoting the burial of organic carbon, punctuated by several very large negative δ^{13}C anomalies associated with the glacial events. There is also a very large negative shift ($\geq$12 ‰ δ^{13}C) in the mid-Ediacaran that is not associated with any known glaciation. A number of geologists have interpreted this event as representing the final oxidation of the deep oceans (Condon et al. 2005; Fike et al. 2006). Other isotopic evidence, however, suggests that the oxidation may have occurred somewhat earlier, perhaps ~600 mya (Canfield et al. 2008; Scott et al. 2008; Shen et al. 2008), and it is quite possible that the oceanic oxidation did not happen synchronously around the world, but instead in a staggered and regionally variable pattern (McFadden et al. 2008), ending near 555 mya. Highly variable carbon isotope ratios persist through the Early Cambrian, with the extreme anomalies gradually declining to levels typical of the post-Cambrian (Halverson et al. 2006).

Geochemical and geologic evidence indicates that environmental conditions during the deglaciation phase were quite severe. Sturtian and Marinoan glacial debris are overlain by unusual carbonate deposits indicative of rapid deposition in highly alkaline seas, probably during an intense climatic greenhouse interval (Crowley et al. 2001; Higgins and Schrag 2003; Corsetti et al. 2006). The various eukaryotic lineages that survived these glaciations must have been able to persist in refugia in the face of the extensive glacial episodes and the harsh postglacial phases. The Gaskiers event at ~580 mya seems less likely to have had a severe impact on diversity.

Earliest Fossil Evidence of Metazoans

Preserved biomolecules, or biomarkers, provide the earliest evidence for metazoans. Pre-Marionoan deposits from Oman contain the degraded remains of C^{30} sterols, a sterol found today only in demosponges (Love et al. 2009). There is no evidence of such biomarkers near the Sturtian glaciation, but this evidence suggests that demosponges must have evolved before 635 mya.

The Doushantuo Biota

The Doushantuo Formation in southern China contains a diverse and exquisitely preserved suite of algae, organic-walled microfossils, and metazoan embryos, providing the best fossil evidence of early metazoan diversification (Fig. 3) (Xiao and Knoll 1999, 2000; Chen et al. 2000, 2002, 2004, 2009a,b; Xiao 2002; Hagadorn et al. 2006). These fossils reveal a wealth of cellular and often subcellular structures and a variety of different cell numbers (up to 2600 cells). The oldest described probable metazoan embryo is *Tianzhushania*, found just above the Marinoan glaciation but continuing through most of the younger parts of the Doushantuo formation (Yin et al. 2007). Assuming that the older specimens, which are known only from their external structure, are internally multicellular as clearly are the younger forms, these would be the oldest fossil metazoans so far unearthed. Given the harsh environmental conditions of the early Ediacaran, the presence of protecting structures surrounding metazoan embryos is not surprising, and indeed, this type of fossil embryo disappears from the fossil record ~550 mya, coincident with the oxygenation of shelf waters (Cohen et al. 2009). As with many fossil assemblages, preservational problems pose many difficulties in distinguishing the original morphology from subsequent alteration (Xiao and Knoll 2000; Dornbos et al. 2006). That the Doushantuo microfossils represent animal embryos is certain, but preservational problems do pose challenges in establishing their phylogenetic affinities.

The Doushantuo microfossil assemblage is unusual in that the preservation conditions allowed fossilization of soft cellular structures. The major feature of this assemblage is the diversity of probable eumetazoan forms implied by even the limited amount of evidence so far available. Figure 3 provides some examples. Here we see, for example, well-preserved adult forms, albeit microscopic in scale, of two different cnidarian clades: a chambered coral similar to some known from the Cambrian and a possible hydroid-like organism (Chen et al. 2002). Figure 3C shows a bilaterally organized, probably coelomate animal fossil *Vernanimalcula*, that is only ~200 μm long but appears to be triploblastic (Chen et al. 2004a,b). Although this form is only known from sections, a number of additional specimens have now been recovered (D.J. Bottjer, unpubl.). The largest amount of evidence relevant to our considerations here is from synchrotron X-ray tomographic (SXRT) studies of fossilized embryos from the Doushantuo formation (Chen et al. 2006, 2009a,b; Hagedorn et al. 2006). Many specimens of near-identical morphology and dimensions have been reported for each of the embryonic forms illustrated in Figure 3. Figure 3D and D2 show computational sections of a frequently occurring, probably noneumetazoan, chorionated embryo that has a unique cleavage pattern similar to that of some modern sponge eggs (Chen et al. 2009a). A cleavage-stage embryo that forms polar lobes, as do a variety of modern protostome eggs, is reproduced in Figure 3E (Chen et al. 2006), and another unique cleavage form known today in acoel worms is illustrated in Figure 3F (Chen et al. 2009a). The successive computational SXRT sections display, from the "vegetal" side, two macromeres, and orthogonally arranged on the opposite side are four micromeres. Figure 3G shows a hollow gastrulating form, possibly cnidarian (Chen et al. 2009a). Finally, among many other forms that could have been included, Figure 3H shows a complex, later-stage bilaterally organized embryo with clearly diverse cell types, including macromeres, micromeres, and a central cord of possibly endodermal cells (Chen et al. 2009b). Clearly, what is thus far missing are the adult forms that produced all of these and the many further types of

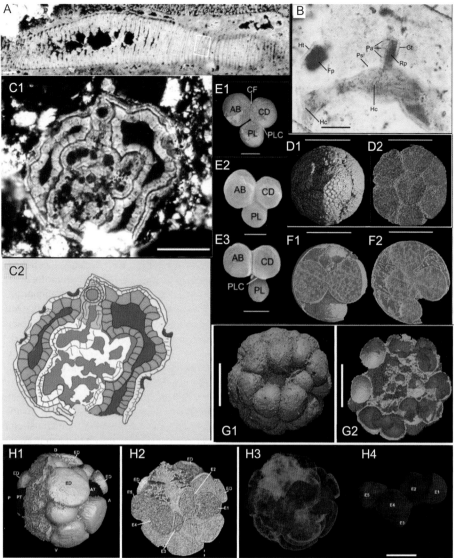

Figure 3. Fossil metazoans from the Doushantuo Formation, China. (*A–C*) Visualizations of stereomicroscope sections. (*A*) Colony of *Sinocyclocylicus guizhouensis*, a tabulate coral-like animal. Bar, 50 μm. (*B*) Hydrozoan-like animal. Features: (Gt) Gonotheca, (Pa) Perisarc annuli, (Ht) hydrotheca, (Ps) perisarc, (Rp) possible reproductive polyp, (Fp) possible feeding polyp, (Hc) hydrocaulus. Bar, 100 μm. (*C1*) Specimen of *Vernanimalcula guizhouena*; digital image of 50-μm-thick section. Bar, 40 μm. (*C2*) Interpretation: (Orange) Mesodermal coelomic layer, (pink) pharyngeal structure, (green) mouth, (tan) endodermal gut with inclusions, (yellow) ectodermal layer, (red) surface pits. (*D–H*) SEM (scanning electron microscope) and SXRT (synchrotron X-ray tomography) computational sections. (*D1*) External SEM of chorionated embryo; note cellular imprints each bearing a pit, possibly site of a cilium, with chorion partly broken to reveal surface of embryo below. (*D2*) Same embryo seen in medial SXRT section, revealing a central blastomere surrounded by six external blastomeres (16-cell stage). Bar, 400 μm. (*E1–E3*) Individual "trefoil stage" polar lobe cleavage-stage embryos, viewed by SEM. Bar, 250 μm. (*F1,F2*) Successive SXRT sections of a chorionated 6-cell cleavage-stage embryo, viewed from "vegetal" or large macromere duet end. Note prominent inclusions in macromeres that probably represent yolk platelets. Bar, 400 μm. (*G1*) External SXRT view of advanced embryo, polar view of site of invagination, (*G2*) medial section of same embryo showing blastocoelar cavity containing individual ingressed cells. Bar, 250 μm. (*H*) Complex later-stage embryo; (A) anterior, (P) posterior, (E) endodermal cell, (V) ventral, (D) dorsal, (ED) ectodermal cell. (*H1*) External view, from right side, (*H2*) SXRT section 43% in from right surface, showing distinct putative endodermal cell types, (*H3*) transparent SXRT view from right side, endodermal cord colored in red, (*H4*) computationally isolated endodermal cord. Bar, 250 μm. (*A,B*, Reprinted, with permission, from Chen et al. 2002 [© Elsevier]; *C*, reprinted, with permission, from Chen et al. 2004a [© AAAS]; *D,F,G*, reprinted, with permission, from Chen et al. 2009a [© Elsevier]; *E*, reprinted, with permission, from Chen et al. 2006 [© AAAS]; *H*, reprinted, with permission, from Chen et al. 2009b [© National Academy of Sciences].)

embryo, but even so, their diversity is incontrovertible, and their topologies are known today only in eumetazoans. Overall, the evidence suggests strongly that animal life of several eumetazoan clades was already extant.

The base of the Doushantuo formation is well dated at 635 mya, and radiometric dates and fossils establish a Cambrian age for the uppermost rocks. The age of the layers of the Doushantuo rich in fossil embryos is less certain, however. Integration of U-Pb geochronology from China, Oman, and Namibia, stratigraphic correlations in south China, as well as $\delta^{13}C$ data suggest that the Doushantuo fossils are younger than the Gaskiers glaciation at

580 mya, and thus broadly correlative with the soft-bodied Ediacaran biota (Condon et al. 2005). These results conflict with earlier reports of Lu-Hf and Pb-Pb dates from phosphorites of the Doushantuo Formation that yielded dates of 602 ± 48 and 599 ± 4 mya, respectively (Barfod et al. 2002). However, these analyses are whole-rock analyses of diagenetically altered phosphorites and use systems that can be less reliable than U-Pb radiometric dates. Black shales between the fossiliferous upper and lower phosphorite beds were recently dated by Pb-Pb methods to 572 ± 36 mya (Chen et al. 2009c). This discrepancy has not been resolved and we simply note two alternatives: (1) dates of 600–590 mya for the Doushantuo fossils or (2) post-580 mya (see Fig. 2).

The Ediacaran Biota

A morphologically diverse suite of macroscopic soft-bodied fossils is found in rocks from many parts of the world dating from 579 to 542 mya (Figs. 2 and 4). These fossils consist of a variety of discs, fronds, and more complex forms, some exhibiting apparently bilateral symmetry (such as *Dickinsonia* and *Yorgia*, illustrated in Fig. 4) (Gehling et al. 2005; Narbonne 2005; Fedonkin et al. 2007a; Xiao and Laflamme 2008). None of the published fossils exhibit evidence of a mouth, appendages, or other morphological structures indicative of phylogenetic affinities of protostomes or deuterostomes. Indeed, recent comparative developmental studies are consistent with all of these fossils representing clades between sponges, at the origin of metazoan, and the acoels (Erwin 2009).

Kimberella, found in rocks dating to 555 mya (Martin et al. 2000), is the notable exception. *Kimberella* is an oval fossil, ~5 cm in length, with an apparently muscular foot surrounded by crenulated depressions that likely represent some sort of a frill (Fig. 4D). A number of specimens preserve signs of a proboscis, plausibly in the anterior end of the animal. Specimens of *Kimberella* are often associated with radiating, parallel scratch marks that have been interpreted as indications of the animal feeding on a microbial mat (Fedonkin and Waggoner 1997; Fedonkin et al. 2007b). Although many of these features are similar to those of a mollusk (Fedonkin and Waggoner 1997), it is probably premature to assign *Kimberella* to the Mollusca. But this is convincing fossil evidence that the protostome–deuterostome ancestor likely predates 555 mya.

Ediacaran Trace Fossils

Trails and burrows are another form of fossil, in addition to the molecular fossils and body fossils already described. There is a rapid increase in the diversity and complexity of trace fossils after ~560 mya, in the latest Ediacaran and Early Cambrian. Most Ediacaran trace fossils are poorly organized, meandering, horizontal forms laid down on the surface of the sediment. A more complex form, *Helminthorhaphe* from South Australia, has a relatively tight spiral meander connected to a more random trail. The complex spiral suggests a moderately well-developed sensory system in a bilaterian animal. Although many papers have described a variety of Ediacaran trace fossils, some of which appear to have been generated by bilaterians, recent reanalyses indicate that the oldest valid metazoan trace fossils date to 560–555 mya (Jensen et al. 2005, 2006), roughly coincident with the first appearance of *Kimberella*. Not until the very end of the Ediacaran do we find any vertical, penetrating burrows, indicating the first appearance of organisms with a hydrostatically resistant coelom. Numerous paleontologists have used these burrows as an indication of the presence of the organisms of the grade of modern protostomes. Because acoel flatworms are bilaterians, primitive bilaterians were probably small and lacked a coelom. In addition, paleontologists have recently recognized a suite of tubes that represent body fossils, rather than trace fossils, and which appear to be bilaterian, but the phylogenetic affinities of the organisms that produced these tubes are

Figure 4. Ediacaran soft-bodied fossils. (*A*) *Pteridinium* from southern Namibia; image is ~1 foot across. (*B*) Portion of a frond of *Rangea* from southern Namibia; specimen is ~3.5 inches long. (*C*) *Fractofucus*, a rangeamorph from Mistaken Point, Newfoundland; specimen is ~4 inches long. (*D*) *Kimberella*, probably bilaterian, from the White Sea, northwestern Russia, anterior to the right. (*E*) *Dickinsonia* from South Australia, ~2 inches long. (*F*) *Yorgia*, from the White Sea, Russia; the asymmetric anterior shield is in the upper part of the image, ~4 inches across. Photographs by DHE.

unknown. Because discrete trace fossils less than ~1 cm in diameter do not necessarily preserve well, their record suggests that benthic bilaterian animals larger than this size were not present until after ~560 mya, although smaller bilaterians could well have been present ealier.

In summary, the fossil record demonstrates that the origin of metazoans predates 635 mya, consistent with evidence from molecular clocks. The Ediacaran-age Duoshantuo Formation from South China contains well-preserved fossil embryos of various stages, consistent with a variety of animals above cnidarians on the phylogenetic tree. We cannot yet be certain whether the diverse metazoan Doushantuo embryo assemblages date from 600–590 mya or are younger than 580 mya. The phylogenetic affinities of the soft-bodied Ediacaran biota (579–542 mya) are complex and represent a variety of lineages. We view them as a series of clades arising independently along the metazoan backbone. Although the morphology of several of the clades, particularly the Dickinsoniamorphs, seem to be superficially bilaterian, so far only *Kimberella* shows convincing bilaterian characters. Geochemical measurements indicate that the deep oceans had become well oxygenated ~555 mya, and the appearance of *Kimberella* and the oldest metazoan traces at this time is unlikely to be a coincidence.

INTEGRATION OF FOSSIL AND DEVELOPMENTAL DATA WITHIN THE PHYLOGENETIC CONTEXT

Phylogenetic and paleontological evidence provides a new perspective on the genetic regulatory structure of deep time eumetazoan evolution and the underlying bilaterian developmental process. Here, we view eumetazoan evolution through the lens of the structure of the developmental GRNs (dGRNs) that have recently been solved. We then attempt to integrate the conclusions from this kind of information with those emerging from the preceding discussion.

The Developmental Perspective

Although the body of evidence is yet slim, comparison of the few relatively well-known and extensive dGRNs for which we have embryonic development reveals certain common structural properties. As examples, this pertains to the dGRN for the *Drosophila* dorsoventral system (Stathopoulos and Levine 2005); the sea urchin embryo skeletogenic mesoderm dGRN (Oliveri et al. 2008) and anterior and posterior endoderm dGRNs (Peter and Davidson 2009a,b); the dGRN for *Xenopus* dorsoventral specification (Koide et al. 2005); and gut-lineage specification in *Caenorhabditis elegans* (Owraghi et al. 2009). Each of these dGRNs consist exclusively of genes encoding transcription factors and signaling molecules and the functional *cis*-regulatory linkages among them. Their general characteristics include the following:

1. They are many layers deep, i.e., they extend from the initial spatial inputs, which in each case initiate the zygotic cascade of transcriptional expressions, to terminal states of cellular specialization in embryonic space.

2. Their depth follows from the different kinds of subcircuits they include, viz. subcircuits that install the initial state of specification in the respective spatial domains, subcircuits that then lock down the specification state, subcircuits that exclude other specification states, subcircuits that operate intercell signaling systems, and finally, subcircuits that run differentiation gene batteries (for review, see Davidson 2006; Peter and Davidson 2009b).

3. They are hierarchical and determinate, in that at each stage, the upstream subcircuits determine the activity (or silence) of those in the next step downstream.

4. They involve, for each "component" of the developmental process, on the order of 20–50 different regulatory genes.

5. But these same regulatory genes are usually found wired into other dGRNs in the same genome. Regulatory genes operate at multiple times and places where they use entirely different input connections; like signaling systems, transcription factors are used continuously in the life cycle and almost none are dedicated to single developmental events. dGRNs control the formation of very disparate kinds of structure. Development by the pathways used in all well-known examples requires execution of the variety of regulatory "jobs" that dGRN subcircuits do. A conclusion that is unlikely to be far wrong is that the development of any eumetazoan embryo or postembryonic body part is likely to be controlled by dGRN components of similar depth, complexity, and subcircuit diversity. The difference between is often called a "more complex animal" and a "less complex animal," to the extent that if there is any difference in the underlying dGRN structures, it probably lies in the number of dGRN components required to build the body plan of the animal. This will depend on the number of qualitatively different body parts, stages, and morphological features for which the development of the body plan has to account. But each dGRN component will have the depth and other characteristics enumerated above, i.e., in comparing diverse animals, the depth per dGRN component will be similar, but the number of these, or the breadth, will vary.

The general quality of eumetazoan dGRN structure requires that there must have been preceding stages in dGRN evolution, in which the dGRN complexity was lower, the depth shallower, the variety of subcircuits less, and hierarchy less dominant a feature. In the following sections, we expand on the subject of dGRN evolution per se. Our point here is that if eumetazoan body plans require deep dGRN wiring, developmental regulatory programs structured in this way must have preceded the divergence of the eumetazoa. The obverse is that animals lacking these program characteristics were not eumetazoans.

Bilaterian Evolution

The "Bilateria" have traditionally been thought of as the clade of animals descended from the last common protostome–deuterostome ancestor. Aside from the issue of whether the protostomes are really a monophyletic clade, the phylogeny discussed above destroys this equivalence: The Bilateria must encompass the acoels, yet the acoels lie outside the clade composed of protostomes plus deuterostomes. This leads to the realization that these characters appeared in evolution piecemeal, in organisms that also display nonbilaterian characters (Fig. 5). The cnidarians are the great example: As has been pointed out (Technau 2001; Finnerty et al. 2004; Martindale et al. 2004; Technau et al. 2005; Putnam et al. 2007; Hejnol and Martindale 2008), in development of the most basal cnidarian clade, anthozoans, regulatory genes are expressed in bilateral axial patterns that indicate an underlying set of bilateral regulatory states. Furthermore, these patterns include examples of regulatory genes expressed in relative positions reminiscent of the patterns of expression of the same genes in bilaterian development. Cnidarians seem to possess almost the complete bilaterian regulatory gene tool kit, including anterior and posterior *hox* genes (Finnerty et al. 2004; Putnam et al. 2007) and generate many of the ectodermal, neuronal, endodermal, and mesodermal differentiated cell types also found in bilaterians. Yet, the same genomic regulatory systems encode the nonbilaterian features of anthozoan larvae and adults as well. Similarly, the acoels have many features of other bilaterians, but unlike the basal body plans of protostomes and deuterostomes, they use a blind rather than a through gut and have other special characters (Baguña et al. 2008).

Thinking of the bilaterian character suite as the result of a gradual assembly process changes our expectations of the phylogenetic breadth of Ediacaran assemblages. As noted above, this first macroscopic animal fossil assemblage is very diverse. We might expect it to include then current representatives of all the eumatazoan clades: cnidarians, acoel grade bilaterians, protostomes, and deuterostomes, and perhaps now-extinct clades that possessed some but not all bilaterian characters. There is a potent underlying mechanism devolving directly from the nature of eumetazoan dGRNs that will result in the appearance and persistence of given diverse clades (instead of a continuous array of morphotypes). This is the canalization of developmental regulatory process in each lineage, i.e., once the upper-level subcircuits in the hierarchical dGRN have emplaced regulatory states in respective portions of the organism, fundamental aspects of the body plan are fixed in each lineage.

Eumetazoan Lineages and dGRN Kernels

Canalization of developmental processes in eumetazoan lineages means that the responsible upper-level subcircuits will be evolutionarily conserved in all descendants of the ancestor of that lineage. Upper-level subcircuit conservation is the other side of the coin of canalization. Our dGRN kernel concept (Davidson 2006; Davidson and Erwin 2006; Erwin and Davidson 2009) originated in considerations of extremely conserved dGRN subcircuits, which in the development of modern animals determine the early regulatory states for given progenitor fields, i.e., kernels cause expression of regulatory genes that together define a previously undetermined area of cells as the domain where a certain developmental outcome will ensue. For example, two cases that we discussed in earlier work concerned an extremely conserved kernel that is involved in setting up endoderm specification in very distant members of the echinoderm phylum (Hinman et al. 2003) and a kernel at the top of the dGRN for heart specification that is very similar from *Drosophila* to vertebrates (Davidson 2006). Another echinoderm-specific kernel has been discovered that underlies embryonic mesoderm specification (McCauley et al. 2009). Consideration of the evolutionarily canalizing upper-level subcircuits of dGRNs leads to the idea of dGRN kernels, just as does the observation of real, unusually conserved upper-level dGRN subcircuits in modern animals. Thus, the discrete forms of eumetazoan body plan must be built by circuits (i.e., kernels) high up in the structure of the dGRN that are the same in all members of each clade, because they all share the developmental outcome of the clade-specific body plan. It is interesting that in the (few) cases known, the several genes of the kernel subcircuit have acquired multiple internal feedback linkages ("recursive wiring"; Erwin and Davidson 2009), so that interference with expression of any of them by mutation or experimental manipulation has severe effects on the phase of development that they initiate. This accentuates the selective conservation of the whole subcircuit, on pain of developmental catastrophe.

We must emphasize that as we envision it, there is nothing unusual about the initial formation of kernels as opposed to other subcircuits of the eumetazoan dGRN; what is unusual is the role that they have downstream because of their hierarchical position in the dGRN once they are formed. This role has strong phylogenetic implications. By canalizing the possibilities of development downstream, kernels essentially define the "developmental morphospace" within which developmental variation is allowed. If, for example, a kernel sets up the initial regulatory state leading to endoderm specification in a given patch of cells, development is canalized in the sense that the only fates allowable in that patch of cells are endoder-

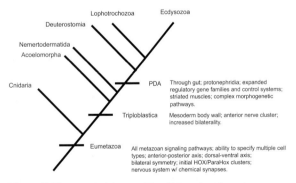

Figure 5. The sequential assembly of bilaterian characters was a gradual process. Both the developmental tools and the morphologic characteristics of Bilateria were sequentially adopted during early metazoan evolution. Key features are shown at relevant nodes.

mal, and the morphospace for future variations in the structure of the gut is defined. Taxonomic clades are defined by the structures emerging from these same spatial elements of the body plan. Therefore, these clades must reflect the assignment of specification states for spatial elements of the body plan by dGRNs (Davidson and Erwin 2006). dGRNs represent the deep structure of developmental systems, and it is their hierarchical structure that is imperfectly reflected in the early 19th century hierarchical Linnaean concept of animal taxonomy. Categories of the Linnean taxonomic hierarchy are not always monophyletic clades (although they should be), but we might predict that the entities recognized as superphyla and phyla should be defined by kernels responsible for their morphological attributes. Similarly, class-specific kernels and perhaps lower-level kernels should exist as well (Fig. 6). Differences defining genera and species are smaller, i.e., they occur as the result of variations in dGRN wiring at lower levels of the hierarchy. These variations are at the level of deployment and nature of differentiation gene batteries and the deployment of signal-driven and other switch systems (Davidson and Erwin 2006). Because almost all variation in body plans since the Early Cambrian have been at the subphylum level, phylum- and superphylum-level kernels must have originated in the Late Neoproterozoic and perhaps Early Cambrian. By this argument, generation of class-level kernels must have continued at the least into the Ordovician.

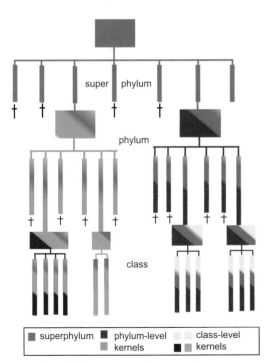

Figure 6. Kernels determining characters of the body plan at superphylum, phylum, and class levels. Each clade is represented by a vertical line. Each has perhaps the same number of total characters, but only those characters specified within the morphospace defined by the respective dGRN kernels are indicated by the color coding.

BEFORE THE EUMETAZOA: INFERRED STAGES OF EVOLUTION OF DEVELOPMENTAL GRNS

To imagine how the hierarchical eumetazoan dGRN with its clade-specific kernels at the top could have arisen, we need to take a conceptual journey backward and deconstruct the eumetazoan dGRN, considering the origins of each of its component types of circuitry.

Initial Steps of dGRN Evolution

We begin with the earliest dGRN subcircuits in evolutionary terms. These are differentiation gene batteries. The most basal extant metazoans, sponges (see Fig. 1), deploy a number of differentiated cell types. In terms of genetic regulatory circuitry, the simultaneous expression of diverse differentiation gene batteries in temporally coeval but spatially separate cellular domains is the fundamental property of metazoans (in contrast, for example, to single-celled organisms that express different downstream genes at different times in their life cycle or in response to different external cues). Differential expression of downstream gene batteries is the simplest regulatory structure that can produce an organism composed of simultaneously present, diverse specialized cell types. The fundamental dividing line is that metazoans must execute spatial developmental gene regulation in order to direct the construction of regulatory states in different morphological compartments that must have a genetically specified geometrical relation to one another. We know enough about the control structure of differentiation gene batteries to infer the minimum regulatory requirements. Unlike the regulatory states produced by the deep dGRNs of eumetazoans, the regulatory states needed just to run differentiation gene batteries consist of only a very small number of transcriptional regulators that together drive the activity of all the downstream protein-coding genes of the battery (for review, see Davidson 2006). An additional requirement for the earliest developmental process deploying differentiation gene batteries is some form of developmental "address" that would cause activation of diverse driver regulators in particular domains of the multicellular structure. The address might consist, for instance, of spatial polarizations in distribution of molecules with gene regulatory activity. The many examples we have of polarizations of regulatory significance in eggs provide models of what we might expect to have been used for this purpose. To summarize, as indicated in Figure 7A, the minimal GRN for the construction of a multicellular animal expressing different differentiation gene batteries in different cellular domains is a relatively shallow structure, similar to the differentiation gene batteries that lie at the periphery of all modern dGRNs.

Signaling, and the Augmentation of dGRN Architecture

All modern bilaterian dGRNs deploy intercellular signaling in a variety of ways. Developmental signaling depends

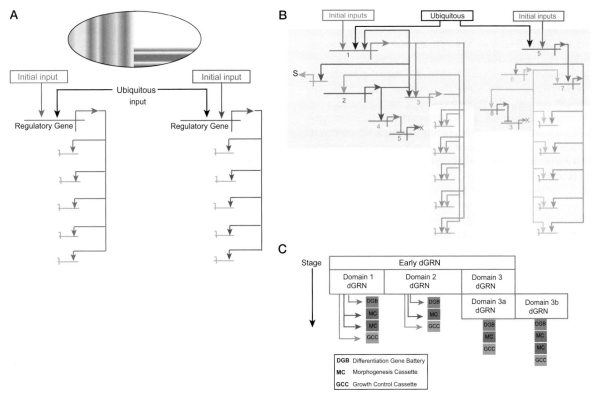

Figure 7. Stages in early dGRN evolution and downstream functional cassettes. (*A*) Initial stage (see text): Spatial inputs in a simple metazoan are used to activate transcriptional drivers of two different differentiation gene batteries. (*B*) Later stage, illustrating intercalary network evolution. Intercalated circuitry is shown on a tan background and the same differentiated gene batteries as in *A* on a lavender background. Genes 1 and 5 have the specific role of transducing the localized initial inputs that are the same as in *A*. Genes 2 and 3 and 6 and 7 set up feedback lockdown subcircuits stabilizing the regulatory states and providing for a stable generation of the differentiation gene batteries. Genes 4 and 8 repress key genes in the alternate regulatory state, i.e., genes 3 and 5, providing additional robustness of state. (*C*) Downstream functional cassettes deployed due to regulatory inputs from the respective spatially active dGRNs: (blue) diverse differentiation gene batteries, (red) diverse morphogenesis cassettes, (green) cell cycle control cassettes.

mainly on a small set of continually reused signal ligand–receptor systems, each of which operates a signal transduction system the relevant biochemical targets of which are transcription factors. Signal systems affect development because they decisively alter the functional properties of their target transcription factors and thereby affect spatial expression of regulatory genes. A whole set of developmental signal systems are used for the development of every known bilaterian, with a few possible exceptions. Even choanoflagellates possess some of these signaling systems (King et al. 2003, 2008; Erwin 2009; Marshall and Valentine 2009) as do sponges, and they are all found in cnidarian genomes (Matus et al. 2007). This can be considered a "preadaptation" to the subsequent evolution of dGRNs (Marshall and Valentine 2009). Signaling systems may have been incorporated in dGRNs very early in their evolution. They are very useful, and it is interesting that the following list of their developmental uses could apply to the simplest levels of dGRN architecture:

1. Inductive signaling: Sending cells alter the regulatory state of receiving cells.

2. Global spatial control: Cells receiving the signal are allowed to express signal system target genes, whereas in all other cells, these same target genes are default-repressed by the same transcription factor that transduces the signal.

3. Community effect signaling: Cells within a territory express a regulatory state signal among themselves in order to uniformly maintain that regulatory state (Bolouri and Davidson 2009).

Figure 7B represents a later stage of dGRN evolution in which several subcircuit devices have been inserted that greatly increase the robustness and accuracy of the differential control system in Figure 7A. These include a feedback subcircuit that stabilizes the expression of the differentiation driver genes, a subcircuit that is dedicated to interpretation of the initial spatial inputs, which because of the feedback circuit may now be present only transiently, and a subcircuit that represses a key gene of the alternative territory. In addition, within each domain, community effect signaling could be operating (not shown), and each domain might express an inductive signal that can be used to specify an adjacent territory. We now have a more hierarchical, multifunctional, nascent dGRN. Note that the additional circuitry has been inserted between the differentiation genes and the initial embryological address ("intercalary" evolution; Gehring and Ikeo 1999). This circuitry controls two stages of development: what we would call specification, i.e., setting up the regulatory state, and what we would call differentiation. It also carries within it the mechanistic possibility of inserting further stages and of further subdividing territories by deployment of signal systems using the same kind of subcircuit devices. Such continued processes can account for the evolution of the modern eumetazoan dGRN. An implication is that there could be a fixed repertoire of subcircuit topologies that have been incorporated over and over and over again as dGRNs evolve in depth and breadth. There is indeed beginning to emerge strong evidence for this from the comparative study of modern dGRNs (Levine and Davidson 2005; Davidson 2009; Peter and Davidson 2009a,b).

Something is still missing, however, and that is the control of morphogenetic functions. We know little of how the morphogenetic cassettes that generate cell movement, epithelial sheets, tubes, columns of cells, outgrowths, branching structures, invagination, etc., are structured or controlled. But clearly, they are linked to the upstream dGRNs (Christiaen et al. 2008), as are cell division control systems (for review, see Davidson 2006), because it is these dGRNs that determine their deployment and thus the morphology of the organism. The organism also has to operate physiological GRNs, for example, those that control its responses to immune challenge and environmental stress (Amit et al. 2009). Thus, in the end, the dGRN is the "brain" that determines the sequence of activation, location, and identity of the effector subcircuits of downstream genes that build the body plan of the organism (Fig. 7C).

Finally, in considering the assembly of the eumetazoan dGRN, a prominent feature of postgastrular development in the macroscopic modern bilaterians is the use of "vectorial" regulatory systems that act as switches to initiate or repress the activity of dGRN subcircuits in different regions of the body plan. Two very different kinds of these switches are diffusible molecules that affect gene expression, such as retinoic acid along the anteroposterior neuraxis and the *hox* gene system. The latter is used and reused as a set of vectorial patterning switches (for review, see Davidson 2001; Davidson and Erwin 2006) These systems are useful for controlling differences in regulatory state in serially reiterated domains and in organizing regulatory states in nested or contiguous spatial patterns. From the viewpoint of the dGRN, *hox* gene functions are in many of the most famous cases equivalent to what we have termed "input/output switches" with respect to those subcircuits of the dGRN that actually create pattern and deploy downstream morphogenesis cassettes and differentiation gene batteries. This role of *hox* genes is probably not to be considered an early, sine qua non evolutionary invention in the long history leading to the eumetazoan dGRN. For example, the sea urchin larva, a small, bilateral free-living organism, develops without ever deploying its *hox* gene complex (Arenas-Mena et al. 1998), which is used only later in generating the macroscopic adult form.

MECHANISTIC ASPECTS OF dGRN EVOLUTION: CHANGE AT THE DNA LEVEL

The basic mechanism of invention, and of every kind of structural change in dGRNS, is *cis*-regulatory alteration of the control systems of regulatory genes. These alterations either provide genes with new target sites and thus new inputs (i.e., forming new GRN linkages) or destroy preexisting target sites (i.e., breaking prior GRN linkages). As we have discussed recently elsewhere (Davidson and Erwin 2009), this kind of change converges on the process underlying the evolutionary phenomenon of "cooption," which denotes redeployment of regulatory gene expression to a novel spatial and/or temporal locus in development. In considering formulation of new dGRN circuitry, as in the progression from the grade of organization in Figure 7A to that in Figure 7B, the whole process boils down to acquisition of new *cis*-regulatory inputs in preexisting regulatory genes. A hidden presumption here is that the regulatory gene tool kit per se is preexistent, and this is now clearly supported by the results of genome projects and other data (Erwin 2009; Marshall and Valentine 2009). There has long been a conceptual disconnect between the continuous mechanism of evolution inferred from traditional protein evolution population genetics, in which phenotypic change is attained when a genetic alteration becomes homozygous, and the kind of process that follows from regulatory gene cooption. In the first place, formation of a new GRN linkage can obviously produce large regulatory, i.e., developmental, effects (or no effects; Davidson and Erwin 2009). But in addition, cooptive change is essentially a regulatory gain-of-function event, and this has profound consequences. As first pointed out by Ruvkun et al. (1991) and further discussed by ourselves (Davidson and Erwin 2009), extensive laboratory experiments show that regulatory gene gain of function is usually a haplodominant event. Particularly in embryonic development, expression of a single copy of a regulatory gene usually suffices to support downstream dGRN function (or else we would not have all the regulatory genes discovered in haploid mutant screens!). The evolutionary significance is that regulatory change in dGRN structure suddenly becomes a likely rather than an impossibly unlikely event, judging from the high rate of *cis*-regulatory change we observe by comparing related modern organisms. Each lineage of animals descended from a founder bearing a haploid regulatory gain of function will likewise express that function, which will thus be available to be combined with such further changes.

Subcircuit Assembly: How Do dGRNs Get Built?

In considering this question, the functional character of modern dGRNs must be our primary guide. Here, the most important feature is that dGRNs are modular in structure: The unit "jobs" of development are performed by individual subcircuits, as discussed above. The form of a dGRN consists of the sum of its particular subcircuits plus the morphology of the switches and other connections among the subcircuits. So our problem is to establish how subcircuits can be assembled by a process of (haplodominant) regulatory gene cooption. Typical subcircuits consist of three to six genes with many more linkages, because each gene has multiple inputs. In the process of subcircuit assembly, preexistent linkages and preexistent *cis*-regulatory modules or enhancers may often be used (sometimes after duplication), but at a different address due to insertion or mutational creation of new target sites. *Cis*-regulatory modules have many sites for ubiquitously present factors that perform mechanistic functions other than determining when and where the module will be active (e.g., mediating intermodule interaction, looping to the basal promoter, and amplifying regulatory output, etc; see, e.g., Yuh et al. 2001). Thus, addition of a new site is likely to be functional in the proximity of, or within, a preexistent *cis*-regulatory complex. But another evolutionary aspect to consider is at the level of the whole genome. As different developmentally localized dGRNs for various aspects of development accumulate in the genome, providing increasing overall dGRN breadth, the genome contains a larger and larger repertoire of already extant subcircuits that can be "highjacked" and coopted to a new function by relatively small changes that alter the deployment of the subcircuit as a whole. This predicts that as the global regulatory system becomes more complex, the possibilities of dGRN change will increase sharply.

An example from comparison of modern dGRNs will illustrate some of these points. In the evolution of echinoids (sea urchins), the euechinoid and cidaroid lineages diverged perhaps 275 mya (Littlewood and Smith 1995). The euechinoid lineage makes an embryonic skeleton from a precociously specified embryonic cell lineage and the cidaroid lineage does not. The euechinoid dGRN controlling specification, development, and differentiation of the skeletogenic lineage is well known (Oliveri et al. 2008). This very complex dGRN, which includes ~25 regulatory genes, must have appeared in the genomic control system for embryogenesis since divergence of these echinoid clades. In a effort to determine its evolutionary origin, we discovered that the entire skeletogenic network was highjacked from the portion of the genomic regulatory system that controls adult skeletogenesis (spines and body wall plates; Gao and Davidson 2008). This network was grafted onto the initial specification apparatus that defines the initial regulatory state of the skeletogenic founder cell lineage. In addition, four regulatory genes not used in adult skeletogenesis were coopted for participation in the embryonic skeletogenic system. All that might have been required to accomplish highjacking of the adult skeletogenic subcircuit was insertion of sites for a single (repressive) transcription factor in the *cis*-regulatory modules of three genes at the top of the adult skeletogenic network.

Rates of Morphological Change

The evolution of eumetazoan body plans by the kinds of alterations in dGRNs discussed here implies that, over time, rates of change in morphology will vary greatly. This is, in fact, just the pattern of change seen in studies of morphological diversity (disparity) in the fossil record (for review, see Erwin 2007). The maximum rates of increase in disparity generally occur near the origin of a clade, with much of subsequent evolution largely "filling in" the morphospace defined by the initial diversification. Part of the explanation must be ecologic, and during the initial diversification of the eumetazoan clades, positive feedback must have occurred between organisms and their environment. In a word, as animals diversified, they altered the environment, producing more selective opportunities for more animal diversification (Erwin 2008). As with any positive feedback, the expected consequence is sharp acceleration of the process. Turning to the nature of change within dGRNs, a primary mechanism for rate variation derives from the developmental morphospace concept (see above). In the initial phase, as morphospace downstream from a newly evolved spatial specification system is filled in, all the variety of allowed solutions will appear, if selectively viable, but thereafter, dynamic stasis will ensue. The expected result is indeed similar to the rapid variation in morphology followed by long stability canonically observed in the fossil record. We have also suggested that there is an additional feature of crown group dGRNs that contributes to morphological stasis, and that is their deeply "overwired" regulatory circuitry. Such circuits prevent any deviation from the correct developmental outcome, by means of multiple regulatory devices that not only produce a given developmental result, but also actively prohibit alternative possibilities (Davidson and Erwin 2009; Peter and Davidson 2009a; Smith and Davidson 2009). In general, the evolutionary consequence would be to buffer these systems against further change. But at the same time, this also means that during the early diversification of eumetazoans, the dGRNs would have lacked some of the fail-safe devices we now see and they would have been more flexible, allowing the generation of a greater variety of morphologic novelties.

In contrast to large-scale morphological evolutionary change, traditionally recognized as the origination of new phyla, classes, and orders of animals, changes at the periphery of the dGRN can happen rapidly. There are now several justly famous cases where dramatic morphological changes that have occurred very recently have been shown to be due to alteration in regulation of single genes, encoding either signaling ligands (Abzhanov et al. 2006) or transcription factors (Cretekos et al. 2008; Chan et al. 2009; Rebeiz et al. 2009). The periphery of a dGRN is a different place from its interior, where multigenic subcircuits do the work, embedded in regulatory linkages upstream and downstream, and frequently interlaced with feedbacks. As we have discussed elsewhere (Erwin and Davidson 2009), change at lower taxonomic levels occurs by redeployment of switches, often signaling switches, and by redeployment of differentiation gene batteries and change of protein-coding genes within such batteries. These kinds of events affect the periphery of the dGRN and late processes in the morphological development of the animal. In a regulatory sense, the underlying DNA-level changes are located far from the constraints that for hundreds of millions of years have defined the clade-specific morphospace of modern bilaterian lineages at higher taxonomic levels.

In conclusion, evolution of eumetazoan body plans has not been a uniform process. The nature of allowable genetic change in dGRNs during the early diversification of eumetazoans channeled the construction of the highly structured regulatory networks responsible for key developmental events. Today, much evolution involves the peripheral modification of these networks, not their construction, and is no more similar to early dGRN evolution than is the modern environment to Neoproterozoic environments. Although we still have much to learn about the timing of the earliest phases of eumetazoan evolution, and we look forward to resolving the phylogenetic position of ctenophores and other groups, enough information is available to sketch the relationship between these events and environmental changes in the late Neoproterozoic. Figure 8A illustrates minimal divergence times based exclusively on a strict reading of the fossil record, assuming the later dates for the Doushantuo, showing most metazoan evolution occurring during the second half of the Ediacaran, after the Gaskiers glaciation. Figure 8B scales the divergences to the results of phylogenetic molecular clock studies for the origin of metazoa, cnidarians, and the protostome–deuterostome ancestor. Here, much of metazoan divergence appears closely tied to the Marinoan glaciation, although with the broad uncertainties inherent in molecular clock studies, we can make no claims for any direct relationship to this major paleoecological event. All the divergences shown in Figure 8B occur well before the appearance of these lineages in the currently known fossil record, even assuming the earlier dates for the Doushantuo.

In studying the developmental basis of these major evolutionary events, we cannot rely on looking at small changes. We can, however, look forward to a more direct approach: predictive, experimental alteration of morphology in the developmental gene regulation laboratory.

SUMMARY

The following are the major conclusions from the arguments and evidence we traversed in this paper:

1. Integration of fossil, phylogenetic, and developmental evidence indicates that "bilaterians" evolved progressively from cnidarians through acoels to the protostome–deuterostome divergence. Consequently, there is not a single node associated with the "origin of the bilateria."

2. A variety of eumetazoan clades were present by the time of the Doushantuo fossil embryos in the Ediacaran Period, including cnidarians and possibly stem group acoels; some also exhibit bilaterian characteristics.

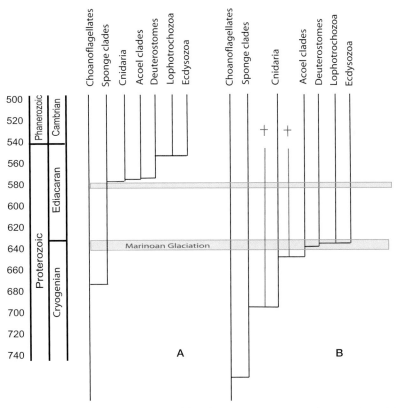

Figure 8. Alternative models for the timing of origination of bilaterian animals. (*A*) A strict interpretation of the origin of metazoan lineages that is consistent with the fossil record, assuming that sponges appear before the Marinoan glaciation, that the Doushantuo Formation postdates the Gaskiers glaciation, and that *Kimberella* marks the first appearance of a lophotrochozoan lineage. (*B*) Timing of the appearances of metazoan lineages in relationship to geological events, with divergences scaled to molecular clock results of Douzery et al. (2004) and Peterson et al. (2008). Uncertainties on the divergence points are not shown, but they are generally 10 million years or more. We have also added extinct lineages (denoted by crosses) representing the possible positions of various clades of the Ediacaran fauna.

3. Although the soft-bodied Ediacaran fossils of 579–542 mya include a number of independent metazoan clades, only one (*Kimberella*) clearly lies above the protostome–deuterostome divergence. Some of these clades could represent lineages between the cnidarians and the acoels or Nemertodermatida.

4. Geochemical evidence, the appearance of *Kimberella*, and the oldest definitive macroscopic eumetazoan traces all occur near 555 mya, consistent with the development of sufficient oxygen levels to sustain large complex animals of eumetazoan grade near this time.

5. The increasing morphologic complexity of eumetazoan lineages likely reflects the number, depth, and subcircuit diversity of the dGRNs required to build the body plan of the animal.

6. From this, it follows that previous stages in metazoan evolution were associated with dGRNs of less complexity, shallower depth, and fewer subcircuits, and with a less hierarchically structured regulatory network.

7. The canalization of developmental processes in eumetazoan lineages is due mechanistically to dGRN hierarchy. This means that the upper-level subcircuits (kernels) will be evolutionarily conserved in all descendants of the lineage. These recursively wired subcircuits provide a developmental explanation for the long-term stability of eumetazoan body plans through the Phanerozoic.

8. Once they had appeared and became responsible for regional patterning in embryonic development, the collective suite of kernels and other upstream dGRN linkages essentially defined the possible "developmental morphospace" of each clade.

9. Linnean taxonomic hierarchy is an imperfect but relevant mirror of the patterning functions established by kernels and other elements of dGRNs in building the body plan.

10. The evolution of dGRNs involves the insertion of additional subcircuit devices through intercalary evolution. Continuation of the process results in expansion of network structure. Some subcircuits are formed that act as switches and are frequently reused in the evolution of development.

11. The downstream processes of morphogenesis are controlled by functional dGRN regulatory linkages through differentiation gene batteries, cell biology, and morphogenetic gene cassettes. These directly deploy morphogenetic cellular functions, and the evolution of morphology has occurred by changes in the circuitry controlling their deployment.

ACKNOWLEDGMENTS

E.H.D. acknowledges support from National Science Foundation grant IOS-0641398. D.H.E. acknowledges support from NASA's National Astrobiology Institute.

REFERENCES

Abzhanov A, Kuo WP, Hartmann C, Grant BR, Grant PR, Tabin CJ. 2006. The calmodulin pathway and evolution of elongated beak morphology in Darwin's finches. *Nature* 442: 563–567.

Amit I, Garber M, Chevrier N, Leite AP, Donner Y, Eixenhaure T, Guttman M, Grenier JK, Li WG, Zuk O, et al. 2009. Unbiased reconstruction of a mammalian transcriptional network mediating pathogen responses. *Science* 325: 257–263.

Arenas-Mena C, Martinez C, Cameron AR, Davidson EH. 1998. Expression of the *Hox* gene complex in the indirect development of a sea urchin. *Proc Natl Acad Sci* 95: 13062–13067.

Arendt D, Wittbrodt J. 2001. Reconstructing the eyes of *Urbilateria*. *Philos Trans R Soc Lond B Biol Sci* 356: 1545–1563.

Aris-Brosou S, Yang Z. 2003. Bayesian models of episodic evolution support a late precambrian explosive diversification of the metazoa. *Mol Biol Evol* 20: 1947–1954.

Baguñà J, Martinez P, Paps J, Riutort M. 2008. Back in time: A new systematic proposal for the *Bilateria*. *Philos Trans R Soc Lond B Biol Sci* 363: 1481–1491.

Balavoine G, Adoutte A. 2003. The segmented *Urbilateria*: A testable scenario. *Int Comp Biol* 43: 137–147.

Barfod GH, Albarede F, Knoll AH, Xiao SH, Telouk P, Frei R, Baker J. 2002. New Lu-Hf and Pb-Pb age constraints on the earliest animal fossils. *Earth Planet Sci Lett* 201: 203–212.

Bolouri H, Davidson EH. 2009. The gene regulatory network basis of the "community effect," and analysis of a sea urchin embryo example. *Dev Biol* (in press).

Butterfield NJ. 2009. Oxygen, animals and oceanic ventilation: An alternative view. *Geobiology* 7: 1–7.

Canfield DE, Poulton SW, Knoll AH, Narbonne GM, Ross G, Goldberg T, Strauss H. 2008. Ferruginous conditions dominated later Neoproterozoic deep-water chemistry. *Science* 321: 949–952.

Catling DC, Glein CR, Zahnle KJ, McKay CP. 2005. Why O_2 is required by complex life on habitable planets and the concept of planetary "oxygenation time." *Astrobiology* 5: 415–438.

Chan YF, Marks ME, Jones FC, Villarreal G Jr, Shapiro MD, Brady SD, Southwick AM, Absher DM, Grimwood J, Schmutz J, et al. 2009. Adaptive evolution of pelvic reduction in sticklebacks by recurrent deletion of a Pitx1 enhancer. *Science* 327: 302–305.

Chen JY, Oliveri P, Li C-W, Zhou G-Q, Gao F, Hagadorn JW, Peterson KJ, Davidson EH. 2000. Precambrian animal diversity: Putative phosphatized embryos from the Doushantuo Formation of China. *Proc Natl Acad Sci* 97: 4457–4462.

Chen JY, Oliveri P, Gao F, Dornbos SQ, Li CW, Bottjer DJ, Davidson EH. 2002. Precambrian animal life: Probable developmental and adult cnidarian forms from Southwest China. *Dev Biol* 248: 182–196.

Chen JY, Bottjer DJ, Oliveri P, Dornbos SQ, Gao F, Ruffins S, Chi H, Li CW, Davidson EH. 2004a. Small bilaterian fossils from 40 to 55 million years before the cambrian. *Science* 305: 218–222.

Chen JY, Oliveri P, Davidson EH, Bottjer DJ. 2004b. Response to comment on "Small bilaterian fossils from 40 to 55 million years before the Cambrian." *Science* 306: 1291b.

Chen JY, Bottjer DJ, Davidson EH, Dornbos SQ, Gao X, Yang YH, Li C-W, Li G, Wang X-Q, Xian D-C, et al. 2006. Phosphatized polar lobe-forming embryos from the Precambrian of Southwest China. *Science* 312: 1644–1646.

Chen JY, Bottjer DJ, Davidson EH, Li G, Gao F, Cameron RA, Hadfield MG, Xian DC, Tafforeau P, Jia QJ, et al. 2009a. Phase contrast synchrotron X-ray microtomography of Ediacaran (Doushantuo) metazoan microfossils: Phylogenetic diversity and evolutionary implications. *Precambrian Res* 173: 191–200.

Chen JY, Bottjer DJ, Li G, Hadfield MG, Gao F, Cameron AR, Zhang CY, Xian DC, Tafforeau P, Liao X, et al. 2009b. Complex embryos displaying bilaterian characters from Precambrian Doushantuo phosphate deposits, Weng'an, Guizhou, China. *Proc Natl Acad Sci* 106: 19056–19060.

Chen YQ, Jiang SY, Ling HF, Yang JH. 2009c. Pb-Pb dating of black shales from the lower Cambrian and Neoproterozoic strata, South China. *Chemie der Erde* 69: 183–189.

Christiaen L, Davidson B, Kawashima T, Powell W, Nolla H,

Vranizan K, Levine M. 2008. The transcription/migration interface in heart precursors of *Ciona intestinalis*. *Science* **320**: 1349–1352.

Cohen PA, Knoll AH, Kodner RB. 2009. Large spinose microfossils in Ediacaran rocks as resting stages of early animals. *Proc Natl Acad Sci* **106**: 6519–6524.

Condon D, Zhu M, Bowring S, Wang W, Yang A, Jin Y. 2005. U-Pb ages from the Neoproterozoic Doushantuo Formation, China. *Science* **308**: 95–98.

Corsetti FA, Olcott AN, Bakermans C. 2006. The biotic response to Neoproterozoic snowball Earth. *Paleaeogeogr Palaeoclimatol Palaeoecol* **232**: 114–130.

Cretekos CJ, Wang Y, Green ED, Martin JF, Rasweiler JJ IV, Behringer RR. 2008. Regulatory divergence modifies limb length between mammals. *Genes Dev.* **22**: 141–151.

Crowley TJ, Hyde WT, Peltier WR. 2001. CO_2 levels required for deglaciation of a "near-snowball" Earth. *Geophys Res Lett* **28**: 283–286.

Davidson EH. 2001. *Genomic regulatory systems*. Academic, San Diego.

Davidson EH. 2006. *The regulatory genome*. Academic, San Diego.

Davidson EH. 2009. Network design principles from the sea urchin embryo. *Curr Opin Genet Dev* **19**: 535–540.

Davidson EH, Erwin DH. 2006. Gene regulatory networks and the evolution of animal body plans. *Science* **311**: 796–800.

Davidson EH, Erwin DH. 2009. Evolutionary innovation and stability in animal gene networks. *J Exp Zool B Mol Dev Evol* **312**: (in press).

Dornbos SQ, Bottjer DJ, Chen JY, Gao F, Oliveri P, Li C-W. 2006. Environmental controls on the taphonomy of phosphatized animals and animal embryos from the Neoproterozoic Doushantuo Formation, Southwest China. *Palaios* **21**: 3–14.

Douzery E, Snell E, Bapteste E, Delsuc F, Philippe H. 2004. The timing of eukaryotic evolution: Does a relaxed molecular clock reconcile proteins and fossils? *Proc Natl Acad Sci* **101**: 15386–15391.

Dunn EF, Moy VN, Angerer LM, Angerer RC, Morris RL, Peterson KJ. 2007. Molecular paleontology: Using gene regulatory analysis to address the origins of complex life cycles in the late Precambrian. *Evol Dev* **9**: 10–24.

Dunn CW, Hejnol A, Matus DQ, Pang K, Browne WE, Smith SA, Seaver E, Rouse GW, Obst M, Edgecombe GD, et al. 2008. Broad phylogenomic sampling improves resolution of the animal tree of life. *Nature* **452**: 745–749.

Erwin DH. 2007. Disparity: Morphological pattern and developmental context. *Palaeontology* **50**: 57–73.

Erwin DH. 2008. Macroevolution of ecosystem engineering, niche construction and diversity. *Trends Ecol Evol* **23**: 304–310.

Erwin DH. 2009. Early origin of the bilaterian developmental toolkit. *Philos Trans R Soc Lond B Biol Sci* **364**: 2253–2261.

Erwin DH, Davidson EH. 2009. The evolution of hierarchical gene regulatory networks. *Nat Rev Genet* **10**: 141–148.

Fedonkin MA, Waggoner BM. 1997. The late Precambrian fossil *Kimberella* is a mollusc-like bilaterian organism. *Nature* **388**: 868.

Fedonkin MA, Gehling JG, Grey K, Narbonne GM, Vickers-Rich P. 2007a. *The rise of animals*. Johns Hopkins University Press, Baltimore.

Fedonkin MA, Simonetta A, Ivantsov AY. 2007b. New data on *Kimberella*, the Vendian mollusc-like organism (White Sea region, Russia): Paleontological and evolutionary implications. In *The rise and fall of the Ediacaran Biota* (ed. P Vickers-Rich and P Komarower), pp. 157–179. Geological Society, London.

Fike DA, Grotzinger JP, Pratt LM, Summons RE. 2006. Oxidation of the Ediacaran Ocean. *Nature* **444**: 744–747.

Finnerty JR. 2005. Did internal transport, rather than directed locomotion, favor the evolution of bilateral symmetry in animals? *BioEssays* **27**: 1174–1180.

Finnerty JR, Pang K, Burton P, Paulson D, Martindale MQ. 2004. Origins of bilateral symmetry: *Hox* and *dpp* expression in a sea anemone. *Science* **304**: 1335–1337.

Gao F, Davidson EH. 2008. Transfer of a large gene regulatory apparatus to a new developmental address in echinoid evolution. *Proc Natl Acad Sci* **105**: 6091–6096.

Gehling JG, Droser ML, Jensen SR, Runnegar BN. 2005. Ediacara organisms: Relating form to function. In *Form and function: Fossils and development* (ed. DEG Briggs), pp. 43–66. Peabody Museum of Natural History, Yale University, New Haven, CT.

Gehring WJ, Ikeo K. 1999. Pax 6. Mastering eye morphogenesis and eye evolution. *Trends Genet* **15**: 371–375.

Hagadorn JW, Xiao S, Donoghue PC, Bengtson S, Gostling NJ, Pawlowska M, Raff EC, Raff RA, Turner FR, Chongyu Y, et al. 2006. Cellular and subcellular structure of Neoproterozoic animal embryos. *Science* **314**: 291–294.

Halverson GP, Hoffman PF, Schrag DP, Maloof AC, Rice AHN. 2006. Toward a Neoproterozoic composite carbon isotope record. *GSA Bull* **117**: 1181–1207.

Hejnol A, Martindale MQ. 2008. Acoel development supports a simple planula-like urbilaterian. *Philos Trans R Soc Lond B Biol Sci* **363**: 1493–1501.

Higgins JA, Schrag DP. 2003. Aftermath of a snowball Earth. *Geochem Geophys Geosys* **4**: 1028.

Hinman VF, Nguyen AT, Cameron RA, Davidson EH. 2003. Developmental gene regulatory network architecture across 500 million years of echinoderm evolution. *Proc Natl Acad Sci* **100**: 13356–13361.

Hoffman PF, Schrag DP. 2002. The snowball Earth hypothesis: Testing the limits of global change. *Terra Nova* **14**: 129–155.

Hoffman PF, Kaufman AJ, Halverson GP, Schrag DP. 1998. A Neoproterozoic snowball Earth. *Science* **281**: 1342–1346.

Jensen S, Droser ML, Gehling JG. 2005. Trace fossil preservation and the early evolution of animals. *Palaeogeogr Palaeoclimatol Palaeoecol* **220**: 19–29.

Jensen S, Droser ML, Gehling JG. 2006. A critical look at the Ediacaran trace fossil record. In *Neoproterozoic geobiology and paleobiology* (ed. S Xiao and AJ Kaufman), pp. 115–157. Springer, Berlin.

Kaufman AJ, Knoll AH, Narbonne GM. 1997. Isotopes, ice ages, and terminal Proterozoic earth history. *Proc Natl Acad Sci* **94**: 6600–6605.

Kimmel CB. 1996. Was *Urbilateria* segmented? *Trends Genet* **12**: 329–332.

King N, Hittinger CT, Carroll SB. 2003. Evolution of key cell signaling and adhesion protein families predates animal origins. *Science* **301**: 361–363.

King N, Westbrook MJ, Young SL, Kuo A, Abedin M, Chapman J, Fairclough S, Hellsten U, Isogai Y, Letunic I, et al. 2008. The genome of the choanoflagellate *Monosiga brevicollis* and the origin of metazoans. *Nature* **451**: 783–788.

Koide T, Hayata T, Cho KWY. 2005. *Xenopus* as a model system to study transcriptional regulatory networks. *Proc Natl Acad Sci* **102**: 4943–4948.

Levine M, Davidson EH. 2005. Gene regulatory networks for development. *Proc Natl Acad Sci* **102**: 4936–4942.

Littlewood DTJ, Smith AB. 1995. A combined morphological and molecular phylogeny for sea urchins. *Philos Trans R Soc Lond B Biol Sci* **347**: 213–234.

Love GD, Grosjean E, Stalvies C, Fike DA, Grotzinger JP, Bradley AS, Kelly AE, Bhatia M, Meredith W, Snape CE, et al. 2009. Fossil steroids record the appearance of Demospongiae during the Cryogenian period. *Nature* **457**: 718–721.

Marshall CR, Valentine JW. 2009. The importance of preadapted genomes in the origin of the animal bodyplans and the Cambrian explosion. *Evolution* **9999**: 999A.

Martin MW, Grazhdankin DV, Bowring SA, Evans DA, Fedonkin MA, Kirschvink JL. 2000. Age of Neoproterozoic bilaterian body and trace fossils, White Sea, Russia: Implications for metazoan evolution. *Science* **288**: 841–845.

Martindale MQ, Hejnol A. 2009. A developmental perspective: Changes in the position of the blastopore during bilaterian evolution. *Dev Cell* **17**: 162–174.

Martindale MQ, Finnerty JR, Henry JQ. 2002. The Radiata and the evolutionary origins of the bilaterian body plan. *Mol Phylogenet Evol* **24**: 358–365.

Martindale MQ, Pang K, Finnerty JR. 2004. Investigating the origins of tripoblasty: 'Mesoderma' gene expression in a diploblastic animal, the sea anemone *Nematostella vectensis* (Phylum, Cnidaria; class Anthozoa). *Development* **131:** 2463–2474.

Matus DQ, Thomsen GH, Martindale MQ. 2007. FGF signaling in gastrulation and neural development in *Nematostella vectensis*, an anthozoan cnidarian. *Dev Genes Evol* **217:** 137–148.

McCauley BS, Weideman EP, Hinman VF. 2009. A conserved gene regulatory network subcircuit drives different developmental fates in the vegetal pole of highly divergent echinoderm embryos. *Dev Biol* (in press).

McFadden KA, Huang J, Chu X, Jiang G, Kaufman AJ, Zhou C, Yuan X, Xiao S. 2008. Pulsed oxidation and biological evolution in the Ediacaran Doushantuo Formation. *Proc Natl Acad Sci* **105:** 3197–3202.

Minelli A. 2009. *Perspectives in animal phylogeny and evolution.* Oxford University Press, Oxford.

Narbonne GM. 2005. The Ediacara biota: Neoproterozoic origin of animals and their ecosystems. *Annu Rev Earth Planet Sci* **33:** 421–442.

Nielsen C. 2008. Six major steps in animal evolution: Are we derived sponge larvae? *Evol Dev* **10:** 241–257.

Oliveri P, Tu Q, Davidson EH. 2008. Global regulatory logic for specification of an embryonic cell lineage. *Proc Natl Acad Sci* **105:** 5955–5962.

Owraghi M, Broitman-Maduro G, Luu T, Roberson H, Maduro MF. 2009. Roles of the Wnt effector POP-1/TCT in the *C. elegans* endomesoderm specification gene network. *Dev Biol* (in press).

Paps J, Baguñà J, Ruitort M. 2009. Lophotrochozoa internal phylogeny: New insights from an up-to-date analysis of nuclear ribosomal genes. *Proc Biol Sci* **276:** 1245–1254.

Peter IS, Davidson EH. 2009a. The endoderm gene regulatory network in sea urchin embryos up to the mid-blastula stage. *Dev Biol* (in press).

Peter IS, Davidson EH. 2009b. Modularity and design principles in the sea urchin embryo gene regulatory network. *FEBS Lett* **583:** 3948–3958.

Peterson KJ, Butterfield NJ. 2005. Origin of the Eumetazoa: Testing ecological predictions of molecular clocks against the Proterozoic fossil record. *Proc Natl Acad Sci* **102:** 9547–9552.

Peterson KJ, Cotton JA, Gehling JG, Pisani D. 2008. The Ediacaran emergence of bilaterians: Congruence between the genetic and the geological fossil records. *Philos Trans R Soc Lond B Biol Sci* **363:** 1435–1443.

Philippe H, Derelle R, Lopez P, Pick K, Borchiellini C, Boury-Esnault N, Vacelet J, Renard E, Houliston E, Queinnec E, et al. 2009. Phylogenomics revives traditional views on deep animal relationships. *Curr Biol* **19:** 706–712.

Putnam NH, Srivastava M, Hellsten U, Dirks B, Chapman J, Salamov A, Terry A, Shapiro H, Lindquist E, Kapitonov VV, et al. 2007. Sea anemone genome reveals ancestral eumetazoan gene repertoire and genomic organization. *Science* **317:** 86–94.

Rebeiz M, Pool JE, Kassner VA, Aquadro CF, Carroll SB. 2009. Stepwise modification of a modular enhancer underlies adaptation in a *Drosophila* population. *Science* **326:** 1663–1667.

Ruvkun G, Wightman B, Burglin T, Arasu P. 1991. Dominant gain-of-function mutations that lead to misregulation of the *C. elegans* heterochronic gene *lin-14*, and the evolutionary implications of dominant mutations in pattern-formation genes. *Dev Suppl* **1:** 47–54.

Schierwater B, Eitel M, Jakob W, Osigus HJ, Hadrys H, Dellaporta SL, Kolokotronis SO, Desalle R. 2009. Concatenated analysis sheds light on early metazoan evolution and fuels a modern "urmetazoon" hypothesis. *PLoS Biol* **7:** e20.

Scott C, Lyons TW, Bekker A, Shen Y, Poulton SW, Chu X, Anbar AD. 2008. Tracing the stepwise oxygenation of the Proterozoic ocean. *Nature* **452:** 456–459.

Shen Y, Zhang TG, Hoffman PF. 2008. On the coevolution of Ediacaran oceans and animals. *Proc Natl Acad Sci* **105:** 7376–7381.

Smith J, Davidson EH. 2009. Regulative recovery in the sea urchin embryo and the stabilizing role of fail-safe gene network wiring. *Proc Natl Acad Sci* **106:** 18291–18296.

Sperling EA, Pisani D, Peterson KJ. 2007. Poriferan paraphyly and its implications for Precambrian palaeobiology. In *The rise and fall of the Ediacaran biota* (ed. P Vickers-Rich and P Komarower), pp. 355–368. Geological Society, London.

Sperling EA, Peterson KJ, Pisani D. 2009a. Phylogenetic-signal dissection of nuclear housekeeping genes supports the paraphyly of sponges and the monophyly of Eumetazoa. *Mol Biol Evol* **26:** 2261–2274.

Sperling EA, Robinson JM, Pisani D, Peterson KJ. 2009b. Where is the glass? Biomarkers, molecular clocks, and microRNAs suggest a 200-Myr missing Precambrian fossil record of siliceous sponge spicules. *Geobiology* **8:** 24–36.

Srivastava M, Begovic E, Chapman J, Putnam NH, Hellsten U, Kawashima T, Kuo A, Mitros T, Salamov A, Carpenter ML, et al. 2008. The Trichoplax genome and the nature of placozoans. *Nature* **454:** 955–960.

Stathopoulos A, Levine M. 2005. Genomic regulatory networks and animal development. *Dev Cell* **9:** 449–462.

Technau U. 2001. *Brachyury*, the blastopore and the evolution of the mesoderm. *BioEssays* **23:** 788–794.

Technau U, Rudd S, Maxwell P, Gordon PM, Saina M, Grasso LC, Hayward DC, Sensen CW, Saint R, Holstein TW, et al. 2005. Maintenance of ancestral complexity and non-metazoan genes in two basal cnidarians. *Trends Genet* **21:** 633–639.

Thompson MD, Bowring SA. 2000. Age of the Squantum "tillite" Boston Basin, Massachusetts: U-Pb zirco constraints on terminal Neoproterozoic glaciation. *Am J Sci* **300:** 630–655.

Towe KM. 1970. Oxygen-collagen priority and the early metazoan fossil record. *Proc Natl Acad Sci* **65:** 781–788.

Wallberg A, Curini-Galletti M, Ahmadzadeh A, Jondelius U. 2007. Dismissal of Acoelomorpha: Acoela and Nemertodermatida are separate early bilaterian clades. *Zool Scripta* **36:** 509–523.

Xiao SH. 2002. Mitotic topologies and mechanics of Neoproterozoic algae and animal embryos. *Paleobiology* **28:** 244–250.

Xiao SH, Knoll AH. 1999. Fossil preservation in the Neoproterozoic Douchantuo phosphorite Lagerstatte, South China. *Lethaia* **32:** 219–240.

Xiao SH, Knoll AH. 2000. Phosphatized animal embryos from the Neoproterozoic Doushantuo Formation at Weng'an, Guizhou, South China. *J Paleontol* **74:** 767–788.

Xiao SH, Laflamme M. 2008. On the eve of animal radiation: Phylogeny, ecology and evolution of the Ediacara biota. *Trends Ecol Evol* **24:** 31–40.

Yin L, Zhu M, Knoll AH, Yuan X, Zhang J, Hu J. 2007. Doushantuo embryos preserved inside diapause egg cysts. *Nature* **446:** 661–663.

Yuh CH, Bolouri H, Davidson EH. 2001. *Cis*-regulatory logic in the *endo16* gene: Switching from a specification to a differentiation mode of control. *Development* **128:** 617–629.

The Dawn of Developmental Signaling in the Metazoa

G.S. RICHARDS AND B.M. DEGNAN

School of Biological Sciences, University of Queensland, Brisbane, Australia 4072
Correspondence: b.degnan@uq.edu.au

Intercellular signaling underpins metazoan development by mediating the induction, organization, and cooperation of cells, tissues, and organs. Herein, the origins of the four major signaling pathways used during animal development and differentiation—Wnt, Notch, transforming growth factor-β (TGF-β), and Hedgehog—are assessed by comparative analysis of genomes from bilaterians, early branching metazoan phyla (poriferans, placozoans, and cnidarians), and the holozoan sister clade to the animal kingdom, the choanoflagellates. On the basis of the incidence and domain architectures of core pathway ligands, receptors, signal transducers, and transcription factors in representative species of these lineages, it appears that the Notch, Wnt, and TGF-β pathways are metazoan synapomorphies, whereas the Hedgehog pathway arose in the protoeumetazoan lineage, after its divergence from poriferan and placozoan lineages. Examination of the binding domains and motifs present in signaling pathway components of nonbilaterians reveals cases in which signaling interactions are unlikely to be operating in accordance with bilaterian canons. Overall, this study highlights the stability and antiquity of the core cytosolic components of each pathway, juxtaposed with the more variable and recently evolved molecular interactions taking place at the cell surface.

Developmental signaling pathways embody two key quintessential aspects of the multicellular lifestyle—that of determinate, epigenetic interactions between cells and the entrainment of a developmental program to enable reproducible morphogenetic outcomes to be achieved from specific genotypes.

Animal signaling pathways are characterized by core suites of molecules that transmit a signal from one cell to another. Classically, this involves a secreted or cell-surface-associated ligand being bound by a surface receptor of a surrounding cell. With this binding, a cascade of molecular interactions is initiated within the receiving cell, culminating in a change in its transcriptional activity. In such a manner, molecules released by one cell can initiate a new program of gene expression in another.

A handful of signaling pathways—Wnt, Notch, TGF-β, and Hedgehog—are the key arbitrators of bilaterian development, with any perturbation of their signaling activity having severe consequences on developmental outcomes (Gerhart 1999). Accordingly, variations in signaling pathway evolution and deployment can result in macroevolutionary changes between metazoan lineages, with the modulation and co-option of signaling pathways into the development of novel characters emerging as a common evolutionary theme (for review, see Pires-daSilva and Sommer 2003). It has become clear that these four major developmental pathways were already present in the last common ancestor (LCA) to all living bilaterians, with the majority of their components being highly conserved in both sequence and functionality across model species in a number of bilaterian phyla. Less clear, however, is how and when in prebilaterian evolution these signaling cascades came about.

The advent of genome sequencing beyond model bilaterians has shown that many of these signaling components are in fact not unique to Bilateria or even to multicellular animals (Putnam et al. 2007; King et al. 2008; Srivastava et al. 2008 and in prep.) In particular, studies in cnidarians have characterized the expression of key signaling molecules and, in some cases, further confirmed the conservation of their functional interactions (Hobmayer et al. 2000; Samuel et al. 2001; Wikramanayake et al. 2003; Broun et al. 2005; Kusserow et al. 2005; Matus et al. 2006, 2008; Käsbauer et al. 2007; Khalturin et al. 2007; Lee et al. 2007; Momose et al. 2008). More currently, data have become available from an increasingly broad range of nonbilaterian metazoans and holozoans; the genomes of the placozoan *Trichoplax adherans* and the demosponge *Amphimedon queenslandica* have been fully sequenced, as well as a genome from the metazoan sister-group, the choanoflagellate *Monosiga brevicollis* (King et al. 2008; Srivastava et al. 2008 and in prep.). These genomes allow us to document the presence and absence of the full repertoires of molecules that comprise these signaling cascades, recognizing the caveats associated with lineage-specific gene loss and the limited sampling of genomes from these phyla. Using these data, we can annotate orthologous genes of members of bilaterian signaling pathways and confirm the presence, absence, and internal organization of specific functional domains and motifs encoded in those genes that determine the molecular interactions and regulation of each pathway. Because domain architecture is likely to be critical to the overall functional capacity of signaling pathway factors, any robust assignment of gene orthology requires an analysis of encoded domain configurations in full-length coding sequences. Expressed sequence tag (EST) analyses, in which partial sequence reads are annotated on the basis of BLAST similarity searches alone, can thus contribute limited information toward understanding the evolution of these and other biological pathways.

Herein we present a synthesis of current genomic data highlighting the distribution of the molecules that comprise the Wnt, Notch, TGF-β, and Hedgehog developmental signaling pathways in nonbilaterian holozoans. Our aim is to document the piecing together of these pathways through protometazoan and early metazoan evolution. Importantly, we investigate whether all attributes of pathway functionality—as described from bilaterian model systems—can be observed in the genomes of nonbilaterians. Such an approach enables us to infer at which nodes on the holozoan tree (Fig. 1) (Philippe et al. 2009) each pathway could be considered as functionally operable according to the bilaterian ideal. As stated above, this requires not only the phylogenetic analysis of conserved gene sequences, but also an examination of domain architectures and binding motifs.

NOTCH PATHWAY

Notch signaling is classically implicated in mediating binary cell-fate decisions and is unusual in that both the ligand and receptor molecules are membrane-bound such that a signal can only be propagated between directly neighboring cells (for review, see Fortini 2009; Kopan and Ilagan 2009). The ligands (Delta and Jagged/Serrate) and receptor (Notch) are multidomain transmembrane proteins, and although the majority of these domains (epidermal growth factor [EGF]; Ankyrin [ANK]; Notch/Lin repeat [NLR]) are not restricted to the Metazoa, the configurations of the domains are thought to be metazoan specific (King et al. 2008). The ligand molecules bear a receptor-binding Delta/Serrate/Lag (DSL) domain, which is a metazoan innovation and apparently a derivation of the pan-eukaryotic EGF domain. Delta-type ligands are present in the *Amphimedon* genome, but Jagged/Serrate-type ligands (which have an additional von Willebrand factor C [vWFC] domain and expanded EGF region) appear later, in the Eumetazoa (Gazave et al. 2009). Notch receptors in bilaterians contain two additional domains, Nod and Nodp, that are variously present in some nonbilaterians (both in *Nematostella*; Nod only in *Hydra* and *Trichoplax*) but absent from *Amphimedon* (Käsbauer et al. 2007; Putnam et al. 2007; Richards et al. 2008; Srivastava et al. 2008). These domains have no described function, but their location within bilaterian Notch receptors (adjacent to cleavage sites in the transmembrane region) suggests that they may be implicit in pathway activity. As such, the lack of Nod and/or Nodp in nonbilaterians may have consequence with regard to receptor functionality. Of interest, *Monosiga* possesses a gene model containing the same domain configuration as *Amphimedon* Notch, albeit with a greatly reduced number of EGF repeats (Gazave et al. 2009). If metazoan Notch receptors are related to this gene, which seems likely, then Notch emerged through the expansion of the EGF region in the metazoan stem, with the evolution of the EGF-NLR-ANK domain configuration occurring before the divergence of choanoflagellate and metazoan lineages.

Both the Notch receptor and ligands undergo glycosylation before reaching the cell membrane; the degree of glycosylation affects ligand/receptor-binding properties and is a key regulatory aspect of the pathway (for review, see Stanley 2007). Glycosylation occurs in the Golgi apparatus through the actions of a holozoan-specific *O*-fucosyltransferase (*O*-fut) and the metazoan-specific Fringe proteins that belong to the pan-eukaryotic superfamily of β-3-glycosyltransferases (B3GLT) (Panin et al. 2002; M Srivastava et al., in prep.). Fringe genes have undergone an independent expansion in the Porifera, with six Fringes identified in the *Amphimedon* genome (compared to one in *Trichoplax*) (Gazave et al. 2009). A corresponding expansion of Delta ligands in *Amphimedon* (five, compared to one in *Trichoplax*) hints at an unexpectedly diverse level of glycosyl-mediated ligand/receptor regulation in this species. Fringe-related genes have also been identified in plants, but they are not close relatives; they belong to the B3GLT superfamily but do not clade with metazoan Fringes in phylogenetic analyses (M Srivastava et al., in prep.).

The Notch receptor undergoes three proteolytic cleavages in its life. The first cleavage (S1) of Notch, by the subtilisin-like prohormone convertase Furin, occurs in the Golgi and results in the assembly of a heterodimer before Notch reaches the cell surface (Blaumueller et al. 1997; Logeat et al. 1998). The second (S2) and third (S3) occur upon ligand binding, such that the Notch signal is propagated via regulated intramembrane proteolysis. This ligand-induced processing of the Notch receptor results in a region of the Notch carboxyl terminus (Notch intercellular domain [NICD]) cleaving off and translocating to the nucleus, where it affects a transcriptional response. The S2 cleavage is performed by members of the ADAM family of proteins (ADAM 10/17) and releases the extracellular region of the receptor (Brou et al. 2000; Lieber et al. 2002). S3 releases the NICD signaling fragment and occurs via the action of the transmembrane γ-secretase complex (Mumm et al. 2000). These proteolytic proteins are common to all metazoans and many of them have ancient origins (e.g., Pan-Eukaryota: Presenilin, Nicastrin) (Gazave et al. 2009). Beyond the Bilateria, it has been shown that the Notch receptor of the cnidarian *Hydra vulgaris* is similarly cleaved, with the carboxyl terminus of HvNotch also displaying nuclear localization (Käsbauer et al. 2007). Furthermore, this localization can be inhibited by treating the animals with drugs that knock down γ-secretase activity, thereby blocking the S3 cleavage event (Käsbauer et al. 2007).

On reaching the nucleus, NICD forms a transcriptional

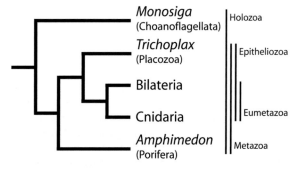

Figure 1. Phylogenetic basis for analyses of gene origin.

complex with the metazoan-specific CBF1/Suppressor of hairless/Lag1 (CSL) DNA-binding protein, which transforms CSL's activity from transcriptional repression to activation. This implicates a number of other nuclear cofactors including the pan-metazoan histone deacetylases (HDACs), core-binding factor-β (CBF-β), and CREB-binding protein (CBP) (for review, see Bray 2006). The cofactor Mastermind is an integral part of this complex in vertebrates and *Drosophila* (Wilson and Kovall 2006), but it has apparently been lost in a number of other bilaterian lineages (Gazave et al. 2009). Although it is present in Cnidaria, Mastermind has not been identified in *Amphimedon* or *Trichoplax*, suggesting that a response to Notch signaling may be elicited via different molecules in poriferans and placozoans (M Srivastava et al., in prep.). The phyletic distribution of Mastermind may highlight variability in the transcriptional activity that a Notch signal can induce, although the high level of sequence divergence between identified Masterminds in vertebrates and *Drosophila* may indicate that homologs in other species will be difficult to find through similarity searching alone.

A well-described inhibitor of the Notch pathway is Numb. Asymmetric subcellular localization of Numb drives Notch-mediated differential acquisition of cell fates between daughter cells (Guo et al. 1996; Spana and Doe 1996; Zhong et al. 1996). Numb has not been identified outside the Bilateria (Gazave et al. 2009; M Srivastava et al., in prep.), suggesting that Numb regulation of Notch signaling is a later addition to pathway regulation that did not evolve until the protobilaterian lineage.

The core members of the Notch pathway can be identified across all metazoans, indicating that this signaling system arose on the protometazoan stem (Fig. 2). Although many of the cytosolic components are premetazoan, the ligand and receptor molecules are animal specific. The Notch-like molecule of *Monosiga* may indicate that before the evolution of Delta/Jagged/Serrate ligands, a proto-Notch receptor was present at the cell surface, perhaps involved in sensing environmental stimuli. A response to such stimuli may then have been relayed into the cell by methods similar to those used by canonical Notch signaling in metazoans—through proteolytic processing of Notch by ADAMs and the γ-secretase complex. CSL is not present in *Monosiga*, but the genome does contain a predicted gene with high similarity to both the Lag1 DNA-binding domain (NTD) and the immunoglobulin fold (CTD) of CSL proteins. Domain prediction software does not recognize a β-trefoil domain in this gene—the major site of interaction with the Notch receptor (Nam et al. 2006; Wilson and Kovall 2006)—but alignments show some sequence similarity in the relevant region. Nonetheless, this gene could still be a binding partner for the Ankyrin repeats of the *Monosiga* Notch-like factor via its CTD and could be acting as a transcriptional regulator via its NTD. Further experimental work in *Monosiga* would help to address whether its Notch-like receptor does undergo proteolysis and interact with the CSL-like gene. If so, this interaction may reflect a functional module that existed in the holozoan LCA before being co-opted for cell–cell signaling in metazoans.

WNT PATHWAY

The Wnt pathway has been well documented as a key factor in determining early axial development in eumetazoans (see, e.g., Wikramanayake et al. 2003; Broun et al. 2005; Lee et al. 2007; Momose et al. 2008) and polar expression of Wnt pathway components in *Amphimedon* embryos may indicate a pan-metazoan distribution of this function (Adamska et al. 2007a and in prep.). Wnt pathway ligands are secreted glycoproteins containing a conserved sequence of cysteine residues. Wnts are only found in metazoans, although the number of Wnts present in the metazoan LCA is unclear. *Amphimedon* has three Wnt family genes, but these cannot be confidently assigned to the defined bilaterian orthology groups, nor do they appear to represent a lineage-specific expansion (M Adamska et al., in prep.). Nonetheless, dramatic expansion of the Wnt family has occurred after poriferan divergence, with eumetazoans possessing 12–15 Wnt genes (Kusserow et al. 2005).

Wnt ligands stimulate the association of a heteromeric receptor complex on the surface of a receiving cell comprising a Frizzled (Fzd) receptor and a low-density lipoprotein receptor-related protein (LRP5/6) (for review, see Cadigan and Liu 2006; Gordon and Nusse 2006). Fzds are a family of 7-transmembrane proteins that are present in all metazoans and carry a cysteine-rich amino-terminal domain (CRD) that is sufficient for Wnt binding (Lin et al. 1997). The *Amphimedon* genome has two Fzd genes that can be assigned to bilaterian orthology groups (M Adamska et al., in prep.). Fzd-related genes have also been described from amebozoans (Grimson et al. 2000), yet no other Fzd-like genes are present in any other nonmetazoan organisms, making the ancestry of this family unclear. LRP5/6s are single-pass multidomain transmembrane proteins with no discernible homologs outside the Metazoa (M Srivastava et al., in prep.). Currently, it is proposed that the role of Wnt ligands is to unite Fzd and LRP5/6 receptors on the plasma membrane, thereby bringing their carboxy-terminal intracellular tails into close proximity (see, e.g., Tolwinski et al. 2003). It is the interaction of these regions that stimulates the signaling cascade.

Dual processes are initiated as a result of ligand binding to Fzd and LRP5/6. Dishevelled (Dsh) interacts with Fzd (putatively via phosphorylation), whereas Axin is bound by LRP5/6 following LRP5/6's phosphorylation by caseine kinase 1 (CK1) and glycogen synthase kinase (GSK) (Cadigan and Liu 2006). The sum effect of these interactions is the dissolution of the core cytosolic component in Wnt signaling, the so-called destruction complex. Composed of Axin, adenomatous polypotis coli (APC), and GSK, in the absence of a Wnt signal the complex phosphorylates, and subsequently degrades, cytosolic β-catenin. GSK is a pan-eukaryotic kinase, but APC and Axin are not found outside the Metazoa (M Srivastava et al., in prep.). When considering the makeup of the destruction complex, it is evident that the nonbilaterian Axins and APCs are lacking specific protein–protein-binding motifs that are required for the correct formation of the destruction complex in their bilaterian counterparts (M Adamska et al., in prep.). What remains to be seen is whether a lack

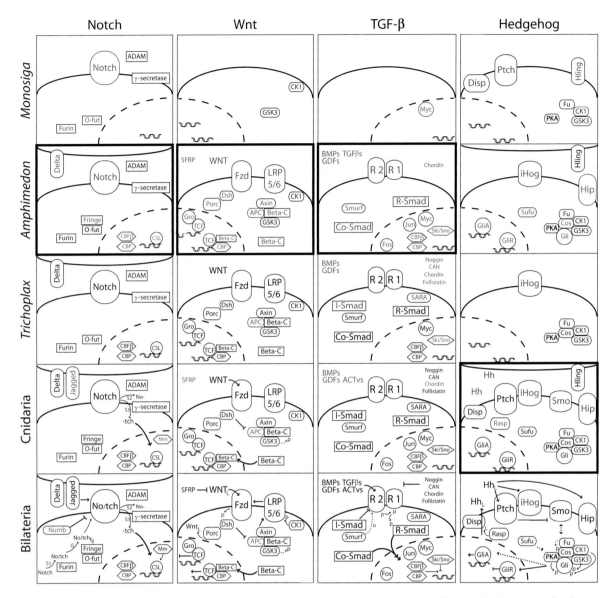

Figure 2. Genomic inventories of developmental signaling pathways across holozoans. Functional interactions between molecules are annotated only when confirmed by experimental evidence. Red indicates earliest appearance of a canonical molecule, and bolded borders indicate the minimal molecular complement that could support signaling based on bilaterian canons. Genes in gray have queries with regard to assignment/function (see text for details).

of detection of binding motifs via similarity-based searches truly reflects a lack of interaction between these proteins. For example, although APC lacks recognizable Axin-binding domains, APC-binding domains are detected in Axin, suggesting that the molecules can interact at some level. Because GSK can phosphorylate β-catenin on its own (in vitro) (Yost et al. 1996), it may be that the destruction complex is a more evolutionary labile element of the Wnt pathway than initially envisioned (M Adamska et al., in prep.).

The target of the destruction complex is another panmetazoan protein, β-catenin. The release of β-catenin from the destruction complex results in its translocation to the nucleus where it interacts with T-cell-specific transcription factor/lymphoid enhancer-binding factor 1 (Tcf/Lef) proteins (for review, see Hoppler and Kavanagh 2007). In the absence of nuclear β-catenin, Tcf/Lef forms a transcriptional repression complex by recruiting the corepressor Groucho and HDACs. When Wnt signaling stimulates the nuclear accumulation of β-catenin, Groucho is displaced and a transcriptional activation complex of β-catenin, Tcf/Lef, and the histone acetylase CBP is formed instead (Bienz and Clevers 2003). Tcf/Lef and Groucho, like the majority of Wnt pathway components, also likely arose on the metazoan stem (M Srivastava et al., in prep.).

Inhibition of the Wnt pathway can occur via ligand traps, which are molecules that bind ligands in the extracellular space and prevent them from interacting with pathway receptors. One such group is the secreted frizzled receptor proteins (SFRPs), which share the Wnt-binding CRD of Fzd receptors, but rather than being membrane bound, they possess a carboxy-terminal Netrin domain that is proposed to anchor them in the extracellular matrix (Rattner et al. 1997). As discussed by M. Adamska et al. (in prep.), the origins and emergence of SFRP genes are not straightforward. The placozoan *Trichoplax* has no obvious SFRP; however, both *Amphimedon* and *Nematostella* possess a number of SFRP-like genes, with and without the addition of Netrin domains (M Srivastava et al., in prep.). Phylogenetic analysis of these genes suggests an independent diversification of SFRPs in multiple metazoan lineages, with multiple convergent acquisitions of Netrin domains (M Adamska et al., in prep.).

A β-catenin-related gene (*Aardvark*) is present in amebozoans, but Aardvark is more akin to certain plant proteins, that also share armadillo repeats, in both sequence similarity and domain composition (Grimson et al. 2000; Coates 2003). Aardvark is implicated in signaling in *Dictyostelium*, and this occurs in a GSK-dependent manner (Grimson et al. 2000). In contrast to metazoan Wnt signaling, GSK activity appears to activate β-catenin-dependent gene expression, instead of down-regulating it via a destruction mechanism. Nonetheless, the coupling of a β-catenin-like protein with GSK to affect transcriptional outcomes in *Dictyostelium* suggests that this module has arisen multiple times in evolution and implies a robust utility to the interaction.

TGF-β SIGNALING PATHWAY

TGF-β signaling is restricted to the Metazoa with neither ligand nor receptor molecules being found outside the animal kingdom (Fig. 2). Implicated in countless cell signaling events, the iconic role of TGF-β developmental signaling is in dorsoventral patterning and the induction of the Spemann organizer in amphibians (for review, see Massagué 1998; De Robertis and Kuroda 2004).

TGF-β ligands are composed of a carboxy-terminal signaling domain and an amino-terminal propeptide domain that is cleaved before ligand release, with both domains having emerged on the protometazoan stem (M Srivastava et al., in prep.). Two major clades of ligands are generally recognized: the TGF-β *sensu stricto*/TGF-β related (e.g., Activins, Leftys, and GDF8s) and bone morphogenetic protein (BMP) related (e.g., BMPs and Nodals) (Matus et al. 2008). Of the eight pathway ligands in *Amphimedon*, one lies outside these clades, along with other divergent ligands such as GDF9/15 (M Srivastava et al., in prep.). A further five *Amphimedon* genes group together to the exclusion of all other ligands, suggesting paralogy and an independent expansion event in this lineage (M Srivastava et al., in prep.). Within these five genes, there are two instances in which two genes are highly similar and are separated by <5 kb on genomic contigs with no intervening coding sequences, suggesting a pair of recent tandem duplication events. The remaining two *Amphimedon* genes are nested within the TGF-β-related clade, as sister to the TGF-β *sensu stricto* subclass (M Srivastava et al., in prep.). This placement of *Amphimedon* ligands warrants further investigation, because to date, TGF-β *sensu stricto* ligands have only been identified in deuterostomes, with no members found in genomic screens of *Caenorhabditis elegans*, *Drosophila*, or *Nematostella* (Matus et al. 2006).

TGF-β pathway receptors are serine threonine kinases (STKRs) that are divided into two subfamilies based on their structural and functional properties: type I and type II (for review, see Massagué 1998). Receptor kinase proteins certainly play a part in unicellular signaling systems, for example, *Monosiga* tyrosine kinases (TKs) and *Dictyostelium* tyrosine-kinase-like (TKLs), but these are quite distinct from the TGF-β STKR family (Manning et al. 2008). Both type I and type II TGF-β receptors have been identified in all metazoans; however, the *Amphimedon* receptors cannot be assigned to eumetazoan subfamilies within the type I and II groupings (compare *Trichoplax* receptors) (M Srivastava et al., in prep.). A TGF-β signal is propagated by ligands binding to the constitutively phosphorylated type II receptors, which stimulates the recruitment of type I receptors and the formation of a heterodimeric receptor complex. The type I receptors are then transphosphorylated by the type II receptors at a conserved "GS" domain containing a characteristic glycine-serine repeat sequence (for review, see Massagué 1998). Conveyance of the TGF-β signal from the membrane to the nucleus occurs via Smad family proteins, another metazoan innovation (M Srivastava et al., in prep.). *Monosiga* does have a Smad-like MH2 domain, but this is coupled with a C2H2 zinc finger, in contrast to the metazoan Smads, which comprise an MH1 and MH2 domain only (M Srivastava et al., in prep). Type I receptors recruit and phosphorylate receptor-regulated Smads (R-Smads; Smad1/5, Smad2/3) that form multisubunit complexes with common partner Smads (Co-Smads; Smad4) before entering the nucleus to affect a response (for review, see Massagué et al. 2005). Both R-Smads and Co-Smads are found in all metazoans (M Srivastava et al., in prep.). A further category of Smads, inhibitory Smads (I-Smads; Smad6/7), have an inhibitory role in the pathway by interfering with the phosphorylation of R-Smads and stimulating receptor degradation or by competing with R-Smads in the formation of complexes with Smad4 (for review, see Massagué et al. 2005). I-Smads have not been located outside the Epitheliozoa, suggesting that the regulatory activity of I-Smads did not evolve until after the divergence of the poriferan lineage. Another epitheliozoan invention is SARA (Smad anchor for receptor activation), a FYVE (Fap1p/ YOTP/Vac1/EEA1) zinc-finger-binding protein that mediates TGF-β receptor/Smad interactions by recruiting R-Smads to the cell membrane and thereby bringing them into the proximity of the TGF-β receptors (Tsukazaki et al. 1998).

TGF-β signaling relies on the activation of transmembrane receptor kinases leading to the nuclear accumulation of Smad complexes. Smads themselves are not highly specific in their binding to DNA; their role is that of comodulators acting in concert with transcription factors (pan-metazoan Fos/Jun [bZIP] and Myc [bHLH]), transcription coactivators (pan-metazoan CBP and CBF-β), and transcription corepressors (Ski/Sno) to recruit basal transcription machinery (for review, see Attisano and Wrana 2000). The activity of Smad complexes in undertaking a wide range of cooperative interactions enables TGF-β signaling to inspire a diversity of outputs, depending on the nuclear population of the receiving cell. The corepressor Ski/Sno is common to epitheliozoans, but in *Amphimedon*, the most similar protein lacks a DNA-binding domain (M Srivastava et al., in prep.). This protein does possess a Co-Smad-binding region, however, which hints at some level of involvement in TGF-β signaling. In such cases, it may be that the defined DNA-binding domain, as characterized in Bilateria, has divergently evolved in the poriferan lineage to a point where it is no longer recognizable based on bilaterian characteristics.

Similarly to Wnt signaling, inhibition of the TGF-β pathway can occur via the action of ligand-trapping molecules. This functionality is of special importance when setting up morphogenetic gradients of ligand localization in developing embryos. Four major ligand traps have been characterized—Noggin, Chordin, Follistatin, and the CAN family (Cerebrus/DAN/Gremlin) (for review, see Balemans and Van Hul 2002). Follistatin, CAN, and Noggin are unique to epitheliozoans (M Srivastava et al., in prep.). True Chordin genes, with CHRD domains as well as CR regions, are restricted to the Eumetazoa; however, Chordin-like genes (with CR domains only) are present in *Trichoplax* and *Amphimedon* (M Srivastava et al., in prep.). CR domains can be found in combination with various other domains (e.g., thrombospondins, von Willebrand factor D [vWFD], cysteine knots, and insulin growth factor binding [IGFB]) in a number of disparate proteins that are also capable of binding TGF-β ligands (e.g., Kielin, CRIM, and Cross-veinless) (Garcia-Abreu et al. 2002). It seems likely that Chordin arose via the incorporation of CHRD domains into a CR domain containing a Chordin-like gene that could already have been functional in TGF-β ligand trapping. The likely additional functions conferred by the CHRD domain are in positioning the CR domains for ligand binding, as well as interaction with the Chordin inhibitor Tolloid (Hyvonen 2003). Of interest, Chordin's iconic role is in the patterning of a secondary developmental axis (Sasai et al. 1994), an attribute that has been described in cnidarians (Matus et al. 2006) but is not evident in placozoans or poriferans. The emergence of Chordin in protoeumetazoans may thus have been implicit in the development of bilateral symmetry in this lineage.

Other important negative regulators of TGF-β signaling are the E3 ubiquitin ligases of the Smurf (Smad ubiquitin regulatory factor) family. These represent an animal-specific subfamily of E3 ligases that target R-Smads for degradation in the cytosol and are also recruited by I-Smads to the membrane where they cause the degradation of TGF-β receptors (Ebisawa et al. 2001). Although the *Amphimedon* genome does encode a Smurf ortholog, it lacks an I-Smad (M Srivastava et al., in prep.), so whether Smurf can be recruited for TGF-β receptor inhibition in *Amphimedon* remains to be seen. If not, the co-option of Smurf into this role occurred after the divergence of the Porifera, in the protoepitheliozoan lineage.

Overall, these data indicate that the main players in the TGF-β pathway (ligands, receptors, and Smads) have emerged in the metazoan stem lineage, before poriferan and placozoan divergence, with no discernible precursors present in choanoflagellates. Furthermore, a distinction can be made between the potential functionality of the TGF-β pathway in the Porifera and the Epitheliozoa, with the addition of I-Smads, multiple ligand traps, and SARA occurring after the divergence of sponges.

HEDGEHOG PATHWAY

Hedgehog (Hh) signaling has key roles in delineating developmental boundaries and was discovered for its part in establishing segment polarity in *Drosophila* embryos (Nüsslein-Volhard and Wieschaus 1980). In brief, canonical Hh signaling is activated by the binding of a secreted ligand from the Hh family to the multitransmembrane receptor Patched (Ptch), which releases a second transmembrane protein Smoothened (Smo) from Patched repression. Smo regulates the activity of transcription factors belonging to the Gli/Ci group of bifunctional zinc-finger-binding proteins through the actions of the serine-threonine kinase Fused, Suppressor of Fused (Sufu), protein kinase A (PKA), casein kinase 1 (CK1), and GSK (for review, see Hooper and Scott 2005; Ingram 2008).

Strikingly, no evidence of Hh ligands has been found outside the Eumetazoa. Although the *Monosiga* and *Amphimedon* genomes do possess Hh amino-terminal signaling domains (HhN), they are located within the large membrane-bound Hedgling (Hling) proteins instead of linked to an autocatalytic intein domain as in the Eumetazoa (Adamska et al. 2007b). This has led to hypotheses of Hh signaling occurring through Hling instead of Hh in *Amphimedon*, implying that the Hh pathway originated as a short-range cell–cell mechanism with the evolution of the diffusible ligand occurring later in the protoeumetazoan stem (Adamska et al. 2007b). But importantly, the *Amphimedon* genome does not encode a Ptch receptor so it is not feasible for Hh signaling to be operating in *Amphimedon* as defined by the bilaterian canon. Although there are two Ptch-like proteins in *Monosiga*, the most similar gene model in *Amphimedon* is a member of the Ptch-related Niemann Pick-C family of sterol-sensing receptors (M Srivastava et al., in prep.). *Amphimedon* also lacks the transmembrane protein Dispatched (Disp), which transports Hh ligands across the membranes of signaling cells, although yet again, there is a Disp-like molecule in *Monosiga* (King et al. 2008; M Srivastava et al., in prep.). In lieu of Ptch, a potential candidate receptor for Hling is the Ihog/CDON family of IgCAM proteins (Yao et al. 2006). This family cooperatively binds the Hh ligand in vertebrates and *Drosophila*, and gene models with similar domain configurations to Ihogs/CDONs are present in the *Amphimedon* genome, although not in *Monosiga* (M Srivastava et al., in prep.).

Of note, Ptch and Disp (as well as Niemann Pick-C) display homology with the resistance-nodulation-division (RND) family of prokaryotic permeases (Davies et al. 2000; Taipale et al. 2002). RND proteins pump molecules across plasma membranes, a feature akin to that of Disp (export of Hh) and one that is proposed to underpin Smo inhibition by Ptch, by transporting Smo agonists and/or antagonists across the cell membrane (Taipale et al. 2002). Accordingly, recent evolutionary hypotheses have posited that the origins of Hh signaling lie in the co-option of a preexisting lipid transport/homeostasis pathway in which Ptch and Smo were functionally coupled (Hausmann et al. 2009). At odds with this scenario is the antiquity of the Hling molecule. Hausmann et al. (2009) propose that Ptch-Hh binding was originally mediated solely by the sterol conjugate of Hh, with a protein–protein-binding interface between Ptch and Hh evolving as a consequence of this interaction. However, the phyletic distribution of Hling suggests that the HhN signaling domain predates the Hh–Ptch interaction and therefore could not have arisen in consequence of it. Nonetheless, the presentation of these tangible and testable hypotheses provides an exciting field of study by which to unravel the provenance of the interactions underpining Hedgehog signaling.

The G-protein-coupled receptor (GPCR)-related receptor Smo is also not found outside Eumetazoa (M Srivastava et al., in prep.); thus, the initiation of Hh signaling through Hh–Ptch–Smo interactions is a Eumetazoan invention. A further invention is the Hedgehog interference protein (Hhip), which can also bind Hh ligands and in doing so regulates the availability of Hh to the Ptch receptor (Chuang and McMahon 1999). In vertebrates, Hhip proteins possess several EGF domains in the intracellular carboxyl terminus along with the amino-terminal folate receptor and Gluc-dehydrogenase domains. This domain configuration is not recovered in searches for Hhip proteins in nonvertebrates, which lack the EGF repeats. However, because the EGF domains have not been implicated in the Hh inhibitory function of Hhip (Ochi et al. 2006), these Hhip-like proteins may still be binding Hh extracellularly and may thus still be involved in the regulation of Hh signaling.

Signal transduction downstream from Smo is not wholly conserved between *Drosophila* and vertebrates (for review, see Huangfu and Anderson 2006). Although both systems make use of Sufu, CK1, GSK, and PKA (all of which are pan-metazoan proteins), it remains unresolved whether Fused and the kinesin Cos2 are also common components. Vertebrate Hh signaling is reliant on the transport of pathway components into cilia. This is not observed in *Drosophila*, in which Cos2 provides a scaffold to which pathway components are recruited. The carboxy-terminal tail of Smo is highly divergent between vertebrates and *Drosophila*, and it is suggested that this sequence divergence underlies the differences in Hh signaling that occurs downstream from Smo between these phyla (Huangfu and Anderson 2006). Because the carboxy-terminal tail of *Nematostella* Smo1 is truncated, and *Nematostella* Smo2 shows no similarity to either the vertebrate or fly proteins (Matus et al. 2008), it will be important to investigate how a Hh signal is propagated in cnidarians and determine what mode of transduction occurred downstream from Smo in the eumetazoan LCA.

The outcome of canonical Hh signaling is the regulation of Gli/Ci transcription factors, which are common to all metazoans and represent an animal specific subfamily of zinc-finger-binding proteins (M Srivastava et al., in prep.). A multiprotein complex, comprising the pan-eukaryotic Fused, CK1, GSK, and PKA, functions in the phosphorylation and processing of Gli/Ci, along with the metazoan-specific protein Sufu (for review, see Aikin et al. 2008). In the absence of a Hh signal, Gli/Ci is phosphorylated and cleaved to produce a truncated protein that acts as a transcriptional repressor. In the presence of a Hh signal, changes in the confirmation and phosphorylated state of the multiprotein complex results in Gli/Ci no longer being cleaved and instead moving into the nucleus as a full-length transcriptional activator (Alexandre et al. 1996; Aza-Blanc et al. 1997).

The cytosolic components of the Hh pathway are common to all metazoans; however, the absence of Hh and Smo from noneumetazoans suggests that these factors, and canonical Hh signaling, are eumetazoan synapomorphies (Fig. 2). Of note, the Hh complement of *Trichoplax* is reduced in comparison to both the *Amphimedon* and cnidarian genomes, suggesting lineage-specific losses of Gli, Sufu, Hling, and Hhip-like in the Placozoa and making it very unlikely that any approximation of Hh signaling could be occurring in this phyla (Srivastava et al. 2008). A subject that requires further investigation is the presence of Ptch- and Disp-like genes in *Monosiga*. If these are true orthologs, then these subfamilies of transmembrane receptors evolved in the holozoan stem lineage and were subsequently lost from both *Amphimedon* and *Trichoplax*, while being maintained in protoeumetazoans and co-opted into the Hh signaling pathway. Alternatively, if the *Monosiga* Ptch-like receptors are shown to bind the HhN domains of *Monosiga* Hling, this would have important consequences for interpreting the origins and emergence of Hh signaling.

INNOVATION AT THE CELL SURFACE: RECEPTOR AND LIGAND EVOLUTION

When comparing the genome complements of the Wnt, Notch, TGF-β, and Hh pathways, the evolvability of pathway ligands becomes apparent. For example, the Notch-binding DSL domains of Delta and Jagged/Serrate are a metazoan innovation, and both genes are proposed to have been single-copy genes in the bilaterian LCA (Lissemore and Starmer 1999). Within the Bilateria, various lineages have expanded their repertoire of DSL proteins. *C. elegans* notably contains 10 genes in possession of a DSL domain, as well as a number of DOS-motif-containing genes that do not contain a DSL but nonetheless show similarity to Delta and Jagged/Serrate ligands and have been implicated in Notch signaling (Komatsu et al. 2008). In *Amphimedon*, five DSL-domain-containing genes are present, all of which are developmentally expressed in cell-type-specific patterns, and these appear to be the result of an independent expansion event (GS Richards and BM Degnan, in prep.). *Trichoplax* has a more conventional DSL complement, with the genome encoding only a single Delta-like gene; however, *Nematostella*, along with Delta and Jagged/Serrate-like genes, possesses a number of genes in which multiple DSL domains are strung together, in several cases, more than 10 in a row (GS Richards, unpubl.). Assuming that these gene predictions are not spurious, it will be of great interest to see how they may interact with cnidarian Notch receptors and further shape our understanding of Notch pathway flexibility.

The Hh ligands belong to a family of Hh-related proteins, all of which share a carboxy-terminal Hog (intein +

sterol recognition region) domain. Canonical Hh ligands are composed of an HhN signaling domain and a Hog domain, and as already stated, these ligands are not present in noneumetazoans, which instead have an HhN domain conjugated to a large membrane-bound protein (Hedgling) (Adamska et al. 2007b). Additionally, Hh ligands are absent from some nematodes. These species do contain Hog domains related to Hh, but these Hogs are conjugated to secreted amino-terminal regions that are distinct from the HhN: "Wart-hog," "Qua-hog," and "Ground-hog" (Aspöck et al. 1999). Similarly, although the lophotrochozoans *Capitella* and *Lottia* both contain canonical Hh genes, they also contain genes with Hog domains conjugated to unique secreted amino-terminal regions: "Lopho-hogs" (Bürglin 2008). The possession of a Hog domain enables a protein precursor to be cleaved (by the intein module) and modified by sterols (due to ligation of the flanking sterol recognition region), resulting in the production of small lipid-modified signaling molecules for secretion. The diversity of Hh-related genes across nonmodel Bilateria therefore deserves further investigation. It may be that the Hh signaling pathway as we know it represents only one facet of the intercellular signaling interactions that are facilitated by the Hog domain.

Wnt and TGF-β ligands are less labile than Notch and Hh ligands at the level of domain composition; however, notable expansions of these families have occurred during metazoan evolution. *Nematostella* possesses six TGF-β ligands and 14 Wnts that generally clade with recognized orthology groups in bilaterians (Kusserow et al. 2005; Matus et al. 2006), indicating a minimal complement of six TGF-β and 12 Wnt genes present in the eumetazoan LCA. The *Trichoplax* genome has five genes with similarity to TGF-β ligands (although two of the models are incomplete predictions) and three Wnts, and *Amphimedon* has seven TGF-β genes, most of which appear to be the result of recent tandem duplication events, and three Wnt genes. Phylogenetic methods fail to resolve the majority of the *Trichoplax* and *Amphimedon* genes into recognized eumetazoan orthology groups, suggesting either multiple expansions in each phylum or high levels of divergence following a series of ancient protometazoan expansions (see, e.g., Srivistava et al. 2008 and in prep). Either way, these data indicate that in the evolution of signal transduction, it may be selectively advantageous to develop and diversify the ligand repertoire in these pathways.

Both Wnt and Hh ligands are lipid modified before release from the cell, causing the molecules to be far more hydrophobic than would be predicted from their sequences; it has been proposed that this makes them more targeted toward cell membranes and has important implications for their mode of transport between cells (Pepinsky et al. 1998; Willert et al. 2003). In Hh signaling, the addition of lipid is performed by Rasp (also known as Skinny Hedgehog) (Micchelli et al. 2002), and in accordance with the origins of the Hh ligand, Rasp is also only found in cnidarians and bilaterians, indicating a close evolutionary relationship between the ligand and its modifier. Similarly, Porcupine (Porc), the predicted modifier of Wnt ligands (Tanaka et al. 2002), is found throughout all Metazoa, as is the Wnt ligand.

Finally, when considering the antiquity of the receptor and ligand combinations, it is intriguing to note that in general, the receptors and their domains are recognizable as variations on molecules that exist beyond the Metazoa (TGF-βR: receptor kinases; Fzd: 7TM serpentine receptors; Ptch: RND translocators; Notch: EGFs, NLRs, and ANKs), whereas the ligands (with the exception of Hh) generally present novel metazoan-specific domains (Delta/Serrate/Jagged: DSL; TGF-β: TGF-signaling domain and preprotein domain; Wnt: Wnt domain). This trend is consistent with these receptors being co-opted into metazoan-signaling pathways from preexisting ancestral roles in the reception of environmental stimuli. In contrast, the provenance of signaling pathway ligands, which are synthesized and secreted specifically for intercellular communication, is less tractable. Nonetheless, it is logical to infer that the emergence of the protoligands in the metazoan stem instigated the co-option of the receptors into cell–cell signaling roles.

CONCLUSIONS

From the comparison of representative choanoflagellate, sponge, placozoan, cnidarian, and bilaterian genomes, it is evident that the Wnt, Notch, Hh, and TGF-β signaling pathways did not arise completely de novo after the divergence of the animal and choanoflagellate lineages. Furthermore, we can identify preexisting functional modules within these pathways that may have provided a foundation for their emergence in the metazoan stem. Although the origins of many of the molecular families and domains involved in developmental signaling lie beyond the Metazoa, a significant amount of innovation has occurred in the protometazoan lineage, coincident with the emergence of multicellularity (Fig. 2). These innovations have occurred via domain shuffling, the emergence and expansion of metazoan-specific subfamilies, and the appearance of hitherto unseen domains and molecules.

Regarding the antiquity of each signaling cascade, the complement of TGF-β and Notch molecules in *Amphimedon* indicates that both of these pathways have the potential to be functioning in poriferans, as we know them to behave in bilaterians. These signal transduction cascades thus emerged in the protometazoan lineage and were likely operable in the metazoan LCA. Notable regulatory additions to these pathways have occurred via the incorporation of I-Smads and ligand traps (Noggin, CAN, and Follistatin) into TGF-β signaling, as well as Numb inhibition of Notch signaling. In contrast, the Hh pathway cannot be considered a metazoan synapomorphy. The Cnidaria are the earliest phylum in which the minimal complement of pathway molecules are present; however, the interactions that occur downstream from Smo require further investigation due to the divergence of Smo intracellular tails across the three major eumetazoan lineages. The Wnt pathway provides a similar case; although all core components are present across the Metazoa, the domain configurations of key genes in the destruction complex pose questions as to whether its functionality in nonbilaterian metazoans is the same as that in the Bilateria.

Overall, there is an apparent lack of significant elaboration of these pathways once their core members have emerged, with their ongoing evolution primarily occurring through the expansion of existing molecular families, rather than the addition of numerous new players. This is most plainly seen when considering the trends occurring at the cell surface and in the nucleus. All ligand and receptor classes display continued expansion, with many independent expansion events occurring in specific lineages. A parallel expansion of developmental transcription factor families throughout metazoan cladogenesis has also been well documented (Larroux et al. 2008). It seems that flexibility in signaling has evolved as a product of producing multiple and varying inputs, transducing them through a robust cytosolic module, and then providing a diverse range of options in the nucleus with which to effect a cellular response. In this manner, relatively few signaling pathways have been able to orchestrate the broad display of morphological diversity that characterizes the animal kingdom.

ACKNOWLEDGMENTS

We acknowledge the significant contribution of Dan Rokshar and the U.S. Department of Energy Joint Genome Institute for the provision of genomic sequence data and resources. We thank members of the Degnan and Adamska labs for sharing unpublished sequence data and analyses. This research is supported by Australian Research Council grants to B.M.D.

REFERENCES

Adamska M, Larroux C, Degnan SM, Green KM, Adamski M, Craigie A, Degnan BM. 2007a. Wnt and TGF-β expression in the sponge *Amphimedon queenslandica* and the origin of metazoan embryonic patterning. *PLoS ONE* **2**: e1031.

Adamska M, Matus DQ, Adamski M, Green K, Rokhsar DS, Martindale MQ, Degnan BM. 2007b. The evolutionary origin of hedgehog proteins. *Curr Biol* **17**: R836–R837.

Aikin RA, Ayers KL, Therond PP. 2008. The role of kinases in the Hedgehog signaling pathway. *EMBO Rep* **9**: 330–336.

Alexandre C, Jacinto A, Ingham PW. 1996. Transcriptional activation of Hedgehog target genes in *Drosophila* is mediated directly by the Cubitus interruptus protein, a member of the GLI family of the zinc finger DNA-binding proteins. *Genes Dev* **10**: 2003–2013.

Aspöck G, Kagoshima H, Niklaus G, Bürglin TR. 1999. *Caenorhabditis elegans* has scores of hedgehog-related genes: Sequence and expression analysis. *Genome Res* **9**: 909–923.

Attisano L, Wrana JL. 2000. Smads as transcriptional co-modulators. *Curr Opin Cell Biol* **12**: 235–243.

Aza-Blanc P, Ramírez-Weber FA, Laget MP, Schwartz C, Kornberg TB. 1997. Proteolysis that is inhibited by hedgehog targets Cubitus interruptus protein to the nucleus and converts it to a repressor. *Cell* **89**: 1043–1053.

Balemans W, Van Hul W. 2002. Extracellular regulation of BMP signaling in vertebrates: A cocktail of modulators. *Dev Biol* **250**: 231–250.

Bienz M, Clevers H. 2003. Armadillo/β-catenin signals in the nucleus: Proof beyond a reasonable doubt? *Nat Cell Biol* **5**: 179–182.

Blaumueller CM, Qi H, Zagouras P, Artavanis-Tsakonas S. 1997. Intracellular cleavage of Notch leads to a heterodimeric receptor on the plasma membrane. *Cell* **90**: 281–291.

Bray SJ. 2006. Notch signaling: A simple pathway becomes complex. *Nat Rev Mol Cell Biol* **7**: 678–689.

Brou C, Logeat F, Gupta N, Bessia C, LeBail O, Doedens JR, Cumano A, Roux P, Black RA, Israel A. 2000. A novel proteolytic cleavage involved in Notch signaling: The role of the disintegrin-metalloprotease TACE. *Mol Cell* **5**: 207–216.

Broun M, Gee L, Reinhardt B, Bode H. 2005. Formation of the head organizer in *Hydra* involves the canonical Wnt pathway. *Development* **132**: 2907–2916.

Bürglin TR. 2008. The Hedgehog protein family. *Genome Biol* **9**: 241.

Cadigan KM, Liu YI. 2006. Wnt signaling: Complexity at the surface. *J Cell Sci* **119**: 395–402.

Coates JC. 2003. Armadillo repeat proteins: Beyond the animal kingdom. *Trends Cell Biol* **13**: 463–471.

Chuang P-T, McMahon AP. 1999. Vertebrate Hedgehog signaling modulated by induction of a Hedgehog-binding protein. *Nature* **397**: 617–621.

Davies JP, Chen FW, Ioannou YA. 2000. Transmembrane molecular pump activity of Niemann-Pick C1 protein. *Science* **290**: 2295–2298.

De Robertis EM, Kuroda H. 2004. Dorsal-ventral patterning and neural induction in *Xenopus* embryos. *Annu Rev Cell Dev Biol* **20**: 285–308.

Ebisawa T, Fukuchi M, Murakami G, Chiba T, Tanaka K, Imamura T, Miyazono K. 2001. Smurf1 interacts with transforming growth factor-β type I receptor through Smad7 and induces receptor degradation. *J Biol Chem* **276**: 12477–12480.

Fortini ME. 2009. Notch signaling: The core pathway and its post-translational regulation. *Dev Cell* **16**: 633–647.

Garcia-Abreu J, Coffinier C, Larrain J, Oelgeschläger M, De Robertis EM. 2002. Chordin-like CR domains and the regulation of evolutionarily conserved extracellular signaling systems. *Gene* **287**: 39–47.

Gazave E, Lapebie P, Richards GS, Brunet F, Ereskovsky AV, Degnan BM, Borchiellini C, Vervoort M, Renard E. 2009. Origin and evolution of the Notch signaling pathway: An overview from eukaryotic genomes. *BMC Evol Biol* **9**: 249.

Gerhart J. 1999. 1998 Warkany Lecture: Signaling pathways in development. *Teratology* **60**: 226–239.

Gordon MD, Nusse R. 2006. Wnt signaling: Multiple pathways, multiple receptors, and multiple transcription factors. *J Biol Chem* **281**: 22429–22433.

Grimson MJ, Coates JC, Reynolds JP, Shipman M, Blanton RL, Harwood AJ. 2000. Adherens junctions and β-catenin-mediated cell signalling in a non-metazoan organism. *Nature* **408**: 727–731.

Guo M, Jan LY, Jan YN. 1996. Control of daughter cell fates during asymmetric division: Interaction of Numb and Notch. *Neuron* **17**: 27–41.

Hausmann G, von Mering C, Basler K. 2009. The Hedgehog signaling pathway: Where did it come from? *PLoS Biol* **7**: e10000146.

Hobmayer B, Rentzsch F, Kuhn K, Happel CM, von Laue C, Snyder P, Rothbächer U, Holstein TW. 2000. WNT signaling molecules act in axis formation in the diploblastic metazoan *Hydra*. *Nature* **407**: 186–189.

Hooper JE, Scott MP. 2005. Communicating with Hedgehogs. *Nat Rev Mol Cell Biol* **6**: 306–317.

Hoppler S, Kavanagh CL. 2007. Wnt signaling: Variety at the core. *J Cell Sci* **120**: 385–393.

Huangfu D, Anderson KV. 2006. Signalling from Smo to Ci/Gli: Conservation and divergence of Hedgehog pathways from *Drosophila* to vertebrates. *Development* **133**: 3–14.

Hyvonen M. 2003. CHRD, a novel domain in the BMP inhibitor Chordin, is also found in microbial proteins. *Trends Biol Sci* **28**: 470–473

Ingram P. 2008. Hedgehog signaling. *Curr Biol* **18**: R238–R241.

Käsbauer T, Towb P, Alexandrova O, David CN, Dall'armi E, Staudigl A, Stiening B, Böttger A. 2007. The Notch signaling pathway in the cnidarian *Hydra*. *Dev Biol* **303**: 376–390.

Khalturin K, Anton-Erxleben F, Milde S, Plötz C, Wittlieb J, Hemmrich G, Bosch TC. 2007. Transgenic stem cells in *Hydra* reveal an early evolutionary origin for key elements controlling self-renewal and differentiation. *Dev Biol* **309**: 32–44.

King N, Westbrook MJ, Young SL, Kuo A, Abedin M, Chapman J, Fairclough S, Hellsten U, Isogai1 Y, Letunic I, et al. 2008. The genome of the choanoflagellate *Monosiga brevicollis* and the origin of metazoans. *Nature* **451**: 783–788.

Komatsu H, Chao MY, Larkins-Ford J, Corkins ME, Somers GA, Tucey T, Dionne HM, White JQ, Wani K, Boxem M, Hart AC. 2008. OSM-11 facilitates LIN-12 Notch signaling during *Caenorhabditis elegans* vulval development. *PloS Biol* **6:** e196.

Kopan R, Ilagan MXG. 2009. The canonical Notch signaling pathway: Unfolding the activation mechanism. *Cell* **137:** 216–233.

Kusserow A, Pang K, Sturm C, Hrouda M, Lentfer J, Schmidt HA, Technau U, von Haeseler A, Hobmayer B, Martindale MQ, Holstein TW. 2005. Unexpected complexity of the *Wnt* gene family in a sea anemone. *Nature* **433:** 156–160.

Larroux C, Luke GN, Koopman P, Rokhsar DS, Shimeld SM, Degnan BM. 2008. Genesis and expansion of metazoan transcription factor gene classes. *Mol Biol Evol* **25:** 980–996.

Lee PN, Kumburegama S, Marlow HQ, Martindale MQ, Wikramanayake AH. 2007. Asymmetric developmental potential along the animal-vegetal axis in the anthozoan cnidarian, *Nematostella vectensis*, is mediated by Dishevelled. *Dev Biol* **310:** 169–186.

Lieber T, Kidd S, Young MW. 2002. *kuzbanian*-mediated cleavage of *Drosophila* Notch. *Genes Dev* **16:** 209–221.

Lin K, Wang S, Julius MA, Kitajewski J, Moos M Jr, Luyten F.P. 1997. The cysteine-rich frizzled domain of Frzb-1 is required and sufficient for modulation of Wnt signaling. *Proc Natl Acad Sci* **94:** 11196–11200.

Lissemore JL, Starmer WT. 1999. Phylogenetic analysis of vertebrate and invertebrate Delta/Serrate/Lag-2 (DSL) proteins. *Mol Phylogenet Evol* **11:** 308–319.

Logeat F, Bessia C, Brou C, LeBail O, Jarriault S, Seidah NG, Israel A. 1998. The Notch1 receptor is cleaved constitutively by a furin-like convertase. *Proc Natl Acad Sci* **95:** 8108–8112.

Manning G, Young SL, Miller WT, Zhai Y. 2008. The protist, *Monosiga brevicollis*, has a tyrosine signaling network more elaborate and diverse than found in any known metazoan. *Proc Natl Acad Sci* **105:** 9674–9679.

Massagué J. 1998. TGF-β signal transduction. *Annu Rev Biochem* **67:** 753–791.

Massagué J, Seoane J, Wotton D. 2005. Smad transcription factors. *Genes Dev* **19:** 2783–2810.

Matus DQ, Thomsen GH, Martindale MQ. 2006. Dorso/ventral genes are asymmetrically expressed and involved in germ-layer demarcation during cnidarian gastrulation. *Curr Biol* **16:** 499–505.

Matus DQ, Magie CR, Pang K, Martindale MQ, Thomsen GH. 2008. The Hedgehog gene family of the cnidarian, *Nematostella vectensis,* and implications for understanding metazoan Hedgehog pathway evolution. *Dev Biol* **313:** 501–518.

Micchelli CA, The I, Selva E, Mogila V, Perrimon N. 2002. Rasp, a putative transmembrane acyltransferase, is required for Hedgehog signaling. *Development* **129:** 843–851.

Momose T, Derelle R, Houliston E. 2008. A maternally localised Wnt ligand required for axial patterning in the cnidarian *Clytia hemisphaerica*. *Development* **135:** 2105–2113.

Mumm JS, Schroeter EH, Saxena MT, Griesemer A, Tian X, Pan DJ, Ray WJ, Kopan R. 2000. A ligand-induced extracellular cleavage regulates γ-secretase-like proteolytic activation of Notch1. *Mol Cell* **5:** 197–206.

Nam Y, Sliz P, Song L, Aster JC, Blacklow SC. 2006. Structural basis for cooperativity in recruitment of MAML coactivators to Notch transcription complexes. *Cell* **124:** 973–983.

Nüsslein-Volhard C, Wieschaus E. 1980. Mutations affecting segment number and polarity in *Drosophila*. *Nature* **287:** 795–801.

Ochi H, Pearson BJ, Chuang P-T, Hammerschmidt M, Westerfield M. 2006. Hhip regulates zebrafish muscle development by both sequestering Hedgehog and modulating localization of Smoothened. *Dev Biol* **297:** 127–140.

Panin VM, Shao L, Lei L, Moloney DJ, Irvine KD, Haltiwanger RS. 2002. Notch ligands are substrates for protein O-fucosyltransferase-1 and Fringe. *J Biol Chem* **277:** 29945–29952.

Pepinsky RB, Zeng C, Wen D, Rayhorn P, Baker DP, Williams KP, Bixler SA, Ambrose CM, Garber EA, Miatkowski K, et al. 1998. Identification of a palmitic acid-modified form of human Sonic hedgehog. *J Biol Chem* **273:** 14037–14045.

Philippe H, Derelle R, Lopez P, Pick K, Borchiellini C, Boury-Esnault N, Vacelet J, Renard E, Houliston E, Quéinnic E, et al. 2009. Phylogenomics revives traditional views on deep animal relationships. *Curr Biol* **19:** 706–712.

Pires-daSilva A, Sommer RJ. 2003. The evolution of signaling pathways in animal development. *Nat Rev Genet* **4:** 39–49.

Putnam NH, Srivastava M, Hellsten U, Dirks B, Chapman J, Salamov A, Terry A, Shapiro H, Lindquist E, Kapitonov VV, et al. 2007. Sea anemone genome reveals ancestral eumetazoan gene repertoire and genomic organization. *Science* **317:** 86–94.

Rattner A, Hsieh JC, Smallwood PM, Gilbert DJ, Copeland NG, Jenkins NA, Nathans J. 1997. A family of secreted proteins contains homology to the cysteine-rich ligand-binding domain of Frizzled receptors. *Proc Natl Acad Sci* **94:** 2859–2863.

Richards GS, Simionato E, Perrron M, Adamska M, Vervoort M, Degnan BM. 2008. Sponge genes provide new insight into the evolutionary origin of the neurogenic circuit. *Curr Biol* **18:** 1156–1161.

Samuel G, Miller D, Saint R. 2001. Conservation of a DPP/BMP signaling pathway in the nonbilaterial cnidarian *Acropora millepora*. *Evol Dev* **3:** 241–250.

Sasai Y, Lu B, Steinbeisser H, Geissert D, Gont LK, De Robertis EM. 1994. *Xenopus* chordin: A novel dorsalizing factor activated by organizer-specific homeobox genes. *Cell* **79:** 779–790.

Spana EP, Doe CQ. 1996. Numb antagonizes Notch signaling to specify sibling neuron cell fates. *Neuron* **17:** 21–26.

Srivastava M, Begovic E, Chapman J, Putnam NH, Hellsten U, Kawashima T, Kuo A, Mitros T, Salamov A, Carpenter ML, et al. 2008. The *Trichoplax* genome and the nature of placozoans. *Nature* **454:** 955–961.

Stanley P. 2007. Regulation of Notch signaling by glycosylation. *Curr Opin Struct Biol* **17:** 530–535.

Taipale J, Cooper MK, Maiti T, Beachy PA. 2002. Patched acts catalytically to suppress the activity of Smoothened. *Nature* **418:** 892–897.

Tanaka K, Kitagawa Y, Kadowaki T. 2002. *Drosophila* segment polarity gene product porcupine stimulates the posttranslational N-glycosylation of wingless in the endoplasmic reticulum. *J Biol Chem* **277:** 12816–12823.

Tolwinski NS, Wehrli M, Rives A, Erdeniz N, DiNardo S, Wieschaus E. 2003. Wg/Wnt signal can be transmitted through arrow/LRP5,6 and Axin independently of Zw3/Gsk3β activity. *Dev Cell* **4:** 407–418.

Tsukazaki T, Chiang TA, Davison AF, Attisano L, Wrana JL. 1998. SARA, a FYVE domain protein that recruits Smad2 to the TGF-β receptor. *Cell* **95:** 779–791.

Wikramanayake AH, Hong M, Lee PN, Pang K, Byrum CA, Bince JM, Xu R, Martindale MQ. 2003. An ancient role for nuclear β-catenin in the evolution of axial polarity and germ layer segregation. *Nature* **27:** 446–450.

Willert K, Brown JD, Danenberg E, Duncan AW, Weissman IL, Reya T, Yates JR, Nusse R. 2003. Wnt proteins are lipid-modified and can act as stem cell growth factors. *Nature* **423:** 448–452.

Wilson JJ, Kovall RA. 2006. Crystal structure of the CSL-Notch-Mastermind ternary complex bound to DNA. *Cell* **124:** 985–996.

Yao S, Lum L, Beachy P. 2006. The Ihog cell-surface proteins bind Hedgehog and mediate pathway activation. *Cell* **125:** 343–357.

Yost C, Torres M, Miller JR, Huang E, Kimelman D, Moon RT. 1996. The axis-inducing activity, stability, and subcellular distribution of β-catenin is regulated in *Xenopus* embryos by glycogen synthase kinase 3. *Genes Dev* **10:** 1443–1454.

Zhong W, Feder JN, Jiang M-M, Jan LY, Jan YN. 1996. Asymmetric localization of a mammalian Numb homolog during mouse cortical neurogenesis. *Neuron* **17:** 43–53.

Evolving Views of DNA Replication (In)Fidelity

T.A. KUNKEL

Laboratory of Molecular Genetics and Laboratory of Structural Biology, National Institute of Environmental Health Sciences, NIH, DHHS, Research Triangle Park, North Carolina 27709
Correspondence: kunkel@niehs.nih.gov

"It has not escaped our notice that the specific pairing we have postulated immediately suggests a possible copying mechanism for the genetic material" (Watson and Crick 1953).

In the years since this remarkable understatement, we have come to realize the enormous complexity of the cellular machinery devoted to replicating DNA with the accuracy needed to maintain genetic information over many generations, balanced by the emergence of mutations on which selection can act. This complexity is partly based on the need to remove or tolerate cytotoxic and mutagenic lesions in DNA generated by environmental stress. Considered here is the fidelity with which undamaged and damaged DNA is replicated by the many DNA polymerases now known to exist. Some of these seriously violate Watson–Crick base-pairing rules such that, depending on the polymerase, the composition and location of the error, and the ability to correct errors (or not), DNA synthesis error rates can vary by more than a millionfold. This offers the potential to modulate rates of point mutations over a wide range, with consequences that can be either deleterious or beneficial.

In organisms from viruses to man, the fidelity with which genetic information is replicated depends on the ability of polymerases to select correct nucleotides—rather than incorrect and/or damaged nucleotides—for incorporation without adding or deleting nucleotides. Polymerase selectivity is the prime determinant of fidelity both at the replication fork and during synthesis to repair DNA damage generated by endogenous cellular metabolism or exposure to the environment (Friedberg et al. 2006). In many organisms, fidelity can be increased by exonucleolytic proofreading of mismatches during replication and by DNA mismatch repair (MMR) (for review, see Kunkel and Erie 2005; Iyer et al. 2006; Hsieh and Yamane 2008). Certain proteins involved in MMR also can also signal for DNA-damage responses, prevent homologous recombination, promote meiotic recombination, modulate somatic hypermutation of immunoglobulin genes, or even stabilize certain misaligned repetitive DNA sequences. When DNA damage is not removed before replication, helix-distorting lesions can impede replication fork progression. In such circumstances, cell survival can be enhanced by specialized DNA transactions, some of which can be mutagenic via translesion DNA synthesis (Jansen et al. 2007; Yang and Woodgate 2007; Chang and Cimprich 2009). Considered here is the amazing diversity of evolutionarily conserved DNA polymerases involved in these transactions, many of which have been discovered relatively recently. Emphasis is on their fidelity and on the contributions of proofreading and MMR to replication fidelity, which can vary over a much wider range than was appreciated even a decade ago.

MULTIPLE POLYMERASES WITH MULTIPLE OVERLAPPING FUNCTIONS

DNA polymerases were first discovered using assays for polymerization activity (Bessman et al. 1956). This approach revealed that bacteria and eukaryotes harbor multiple polymerases (Kornberg and Baker 1992). However, just how many only came to light more recently when sequence alignments and recombinant DNA technology were used to find low-activity low-abundance polymerases. Sequence alignments now permit classification of DNA polymerases into several different families, with most organisms encoding more than one (Shcherbakova et al. 2003a; Bebenek and Kunkel 2004; Loeb and Monnat 2008). For example, *Escherichia coli* encodes five polymerases (Friedberg et al. 2005), one each from families A, B, and C and two from different subfamilies of family Y, each with important but somewhat different functions. The human genome encodes even more (Table 1) from families A (three polymerases), B (four polymerases), X (four polymerases), Y (four polymerases), and reverse transcriptase (RT) (telomerase). Because polymerases can have multiple functions (Table 1) and can sometimes compensate one for another, it is a continuing challenge to understand exactly where and when each polymerase operates in vivo.

Despite differences in primary sequence, DNA polymerases in different families share a common general structure for the polymerase domain (Ollis et al. 1985), which is composed of fingers, thumb, and palm subdomains (colored blue, green, and red, respectively, in Fig. 1A). The palm contains three highly conserved carboxy-

Table 1. Human DNA-template-dependent DNA Polymerases

Polymerase	Family	Mass (kDa)	Gene (alias)	Associated activities	Proposed functions
α	B	165	POLA	RNA primase	nuclear genome replication, S-phase checkpoint
β	X	39	POLB	dRP lyase AP lyase	BER MMR
γ	A	140	POLG	3′→5′ exonuclease dRP lyase	mitochondrial genome replication and BER
δ	B	125	POLD1	3′→5′ exonuclease	nuclear genome replication NER, BER, MMR, DSB repair
ε	B	255	POLE	3′→5′ exonuclease	nuclear genome replication NER, BER, MMR, DSB repair S-phase checkpoint
ζ	B	353	POLZ (REV3)	–	TLS, DSB repair, ICL repair, SHM
η	Y	78	POLH (RAD30, RAD30A, XPV)	–	TLS, SHM, BER? recombination repair
θ	A	198	POLQ	helicase motifs	ICL repair? TLS, SHM?
ι	Y	80	POLI (RAD30B)	dRP lyase	TLS, BER SHM?
κ	Y	76	POLK (DINB1)	–	TLS NER
λ	X	66	POLL	dRP lyase	V(D)J recombination NHEJ, BER
μ	X	55	POLM	terminal transferase	V(D)J recombination NHEJ
ν	A	100	POLN	–	TLS?
σ	X	60	POLS (TRF4-1)	3′→5′ exonuclease	sister chromatid cohesion
TdT	X	58	TdT	–	V(D)J recombination
REV1	Y	138	REV1	dCTP incorporation	TLS

(BER) Base excision repair, (NER) nucleotide excision repair, (MMR) mismatch repair, (DSB repair) double-strand break repair, (TLS) translesion synthesis, (SHM) somatic hypermutation, (ICL repair) interstrand cross-link repair, (NHEJ) nonhomologous end-joining of double-strand breaks.

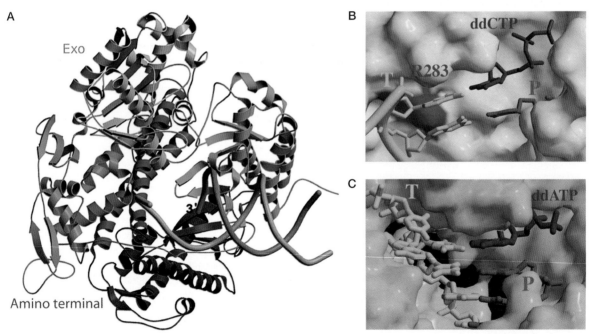

Figure 1. X-ray crystal structures of DNA polymerases. (*A*) Shown is the structure of a representative replicative DNA polymerase from bacteriophage RB69 (family B). Polymerase domains share three common subdomains: designated fingers (blue), palm (red), and thumb (green). Other domains for specialized functions are shown in purple and yellow. (*B*) The active site of human DNA polymerase β. The surface of Arg-283 is highlighted in pink to emphasize the importance to fidelity of polymerase interactions with the DNA minor groove. (*C*) The more open and solvent-accessible active site of low-fidelity *Sulfolobus sulfataricus* Dpo4. See text for further descriptions. (*A*, Prepared by Miguel Garcia-Diaz, using the structure in Franklin et al. [2001]; *B* and *C*, reprinted, with permission, from Kunkel et al. 2003.)

lates that bind two divalent metal ions required for catalysis via an in-line nucleophilic attack of the 3′-OH on the α-phosphate of the incoming dNTP. This mechanism is thought to be common to all DNA polymerases (Steitz 1993), yet it appears to have resulted from convergent evolution, because some polymerase families have the active site carboxylates in a "right-handed" configuration and others (families X and C) have it in a "left-handed" configuration (see, e.g., Wing et al. 2008 and references therein).

The polymerase domains are usually attached to other domains needed for the variable functions of these proteins. For example, polymerases that perform the bulk of genome replication often have a domain harboring 3′ exonuclease activity that proofreads replication errors (Fig. 1A). Nonetheless, most DNA polymerases lack an intrinsic 3′ exonuclease activity (Table 1), which is interesting given the importance of proofreading to genome stability (see below). Other specialized domains include a "little finger" domain (Yang and Woodgate 2007) unique to family-Y members involved in translesion DNA synthesis, and an 8-kDa domain unique to family-X polymerases (Moon et al. 2007) that assists in filling small gaps during DNA repair and that, in polymerases β and λ, harbors a dRP (deoxyribose phosphate) lyase activity needed for repair (Table 1). Still other domains include the BRCT (BRCA1 carboxy terminal) domains of family-X polymerases involved in nonhomologous end-joining of double-stranded DNA breaks and amino- or carboxy-terminal regions of polymerase catalytic subunits that are involved in cellular responses to DNA damage, including via partnerships with other proteins. In fact, DNA polymerases typically operate in DNA transactions that require coordinated interactions with many other proteins (e.g., noncatalytic accessory subunits, processivity clamps, and single-stranded DNA-binding proteins), whose properties and functions are subjects of continuing interest (see, e.g., Shcherbakova et al. 2003a; Bebenek and Kunkel 2004; Friedberg et al. 2005; Jansen et al. 2007; Loeb and Monnat 2008; Burgers 2009; Chang and Cimprich 2009 and references therein).

THE FIDELITY OF DNA SYNTHESIS

Measurements of the fidelity of DNA synthesis in vitro by purified DNA polymerases reveal a remarkable variation in error rates for the two major types of errors that polymerases generate: single-base-pair substitutions and single-base deletions (Fig. 2). These error rates reflect the contribution of nucleotide selectivity at the polymerase active site and proofreading by those polymerases harboring an associated 3′ exonuclease.

Major Replicative Polymerases

To maintain species identity, the accuracy of genomic replication is expected to be high. Consistent with this expectation, the major replicative polymerases nearly always insert correct dNTPs onto properly aligned primer-templates (exemplified in Fig. 2 by polymerases α, δ, ε, and γ, but also true for replicative polymerases from other organisms). High nucleotide selectivity at the polymerase active site is illustrated by the relatively low base substitution and indel error rates ($\sim 10^{-4}$) of polymerase α, which naturally lacks proofreading activity. Similarly, low error rates are seen for polymerases δ, ε, and γ when their intrinsic proofreading exonucleases are inactivated (Longley et al. 2001; Shcherbakova et al. 2003b; Fortune et al. 2005).

Nucleotide Selectivity

What determines the high nucleotide selectivity of accurate DNA polymerases? Hydrogen bonding between template bases and incoming dNTPs is clearly important for replication fidelity. However, this alone is unlikely to explain high selectivity because the free-energy difference between correct and incorrect base pairs in solution accounts for error rates of ~1:100 (Loeb and Kunkel 1982). Thus, other ideas have been put forth to account for the higher selectivity of accurate polymerases. For example, for the incoming dNTP to hydrogen bond to a template base, water molecules that are hydrogen-bonded to the base of the incoming dNTP must be removed,

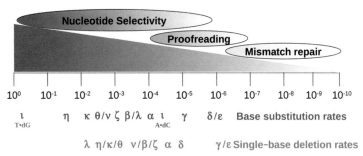

Figure 2. Polymerase error rates and the contributions of each fidelity process to mutation rate. The image illustrates the wide ranges over which polymerase nucleotide selectivity, exonucleolytic proofreading, and mismatch repair contribute to spontaneous mutation rates of organisms. Also depicted are the average rates at which purified eukaryotic DNA polymerases generate single-base substitution and single-base deletion errors when performing gap-filling DNA synthesis in vitro. See text for further descriptions. For details on the source and composition of the polymerases used and on their error specificity, see McCulloch and Kunkel (2008 and references therein).

thereby decreasing the entropy of the system. This magnifies the contribution of enthalpy to the free-energy difference (Petruska and Goodman 1995), thereby increasing nucleotide selectivity. Another idea supported by substantial evidence (for review, see Kunkel and Bebenek 2000; Kool 2002; Beard and Wilson 2003; Kim et al. 2005) is that high nucleotide selectivity partly results from the shape complementarity in the nascent base-pair-binding pocket. The four canonical Watson–Crick base pairs are nearly identical in size and shape. Structural studies reveal that correct base pairs fit within the nascent base-pair-binding pocket without steric clashes. Particularly important to fidelity are amino acid side chains (e.g., Fig. 1B, Arg-283 [purple] in polymerase β) that interact with the O2 atom of pyrimidines and the N3 atom of purines, which are isosteric in the four correct Watson–Crick base pairs. This is illustrated in Figure 1B, which shows the active site of DNA polymerase β, a relatively accurate repair enzyme, with a correct base pair poised for catalysis. The correct pair fits snugly, whereas mismatches with different and variable geometries are predicted to have steric clashes that would reduce incorrect dNTP-binding affinity, affect subsequent conformational changes needed to set up the proper geometry for catalysis, and/or reduce the rate of phosphodiester bond formation.

Insertion–Deletion Errors (Indels)

DNA polymerases also insert and delete nucleotides during DNA synthesis. These errors result from strand misalignments that generate unpaired bases in the primer strand, leading to additions, or in the template strand, leading to deletions. Ideas to account for how these misalignments initiate and are stabilized for continued synthesis include classical DNA strand slippage, misinsertion followed by primer relocation, and misalignment of a nucleotide at the polymerase active site. Biochemical and structural support exists for all three models (for review, see Bebenek and Kunkel 2000; Garcia-Diaz and Kunkel 2006). Replicative DNA polymerases generate single-base deletions at rates that are similar to those for single-base substitutions (Fig. 2). Single-base deletion error rates are usually higher than single-base addition error rates or rates for indels involving large numbers of nucleotides, with possible explanations considered elsewhere (Bebenek and Kunkel 2000; Garcia-Diaz and Kunkel 2006). Importantly, the single-base substitution and deletion error rates in Figure 2 are average values, with wide variations observed depending on the type of mismatch and the sequence context in which the mismatch is located (Kunkel and Bebenek 2000). Prime examples of such variability among lower fidelity polymerases involved in DNA repair and translesion synthesis are considered below.

Proofreading by Replicative DNA Polymerases

Average base substitution error rates of proofreading-proficient replicative DNA polymerases are typically $\geq 10^{-6}$. Their exonuclease-deficient derivatives are considerably less accurate, indicating that on average, proofreading improves replication fidelity by about ~10-fold to 100-fold (Fig. 2). The energetic cost of improving fidelity by more than this could be unacceptable due to excessive excision of correctly paired bases (Fersht et al. 1982). The biological importance of proofreading is illustrated by studies showing that when highly conserved residues near the active sites of *Saccharomyces cerevisiae* replicative polymerases are replaced with nonconservative amino acids, the mutant enzymes have decreased DNA synthesis fidelity in vitro (Longley et al. 2001; Shcherbakova et al. 2003b; Fortune et al. 2005) and generate mutator phenotypes in vivo (Morrison and Sugino 1994). Moreover, mice with homologous replacements in polymerase δ have decreased genomic stability and accelerated tumorigenesis (Goldsby et al. 2001).

The key to proofreading efficiency is the balance between polymerization and excision at a growing primer terminus (Fig. 3A). Under normal circumstances, correct incorporation allows subsequent incorporations to occur rapidly with little opportunity for proofreading (line 1). However, misinsertion generates a mismatched primer terminus that is more difficult to extend. This slows polymerization, allowing the primer terminus to fray and move single-stranded DNA into the exonuclease active site for excision of the error (line 2). On the basis of early work (for review, see Kornberg and Baker 1992) and on more recent studies, we now realize that there are several ways to influence this critical balance between polymerization and excision (Table 2). Proofreading can be inactivated by amino acid substitutions in the exonuclease active site, or exonuclease activity can be inhibited if the end product of excision, a dNMP, binds to the exonuclease active site. Proofreading can be reduced by amino acid substitutions in replicative polymerases that prevent movement of the frayed primer terminus to the exonuclease active site (so-called "switching mutants") (see, e.g., Jin et al. 2005 and references therein) or by amino acid substitutions in the polymerase active site that promote mismatch extension (see, e.g., Nick McElhinny et al. 2008 and references therein). Proofreading can be suppressed by increasing the concentration of the next correct nucleotide to be incorporated after a misinsertion (dCTP for the examples in Fig. 3A), thereby promoting mismatch extension at the expense of excision (Ninio 1975). Finally, under some circumstances, mismatches escape proofreading by tricking the replicative polymerases. A well-known example involves 8-oxo-guanine, a common lesion resulting from oxidative stress. Replication of template 8-oxo-G can generate 8-oxoG•dA mismatches whose geometry is similar to that of a correct base pair, such that the mismatch largely escapes proofreading, e.g., by replicative T7 DNA polymerase (see, e.g., Brieba et al. 2004). Another example with high biological relevance involves proofreading of insertion–deletion mismatches during replication of repetitive sequences (Fig. 3A, line 3). Proofreading does correct misaligned intermediates containing extra bases in one strand or the other near the primer terminus, as illustrated by the higher indel error rates of exonuclease-deficient polymerases δ, ε, and γ when compared to their proofreading-proficient

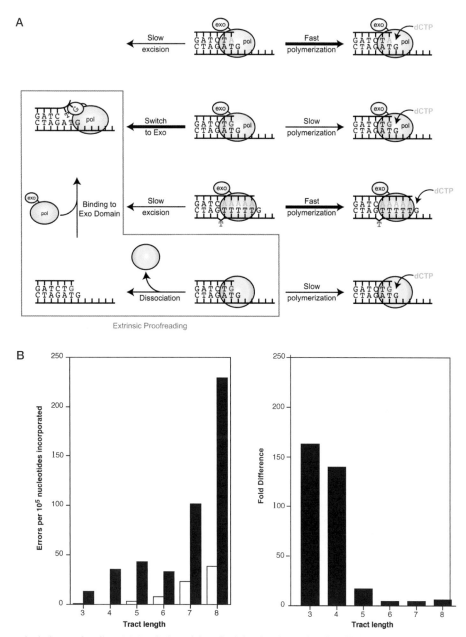

Figure 3. Exonucleolytic proofreading. (*A*) Depiction of the principles that determine the efficiency of proofreading. A proofreading-proficient polymerase (blue) harbors its polymerase and exonuclease activities in separate domains (e.g., see RB69 polymerase in Fig. 1A), depicted as large and small ovals, respectively. The partitioning between these two activities determines the efficiency of proofreading. Also shown is the possibility that errors made by an exonuclease-deficient polymerase (yellow) may be proofread by a separate exonuclease, either that of a proofreading-proficient polymerase (as shown) or present in another protein. See text for a further description and Nick McElhinny et al. (2006 and references therein) for additional discussion and information. (*B, left panel*) Single-base deletion error rates of proofreading-proficient (open bars) and proofreading-deficient (closed bars) T7 DNA polymerase when copying tracks of three to eight consecutive template Ts. (*Right panel*) Ratio of the error rates of the two polymerases to illustrate the decreasing efficiency of proofreading as a function of increasing repetitive sequence track length. (Reprinted, with permission, from Kroutil and Kunkel 1998.)

counterparts (Longley et al. 2001; Shcherbakova et al. 2003b; Fortune et al. 2005). However, the efficiency of indel proofreading decreases as the length of a repetitive sequence increases (e.g., Fig. 3B). This is because in a long repetitive sequence, the mismatch generated by strand slippage (i.e., the unpaired base) is likely to be located upstream of the polymerase active site, such that it does not strongly reduce the rate of polymerization (Fig. 3A, line 3). Such diminished proofreading, in conjunction with a higher rate of strand slippage by polymerases (e.g., left panel in Fig. 3B; see Garcia-Diaz and Kunkel 2006), contributes to the observation that long repetitive sequences are at risk for a high rate of replication slippage errors, as evidenced by the well-known "microsatellite

Table 2. Variables that Can Modulate the Efficiency of Correcting Replication Errors

Reduced proofreading by
  Mutational inactivation of 3′ exonuclease
  Inhibiting 3′ exonuclease activity—dNMPs
  Suppressing proofreading by
      Reducing switching from polymerase to exonuclease active site
  Promoting MM extension by
      Polymerase active site mutations
      High concentration of next correct dNTPs
  Mismatch mimicry of correct base pairing
  Internalizing a mismatch in a repetitive sequence

Reduced mismatch repair by
  Mutational inactivation of MMR proteins
  Cadmium inhibition of MMR in *S. cerevisiae*
  Promoter hypermethylation to silence expression of human Mlh1
  Saturation of MMR
      Rapid replication in proofreading-defective *E. coli* (MutD)
      DNA damage
  Imbalanced expression of MMR proteins
      Human Msh3
      S. cerevisiae Mlh1

instability" phenotype of eukaryotic cells defective in DNA mismatch (see below). On the basis of the logic in Figure 3 and the parameters in Table 2, it is now very clear that just as for nucleotide selectivity, the contribution of proofreading to replication fidelity can vary over a wide range (Fig. 2), from almost none (8-oxoG•dA mismatches) to several hundredfold (e.g., for bacteriophage T7 replication; see, e.g., Donlin et al. 1991).

"Extrinsic" Proofreading May Also Contribute to Genome Stability

Interestingly, among many mammalian DNA polymerases, only those responsible for the bulk of chain elongation during replication (δ, ε, and γ) contain intrinsic 3′ exonucleolytic proofreading activity. Nonetheless, the exonuclease-deficient polymerases have very important roles in maintaining genome stability (Table 1). Are errors made by exonuclease-deficient polymerases subject to "extrinsic" proofreading by a separate exonuclease? The idea (Fig. 3A, line 4) is that, after making a mismatch, the polymerase would dissociate, allowing the exonuclease activity of another protein to excise the mismatch. Indeed, the major *E. coli* replicative polymerase, DNA polymerase III, harbors its polymerase and exonuclease activities in two different subunits (the α and ε subunits, respectively), and these two proteins work in concert to achieve high replication fidelity. Proofreading by a separate protein may also occur in eukaryotes. For example, yeast DNA polymerase α lacks its own proofreading activity yet synthesizes perhaps 10% of each Okazaki fragment on the lagging strand, i.e., ~5% of the human genome. Given a base substitution error rate of ~10^{-4} (Fig. 2), this amount of replication would generate 30,000 mismatches during each replication cycle. Can polymerase α errors be proofread by a separate exonuclease? This possibility was recently examined in a genetic study of yeast polymerase α with a Leu-868Met (L868M) substitution at the polymerase active site (Pavlov et al. 2006). L868M polymerase α copies DNA in vitro with normal activity and processivity but with reduced fidelity. In vivo, the *pol1-L868M* allele confers a mutator phenotype, which is strongly increased upon inactivation of the 3′ exonuclease of polymerase δ but not that of ε. Among several possible (nonexclusive) explanations, the results support the hypothesis that the 3′ exonuclease of polymerase δ proofreads errors generated by polymerase α during initiation of Okazaki fragments. Given the existence of many other specialized, naturally proofreading-deficient DNA polymerases with even lower fidelity than polymerase α, intrinsic proofreading could be relevant to other DNA transactions that control genome stability (for review, see Nick McElhinny et al. 2006), such as base excision repair and possibly translesion synthesis by polymerase η (see below).

REPLICATION ASYMMETRY AND FIDELITY

The two strands of duplex DNA are oriented antiparallel to each other, and DNA polymerases copy DNA in only the 5′ to 3′ direction. Thus, replication of duplex DNA is intrinsically asymmetric. This asymmetry is illustrated by the simple model of a eukaryotic replication fork shown in Figure 4A, left. Recent evidence in budding yeast suggests that the leading strand is primarily replicated by polymerase ε (Pursell et al. 2007), whereas Okazaki fragments on the lagging strand are initiated by polymerase α-primase and then primarily completed by polymerase δ (Nick McElhinny et al. 2008). These enzymes differ from each other in primary sequence, subunit composition, interactions with other proteins, and several biochemical properties, including processivity, proofreading capacity, fidelity, and error specificity. It is therefore possible that the fidelity of leading- and lagging-strand replication may differ, perhaps even more so under nonstandard replication conditions arising under stress (Fig. 4A, right), either environmental (DNA lesions) or genetic (mutations in key genes). Evidence for differences in leading- and lagging-strand replication fidelity have been reported in *E. coli* (Fijalkowska et al. 1998), where both strands are replicated by the same polymerase acting as a multisubunit dimer, DNA polymerase III holoenzyme.

FIDELITY OF DNA REPAIR POLYMERASES

Efficient and accurate replication requires clean substrates, such that many organisms devote great attention and energy to repairing DNA lesions that can result from endogenous metabolic processes and from exposure to physical and chemical agents in the external environment. Many different repair processes exist and can be distinguished by lesion specificity, the enzymes involved, and when they operate (for review, see Friedberg et al. 2006). For many of these repair pathways, e.g., BER, NER, NHEJ,

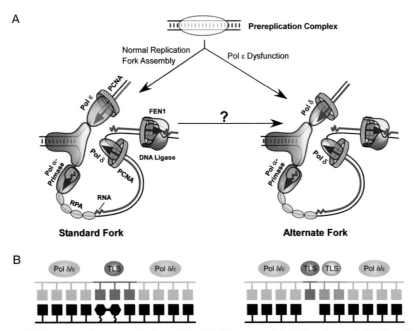

Figure 4. Replication fork and translesion synthesis models. (*A*) Current model of the eukaryotic DNA replication fork (left), with polymerase ε replicating the leading-strand template and polymerases α and δ replicating the lagging-strand template. On the right is an "alternative fork" that might result from stress. For further discussion, see text and Kunkel and Burgers (2008). (*B*) "One TLS polymerase" (left) and "two TLS polymerase" (right) models for translesion DNA synthesis. See text for further discussion. (*A*, Reprinted, with permission, from Kunkel and Burgers 2008.)

MMR, and ICL repair, excision of a lesion is followed by gap-filling DNA synthesis and ligation (Friedberg et al. 2006). The gap filling is conducted by polymerases that are highly accurate (e.g., polymerase δ for filling long gaps during NER and MMR) or moderately accurate (polymerase β for filling short gaps during BER). However, certain gap-filling transactions in cells may involve inaccurate DNA polymerases, e.g., κ in NER, ζ/θ in ICL repair, and μ in NHEJ, such that DNA synthesis errors occurring during repair may contribute to mutagenesis. Perhaps the best examples are for polymerases ζ and η, both of which are implicated in somatic hypermutation of immunoglobulin genes, a process that involves processing uracil in DNA generated by cytosine deamination catalyzed by activation-induced cytosine deaminase (Diaz and Lawrence 2005).

Fidelity of Translesion Synthesis Polymerases

Replication forks can stall upon encountering lesions that distort helix geometry. Among several possible solutions that allow complete genome replication, one is translesion synthesis (TLS) by DNA polymerases. Two general models have been put forth for TLS, one involving a single TLS polymerase for lesion bypass (Fig. 4B, left) and another involving two TLS polymerases, one for insertion opposite a lesion and another for extending aberrant primer termini (Fig. 4B, right). Several specialized TLS polymerases have been discovered in the past decade, and these are evolutionarily conserved (Ohmori et al. 2001; Prakash and Prakash 2002; Yang and Woodgate 2007). In mammals, they include family-B polymerase ζ, family-Y polymerases η, κ, and ι, and possibly family-A polymerases θ and ν. These are the least accurate of polymerases (Fig. 2). They all lack proofreading activity and also have lower nucleotide selectivity than the major replicative polymerases, as indicated by their higher error rates for base substitutions and indels (Fig. 2). The extreme case is for polymerase ι, a conserved family-Y member that rarely generates certain mismatches (e.g., A•dC) but can preferentially misincorporate dG as compared to dA opposite template T (Fig. 2) (see, e.g., Bebenek et al. 2001). This rather amazing violation of Watson–Crick base-pairing dogma leads one to wonder what the true physiological substrates and functions of polymerase ι might be in vivo, whether in TLS (Dumstorf et al. 2006) or in yet to be discovered DNA transactions. Structural and biochemical studies suggest that the low fidelity of family-Y enzymes is partly due to relaxed geometric selectivity in the nascent base-pair-binding pocket, which is more open and solvent accessible than those of more accurate DNA polymerases. An example is shown in Figure 1C, which depicts the active site of a bacterial Y-family polymerase: S*so* Dpo4. Indeed, much of the seminal work on family-Y polymerases has been performed using bacterial enzymes, which include two DNA-damage-induced *E. coli* DNA polymerases: IV and V (Fuchs et al. 2004). Another TLS polymerase is the B-family member ζ. When copying undamaged DNA, ζ has somewhat higher fidelity than the Y-family polymerases, but lower fidelity than the other B-family members (Fig. 2). The ability of ζ to generate both base substitutions and indels at relatively high rates is consistent with its known participation in a large majority of spontaneous mutations, as well as in mutagenesis induced by a variety of DNA-damaging agents (Lawrence 2002). Polymerase ζ's high base substitution error rate clearly demonstrates that it has relatively low nucleotide selectivity, consistent with a possible direct role in mutagenic misinsertion of dNTPs in vivo. Polymerase ζ also efficiently extends terminal mismatches when copying undamaged DNA, as well as efficiently extending damaged termini, the latter being consistent with a role for ζ in the extension step of TLS in the two-polymerase model (Fig. 4B). A similar role has also been proposed for polymerase κ, which, like ζ, is promiscuous for mismatch extension (for review, see Prakash and Prakash 2002). During DNA synthesis in vitro, ζ also generates "complex" mutations that contain multiple substitutions and indels within a short tract of DNA (Sakamoto et al. 2007; Stone et al. 2009). Consistent with this property, ζ also generates complex errors in vivo, which could be significant from an evolutionary perspective. The biological relevance of TLS is perhaps best illustrated by the role of polymerase V in the mutagenic SOS response in *E. coli,* and the fact that loss of polymerase η function in humans and in mice results in sensitivity to sunlight, predisposition to skin cancer, and altered specificity of somatic hypermutation of immunoglobulin genes. The topics and the TLS ability and fidelity of various polymerases when encountering a wide range of structurally diverse lesions have been described in great detail elsewhere (see, e.g., Prakash and Prakash 2002; Fuchs et al. 2004; Diaz and Lawrence 2005; Friedberg et al. 2005; Yang and Woodgate 2007).

DNA Mismatch Repair

Replication errors are corrected by DNA mismatch repair (MMR) (for review, see Kunkel and Erie 2005; Iyer et al. 2006; Hsieh and Yamane 2008). The reactions and proteins catalyzing MMR are evolutionarily conserved from *E. coli* (Fig. 5A) through humans (Fig. 5B). MMR requires initial recognition of mismatches by bacterial MutS protein or its eukaryotic homologs (Msh2-Msh6 or Msh2-Msh3). This is followed by binding of a second protein, MutL, or its eukaryotic homologs Mlh1-Pms1 (Pms2 in humans), Mlh1-Mlh2, or Mlh1-Mlh3. These MutS and MutL proteins bind and hydrolyze ATP, and in so doing, these complexes undergo conformational changes that help to coordinate the multiple protein partnerships and reactions needed to find the strand-discrimination signal, incise the nascent strand, excise the replication error, correctly synthesize new DNA, and then ligate the nascent strand. In addition to their functions in repairing replication errors, some MMR proteins also participate in other DNA transactions, including critical environmental stress-response pathways such as repair of double-stranded DNA breaks and DNA-damage surveillance to signal apoptosis. As a consequence, loss of MMR is associated with elevated mutation rates and altered survival in response to DNA damage. These in turn can give rise to microbial populations with increased fitness to survive adverse environmental conditions and to somatic cells with increased resistance to chemotherapeutic agents and increased probability of tumorigenesis. MMR proteins also prevent recombination between DNA sequences with imperfect homology, thereby influencing speciation. Some MMR proteins participate in meiotic recombination, such that loss of MMR results in infertility. MMR also modulates somatic hypermutation of immunoglobulin genes, such that loss of MMR protein functions alters the specificity of somatic hypermutation (SHM).

Modulating MMR Efficiency

Given these many functions and biological effects, intensive studies of MMR have revealed several ways to modulate MMR activity (Table 2). These include partial or complete inactivation by mutations in various MMR genes (for review, see Kunkel and Erie 2005; Iyer et al. 2006; Hsieh and Yamane 2008), cadmium inhibition of MMR in budding yeast (Jin et al. 2003), silencing Mlh1 expression by promoter hypermethylation (see, e.g., Herman et al. 1998), saturating MMR repair capacity under conditions of stress (Schaaper and Radman 1989), and imbalanced expression of certain MMR proteins that can reduce MMR efficiency (see Drummond et al. 1997; Shcherbakova et al. 2001 and references therein). Indeed, just as for nucleotide selectivity and proofreading, the contribution of MMR to replication fidelity can vary over a wide range (Fig. 2). On average, complete loss of bacterial MutS-dependent MMR or eukaryotic Msh2-dependent MMR elevates point mutation rates ~100-fold to 1000-fold. On the edges of the MMR efficiency continuum are some mismatches that are poorly corrected, e.g., about fivefold for 8-oxo-G•A mismatches (Pavlov et al. 2003), and others that are repaired incredibly efficiently, e.g., exceeding 10,000-fold for single-base deletion mismatches in long homo-nucleotide runs. The latter illustrate that MMR is the major guardian against the instability of repetitive sequence elements, which, as explained above, are prone to slippage and poorly proofread.

CONCLUSIONS

In the half century since the DNA double helix was described, we have come to more fully, but still incompletely, appreciate the elegant complexity of the DNA transactions required to replicate genomes with the fidelity needed to maintain genetic identity in the face of environmental insults, coupled with the advantage of some promiscuity for survival and evolution. This flexibility stems from the wide variability in the contributions of the three processes that determine DNA synthesis fidelity (Fig. 2) and the fact that proofreading and MMR, the two main replication-error correction mechanisms that ensure genome stability, are not essential for cellular survival. Thus, neither proofreading nor MMR contributes to replication of RNA viruses, and these viruses have very high mutation rates that fit their lifestyle (Drake and Holland 1999). Some DNA viruses use proofreading to achieve lower mutation rates but do not take advantage of their host's MMR machinery (e.g., bacte-

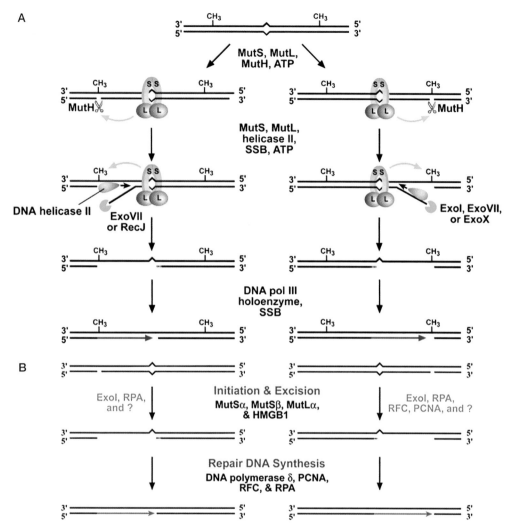

Figure 5. Models for DNA mismatch repair in *E. coli* (*A*) and eukaryotes (*B*). See text for description. (Reprinted, with permission, from Iyer et al. 2006.)

riophage T4) (Santos and Drake 1994). As a consequence, they have mutation rates per base pair that are higher than organisms that do use MMR. Interestingly, given the differences in genome size, the mutation rate per genome is relatively constant among DNA-based organisms, at 0.003 (Drakes rule; Drake 1991, 1999). Also of interest is the fact that the genomes of certain bacteria do not encode obvious homologs of the major MMR genes. This leads one to wonder whether they forego MMR altogether or correct replication errors in a manner yet undiscovered.

ACKNOWLEDGMENTS

I thank Alan Clark and Mercedes Arana for thoughtful comments on the manuscript. The research conducted in the author's laboratory is supported by the Intramural Research Program of the National Institutes of Health, National Institute of Environmental Health Sciences (Projects Z01 ES065070 and Z01 ES065089).

REFERENCES

Beard WA, Wilson SH. 2003. Structural insights into the origins of DNA polymerase fidelity. *Structure* **11:** 489–496.

Bebenek K, Kunkel TA. 2000. Streisinger revisited: DNA synthesis errors mediated by substrate misalignments. *Cold Spring Harbor Symp Quant Biol* **65:** 81–91.

Bebenek K, Kunkel TA. 2004. Functions of DNA. *Adv Protein Chem* **69:** 137–165.

Bebenek K, Tissier A, Frank EG, McDonald JP, Prasad R, Wilson SH, Woodgate R, Kunkel TA. 2001. 5′-Deoxyribose phosphate lyase activity of human DNA polymerase ι in vitro. *Science* **291:** 2156–2159.

Bessman MJ, Kornberg A, Lehman IR, Simms ES. 1956. Enzymic synthesis of deoxyribonucleic acid. *Biochim Biophys Acta* **21:** 197–198.

Brieba LG, Eichman BF, Kokoska RJ, Doubliâe S, Kunkel TA, Ellenberger T. 2004. Structural basis for the dual coding potential of 8-oxoguanosine by a high-fidelity DNA polymerase. *EMBO J* **23:** 3452–3461.

Burgers PM. 2009. Polymerase dynamics at the eukaryotic DNA replication fork. *J Biol Chem* **284:** 4041–4045.

Chang DJ, Cimprich KA. 2009. DNA damage tolerance: When it's OK to make mistakes. *Nat Chem Biol* **5:** 82–90.

Diaz M, Lawrence C. 2005. An update on the role of translesion synthesis DNA polymerases in Ig hypermutation. *Trends Immunol* **26:** 215–220.

Donlin MJ, Patel SS, Johnson KA. 1991. Kinetic partitioning between the exonuclease and polymerase sites in DNA error correction. *Biochemistry* **30:** 538–546.

Drake JW. 1991. A constant rate of spontaneous mutation in DNA-based microbes. *Proc Natl Acad Sci* **88:** 7160–7164.

Drake JW. 1999. The distribution of rates of spontaneous mutation over viruses, prokaryotes, and eukaryotes. *Ann NY Acad Sci* **870:** 100–107.

Drake JW, Holland JJ. 1999. Mutation rates among RNA viruses. *Proc Natl Acad Sci* **96:** 13910–13913.

Drummond JT, Genschel J, Wolf E, Modrich P. 1997. DHFR/MSH3 amplification in methotrexate-resistant cells alters the hMutSα/hMutSβ ratio and reduces the efficiency of base-base mismatch repair. *Proc Natl Acad Sci* **94:** 10144–10149.

Dumstorf CA, Clark AB, Lin Q, Kissling GE, Yuan T, Kucherlapati R, McGregor WG, Kunkel TA. 2006. Participation of mouse DNA polymerase ι in strand-biased mutagenic bypass of UV photoproducts and suppression of skin cancer. *Proc Natl Acad Sci* **103:** 18083–18088.

Fersht AR, Knill-Jones JW, Tsui WC. 1982. Kinetic basis of spontaneous mutation. Misinsertion frequencies, proofreading specificities and cost of proofreading by DNA polymerases of *Escherichia coli*. *J Mol Biol* **156:** 37–51.

Fijalkowska IJ, Jonczyk P, Tkaczyk MM, Bialoskorska M, Schaaper RM. 1998. Unequal fidelity of leading strand and lagging strand DNA replication on the *Escherichia coli* chromosome. *Proc Natl Acad Sci* **95:** 10020–10025.

Fortune JM, Pavlov YI, Welch CM, Johansson E, Burgers PM, Kunkel TA. 2005. *Saccharomyces cerevisiae* DNA polymerase δ: High fidelity for base substitutions but lower fidelity for single- and multi-base deletions. *J Biol Chem* **280:** 29980–29987.

Franklin MC, Wang J, Steitz TA. 2001. Structure of the replicating complex of a pol α family DNA polymerase. *Cell* **105:** 657–667.

Friedberg EC, Lehmann AR, Fuchs RP. 2005. Trading places: How do DNA polymerases switch during translesion DNA synthesis? *Mol Cell* **18:** 499–505.

Friedberg EC, Walker GC, Siede W, Wood RD, Schultz RA, Ellenberger T. 2006. *DNA repair and mutagenesis*, 2nd ed. ASM Press, Washington, D.C.

Fuchs RP, Fujii S, Wagner J. 2004. Properties and functions of *Escherichia coli:* Pol IV and Pol V. *Adv Protein Chem* **69:** 229–264.

Garcia-Diaz M, Kunkel TA. 2006. Mechanism of a genetic glissando: Structural biology of indel mutations. *Trends Biochem Sci* **31:** 206–214.

Goldsby RE, Lawrence NA, Hays LE, Olmsted EA, Chen X, Singh M, Preston BD. 2001. Defective DNA polymerase-δ proofreading causes cancer susceptibility in mice. *Nat Med* **7:** 638–639.

Herman JG, Umar A, Polyak K, Graff JR, Ahuja N, Issa JP, Markowitz S, Willson JK, Hamilton SR, Kinzler KW, et al. 1998. Incidence and functional consequences of hMLH1 promoter hypermethylation in colorectal carcinoma. *Proc Natl Acad Sci* **95:** 6870–6875.

Hsieh P, Yamane K. 2008. DNA mismatch repair: Molecular mechanism, cancer, and ageing. *Mech Ageing Dev* **129:** 391–407.

Iyer RR, Pluciennik A, Burdett V, Modrich PL. 2006. DNA mismatch repair: Functions and mechanisms. *Chem Rev* **106:** 302–323.

Jansen JG, Fousteri MI, de Wind N. 2007. Send in the clamps: Control of DNA translesion synthesis in eukaryotes. *Mol Cell* **28:** 522–529.

Jin YH, Clark AB, Slebos RJ, Al-Refai H, Taylor JA, Kunkel TA, Resnick MA, Gordenin DA. 2003. Cadmium is a mutagen that acts by inhibiting mismatch repair. *Nat Genet* **34:** 326–329.

Jin YH, Garg P, Stith CM, Al-Refai H, Sterling J, Murray LJ, Kunkel TA, Resnick MA, Burgers PM, Gordenin DA. 2005. The multiple biological roles of the 3′→5′ exonuclease of *Saccharomyces cerevisiae* DNA polymerase δ require switching between the polymerase and exonuclease domains. *Mol Cell Biol* **25:** 461–471.

Kim TW, Delaney JC, Essigmann JM, Kool ET. 2005. Probing the active site tightness of DNA polymerase in subangstrom increments. *Proc Natl Acad Sci* **102:** 15803–15808.

Kool ET. 2002. Active site tightness and substrate fit in DNA replication. *Annu Rev Biochem* **71:** 191–219.

Kornberg A, Baker T. 1992. *DNA replication,* 2nd ed. Freeman, New York.

Kroutil LC, Kunkel TA. 1998. DNA replication errors involving strand misalignments. In *Genetic instabilities and hereditary neurological diseases* (ed. RD Wells and ST Warren), pp. 699–716. Academic, San Diego.

Kunkel TA, Bebenek K. 2000. DNA replication fidelity. *Annu Rev Biochem* **69:** 497–529.

Kunkel TA, Burgers PM. 2008. Dividing the workload at a eukaryotic replication fork. *Trends Cell Biol* **18:** 521–527.

Kunkel TA, Erie DA. 2005. DNA mismatch repair. *Annu Rev Biochem* **74:** 681–710.

Kunkel TA, Pavlov YI, Bebenek K. 2003. Functions of human DNA polymerases η, κ and ι suggested by their properties, including fidelity with undamaged DNA templates. *DNA Repair* **2:** 135–149.

Lawrence CW. 2002. Cellular roles of DNA polymerase ζ and Rev1 protein. *DNA Repair* **1:** 425–435.

Loeb LA, Kunkel TA. 1982. Fidelity of DNA synthesis. *Annu Rev Biochem* **51:** 429–457.

Loeb LA, Monnat RJ Jr. 2008. DNA polymerases and human disease. *Nat Rev Genet* **9:** 594–604.

Longley MJ, Nguyen D, Kunkel TA, Copeland WC. 2001. The fidelity of human DNA polymerase γ with and without exonucleolytic proofreading and the p55 accessory subunit. *J Biol Chem* **276:** 38555–38562.

McCulloch SD, Kunkel TA. 2008. The fidelity of DNA synthesis by eukaryotic replicative and translesion synthesis polymerases. *Cell Res* **18:** 148–161.

Moon AF, Garcia-Diaz M, Bebenek K, Davis BJ, Zhong X, Ramsden DA, Kunkel TA, Pedersen LC. 2007. Structural insight into the substrate specificity of DNA polymerase μ. *Nat Struct Mol Biol* **14:** 45–53.

Morrison A, Sugino A. 1994. The 3′→5′ exonucleases of both DNA polymerases δ and ε participate in correcting errors of DNA replication in *Saccharomyces cerevisiae*. *Mol Gen Genet* **242:** 289–296.

Nick McElhinny SA, Pavlov YI, Kunkel TA. 2006. Evidence for extrinsic exonucleolytic proofreading. *Cell Cycle* **5:** 958–962.

Nick McElhinny SA, Gordenin DA, Stith CM, Burgers PM, Kunkel TA. 2008. Division of labor at the eukaryotic replication fork. *Mol Cell* **30:** 137–144.

Ninio J. 1975. Kinetic amplification of enzyme discrimination. *Biochimie* **57:** 587–595.

Ohmori H, Friedberg EC, Fuchs RP, Goodman MF, Hanaoka F, Hinkle D, Kunkel TA, Lawrence CW, Livneh Z, Nohmi T, et al. 2001. The Y-family of DNA polymerases. *Mol Cell* **8:** 7–8.

Ollis DL, Brick P, Hamlin R, Xuong NG, Steitz TA. 1985. Structure of large fragment of *Escherichia coli* DNA polymerase I complexed with dTMP. *Nature* **313:** 762–766.

Pavlov YI, Mian IM, Kunkel TA. 2003. Evidence for preferential mismatch repair of lagging strand DNA replication errors in yeast. *Curr Biol* **13:** 744–748.

Pavlov YI, Frahm C, Nick McElhinny SA, Niimi A, Suzuki M, Kunkel TA. 2006. Evidence that errors made by DNA polymerase α are corrected by DNA polymerase δ. *Curr Biol* **16:** 202–207.

Petruska J, Goodman MF. 1995. Enthalpy-entropy compensation in DNA melting thermodynamics. *J Biol Chem* **270:** 746–750.

Prakash S, Prakash L. 2002. Translesion DNA synthesis in eukaryotes: A one- or two-polymerase affair. *Genes Dev* **16:** 1872–1883.

Pursell ZF, Isoz I, Lundstrom EB, Johansson E, Kunkel TA. 2007. Yeast DNA polymerase ε participates in leading-strand DNA replication. *Science* **317:** 127–130.

Sakamoto AN, Stone JE, Kissling GE, McCulloch SD, Pavlov

YI, Kunkel TA. 2007. Mutator alleles of yeast DNA polymerase ζ. *DNA Repair* **6:** 1829–1838.

Santos ME, Drake JW. 1994. Rates of spontaneous mutation in bacteriophage T4 are independent of host fidelity determinants. *Genetics* **138:** 553–564.

Schaaper RM, Radman M. 1989. The extreme mutator effect of *Escherichia coli* mutD5 results from saturation of mismatch repair by excessive DNA replication errors. *EMBO J* **8:** 3511–3516.

Shcherbakova PV, Hall MC, Lewis MS, Bennett SE, Martin KJ, Bushel PR, Afshari CA, Kunkel TA. 2001. Inactivation of DNA mismatch repair by increased expression of yeast MLH1. *Mol Cell Biol* **21:** 940–951.

Shcherbakova PV, Bebenek K, Kunkel TA. 2003a. Functions of eukaryotic DNA polymerases. *Sci Aging Knowledge Environ* **2003:** RE3.

Shcherbakova PV, Pavlov YI, Chilkova O, Rogozin IB, Johansson E, Kunkel TA. 2003b. Unique error signature of the four-subunit yeast DNA polymerase ε. *J Biol Chem* **278:** 43770–43780.

Steitz TA. 1993. DNA- and RNA-dependent DNA polymerases. *Curr Opin Struct Biol* **3:** 31–38.

Stone JE, Kissling GE, Lujan SA, Rogozin IB, Stith CM, Burgers PM, Kunkel TA. 2009. Low-fidelity DNA synthesis by the L979F mutator derivative of *Saccharomyces cerevisiae* DNA polymerase ζ. *Nucleic Acids Res* **37:** 3774–3787.

Watson JD, Crick FH. 1953. Molecular structure of nucleic acids; a structure for deoxyribose nucleic acid. *Nature* **171:** 737–738.

Wing RA, Bailey S, Steitz TA. 2008. Insights into the replisome from the structure of a ternary complex of the DNA polymerase III α-subunit. *J Mol Biol* **382:** 859–869.

Yang W, Woodgate R. 2007. What a difference a decade makes: Insights into translesion DNA synthesis. *Proc Natl Acad Sci* **104:** 15591–15598.

Protein Folding Sculpting Evolutionary Change

S. LINDQUIST

Whitehead Institute for Biomedical Research, Massachusetts Institute of Technology, Cambridge, Massachusetts 02142

Correspondence: lindquist_admin@wi.mit.edu

Our work suggests that the forces that govern protein folding exert a profound effect on how genotypes are translated into phenotypes and that this in turn has strong effects on evolutionary processes. Molecular chaperones, also known as "heat-shock proteins" (Hsps), promote the correct folding and maturation of many other proteins in the cell. Hsp90 is an abundant and highly specialized chaperone that works on a particularly interesting group of client proteins: metastable signal transducers that are key regulators of a broad spectrum of biological processes. Such proteins often have evolved to finish folding only when they have received a specific signal, such as the binding of a ligand or a posttranslational modification. Importantly, the folding of Hsp90 clients is particularly sensitive to changes in the external and internal environment of the cell. Therefore, Hsp90 is uniquely positioned to couple environmental contingencies to the evolution of new traits. Our work has helped to define two mechanisms by which Hsp90 might influence the acquisition of new phenotypes. First, by robustly maintaining signaling pathways, Hsp90 can buffer the effects of mutations in those pathways, allowing the storage of cryptic genetic variation that is released by stress. In this case, when the Hsp90 buffer is compromised by environmental stress, new traits appear. These traits can also be assimilated, so that they become manifest even in the absence of stress, when genetic recombination and selection enrich causative variants in subsequent generations. Second, Hsp90 can potentiate the effects of genetic variation, allowing new mutations to produce immediate phenotypes. In this case, when Hsp90 function is compromised, new traits are lost. These traits can also be assimilated, so that they are maintained under environmental stress, but this is achieved through new mutations. We have discovered these powerful evolutionary mechanisms in fruit flies, mustard plants, and fungi, but expect them to operate in all eukaryotes. Another line of work relating protein folding to the evolution of new traits involves protein-based hereditary elements known as prions. These produce changes in phenotype through heritable, self-perpetuating changes in protein conformation. Because changes in protein homeostasis occur with environmental stress, prions can be cured or induced by stress, creating heritable new phenotypes that depend on the genetic variation present in the organism. Both prions and Hsp90 provide plausible mechanisms for allowing genetic diversity and fluctuating environments to fuel the pace of evolutionary change. The multiple mechanisms by which protein folding can influence the evolution of new traits provide both a new paradigm for understanding rapid, stepwise evolution and a framework for targeted therapeutic interventions.

Proteins are synthesized as long, linear chains of amino acids. To function, they must fold into intricate, stereotyped, three-dimensional structures. Moreover, individual proteins must do so in a highly chaotic environment. Inside a living cell, the protein concentration reaches 300 mg/ml, and those many proteins possess a tremendous amount of kinetic energy. Individual proteins therefore frequently encounter folding problems due to macromolecular crowding. In fact, "off-pathway" folding events are a constant problem. Individual cells, and even entire organisms, live poised on the precipice of a protein-folding crisis, a situation that is worsened by environmental stress.

In the last 15 years, two different areas of research in my laboratory have pointed to a pivotal role for protein folding in evolutionary processes. One of these involves the action of chaperone proteins, which facilitate protein folding in the complex cellular environment, thereby helping proteins to assume their normal functions. The second closely related line of work involves a type of hereditary element, a prion, that produces a change in phenotype through a heritable, self-perpetuating change in protein conformation. Reflecting time constraints of the Symposium, I focus my presentation on the role of the chaperone protein Hsp90 and only touch on yeast prions very briefly.

HEAT-SHOCK PROTEINS

Nearly 5% of the protein mass of the cell is devoted to helping other proteins fold properly and to correcting problems that occur when they misfold. An important class of helper proteins (known as chaperones) is the heat-shock proteins, so called because they can be induced by a shift from a normal to a slightly higher growth temperature. We concentrate here on the abundant chaperone Hsp90 and its role in facilitating the rapid evolution of novel traits. Hsp90 is essential for life in eukaryotic cells. Only small amounts of Hsp90 are required for growth and development under normal conditions, but much more is required under stressful conditions. Unlike most chaperones, it is made at much higher levels than needed and is present in excess under normal conditions and therefore acts as a protein-folding buffer. This enables it to facilitate protein folding even during sudden, unexpected stresses.

Also unlike most other Hsps, Hsp90 is involved in the folding of specific regulatory proteins positioned strategically in major signal transduction networks, comprising perhaps only ~2%–5% of the total protein in the cell. Hsp90's client proteins are characteristically metastable proteins that do not finish folding into their final stable conformations until something else has activated them, for example, the binding of a ligand, a phosphorylation event, or translocation of the protein to the membrane. Protein complexes of Hsp90 with its clients are therefore extremely dynamic and strongly affected by stress. Their metastable nature means that much higher concentrations of Hsp90 are then required to drive its client proteins into appropriate complexes. Stressful conditions also cause additional cellular proteins to become unfolded (thereby creating increased need for Hsp90 and further depleting the buffer). Together, these properties of Hsp90–client interactions have a strong influence on how genetic variation is translated into phenotypic novelty and diversity.

HSP90 BUFFERS POLYMORPHISMS

Hsp90 may affect evolutionary change by buffering genetic polymorphisms. The excess Hsp90 chaperone capacity allows a variety of different polymorphisms throughout the genome to accumulate without phenotypic consequence. Within a certain range, it keeps these polymorphisms silent because it is able to maintain the normal folding and functioning of the signal transduction pathways controlled by its client proteins. But when stress depletes the Hsp90 buffer, phenotypes associated with these accumulated polymorphisms are revealed suddenly and in a combinatorial fashion. In this way, Hsp90 acts as a "capacitor," allowing the storage and controlled release of genetic variation. This can have enormous functional consequences for the organism.

Our first experiments suggesting this mechanism involved some Hsp90 mutants in *Drosophila* lines that we obtained from the Rubin and Hoffmann laboratories (Rutherford and Lindquist 1998). We confirmed their findings that homozygotes (completely lacking Hsp90 function) died, and heterozygotes (with reduced Hsp90 function) were normal. However, we noticed that a small number of them had astonishing, and diverse, developmental anomalies (Fig. 1A–F). Interestingly, these rare developmental anomalies were generally specific to the genetic background into which the mutations were transferred. These observations suggested that reducing Hsp90 function does not just destabilize development; instead, it reveals the effects of previously hidden genetic variation that can affect virtually every trait in the fly.

Further experiments suggested that populations of these flies contained a great diversity of different genetic polymorphisms. Some rare individuals exhibited several novel and unusual traits. This itself was a big surprise, because most geneticists believed that evolution worked through small incremental mutations. When we reduced Hsp90 function by genetic mutation, or grew flies at a higher temperature, we observed novel phenotypes in those flies more frequently. When we crossed two of the rare flies together and selected for that novel phenotype, we enriched subsequent generations for that underlying genetic variation to the point where those phenotypes were maintained, i.e., variation was assimilated, even when Hsp90 function was normal (Rutherford and Lindquist 1998). We hypothesized that this interface between environmental stress and protein homeostasis could have a large effect on evolution.

We then turned to the mustard plant *Arabidopsis thaliana*, an inbreeding organism with a sessile lifestyle and considerably less heterozygosity than *D. melanogaster*. We asked, again, what happened when Hsp90 function was reduced in different accession lines of *Arabidopsis*. We observed a stunning array of phenotypic abnormalities that were specific to the individual lines: For example, leaves coming out in whorl shapes, plants growing upside down, hairy roots, or vine-like growth (Fig. 2). We also obtained the same array of phenotypes by growing the plants at elevated temperatures (Queitsch et al. 2002; Sangster et al. 2007). Again, these experiments revealed hidden genetic variation when the Hsp90 buffer was reduced by stress.

We next examined the progeny of recombinant inbred plants from different ecotypes. We studied a wide variety of normal life history traits in *Arabidopsis*, such as flowering time and number of seeds that were set. Every single trait we studied in these lines had some hidden genetic variation that was exposed when Hsp90 function was reduced, and we were able to build a crude map of quantitative trait loci

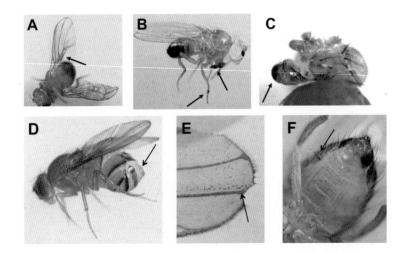

Figure 1. Examples of developmental abnormalities associated with Hsp90 deficits in *Drosophila*. (*A*) Thickened wing veins, (*B*) transformed second leg with an ectopic sex comb, (*C*) deformed eye with an extra antenna, (*D*) disorganized abdominal tergites, (*E*) notched wings, (*F*) extraneous tissue growing out of tracheal pit. (Reprinted, with permission, from Rutherford and Lindquist 1998 [© Nature Publishing Group].)

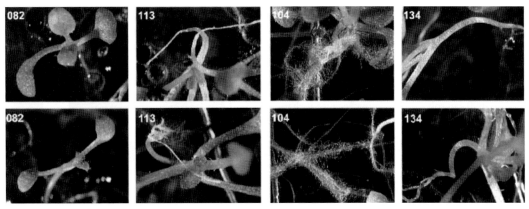

Figure 2. The same recombinant inbred line (RIL)-specific phenotypes are uncovered when Hsp90 buffering capacity is challenged by GdA (*top*) or by growth at 27°C (*bottom*). (Line 082 seedlings) S-shaped rosettes with vertically oriented leaf blades. (Line 113 seedlings) Extreme hypocotyl curls and roots partially extended into air. (Line 104 seedlings) Abundant root hair growth. (Line 134 seedlings) Bent hypocotyls with rosette touching the medium surface. (Reprinted, with permission, from Queitsch et al. 2002 [© Nature Publishing Group].)

(QTL) involved in these traits by linkage analysis (Sangster et al. 2008a,b). This suggested that most of the underlying variation was genetic rather than epigenetic. In addition to examining Hsp90 inhibition with chemical compounds, we used RNA interference (RNAi) inhibition, establishing that the changes in phenotype were absolutely due to the interface between Hsp90 and genetic variation. That Hsp90 has such a massive influence on the genotype-to-phenotype map in organisms with profoundly different lifestyles, separated by at least 1 billion years of evolution, suggests that this mechanism has influenced evolutionary processes very broadly.

HSP90 POTENTIATES POLYMORPHISMS

Another way that Hsp90 translates genetic variation into new phenotypes is by directly chaperoning mutated proteins. Some mutations can create new activities in proteins, but often the mutated proteins are unable to fold well on their own. They therefore benefit from the Hsp90 buffer present in the cell to enable them to fold correctly into a functional state and create new phenotypes. In this way, Hsp90 acts as a potentiator of new variation rather than as a capacitor.

The first class of proteins in which we observed this phenomenon was the oncogenic kinases such as v-Src (other kinases have the same basic phenomena). Early experiments on Hsp90 were aimed at determining its basic function and that of the other heat-shock proteins. At that time, we had recently developed techniques to reduce Hsp90 levels in yeast, and we introduced v-Src into those cells to model Hsp90's effects on the protein. Several different laboratories had found that oncogenic kinases were inactive when complexed with Hsp90, leading to speculation that it was a "protein repressor." We found small differences in levels of v-Src accumulation in cells with high versus low levels of Hsp90, but the *activity* of v-Src in the two cell types was profoundly different (Fig. 3A) (Xu and Lindquist 1993). Surprisingly, rather than being repressed, v-Src activity was much higher in the cells with high lev-

els of Hsp90. In contrast, cells with low levels of Hsp90 were able to grow normally but unable to functionally mature the mutated, dysregulated v-Src kinase (Xu et al. 1999). We also found that Hsp90 levels do not make nearly as much difference to the activity of c-Src, a nonmutated and much less active form of the kinase (Fig. 3B). This observation has held true for a host of oncogenic kinases and has been extended to several other types of oncogenes as well. Indeed, studies are under way that suggest that reducing Hsp90 function may offer great promise as a novel therapeutic strategy for treating cancer (Dai and Whitesell 2005; Whitesell and Lindquist 2005; Cullinan and Whitesell 2006).

In any case, these studies suggested that the excess chaperone activity of Hsp90 is a latent reservoir that can be used to fold newly mutated proteins and provide new functions. If such mutations are ever beneficial to the organism, they too could enhance the rate of evolution.

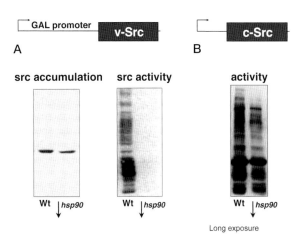

Figure 3. Maturation and activity of the mutated tyrosine kinase v-Src depend on hsp90. (*A*) v-Src accumulates at the same rate when *hsp90* is inhibited, but its activity is reduced. (*B*) c-Src is much less active in general than v-Src, and its activity is much less dependent on hsp90 levels.

Cowen LE. 2008. The evolution of fungal drug resistance: Modulating the trajectory from genotype to phenotype. *Nat Rev Microbiol* **6:** 187–198.

Cowen LE, Lindquist SL. 2005. Hsp90 potentiates the rapid evolution of new traits: Drug resistance in diverse fungi. *Science* **309:** 2185–2189.

Cowen LE, Singh SD, Kohler JR, Collins C, Zaas AK, Schell WA, Aziz H, Mylonakis E, Perfect JR, Whitesell L, et al. 2009. Harnessing Hsp90 function as a powerful, broadly effective therapeutic strategy for fungal infectious disease. *Proc Natl Acad Sci* **106:** 2818–2823.

Cruz MC, Goldstein AL, Blankenship JR, Del Poeta M, Davis D, Cardenas ME, Perfect JR, McCusker JH, Heitman J. 2002. Calcineurin is essential for survival during membrane stress in *Candida albicans*. *EMBO J* **21:** 546–559.

Cullinan SB, Whitesell L. 2006. Heat shock protein 90: A unique chemotherapeutic target. *Semin Oncol* **33:** 457–465.

Dai C, Whitesell L. 2005. HSP90: A rising star on the horizon of anticancer targets. *Future Oncol* **1:** 529–540.

Fox DS, Heitman J. 2002. Good fungi gone bad: The corruption of calcineurin. *Bioessays* **24:** 894–903.

Imai J, Yahara I. 2000. Role of HSP90 in salt stress tolerance via stabilization and regulation of calcineurin. *Mol Cell Biol* **20:** 9262–9270.

Namy O, Galopier A, Martini C, Matsufuji S, Fabret C, Rousset JP. 2008. Epigenetic control of polyamines by the prion [PSI+]. *Nat Cell Biol* **10:** 1069–1075.

Queitsch C, Sangster TA, Lindquist SL. 2002. Hsp90 as a capacitor of phenotypic variation. *Nature* **417:** 618–624.

Rutherford SL, Lindquist SL. 1998. Hsp90 as a capacitor for morphological evolution. *Nature* **396:** 336–342.

Sangster TA, Bahrami A, Wilczek A, Watanabe E, Schellenberg K, McLellan C, Kelley A., Kong SW, Queitsch C, Lindquist SL. 2007. Phenotypic diversity and altered environmental plasticity in *Arabidopsis thaliana* with reduced Hsp90 levels. *PLoS ONE* **2:** e648.

Sangster TA, Salathia N, Undurraga S, Milo R, Schellenberg K, Lindquist S, Queitsch C. 2008a. HSP90 affects the expression of genetic variation and developmental stability in quantitative traits. *Proc Natl Acad Sci* **105:** 2963–2968.

Sangster TA, Salathia N, Lee HN, Watanabe E, Schellenberg K, Morneau K, Wang H, Undurraga S, Queitsch C, Lindquist S. 2008b. HSP90-buffered genetic variation is common in *Arabidopsis thaliana*. *Proc Natl Acad Sci* **105:** 2969–2974.

Shorter J, Lindquist SL. 2005. Prions as adaptive conduits of memory and inheritance. *Nat Rev Genet* **6:** 435–450.

True HL, Lindquist SL. 2000. A yeast prion provides a mechanism for genetic variation and phenotypic diversity. *Nature* **407:** 477–483.

Tyedmers J, Madariaga ML, Lindquist S. 2008. Prion switching in response to environmental stress. *PLoS Biol* **6:** e294.

Whitesell L, Lindquist SL. 2005. HSP90 and the chaperoning of cancer. *Nat Rev Cancer* **5:** 761–772.

Xu Y, Lindquist SL. 1993. Heat-shock protein hsp90 governs the activity of pp60v-src kinase. *Proc Natl Acad Sci* **90:** 7074–7078.

Xu Y, Singer MA, Lindquist S. 1999. Maturation of the tyrosine kinase c-src as a kinase and as a substrate depends on the molecular chaperone Hsp90. *Proc Natl Acad Sci* **96:** 109–114.

Predicting Virus Evolution: The Relationship between Genetic Robustness and Evolvability of Thermotolerance

C.B. Ogbunugafor, R.C. McBride, and P.E. Turner

Department of Ecology and Evolutionary Biology, Yale University, New Haven, Connecticut 06520-8106
Correspondence: paul.turner@yale.edu

Evolutionary biologists often seek to infer historical patterns of relatedness among organisms using phylogenetic methods and to gauge the evolutionary processes that determine variation among individuals in extant populations. But relatively less effort is devoted to making evolutionary biology a truly predictive science, where future evolutionary events are precisely foreseen. Accurate predictions of evolvability would be particularly useful in the evolution of infectious diseases, such as the ability to preemptively address the challenge of pathogens newly emerging in humans and other host populations. Experimental evolution of microbes allows the possibility to rigorously test hypotheses regarding pathogen evolvability. Here, we review how genetic robustness was a useful predictor in gauging which variants of RNA virus φ6 should evolve faster in a novel high-temperature environment. We also present new data on the relative survival of robust and brittle viruses across elevated temperatures and durations of ultraviolet exposure, to infer a possible mechanism for robustness. Our work suggests that virus adaptability in a new environment can be predicted given knowledge of virus canalization in the face of mutational input. These results hint that accurate predictions of virus evolvability are a realistic possibility, at least under circumstances of adaptive thermotolerance.

As a naturalist, Charles Darwin was understandably fascinated by the species diversity visible in the natural world. However, he understood little of the invisible realm of genes—the units of inheritance that are passed across generations and contribute to the phenotypic variation evident in biological populations. Yet Darwin showed remarkable insight when formulating his ideas on evolution via natural selection, as a process whereby the environment determines which variants in a population are favored to contribute their traits to succeeding generations, leading to microevolutionary changes over short timescales. Furthermore, Darwin perceptively understood that natural selection is a process that can also primarily account for nature's vast biodiversity—resulting from macroevolution occurring over long timescales—which Darwin described as "endless forms most wonderful" (Darwin 1859).

Many modern-day evolutionary biologists are concerned with elucidating patterns of extant diversity, seeking to understand how descent with modification from common ancestry accounts for Earth's myriad forms that have evolved since life arose billions of years ago. In contrast, other evolutionary biologists are largely concerned with the processes of evolutionary change that underlie these patterns, examining how mechanisms such as selection and drift can lead to altered phenotypes and genotypes across relatively few generations. Both camps of evolutionary biologists have increasingly powerful and sophisticated approaches for studying patterns and processes of evolution, allowing more accurate glimpses into the mysterious "black boxes" of past and ongoing evolutionary events (see, e.g., Lenski et al. 2003; Thornton et al. 2003; Vrba and DeGusta 2004; Weinreich et al. 2006).

Arguably, evolutionary biology mostly concerns scrutiny of past and present events, rather than forthcoming events, i.e., evolutionary biologists tend to resist the temptation to study and predict how evolution will play out in the future. Such predictive efforts are necessarily more difficult than retrospective analyses, because foretelling the future is inherently less precise than resolving the past. For this reason, relatively less energy has been placed into developing evolutionary biology into a truly predictive science. Of course, like other scientists, evolutionary biologists often make explicit predictions, referred to as hypothesis testing. But there is a distinct difference between formulating a hypothesis that concerns events that have already occurred and formulating a hypothesis that *predicts* events yet to occur.

EVOLVABILITY STUDIED USING MATHEMATICAL THEORY AND BIOLOGICAL EXPERIMENTS

Mathematical studies in evolutionary biology are sometimes future-minded. For example, John Maynard Smith (1982) famously popularized evolutionary game theory, a discipline that often seeks to mathematically determine whether a population can evolve an unbeatable strategy when dealing with competition for finite resources, such as food, mates, or nesting sites. If this evolved strategy is truly unbeatable, it is deemed an evolutionarily stable strategy (ESS) whereby all members of the population should adopt the strategy, preventing the population's invasion by any competing strategy in the future. However, evolutionary game theory usefully acknowledges that the strategy may fail to be an ESS should the environment change unexpectedly, because the rules of engagement may summarily change as well. Because the future may be uncertain, evolutionary games such as tit for tat allow players to change their strategies through time, depending on each other's actions and similarly variable conditions (Turner and Chao 2003; Cressman et al. 2004; Wagner 2006).

Many other examples of future-minded mathematical theory exist in evolutionary biology, centering on the topic of evolvability. Here, the general goal is to predict which genotypes should be advantaged to give rise to evolutionarily successful populations, even when precise environmental changes are difficult to foresee. Evolvability can be defined as the ability to adapt through natural selection (Wagner and Altenberg 1996; Kirschner and Gerhart 1998; Brookfield 2009). Mathematical treatments of evolvability are highly useful, but like all mathematical models in biology, they necessarily must work within confined frameworks reflecting a subset of real-world conditions, i.e., all relevant biological parameters cannot possibly be incorporated into a mathematical model or else the model quickly becomes mathematically intractable and too complicated to be useful. A helpful goal is to identify biological factors deemed most important for making the evolvability prediction and then use these to build the model.

The preparation of flu vaccines is one example of the need to apply evolvability thinking to a practical problem. How can we best predict which variants of influenza virus will be most prevalent during the next flu season, given that stocks of the flu vaccine must be prepared and distributed to medical workers long before the flu season actually strikes? Mathematical and genomic approaches developed for evolutionary phylogenetics have proven to be highly useful for this vital medical task (Bush et al. 1999; Koelle et al. 2006; Coburn et al. 2009). Essentially, one can look retrospectively to determine which virus variants have been recent evolutionary successes and then apply an algorithm to deduce which of these viruses will be prevalent in the near future to warrant choosing them as vaccine targets. Of course, it is well known that mass-produced vaccines are not always highly effective in protecting against the flu, and entirely new variants of influenza virus may emerge unexpectedly to threaten human health (e.g., H1N1 variants identified during a 2009 outbreak in Mexico). These observations aptly demonstrate our sometimes limited capability to accurately predict the course of future evolutionary events. Do such findings indicate that studies of evolvability and other topics in evolutionary prediction should remain locked into the realm of mathematical theory?

Not necessarily. Experimental evolution is one empirical discipline in evolutionary biology where researchers are willing to let their science be forward-looking. Experimental evolution is the study of evolution in action, often involving biological populations placed in controlled laboratory environments where researchers scrutinize phenotypic and molecular changes occurring across generations. Because experimental evolution allows the study of evolutionary change as it unfolds, such studies span the past (a founding ancestor) through the present (evolved descendants). Furthermore, if the study system can be archived (stored under suspended animation), genotypes from different evolutionary time periods may be directly compared, such as through head-to-head competition. Microbes such as viruses, bacteria, and yeast often allow this intense scrutiny because they may be stored indefinitely under ultralow temperature and then revived at a later time. Thus, experimental evolution of microbes allows evolving populations to march forward in time, and the future conveniently becomes the present as researchers examine changes taking place across this completely accessible "fossil record."

Here, we review our studies in experimental evolution of viruses, emphasizing how this work can be used to study evolvability. We deemed that it was useful to first identify whether viral genomes evolved to differ in genetic robustness: relative ability to maintain a constant phenotype despite random mutational input (de Visser et al. 2003). This link between phenotype and genotype is crucial for evolution, because natural selection is fueled by phenotypically expressed genetic variation. As we discuss below, robustness is a trait that dictates the translation of genotype to phenotype, and knowledge of virus robustness may be used to predict whether viruses should be relatively advantaged in their evolvability when encountering future environmental challenges.

ROBUSTNESS DESCRIBES THE LINK BETWEEN GENOTYPE AND PHENOTYPE

Evolution is defined as the change in genetic makeup of a population through time. This change can be driven by the process of natural selection, which leads indirectly to changes in gene frequencies by acting on the phenotypes that genes produce. Thus, the translation of phenotype from genotype is vital to the natural selection process. Robustness and brittleness are terms used for measuring and describing the relative accuracy of this genotype-to-phenotype translation (Fig. 1). If a mutation changes a phenotype, we consider the gene—or, more broadly, the genome—to be relatively nonrobust (brittle) against mutational input. However, if a genome changes but its phenotype remains unaffected, the genome is considered genetically robust. Because other types of robustness are of interest, especially environmental robustness that describes whether a phenotype persists when an environment changes, it is crucial when examining robustness to define the specific phenotype and perturbation being measured and discussed. As we describe below, to better understand how natural selection shapes the adaptive trajectory of a population, it would be useful to know the average effectiveness of genotype-to-phenotype translation (genetic robustness) for individuals in the population when adaptive challenges are encountered.

Despite a long-standing interest in evolution of robustness and its associated mathematics, evolutionary biologists have generated few empirical data on this subject. Some empirical studies have allowed organisms to evolve for long periods in a constant environment (see, e.g., Lenski and Travisano 1994), and one might expect that these lineages would be ideal candidates to study evolution of genetic robustness; at evolutionary equilibrium—mutation-selecti-

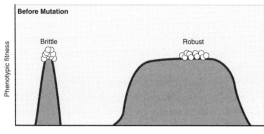
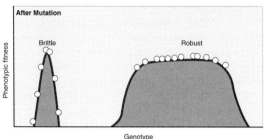

Figure 1. Brittle and robust organisms can be defined by their fitness response to mutational change, using the metaphor of fitness landscapes. Fitness is vertical height on the landscape. Mutation causes genotypes to move away from their original position on the horizontal axis. After mutation, brittle individuals experience large changes in fitness as they are "pushed off" the narrow fitness peak. Because robust individuals reside on flatter portions of the landscape, they are phenotypically buffered against mutational change. (Modified, with permission, from Wilke and Adami 2003 [© Elsevier].)

on balance—any mutation entering the population should be either neutral or deleterious, perhaps causing strong selection for genetic robustness that protects the phenotype against mutations. However, theory suggests that such selection may be weak or even absent (Wagner et al. 1997) due to opposing effects of stabilizing selection that increases the fitness effect of the mutations present in a population while simultaneously reducing the frequency of mutants in the equilibrium population. In contrast, because spontaneous mutations are believed typically to be deleterious, selection favoring evolution of robustness should be especially strong if mutation rates are elevated, even when populations are away from equilibrium (Wilke et al. 2001).

Some convincing data on evolution of robustness stem from studies looking at virtual organisms, namely, self-replicating computer programs that change randomly and thus have the potential to evolve. According to one such study (Wilke et al. 2001), elevated mutation rates can cause robust genotypes to be selectively favored over their brittle counterparts, even though robustness against mutations coincided with lower reproductive fitness. Thus, the fittest gave way to the "flattest" with selection favoring those variants having the greatest phenotypic constancy and residing on flat regions of the fitness landscape (Fig. 1) (Wilke et al. 2001). Other studies successfully examined robustness by following changes in proteins that are evolved in vitro (Bloom et al. 2007; Bloom and Arnold 2009).

Microbial populations are logical candidates for examining the evolution of robustness due to their rapid generation times and large population sizes. Similar to digital organism experiments, elevated mutation rates should be a key prerequisite for microbial populations to adapt by altering their genetic (mutational) robustness (Wilke et al. 2001). RNA viruses seem to be particularly appropriate for examining genetic robustness because their mutation rates generally exceed those of other organisms, including DNA viruses, by at least one order of magnitude (Drake et al. 1998).

DEMONSTRATING EVOLUTION OF ROBUSTNESS IN RNA VIRUSES

A handful of experimental evolution studies with RNA viruses have been used to show that genetic robustness can increase or decrease, depending on the particulars of the selective environment. In experiments with RNA viroids of plants and with vesicular stomatitis virus, an RNA virus that infects mammals and insect vectors, populations evolved in treatments causing elevated mutation rates were observed to undergo selection for increased genetic robustness (Codoñer et al. 2006; Sanjuán et al. 2007). These results echo the conclusions drawn from digital organism experiments where robustness evolves in response to mutation pressure (Wilke et al. 2001).

In our experiments with the RNA virus φ6, we examined a different selection pressure where evolved changes in robustness were expected (Montville et al. 2005; see also Turner and Chao 1998). We used a single genotype of the lytic RNA bacteriophage (phage) φ6 to found six populations. Three of these populations were allowed to adapt to growth on the bacterium *Pseudomonas syringae* pathovar *phaseolicola* at very low multiplicity of infection (moi; ratio of viruses to bacteria); when moi $\ll 1.0$, the vast majority of infection events are expected to be single phage particles infecting individual host cells to produce viral progeny (Turner et al. 1999). In parallel, the remaining three populations evolved at a high multiplicity (moi ≈ 5), where two to three phage particles on average should coinfect an individual host cell (Turner et al. 1999). This experimental evolution study lasted for 300 phage generations (60 days, five generations per day) (Fig. 2).

A key difference between the two experimental treatments was that high-multiplicity viruses could experience complementation, a mechanism that can effectively allow viruses to experience robustness against mutations (Froissart et al. 2004). In theory, adaptive robustness assumes that phenotypic expression results solely from the underlying genotype. However, this assumption should not hold true for viruses that can experience complementation, because virus genotypes typically have no ownership rights over the proteins encoded by their genomes. For example, during coinfection, a low-fitness viral genotype can phenotypically profit from intracellular proteins made by a coinfecting virus of higher fitness, causing the low-fitness virus to be overrepresented among the viral progeny exiting the cell (Froissart et al. 2004). Therefore, complementation during coinfection can automatically cause viral phenotypes to be beneficially buffered against mutations. By analogy, viral complementation has a role similar to that of gene duplication and diploidy, genetic mechanisms that may provide robustness in cellular or multicellular organisms.

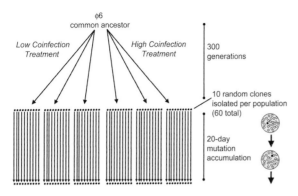

Figure 2. Design of an experiment where a wild-type phage φ6 ancestor founded three lineages evolved under a low level of coinfection, and three lineages evolved at high coinfection. After 60 days (300 generations), 10 clones were isolated from each population and used to found lineages subjected to mutation accumulation (MA). The MA lineages were then analyzed to see whether prior ecological history (low vs. high coinfection) affected robustness: maintenance of constant phenotype (fitness) despite mutational change. (Modified from McBride et al. 2008.)

Because complementation buffers mutational effects, it offers a built-in robustness mechanism for populations of phage φ6 that experience coinfection. By this logic, conditions fostering complementation (and hence, coinfection) may weaken selection for phage φ6 genomes to maintain their individual-level robustness, because coinfection provides mutational buffering. We therefore predicted that the degree of coinfection—high multiplicity versus low multiplicity—should influence evolution of robustness in phage φ6 populations (Montville et al. 2005). More specifically, we hypothesized that selection for robustness should be greatly relaxed in the phage φ6 populations that evolved for 300 generations under high levels of coinfection. If true, this would mean that the high-multiplicity populations should be dominated by φ6 genotypes that are relatively less robust to mutations.

To test the idea, we isolated 10 phage clones at random from each of the three low- and high-multiplicity populations (60 clones total). We then used each clone to found an independent population. These 60 populations were then subjected to a mutation accumulation experiment, where a population experiences an extreme daily bottleneck of a single virus particle (Fig. 2). To do so, we plated a dilution of each phage φ6 population on a host lawn and the next day chose a plaque at random for the next round of population growth. Thus, the population is propagated via daily plaque-to-plaque transfers imposing an extremely small population size, allowing the effects of genetic drift to be relatively more important than natural selection in dictating evolutionary change. This propagation scheme enables a phage population to accumulate nonlethal mutations at random (Chao 1990; Burch and Chao 2004). Because random mutations are expected to be deleterious on average, mean fitness of a virus population is expected to decline through time as random mutations accumulate.

The 60 phage populations were subjected to 20 consecutive days of extreme bottlenecking, providing the opportunity for ~1.3 mutations to fix in each population, based on an estimated rate of 0.067 mutations per generation in phage φ6 (Burch and Chao 2004). To examine how the amassed mutations affected phenotypic fitness (W, relative growth rate on the host bacteria), we measured $\log_{10}W$ of each prebottleneck and postbottleneck population. The difference between these two values, $\Delta\log_{10}W$, revealed the sensitivity of the population to phenotypic effects of the accumulated mutation(s) (Fig. 3). Support for the hypothesis would be that the 30 populations initiated by clones that were historically evolved under high multiplicity (frequent coinfection) would show greater variance in $\Delta\log_{10}W$ values, owing to weakened selection for them to maintain robustness as an individual trait. Additional support would come from a greater mean magnitude of $\Delta\log_{10}W$ values for the high-multiplicity populations, indicating that they suffered a greater drop in fitness on average, because they are less able to withstand the deleterious effects of mutations. These hypothesized outcomes are depicted in Figure 3.

The data supported the general predictions (Fig. 4), confirming that selection to maintain mutational robustness is weaker with viral coinfections—viruses that were historically evolved under low-coinfection conditions were measurably advantaged in robustness relative to their high-coinfection-evolved counterparts, i.e., results demonstrated that viruses evolved under high moi showed relatively greater mean magnitude and variance in the fitness changes generated by addition of random mutations (Fig. 4). From these results, we concluded that the low-coinfection populations could henceforth be defined as relatively genetically robust, whereas the high-coinfection populations were considered to be relatively brittle.

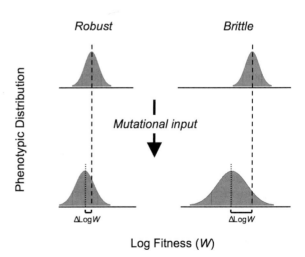

Figure 3. Hypothetical diagram showing how relative robustness of organisms may be revealed through measurements of fitness change following mutational input. Because the population on the right presents greater variance in fitness and lower mean fitness following mutational change, it can be defined as relatively less robust (i.e., brittle).

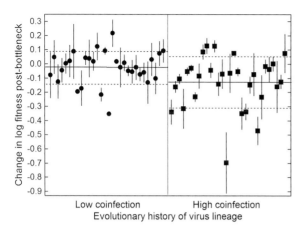

Figure 4. Phage ϕ6 strains that evolved under low coinfection are more robust than those evolved under high coinfection, owing to their lower variance in fitness and higher mean fitness following mutation accumulation. Each point is the mean change in $\log_{10}$ fitness resulting from mutation accumulation, for an independent lineage founded by a virus clone previously evolved under a low level of coinfection (closed circles) or a high level of coinfection (closed squares). (Horizontal lines) Grand means among lineages within a treatment, (dashed lines) one standard deviation away from the mean. (Modified from Montville et al. 2005.)

So far, the exact molecular mechanism responsible for robustness in phage ϕ6 has not been determined. But some clues regarding the mechanism stem from additional sets of experiments examining the relationship between genetic robustness and evolvability and between genetic robustness and generalized environmental robustness.

HOW DOES ROBUSTNESS IN THE CURRENT ENVIRONMENT IMPACT ADAPTATION SHOULD THE ENVIRONMENT CHANGE?

On the one hand, constancy in the face of environmental and mutational changes provides obvious benefits to an organism during growth and reproduction. Robustness buffers organisms against such perturbations, affording constancy in terms of cellular function, development, and offspring production, i.e., robustness provides reliability in the very currencies by which natural selection judges phenotypes. On the other hand, rigidity in the face of change may pose problems. For example, if organisms are steadfast under environmental change, how can they possibly adapt to new conditions? Because natural selection acts on phenotypic variation, robustness that buffers this variation could impede evolution.

These conflicting necessities force organisms to strike a balance between withstanding some changes and maintaining an ability to adapt to new circumstances. This compromise is the balance between robustness and evolvability, the capacity to adapt. It is conceivable that some variants are selectively favored over others in striking this balance, such that these favored genotypes are more likely to give rise to progeny that dominate the population should a novel environmental challenge be encountered. Thus, by examining this balancing act, we may learn whether evolvability can itself evolve. Furthermore, we can explore the intriguing—and contentious—idea that natural selection shapes evolution itself (Wagner and Altenberg 1996; Kirschner and Gerhart 1998; Wagner 2005)

Where does one begin in addressing issues of such high stakes in evolutionary biology? Some prior theoretical and empirical studies in biology offer clues; evolvability has been explored in disciplines ranging from protein biophysics to yeast genetics (Bloom et al. 2004, 2006; Tanay et al. 2005; Tokuriki and Tawfik 2009). These varying contexts often share a common theme: Prior evolution of protein-folding stability seems to be beneficial for future evolutionary innovation. When faced with new environmental challenges, robust proteins (those that maintain their folding stability despite underlying genetic changes) seem to be better capable of experiencing mutations that permit innovation without compromising proper folding. Thus, populations composed of individuals with relatively robust proteins are expected to have an adaptive advantage in novel environments.

Although many environmental stressors can harm organisms by negatively impacting protein folding, heat shock is an especially direct proxy for protein stability, as several studies in protein thermostability have contributed to our broader understanding of protein structure and stability (Kumar et al. 2001; Razvi and Scholtz 2006). This notion is supported by the observation that microbial species thriving in extremely high-temperature environments typically possess proteins with unusually stable folding patterns that facilitate growth and survival under thermodynamic stress at elevated temperatures (Huber and Stetter 2001; Sterner and Liebl 2001; England et al. 2003). The widespread existence of heat shock protein chaperones in biological systems offers further evidence that protein stability under heat shock is a target of natural selection (e.g., in yeast species) (Craig et al. 1993; Parsell et al. 1993) .

For these reasons, we can infer that phage ϕ6 genotypes which differ in genetic robustness may also differ in survival under heat shock. In addition, evolution of the ability to tolerate survival under high temperatures provides a logical choice for interrogating whether relatively robust genotypes of the virus are advantaged when founding populations that need to adapt under novel conditions. Below, we summarize how we used phage ϕ6 as a biological system to examine the potential positive relationship between genetic robustness and evolvability.

EXAMINING THE LINK BETWEEN GENETIC ROBUSTNESS AND EVOLVABILITY IN RNA VIRUS ϕ6

We have used experiments with phage ϕ6 to address whether robustness promotes or hinders evolvability (McBride et al. 2008; see also McBride and Turner 2008). In the wild, phage ϕ6 and other members of the *Cystoviridae* virus family seem to specifically attack plant pathogenic *Pseudomonas* species that colonize plant surfaces (Silander et al. 2005). Thus, phage ϕ6 is typically cultured in the laboratory at the benign temperature of 25°C. We observed that when bacteria-free lysates of wild-type ϕ6 were exposed to 5-min heat shock at temperatures above 40°C, increasingly fewer virus particles remained viable—able to successfully infect the typical host *P. syringae* pathovar *phaseolicola* (McBride et al. 2008). At heat shock of 45°C, ~80% of wild-type virus particles became nonviable. Presumably, the 45°C heat shock severely damages viral proteins, such as the P3 attachment protein of the virus that is needed for cell attachment during the early stages of infection.

From the collection of 60 virus clones that were isolated from populations which experienced 300 generations of viral evolution under high versus low moi (Montville et al. 2005), we chose a subset of 24 clones to study the relationship between robustness and evolvability. These 24 clones (12 robust, 12 brittle) were first examined for their survival under 45°C heat shock. Results showed that the two groups of viruses were similarly sensitive in terms of average percent survival (%S), with mean survival for both robust and brittle genotypes of only 14% (McBride et al. 2008). Because both groups of φ6 viruses were earlier subjected to 300 generations of evolution with daily incubation at 25°C, we did not expect the robust and brittle clones to differ in their sensitivity to high temperature. However, this information confirmed that we could conduct an evolution experiment examining whether robustness enhanced or suppressed evolvability under heat shock conditions, without biasing in favor of one group of the viruses.

Our preliminary experiments also addressed other potential confounders. In particular, we examined whether the two groups of virus populations (robust and brittle) differed in mutation rate. Regardless of robustness differences, a higher mutation rate could foster more rapid adaptation in a new environment because a population can more easily access rare beneficial mutations. Such a difference could then lead to spurious conclusions regarding the link between robustness and evolvability in a new habitat. However, two separate methods for gauging spontaneous frequencies of viral mutants did not suggest that robust and brittle viruses consistently differed in mutation rate (McBride et al. 2008). A second potential confounder was whether one set of viruses was better fit under low-moi conditions, because we intended to gauge evolvability in the absence of coinfection. The robust and brittle viruses had historically evolved under low- versus high-moi conditions, respectively (Turner and Chao 1998). But our assays showed that after 300 generations of virus evolution, the fitness of robust and brittle clones was equivalent at low moi (McBride et al. 2008), even though performance trade-offs across these moi environments were evident earlier in the evolutionary process (Turner and Chao 1999). Because these trade-offs had eroded by the time of our evolvability study, we deemed that any observed differences in the rate of adaptation to 45°C heat shock would be due to differences in intrinsic evolvability of the viruses.

To conduct the experimental evolution under heat shock, we used each of the 12 robust and 12 brittle clones to found an independent test population. These 24 test populations were allowed to undergo 50 generations (10 days) of evolution, where each population experienced growth on *P. phaseolicola* at 25°C, with periodic (every fifth generation) exposure to 45°C heat shock (Fig. 5, top). At the end of the study, we measured mean %S at 45°C for each founding clone and its derived endpoint population. These replicated measures were used to estimate mean Δ%S, the average change in percent survival after 50 generations of selection to resist damaging heat shock. The results showed that the populations founded by robust genotypes were more evolvable (Fig. 5, bottom). From these data, we concluded that robustness promoted evolv-

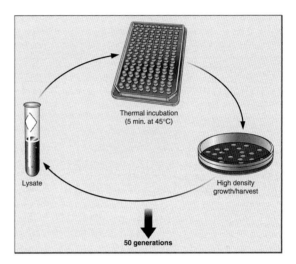

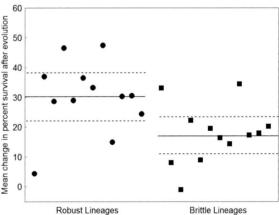

Figure 5. (*Top*) Design for an evolution experiment where phage φ6 lineages were selected to withstand damaging effects of heat shock. A phage lysate was exposed to 45°C incubation for 5 min, and a dilution of the surviving progeny was plated on a lawn of *P. phaseolicola* bacteria. Overnight plaque formation at 25°C corresponded to five generations of phage evolution. The plaques were then harvested, and the process was repeated for 10 days, equivalent to 50 phage generations. (*Bottom*) Mean change in %S after selection with heat shock (45°C) is greater for virus lineages founded by 12 robust strains (closed circles), relative to lineages initiated by 12 brittle strains (closed squares). All populations were subjected to 10 days (50 generations) of periodic heat shocks. (Solid line) Grand mean for the group, (dashed lines) 95% confidence intervals. (*Top*, Modified, with permission, from McBride and Turner 2008 [© ASM]; *Bottom*, modified from McBride et al. 2008.)

ability in phage φ6, at least when viruses underwent adaptation to resist the damaging effects of survival at high temperature. Below, we present new data relating to a potential mechanism for robustness in phage φ6.

NEW DATA ON A POSSIBLE UNDERLYING MECHANISM FOR VIRUS ROBUSTNESS

Data from in vitro protein studies suggest that less-sensitive (relatively robust) proteins are more likely to maintain their function in a new environment where innovation is needed (Bloom et al. 2006; Ortlund et al. 2007). A similar mechanism may explain differences in robustness among phage φ6 strains and may account for the link between robustness and evolvability of thermotolerance (McBride et al. 2008), i.e., robust strains may contain proteins that tend to tolerate mutations while maintaining proper folding. Under heat shock selection, evolution of greater protein thermostability would be advantageous. We infer that robust viruses featured one or more proteins that were better able to maintain proper folding while allowing the input of spontaneous mutation(s), which led to exploration of novel thermostable genotypes. This beneficial combination would explain the increased ability of the robust viruses to adapt under heat stress. In contrast, the brittle viruses were perhaps constrained in their ability to adapt because their proteins encountered mutations that increased thermostability while compromising the ability to produce viral progeny efficiently. Better thermotolerance of the robust viruses would support existing arguments for the connection between robustness and thermostability (Bornberg-Bauer 1997; Bornberg-Bauer and Chan 1999).

We performed additional experiments to further examine the protein stability mechanism that links phage φ6 robustness to evolvability. Empirical support for the mechanism would be increased survivability of robust strains at less-extreme temperatures than the 45°C selective environment, where survival of robust and brittle viruses was shown to be equally poor (McBride et al. 2008). Thus, we hypothesized that robust viruses should show better average survival under moderate-temperature heat shock, relative to brittle viruses. To test this idea, we compared survival of 12 phage clones (six robust, six brittle) that constitute a subset of the strains used to found populations in our evolvability experiment. In particular, we chose six phage clones from the group of viruses historically evolved under low coinfection, previously described as relatively robust and better evolvable under 45°C heat shock: L1.7, L1.9, L2.4, L2.6, L3.2, and L3.8. The remaining six phage clones were from the group evolved under high coinfection, described as comparatively brittle and less adaptable to heat shock: H1.1, H1.3, H2.10, H2.4, H3.5, and H3.9.

Each of the 12 phage strains was used to obtain a high-titer lysate, which was diluted to a concentration of ~2.5 × 10^3 plaque-forming units (pfu)/ml. A 30-µL volume from each lysate (~750 pfu) was placed in a sterile PCR (polymerase chain reaction) tube and immediately sampled onto a lawn of the *P. phaseolicola* host; plaques that formed during overnight incubation were used to accurately estimate initial virus titer (N_i). The PCR tube was then incubated for 5 min in an Eppendorf thermocycler, preheated to an assay temperature of 25°C, 37.5°C, 40°C, 42.5°C, or 45°C. After this temperature shock, 10 µL of the test lysate was sampled onto a host lawn of bacteria to estimate final virus titer (N_f). %S was calculated using the formula %S = (N_f/N_i) × 100. Three independent replicate measurements of %S were obtained for each virus clone at each of the five assay temperatures, and we calculated the mean %S for each clone at a given temperature. Grand mean %S values for the two groups of viruses (robust, brittle) at a given temperature were obtained by averaging the clone means within each group.

The results showed that mean survival of robust and brittle viruses was not statistically different at the standard growth temperature of 25°C (*t*-test with $t = 0.067$, $df = 10$, $P = 0.947$). Figure 6 plots the reaction norm for the grand mean %S of robust viruses and of brittle viruses, across the four elevated heat shock environments (37.5°C, 40°C, 42.5°C, 45°C). Visual inspection suggests that the reaction norm of the brittle viruses is linear, whereas that of the robust viruses is curvilinear, concave downward. A statistical analysis confirmed that the data for the brittle viruses were best described by a simple linear regression ($P = 0.007$). In contrast, a quadratic regression ($P = 0.071$) provided a better fit to the data for the robust viruses than did a linear regression ($P = 0.083$). However, we noted that both of these fits to the data for the robust viruses were marginally significant, indicating variation in %S among the robust strains (data not shown). We also conducted a Kaplan–Meier survival estimator to analyze differences in the survival functions of the two virus groups. A log-rank comparison test showed that the robust viruses had a survival function that significantly differed from that of the brittle viruses ($\alpha = 0.05$; $P = 0.015$). From these analyses, we concluded that the robust viruses showed better average survival at moderately warm temperatures. This outcome is consistent with

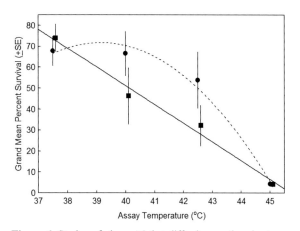

Figure 6. Strains of phage φ6 that differ in genetic robustness also differ in survival at moderately high temperatures. Brittle (closed circles, dashed line) and robust (closed squares, solid line) viruses were exposed to heat shock at a range of temperatures, and the two groups of viruses differed in survival as a function of temperature. Data points are offset for clarity.

the protein stability mechanism as a possible explanation for the robustness and evolvability advantage of viruses evolved under low-moi conditions.

We sought to test whether the observed survival difference between robust and brittle viruses under moderately elevated temperatures carried over to a separate environmental perturbation known to cause virus damage and degradation. If so, the results would suggest that the genetically robust viruses are generally advantaged in terms of environmental robustness as well. Some theory predicts that genetic robustness and generalized environmental robustness should coincide, whereas other theory suggests no such relationship should be expected (de Visser et al. 2003). If equivalent survival of robust and brittle viruses was observed in an environmental perturbation other than thermal stress, it would further suggest that protein stability under heat shock may be the causal mechanism explaining robustness differences in phage ɸ6.

We chose survival under ultraviolet (UV) radiation as an additional challenge environment to examine relative survival of genetically robust and brittle phage strains. UV radiation can be a mutagen that increases the number of mutational errors occurring during DNA or RNA replication. However, our preliminary results (data not shown) found no difference in the frequency of reversion mutants in experiments where genetically marked phages were and were not exposed to UV-C (254-nm wavelength) radiation. Consistent with this observation, prior work shows that UV alone is not highly mutagenic to phage ɸ6, but that UV is effective at causing damage to ɸ6 virions presumably due to protein degradation (Lytle et al. 1993)

Using the same 12 phage strains (six robust, six brittle) for which we measured thermal reaction norms, we tested relative ability for viruses to survive exposure to UV-C (254 nm) radiation. Each strain was grown to a high-titer lysate as described above. Then, 200-µL aliquots containing a known initial quantity of virus (N_i) were randomly distributed among wells of a flat-bottomed polystyrene 96-well plate. A UV illuminator (Spectroline Long Life Filter) was placed over the 96-well plate, exposing the virus samples to UV-C emission. Plates were exposed to UV radiation for durations of 5, 7.5, 10, or 15 min. Following exposure, viruses in the individual wells were titered using standard plating procedures to estimate the post-UV exposure number of surviving virus particles (N_f). %S was calculated similarly to the temperature reaction-norm experiments ($[N_f/N_i] \times 100$). Between three and six replicate measurements were obtained for each phage strain at each of the UV-exposure durations.

The results showed that the robust and brittle viruses were similarly damaged by UV-C radiation, presenting "death curves" characterized by decreasing grand mean survival following longer durations of UV exposure (Fig. 7). Although visual inspection suggested that grand mean %S was saturating with increasing UV exposure, for both groups of viruses, a simple linear regression provided an adequate fit to the data that was not improved through a quadratic model (robust strains: $P < 0.001$; brittle strains: $P < 0.001$). The observations (Fig. 7) hinted that robust strains may be advantaged in relative survival under inter-

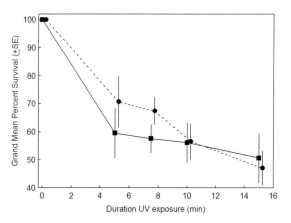

Figure 7. Strains of phage ɸ6 that differ in genetic robustness do not differ in survival following varying dosages of UV-C radiation. Robust (closed circles, dashed line) and brittle (closed squares, solid line) viruses were exposed to UV radiation for different durations, and the two groups of viruses showed equivalent survival as a function of UV dosage. Data points are offset for clarity.

mediate durations of UV exposure (7.5, 10 min), echoing their greater survival at intermediate elevated temperatures. However, ANOVA showed that the interactive effect of treatment history (low- versus high-moi evolution) and duration of UV exposure on the dependent variable %S was not statistically significant ($P = 0.686$). Furthermore, a Kaplan–Meier estimator with a log-rank test did not demonstrate a significant difference in survival function for the two groups of viruses over the range of UV dosages examined ($P = 0.674$). Thus, our results did not suggest that genetically robust strains were relatively advantaged in survival under UV radiation, compared with their brittle counterparts.

Overall, the data hint that survival differences under moderate temperatures (but not under UV radiation) point to protein thermostability as a mechanism explaining relative differences in robustness among phage ɸ6 genotypes. Although survival of robust and brittle genotypes is equally poor at the elevated 45°C temperature, the general thermostability advantage of robust strains might have allowed lineages founded by these genotypes to more easily adapt when survival at 45°C was experienced as a selection pressure (McBride et al. 2008). These observations are consistent with the idea that robust organisms can accumulate genetic variation that is neutral in the current environment but which can be used for adaptive innovation when the environment changes. Thus, our work suggests that robust strains of the virus may be preadapted to better survive temperatures approaching (but not including) 45°C, provided survival is predicated through adaptive change via natural selection rather than simply immediate survival upon entering the novel environment. We have cautioned that this observed adaptive advantage of robust genotypes may not generally carry over to other environmental challenges, such as viral adaptation to a novel host type that is important in virus emergence (McBride et al. 2008). Such generalized evolvability questions may be addressed using phage ɸ6 or other experimental systems where relative differences in robustness among genotypes have been observed.

CONCLUDING REMARKS

Our studies with phage φ6 provided a unique collection of virus genotypes that were shown to differ in their evolved genetic robustness (Montville et al. 2005). With this known difference in hand, we were able to capitalize on the collection of viruses to test whether a positive relationship exists between robustness and evolvability, as suggested by theory but rarely examined experimentally (McBride et al. 2008). This combination of rapidly evolving study system, observed differences in robustness, and existing theory on robustness and evolvability was admittedly fortuitous. Undoubtedly, these circumstances aided our ability to harness a biological system to test a hypothesis that predicts the course of future evolutionary events.

Traits other than genetic robustness, however, have been suggested to influence evolvability. One example is the presence of tandem repeats in promoter sequences of the yeast *Saccharomyces cerevisiae*, which are associated with higher rates of transcriptional divergence that indicate increased evolvability (Vinces et al. 2009). Other traits that have been associated with evolvability include variation in endocrine signaling pathways in social insects (Amdam et al. 2007) and protein dynamism (Tokuriki and Tawfik 2009). In addition, the molecular chaperone Hsp90 has been identified as a possible regulator of evolvability (Rutherford and Lindquist 1998) as has the yeast prion PSI (True and Lindquist 2000). These various biological systems could be harnessed to examine the proposed link between robustness and evolvability, in ways similar to our approach using phage φ6.

In addition, avoidance of extinction is a concept that relates to the propensity to adapt. Some studies have examined whether lineages or groups of organisms tend to avoid extinction better than others. A well-known result used fossil evidence to conclude that increased geographic range (i.e., niche breadth) decreased extinction probability in species of marine bivalves and gastropods (Jablonski 1986). This study supports the idea that greater habitat use should increase the likelihood that a species locates a habitat suitable for survival. One intriguing possibility for future research is to examine whether niche breadth in organisms such as pathogenic microbes similarly promotes avoidance of extinction. Resource generalization is often assumed to be costly, fostering arguments that natural species biodiversity can be explained by a tendency for niche specialists to be favored over niche generalists (Finke and Snyder 2008). However, many experimental evolution studies in microbes such as RNA viruses have shown that genotypes that evolve to use multiple hosts are not necessarily fitness disadvantaged relative to their more specialized counterparts (Novella et al. 1999; Turner and Elena 2000; Remold et al. 2008). These collections of microbes could be harnessed to conduct studies examining whether an evolved broad host range is superior to a narrow host range, when avoidance of extinction is the selective challenge.

Would the full elucidation of mechanisms for evolvability serve as a "Rosetta stone" for predictive evolution? Perhaps—but the very real challenge is to identify the myriad ways that an evolvability advantage might be obtained by biological systems. As suggested in studies exploring robustness and epistasis (Elena et al. 2006; Sanjuán and Elena 2006), it is quite possible that evolvability is governed by principles that are highly contextual and organism specific. That said, our proven ability to identify evolved robustness as a distinguishing trait in phage φ6, and to successfully use this knowledge to connect the trait to predictions regarding evolvability, reveals that such efforts are far from hopeless and that further experiments are warranted. If evolution can be truly transformed into a predictive science with a falsifiable set of governing principles, the implications would be enormous for basic research as well as biological applications. For example, public health would benefit by our increased ability to harness evolutionary predictions to prepare against emerging pathogens, and pharmacologists could be better informed in designing therapies that are resilient against evolutionary escape by targeted infectious agents. Overall, evolutionary biology could experience its next great paradigm shift, where the science becomes transformed from a discipline mostly concerned with elucidating the past to one that also boldly makes predictions about biological events that have yet to transpire.

ACKNOWLEDGMENTS

We thank Jeremy Draghi, James Pease, and Sonia Singhal for assistance with some of the experiments. This work was partially funded by a graduate dissertation fellowship to C.B.O. from UNCF-Merck, by National Institutes of Health training grant T32GM07205, and by grant DEB-0452163 to P.E.T. from the National Science Foundation.

REFERENCES

Amdam GV, Nilsen KA, Norberg K, Fondrk MK, Hartfelder K. 2007. Variation in endocrine signaling underlies variation in social life history. *Am Nat* **170:** 37–46.

Bloom JD, Arnold FH. 2009. In the light of directed evolution: Pathways of adaptive protein evolution. *Proc Natl Acad Sci* (suppl. 1) **106:** 9995–10000.

Bloom JD, Wilke CO, Arnold FH, Adami C. 2004. Stability and the evolvability of function in a model protein. *Biophys J* **86:** 2758–2764.

Bloom JD, Labthavikul ST, Otey CR, Arnold FH. 2006. Protein stability promotes evolvability. *Proc Natl Acad Sci* **103:** 5869–5874.

Bloom JD, Lu Z, Chen D, Raval A, Venturelli OS, Arnold FH. 2007. Evolution favors protein mutational robustness in sufficiently large populations. *BMC Biol* **5:** 29.

Brookfield JF. 2009. Evolution and evolvability: Celebrating Darwin 200. *Biol Lett* **5:** 44–46.

Bornberg-Bauer E. 1997. How are model protein structures distributed in sequence space? *Biophys J* **73:** 2393–2403.

Bornberg-Bauer E, Chan HS. 1999. Modeling evolutionary landscapes: Mutational stability, topology, and superfunnels in sequence space. *Proc Natl Acad Sci* **96:** 10689–10694.

Burch CL, Chao L. 2004. Epistasis and its relationship to canalization in the RNA virus φ6. *Genetics* **167:** 559–567.

Bush RM, Bender CA, Subbarao K, Cox NJ, Fitch WM. 1999.

Predicting the evolution of human influenza A. *Science* **286**: 1921–1925.

Chao L. 1990. Fitness of RNA virus decreased by Muller's ratchet. *Nature* **348**: 454–455.

Coburn BJ, Wagner BG, Blower S. 2009. Modeling influenza epidemics and pandemics: Insights into the future of swine flu (H1N1). *BMC Med* **7**: 30.

Codoñer FM, Darós JA, Solé RV, Elena SF. 2006. The fittest versus the flattest: Experimental confirmation of the quasispecies effect with subviral pathogens. *PLoS Pathog* **2**:e136.

Craig EA, Gambill BD, Nelson RJ. 1993. Heat shock proteins: Molecular chaperones of protein biogenesis. *Microbiol Rev* **57**: 402–414.

Cressman R, Krivan V, Garay J. 2004. Ideal free distributions, evolutionary games, and population dynamics in multiple-species environments. *Am Nat* **164**: 473–489.

Darwin C. 1859. *On the origin of species by means of natural selection, or the preservation of favoured races in the struggle for life*. Murray, London.

de Visser JA, Hermisson J, Wagner GP, Ancel Meyers L, Bagheri-Chaichian H, Blanchard JL, Chao L, Cheverud JM, Elena SF, Fontana W, et al. 2003. Perspective: Evolution and detection of genetic robustness. *Evolution* **57**: 1959–1972.

Drake JW, Charlesworth B, Charlesworth D, Crow JF. 1998. Rates of spontaneous mutation. *Genetics* **148**: 1667–1686.

Elena SF, Carrasco P, Darós JA, Sanjuán R. 2006. Mechanisms of genetic robustness in RNA viruses. *EMBO Rep* **7**: 168–173.

England JL, Shakhnovich BE, Shakhnovich EI. 2003. Natural selection of more designable folds: A mechanism for thermophilic adaptation. *Proc Natl Acad Sci* **100**: 8727–8731.

Finke DL, Snyder WE. 2008. Niche partitioning increases resource exploitation by diverse communities. *Science* **321**: 1488–1490.

Froissart R, Wilke CO, Montville R, Remold SK, Chao L, Turner PE. 2004. Co-infection weakens selection against epistatic mutations in RNA viruses. *Genetics* **168**: 9–19.

Huber R, Stetter KO. 2001. Discovery of hyperthermophilic microorganisms. *Methods Enzymol* **330**: 11–24.

Jablonski D. 1986. Background and mass extinctions: The alternation of macroevolutionary regimes. *Science* **231**: 129–133.

Kirschner M, Gerhart J. 1998. Evolvability. *Proc Natl Acad Sci* **95**: 8420–8427.

Koelle K, Cobey S, Grenfell B, Pascual M. 2006. Epochal evolution shapes the phylodynamics of interpandemic influenza A (H3N2) in humans. *Science* **314**: 1898–1903.

Kumar S, Tsai CJ, Nussinov R. 2001. Thermodynamic differences among homologous thermophilic and mesophilic proteins. *Biochemistry* **40**: 14152–14165.

Lenski RE, Travisano M. 1994. Dynamics of adaptation and diversification: A 10,000-generation experiment with bacterial populations. *Proc Natl Acad Sci* **91**: 6808–6814.

Lenski RE, Ofria C, Pennock RT, Adami C. 2003. The evolutionary origin of complex features. *Nature* **423**: 139–144.

Lytle CD, Wagner SJ, Prodouz NJ. 1993. Antiviral activity of gilvocarcin V plus UVA radiation. *Photochem Photobiol* **58**: 818–821.

Maynard Smith J. 1982. *Evolution and the theory of games*. Cambridge University Press, Cambridge.

McBride RC, Turner PE. 2008. Genetic robustness and adaptability of viruses. *Microbe* **3**: 409–415.

McBride RC, Ogbunugafor CB, Turner PE. 2008. Robustness promotes evolvability of thermotolerance in an RNA virus. *BMC Evol Biol* **8**: 231.

Montville R, Froissart R, Remold SK, Tenaillon O, Turner PE. 2005. Evolution of mutational robustness in an RNA virus. *PLoS Biol* **3**: 1939–1945.

Novella IS, Hershey CL, Escarmis C, Domingo E, Holland JJ. 1999. Lack of evolutionary stasis during alternating replication of an arbovirus in insect and mammalian cells. *J Mol Biol* **287**: 459–465.

Ortlund EA, Bridgham JT, Redinbo MR, Thornton RW. 2007. Crystal structure of an ancient protein: Evolution by conformational epistasis. *Science* **317**: 1544–1548.

Parsell DA, Taulien J, Lindquist S. 1993. The role of heat-shock proteins in thermotolerance. *Philos Trans R Soc Lond B Biol Sci* **339**: 279–285

Razvi A, Scholtz JM. 2006. Lessons in stability from thermophilic proteins. *Protein Sci* **15**: 1569–1578.

Remold SK, Rambaut A, Turner PE. 2008. Evolutionary genomics of host adaptation in vesicular stomatitis virus. *Mol Biol Evol* **25**: 1138–1147.

Rutherford SL, Lindquist S. 1998. Hsp90 as a capacitor for morphological evolution. *Nature* **396**: 336–342

Sanjuán R, Elena SF. 2006. Epistasis correlates to genomic complexity. *Proc Natl Acad Sci* **103**: 14402–14405.

Sanjuán R, Cuevas JM, Furió V, Holmes EC, Moya A. 2007. Selection for robustness in mutagenized RNA viruses. *PLoS Genet* **3**: 939–946.

Silander OK, Weinreich DM, Wright KM, O'Keefe KJ, Rang CU, Turner PE, Chao L. 2005. Widespread genetic exchange among terrestrial bacteriophages. *Proc Natl Acad Sci* **102**: 19009–19014.

Sterner R, Liebl W. 2001. Thermophilic adaptation of proteins. *Crit Rev Biochem Mol Biol* **36**: 39–106.

Tanay A, Regev A, Shamir R. 2005. Conservation and evolvability in regulatory networks: The evolution of ribosomal regulation in yeast. *Proc Natl Acad Sci* **102**: 7203–7208.

Thornton JW, Need E, Crews D. 2003. Resurrecting the ancestral steroid receptor: Ancient origin of estrogen signaling. *Science* **301**: 1714–1717.

Tokuriki N, Tawfik DS. 2009. Protein dynamism and evolvability. *Science* **324**: 203–207.

True HL, Lindquist SL. 2000. A yeast prion provides a mechanism for genetic variation and phenotypic diversity. *Nature* **407**: 477–483.

Turner PE, Chao L. 1998. Sex and the evolution of intrahost competition in RNA virus φ6. *Genetics* **150**: 523–532.

Turner PE, Chao L. 1999. Prisoner's dilemma in an RNA virus. *Nature* **398**: 441–443.

Turner PE, Chao L. 2003. Escape from prisoner's dilemma in RNA phage φ6. *Am Nat* **161**: 497–505.

Turner PE, Elena SF. 2000. Cost of host radiation in an RNA virus. *Genetics* **156**: 1465–1470.

Turner PE, Burch CL, Hanley KA, Chao L. 1999. Hybrid frequencies confirm limit to coinfection in the RNA bacteriophage φ6. *J Virol* **73**: 2420–2424.

Vinces MD, Legendre M, Caldara M, Hagihara M, Verstrepen KJ. 2009. Unstable tandem repeats in promoters confer transcriptional evolvability. *Science* **324**: 1213–1216.

Vrba ES, DeGusta D. 2004. Do species populations really start small? New perspectives from the Late Neogene fossil record of African mammals. *Philos Trans R Soc Lond B Biol Sci* **359**: 285–293.

Wagner A. 2005. Robustness, evolvability, and neutrality. *FEBS Lett* **579**: 1772–1778.

Wagner A. 2006. Cooperation is fleeting in the world of transposable elements. *PLoS Comput Biol* **2**: e162.

Wagner GP, Altenberg L. 1996. Complex adaptations and the evolution of evolvability. *Evolution* **50**: 967–976.

Wagner GP, Booth G, Bagheri-Chaichian H. 1997. A population genetic theory of canalization. *Evolution* **51**: 329–347.

Weinreich DM, Delaney NF, DePristo MA, Hartl DL. 2006. Darwinian evolution can follow only very few mutational paths to fitter proteins. *Science* **312**: 111–114.

Wilke CO, Adami C. 2003. Evolution of mutational robustness. *Mutat Res* **522**: 3–11.

Wilke CO, Wang JL, Ofria C, Lenski RE, Adami C. 2001. Evolution of digital organisms at high mutation rates leads to survival of the flattest. *Nature* **412**: 331–333.

Genome-wide Mutational Diversity in an Evolving Population of *Escherichia coli*

J.E. BARRICK AND R.E. LENSKI

Department of Microbiology and Molecular Genetics, Michigan State University, East Lansing, Michigan 48824
Correspondence: lenski@msu.edu

The level of genetic variation in a population is the result of a dynamic tension between evolutionary forces. Mutations create variation, certain frequency-dependent interactions may preserve diversity, and natural selection purges variation. New sequencing technologies offer unprecedented opportunities to discover and characterize the diversity present in evolving microbial populations on a whole-genome scale. By sequencing mixed-population samples, we have identified single-nucleotide polymorphisms (SNPs) present at various points in the history of an *Escherichia coli* population that has evolved for almost 20 years from a founding clone. With 50-fold genome coverage, we were able to catch beneficial mutations as they swept to fixation, discover contending beneficial alleles that were eliminated by clonal interference, and detect other minor variants possibly adapted to a new ecological niche. Additionally, there was a dramatic increase in genetic diversity late in the experiment after a mutator phenotype evolved. Still finer-resolution details of the structure of genetic variation and how it changes over time in microbial evolution experiments will enable new applications and quantitative tests of population genetic theory.

Several next-generation platforms capable of sequencing more than 1 billion DNA bases in a single run have recently become commercially available (Mardis 2008), and more are under development (Gupta 2008). The compact genomes of microorganisms put them at the forefront of efforts to open new windows on the study of genetic diversity and evolution using the massive throughput of these technologies. Metagenomic surveys that profile the species abundance and metabolic composition of microbial communities by sampling environmental DNA (Vieites et al. 2009) can also be used to infer some population genetic parameters in nature (Johnson and Slatkin 2006). Population genomic approaches have begun to fill in our knowledge of sequence diversity among isolates of a single species and between closely related species. For example, one study characterized the patterns of genome-wide mutational variation in yeasts and reconstructed details of their life history since domestication from next-generation sequencing data (Liti et al. 2009). Ultradeep sequencing of genes from viruses such as human immunodeficiency virus (HIV) has even begun to reveal patterns of within-host diversity during infections, including subpopulations with drug resistance mutations (Wang et al. 2007; Eriksson et al. 2008).

We are interested in how these new technologies can be used to better understand evolutionary processes and advance population genetic theory in the context of experiments with microorganisms (Elena and Lenski 2003; Kassen and Rainey 2004; Buckling et al. 2009). Evolution experiments have the advantage over studies of natural and clinical populations in that they take place under laboratory conditions where environmental conditions and sampling regimens are rigorously controlled. To date, whole-genome resequencing has mainly been used to find the beneficial mutations in "winning" clones isolated at the end of bacterial evolution experiments. For example, next-generation sequencing platforms were used to identify a single mutation responsible for the reacquisition of social swarming in a population of *Myxococcus xanthus* that previously lost that capacity (Fiegna et al. 2006) and several mutations that improve the growth of *E. coli* in a glycerol-based medium after 44 days of continuous culture (Herring et al. 2006). Next-generation platforms have not yet been used to examine the genetic variation present in these populations and how that diversity changes over time.

The frozen "fossil record" of a long-term experiment with *E. coli* spanning almost 20 years and 40,000 generations of evolution provides a unique opportunity for exploring these issues (Lenski and Travisano 1994). In this experiment, 12 populations of *E. coli* were founded from the same ancestral strain and maintained by the daily transfer of 1% of each culture into fresh glucose minimal media. After years of intensive study, a great deal is known about the fitness trajectories, phenotypic changes, and beneficial mutations that occur in this environment (Lenski 2004; Philippe et al. 2007). We have recently resequenced *E. coli* clones isolated at different time points from one of these populations to examine the coupling between the rates of genomic evolution and adaptation to the environment (JE Barrick et al. 2009). We found that fitness increased dramatically during the first 20,000 generations of the experiment but that new mutations accumulated at a near-constant rate during this time. This result was surprising because a clock-like accumulation of mutations is usually taken as a signature of neutral evolution, yet several lines of evidence indicate that most of these mutations are beneficial.

These clone genomes offer only a fragmentary picture of the history of this *E. coli* population. Knowledge of the details of the ebb and flow of genetic diversity could potentially reveal how the population's adaptive trajectory was influenced by selective sweeps driving mutations to fixation, beneficial mutations that transiently accumulated but were ultimately unsuccessful, changes in mutation rates, and ecological interactions between divergent lineages. Here, we show that it is possible to identify mutational variants in mixed bacterial population samples and to follow these processes on a genome-wide scale with current DNA sequencing technologies.

MATERIALS AND METHODS

DNA samples. During the long-term *E. coli* evolution experiment, mixed-population samples (M) and clones (C) isolated from these populations were periodically frozen at –80°C in 15% (w/v) glycerol. We revived clones from traces of frozen cultures by growing them overnight (16–24 h) as 10-mL cultures in 50-mL Erlenmeyer flasks at 37°C with shaking at 120 rpm in LB media, and mixed populations from 100 μL of frozen samples under the same conditions except in Davis minimal medium supplemented with 2 mg/L glucose. Genomic DNA was harvested and purified from several milliliters of each culture using a Qiagen Genomic-tip 100/G kit. Any deviations in the frequencies of mutations in DNA samples from the source populations due to these revival and reculturing steps appear to be minor.

Genome resequencing. We sequenced DNA fragment libraries derived from these samples on Genome Analyzer systems (Illumina, San Diego, California) in two separate runs. Clones were sequenced as paired-end libraries on a Genome Analyzer 1G machine by Macrogen (Seoul, South Korea), and mixed-population samples were sequenced in an unpaired library format using a Genome Analyzer 2G system by the Research Technology Support Facility at Michigan State University (East Lansing). Both devices generate 36-base reads, and the overall number of bases obtained for each sample was roughly equivalent. Per-base quality scores were calibrated on the basis of alignments to the genome sequence of the ancestral *E. coli* B strain REL606 using the standard Illumina resequencing pipeline.

Read alignment. We aligned reads from each data set to the ancestral genome using MUMmer v3.20 (Kurtz et al. 2004). Only reads where the entire 36-base length mapped with no base insertions or deletions and at most one base mismatch with respect to the reference genome were analyzed, because reads with indels or multiple discrepancies are more likely to be mapped incorrectly and may exhibit different error signatures. We also restricted our attention to "unique-only" reference genome positions, where all of the bases that mapped to a given position came from reads matching only a single site in the reference genome. It is not as straightforward to predict and interpret polymorphisms in repeat regions and sites on their periphery, where a mixture of reads uniquely and degenerately map to a position.

Base error model. For each data set, we created a null model that gives the probability of observing each of the four possible bases in a read, given the quality score assigned to that base and the identity of the base that was sequenced in the reference genome. We estimated the values in this series of 4 × 4 matrices directly from the observed counts of base discrepancies. This simple empirical strategy, which neglects the presence of real mutations, is justified because the vast majority of base mismatches are due to sequencing errors. The maximum number of point mutations found in the clone samples was 627 at generation 40,000 (JE Barrick et al. 2009). Assuming average coverage of these sites, this means that only 2.4% of the mismatches in this data set are due to consensus mutations. If the same number of base mismatches were due to mutations in the 40,000-generation mixed-population sample and they all had the highest quality score typically found in this data set, they would still raise the inferred error rate by only twofold. As the sensitivity analysis shows (see below), a change in the error rate of this magnitude barely alters our ability to discover SNPs. Thus, this simple error-rate estimate should suffice. For a given quality score and reference base, there is typically at most twofold or threefold variation among the inferred rates of each of the three possible base errors.

SNP prediction. To predict SNPs, we began by identifying the two most common bases aligned to each reference position. We then used a likelihood ratio test to decide between the null hypothesis, that the underlying population had only the most common base at this position, and the alternative hypothesis, that the population contained a mixture of alleles at this site. The probability of the observed data given the null hypothesis was calculated from the identities of the aligned bases, their quality scores, and the error model. The maximum likelihood of the alternative hypothesis was determined by scanning prospective mixtures of the two bases at 0.1% intervals. Twice the negative logarithm of the ratio between the two likelihoods was then compared to a χ^2 distribution with 1 degree of freedom to calculate a *p*-value for rejection of the null hypothesis of no polymorphism. Finally, we multiplied each *p*-value by the number of sites with unique-only coverage in the genome to obtain an *E*-value that reflects the number of SNP predictions expected to have this level of significance by chance.

For individually significant SNPs, we estimated the maximum likelihood frequency of the mutated allele in the underlying population. For this calculation, we only considered observations of the two most frequent bases. We performed 10^5 simulations of each alignment column at underlying base compositions in 0.1% intervals and recorded the underlying allele frequency that generated the actual mixture of bases at each site with the maximum likelihood, taking into account the chances of sequencing errors between these two bases.

When applied to the clone genomes, this procedure discovered all known point mutations outside of repeat regions as sites where a majority of the observed bases correspond to the new allele and there is no prediction of a polymorphism. Some spurious SNP predictions arose because the underlying genomic structure had changed due to large deletions or new mobile element insertions, but whole reads reflecting this changed sequence still aligned to the reference genome with one or fewer mismatches. We therefore manually eliminated, from all samples, SNP predictions adjacent to examples of those changes identified previously in clone genomes.

Bias filtering. We further rejected SNP predictions when bases supporting the mutant allele had consistently low quality scores or reads supporting the mutant allele showed a strand bias. To test for quality score bias, we performed a Kolmogorov–Smirnov test for the one-sided hypothesis that the quality scores supporting the mutant allele were lower than those supporting the ancestral base at that position. To test for strand bias, we performed Fisher's exact test on the two-tailed hypothesis that the distributions of mutant and reference base observations in reads on each genomic strand were different. Finally, we combined the *p*-values from these two independent tests using Fisher's method, and we rejected those predicted SNPs for which there was <5% chance of observing these differences in signature between the mutated and ancestral alleles by chance alone.

Sensitivity analysis. We performed the SNP prediction procedure without bias filtering on two sets of simulated data in order to estimate the chances of discovering SNPs at various frequencies. In the first set, we fixed the proportion of the mutant allele over a range of values and then resampled alignment columns according to the coverage and quality score distributions observed in the 2000-generation mixed-population data. Observations were also subject to a simplified error model where the overall chances of error for each quality score were the same as in the sample, but the probabilities of all base errors were made equal. In the second set of simulations that explores limits on rare SNP detection, all resampled alignments were uniformly assigned the same coverage and all bases had the same error rate.

RESULTS

Expectations of Diversity in an Evolving Bacterial Population

The *E. coli* populations in the long-term evolution experiment were each founded from a genetically homogeneous clone. No plasmids, viruses, or other mechanisms for horizontal gene exchange are present in the populations, and they evolve in a strictly clonal (asexual) manner. How quickly do we expect measurable diversity to accumulate in one of these populations, and what evolutionary forces will impact the patterns of variation within a population over time?

Figure 1 shows a numerical simulation of the spread of beneficial mutations in a population with parameters similar to that of the *E. coli* long-term evolution experiment (Woods 2005). Many beneficial mutations will be lost to

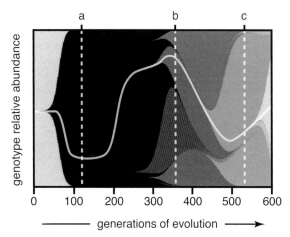

Figure 1. Expected dynamics in an evolving bacterial population. Lineages with new beneficial mutations are depicted as shaded wedges that originate in a previous genetic background and rise in frequency as they outcompete their ancestor and other lineages (Muller 1932). The same shading indicates that lineages have equivalent fitnesses, and the light gray curve highlights the path to the final dominant genotype containing five mutations. This figure was produced using a simulation with population size and mutation parameters meant to model the first 600 generations of the *E. coli* long-term evolution experiment (Woods 2005). Note how the level of genetic diversity changes over time. Early on, a new beneficial mutation sweeps to fixation and the population has little diversity (*a*). Later, four lineages with different mutations coexist at appreciable frequencies for a time (*b*) before the descendants of one lineage become a majority (*c*).

drift when they are rare (Lenski et al. 1991), with only those that achieve substantial frequency visible in the evolving frequency distribution. Among those that survive drift, some will eventually fix, but others will be eliminated by clonal interference (Gerrish and Lenski 1998). In some cases, the winning lineage may be decided not by the effects of single beneficial mutations, but rather by the combined effects of multiple mutations that accumulate before any one lineage is fixed (Fogle et al. 2008). Note that diversity does not increase monotonically, but rather it waxes and wanes as new beneficial mutations take hold and their fates are resolved by selection.

As beneficial mutations spread, they will perturb the frequencies of neutral and deleterious mutations owing to linkage disequilibrium in an asexual population. These perturbations are the classic signature of periodic selection (Atwood et al. 1951). However, we do not expect to detect many neutral or deleterious mutations in the evolution experiment. Under a pure drift process, the number of generations required for a neutral mutation to drift to fixation is on the order of the population size (Kimura 1983), which is many millions even after accounting for the bottlenecks during serial transfers (Lenski et al. 1991). Selective sweeps, however, reduce the effective size so that a rare neutral mutation may hitchhike to fixation much faster than it can spread by pure drift, although the vast majority of neutral mutations will be purged by these sweeps. In this case, the expected number of neutral mutations that fix equals the product of the genomic mutation rate, the proportion of neutral sites, and the number of generations (Kimura 1983). For *E. coli*, the genomic mutation rate is on the order of 10^{-4} to 10^{-3} per generation (Lenski et al. 2003), and perhaps 50% of sites are neutral, so we expect it to take on the order of 2,000–20,000 generations for even one neutral mutation to fix in the population. It is therefore unlikely that many neutral mutations would reach high frequency in the first few thousand generations of evolution. Deleterious mutations will fare even worse. Mutations causing extreme fitness defects will be rapidly lost from the population. A pool of many slightly deleterious mutations may accumulate and persist at mutation-selection balance, but these mutations are even less likely to fix or reach high frequency than neutral ones.

If the genomic mutation rate were much higher, however, neutral and weakly deleterious alleles could spread more easily and more would potentially reach high frequency. Several populations in the long-term experiment evolved mutator phenotypes, leading to mutation rates roughly two orders of magnitude higher than the ancestral rate (Sniegowski et al. 1997). In fact, a *mutT* mutator phenotype evolved in the population studied here, making its first appearance by generation 26,500 and becoming numerically dominant by generation 29,000 (JE Barrick et al. 2009).

Other polymorphisms may evolve and be maintained by negative-frequency-dependent interactions, in which some genotype has a selective advantage when rare but is disadvantaged at high frequency. Acetate and short-chain fatty acids are by-products of glucose fermentation by *E. coli*. These compounds are normally excreted during growth on glucose and then reabsorbed and used after the glucose is depleted. Mutants that are better competitors for acetate have been observed to evolve and persist via frequency-dependent selection in some chemostat experiments with *E. coli* (Rosenzweig et al. 1994). The low concentration of glucose and the serial-transfer regime used in the long-term experiment lead to low cell densities and correspondingly low levels of excreted metabolites, so that cross-feeding genotypes should be rarer and harder to detect. Indeed, a sustained cross-feeding interaction is only known to have evolved in one of the long-term populations (Rozen and Lenski 2000; Rozen et al. 2005, 2009), and not the one that is the focus of our study, although there appear to be weaker frequency-dependent interactions in some other populations (Elena and Lenski 1997).

Mixed-population Sequence Data Sets

We examined the genetic diversity over time in one experimental line from the long-term *E. coli* evolution experiment (designated Ara-1) by sequencing whole-population samples from 2,000 (M2K), 5,000 (M5K), 10,000 (M10K), 15,000 (M15K), 20,000 (M20K), 30,000 (M30K), and 40,000 (M40K) generation time points. Clones that were the subject of a previous study (JE Barrick et al. 2009) serve as controls for the mixed-population analysis. These clones were isolated from the same population at 2,000 (C2K), 5,000 (C5K), 10,000 (C10K), 15,000 (C15K), 20,0000 (C20K), and 40,000 (C40K) generations. We also include the founder of the Ara+1 experimental population that differs by two point mutations from the ancestor of the Ara-1 line (C0K).

Clones and mixed-population samples were sequenced, one per lane, in two separate runs of the Genome Analyzer system. Alignment of the resulting 36-base reads to the ancestral sequence yielded 40- to 60-fold average coverage outside of repeat regions for each genome (Table 1). Positions with zero coverage are not counted in these estimates, because they almost always proved to represent true

Table 1. Data-set Statistics and SNP Prediction Summary

Sample	Unique-only positions	Coverage μ	σ^2/μ	Base errors	SNP predictions (dN/dS) E-value ≤1		bias filtered	
Clone samples								
C0K	4,475,960	40.2	3.1	1,282,378	22	(17/3)	8	(6/2)
C2K	4,469,732	45.1	3.2	1,332,415	29	(18/5)	5	(3/0)
C5K	4,468,358	53.5	3.6	1,497,884	38	(25/9)	7	(4/3)
C10K	4,441,204	51.8	3.3	1,366,094	47	(37/4)	5	(5/0)
C15K	4,442,023	52.3	3.4	1,248,779	51	(38/11)	2	(1/1)
C20K	4,441,245	48.0	3.2	1,252,909	42	(30/9)	7	(5/1)
C40K	4,411,765	53.8	3.4	1,394,982	53	(33/13)	8	(4/2)
Mixed-population samples								
M2K	4,476,202	54.2	4.1	702,687	59	(51/3)	7	(7/0)
M5K	4,476,629	56.7	3.6	716,580	70	(46/10)	5	(4/1)
M10K	4,465,577	57.4	4.3	688,297	117	(71/25)	20	(10/1)
M15K	4,469,198	58.7	4.2	667,182	115	(72/24)	19	(12/0)
M20K	4,443,151	40.6	4.4	776,788	34	(22/3)	6	(4/1)
M30K	4,444,156	52.1	4.5	720,115	415	(314/50)	364	(270/46)
M40K	4,449,187	58.3	5.5	611,554	1150	(817/167)	1062	(754/148)

Bacterial clone (C) and mixed-population samples (M) from different generations of the Ara-1 population of the long-term *E. coli* evolution experiment were sequenced on Genome Analyzer systems. Each data set had the specified number of positions with coverage only from reads with unique best matches to the ancestral genome, mean (μ) and index of dispersion (σ^2/μ) for the distribution of read coverage depth at these unique-only positions, and number of base mismatch errors in reads with a unique best alignment to the ancestral genome. The numbers of SNPs predicted by our procedure using an E-value cutoff of 1 and after further filtering out predictions with biased base quality score and strand distributions are also reported for each sample. Ratios of nonsynonymous to synonymous substitutions are shown in parentheses.

deletions relative to the ancestral sequence in clone genomes (JE Barrick et al. 2009). The ancestral clone (C0K) has at least 10-fold read coverage at 99.9% of the positions in the reference genome. The decrease in the number of sites with coverage at later generations in both the clone and mixed-population samples is consistent with sizable deletions becoming fixed in the population.

If the sampling of reads from different locations in the genome were perfectly random, the number of sites with a given coverage depth would fit a Poisson distribution with equal mean and variance. We find that the index of dispersion (the variance divided by the mean) for the coverage distribution is much greater than unity in these genomes, ranging from 3.1 to 5.5 (Table 1), with the mixed samples showing slightly more dispersion. A maximum likelihood fit to a negative binomial distribution, which is commonly used to model overdispersed count data, reproduces most of the observed coverage structure (Fig. 2A). Higher coverage within GC-rich regions has been reported for Genome Analyzer sequence data, possibly due to more efficient processing of these fragments during library preparation on account of their greater duplex stability (Dohm et al. 2008). This bias may contribute to the overdispersion we observe and could systematically affect the recovery of polymorphisms in specific chromosomal regions.

There are hundreds of thousands to millions of base mismatches in the reads with unique best alignments to the ancestral genome in each data set (Table 1). When constructing a model for the base error rate, we verified that bases assigned high quality scores by the resequencing analysis software usually have fewer mismatches to the ancestral sequence (Fig. 2B). A majority of the bases in each run were assigned high quality scores. However, there were fewer overall errors, and bases with higher quality scores had fewer errors, in the mixed-population data sets. For example, 50% of the base calls have quality scores corresponding to error rates of about 0.02% or lower per base, and 75% have error rates below 0.04%, in the 2K mixed-population sample. By comparison, 68% of bases in the 2K clone data have quality scores with error rates below 0.04%, but only about 1% have error rates below 0.02%.

Distinguishing SNPs from Sequencing Errors

Our aim is to determine what diversity in a set of whole-population genome sequences is due to biological variation, as opposed to confounding mechanical errors and biases introduced during DNA preparation and sequencing. We restrict our analysis here to SNPs, representing new mutations that have risen to a measurable frequency, but not fixed, in a bacterial population at the time of sampling. Although there is information about deletions, insertions, and rearrangements in genome resequencing data, it is more difficult to interpret in terms of population frequencies, and so we have not yet attempted to analyze these other polymorphisms. Approximately two-thirds of the changes found in a more detailed analysis of the 20K clone from this population were point mutations (JE Barrick et al. 2009).

After aligning the reads in each data set to the reference genome, we used a likelihood ratio test to determine whether there was evidence of a SNP at each site. This test compares the likelihood of observing the collection of bases at a site under the null hypothesis of no genetic variation (i.e., all mismatches due to sequencing errors) to the maximum likelihood possible under the alternative hypothesis that there is a mixture of two alleles in the population. A much greater probability of the data given the alternative hypothesis indicates that the population from which DNA fragments were sampled consisted of subpopulations with different bases at this position. We report an E-value for each SNP prediction that is an estimate of its genome-wide significance, i.e., the likelihood ratio test p-value at a given site corrected for multiple testing. An E-value thus also represents the approximate number of false-positive predictions expected in a genome at a given significance level by chance.

Owing to the stochastic nature of both sequencing errors and sampling DNA fragments from different individuals, a

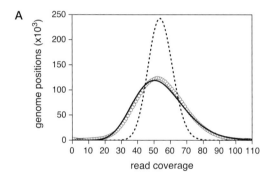
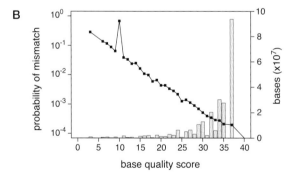

Figure 2. Example coverage distribution and base error rates. The 2K mixed-population sample is displayed as representative of the sequence datasets. (*A*) The distribution of the number of ancestral genomic positions with a given read coverage depth (open circles) is overdispersed relative to a Poisson model (dashed line) but is fit reasonably well by a negative binomial model (solid line). Repeat regions were excluded from this analysis. (*B*) The probability of a base error at a given quality score estimated from the number of observed mismatches in reads aligned to the reference genome usually decreases as a higher quality score is assigned to a base. Bases assigned a quality score of 10 had an anomalously high error rate in this data set. The accompanying histogram shows that most bases in the data set had high quality scores. Bases assigned a quality score of 40 do not appear on the log scale because they had zero errors.

true polymorphism that has a 50% frequency in the population is far more likely to achieve a significant E-value than one at 5%. We used simulated data with the same coverage and quality score distributions as the 2K mixed-population sample to estimate the chances of discovering polymorphisms at various frequencies in the population by our procedure (Fig. 3A). At an E-value threshold of 1, we expect to recover nearly all of the polymorphisms with frequencies of 20%–80%, ~50% of the polymorphisms present in 5% of the individuals, and only 1.6% of the polymorphisms at a frequency of 1%. Lowering the E-value cutoff to 0.01 reduces the sensitivity by factors of 1.6 and 5.4 for finding polymorphisms at frequencies of 5% and 1%, respectively.

In light of ever-improving technologies, we also investigated how better coverage and error rates would affect the discovery of SNPs at very low frequencies in a population (Fig. 3B). We performed further simulations to address this issue, with a simplified model that assumes uniform coverage and the same rate for all base errors (i.e., no differences in base quality). At an E-value cutoff of 1, the threshold frequency for a 50% chance of SNP discovery drops from 8.9% to 0.63% as coverage increases from 30- to 1000-fold. Reducing the error rate by an order of magnitude to 0.01% does not affect the recovery of SNPs at 30-fold coverage and only slightly improves the frequency for 50% detection probability to 0.36% at 1000-fold coverage. This sensitivity analysis therefore predicts that increasing coverage would be more effective for improving rare SNP detection than reducing the base error rate by a similar factor.

SNP Predictions

We chose to examine SNP predictions below a relatively permissive E-value cutoff of 1 in hopes of identifying real polymorphisms that were at low frequencies in the mixed-population samples. We first discovered that there were many more SNP predictions at this significance level than the average of 1 expected in each of the clone data sets, with the values ranging from 22 to 53 per clone (Table 1). Many of these predictions appear highly significant: 61 have E-values ≤0.01. During the outgrowth of a single cell, it is highly unlikely that even a single polymorphism will reach a frequency of >1%, because a mutation would have had to occur within the first seven generations (i.e., 2^7 = 128 cell divisions). Furthermore, if mutations that arose while culturing these samples after picking a clone were responsible for these SNP predictions, we would expect many more in the 40K clone because it is a mutator with a ~100-fold elevated mutation rate (Barrick et al. 2009), yet we see about the same number in this clone as in any other.

Instead, the unexpectedly high rate of false-positive predictions in the clones appears to result from sequencing or alignment errors that are outside the scope of our statistical model. Certain genomic sites appear to be especially prone to these errors, because many of the exact same SNPs are predicted in multiple samples and in sequence data sets from both the clone and mixed-population runs. Fortunately, many of these spurious predictions can be recognized by two kinds of biases. Base calls supporting the putative mutated base often have consistently lower qualities than those supporting the reference base for these polymorphisms, and reads supporting these SNPs are often derived largely or even exclusively from one strand of the genomic sequence. We developed a bias filtering step to reject putative SNPs with these error signatures. It reduces the number of predictions in clones to at most eight per genome (Table 1) and removes all but five clone predictions with E-values ≤0.01. This filter does not, however, eliminate any SNP predictions in the mixed-population samples thought to be real (see below).

There are many more highly significant predictions in the mixed-population samples than in the clones after bias filtering (Fig. 4). Every population data set has at least one predicted SNP with an E-value <10^{-5}, whereas the best prediction in any clone has an E-value almost two orders of magnitude higher. Even at 20K, where the mixed sam-

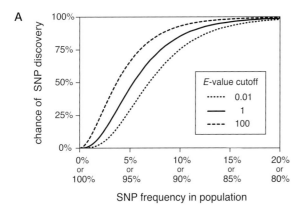

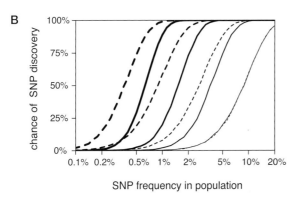

Figure 3. Sensitivity of SNP prediction procedure. (*A*) Estimates of the probability that our statistical procedure would detect SNPs present at various frequencies in a mixed-population sample at different E-value cutoffs. For these calculations, the coverage and quality score distributions were those of the mixed-population 2K sample. (*B*) Estimates of sensitivity improvements possible by increasing sequencing coverage and by reducing the rate of base errors. For these calculations, all sites had uniform coverage and the same error rate for all bases.

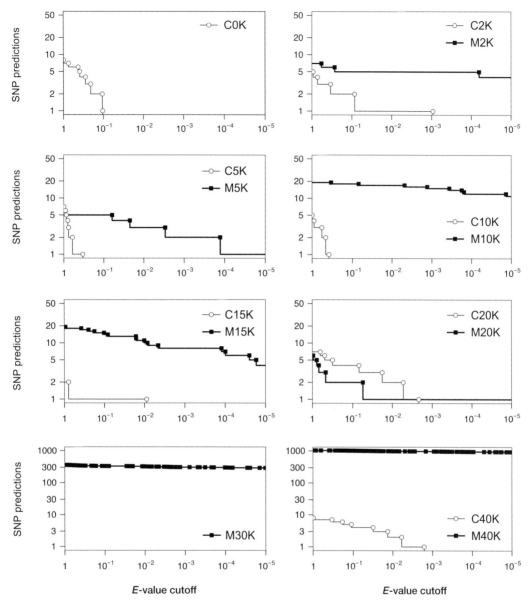

Figure 4. SNP predictions. The cumulative distributions of predictions below a given E-value threshold that also passed the bias filtering step are plotted for each data set. Each panel contains a generation-paired mixed-population sample (squares) and clone (circles), except there is only a clone at 0K and only a mixed population at 30K.

ple has fewer predicted SNPs than the paired clone, one of the mixed-population SNPs is very highly significant. Our detailed knowledge of the long-term experiment allows us to further evaluate these SNP predictions. We believe that 49 of the 57 predicted SNPs in the 2K to 20K population samples (Table 2) are probably both accurate and biologically important for the reasons presented below.

Elevated dN/dS ratio. We expect that most alleles that reach a high enough frequency in the population to be detected as SNPs during the first 20,000 generations will be beneficial mutations. Synonymous substitutions are likely to be neutral, and so an elevated ratio of nonsynonymous to synonymous mutations, dN/dS, provides evidence of positive selection. In the ancestral genome, there is a 20.4% chance that a random base substitution in a protein-coding region is synonymous. There are 37 nonsynonymous and 3 synonymous mutations in the predicted SNPs from the pooled set of 2K to 20K mixed-population samples, which is a significantly higher dN/dS ratio than expected by chance (one-tailed binomial test, $p = 0.03$). In contrast, taking all putative SNPs in the seven clone data sets together, there is no evidence that their dN/dS ratio is elevated ($p = 0.79$).

Mutator phenotype. We expect an increase in the amount of genetic variation in this population after a mutator phenotype evolved. Indeed, there is a dramatic increase

Table 2. SNPs of Particular Biological Interest

$-\log_{10} E$	Freq (%)	Gene	Notes
2K Mixed-population sample			
16.7	10.0	*mrdB*	cell wall
12.2	12.8	*mreB*	cell wall
12.2	14.8	*yegI*	M5K+ C5K+
6.4	6.9	*pykF*	new allele
4.2	10.1	*hslU*	M5K+ C5K+
0.6	4.4	*iclR*	new allele
5K Mixed-population sample			
5.1	94.4	*infB*	M10K+ C5K+
3.9	92.9	*malT*	M10K+ C5K+
2.5	6.8	*atoC*	SCFA
1.7	94.9	*spoT*	*M2K+ C2K+
10K Mixed-population sample			
80.8	45.6	*yghJ*	M20K+ C10K+
77.0	33.1	*hsdM*	C10K
72.8	61.6	*rpsM*	M20K+ C10K+
69.9	43.2	*araJ*	M20K+ C10K+
65.2	53.8	*yhdG/fis*	M20K+ C10K+
64.4	34.1	*acs/nrfA*	acetate
60.1	57.3	*rpsA*	ribosome
49.0	39.0	*yedW/yedX*	M20K+ C10K+
44.3	33.1	*maeB/talA*	C10K
23.4	16.2	*nuoM*	respiration
20.7	13.6	*nuoG*	respiration
20.3	23.0	*elaD*	synonymous
4.9	9.8	*ompF/asnS*	new allele
3.8	8.1	*nadR*	M15K+ C15K+
3.8	10.3	*ompF/asnS*	new allele
3.4	9.8	*iclR/metH*	new allele
2.9	8.7	*leuO/ilvI*	regulation
2.3	5.5	*atoS*	SCFA
15K Mixed-population sample			
31.3	82.8	*iclR*	M20K+ C15K+
10.5	90.6	*rpsM*	M20K+ C10K+
6.8	91.2	*pcnB*	M20K+ C15K+
5.2	90.3	*arcB*	M20K+ C15K+
4.8	94.7	*infB*	*M10K+ C5K+
4.6	91.9	*dhaM*	M20K+ C15K+
4.0	94.5	*araJ*	M20K+ C10K+
3.9	91.3	*narI/ychS*	M20K+ C15K+
2.3	4.1	*yaaH*	acetate
2.1	95.8	*yghJ*	M20K+ C10K+
2.0	7.7	*ydiV/nlpC*	regulation
1.8	91.2	*ompF/asnS*	M20K+ C15K+
1.8	7.6	*ycbX/ycbY*	ribosome
1.1	93.9	*yhdG/fis*	M20K+ C10K+
1.0	95.9	*yedW/yedX*	M20K+ C10K+
0.8	5.9	*gyrB*	regulation
0.6	95.2	*yegI*	*M5K+ C5K+
0.4	4.6	*rspA/ynfA*	regulation
0.0	95.6	*ebgR*	M20K+ C15K+
20K Mixed-population sample			
40.0	35.3	*hypF*	C20K+
1.3	5.6	*mgrB/yobH*	regulation

For selected SNPs in the mixed-population samples, the negative base-10 logarithm of the E-value, maximum likelihood prediction of the frequency of the derived allele in the population, and gene (e.g., *araJ*) or intergenic region (e.g., *rspA/ynfA*) containing the SNP are shown. Samples where the same mutation was fixed (within statistical resolution) in a mixed-population sample (M) or present in a sequenced clone (C) at a given generation are marked in the notes column, with a plus sign further indicating that the mutation was also found in all later samples, and asterisks marking a few SNPs that appeared (erroneously, owing to statistical uncertainty) to have been fixed in earlier population samples. Other mutations are likely to be beneficial because they are in the same gene or promoter region as mutations that later swept to fixation in this population (new allele), probably affect cellular processes known to be targets of selection in this experiment (cell wall, respiration, ribosome, regulation), or possibly improve growth on metabolic byproducts (acetate, SCFA). A complete list of all predicted SNPs that includes further details is available on the author's website (http://myxo.css.msu.edu/papers/).

in the number of predicted SNPs, from an average of 11.4 in the 2K to 20K population samples to 364 and 1062 in the 30K and 40K samples, respectively. The *mutT* defect that evolved specifically elevates the rate of A·T → C·G transversions, and so we also expect the SNPs in the 30K and 40K samples to exhibit almost exclusively this sequence signature. In the 2K to 20K mixed-population samples, 26.3% of the predicted SNPs are A → C or T → G changes. As expected, there is an extremely significant shift in this fraction to 98.1% (one-tailed Fisher's exact test, $p = 1.0 \times 10^{-37}$) and 91.5% ($p = 1.3 \times 10^{-29}$) in the 30K and 40K population samples, respectively.

Selective sweeps reaching fixation. If our procedure finds true polymorphisms, we would expect some predicted SNPs to be mutations that were rising in frequency during a selective sweep that would ultimately reach fixation. In fact, nearly half (25/57) of the predicted SNPs in the 2K to 20K population samples are mutations that were later found at 100% frequency in the population. In contrast, none of the suspect SNP predictions from clonal samples corresponds to mutations that were fixed in the population or observed in other clones.

Unsuccessful mutations in genes where other alleles fixed. If our procedure for SNP discovery is accurate, we expect to find evidence for selective sweeps that failed due to clonal interference. Consistent with that expectation, five predicted SNPs in the 2K to 15K population samples are in genes where a different mutation fixed by 20,000 generations (Table 2). These transient polymorphisms probably represent alternative beneficial alleles at genes under strong selection in the long-term experiment. Given that *E. coli* has ~4000 genes and that 26 predicted SNPs from this period did not reach fixation, there is only a small chance (one-tailed binomial test, $p = 8 \times 10^{-7}$) of picking five or more SNPs at random in the 27 genes with mutations that later fixed. Of the putative SNPs in clones, only 1 in 42 impinges on this same set of 27 genes, which is not unlikely by chance ($p = 0.25$).

Among these unsuccessful mutations, the 10K mixed sample included two different SNPs at adjacent bases upstream of the *ompF* gene. The *ompF* allele that eventually fixed is also in the promoter region and appears as a SNP at 15K. Surprisingly, it has a highly deleterious effect when moved alone into the ancestral chromosome (Barrick et al. 2009). The finding of two similar contending mutations provides compelling evidence that these *ompF* mutations are actually beneficial in a genetic background that had become common by 10,000 generations. There are also transient SNPs affecting the *pykF* and *iclR* genes, each of which eventually fixed a different allele.

Other unsuccessful beneficial mutations. We would also expect some unsuccessful lineages to have beneficial mutations in other genes. Two predicted SNPs in the 10K population (*hsdM* and *maeB/talA*) are clearly real because they were also present in the 10K clone genome, although this clone was off the main line of descent. Although we have no direct evidence that other transient SNPs are biologically significant, it seems plausible that at least 15 of them are beneficial in this environment. Eleven transient SNPs in the 2K to 20K mixed-population samples occur in genes involved in processes thought to be key targets of selection (Table 2) including cell wall synthesis, respiration, ribosomal function, and gene regulation (Philippe et al. 2007, 2009). For example, *mrdA* and *mrdB* are two genes in the same operon involved in cell wall synthesis; a transient SNP in *mrdB* occurs in the 2K population sample, and a mutation in *mrdA* was fixed in every population sample from 5K onward.

Cross-feeding adaptations. The four remaining transient SNPs in the 2K to 20K mixed-population samples, which occur in genes related to acetate and short-chain fatty acid (SCFA) metabolism (Table 2), may be cross-feeding adaptations. Three of these mutations were rare, and all of them were ultimately lost from this population. One is in the promoter region of the *acs* gene, which encodes an enzyme for acetate utilization. A second is a nonsynonymous change in *yaaH*, which is predicted to have a role in acetate transport. The third and fourth cause amino acid substitutions in *atoS* and *atoC*, which together regulate an operon involved in SCFA degradation. Two early transient SNPs in the *iclR* gene, which encodes a repressor for glyoxylate bypass enzymes that are induced when *E. coli* grows on acetate or SCFAs, may also promote cross-feeding interactions, even though a different mutation in *iclR* was eventually adopted by the dominant lineage and fixed in the population by 20,000 generations.

It is possible that cross-feeding genotypes off the main line of descent may have persisted in this population at low levels below our detection limit for SNPs, with occasional increases in frequency, perhaps in association with other beneficial mutations. The presence of such cross-feeding genotypes could explain the weaker frequency-dependent interactions observed in populations other than the one with the stable polymorphism (Elena and Lenski 1997). None of the potential cross-feeding alleles detected as SNPs remain at detectable frequencies in successive samples, so we suspect that they were evolutionary dead ends in this population. However, new cross-feeding genotypes could periodically reevolve from the main lineage to exploit that niche and, in turn, later become extinct (Rozen et al. 2005).

Changes in Genetic Diversity Over Time

Figure 5 summarizes the genetic diversity observed in mixed-population samples over time and the tempo with which mutations were fixed in the population. The top panel shows the origin and eventual fate of all of the point mutations discovered in the 2K to 40K population samples, and the bottom panel provides a visual summary of the main patterns in these data.

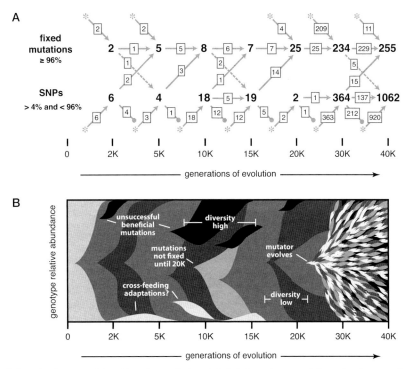

Figure 5. Mutational diversity in an evolving *E. coli* population. (*A*) Origin and eventual fate of point mutations in the 2K to 40K mixed-population samples. New mutations that first appear as SNPs or fixed alleles are shown as asterisks along the bottom or top, respectively, with arrows leading to the corresponding pools of SNPs and fixed mutations. Transient SNPs that were lost from the population are shown by descending lines ending in closed circles. Note that we only detect SNPs when they are between ~4% and 96% frequency in the population and that we only recover ~50% of the SNPs at 5% frequency. Only the 49 SNP predictions in Table 2 were included for the 2K to 20K samples. (*B*) Stylized summary of the mixed-population SNP analysis. Shaded wedges represent subpopulations containing new mutations relative to the previous genetic background. Mutations are grouped to highlight their eventual fates, but we do not always have linkage information to resolve which SNPs occurred together. Labeled features are explained in the text.

this technology due to the fact that they are not caused by expression level differences or that they are so spatially restricted that their magnitude would not stand out against the background of activity elsewhere. Despite all of these caveats, expression profiles provide a rough sketch of the differences underlying selection effects.

Most laboratory selection experiments with *Drosophila* are short term and performed on small populations, on the order of 20–30 generations selecting the top ~10% of a population of ~200 (Greenspan 2004a). Expression profiles have been done on RNA extracted from fly heads (mostly brain) of *D. melanogaster* strains differentially selected for mating speed (Mackay et al. 2005), aggression (Fig. 1) (see Dierick and Greenspan 2006; Edwards et al. 2006), and locomotor activity (Jordan et al. 2007). All have yielded a bewilderingly wide range of genes, as is common with microarray results. Two notable attributes are shared by all of these studies. First, the expression differences found for those genes significantly differing between strains are relatively small, almost all less than twofold in magnitude. One might attribute this feature to the likelihood that most natural variants are relatively mild. The second feature is the lack of any overlap between these genes and any that had previously been identified through classical, Benzerian mutant analysis (Hall 1982, 1994a; Kyriacou and Hall 1994; Wilson et al. 2008). This is all the more surprising given the number and range of genes that differ in these experiments. The question of how well expression analysis identifies the pertinent genes is addressed below.

Few selection experiments, behavioral or otherwise, go on for more than 30 generations, and very few of these have been analyzed molecularly. One of the rarities is a medium-term selection experiment for divergence in wing morphology that was conducted during a time frame more than twice as long as usual. An initial 20 generations of selection produced divergence in wing shape as measured by the ratio of a transect in the distal wing to a transect in the proximal wing (Fig. 2, left) (see Weber 1990). At generation 21 (Fig. 2, right), the two divergent lines were mixed and allowed to interbreed for 34 generations, at which time they were reselected for the same wing parameter divergence—four replicates in each direction—for another 25 generations (Weber et al. 2008).

Expression profiles were done on wing disc tissue dissected from these strains. The magnitude of the differences was somewhat higher than in the shorter-term selection experiments, with many more of them exceeding twofold (Weber et al. 2008). Much more striking, however, is the continued lack of overlap between these differentially expressed genes and any of the more than 200 genes identified over the years as affecting wing development (Brody 1999). Any concern from the earlier analyses that the phenotypes may have been undersampled in the mutagenesis studies is adequately allayed here.

The final example of a selection experiment subjected to expression analysis is perhaps the record holder for long-term laboratory selection in a metazoan. The indoor record for selection in any organism is 10,000 generations for *Escherichia coli* (Lenski and Travisano 1994), but in 1958, Jerry Hirsch began selecting *Drosophila* for divergence in geotaxis behavior and continued doing so for more than 600 generations until the lines stabilized spontaneously in 1983 (Fig. 3, left) (see Hirsch 1959; Ricker and Hirsch 1985). When tested again in 2000, they still displayed the appropriate phenotypic difference (Fig. 3, right) (see Toma et al. 2002).

Gene expression profiles were obtained for RNA from heads of the divergent strains, and, as usual, many genes differed. The magnitude of expression difference in these flies was considerably greater than that in any of the previous examples, with many showing fourfold to eightfold differences and a few even higher (Toma et al. 2002). And as in the earlier examples, the differing genes showed no overlap with any of the genes identified previously as geotaxis mutants (Armstrong et al. 2006).

Two trends stand out from this admittedly small set of examples. First, none of the genes match those found in standard mutant searches, and second, the magnitude of the genetic variation between strains, as measured by expression differences, increases with longer selection regimes.

The discrepancy between the two sets of genes may have several possible explanations. Many of the expression differences that diverge in these strains will be irrelevant to the divergent phenotypes. Some will be due to hitchhiking of variants that happen to be linked to those that are relevant, and others will be due to alterations in branches of

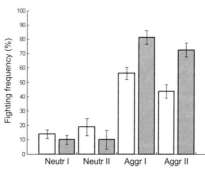

Figure 1. (*Left*) Fly aggression. (*Right*) Increased aggression with 20 generations of selection. (*Right panel,* Reprinted, with permission, from Dierick and Greenspan 2006 [©Nature Publishing Group].)

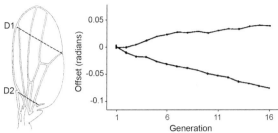

Figure 2. (*Left*) Wing parameters subjected to selection. (*Right*) Divergence during selection on wing parameters. (Reprinted, with permission, from Weber 1990 [©Genetics Society of America].)

transcriptional hierarchies that are not related to the selected phenotype. Genetic polymorphisms do not necessarily alter expression level; they can also alter timing, placement, or protein sequence. And even if they do, the effect could be so localized that it would not be detectable when averaging over a whole head or wing disc. On the other hand, it is well known that a great many mutations, including P-element mutations (see, e.g., Lerman et al. 2003) and natural variants (see, e.g., Osborne et al. 1997) in *Drosophila*, do actually alter levels of transcription.

The total lack of overlap is, however, rather striking, particularly in the case of wing morphology, where the tissue assayed was quite restricted (i.e., third instar larval wing discs), and the battery of genes known to affect wing development at that stage is extensive. Even if many of the expression differences are irrelevant, one might imagine that *some* of them would correspond to known mutants. One explanation is that there are many more ways of affecting a given phenotype beyond the "core" set of genes identified by standard mutagenesis.

MORE THAN ONE WAY TO SKIN A CAT

Support for the idea of a broad palette of available genes comes from several quarters. In a previous study of 50 randomly generated, homozygous viable P-element insertions, Weber et al. (2005) found 11 that affected the same wing morphology phenotype, and none overlapped with the known set of wing development genes. In a much broader screen of more than 2800 P-element insertions for alterations in bristle number, a similar proportion (~22%) was found to produce significant effects, and these covered a much wider range of genes than those previously identified in mutant screens (Norga et al. 2003). Another, much larger, screen for randomly generated homozygous viable P-element insertions yielded 263 that were then tested for activity levels of 14 metabolic enzymes (Clark et al. 1995). More than 50% of them (153) produced significant alterations in activity for at least one enzyme, and 15%–20% of them affected two or three enzymes. These are the so-called "housekeeping" enzymes that are not supposed to change. And finally, a recent search for P-element mutants affecting the startle response yielded 267 of 720 lines screened, including many with no obvious relationship to neural mechanisms (Yamamoto et al. 2008).

Coming to the same conclusion from a different origin, genome-wide screens for RNA interference (RNAi) effects have also revealed a much wider set of genes and mechanisms capable of significantly altering a phenotype than predicted from previous mutant studies, whether conducted in cell culture (Friedman and Perrimon 2007) or in the intact animal (Byrne et al. 2007). And from yet another direction, analysis of differences in natural variants between two different fly strains, and the effects on global gene expression of those strain differences, implicates a wide range of pleiotropic genes in many behaviors (Ayroles et al. 2009; Edwards et al. 2009; Harbison et al. 2009).

The second trend seen with increasing length of selection is the greater magnitude of gene expression differences. These could be either due to changes at the relevant loci themselves or the result of selection for gene interactions that affect expression. It has previously been suggested that new mutations could arise and spread after as few as 20 generations (Robertson 1980) and that combinations of interacting genes would increase over successive generations of selection (Wright 1963). The question of gene interactions resulting from selection bears further examination.

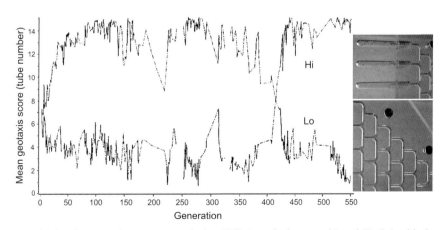

Figure 3. Long-term selection for geotaxis response producing "Hi" (negatively geotaxic) and "Lo" (positively geotaxic) strains. (Reprinted, with permission, from Ricker and Hirsch 1985 [©APA].) *Right upper panel:* Topmost tubes of geotaxis apparatus showing "Hi" flies at the end of their run (courtesy of R.J. Greenspan). *Right lower panel:* Entry of "Hi" flies into the maze (courtesy of R.J. Greenspan).

WEAVING THE NETWORK OF GENE INTERACTIONS

The importance of gene interactions in evolution has been a matter of controversy since the early days of population genetics (Fisher 1930; Wright 1931) and it goes on to this day (Coyne et al. 1997; Wade and Goodnight 1998; Desai et al. 2007). The original protagonists in this matter were R.A. Fisher and Sewall Wright. Their respective viewpoints are summed up in Figure 4, taken from one of Wright's last papers (Wright 1982), along with a third relevant viewpoint from J.B.S. Haldane. Fisher considered selected traits to be the result of the action of a large number of genes, each one of small effect and acting independently, summing to produce the final outcome. Wright also conceived of the process as multigenic, but rather than acting independently and summing, he saw the genes as interacting extensively in nonadditive (epistatic) ways. Haldane, in contrast, imagined single large-effect variants as responsible for phenotypic change. How do these formulations square with the findings from various selection experiments?

On the basis of statistical analyses of the phenotypes of F_1, F_2, and backcross progeny, short-term selection seems to result mainly from additive interactions. For two of the selection experiments discussed here, additivity was found in the initial stages (generation 20) of Weber's wing morphology selection (Weber et al. 1999, 2001) and in the initial stages (generation 28) of Hirsch's geotaxis selection (Hirsch and Erlenmeyer-Kimling 1962). Further examples of short-term selection that found a similar lack of epistasis include knockdown resistance to ethanol (Cohan et al. 1989) and male mating speed (Caseres et al. 1993). Some short-term experiments did show epistasis as well as additivity: central excitatory state (Vargo and Hirsch 1986), female remating speed (Fukui and Gromko 1991), and locomotor activity (Mackay et al. 2005).

When selection goes on for longer times, it apparently produces greater epistasis, as illustrated in Hirsch's geotaxis experiment at generation 133 (McGuire 1992) and at generation 566 (Ricker and Hirsch 1988). A similar conclusion can be drawn by calculating interlocus epistasis in the data from a selection experiment for locomotion against a wind current performed by Weber (1996). Using the means of hybrids and parents as a measure, where $4F2 - 2F1 - P1 - P2 \neq 0$ if there is interlocus epistasis (Mather and Jinks 1977), significant epistasis begins to be detectable at generation 98.

LARGE-EFFECT MUTATIONS: ARTIFICAL OR NATURAL?

The essence of single-gene mutant studies, and the principal criterion applied since Benzer inaugurated the strategy (Benzer 1967) and which has continued through the many screens for all kinds of phenotypes since then, is to isolate strong-effect mutations with little or no effect on other phenotypes. As regards the specificity of these mutations, this criterion is more honored in the breach than in the observance, as documented occasionally through the past few decades (Hall 1982, 1994b; Greenspan 2001). But by and large, the general idea holds, especially for the criterion of large effects.

Natural variants, in contrast, generally tend to be small in their effects (see, e.g., Hill and Caballero 1992; Mackay 2001), with rare exceptions (see below). This observation fits with the Fisher and Wright models for the polymorphisms that are acted upon by selection (Fig. 4, left and middle). As for the Haldane model (Fig. 4, right), is there any evidence for it? The notable exceptions to the small-effect rule are single-gene polymorphisms that have been shown to account for most of the variance affecting central excitatory state in the blowfly *Phormia* (Tully and Hirsch 1982), larval foraging behavior in *D. melanogaster* (Sokolowski 1980), and aggregation behavior in *Caenorhabditis elegans* (de Bono and Bargmann 1998).

Among the selection experiments discussed above, Hirsch's long-term geotaxis selection exhibits two features that have more of a Haldanish quality: larger expression differences (Toma et al. 2002) and an ability of induced mutations in a single one of the differing genes, the neuropeptide *Pdf* pigment-dispersing factor (Renn et al. 1999), to recapitulate the strength and range of the selected phenotype when driven toward either nullness or overexpression (Toma et al. 2002). This degree of recapitulation is unmatched in any of the other selection experiments for which such tests have been made (Dierick and Greenspan 2006; Edwards et al. 2006) and constitutes a "near miss" in finding a gene by selection that could also have been found by a classical screen for single-gene mutants. Further evidence for the potency of this gene subsequently came from isolation of mutants in the neuropeptide's receptor, which also exhibited a negative geotaxis response as strong as that of Hirsch's "Hi" line (Mertens et al. 2005). Although far from definitive, these observations nonetheless suggest that long-term selection may be more likely to accumulate, and perhaps be more permissive for, large-effect variants.

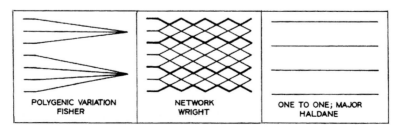

Figure 4. Three classic models of selection. (Reprinted, with permission, from Wright 1982 [©Wiley-Blackwell].)

THE SEASONS OF SELECTION

The foregoing discussion may offer a way of reconciling the differing points of view on modes of selection illustrated in Figure 4. As often happens in biological controversies, opposing viewpoints may simply represent valid pictures of what goes on at different times. Short-term selection appears Fisherian with its multiple small, additive effect variants. Medium-term selection appears Wrightian with its multiple, increasingly epistatic interactions. And long-term selection appears Haldanish, with a strong Wrightian streak, in its larger-effect variants and more extensive epistatic interactions. Mechanistically, these stages are feasible from the standpoint of what is most likely to be available to work with initially, leading to a more extensive intertwining during a longer time period and eventually allowing for the appearance and fixation of new, larger-effect variants.

THE DISCREPANCIES REMAIN

The quandary remains, however, as to the apparent lack of concordance between selected variants and induced mutants. As summarized above, there is ample evidence that many more genes can affect phenotypes than are generally found in mutant screens, but what is the relationship between these two sets?

One interesting possibility, for which there is increasing evidence, is that the gene networks that subserve any phenotype are much wider ranging than previously suspected (Greenspan 2001, 2004b, and unpubl.). If it is the case that gene networks are wider ranging than we had thought, then it almost certainly must be due to extensive gene interactions. Can we also find evidence for this?

The most comprehensive approach of this sort has been global gene interaction studies in yeast (Tong et al. 2004; Roguev et al. 2008), demonstrating ~4000 interactions among ~1000 genes, with interactions/gene ranging from 1 to 146 (μ = 34). More extensive evidence for the breadth of gene action can be found in the literature on suppressor and enhancer screens. A representative example is the now classic studies done on the *sevenless* gene of *Drosophila* and its role in cell-fate determination in the retina. The standard protocol was to start with a moderate allele of the gene, with a rough eye phenotype that allowed for exacerbation by enhancers and amelioration by suppressors. Out of such screens, the *sevenless* pathway was assembled and its homology with mammalian oncogene tyrosine kinase/Ras/MAP kinase signal transduction was clearly shown (Brennan and Moses 2000). If one looks more closely at the products of these suppressor/enhancer screens, one finds that a very wide range of interacting genes was identified (see, e.g., Simon et al. 1991), but not followed up, presumably due to the lack of clear involvement in the Ras pathway. These genes were, nonetheless, capable of modifying the *sevenless* phenotype and were thus sampling the wider space of relevant genes, now largely lost to the world.

THE FLEXIBILITY OF GENE NETWORKS

It has generally been assumed that the relationships among the elements of a pathway or network are stable. We have tested this assumption in an analysis of interactions among a set of genes affecting loss of coordination in *Drosophila*. Unlike most studies of gene networks, we defined a network in terms of functional interactions (rather than common phenotype), common biochemical function, or covariation of gene expression patterns. A classical analysis of epistasis was then performed on 16 mutations isolated on the basis of their interaction with a mutation of *Syntaxin-1A* (*Syx1A*), a component of the machinery of secretion and synaptic transmission (Richmond and Broadie 2002). We treated the temperature-sensitive induction of uncoordination in $Syx1A^{3-69}$ mutants (Littleton et al. 1998) as a quantitative phenotype and tested their interactions based on statistical alteration of its time course of onset (Fig. 5, left).

Using epistatic (nonlinear) interactions between two genes as an indicator of a more intimate network relationship than additivity, we measured the relationships among all 16 suppressor/enhancer genes on an otherwise similar wild-type background (Fig. 5, middle). We then measured these interactions on the same background with the reintroduction of the original *Syx1A* mutation, against which the suppressor/enhancers were isolated (Fig. 5, right). The resulting set of interactions defines a series of relationships among the genes.

The unexpected result is that the functional relationships among the 16 genes vary with genetic context: They

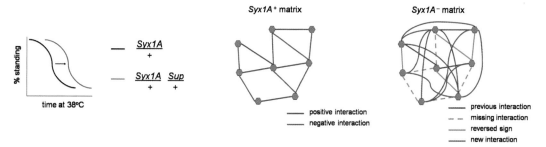

Figure 5. (*Left*) Test for epistasis among *Syx-1A* suppressors/enhancers. (*Middle*) Network of interactions among *Syx-1A* suppressor/enhancers in the absence $Syx1A^+$ and (*right*) presence of the original *Syx-1A* mutation used to identify them. (Modified, with permission, from Greenspan 2001 ©Nature Publishing Group].)

change dramatically depending on the presence or absence of the *Syx1A* mutation. This indicates a potential for network flexibility beyond that predicted by standard molecular biological models of gene interactions and implies that the network takes on different configurations under different conditions.

FLEXIBILITY, ROBUSTNESS, AND DEGENERACY

Flexible relationships among elements of a network are likely to be a major source of robustness as well as a source for the emergence of new properties. In contrast to the conventionally invoked mechanisms of local feedback or redundancy to account for such properties (Hartman et al. 2001; Davidson et al. 2002), the more far-flung interactions that we have uncovered may be better attributed to degeneracy, the wide-ranging ability of a system to produce the same output by different strategies. The finding that many different genotypes produce nearly identical behavioral scores exemplifies this property (cf. Hirsch 1963).

Degeneracy is a signature feature of biological systems in general (Edelman and Gally 2001) and of gene networks in particular (Greenspan 2001). Degenerate biological systems have many nonidentical elements (e.g., genes) that are extensively interconnected but that have nonuniform patterns of connectivity. The effective range of each gene is further enhanced by pleiotropy (see discussion above and Wright 1968). These properties endow biological systems with the ability to compensate for perturbations that may never have been encountered before. An important consequence of this property is that there is a great deal more latent potential in gene networks than has previously been revealed either by classical quantitative genetics, where the identities of interacting genes were not known, or by classical mutant analysis, where the scope of interaction was relatively narrow.

A SHIFTING GENETIC LANDSCAPE

An implication of our findings and of these ideas for mechanistic studies of gene action then arises: If the functional state of a gene network is perturbed when one of its elements is changed, then caution must be exercised in extrapolating back to a "normal" system state from mutant data. Moreover, it casts a shadow over the very concept of a normal system state, in much the same way that population geneticists have long questioned the notion of a "normal" individual (Hirsch 1963; Lewontin 1974).

This property has its counterpart in the nervous system where the functional connections shift under different conditions of stimulation (Marder and Thirumalai 2002). Unlike the nervous system, gene networks lack a constant underlying physical anatomy of connections to map. Thus, interaction mapping studies could potentially go on forever without ever producing a coherent system-wide set of relationships. Each result would be valid in its own context but not necessarily in many others.

IMPLICATIONS: A PATH FROM MICRO TO MACRO

What are the implications of all of this for understanding the process of evolution? If we take into account (1) the broad palette of networked genes available to affect any phenotype, (2) the increase in both epistasis and magnitude of gene effects with prolonged selection, and (3) the flexibility of network configurations with the introduction of new genetic variants, then a mechanism for the transition from small-effect to large-effect genetic changes during evolution may be discerned (Fig. 6).

In the short-term, selection gathers together the small-effect variants that are segregating in the population to achieve a particular phenotypic outcome. As more time passes under selection, more intricate combinations of variants are woven together to produce a more potent mixture based on epistatic interactions. These interactions, in turn, mold the network into a new shape, commensurate with the selected phenotype. With still longer times, more extensive intertwining and network sculpting occurs, creating a genetic environment more tolerant of a newly arising large-effect variant that reinforces the phenotype without causing disruptions so drastic that the whole system crashes. Should such a variant arise, not only would it reinforce the existing network configuration, but it would also be far easier to inherit than the web of many variants required for the phenotypic effect up to that point. In time, the many small-effect variants could then scatter, leaving the large-effect change to perpetuate the network configuration. During that period, finer honing

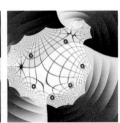

Figure 6. Stages in gene network alteration during selection for a new trait. (*First panel*) Topography of a phenotypic landscape. (*Second panel*) Wide-ranging gene network capable of affecting a given phenotype. (*Third panel*) Accumulation of multiple small-effect variants and consequent distortion of the network relationships, creating an environment capable of tolerating a large-effect variant (*fourth panel*) that reinforces the same network distortion. (*Fifth panel*) Subsequent loss of the small variants and persistence of the large variant.

of the large variant, such as further nucleotide changes to a *cis*-acting sequence, could reduce deleterious effects or narrow the scope of the variant's effect.

A scenario of this sort could account for the relatively specific, relatively large-effect single-gene differences we find today as the remnants of species differences (see, e.g., Carroll et al. 1995). It could also provide a mechanism to account for the transition from microevolutionary to macro-evolutionary transformations.

REFERENCES

Armstrong JD, Texada MJ, Munjaal R, Baker DA, Beckingham KM. 2006. Gravitaxis in *Drosophila melanogaster:* A forward genetic screen. *Genes Brain Behav* **5:** 222–239.

Ayroles JF, Carbone MA, Stone EA, Jordan KW, Lyman RF, Magwire MM, Rollmann SM, Duncan LH, Faye Lawrence F, Anholt RH, Mackay TFC. 2009. Systems genetics of complex traits in *Drosophila melanogaster*. *Nat Genet* **41:** 299–307

Benzer S. 1967. Behavioral mutants of *Drosophila* isolated by countercurrent distribution. *Proc Natl Acad Sci* **58:** 1112–1119.

Brennan CA, Moses K. 2000. Determination of *Drosophila* photoreceptors: Timing is everything. *Cell Mol Life Sci* **57:** 195–214.

Brody T. 1999. The interactive fly: Gene networks, development and the internet. *Trends Genet* **15:** 333–334.

Byrne AB, Weirauch MT, Wong V, Koeva M, Dixon SJ, Stuart JM, Roy PJ. 2007. A global analysis of genetic interactions in *Caenorhabditis elegans*. *J Biol* **6:** 8.

Carroll SD, Weatherbee SB, Langeland JA. 1995. Homeotic genes and the regulation and evolution of insect wing number. *Nature* **375:** 58–61.

Casares P, Carracedo MC, San Miguel E, Piñeiro R, García-Flórez L. 1993. Male mating speed in *Drosophila melanogaster*: Differences in genetic architecture and in relative performance according to female genotype. *Behav Genet* **23:** 349–358.

Clark AG, Wang L, Hulleberg T. 1995. P-element-induced variation in metabolic regulation in *Drosophila*. *Genetics* **139:** 337–438.

Cohan FM, Hoffmann AA, Gayley TW. 1989. A test of the role of epistasis in divergence under uniform selection. *Evolution* **43:** 766–774.

Coyne JA, Barton NH, Turelli M. 1997. Perspective: A critique of Sewall Wright's shifting balance theory of evolution. *Evolution* **51:** 643–671.

Davidson EH, Rast JP, Oliveri P, Ransick A, Calestani C, Yuh CH, Minokawa T, Amore G, Hinman V, Arenas-Mena C, et al. 2002. A genomic regulatory network for development. *Science* **295:** 1669–1678.

de Bono M, Bargmann CI. 1998. Natural variation in a neuropeptide Y receptor homolog modifies social behavior and food response in *C. elegans*. *Cell* **94:** 679–689.

Desai MM, Weissman D, Feldman MW. 2007. Evolution can favor antagonistic epistasis. *Genetics* **177:** 1001–1010.

Dierick HA, Greenspan RJ. 2006. Molecular analysis of flies selected for aggressive behavior. *Nat Genet* **38:** 1023–1031.

Edelman GM, Gally JA. 2001. Degeneracy and complexity in biological systems. *Proc Natl Acad Sci* **98:** 13763–13768.

Edwards AC, Rollmann SM, Morgan TJ, Mackay TF. 2006. Quantitative genomics of aggressive behavior in *Drosophila melanogaster*. *PLoS Genet* **2:** e154.

Edwards AC, Ayroles JF, Stone EA, Carbone MA, Lyman RF, Mackay TF. 2009. A transcriptional network associated with natural variation in *Drosophila* aggressive behavior. *Genome Biol* **10:** R76.

Fisher RA. 1930. *The genetical theory of natural selection*. Oxford University Press, New York.

Friedman A, Perrimon N. 2007. Genetic screening for signal transduction in the era of network biology. *Cell* **128:** 225–231.

Fukui HH, Gromko MH. 1991. Genetic basis for remating in *Drosophila melanogaster*. IV. A chromosome substitution analysis. *Behav Genet* **21:** 169–181.

Greenspan RJ. 1997. A kinder, gentler genetic analysis of behavior. *Curr Opin Neurobiol* **7:** 805–811.

Greenspan RJ. 2001. The flexible genome. *Nat Rev Genet* **2:** 383–387.

Greenspan RJ. 2004a. The varieties of selectional experience in behavioral genetics. *J Neurogenet* **17:** 241–270.

Greenspan RJ. 2004b. E pluribus unum, ex uno plura: Quantitative- and single-gene perspectives on the study of behavior. *Annu Rev Neurosci* **27:** 79–105.

Greenspan RJ. 2008. The origins of behavioral genetics. *Curr Biol* **18:** R192–R198.

Hall JC. 1982. Genetics of the nervous system in *Drosophila*. *Q Rev Biophys* **15:** 223–479.

Hall JC. 1994a. The mating of a fly. *Science* **264:** 1702–1714.

Hall JC. 1994b. Pleiotropy of behavioral genes. In *Flexibility and constraint in behavioral systems* (ed. RJ Greenspan and CP Kyriacou), pp. 15–27. Dahlem Konferenzen Publications, Berlin.

Harbison ST, Carbone MA, Ayroles JA, Stone EA, Lyman RF, Mackay TFC. 2009. Co-regulated transcriptional networks contribute to natural genetic variation in *Drosophila* sleep. *Nat Genet* **41:** 371–375.

Hartman JL IV, Garvik B, Hartwell L. 2001. Principles for buffering of genetic variation. *Science* **291:** 1001–1004.

Hill WB, Caballero A. 1992. Artificial selection experiments. *Annu Rev Ecol Syst* **23:** 287–310.

Hirsch J. 1959. Studies in experimental behavior genetics. *J Comp Physiol Psychol* **52:** 304–308.

Hirsch J. 1963. Behavior genetics and individuality understood. *Science* **142:** 1436–1442.

Hirsch J, Erlenmeyer-Kimling LF. 1962. Studies in experimental behavior genetics. IV. Chromosome analyses for geotaxis. *J Comp Physiol Psychol* **55:** 732–739.

Hoyle G. 1974. Neural machinery underlying behavior in insects. In *The neurosciences: Third study program* (ed. FO Schmitt and FG Worden), pp. 397–410. MIT Press, Cambridge, MA.

Jordan KW, Carbone MA, Yamamoto A, Morgan TJ, Mackay TFC. 2007. Quantitative genomics of locomotor behavior in *Drosophila melanogaster*. *Genome Biol* **8:** R172.

Kyriacou CP, Hall JC. 1994. Genetic and molecular analysis of *Drosophila* behavior. *Adv Genet* **31:** 139–186.

Lenski RE, Travisano M. 1994. Dynamics of adaptation and diversification: A 10,000-generation experiment with bacterial populations. *Proc Natl Acad Sci* **91:** 6808–6814.

Lerman DN, Michalak P, Helin AB, Bettencourt BR, Feder ME. 2003. Modification of heat-shock gene expression in *Drosophila melanogaster* populations via transposable elements. *Mol Biol Evol* **20:** 135–144.

Lewontin RC. 1974. *The genetic basis of evolutionary change*. Columbia University Press, New York.

Littleton JT, Chapman ER, Kreber R, Garment MB, Carlson SD, Ganetzky B. 1998. Temperature-sensitive paralytic mutations demonstrate that synaptic exocytosis requires SNARE complex assembly and disassembly. *Neuron* **21:** 401–413.

Mackay TF. 2001. Quantitative trait loci in *Drosophila*. *Nat Rev Genet* **2:** 11–20.

Mackay TF, Heinsohn SL, Lyman RF, Moehring AJ, Morgan TJ, Rollmann SM. 2005. Genetics and genomics of *Drosophila* mating behavior. *Proc Natl Acad Sci* (suppl. 1) **102:** 6622–6629.

Marder E, Thirumalai V. 2002. Cellular, synaptic and network effects of neuromodulation. *Neural Netw* **15:** 479–493.

Mather K, Jinks JL. 1977. *Introduction to biometrical genetics*. Cornell University Press, Ithaca, NY.

McGuire TR. 1992. A biometrical genetic approach to chromosome analysis in *Drosophila:* Detection of epistatic interactions in geotaxis. *Behav Genet* **22:** 453–467.

Mertens I, Vandingenen A, Johnson EC, Shafer OT, Li W, Trigg JS, De Loof A, Schoofs L, Taghert PH. 2005. PDF receptor signaling in *Drosophila* contributes to both circadian and geotactic behaviors. *Neuron* **48:** 213–219.

Norga KK, Gurganus MC, Dilda CL, Yamamoto A, Lyman RF, Patel PH, Rubin GM, Hoskins RA, Mackay TF, Bellen HJ.

2003. quantitative analysis of bristle number in *Drosophila* mutants identifies genes involved in neural development. *Curr Biol* **13**: 1388–1396.

Osborne KA, Robichon A, Burgess E, Butland S, Shaw RA, Coulthard A, Pererira HS, Greenspan RJ, Sokolowski MB. 1997. Natural behavior polymorphism due to a cGMP-dependent protein kinase of *Drosophila*. *Science* **277**: 834–836.

Purves D, Lichtman JW. 1985. *Principles of neural development*. Sinauer, Sunderland, MA.

Renn SC, Park JH, Rosbash M, Hall JC, Taghert PH. 1999. A *pdf* neuropeptide gene mutation and ablation of pdf neurons each cause severe abnormalities of behavioral circadian rhythms in *Drosophila*. *Cell* **99**: 791–802.

Richmond JE, Broadie KS. 2002. The synaptic vesicle cycle: Exocytosis and endocytosis in *Drosophila* and *C. elegans*. *Curr Opin Neurobiol* **12**: 499–507.

Ricker JP, Hirsch J. 1985. Evolution of an instinct under long-term divergent selection for geotaxis in domesticated populations of *Drosophila melanogaster*. *J Comp Psychol* **99**: 380–390.

Ricker JP, Hirsch J. 1988. Genetic changes occurring over 500 generations in lines of *Drosophila melanogaster* selected divergently for geotaxis. *Behav Genet* **18**: 13–25.

Robertson A, ed. 1980. *Selection experiments in laboratory and domestic animals*. Commonwealth Agricultural Bureau, Slough, United Kingdom.

Roguev A, Bandyopadhyay S, Zofall M, Zhang K, Fischer T, Collins SR, Qu H, Shales M, Park HO, Hayles J, et al. 2008. Conservation and rewiring of functional modules revealed by an epistasis map in fission yeast. *Science* **322**: 405–410.

Simon MA, Botwell DL, Dodson GS, Laverty TR, Rubin GM. 1991. Ras1 and a putative guanine nucleotide exchange factor perform crucial steps in signalling by the sevenless protein tyrosine kinase. *Cell* **67**: 701–716.

Sokolowski MB. 1980. Foraging strategies of *Drosophila melanogaster*: A chromosomal analysis. *Behav Genet* **10**: 291–302.

Stent G. 1981. Strength and weakness of the genetic approach to the development of the nervous system. *Annu Rev Neurosci* **4**: 163–194.

Toma DP, White KP, Hirsch J, Greenspan RJ. 2002. Identification of genes involved in *Drosophila melanogaster* geotaxis, a complex behavioral trait. *Nat Genet* **31**: 349–353.

Tong AH, Lesage G, Bader GD, Ding H, Xu H, Xin X, Young J, Berriz GF, Brost RL, Chang M, et al. 2004. Global mapping of the yeast genetic interaction network. *Science* **303**: 808–813.

Tully T, Hirsch J. 1982. Behavior-genetic analysis of *Phormia regina*. I. Isolation of pure-breeding lines for high and low levels of the central excitatory state (CES) from an unselected population. *Behav Genet* **12**: 395–415.

Vargo M, Hirsch J. 1986. Biometrical and chromosome analyses of lines of *Drosophila melanogaster* selected for central excitation. *Heredity* **56**: 19–24.

Wade MJ, Goodnight CJ. 1998. Perspective: The theories of Fisher and Wright in the context of metapopulations: When nature does many small experiments. *Evolution* **52**: 1537–1553.

Weber KE. 1990. Selection on wing allometry in *Drosophila melanogaster*. *Genetics* **126**: 975–989.

Weber K. 1996. Large genetic change at small fitness cost in large populations of *Drosophila melanogaster* selected for wind tunnel flight: Rethinking fitness surfaces. *Genetics* **144**: 205–213.

Weber K, Eisman R, Morey L, Patty A, Sparks J, Tausek M, Zeng ZB. 1999. An analysis of polygenes affecting wing shape on chromosome 3 in *Drosophila melanogaster*. *Genetics* **153**: 773–786.

Weber K, Eisman R, Higgins S, Morey L, Patty A, Tausek M, Zeng ZB. 2001. An analysis of polygenes affecting wing shape on chromosome 2 in *Drosophila melanogaster*. *Genetics* **159**: 1045–1057.

Weber K, Johnson N, Champlin D, Patty A. 2005. Many P-element insertions affect wing shape in *Drosophila melanogaster*. *Genetics* **169**: 1461–1475.

Weber KE, Greenspan RJ, Chicoine DR, Fiorentino K, Thomas MH, Knight TL. 2008. Microarray analysis of replicate populations selected against a wing-shape correlation in *Drosophila melanogaster*. *Genetics* **178**: 1093–1108.

Wilson RJ, Goodman JL, Strelets VB, FlyBase Consortium. 2008. FlyBase: Integration and improvements to query tools. *Nucleic Acids Res* **36**: D588–D593.

Wright S. 1931. Evolution in Mendelian populations. *Genetics* **16**: 97–159.

Wright S. 1963. Genic interaction. In *Methods in mammalian genetics* (ed. WJ Burdette), pp. 159–192. Holden-Day, San Francisco.

Wright S. 1968. *Evolution and the genetics of populations*, Vol. 1. University of Chicago Press, Chicago, IL.

Wright S. 1982. Character change, speciation, and the higher taxa. *Evolution* **36**: 427–443.

Yamamoto A, Zwarts L, Callaerts P, Norga K, Mackay TFC, Anholt RRH. 2008. Neurogenetic networks for startle-induced locomotion in *Drosophila melanogaster*. *Proc Natl Acad Sci* **105**: 12393–12398.

The Oligogenic View of Adaptation

G. BELL

Department of Biology, McGill University, Montréal, Québec, Canada H3A 1B1
Correspondence: graham.bell@mcgill.ca

The traditional view is that evolution proceeds very slowly, over immense periods of time, driven by weak selection acting on innumerable genes of small effect. Recent studies of rapid evolution, in the laboratory and in the field, have given a radically different picture. Although beneficial mutations tend to be small in effect when they first appear, those that survive to spread and become fixed are usually among the minority with large effect. Hence, although hundreds of loci of small effect may contribute to variation in character state, adaptation is predominantly caused by alleles of large effect. This leads to the hope that the particular mutations responsible for adaptation to altered conditions of life can be identified and characterized. This has been achieved in some cases and may soon become routine. Furthermore, it raises the possibility that adaptive change can be predicted from a knowledge of genetics and ecology. Experimental evolution suggests that any given selection line that is adapting to changed conditions will follow one of a few themes (broadly speaking, loci), each of which may have many variations (mutations within the locus producing similar phenotypes). Hence, evolutionary change can be predicted only within limits, even in principle. Nevertheless, recent attempts to predict how very simple genomes change have been surprisingly successful, and we may be close to a new predictive understanding of the genetic basis of adaptation.

History is full of old battles, bitterly contested, leading to an outcome, decisive at the time, which is later quietly reversed. The struggle between saltationist and gradualist views of evolution that spanned the rediscovery of Mendelian genetics is the leading example in our own field. Darwin established the gradualist view at a very early period in the history of evolutionary biology: Natural selection is a weak force, acting on slight inappreciable variation, that leads to adaptation only after the lapse of vast periods of time. Although there were some dissenting voices (including T. Huxley and the codiscoverer A.R. Wallace), this view prevailed up to the debates about the nature of variation around the turn of the 20th century. The convoluted arguments between the biometrical and Mendelian schools have been reviewed by Provine (1971). The outcome was a delicate compromise between the apparently continuous nature of most variation, especially morphological variation, and the discrete hereditary units of Mendelism. It was achieved by supposing that many genes each make a small and equal contribution to the value of most characters, which therefore appear to be very nearly continuously distributed.

The physical basis for this theory lay in a distinction between major genes and "polygenes" (Mather 1941, 1949). Major genes are essential for normal function and development, and mutations are severely deleterious. Polygenes are more or less interchangeable, and the effect of mutation is to cause some trifling alteration of character value. Major genes are responsible for occasional "sports" differing strongly from the ancestral type. Although these figured prominently in early discussions of evolutionary genetics, especially in the Mutation Theory of de Vries (1900), they became discredited as the Mendelian interpretation of continuous variation gained ground. When they were reintroduced by Goldschmidt (1940), the phrase "hopeful monster" became a term of ridicule.

The theoretical interpretation of polygenic variation is the infinitesimal model, which was introduced by Fisher (1918) in the course of his demonstration that Mendelian inheritance necessarily gives rise to the observed pattern of correlation among relatives. Continuous characters are held to be influenced by many genes of small and equal effect, and evolutionary change can be represented in terms of the flux in frequency of these polygenes. This is a mathematical convenience that made it possible to develop Normal theory, in which the mean value of a character can change under selection, while allele frequencies remain almost unchanged. Fisher made it clear, however, that he expected this theoretical extreme to correspond quite closely to the physical basis of variation. His reason is ingenious: Alleles of large effect will be rapidly fixed or lost, leaving only those of small effect segregating in the population. This is unassailable, provided that selection acts consistently in the same direction over long periods of time.

During the half-century following the resolution of the mechanism of inheritance, the gradualist view that adaptation usually involves very slow transformation driven by weak directional selection over long periods of time became firmly entrenched. This had some unfortunate consequences. Field studies were discouraged because they would be unlikely to detect any measurable change within a research program of reasonable duration. Experimental studies of natural selection failed to flourish for the same reason. Although there were some prominent exceptions to these broad statements, the situation began to change decisively only about the time of Cold Spring Harbor Symposium XXIV in 1959.

This work is derived in part from previously published material (Bell 2008) by kind permission of Oxford University Press.

THE BASIS OF QUANTITATIVE VARIATION

It proved to be very difficult to decide whether quantitative variation is usually attributable to many genes, each with small and nearly equal effects (Mather 1941), or to a few genes of large effect that are de-Mendelized by small-effect genes or environmental variation (Robertson and Reeve 1952). A number of hypothetical "polygenes" were identified through linkage with major genes, but it was only after 1980 that it was possible to use cheap, highly polymorphic marker systems to locate the genes responsible for quantitative variation routinely and with reasonable precision. They were redubbed "quantitative trait loci," given the acronym QTLs, and thereupon became a fashionable research theme. Identifying QTLs can in principle estimate the number of genes responsible for quantitative variation and the size of their effects, although in practice, these estimates may be biased in several ways (see Erickson et al. 2004). Kearsey and Farquhar (1998) reviewed the literature and found that the average number of QTLs reported was four, with almost all studies reporting eight or fewer. They explained an average of ~50% of the variance of the character, independently of the number of QTLs. A typical QTL is thus associated with ~10% of the variance, showing that the variation of quantitative characters is often attributable to a few genes with rather large effects.

In large experiments, the distribution of QTL effects is highly skewed, with a long tail containing a few QTLs of large effect (Edwards et al. 1987 for maize; Mackay 1996 for *Drosophila;* Hayes and Goddard 2001 for livestock; Xu 2003 for barley). There may be several uninteresting reasons for this (Bost et al. 2001). QTL effects are often reported in terms of the fraction of phenotypic variance explained, which will be proportional to the square of the genetic effect, and the size of effect is biased upward because small studies will only discover large effects (and because studies that fail to detect QTLs might not be published at all). QTLs are not really loci, but rather long and variable chromosome segments between flanking markers; thus, QTLs of large effect might be long segments containing many genes of small effect. Nevertheless, orthologous QTLs are often detected in related species, showing that their effects have been accurately estimated, and the effect of QTLs is unrelated to its length. Effect size can be fitted to negative exponential (Otto and Jones 2000) or γ (Xu 2003) distributions. In short, the evidence supports Robertson's view that gene effects will be roughly exponentially distributed, with a few genes of large effect and a much greater number with small effects (Robertson 1967).

THE EFFECT OF BENEFICIAL MUTATIONS

The genes responsible for adaptation to novel conditions of growth are not necessarily representative of those contributing to the variation of the characters involved. Any character may be influenced, if only to a very small extent, by many genes—most of the genome, perhaps, if sufficiently detailed observations were available. Very few of these are likely to respond to changed conditions of growth. The reason is that a beneficial allele borne by only a few individuals is likely to be lost by drift before increasing to a frequency that will permit it to spread nearly deterministically. It was realized very early in the history of population genetics that the probability of fixation of a novel beneficial mutation is proportional to its selective advantage. Even if the great majority of novel mutations have small effects on fitness, most will soon become extinct. The mutations most likely to be fixed are those of moderate effect, whose rarity is compensated by their greater probability of survival. Consequently, the distribution of effects for the first mutation to be fixed is likely to be modal, with a peak at intermediate values (Kimura 1983).

An elegant interpretation of the initial stages of adaptation can be developed from the assumption that the wild type is likely to be very well adapted, even to conditions that have recently changed. The distribution of effect among all mutations is unknown, but the great majority are deleterious and will not contribute to adaptation. The distribution of effect among the very small minority of beneficial mutations that is alone relevant to adaptation is then predicted from general extreme-value theory, an approach pioneered by Gillespie (1984) and developed more fully by Orr (2003). Suppose that we rank all possible beneficial mutations from 1 (the fittest) to λ (the current wild-type allele) and that the difference in fitness between the top-ranked and second-ranked alleles is Δ. Then it can be shown that the average rank of the first mutation fixed is $(\lambda - 2)/4$, and the expected increase in fitness is about 2Δ. Hence, the initial step in adaptation is likely to be the substitution of a beneficial mutation of rather large effect.

This theory predicts that the distribution of the effects of beneficial mutations will shift from exponential when they first appear to modal among those that have become fixed. Kassen and Bataillon (2006) isolated a set of single mutations of *Escherichia coli* resistant to nalidixic acid and then tested them against the ancestor in medium lacking nalidixic acid. They found that 28/665 mutations increased fitness, and their effects were consistent with an exponential distribution. It is noteworthy that so large a fraction of mutations was beneficial, which reinforces the impression that in some circumstances, beneficial mutation is not very rare. Modal distributions among fixed mutations have been described for *E. coli* in minimal glucose medium (Rozen et al. 2002) and for *Pseudomonas* in serine medium (Barrett et al. 2006) by trapping mutations at or near the end of selective sweeps. In the *Pseudomonas* study, the average effect of a fixed mutation was a doubling of wild-type fitness. These experimental results, although still rather meager, seem to show quite clearly how the predominance of nascent mutations of small effect is translated into fixed mutations of generally much greater effect. It is the latter which imply that selection in novel environments will initially involve large increases in fitness.

The obvious objection to this conclusion is Fisherian: If mutations of large effect are rapidly substituted, only polygenes will remain as a source of variation. The response is that conditions change so frequently that a new series of beneficial mutations is continuously recreated. The evidence for this assertion lies beyond this review and is summarized elsewhere (Bell 2008, 2010).

IDENTIFYING THE GENES RESPONSIBLE FOR ADAPTATION

These theoretical and experimental results suggest that an oligogenic theory of adaptation may often be more appropriate than the infinitesimal model. This is not to suggest that mutation guides the course of evolution, as de Vries (1900) claimed, but rather that adaptation often involves a few discrete steps, each driven by strong selection. One of the most desirable consequences of this view is that the genes contributing to any particular episode of adaptation can be identified, because there are so few of them, and the way in which they are modified can be determined.

A start on this research project was made 30 years ago by a redoutable group of geneticists and biochemists whose work is exemplified in the volume edited by Mortlock (1984). The selection of new amidases in *Pseudomonas*, for example, is an elegant example of cumulative adaptation to refractory substrates (see Clarke 1984). The ancestral strain could metabolize only the simplest two- and three-carbon amides, acetamide and propionomide. It grows only very slowly on the four-carbon butyramide, because the native amidase is inefficient and is not induced by the new substrate. Adaptation involved the appearance of constitutive mutants that overproduced the amidase, then further modification of gene regulation, and finally a mutation in the amidase structural gene. Further selection led to strains that could grow on more complex amides, including those containing an aromatic ring. There are many similar examples documenting how bacteria adapt to exotic substrates (see Bell 2008). In many cases, the adaptive walk seems to follow a predictable succession of events involving exaptation, deregulation, amplification, and modification (the EDAM model). The particular route toward adaptation is often unique, because of the historical nature of successive substitution. However, it almost always involves a small number of mutations that have a large effect on fitness.

The development of rapid sequencing technology has now made it possible to track evolutionary change in great detail. For a detailed description of the mutations responsible for enhanced glucose uptake after ~300 generations of growth in gluose-limited chemostats, see, e.g., Notley-McRobb and Ferenci (1999a,b). In some cases, phosphotransferase activity was elevated, apparently because of loss-of-function mutations at *mlc*, a gene that regulates sugar transport. The second inner membrane system Mgl is more important at low glucose concentration and was overexpressed in almost all lines, leading to very large increases in glucose uptake. The underlying genetic changes were substitutions, frameshifts, and short insertions/deletions in both the *mgl* operator and the MglD repressor protein. At micromolar glucose concentrations, uptake at the outer membrane is undertaken mainly by the LamB glycoporin, which is regulated by *mal*, which is in turn regulated by the global repressor *mlc*. LamB activity and *mal* expression were elevated in almost all lines, as the consequence of point mutations in the *mal* structural gene and by mutations in *mlc*. Thus, the basis of adaptation to glucose-limited chemostat conditions was constitutive production of the LamB protein on the outer membrane and the Mgl proteins on the inner membrane, causing greatly increased uptake of glucose. These mutations lead to an increase in the rate of glucose transport by factors of 8–15. In glucose-limited chemostats, fitness is linearly related to glucose flux, which will depend primarily on uptake. Consequently, the first beneficial mutations fixed will often have a large effect on fitness. Dykhuizen and Hartl (1981) found that competitive fitness increased by ~13% in the first 40 generations of culture: This is a minimal estimate of the fitness effect of the first beneficial mutation to be fixed.

For organisms with small genomes, a complete accounting of the genetic basis of adaptation is now feasible. For example, a single lineage of phage φX174 typically adapts to high temperature through mutations in 10–20 nucleotides of the 5400 in its genome. Replicate lines evolve similar levels of fitness through a large number of possible beneficial mutations, some of which are unique to a particular line, whereas others recur in two or more lines. Pairs of replicate lines selected at high temperature on the same host species shared on average 20% of their beneficial mutations, and ~50% of all mutations were found in two or more of five replicate lines (Bull et al. 1997). When two replicate lines of φX174 were cultured on a novel host bacterium at high temperature, 22 mutations were fixed, of which 7 occurred in both (Wichman et al. 1999). These shared mutations were substituted in a completely different order in the two populations, however, suggesting that they act independently. In contrast, adaptation of phage T4 to high temperature consistently involved the same point mutations substituted in the same order, and go-back experiments in which stored intermediate sequences were rerun confirmed the repeatability of mutational order (Holder and Bull 2001).

As a very broad-brush conclusion, bacterial or viral evolution in a simple microcosm often seems to involve a few themes and countless variations. The few themes are the major genes where beneficial mutations can occur. The course of adaptation can often be predicted, in terms of the types of genes and proteins likely to be responsible for improvements in growth and fitness, because the number of themes is limited. It cannot be completely predicted, however, because there is usually more than one theme, and this gives rise to genetic differences between lines. The variations are the alleles of the major genes, which may be exceedingly numerous and give rise to genetic diversity within lines. At this level, the course of adaptation is scarcely predictable at all. In the initial stage of adaptation, replicate lines will discover a few broad themes and will then build on these in subsequent evolution.

Nevertheless, the ability to identify the precise mutational steps responsible for adaptation leads to the even more enticing prospect of predicting the course of adaptation from a knowledge of biochemistry and genetics. For example, most gene expression by phage T7 requires the RNA polymerase (RNAP) gene. If this is deleted, the equivalent gene from phage T3 can be used, but it is much less efficient. A single base-pair change (G→C) at a particular position (–11) in the T7 promoter produces a marked elevation in expression. Hence, Bull et al. (2007) predicted that RNAP deletion would be compensated by G→C mutations at –11 in T7 promoters when the T3 gene product was supplied in *trans*. The outcome was only partly consistent with this prediction. There was indeed a large increase in fitness caused by compensatory mutations, although only about half the promoters were modified. Furthermore, although position –11 was often modified, the most common change was G→A rather than G→C. Reviewing several similar experiments, Bull and Molineux (2008) concluded that only about one-third to one-half of fixed mutations were successfully predicted, or at least rationalized, from a knowledge of T7 biochemistry. This is scarcely failure: The phage work clearly points the way to a predictive model for population genetics quite different from the black box of polygenes.

ADAPTATION IN NATURAL POPULATIONS

It has taken half a century for microbial experimental evolution to edge into the mainstream of evolutionary biology, perhaps because many have been reluctant to believe that the relatively simple genomes of phage and bacteria could be used to understand events in complex multicellular organisms. Nevertheless, some of the classical examples of natural selection in the field that began to be systematically investigated in the 1950s, such as industrial melanism in moths and color pattern in snails, were clearly based on the effects of major genes. More recently, mutations in specified major genes have been implicated in other well-studied situations involving continuous or semicontinuous variation.

The three-spined stickleback *Gasterosteus aculeatus* is a small fish widely distributed in the colder waters of the northern hemisphere. The marine form, living in coastal and brackish water, is heavily armored with dorsal and pelvic spines capable of locking into place and connected by a series of bony plates that extend to the root of the tail. The spines and plates protect sticklebacks from predators by making them difficult to swallow, protecting them from injury, and facilitating their escape (Reimchen 2000). These marine populations have repeatedly invaded streams since the last glacial retreat 10,000–15,000 years ago, giving rise to a very large number of independently derived freshwater populations. Freshwater sticklebacks are often different in appearance from their marine ancestors, and because these differences must have evolved within a few thousand years, or perhaps much less, they have been intensively studied as examples of rapid diversification and even speciation (Hagen and Gilbertson 1973; Bell and Foster 1994; McKinnon and Rundle 2002).

The heavy armor of marine sticklebacks is strongly reduced in populations that have adapted to living in streams. Armor development is controlled largely by a single QTL (Colosimo et al. 2004) that maps to the *Ectdysoplasin* (*Eda*) gene (Colosimo et al. 2005). The product of this gene is a signal molecule that is required for normal scale development in other fish and for the development of ectodermal structures such as hair and teeth in mammals. *Eda* alleles from different populations have common ancestry, suggesting that the invasion of freshwater was accompanied by sorting low-plated alleles from the ancestral marine population. The large pelvic spines of marine sticklebacks are also reduced or completely lost in freshwater populations. Pelvic spine reduction is a Mendelian character involving a single QTL that appears to represent a regulatory mutation in *Pitx1*, a gene whose homolog is necessary for normal hindlimb development in mice (Shapiro et al. 2004). Plates and spines are reduced very rapidly in freshwater. Bell (2001) cites several cases in which marine populations have moved into newly created or newly vacated freshwater sites and have evolved greatly reduced armor and a deeper body within 10 generations or so. The agent of selection has not been identified, although deep-bodied weakly armored individuals may be more maneuverable and thus more proficient in capturing benthic invertebrates. The strong, repeatable natural selection based primarily on a few genes of large effect echoes the outcome of experimental evolution in laboratory microcosms.

A famous example of an historical process of selection driven by a known selective agent is the change of beak shape in the large ground finch (Darwin's finch) *Geospiza fortis* on the island of Daphne Major in the Galapagos. A prolonged drought in 1976–1977 caused a change in the composition of the vegetation by favoring plants with large tough-shelled seeds. These could be consumed only by finches with unusually large and powerful beaks, and between 1976 and 1978, beak depth increased at a rate of 26.1 kDar (0.66 Hal) (Boag and Grant 1981). Heavy rain in 1983 reversed the trend in the vegetation by favoring plants with smaller softer seeds that germinated more readily and thereby favored birds with smaller beaks that were more adept at processing them (Gibbs and Grant 1987). Within a few years, the response to reversed selection at a rate of 8.8 kDar (0.37 Hal) had more or less restored the status quo (Grant and Grant 1995). This study has become a classic example of selection in a nearly pristine environment, the thoroughness of the fieldwork being buttressed by detailed knowledge of the ecology of the populations and the genetics of beak shape. Beak shape is modulated by *Bmp4*, whose product is a bone morphogen, which is strongly expressed early in the development of *Geospiza* species with deep beaks but not in those with long thin beaks (Abzhanov et al. 2004; Grant et al. 2006). Thus, selection on this quantitative character may act primarily through alleles of a single gene to produce adaptation.

A third example is provided by the parallel adaptation of populations of the wood mouse *Peromyscus* to a dune habitat, principally through the evolution of paler pelage. This involves the fixation of mutations in three genes, apparently in a predictable sequence. This case is described in detail by Hoekstra (2009) in this volume.

THE OLIGOGENIC VIEW OF ADAPTATION

One of the most important advances in evolutionary biology since the last CSHL Symposium on evolutionary biology has been the ability to identify and characterize precisely the genetic changes responsible for adaptation. In many cases, it is clear that rapid evolution has been driven by strong selection acting on mutations at one or a few loci. The classical gradualist view that dominated the first century of Darwinism has become supplemented by an oligogenic interpretation that is supported by theory, laboratory experiments, and detailed analyses of selection in natural populations. One welcome outcome of this development is that we are now able to identify precisely the genetic changes that underlie adaptation. It even seems likely that we can begin to build a predictive theory of how populations will evolve in response to some defined stress. This will have very important consequences not only for the study of evolution, but also for our ability to apply evolutionary principles to situations of social and economic concern.

REFERENCES

Abzhanov A, Protas M, Grant BR, Grant PR, Tabin CJ. 2004. Bmp4 and morphological variation of beaks in Darwin's finches. *Science* **305:** 1462–1465.

Barrett RDH, MacLean RC, Bell G. 2006. Mutations of intermediate effect are responsible for adaptation in evolving *Pseudomonas fluorescens* populations. *Biol Lett* **2:** 236–238.

Bell MA. 2001. Lateral plate evolution in the threespine stickleback: Getting nowhere fast. *Genetica* **112–113:** 445–461.

Bell G. 2008. *Selection: The mechanism of evolution*, 2nd ed. Oxford University Press, Oxford.

Bell G. 2010. Fluctuating selection: The perpetual renewal of adaptation in variable environments. *Proc R Soc Lond B Biol Sci* (in press).

Bell MA, Foster SA. 1994. *The evolutionary biology of the threespine stickleback*. Oxford University Press, Oxford.

Boag PT, Grant PR. 1981. Intense natural selection in a population of Darwin's finches (Geospizinae) in the Galapagos. *Science* **214:** 82–85.

Bost B, de Vienne D, Hospital F, Moreas L, Dillmann C. 2001. Genetic and nongenetic bases for the L-shaped distribution of quantitative trait loci effects. *Genetics* **157:** 1773–1787.

Bull JJ, Molineux IJ. 2008. Predicting evolution from genomics: Experimental evolution of bacteriophage T7. *Heredity* **100:** 453–463.

Bull JJ, Badgett MR, Wichman HA, Huelsenbeck JP, Hillis DM, Gulati A, Ho C, Molineux IJ. 1997. Exceptional convergent evolution in a virus. *Genetics* **147:** 1497–1507.

Bull JJ, Springman R, Molineux IJ. 2007. Compensatory evolution in response to a novel RNA polymerase: Orthologous replacement of a central network gene. *Mol Biol Evol* **24:** 900–908.

Clarke PH. 1984. Amidases of *Pseudomonas aeruginosa*. In *Microorganisms as model systems for studying evolution* (ed. RP Mortlock), pp. 187–232. Plenum, New York.

Colosimo PF, Peichel CL, Nereng K, Blackman BK, Shapiro MD, Schluter D, Kingsley DM. 2004. The genetic architecture of parallel and armor plate reduction in threespine sticklebacks. *PLoS Biol* **2:** E109.

Colosimo PF, Hosemann KE, Balabhadra S, Villarreal G Jr, Dickson M, Grimwood J, Schmutz J, Myers RM, Schluter D, Kingsley DM. 2005. Widespread parallel evolution in sticklebacks by repeated fixation of Ectodysplasin alleles. *Science* **307:** 1928–1933.

de Vries H. 1900. *The mutation theory* (transl. JB Farmer and AD Darbishire). Open Court, Chicago, Illinois.

Dykhuizen DE, Hartl DL. 1981. Evolution of competitive ability in *Escherichia coli*. *Evolution* **35:** 581–594.

Edwards MD, Stuber CW, Wendel JF. 1987. Molecular-marker-facilitated investigations of quantitative trait loci in maize. I. Numbers, genomic distribution and type of gene action. *Genetics* **116:** 113–125.

Erickson DL, Fenster CB, Stenøien HK, Price D. 2004. Quantitative trait locus analysis and the study of evolutionary processes. *Mol Ecol* **13:** 2505–2522.

Fisher RA. 1918. The correlation between relatives on the supposition of Mendelian inheritance. *Trans R Soc Edinb* **52:** 399–433.

Gibbs HL, Grant PR. 1987. Ecological consequences of an exceptionally strong El Nino event on Darwin's finches. *Evolution* **68:** 1735–1746.

Gillespie JH. 1984. Molecular evolution over the mutational landscape. *Evolution* **38:** 1116–1129.

Goldschmidt R. 1940. *The material basis of evolution*. Yale University Press, New Haven.

Grant PR, Grant BR. 1995. Predicting microevolutionary responses to directional selection on heritable variation. *Evolution* **49:** 241–251.

Grant PR, Grant BR, Abzhanov A. 2006. A developing paradigm for the development of bird beaks. *Biol J Linnean Soc* **88:** 17–22.

Hagen DW, Gilbertson LG. 1973. Selective predation and the intensity of selection acting on the lateral plates of three-spine sticklebacks. *Heredity* **30:** 273–287.

Hayes B, Goddard ME. 2001. The distribution of the effects of genes affecting quantitative traits in livestock. *Genet Sel Evol* **33:** 209–229.

Holder KK, Bull JJ. 2001. Profiles of adaptation in two similar viruses. *Genetics* **159:** 1393–1404.

Kassen R, Bataillon T. 2006. Distribution of fitness effects among beneficial mutations before selection in experimental populations of *Pseudomonas fluorescens*. *Nat Genet* **38:** 484–488.

Kearsey MJ, Farquhar AGL. 1998. QTL analysis in plants: Where are we now? *Heredity* **80:** 137–142.

Kimura M. 1983. *The neutral theory of molecular evolution*. Cambridge University Press, Cambridge.

Mackay TFC. 1996. The nature of quantitative genetic variation revisited: Lessons from *Drosophila* bristles. *BioEssays* **18:** 113–121.

Mather K. 1941. Variation and selection of polygenic characters. *J Genet* **41:** 159–193.

Mather K. 1949. *Biometrical genetics*. Methuen, London.

McKinnon JS, Rundle HD. 2002. Speciation in nature: The threespine stickleback model systems. *Trends Ecol Evol* **17:** 480–488.

Mortlock RP, Ed. 1984. *Microorganisms as model systems for studying evolution*. Plenum, New York.

Notley-McRobb L, Ferenci T. 1999a. Adaptive *mgl*-regulatory mutations and genetic diversity evolving in glucose-limited *Escherichia coli* populations. *Environ Microbiol* **1:** 33–43.

Notley-McRobb L, Ferenci T. 1999b. The generation of multiple co-existing *mal*-regulatory mutations through polygenic evolution in glucose-limited populations of *Escherichia coli*. *Environ Microbiol* **1:** 45–52.

Orr HA. 2003. The distribution of fitness effects among beneficial mutations. *Genetics* **163:** 1519–1526.

Otto SP, Jones CD. 2000. Detecting the undetected: Estimating the total number of loci underlying a quantitative trait. *Genetics* **156:** 2093–2017.

Provine WB. 1971. *The origins of theoretical population genetics*. University of Chicago Press, Chicago, Illinois.

Reimchen TE. 2000. Predator handling failures of lateral plate morphs in *Gasterosteus aculeatus*: Functional implications for the ancestral plate condition. *Behaviour* **137:** 1081–1096.

Robertson A. 1967. The nature of quantitative genetic variation. In *Heritage from Mendel* (ed. A Brink), pp. 265–280. University of Wisconsin Press, Madison.

Robertson FW, Reeve E. 1952. Studies in quantitative inheritance. I. Effects of selection of wing and thorax length in *Drosophila melanogaster*. *J Genet* **50:** 414–448.

Rozen D, de Visser JA, Gerrish P. 2002. Fitness effects of fixed beneficial mutations in microbial populations. *Curr Biol* **12:** 1040–1045.

Shapiro MD, Marks ME, Peichel CL, Blackman BK, Nereng KS, Jónsson B, Schluter D, Kingsley DM. 2004. Genetic and developmental basis of evolutionary pelvic reduction in threespine sticklebacks. *Nature* **428:** 717–723.

Wichman HA, Badgett MR, Scott LA, Boulianne CM, Bull JJ. 1999. Different trajectories of parallel evolution during viral adaptation. *Science* **285:** 422–424.

Xu S. 2003. Estimating polygenic effects using markers of the entire genome. *Genetics* **163:** 789–801.

Genetic Dissection of Complex Traits in Yeast: Insights from Studies of Gene Expression and Other Phenotypes in the BYxRM Cross

I.M. Ehrenreich,[1–4] J.P. Gerke,[1–4] and L. Kruglyak[1–3]

[1]*Lewis-Sigler Institute for Integrative Genomics, Princeton University, Princeton, New Jersey 08544;*
[2]*Howard Hughes Medical Institute, Princeton University, Princeton, New Jersey 08544;* [3]*Department of Ecology and Evolutionary Biology, Princeton University, Princeton, New Jersey 08544*

Correspondence: leonid@genomics.princeton.edu

The genetic basis of many phenotypes of biological and medical interest, including susceptibility to common human diseases, is complex, involving multiple genes that interact with one another and the environment. Despite decades of effort, we possess neither a full grasp of the general rules that govern complex trait genetics nor a detailed understanding of the genetic basis of specific complex traits. We have used a cross between two yeast strains, BY and RM, to systematically investigate the genetic complexity underlying differences in global gene expression and other traits. The number and diversity of traits dissected to the locus, gene, and nucleotide levels in the BYxRM cross make it arguably the most extensively characterized system with regard to causal effects of genetic variation on phenotype. We summarize the insights obtained to date into the genetics of complex traits in yeast, with an emphasis on the BYxRM cross. We then highlight the central outstanding questions about the genetics of complex traits and discuss how to answer them using yeast as a model system.

Many phenotypes in nature display continuous variation and complex genetic inheritance (Falconer and Mackay 1996). Mapping the genetic basis of such traits is difficult due to the combined effects of environmental variation and multiple genetic loci (Lynch and Walsh 1998). This complexity poses a challenge for answering questions about the number of loci governing a trait, the allele frequencies and distribution of effects of these loci, and the prevalence of genetic and gene–environment interactions.

Because of the historical challenge of mapping complex traits to their multiple underlying loci, it is unknown whether general rules exist for how complex traits are specified at the genetic level. Theory predicts that a wide range of genetic architectures can exist, depending on the fitness effects of the underlying loci and the selective pressures experienced by a population (Orr 2005). The current challenge lies in empirically defining the distribution of genetic architectures that actually exist, which can only be achieved through the genetic dissection of a large number of complex traits.

The identification of the individual genes and polymorphisms underlying complex traits is essential for a full understanding of genetic architecture. For example, if a single genomic region causes variation in five traits, then only the identification of the causative gene(s) underlying a quantitative trait locus (QTL) will determine whether this architecture is due to five genes with single effects or one gene with effects on five traits. Furthermore, the identification of the specific DNA sequence polymorphisms that cause trait differences makes it possible to determine how trait variation arises at the molecular level, as well as how such causal polymorphisms are distributed across a population. Obtaining molecular resolution of complex traits remains a great challenge. However, this challenge is not insurmountable, and upcoming research holds great promise for uncovering the basic principles of trait variation.

The single-celled brewer's yeast *Saccharomyces cerevisiae* has many qualities that make it the organism of choice for the study of a large number of quantitative traits at molecular resolution. Its rapid generation time and ease of growth and maintenance, combined with the ability to keep frozen stocks indefinitely, mean that entire populations of strains can be repeatedly tested for different phenotypes under a variety of conditions. Once QTLs are identified, the small genome, high recombination rate, and the ease of site-directed mutagenesis of yeast (Storici et al. 2001) enable fine mapping and precise functional tests to determine causative genes and nucleotides.

Our lab has used a cross of two yeast strains to systematically dissect the genetic architecture of complex traits and in some cases identify causative alleles at gene and nucleotide resolution. The parents of this cross are BY4716, a common laboratory strain, and RM-11, a strain isolated from a California vineyard. These strains differ on average at 1 nucleotide every 200 base pairs (Ruderfer et al. 2006). More than 100 haploid meiotic segregants from the BYxRM F$_1$ hybrid provide independent recombinations of the two parental genomes (Fig. 1). These segregants have been genotyped genome-wide at thousands of markers and assayed for global gene expression and many

---

[4]These authors contributed equally to the writing of this paper.

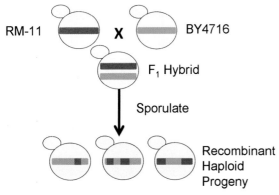

Figure 1. Crossing scheme used to create the BY×RM mapping population. Haploid parental strains are crossed to form a heterozygous diploid ("hybrid") that is then sporulated to produce recombinant haploid offspring. More than 100 segregants were sampled from the BY×RM cross.

other phenotypes. These data have provided general insights into the genetic architectures underlying complex traits. Here, we review the main lessons learned so far and examine how these lessons can guide future research on the genotype–phenotype relationship in yeast.

INSIGHTS FROM THE BY×RM CROSS

Using the BY×RM cross, our lab was the first to combine classical linkage analysis with DNA microarrays to dissect the genetic basis of global gene expression (Brem et al. 2002). Measuring gene expression across the genome enables the analysis of thousands of quantitative traits in parallel, thereby providing a route for the systematic study of genetic architecture (Fig. 2). In addition, gene expression provides a direct connection between DNA sequence and phenotype, and specific tests can determine if a gene's own sequence causes variation in its expression.

Once the basic principles of the genetics of gene expression were understood in yeast, this new approach to studying quantitative genetic variation was quickly expanded to other organisms (for review, see Rockman and Kruglyak 2006). Mapping the loci that cause gene expression variation in the BY×RM segregants has revealed a number of general insights into the genetic basis of gene expression variation and complex traits in general. Studying traits other than gene expression levels in the BY×RM cross has made it possible to connect expression variation to other phenotypes and build a more comprehensive picture of the relationship between genotype and phenotype.

Gene Expression Traits Show Complex Inheritance

A central finding of the BY×RM experiments is that most gene expression traits are complex, influenced by both multiple interacting genetic factors and the environment. Gene expression variation in the BY×RM mapping population is highly heritable, with a median heritability of 84% (Brem et al. 2002). More than 75% of all transcripts map to at least one QTL in the environments studied so far (Brem et al.

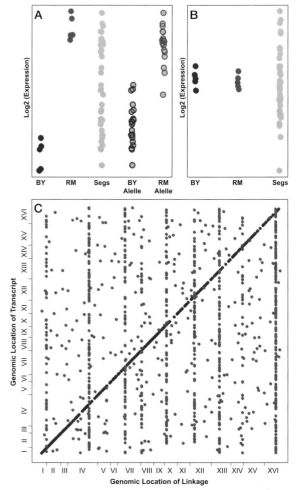

Figure 2. Primer on the genetics of global gene expression. (*A*) Example of a transcript where BY and RM differ, and a single locus has a major effect on the expression level of the segregants. (*B*) Example of a transcript showing transgressive segregation, where BY and RM have comparable expression levels but the segregants exhibit much more variation than is seen in the parents. (*C*) Plot of local (blue) and distant (red) linkages across the entire genome. A number of locations in the genome show linkage to many distant transcripts (vertical bands). These loci are referred to as hot spots. (*A–C* are illustrations and do not represent actual data.)

2002; Smith and Kruglyak 2008). However, the combined effects of the detected linkages explain less than 30% of the genetic variance for more than half of the transcripts in the genome, implying that most of the causal loci remain undetected (Brem and Kruglyak 2005). This assertion is strengthened by two observations: (1) transgressive segregation is rampant in BY×RM segregants, meaning that multiple alleles of opposite effect are present in BY and RM (Fig. 2B), and (2) many highly heritable transcripts show no linkages, suggesting that we do not have the power to detect the causal genetic basis of many expression differences (Brem and Kruglyak 2005). Together, these results indicate that gene expression variation is primarily under genetic control, but the genetic basis of this control is due to a potentially large number of loci (Brem et al. 2002; Brem and

Kruglyak 2005; Smith and Kruglyak 2008). Modeling of the transcript variation found in this cross supports a complex genetic basis for gene expression variation, suggesting that 50% of all transcripts have at least five additive QTLs and 20% of all transcripts have at least 10 additive QTLs (Brem and Kruglyak 2005).

Additional studies of the BYxRM cross have shown that both genetic and gene–environment interactions are prominent in shaping gene expression variation. Purely synthetic interactions, in which polymorphisms in multiple genes cause a phenotypic change without any of these genes having additive effects on their own, appear to be rare in the BYxRM cross (Storey et al. 2005). Nonetheless, genetic interactions are common in these data, with more than 57% of all transcripts influenced by a genetic interaction (Brem et al. 2005; Storey et al. 2005). In the majority of these interactions, one of the loci showed an additive effect, whereas the other had an effect too modest to be detected without accounting for the interaction (Brem et al. 2005; Storey et al. 2005). Recent evidence suggests that genetic interactions affecting gene expression can involve more than two loci (Litvin et al. 2009).

In addition to genetic interactions, many transcripts exhibit gene–environment interactions. Global gene expression was recently measured across the BYxRM segregants in two different carbon sources: glucose and ethanol. Nearly half of all transcripts (47%) were found to have a gene–environment effect (Smith and Kruglyak 2008). Many different types of gene–environment interactions were detected, including loci with effects only under a single condition, loci with effects in the same direction but of different magnitudes under the two conditions, and loci with effects in the opposite direction across the two conditions (Smith and Kruglyak 2008). Taken together, these results show that gene expression variation, like most complex trait variation, is due to a mixture of additive, nonadditive, and genotype-by-environment effects.

The Molecular Mechanisms of Gene Expression Variation

Because transcripts are transcribed from genes with specific genomic locations, linkages for expression traits can be separated into two classes: (1) local linkages that arise from polymorphisms in or near the differentially expressed genes themselves and (2) distant linkages that give rise to expression variation at unlinked transcripts (Fig. 2C). Local linkages are common in the BYxRM cross, with more than 25% of all genes in the yeast genome showing local linkage (Ronald et al. 2005). Classic *cis-trans* tests using allele-specific expression assays in a diploid hybrid of BY and RM demonstrated that between 52% and 78% of local linkages were due to *cis*-regulatory polymorphisms at the differentially expressed genes themselves (Ronald et al. 2005). Not all *cis* effects could be explained by polymorphisms in transcription-factor-binding sites, suggesting that variants in a number of sequence classes, including within transcripts and 3'UTRs (untranslated regions), can give rise to *cis* effects. Ronald et al. (2005) also showed that local linkages can act in *trans*, because allele-specific measurements for multiple transcripts were inconsistent with *cis* effects on expression. In addition, an amino acid polymorphism in *AMN1* was shown to cause an increase in *AMN1*'s expression through a feedback loop. Population genetic modeling based on the *cis* linkages in the BYxRM cross suggests that if gene expression were mapped across a wider range of *S. cerevisiae* isolates, every gene in the genome might be found to have a *cis*-regulatory polymorphism (Ronald and Akey 2007).

Consistent with the complex genetic architecture of gene expression variants, extensive distant linkages are observed in the BYxRM cross. Many of these *trans* effects stem from a small number of hot spots that influence the transcript abundance of tens to hundreds of genes (Fig. 3) (Brem et al. 2002; Yvert et al. 2003; Smith and Kruglyak 2008). Eight hot spots, each being linked to 7–94 genes of related function, were observed in the initial study of this cross (Brem et al. 2002), and subsequent studies identified additional hot spots (Smith and Kruglyak 2008). Positional cloning and bioinformatic analyses of distant linkages at both *trans* hot spots and other loci have shown that these effects are often not due to transcription factor polymorphisms (Yvert et al. 2003).

Analysis of global gene expression in BYxRM segregants in both glucose and ethanol showed that distant and local linkages respond very differently across environments. Local effects were more consistent across environments, whereas distant effects exhibited greater sensitivity to environment. Thirteen *trans* hot spots were found in each environment, but only seven were common between the two environments, illustrating the potential importance of distant linkages and, more specifically, *trans* hot spots in mediating transcriptional responses to the environment (Smith and Kruglyak 2008). Strikingly, 78% of the time, distant linkages had opposite effects in glucose and ethanol, suggesting that distant linkages frequently display gene–environment interactions (Smith and Kruglyak 2008).

Genetic Dissection of Other Types of Traits in the BYxRM Cross

The BYxRM segregants have been used to map a variety of higher-order phenotypes reflecting different aspects of yeast cell biology and physiology. Along and in combination with the genetics of gene expression, the study of these traits has shed light on basic principles regarding the genetics of complex traits.

Gene expression differences should affect cellular phenotypes through changes in protein abundance. Measuring genetic variation in protein abundance is therefore a logical next step in the analysis of complex traits in yeast. Studying protein abundance, like gene expression, also allows the comparison of the effects of local and distant linkages. Mass spectrometry enables the analysis of many proteins in parallel. Foss et al. (2007) used this technique to measure variation in the abundance of ~200 proteins among the BYxRM segregants. The detected protein abundance linkages mirror the gene expression linkages in

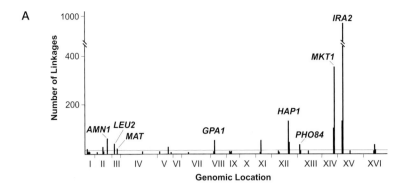

Figure 3. Compendium of the hot spots observed in the BYxRM cross. The locations (*A*) and descriptions (*B*) of a subset of the hot spots in the genome. This plot is a modified version of a plot in Smith and Kruglyak (2008). Causal genes for each of these hot spots are named, although in some cases it is unclear whether coincident hot spots across traits are due to different closely linked genes or to the pleiotropic effects of a single gene. For instance, the *AMN1* gene expression hot spot is likely due to multiple genes with effects on different transcripts (Yvert et al. 2003). Red horizontal line in *A* indicates the threshold for calling a locus a hot spot. References for *B* are provided in the text.

several ways. As with variation in gene expression, variation in protein abundance shows high heritability. The average heritability in protein abundance was 0.62, and ~100 linkages were identified. Despite high heritability, 38% of the peptides analyzed showed no significant linkages. Because most traits under the control of a single locus should show linkage with the statistical power available, this result means that like gene expression variation, variation in protein abundance likely has a complex genetic basis. Also consistent with previous results in the BYxRM cross, there are clear examples of transgressive segregation in protein abundance, and four hot spots were observed in which the abundance of at least five proteins links to the same genomic location.

The results of Foss et al. (2007) provide interesting comparisons between variation in transcript levels and protein abundance. There is significant overlap between genes that varied in both the transcript and protein data sets. But this overlap is not complete, because only 43% of genes with protein abundance differences between BY and RM showed detectable differences in gene expression. In addition, only three of the gene expression hot spots resulted in protein abundance hot spots, and one protein abundance hot spot did not have a corresponding hot spot for gene expression. These results show that there are many cases where a gene expression difference does not cause a change in protein levels, and there are also cases where differences in protein abundance arise only from posttranscriptional effects.

Another way to analyze a large number of traits at once is through the application of libraries of small molecules. Perlstein et al. (2007) measured differences in the growth of the BYxRM segregants in response to a panel of 100 molecules that included various classes of compounds and some FDA-approved medications. Linkage analysis identified 124 QTLs involved in small-molecule response. Resistance to small molecules shows a substantial genetic component, because more than half of the compounds link to at least one QTL. Many compounds link to multiple QTLs, indicating a complex genetic basis. There are clear examples of transgressive segregation, and there is striking overlap between hot spots for small-molecule resistance and gene expression variation. Eight regions of the genome link significantly to at least four small molecules, and seven of these regions overlap with hot spots for gene expression variation. These results suggest that gene expression variation, protein abundance, and small-molecule resistance have similarly complex genetic architectures, and in some cases they may share a common functional basis. Accordingly, a missense mutation in *PHO84*, a phosphate transporter, underlies one of the small-molecule hot spots and may explain gene expression variation linking to this region (Perlstein et al. 2007).

Other traits studied in the BYxRM cross include telomere length (Gatbonton et al. 2006), morphological variation (Nogami et al. 2007), noise in gene expression (Ansel et al. 2008), and sensitivity to DNA-damaging agents (Demogines et al. 2008). In all cases, the genetic basis of variation is complex. This observation is consistent with results from additional crosses in other strain backgrounds, which have identified complex genetic variation in ethanol production (Hu et al. 2007), high-temperature growth (Steinmetz et al. 2002; Sinha et al. 2008), and sporulation efficiency (Deutschbauer and Davis 2005; Gerke et al. 2009). Just as with multicellular organisms, the genotype–phenotype relationship in yeast is often due to allelic variation at multiple genes that exhibit pleiotropy, gene–gene interaction, and gene–environment interaction.

How Prevalent Are Hot Spots?

One of the most striking observations across the many data sets generated for the BYxRM cross is the importance of the QTL hot spots. Seven of eight hot spots governing small-molecule resistance overlap with hot spots for gene expression variation, as do three of the four protein abundance hot spots. Additionally, causative alleles in hot-spot regions are known to affect small-molecule resistance, sensitivity to DNA-damaging agents, 16 cell size and shape parameters, sporulation efficiency, and high-temperature growth (Fig. 3). These results strongly suggest that some complex traits in the BYxRM cross share a functional basis due to a set of polymorphisms with widespread effects.

To date, most of the hot spots in the BYxRM cross have been found to be due to derived alleles only present in the BY background. This suggests that the polymorphisms that cause hot spots may be at low frequency in general, potentially because they are deleterious under most conditions (Ronald and Akey 2007). This hypothesis is consistent with comparisons of gene expression variation with sequence divergence in multiple species, which indicate that gene expression variation is subject to stabilizing selection (Lemos et al. 2005; Bedford and Hartl 2009). This trend is also seen in mutation-accumulation experiments, which have shown that random mutations affect gene expression to a greater degree than polymorphisms that reach fixation between species (Rifkin et al. 2005). Together, these studies suggest that mutations with large effects on gene expression are typically culled from populations. Perhaps some hot spots have been maintained because of lineage-specific positive selection that negates or outweighs any deleterious effects of having a major regulatory change (Ronald and Akey 2007; Lang et al. 2009). This scenario is especially plausible when considering the BY strain, which has been suggested to have undergone elevated rates of evolution due to manipulation in the lab (Gu et al. 2005). Because there have been no studies to date to examine genetic architecture across a large set of traits in any yeast cross besides BYxRM, the true prevalence of hot spots in *S. cerevisiae* remains to be empirically determined.

Summary

Important quantitative genetic principles and molecular mechanisms underlying trait variation have been identified through the genetic dissection of global gene expression and other trait variation in the BYxRM cross. Because complex traits in yeast resemble those of higher organisms, we expect yeast to continue to be a valuable model system for quantitative and evolutionary genetics. However, at present, we have only scratched the surface in terms of using yeast to understand complex traits. In the following sections, we discuss more specifically the enigmas that remain and what resources and methods will be necessary to elucidate them.

LOOKING FORWARD

The BYxRM cross has provided many insights into the genetic architecture of trait variation between two strains. One challenge now is to determine how genetic architectures vary across an entire population (Fig. 4). For example, in a set of parallel crosses between multiple strains, would we observe the same QTLs governing a trait or do different loci arise in each case? Do traits tend to be governed by rare or common polymorphisms? To what extent do the answers to these questions depend on population structure? Understanding how the basis of complex traits varies in a population is central to improving the methods used to dissect the genetic basis of common human diseases.

The yeast population provides a rich resource of variation among strains. *S. cerevisiae* strains have been isolated from a wide variety of habitats such as vineyards, palm wine, rotting fruit, oak trees, sake fermentations, Bertram palms, and immunocompromised humans. An important feature of the yeast population is that the strains from some of these habitats, including vineyard and sake strains, appear to be genetically isolated (Fay and Benavides 2005). Others, such as the clinical isolates, appear to derive from various sources but are likely adapted to a common niche (Schacherer et al. 2009). Recently, comprehensive surveys of sequence variation were performed for many yeast strains (Liti et al. 2009; Schacherer et al. 2009).

The environments in which yeast can be found impose unique stresses and selective pressures, and this variation is reflected in the phenotypic diversity among yeast isolates from different sources. For example, vineyard isolates have likely experienced selection for resistance to the antifungal agent copper sulfate (Fay et al. 2004). Genetic evidence suggests that oak tree strains exhibit high sporulation efficiency and freeze-tolerance relative to other strains due to selective pressures in woodland environments (Gerke et al. 2006; Kvitek et al. 2008). A systematic study across 52 yeast strains demonstrated a high degree of phenotypic variation in both gene expression and response to stresses (Kvitek et al. 2008). *S. cerevisiae* provides an opportunity to link molecular genetic variation, phenotypic variation, natural selection, and population structure across a variety of strains and environments.

Other yeast species offer alternative demographic scenarios not observed in *S. cerevisiae*. *S. paradoxus* is often found in the same soil and oak tree samples as *S. cerevisiae* (Sniegowski et al. 2002; JP Gerke, unpubl.). Unlike *S. cerevisiae*, however, *S. paradoxus* has not been the subject of domestication by humans, and it displays a fundamentally different population structure. Whereas *S. cerevisiae* displays a population structure that correlates primarily with habitat (Fay and Benavides 2005; Schacherer et al. 2009), genetic differentiation in *S. paradoxus* correlates prima-

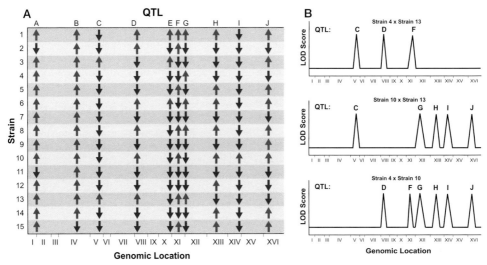

Figure 4. Genetic architecture across a population. (*A*) QTLs causing variation in a trait segregate with varying allele frequencies across 15 strains in a population. The upward red arrows and downward blue arrows are opposite effect alleles that cause relative increases or decreases in the trait value of an individual. (*B*) As a result of the large number of loci affecting the trait in the population, in an intercross, both the number and locations of QTLs will depend on the two parental strains chosen for the cross. In the three intercrosses chosen here, QTLs "C" and "I" appear to be common, but identifying the causal polymorphisms in the other 12 strains would reveal that the minor alleles are actually rare.

rily with geographical distance (Liti et al. 2009). Some isolates of *S. paradoxus* display increased reproductive isolation (Sniegowski et al. 2002). The attributes that make *S. cerevisiae* an ideal model for molecular quantitative genetics—a small genome and efficient homologous recombination—are also present in *S. paradoxus*. Studies of genetic and phenotypic variation among the geographically diverged *S. paradoxus* strains will provide a useful complement to research in *S. cerevisiae*.

APPROACHES TO DISSECTING COMPLEX TRAITS IN YEAST

Yeast represents a powerful system for studying complex traits because it permits the use of all conventional mapping methods and is especially well suited to high-throughput genotyping and phenotyping. High recombination rate, small genome, and the ease of gene replacement enable rapid dissection of QTLs to individual genes. We present a brief summary of the approaches used to map QTLs in yeast, along with their respective advantages and disadvantages.

Methods for Mapping Complex Traits to Genomic Loci

Linkage mapping. As described earlier in the discussion of the BYxRM cross, linkage mapping is a powerful approach for identifying QTLs. Linkage mapping uses related individuals with known pedigrees to identify causal genomic loci (Lynch and Walsh 1998). Within the context of laboratory experiments, two inbred strains are typically crossed to produce segregating progeny. Most commonly, interval mapping (Lander and Botstein 1989), which defines regions of the genome that exhibit linkage to the trait of interest, is used to identify QTLs in a mapping population of segregants. The limitations of linkage mapping are that only a small number of parental genomes (typically two) are compared, and that identified QTLs often span a large number of genes. Indeed, even in yeast, QTLs typically span dozens of genes or more, making it difficult to localize the QTLs to the underlying quantitative trait genes (QTGs) through this approach alone.

Association mapping. Association mapping identifies correlations between genetic markers and traits using panels of individuals sampled from natural populations. Linkage disequilibrium (LD) determines the extent of the associated genomic region. LD decays rapidly in yeast (Schacherer et al. 2009), meaning that this approach should have a very high mapping resolution. Association mapping samples a much larger fraction of the genetic variation in a species than typical linkage mapping experiments. The major limitation of association mapping is its high false-positive rate, which arises from noncausal genotype–phenotype correlations due to population structure. Numerous methods exist to control for population structure while performing statistical tests for association (Price et al. 2006; Yu et al. 2006), but these structured association mapping techniques also cause a high rate of false negatives, often eliminating the signal of association from true positive loci that are strongly correlated with population structure (Zhao et al. 2007).

Hybrid linkage/association approaches. The respective advantages of linkage and association mapping can be combined by creating multiparent inbred line populations (Churchill et al. [The Complex Traits Consortium]

2004; Macdonald and Long 2007; Kover et al. 2009) or multiple two-parent mapping populations that share a common parent (Yu et al. 2008). In such populations, linkage mapping can be used to identify QTLs, and association mapping can be used to refine the QTLs to smaller genomic intervals. In yeast, which has an innately high recombination rate, it may be that multiparent mapping populations will allow causal loci to be mapped to a high resolution, perhaps to specific genes.

Bulk segregant analysis. A final, underused approach to mapping QTLs is bulk segregant analysis (BSA). In BSA, segregants with extreme phenotypes are genotyped collectively, and loci with significantly biased allele frequencies represent candidate QTLs (Michelmore et al. 1991). This approach has been previously used in yeast to map an auxotophy, a growth defect on acetate, a locus involved in flocculation (Brauer et al. 2006), loci involved in adaptation to fluctuating carbon sources (Segre et al. 2006), resistance to DNA-damaging agents (Demogines et al. 2008), and a locus involved in resistance to leucine starvation (Boer et al. 2008). A major limitation in this approach is that it requires very precise, quantitative genotyping for a large number of markers. Previously, this had been a challenge; however, with easy custom array design and next-generation sequencing, there are now multiple options to measure allele frequencies in DNA pools. The advantage of this approach is that unlike other approaches, it permits a much larger number of individuals to be surveyed, which means that it is likely to be more powerful in detecting the multiple loci with modest effects that underlie complex traits. Despite its promise, BSA has yet to be used effectively to map multiple loci underlying a complex trait.

Methods for refining QTLs to Genes or Causal Polymorphisms

Reciprocal hemizygosity. In this approach, two nearly identical strains are compared. Both strains are diploids descended from the same two parental strains and are hemizygous for a putative QTG. The only difference between the strains is which parent's allele of the putative QTG they possess. Comparison of these two hemizygous strains makes it possible to test the effects of the two alleles in genetic backgrounds that are otherwise isogenic. Deletion collections exist in yeast in which there is a DNA bar-coded single-gene knockout strain for nearly every nonessential gene in the genome (Winzeler et al. 1999). These deletion collections allow reciprocal hemizygosity tests to be conducted on a genome-wide scale, either by screening with individual knockout strains en masse or by competing all knockout strains in heterozygous diploid pools (Steinmetz and Davis 2004). One drawback of this approach is that it does not provide nucleotide-level resolution.

Allele swaps. Perhaps the greatest advantage of yeast is the ease with which targeted gene replacement can be done. Homologous recombination can be used to create lines that differ at a specific site but are otherwise isogenic. If necessary, allele swap lines can be generated for all genes in a QTL and used to delineate the causal gene(s) underlying the QTL. Various allele replacement lines can then be intercrossed to directly measure the effects of allelic interactions. A drawback of this approach is that if genetic interactions exist, allele replacements may need to be done in a specific genetic background.

Introgression and targeted recombination. An additional approach to isolating QTLs is by recurrent crossing into a different background, using either phenotypic or marker-assisted selection. This approach can increase the power to identify small-effect QTLs by eliminating other loci with effects and by isolating a QTL allele from one parent in a genetic background that is largely composed of the other parent. This iterative crossing approach can be used to break down the amount of flanking parent-of-origin sequence, whittling down a QTL to a smaller genomic region. Either independently or in concert with introgression, selectable flanking markers can be used to identify segregants with recombination events specifically in a QTL. These targeted recombinants can be used to rapidly fine map a QTL.

NEEDED RESOURCES FOR THE FUTURE

Attaining a more comprehensive understanding of complex traits will require identifying a large fraction of the polymorphisms that encode trait differences within and between yeast species. Accomplishing this will require new mapping populations and high-throughput genotyping and phenotyping approaches.

A prerequisite to creating new mapping populations is the identification of a large number of genome-wide markers. Obtaining such markers is now simple: It can be done by resequencing 10 or so strains in a single lane of an Illumina Genome Analyzer 2 run. In addition, an already published population genomics project has produced low-coverage genome sequences for a number of strains for both *S. cerevisiae* and *S. paradoxus* (Liti et al. 2009). Exisiting genome-wide polymorphism data provide a foundation for understanding the genetic diversity and population structure of these species (Liti et al. 2009; Schacherer et al. 2009). Additionally, next-generation sequencing technologies can be used for rapid targeted resequencing of strains with phenotypes of interest or genotyping of new mapping populations.

A central consideration going forward is whether yeast needs an infrastructure to promote community collaboration in the dissection of complex traits, as exists for *Arabidopsis thaliana* (Weigel and Mott 2009), *Drosophila* (Mackay et al. 2008), and mouse (Churchill et al. [The Complex Traits Consortium] 2004). In a sense, the simplicity of yeast diminishes the need for collaboration. At the same time, a concerted community effort to dissect the genetic basis of many traits to as many causal polymorphisms as possible in as many backgrounds as is feasible could result in a basic understanding of complex traits that may be impossible to achieve in any other organism.

The cloning of individual QTGs has been accomplished in a wide range of organisms, and the challenge now is to advance to a new level of understanding of complex traits. To achieve this, we must identify all QTGs for many genetically complex traits across a diverse set of strains and determine the underlying variants at the nucleotide level. Such a comprehensive undertaking is essential if we are to understand mechanistically the genetic architecture and evolutionary implications of complex traits, and if we are to harness natural variation for applied purposes.

ACKNOWLEDGMENTS

The authors thank Rachel Brem, Gael Yvert, Erin Smith, Eric Foss, James Ronald, Josh Akey, Ethan Perlstein, John Storey, Jackie Whittle, and other Kruglyak lab members and collaborators for their contributions to the research based on the BYxRM cross. This work was supported by National Institutes of Health grant MH059520 and a James S. McDonnell Foundation Centennial Fellowship to L.K., National Institutes of Health grant GM071508 to the Lewis-Sigler Institute, and the Howard Hughes Medical Institute.

REFERENCES

Ansel J, Bottin H, Rodriguez-Beltran C, Damon C, Nagarajan M, Fehrmann S, Francois J, Yvert G. 2008. Cell-to-cell stochastic variation in gene expression is a complex genetic trait. *PLoS Genet* **4:** e1000049.

Bedford T, Hartl DL. 2009. Optimization of gene expression by natural selection. *Proc Natl Acad Sci* **106:** 1133–1138.

Boer VM, Amini S, Botstein D. 2008. Influence of genotype and nutrition on survival and metabolism of starving yeast. *Proc Natl Acad Sci* **105:** 6930–6935.

Brauer MJ, Christianson CM, Pai DA, Dunham MJ. 2006. Mapping novel traits by array-assisted bulk segregant analysis in *Saccharomyces cerevisiae*. *Genetics* **173:** 1813–1816.

Brem RB, Kruglyak L. 2005. The landscape of genetic complexity across 5,700 gene expression traits in yeast. *Proc Natl Acad Sci* **102:** 1572–1577.

Brem RB, Yvert G, Clinton R, Kruglyak, L. 2002. Genetic dissection of transcriptional regulation in budding yeast. *Science* **296:** 752–755.

Brem RB, Storey JD, Whittle J, Kruglyak L. 2005. Genetic interactions between polymorphisms that affect gene expression in yeast. *Nature* **436:** 701–703.

Churchill GA, Airey DC, Allayee H, Angel JM, Attie AD, Beatty J, Beavis WD, Belknap JK, Bennett B, Berrettini W, et al. (The Complex Traits Consortium) 2004. The Collaborative Cross, a community resource for the genetic analysis of complex traits. *Nat Genet* **36:** 1133–1137.

Demogines A, Smith E, Kruglyak L, Alani E. 2008. Identification and dissection of a complex DNA repair sensitivity phenotype in baker's yeast. *PLoS Genet* **4:** e1000123.

Deutschbauer AM, Davis RW. 2005. Quantitative trait loci mapped to single-nucleotide resolution in yeast. *Nat Genet* **37:** 1333–1340.

Falconer D, Mackay T. 1996. *Introduction to quantitative genetics*. Pearson Prentice Hall, Harlow, England.

Fay JC, Benavides JA. 2005. Evidence for domesticated and wild populations of *Saccharomyces cerevisiae*. *PLoS Genet* **1:** 66–71.

Fay JC, McCullough HL, Sniegowski PD, Eisen MB. 2004. Population genetic variation in gene expression is associated with phenotypic variation in *Saccharomyces cerevisiae*. *Genome Biol* **5:** R26.

Foss EJ, Radulovic D, Shaffer SA, Ruderfer DM, Bedalov A, Goodlett DR, Kruglyak L. 2007. Genetic basis of proteome variation in yeast. *Nat Genet* **39:** 1369–1375.

Gatbonton T, Imbesi M, Nelson M, Akey JM, Ruderfer DM, Kruglyak L, Simon JA, Bedalov A. 2006. Telomere length as a quantitative trait: Genome-wide survey and genetic mapping of telomere length-control genes in yeast. *PLoS Genet* **2:** e35.

Gerke JP, Chen CT, Cohen BA. 2006. Natural isolates of *Saccharomyces cerevisiae* display complex genetic variation in sporulation efficiency. *Genetics* **174:** 985–997.

Gerke J, Lorenz K, Cohen B. 2009. Genetic interactions between transcription factors cause natural variation in yeast. *Science* **323:** 498–501.

Gu Z, David L, Petrov D, Jones T, Davis RW, Steinmetz LM. 2005. Elevated evolutionary rates in the laboratory strain of *Saccharomyces cerevisiae*. *Proc Natl Acad Sci* **102:** 1092–1097.

Hu XH, Wang MH, Tan T, Li JR, Yang H, Leach L, Zhang RM, Luo ZW. 2007. Genetic dissection of ethanol tolerance in the budding yeast *Saccharomyces cerevisiae*. *Genetics* **175:** 1479–1487.

Kover PX, Valdar W, Trakalo J, Scarcelli N, Ehrenreich IM, Purugganan MD, Durrant C, Mott R. 2009. A multiparent advanced generation inter-cross to fine-map quantitative traits in *Arabidopsis thaliana*. *PLoS Genet* **5:** e1000551.

Kvitek DJ, Will JL, Gasch AP. 2008. Variations in stress sensitivity and genomic expression in diverse *S. cerevisiae* isolates. *PLoS Genet* **4:** e1000223.

Lander ES, Botstein D. 1989. Mapping Mendelian factors underlying quantitative traits using RFLP linkage maps. *Genetics* **121:** 185–199.

Lang GI, Murray AW, Botstein D. 2009. The cost of gene expression underlies a fitness trade-off in yeast. *Proc Natl Acad Sci* **106:** 5755–5760.

Lemos B, Meiklejohn CD, Caceres M, Hartl DL. 2005. Rates of divergence in gene expression profiles of primates, mice, and flies: Stabilizing selection and variability among functional categories. *Evolution* **59:** 126–137.

Liti G, Carter DM, Moses AM, Warringer J, Parts L, James SA, Davey RP, Roberts IN, Burt A, Koufopanou V, et al. 2009. Population genomics of domestic and wild yeasts. *Nature* **458:** 337–341.

Litvin O, Causton HC, Chen BJ, Pe'er D. 2009. Modularity and interactions in the genetics of gene expression. *Proc Natl Acad Sci* **106:** 6441–6446.

Lynch M, Walsh B. 1998. *Genetics and analysis of quantitative traits*. Sinauer, Sunderland, MA.

Macdonald SJ, Long AD. 2007. Joint estimates of quantitative trait locus effect and frequency using synthetic recombinant populations of *Drosophila melanogaster*. *Genetics* **176:** 1261–1281.

Mackay TF, Richards S, Gibbs R. 2008. Proposal to sequence a *Drosophila* genetic reference panel: A community resource for the study of genotypic and phenotypic variation. *White Paper*.

Michelmore RW, Paran I, Kesseli RV. 1991. Identification of markers linked to disease-resistance genes by bulked segregant analysis: A rapid method to detect markers in specific genomic regions by using segregating populations. *Proc Natl Acad Sci* **88:** 9828–9832.

Nogami S, Ohya Y, Yvert G. 2007. Genetic complexity and quantitative trait loci mapping of yeast morphological traits. *PLoS Genet* **3:** e31.

Orr HA. 2005. The genetic theory of adaptation: A brief history. *Nat Rev Genet* **6:** 119–127.

Perlstein EO, Ruderfer DM, Roberts DC, Schreiber SL, Kruglyak L. 2007. Genetic basis of individual differences in the response to small-molecule drugs in yeast. *Nat Genet* **39:** 496–502.

Price AL, Patterson NJ, Plenge RM, Weinblatt ME, Shadick NA, Reich D. 2006. Principal components analysis corrects for stratification in genome-wide association studies. *Nat Genet* **38:** 904–909.

Rifkin SA, Houle D, Kim J, White KP. 2005. A mutation accumulation assay reveals a broad capacity for rapid evolution of gene expression. *Nature* **438:** 220–223.

Rockman MV, Kruglyak L. 2006. Genetics of global gene expression. *Nat Rev Genet* **7:** 862–872.

Ronald J, Akey JM. 2007. The evolution of gene expression QTL in *Saccharomyces cerevisiae*. *PLoS ONE* **2:** e678.

Ronald J, Brem RB, Whittle J, Kruglyak L. 2005. Local regulatory variation in *Saccharomyces cerevisiae*. *PLoS Genet* **1:** e25.

Ruderfer DM, Pratt SC, Seidel HS, Kruglyak L. 2006. Population genomic analysis of outcrossing and recombination in yeast. *Nat Genet* **38:** 1077–1081.

Schacherer J, Shapiro JA, Ruderfer DM, Kruglyak L. 2009. Comprehensive polymorphism survey elucidates population structure of *Saccharomyces cerevisiae*. *Nature* **458:** 342–345.

Segre AV, Murray AW, Leu JY. 2006. High-resolution mutation mapping reveals parallel experimental evolution in yeast. *PLoS Biol* **4:** e256.

Sinha H, David L, Pascon RC, Clauder-Munster S, Krishnakumar S, Nguyen M, Shi G, Dean J, Davis RW, Oefner PJ, et al. 2008. Sequential elimination of major-effect contributors identifies additional quantitative trait loci conditioning high-temperature growth in yeast. *Genetics* **180:** 1661–1670.

Smith EN, Kruglyak L. 2008. Gene-environment interaction in yeast gene expression. *PLoS Biol* **6:** e83.

Sniegowski PD, Dombrowski PG, Fingerman E. 2002. *Saccharomyces cerevisiae* and *Saccharomyces paradoxus* coexist in a natural woodland site in North America and display different levels of reproductive isolation from European conspecifics. *FEMS Yeast Res* **1:** 299–306.

Steinmetz LM, Davis RW. 2004. Maximizing the potential of functional genomics. *Nat Rev Genet* **5:** 190–201.

Steinmetz LM, Sinha H, Richards DR, Spiegelman JI, Oefner PJ, McCusker JH, Davis RW. 2002. Dissecting the architecture of a quantitative trait locus in yeast. *Nature* **416:** 326–330.

Storey JD, Akey JM, Kruglyak L. 2005. Multiple locus linkage analysis of genomewide expression in yeast. *PLoS Biol* **3:** e267.

Storici F, Lewis LK, Resnick MA. 2001. In vivo site-directed mutagenesis using oligonucleotides. *Nat Biotechnol* **19:** 773–776.

Weigel D, Mott R. 2009. The 1001 Genomes project for *Arabidopsis thaliana*. *Genome Biol* **10:** 107.

Winzeler EA, Shoemaker DD, Astromoff A, Liang H, Anderson K, Andre B, Bangham R, Benito R, Boeke JD, Bussey H, et al. 1999. Functional characterization of the *S. cerevisiae* genome by gene deletion and parallel analysis. *Science* **285:** 901–906.

Yu J, Pressoir G, Briggs WH, Vroh Bi I, Yamasaki M, Doebley JF, McMullen MD, Gaut BS, Nielsen DM, Holland JB, et al. 2006. A unified mixed-model method for association mapping that accounts for multiple levels of relatedness. *Nat Genet* **38:** 203–208.

Yu J, Holland JB, McMullen MD, Buckler ES. 2008. Genetic design and statistical power of nested association mapping in maize. *Genetics* **178:** 539–551.

Yvert G, Brem RB, Whittle J, Akey JM, Foss E, Smith EN, Mackelprang R, Kruglyak L. 2003. *Trans*-acting regulatory variation in *Saccharomyces cerevisiae* and the role of transcription factors. *Nat Genet* **35:** 57–64.

Zhao K, Aranzana MJ, Kim S, Lister C, Shindo C, Tang C, Toomajian C, Zheng H, Dean C, Marjoram P, Nordborg M. 2007. An *Arabidopsis* example of association mapping in structured samples. *PLoS Genet* **3:** e4.

Measuring Natural Selection on Genotypes and Phenotypes in the Wild

C.R. LINNEN AND H.E. HOEKSTRA
Department of Organismic and Evolutionary Biology and Museum of Comparative Zoology, Harvard University, Cambridge, Massachusetts 02138
Correspondence: hoekstra@oeb.harvard.edu

A complete understanding of the role of natural selection in driving evolutionary change requires accurate estimates of the strength of selection acting in the wild. Accordingly, several approaches using a variety of data—including patterns of DNA variability, spatial and temporal changes in allele frequencies, and fitness estimates—have been developed to identify and quantify selection on both genotypes and phenotypes. Here, we review these approaches, drawing on both recent and classic examples to illustrate their utility and limitations. We then argue that by combining estimates of selection at multiple levels—from individual mutations to phenotypes—and at multiple timescales—from ecological to evolutionary—with experiments that demonstrate *why* traits are under selection, we can gain a much more complete picture of the adaptive process.

To account for adaptation—the remarkably precise fit between organisms and their environments—Darwin (1859) and Wallace (1858) independently proposed the theory of evolution by natural selection. One hundred and fifty years later, we have amassed a large body of theoretical and empirical work that extensively describes and documents natural selection, yet we are still largely ignorant as to how, exactly, natural selection acting on beneficial mutations leads to adaptation (Orr 2005). There are many fundamental questions that remain largely unanswered. These include: Where do most adaptive genetic variants come from—ancestral variation or de novo mutation? How strong is selection, on average, and does its strength vary for different types of phenotypic traits? Does the strength of selection change in a predictable way as mutations are fixed and populations approach phenotypic optima? Do most adaptations involve a small number of genes with large phenotypic effects or many genes of small effect? To what extent do competing ecological demands, genetic linkage, and pleiotropy constrain adaptation? Finally, how often does natural selection rely on the same genes and/or mutations to drive convergent evolution? Answering these questions is challenging because it requires knowing the precise phenotypic targets of selection, identifying the genetic loci contributing to those adaptive traits, and measuring the strength of selection acting on both phenotypes and genotypes.

In a now classic book, Endler (1986) compiled evidence for selection in natural populations from a diversity of species measured using a variety of approaches. This and other widely used references (see, e.g., Hartl and Clark 2007) classify approaches to estimating selection into two broad categories: those that can be applied to discrete polymorphisms and those for continuous characters. These approaches have developed largely in isolation (but see Kimura and Crow 1978; Milkman 1978); however, with increased power to link genotype to phenotypes in natural populations, it is now possible to estimate selection at multiple levels of biological organization simultaneously and thus use a diverse set of complementary methods to better understand adaptation in the wild.

Here, we discuss approaches used to estimate the strength of selection, drawing on both classic and recent examples to illustrate their utility. Because of its association with adaptation and the evolution of novel forms and functions, we focus on positive selection. In the first four sections, we describe methods for estimating the strength of selection on genotypes and phenotypes using four distinct types of data: (1) within-generation fitness estimates from individuals bearing different genotypes and/or phenotypes, (2) changes in allele frequencies or phenotypic means over multiple generations, (3) changes in allele frequencies or phenotypic means in space, and (4) DNA sequence data from genes that contribute to phenotypic differences in natural populations. For clarity, we divide the first two sections by the type of trait variation (genotypic or phenotypic) under study. These diverse approaches allow us to measure selection at multiple levels—from single-nucleotide changes to quantitative traits—and at multiple timescales—from ecological to evolutionary. However, none can tell us *why* particular traits are under selection; this issue is discussed in a fifth section. Finally, we conclude by discussing how applying multiple methods to traits for which we can make genotype–phenotype and phenotype–fitness links can address fundamental questions regarding the genetics of adaptation in natural populations.

FITNESS DIFFERENCES BETWEEN GENOTYPES OR PHENOTYPES (WITHIN A GENERATION)

Natural selection need not result in evolution—a change in allele frequencies over time. For example, when phenotypes do not differ in their underlying genotypes or when selection favors heterozygotes, allele frequencies may remain the same from one generation to the next. Thus, the most straightforward way to estimate the strength of selection is to focus on a single generation and compare the success of different phenotypes or genotypes at survival and reproduction. This approach requires information regarding the fitness of individuals (or classes of individuals) in a population. Unfortunately, "fitness" is notoriously difficult to define and measure (see Endler 1986; De Jong 1994; McGraw and Caswell 1996; Orr 2009). Therefore, most studies measure components of fitness (e.g., survival to sexual maturity, survival following an environmental change, number of mates, or number of offspring produced) as surrogates for total fitness. Because selection may differ over the course of a single generation or across years (Schemske and Horvitz 1989; Hoekstra et al. 2001; Siepielski et al. 2009), the most comprehensive studies estimate fitness across multiple life stages and are repeated across multiple years; two such examples are described below.

Genotypes

Selection on a single Mendelian locus, or a discrete polymorphism presumed to be under the control of a single gene, can be quantified using a selection coefficient (s) that describes the intensity of selection against genotypes (or alleles). Specifically, s for a given genotype is equal to $1-w$, where w is the relative fitness of that genotype. The relative fitness for each genotypic class is calculated by dividing the absolute fitness for that class (estimated from survival rates or number of offspring produced, see above) by the highest absolute fitness across all genotypic classes. Traditionally, s has been calculated for discrete morphological characters (see, e.g., Schemske and Bierzychudek 2001) or enzyme polymorphisms (see, e.g., Eanes 1999). More recently, however, advances in genomic technologies and statistical methods for analyzing genotypic and phenotypic data have enabled us to identify genes contributing to variation in quantitative traits (for review, see Mackay 2001; Feder and Mitchell-Olds 2003; Luikart et al. 2003; Erickson et al. 2004; Vasemagi and Primmer 2005; Ehrenreich and Purugganan 2006; Ellegren and Sheldon 2008; Hoffmann and Willi 2008; Naish and Hard 2008; Pavlidis et al. 2008; Stinchcombe and Hoekstra 2008; Mackay et al. 2009; Slate et al. 2009). Thus, it is now possible to estimate selection coefficients for individual quantitative trait loci (QTLs) (see, e.g., Schemske and Bradshaw 1999; Rieseberg and Burke 2001; Lexer et al. 2003a; Mullen and Hoekstra 2008). One such analysis, described below, was performed on a major-effect QTL contributing to adaptive morphological variation in threespine sticklebacks (*Gasterosteus aculeatus*).

Following the last ice age, marine threespine sticklebacks repeatedly colonized freshwater environments and underwent concomitant losses in bony armor plating (Fig. 1A) (Bell and Foster 1994). Repeated loss in similar environments suggests that selection is responsible for these morphological differences. QTL mapping followed by functional verification via transgenic studies have implicated the gene *Ectodysplasin* (*Eda*), which explains more than 75% of the variation in plate number between marine and freshwater fish (Colosimo et al. 2004, 2005). Most marine fish are homozygous for the "complete" *Eda* allele (CC) and have 30–36 plates, freshwater fish are often homozygous for the "low" allele (LL) and have 0–9 plates, and heterozygotes (CL), rare in both habitats, have an intermediate number of plates. The low allele, estimated to be 2 million years old, is present in low frequencies (~1%) in marine populations, suggesting that freshwater populations, which form a monophyletic group at *Eda*, used standing genetic variation to adapt to the novel lake habitat (Colosimo et al. 2005).

To mimic selection pressures experienced by threespine stickleback upon invasion of novel freshwater habitats, Barrett, Rogers, and Schluter (2008) created four experimental ponds into which they introduced marine fish known to be heterozygous for the low allele (CL). They then sampled and genotyped the F_1 progeny of these fish (in which all three genotypic classes were present) at 10 time points over the course of 1 year. Selection coefficients were calculated from changes in genotype and allele frequencies. As expected, individuals with at least one copy of the low allele had increased growth rates and higher overwinter survival (October–July; $s = 0.52$ against the C allele). However, contrary to expectations, the frequency of the low allele actually decreased during the summer (July–October; $s = 0.50$ against the L allele), before the development of armor plates, which suggests conflicting selection pressures on *Eda* or a linked gene (Fig. 1A) (Barrett et al. 2008). At present, it is unclear why complete alleles are favored in the summer, but complementary studies focusing on phenotypic change (see below) might provide insight into the precise targets of selection at different times during the year. Nevertheless, had allele frequencies only been measured at the start and end of the year, changes in the direction and intensity of selection acting on the low allele would have been missed.

Phenotypes

Because selection acts at the phenotypic level, we can estimate its strength even in the absence of any knowledge about the genetic basis of fitness-related traits. A series of papers starting in the late 1970s laid the foundation for the analysis of selection on continuously varying phenotypic traits (Lande 1979; Lande and Arnold 1983; Arnold and Wade 1984a,b). Under this quantitative genetic approach (also referred to as the "Chicago School" approach), the mode and intensity of natural selection are estimated by regressing relative fitness (for continuous traits, fitness is typically calculated relative to the population mean) onto phenotypic values. Directional selection is characterized by a linear relationship between fitness and phenotype, and the slope of this relationship, calculated using linear regression, estimates the strength of selection (i.e., change in phenotypic means that is due to selection). When phenotypic values are standardized by subtracting the population mean and dividing by the population standard deviation, the slope of the phenotype–fitness regression equals the selection differential (S), which is defined as the covariance between fitness and the trait ($Cov[w,z]$) (Price 1970).

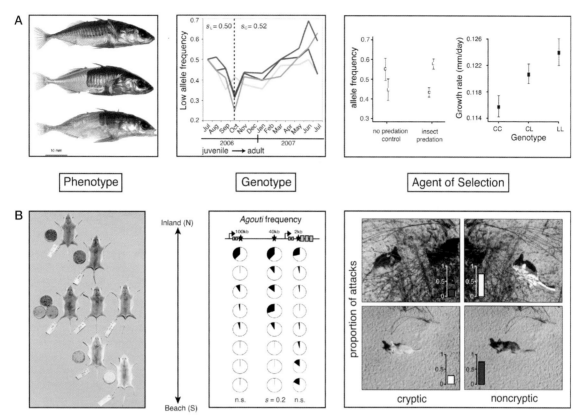

Figure 1. Two examples for which both the targets (phenotypic *and* genotypic) and agents of natural selection have been identified. (*A*) Selection on body armor in the threespine stickleback, *Gasterosteus aculeatus*. (*Left panel*) Complete (*top*), partial (*middle*), and low (*bottom*) lateral plate morphs. (*Middle panel*) Changes in low *Eda* allele frequency within a single generation in four replicate ponds (colored lines). Selection coefficients are given for selection against the low allele from July to October (s_L) and from selection against the complete allele from October to July (s_C). (*Right panel*) Relative to the complete *Eda* allele (C), individuals carrying the low *Eda* allele (L) enjoy decreased predation by insects (*left*) and increased growth rates in fresh water (*right*). (*B*) Selection on coat color in the oldfield mouse *Peromyscus polionotus*. (*Left panel*) Representative mice and soil sampled from collection sites along a 150-km transect from northwestern Florida (beach) to southeastern Alabama (inland). (*Middle panel*) Allele frequencies at three polymorphic sites (stars) within the pigmentation gene *Agouti* (large boxes: coding exons; small boxes: untranslated exons) sampled from eight populations along the same N-S transect. Pie charts and mice are arranged N (*top*) to S (*bottom*) and light allele frequencies are indicated in white. One of the three single-nucleotide polymorphisms (SNPs) (40 kb), but not the others, varies clinally. The selection coefficient is given for this SNP. (*Right panel*) Increased attack rates on noncryptic clay models relative to cryptic clay models on both light (beach) and dark (inland) soils demonstrate that visually hunting predators are an important selective agent targeting color variation within and between *P. polionotus* populations. (*A* [*left* and *middle panels*], Reprinted, with permission, from Barrett et al. 2008 [© AAAS]; *A [right panel, left]*, reprinted, with permission, from Marchinko 2009; *A [right panel, right]*, modified, with permission, from Barrett et al. 2009; *B [left panel]*, reprinted, with permission, from Mullen and Hoekstra 2008; *B [middle panel]*, modified, with permission, from Mullen and Hoekstra 2008; *B [right panel]*, modified, with permission, from Vignieri et al. 2010 [all © Wiley].)

Standardizing data also provides selection estimates that are comparable across different traits and organisms (Kingsolver et al. 2001).

Selection differentials measure total selection on a given trait. However, phenotypes can be correlated with fitness either because they impact fitness directly (direct selection) or because they are correlated with other traits that affect fitness (indirect selection). Selection gradients (β), in contrast, are calculated using multiple regression to control for indirect selection, thereby estimating direct selection on a trait (Lande and Arnold 1983). Selection gradient analysis has now been applied to a wide range of plant and animal taxa (compiled in Endler 1986; Hoekstra et al. 2001; Kingsolver et al. 2001; Siepielski et al. 2009). In most cases studied to date, estimates of β and S are similar, suggesting that, for the traits that were investigated, indirect selection is usually small relative to direct selection (Kingsolver et al. 2001). This observation, however, does not suggest that indirect selection is low or unimportant, but rather that we tend to focus on traits for which we have a priori reasons to believe are targets of selection. In fact, strong indirect selection can overcome direct selection in an opposing direction, and selection gradients and differentials will have opposite signs. These cases—the most famous of which involves Darwin's finches—illustrate the importance of measuring multiple traits and estimating both direct and total selection to gain an accurate picture of adaptation and evolutionary constraint in natural populations.

Every year since 1973, Peter and Rosemary Grant and colleagues have measured survival, reproduction, and phenotypes of marked individuals of *Geospiza fortis* (medium ground finch) living on the Galápagos of

Daphne Major (Grant and Grant 2002; for review, see Grant 2003; Grant and Grant 2008). From 1976 to 1977, a severe drought decimated seed supplies on the island, resulting in no reproduction and high adult mortality. During this time, virtually the only type of food available was large, hard seeds, which are most efficiently handled by large birds with deep beaks (Boag and Grant 1981; Grant 1981; Price et al. 1984). Estimates of the selection differential (S) obtained by comparing phenotypic means before and after the drought confirmed that selection indeed favored birds with large bodies ($S = 0.74$) and beaks ($S = 0.53$ to 0.63, depending on beak trait). However, because phenotypic correlations were taken into account, they found that although direct selection (β) favored a decrease in beak length and width ($\beta = -0.14$ and -0.45, respectively), these trait values nevertheless increased due to strong positive correlations with beak depth (Fig. 2A) (Price et al. 1984; Grant and Grant 1995). A second drought from 1984 to 1986 also resulted in decreased food supplies and high adult mortality; however, this drought followed an exceptionally wet season that resulted in an increased abundance of small, soft seeds. During this environmental perturbation, selection differentials for body ($S = -0.11$) and beak ($S = -0.03$ to -0.17) size were uniformly negative, even though direct selection favored an increase in beak length ($\beta = 0.25$) (Fig. 2A) (Gibbs and Grant 1987; Grant and Grant 1995). Thus, in *G. fortis*, the strength and direction of selection can vary greatly from one year to the next, and due to phenotypic correlations, the direct targets of selection need not always change in the expected direction. Such long-term field studies are rare, but they are invaluable for understanding changes in phenotype over time.

CHANGES IN ALLELE FREQUENCIES OR TRAIT MEANS OVER TIME (BETWEEN GENERATIONS)

The magnitude of the phenotypic response to selection depends on both the heritability and the strength of selection. Therefore, observed changes in allele frequencies or phenotypic means over multiple generations, when coupled with information on the relationship between genotype and phenotype, can be used to estimate the strength of selection. This approach is especially useful for organ-

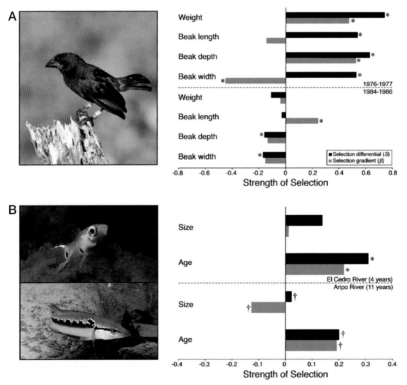

Figure 2. Strength of selection can vary in time and space, and indirect selection (via genetic correlations) can oppose direct selection on traits, as shown in Darwin's finches and Trinidadian guppies. (*A*) Selection differentials (S) and gradients (β) for four traits in medium ground finches (*Geospiza fortis*) calculated from changes in phenotypic means (within generation) following two bouts of selection (1976–1977 and 1984–1986 droughts). Asterisks indicate S and β that are significantly different from zero ($p < 0.05$). (*B*) Selection differentials (S) and gradients (β) for age and size at maturity in male guppies (*Poecilia reticulata*) calculated from response to selection (between generations) under low-predation conditions in two streams (El Cedro River and Aripo River). Results are from bivariate analyses, and asterisks indicate values significantly different from zero. The significance of bivariate Aripo River estimates (denoted by †) could not be evaluated, but these estimates were significantly different from zero in a univariate analysis. Photos are a male guppy (*top*) and its chief predator, the pike cichlid (*Crenicichla alta*) (*bottom*). (*A*, Data from Grant and Grant 1995 [photo by B.R. Grant]; *B*, data from Reznick et al. 1997 [photos by P. Bentzen].)

isms for which it is difficult or impossible to track differences in survival and/or reproduction among a cohort of individuals. For example, although we usually cannot measure fitness in long-dead organisms preserved as ancient DNA samples or in natural history collections, these resources provide us with invaluable time-series data on allele frequencies and/or phenotypic means that can be used to estimate selection.

Genotypes

Simple population genetic models predict changes in allele frequencies after one generation of selection. If p and q are the frequencies of alleles A and B; w_{AA}, w_{AB}, and w_{BB} are the relative fitnesses of genotypes AA, AB, and BB; and $\bar{w}$ is the mean fitness of the population, then change in the frequency of the A allele ($\Delta p = p' - p$), is given by

$$\Delta p = \frac{pq[p(w_{AA} - w_{AB}) + q(w_{AB} - w_{BB})]}{\bar{w}}. \quad (1)$$

Therefore, if we know allele frequencies in two consecutive generations and the degree of dominance, we can calculate the selection coefficient. For example, assuming $w_{AA} = 1$ and w_{AB} and w_{BB} both equal $1-s$ (i.e., A is recessive and AA is the most fit genotype), the selection coefficient is given by

$$s = \frac{\Delta p}{p'(1-p^2) - pq}. \quad (2)$$

Over multiple generations, selection generates a sigmoidal response curve of allele frequency change that can be used to estimate s (Haldane 1924). Alternatively, s can be estimated by running a recurrence equation over n generations under alternative values of s, choosing the value that best explains the observed allele frequencies (Cook 2003). These approaches were used by Haldane (1924) and others (for review, see Cook 2003; see below) to estimate selection on melanic and typical forms of the iconic peppered moth *Biston betularia*.

More recently, Bollback et al. (2008) developed a maximum likelihood (ML) approach that uses the diffusion process (see Ewens 2004) to estimate effective population size (N_e) and selection (s) from time-series data of allele frequencies. They applied this method to frequency data for the human CCR5-Δ32 mutation, which confers resistance to HIV (human immunodeficiency virus) infection, from European samples gathered from 2900 years ago to the present (Hummel et al. 2005). Contrary to previous work that reported large selection coefficients (Novembre et al. 2005), Bollback et al. (2008) found an ML estimate of s near zero, which suggests that before the origin of HIV, this mutation was neutral or nearly neutral. In another ancient DNA study, Ludwig et al. (2009) applied this method to Eurasian horse samples dating from 20,000 years ago to the present. They typed fossils at six loci known to contribute to color variation in horses and found both a dramatic increase in horse coat-color variation coinciding with domestication (~5500 years ago;

Outram et al. 2009), and, for two of the six loci, selection coefficients were significantly different from zero (agouti signaling protein [*Agouti*], $s = 0.0007$, and melanocortin-1 receptor [*Mc1r*], $s = 0.0019$). These results suggest that domestication and selective breeding contributed to changes in horse coloration.

Phenotypes

Estimating the strength of phenotypic selection from multigenerational data requires information on the genetic basis of traits under selection. For quantitative traits, the standard equation for predicting evolutionary response to selection is the breeder's equation $R = h^2 S$, where R is the response to selection (i.e., change in trait mean), h^2 is narrow-sense heritability (i.e., the fraction of total phenotypic variation that is additive), and S is the selection differential (Falconer and Mackay 1996). Because phenotypes are often genetically correlated, Lande (1979) developed a multivariate version of this equation that predicts evolutionary response on a trait by accounting for selection on correlated traits (and their corresponding genetic covariances) in addition to direct selection (and its genetic variance). The multivariate analog of $R = h^2 S$ is $\mathbf{R} = \mathbf{G}\boldsymbol{\beta}$, where $\mathbf{G}$ is the genetic variance–covariance matrix (i.e., the $\mathbf{G}$ matrix, in which diagonal elements are additive genetic variances for n traits, and off-diagonal elements are additive genetic covariances between traits) and $\boldsymbol{\beta}$ is a vector of selection gradients (Lande 1979; Lande and Arnold 1983).

This eponymous equation is most commonly used by breeders to either predict phenotypic response to selection or to estimate heritability, but it can also be used to estimate phenotypic selection in nature (rearranging, $S = R/h^2$ and $\boldsymbol{\beta} = \mathbf{R}\mathbf{G}^{-1}$) (Thompson 2008). Two advantages of this approach are that it allows one to estimate selection when within-generation fitness estimates are unattainable, and it takes into account the impact of trait variation on total lifetime fitness, not just a single fitness component. An important disadvantage, however, is that its accuracy is sensitive to errors in estimates of genetic variance, failure to measure all traits under selection, and/or fluctuating environmental conditions (Kruuk et al. 2008). Nonetheless, this equation has been shown to accurately predict evolutionary response, particularly when trait heritabilities are high and have narrow confidence intervals (see, e.g., Grant and Grant 1995, 2002), as is the case for life-history traits in Trinidadian guppies.

Guppies (*Poecilia reticulata*) living in Trinidad occur in two habitats: high-predation habitats, defined by presence of pike predator (*Crenicichla alta*) that selectively prey on large mature guppies, and low-predation habitats that lack this cichlid (Fig. 2B). Fish from low-predation populations mature later and at a larger size than their low-predation counterparts, and these differences are genetically based (Reznick 1982; Reznick and Endler 1982). To test if these differences are caused by selection, Reznick et al. (1997) moved guppies from two high-predation to two low-predation streams. After 4–11 years of selection, they quantified phenotypic response (change in mean size and age at maturity), heritabilities, and **G** matrices for fish in each of the two streams and used these values to calculate S and β. As predicted, both size and age at reproductive maturity increased (size, $S = 0.023$ and 0.138; age, $S = 0.201$ and 0.310, for Aripo and El Cedro, respectively) once fish were freed from predation pressure. In both streams, there was strong direct selection for increased age at maturity, and selection on age exceeded selection on size (size, $\beta = -0.127$ and 0.013; age, $\beta = 0.193$ and 0.220, for Aripo and El Cedro, respectively). However, in the Aripo River, but not in the El Cedro River, there was selection for a decrease in size at maturity (Fig. 2B). These differences likely stem from the different genetic compositions of the starting populations and/or ecological differences between the two streams. Like the threespine stickleback example, this work demonstrates the importance of replication to detect unique and shared responses to novel selection pressures; like the Darwin's finch example, this work highlights the importance of measuring both direct and total (direct + indirect) phenotypic selection.

SPATIAL PATTERNS OF ALLELE FREQUENCIES AND PHENOTYPE MEANS

The spatial distribution of genotypes and phenotypes can tell us a great deal about the strength of natural selection maintaining patterns of geographic variation (Haldane 1948; Slatkin 1973, 1975; Endler 1977; Barton 1979a,b, 1983; Barton and Hewitt 1985). For example, sharp transitions in allele frequencies or trait means, particularly if they coincide with ecotones (transition areas between adjacent ecological communities), suggest a role for selection because in the absence of selection, gene flow will homogenize populations (Haldane 1930; Lenormand 2002; Slatkin 2003). Thus, the frequency of alleles in adjacent populations, and the sharpness of the transition between them, is determined by both the amount of gene flow and the strength of selection against deleterious alleles in each habitat. When gene flow and selection reach an equilibrium, the populations are considered to be in migration-selection balance. A mathematical model of migration-selection balance was proposed by Haldane (1930) and Wright (1931), who showed that the change in the deleterious (immigrant) allele frequency in a population (Δq) is given by

$$\Delta q = \frac{-spq[q + h(p-q)]}{1 - sq(2hp + q)} + mQ - Mq, \quad (3)$$

in which s is the selection coefficient against the deleterious allele, q and p are allele frequencies of the deleterious and nondeleterious alleles, h is the dominance coefficient, m is the migration rate into the population, Q is the frequency of the deleterious allele outside the population, and M is the emigration rate. Using this approach, Hoekstra et al. (2004) estimated the strength of selection acting on allelic variation at a single locus (*Mc1r*) that determines coat color in pocket mice (*Chaetodipus intermedius*) living on light- and dark-colored rocks in the southwestern United States. They reported strong selection for background matching but found that selection estimates were not symmetrical—light mice were more strongly selected against on dark rock ($s = 0.013$–0.390) than dark mice on light rock ($s = 0.0002$–0.020). This is consistent with visual perception abilities of avian predators, with selection on light/melanic forms of peppered moths (Cook and Mani 1980), and with the direction of evolutionary change (i.e., light mice colonized newly formed lava flows, and strong selection against these mismatched mice favored the evolution of the novel melanic form).

As shown, Equation 3 can be used for populations sampled from two distinct habitats. Alternatively, when multiple populations are sampled along an ecotone, the distance over which allele frequencies or trait means change (i.e., the cline width [w]) can be used to estimate selection (s) because w is proportional to $\sigma/\sqrt{s}$, where σ is the standard deviation of the adult–offspring dispersal distance (Haldane 1948; Slatkin 1973). Cline width is estimated by fitting a sigmoidal curve to allele frequencies or population trait means plotted as a function of geographic distance. The cline width is defined as the inverse of the maximum slope of this curve for allele frequencies; for quantitative traits, cline width is equal to $\Delta z/(\delta z/\delta x)$, where Δz is the difference in population means on either side of the cline and $\delta z/\delta x$ is the maximum slope of phenotypic change over distance x (Slatkin 1978; Barton and Gale 1993). This approach has been used widely to analyze clines in allele frequencies (see, e.g., Mallet et al. 1990) and quantitative traits (see, e.g., Nurnberger et al. 1995), but only rarely are genotypes and phenotypes analyzed together. One notable exception is the work of Mullen and Hoekstra (2008), who took advantage of knowing the genetic basis of pigment variation to analyze selection acting to maintain both a cline in pigmentation and its underlying genes (Fig. 1B).

In the southeastern United States, there is a sharp transition in soil color from the white sandy beaches of Florida to darker inland soils. In the 1920s, Francis Sumner sampled oldfield mice (*Peromyscus polionotus*) along a 150-km transect and found that as he moved inland and the soil got darker, pale-colored mice were replaced by a darker form more typical of the genus (Sumner 1929a,b). Using Sumner's original museum specimens, Mullen and Hoekstra (2008) quantified brightness over multiple body regions with a spectrophotometer. From these phenotypic data and estimates of dispersal distances in *P. polionotus*, they calculated the width of the cline and strength of selection; selection on coat color was strong (assuming an ecotonal model, $s = 0.07$–0.21, depending on body region measured). In addition, because previous work had identified two genes (*Mc1r* and *Agouti*) that contribute to pigment differences in these mice (Hoekstra et al. 2006; Steiner et al. 2007), they were able to estimate cline widths and selection strengths for these alleles (Fig. 1B). Although a single molecular marker in the *Agouti* locus showed clinal variation similar to that observed for phenotypes (cline width, and therefore selection strength, was statistically indistinguishable from the phenotypic cline, $s = \sim 0.2$), allelic variation at *Mc1r* showed a surprising lack of clinal variation. This was due to dark mice harboring "light" *Mc1r* alleles in the northernmost populations. One explanation for this pattern is that epistatically interacting alleles (e.g., *Agouti*) (Steiner et al. 2007) mask the effects of *Mc1r* and therefore relax selection against the light *Mc1r* allele when on a dark *Agouti* genetic background (common in dark soil habitats). Additional data are needed to test this hypothesis, but these results clearly show how selection on both phenotypes and genotypes may vary in strength and direction across different environments and sometimes in complex ways.

SELECTION ESTIMATES BASED ON DNA SEQUENCE DATA

With recent advances in genomic technologies and powerful new statistical methods for linking genotype to phenotype in natural populations has come an explosion of methods for detecting natural selection at the molecular level (for review, see Nielsen 2005; Biswas and Akey 2006; Eyre-Walker 2006; Sabeti et al. 2006; Jensen et al. 2007b; Thornton et al. 2007; Grossman et al. 2010). Natural selection shapes the distribution of alleles within and between populations and species; thus, both population-genetic and comparative data—analyzed jointly or in isolation—can be used to infer selection. Two general approaches use these data. Many recent studies have taken a bottom-up approach in which genome-scale sequence data are screened for signatures of selection, either to estimate the proportion of the genome affected by selection (for review, see Eyre-Walker 2006; Sella et al. 2009) or to identify promising loci for future functional (and ecological) verification (see, e.g., Nielsen et al. 2005; Williamson et al. 2007; Grossman et al. 2010). Here, we focus on the top-down approach, in which the evidence for selection is evaluated for a candidate gene (or mutation) chosen a priori based on its known effects on an individual's phenotype (and, ideally, fitness) (see, e.g., Bersaglieri et al. 2004; Olsen et al. 2007; Pool and Aquadro 2007; Linnen et al. 2009). Because methods for detecting selection have been recently and thoroughly reviewed, we give a brief overview of some more widely used approaches, with particular emphasis on how these methods can be used to obtain quantitative estimates of s (or related parameters).

When a novel mutation is fixed in a population by natural selection, linked neutral variation is carried along with it. Numerous methods for detecting and measuring selection are based on the characteristic patterns of variation created by this "hitchhiking" effect (Maynard Smith and Haigh 1974). Specifically, selective sweeps are expected to reduce heterozygosity surrounding a selected site while producing an excess of low- and high-frequency-derived alleles surrounding the target of selection (i.e., a U-shaped site-frequency spectrum [SFS]) (Tajima 1989; Fu and Li 1993; Braverman et al. 1995; Fu 1997; Fay and Wu 2000). On the basis of these predictions, Kim and Stephan (2002) developed a model-based approach that uses a composite likelihood ratio (CLR) test to compare the likelihood of polymorphism data under the standard neutral model to the likelihood of the data under a hitchhiking model. Because sweeps are also expected to affect patterns of linkage disequilibrium (LD)—for example, strong LD is expected on either side of a beneficial mutation but not across the two sides—Kim and Nielsen (2004) later extended this method to include information regarding LD (see also Sabeti et al. 2002; Stephan et al. 2006; Jensen et al. 2007a). Both approaches generate ML estimates of the strength and target of selection under the hitchhiking model. In addition, although the CLR test assumes that the swept allele has gone to fixation, Meiklejohn et al. (2004) demonstrated that this test can also be applied to incomplete sweeps by analyzing only those chromosomes that carry the beneficial allele. Linnen et al. (2009) used this strategy to estimate the strength of selection acting on a partially swept allele contributing to adaptive coloration in deer mice (*Peromyscus maniculatus*) living on the light soils of the Nebraska Sand Hills (Fig. 3). On the basis of this analysis, they concluded that selection for light color is relatively strong (s = 0.006), comparable in magnitude to estimates of s obtained for color polymorphisms in other organisms (compiled by Hoekstra et al. 2004).

Whereas polymorphism-based methods such as the CLR test estimate the strength of very recent selection acting on individual mutations (i.e., ~0.4N generations or less since the fixation of an allele) (see Kim and Stephan 2002; Przeworski 2002), methods that use comparative (between-species) data, either in addition to or in the absence of within-species data, calculate the average strength of selection acting on a particular locus (or site; see Fitch et al. 1997; Nielsen and Yang 1998; Suzuki and Gojobori 1999) over longer periods of evolutionary time. The most widely used of these are the Hudson–Kreitman–Aguade (HKA) test (Hudson et al. 1987), the McDonald–Kreitman (MK) test (McDonald and Kreitman 1991), and d_N/d_S-based tests (Kimura 1977; Yang and Bielawski 2000). The HKA test is based on the expectation that, under neutrality, the ratio of intraspecific polymorphism to interspecific divergence will be equal across loci (Kimura 1983). Selection is inferred when there are significant differences in polymorphism-to-divergence ratios among loci. As it is typically implemented, the HKA test does not provide estimates for the strength of selection. Quantitative estimates can be obtained, however, using an ML alternative to the standard HKA test that was developed by Wright and Charlesworth (2004).

The MK test also uses interspecific and intraspecific data, but it differs from the HKA test in that it partitions data into functional classes and can be applied to data from a single locus. This test is most often applied to protein-coding regions (but see Andolfatto 2005; Pollard et al. 2006; Hahn 2007), for which the ratio of nonsynonymous to synonymous polymorphisms within a species is compared to the ratio of nonsynonymous to synonymous differences between species; a significant difference in these ratios implicates selection. Sawyer and Hartl (1992) developed an explicit mathematical framework—the Poisson Random Field—that provides an estimate of the average selection coefficient for a locus from an MK table (see also Bustamante et al. 2002, 2005; Sawyer et al. 2003,

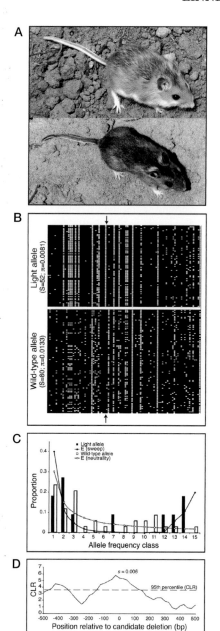

Figure 3. Molecular evidence of a partial selective sweep on the light-color allele in deer mice *Peromyscus maniculatus*. (*A*) Deer mice living on the pale soils of the Nebraska Sand Hills are lighter (*top*) than deer mice from darker surrounding areas (*bottom*). Mice are shown on contrasting soil backgrounds (*bottom*: Sand Hills soil; *top*: soil from outside the Sand Hills). (*B*) Consistent with patterns expected under recent selection on a de novo mutation, variation among light *Agouti* haplotypes is reduced compared to variation among wild-type haplotypes. (*C*) Site-frequency spectrum (SFS) for the light *Agouti* allele matches patterns expected under a selective sweep (black circles and solid line), whereas the SFS for the wild-type allele matches neutral expectations (white circles/dashed line). (*D*) Composite likelihood ratio (CLR) as a function of distance from candidate polymorphism within *Agouti* for the light allele. Values above the line reject the neutral model, and the maximum CLR value, which indicates the putative target of selection, is near the candidate deletion. An estimate of the selection coefficient (*s*), obtained by maximization of the composite likelihood function, is given. (*A*, Photos by E. Kay; *A*, *B*, reprinted, with permission, from Linnen et al. 2009 [© AAAS]; *C*, data from Linnen et al. 2009; *D*, modified from Linnen et al. 2009.)

2007; Eyre-Walker 2006; Andolfatto 2007). Like the MK test, d_N/d_S tests partition data into functional classes. Specifically, these tests compare the rate of nonsynonymous substitutions (d_N) to the rate of synonymous substitutions (d_S), with the expectation that $d_N/d_S = 1$ under neutrality, $d_N/d_S > 1$ under positive selection, and $d_N/d_S < 1$ under negative selection. This approach differs from the MK test in that only comparative data (fixed differences between species) are used. Using an approach similar to that of Sawyer and Hartl (1992), Nielsen and Yang (2003) devised a method to estimate the distribution of selection coefficients *s* for a given locus from d_N/d_S data.

One important consideration for applying molecular tests of selection is that in some cases, observed deviations from the standard neutral model may be due to demography, not selection. Fortunately, divergence-based methods are expected to be relatively insensitive to demographic assumptions (Nielsen 2005; Garrigan et al. 2010; but see Eyre-Walker 2002; Ingvarsson 2004). In contrast, a variety of demographic scenarios can replicate patterns of genetic variation expected under hitchhiking (Tajima 1989; Fu and Li 1993; Wakeley and Aliacar 2001; Jensen et al. 2005, 2007b; Thornton and Andolfatto 2006; Thornton et al. 2007). For example, Jensen et al. (2005) demonstrated that the CLR test produces many false positives (up to 90%) when there is population structure and/or a recent population bottleneck. They therefore developed a goodness-of-fit test that can be used in conjunction with the CLR test to reduce the number of false positives. An alternative approach is to estimate demographic parameters and incorporate these into comparisons between neutral and selected models (Thornton and Andolfatto 2006; Thornton and Jensen 2007; Nielsen et al. 2009). In addition to improving the false-positive rate, this approach also should yield more accurate estimates of *s*. A final consideration is that deviations from the classic hitchhiking model (Maynard Smith and Haigh 1974; Kaplan et al. 1989), such as selection on standing genetic variation (see, e.g., Hermisson and Pennings 2005; Przeworski et al. 2005), recurrent mutation (see, e.g., Pennings and Hermisson 2006a,b), or recurrent selective sweeps (see, e.g., Kim 2006; Jensen et al. 2008), also will impact our ability to detect and measure selection using population-genetic data. Nonetheless, as statistical methods continue to improve, we will be able to estimate selection under a broader range of demographic and selective scenarios.

IDENTIFYING THE AGENT OF SELECTION

Estimating the strength of selection acting on phenotypic and genotypic "targets" can clearly tell us a great deal about adaptation. However, a complete understanding of this process also requires that we determine *why* phenotypes and genotypes are under selection; in other words, we must identify the "agents" of natural selection in addition to its "targets" (Endler 1986; Conner 1996; Conner and Hartl 2004). When a significant relationship between phenotype (or genotype) and fitness is observed, it is always possible that this relationship is due (partially or completely) to correlation with an unmeasured character. Although measuring multiple traits may increase our confidence that such a scenario is unlikely, determining causation ultimately requires that we generate and test adaptive hypotheses. For example, recent experimental work reveals two reasons why selection favors the *Eda* low allele when threespine stickleback invade freshwater habitats (see Fig. 1A). First, due to a tradeoff between armoring and growth in freshwater, fish carrying the low allele experience increased growth rates, which leads to higher overwinter survival and reproductive success (Marchinko and Schluter 2007; Barrett et al. 2008, 2009). Second, juvenile fish carrying the low allele enjoy reduced predation by insects, possibly because these fish have shorter dorsal spines, thereby reducing the ability of insects to hold and consume them (Marchinko 2009). Together, this work suggests that multiple agents of selection can favor the same genetic target.

In some cases, experimental manipulation also can yield direct estimates of the strength of selection. As discussed above, the striking match between coat color and local soil color in *Peromyscus* populations has long been hypothesized to be the result of selection for crypsis. In a classic experiment, Dice (1947) released equal frequencies of lab-reared deer mice with light or dark coats into enclosures that varied in substrate color and then subjected them to owl predation. He found that, as predicted, conspicuous mice were captured at much higher rates. He also devised a selection index (SI) to describe the relative survival of two equally abundant phenotypes:

$$SI = \frac{(a-b)}{(a+b)}, \quad (4)$$

in which a and b are the number of attacked individuals in each phenotypic class. The significance of the SI can be tested using a χ^2 test, as described by Dice (1947, 1949). Although Dice's experiments imply that predation may be an important agent of selection and that color is a target, his SI estimates (0.24–0.29) are probably overestimates and not directly comparable to estimates from natural populations because both predator and prey densities are inflated in his experimental enclosures.

To estimate the magnitude of selection for crypsis in nature—and to control for possible selection on correlated traits, such as odor, activity level, or escape behavior—Vignieri et al. (2010) constructed clay models of *P. polionotus* and painted these to resemble either the dark oldfield mouse (*P. p. subgriseus*) or the light Santa Rosa Island beach mouse (*P. p. leucocephalus*). They then deployed these models in beach (light) and inland (dark) habitats known to be occupied by *P. polionotus* and recorded the number of attacks—inferred from the presence of predatory marks, such as tooth or beak marks from mammals and birds—on each model type in each habitat. Across both habitats, conspicuous models were more than three times more likely to be attacked than cryptic ones (SI = 0.5, a value even higher than that in Dice's experiments), demonstrating that both the agent (visually hunting predators) and the target (cryptic coloration) of selection had been correctly identified (Fig. 1B). Still another way to control for selection on traits other than color that may differ between beach and inland mice would be to take advantage of our ability to cross these subspecies and thus to introgress pigment alleles onto a common genetic background. If these hybrid mice were released in natural enclosures, one could estimate and compare selection estimates on individual alleles (e.g., similar to the pond experiments conducted in sticklebacks), combinations of alleles, and phenotypes.

INTEGRATION ACROSS LEVELS AND TIMESCALES OF SELECTION

The methods described here use different types of data and rely on different biological assumptions. These studies generate estimates of selection at multiple levels—from phenotypes to genotypes, and in some cases on single-nucleotide changes (Fig. 4A). And they are applicable at different timescales—estimates based on fitness or allele frequency data correspond to ecological time (one to tens of generations), methods that use neutral polymorphism data (e.g., CLR test) detect ongoing or recent selective sweeps, and comparative methods detect repeated bouts of selection over long periods of evolutionary time (Fig. 4B). Thus, we can make three types of comparisons among these methods: (1) between estimates calculated at the same level and timescale, but using different data and methods, (2) across different levels of selection (phenotypes and underlying genotypes), and (3) across different

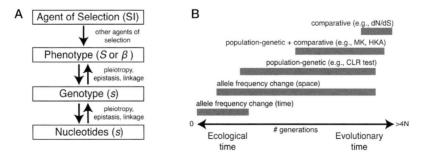

Figure 4. Comparison of selection estimates at different levels and at different timescales. (*A*) Levels at which one can estimate selection are given in boxes with corresponding symbols in parentheses. Processes that can create discord between different levels are indicated beside arrows. (*B*) Approximate timescales (gray boxes) over which different types of data can inform selection studies.

timescales. Such comparisons have the potential to reveal not only strengths and weaknesses of individual methods, but also novel insights into the adaptive process.

First, for a particular trait and organism, it can be advantageous to apply multiple methods to independent data sets to determine whether selection estimates are concordant; if not, these data may reveal why. This point is illustrated by research on one of the most intensively studied adaptive polymorphisms to date: melanic coloration in the peppered moth *Biston betularia* (for review, see Cook 2003; Saccheri et al. 2008). Following the onset of industrialization in 19th-century England, there was a rapid increase in the frequency of a melanic morph (*carbonaria*) in the peppered moth, an exemplar of strong selection (Haldane 1924, 1932). This trend continued until legislation to improve air quality was introduced in the 1960s, and the frequency of the *carbonaria* morph began to decline (Cook 2003). These changes appear to be driven by selection for crypsis to avoid avian predation: *carbonaria* is more cryptic on dark backgrounds (i.e., soot-covered trees), whereas the nonmelanic (*typical*) morph is more cryptic on lighter, unpolluted backgrounds (Cook 2003). Three types of data have been used to estimate selection coefficients against the *carbonaria* form following the decline in air pollution (and therefore dark backgrounds to rest upon): changes in allele frequencies over time, changes in allele frequencies over space, and both field and laboratory experiments involving predators. Although estimates of selection obtained from time-series and cline data are largely concordant ($s = \sim 0.1$), predation experiments suggest that selection against *carbonaria* is much stronger than the allele frequency data imply. These discrepancies led some to hypothesize a nonvisual advantage to melanics that partially counteracts predator-driven selection (Bishop 1972; Mani 1982). However, in a recent study that combines spatial and temporal data into a single analysis, Saccheri et al. (2008) found that gene flow was much higher than previously appreciated and that estimates of selection against the *carbonaria* form were consistent with predation experiments ($s = \sim 0.2$), thereby suggesting that nonvisual selection was less important than had been supposed.

It is worth noting that, despite more than 80 years of work to estimate selection in *B. betularia*, the gene responsible for melanism has yet to be identified. In contrast, there are a small but growing number of ecologically relevant traits for which researchers have pinpointed the precise genetic changes responsible for phenotypic variation, thereby allowing comparisons between genotypic and phenotypic selection estimates (Fig. 4A). A particularly informative comparison comes from recent work on wild sunflower (*Helianthus*) hybrids (Lexer et al. 2003a,b). Molecular phylogenetic work has shown that two widespread annual sunflowers, *Helianthus annuus* and *H. petiolarus*, have given rise to three diploid hybrid species (Rieseberg et al. 1990; Rieseberg 1991). Each of these hybrid species occurs in "extreme" environments uninhabitable by either parental species. The prevailing explanation for this pattern is that genetic contributions from both parental species give rise to traits present in neither (i.e., transgressive traits) (Rieseberg et al. 1999). Lexer et al. (2003a,b) confirmed this hypothesis by mimicking the events that led to the formation of one hybrid species by crossing the parental forms and placing early-generation hybrids in the salt marsh habitat of the hybrid species *H. paradoxus*. In these *paradoxus*-like hybrids, these authors first measured the strength of directional selection acting on several transgressive characters, including mineral ion (e.g., Ca^{2+} and Na^+) uptake (Lexer et al. 2003b). Next, they mapped multiple QTLs contributing to ion uptake and survivorship and measured the strength of selection acting on these QTLs in the salt marsh habitat (Lexer et al. 2003a). Comparison of these selection estimates reveals qualitative concordance across genotypic and phenotypic levels; QTLs contributing to phenotypic traits under selection were also selected in the same direction (Table 1). However, selection on each QTL was much higher than expected based on simple predictive equations (see Table 1) (Rieseberg and Burke 2001). This suggests that pleiotropy and/or genetic linkage is pervasive and increases selection on an individual QTL. Consistent with this hypothesis, extensive genetic correlations were observed both in this study and in a larger, greenhouse-based QTL study using the same cross design (Rieseberg et al. 2003; Lexer et al. 2005). These results also suggest that several loci with intermediate effects on fitness—not a large number of mutations of small effect or a small number of mutations of large effect—contributed to adaptation in salt marsh habitats. Thus, adaptation from standing genetic variation (a scenario for which hybrid species formation represents an extreme case) may be characterized by different patterns than adaptation from de novo mutation (Hermisson and Pennings 2005; Orr 2005; Przeworski et al. 2005).

Finally, as a recent review on the temporal dynamics of selection vividly illustrates, a third type of comparison we can make is among selection estimates obtained at different points in time (Fig. 4B). Siepielski et al. (2009) compiled a large database of temporally replicated studies of selection in natural populations and found (1) considerable variation in the strength of selection from year to year, (2) frequent reversals in the direction of selection, and (3) apparent changes in the form of selection (i.e., directional to stabilizing). These findings highlight the importance of field studies of selection that extend over multiple years (see, e.g., Grant 2003). They also raise questions regarding the trajectory of evolutionary change, which can be addressed by estimating selection acting on a particular mutation or gene at different temporal scales. For example, within-generation studies of selection on *Eda* (Barrett et al. 2008) could be complemented both by multigeneration studies in experimental ponds and by studies estimating the strength of selection driving the low armor-plate allele to fixation (or near fixation) in natural ponds. Similarly, if recent selection has been documented for a single mutation, comparative methods can be used to investigate the role of that gene in driving adaptive change over longer periods of evolutionary time. Comparing the average strength of selection on different genes may then reveal the degree to which they are constrained. For example, despite both genes being involved in the pigmentation pathway and capable of producing similar effects on color, mutations in *Mc1r* are more commonly

Table 1. Phenotypic and Genotypic Selection Estimates

Trait	Linkage group	PVE (%)[a]	Phenotypic selection		Genotypic selection	
			S[b]	β[b]	Observed s[c]	Expected s[d]
Ca	1	32	0.062	0.237	0.126	0.010
Na	4	15	−0.250	−0.289	−0.084	−0.019
Na	17b	18	−0.250	−0.289	−0.094	−0.023
Mg	4	17	−0.250	−0.289	−0.084	−0.021

These estimates are for mineral uptake traits and corresponding QTLs in a second-generation backcross population (BC$_2$) of *Helianthus annuus* × *H. petiolarus* in the natural habitat of the hybrid species *H. paradoxus*.
[a]Percent phenotypic variance explained.
[b]Data from Lexer et al. 2003b.
[c]Data from Lexer et al. 2003a.
[d]Expected s for a given QTL was calculated by multiplying the selection differential (S) for the trait by the PVE and then dividing by two to account for diploidy (following Rieseberg and Burke 2001).

associated with natural color variation than mutations in *Agouti* (Hoekstra 2006; but see Kingsley et al. 2009). One explanation is that *Agouti* evolution is constrained because mutations in this gene have greater negative pleiotropic consequences on fitness (e.g., embryonic lethality, increased tumor susceptibility, diabetes, hyperphagia, and obesity) (Duhl et al. 1994; Siracusa 1994; Miltenberger et al. 2002), on average, than mutations to *Mc1r*. This hypothesis predicts that, over evolutionary time, *Mc1r* may be the target of positive selection more often than *Agouti*; average per-locus selection coefficients are therefore expected to be higher for *Mc1r* than for *Agouti*. These predictions can be tested at multiple levels of divergence using MK, HKA, and d_N/d_S tests (see, e.g., Mundy and Kelly 2006).

CONCLUSIONS

This is an exciting time in evolutionary biology. As an increasing number of genes underlying adaptive phenotypes are identified and statistical methods for estimating selection at the molecular level continue to improve, we will undoubtedly accumulate more precise estimates of selection acting on individual genes and mutations. In contrast, although the methods for estimating phenotypic selection are not new, they continue to provide important insights into how natural selection shapes the distribution of phenotypic variation in space and time. Although estimates of genotypic and phenotypic selection are each informative in their own right, comparisons across both levels, when coupled with identification of the agent(s) of selection, allow us to link genotype, phenotype, and the environment. At present, such studies are rare, but we suspect that comparisons among selection estimates—measured with different data and using distinct approaches—will ultimately provide a more complete picture of the adaptive process.

ACKNOWLEDGMENTS

We thank R. Barrett, P. Grant, R. Grant, J. Jensen, C. Lexer, J. Mallet, and D. Reznick for comments that improved this manuscript. R. Grant and D. Reznick kindly provided photographs. C.R.L. was supported by a Ruth Kirschstein National Research Service Award. Work from the Hoekstra Lab reported here was largely funded by the National Science Foundation.

REFERENCES

Andolfatto P. 2005. Adaptive evolution of non-coding DNA in *Drosophila*. *Nature* **437:** 1149–1152.
Andolfatto P. 2007. Hitchhiking effects of recurrent beneficial amino acid substitutions in the *Drosophila melanogaster* genome. *Genome Res* **17:** 1755–1762.
Arnold SJ, Wade MJ. 1984a. On the measurement of natural and sexual selection: Applications. *Evolution* **38:** 720–734.
Arnold SJ, Wade MJ. 1984b. On the measurement of natural and sexual selection: Theory. *Evolution* **38:** 709–719.
Barrett RDH, Rogers SM, Schluter D. 2008. Natural selection on a major armor gene in threespine stickleback. *Science* **322:** 255–257.
Barrett RDH, Rogers SM, Schluter D, Peichel C. 2009. Environment specific pleiotropy facilitates divergence at the *Ectodysplasin* locus in threespine stickleback. *Evolution* **63:** 2831–2837.
Barton NH. 1979a. Gene flow past a cline. *Heredity* **43:** 333–339.
Barton NH. 1979b. The dynamics of hybrid zones. *Heredity* **43:** 341–359.
Barton NH. 1983. Multilocus clines. *Evolution* **37:** 454–471.
Barton NH, Gale KS. 1993. Genetic analysis of hybrid zones. In *Hybrid zones and the evolutionary process* (ed. RG Harrison), pp. 13–45. Oxford University Press, New York.
Barton NH, Hewitt GM. 1985. Analysis of hybrid zones. *Annu Rev Ecol Syst* **16:** 113–148.
Bell MA, Foster SA. 1994. *The evolutionary biology of the threespine stickleback*. Oxford University Press, Oxford.
Bersaglieri T, Sabeti PC, Patterson N, Vanderploeg T, Schaffner SF, Drake JA, Rhodes M, Reich DE, Hirschhorn JN. 2004. Genetic signatures of strong recent positive selection at the lactase gene. *Am J Hum Genet* **74:** 1111–1120.
Bishop JA. 1972. An experimental study of the cline of industrial melanism in *Biston betularia* (L.) (Lepidoptera) between urban Liverpool and rural north Wales. *J Anim Ecol* **41:** 209–243.
Biswas S, Akey JM. 2006. Genomic insights into positive selection. *Trends Genet* **22:** 437–446.
Boag PT, Grant PR. 1981. Intense natural selection in a population of Darwin's finches (Geospizinae) in the Galapagos. *Science* **214:** 82–85.
Bollback JP, York TL, Nielsen R. 2008. Estimation of $2N_e s$ from temporal allele frequency data. *Genetics* **179:** 497–502.
Braverman JM, Hudson RR, Kaplan NL, Langley CH, Stephan W. 1995. The hitchhiking effect on the site frequency spectrum of DNA polymorphisms. *Genetics* **140:** 783–796.
Bustamante CD, Nielsen R, Sawyer SA, Olsen KM, Purugganan MD, Hartl DL. 2002. The cost of inbreeding in *Arabidopsis*. *Nature* **416:** 531–534.
Bustamante CD, Fledel-Alon A, Williamson S, Nielsen R, Hubisz MT, Glanowski S, Tanenbaum DM, White TJ, Sninsky JJ, Hernandez RD, et al. 2005. Natural selection on protein-coding genes in the human genome. *Nature* **437:** 1153–1157.
Colosimo PF, Peichel CL, Nereng K, Blackman BK, Shapiro MD, Schluter D, Kingsley DM. 2004. The genetic architecture of par-

allel armor plate reduction in threespine sticklebacks. *PLoS Biol* **2:** e109.
Colosimo PF, Hosemann KE, Balabhadra S, Villarreal G, Dickson M, Grimwood J, Schmutz J, Myers RM, Schluter D, Kingsley DM. 2005. Widespread parallel evolution in sticklebacks by repeated fixation of *ectodysplasin* alleles. *Science* **307:** 1928–1933.
Conner JK. 1996. Understanding natural selection: An approach integrating selection gradients, multiplicative fitness components, and path analysis. *Ethol Ecol Evol* **8:** 387–397.
Conner JK, Hartl DL. 2004. *A primer of ecological genetics*. Sinauer, Sunderland, MA.
Cook LM. 2003. The rise and fall of the *carbonaria* form of the peppered moth. *Q Rev Biol* **78:** 399–417.
Cook LM, Mani GS. 1980. A migration-selection model for the morph frequency variation in the peppered moth over England and Wales. *Biol J Linn Soc* **13:** 179–198.
Darwin C. 1859. *On the origin of species by means of natural selection, or the preservation of favoured races in the struggle for life*. John Murray, London.
De Jong G. 1994. The fitness of fitness concepts and the description of natural selection. *Q Rev Biol* **69:** 3–29.
Dice LR. 1947. Effectiveness of selection by owls of deer-mice (*Peromyscus maniculatus*) which contrast in color with their background. *Contrib Lab Vertebrate Biol Univ Michigan* **34:** 1–20.
Dice LR. 1949. The selection index and its test of significance. *Evolution* **3:** 262–265.
Duhl DM, Vrieling H, Miller KA, Wolff GL, Barsh GS. 1994. Neomorphic *agouti* mutations in obese yellow mice. *Nat Genet* **8:** 59–65.
Eanes WF. 1999. Analysis of selection on enzyme polymorphisms. *Annu Rev Ecol Syst* **30:** 301–326.
Ehrenreich IM, Purugganan MD. 2006. The molecular genetic basis of plant adaptation. *Am J Bot* **93:** 953–962.
Ellegren H, Sheldon BC. 2008. Genetic basis of fitness differences in natural populations. *Nature* **452:** 169–175.
Endler JA. 1977. *Geographic variation, speciation, and clines*. Princeton University Press, Princeton, NJ.
Endler JA. 1986. *Natural selection in the wild* (Monographs in Population Biology). Princeton University Press, Princeton, NJ.
Erickson DL, Fenster CB, Stenoien HK, Price D. 2004. Quantitative trait locus analyses and the study of evolutionary process. *Mol Ecol* **13:** 2505–2522.
Ewens WJ. 2004. *Mathematical population genetics*. Springer, New York.
Eyre-Walker A. 2002. Changing effective population size and the McDonald-Kreitman test. *Genetics* **162:** 2017–2024.
Eyre-Walker A. 2006. The genomic rate of adaptive evolution. *Trends Ecol Evol* **21:** 569.
Falconer DS, Mackay TFC. 1996. *Introduction to quantitative genetics*. Longman, Essex, England.
Fay JC, Wu C-I. 2000. Hitchhiking under positive Darwinian selection. *Genetics* **155:** 1405–1413.
Feder ME, Mitchell-Olds T. 2003. Evolutionary and ecological functional genomics. *Nat Rev Genet* **4:** 651–657.
Fitch WM, Bush RM, Bender CA, Cox NJ. 1997. Long term trends in the evolution of H(3) HA1 human influenza type A. *Proc Natl Acad Sci* **94:** 7712–7718.
Fu YX. 1997. Statistical tests of neutrality of mutations against population growth, hitchhiking and background selection. *Genetics* **147:** 915–925.
Fu YX, Li WH. 1993. Statistical tests of neutrality of mutations. *Genetics* **133:** 693–709.
Garrigan D, Lewontin R, Wakeley J. 2010. Measuring the sensitivity of single-locus "neutrality tests" using a direct perturbation approach. *Mol Biol Evol* **27:** 73–89.
Gibbs HL, Grant PR. 1987. Oscillating selection on Darwin's finches. *Nature* **327:** 511–513.
Grant PR. 1981. The feeding of Darwin's finches on *Tribulus cistoides* (L.) seeds. *Anim Behav* **29:** 785–793.
Grant BR. 2003. Evolution in Darwin's finches: A review of a study on Isla Daphne Major in the Galapagos archipelago. *Zoology* **106:** 255–259.
Grant PR, Grant BR. 1995. Predicting microevolutionary responses to directional selection on heritable variation. *Evolution* **49:** 241–251.
Grant PR, Grant BR. 2002. Unpredictable evolution in a 30-year study of Darwin's finches. *Science* **296:** 707–711.
Grant PR, Grant BR. 2008. *How and why species multiply: The radiation of Darwin's finches*. Princeton University Press, Princeton, NJ.
Grossman SR, Shylakhter I, Karlsson EK, Byrne EH, Morales S, Frieden G, Hostetter E, Angelino E, Garber M, Zuk O, et al. 2010. A composite of multiple signals distinguishes causal variants in regions of positive selection. *Science* **327:** 883–886.
Hahn M. 2007. Detecting natural selection on *cis*-regulatory DNA. *Genetica* **129:** 7–18.
Haldane JBS. 1924. A mathematical theory of natural and artificial selection. *Trans Camb Philos Soc* **23:** 19–41.
Haldane JBS. 1930. A mathematical theory of natural and artificial selection. Part VI. Isolation. *Proc Camb Philos Soc* **26:** 220–230.
Haldane JBS. 1932. *The causes of evolution*. Longmans, London.
Haldane JBS. 1948. The theory of a cline. *J Genet* **48:** 277–284.
Hartl DL, Clark AG. 2007. *Principles of population genetics*. Sinauer, Sunderland, MA.
Hermisson J, Pennings PS. 2005. Soft sweeps: Molecular population genetics of adaptation from standing genetic variation. *Genetics* **169:** 2335–2352.
Hoekstra HE. 2006. Genetics, development and evolution of adaptive pigmentation in vertebrates. *Heredity* **97:** 222–234.
Hoekstra HE, Hoekstra JM, Berrigan D, Vignieri SN, Hoang A, Hill CE, Beerli P, Kingsolver JG. 2001. Strength and tempo of directional selection in the wild. *Proc Natl Acad Sci* **98:** 9157–9160.
Hoekstra HE, Drumm KE, Nachman MW. 2004. Ecological genetics of adaptive color polymorphism in pocket mice: Geographic variation in selected and neutral genes. *Evolution* **58:** 1329–1341.
Hoekstra HE, Hirschmann RJ, Bundey RA, Insel PA, Crossland JP. 2006. A single amino acid mutation contributes to adaptive beach mouse color pattern. *Science* **313:** 101–104.
Hoffmann AA, Willi Y. 2008. Detecting genetic responses to environmental change. *Nat Rev Genet* **9:** 421–432.
Hudson RR, Kreitman M, Aguade M. 1987. A test of neutral molecular evolution based on nucleotide data. *Genetics* **116:** 153–159.
Hummel S, Schmidt D, Kremeyer B, Herrmann B, Oppermann M. 2005. Detection of the *CCR5-Δ32* HIV resistance gene in Bronze Age skeletons. *Genes Immun* **6:** 371–374.
Ingvarsson P. 2004. Population subdivision and the Hudson-Kreitman-Aguade test: Testing for deviations from the neutral model in organelle genomes. *Genet Res* **83:** 31–39.
Jensen JD, Kim Y, DuMont VB, Aquadro CF, Bustamante CD. 2005. Distinguishing between selective sweeps and demography using DNA polymorphism data. *Genetics* **170:** 1401–1410.
Jensen JD, Thornton KR, Bustamante CD, Aquadro CF. 2007a. On the utility of linkage disequilibrium as a statistic for identifying targets of positive selection in nonequilibrium populations. *Genetics* **176:** 2371–2379.
Jensen JD, Wong A, Aquadro CF. 2007b. Approaches for identifying targets of positive selection. *Trends Genet* **23:** 568–577.
Jensen JD, Thornton KR, Andolfatto P. 2008. An approximate Bayesian estimator suggests strong, recurrent selective sweeps in *Drosophila*. *PLoS Genet* **4:** e1000198.
Kaplan NL, Hudson RR, Langley CH. 1989. The "hitchhiking effect" revisited. *Genetics* **123:** 887–899.
Kim Y. 2006. Allele frequency distribution under recurrent selective sweeps. *Genetics* **172:** 1967–1978.
Kim Y, Nielsen R. 2004. Linkage disequilibrium as a signature of selective sweeps. *Genetics* **167:** 1513–1524.
Kim Y, Stephan W. 2002. Detecting a local signature of genetic hitchhiking along a recombining chromosome. *Genetics* **160:** 765–777.
Kimura M. 1977. Preponderance of synonymous changes as evidence for the neutral theory of molecular evolution. *Nature* **267:** 275–276.

Kimura M. 1983. *The neutral theory of molecular evolution.* Cambridge University Press, Cambridge.

Kimura M, Crow JF. 1978. Effect of overall phenotypic selection on genetic change at individual loci. *Proc Natl Acad Sci* **75**: 6168–6171.

Kingsley EP, Manceau M, Wiley CD, Hoekstra HE. 2009. Melanism in *Peromyscus* is caused by independent mutations in *Agouti*. *PLoS ONE* **4**: e6435.

Kingsolver JG, Hoekstra HE, Hoekstra JM, Berrigan D, Vignieri SN, Hill CE, Hoang A, Gibert P, Beerli P. 2001. The strength of phenotypic selection in natural populations. *Am Nat* **157**: 245–261.

Kruuk LEB, Slate J, Wilson AJ. 2008. New answers for old questions: The evolutionary quantitative genetics of wild animal populations. *Annu Rev Ecol Evol Syst* **39**: 525–548.

Lande R. 1979. Quantitative genetic analysis of multivariate evolution, applied to brain: Body size allometry. *Evolution* **33**: 402–416.

Lande R, Arnold SJ. 1983. The measurement of selection on correlated characters. *Evolution* **37**: 1210–1226.

Lenormand T. 2002. Gene flow and the limits to natural selection. *Trends Ecol Evol* **17**: 183–189.

Lexer C, Welch ME, Durphy JL, Rieseberg LH. 2003a. Natural selection for salt tolerance quantitative trait loci (QTLs) in wild sunflower hybrids: Implications for the origin of *Helianthus paradoxus*, a diploid hybrid species. *Mol Ecol* **12**: 1225–1235.

Lexer C, Welch ME, Raymond O, Rieseberg LH, Nason J. 2003b. The origin of ecological divergence in *Helianthus paradoxus* (Asteraceae): Selection on transgressive characters in a novel hybrid habitat. *Evolution* **57**: 1989–2000.

Lexer C, Rosenthal DM, Raymond O, Donovan LA, Rieseberg LH. 2005. Genetics of species differences in the wild annual sunflowers, *Helianthus annuus* and *H. petiolaris*. *Genetics* **169**: 2225–2239.

Linnen CR, Kingsley EP, Jensen JD, Hoekstra HE. 2009. On the origin and spread of an adaptive allele in deer mice. *Science* **325**: 1095–1098.

Ludwig A, Pruvost M, Reissmann M, Benecke N, Brockmann GA, Castanos P, Cieslak M, Lippold S, Llorente L, Malaspinas A-S, et al. 2009. Coat color variation at the beginning of horse domestication. *Science* **324**: 485.

Luikart G, England PR, Tallmon D, Jordan S, Taberlet P. 2003. The power and promise of population genomics: From genotyping to genome typing. *Nat Rev Genet* **4**: 981–994.

Mackay TFC. 2001. Quantitative trait loci in *Drosophila*. *Nat Rev Genet* **2**: 11–20.

Mackay TFC, Stone EA, Ayroles JF. 2009. The genetics of quantitative traits: Challenges and prospects. *Nat Rev Genet* **10**: 565–577.

Mallet J, Barton N, Gerardo LM, Jose SC, Manuel MM, Eeley H. 1990. Estimates of selection and gene flow from measures of cline width and linkage disequilibrium in *Heliconius* hybrid zones. *Genetics* **124**: 921–936.

Mani GS. 1982. A theoretical analysis of the morph frequency variation in the peppered moth over England and Wales. *Biol J Linn Soc* **17**: 259–267.

Marchinko KB. 2009. Predation's role in repeated phenotypic and genetic divergence of armor in threespine stickleback. *Evolution* **63**: 127–138.

Marchinko KB, Schluter D. 2007. Parallel evolution by correlated response: Lateral plate reduction in threespine stickleback. *Evolution* **61**: 1084–1090.

Maynard Smith J, Haigh J. 1974. The hitch-hiking effect of a favourable gene. *Genet Res* **23**: 23–35.

McDonald JH, Kreitman M. 1991. Adaptive protein evolution at the *Adh* locus in *Drosophila*. *Nature* **351**: 652–654.

McGraw JB, Caswell H. 1996. Estimation of individual fitness from life-history data. *Am Nat* **147**: 47–64.

Meiklejohn C, Kim Y, Hartl D, Parsch J. 2004. Identification of a locus under complex positive selection in *Drosophila simulans* by haplotype mapping and composite-likelihood estimation. *Genetics* **168**: 265–279.

Milkman R. 1978. Selection differentials and selection coefficients. *Genetics* **88**: 391–403.

Miltenberger RJ, Wakamatsu K, Ito S, Woychik RP, Russell LB, Michaud EJ. 2002. Molecular and phenotypic analysis of 25 recessive, homozygous-viable alleles at the mouse *Agouti* locus. *Genetics* **160**: 659–674.

Mullen LM, Hoekstra HE. 2008. Natural selection along an environmental gradient: A classic cline in mouse pigmentation. *Evolution* **62**: 1555–1569.

Mundy NI, Kelly J. 2006. Investigation of the role of the agouti signaling protein gene (*ASIP*) in coat color evolution in primates. *Mamm Genome* **17**: 1205–1213.

Naish KA, Hard JJ. 2008. Bridging the gap between the genotype and the phenotype: Linking genetic variation, selection and adaptation in fishes. *Fish Fisheries* **9**: 396–422.

Nielsen R. 2005. Molecular signatures of natural selection. *Annu Rev Genet* **39**: 197–218.

Nielsen R, Yang Z. 1998. Likelihood models for detecting positively selected amino acid sites and applications to the HIV-1 envelope gene. *Genetics* **148**: 929–936.

Nielsen R, Yang Z. 2003. Estimating the distribution of selection coefficients from phylogenetic data with applications to mitochondrial and viral DNA. *Mol Biol Evol* **20**: 1231–1239.

Nielsen R, Bustamante C, Clark AG, Glanowski S, Sackton TB, Hubisz MJ, Fledel-Alon A, Tanenbaum DM, Civello D, White TJ, et al. 2005. A scan for positively selected genes in the genomes of humans and chimpanzees. *PLoS Biol* **3**: e170.

Nielsen R, Hubisz MJ, Hellmann I, Torgerson D, Andrés AM, Albrechtsen A, Gutenkunst R, Adams MD, Cargill M, Boyko A, et al. 2009. Darwinian and demographic forces affecting human protein coding genes. *Genome Res* **19**: 838–849.

Novembre J, Galvani AP, Slatkin M. 2005. The geographic spread of the CCR5 Δ32 HIV-resistance allele. *PLoS Biol* **3**: e339.

Nurnberger B, Barton N, MacCallum C, Jason G, Appleby M. 1995. Natural selection on quantitative traits in the *Bombina* hybrid zone. *Evolution* **49**: 1224–1238.

Olsen K, Sutherland B, Small L. 2007. Molecular evolution of the *Li/li* chemical defence polymorphism in white clover (*Trifolium repens* L.). *Mol Ecol* **16**: 4180–4193.

Orr HA. 2005. The genetic theory of adaptation: A brief history. *Nat Rev Genet* **6**: 119–127.

Orr HA. 2009. Fitness and its role in evolutionary genetics. *Nat Rev Genet* **10**: 531–539.

Outram AK, Stear NA, Bendrey R, Olsen S, Kasparov A, Zaibert V, Thorpe N, Evershed RP. 2009. The earliest horse harnessing and milking. *Science* **323**: 1332–1335.

Pavlidis P, Hutter S, Stephan W. 2008. A population genomic approach to map recent positive selection in model species. *Mol Ecol* **17**: 3585–3598.

Pennings PS, Hermisson J. 2006a. Soft sweeps II: Molecular population genetics of adaptation from recurrent mutation or migration. *Mol Biol Evol* **23**: 1076–1084.

Pennings PS, Hermisson J. 2006b. Soft sweeps III: The signature of positive selection from recurrent mutation. *PLoS Genet* **2**: e186.

Pollard KS, Salama SR, Lambert N, Lambot M-A, Coppens S, Pedersen JS, Katzman S, King B, Onodera C, Siepel A, et al. 2006. An RNA gene expressed during cortical development evolved rapidly in humans. *Nature* **443**: 167–172.

Pool JE, Aquadro CF. 2007. The genetic basis of adaptive pigmentation variation in *Drosophila melanogaster*. *Mol Ecol* **16**: 2844–2851.

Price GR. 1970. Selection and covariance. *Nature* **227**: 520–521.

Price TD, Grant PR, Gibbs HL, Boag PT. 1984. Recurrent patterns of natural selection in a population of Darwin's finches. *Nature* **309**: 787–789.

Przeworski M. 2002. The signature of positive selection at randomly chosen loci. *Genetics* **160**: 1179–1189.

Przeworski M, Coop G, Wall J. 2005. The signature of positive selection on standing genetic variation. *Evolution* **59**: 2312–2323.

Reznick D. 1982. The impact of predation on life history evolution in Trinidadian guppies: Genetic basis of observed life history patterns. *Evolution* **36**: 1236–1250.

Reznick D, Endler JA. 1982. The impact of predation on life history evolution in Trinidadian guppies (*Poecilia reticulata*).

Evolution **36:** 160–177.
Reznick DN, Shaw FH, Rodd FH, Shaw RG. 1997. Evaluation of the rate of evolution in natural populations of guppies (*Poecilia reticulata*). *Science* **275:** 1934–1937.
Rieseberg LH. 1991. Homoploid reticulate evolution in *Helianthus* (Asteraceae): Evidence from ribosomal genes. *Am J Bot* **78:** 1218–1237.
Rieseberg LH, Burke JM. 2001. The biological reality of species: Gene flow, selection, and collective evolution. *Taxon* **50:** 47–67.
Rieseberg LH, Carter R, Zona S. 1990. Molecular tests of the hypothesized hybrid origin of two diploid *Helianthus* species (Asteraceae). *Evolution* **44:** 1498–1511.
Rieseberg LH, Archer MA, Wayne RK. 1999. Transgressive segregation, adaptation and speciation. *Heredity* **83:** 363–372.
Rieseberg LH, Raymond O, Rosenthal DM, Lai Z, Livingstone K, Nakazato T, Durphy JL, Schwarzbach AE, Donovan LA, Lexer C. 2003. Major ecological transitions in wild sunflowers facilitated by hybridization. *Science* **301:** 1211–1216.
Sabeti PC, Reich DE, Higgins JM, Levine HZ, Richter DJ, Schaffner SF, Gabriel SB, Platko JV, Patterson NJ, McDonald GJ, et al. 2002. Detecting recent positive selection in the human genome from haplotype structure. *Nature* **419:** 832–837.
Sabeti PC, Schaffner SF, Fry B, Lohmueller J, Varilly P, Shamovsky O, Palma A, Mikkelsen TS, Altshuler D, Lander ES. 2006. Positive natural selection in the human lineage. *Science* **312:** 1614–1620.
Saccheri IJ, Rousset F, Watts PC, Brakefield PM, Cook LM. 2008. Selection and gene flow on a diminishing cline of melanic peppered moths. *Proc Natl Acad Sci* **105:** 16212–16217.
Sawyer SA, Hartl DL. 1992. Population genetics of polymorphism and divergence. *Genetics* **132:** 1161–1176.
Sawyer S, Kulathinal R, Bustamante C, Hartl D. 2003. Bayesian analysis suggests that most amino acid replacements in *Drosophila* are driven by positive selection. *J Mol Evol* **57:** S154–S164.
Sawyer SA, Parsch J, Zhang Z, Hartl DL. 2007. Prevalence of positive selection among nearly neutral amino acid replacements in *Drosophila*. *Proc Natl Acad Sci* **104:** 6504–6510.
Schemske DW, Bierzychudek P. 2001. Evolution of flower color in the desert annual *Linanthus parryae*: Wright revisited. *Evolution* **55:** 1269–1282.
Schemske DW, Bradshaw HD. 1999. Pollinator preference and the evolution of floral traits in monkeyflowers (*Mimulus*). *Proc Natl Acad Sci* **96:** 11910–11915.
Schemske DW, Horvitz CC. 1989. Temporal variation in selection on a floral character. *Evolution* **43:** 461–465.
Sella G, Petrov DA, Przeworski M, Andolfatto P. 2009. Pervasive natural selection in the *Drosophila* genome? *PLoS Genet* **5:** e1000495.
Siepielski AM, DiBattista JD, Carlson SM. 2009. It's about time: The temporal dynamics of phenotypic selection in the wild. *Ecol Lett* **12:** 1261–1276.
Siracusa LD. 1994. The *agouti* gene: Turned on to yellow. *Trends Genet* **10:** 423–428.
Slate J, Gratten J, Beraldi D, Stapley J, Hale M, Pemberton JM. 2009. Gene mapping in the wild with SNPs: Guidelines and future directions. *Genetica* **136:** 97–107.
Slatkin M. 1973. Gene flow and selection in a cline. *Genetics* **75:** 733–756.
Slatkin M. 1975. Gene flow and selection in a two-locus system. *Genetics* **81:** 787–802.
Slatkin M. 1978. Spatial patterns in the distributions of polygenic characters. *J Theor Biol* **70:** 213–228.
Slatkin M. 2003. Gene flow in natural populations. *Annu Rev Ecol Syst* **16:** 393–430.
Steiner CC, Weber JN, Hoekstra HE. 2007. Adaptive variation in beach mice produced by two interacting pigmentation genes. *PLoS Biol* **5:** e219.
Stephan W, Song Y, Langley C. 2006. The hitchhiking effect on linkage disequilibrium between linked neutral loci. *Genetics* **172:** 2647–2663.
Stinchcombe JR, Hoekstra HE. 2008. Combining population genomics and quantitative genetics: Finding the genes underlying ecologically important traits. *Heredity* **100:** 158–170.
Sumner FB. 1929a. The analysis of a concrete case of intergradation between two subspecies. *Proc Natl Acad Sci* **15:** 110–120.
Sumner FB. 1929b. The analysis of a concrete case of intergradation between two subspecies. II Additional data and interpretations. *Proc Natl Acad Sci* **15:** 481–493.
Suzuki Y, Gojobori T. 1999. A method for detecting positive selection at single amino acid sites. *Mol Biol Evol* **16:** 1315–1328.
Tajima F. 1989. Statistical method for testing the neutral mutation hypothesis by DNA polymorphism. *Genetics* **123:** 585–595.
Thompson R. 2008. Estimation of quantitative genetic parameters. *Proc R Soc Lond B Biol Sci* **275:** 679–686.
Thornton K, Andolfatto P. 2006. Approximate bayesian inference reveals evidence for a recent, severe bottleneck in a Netherlands population of *Drosophila melanogaster*. *Genetics* **172:** 1607–1619.
Thornton KR, Jensen JD. 2007. Controlling the false-positive rate in multilocus genome scans for selection. *Genetics* **175:** 737–750.
Thornton KR, Jensen JD, Becquet C, Andolfatto P. 2007. Progress and prospects in mapping recent selection in the genome. *Heredity* **98:** 340–348.
Vasemagi A, Primmer CR. 2005. Challenges for identifying functionally important genetic variation: The promise of combining complementary research strategies. *Mol Ecol* **14:** 3623–3642.
Vignieri SN, Larson JG, Hoekstra HE. 2010. The selective advantage of crypsis in mice. *Evolution* (in press).
Wakeley J, Aliacar N. 2001. Gene genealogies in a metapopulation. *Genetics* **159:** 893–905.
Wallace AR. 1858. On the tendency of varieties to depart indefinitely from the original type. *J Proc Linn Soc* **3:** 53–62.
Williamson SH, Hubisz MJ, Clark AG, Payseur BA, Bustamante CD, Nielsen R. 2007. Localizing recent adaptive evolution in the human genome. *PLoS Genet* **3:** e90.
Wright SJ. 1931. Evolution in Mendelian populations. *Genetics* **16:** 97–159.
Wright SI, Charlesworth B. 2004. The HKA test revisited: A maximum-likelihood-ratio test of the standard neutral model. *Genetics* **168:** 1071–1076.
Yang Z, Bielawski JP. 2000. Statistical methods for detecting molecular adaptation. *Trends Ecol Evol* **15:** 496–503.

Exploring the Molecular Landscape of Host–Parasite Coevolution

D.E. ALLEN AND T.J. LITTLE

Institute of Evolutionary Biology, University of Edinburgh, Edinburgh EH9 3JT, Scotland, United Kingdom
Correspondence: desiree.allen@ed.ac.uk

Host–parasite coevolution is a dynamic process that can be studied at the phenotypic, genetic, and molecular levels. Although much of what we currently know about coevolution has been learned through phenotypic measures, recent advances in molecular techniques have provided tools to greatly deepen this research. Both the availability of full-genome sequences and the increasing feasibility of high-throughput gene expression profiling are leading to the discovery of genes that have a key role in antagonistic interactions between naturally coevolving species. Identification of such genes can enable direct observation, rather than inference, of the host–parasite coevolutionary dynamic. The *Daphnia magna–Pasteuria ramosa* host–parasite model is a prime example of an interaction that has been well studied at the population and whole-organism levels, and much is known about genotype- and environment-specific interactions from a phenotypic perspective. Now, with the recent completion of genome sequences for two *Daphnia* species, and a transcriptomics project under way, coevolution between these two enemies is being investigated directly at the level of interacting genes.

Charles Darwin appears to have had little appreciation for parasites. His neglect was probably symptomatic of the mood of naturalists of his time: Parasites were viewed as degenerate and generally unworthy of the attention of sensible Victorian naturalists. This is unfortunate, because the field of parasitology offers some of the most striking and elegant examples of adaptation that could no doubt have aided Darwin's efforts to have his theory of natural selection gain acceptance. Indeed, it is now clear that parasites and pathogens offer paradigmatic examples of evolution in action: Due to their often short generation time, they can be observed evolving in the laboratory (Rainey 2004; Morgan et al. 2005) and even in the real world, as evidenced by the evolution of drug resistance (Marchese et al. 2000), emerging diseases (Morse 1994), and vaccine escape mutants (Bangham et al. 1999). Of course, parasite adaptation is only half the story, and host defense systems offer equally striking, indeed strikingly complex, examples of adaptation. Thus, the host–parasite interaction is a dynamic *coevolutionary* relationship. Darwin was clearly aware of the importance of some coevolutionary scenarios (in particular, pollination), but the ramifications of host–parasite coevolution were not among them. Today, there remains considerable need to understand the nature and full significance of antagonistic parasitic interactions.

For example, dynamic ongoing adaptation and counter-adaptation between hosts and parasites may foster diversity within species and influence their mode of reproduction. Computer simulations have confirmed that parasitic interactions should promote diversity (Clarke 1979; Seger 1988; Hamilton et al. 1990; Frank 1993), and indeed, immune system genes may show striking levels of polymorphism (Hill et al. 1992; Hedrick 1994, 1998; Bishop et al. 2000; Obbard et al. 2008). Therefore, an explanation for genetic differences in susceptibility, which are pervasive in parasitic interactions (Little 2002; Woolhouse et al. 2002), is that a dynamic coevolutionary process has promoted these differences and will determine the fate of genetic variants. These dynamics can be important on medically relevant timescales (e.g., in vector control programs), but, even in longer-lived hosts, genetic data need to be interpreted in terms of host–parasite coevolution (e.g., the genetic polymorphism that underlies sickle cell anemia) (Hill et al. 1991; Gilbert et al. 1998). They are also of broad evolutionary importance: Frequency-dependent dynamics of host and parasite genes may select for recombination, and thus parasitism offers an explanation for why sexual reproduction predominates in metazoan organisms (The Red Queen Hypothesis) (Hamilton 1980; Lively 1993; Salathé et al. 2008).

Despite our knowledge of the occurrence of genetic variation for disease-related traits, we lack a thorough understanding of the coevolutionary dynamic. For instance, the predominant selective force during coevolutionary interactions could be selective sweeps, where genetic polymorphism is transient, observable only as one genotype rises to fixation, leaving populations in a single state that remains until the next mutation arises. Alternatively, there could be frequency-dependent selection where common genotypes are disfavored, such that no single genotype can go to fixation and rare genotypes never become extinct. An additional nuance is that the fixation of successful genotypes may be prevented by trade-offs, for example, if a resistance allele confers poor fitness in parasite-free environments. A more general form of the trade-off hypothesis is that polymorphism might further be maintained by pervasive genotype-by-environment interactions, where alternative host genotypes are favored under different environmental conditions. Without an understanding of these mechanisms, we lack the capacity to determine the rate at which evolution occurs in host–parasite interactions and the likely outcomes for the health of host populations.

Modern molecular approaches hold the key to unraveling the intricacies of the coevolutionary process. Molecular techniques have successfully identified immune system genes that control infection, and this explosion of genetic information is fertile ground for advancing our understanding of coevolution. Much work in the molecular study of immune systems has been confined to organisms of obvious medical relevance (e.g., humans) or those that are standard models for genetic study (e.g., mice and *Drosophila*). However, these organisms are either too long-lived to permit observations of genetic change within reasonable time frames, there is a lack of suitable knowledge regarding their natural parasites, or they are not amenable to experimentation. Thus, the study of evolution at immune genes remains incomplete, and the contribution of genomics to the understanding of coevolution is less than it could be.

Study systems are required that can provide a comprehensive view on host–parasite interactions, i.e., systems that are amenable to (1) gene discovery in both host and parasite, (2) experimental manipulation including whole-organism experimentation with parasites/pathogens, and hence, measurement of fitness and phenotypes, and (3) epidemiological or evolutionary studies in the field. Ideally, all such studies will concern naturally coevolving interactors as opposed to artificial interactions or the use of pathogen mimics (although these methods certainly have merits for some questions). Biology's major metazoan models (e.g., *Caenorhabditis* and *Drosophila*) tend to fulfill criterion 1 but fall short on 2 and 3, especially with respect to the use of naturally coevolving interactors (but see Bangham et al. 2008). In contrast, the crustacean *Daphnia* has historically strong credentials in the second two categories, and it is developing rapidly in terms of gene discovery projects. This chapter briefly discusses the development of molecular techniques that can be used to elucidate the genetic underpinnings of host–parasite interactions and then describes how these tools can and have been used in the context of *Daphnia* and its parasites, in conjunction with more traditional phenomenological-based approaches, to provide a unique opportunity to study host–parasite coevolution in a multifaceted way.

METHODS OF GENE DISCOVERY: PHYLOGENETIC CANDIDATE GENE APPROACHES

In the absence of any a priori information on immune system genes in a particular species, the best available option for locating genes of interest is de novo sequencing using degenerative primers designed from genes in a closely related species. Once the gene of interest is sequenced, inference as to its associated function can be made and phylogenetic comparisons with other closely related species can elucidate the evolutionary context. Depending on the genetic distance between the focal species and the species used to design primers, this method will have variable success. Moreover, it is constrained as to the information that it can reveal about the immune system of the focal species by the immune system of the related species from which the original sequence information was obtained; novel immune mechanisms will be overlooked. Finally, this method is limited in its utility except for highly conserved immune/infection-related pathways and may be of little use for rapidly evolving genes, although even highly conserved genes may have regions within them that evolve rapidly (Little et al. 2004; Little and Cobbe 2005). However, as the only option for gene discovery in some species, this is often a useful first pass.

METHODS OF GENE DISCOVERY: QUANTITATIVE TRAIT LOCI

Quantitative trait locus (QTL) or association mapping is a means of identifying chromosomal regions containing genes underlying a phenotype of interest, such as those produced during host–parasite interactions. This is achieved via crossing, phenotyping, and genotyping progeny in the laboratory. Specifically, this is a labor-intensive method requiring a linkage map, mapping populations consisting of genetic crosses between highly homozygous lines (often requiring generations of inbreeding or backcrosses) or a well-established pedigree. Additionally, a large number of microsatellite or single-nucleotide polymorphism (SNP) markers are required in order to have the statistical power to locate the genomic regions conferring the phenotype. Although quite involved, and not suitable for all species (especially those that reproduce poorly in the laboratory), QTL and association mapping have been a highly effective means of identifying genes involved in host–parasite interactions (see, e.g., Niaré et al. 2002; Lazzaro et al. 2004). QTL analyses have been put to particularly good use in plants: Due to their tractability to genetic crossing and the production of large sample sizes, the genetic bases of plant–pathogen interactions have been widely elucidated through this technique (for references, see Kover and Caicedo 2001; Wilfert and Schmid-Hempel 2008).

Fine-scale genetic mapping via the QTL approach is useful for gene discovery even without the advantage of a whole-genome sequence. If a chromosomal region can be narrowed sufficiently, through sequential crosses or backcrosses, de novo sequencing can identify genes within the region. This provides the material for comparison against known sequences from other taxa and may lead to the actual gene or genes responsible for the phenotype of interest. Obviously, with the availability of a genome sequence, identifying the genes within the QTL region becomes simpler. Even in this case, however, the genes within a located QTL region will not necessarily be identifiable as underlying the phenotype of interest, and it is important to bear in mind that wholly novel (i.e., genes that have not been characterized as immunity genes in any organism) genes could be the key genes. To overcome such limitations, it can be valuable to look at the expression pattern of all genes, and putative genes, located within the identified region. Using microarray, quantitative polymerase chain reaction (Q-PCR), or next-generation sequencing approaches (detailed below), any changes in expression patterns during exposure to experimental conditions can be monitored and the gene or genes responding to the challenge can be identified (Brown et al. 2005).

In addition to providing information on the location of chromosomal regions, and potentially the actual genes underlying a particular phenotype, QTL mapping can provide information on the way genes interact to produce a phenotype. The field of quantitative genetics has well-established methods for determining the additive, dominance, and epistatic interactions of alleles and loci (Lynch and Walsh 1998). Indeed, in a recent review of QTL studies of parasite/pathogen susceptibility, Wilfert and Schmid-Hempel (2008) were able to extract details about the number and effect of genes involved, concluding that resistance is based on few loci, and both additive and epistatic effects are important. These researchers also assessed the implications for evolutionary puzzles such as the evolution of sex and suggest that epistasis in resistance may have a major impact on the evolution and maintenance of meiotic segregation and recombination.

METHODS OF GENE DISCOVERY: GENE EXPRESSION

The development of microarrays, pyrosequencing technology, and next-generation massively parallel sequencing methods has introduced high-throughput means of discovering genes and gene networks underlying host and parasite traits and interactions (Hill et al. 2005; Keeler et al. 2007; Baton et al. 2008). These technologies have enabled the field to move from the previously described steps of inferring gene involvement through phylogenetic relationships or associating quantitative traits with genetic regions to directly measuring transcriptional changes in potentially hundreds of different genes during the production of a particular phenotype. As with most new molecular technology, these techniques were originally only cost-effective for well-established genetic systems. However, technology has now advanced to the point where these techniques are readily available for essentially any organism of interest. The main difference in the utility of transcriptional profiling compared to QTL analyses is that the latter can locate genes that do not show differential regulation during infection (e.g., vertebrate major histocompatibility genes are important, but only constitutively expressed), whereas the former can uncover up-regulation or down-regulation of a gene and as such might overlook parts of the immune system.

Transcriptional profiling, either via microarray or sequencing of cDNAs, can be used both to identify genes or pathways underlying the trait of interest or to specifically look at transcriptional profiles of target genes. Tissue-specific gene expression arrays are often preferred in medical and veterinary research, but where little is known about the genetic basis of the trait of interest, whole-genome microarrays or cDNA sequencing can be used. This process is effective for detecting the genetic basis of both host immune responses and parasite infection strategies. Differential gene expression between exposed and unexposed individuals has been used to simultaneously identify hundreds of genes underlying resistance phenotypes in a diverse range of organisms, such as cattle responding to intestinal nematodes (Araujo et al. 2009), *Anopheles* as a vector responding to *Plasmodium* (Vlachou et al. 2005), and shrimp exposed to white spot syndrome virus (Lan et al. 2006). In a similar manner, differential expression of genes underlying infection traits, such as the fungal pathogen *Metarhizium*'s response to different insect hosts' cuticles (Freimoser et al. 2005), is key to identifying genes determining a parasite's coevolutionary trajectory.

In addition to its use in de novo gene identification, transcriptional profiling can be used on a narrower range of genes, such as those located within a QTL region or previously identified as underlying a phenotype of interest. When used in combination with QTL mapping, not only can gene expression changes be used to locate the specific genes within these regions that are underlying the trait, but the expression profile itself can subsequently be treated as a quantitative trait and the heritability and genetic architecture of gene expression identified (Brem et al. 2002).

Once genes are identified via any of the described methods, further gene expression measurement can be used to verify their involvement in the phenotype of interest, and this is probably best done on individual genes using quantitative reverse transcription PCR (qrtPCR) as opposed to further use of large-scale methods. However, the true power of transcriptional profiling for illuminating host–parasite coevolution is that it can be used in an experimental context, particularly once target genes are identified. Understanding how specific genes respond to genetic and environmental variation is fundamental to our understanding of coevolution. Thus, using appropriately designed experiments, questions can be asked of any system and the differential expression of genes can be measured, rather than the associated phenotype.

DAPHNIA

Large swathes of data are rarely of use unless they can be contextualized. With the advent of technology enabling the production of information such as whole-genome transcriptional profiling, the onus is on us to develop ways to apply these data to questions of broader relevance. Medical, veterinary, and agricultural areas of research use these technologies for purposes such as the development of vaccines and the breeding of animals or plants more resistant to specific parasites or pathogens. However, evolutionary biologists have a different agenda, i.e., to elucidate the mechanisms underlying evolutionary change. With our work on the crustacean *Daphnia*, we are specifically interested in host–parasite coevolutionary change, and thus our goal is to apply these molecular techniques to questions more traditionally addressed using phenomenological, whole-organism studies. By integrating these two approaches in the *Daphnia* host–parasite system, we can hope to address not only the genetic basis of host–parasite interactions, but also the role of genetic and environmental variations underlying coevolution in general.

Daphnia and the Potential for Coevolution

Daphnia are small (~2 mm) ubiquitous crustaceans that have been the focus of more than a century of intense and diverse study, including toxicology, life history, physiology, nutrition, and parasitology. Critically, the distribution and virulence of the prevalent bacterial, microsporidian, and fungal parasites of *Daphnia* have been characterized, and easy to perform experimental manipulations on these naturally coevolving parasites have revealed extensive genetic variation for susceptibility (Ebert et al. 1998; Little and Ebert 1999, 2000b, 2001; Carius et al. 2001). Especially useful is the fact that *Daphnia* are facultative parthenogens; thus, they can be maintained clonally in the laboratory, enabling precise comparison of genetic backgrounds or the study of different environments on replicates of the same genetic background. They can also reproduce sexually, which permits traditional crossing experiments, with the caveat that sexual reproduction is mediated by environmental conditions to which genotypes vary in their sensitivities.

The short generation time of *Daphnia* (~10 d) enables the study of real-time evolutionary responses to parasites (Little and Ebert 1999). Gathering epidemiological data is also straightforward: The clear carapace of *Daphnia* makes infections easy to identify in the field, and infection dynamics can be tracked. Epidemics are common and severe but highly variable in space and time (see, e.g., Duncan et al. 2006; Lass and Ebert 2006; Duffy and Sivars-Becker 2007; Duncan and Little 2007; Wolinska et al. 2007). Past work has shown that *Daphnia* genotype frequencies (identified by neutral molecular markers) fluctuate wildly (Hebert 1974) and that these dynamics are linked to infection (Little and Ebert 1999; Mitchell et al. 2004; Duncan and Little 2007). Such studies were a reasonable, if rough, approximation of parasite-driven dynamics. In the postgenomic era, however, we can do much better through the identification and tracking of genes that specifically influence infection outcomes.

Simultaneous comparison of both host and parasite genetic backgrounds has lead to the discovery of genetic specificity in the interaction between *Daphnia magna* and its bacterial pathogen *Pasteuria ramosa* (Carius et al. 2001). Genetic specificity should not be confused with immunological specificity of the sort that is commonly generated by vertebrate immune systems and possibly by some invertebrates as well (Little et al. 2003; Sadd and Schmid-Hempel 2006). Immunological specificity is the case where the immune system learns to recognize and rapidly respond to a particular pathogen genotype (more specifically, a particular antigen) and does not explicitly take account of host genetic variation. Genetic specificity is about simultaneous genetic variation in hosts and parasites and, in particular, when the ability to resist parasites is tightly dependent on which parasite genotype is encountered (irrespective of previous encounters), whereas the ability of a particular parasite genotype to establish infection is tightly dependent on which host genotype it encounters (again, irrespective of infection history). Thus, genetic specificity is defined as host genotype by parasite genotype interactions (hereafter, referred to as $G_H \times G_P$).

Genetic specificity indicates the potential for frequency-dependent coevolution. Many studies have determined genes that have a key role in resisting parasites or pathogens (e.g., in *Anopheles Gambiae*; see, e.g., Osta et al. 2004), whereas fewer studies have used variable hosts as a starting point to detail genes that have a role in genetic variability for infection. No studies that we are aware of on animals (the plant literature is, however, comparatively rich; see, e.g., Thompson and Burdon 1992; Stahl and Bishop 2000) have used host and parasite combinations that show tight patterns of genetic specificity as a starting point for gene discovery. This, however, would be the ideal scenario given the relationship between genetic specificities and the coevolutionary process.

Daphnia Genomics: Characterizing the Immunome

Any gene discovery research program finds a strong foundation on a full-genome sequencing project, with *Daphnia* being no exception. The first crustacean genome to be fully sequenced was that of *Daphnia pulex*. Using sequence homology with the immune systems of other arthropods, 82 genes and 21 gene families with putative immune function were identified in *D. pulex* (McTaggart et al. 2009). This study identified pathways such as the TOLL pathway that are well conserved across invertebrate taxa and characterized areas of the *Daphnia* immune system that may be missing or evolving rapidly relative to other arthropods, such as antimicrobial peptides and antiviral RNA interference (RNAi) genes, respectively.

Although *D. pulex* is not suited for experimental host–parasite studies, its full-genome sequence has enabled the characterization of its innate immune system and provided a first set of candidate genes for study in a species well established for the study of host–parasite interactions, *D. magna*. Several genes identified from this genome sequence have been used for subsequent analysis in *D. magna*. Full cDNAs for one prophenyloxidase and two nitric oxide synthetase genes were acquired for *D. magna* (Labbé and Little 2009; Labbé et al. 2009). Using qrtPCR, changes in host-gene expression levels were measured in response to exposure to the bacterial parasite *P. ramosa*. The host and parasite genotypes used in these expression experiments have been studied previously (Carius et al. 2001), with this particular parasite genotype eliciting a high rate of infection in some genotypes but not in others. Despite this, there was no significant upregulation of any of the three genes tested, suggesting that they may not be directly involved in response to infection. This result was somewhat surprising, given that the genes studied almost invariably have a role in immunity in other taxa (Soderhall and Cerenius 1998; Rivero 2006). However, further testing is needed to draw conclusions about their potential role in generating resistant phenotypes in *Daphnia*. In particular, given the large variation typically observed among *Daphnia* genotypes (Little and

Ebert 2000b; Carius et al. 2001), it will be crucial to compare expression levels of a larger number of genotypes that vary in their resistance capabilities.

A major limitation of this type of gene discovery approach is that any immune system genes novel to the focal species will not be identified. Although the full-genome sequence of *D. magna* has now also been completed (see http://wfleabase.org/), any genes identified via phylogenetic comparison will still be restricted to those already known in other taxa. Furthermore, this approach assumes that if a gene has an immune function in one species (typically a well-characterized species such as *Drosophila melanogaster*), it will also be immunity related in the focal species. Although this assumption may be reasonable for genes that appear to have a universal role in immunity (e.g., prophenoloxidase, nitric oxide synthethase, α-2-macroglobulins), it is unlikely to always hold true. Thus, our current research is directed toward finding genes underlying coevolutionary traits specifically in *D. magna*. Toward this end, a transcriptomics project designed to identify genes specifically involved in the *D. magna*–*P. ramosa* interaction has been initiated.

Rather than sequencing a single genotype, an experimental design was used that enables both gene identification and experimental comparison. Two host genotypes, one resistant and one susceptible, and a single parasite strain were used in a 2 × 2 design whereby each of the two host genotypes were used to produce an unexposed control sample and an exposed treatment sample. The gene expression profiles of these four samples will produce (1) a comparison between the standing level of gene expression in susceptible and resistant individuals, (2) a comparison between the susceptible and resistant individuals' gene expression levels during exposure to a parasite, and (3) changes in gene expression between exposed and unexposed individuals of the same genotype. The differential regulation of genes in the two samples exposed to *P. ramosa* relative to their controls should provide a fairly comprehensive list of genes potentially underlying the innate immune response of *D. magna*. However, it will be desirable to expand this work to include additional genotypes.

FROM GENE DISCOVERY TO COEVOLUTION

Although identifying host genes that respond to parasite exposure is a major advance in understanding the *Daphnia* immune system, it is simply a first step toward our understanding of coevolutionary processes. Coevolution is clearly a dynamic, ongoing process, and thus, the genes involved must be characterized in terms of past selection pressure, current standing variation, and interaction with factors affecting their potential for evolutionary change. To this end, we are interested in applying standard molecular evolution analyses (e.g., McDonald Kreitman tests; McDonald and Kreitman 1991), laboratory experimental manipulations, and field-based samples and experiments to the set of genes identified from the transcriptome and to the set identified via bioinformatics comparison (McTaggart et al. 2009).

MOLECULAR EVOLUTION

Once candidate genes have been identified through differential gene expression, they can be subject to standard analyses of molecular evolution in order to examine the evolutionary history and signatures of selection on these genes. This information can be gained through the study of DNA polymorphism and divergence. It is, for example, possible to identify genes that are evolving rapidly, and thus possibly engaged in a host–parasite arms race (Yang and Bielawski 2000; Ford 2002). Strong directional selection, such as that which occurs during a host–parasite arms race, is expected to increase the rate of amino acid substitution among species, and the concomitant spread of new advantageous alleles will reduce the level of within-species genetic diversity around the selected locus. If, on the other hand, there is negative frequency-dependent selection, where rare alleles are at a selective advantage, genetic diversity will be maintained for extended periods of time and divergence between haplotypes can become the extreme, as is seen, for example, at vertebrate major histocompatibility complex (MHC) loci (see, e.g., Hedrick 1998) and in plant resistance genes (Stahl et al. 1999).

The detection of arms races requires an outgroup species that is neither too distantly related to the focal species (such that there is saturation of nucleotide divergence) nor so closely related that nucleotide differences are too scarce for analyses. This problem, for example, is particularly acute in the study of mosquitoes, which tend to form species flocks with little diversity within flocks but substantial (too substantial for outgroup comparisons) divergence between flocks (Obbard et al. 2007, 2009). The number and average genetic distance separating *Daphnia* species suggest that this is unlikely to be a problem in this genus (Colbourne et al. 1998). Indeed, we have found *D. pulex*–*D. parvula* comparisons to be effective for the detection of arms races (S McTaggart and T Little, in prep.), although similar efforts on *D. magna* have yet to be undertaken. The detection of negative frequency-dependent selection or other forms of balancing selection faces different but equally challenging issues. In particular, it can be difficult to distinguish balancing selection from relaxed constraint, because both forces can lead to very high polymorphism. Data on linkage disequilibrium can to a considerable degree resolve the issue, but even with this in hand, other factors, such as gene conversion, can potentially derail interpretation (Obbard et al. 2008). Despite these obstacles, there are many instances where analyses of molecular evolution have yielded insight into the evolutionary and coevolutionary process (see, e.g., Hedrick 1998; Hurst and Smith 1999; Stahl et al. 1999; Lazzaro and Clark 2001, 2003; Schlenke and Begun 2003; Lazzaro et al. 2004; Jiggins and Kim 2006; Obbard et al. 2006)

EXPERIMENTAL GENOMICS

The utility of identifying genes underlying traits involved in coevolution is not limited to elucidating historical patterns of natural selection via molecular evolution approaches but rather can be extended to actual laboratory and field experimental applications. Much of our understanding of *Daphnia* coevolution has come from whole-organism empirical work revealing the impact of both host and parasite genotypes and various aspects of the environment on this interaction. Now, with candidate genes in hand, we can rework these experiments and identify the genetic changes or dynamics that result in phenotypic variation due to $G_H \times G_P$ or G × E interactions. For example, host by parasite genetic interactions have been experimentally identified as being a key determinant of infection in the *D. magna–P. ramosa* system (Carius et al. 2001). It is now possible to repeat this type of laboratory interaction experiment and directly associate the phenotypic response with allelic or nucleotide variation in genes previously identified as underlying resistance or susceptibility. This information is key to determining which genes may be ongoing targets of selection during coevolution.

Extending this, a unique power of the *Daphnia* system is the capacity to verify the "real world" relevance and evolution of any genes highlighted by QTL or expression studies. Comparing the genotype(s) of naturally infected individuals to that of naturally uninfected individuals can reveal whether these genetic variants determine which individuals become infected and which do not in nature. Normally, verification might be accomplished with experimental knockout lines in the laboratory, but a disease association in natural populations, where the focal gene is found within a huge range of genetic backgrounds, is a different but perhaps more relevant form of verification. Moreover, these data are crucial because they also check for the phenotypic consequence of a gene when placed in a range of environments. Previous work on *Daphnia* has shown the influence of environmental variables that effect certain genotypes more than others, i.e., there are strong genotype by environment (G × E) interactions in the lab (Mitchell et al. 2005; Vale et al. 2008), as have studies on other parasitic interactions (see, e.g., Blanford et al. 2003). Genotype by environment effects are not limited to invertebrates. For example, the widely used mouse strains BALB/c and C57BL/6 clearly differ in susceptibility to parasites under tightly controlled conditions, but they may show no such differences in a natural arena (Scott 1991). Thus, in any taxon, strong G × E interactions could generate a gene infection association in one environment, such as the lab where the QTL or expression work occurred, but no association in other environments such as the field. We stress that the clearest measure of a gene's significance in the host–parasite interaction is that its effects penetrate noisy natural environments in a range of genetic backgrounds.

In this sense, this information is interesting simply from the point of view of understanding how immune effector systems variably respond under genetic and environmental variation. We see a need for the field of immunology to expand to systems tractable for the study of variation or to incorporate variation into existing systems. The experimental designs traditionally used in the field of immunology are simplified: Most experiments are performed in the absence of pathogens, under ideal laboratory conditions, and in homogeneous inbred genetic backgrounds. Thus, although it provides the necessary mechanistic backbone for studying infection, immunology has not typically addressed variation in natural populations, despite the observation that the impact of genetic and environmental variation on infection is likely large (Boulinier et al. 1997; Sorci et al. 1997; Coltman et al. 1999; Little 2002; Thomas and Blanford 2003; Bedhomme et al. 2004; Mitchell et al. 2005; Vale et al. 2008), as are G × E effects (references above). It remains a salient challenge for both immunologists and ecologists to link variation in host–parasite outcomes (as determined by genetic and environmental factors) to immunological mechanisms.

Associations between an immune-related gene and infection in field populations is valuable information, but if possible should be confirmed with phenotypic assays in the laboratory. Here, organisms, such as *Daphnia*, that reproduce clonally hold great promise because individuals can be brought to the laboratory and maintained as clonal lineages for genotyping and experimentation. Even field-collected infected (and sterilized) individuals can be cured (Little and Ebert 2000a) and kept clonally in the laboratory. By taking a sample of clones (potentially different genotypes) from natural populations, it is possible to effectively establish a snapshot of the population at the time of collection that can then be studied at leisure in the lab. Because many *Daphnia* parasites can be frozen, snapshots of them can also be taken. These sampling possibilities are relevant because it is important not only to observe allele frequency differences, but ultimately to know how and why they vary, and thus determine the evolutionary consequences for the population. For example, a change in allele frequencies might be accompanied by a decline in parasite prevalence, but that decline might also be the consequence of a change in temperature. By bringing samples of hosts from appropriate time points into the laboratory and testing their phenotypic resistance under controlled conditions, it is possible to verify the cause of gene frequency changes observed in nature.

Such a complete level of information will be difficult to obtain with other prominent host–parasite systems (e.g., *Anopheles–Plasmodium*), but this level is required when the goal is to elucidate general evolutionary principles. When complete, something specific about the immune pathways of *Daphnia* will have been uncovered, but more importantly, this research program should also enhance general understanding about the impact of parasitism: How quickly do host genotype frequencies change, how quickly can parasite populations respond, and how will this affect parasite prevalence? The mechanisms by which this occurs are key: Are selective sweeps and transient polymorphisms the norm or does frequency-dependent selection maintain polymorphism? Or maybe it is neither of these. Perhaps polymorphism is maintained by pervasive genotype by environment interactions or epistasis. These are salient questions if we are to understand how coevolution influences the health of populations.

ACKNOWLEDGMENTS

T.J.L. and D.E.A. are supported by The Wellcome Trust, United Kingdom.

REFERENCES

Araujo RN, Padilha T, Zarlenga D, Sonstegard T, Connor EE, Van Tassel C, Lima WS, Nascimento E, Gasbarre LC. 2009. Use of a candidate gene array to delineate gene expression patterns in cattle selected for resistance or susceptibility to intestinal nematodes. *Vet Parasitol* **162:** 106–115.

Bangham C, Anderson RM, Baquero F, Bax R, Hastings I, Koella JC, Lipsitch M, Mclean A, Smith T, Taddei F, et al. 1999. Evolution of infectious diseases: The impact of vaccines, drugs and social factors. In *Evolution in health and disease* (ed. SC Stearns), pp. 152–160. Oxford University Press, New York.

Bangham J, Kim KW, Webster C, Jiggins FM. 2008. Genetic variation affecting host-parasite interactions: Different genes affect different aspects of sigma virus replication and transmission in *Drosophila melanogaster*. *Genetics* **178:** 2191–2199.

Baton LA, Garver L, Xi Z, Dimopoulos G. 2008. Functional genomics studies on the innate immunity of disease vectors. *Insect Sci* **15:** 15–27.

Bedhomme S, Agnew P, Sidobre C, Michalakis Y. 2004. Virulence reaction norms across a food gradient. *Proc R Soc Lond B Biol Sci* **271:** 739–744.

Bishop JG, Dean AM, Mitchell-Olds T. 2000. Rapid evolution in plant chitinases: Molecular targets of selection in plant-pathogen coevolution. *Proc Natl Acad Sci* **97:** 5322–5327.

Blanford S, Thomas MB, Pugh C, Pell JK. 2003. Temperature checks the Red Queen? Resistance and virulence in a fluctuating environment. *Ecol Lett* **6:** 2–5.

Boulinier T, Sorci G, Monnat JY, Danchin E. 1997. Parent-offspring regression suggests heritable susceptibility to ectoparasites in a natural population of kittiwake *Rissa tridactyla*. *J Evol Biol* **10:** 77–85.

Brem RB, Yvert G, Clinton R, Kruglyak L. 2002. Genetic dissection of transcriptional regulation in budding yeast. *Science* **296:** 752–755.

Brown A, Olver W, Donnelly C, May M, Naggert J, Shaffer D, Roopenian D. 2005. Searching QTL by gene expression: Analysis of diabesity. *BMC Genet* **6:** 12.

Carius HJ, Little TJ, Ebert D. 2001. Genetic variation in a host–parasite association: Potential for coevolution and frequency dependent selection. *Evolution* **55:** 1136–1145.

Clarke BC. 1979. The evolution of genetic diversity. *Proc R Soc Lond B Biol Sci* **205:** 453–474.

Colbourne JK, Crease TJ, Weider LJ, Hebert PDN, Dufresne F, Hobaek A. 1998. Phylogenetics and evolution of a circumarctic species complex (Cladocera: *Daphnia pulex*). *Biol J Linn Soc* **65:** 347–365.

Coltman DW, Pilkington JG, Smith JA, Pemberton JM. 1999. Parasite-mediated selection against inbred Soay sheep in a free-living, island population. *Evolution* **53:** 1259–1267.

Duffy MA, Sivars-Becker L. 2007. Rapid evolution and ecological host-parasite dynamics. *Ecol Lett* **10:** 44–53.

Duncan AB, Little TJ. 2007. Parasite-driven genetic change in a natural population of *Daphnia magna*. *Evolution* **61:** 796–803.

Duncan A, Mitchell SE, Little TJ. 2006. Parasite-mediated selection in *Daphnia*: The role of sex and diapause. *J Evol Biol* **19:** 1183–1189.

Ebert D, Zschokke-Rohringer CD, Carius HJ. 1998. Within- and between-population variation for resistance of *Daphnia magna* to the bacterial endoparasite *Pasteuria ramosa*. *Proc R Soc Lond B Biol Sci* **265:** 2127–2134.

Ford MJ. 2002. Applications of selective neutrality tests to molecular ecology. *Mol Ecol* **11:** 1245–1262.

Frank SA. 1993. Evolution of host-parasite diversity. *Evolution* **47:** 1721–1732.

Freimoser FM, Hu G, St Leger RJ. 2005. Variation in gene expression patterns as the insect pathogen *Metarhizium anisopliae* adapts to different host cuticles or nutrient deprivation in vitro. *Microbiology* **151:** 361–371.

Gilbert SC, Plebanski M, Gupta S, Morris J, Cox M, Aidoo M, Kwiatkowski D, Greenwood BM, Whittle HC, Hill AV. 1998. Association of malaria parasite population structure, HLA, and immunological antagonism. *Science* **279:** 1173–1177.

Hamilton WD. 1980. Sex versus non-sex versus parasite. *Oikos* **35:** 282–290.

Hamilton WD, Axelrod R, Tanese R. 1990. Sexual reproduction as an adaptation to resist parasites. *Proc Natl Acad Sci* **87:** 3566–3573.

Hebert PD. 1974. Enzyme variability in natural populations of *Daphnia magna*. II. Genotypic frequencies in permanent populations. *Genetics* **77:** 323–334.

Hedrick PW. 1994. Evolutionary genetics of the major histocompatibility complex. *Am Nat* **143:** 945–964.

Hedrick PW. 1998. Balancing selection and MHC. *Genetica* **104:** 207–214.

Hill AVS, Allsopp CEM, Kwiatkowski D, Anstey NM, Twumasi P, Rowe PA, Bennett S, Brewster D, McMichael AJ, Greenwood BM. 1991. Common West African HLA antigens are associated with protection from severe malaria. *Nature* **352:** 595–600.

Hill AVS, Kwiatkowski D, McMichael AJ, Greenwood BM, Bennett S. 1992. Maintenance of Mhc polymorphism: Reply. *Nature* **355:** 403–403.

Hill CA, Kafatos FC, Stansfield SK, Collins FH. 2005. Arthropod-borne diseases: Vector control in the genomics era. *Nat Rev Microbiol* **3:** 262–268.

Hurst LD, Smith NGC. 1999. Do essential genes evolve slowly? *Curr Biol* **9:** 747–750.

Jiggins FM, Kim K-W. 2006. Contrasting evolutionary patterns in *Drosophila* immune receptors. *J Mol Evol* **63:** 769–780.

Keeler CL Jr, Bliss TW, Lavric M, Maughan MN. 2007. A functional genomics approach to the study of avian innate immunity. *Cytogenet Genome Res* **117:** 139–145.

Kover PX, Caicedo AL. 2001. The genetic architecture of disease resistance in plants and the maintenance of recombination by parasites. *Mol Ecol* **10:** 1–16.

Labbé P, Little TJ. 2009. ProPhenolOxidase in *Daphnia magna*: cDNA sequencing and expression in relation to resistance to pathogens. *Dev Comp Immunol* **33:** 674–680.

Labbé P, McTaggart SJ, Little TJ. 2009. An ancient immunity gene duplication in *Daphnia magna*: RNA expression and sequence analysis of two nitric oxide synthase genes. *Dev Comp Immunol* **33:** 1000–1010.

Lan Y, Xu X, Yang F, Zhang X. 2006. Transcriptional profile of shrimp white spot syndrome virus (WSSV) genes with DNA microarray. *Arch Virol* **151:** 1723–1733.

Lass S, Ebert D. 2006. Apparent seasonality of parasite dynamics: Analysis of cyclic prevalence patterns. *Proc R Soc Lond B Biol Sci* **273:** 199–206.

Lazzaro BP, Clark AG. 2001. Evidence for recurrent paralogous gene conversion and exceptional allelic divergence in the Attacin genes of *Drosophila melanogaster*. *Genetics* **159:** 659–671.

Lazzaro BP, Clark AG. 2003. Molecular population genetics of inducible antibacterial peptide genes in *Drosophila melanogaster*. *Mol Biol Evol* **20:** 914–923.

Lazzaro BP, Sceurman BK, Clark AG. 2004. Genetic basis of natural variation in *D. melanogaster* antibacterial immunity. *Science* **303:** 1873–1876.

Little TJ. 2002. The evolutionary significance of parasitism: Do parasite-driven genetic dynamics occur *ex silico*? *J Evol Biol* **15:** 1–9.

Little TJ, Cobbe N. 2005. The evolution of immune-related genes from disease carrying mosquitoes: Diversity in a peptidoglycan- and a thioester-recognising protein. *Insect Mol Biol* **14:** 599–605.

Little TJ, Ebert D. 1999. Associations between parasitism and host genotype in natural populations of *Daphnia* (Crustacea: Cladocera). *J Anim Ecol* **68:** 134–149.

Little TJ, Ebert D. 2000a. Sex, linkage disequilibrium and patterns of parasitism in three species of cyclically parthenogenic *Daphnia* (Crustacea: Cladocera). *Heredity* **85**: 257–265.

Little TJ, Ebert D. 2000b. The cause of parasitic infection in natural populations of *Daphnia* (Crustacea: Cladocera): The role of host genetics. *Proc R Soc Lond B Biol Sci* **267**: 2037–2042.

Little TJ, Ebert D. 2001. Temporal patterns of genetic variation for resistance and infectivity in a *Daphnia*-microparasite system. *Evolution* **55**: 1146–1152.

Little TJ, O'Connor B, Colegrave N, Watt K, Read AF. 2003. Maternal transfer of strain-specific immunity in an invertebrate. *Curr Biol* **13**: 489–492.

Little TJ, Colbourne JK, Crease TJ. 2004. Molecular evolution of *Daphnia* immunity genes: Polymorphism in a gram negative binding protein and an α-2-macroglobulin. *J Mol Evol* **59**: 498–506.

Lively CM. 1993. Rapid evolution by biological enemies. *Trends Ecol Evol* **8**: 345–346.

Lynch M, Walsh B. 1998. *Genetics and analysis of quantitative traits*. Sinauer, Sunderland, MA.

Marchese A, Schito GC, Debbia EA. 2000. Evolution of antibiotic resistance in gram-positive pathogens. *J Chemother* **12**: 459–462.

McDonald JH, Kreitman M. 1991. Adaptive protein evolution at the *Adh* locus in *Drosophila*. *Nature* **351**: 652–654.

McTaggart S, Conlon C, Colbourne J, Blaxter M, Little T. 2009. The components of the *Daphnia pulex* immune system as revealed by complete genome sequencing. *BMC Genomics* **10**: 175.

Mitchell SE, Read AF, Little TJ. 2004. The effect of a pathogen epidemic on the susceptibility to infection, reproductive investment and genetic structure of the cyclical parthenogen *Daphnia magna*. *Ecol Lett* **7**: 848–858.

Mitchell SE, Rogers ES, Little TJ, Read AF. 2005. Host-parasite and genotype-by-environment interactions: Temperature modifies potential for selection by a sterilizing pathogen. *Evolution* **59**: 70–80.

Morgan AD, Gandon S, Buckling A. 2005. The effect of migration on local adaptation in a coevolving host-parasite system. *Nature* **437**: 253–256.

Morse SS. 1994. *The evolutionary biology of viruses*. Raven, New York.

Niaré O, Markianos K, Volz J, Oduol F, Touré A, Bagayoko M, Sangaré D, Traoré SF, Wang R, Blass C, et al. 2002. Genetic loci affecting resistance to human malaria parasites in a West African mosquito vector population. *Science* **298**: 213–216.

Obbard DJ, Jiggins FM, Little TJ. 2006. Rapid evolution of antiviral RNAi genes. *Curr Biol* **16**: 580–585.

Obbard D, Linton Y, Jiggins F, Yan G, Little T. 2007. Population genetics of *Plasmodium* resistance genes in *Anopheles gambiae*: No evidence for strong selection. *Mol Ecol* **16**: 3497–3510.

Obbard DJ, Callister DM, Jiggins FM, Soares DC, Yan G, Little TJ. 2008. The evolution of TEP1, an exceptionally polymorphic immunity gene in *Anopheles gambiae*. *BMC Evol Biol* **8**: 274.

Obbard DJ, Welch JJ, Little TJ. 2009. Inferring selection in the *Anopheles gambiae* species complex: An example from immune-related serine protease inhibitors. *Malar J* **8**: 117.

Osta MA, Christophides GK, Kafatos FC. 2004. Effects of mosquito genes on *Plasmodium* development. *Science* **303**: 2030–2032.

Rainey P. 2004. Bacterial populations adapt genetically by natural selection even in the lab! *Microbiol Today* **31**: 160–162.

Rivero A. 2006. Nitric oxide: An antiparasitic molecule of invertebrates. *Trends Parasitol* **22**: 219–225.

Sadd BM, Schmid-Hempel R. 2006. Insect immunity shows specificity in protection on secondary pathogen exposure. *Curr Biol* **16**: 1206–1210.

Salathé M, Kouyos RD, Bonhoeffer S. 2008. The state of affairs in the kingdom of the Red Queen. *Trends Ecol Evol* **23**: 439–445.

Schlenke TA, Begun DJ. 2003. Natural selection drives *Drosophila* immune system evolution. *Genetics* **164**: 1471–1480.

Scott ME. 1991. *Heligmosomoides polygrus* (Nematoda): Susceptible and resistant strains are indistinguishable following natural infection. *Parasitology* **103**: 429–438.

Seger J. 1988. Dynamics of some simple host-parasite models with more than two genotypes in each species. *Philos Trans R Soc Lond B Biol Sci* **319**: 541–555.

Soderhall K, Cerenius L. 1998. Role of the prophenoloxidase-activating system in invertebrate immunity. *Curr Opin Immunol* **10**: 23–28.

Sorci G, Moller AP, Boulinier T. 1997. Genetics of host-parasite interactions. *Trends Ecol Evol* **12**: 196–200.

Stahl EA, Bishop JG. 2000. Plant-pathogen arms races at the molecular level. *Curr Opin Plant Biol* **3**: 299–304.

Stahl EA, Dwyer G, Mauricio R, Kreitman M, Bergelson J. 1999. Dynamics of disease resistance polymorphism at the *Rpm1* locus of *Arabidopsis*. *Nature* **400**: 667–671.

Thomas MB, Blanford S. 2003. Thermal biology in insect-parasite interactions. *Trends Ecol Evol* **18**: 344–350.

Thompson JN, Burdon JJ. 1992. Gene-for-gene coevolution between plants and parasites. *Nature* **360**: 121–125.

Vale PF, Stjernman M, Little TJ. 2008. Temperature-dependent costs of parasitism and maintenance of polymorphism under genotype-by-environment interactions. *J Evol Biol* **21**: 1418–1427.

Vlachou D, Schlegelmilch T, Christophides GK, Kafatos FC. 2005. Functional genomic analysis of midgut epithelial responses in *Anopheles* during *Plasmodium* invasion. *Curr Biol* **15**: 1185–1195.

Wilfert L, Schmid-Hempel P. 2008. The genetic architecture of susceptibility to parasites. *BMC Evol Biol* **8**: 187.

Wolinska J, Keller B, Manca M, Spaak P. 2007. Parasite survey of a *Daphnia* hybrid complex: Host-specificity and environment determine infection. *J Anim Ecol* **76**: 191–200.

Woolhouse ME, Webster JP, Domingo E, Charlesworth B, Levin BR. 2002. Biological and biomedical implications of the coevolution of pathogens and their hosts. *Nat Genet* **32**: 569–577.

Yang Z, Bielawski JP. 2000. Statistical methods for detecting molecular adaptation. *Trends Ecol Evol* **15**: 496–502.

Genetic Recombination and Molecular Evolution

B. Charlesworth,[1] A.J. Betancourt,[1] V.B. Kaiser,[1] and I. Gordo[2]

[1]*Institute of Evolutionary Biology, School of Biological Sciences, University of Edinburgh, Edinburgh, EH9 3JT, United Kingdom;* [2]*Instituto Gulbenkian de Ciência, 2780-156 Oeiras, Portugal*

Correspondence: Brian.Charlesworth@ed.ac.uk

Reduced rates of genetic recombination are often associated with reduced genetic variability and levels of adaptation. Several different evolutionary processes, collectively known as Hill–Robertson (HR) effects, have been proposed as causes of these correlates of recombination. Here, we use DNA sequence polymorphism and divergence data from the noncrossing over dot chromosome of *Drosophila* to discriminate between two of the major forms of HR effects: selective sweeps and background selection. This chromosome shows reduced levels of silent variability and reduced effectiveness of selection. We show that neither model fits the data on variability. We propose that, in large genomic regions with restricted recombination, HR effects among nonsynonymous mutations undermine the effective strength of selection, so that their background selection effects are weakened. This modified model fits the data on variability and also explains why variability in very large nonrecombining genomes is not completely wiped out. We also show that HR effects of this type can produce an individual selection advantage to recombination, as well as greatly reduce the mean fitness of nonrecombining genomes and genomic regions.

In eukaryotes, the disjunction of homologous centromeres in the first division of meiosis results in the independent assortment of genes on different chromosomes; recombinational exchange (gene conversion and reciprocal crossing over) reshuffles the genetic material between the homologous chromosomes contributed by the two parents. These processes cause different sites in the genome to have more or less distinct ancestries, unless recombination is absent or ineffective. Recombination can therefore have important consequences for the effectiveness of selection (Barton, this volume). In particular, Fisher (1930, p.103) pointed out that there may be an evolutionary cost to recombination. When two loci are each polymorphic for two alleles with epistatic fitness effects that create linkage disequilibrium (LD), recombination reduces the frequencies of the selectively favorable combinations of alleles. He concluded that if there is genetic variability in the frequency of recombination, this "…will always tend to diminish recombination, and therefore to increase the intensity of linkage in the chromosomes.… ." Subsequent theoretical work has put this verbal argument on a firm theoretical basis (Zhivotovsky et al. 1994; Otto and Lenormand 2002).

Why, therefore, does the genome not "congeal" to a state of zero recombination (Turner 1967)? There is evidence pointing to a countervailing selective advantage to recombination; for example, several ecological factors correlate with rates of crossing over. Mammalian species with long development times tend to have higher rates of crossing over per chromosome than fast-developing species (Burt and Bell 1987; Sharp and Hayman 1988), and highly self-fertilizing species of plants tend to have higher rates of cytologically detectable crossovers than related outcrossing species (Roze and Lenormand 2005). Recombination rates therefore appear to vary in response to selective pressures, so that we need to search for population genetic processes that relate recombination to higher fitness. Several credible candidates for this have been identified (Barton, this volume). The challenge is to identify biological patterns that indicate that the level of recombination has evolutionary consequences and that also shed light on which processes may be involved.

We argue that a variety of lines of evidence suggest that, as proposed by Felsenstein (1974), Hill–Robertson (HR) effects have a major role in causing the evolutionary effects of recombination. Hill and Robertson (1966) showed that selection at one site in the genome impedes the action of selection at another site, especially when recombination between them is rare or absent (Fig. 1). This is because a finite population cannot contain all possible combinations of variants at different sites. If mutations arise in different individuals, a favorable variant at one site will generally be present in a genotype with a deleterious variant at another site, i.e., negative LD exists among selectively favorable variants. Because recombination breaks down this LD, it enhances the population's ability to respond to selection. Table 1 lists the main types of HR effects.

A useful way to understand HR effects is to consider them in terms of the effective population size N_e. This is essentially the number of individuals in the population that successfully transmit genes to the next generation and is often much smaller than the number of individuals of breeding age (Wright 1931; Charlesworth 2009). Selection implies the existence of heritable variance in fitness among individuals; this reduces N_e because genes are preferentially transmitted through the fittest members of the population (Robertson 1961). A nucleotide site that is closely linked to another site with variants that are under selection experiences an especially large effect,

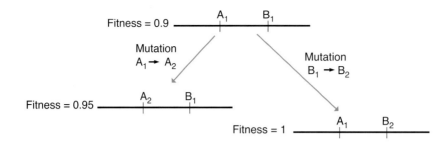

Figure 1. Hill–Robertson interference between the action of selection for advantageous mutations at two sites in a genome when recombination between them is rare or absent. The diagram shows two loci or sites, A and B, at which advantageous mutations arise in a haploid population, each of which increases fitness. The highest fitness is achieved when both advantageous mutations are present in an individual.

because the influence of the variance in fitness at one site on the behavior of closely linked sites is maintained for many generations (Santiago and Caballero 1998; Barton, this volume).

The equilibrium level of diversity at neutral nucleotide sites is equal to the product of $4N_e$ and the mutation rate u, provided that $4N_e u \ll 1$ (Kimura 1971). In addition, the probability that genetic drift fixes a deleterious mutation that reduces the fitness of its homozygous carriers by s (the selection coefficient) is close to the value for a neutral mutation when $N_e s < 1$ but is negligible when $N_e s \gg 1$ (Fisher 1930, ch. 5; Kimura 1962). Similarly, the chance that a selectively favorable mutation with selective advantage s becomes established in a population is close to the neutral value when $N_e s \ll 1$ but approaches the value for an infinitely large population when $N_e s > 1$. Not all features of HR effects can be understood simply in terms of a reduction in N_e (Comeron et al. 2008; Kaiser and Charlesworth 2009), but it nevertheless provides a useful heuristic for their interpretation.

Reduced N_e caused by HR effects is thus expected to cause a reduction in the level of variability with respect to neutral or nearly neutral nucleotide variants. It will also cause loci to accumulate more slightly deleterious mutations, and fix fewer advantageous ones, than when HR effects are absent. The following patterns that are

Table 1. Main Categories of HR Effects

1. Interference by Favorable Mutations (Selective Sweeps)

The spread of a favorable mutation drags to fixation any closely linked neutral or deleterious mutant alleles initially associated with it, so that successive adaptive substitutions in a low recombination region of the genome can lead to a loss in neutral or nearly neutral variability and the fixation of slightly deleterious mutations at many loci. In addition, the spread of a favorable mutation at one locus can prevent the spread of a favorable mutation at another, closely linked locus.

2. Interference by Deleterious Mutations (Background Selection)

Deleterious mutations are assumed to enter the population at sites distributed over the genomic region in question and to be removed by selection with near certainty. A neutral or weakly selected mutation that arises in a nonrecombining section of the genome has a nonzero chance of survival only if it arises on a chromosome free of these mutations. This accelerates the fixation of weakly deleterious mutations and retards the fixation of advantageous mutations.

3. Muller's Ratchet

This involves the stochastic loss from a finite population of the class of chromosomes carrying the fewest deleterious mutations. In the absence of recombination and back mutation, this class of chromosome cannot be restored. The next best class then replaces it and is in turn lost in a process of successive irreversible steps. Each such loss is quickly followed by fixation of a deleterious mutation on the chromosome. Mutations at most sites remain close to their equilibrium frequencies.

4. Mutual Interference among Weakly Selected Sites (Weak HR Effects)

With a very large number of closely linked sites, subject to reversible mutation between favored and disfavored alleles, the mean level of adaptation can be strongly reduced in nonrecombining regions. This is because a mutual interference exists among the different sites under selection, allowing selectively deleterious variants to become much more frequent than expected under mutation-selection equilibrium frequencies, as a result of genetic drift.

For further details, see Otto and Lenormand (2002), Comeron et al. (2008), Charlesworth and Charlesworth (2010, ch. 10), and Barton (this volume).

consistent with these expectations have been uncovered.

1. Regions of the genome with low levels of genetic recombination often show low levels of genetic diversity (see, e.g., Presgraves 2005).
2. Species with low levels of genome-wide recombination, such as highly self-fertilizing species of animals and plants, also show reduced genetic diversity (see, e.g., Charlesworth 2003).
3. These reductions in diversity are often associated with reduced levels of adaptation at the molecular level (see, e.g., Presgraves 2005; Moran et al. 2008).

We examine in detail here some examples of such evidence from our recent work with *Drosophila*. In addition, we describe recent theoretical work that helps to resolve some contradictions between the theoretical predictions and the data

THE RELATION OF RECOMBINATION TO GENETIC VARIATION

Previous Work with *Drosophila*

About 20 years ago, it was found that within-population variability was unusually low in regions of the *Drosophila* genome with low levels of crossing over (Aguadé et al. 1989; Stephan and Langley 1989). Begun and Aquadro (1992) showed that a high correlation exists between the estimated level of variability in a gene and the local rate of recombination determined from the standard genetic map, whereas divergence at silent sites showed no relation to recombination rates. These observations have been replicated in more recent studies, for example, by Presgraves (2005).

Some form of HR effect seems to be the only credible explanation of these patterns. Begun and Aquadro (1992) favored hitchhiking effects caused by the spread of favorable mutations (Maynard Smith and Haigh 1974), often now referred to as "selective sweeps" (Berry et al. 1991). Stephan (1995) showed that it is possible to fit the observed relation between recombination rate and level of variability by this model; this has recently been extended to large *Drosophila melanogaster* polymorphism data sets (Andolfatto 2007). However, the alternative mode of hitchhiking by selection against recurrent deleterious mutations ("background selection," Table 1) also fits the data on the relation between levels of variability and local recombination rates (Charlesworth 1996). Attempts to discriminate between selective sweeps and background selection have largely been inconclusive.

Genetic Diversity on the Dot Chromosome of *Drosophila americana*

We have recently revisited this question (Betancourt et al. 2009), using the close relative of *D. virilis*, *D. americana*, for a survey of within-species variation and divergence among species at 14 genes on the small "dot" chromosome (Muller's element F), a chromosome that shows highly reduced levels of crossing over (Ashburner et al. 2005). A data set on variability at 18 genes on other chromosomes is available for the purpose of comparison (Maside and Charlesworth 2007). Genetic data show that there is little population subdivision in *D. americana* and little evidence for demographic effects such as population expansion that complicate the interpretation of population genetic data (McAllister 2002; Maside and Charlesworth 2007); this makes it appropriate material for population genetic studies.

Silent nucleotide site diversity on the *D. americana* dot chromosome is about 17 times lower than the genome-wide average, a reduction in variability similar to that in other *Drosophila* species (Berry et al. 1991; Jensen et al. 2002; Wang et al. 2002, 2004; Sheldahl et al. 2003). As expected, the polymorphism data show that dot chromosome loci have a very low, but nonzero, recombination rate, as was seen in the other species. These recombination events are probably due to gene conversion rather than crossovers (Langley et al. 2000; Gay et al. 2007).

Interpretation of the Results

Coalescent simulations show that a recent selective sweep is not compatible with the observed distribution of frequencies of nucleotide site variants on the *D. americana* dot chromosome (Betancourt et al. 2009): There are too many intermediate-frequency variants compared with what is expected after a selective sweep on a nonrecombining chromosome (Braverman et al. 1995; Simonsen et al. 1995). Can background selection explain these data? The expected reduction in diversity for the dot chromosome can be determined from the classical background selection equation (Hudson and Kaplan 1995; Nordborg et al. 1996). For this purpose, we need to know the distribution of selection coefficients against deleterious mutations. Estimates of the parameters of this distribution for nonsynonymous mutations can be obtained from polymorphism data (Loewe and Charlesworth 2006; Loewe et al. 2006; Keightley and Eyre-Walker 2007; Sawyer et al. 2007). These data show that there is a wide distribution of the selective effects of deleterious mutations but the mean selection coefficient against a segregating amino acid mutation is extremely small (of the order of 10^{-5} to 10^{-4}). The classical background selection model with these estimates predicts that the dot chromosome should have ~1000-fold lower variation than the other autosomes (Loewe and Charlesworth 2007), rather than the 17-fold lower value in *D. americana*, and similar values for the other cases cited above.

EXPLAINING THE OBSERVED PATTERNS OF VARIABILITY

Reformulating the Background Selection Model

Both of the standard explanations for reduced variability in regions with low levels of recombination appear to be incompatible with our data on patterns of variability on the dot chromosome. To try and resolve this paradox, we have reexamined the theory of background selection in a large, low recombination genomic region. The standard model assumes that the frequencies of deleterious mutant variants involved are close to those expected under mutation-selection balance equilibrium in an infinite population (Charlesworth et al. 1993; Hudson and Kaplan 1995; Nordborg et al. 1996). When recombination rates are extremely low, however, the model predicts a larger reduction in variability than is found in simulations (Charlesworth et al. 1993; Nordborg et al. 1996; Gordo et al. 2002). This suggests that low recombination may cause HR interference among the sites involved (for which $N_e s > 1$ when there are no HR effects), undermining the effectiveness of selection on these sites and causing the frequencies of deleterious mutations to drift up to much higher values than with mutation-selection balance. HR interference of this type has been studied previously by Monte Carlo simulations of selection at many linked sites (for review, see Comeron et al. 2008). These studies used fixed selection coefficients at each site and were designed primarily to model weak selection on codon usage (except for Tachida 2000).

In view of the recent increase in our knowledge of the intensity of purifying selection on nonsynonymous mutations (see above), it seemed important to model HR effects with realistic selective effects to see whether they can explain the *Drosophila* data on variability in genomic regions that lack crossing over. We have performed Monte Carlo simulations of randomly mating populations with either normal or reduced rates of recombination (Kaiser and Charlesworth 2009). Haploid populations of 1000 individuals (equivalent to $N = 500$ diploids) were modeled. The state of a site under selection is represented as 1 or 0, where 0 is wild type and 1 is a deleterious alternative variant (Fig. 2). The fitness effect of a mutation at the ith chromosomal site under selection is denoted by s_i; the fitness, w, of an individual carrying a set of mutations is given by the standard multiplicative model (Haldane 1937), such that $\ln(w) = \Sigma_i \ln(1 - s_i)$.

Pairs of adjacent selected sites followed by neutral sites were distributed along the chromosome, with the total number of sites (L) varying between simulations. The selected sites correspond to first and second codon positions, where all mutations were assumed to be nonsynonymous; the neutral sites correspond to third codon positions that experience only synonymous mutations. Mutations at both types of sites arise at a rate u per base pair in each direction. This reversible mutation model applies to nucleotide mutations and allows the population to reach statistical equilibrium between drift, mutation, and selection (McVean and Charlesworth 2000). When the evolutionary forces are all weak, measures of the deterministic forces scaled by multiplying their values by N_e completely describe the system if time is measured in units of N_e generations (Ewens 2004). We can thus infer the behavior of large natural populations from our small simulated populations by using these scaled parameter values (McVean and Charlesworth 2000).

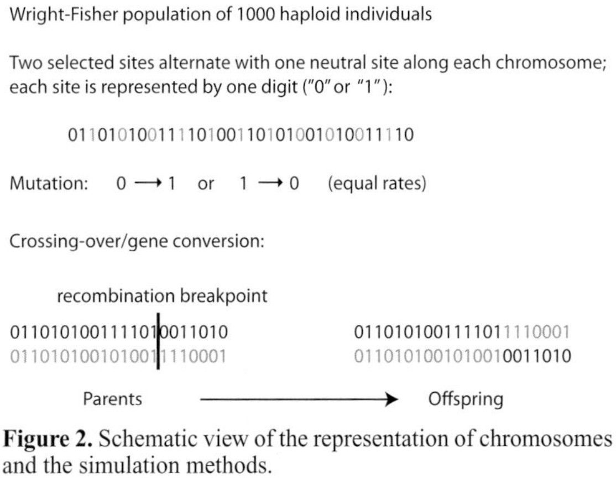

We chose mutation rates, recombination rates between adjacent sites, and a distribution of selection coefficients such that the products of N and the relevant parameter values are similar to those for genes in regions of normal recombination in a *Drosophila* population (Loewe and Charlesworth 2007). As a measure of the effectiveness of background selection, we used the ratio B of the mean pairwise diversity at the simulated neutral sites, relative to the theoretical equilibrium value for a population free of HR effects ($\pi = 4Nu$; Kimura 1971). Figure 3 (top) shows the effects of selection at the "nonsynonymous" sites on B. For noncrossover regions, there is a rapid initial decline of B with L, but B is always much larger than predicted by the classical background selection formula (Fig. 3, bottom; Hudson and Kaplan 1995; Nordborg et al. 1996). Importantly, when there is no crossing over, B levels off at a value of ~0.015 for $L > 640,000$ sites. This suggests that the HR effects between sites under selection progressively undermine the effectiveness of selection as more selected sites are packed into a region where crossing over is absent, so that additional selected sites eventually have no further effect on variability at the linked neutral sites.

Selection also distorts the gene genealogies at linked sites, especially when a large number of sites are under selection (Gordo et al. 2002; Williamson and Orive 2002), so that the reduction in N_e is not a complete descriptor of HR effects, as mentioned earlier. This distortion can be examined using Tajima's D_T statistic, which measures the difference between the estimate of variability from the mean number of sequence differences between all pairs of alleles in a sample and the estimate from the number of segregating sites in the sample (Tajima 1989). This has been used to test for selective sweeps, because these are expected to cause negative D_T values, reflecting an excess of rare variants (Braverman et al. 1995; Simonsen et al. 1995). Our simulations show that D_T for neutral sites is negative and increases in magnitude with L; with no crossing over, it approaches its maximum value with the largest L values that we have simulated (Kaiser and Charlesworth 2009).

How do the simulation results relate to observations on genomic regions with low levels of recombination? As described in the previous section, the observed mean diver-

Figure 2. Schematic view of the representation of chromosomes and the simulation methods.

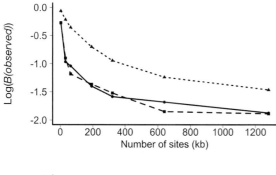

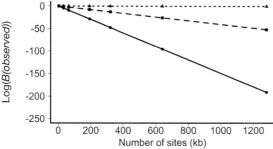

Figure 3. Effects of Hill–Robertson interference among strongly selected mutations on levels of diversity at linked neutral sites. (Dotted line) Free recombination, (dashed line) gene conversion only, (solid line) no recombination. (*Top*) The logarithm to base 10 of $B(observed) = \pi_s/(4Nu)$ plotted against the number of sites. $B(observed)$ decreases with an increasing number of sites. An asymptotic value of ~1.5% is reached for large values of L. (*Bottom*) The logarithm of $B(expected)$ as a function of L, where $B(expected)$ is calculated from the standard background selection formula. With an increasing number of sites, neutral diversity is expected to decline exponentially, and the rate of decline is greater if recombination rates are low.

sity value on the dot chromosome in several *Drosophila* species is ~5% of the genome-wide average, close to the value of 6.5% in our simulations of a noncrossing over region of this size (Fig. 3, top). In addition, in *D. americana*, the observed value of D_T is not significantly different from the predicted value for a noncrossover region of this size (Kaiser and Charlesworth 2009).

Similarly, the neo-Y chromosome of *D. miranda* completely lacks recombination, and it contains ~3.7 Mb of coding sequence, of which approximately one-half are nonfunctional and thus unlikely to cause HR effects (Bachtrog et al. 2008). The mean average silent-site diversity for genes on the *D. miranda* neo-Y is ~1% of the value for their homologs on the recombining neo-X chromosome (Bartolomé and Charlesworth 2006), which is quite close to the predicted value with large L (Kaiser and Charlesworth 2009). There is a large negative D_T for the neo-Y (Bartolomé and Charlesworth 2006), which has been interpreted as having been caused by a recent selective sweep (Bachtrog 2004). However, this D_T value is also consistent with our modified model of background selection (Kaiser and Charlesworth 2009).

The simulations also show that the frequencies of sites fixed for deleterious nonsynonymous variants are greatly increased in large regions with reduced recombination, with a corresponding reduction in the mean fitness of the population (Table 2), as found previously in HR models with weak selection and fixed selection coefficients (McVean and Charlesworth 2000; Comeron et al. 2008). Even a small nonrecombining genomic region can experience a noticeable reduction in its mean fitness, as illustrated in Figure 4, where a population with a nonrecombining chromosome of only 32 kb in length (21,333 sites under selection) equilibrates at a natural logarithm of mean fitness of –7.78. If we retain the same $N_e s$ values but rescale the population size to 1 million, which is reasonable for a *Drosophila* population, most selection coefficients are very small, so that the expression for ln(w) given above is well approximated by $-\Sigma_i s_i$. Using this simplification, the mean of log fitness in the simulations corresponds to a log mean fitness for *Drosophila* of $-7.78 \times 500/1000000 = -0.0039$, whereas the equilibrium log mean fitness of a freely recombining population with the same parameters is –0.0001, using the for-

Table 2. Effect of Zero Recombination on the Proportion of Selected Sites with Deleterious Mutations and on the Mean Fitness of the Population

Length of chromosome (kb)	Proportion of sites carrying a deleterious mutation[a]	Mean s at fixed sites[b] ($\times 10^5$)	Log fitness[c]	Relative reduction in log fitness[d]
3.2	0.00104	0.16	0	0
32	0.0168	0.63	–0.002	0.002
64	0.0254	0.83	–0.009	0.009
192	0.0377	1.19	–0.057	0.056
320	0.0464	1.35	–0.134	0.131
640	0.0533	1.58	–0.360	0.358
1280	0.0620	1.80	–0.952	0.948

[a]This is for a chromosome chosen randomly from the population after 10,000 generations.

[b]Mean s is scaled to a population size of 10^6 (the simulation value is multiplied by $500/10^6$).

[c]This is the negative of the number of nonsynonymous sites (two-thirds of column one) multiplied by the product of columns two and three. It provides a lower bound to the equilibrium reduction in fitness per nonsynonymous site, because the mean selection coefficient at sites that are fixed is lower than average.

[d]This is the difference in the estimated log fitness for the nonrecombining chromosome and the equilibrium log fitness for a freely recombining chromosome. The latter is calculated from the product of the mutation rate per site and the number of nonsynonymous sites (Haldane 1937).

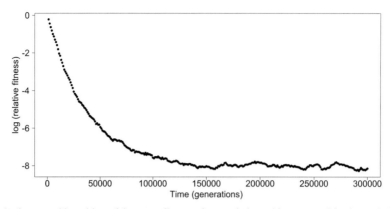

Figure 4. The decline in the natural logarithm of the mean fitness of a population without recombination, relative to the expected fitness under free recombination, plotted against the number of generations since the start (when the population was at equilibrium), for simulations without recombination and a chromosome length of 32 kb. The initial decline is linear, similar to what is found with Muller's ratchet, but eventually statistical equilibrium is reached.

mula of Haldane (1937). The relative reduction in mean fitness is thus ~0.0038.

It is harder to estimate this reduction for cases with large numbers of sites, because the time to reach equilibrium with respect to mean fitness becomes very long. Table 2 shows some estimates of this reduction. With large L, there is a large reduction in mean fitness for a nonrecombining population, with the mean fitness approaching 39% of the free recombination value. The equilibrium mean fitness of a completely asexual population of a higher eukaryote, where a whole genome of more than 1 million nonsynonymous sites is nonrecombining, must therefore be extremely small, so that this effect may contribute to the apparent long-term evolutionary disadvantages of asexual lineages (Maynard Smith 1978; Bell 1982; Otto and Lenormand 2002; Normark et al. 2003).

EVIDENCE FOR A REDUCED EFFECTIVENESS OF SELECTION IN LOW RECOMBINATION GENOMIC REGIONS

These results lead naturally to the question of the extent to which levels of adaptation at the protein and DNA sequence level are indeed reduced in low recombination genomes or genomic regions. A severe reduction in effective population size should be reflected in a reduction in the efficacy of both purifying selection and positive selection, as well as a reduction in diversity. Various studies have found evidence for such effects, primarily in nonrecombining genomes or chromosomes. For example, codon usage bias is reduced in regions of the *Drosophila* genome that lack crossing over, especially the dot chromosome (Kliman and Hey 1993, 2003; Marais et al. 2003; Haddrill et al. 2007), as we also found for the *D. americana* dot chromosome (Betancourt et al. 2009).

Studies of the relationships between local recombination rate and protein sequence variation and between-species divergence in *Drosophila* (Betancourt and Presgraves 2002; Presgraves 2005; Haddrill et al. 2007) also suggest that recombination affects the efficiency of selection on amino acid sequences, with reduced rates of adaptive evolution in regions of low recombination and relaxed selection against deleterious mutations. We used our *D. americana* data to ask if dot chromosome loci experience relaxed purifying selection on amino acid mutations. As expected, protein sequence divergence between *D. americana* and two related species, *D. virilis* and *D. ezoana*, is elevated on the dot chromosome (Betancourt et al. 2009).

This pattern can be explained by relaxed purifying selection on the dot chromosome causing the fixation of slightly deleterious amino acid variants, but it is also possible that it is due to a higher rate of fixation of beneficial mutations, although this is contrary to theoretical expectations. We can distinguish between these alternatives using the *D. americana* polymorphism data, because elevated levels of protein polymorphism relative to silent polymorphism indicate relaxed purifying selection: The *D. americana* dot chromosome loci indeed show such an effect (Betancourt et al. 2009). The average strength of purifying selection acting on the heterozygous carriers of segregating amino acid variants was estimated to be $N_e s \approx 4$ for dot loci and $N_e s \approx 29$ for nondot autosomal loci. Provided the two sets of loci experience similar levels of functional constraints, these results suggest that selection against amino acid variants is less effective in low recombination regions.

We also investigated whether adaptive evolution is similarly compromised on the *D. americana* dot chromosome. We used the combination of polymorphism and divergence data to estimate α, the proportion of nonsynonymous between-species differences caused by positive selection (Fay et al. 2002; Smith and Eyre-Walker 2002; Bierne and Eyre-Walker 2004). Using a maximum likelihood implementation of this method (Welch 2006), we found a significantly lower α estimate for the dot chromosome loci than for the other genes. This not only suggests that adaptive evolution may be compromised on the dot chromosome, but also excludes the possibility that the elevated protein sequence for dot chromosome loci is due to more frequent selectively driven substitutions.

DISCUSSION

The data that we have presented on sequence divergence and polymorphism on the dot chromosome of *D. americana* show that both levels of variability and the effectiveness of selection on protein sequences and codon usage are significantly reduced. These findings are in agreement with those previously reported for this small (~80 genes), noncrossing over component of the *Drosophila* genome, but ours is the most comprehensive study that combines both polymorphism and divergence data for this chromosome. We observed an apparent absence of amino acid sequence differences that have been fixed by positive selection, in contrast to the ~60% fraction for genes on other chromosomes. This agrees with the proposal of Betancourt and Presgraves (2002) and Presgraves (2005) that recombination accelerates adaptive protein sequence evolution.

The main caveat is that we cannot exclude the possibility that the two sets of genes which we have studied differ in properties that affect the rate of protein sequence. We note, however, that the major determinant of the rate of protein sequence evolution is the level of gene expression, with low expression genes showing higher rates of nonsynonymous substitutions (Drummond and Wilke 2008). There is, however, no significant difference in expression levels between the dot chromosome genes and the rest of the genome in *D. virilis* (Betancourt et al. 2009), so that this factor can be ruled out. Indeed, Haddrill et al. (2008) found significantly *higher* expression levels for genes on the *D. melanogaster* dot chromosome compared with the rest of the genome, and they suggested that this might reflect an adaptation to compensate for the lower functionality of protein sequences on this chromosome. It is interesting to note that a protein Painting of fourth (POF) has been characterized that binds specifically to the dot chromosome in *D. melanogaster* and appears to increase the expression of genes on this chromosome (Larsson et al. 2004; Johannson et al. 2007).

The only way of definitively dealing with this difficulty is to exploit systems in which the same genes can be compared in different recombinational environments. This is possible for homologous genes located on the two different sex chromosomes, when the Y or W chromosome does not recombine in the heterogametic sex. In the case of the *D. miranda* neo-Y and neo-X chromosomes mentioned above, there is evidence for accelerated protein sequence evolution associated with relaxed purifying selection on the nonrecombining neo-Y chromosome (Bartolomé and Charlesworth 2006; Bachtrog et al. 2008). The reduced levels of gene expression on this chromosome, and losses of gene function due to major mutational lesions, do not account for this effect (Bachtrog 2006). Similarly, accelerated protein sequence evolution and reduced variability have been observed on the W chromosome of birds (Berlin and Ellegren 2006) and the recently evolved Y chromosome of the white campion *Silene latifolia* (Marais et al. 2008). In all of these cases, the same genes are being compared between the two recombinational environments.

This strongly suggests that selective forces in a low recombination environment do indeed reduce N_e, leading to an impaired effectiveness of selection. It is difficult to be sure which of the factors listed in Table 1 is likely to be the most important cause. However, our study of the *D. americana* dot chromosome shows that it is apparently impossible to account for its reduced N_e by a recent selective sweep—we observe too many variants at intermediate frequencies to be consistent with such an event. As mentioned earlier, the levels of variability on both the dot chromosome and the *D. miranda* neo-Y chromosome are also inconsistent with the classical background selection model.

We propose that this paradox can be resolved by invoking HR interference among the sites subject to reversible mutation and purifying selection, when a large number of such sites are included in a low recombination genomic region. Our simulation results show that this produces weakening of the effective strength of selection on nonsynonymous variants. The HR effects mean that these variants are more likely to drift to intermediate frequencies and can become fixed more easily than with normal levels of recombination. This reduction in the effectiveness of selection means that the mutations in question have reduced effects on variability at linked neutral sites. As we discussed above, it seems likely that an asymptotic state is reached as the extent of a nonrecombining region increases, whereby adding more selected sites into a low recombination region has little or no effect on levels of variability. This may well account for the observation that the *D. miranda* neo-Y has a silent-site diversity value that is 1% of the neo-X value, despite the very large number of functional genes that it carries (Bachtrog et al. 2008).

We can also ask whether this type of process may provide a selective advantage to recombination at the individual level (see Barton, this volume), i.e., will selection resist the invasion of a freely recombining population by a genetic factor or chromosome rearrangement that reduces recombination or favor the invasion of a low recombination population by a modifier that increases it? Keightley and Otto (2006) investigated a similar model, but assumed unidirectional mutation from wild-type to deleterious alleles. This type of system cannot reach an equilibrium, and so is more similar to Muller's ratchet (Table 1) than ours (Gordo and Campos 2008).

We have therefore explored the possibility that HR interference among relatively strongly selected mutations may cause individual selection on modifiers of recombination, using simulations similar to those of Keightley and Otto (2006), but with populations at statistical equilibrium under reversible mutation, selection, and drift, similar to those described above. We introduced a single copy of a modifier allele that either completely suppresses recombination (if the initial population has a normal level of recombination) or increases it (if the population has zero recombination). To save computer time, a fixed selection coefficient was assigned to each site. The results are shown in Table 3. The advantage of increased recombination and disadvantage of decreased recombination tend to level off as the number of selected

Table 3. Fates of Modifiers of Recombination with Hill–Robertson Effects

2N	2Nu	2Ns	L	P_{fix}/P_{neu}	2 × s.e.
A. Initial population with map length 90 cM ($2Nr = 450$ for the whole chromosome); a modifier that completely suppresses recombination is introduced at equilibrium.					
500	0.01	10	10,000	0.75	0.27
500	0.01	10	20,000	0.80	0.28
500	0.01	10	50,000	0.28	0.17
500	0.01	10	100,000	0.08	0.09
500	0.01	10	200,000	0.15	0.12
B. Initial population with no recombination; a modifier that increases map length to 90 cM is introduced at equilibrium. (The modifier is located at one end of the chromosome.)					
500	0.01	10	1,000	1.25	0.50
500	0.01	10	5,000	1.50	0.77
500	0.01	10	10,000	3.00	1.09
500	0.01	10	20,000	6.00	2.43
500	0.01	10	40,000	5.25	2.28
500	0.01	10	50,000	6.56	1.19
500	0.01	10	100,000	9.21	2.04
500	0.01	10	200,000	10.27	2.58
500	0.01	10	300,000	13.09	2.51
500	0.01	10	400,000	11.20	2.09

10,000 to 20,000 simulations were run for each parameter set. P_{fix} and P_{neu} are the fixation probabilities for a recombination modifier and a neutral mutation, respectively.

sites increases, similar to the effects on neutral diversity that we have described. The results show that HR effects among nonsynonymous mutations can have a significant influence on individual-level selection for recombination.

This raises the question of why some regions of the genome have low frequencies of recombination if selection generally favors increased recombination. The answer must lie in a selective advantage to recombination suppression that is sufficient to overcome selection in favor of recombination. In the case of Y chromosomes, a plausible scenario is that suppression of crossing over between incipient X and Y chromosomes is advantageous because it prevents alleles that are favored in one sex, but deleterious in the other, recombining into the "wrong" sex (Charlesworth et al. 2005). It is less clear why certain regions of the genome, such as centromeres and telomeres, are associated with suppression of crossing over (Sherman and Stack 1995; Gerton et al. 2000; Ashburner et al. 2005). (The lack of crossing over on the dot chromosome probably reflects the fact that its small size means that its euchromatin is adjacent to both the telomere and the centromere.)

Centromeres in most species are compound structures, containing repetitive sequences around which the structure that binds the spindle fiber to microtubules forms (Charlesworth et al. 1986; Ashburner et al. 2005). Telomeres are also made up of repetitive units, of a different type from the centromere (Chan and Blackburn 2004). Unequal crossing over between the repeat units would produce aberrant numbers of repeats, which could lead to aberrant chromosome segregation in mitosis and meiosis, resulting in aneuploid cells and reduced fitness (Charlesworth et al. 1986). This could result in a selective advantage to reduced crossing over near centromeres and telomeres (Charlesworth et al. 1986). Additionally, exchanges in or near centromeres (even without unequal crossing over) may directly interfere with centromere disjunction in meiosis, again leading to aneuploidy. There is evidence for this from the smut fungus *Microbotryum violaceum* (Cattrall et al. 1978), yeast, *Drosophila*, and humans (Rockmill et al. 2006). The suppression of crossing over in these genomic regions is thus likely to have an adaptive basis.

ACKNOWLEDGMENTS

A.J.B. was supported by a research grant from the UK Biotechnology and Biological Sciences Research Council, B.C. by the Royal Society, I.G. by the Fundação para a Ciência e Tecnologia, Portugal, and V.B.K. by the School of Biological Sciences, University of Edinburgh.

REFERENCES

Aguadé M, Miyashita N, Langley CH. 1989. Reduced variation in the *yellow-achaete-scute* region in natural populations of *Drosophila melanogaster*. *Genetics* **122**: 607–615.

Andolfatto P. 2007. Hitchhiking effects of recurrent beneficial amino acid substitutions in the *Drosophila melanogaster* genome. *Genome Res* **17**: 1755–1762.

Ashburner M, Golic KG, Hawley RS. 2005. Drosophila: *A laboratory handbook*. Cold Spring Harbor Laboratory Press, Cold Spring Harbor, NY.

Bachtrog D. 2004. Evidence that positive selection drives Y-chromosome degeneration in *Drosophila miranda*. *Nat Genet* **36**: 518–522.

Bachtrog D. 2006. Expression profile of a degenerating neo-Y chromosome in *Drosophila*. *Curr Biol* **16**: 1694–1699.

Bachtrog D, Hom E, Wong KM, Maside X, De Jong P. 2008. Genomic degradation of a young Y chromosome in *Drosophila miranda*. *Genome Biol* **9**: R30.

Bartolomé C, Charlesworth B. 2006. Evolution of amino-acid sequences and codon usage on the *Drosophila miranda* neo-sex chromosomes. *Genetics* **174**: 2033–2044.

Begun DJ, Aquadro CF. 1992. Levels of naturally occurring DNA polymorphism correlate with recombination rate in *Drosophila melanogaster*. *Nature* **356**: 519–520.

Bell G. 1982. *The masterpiece of nature*. Croom-Helm, London.

Berlin S, Ellegren H. 2006. Fast accumulation of nonsynonymous mutations on the female-specific W chromosome in birds. *J Mol Evol* **62**: 66–72.

Berry AJ, Ajioka JW, Kreitman M. 1991. Lack of polymorphism on the *Drosophila* fourth chromosome resulting from selection. *Genetics* **129**: 1111–1117.

Betancourt AJ, Presgraves DC. 2002. Linkage limits the power of natural selection. *Proc Natl Acad Sci* **99**: 13616–13620.

Betancourt AJ, Welch JJ, Charlesworth B. 2009. Reduced effectiveness of selection caused by lack of recombination. *Curr Biol* **19**: 655–660.

Bierne N, Eyre-Walker A. 2004. The genomic rate of adaptive amino-acid substitutions in *Drosophila*. *Mol Biol Evol* **21**: 1350–1360.

Braverman JM, Hudson RR, Kaplan NL, Langley CH, Stephan W. 1995. The hitchhiking effect on the site frequency spectrum of DNA polymorphism. *Genetics* **140**: 783–796.

Burt A, Bell G. 1987. Mammalian chiasma frequencies as a test of two theories of recombination. *Nature* **326**: 803–805.

Cattrall ME, Baird ML, Garber ED. 1978. Genetics of *Ustilago violacea*. III. Crossing over and nondisjunction. *Bot Gaz* **13**: 266–270.

Chan SR, Blackburn EH. 2004. Telomeres and telomerase. *Philos Trans R Soc Lond B Biol Sci* **359**: 109–121.

Charlesworth B. 1996. Background selection and patterns of

genetic diversity in *Drosophila melanogaster. Genet Res* **68**: 131–150.
Charlesworth D. 2003. Effects of inbreeding on the genetic diversity of plant populations. *Philos Trans R Soc Lond B Biol Sci* **358**: 1051–1070.
Charlesworth B. 2009. Effective population size and patterns of molecular evolution and variation. *Nat Rev Genet* **10**: 195–205.
Charlesworth B, Charlesworth D. 2010. *Elements of evolutionary genetics*. Roberts, Greenwood Village, CO.
Charlesworth B, Langley CH, Stephan W. 1986. The evolution of restricted recombination and the accumulation of repeated DNA sequences. *Genetics* **112**: 947–962.
Charlesworth B, Morgan MT, Charlesworth D. 1993. The effect of deleterious mutations on neutral molecular variation. *Genetics* **134**: 1289–1303.
Charlesworth D, Charlesworth B, Marais G. 2005. Steps in the evolution of heteromorphic sex chromosomes. *Heredity* **95**: 118–128.
Comeron JM, Williford A, Kliman RM. 2008. The Hill-Robertson effect: Evolutionary consequences of weak selection in finite populations. *Heredity* **100**: 19–31.
Drummond DA, Wilke CO. 2008. Mistranslation-induced protein misfolding as a dominant constraint on coding-sequence evolution. *Cell* **13**: 341–352.
Ewens WJ. 2004. *Mathematical population genetics. 1. Theoretical introduction*. Springer, New York.
Fay J, Wyckhoff GJ, Wu C-I. 2002. Testing the neutral theory of molecular evolution with genomic data from *Drosophila. Nature* **415**: 1024–1026.
Felsenstein J. 1974. The evolutionary advantage of recombination. *Genetics* **78**: 737–756.
Fisher RA. 1930. *The genetical theory of natural selection*. Oxford University Press, Oxford.
Gay J, Myers S, McVean G. 2007. Estimating meiotic gene conversion rates from population genetic data. *Genetics* **177**: 881–894.
Gerton JL, DeRisi J, Shroff R, Lichten M, Brown PO, Petes TD. 2000. Global mapping of meiotic recombination hotspots and coldspots in the yeast *Saccharomyces cerevisiae. Proc Natl Acad Sci* **97**: 11383–11390.
Gordo I, Campos RA. 2008. Sex and deleterious mutations. *Genetics* **179**: 621–626.
Gordo I, Navarro A, Charlesworth B. 2002. Muller's ratchet and the pattern of variation at a neutral locus. *Genetics* **161**: 835–848.
Haddrill PR, Halligan DL, Tomaras D, Charlesworth B. 2007. Reduced efficacy of selection in regions of the *Drosophila* genome that lack crossing over. *Genome Biol* **8**: R18.
Haddrill PR, Waldron FM, Charlesworth B. 2008. Elevated levels of expression associated with regions of the *Drosophila* genome that lack crossing over. *Biol Lett* **4**: 758–761.
Haldane JBS. 1937. The effect of variation on fitness. *Am Nat* **71**: 337–349.
Hill WG, Robertson A. 1966. The effect of linkage on limits to artificial selection. *Genet Res* **8**: 269–294.
Hudson RR, Kaplan NL. 1995. Deleterious background selection with recombination. *Genetics* **141**: 1605–1617.
Jensen MA, Charlesworth B, Kreitman M. 2002. Patterns of genetic variation at a chromosome 4 locus of *Drosophila melanogaster* and *D. simulans. Genetics* **160**: 493–507.
Johannson A-M, Stenberg P, Berhhardsson C, Larsson J. 2007. Painting of fourth and chromosome-wide regulation of the 4th chromosome in *Drosophila melanogaster. EMBO J* **26**: 2307–2316.
Kaiser VB, Charlesworth B. 2009. The effects of deleterious mutations on evolution in non-recombining genomes. *Trends Genet* **25**: 9–12.
Keightley PD, Eyre-Walker A. 2007. Joint inference of the distribution of fitness effects of deleterious mutations and population demography based on nucleotide polymorphism frequencies. *Genetics* **177**: 2251–2261.
Keightley PD, Otto SP. 2006. Interference among deleterious mutations favours sex and recombination in finite populations. *Nature* **443**: 89–92.
Kimura M. 1962. On the probability of fixation of a mutant gene in a population. *Genetics* **47**: 713–719.
Kimura M. 1971. Theoretical foundations of population genetics at the molecular level. *Theor Popul Biol* **2**: 174–208.
Kliman RM, Hey J. 1993. Reduced natural selection associated with low recombination in *Drosophila melanogaster. Mol Biol Evol* **10**: 1239–1258.
Kliman RM, Hey J. 2003. Hill-Robertson interference in *Drosophila melanogaster:* Reply to Marais, Mouchiroud and Duret. *Genet Res* **81**: 89–90.
Langley CH, Lazzaro BP, Phillips W, Heikkinen E, Braverman JM. 2000. Linkage disequilibria and the site frequency spectra in the *su(s)* and *su(w^a)* regions of the *Drosophila melanogaster X* chromosome. *Genetics* **156**: 1837–1852.
Larsson J, Svensson MJ, Stenberg P, Mäkitalo M. 2004. Painting of fourth in genus *Drosophila* suggests autosome-specific gene regulation. *Proc Natl Acad Sci* **101**: 9278–9733.
Loewe L, Charlesworth B. 2006. Inferring the distribution of mutational effects on fitness in *Drosophila. Biol Lett* **2**: 426–430.
Loewe L, Charlesworth B. 2007. Background selection in single genes may explain patterns of codon bias. *Genetics* **175**: 1381–1393.
Loewe L, Charlesworth B, Bartolomé C, Nöel V. 2006. Estimating selection on nonsynonymous mutations. *Genetics* **172**: 1079–1092.
Marais G, Mouchiroud D, Duret L. 2003. Neutral effect of recombination on base composition in *Drosophila. Genet Res* **81**: 79–87.
Marais GAB, Nicolas M, Bergero R, Chambrier P, Kejnovsky E, Moneger F, Hobza R, Widmer A, Charlesworth D. 2008. Evidence for degeneration of the Y chromosome in the dioecious plant *Silene latifolia. Curr Biol* **18**: 545–549.
Maside X, Charlesworth B. 2007. Patterns of molecular variation and evolution in *Drosophila americana* and its relatives. *Genetics* **176**: 2293–2305.
Maynard Smith J. 1978. *The evolution of sex*. Cambridge University Press, Cambridge.
Maynard Smith J, Haigh J. 1974. The hitch-hiking effect of a favourable gene. *Genet Res* **23**: 23–35.
McAllister BF. 2002. Chromosomal and allelic variation in *Drosophila americana:* Selective maintenance of a chromosomal cline. *Genome* **45**: 13–21.
McVean GAT, Charlesworth B. 2000. The effects of Hill-Robertson interference between weakly selected mutations on patterns of molecular evolution and variation. *Genetics* **155**: 929–944.
Moran NA, McCutcheon JP, Nakabatchi A. 2008. Genomics and evolution of heritable bacterial symbionts. *Annu Rev Genet* **42**: 165–190.
Nordborg M, Charlesworth B, Charlesworth D. 1996. The effect of recombination on background selection. *Genet Res* **67**: 159–174.
Normark BB, Judson OP, Moran NA. 2003. Genomic signatures of ancient asexual lineages. *Biol J Linn Soc* **79**: 69–84.
Otto SP, Lenormand T. 2002. Resolving the paradox of sex and recombination. *Nat Rev Genet* **3**: 256–261.
Presgraves D. 2005. Recombination enhances protein adaptation in *Drosophila melanogaster. Curr Biol* **15**: 1651–1656.
Robertson A. 1961. Inbreeding in artificial selection programmes. *Genet Res* **2**: 189–194.
Rockmill B, Voelkel-Meiman K, Roeder GS. 2006. Centromere-proximal crossovers are associated with precocious separation of sister chromatids during meiosis in *Saccharomyces cerevisiae. Genetics* **174**: 1745–1754.
Roze D, Lenormand T. 2005. Self-fertilization and the evolution of recombination. *Genetics* **170**: 840–857.
Santiago E, Caballero A. 1998. Effective size and polymorphism of linked neutral loci in populations under selection. *Genetics* **149**: 2105–2117.
Sawyer SA, Parsch J, Zhang Z, Hartl DL. 2007. Prevalence of positive selection among nearly neutral amino acid replacements in *Drosophila. Proc Natl Acad Sci* **104**: 6504–6510.

Sharp PJ, Hayman DL. 1988. An examination of the role of chiasma frequency in the genetic system of marsupials. *Heredity* **60:** 77–85.

Sheldahl LE, Weinreich DM, Rand DM. 2003. Recombination, dominance and selection on amino-acid polymorphisms in the *Drosophila* genome: Contrasting patterns on the X and fourth chromosomes. *Genetics* **165:** 1195–1208.

Sherman JD, Stack SM. 1995. Two-dimensional spreads of synaptonemal complexes from solanaceous plants. VI. High-resolution recombination map for tomato (*Lycopersicon esculentum*). *Genetics* **141:** 683–708.

Simonsen KL, Churchill GA, Aquadro CF. 1995. Properties of statistical tests of neutrality for DNA polymorphism data. *Genetics* **141:** 413–429.

Smith NGC, Eyre-Walker A. 2002. Adaptive protein evolution in *Drosophila*. *Nature* **415:** 1022–1024.

Stephan W. 1995. An improved method for estimating the rate of fixation of favorable mutations based on DNA polymorphism data. *Mol Biol Evol* **12:** 959–962.

Stephan W, Langley CH. 1989. Molecular genetic variation in the centromeric region of the X chromosome in three *Drosophila ananassae* populations. I. Contrasts between the *vermilion* and *forked* loci. *Genetics* **121:** 89–99.

Tachida H. 2000. DNA evolution under weak selection. *Gene* **261:** 3–9.

Tajima F. 1989. Statistical method for testing the neutral mutation hypothesis. *Genetics* **123:** 585–595.

Turner JRG. 1967. Why does the genome not congeal? *Evolution* **21:** 645–656.

Wang W, Thornton K, Berry A, Long M. 2002. Nucleotide variation along the *Drosophila melanogaster* fourth chromosome. *Science* **295:** 134–137.

Wang W, Thornton K, Emerson JJ, Long M. 2004. Nucleotide variation along the *Drosophila simulans* fourth chromosome. *Genetics* **166:** 1783–1794.

Welch JJ. 2006. Estimating the genomewide rate of adaptive protein evolution in *Drosophila*. *Genetics* **173:** 821–827.

Williamson SM, Orive ME. 2002. The genealogy of a sequence subject to purifying selection at multiple sites. *Mol Biol Evol* **19:** 1376–1384.

Wright S. 1931. Evolution in Mendelian populations. *Genetics* **16:** 97–159.

Zhivotovsky LA, Feldman MW, Christiansen FB. 1994. Evolution of recombination among multiple selected loci: A generalized reduction principle. *Proc Natl Acad Sci* **91:** 1079–1093.

Why Sex and Recombination?

N.H. BARTON

Institute of Science and Technology, A-3400 Klosterneuburg, Austria
Correspondence: nick.barton@ist-austria.ac.at

Sex and recombination have long been seen as adaptations that facilitate natural selection by generating favorable variations. If recombination is to aid selection, there must be negative linkage disequilibria—favorable alleles must be found together less often than expected by chance. These negative linkage disequilibria can be generated directly by selection, but this must involve negative epistasis of just the right strength, which is not expected, from either experiment or theory. Random drift provides a more general source of negative associations: Favorable mutations almost always arise on different genomes, and negative associations tend to persist, precisely because they shield variation from selection.

We can understand how recombination aids adaptation by determining the maximum possible rate of adaptation. With unlinked loci, this rate increases only logarithmically with the influx of favorable mutations. With a linear genome, a scaling argument shows that in a large population, the rate of adaptive substitution depends only on the expected rate in the absence of interference, divided by the total rate of recombination. A two-locus approximation predicts an upper bound on the rate of substitution, proportional to recombination rate.

If associations between linked loci do impede adaptation, there can be substantial selection for modifiers that increase recombination. Whether this can account for the maintenance of high rates of sex and recombination depends on the extent of selection. It is clear that the rate of species-wide substitutions is typically far too low to generate appreciable selection for recombination. However, local sweeps within a subdivided population may be effective.

Why are sex and recombination so widespread? In bacteria and archaea, it is arguable that recombination is an incidental by-product of other processes. However, the biology of eukaryotes is dominated by sexual reproduction. Meiosis, with its elaborate molecular machinery, is a key characteristic of eukaryotes, and sex has all kinds of consequences: differentiation of male and female sex chromosomes, the physiological mechanisms for producing male and female gametes, everything involved in finding and attracting mates, and so on. Much of Darwin's work was concerned with sex and its consequences—sexual selection, studies of inbreeding, cross-pollination in plants, etc. (see, e.g., Darwin 1871, 1876).

So, why sex and recombination? We have known the answer to this question for a long time: sexual reproduction is an adaptation that facilitates natural selection by bringing together new, favorable combinations of genes. To use a modern term, recombination is an adaptation for "evolvability," for generating variation useful to selection (Wagner and Altenberg 1996; Kirschner and Gerhart 1998; Sniegowski and Murphy 2006). August Weismann (1889), one of the few champions of natural selection in the late 19th century, put this quite clearly:

> [T]he communication of fresh [genes] to the germplasm implies an augmentation of the variational tendencies, and thus an increase of the power of adaptation.

The necessity of sexual reproduction for successful selection is shown in nature by the degeneration of non-recombining sex chromosomes (Charlesworth et al.; Bellott and Page, both this volume) and by the relatively short life of asexual species (Maynard Smith 1978; Butlin 2002). The importance of outcrossing has been clear to plant and animal breeders long before it was understood genetically. Arnold (this volume) described the selection in vitro of novel enzyme activities, which could not have been designed from first principles. A less familiar example comes from evolutionary computation, where programs compete against one another, so that selection can generate better algorithms. For example, Koza et al. (2002) used selection to design efficient low-pass filters and found several designs ingenious enough to have previously been patented. There is a great deal of activity in "evolutionary computation," with a wide variety of approaches, but almost all depend on some form of recombination (Mitchell 1998). In all these examples, recombination and selection together are more effective than intelligent design.

It has been surprisingly difficult to show exactly why selection requires recombination to be effective: We still do not have a compelling general explanation for why so many organisms go to so much trouble to mix their genomes with others. The intuitive idea, that sex is good for the species, was accepted until the 1970s, when it was realized that there has to be an advantage to individual genes that increase outcrossing and recombination and, moreover, that this advantage has to be strong enough to outweigh all of the obvious costs (Williams 1975; Maynard Smith 1978). For the past 35 years, much effort has focused on this question, and there has been corresponding progress in clearing the theoretical fog. I argue that although the basic theoretical framework is clear, we still do not know whether selection is generally strong enough, and has the right form, to give a general advantage to sex and recombination.

RECOMBINATION INCREASES ADDITIVE GENETIC VARIANCE IF THERE ARE NEGATIVE LINKAGE DISEQUILIBRIA

We start with Fisher's (1930) "Fundamental Theorem of Natural Selection": the increase in mean fitness due to selection on allele frequencies equals the additive genetic variance in fitness (for explanations of Fisher's theorem, see Price [1972] and Edwards [2002]). So, if recombination is to speed up adaptation, it must increase the additive variance in fitness. Moreover, if a modifier allele increases recombination, it will to some extent be associated with the favorable variation that it generates, and will itself increase. So, in principle, an increase in additive genetic variance gives an advantage to both the group and the individual.

Crucially, recombination does not necessarily either increase or decrease the variance in fitness. If genes are already randomly associated in the population, shuffling them will make no difference. Recombination can only have an effect if there are nonrandom associations between genes (i.e., if there is "linkage disequilibrium"). Moreover, if there are positive associations between favorable genes, recombination will *reduce* the variance in fitness, and so slow down adaptation. For recombination to gain an advantage by increasing the variance in fitness, there must be negative associations: + with −, and − with +. Understanding the advantage of recombination comes down to understanding why associations should be predominantly negative—why they should perversely tend to impede selection.

Negative associations can be produced systematically and deterministically by selection, i.e., by negative epistasis. For example, stabilizing selection against extremes of a quantitative trait tends to reduce its variance and generates negative linkage disequilibrium; this slows the response to directional selection (Charlesworth 1993; Burger 1999; Waxman and Peck 1999). The effect of epistasis can be understood, quite generally, from the effect of recombination on the distribution of log fitness (Fig. 1; Barton 1995a). In each generation, selection will increase the mean by precisely the total genotypic variance: If epistasis is generally negative, selection will also reduce the variance. Recombination then has two effects (Fig. 1). First, it will reduce the mean log fitness by breaking up combinations of genes that have just been built up by selection. (If such associations were completely dissipated, the net increase in the mean would equal the additive genetic variance in fitness, which is smaller than the total genotypic variance.) This immediate recombination load will be seen whether epistasis is negative or positive. If epistasis is negative, recombination will also increase the variance, which will give an advantage in the future by increasing the response to selection. This is a very general result that does not depend on exactly how selection acts or on how many genes are involved. Moreover, it involves quantities that in principle can be measured (Charlesworth and Barton 1996; Burt 2000).

Unfortunately, this deterministic explanation seems unlikely. First, it requires that epistasis be negative, but not too negative, and that it should not vary too much

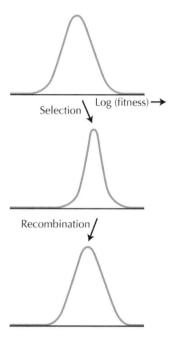

Figure 1. Selection increases the mean log fitness by an amount equal to the total genotypic variance in fitness. If selection favors negative linkage disequilibria, it will also reduce the variance in log fitness. Recombination causes an immediate reduction in mean fitness by breaking up favorable gene combinations, but it may help future adaptation by breaking up negative associations, thus increasing the variance in log fitness.

between genotypes (Otto and Feldman 1997; Kouyos et al. 2006). Otherwise, the immediate recombination load, due to the breakup of well-adapted combinations, would be too high. Second, there is no evidence that epistasis is usually negative (see, e.g., Elena and Lenski 1997, 2001; Bonhoeffer et al. 2004; de Visser and Elena 2007; Kouyos et al. 2007) and no theoretical reason to expect it to be (but see Szathmary 1993; de Visser and Elena 2007).

DRIFT, RECOMBINATION, AND SELECTION: THE HILL–ROBERTSON EFFECT

A more general, and more compelling, explanation is that negative associations are produced by random genetic drift—the random fluctuations in genotype frequencies that are inevitable in finite populations—or when a new mutation increases from a single copy. This process, by which random linkage disequilibria tend to be negative, and so to interfere with selection, is known as the Hill–Robertson effect and can be understood in several ways (Hill and Robertson 1966; Maynard Smith 1978). Random drift generates random associations, but by itself, does not to produce positive or negative associations: by chance, + + or − − genotypes may do better or worse, but on average, there is no systematic bias. It is the interaction between random drift and directional selection that is important: + + and − − combinations will be strongly selected, and will be quickly swept out of the population, whereas negative associations (+ − and − +) will have intermediate fitness, and so will persist.

Another way to think about the process is to look at the extreme case of asexual reproduction. R.A. Fisher (1930) and Hermann Muller (1932) both pointed out that favorable alleles that arise at about the same time in such a population will compete with one another; in the absence of recombination, only one of them can be fixed (Fig. 2). So, there is a strong advantage to recombination, which can bring different favorable mutations together in the same individual. Here, the randomness is in the origin of new mutations in a single copy. In an extremely large population, with recurrent mutation, the double mutant would arise, there would be no linkage disequilibrium, and recombination would have no effect.

Fisher and Muller described an extreme case, but even when there is some recombination, favorable alleles still compete with one another. Hill and Robertson (1966) first quantified this process, and they gave us yet another way to understand it. From the point of view of a particular allele, selection at linked loci will be experienced as random drift. If the allele happens to be on a fit background, it will increase, whereas if it happens to be on an unfit background, it will decrease. Moreover, fluctuations are especially powerful, because if they are due to linked loci, they will tend to persist for many generations and will cause a large inflation of random drift. Any kind of selection can contribute to Hill–Robertson interference: Elimination of deleterious mutations (known as "background selection"; Charlesworth et al., this volume), fluctuating selection, or positively selected substitutions will inflate the rate of random drift and so interfere with selection. Several theoretical studies suggest that random drift has a stronger effect on the evolution of recombination than does negative epistasis, even in large populations. Otto and Barton (2001) use simulations of directional selection on standing variation to show that selection for recombination is primarily due to the Hill–Robertson effect, in populations of up to several thousands, even when compared with an optimal level of epistasis. Iles et al. (2003) use similar simulations to show that as the number of loci increases, the effect of drift predominates in ever larger populations. Keightley and Otto (2006) examine selection against deleterious mutations and show that recombination gains a substantial advantage through random drift, which is insensitive to the form of epistasis.

MAXIMUM RATE OF ADAPTATION

The effects of random drift on the evolution of recombination are harder to analyze or to simulate than the effect of epistasis—the problem is stochastic, and strong effects emerge only when very many loci are involved. So, I concentrate here on a basic question that can be answered in a fairly simple way: How fast, in principle, can a population adapt? Of course, this question is interesting in itself: Just how effective, in principle, can selection be?

Suppose that beneficial mutations arise within a very large population, at a rate U per genome per generation. Mutations act independently (i.e., there is no epistasis), and each of them multiplies fitness by a factor $(1 + s)$. It is convenient to think of a trait, the log fitness, z, with each mutation adding to the trait, and fitness being $\exp(z)$. Then, if z is normally distributed with variance v, the variance in fitness will be $V = e^v$. (Note that the trait z is the genotypic value for log fitness; noninherited fitness variance can be included in the distribution of offspring number, conditioned on z.)

If we could ignore linkage, the chance that any allele will fix is just $2s$, so there would be a baseline rate of substitution, in the absence of Hill–Robertson interference, of $\Lambda_0 = 2sNU$. However, selection across the genome reduces the chance to $P < 2s$ that any one mutation will fix. This probability of fixation will be a decreasing function of the rate of sweeps occurring at other loci. So, $\Lambda = \Lambda_0 P(\Lambda)/2s$. Thus, we can work out the net rate of sweeps if we know by how much that rate will reduce the chance of fixation, $P(\Lambda)$, of any new mutation. This is a huge simplification, because we do not need to worry about how new alleles increase all the way to fixation. Instead, we just focus on whether they can increase from one copy up to a large enough number to avoid random loss, although this is still rare overall.

The fate of a single gene introduced into a very large population can be found by assuming that all of its offspring reproduce independently of one another. Then, it follows a "branching process" that can be analyzed even when the gene may find itself in a variety of locations or genetic backgrounds. I first lay out some general results

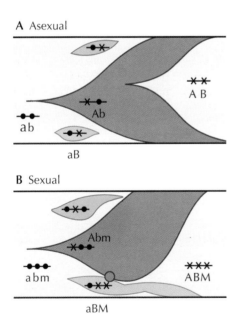

Figure 2. Favorable mutations must be established sequentially in an asexual population (*A*). For example, if allele A is destined to replace a, then any favorable alleles that occur at other loci (B, for instance) can only be fixed if they occur within a genome that carries A. (*B*) With sexual reproduction, favorable mutations at different loci can be combined; this leads to an advantage to modifiers that causes sex and recombination. A favorable allele B that occurs with the unfavorable allele a can be fixed if it can recombine into association with A (red circle); if this requires that a modifier allele M be present, then allele M will also tend to increase by hitchhiking.

for branching processes, showing how fixation probability is related to the classical concept of "reproductive value." I then summarize results for unlinked genes and for a linear genetic map.

First, consider a single gene in an unstructured population. Suppose that it has a probability Ψ_k of having k offspring. Then, the chance that a copy at time t will ultimately be lost, $1 - P_t$, is just the sum over the chance that it leaves k offspring and that all k of these are lost $(1 - P_{t+1})^k$:

$$1 - P_t = \sum_k \Psi_k (1 - P_{t+1})^k. \quad (1)$$

If the allele has a small selective advantage s and therefore a small chance of being fixed ($P \sim s \ll 1$), we can expand Equation 1 as a Taylor series:

$$1 - P_t = \sum_k \Psi_k \left(1 - k P_{t+1} + \frac{k(k-1)}{2} P_{t+1}^2 \ldots \right). \quad (2)$$

If the average number of offspring is $E[k] = 1 + s$, then we have (approximating to continuous time for $s \ll 1$)

$$-\partial_t P = sP - \frac{V}{2} P^2 + O(s^3), \quad (3)$$

where V is the variance in fitness. This equation applies with selection that changes through time, but if it is constant, we simply have $P = 2s/V$: Fixation probability is reduced in proportion to the variance in fitness of individual genes. Note that the effective population size (defined via the rate of drift of neutral alleles) is reduced in the same way and so we could also write the more familiar form $P = 2s(N_e/N)$ (Whitlock and Gomulkiewicz 2005).

This derivation extends to structured populations, in which genes can be in different states, labeled i. These might represent spatial locations or, for our purposes, different genetic backgrounds. We will see that the overall fixation probability, averaged over states, is reduced by the variance in reproductive value across states, in much the same way that random variation in fitness between genes reduces it, as just discussed. Making the same approximation as before, we find that

$$P_{i,t} = (1 + s)\tilde{P}_{i,t} - \frac{\tilde{P}_{i,t}^2}{2} + O(s^3), \quad (4)$$

where $\tilde{P}_{i,t} = \sum_j M_{i,j} P_{j,t+1}$ is the expected fixation probability of a gene in an *offspring* from a parent in state i. This is an average over the different states j in which the gene might be found in the next generation; $M_{i,j}$ is the chance that an offspring from a parent in i will be in j, which here represents recombination between genetic backgrounds. Because we assume that selection is weak ($s \ll 1$), we can again take time to be approximately continuous. Then, Equation 3 extends to

$$-\partial_t P_i = s P_i + \sum_j M^*_{i,j} P_j - \frac{\tilde{P}_i^2}{2} + O(s^3), \quad (5)$$

where $M^* = M - I$ is the rate of movement between states.

A key concept is the "reproductive value" of a gene in state i, v_i, which is defined as the expected number of copies that the gene will leave in the distant future, normalized so that $E[v_i]$. For a weakly selected allele ($s \ll M$), the fixation probability is proportional to the reproductive value (i.e., $P_i \sim v_i$). This is because if genes move between states much faster than the rate of selection (e.g., if recombination is much faster than selection), the relative contribution of a gene in state i to any state j will be proportional to its reproductive value after a long enough time; the probability that it will be picked up by selection is therefore also proportional to the reproductive value v_i.

Finally, we need to know the absolute fixation probability, averaged over states. Averaging Equation 4 over states leads to

$$-\partial_t \bar{P} = s\bar{P} - \frac{E[P^2]}{2}. \quad (6)$$

Because $E[P^2] > \bar{P}^2$, we see that variation in fixation probability across states necessarily reduces the average fixation probability. If the population is at equilibrium, we can write $E[P^2] > \bar{P}^2(1 + CV)$, where CV is the coefficient of variation of P across states, and so $\bar{P} = 2s/(1 + CV)$. This is consistent with Maruyama's (1971) invariance principle that the probability of fixation of an advantageous allele is equal to $2s$ even in a structured population, provided gene flow does not alter allele frequencies. In that case, $P_i = 2s$ for all locations i and is therefore not reduced below the panmictic value of $2s$.

If mixing across states is rapid, then P is proportional to reproductive value, and we immediately find that $\bar{P} = 2s/E[v^2]$—a remarkably simple result. We see that fixation probability, and hence the rate of adaptation, is reduced by a factor $1/E[v^2]$. Therefore, recombination gives an advantage by shuffling genes across backgrounds and so reducing the variance in reproductive value v between backgrounds. This is consistent with Hill and Robertson's (1966) account, in which selected loci interfere with one another by, in effect, inflating the rate of random drift. We can now apply these ideas to two specific models: a large number of unlinked loci or a linear genetic map.

THE INFINITESIMAL MODEL

In the very simplest case, suppose that fitness variation is not inherited. Uncorrelated variance in fitness V will nevertheless reduce the fixation probability by a factor $1/V = e^{-v}$. If the trait is heritable, being determined by a very large number of unlinked loci, fluctuations in fitness will be correlated across generations and so will cause much more random drift. Robertson (1961) argued that a gene in a background with value δz above the trait mean will in the next generation find itself in a background with average value $\delta z/2$. Thus, the excess, summed over generations, is expected to be $\delta z (1 + \frac{1}{2} + \frac{1}{4} \ldots) = 2\delta z$. Therefore, the cumulative variance in the trait value will be multiplied by $2^2 = 4$. This heuristic argument suggests that heritable variance in log fitness will reduce fixation probability by a factor e^{-4v}, rather than e^{-v}. This is confirmed by solving Equation 5. The reproductive value of a gene in a background with value z is e^{2z-v}, which is proportional to the square of the fitness and has expectation 1. The mean square reproductive value is e^{4v}, and so fixation probability is indeed reduced by a factor e^{-4v}. If the rate of beneficial mutation is extremely high, all of the genetic variance is due to selective sweeps, and in a haploid population, $v = \Lambda \bar{s}$, where $\bar{s} = E[s^2]/E[s]$ is the average selection. Solving $\Lambda = \Lambda_0 e^{-4v}$, we find that the rate of adaptive substitutions only increases logarithmically with the input of favorable mutations (Fig. 3, middle line). This argument assumes that every offspring comes from randomly chosen parents; if instead parents mate for life, the effect of inherited fitness variation is much greater, because it is correlated across siblings. The reduction in fixation probability is then e^{-9v}, rather than e^{-4v} (Fig. 3, lower line).

UNLINKED LOCI: SEXUAL VERSUS ASEXUAL REPRODUCTION

We cannot use the infinitesimal model to follow the evolution of recombination because all loci are unlinked, and so all recombination rates are fixed at 1/2. However, we can model facultative sex by assuming that a fraction α reproduce asexually and produce identical offspring.

In an infinite population, under purely directional selection, both sexual and asexual reproduction lead to the same genetic variance and hence the same rate of adaptation. However, the reproductive value depends on the trait in a very different way for sexuals and asexuals: With asexual reproduction, the current trait value stays tied to the genotype indefinitely, and so reproductive value depends extremely steeply on the current trait value. We have seen that in a sexual population, with variance in log fitness v, the mean increases by v in each generation, and the reproductive value in any generation is proportional to e^{2z}. In contrast, the reproductive value of an asexual diverges and increases more and more steeply with z: The number of offspring from an individual with trait value z is proportional to e^z after one generation, e^{2z} after two generations, and e^{tz} after t generations. However, the absolute number of offspring will always fall behind the number of sexuals, because the latter are steadily increasing their mean and must eventually overtake any given asexual lineage. In generation t, an asexual genotype z has absolute fitness e^{z-vt}, and multiplying across generations, the number of asexuals is $\Pi_{r=1}^{t} e^{z-vt} = \exp(zt - vt[t-1]/2)$. This has a maximum at $\exp([z + v/2]^2/[2v])$, which is rather more than $\exp(k^2/2)$ if $z = k\sqrt{v}$ standard deviations. Thus, an asexual lineage can increase to very large numbers in the short term (up to $t \sim [z/v]$ generations), but it must eventually fall behind a sexual population that can continually replenish its genetic variance by recombination (Fig. 4).

This contrast between selection on sexual and asexual populations was understood a century ago in the debate regarding the effectiveness of selection on Mendelian variation. Johanssen promoted the view that selection could only pick out the best genotype available in the F_2 population, which would be the case if subsequent reproduction were asexual (Provine 1971). However, Castle's experiments, and many since, have shown that artificial selection can change the mean by many tens of standard deviations, using only standing variation (Barton and Keightley 2002).

Surprisingly, the same divergence of reproductive value is seen even if there is only a small rate α of asexual reproduction. Then, the reproductive value increases more steeply than e^{2z}, and for sufficiently large z, it diverges over time; this is because the number of offspring produced entirely by asexual reproduction keeps increasing if $\alpha e^z > 1$, so that their very high intrinsic fitness e^z outweighs the loss due to mating back to the sexual population. This divergence implies that in a truly infinite population, the mean square reproductive value would diverge, and the fixation probability would tend to zero, even if individuals reproduced asexually only a small fraction of the time (Fig. 5).

In fact, even in a very large population, extremely fit individuals, many standard deviations above the mean, do not exist. Thus, deterministic arguments that require their existence will fail even in very large populations. (I ignore, for simplicity, the additional biological constraint on the maximum number of offspring.) Simulations suggest that indeed, a moderate rate of asexual reproduction slows the response to selection appreciably, because very fit lineages are lost by chance from the leading edge, causing a reduced variance (i.e., negative linkage disequilibria). Moreover, an unlinked modifier allele that causes a low rate of asexual reproduction can have a substantial

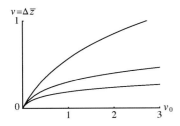

Figure 3. The rate of increase of log mean fitness $\Delta \bar{z} = v$, plotted against the baseline rate $V_0 = 2NU\overline{s^2}$, expected in the absence of interference between sweeps. (*Top*) Uncorrelated fitness fluctuations, (*middle*) unlinked loci, polygamy, (*bottom*) unlinked loci, monogamy.

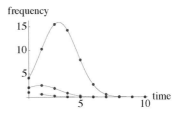

Figure 4. Frequency of an asexual clone that initially has trait value $z = 0$, 1, and 2 standard deviations above the mean of a sexual population, with variance in log fitness $v = 0.5$. Because the log fitness of the sexual population grows as $z = vt$, the asexual population must eventually become extinct. (Generations are discrete; the lines connecting the dots are only for clarity.)

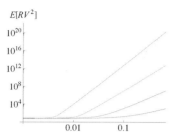

Figure 5. Mean square reproductive value, plotted against the probability of reproducing asexually α, assuming the infinitesimal model and an infinite population. Values are plotted at $t = 4, 6, 8, 10$ generations (*bottom to top*). The variance in reproductive value diverges over time, because very fit individuals contribute an extremely large number of offspring.

disadvantage, even in large populations (Fig. 6). This is surprising, because selection on such a modifier is expected to be weak when it only stays associated with its effects for a short time (Barton 1995a).

This analysis of the infinitesimal model has depended only on standing variation and has not involved mutation, although the heritable variation in fitness must ultimately be sustained by mutation and might be due to the immediate effect of deleterious or beneficial mutations. All that matters in these models of loosely linked loci is the current standing variation. However, because the variance in reproductive value diverges, the fixation probability of new favorable mutations will be strongly reduced by selection, even if the rate of asexual reproduction is low.

A LINEAR GENETIC MAP

By how much do random mutations, scattered over a linear genetic map, reduce fixation probability? For any particular model of selection, this can be answered using Equation 4; now, the states i correspond to the possible genetic backgrounds. For example, Figure 7 shows how the fixation probability depends on when and where on the map a new favorable mutation occurs, given that four sweeps are in progress at linked loci. Its chances are reduced below the baseline level of $2s$ if it occurs just before a sweep or is closely linked to it. However, such calculations are not feasible for more than a few loci, because the number of different genetic backgrounds

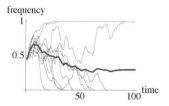

Figure 6. Frequency of a modifier that increases the rate of asexual reproduction from zero to α = 20%. Ten replicate simulations are shown of a population of 1000 haploid individuals. Although individual outcomes are highly variable, the average decline in frequency indicates a selective disadvantage of –3.0% (thick line). Selection acts according to the infinitesimal model, with variance in log fitness $v = 1$.

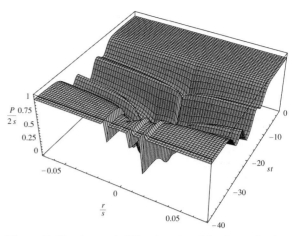

Figure 7. Fixation probability is reduced by four selective sweeps. The fixation probability $\frac{P}{2s}$, measured relative to its baseline value $2s$, is plotted against the map position of the new mutation (r/s) and its time of occurrence (st). The four sweeps occur at evenly spaced map positions $r = 0.01s$ apart; mutations were introduced at times $st = -5, -10, 0, -5$, as single copies in a population of $N = 10^6$.

increases so quickly. Moreover, we would really like to know the loss of fixation probability (i.e., the volume below the flat surface in Fig. 7), averaged over randomly scattered sweeps.

We can, however, make a scaling argument that shows how average fixation probability depends on selection strength, population size, and map length (s, N, R). The key assumption is that N is so large that the establishment of a new mutation is influenced by sweeps that are already established at high frequency and, if it survives, will itself influence mutations only far into the future. Thus, population size does not by itself directly affect the results. (N must be large enough that negative linkage disequilibria generated while it is rare have dissipated, but this effect is weak.) We can also assume that the map is so long that loosely linked genes have little effect; then, what matters is the density of substitutions per generation and per map length Λ/R. (Large numbers of completely unlinked genes, on other chromosomes, can be analyzed by the infinitesimal model, as above.) Finally, the strength of selection has two opposing effects: The length of map affected is approximately the strength of selection on the sweep S, whereas a new rare allele, with advantage s, is vulnerable for $\sim 1/s$ generations. Averaging over a uniform density of times and positions on the map, these two effects cancel: What matters is the ratio S/s between selection on the established sweep and on the new mutation. Overall, we find that density of sweeps Λ/R is a decreasing function $f(\Lambda_0/R)$ of that rate. The dimensionless function f depends on the shape of the distribution of selection but not on its absolute strength.

To obtain a concrete form for this function, we must make an approximation. If the established sweeps are much more strongly selected than the alleles that they chase out of the population, $S >> s$, the fixation probability is closely approximated by $P = 2(s - s^*)$, where s^* is a critical selection coefficient proportional to the additive genetic variance per map length (Barton 1994). In a large population, subject to recurrent sweeps, a weakly selected allele has a negligible chance of fixation if its advantage is less than this critical value (more precisely, P tends to zero as N tends to infinity). This critical value is easily understood: It just counterbalances the expected rate of decrease of a new allele, due to sweeps that will almost always occur on another background (Fig. 8).

Using this approximation, we find that if all mutations have the same effect $(\Lambda/R) \sim (\Lambda_0/R)/(1 + 1.94\,[\Lambda_0/R])$. This approximation works well for moderate rates of substitution but somewhat overestimates the degree of Hill–Robertson interference for high rates. Crucially, however, the scaling argument matches simulations well: Values converge as population size increases to a value that is independent of s and R (Fig. 9).

These arguments show that the maximum rate at which favorable mutations can be accumulated by selection is limited by recombination, via the Hill–Robertson effect. It is unlikely that most hitchhiking is due to selective sweeps, as was assumed here; deleterious mutations must also contribute. Hudson and Kaplan (1995) and Nordborg et al. (1996) showed how such "background selection" reduces neutral diversity, and Barton (1995b) showed that the effect on fixation probability is similar: For deleterious mutations scattered evenly over a linear genetic map, both are reduced by $\sim\exp(-U/R)$, independent of the strength of selection. This dimensionless ratio suggests that a similar scaling argument to that given above could be used to find the joint effects of many genes. (However, this result is essentially an extrapolation from the two-locus theory and, for multiple loci, is only an approximation; Barton 1995b.) Surprisingly, however, analysis of a population at mutation-selection balance is harder than that for multiple overlapping sweeps. This is because we must consider linkage disequilibria among common genotypes that typically carry many deleterious mutations. The analysis of multiple sweeps given above was greatly simplified, because under multiplicative selection in a large population, new alleles sweep through at close to linkage equilibrium.

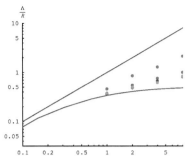

Figure 9. Rate of sweeps per unit map length, plotted against the baseline rate $\Lambda_0/R = 2\,sNU/R$. The series of dots are for population size $N = 10^2, 10^3, \ldots 10^5$ (moving down). (Upper line) $\Lambda = \Lambda_0$, (lower curve) the approximation $(\Lambda_0/R)/(1 + 1.94\,Z[\Lambda_0/r])$. $s = 0.05$, $R = 1$; values are the mean of 10 replicates.

DISCUSSION

The most plausible explanation for the prevalence of sex and recombination is that it facilitates natural selection by maintaining high additive genetic variance in fitness. This requires that there are typically negative associations among favorable alleles, which hinder selection and must be broken up by recombination. Several simulation studies have shown that substantial negative associations can be generated by random drift, even in large populations (Otto and Barton 2001; Iles et al. 2003; Keightley and Otto 2006). This can be understood by realizing that even in a fairly large population, rare alleles—and in particular, new favorable mutations—are vulnerable to random loss, and their chances of fixation can be greatly increased by recombination, which frees them from random, and typically negative, association with their genetic background. Simple arguments, based on branching processes, show how this Hill–Robertson effect depends on the heritable variance in fitness V, on the density of deleterious mutations on the genetic map U/R, and on the density of selective sweeps Λ/R.

Although not discussed in detail here, this analysis extends to show how modifiers that increase recombination can gain an advantage if they are linked to adaptive variation. This can be done both for alleles that are at appreciable frequency, with random drift being treated as a perturbation (Barton and Otto 2005), and for new mutations that follow a branching process, as discussed above (Roze and Barton 2006). In both cases, although the analysis of two or three loci is fairly straightforward, extending results to large numbers of selected genes remains an important open question: It is just this realistic case that simulations suggest will most readily select for recombination (Iles et al. 2003; Keightley and Otto 2006).

How important is Hill–Robertson interference in nature? For unlinked loci, the key parameter is the heritable variance in fitness—the best measure of the total amount of selection that acts. In principle, the genetic variance of fitness could be estimated from the number of offspring in successive generations, requiring simply a large pedigree. Unfortunately, it has been extremely difficult to measure this fundamental quantity in any organism. Large, long-term surveys are needed to estimate the

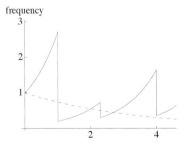

Figure 8. A rare allele will experience selective sweeps at linked loci as a series of random catastrophes that suddenly knock it back to a fraction of its previous frequency. If the rate of sweeps is above a threshold, a rare allele will on average decline and can never fix (dashed line). (Actually, there is a very small chance that the sweep will occur on coupling with the rare allele, giving it a very large boost. However, this can be neglected in large populations.)

genetic variance of any quantitative trait, and fitness is especially problematic because it involves the whole life cycle and is perturbed by noninherited variation (Houle 1992). Thus, although the heritability of fitness (the ratio of the genetic variance to the total) is low, the heritable variance itself may well be high—high enough to reduce the rate of adaptation by more than an order of magnitude. (For example, if $V = e^v = 2$, fixation probability would be reduced by a factor $e^{-4v} = V^{-4} = 0.0625$ in a polygamous species and by $e^{-9v} = V^{-9} = 0.002$ in a monogamous species.) (For reviews, see Burt [2000]; Merila and Sheldon [2000].)

Recently, it has become feasible to measure the total mutation rate directly by comparing DNA sequences across several generations (Haag-Liautard et al. 2007; Lynch et al. 2008)—a great improvement on previous indirect techniques (Drake et al. 1998). It is also possible to estimate the fraction of sequence that is constrained by selection, by comparing rates of molecular evolution with presumed neutral sequence, such as pseudogenes. These recent estimates suggest that the rate of deleterious mutation may be large ($U > 1$, say) in organisms with large functional genomes, such as ourselves. Thus, deleterious mutation may be maintaining high recombination rates in such organisms and clearly does interfere with selection in genomic regions of low recombination (e.g., in Y chromosomes; Charlesworth et al., this volume). However, it does not seem to provide an explanation for high rates of recombination in eukaryotes in general, for which U may be much lower (Lynch et al. 2008).

We now have good estimates of the rate of divergence between species and the fraction that is due to positive selection (see, e.g., Smith and Eyre-Walker 2002). Although a large number of selected substitutions distinguish sister species, the rate of species-wide sweeps per generation must be very low—much too low to cause significant selection for recombination (Roze and Barton 2006). We know that episodes of intense selection can select for higher recombination—recombination rates increase significantly following artificial selection (for review, see Otto and Barton 2001). It may be that such selection is widespread in nature: Local selective sweeps could be frequent enough to maintain high rates of recombination (Martin et al. 2006; Roze 2009) and yet need not contribute significantly to the rate of species-wide sweeps. Genome-wide surveys of geographic differentiation may soon settle this question. For example, Pickrell et al. (2009) use human single-nucleotide polymorphism (SNP) data from 53 populations to identify several examples of loci that have undergone recent positive selection in limited geographic areas; however, Coop et al. (2009) note that there are surprisingly few examples where different alleles are fixed or nearly fixed in diferent human populations. Although the pattern is as yet unclear, such data do have the potential to tell us the rate at which alleles sweep through local populations.

ACKNOWLEDGMENTS

I thank the Royal Society and the Engineering and Physical Sciences for support (GR/ T11753/01). Jonathan Coe provided the simulations for Figure 6.

REFERENCES

Barton NH. 1994. The reduction in fixation probability caused by substitutions at linked loci. *Genet Res* **64:** 199–208.
Barton NH. 1995a. A general model for the evolution of recombination. *Genet Res* **65:** 123–144.
Barton NH. 1995b. Linkage and the limits to natural selection. *Genetics* **140:** 821–841.
Barton NH, Keightley PD. 2002. Understanding quantitative genetic variation. *Nat Rev Genet* **3:** 11–21.
Barton NH, Otto SP. 2005. Evolution of recombination due to random drift. *Genetics* **169:** 2353–2370.
Bonhoeffer S, Chappey C, Parkin NT, Whitcomb JM, Petropoulos CJ. 2004. Evidence for positive epistasis in HIV-1. *Science* **306:** 1547–1551.
Burger R. 1999. Evolution of genetic variability and the advantage of sex and recombination in changing environments. *Genetics* **153:** 1055–1069.
Burt A. 2000. Sex, recombination and the efficacy of selection—Was Weissman right? *Evolution* **54:** 337–351.
Butlin R. 2002. The costs and benefits of sex: New insights from old asexual lineages. *Nat Rev Genet* **3:** 311–317.
Charlesworth B. 1993. Directional selection and the evolution of sex and recombination. *Genet Res* **61:** 205–224.
Charlesworth B, Barton NH. 1996. Recombination load associated with selection for increased recombination. *Genet Res* **67:** 27–41.
Coop G, Pickrell JK, Novembre J, Kudaravalli S, Li J, Absher D, Myers RM, Cavalli-Sforza LL, Feldman MW, Pritchard JK. 2009. The role of geography in human adaptation. *PLoS Genet* **5:** e1000500.
Darwin C. 1871. *The descent of man, and selection in relation to sex*. Murray, London.
Darwin CR. 1876. *The effects of cross and self fertilisation in the vegetable kingdom*. Murray, London.
de Visser JA, Elena SF. 2007. The evolution of sex: Empirical insights into the roles of epistasis and drift. *Nat Rev Genet* **8:** 139–149.
Drake JW, Charlesworth B, Charlesworth D, Crow JF. 1998. Rates of spontaneous mutation. *Genetics* **148:** 1667–1686.
Edwards AWF. 2002. The fundamental theorem of natural selection. *Theor Popul Biol* **61:** 335–338.
Elena SF, Lenski RE. 1997. Test of synergistic interactions among deleterious mutations in bacteria. *Nature* **390:** 395–298.
Elena SF, Lenski RE. 2001. Epistasis between new mutations and genetic background and a test of genetic canalization. *Evolution* **55:** 1746–1752.
Fisher RA. 1930. *The genetical theory of natural selection*. Oxford University Press, Oxford.
Haag-Liautard C, Dorris M, Maside X, Macaskill S, Halligan DL, Houle D, Charlesworth B, Keightley PD. 2007. Direct estimation of per nucleotide and genomic deleterious mutation rates in *Drosophila*. *Nature* **445:** 82–85.
Hill WG, Robertson A. 1966. The effect of linkage on limits to artificial selection. *Genet Res* **8:** 269–294.
Houle D. 1992. Comparing evolvability and variability of quantitative traits. *Genetics* **130:** 195–204.
Hudson RR, Kaplan NL. 1995. Deleterious background selection with recombination. *Genetics* **141:** 1605–1617.
Iles MM, Walters JR, Cannings C. 2003. Selection for recombination in large genomes. *Genetics* **165:** 2249–2258.
Keightley PD, Otto SP. 2006. Interference among deleterious mutations favours sex and recombination in finite populations. *Nature* **443:** 89–92.
Kirschner M, Gerhart J. 1998. Evolvability. *Proc Natl Acad Sci* **95:** 8420–8427.
Kouyos RD, Otto SP, Bonhoeffer S. 2006. Effect of varying epistasis on the evolution of recombination. *Genetics* **173:** 589–597.
Kouyos RD, Silander OK, Bonhoeffer S. 2007. Epistasis between deleterious mutations and the evolution of recombination. *Trends Ecol Evol* **22:** 308–315.
Koza JR, Bennett FH, Andre D, Keane MA. 2002. Genetic programming: Biologically inspired computation that creatively solves non-trivial problems. In *Evolution as computation* (ed.

LF Landweber and E Winfree), pp. 95–124. Springer, Berlin.
Lynch M, Sung W, Morris K, Coffey N, Landry CR, Dopman EB, Dickinson WJ, Okamoto K, Kulkarni S, Hartl DL, Thomas WK. 2008. A genome-wide view of the spectrum of spontaneous mutations in yeast. *Proc Natl Acad Sci* **105:** 9272–9277.
Martin G, Otto SP, Lenormand T. 2006. Selection for recombination in structured populations. *Genetics* **172:** 593–609.
Maruyama T. 1971. An invariant property of a structured population. *Genet Res* **18:** 81–84.
Maynard Smith J. 1978. *The evolution of sex*. Cambridge University Press, Cambridge.
Merila J, Sheldon B. 2000. Lifetime reproductive success and heritability in nature. *Am Nat* **155:** 301–310.
Mitchell M. 1998. *An introduction to genetic algorithms*. MIT Press, Cambridge, MA.
Muller HJ. 1932. Some genetic aspects of sex. *Am Nat* **66:** 118–138.
Nordborg M, Charlesworth B, Charlesworth D. 1996. The effect of recombination on background selection. *Genet Res* **67:** 159–174.
Otto SP, Barton NH. 2001. Selection for recombination in small populations. *Evolution* **55:** 1921–1931.
Otto SP, Feldman MW. 1997. Deleterious mutations, variable epistatic interactions, and the evolution of recombination. *Theor Popul Biol* **51:** 134–147.
Pickrell JK, Coop G, Novembre J, Kudaravalli S, Li JZ, Absher D, Srinivasan BS, Barsh GS, Myers RM, Feldman MW, Pritchard JK. 2009. Signals of recent positive selection in a worldwide sample of human populations. *Genome Res* **19:** 826–837.
Price GR. 1972. Fisher's "fundamental theorem" made clear. *Ann. Hum. Genet* **36:** 129–140.
Provine W. 1971. *The origins of theoretical population genetics*. University of Chicago Press, Chicago, IL.
Robertson A. 1961. Inbreeding in artificial selection programmes. *Genet Res* **2:** 189–194.
Roze D. 2009. Diploidy, population structure and the evolution of recombination. *Am Nat* **174:** S79–S94.
Roze D, Barton NH. 2006. The Hill-Robertson effect and the evolution of recombination. *Genetics* **173:** 1793–1811.
Smith NGC, Eyre-Walker A. 2002. Adaptive protein evolution in *Drosophila*. *Nature* **415:** 1022–1024.
Sniegowski PD, Murphy HA. 2006. Evolvability. *Curr Biol* **16:** R831–R834.
Szathmary E. 1993. Do deletion mutations act synergistically? Metabolic control theory provides a partial answer. *Genetics* **133:** 127–132.
Wagner GP, Altenberg L. 1996. Complex adaptations and the evolution of evolvability. *Evolution* **50:** 967–976.
Waxman D, Peck JR. 1999. Sex and adaptation in a changing environment. *Genetics* **153:** 1041–1053.
Weismann A. 1889. The significance of sexual reproduction in the theory of natural selection. In *Essays upon heredity and kindred biological problems* (ed. EB Poulton et al.). Clarendon, Oxford.
Whitlock MC, Gomulkiewicz R. 2005. Probability of fixation in a heterogeneous environment. *Genetics* **171:** 1407–1417.
Williams GC. 1975. *Sex and evolution*. Princeton University Press, Princeton, NJ.

Eradicating Typological Thinking in Prokaryotic Systematics and Evolution

W.F. DOOLITTLE

Biochemistry and Molecular Biology, Dalhousie University, Nova Scotia, Canada B3H 1X5
Correspondence: ford@dal.ca

In 1982, John Maynard Smith called for an evolutionary "New Synthesis" specific for prokaryotes, observing that *"population thinking has been well developed for fully half a century, but has yet to be adopted by microbiology"* (Maynard Smith 1982). Twenty-seven years later, typological thinking (population thinking's antithesis) still dominates the field. Evidence for this includes the continuing debates on the reality of prokaryotic species, the value of the term "prokaryote," and the significance of the tree of life (TOL). In each case, the unexpected prevalence of interlineage transfer of genetic information has been (or *should* now be) the catalyst for the final Darwinization of our discipline. With examples from phylogenomics, I argue that "species," "domains," and the "TOL" are reifications that we can do without, especially as genomics dissolves into metagenomics.

Much progress has been made in the quarter century since John Maynard Smith wrote the words quoted, in no small part thanks to him. But the called-for synthesis is still unfinished. That we microbiologists remain half-committed to pre-Darwinian typological and essentialist thinking and are (most of us) still realists about categories that might more realistically be understood as nominal—matters of convention only—is one of the two central points I hope to make in this chapter. We seem to believe that **species** and **domains** and the **TOL** exist independently of our minds, that we can discover facts about them by experiment, and that ongoing debates about the use of these terms are scientific in character (being about what *is* the case), not *normative* (about what *should* be our practice). That we ought not to believe these things is my second point.

IMPORTING CONCEPTS FROM THE NONMICROBIAL LITERATURE

Prokaryotic and multicellular eukaryotic systematic and evolutionary discourses are largely disconnected, and the latter, with more input from philosophers, has the richer vocabulary. Some of this I mean to import. For instance, **realism** in systematics traditionally (even before Darwin) entails the belief that individual *species taxa* (*Homo sapiens* or *Escherichia coli*, for instance) exist extramentally ("out there"), are not artifacts of human ways of thinking, and have verifiable characteristics. Before Darwin, realism generally extended to the claim that the *species category* (made up of all properly designated species taxa) is also real, because all species share some trait(s) that makes them species and not varieties or genera, not least their being specially and individually created by God (Hull 1965; Sober 1980; Stamos 2005).

Darwin's position, Beatty (1992) and Ereshefsky (2009) suggest, combined **species taxon realism** with **species category antirealism**. In the *Origin*, Darwin wrote

> To sum up, I believe that species come to be tolerably well-defined objects, and do not at any one period present an inextricable chaos of varying and intermediate links (Darwin 1859, p. 177).

but also, and at first blush contradictorily,

> In short, we shall have to treat species in the same manner as those naturalists treat genera, who admit that genera are merely artificial combinations made for convenience. This may not be a cheering prospect; but we shall be freed from the vain search for the undiscovered and undiscoverable essence of the term species (Darwin 1859, p. 282).

Species taxon realism (as in the first passage) holds that individual recognized species can be *real*—variously well-defined spatiotemporally and cohesive in their properties. These are not arbitrary social constructs: Humans untrained in Western systematics supposedly do (Diamond and Bishop 1999)—and extraterrestrials presumably would—identify the same groupings. Individual organisms in a real species share one or a combination of traits that unite them as members of that species and distinguish them from members of other species. A pre- or post-Darwinian realist who (unlike the man himself) subscribed to species **essentialism** would further hold that there is one or a certain fixed combination of traits necessary and sufficient to define each species taxon uniquely. (The alternative would be to define species by a family resemblance or cluster concept, some number but no specific ones of a larger set of possible traits being required, as in Wittgenstein's famous example of games [Pigliucci 2003].) For pre-Darwinian (i.e., creationist) systematists, species essentialism came associated with a belief, shared by today's creationists, in the fixity of species' essences. It is not surprising that "transmutation," with its alchemical, saltationist, and miraculous connotations, was preferred to "evolution" as a description of such a radical and disallowed transformation.

Species category antirealism has two aspects. First, various entities called species are just not the same kind of thing. Genotypic/phenotypic cohesiveness and mechanisms thought to maintain a species differ widely in degree and kind between *H. sapiens* and *E. coli*, for instance. Only because both are called species do we believe them to be similar. To be sure, organisms that are members of one species are more like one another than like organisms that are members of a sister species, but so too are organisms that are members of varieties and genera more like one another than like organisms that are members of sister varieties or genera. Indeed, that the varieties of what is now called a species may in time become the species of what will be called a genus is the essence of Darwin's theory and the antithesis of special creation.

Ernst Mayr claimed to have restored reality to the species category and nonarbitrariness to classification by replacing essentialism, which he called **typological thinking**, with what he called **population thinking** (Mayr 1975). Although it is not always clear how to apply this dichotomy throughout biology (as Mayr would have had us do), the instantiation of the latter in his Biological Species Concept (BSC) was a signal contribution to systematics. The BSC promised an essence-free approach to species taxa while providing a kind of process-based essence for the species category, one that entails a distinction between varieties, species, and genera. Biological Species are all maximally inclusive groupings (populations or dispersed metapopulations) whose members *can* produce fertile offspring through mating, and any group that can do this is one: "species-ness" is thus redefined.

The hegemony of the BSC notwithstanding, it has always had problems: Members need not look much alike, so it will sometimes be at odds with more intuitive morphological species concepts; interbreeding groups can be paraphyletic or even polyphyletic; interbreeding among subpopulations (for instance, of "ring species") need not be a transitive property; and asexuals (arguably most of the biosphere if we think that prokaryotes are not properly sexual) are excluded—excluded by definition! In coming up with a species concept that can cope better with evolutionary theory then did typology, Mayr sacrificed a general role for the category in systematics: "Species" describes a type of population genetic behavior, not a taxonomic rank (Mayr 1996).

One work-around, suggested, for instance, by Hull (1965), is to define species *disjunctively*, by the BSC for sexuals, by ecological clustering for asexuals, and perhaps by alternative species concepts for other groups. This could satisfy the needs of systematists, if we require that there be some concept, however thrown together, in place for *all* individual organisms. But there are other, primarily ecological, questions that we want to ask of species as a category—for instance, how many species are there, and what fraction of them are sexual?—that cannot be answered except arbitrarily if the two or more sorts of species are disjunctively (and incommensurably) defined. All we could ask then is, "Among sexuals, how many are called species by one criterion and among asexuals, how many are called species by another, incompatible, criterion?"—a question about systematics, not about Nature. The situation is worse than that of apples and oranges because these share an essential fruitiness, which can be used to define a larger category to which both belong. There is no comparable definition of a larger category embracing sexuals and asexuals, other than that they comprise organisms and are *called* species. In my view and that of Ereshefsky (1998), disjunctive definitions are antirealistic.

Tree thinking has now or will soon replace both typological and population thinking in many systematics applications (O'Hara 1998). The success of molecular data and sophisticated phylogenetic algorithms in placing many animal and plant species in unique relationships to one another on what seems a universal TOL encourages us to look at them primarily as its twigs and to abandon altogether any attempt to circumscribe and name more inclusive (Linnaean) groupings. Such an agnostic philosophy is embraced by the DNA bar-coding and Phylocode efforts (Cantino and De Queiroz 2006). Indeed, Darwin believed that such a tree might be a discoverable *historical* entity: He was a **tree realist**.

Tree thinking is the basis of De Queiroz's currently popular *metapopulation lineage concept* for species. He reasons that

> Alternative species concepts agree in treating existence as a separately evolving metapopulation lineage as the primary defining property of the species category, but they disagree in adopting different properties acquired by lineages during the course of divergence (e.g., intrinsic reproductive isolation, diagnosability, monophyly) as secondary defining properties (secondary species criteria). A unified species concept can be achieved by treating existence as a separately evolving metapopulation lineage as the only necessary property of species and the former secondary species criteria (e.g., intrinsic reproductive isolation, diagnosability, monophyly) as different lines of evidence (operational criteria) relevant to assessing lineage separation (De Queiroz 2007).

It is true that no biologist would now accept groups that are *not* "separately evolving metapopulation lineages" as species. But the concept leaves us dependent on incommensurable "secondary criteria" to answer ecological questions about global species diversity, richness, or distribution. Except for those groups (a minority) amenable to the BSC, we still have no principled way to distinguish varieties from species from genera—as we must to answer such questions. And we cannot accommodate reticulation in the tree (introgression, hybridization, or lateral gene transfer [LGT]) without again reifying higher taxa or reintroducing a new sort of essentialism that justifies ignoring much of the data, for instance, privileging ribosomal RNA (rRNA) genes for defining lineage identity. In endorsing rRNA phylogeny as the backbone of universal prokaryotic classification (Guerrero 2001), the publishers of *Bergey's Manual of Systematic Bacteriology* have legitimized what in a later section I call **ribo-essentialism**.

Species, for Prokaryotes

Although divided as to what species might be, many microbiologists clearly believe that they do exist. Cohan (2002) writes, "Bacterial species exist—on this much bacteriologists can agree. Bacteriologists widely recognize that bacterial diversity is organized into discrete phenotypic and genetic clusters, which are separated by large phenotypic and genetic gaps, and these clusters are recognized as species" (Cohan 2002). Ward et al. (2008) hypothesize that there are "such ecologically distinct, species-like groups of bacteria" and believe that "it is essential to identify and define these populations if we are to develop a predictive understanding of microbial community composition, structure and function." That species must exist because we need them is also implied by Fraser et al. (2009): "Evolutionary theory should be able to explain why species exist at all levels of the tree of life, and we need to be able to define species for practical applications in industry, agriculture and medicine." And in asserting that "… it is important to note there is little, if any, substantive data to support the conclusion that bacterial species do not exist," Riley and Lizotte-Waniewski (2009) seem to be laying claim to the null hypothesis, putting the burden of proof on species antirealists.

It would be presumptuous to guess what ontologies underlie these expressions of species category realism: whether their authors believe that microbes *must* form more-or-less discrete phenotypic clusters because species are a universal fact of Nature or instead believe that there exist somewhere extensive meta-analyses showing this to be generally the case, for bacteria. But in either, they would be mistaken. As Hanage et al. (2006) summarize the situation, "To many microbiologists, bacterial species are real entities that can be recognized as clusters of genotypes which are clearly resolved from similar clusters. In fact, there are almost no data that address this assertion, which in essence is a statement of belief." And Konstantinidis et al. (2006), although generally species-friendly, admit that "An important issue that remains unresolved is whether bacteria exhibit a genetic continuum in nature … . It is possible that some environments support clusters, whereas others do not."

Resolving this important issue will be extraordinarily difficult. One approach, common in environmental studies, is culture-independent polymerase chain reaction (PCR) amplification and sequencing of individual marker genes from environmental samples, the signal for significant clustering being **sequence microdiversity** (an excess of closely related sequences). But random birth and death models will also produce clusters (Zhaxybayeva and Gogarten 2004), so any signal denoting the presence of species must be in excess of this random expectation. Acinas et al. (1999) do find such a signal in coastal *Vibrio* populations and Koeppel et al. (2008) claim to have done so for *Bacillus* populations in Israel's "Evolution Canyons." But the great majority of microdiversity studies have been anecdotal and nonquantitative.

Another approach for finding clusters is MLST (multilocus sequence typing) applied to isolates of a described species or species cluster (most often of clinically relevant bacteria). Seven or so unlinked housekeeping genes are sequenced, used to identify sequence types (effectively, haplotypes) and detect recombination between them, and—when concatenated—to construct strain phylogenies. Often, but not always, clusters corresponding to named species are obtained, even though individual loci will give incongruent trees and cluster differently, because of within- and between-species recombination. Such species are variously "fuzzy" and, as Hanage et al. (2005) admit, "The point at which such a group is described as a species is a matter more of human interest and attention than any intrinsic evolutionary process."

It is, moreover, inevitable that with either method, more extensive sampling of the same or different sites (or hosts) and at different seasons (or states of health) will begin to fill in the "phenotypic and genetic gaps" between observed clusters. After all, it cannot logically widen them! And with either, sampling biases (PCR primer choice, selection for specific phenotypes, and inevitably for "culturability") will create the appearance of clusters even where there is a continuum. We simply do not know the real landscape.

That said, there are two appealing models that would account for the formation, maintenance, and divergence of cohesive clusters, if indeed they do exist (see Fig. 1). The older ecotype model (Cohan 2002) sees periodic selection as a force that, by purging genomic diversity each time "fitter types" arise within different niche-defined subpopulations, ensures divergence between and cohesion within them ("speciation"). Intersubpopulation recombination at loci not under selection will frustrate such a process and is in fact the basis of the second and increasingly popular model, patterned on Mayr's BSC (Fraser et al. 2009). In this, it is frequent within-population homologous recombination, coupled with increasingly infrequent recombination between populations as they diverge, that maintains and also creates species. The analogy to animal species is obvious but imperfect, because interspecies-level and (by LGT) intergenus, interphylum, and interdomain exchange are much more frequent for prokaryotes.

Both ecotype and BSC-type speciation processes have been modeled extensively. Under some (not necessarily realistic) parameter settings, either *can* create genomically cohesive clusters, insofar as the "core" of genes shared by all strains of a putative species is concerned. But it is not enough to show that under *some* conditions, one or the other process *can* produce species-like clusters. For any robust species concept useful in systematics and ecology, it is necessary to show that they *must* do so, under all realistic conditions—so that no organisms are left out of species and not too many are relegated to some limbo of incipient speciation. No modeler would claim that this has been shown, or likely believes that it is true. And, if some species arise as ecotypes and some via the BSC, we also have the same kind of disjunctive definition discussed earlier, with sexual and asexual animal species in mind. Systematists might be satisfied, but ecologists would still be unable to answer general questions about species.

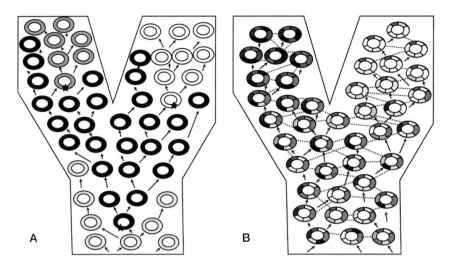

Figure 1. Two models for bacterial speciation. (*A*) The ecotype model, in which fitter-type mutations (stars) initiate periodic selective sweeps, purging diversity at all loci. The model assumes little if any recombination. Speciation can be initiated by a physical barrier, or sympatrically, if subpopulations acquire nonoverlapping adaptations (use of different substrates, for instance). (*B*) The Biological Species Concept for bacteria. The model assumes near-linkage equilibrium, so that fixation of favored mutations does not purge diversity at most loci. Speciation likely requires allopatry or some restriction to gene exchange (resistance to transducing phages or presence of differing restriction-modification systems, for instance.) (Arrows) Cell replications, (dotted lines) recombination.

In short, no known or likely to be discovered law of Nature will constrain known processes working separately or together to always (or perhaps even often) produce "discrete phenotypic and genetic clusters, which are separated by large phenotypic and genetic gaps" (Cohan 2002), unless "discrete" and "large" are allowed to take a very wide range of values. After all, data sets will inevitably show *some* nonrandomness. But to accept just any degree of nonrandomness as evidence for species category realism is to admit that we have no testable hypothesis about what species might be, only faith that they exist.

In any case, there would be the disconnectedness of phenome and genome resulting from LGT. Indeed, it was evidence for LGT's frequent occurrence that first caused many of us to question the notion of prokaryotic species. As noted, the ecotype and BSC models address only the core of genes shared among closely related strains and the extent to which they show genetic cohesion. For most bacteria (and probably most archaea), this core is less than half a genome. The rest is made up of genes present in only *some* "conspecifics." Among 20 recently sequenced *Escherichia coli* strains, for instance, 18,000 families of orthologous genes were found, but only 2000 of these are common to all strains, where they make up on average 42% of the 4700 genes in the average genome (Touchon et al. 2009). Most (at least three-quarters) of the noncore (auxiliary) genes are present in only a few (one to five) genomes. Only the most closely related *E. coli* strains (defined by the similarity of shared core genes) have similar repertoires of auxiliary genes: The latter must turn over rapidly (mostly by LGT balanced by gene loss) and exhibit little deep "phylogenetic signal." Auxiliary genes are often key determinants of relevant strain-specific properties (Touchon et al. 2009): Thus, the phenome and the lineage-defining genomic core are evolutionarily uncoupled, even within and certainly between so-called species, in a manner unfriendly to any but the most liberal sort of category realism.

Domains

Species category realism is an element of microbiology's pre-Darwinian (typological) heritage, a presumption that the species category must exist as part of the natural order. We have simply continued to modify our definitions and criteria for identification to keep ourselves convinced that this is true, despite Darwin's demurral. A similarly persistent but differently articulated sort of category realism or essentialism marks the ongoing wrangle regarding the term "prokaryote" (see Fig. 2).

The contested history of the prokaryote/eukaryote dichotomy and the role of Stanier and van Niel (1962) in popularizing it have been reviewed by Sapp (2006). Norman Pace argues vehemently that this dichotomous scheme must be replaced by the three-domain (Bacteria, Archaea, and Eukarya = Eukaryota) representation of organismal diversity, even though many biologists are comfortable with the former as a distinction between *grades* (types of cellular organization) and the later as one between *clades*. Pace (2006) gives four reasons:

1. Prokaryotes are paraphyletic—eukaryotes are in fact sisters to archaea, so "there is not a single (monophyletic) phylogenetic group on which to hang the tag prokaryote."

2. The pro- and eu- suffixes imply that one group gave rise to the other.

3. The "nuclear line of descent is as ancient as the archaeal line, and not derived from either archaea or bacteria."

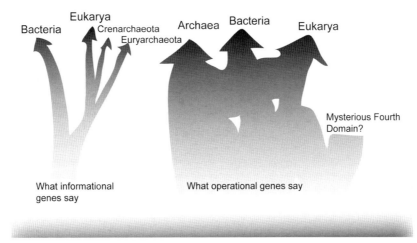

Figure 2. Different perspectives on relationships between domains. Two different stories concerning relationships between domains scales roughly in proportion to the fraction of genes telling each. Many eukaryote-specific genes have no obvious bacterial or archaeal origin: Some researchers have proposed a mysterious fourth domain as their source.

4. If "prokaryotic" means only "not eukaryotic," then this is "a negative and therefore scientifically invalid description."

Each objection is easily countered and indeed has been by W.B. Whitman (2009) in a vigorous defense of the dichotomy. First, the dichotomy was presented by Stanier and van Niel as one between cell types and not monophyletic clades. Ironically, the three-domain tree Pace promotes—with Archaea and Eukarya as sisters—is actually another dichotomy, and by failing to assign a name and domain status to the Archaeal/Eukaryal sisterhood, Pace abandons the same cladistic practices that he claims to defend. Second, if we indeed take "prokaryote" and "eukaryote" only as descriptors of less and more complex cell types, almost certainly the former *did* give rise to the latter. Third, a recent reanalysis (Cox et al. 2008) of much the same data on which Pace bases his conclusion that eukaryotes "are not derived from either archaea or bacteria" in fact shows their derivation from archaea (specifically "eocytes"). And fourth, there *are* key traits that unite prokaryotes. In particular, the coupling of transcription and translation that the lack of a nuclear membrane allows—and that the presence of spliceosomal introns would prevent—is a fundamental and likely ancestral feature that bacteria and archaea share.

It is tree realism (or "historical essentialism; Bapteste and Boucher 2009) that underwrites Pace's confidence in his views. The domains are not for him "merely artificial combinations made for convenience"—as Darwin described species and many before and after him view higher taxonomic ranks—because they appear as monophyletic clades on the rRNA-based TOL. And it is ribo-essentialism that underwrites his confidence in this particular tree, because most nontranslational genes do not support it and many refute it, at many levels of analysis. For instance, in our recent survey of five named *Thermotoga* species (Zhaxybayeva et al. 2009), only a few "core" (mostly ribosomal protein) genes show the expected deep phylogenetic placement and sister relationship to the Aquificales: Almost 20 times as many showed affinity to Firmicutes, tellingly to thermophilic members of the group in particular. And in a phylogenomic study of *Saccharomyces cerevisiae*, Esser et al. (2004) found that "approximately 75% of [850] yeast genes having homologues among the present prokaryotic sample share greater amino acid sequence identity to eubacterial than to archaebacterial homologues," whereas "at high stringency comparisons, only the eubacterial component of the yeast genome is detectable." Eukarya are sisters to Archaea only when the majority of the usable data is discarded!

Ribo-essentialism is problematic in several ways in addition to this frequent need to ignore most of the data (Fig. 2). It privileges (presumed) hosts over (presumed) symbionts in chimeras (such as all the eukaryotes) and thus (in this case) depends on still-contested theories about which is which (Embley and Martin 2006). It relies on the widely believed but not widely tested "complexity hypothesis," according to which translation-related genes are relatively immune to LGT. And, ultimately, most important in the argument regarding "prokaryotes," ribo-essentialism assumes that the concept of monophyly, which surely *must* legitimize any claim for taxon reality at the species level and above (De Queiroz 2007), can be applied in a principled way to evolutionary chimeras. Likely, it cannot (Mindell 1992), and ribo-essentialism simply sweeps this issue under the carpet. The heated rhetoric obscures the fact that no experiment can ever tell us whether Nature holds three domains or two: Such higher taxa exist only in our minds. There remains room to negotiate but nothing to discover, no final truth of this matter.

The Tree of Life

Attacks on and defenses of the TOL are often vague about what this concept is actually meant to be or do. Eric Bapteste and I have argued that the TOL was first presented by Darwin as a hypothesis (the **TOL Hypothesis**), holding that classifications are hierarchical or tree-like (show "groups subordinate to groups") *because* the process responsible for generating the phenotypic similarities and differences on which they were based—descent with modification—is itself tree-like (Doolittle and Bapteste 2007). Hierarchical classification was thus the *explanandum* (that which was to be explained) and evolution the *explanans* (the explanation).

After 1859, systematists such as Haeckel adopted evolution as a legitimating theory, without fundamentally changing their practices. But further elaboration of trees—expansion of the *explanandum*—logically cannot provide proof of the *explanans*, for all that Darwin seemed to want to work it this way. As Eldredge (2005) notes, "It is very much as if Darwin wanted his patterns—the very 'facts' that led him to evolution in the first place—to be retroactively seen as predictions that would necessarily be true if evolution has occurred." Few biologists have owned up to the circularity of this reasoning, and it was not until 1965 that Zuckerkandl and Pauling suggested that congruence between traditional phenotype-based classifications and independently derived molecular phylogenies could provide a much-needed independent proof.

For animals and plants (hybrids aside), such proof arguably has been or soon will be obtained, and these after all were Darwin's main concern. But prokaryotes lack traditional phenotype-based classifications, and the best substitute proof, congruence among gene trees, has failed us. Few would claim that even as many as 10% of prokaryote genes demonstrably have the same tree, all the way back to the last universal common ancestor (LUCA) whose ancient existence the TOL model demands. A LUCA-less population model (Fig. 3) is to be preferred, we suggest (Doolittle and Bapteste 2007). In such a model, (1) modern gene families descend from coalescent last common ancestral genes (LCAGs) that existed in different genomes at different times in the past and (2) modern cells descend from a last common ancestral cell (LCAC) which—although by definition the root of a tree of cells (TOC)—likely contained none of those LCAGs. LCAC cannot be a LUCA in the sense of carrying ancestral versions of all modern genes: a false assumption underlying many attempts to reconstruct "the" ancestral genome (e.g., see Mushegian 2008).

Many molecular phylogeneticists nevertheless consider a universal TOC to be their real (and realistic) target. They pursue it either through (1) phylogenies of the few "informational" (mostly translation-related) genes that are shared by all genomes, concatenating them because individually such genes have weak and contradictory signals, or (2) "genome trees," which assess the evolutionary distances between pairs of genomes using all their shared genes. Efforts of the first sort can be criticized for being based on so few data (one prominent result has been called "The Tree of One Percent") and depending absolutely on the assumption that core genes are seldom transferred (Dagan and Martin 2006). Genome trees are also problematic, because they will track cell history only if LGT (which affects the majority of the genes analyzed)

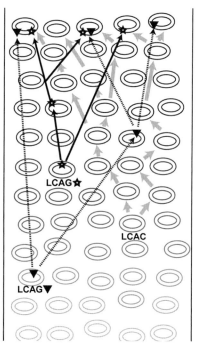

Figure 3. Population model for prokaryote genome evolution. Each contemporary gene family (two shown here as stars and triangles) is derived from a last common ancestral version (LCAG) in the population's past, that population being all prokaryotes, and the past extending back at least 3 billion years. Different gene's ancestors will have existed in different genomes, at different times. LCAC is the cell from which all contemporary cells can be considered descended. It will have existed at some still different time and need not have harbored any LCAGs.

is "random," its effects "cancelling out." Strong nonrandom biases in exchange, as revealed by the *Thermotoga* data just discussed, are, however, surely more the rule than the exception.

Any TOC, however obtained, may tell us very little about the majority of an organism's phenotype-determining genes and cannot be the *explanans* for Darwin's *explanandum*. A TOC might of course form the scaffold for a more complex network-like or web-like scheme for explaining similarities and differences in the sorts of phenotypic traits on which traditional "organismal" phylogenies have been based (Bapteste and Boucher 2009). Still, if trait evolution is web-like, not tree-like, Darwin's original TOL hypothesis founders.

Reluctance to admit this has several sources. One is fear of the creationist opposition, who will see such admission as weakness. But we now have so much other evidence bearing on the genetic and ecological processes that make evolution inevitable that we do not need the TOL hypothesis to be true in order to convince us that evolution has occurred. Another is denial that an evolutionary underpinning for the TOL ever really was a hypothesis, a belief that instead it was simply a truth of Nature, perceived inductively by Darwin. This would thumb a collective nose at how most of us were taught to think science works and ally us with pre-Darwinian systematists for whom Nature's pattern of organized diversity was also a directly perceivable product of a divine intelligence. Few of us now endorse this latter explanation. But tree thinking is surely a form of typological thinking writ large, and the extramental existence of the historical entity that is the universal TOL is surely a concept in which we are predisposed to believe by 2000 years of essentialist philosophy (Hull 1965; Panchen 1992). Our attempts to reconstruct it—despite evidence that the great majority of genes in the (mostly prokaryotic) majority of genomes have through most of Life's (mostly prokaryotic) history evolved in a web-like fashion—seems to testify to the strength of that predisposition.

Consequences

Too much attention to philosophical niceties can no doubt inhibit biological practice. But the ontological status of species, domains, and the TOL all matter, in that what we think about them will variously determine what we do.

If the species category is, as I argue, only nominal, we should bravely give up our addiction to what Darwin called "the vain search for the undiscovered and undiscoverable essence of the term species." And as a practical matter, many questions that microbial ecologists ask about the numbers, diversity, distribution, and community functions of species will have to be reformulated in terms of genes, genomes, and individual cells, at one end of the explanatory spectrum and of loosely organized and likely unstable communities at the other. Metagenomics will force us to start thinking that way anyway, and already there are excellent publications using species-free approaches for correlating genomic diversity and niche diversity (e.g., see Hunt et al. 2008; Martiny et al. 2009). That species taxa remain real is a comfort, though: Clinical, industrial, and environmental microbiologists investigating the activities of microbes in those settings can usefully continue to use these designations. Of course, they will need constant reminding that genes that distinguish beneficial and useful strains from harmful and even deadly ones will often be part of the species' highly variable strain-specific auxiliary gene repertoire (Lapierre and Gogarten 2009; Touchon et al. 2009)

The debate regarding "prokaryotes" needs to stop altogether: Grades and clades are both useful social constructs, but neither is more than that (for either microbes or "macrobes"). There are no scientific questions about what actually exists to be settled here, no facts to be discovered. There are issues of systematic theory (how should cladistics treat chimeras) and perhaps about language to be hammered out, but as between politicians or lawyers. We need to remember that although taxonomy employs science, it is not science.

That the TOL might be a pattern we impose on Nature, not one we discover there, matters most. Even if a TOC could reliably be constructed, Figure 3 tells us that its root was not LUCA, if we mean by LUCA a cell whose genome harbored the last common ancestral versions of all extant contemporary genes. Still, one often sees papers devoted to the in silico reconstruction of such an impossible entity (Mushegian 2008). And we must be cautious about inferring the properties of ancestral lineages, for instance, the biochemical attributes of the α-proteobacteria that gave rise to mitochondria (Esser et al. 2007). Much of the field of paleobiogeochemistry assumes that chemical evidence for the operation of a particular pathway in the ancient past is evidence for the existence then of populations of the bacterial or archaeal group(s) resembling those that harbor that pathway now. There is no guarantee of this.

Pattern Is Prior to Process

Bapteste and Boucher (2009) link what is happening in microbiology to a shift in academic philosophy toward "process philosophy," in which "natural existence consists in and is best understood in terms of processes rather than things" (Rescher 2002), and in which context, for instance, "taxa can be seen as ontologically secondary." In my own view, the history of the science (and biology generally) has involved (1) conceptualization (and often reification) of patterns in Nature and (2) identification and investigation of processes responsible for such patterns, leading to (3) recognition that these processes are not constrained to generate the patterns we thought we saw, after all. We are now just at the edge of (4) understanding that process is prior to pattern, and more "real." I believe that Woese and Goldenfeld (2009) might be saying something such as this too, albeit as part of their own idiosyncratic reading of the history of the discipline.

Bapteste and Boucher (2008, 2009) favor an expanded and more even-handed application of the **eliminative pluralism** of Marc Ereshefsky (1992), in which alternative (genetic, ecological, phylogenetic) evolutionary process-based classifications are seen as simultaneously and equally valid, with the inevitable result that individuals that belong to one taxon by one scheme can be separated in another. I believe that this pluralistic approach—although appealing as ontology—either goes too far (resulting in an ambiguity or flexibility that does not serve many of the traditional or practical reasons for which humans classify) or not far enough (by "classifying" at all). In the brave new metagenomic world that I see over the horizon, we will concern ourselves only with how biological activities and capacities and their genetic determinants are distributed in space and time, and how from knowing some properties of an organism or a community we can predict other properties. As humans, we may always need to make classifications: Almost certainly, this is part of our evolved psychology. But as scientists, we do not have to believe in them.

ACKNOWLEDGMENTS

I thank Marc Ereshefsky, Eric Bapteste, Maureen O'Malley, and Olga Zhaxybayeva for critical reading of and valuable suggestions concerning this manuscript and the Canadian Institutes for Health Research and the Canadian Institute for Advanced Research for support. This essay is dedicated to the memory of Meg Muir.

REFERENCES

Acinas SG, Klepac-Ceraj V, Hunt DE, Pharino C, Ceraj I, Distel DL, Polz MF. 1999. Fine-scale phylogenetic architecture of a complex bacterial community. *Nature* **430:** 551–554.

Bapteste E, Boucher Y. 2008. Lateral gene transfer challenges principles of microbial systematics. *Trends Microbiol* **16:** 200–207.

Bapteste E, Boucher Y. 2009. Epistemological impacts of horizontal gene transfer on classification in microbiology. *Methods Mol Biol* **532:** 55–72.

Beatty J. 1992. Speaking of species: Darwin's strategy. In *The units of evolution* (ed. M Ershefsky), pp. 227–246. MIT Press, Cambridge, MA.

Cantino P, de Queiroz K. 2006. PhyloCode: A phylogenetic code of biological nomenclature version 3a (http//ohiou.edu/PhyloCode).

Cohan FM. 2002. What are bacterial species? *Annu Rev Microbiol* **56:** 457–487.

Cox CJ, Foster PG, Hirt RP, Harris SR, Embley TM. 2008. The archaebacterial origin of eukaryotes. *Proc Natl Acad Sci* **105:** 20356–20361.

Dagan T, Martin W. 2006. The tree of one percent. *Genome Biol* **7:** 118.

Darwin C. 1859. *On the origin of species by means of natural selection*, 1st ed. Murray, London.

De Queiroz K. 2007. Species concepts and species delimitation. *Syst Biol* **56:** 879–886.

Diamond J, Bishop KD. 1999. Ethno-ornithology of the Ketengban People, Indonesian New Guinea. In *Folkbiology* (ed. DL Medin and S Atran), pp. 17–45. Bradford Books, MIT Press, Cambridge, MA.

Doolittle WF, Bapteste E. 2007. Pattern pluralism and the Tree of Life hypothesis. *Proc Natl Acad Sci* **104:** 2043–2049.

Eldredge N. 2005. *Darwin: Discovering the tree of life.* Norton, New York.

Embley TM, Martin W. 2006. Eukaryotic evolution, changes and challenges. *Nature* **440:** 623–630.

Ereshefsky M. 1992. Eliminative pluralism. *Philos Sci* **59:** 671–690.

Ereshefsky M. 1998. Species pluralism and antirealism. *Philos Sci* **65:** 103–120.

Ereshefsky M. 2009. Darwin's solution to the species problem. *Synthese* (in press).

Esser C, Ahmadinejad N, Wiegand C, Rotte C, Sebastiani F, Gelius-Dietrich G, Henze K, Kretschmann E, Richly E, Leister D, et al. 2004. A genome phylogeny for mitochondria among α-proteobacteria and a predominantly eubacterial ancestry of yeast nuclear genes. *Mol Biol Evol* **21:** 1643–1660.

Esser C, Martin W, Dagan T. 2007. The origin of mitochondria in light of a fluid prokaryotic chromosome. *Biol Lett* **3:** 180–184.

Fraser C, Alm EJ, Polz MF, Spratt BG, Hanage WP. 2009. The bacterial species challenge: Making sense of genetic and ecological diversity. *Science* **323:** 741–746.

Guerrero R. 2001. Bergey's manuals and the classification of prokaryotes. *Int Microbiol* **4:** 103–109.

Hanage WP, Fraser C, Spratt BG. 2005. Fuzzy species among recombinogenic bacteria. *BMC Biol* **3:** 6.

Hanage WP, Fraser C, Spratt BG. 2006. Sequences, sequence clusters and bacterial species. *Philos Trans R Soc Lond B Biol Sci* **361:** 1917–1927.

Hull D. 1965. The effect of essentialism on taxonomy: Two thousand years of stasis. *Br J Philos Sci* (Part I) **15:** 314–326, (Part II) **16:** 1–18.

Hunt DE, David LA, Gevers D, Preheim SP, Alm EJ, Polz MF. 2008. Resource partitioning and sympatric differentiation among closely related bacterioplankton. *Science* **320:** 1081–1085.

Koeppel A, Perry EB, Sikorski J, Krizanc D, Warner A, Ward DM, Rooney AP, Brambilla E, Connor N, Ratcliff RM, et al. 2008. Identifying the fundamental units of bacterial diversity: A paradigm shift to incorporate ecology into bacterial systematics. *Proc Natl Acad Sci* **105:** 2504–2509.

Konstantinidis KT, Ramette R, Tiedje JM. 2006. The bacterial species definition in the genomic era. *Philos Trans R Soc Lond B Biol Sci* **361:** 1929–1940.

Lapierre P, Gogarten JP. 2009. Estimating the size of the bacterial pan-genome. *Trends Genet* **25:** 107–110.

Martiny AC, Tai AP, Veneziano D, Primeau F, Chisholm SW. 2009. Taxonomic resolution, ecotypes and the biogeography of *Prochlorococcus*. *Environ Microbiol* **11:** 823–832.

Maynard Smith J, Ed. 1982. *Evolution now: A century after Darwin*. Freeman, San Francisco.

Mayr E. 1975. *Evolution and the diversity of life.* Harvard University Press, Cambridge, MA.

Mayr E. 1996. What is a species and what is not? *Philos Sci* **63:** 262–277.

Mindell DP. 1992. Phylogenetic consequences of symbioses: Eukarya and eubacteria are not monophyletic taxa. *BioSystems* **27:** 58–62.

Mushegian A. 2008. Gene content of LUCA, the last universal common ancestor. *Front Biosci* **13:** 4657–4666.

O'Hara RJ. 1998. Population thinking and tree thinking in systematics. *Zool Scr* **26:** 323–329.

Pace NR. 2006. Time for a change. *Nature* **441:** 289.

Panchen AL. 1992. *Classification, evolution and the nature of biology.* Cambridge University Press, Cambridge.

Pigliucci M. 2003. Species as family resemblance concepts: The (dis-)solution of the species problem? *Bioessays* **25:** 596–602.

Rescher N. 2002. Process philosophy. In *The Stanford encyclopedia of philosophy* (ed. EN Zalta). The Metaphysics Research Lab, Stanford University, Stanford, CA.

Riley MA, Lizotte-Waniewski M. 2009. Population genomics and the bacterial species concept. *Methods Mol Biol* **532:** 367–377.

Sapp J. 2006. Two faces of the prokaryote concept. *Int Microbiol* **9:** 163–172.

Sober E. 1980. Evolution, population thinking, and essentialism. *Philos Sci* **47:** 350–383.

Stamos DN. 2005. Pre-Darwinian taxonomy and essentialism: A reply to Mary Winsor. *Biol Philos* **20:** 79–96.

Stanier RY, van Niel CB. 1962. The concept of a bacterium. *Arch Microbiol* **42:** 17–35.

Touchon M, Hoede C, Tenaillon O, Barbe V, Baeriswyl S, Bidet P, Bingen E, Bonacorsi S, Bouchier C, Bouvet O, et al. 2009. Organised genome dynamics in the *Escherichia coli* species results in highly diverse adaptive paths. *PLoS Genet* **5:** e1000344.

Ward DM, Cohan FM, Bhaya D, Heidelberg JF, Kuhl M, Grossman A. 2008. Genomics, environmental genomics and the issue of microbial species. *Heredity* **100:** 207–219.

Whitman WB. 2009. The modern concept of the procaryote. *J Bacteriol* **191:** 2000–2005.

Woese CR, Goldenfeld N. 2009. How the microbial world saved evolution from the scylla of molecular biology and the charybdis of the modern synthesis. *Microbiol Mol Biol Rev* **73:** 14–21.

Zhaxybayeva O, Gogarten JP. 2004. Cladogenesis, coalescence and the evolution of the three domains of life. *Trends Genet* **20:** 182–187.

Zhaxybayeva O, Swithers KS, Lapierre P, Fournier GP, Bickhart DM, DeBoy RT, Nelson KE, Nesbø CL, Doolittle WF, Gogarten JP, Noll KM. 2009. On the chimeric nature, thermophilic origin, and phylogenetic placement of the Thermotogales. *Proc Natl Acad Sci* **106:** 5865–5870.

Zuckerkandl E, Pauling L. 1965. Evolutionary divergence and convergence in proteins. In *Evolving genes and proteins* (ed. V Bryson and HJ Vogel), pp. 97–166. Academic, New York.

The Phylogenetic Forest and the Quest for the Elusive Tree of Life

E.V. KOONIN, Y.I. WOLF, AND P. PUIGBÒ

National Center for Biotechnology Information, National Library of Medicine, National Institutes of Health, Bethesda, Maryland 20894

Correspondence: koonin@ncbi.nlm.nih.gov

Extensive horizontal gene transfer (HGT) among prokaryotes seems to undermine the tree of life (TOL) concept. However, the possibility remains that the TOL can be salvaged as a statistical central trend in the phylogenetic "forest of life" (FOL). A comprehensive comparative analysis of 6901 phylogenetic trees for prokaryotic genes revealed a signal of vertical inheritance that was particularly strong among the 102 nearly universal trees (NUTs), despite the high topological inconsistency among the trees in the FOL, most likely, caused by HGT. The topologies of the NUTs are similar to the topologies of numerous other trees in the FOL; although the NUTs cannot represent the FOL completely, they reflect a significant central trend. Thus, the original TOL concept becomes obsolete but the idea of a "weak" TOL as the dominant trend in the FOL merits further investigation. The totality of gene trees comprising the FOL appears to be a natural representation of the history of life given the inherent tree-like character of the replication process.

THE TREE OF LIFE CONCEPT IN THE AGE OF GENOMICS

The concept of the TOL introduced by Darwin, captured in the famous single illustration of the *On the Origin of Species* (Darwin 1859), and used by Haeckel as the grand scheme of the history of the actual life-forms is the cornerstone of evolutionary biology and, arguably, of biology in general. For nearly 140 years after the publication of the *Origin*, phylogenetic trees, which were initially constructed using phenotypic characters but, following the seminal work of Zuckerkandl and Pauling (1962, 1965), increasingly relied on molecular sequence comparison, were viewed as a (more or less accurate) depiction of the evolution of the respective organisms. In other words, a tree built for a specific character or a gene was routinely equated with a "species tree." The use of rRNA as the molecule of choice for phylogenetic reconstruction culminated in the now textbook three-domain TOL of Woese and coworkers (Pace et al. 1986; Woese 1987) and was the brilliant culmination of the heroic period of phylogenetics that brought hopes that the detailed, definitive topology of the TOL could be within reach.

Trouble for the TOL concept, however, started even before the advent of genomics as it became clear that common and essential genes of prokaryotes experienced multiple HGTs. So the idea of a "net of life" as a potential replacement for the TOL was proposed (Hilario and Gogarten 1993; Gogarten 1995). Generally, however, in the pregenomic era, HGT was viewed as a minor process of evolution, crucial in some areas such as the spread of antibiotic resistance but secondary for the general scheme of evolution. In the late 1990s, comparative genomics of prokaryotes dramatically changed this picture by showing that the patterns of gene distribution across genomes were typically patchy, whereas the topologies of gene-specific phylogenetic trees were often incongruent. These findings indicated that HGT was extremely common among prokaryotes (bacteria and archaea) (Doolittle 1999a,b, 2000; Martin 1999; Koonin et al. 2001; Gogarten et al. 2002; Koonin and Aravind 2002; Lawrence and Hendrickson 2003; Gogarten and Townsend 2005; Dagan et al. 2008) and could have been important also in the evolution of eukaryotes, especially, as a consequence of endosymbiotic events (Doolittle 1998; Martin and Herrmann 1998; Doolittle et al. 2003; Embley and Martin 2006). Thus, a perfect TOL turned out to be a chimera because extensive HGT prevents any single gene tree from being an accurate representation of the evolution of entire genomes. The realization that HGT among prokaryotes is the dominant rather than an exceptional mode of evolution led to the idea of "uprooting" the TOL, a development that is often interpreted as a paradigm shift in evolutionary biology (Pennisi 1999; Doolittle 2000; O'Malley and Boucher 2005).

Of course, the incongruence of gene phylogenies caused by HGT or other processes cannot alter the fact that all cellular life-forms are linked by a tree of cell divisions (*Omnis cellula e cellula*, according to the famous motto of Rudolf Virchow [1858]) that goes back to the earliest stages of evolution, with the exception of endosymbiotic events that were key to the evolution of eukaryotes but not prokaryotes (Lane and Archibald 2008). The problems with the TOL concept in the era of comparative genomics concern the TOL as it can be derived by the phylogenetic (phylogenomic) analysis of genes and genomes. Thus, the claim that HGT uproots the TOL more accurately means that extensive HGT has the potential to result in complete decoupling of molecular phylogenies from the actual tree of cells. Phylogenetic trees of genes also reflect the evolution of the respective molecular functions, so the phylogenomic analysis has straightforward biological connotations. Thus, the phylogenomic approach and not the abstract tree of cells reveals the actual history of the genetic content of organisms. Accordingly, we examine here the current status of the "phylogenomic TOL."

The views of evolutionary biologists on the status of the TOL in the face of the ubiquitous HGT (O'Malley and Boucher 2005) span the entire range from (1) continued denial of the major role of HGT in the evolution of life (Kurland 2000; Kurland et al. 2003) to (2) "moderate" revision of the TOL concept (Wolf et al. 2002; Zhaxybayeva et al. 2004; Beiko et al. 2005; Ge et al. 2005; Kunin et al. 2005; Galtier and Daubin 2008) to (3) radical uprooting whereby the representation of the evolution of organisms (or genomes) as a TOL is declared meaningless (Bapteste et al. 2005; Doolittle and Bapteste 2007; Koonin 2007). The moderate approach maintains that all of the differences among individual gene trees notwithstanding, the TOL concept remains valid as a central trend that, at least in principle, can be revealed through a comprehensive comparison of gene tree topologies. The radical view counters that the massive HGT obliterates the very distinction between the vertical and horizontal routes of genetic information transmission, so the TOL concept should be abandoned in favor of a (broadly defined) network representation of evolution (Dagan et al. 2008). The TOL conundrum is fittingly emphasized in the recent debate on the "highly resolved tree of life" that was generated from a concatenation of alignments of 31 highly conserved proteins (Ciccarelli et al. 2006), only to be dismissed as a "tree of one percent" (of the genes in any given genome) that does not actually reflect the history of genomes (Dagan and Martin 2006).

We discuss here our recent effort of a comprehensive comparison of the phylogenetic trees for individual genes of prokaryotes. We refer to this set of ~7000 trees as the FOL. We show that there is indeed a central trend in the FOL but the deep splits in this topology cannot be unambiguously resolved, probably, owing to both extensive HGT and methodological problems of tree reconstruction. Nevertheless, computer simulations show that the observed pattern of evolution of archaea and bacteria is better compatible with a compressed cladogenesis (CC) model (Rokas et al. 2005; Rokas and Carroll 2006) than with a "big bang" model that includes non-tree-like phases of evolution (Koonin 2007). These findings are, in principle, compatible with the "TOL as a central trend" concept. However, we argue on more general grounds that the TOL is not a fundamentally necessary concept, and the entire FOL with its different trends could be the most adequate representation of the history of life. We now have the adequate computational methods and tools to identify and analyze these trends.

THE FOREST OF LIFE AND THE NEARLY UNIVERSAL TREES

We analyzed 6901 maximum likelihood phylogenetic trees that were built using clusters of orthologous gene (COG) databases that included a selected representative set of 100 prokaryotes (41 archaea and 59 bacteria) (Tatusov et al. 1997, 2003; Jensen et al. 2008). The majority of these trees include only a small number of species (<20); only 2040 trees included more than 20 species, and only a small set of NUTs included >90% of the analyzed prokaryotes. We sought to identify patterns in the FOL and, in particular, to determine whether there exists a central trend among the trees and whether the topologies of the NUTs reflect such a trend should it exist. To this end, we analyzed the complete, all-against-all matrix of the topological distances between the trees (Puigbò et al. 2007, 2009). This matrix was represented as a network of trees and was subject to classical multidimensional scaling (CMDS) analysis to detect potentially existing distinct clusters of trees. In addition, we introduced a new measure, the inconsistency score (IS), that determines how representative the topology of the given tree is of the entire FOL (IS is the fraction of the times the splits from a given tree are found in all trees of the FOL). Using the IS, we objectively examine trends in the FOL, without relying on the topology of a preselected "species tree" such as a supertree used in the most comprehensive previous study of HGT (Beiko et al. 2005) or a tree of concatenated highly conserved proteins or rRNAs (Mirkin et al. 2003; Ciccarelli et al. 2006; Dagan et al. 2008).

We began the systematic exploration of the FOL from the grove of 102 NUTs, most, although not all, of which, as expected, correspond to genes encoding components of information transmission systems, particularly, translation. The topologies of the NUTs were, in general, highly coherent. Indeed, the inconsistency among the NUTs ranged from 1.4% to 4.3%, whereas the mean value of inconsistency for an equal-sized set (102) of randomly generated trees with the same number of species was ~80% (Fig. 1). In 56% of the NUTs, archaeal and bacterial branches were perfectly separated, whereas the remaining 44% showed indications of HGT between

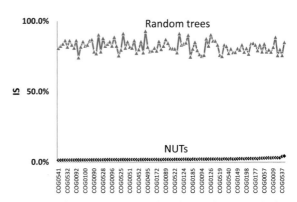

Figure 1. The 102 NUTs have largely consistent topologies. (Black) Inconsistency scores of the 102 NUTs, (gray) IS values for the random trees produced by shuffling the branches in each of the NUTs are represented in black and ordered by increasing IS values. The IS of each NUT was calculated using as the reference set all 102 NUTs, and the IS of each random tree was similarly calculated using as the reference set of all 102 random trees. (Modified from Puigbò et al. 2009.)

archaea and bacteria (13% from archaea to bacteria, 23% from bacteria to archaea, and 8% in both directions). In the rest of the NUTs, interdomain gene transfer was not detected, but there were many probable HGT events within one or both domains (data not shown). We further analyzed the relationships among the 102 NUTs by embedding them into a 30-dimensional tree space using the CMDS procedure and the gap statistics analysis, which revealed a lack of significant clustering among the NUTs in the tree space: All of the NUTs seem to belong to a single unstructured cloud of points scattered around a single centroid (Fig. 2a). This organization of the tree space is best compatible with individual trees randomly deviating from a single dominant topology ("the TOL"), apparently as a result of HGT (but also, possibly, due to random errors of the tree construction procedure).

The overall conclusion on the evolutionary trends among the NUTs is that although the topologies of the NUTs were, for the most part, not identical, so that the NUTs could be separated by their degree of inconsistency (a proxy for the amount of HGT), the overall high consistency level indicated that the NUTs are scattered in the close vicinity of a consensus tree, with the HGT events distributed approximately randomly. These findings are compatible with previous reports on the apparently random distribution of HGT events in the history of highly conserved genes, in particular, those encoding proteins involved in translation (Brochier et al. 2002; Ge et al. 2005).

We further analyzed the structure of the FOL by embedding the 3789 COG trees (the subset of the FOL that included most trees with a large number of organisms) into a 669-dimensional space using the CMDS procedure and found that the optimal partitioning of this set yielded seven clusters of trees; notably, all of the NUTs formed a compact subset of cluster 6 (Fig. 2b). The trees that belonged to different clusters showed considerable differences in the distribution of the trees by the number of species, the partitioning of archaea-only and bacteria-only trees, and the functional classification of the respective COGs (Puigbò et al. 2009). The results of the CMDS clustering support the existence of several distinct "attractors" in the FOL, although trivial separation of the trees by size could substantially contribute to this finding. The key observation is that all of the NUTs occupy a compact and contiguous region of the tree space and, unlike the complete set of the trees, are not partitioned into distinct clusters (compare Fig. 2a,b).

As could be expected, the trees in the FOL show a strong signal of numerous HGT events including interdomain gene transfers. Among the 1473 trees that include at least five archaeal species and at least five bacterial species, perfect separation of archaea and bacteria was seen only in 13%. This is the low bound for the fraction of trees that are free of interdomain HGT because, even for trees with a perfect separation of archaea and bacteria, HGT cannot be ruled out, for instance, in cases when a small compact archaeal branch is embedded within a bacterial lineage (or vice versa).

We constructed a network of all 6901 trees in the FOL and examined the position and the connectivity of the 102 NUTs in this network. At the 50% similarity cutoff and a p value <0.05, the 102 NUTs were connected to 2615 trees (38% of the FOL) (Fig. 3), and the mean similarity of the trees in the FOL to the NUTs was ~50%, with similar distributions of strongly, moderately, and weakly similar trees seen for most of the NUTs (Puigbò et al. 2009). In a sharp contrast, using the same similarity cutoff, 102 randomized NUTs were connected to only 33 trees (~0.5% of the trees in the FOL) and the mean similarity was ~28% to the trees in the FOL. These findings reveal the high and nonrandom topological similarity between the NUTs and a large part of the FOL and show that this similarity is not an artifact of the large number of species in the NUTs.

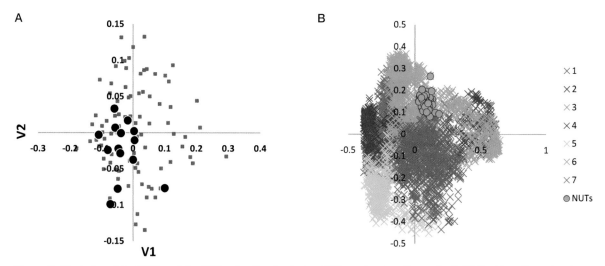

Figure 2. Clustering of the NUTs and the FOL using the classical multidimensional scaling method. (*A*) Best two-dimensional projection of the clustering of the 102 NUTs in a 30-dimensional space. (*B*) Best two-dimensional projection of the clustering of the 3789 COG trees in a 669-dimensional space. The seven clusters are color coded and the NUTs are shown by circles. (Modified from Puigbò et al. 2009.)

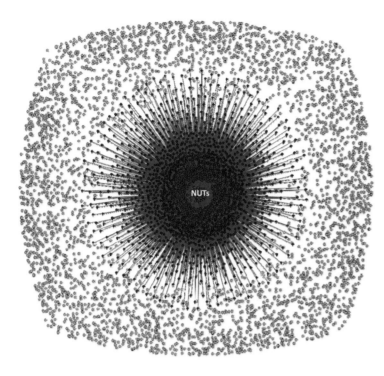

Figure 3. Network of the 6901 trees comprising the FOL. (Red circles) 102 NUTs, (green circles) remainder of the trees. The NUTs are connected to trees with similar topologies (>80% blue, >90% violet, 100% red). (Modified from Puigbò et al. 2009.)

DEPENDENCE OF TREE INCONSISTENCY ON PHYLOGENETIC DEPTH: BIG BANG OR COMPRESSED CLADOGENESIS?

It is well known from many phylogenetic studies and was supported by the examination of a supernetwork of the NUTs (Puigbò et al. 2009) that deep internal nodes in phylogenetic trees tend to be poorly resolved compared to external nodes. Whether there actually is a discernible phylogenetic signal in the deepest nodes of the trees bears on the question of whether there is a central trend in the FOL that potentially could be approximated by the NUTs.

To explore the dependence of the inconsistency between trees on phylogenetic depth quantitatively, we used an ultrametric tree that was produced from the supertree of the 102 NUTs and in which the phylogenetic distances were scaled from 0 to 1 (Puigbò et al. 2009). We found that the inconsistency of the FOL sharply increased, in a phase-transition-like fashion, between the depths of 0.7 and 0.8 (Fig. 4), suggesting that the evolutionary processes which were responsible for the formation of this part of the FOL could be qualitatively distinct from affected lesser phylogenetic depths. We considered two models of early evolution, at the level of archaeal and bacteria phyla: (1) Compressed cladogenesis (CC), under which there is a tree structure even at the deepest levels but the internal branches are extremely short (Rokas and Carroll 2006), and (2) biological big bang (BBB) model, where the early phase of evolution involved horizontal gene exchange so intensive that there is no signal of vertical inheritance in principle (Koonin 2007).

The evolution of the FOL was simulated under each of the two models. We attempted to fit the observed IS-depth dependence (Fig. 4) with the respective curves obtained by simulating the BBB at different phylogenetic depths by randomly shuffling the tree branches at the given depth and modeling the subsequent evolution as a tree-like process with different rates of HGT. The clear-cut result is that only by simulating the BBB at the depth of 0.8, i.e., before the divergence of the major bacterial and archaeal phyla, could a good fit with the empirical data be reached (Fig. 5). In contrast, simulation of the BBB at the critical depth of 0.7 or above, which erases the phylogenetic signal below the phylum level, did not yield a satisfactory fit (Fig. 5) (Puigbò et al. 2009). Thus, the CC model appears to be a more appropriate representation of the early phases of evolution of archaea and bacteria than the BBB model. In other words, the signal of apparent vertical inheritance (a central trend in the FOL) is detectable even for the earliest stages of evolution of each prokaryotic domain, although given the high level of inconsistency, the determination of the correct tree topology of the deepest branches in the tree is problematic at best. This analysis does not rule out a BBB as the generative mechanism underlying the divergence of archaea and bacteria, but this model cannot be tested using the approach described above because of the absence of an outgroup.

THE TRENDS IN THE FOREST OF LIFE

Recent developments in prokaryotic genomics reveal the ubiquity of HGT and overthrow the "strong" TOL concept under which all (or the substantial majority) of the genes would tell a consistent story of genome evolution (the species tree, or the TOL) if analyzed with appropriate methods ("uprooting the tree of life") (Doolittle

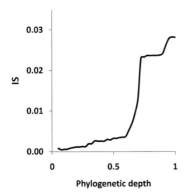

Figure 4. Inconsistency versus phylogenetic depth plot for the 6901 trees in the FOL. The distances (on a 0–1 scale) are from the ultrametric tree that was produced from the supertree of the 102 NUTs. (Modified from Puigbò et al. 2009.)

1999a, 2000; Pennisi 1999; Gogarten et al. 2002; Wolf et al. 2002; Gogarten and Townsend 2005; Doolittle and Bapteste 2007; Koonin 2009a). Is there any hope to salvage the TOL as a statistical central trend (Wolf et al. 2002)? The results of a comprehensive comparative analysis of phylogenetic trees for prokaryotic genes described here suggest that such a trend does exist.

The results of the FOL analysis are twofold. On the one hand, we observed high levels of inconsistency among the trees in the FOL, owing mostly to extensive HGT, as demonstrated more directly by the observations of numerous likely transfers of genes between archaea and bacteria. On the other hand, we also detected a distinct signal of a consensus topology that was particularly strong among the NUTs. Although the NUTs showed a substantial amount of apparent HGT, the transfer events seemed to be distributed randomly and did not obscure the apparent vertical signal. Moreover, the topology of the NUTs was quite similar to those of numerous other trees in the FOL, so although the NUTs certainly cannot represent the FOL completely, this set of largely congruent, nearly universal trees is a reasonable candidate for representing a central trend. However, the opposite side of the coin is that the consistency between the trees in the FOL is high at the external branches of the trees and abruptly drops, almost to the level of random trees, at greater phylogenetic depths that correspond to the radiation of archaeal and bacterial phyla. This observation casts doubt on the reality of a central trend in the FOL and suggests the possibility that the early phases of evolution might have been non-tree-like (a BBB; Koonin 2007). We addressed this problem directly by simulating evolution under the compressed cladogenesis model (Rokas et al. 2005; Rokas and Carroll 2006) and under the BBB model and found that the CC scenario better fits the observed dependence between tree inconsistency and phylogenetic depth. Thus, a consistent phylogenetic signal seems to be discernible throughout the evolution of archaea and bacteria, although, under the CC model, the prospect of unequivocally resolving the relationships between the major archaeal and bacterial clades is bleak.

The detected central trend in the FOL is most likely to represent vertical inheritance permeating the entire history of archaea and bacteria. A contribution from "highways" of HGT (i.e., preferential HGT between certain groups of archaea and bacteria, in particular, those closely related) that could mimic vertical evolution cannot be ruled out (Gogarten et al. 2002). However, the lack of significant clustering within the group of NUTs and the comparable high levels of similarity between the NUTs and different clusters of trees in the FOL suggest that the trend, even if relatively weak, is primarily vertical.

In the following sections, we take a more general, conceptual standpoint to discuss the status of the TOL in light of these findings and additional considerations.

A TREE IS AN ISOMORPHOUS REPRESENTATION OF REPLICATION HISTORY

Replication of nucleic acids with an error rate below the mutational meltdown threshold is both a necessary condition and the direct cause of evolution by random drift and natural selection (Eigen 1971; Koonin and Wolf 2009). Crucially, replication and the ensuing evolution are inherently tree-like processes: A replicating molecule gives rise to two (semiconservative replication of double-stranded DNA that occurs in all cellular organisms and many viruses) or multiple (conservative replication of viruses with single-stranded DNA or single-stranded RNA genomes) copies with errors, resulting in a tree-like process of divergence (Fig. 6). In graph-theoretical terms, such a process can be

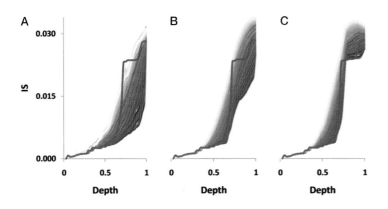

Figure 5. Simulations of a BBB at different phylogenetic depths and with different numbers of HGT events. Each panel is a plot of the mean tree inconsistency versus phylogenetic depth (as in Fig. 4). The empirical dependence is shown by the thick blue line, and the results of simulations with 1 to 200 HGT events are shown by thin lines along a color gradient. (*A*) BBB simulated at depth 0.6, (*B*) BBB simulated at depth 0.7, (*C*) BBB simulated at depth 0.8. (Modified from Puigbò et al. 2009.)

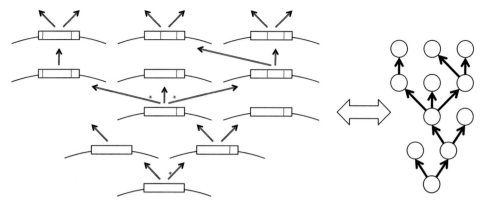

Figure 6. A tree (arborescence) is an isomorphous representation of the error-prone replication process. An idealized scheme of the replication history of a fundamental unit of evolution (FUE) includes both bifurcations and a multifurcation (shown by asterisks). Fixed mutations are shown by red lines. (Reprinted from Koonin and Wolf 2009.)

isomorphously represented with a directed acyclic graph known as arborescence, which is a generalized tree where multifurcations are allowed and all edges are directed away from the root (Fig. 6) (Evans and Minieka 1992). As a result of occasional extinction of one or both progeny molecules, some of the vertices in the resulting graph emit no edges, but this does not violate the definition of an arborescence (Fig. 6) (hereafter, for the sake of simplicity, we speak of trees rather than of arborescences).

A major complication to the tree-like character of evolution is recombination that, if common, would turn the tree-like representation of the history of a replicating lineage into a network. Is there a fundamental "atomic" level of genetic organization at which recombination is negligible? In the case of homologous recombination that is extensive during coreplication of closely related sequences, in particular, in eukaryotes that engage in regular sex, and in "quasi-sexual" prokaryotes (Feil et al. 2001; Spratt et al. 2001; Turner and Feil 2007; Doolittle and Zhaxybayeva 2009), the atomic unit, effectively, is a single base pair which of course is not a level at which any analysis can be conducted. In contrast, homologous recombination between distantly related sequences is impossible, so HGT between diverse prokaryotes involves only nonhomologous (illegitimate) recombination complemented by more specific routes such as dissemination via bacteriophages and plasmids. Unlike the case of homologous recombination, there is a strong preference for evolutionary fixation of nonhomologous recombination events outside genes because preservation of the integrity of a gene after nonhomologous recombination is unlikely. An important exception is fusion and shuffling of domains in multidomain proteins (Basu et al. 2009). Consequently, the evolutionary history of a gene or domain is reticulate on the microscale owing to homologous recombination but is essentially tree-like on the macroscale (Fig. 7).

It was argued that a tree can well describe relationships that have nothing to do with common descent, so "tree thinking" was deemed not to be a priori relevant in biology (Doolittle and Bapteste 2007). Although technically valid, this argument seems to miss the crucial point that a tree is a necessary formal consequence of the descent history of replicating nucleic acids and the ensuing evolution. Therefore, trees cannot be banished from evolutionary biology for the simple reason that they are intrinsic to the evolutionary process. Then, the main pertinent question becomes What are the fundamental units whose evolution should be represented by trees? In the practice of evolutionary biology, trees are most often built for individual genes or for sets of genes that are believed to evolve coherently. However, it is typically stated or implied that the ultimate goal is a species (organismal) tree. In our opinion, the lack of clarity about the basic unit to which tree analysis applies is the source of the entire controversy around the TOL.

FUNDAMENTAL UNITS OF (TREE-LIKE) EVOLUTION

Conceptually, the fundamental unit of evolution (FUE) can be most appropriately defined as the smallest portion of genetic material with a distinct evolutionary identity, i.e., one that replicates independently of other such units. Two distinct classes of FUEs can be defined:

1. Bona fide selfish elements such as viruses, viroids, transposons, and plasmids. All of these elements encode some of the information required for their replication and are united through their ability to promote their evolutionary success by exploiting resources of other organisms (Koonin et al. 2006).

2. Quasi-independent elements that do not encode devices for their own replication but possess distinct selective value and, in that capacity, can be transferred between ensembles of FUEs (genomes) and promote their own replication along with the rest of the genome. Essentially, any functional gene or even a portion of a gene encoding a distinct protein domain with an independent functional role fits this definition.

The concept of FUEs is, in part, derived from the "selfish gene" idea of Dawkins (1976) and the selfish operon hypothesis of Lawrence and Roth (1996; Lawrence 1999). These concepts seem to generate some confusion

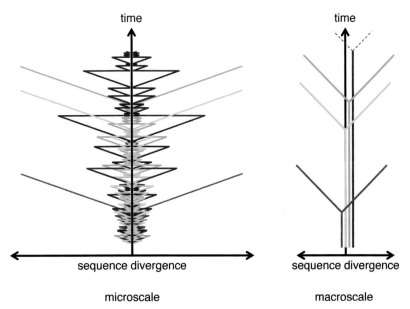

Figure 7. Evolution of an FUE is reticulate on the microscale but tree-like on the macroscale. The scheme schematically shows the evolution of four FUEs designated by different colors. The divergence history of each FUE was simulated under the model of random homologous recombination, with the probability of recombination exponentially decreasing with sequence divergence. At each simulation step, the two daughter FUEs diverge by a constant amount (clock-like divergence) and either undergo homologous recombination (which brings the difference between the two back to zero) or not, preserving the existing state of divergence. After a number of short periods of divergence and recombination, the FUEs stochastically diverge far enough for recombination to become extremely unlikely, after which point they continue diverging without recombination. At a macroscale, this process looks like a simple bifurcation in the tree graph. (Reprinted from Koonin and Wolf 2009.)

by assigning "selfishness" to genetic elements that do not actually contribute to their own replication at the mechanistic level. It seems that the partitioning of FUEs into two distinct classes eliminates this tension, with the understanding that some of the FUEs of the first class (such as large viruses and megaplasmids) could contain multiple FUEs of the second class.

Given the extensive HGT in the prokaryotic world, any gene or a portion of a gene encoding a distinct domain possesses a degree of independence and can be fixed in the recipient population even if the conferred advantage is relatively small, or even neutral (Novozhilov et al. 2005). Therefore, the prokaryotic genetic universe appears to be a consortium of FUEs with varying degrees of independence, some of which form ensembles that evolve as a physical and functional unity during extended time intervals and are more commonly known as genomes of viruses, plasmids, and cellular life forms (Koonin and Wolf 2008).

Additional motivation for the FUE concept comes from theoretical research and simple logical considerations on precellular evolution. It appears inconceivable that the first replicating elements were comparable in size and complexity to those of modern prokaryotic genomes. Evolution of life must have started with ensembles of relatively small FUEs, some of which would provide the means for the replication of others that in turn would provide other benefits, for instance, precursor synthesis, resulting in symbiotic relationships; fully selfish elements would necessarily parasitize on such ensembles. Physical joining of FUEs would be beneficial in many cases, provided sufficient replication fidelity. Qualitative and quantitative models of this early, collective phase in the evolution of life were developed (Szathmary and Demeter 1987; Zintzaras et al. 2002; Koonin and Martin 2005; Wolf and Koonin 2007; Takeuchi et al. 2008).

As we argued previously, this precellular stage of evolution could be considered virus-like in many respects, and the principal classes of extant viruses and other selfish elements, most likely, emerged already at that stage (Koonin et al. 2006; Koonin 2009b). There is an ongoing debate regarding the place of this collective stage of evolution in the history of life, and in particular, whether the last universal cellular ancestor (LUCA) was a typical cell, a cell with a fragmented genome, or a precellular ensemble of genetic elements (Koonin and Martin 2005; Forterre 2006; Glansdorff et al. 2008). Regardless of the ultimate outcome of this debate, in principle, there seems to be no reasonable doubt as to the reality of the collective stage. Furthermore, extensive mixing and matching of FUEs (which may or may not be called HGT, depending on whether this stage is envisaged as cellular) might be not only an inherent feature of this evolutionary stage, but also a prerequisite of a rapid increase in genetic and organizational complexity of life forms (Woese 1998, 2002; Koonin and Martin 2005; Koonin et al. 2006; Koonin 2009b).

Considering the virtual inevitability of an early collective stage of evolution and the extensive HGT that permeates modern prokaryotic world, the entire evolution of prokaryotes can be viewed as a dynamic process that plays out on the network of FUEs, although relatively stable genomes consisting of hundreds and thousands of FUEs, of course, are major components of that network (Koonin and Wolf 2008). Accordingly, FUEs should be construed as fundamental units of evolution, whereas all other levels of genetic organization are more properly viewed as derived.

CONCLUSIONS

Considering that FUEs appear to be fundamental units of evolution and that a tree is a necessary form of description of the evolution of any FUE, the adequate representation of evolution of life as a whole is the full compendium of FUE-specific trees, i.e., the FOL. This being the case, the notion of a species tree becomes if not obsolete at least secondary, being applicable to some phases of evolution of some groups of organisms but not in general. This conclusion does not imply that there is no order in the FOL and that signals of coherence among the trees are not to be sought. Such patterns are indeed discernible (Galtier and Daubin 2008), and as described above, the central trend, even if relatively weak, seems to correspond to the signal of vertical inheritance detectable in the nearly universal trees (Puigbò et al. 2009). Further study of different trends detectable in the FOL is expected to clarify the relationship between vertical and horizontal transmission of genetic material in the evolution of prokaryotes.

ACKNOWLEDGMENTS

The authors thank Liran Carmel (Hebrew University, Israel) for helpful discussions of multidimensional analysis and clustering. E.V.K. thanks Michael Gelfand, Andrei Mironov, and members of the Moscow Seminar on Bioinformatics for an inspiring discussion. The authors' research is supported by the Department of Health and Human Services intramural program (National Institutes of Health, National Library of Medicine).

REFERENCES

Bapteste E, Susko E, Leigh J, MacLeod D, Charlebois RL, Doolittle WF. 2005. Do orthologous gene phylogenies really support tree-thinking? *BMC Evol Biol* **5:** 33.
Basu MK, Poliakov E, Rogozin IB. 2009. Domain mobility in proteins: Functional and evolutionary implications. *Brief Bioinform* **10:** 205–216.
Beiko RG, Harlow TJ, Ragan MA. 2005. Highways of gene sharing in prokaryotes. *Proc Natl Acad Sci* **102:** 14332–14337.
Brochier C, Bapteste E, Moreira D, Philippe H. 2002. Eubacterial phylogeny based on translational apparatus proteins. *Trends Genet* **18:** 1–5.
Ciccarelli FD, Doerks T, von Mering C, Creevey CJ, Snel B, Bork P. 2006. Toward automatic reconstruction of a highly resolved tree of life. *Science* **311:** 1283–1287.
Dagan T, Martin W. 2006. The tree of one percent. *Genome Biol* **7:** 118.
Dagan T, Artzy-Randrup Y, Martin W. 2008. Modular networks and cumulative impact of lateral transfer in prokaryote genome evolution. *Proc Natl Acad Sci* **105:** 10039–10044.
Darwin C. 1859. *On the origin of species by means of natural selection,* 1st ed. Murray, London.
Dawkins R. 1976. *The selfish gene.* Oxford University Press, Oxford.
Doolittle WF. 1998. You are what you eat: A gene transfer ratchet could account for bacterial genes in eukaryotic nuclear genomes. *Trends Genet* **14:** 307–311.
Doolittle WF. 1999a. Phylogenetic classification and the universal tree. *Science* **284:** 2124–2129.
Doolittle WF. 1999b. Lateral genomics. *Trends Cell Biol* **9:** M5–M8.
Doolittle WF. 2000. Uprooting the tree of life. *Sci Am* **282:** 90–95.
Doolittle WF, Bapteste E. 2007. Pattern pluralism and the Tree of Life hypothesis. *Proc Natl Acad Sci* **104:** 2043–2049.
Doolittle WF, Zhaxybayeva O. 2009. On the origin of prokaryotic species. *Genome Res* **19:** 744–756.
Doolittle WF, Boucher Y, Nesbo CL, Douady CJ, Andersson JO, Roger AJ. 2003. How big is the iceberg of which organellar genes in nuclear genomes are but the tip? *Philos Trans R Soc Lond B Biol Sci* **358:** 39–57.
Eigen M. 1971. Selforganization of matter and the evolution of biological macromolecules. *Naturwissenschaften* **58:** 465–523.
Embley TM, Martin W. 2006. Eukaryotic evolution, changes and challenges. *Nature* **440:** 623–630.
Evans JR, Minieka E. 1992. *Optimization algorithms for networks and graphs.* Taylor and Francis, New York.
Feil EJ, Holmes EC, Bessen DE, Chan MS, Day NP, Enright MC, Goldstein R, Hood DW, Kalia A, Moore CE, et al. 2001. Recombination within natural populations of pathogenic bacteria: Short-term empirical estimates and long-term phylogenetic consequences. *Proc Natl Acad Sci* **98:** 182–187.
Forterre P. 2006. Three RNA cells for ribosomal lineages and three DNA viruses to replicate their genomes: A hypothesis for the origin of cellular domain. *Proc Natl Acad Sci* **103:** 3669–3674.
Galtier N, Daubin V. 2008. Dealing with incongruence in phylogenomic analyses. *Philos Trans R Soc Lond B Biol Sci* **363:** 4023–4029.
Ge F, Wang LS, Kim J. 2005. The cobweb of life revealed by genome-scale estimates of horizontal gene transfer. *PLoS Biol* **3:** e316.
Glansdorff N, Xu Y, Labedan B. 2008. The last universal common ancestor: Emergence, constitution and genetic legacy of an elusive forerunner. *Biol Direct* **3:** 29.
Gogarten JP. 1995. The early evolution of cellular life. *Trends Ecol Evol* **10:** 147–151.
Gogarten JP, Townsend JP. 2005. Horizontal gene transfer, genome innovation and evolution. *Nat Rev Microbiol* **3:** 679–687.
Gogarten JP, Doolittle WF, Lawrence JG. 2002. Prokaryotic evolution in light of gene transfer. *Mol Biol Evol* **19:** 2226–2238.
Hilario E, Gogarten JP. 1993. Horizontal transfer of ATPase genes: The tree of life becomes a net of life. *Biosystems* **31:** 111–119.
Jensen LJ, Julien P, Kuhn M, von Mering C, Muller J, Doerks T, Bork P. 2008. eggNOG: Automated construction and annotation of orthologous groups of genes. *Nucleic Acids Res* **36:** D250–D254.
Koonin EV. 2007. The biological big bang model for the major transitions in evolution. *Biol Direct* **2:** 21.
Koonin EV. 2009a. Darwinian evolution in the light of genomics. *Nucleic Acids Res* **37:** 1011–1034.
Koonin EV. 2009b. On the origin of cells and viruses: Primordial virus world scenario. *Ann NY Acad Sci* (in press).
Koonin EV, Aravind L. 2002. Origin and evolution of eukaryotic apoptosis: The bacterial connection. *Cell Death Differ* **9:** 394–404.
Koonin EV, Martin W. 2005. On the origin of genomes and cells within inorganic compartments. *Trends Genet* **21:** 647–654.
Koonin EV, Wolf YI. 2008. Genomics of bacteria and archaea: The emerging dynamic view of the prokaryotic world. *Nucleic Acids Res* **36:** 6688–6719.
Koonin EV, Wolf YI. 2009. The fundamental units, processes and patterns of evolution, and the Tree of Life conundrum. *Biol Direct* (in press).
Koonin EV, Makarova KS, Aravind L. 2001. Horizontal gene transfer in prokaryotes: Quantification and classification. *Annu Rev Microbiol* **55:** 709–742.

Koonin EV, Senkevich TG, Dolja VV. 2006. The ancient virus world and evolution of cells. *Biol Direct* **1:** 29.

Kunin V, Goldovsky L, Darzentas N, Ouzounis CA. 2005. The net of life: Reconstructing the microbial phylogenetic network. *Genome Res* **15:** 954–959.

Kurland CG. 2000. Something for everyone. Horizontal gene transfer in evolution. *EMBO Rep* **1:** 92–95.

Kurland CG, Canback B, Berg OG. 2003. Horizontal gene transfer: A critical view. *Proc Natl Acad Sci* **100:** 9658–9662.

Lane CE, Archibald JM. 2008. The eukaryotic tree of life: Endosymbiosis takes its TOL. *Trends Ecol Evol* **23:** 268–275.

Lawrence J. 1999. Selfish operons: The evolutionary impact of gene clustering in prokaryotes and eukaryotes. *Curr Opin Genet Dev* **9:** 642–648.

Lawrence JG, Hendrickson H. 2003. Lateral gene transfer: When will adolescence end? *Mol Microbiol* **50:** 739–749.

Lawrence JG, Roth JR. 1996. Selfish operons: Horizontal transfer may drive the evolution of gene clusters. *Genetics* **143:** 1843–1860.

Martin W. 1999. Mosaic bacterial chromosomes: A challenge en route to a tree of genomes. *Bioessays* **21:** 99–104.

Martin W, Herrmann RG. 1998. Gene transfer from organelles to the nucleus: How much, what happens, and why? *Plant Physiol* **118:** 9–17.

Mirkin BG, Fenner TI, Galperin MY, Koonin EV. 2003. Algorithms for computing parsimonious evolutionary scenarios for genome evolution, the last universal common ancestor and dominance of horizontal gene transfer in the evolution of prokaryotes. *BMC Evol Biol* **3:** 2.

Novozhilov AS, Karev GP, Koonin EV. 2005. Mathematical modeling of evolution of horizontally transferred genes. *Mol Biol Evol* **22:** 1721–1732.

O'Malley MA, Boucher Y. 2005. Paradigm change in evolutionary microbiology. *Stud Hist Philos Biol Biomed Sci* **36:** 183–208.

Pace NR, Olsen GJ, Woese CR. 1986. Ribosomal RNA phylogeny and the primary lines of evolutionary descent. *Cell* **45:** 325–326.

Pennisi E. 1999. Is it time to uproot the tree of life? *Science* **284:** 1305–1307.

Puigbò P, Garcia-Vallvé S, McInerney JO. 2007. TOPD/FMTS: A new software to compare phylogenetic trees. *Bioinformatics* **23:** 1556–1558.

Puigbò P, Wolf YI, Koonin EV. 2009. Search for a "Tree of Life" in the thicket of the phylogenetic forest. *J Biol* **8:** 59.

Rokas A, Carroll SB. 2006. Bushes in the tree of life. *PLoS Biol* **4:** e352.

Rokas A, Kruger D, Carroll SB. 2005. Animal evolution and the molecular signature of radiations compressed in time. *Science* **310:** 1933–1938.

Spratt BG, Hanage WP, and Feil EJ. 2001. The relative contributions of recombination and point mutation to the diversification of bacterial clones. *Curr Opin Microbiol* **4:** 602–606.

Szathmary E, Demeter L. 1987. Group selection of early replicators and the origin of life. *J Theor Biol* **128:** 463–486.

Takeuchi N, Salazar L, Poole AM, Hogeweg P. 2008. The evolution of strand preference in simulated RNA replicators with strand displacement: Implications for the origin of transcription. *Biol Direct* **3:** 33.

Tatusov RL, Koonin EV, Lipman DJ. 1997. A genomic perspective on protein families. *Science* **278:** 631–637.

Tatusov RL, Fedorova ND, Jackson JD, Jacobs AR, Kiryutin B, Koonin EV, Krylov DM, Mazumder R, Mekhedov SL, Nikolskaya AN, et al. 2003. The COG database: An updated version includes eukaryotes. *BMC Bioinformatics* **4:** 41.

Turner KM, Feil EJ. 2007. The secret life of the multilocus sequence type. *Int J Antimicrob Agents* **29:** 129–135.

Virchow RLK. 1858. *Die Cellularpathologie in ihrer Begründung auf physiologische und pathologische Gewebelehre*. Hirschwald, Berlin.

Woese CR. 1987. Bacterial evolution. *Microbiol Rev* **51:** 221–271.

Woese C. 1998. The universal ancestor. *Proc Natl Acad Sci* **95:** 6854–6859.

Woese CR. 2002. On the evolution of cells. *Proc Natl Acad Sci* **99:** 8742–8747.

Wolf YI, Koonin EV. 2007. On the origin of the translation system and the genetic code in the RNA world by means of natural selection, exaptation, and subfunctionalization. *Biol Direct* **2:** 14.

Wolf YI, Rogozin IB, Grishin NV, Koonin EV. 2002. Genome trees and the tree of life. *Trends Genet* **18:** 472–479.

Zhaxybayeva O, Lapierre P, Gogarten JP. 2004. Genome mosaicism and organismal lineages. *Trends Genet* **20:** 254–260.

Zintzaras E, Santos M, Szathmary E. 2002. "Living" under the challenge of information decay: The stochastic corrector model vs. hypercycles. *J Theor Biol* **217:** 167–181.

Zuckerkandl E, Pauling L. 1962. Molecular evolution. In *Horizons in biochemistry* (ed. M Kasha and B Pullman), pp. 189–225. Academic, New York.

Zuckerkandl E, Pauling L. 1965. Evolutionary divergence and convergence of proteins. In *Evolving gene and proteins* (ed. V Bryson and HJ Vogel), pp. 97–165. Academic, New York.

On the Origins of Species: Does Evolution Repeat Itself in Polyploid Populations of Independent Origin?

D.E. SOLTIS,[1,2] R.J.A. BUGGS,[1] W.B. BARBAZUK,[1,2] P.S. SCHNABLE,[3] AND P.S. SOLTIS[2,4]

[1]*Department of Biology, University of Florida, Gainesville, Florida 32611;* [2]*Genetics Institute, University of Florida, Gainesville, Florida 32610;* [3]*Center for Plant Genomics, Iowa State University, Ames, Iowa 50011;* [4]*Florida Museum of Natural History, University of Florida, Gainesville, Florida 32611*
Correspondence: dsoltis@botany.ufl.edu

Multiple origins of the same polyploid species pose the question: Does evolution repeat itself in these independently formed lineages? *Tragopogon* is a unique evolutionary model for the study of recent and recurrent allopolyploidy. The allotetraploids *T. mirus* (*T. dubius* × *T. porrifolius*) and *T. miscellus* (*T. dubius* × *T. pratensis*) formed repeatedly following the introduction of three diploids to the United States. Concerted evolution has consistently occurred in the same direction (resulting in loss of *T. dubius* rDNA copies). Both allotetraploids exhibit homeolog loss, with the same genes consistently showing loss, and homeologs of *T. dubius* preferentially lost in both allotetraploids. We have also documented repeated patterns of tissue-specific silencing in multiple populations of *T. miscellus*. Hence, some aspects of genome evolution may be "hardwired," although the general pattern of loss is stochastic within any given population. On the basis of the study of F_1 hybrids and synthetics, duplicate gene loss and silencing do not occur immediately following hybridization or polyploidization, but gradually and haphazardly. Genomic approaches permit analysis of hundreds of loci to assess the frequency of homeolog loss and changes in gene expression. This methodology is particularly promising for groups such as *Tragopogon* for which limited genetic and genomic resources are available.

In *On the origin of species*, Darwin (1859) presented a mechanism by which evolution could occur via natural selection. Our understanding of evolution and speciation has obviously improved dramatically in the past 150 years. Of the numerous new insights, perhaps one of the more surprising discoveries is the relatively recent finding that the same species can actually form multiple times. Specifically, in the case of polyploid organisms, the same polyploid species may form not once, but repeatedly in multiple locations where the diploid parents come into contact and hybridize. In fact, the use of molecular techniques has shown that most polyploid species have probably formed more than once—that multiple origins of the same polyploid species is the norm, not the exception, in polyploidy organisms (see, e.g., Soltis and Soltis 1993, 1999). Given the importance and prevalence of polyploidy in some plant groups, particularly ferns and angiosperms, the recurrent formation of the same polyploid species becomes a major evolutionary factor. Numerous examples of recurrent polyploidization have now been proposed for polyploid animals and plants (for review, see Soltis and Soltis 1993, 1999). But, the "repeatability" of polyploid speciation may be best seen on a broad geographic scale in the arctic, where diploid progenitor species come into contact over and over again on a circumpolar scale, hybridizing and subsequently generating the same polyploids again and again (Brochmann et al. 2004; Grundt et al. 2006).

Recurrent formation of the same polyploid species poses intriguing evolutionary questions, a major one being: Does evolution repeat itself in polyploid lineages of independent origin? On a grand scale, evolutionary biologists have pondered this question. Gould (1994), for example, suggested that if the tape of evolution of life on Earth could be replayed, it would play out differently each time—"history involves too much chaos," and too many chance events are involved for the evolutionary process to be repetitive. However, is this true on a finer scale? What about at the level of species and during shorter time frames? That is, are some genetic features of the polyploidization process "hardwired" so that the same genomic/genetic changes will recur in polyploid populations of independent formation? Recent research has suggested that at deep levels across broad clades of life, preservation of duplicated gene copies following genome duplication is far from random, with specific functional categories preferentially retained and reduplicated in subsequent polyploidizations (Seoighe and Gehring 2004; Chapman et al. 2006). Independent whole-genome duplications in the ancestors of *Arabidopsis*, *Oryza* (rice), *Saccharomyces* (yeast), and *Tetraodon* (pufferfish) appear to have been followed by convergent fates of many gene families (Paterson et al. 2006). Collectively, these observations indicate that on a broad scale, there *may* exist certain "principles" that govern the fates of gene and genome duplications. On the basis of these data, perhaps the tape of evolution would replay in the very same or similar way each time at the level of independently formed polyploid lines. Conversely, perhaps stochasticity has a major role, resulting in little repeatability or predictability across populations of independent formation. As one more alternative, perhaps the end result is some place between these two extremes.

With this brief introduction, it is apparent that polyploid plants of independent origin provide an unusual opportunity to address a fundamental question: Does evolution repeat itself? A particularly useful plant system for studying the early phases of polyploidization and addressing this fundamental question is provided by members of the genus *Tragopogon* (goatsbeard) (Asteraceae; sunflower family). As reviewed below, *Tragopogon* is a unique evolutionary model for the study of recent and recurrent polyploidy, providing a superb system for investigating the repeatability of the evolutionary process.

THE *TRAGOPOGON* SYSTEM: A BRIEF HISTORY

Although polyploidy has long been recognized as prevalent in plants (see, e.g., Müntzing 1936; Darlington 1937; Clausen et al. 1945; Stebbins 1947, 1950; Löve and Löve 1949; Grant 1981), genomic data have now revealed that it is an even more significant force than previously proposed (see, e.g., Blanc et al. 2000, 2003; Paterson et al. 2000; Vision et al. 2000; Simillion et al. 2002; Bowers et al. 2003; Blanc and Wolfe 2004; Schlueter et al. 2004; Cui et al. 2006). The question being asked is no longer "what proportion of angiosperms are polyploid?" but "how many episodes of polyploidy characterize any given lineage?"

Despite enormous progress in our understanding of many aspects of polyploidy (see, e.g., Wendel 2000; Tate et al. 2004; Wendel and Doyle 2004; Doyle et al. 2008), the early stages of polyploid evolution remain poorly understood, particularly in natural populations. Although polyploidy is ubiquitous in plants, only a few polyploid species are known to have arisen recently, i.e., within just the past 150–200 years: *Cardamine schulzii* (Urbanska et al. 1997), *Spartina anglica* (Huskins 1931; for review, see Ainouche et al. 2004), *Senecio cambrensis* (Rosser 1955) and *Senecio eboracensis* (Abbott and Lowe 2004), and two species of *Tragopogon*, *T. mirus* and *T. miscellus* (Ownbey 1950; for review, see Soltis et al. 2004). All of these new polyploids provide the opportunity to examine the early stages of polypoidization; all but *C. schulzii* have garnered considerable recent attention (see, e.g., Ainouche et al. 2004; Hegarty et al. 2005, 2006). Of these taxa, *Tragopogon* provides the best system for the study of *recent* and *recurring* polyploidy in natural populations.

Tragopogon consists of ~150 species native to Eurasia, most of which are diploid ($2n = 12$). Three of these diploids, *T. dubius*, *T. pratensis*, and *T. porrifolius*, were introduced into North America as it was settled by Europeans. *T. dubius* and *T. pratensis* were likely introduced accidentally, but *T. porrifolius* (salsify) has an edible root, was planted, and escaped. In the Palouse region of eastern Washington and adjacent Idaho, the three diploids came into close contact, which rarely happens in Europe where they are ecologically and in part geographically isolated. Hybridization occurred and two new allopolyploid species with $2n = 24$ ultimately formed: *T. mirus* (*T. porrifolius* × *T. dubius*) and *T. miscellus* (*T. dubius* × *T. pratensis*) (Fig. 1). The new allotetraploids have not formed in Europe, but they are native to the New

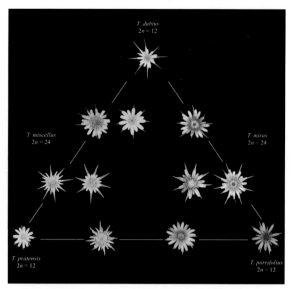

Figure 1. Summary of parentage of tetraploid *Tragopogon* species comparing what we have produced synthetically ("man") and what has occurred in nature ("wild"). The diploid parents (with $2n = 12$) are at the corners of the triangle; polyploids ($2n = 24$) are in between the corners. Synthetic polyploids are to the outside of the triangle; those polyploids forming naturally are to the inside of the triangle. In nature, *T. miscellus* has formed reciprocally, and *T. mirus* has formed only in one direction (with *T. porrifolius* as the maternal parent). However, we have made reciprocal synthetic lines of both and have also made reciprocal polyploids of *T. pratensis* × *T. dubius* ("*T. floridana*"); this polyploid has not formed in nature. Note that populations of *T. miscellus* of reciprocal origin differ in morphology. Those with *T. pratensis* as the maternal parent have short ligules, and those with *T. dubius* as the maternal parent have long ligules. (Photographs contributed by A. Doust and V. Symonds; plate courtesy of J. Tate.)

World, although their parents are aliens. Hybrids formed (and still form) between *T. pratensis* and *T. porrifolius*, but a polyploid has never been detected.

Ownbey first collected *T. mirus* and *T. miscellus* in 1949 and named these new species in 1950. Given his expertise as a systematist, it is likely that he discovered these new species not long after they first formed. Herbarium records indicate that the three diploid parents did not all occur in the Palouse region before 1928. Hence, the polyploids are probably not more than 80 years old. Given that these plants are biennials, the timescale involved since the formation of these two new species is fewer than 40 generations.

MULTIPLE ORIGINS AND A NORTH AMERICAN SUCCESS STORY

The parentage of both polyploids (*T. mirus* and *T. miscellus*) has now been confirmed using multiple approaches (for review, see Soltis et al. 2004). Ownbey (1950) referred to the small populations of the new polyploids as "small and precarious" but indicated that they appeared to be competing successfully with their parents and that it would be important to "follow the two polyploids over time."

Novak et al. (1991) conducted a survey to determine how common the two polyploids had become since Ownbey's discovery. One or both polyploids were found in most towns of the Palouse with populations ranging from small (<100 individuals) to many thousands of individuals. *T. miscellus* had become one of the most common weeds in and around Spokane, Washington, as well as in parts of Moscow, Idaho, and Spangle, Washington. Populations of *T. mirus* and *T. miscellus* often formed dense stands and were, in fact, displacing their parents, particularly *T. pratensis* and *T. porrifolius*.

Genetic markers revealed how the two *Tragopogon* polyploids had so quickly occupied towns across eastern Washington and Idaho. Both polyploids had formed repeatedly. A diverse array of approaches (for review, see Soltis et al. 2004) indicated as many as 21 distinct lineages of separate origin of *T. miscellus* and perhaps 11 lineages of *T. mirus* (Soltis and Soltis 2000; Soltis et al. 2004). Ownbey had in fact suggested multiple formations based on morphology and cytology (Ownbey and McCollum 1953, 1954). Our recent use of microsatellite markers indicates even more extensive multiple formations than initially thought. In several cases, distinct populations of *T. mirus* and *T. miscellus* from the same small town, separated by less than several kilometers, have formed independently (VV Symonds et al., in prep.).

On a larger geographic scale, both polyploids have also formed in Flagstaff, Arizona; *T. miscellus* has formed in Gardiner, Montana (now apparently extinct; DE Soltis, unpubl.) and Sheridan, Wyoming (for review, see Soltis et al. 2004).

SYNTHETIC LINES: MAN VS. WILD

Adding to the utility of the *Tragopogon* system as an evolutionary model is the recent production of multiple synthetic lines of both *T. mirus* and *T. miscellus* (Tate et al. 2009). These lines provide the added opportunity of examining multiple synthetic lines of both polyploids, following polyploidization from its inception. Comparison of the synthetics to natural populations of separate origin adds another important dimension to the "does evolution repeat itself?" question.

In nature, all formations of *T. mirus* have *T. porrifolius* as the maternal parent and *T. dubius* as the paternal parent, but we have synthesized *T. mirus* reciprocally (both combinations). *T. miscellus* has formed reciprocally in nature (resulting in a dramatic change in floral head morphology; Fig. 1), and Tate et al. (2009) also created synthetic lines with *T. dubius* as both the maternal and the paternal parent. Tate et al. (2009) also produced what has not formed in nature, polyploids between *T. pratensis* and *T. porrifolius* (Fig. 1). All of these synthetic lines are now in the second generation and offer the unique opportunity for comparative genetic/genomic study of repeated formations of both natural and synthetic polyploids.

So, with this background to the *Tragopogon* polyploid evolutionary model system, does genome evolution repeat itself in natural populations of separate origin of *T. miscellus* and *T. mirus* and also in multiple synthetic lines of the two species? We examine the data now available in the sections below.

rDNA LOCI AND CONCERTED EVOLUTION

Concerted evolution, which results in the homogenization of gene sequences to one type, is a common feature of ribosomal RNA (rRNA) genes (see, e.g., Zimmer et al. 1980). In F_1 hybrids and in some allopolyploids, the rDNA types of both parented diploids are present. Both parental arrays may be present in an allopolyploid, as in some allopolyploids of *Glycine* (Doyle and Beachy 1985; Rauscher et al. 2004), *Triticum* (Appels and Dvorak 1982), *Paeonia* (Sang et al. 1995), *Krigia* (Kim and Jansen 1994), *Brassica napus* (Bennett and Smith 1991), and *Arabidopsis suecica* (O'Kane et al. 1996). But in some allopolyploids, only one parental type is present, with homogenization to one parental type having occurred as reported in species of *Gossypium* L. (Wendel et al. 1995), *Nicotiana* L. (Volkov et al. 1999; Lim et al. 2000; Kovarik et al. 2004), *Cardamine* L. (Franzke and Mummenhoff 1999), *Triticum* L. (Flavell and O'Dell 1976), *Glycine* (Rauscher et al. 2004), and *Senecio* L. (Abbott and Lowe 2004).

In *T. mirus* and *T. miscellus*, concerted evolution is ongoing, but incomplete; i.e., we have essentially caught it in the act (Kovarik et al. 2005). F_1 hybrids have equal contributions of the diploid parents, as do the new synthetic polyploids and the earliest natural populations of *T. mirus* and *T. miscellus* (based on DNA from herbarium specimens). But in all modern day natural populations examined representing distinct origins, the rDNA type of *T. dubius* is consistently in very low abundance, with either the *T. pratensis* rDNA type (in *T. miscellus*) or *T. porrifolius* rDNA type (in *T mirus*) in much greater abundance. Thus, concerted evolution has consistently occurred in these new polyploid lines of separate origin, and it has repeatedly operated "against" *T. dubius*, homogenizing those copies in the direction of the other parent. This is readily seen on Southern blots (Fig. 2). Thus, in the case of the rDNA cistron, molecular evolution of rDNA *does appear* to have repeated itself in *Tragopogon*. Surprisingly, despite being the least abundant in terms of rDNA gene copy number, *T. dubius* is by far the most abundant transcript in natural polyploidy populations (Matyasek et al. 2007).

HOMEOLOG LOSS AND GENE SILENCING

Tragopogon is an evolutionary model, but not a genetic model organism; hence, genetic resources are not available. As a result, genetic and genomic changes in the newly formed tetraploids have been so far examined using a "one gene at a time" approach (Fig. 3). This is slow tedious work that has required ~5 years to examine ~30 genes in multiple populations of both *T. mirus* and *T. miscellus* (Tate et al. 2006; Buggs et al. 2009; J Koh et al., in prep.). The genes analyzed to date were chosen based on several different approaches. Amplified-fragment-length polymorphism (AFLP)-cDNA display was initially used to screen plants of *T. miscellus* and *T. mirus* and parental diploids to look

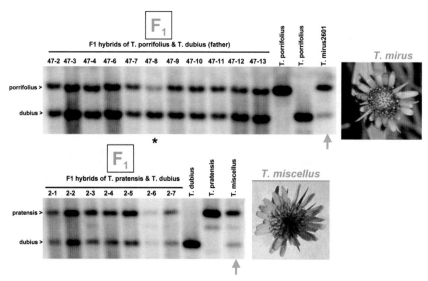

Figure 2. Southern blot hybridization of the ITS region in *Tragopogon* polyploids and parents. (*Top*) Southern blot hybridization of genomic DNAs showing variability in the ITS1 content among populations of separate origin of *T. mirus*. The DNAs were digested with restriction enzymes yielding diagnostic fragments for the diploid parents: *T. porrifolius* (po) and *T. dubius* (do). (*Bottom*) Southern blot hybridization of genomic DNAs showing variability in the ITS1 content among populations of separate origin of *T. miscellus*. Same enzymes and digestion conditions as in top panel.

for promising candidate genes, i.e., fragments that did not show additivity in the allopolyploids as would be expected (Tate et al. 2006; J Koh et al., in prep.). However, additional genes were chosen for survey because they were orthologous to genes that were singletons in *Arabidopsis* (Buggs et al. 2009; J Koh et al., in prep.); the fate of such genes seemed to be of particular interest in new polyploids. Are these "singleton" genes rapidly returned to single-copy status in new *Tragopogon* polyploids?

The results of these gene surveys are presented in detail elsewhere, and we only summarize the major features of those studies here (Tate et al. 2006; Buggs et al. 2009; J Koh et al., in prep.). Some important generalizations have emerged from these studies across polyploid populations of separate origin of both *T. mirus* and *T. miscellus*. Most of the changes observed in populations of both young polyploids are homeolog-loss events; these far outnumber gene-silencing events in these plants. Furthermore, most of the homeolog losses in both polyploids have involved *T. dubius*, the diploid parent shared by both *T. mirus* and *T. miscellus*.

It is also noteworthy that the same suite of genes consistently shows additivity of the parental gene copies (no loss or silencing) in polyploid populations of separate

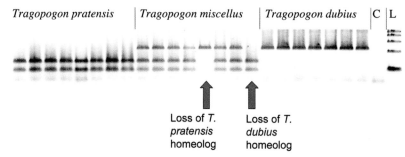

Figure 3. Example of use of genomic CAPS (cleaved amplified polymorphic sequence) markers to examine homeolog loss (see Tate et al. 2006; Buggs et al. 2009). The stained 4% metaphor agarose gel shows homeolog loss in two *T. miscellus* individuals (lanes 13 and 16) from the Oakesdale population for gene D1 (an LRR [leucine-rich repeat] protein kinase). The DNA fragments visualized are products of a restriction enzyme digest on polymerase chain reaction (PCR)-amplified fragments of the gene in individual plants. The first eight lanes of the gel show *T. pratensis* individuals (Soltis and Soltis collection number 2672, individuals 1, 2, 3, 6, 7, 8, 9, and 10, respectively); lanes 9 to 16 show *T. miscellus* individuals (Soltis and Soltis collection number 2671, individuals 2, 3, 4, 5, 7, 8, 10, and 11, respectively); lanes 17 to 23 show *T. dubius* individuals (Soltis and Soltis collection number 2670, individuals 1–7, respectively); lane 24 shows a negative PCR control; and lane 25 shows HyperLadder IV (Bioline, Taunton, Massachusetts) markers from 100–500 bp. (This example modified from Buggs et al. 2009.)

origin, whereas some of the genes analyzed consistently show some evidence of loss across at least some of the populations surveyed. Significantly, at deep levels across eukaryotes, Paterson et al. (2006) showed that genes with some PFam domains may be consistently returned to singleton status following genome-wide duplication, whereas other genes consistently are retained as duplicates across highly divergent lineages. Thus, our observations for *Tragopogon* polyploids are also in agreement with the hypothesis that some underlying "principles" to polyploidization may exist at the genetic/biochemical level.

But there is an element of randomness operating as well in these young polyploids. Although homeolog loss is present in the polyploid populations, the process is ongoing and appears to be stochastic within individual populations. In no population examined has silencing or loss been complete, i.e., observed in all individuals of a population. Furthermore, these losses and gene-silencing events were not detected in F_1 hybrids or early-generation synthetic lines (S_1). Hence, loss of homeologs and gene silencing are not immediate consequences of hybridization or polyploidization in *Tragopogon*, but they appear to occur somewhat gradually and haphazardly for certain genes following polyploidy.

TISSUE-SPECIFIC SILENCING

Several outcomes exist for genes following duplication: (1) Both members of a duplicate gene pair may retain their original function; (2) one copy of a duplicate gene pair may retain the original function, but the other copy may become lost or silenced (Lynch and Conery 2000; Adams et al. 2003; Adams 2007; Sterck et al. 2007); (3) duplicate genes may partition the original gene function (subfunctionalization), with one copy active, for example, in one tissue and the other copy active in another tissue (Lynch and Conery 2000; Lynch and Force 2000); and (4) one copy may retain the original function while the other develops a new function (neofunctionalization) (Drea et al. 2006; Teshima and Innan 2008).

We recently examined (RJA Buggs et al., in press) tissue-specific silencing in 10 individuals from two reciprocally formed natural populations of *T. miscellus* (Asteraceae). Using cleaved amplified polymorphic sequence analysis of 18 homeologs, we found homeolog silencing in 14 genes in at least one individual and, of these, eight genes showed tissue-specific silencing. Patterns of tissue-specific homeolog silencing varied a great deal among individuals, but there was an occasional repetition of pattern, such as repeated silencing in corolla tissue. Patterns of silencing were not determined by the direction of the parental cross. By comparison with synthetic allopolyploids, F_1 hybrids, and parental diploids, we showed that most cases of homeolog silencing have arisen in early generations after whole-genome duplication. Semiquantitative analysis showed greater variance of expression between individuals and tissues in the natural populations compared to the synthetics and hybrids for seven genes. Some silencing was due to genomic loss of duplicate genes. Our data suggest that more homeolog expression changes occur in early generations of new polyploids than do the processes of hybridization and whole-genome duplication and thus contribute to duplicated gene evolution.

TRAGOPOGON GOES GENOMIC

As noted, *Tragopogon* is not a genetic model system. Hence, our investigations to date have been limited to a gene-by-gene approach. Recent advances in high-throughput sequencing technology provide a rapid and cost-effective means to generate sequence data. This and other genomic approaches offer the opportunity to accelerate dramatically our ability to survey gene loss and expression changes in nonmodel species such as *Tragopogon*. Essentially, instead of building a wall one brick at a time, with genomic methods, we can pour it or build it all at once.

We recently (RJA Buggs et al., in press) undertook a hybrid next-generation sequencing approach to identifying single-nucleotide polymorphism (SNP) markers for homologous genes in *Tragopogon miscellus*. This general approach (Fig. 4) makes it possible to rapidly build genetic resources for many "nonmodel" plant systems such as *Tragopogon*. We (RJA Buggs et al., in press) generated reference expressed sequence tags (ESTs) from the transcriptome of *T. dubius* using 454 FLX sequencing. We then generated and aligned Illumina reads from the diploids *T. pratensis* and *T. dubius* to this reference. The resulting alignments generated 7782 SNPs within 2885 contigs between *T. dubius* and *T. pratensis* at high stringency. We then examined these SNPs in a pilot transcriptome profile for *T. miscellus* using Illumina sequence reads. Of the 7782 SNPs, 2064 (27%) appeared to show equal homeolog expression in *T. miscellus*, 671 (9%) showed differential expression in *T. miscellus*, and 254 (3%) showed potential homeolog loss in *T. miscellus*. Most of the potential homeolog losses were of the *T. dubius* homeolog (164/254) with a minority the *T. pratensis* homeolog (90/254), in agreement with results from our gene-by-gene approach. Sequenom analyses confirmed that in a sample of 27 of the SNPs showing potential gene loss from the transcriptome profile, 23 (85%) were cases of genomic homeolog loss.

MECHANISM OF GENE LOSS

Using GISH (genomic in situ hybridization) and FISH (fluorescence in situ hybridzation), we have detected surprising chromosomal variation in natural populations of *T. mirus* and *T. miscellus* including inversions and intergenomic translocations, as well as fertile plants of both polyploids having three copies of one chromosome, but one copy of another (reciprocal trisomy/monosomy) (Lim et al. 2008). These rearrangements provide one possible mechanism for homeolog loss in these plants. Additional chromosomal studies are needed; it is unknown, for example, if similar chromosomal changes are present in polyploidy populations of separate formation.

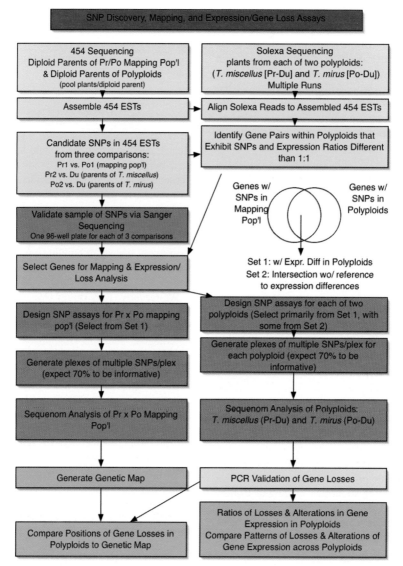

Figure 4. General flow chart illustrating methods that can be used to rapidly develop genomic resources and markers for the *Tragopogon* system.

COMPARING *TRAGOPOGON* TO WELL-STUDIED SYSTEMS

Genome evolution has been well studied in model polyploid systems (e.g., *Gossypium, Triticum, Brassica,* and *Arabidopsis*). Although much more work on *Tragopogon* is needed, at this stage, polyploid evolution in *T. mirus* and *T. miscellus* exhibits important similarities and differences to the better-studied genetic models. *Tragopogon* is noteworthy in that initial studies show that homeolog losses far outnumber true changes in duplicate gene expression (Tate et al. 2006; Buggs et al. 2009; T Koh et al., unpubl.). For example, of the initial 23 genes analyzed in *T. miscellus*, 15 showed homeolog loss in one or more plants from nature, and only eight showed true expression changes; results for *T. mirus* are comparable. In contrast, in synthetic wheat (*Triticum*) and synthetic *Arabidopsis thaliana* and *A. suecica* polyploids, expression changes dominate (Kashkush et al. 2002; Madlung et al. 2005; Wang et al. 2006), although losses do occur in wheat. Genome evolution in *Tragopogon* may be most similar to *Brassica napus*, in which most of the apparent expression changes observed in later generations were due to losses, most likely resulting from genomic rearrangements (Song et al. 1995; Gaeta et al. 2007).

In *Tragopogon*, *T. dubius* appears to be the "loser" diploid genome in both *T. mirus* and *T. miscellus*. Preferential loss of expression, or homeologs, has also been observed in some of the model polyploid systems. For example, across 50 synthetic lines of *B. napus*, genetic changes are equally distributed between the parental diploid genomes (Gaeta et al. 2007). The system is dynamic—some lines become more "*oleracea*-like" and others more "*rapa*-like" in terms of losses and corresponding expression differences. In *A. suecica* allopolyploids, silencing of homeologs from one parent (*A. thaliana*) was observed more frequently than silencing of homeologs from the other parent, *A. arenosa* (Wang et al. 2006). In synthetic polyploids in *Triticale,* the contribution of the rye genome (*Secale cereale*) is preferentially silenced (Ma et al. 2004; Ma and Gustafson 2006). In the allohexaploid *Senecio cambrensis,* gene expression is more similar to that of the tetraploid parent (*S. vulgaris*) than to the diploid parent *S. squalidus* (Hegarty et al. 2006). In cotton, tissue-specific subfunctionalization was observed; there is no global bias in the expression of transcripts from one parental diploid genome over the other (Adams et al. 2003; Adams and Wendel 2004). In both *T. miscellus* and *T. mirus*, homeologs of one diploid genome (*T. dubius*) are more often lost or not expressed; i.e., *T. dubius* is often the "loser genome" based on the set of genes surveyed to date. It remains to be seen whether these patterns are maintained over the entire genome; i.e., are homeologs of the diploid parent (*T. dubius*) shared by both *T. mirus* and *T. miscellus* (Fig. 1) preferentially lost or silenced?

SUMMARY/FUTURE PROSPECTS

Most of what we know about the genetic and genomic consequences of polyploidy has been derived from the study of crops, synthetic polyploids, and models. To understand how polyploidy shapes genome evolution and gene function in nature, we must extend from a few models and synthetics to naturally occurring polyploids. *Tragopogon* provides the unique opportunity to investigate the genetic and genomic changes that occur across a continuum from F_1 hybrids, synthetic allopolyploids, independently formed natural populations of *T. mirus* and *T. miscellus* that are 60–80 years postformation, to older Eurasian polyploids. With this system, we can assess the relative frequency of gene loss and gene expression changes in natural populations of the two allotetraploid species and in *synthetic* lines of the two allotetraploids and determine whether ***evolution repeats itself***.

To date, we have used a gene-by-gene approach to examine gene loss and expression changes. The data so far garnered for ~30 genes indicate that some aspects of genomic evolution may be hardwired—that some features are repeated in polyploid populations of separate origin; that there may be underlying **"rules"** that govern some genomic interactions in polyploids. But we need to extend from a few genes to a better representation of the genome. We are now using a genomics approach; rather than slowly building a wall of genetic data for *Tragopogon* one brick at a time, we are using a genomics approach to build a wall all at once, querying thousands of genes. The proposed methodology has not been previously applied to "nonmodel" species, and our initial data indicate that this general approach is particularly promising for species such as *Tragopogon* for which limited genetic and genomic resources are available.

ACKNOWLEDGMENTS

This work was supported in part by National Science Foundation grants MCB 034637, DEB 0919254, and DEB 0919348.

REFERENCES

Abbott RJ, Lowe AJ. 2004. Origins, establishment and evolution of new polyploid species: *Senecio cambrensis* and *S. eboracensis* in the British Isles. *Biol J Linn Soc* **82:** 467–474.

Adams KL. 2007. Evolution of duplicate gene expression in polyploid and hybrid plants. *J Hered* **98:** 136–141.

Adams KL, Wendel JF. 2004. Exploring the genomic mysteries of polyploidy in cotton. *Biol J Linn Soc* **82:** 573–581.

Adams KL, Cronn R, Percifield R, Wendel JF. 2003. Genes duplicated by polyploidy show unequal contributions to the transcriptome and organ-specific reciprocal silencing. *Proc Natl Acad Sci* **100:** 4649–4654.

Ainouche ML, Salmon A, Baumel A, Yannic G. 2004. Hybridization, polyploidy and speciation in *Spartina* (Poaceae). *New Phytol* **161:** 165–172.

Appels R, Dvorak J. 1982. The wheat ribosomal DNA spacer region: Its structure and variation in populations and among species. *Theor Appl Genet* **63:** 337–348.

Bennett RI, Smith AG. 1991. Use of a genomic clone for ribosomal RNA from *Brassica oleracea* in RFLP analysis of *Brassica* species. *Plant Mol Biol* **16:** 685–688.

Blanc G, Wolfe KH. 2004. Widespread paleopolyploidy in model plant species inferred from age distributions of duplicate genes. *Plant Cell* **16:** 1667–1678.

Blanc G, Barakat A, Guyot R, Cooke R, Delseny M. 2000. Extensive duplication and reshuffling in the *Arabidopsis* genome. *Plant Cell* **12:** 1093–1101.

Blanc G, Hokamp K, Wolfe KH. 2003. A recent polyploidy superimposed on older large-scale duplications in the *Arabidopsis* genome. *Genome Res* **13:** 137–144.

Bowers JE, Chapman BA, Rong JK, Paterson AH. 2003. Unravelling angiosperm genome evolution by phylogenetic analysis of chromosomal duplication events. *Nature* **422:** 433–438.

Brochmann C, Brysting AK, Alsos IG, Borgen L, Grundt HH, Scheen A-C, Elven R. 2004. Polyploidy in arctic plants. *Biol J Linn Soc* **82:** 521–536.

Buggs RJA, Chamala S, Wu W, Gao L, May GD, Schnable PS, Soltis DE, Soltis PS, Barbazuk WB. 2009. Characterization of duplicate gene evolution in the recent natural allopolyploid *Tragopogon miscellus* by next-generation sequencing and Sequenom iPLEX genotyping. *Mol Ecol* (in press).

Buggs RJA, Doust AN, Tate JA, Koh J, Soltis K, Feltus FA, Paterson AH, Soltis PS, Soltis DE. 2009. Gene loss and silencing in *Tragopogon miscellus* (Asteraceae): Comparison of natural and synthetic allotetraploids. *Heredity* **24:** 1–9.

Chapman BA, Bowers JE, Feltus FA, Paterson AH. 2006. Buffering of crucial functions by paleologous duplicated genes may contribute cyclicality to angiosperm genome duplication. *Proc Natl Acad Sci* **103:** 2730–2735.

Clausen J, Keck DD, Hiesey WM. 1945. Experimental studies on the nature of species. II. Plant evolution through amphiploidy and autoploidy, with examples from the Madiinae. *Carnegie Inst Wash Publ,* no. 564.

Cui L, Wall PK, Leebens-Mack JH, Lindsay BG, Soltis DE, Doyle JJ, Soltis PS, Carlson JE, Arumuganathan K, Barakat A, et al. 2006. Widespread genome duplications throughout the history of flowering plants. *Genome Res* **16:** 738–749.

Darlington CD. 1937. *Recent advances in cytology.* Blakiston, Philadelphia.

Darwin C. 1859. *On the origin of species by means of natural selection,* 1st ed. Murray, London.

Doyle JJ, Beachy RN. 1985. Ribosomal gene variation in soybean (*Glycine max*) and its relatives. *Theor Appl Genet* **70:** 369–376.

Doyle JJ, Flagel LE, Paterson AH, Rapp RA, Soltis DE, Soltis PS, Wendel JF. 2008. Evolutionary genetics of genome merger and doubling in plants. *Annu Rev Genet* **42:** 443–461.

Drea SC, Lao NT, Wolfe KH, Kavanagh TA. 2006. Gene duplication, exon gain and neofunctionalization of OEP16-related genes in land plants. *Plant J* **46:** 723–735.

Flavell RB, O'Dell M. 1976. Ribosomal RNA genes in homeologous chromosomes of groups 5 and 6 in hexaploid wheat. *Heredity* **37:** 377–385.

Franzke A, Mummenhoff K. 1999. Recent hybrid speciation in *Cardamine* (Brassicaceae): Conversion of nuclear ribosomal ITS sequences in statu nascendi. *Theor Appl Genet* **98**: 831–834.

Gaeta RT, Pires JC, Iniguez-Luy F, Leon E, Osborn TC. 2007. Genomic changes in resynthesized *Brassica napus* and their effect on gene expression and phenotype. *Plant Cell* **19**: 3403–3417.

Gould SJ. 1994. The evolution of life on earth. *Sci Am* **271**: 85–86.

Grant V. 1981 *Plant speciation*. Columbia University Press, New York.

Grundt HH, Kjølner S, Borgen L, Rieseberg LH, Brochmann C. 2006. High biological species diversity in the arctic flora. *Proc Natl Acad Sci* **103**: 972–975.

Hegarty MJ, Jones JM, Wilson ID, Barker GL, Coghill JA, Sanchez-Barcaldo P, Liu G, Buggs RJA, Abbott RJ, Edwards KJ, Hiscock SJ. 2005. Development of anonymous cDNA microarrays to study changes to the *Senecio* floral transcriptome during hybrid speciation. *Mol Ecol* **14**: 2493–2510.

Hegarty MJ, Barker GL, Wilson ID, Abbott RJ, Edwards KJ, Hiscock SJ. 2006. Transcriptome shock after interspecific hybridization in *Senecio* is ameliorated by genome duplication. *Curr Biol* **16**: 1652–1659.

Huskins CL. 1931. The origin of *Spartina townsendii*. *Genetica* **12**: 531–538.

Kashkush K, Feldman M, Levy AA. 2002. Gene loss, silencing and activation in a newly synthesized wheat allotetraploid. *Genetics* **160**: 1651–1659.

Kim KJ, Jansen RK. 1994. Comparisons of phylogenetic hypotheses among different data sets in dwarf dandelions (*Krigia*): Additional information from internal transcribed spacer sequences of nuclear ribosomal DNA. *Plant Syst Evol* **190**: 157–159.

Kovarik A, Matyasek R, Lim KY, Skalická K, Koukalová B, Knapp S, Chase M, Leitch AR. 2004. Concerted evolution of 18–5.8-26S rDNA repeats in *Nicotiana* allotetraploids. *Biol J Linn Soc* **82**: 615–625.

Kovarik A, Pires JC, Leitch AR, Lim KY, Sherwood A, Matyasek R, Rocca J, Soltis DE, Soltis PS. 2005. Rapid concerted evolution in two allopolyploids of recent and recurrent origin. *Genetics* **169**: 931–944.

Lim KY, Matyasek R, Lichtenstein CP, Leitch AR. 2000. Molecular cytogenetic analyses and phylogenetic studies in the *Nicotiana* section *Tomentosae*. *Chromosoma* **109**: 245–258.

Lim KY, Soltis DE, Soltis PS, Tate J, Matyasek R, Srubarova H, Kovarik A, Pires JC, Xiong Z, Leitch AR. 2008. Rapid chromosome evolution in recently formed polyploids in *Tragopogon* (Asteraceae). *PLos One* **3**: e3353.

Löve A, Löve D. 1949. The geobotanical significance of polyploidy. I. Polyploidy and latitude. *Port Acta Biol Ser A* (special vol) **2**: 273–352.

Lynch M, Conery JS. 2000. The evolutionary fate and consequences of duplicate genes. *Science* **290**: 1151–1155.

Lynch M, Force A. 2000. The probability of duplicate gene preservation by subfunctionalization. *Genetics* **154**: 459–473.

Ma XF, Gustafson JP. 2006. Timing and rate of genome variation in *Triticale* following allopolyploidization. *Genome* **49**: 950–958.

Ma XF, Fang P, Gustafson JP. 2004. Polyploidization-induced genome variation in *Triticale*. *Genome* **47**: 839–848.

Madlung A, Tyagi AP, Watson B, Jiang HM, Kagochi T, Doerge RW, Martienssen R, Comai L. 2005. Genomic changes in synthetic *Arabidopsis* polyploids. *Plant J* **41**: 221–230.

Matyasek R, Tate JA, Lim YK, Srubarova H, Koh J, Leitch AR, Soltis DE, Soltis PS, Kovarik A. 2007. Concerted evolution of rDNA in recently formed *Tropogon* allotetraploids is typically associated with an inverse correlation between gene copy number and expression. *Genetics* **176**: 2509–2519.

Müntzing A. 1936. The evolutionary significance of autopolyploidy. *Hereditas* **21**: 263–378.

Novak SJ, Soltis DE, Soltis PS. 1991. Ownbey *Tragopogons*: 40 Years later. *Am J Bot* **78**: 1586–1600.

O'Kane SL, Schaal BA, Al-Shehbaz IA. 1996. The origins of *Arabidopsis suecica* as indicated by nuclear rDNA sequences. *Syst Bot* **21**: 559–566.

Ownbey M. 1950. Natural hybridization and amphiploidy in the genus *Tragopogon*. *Am J Bot* **37**: 487–499.

Ownbey M, McCollum GD. 1953. Cytoplasmic inheritance and reciprocal amphiploidy in *Tragopogon*. *Amer J Bot* **40**: 788–796.

Ownbey M, McCollum GD. 1954. The chromosomes of *Tragopogon*. *Rhodora* **56**: 7–21.

Paterson AH, Bowers JE, Burow MD, Draye X, Elsik CG, Jiang C-X, Katsar CS, Lan T-H, Lin Y-R, Ming R, Wright RJ. 2000. Comparative genomics of plant chromosomes. *Plant Cell* **12**: 1523–1539.

Paterson AH, Chapman BA, Kissinger JC, Bowers JE, Feltus FA, Estill JC. 2006. Many gene and domain families have convergent fates following independent whole-genome duplication events in *Arabidopsis*, *Oryza*, *Saccharomyces* and *Tetraodon*. *Trends Genet* **22**: 597–602.

Rauscher JT, Doyle JJ, Brown AHD. 2004. Multiple origins and rDNA internal transcribed homoeolog evolution in the *Glycine tomentella* (Leguminose) allopolyploid complex. *Genetics* **166**: 987–998.

Rosser EM. 1955. A new British species of *Senecio*. *Watsonia* **3**: 228–232.

Sang T, Crawford DJ, Stuessy TF. 1995. Documentation of reticulate evolution in peonies (*Paeonia*) using internal transcribed spacer sequences of nuclear ribosomal DNA: Implications for biogeography and concerted evolution. *Proc Natl Acad Sci* **92**: 6813–6817.

Schlueter JA, Dixon P, Granger C, Grant D, Clark L, Doyle JJ, Shoemaker RC. 2004. Mining EST databases to resolve evolutionary events in major crop species. *Genome* **47**: 868–876.

Seoighe C, Gehring C. 2004. Genome duplication led to highly selective expansion of the *Arabidopsis thaliana* proteome. *Trends Genet* **20**: 461–464.

Simillion C, Vandepoele K, Van Montagu MCE, Zabeau M, Van de Peer Y. 2002. The hidden duplication past of *Arabidopsis thaliana*. *Proc Natl Acad Sci* **99**: 13627–13632.

Soltis DE, Soltis PS. 1993. Molecular data facilitate a reevaluation of traditional tenets of polyploid evolution. *Crit Rev Plant Sci* **12**: 243–273.

Soltis DE, Soltis PS. 1999. Polyploidy: Recurrent formation and genome evolution. *Trends Ecol Evol* **14**: 348–352.

Soltis PS, Soltis DE. 2000. The role of genetic and genomic changes in the success of polyploids. *Proc Natl Acad Sci* **97**: 7051–7057.

Soltis DE, Soltis PS, Pires JC, Kovarík A, Tate JA, Mavrodiev EV. 2004. Recent and recurrent polyploidy in *Tragopogon* (Asteraceae): Cytogenetic, genomic, and genetic comparisons. *Biol J Linn Soc* **82**: 485–501.

Song K, Lu P, Tang K, Osborn TC. 1995. Rapid genome change in synthetic polyploids of *Brassica* and its implications for polyploidy evolution. *Proc Natl Acad Sci* **92**: 7719–7723.

Stebbins GL. 1947. Types of polyploids: Their classification and significance. *Adv Genet* **1**: 403–429.

Stebbins GL. 1950. *Variation and evolution in plants*. Columbia University Press, New York.

Sterck L, Rombauts S, Vandepoele K, Rouze P, Van de Peer Y. 2007. How many genes are there in plants (... and why are they there)? *Curr Opin Plant Biol* **10**: 199–203.

Tate JA, Soltis PS, Soltis DE. 2004. Polyploidy in plants. In *The evolution of the genome* (ed. TR Gregory), pp. 372–426. Academic, New York.

Tate JA, Ni ZF, Scheen AC, Koh J, Gilbert CA, Lefkowitz D, Chen ZJ, Soltis PS, Soltis DE. 2006. Evolution and expression of homeologous loci in *Tragopogon miscellus* (Asteraceae), a recent and reciprocally formed allopolyploid. *Genetics* **173**: 1599–1611.

Tate JA, Symonds VV, Doust AN, Buggs RJA, Mavrodiev EV, Soltis PS, and Soltis DE. 2009. Synthetic polyploids of *Tragopogon miscellus* and *T. mirus* (Asteraceae): 60 Years after Ownbey's discovery. *Am J Bot* **96**: 979–988.

Teshima KM, Innan H. 2008. Neofunctionalization of duplicated genes under the pressure of gene conversion. *Genetics* **178:** 1385–1398.

Urbanska KM, Hurka H, Landolt E, Neuffer B, Mummenhoff K. 1997. Hybridization and evolution in *Cardamine* (Brassicaceae) at Urnerboden, Central Switzerland: Biosystematic and molecular evidence. *Plant Syst Evol* **204:** 233–256.

Vision TJ, Brown DG, Tanksley SD. 2000. The origins of genomic duplications in *Arabidopsis*. *Science* **290:** 2114–2117.

Volkov RA, Borisjuk NV, Panchuk II, Schweizer D, Hemleben V. 1999. Elimination and rearrangement of parental rDNA in the allotetraploid *Nicotiana tabacum*. *Mol Biol Evol* **16:** 311–320.

Wang J, Tian L, Lee HS, Wei NE, Jiang H, Watson B, Madlung A, Osborn TC, Doerge RW, Comai L, Chen ZJ. 2006. Genomewide nonadditive gene regulation in *Arabidopsis* allotetraploids. *Genetics* **172:** 507–517.

Wendel JF. 2000. Genome evolution in polyploids. *Plant Mol Biol* **42:** 225–249.

Wendel JF, Doyle JJ. 2004. Polyploidy and evolution in plants. In *Plant diversity and evolution* (ed. R Henry), pp. 97–117. CABI, Cambridge.

Wendel JF, Schnabel A, Seelanan T. 1995. Bidirectional interlocus concerted evolution following allopolyploid speciation in cotton (*Gossypium*). *Proc Natl Acad Sci* **92:** 280–284.

Zimmer EA, Martin SL, Beverley SM, Kan YW, Wilson AC. 1980. Rapid duplication and loss of genes coding for the α chains of hemoglobin. *Proc Natl Acad Sci* **77:** 2158–2162.

Molecular Evolution of piRNA and Transposon Control Pathways in *Drosophila*

C.D. MALONE AND G.J. HANNON

*Watson School of Biological Sciences, Howard Hughes Medical Institute,
Cold Spring Harbor Laboratory, Cold Spring Harbor, New York 11724*
Correspondence: hannon@cshl.edu

The mere prevalence and potential mobilization of transposable elements in eukaryotic genomes present challenges at both the organismal and population levels. Not only is transposition able to alter gene function and chromosomal structure, but loss of control over even a single active element in the germline can create an evolutionary dead end. Despite the dangers of coexistence, transposons and their activity have been shown to drive the evolution of gene function, chromosomal organization, and even population dynamics (Kazazian 2004). This implies that organisms have adopted elaborate means to balance both the positive and detrimental consequences of transposon activity. In this chapter, we focus on the fruit fly to explore some of the molecular clues into the long- and short-term adaptation to transposon colonization and persistence within eukaryotic genomes.

TRANSPOSON OCCUPATION IN EUKARYOTIC GENOMES

Transposable elements, or transposons, are selfish genetic elements that possess the ability to mobilize within a genome, potentially causing serious damage. They can accomplish this through several well-understood mechanisms. Most move via an RNA intermediate (class I: retrotransposons), whereas others colonize new locations through the direct transfer of transposon DNA (class II: DNA transposons) (for review, see Slotkin and Martienssen 2007). In the case of retrotransposons, active elements within the genome are transcribed and translated into functional proteins that subsequently generate an additional DNA copy of the element and catalyze its integration into a new genomic site. DNA transposons are also transcribed and translated into proteins that instead excise a physical DNA copy of an element from the genome and then insert that DNA elsewhere in the genome. This "cut-and-paste" mechanism generates no net increase in copies of the element on an individual level (although small genomic lesions are left behind), whereas retrotransposition actively increases transposon copy number within an individual genome. In both cases, transposon mobilization can have positive and negative consequences for the viability of an organism and its progeny, notably on the organization of its genome.

The potential detrimental effects of unregulated transposon activity are obvious. For example, by inserting into an essential gene, especially one with a sensitive dosage requirement, a transposon could severely alter cell or organismal viability. The impact of landing in a nonessential gene is less dramatic, but such events can still generate clear phenotypic manifestations (Demerec 1926a,b). Transposition into important gene regulatory regions, such as splicing regulators or transcriptional enhancer domains, could significantly change transcriptional and posttranscriptional gene expression programs (White et al. 1994). Moreover, transposons can promote alterations in chromosomal structure by generating double-stranded DNA breaks that can precede the formation of unstable dicentric and acentric chromosomes (McClintock 1950) and induce other spurious chromosomal aberrations. Finally, important chromatin domains, such as insulator and boundary elements, can be interrupted, leading to a general transcriptional misregulation in the surrounding genomic space.

Not all instances of transposon mobilization necessarily lead to such negative consequences for an organism. In fact, transposons are capable of positively influencing genomic content, structure, and evolution (Kazazian 2004). For instance, transposon movement can lead to the gain or loss of introns and exons, generating novel transcriptional output. They can also generate additional coding information in the genome, either being themselves domesticated as components of host transcripts (Miller et al. 1992; Baudry et al. 2009) or by inducing duplication of endogenous genes (Esnault et al. 2000). Transposon-induced reorganization of large chromosomal tracts can drive substantial genome-scale evolution, assuming creation of a positive selective advantage. Likewise, reorganization of transcriptional regulatory regions can drive the evolution of expression programs through either disruption or donation of novel regulatory elements. Finally, transposon-induced chromatin state changes can modify the genomic landscape by producing new broadscale regulatory domains, perhaps creating variation by altering the transcriptional output at particular developmental stages. Each of these types of events can provide substrates for evolutionary selection and thus have profound impacts on the adaptability of the organism.

Transposable Element Diversity and Challenges to Their Control

Transposon occupation in eukaryotic genomes varies at nearly every definable level. First, the total percentage of transposon-derived DNA in a genome can range from just a few percent to more than 90%. Although there is no strict correlation between organismal "complexity" and transposon load, there are certainly dramatic consequences to an organism's genomic complexity and capacity for rapid evolution. Second, element diversity varies enormously, ranging from only a few classes in organisms such as yeast and mouse to upward of 150 in the fruit fly. Again, this does not necessarily relate to phenotypic "complexity," but it could simply reflect waves of transposon challenge and fixation within wild populations. Individual transposon families possess particular transcriptional profiles, honed to maximize their propagation in germ cells and hence expansion in populations. Increased transposon diversity could also reflect the availability of multiple cellular niches, developmental stages, or tissue types that may offer some opportunity to evade control mechanisms. Third, despite their shared imperative to propagate, different transposons, even within the same element class, can be quite dissimilar in their DNA sequence composition. Therefore, to silence the diversity of element structures and sequences, cells must exploit a unified feature of transposons as a means for selective targeting.

The challenge from potentially active elements and their representation versus inactive remnants can also vary dramatically between organisms. In *Drosophila*, some elements have expanded to more than 50 active copies (roo has 58 in the sequenced strain genome), whereas for others, only one or even no active copies may be present (Kaminker et al. 2002). From an adaptive perspective, this may reflect the degree to which an organism has brought a particular element under control. For instance, although there are approximately eight active *I*-element copies in the sequenced *Drosophila* genome, there are roughly three times the amount of diverged, fragmented copies scattered throughout heterochromatic regions of the genome (Kaminker et al. 2002; Brennecke et al. 2008). This is consistent with a hypothesis that an expansion of heterochromatic fragments contributed to the ability of the cell to effectively silence active elements, but it is equally possible that insertions are simply better tolerated in these gene-poor regions. In support of the former hypothesis, transposition of a single, truncated copy of the *Drosophila* P element into a repetitive telomeric locus is sufficient to induce the silencing of any additionally introduced active copies (Ronsseray et al. 1991, 1998; Marin et al. 2000). For many elements, maintenance of heterochromatic copies, even in the absence of active elements, preserves the ability to silence, indicating a selective pressure to retain these often truncated fragments (Pélisson et al. 2007). However, additional work is required to investigate a more generalizable relationship between the control of elements and the number of active and fragmented transposon copies, because they do not appear to correlate universally (Kaminker et al. 2002).

All of these complexities challenge the cell at many levels to build an effective and adaptive silencing program. However, the molecular and evolutionary mechanisms used by the cell to successfully regulate transposon activity have begun to yield a molecular understanding during the past several years.

THE ADAPTATION OF TRANSPOSON CONTROL

On the basis of their variety alone, transposable elements present an imposing threat to a cell and its genome. In *Drosophila*, there are ~150 different element types, each with unique expression, replication, and mobilization strategies. These also possess an astounding array of sequence-level diversity. Therefore, cells must adopt a strategy of control that relies on features shared by all transposons—one being that at some point, all active elements, regardless of their type or abundance, propagate by moving to new locations in the genome. Moreover, all encode proteins essential for their mobility (or parasitize other transposons). Thus, transposon-encoded RNAs (Fig. 1) are a logical target for control by posttranscriptional gene silencing.

Small RNA-based Pathways of Transposon Regulation

Essentially, the RNA interference (RNAi) pathway uses 20–30-nucleotide RNAs, termed small RNAs, as guides to target larger genic, repeat, or virally derived RNA molecules. At the heart of RNAi lies an Argonaute protein that binds small RNAs and uses them as a guide to either "slice" homologous RNA transcripts, bind to and inhibit their translation, or target repression by chromatin-level changes (Hannon 2002; Reinhart and Bartel 2002; Volpe et al. 2002; Malone and Hannon 2009). These silencing effects are often potent and have been shown to be essential for processes such as stem cell maintenance (Cox et al. 1998, 2000) regulating developmental progression (Houwing et al. 2007, 2008), and they are even altered in the development of some types of cancers (Gilbert et al. 2002; Symer et al. 2002; Belgnaoui et al. 2006). To accomplish these tasks, the cell must actively generate the small RNAs used to direct the activity of Argonaute proteins. Two broad biogenesis pathways for small RNAs have emerged from work of the past decade. The best understood uses RNase III–family enzymes, Dicer and/or Drosha, to generate small RNAs from double-strand substrates. However, there are clearly Dicer- and Drosha-independent mechanisms for creating small RNAs that flow into Argonaute proteins.

Germline cells have a special need to protect their genome, because this copy must faithfully transmit genetic information to offspring. Therefore, animals have developed elaborate means to adapt to the ever-changing threats to the integrity of germ-cell genomes. As in other cells of the body, a key to germline transposon regulation comes by way of RNAi (Aravin et al. 2001). However, the

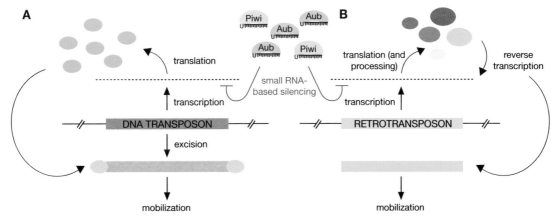

Figure 1. RNA-based targeting of transposable elements. Transposon expression, regardless of type, generates a universal target for the RNAi machinery. (*A*) DNA transposons are transcribed and then translated to generate a transposase enzyme capable of catalyzing the excision and *cis* or *trans* mobilization via this "cut-and-paste" mechanism. The initial transcription of the locus provides a target for silencing. (*B*) Retrotransposons are transcribed both to be translated (non-LTR retrotransposon proteins are further posttranslationally processed) and also to serve as the substrate for reverse transcriptase to manufacture additional transposon copies. Here, either transcription products or reverse transcription substrate molecules serve as the target of the RNAi machinery.

canonical Argonaute protein was shown to be dispensable for this regulation, whereas instead, members of a separate clade of Argonautes, termed Piwi proteins and composed of Piwi, Aubergine, and AGO3 in *Drosophila*, had an essential role (Vagin et al. 2006). Piwi proteins bind a larger (23–32 nucleotides) class of small RNA than do canonical pathway Argonautes. These RNAs were initially termed rasiRNAs (for repeat-associated small interfering RNAs), based on their similarity to repeat elements (Aravin et al. 2001, 2003) and subsequently dubbed Piwi-interacting RNAs, or piRNAs (Aravin et al. 2006; Girard et al. 2006; Lau et al. 2006) based on their protein interaction patterns. Interestingly, piRNA production did not depend on the activity of a dicer enzyme, but instead appeared dependent on the RNA slicing activity of the Piwi proteins themselves (Vagin et al. 2006).

These observations led to a model of piRNA biogenesis that, like dicer-based pathways, requires sense and antisense RNA transcripts. However, these appear to be required in *trans* as single-stranded RNA, and not double-stranded RNA precursors, as seen in canonical RNAi pathways. Here, active transposon RNAs are recognized by a Piwi protein (mainly Aubergine in *Drosophila*), bound with a near-perfect, reverse-complementing antisense piRNA. This Piwi protein then cleaves the transposon RNA 10 nucleotides distal to the 5′ end of its bound piRNA (Elbashir et al. 2001). The sliced transcript is then further processed by an as yet defined machinery to generate the 3′ end of a new piRNA. This sense-piRNA is then loaded into the Argonaute3 (AGO3) protein (in *Drosophila*), which then targets antisense transposon transcripts in the cell for cleavage (Li et al. 2009). After slicing and again further processing, a new antisense piRNA is produced and loaded back into Aubergine to further target active transposon transcripts. This feed forward amplification cycle, called piRNA "ping-pong," shapes the population of piRNAs, optimizing control of expressed elements (Brennecke et al. 2007; Gunawardane et al. 2007).

In considering such a cycle, a major question is the source of the antisense transposon information. Because heterochromatic transposon fragments had been previously linked to active transposon silencing (Simonelig et al. 1988; Dimitri and Bucheton 2005), this seemed to be a logical starting point from which to investigate the origin of antisense transcripts. In fact, many fragments do not exist as isolated truncated transposons, but instead sit within what appear as transposon "graveyards," containing countless nested transposon fragments, most of which have accumulated significant mutations, leaving them otherwise inert (Brennecke et al. 2007). On the basis of sequence divergence, these loci were demonstrated to produce abundant piRNAs and were therefore termed "piRNA clusters." These are typically, although not exclusively, transcribed in both orientations as long RNA transcripts (Malone et al. 2009) that are parsed into small RNAs. In the case of ping-pong, cluster-derived transcripts are proposed to serve as the source of antisense transposon content, after slicing by Piwi proteins harboring sense piRNAs (Brennecke et al. 2007; Gunawardane et al. 2007). Although the transcriptional regulation of piRNA clusters is not well understood, it appears that at least double-strand cluster transcription may be triggered by a specialized HP1 homolog, Rhino (Klattenhoff et al. 2007). Regardless, many interesting questions remain about the selection of RNA transcripts to feed the transposon silencing program.

Coevolution of Transposons and Their Control Pathways

All sexual organisms must transmit an intact and functioning copy of their genome to offspring. To accomplish this, germline cells must be protected from transposable element mobilization that can interrupt normal gene function, compete for essential cellular localization signals (Van De Bor et al. 2005), and generate widespread genomic instability, inducing catastrophic DNA damage, triggering cellular checkpoints (Theurkauf et al. 2006; Klattenhoff et al. 2007) and a dramatic loss in fertility (Kidwell et al. 1977).

The threat posed by transposons comes on many fronts. Specifically, transposons are incredibly diverse in number, composition, and pattern of expression. This allows individual transposon types to fill specific niches, being expressed at different developmental stages and in particular cell types. As an extreme example, in *Drosophila*, the *gypsy* family of LTR (long terminal repeat) retrotransposon limits its expression to ovarian somatic cells that surround the germline compartment. This initially seems at odds with the imperative to expand copy number in the germline genome. However, *gypsy* family elements, unlike others in *Drosophila*, have regained features of their viral past, reconciling somatic expression and germline transmission.

Gypsy elements belong to the class of errantivirus, functionally related to human endogenous retroviruses and defined by their possession of a virally derived envelope gene (Kim et al. 1994). This allows for the packaging of *gypsy* RNA and protein into viral particles that can infect neighboring cells. In *Drosophila*, when somatic cells of the ovary fail to silence these elements, not only can this process be visualized using electron microscopy, but germline transmission and mutagenesis by *gypsy* elements can also be detected (Pélisson et al. 1994; Song et al. 1994, 1997; Lécher et al. 1997). In response to this apparent threat, organisms have two choices: either target the transmitted copies for destruction in the germline or silence these elements during production in the surrounding somatic cells. *Drosophila* appears to have adapted to regulate these elements at their source in the soma.

The piRNA-based pathway that targets *gypsy* in the soma has evolved unique molecular, cellular, and genomic features that distinguish it from the germline pathway (Fig. 2). First, in contrast to germline cells that express all three Piwi proteins, somatic cells use only the Piwi protein itself to combat transposon activity. Second, only one specialized, single-stranded piRNA cluster, *flamenco*, appears to be expressed in this niche. Molecular genetic

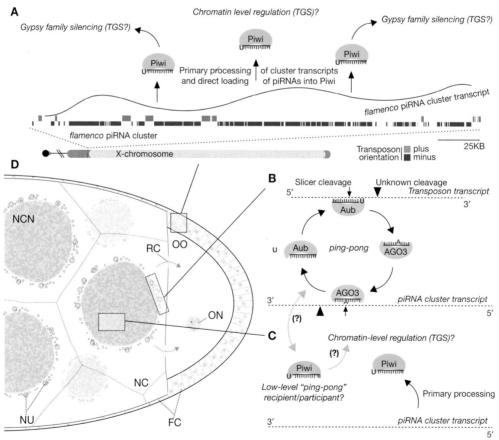

Figure 2. Models of piRNA pathway silencing in distinct tissues of the *Drosophila* ovary. Germline and somatic cells of the *Drosophila* ovary use vastly different mechanisms to combat particular transposon threats. (*A*) In somatic cells of the ovary, *flamenco* cluster transcription precedes its processing to piRNAs that are directly loaded into the Piwi protein, which targets somatically expressed transposons (*gypsy* family) for silencing. Given Piwi's nuclear localization, regulation is likely occurring at the transcriptional level. (TGS) Transcriptional gene silencing. (*B*) The Aub and AGO3 proteins actively cycle in a germline-specific feed forward amplification loop to generate a potent and abundant pool of silencing-capable piRNAs. (*C*) Piwi may act similarly in the germline as in the soma. Here, the Piwi protein may directly bind antisense, cluster-derived piRNAs to silence elements by TGS. Additionally, Piwi may serve as a low-level recipient or participant with AGO3 in the ping-pong cycle. (*D*) Diagram of a mid/late-stage egg chamber of the *Drosophila* ovary. (NC) Nurse cell (germline); (NCN) nurse cell nucleus; (OO) oocyte; (ON) oocyte nucleus; (NU) nuage; (FC) follicular cell (somatic); (RC) ring canal.

studies previously classified this locus a master regulator of *gypsy* family elements (Prud'homme et al. 1995; Mével-Ninio et al. 2007), and *flamenco* was later annotated as a prominent piRNA cluster (Brennecke et al. 2007; Malone et al. 2009). Third, biogenesis of *gypsy* element piRNAs requires no piRNA "ping-pong," but instead involves the processing of long *flamenco* transcripts into primary piRNAs (secondary being "ping-pong"–derived) (Malone et al. 2009). These are directly loaded into the Piwi protein, where they effectively guide silencing of *gypsy* family elements. Here, selective pressures on the orientation of transposon fragments within the cluster allow for the almost exclusive production of antisense piRNAs necessary to silence active transposon transcripts. This feature is not shared with any abundant germline clusters that instead enrich for antisense species in a manner that depends on piRNA "ping-pong."

The mechanisms of piRNA production and silencing by Piwi in somatic cells remain a mystery, although the nuclear localization of Piwi could indicate a role of chromatin-level silencing. This is supported by Piwi association with the chromatin-bound HP1 (Brower-Toland et al. 2007) and links between the Piwi protein and chromatin regulation in other tissues (Pal-Bhadra et al. 2004). Finally, the *flamenco* locus, found within the pericentric heterochromatin of the X chromosome, appears to be under positive selection, because it is identifiable at the same locus in at least two other sequenced Drosophilid species (Malone et al. 2009). Interestingly, although *gypsy* elements remain enriched in these putative *flamencos*, the individual elements present are distinct, indicating that this locus coevolves to combat the changing transposon threat in somatic cells of the *Drosophila* ovary.

Given this apparent pressure for piRNA clusters to maintain and combat the transposons activated in a specific tissue or cell type, it is likely that many or all clusters occupy their own spatial and temporal niches, despite presently being lumped as germline loci. Here, a cluster would become established in the genome to provide either antisense transposon transcripts for the "ping-pong" cycle or antisense primary piRNAs for Piwi. Presumably, a cluster can only act to silence elements with which it is coexpressed, and as is the case for *flamenco*, it also exclusively retains elements that pose a threat within the same transcriptional niche (Fig. 3). Therefore, given the need for piRNA clusters to consume active transposons for subsequent silencing and the propensity for transposons to mobilize into actively transcribed regions of the genome, the molecular coevolution of transposons and their control pathways may be linked at a transcriptional level. In mice, where gonadal maturation occurs in a more synchronized fashion, piRNA cluster expression and their targets are distinct throughout different stages of development (Aravin et al. 2007, 2008).

Long- and Short-term Evolution of Transposon Silencing

One question stemming from this work is whether the integration of a transposable element into a cluster is alone sufficient for its recognition by the pathway and silencing of active elements. By looking at active elements integrated into abundant piRNA clusters, such as the somatic *flamenco* and germline 42AB, it appears that integration leads to the immediate generation of piRNAs, because piRNAs spanning insertion boundaries are detectable (data not shown). However, after insertion within a cluster, there may be a selective pressure to mutate elements away from consensus, meaning that they would lose the capacity to transcribe on their own, mobilize, or generate functional transposon proteins. In fact, most of the elements in clusters are highly diverged from an active form. That being said, there must also be a pressure to restrict element mutation from reaching a point where the sequence similarity required to target active elements is lost. This may well be an example of purifying, or stabilizing, selection, where the extremes (few mutations leaving active transposons or high mutations resulting in a loss of sequence identity) are selected against, and instead a middle ground, where both needs are satisfied, is favored. One such example holds for the *I*-element LINE (long interspersed nucleotide element) retrotransposon, where although a particular lab strain of *Drosophila* contains no active elements, it maintains ancient fragments within a cluster that are the same as those seen in strains that silence colonizing, active *I* elements. This indicates some pressure to diverge but not lose even ancient transposon fragments within a cluster, a point also supported by the conservation of *gypsy* fragments in *flamenco* despite the absence of an active threat (Pélisson et al. 2007). In the case of the *I* element, these fragments alone are insufficient to silence active elements in the lab strain but instead can only act in the presence of a maternally transmitted pool of Piwi-bound piRNAs (Brennecke et al. 2008), a point discussed in more detail below.

This sequence-level conservation of piRNAs lies in stark contrast to the miRNA pathway, which uses individual loci that are under strong selection and are transcribed to generate a single, potent, small RNA to target the regulation of specific mRNAs harboring complementary sites. These individual elements remain conserved throughout eukaryotes, despite otherwise substantial genomic divergences. In contrast, piRNA pathways use abundant piRNA molecules that tile across transposable elements. With many piRNAs diverging significantly from transposon consensus, there is the possibility that drift of sequences within piRNA clusters may be important not only to combat the evolution of transposons that already exist within a genome, but also to provide some measure of protection from horizontal transmission of elements from closely related species (Brennecke et al. 2007).

A key mystery is how the cell selects transposon and cluster transcripts to feed them into the pathway, leaving the majority of other cellular RNAs untouched. For instance, the vast majority of mRNAs, tRNAs, and rRNAs are excluded from metabolism into piRNAs, whereas the vast majority of transposon transcripts, most of which will be translated into protein just as are gene products, are funneled into the pathway. Although upward of 85% of piRNAs are generated from transposons or other repeats, a small fraction is in fact gene derived (Brennecke et al. 2007; Saito et al. 2009). In general, these are from the sense strand, indicating their direct production from the target

sophila species (*melanogaster* and *simulans*) can be mated to each other to generate some viable and fertile offspring (Davis et al. 1996). This effect is not fully penetrant, can show a gender bias in viable offspring, and can be attributed to several processes, including single-gene incompatibilities. However, *D. simulans* contains at least one transposable element, *mariner*, which is absent from *D. melanogaster* (Maruyama and Hartl 1991), further supporting the possibility that transposon differences can contribute, at least at some level, to speciation mating incompatibilities (Rose and Doolittle 1983). Regardless, the phenomenon of hybrid dysgenesis presents an empirically validated phenomenon by which transposon-based mating incompatibilities are apparent within a single generation and likewise may represent a substantial evolutionary force. In this model, the piRNA pathway and its design as a supremely adaptable system is likely critical for maintaining species integrity.

CONCLUSION

Elaborate and specialized small RNA pathways of transposon regulation have developed in diverse organisms and in a multitude of tissues types. However, our understanding of the molecular basis of the acquisition to and adaptation of transposon control mechanisms remains illusive on many levels. First, an organism must have a means of recognizing invading elements as foreign and likewise mark these for silencing. This may very well occur through a single, heritable insertion of a transposon into a piRNA cluster, leading to the initial production of small RNAs and likewise offering the cell a means by which to silence additional elements of the same type and sequence. However, this almost certainly cannot explain the entirety of adaptation to transposon control because there is a clear and essential role for deposited Piwi proteins/piRNAs in kick-starting an effective "ping-pong" biogenesis and silencing program for some elements. Regardless, piRNA clusters no doubt provide a means to capture new transposon threats and to retain a memory of transposon challenge.

One mechanism by which transposon expression and mobilization are combated by the cell would be selecting coexpression of specific piRNA clusters and elements at similar developmental stages (see Fig. 3). Once an element lands in a cluster with which it is coexpressed, that transposon and its control mechanisms will be transcriptionally linked. This implies a diversity of cluster expression patterns that has yet to be investigated. Some support for this model comes from the existence of a potent piRNA silencing pathway in somatic cells of the *Drosophila* ovary, which uses a single piRNA cluster, *flamenco*, in the regulation of *gypsy* family elements (Malone et al. 2009).

An important observation involving the molecular evolution of piRNA pathways involves not the adaptation to control, but simply the conservation of particular loci to combat various invading elements. Here, using the *flamenco* locus as a model of piRNA cluster and heterochromatic selection, it appears that this tract of pericentric heterochromatin has served as a catalog of the specific somatic transposon threat in several distinct species. Interestingly, the type, but not the individual elements occupying these putative *flamenco* clusters, is conserved. This indicates that a heterochromatic domain is selectively maintained by populations with the functional capacity to serve presumably similar roles in preserving a memory of somatic transposon expression. Given the selective advantage to maintain other specified "heterochromatic" loci, such as centromeres and pericentric arrays, it seems feasible that there are diverse and distinct evolutionary pressures acting broadly on both euchromatic and heterochromatic sequences of the genome.

Many questions remain about the relationship between host genomes and colonizing transposable elements, most of which relate to either the silencing mechanisms themselves or the adaptation of populations to control invading elements. Organisms have developed a defense strategy that successfully balances both the deleterious consequences of transposon mobilization and the benefits for genome evolution and adaptive control of future invasions. Either way, small RNA-based pathways represent a key transition by organisms to engage in an elegant and adaptive battle against the selfish needs of transposable elements.

ACKNOWLEDGMENTS

We thank members of the Hannon Laboratory for helpful discussion, especially Julius Brennecke and Ralph Burgess. C.D.M. is a Beckman fellow of the Watson School of Biological Sciences and is supported by a National Science Foundation graduate research fellowship. This work was supported in part by grants from the National Institutes of Health to G.J.H. and a kind gift from Kathryn W. Davis (G.J.H.).

REFERENCES

Aravin AA, Naumova NM, Tulin AV, Vagin VV, Rozovsky YM, Gvozdev VA. 2001. Double-stranded RNA-mediated silencing of genomic tandem repeats and transposable elements in the *D. melanogaster* germline. *Curr Biol* **11:** 1017–1027.

Aravin AA, Lagos-Quintana M, Yalcin A, Zavolan M, Marks D, Snyder B, Gaasterland T, Meyer J, Tuschl T. 2003. The small RNA profile during *Drosophila melanogaster* development. *Dev Cell* **5:** 337–350.

Aravin A, Gaidatzis D, Pfeffer S, Lagos-Quintana M, Landgraf P, Iovino N, Morris P, Brownstein MJ, Kuramochi-Miyagawa S, Nakano T, et al. 2006. A novel class of small RNAs bind to MILI protein in mouse testes. *Nature* **442:** 203–207.

Aravin AA, Sachidanandam R, Girard A, Fejes-Toth K, Hannon GJ. 2007. Developmentally regulated piRNA clusters implicate MILI in transposon control. *Science* **316:** 744–747.

Aravin AA, Sachidanandam R, Bourc'his D, Schaefer C, Pezic D, Toth KF, Hannon GJ. 2008. A piRNA pathway primed by individual transposons is linked to de novo DNA methylation in mice. *Mol Cell* **31:** 785–799.

Baudry C, Malinsky S, Restituito M, Kapusta A, Rosa S, Meyer E, Bétermier M. 2009. PiggyMac, a domesticated *piggyBac* transposase involved in programmed genome rearrangements in the ciliate *Paramecium tetraurelia*. *Genes Dev* **23:** 2478–2483.

Bayes JJ, Malik HS. 2009. Altered heterochromatin binding by a hybrid sterility protein in *Drosophila* sibling species. *Science* **326:** 1538–1541.

Belgnaoui SM, Gosden RG, Semmes OJ, Haoudi A. 2006. Human LINE-1 retrotransposon induces DNA damage and apoptosis in cancer cells. *Cancer Cell Int* **6:** 13.

Blumenstiel JP, Hartl DL. 2005. Evidence for maternally transmitted small interfering RNA in the repression of transposition in *Drosophila virilis*. *Proc Natl Acad Sci* **102:** 15965–15970.

Bregliano JC, Picard G, Bucheton A, Pélisson A, Lavige JM, L'Heritier P. 1980. Hybrid dysgenesis in *Drosophila melanogaster*. *Science* **207:** 606–611.

Brennecke J, Aravin AA, Stark A, Dus M, Kellis M, Sachidanandam R, Hannon GJ. 2007. Discrete small RNA-generating loci as master regulators of transposon activity in *Drosophila*. *Cell* **128:** 1089–1103.

Brennecke J, Malone CD, Aravin AA, Sachidanandam R, Stark A, Hannon GJ. 2008. An epigenetic role for maternally inherited piRNAs in transposon silencing. *Science* **322:** 1387–1392.

Brideau NJ, Flores HA, Wang J, Maheshwari S, Wang X, Barbash DA. 2006. Two Dobzhansky-Muller genes interact to cause hybrid lethality in *Drosophila*. *Science* **314:** 1292–1295.

Brower-Toland B, Findley SD, Jiang L, Liu L, Yin H, Dus M, Zhou P, Elgin SC, Lin H. 2007. *Drosophila* PIWI associates with chromatin and interacts directly with HP1a. *Genes Dev* **21:** 2300–2311.

Bucheton A, Paro R, Sang HM, Pélisson A, Finnegan DJ. 1984. The molecular basis of I-R hybrid dysgenesis in *Drosophila melanogaster*: Identification, cloning, and properties of the I factor. *Cell* **38:** 153–163.

Chambeyron S, Bucheton A. 2005. I elements in *Drosophila*: In vivo retrotransposition and regulation. *Cytogenet Genome Res* **110:** 215–222.

Cox DN, Chao A, Baker J, Chang L, Qiao D, Lin H. 1998. A novel class of evolutionarily conserved genes defined by *piwi* are essential for stem cell self-renewal. *Genes Dev* **12:** 3715–3727.

Cox DN, Chao A, Lin H. 2000. *piwi* encodes a nucleoplasmic factor whose activity modulates the number and division rate of germline stem cells. *Development* **127:** 503–514.

Davis AW, Roote J, Morley T, Sawamura K, Herrmann S, Ashburner M. 1996. Rescue of hybrid sterility in crosses between *D. melanogaster* and *D. simulans*. *Nature* **380:** 157–159.

Demerec M. 1926a. Reddish—A frequently "mutating" character in *Drosophila virilis*. *Proc Natl Acad Sci* **12:** 11–16.

Demerec M. 1926b. Miniature-α—A second frequently mutating character in *Drosophila virilis*. *Proc Natl Acad Sci* **12:** 687–690.

Dimitri P, Bucheton A. 2005. I element distribution in mitotic heterochromatin of *Drosophila melanogaster* reactive strains: Identification of a specific site which is correlated with the reactivity levels. *Cytogenet Genome Res* **110:** 160–164.

Elbashir SM, Lendeckel W, Tuschl T. 2001. RNA interference is mediated by 21- and 22-nucleotide RNAs. *Genes Dev* **15:** 188–200.

Esnault C, Maestre J, Heidmann T. 2000. Human LINE retrotransposons generate processed pseudogenes. *Nat Genet* **24:** 363–367.

Gerke J, Lorenz K, Cohen B. 2009. Genetic interactions between transcription factors cause natural variation in yeast. *Science* **323:** 498–501.

Gilbert N, Lutz-Prigge S, Moran JV. 2002. Genomic deletions created upon LINE-1 retrotransposition. *Cell* **110:** 315–325.

Girard A, Sachidanandam R, Hannon GJ, Carmell MA. 2006. A germline-specific class of small RNAs binds mammalian Piwi proteins. *Nature* **442:** 199–202.

Gunawardane LS, Saito K, Nishida KM, Miyoshi K, Kuwamura Y, Nagami T, Siomi H, Siomi MC. 2007. A slicer-mediated mechanism for repeat-associated siRNA 5′ end formation in *Drosophila*. *Science* **315:** 1587–1590.

Hannon GJ. 2002. RNA interference. *Nature* **418:** 244–251.

Houwing S, Kamminga LM, Berezikov E, Cronembold D, Girard A, van den Elst H, Filippov DV, Blaser H, Raz E, Moens CB, et al. 2007. A role for Piwi and piRNAs in germ cell maintenance and transposon silencing in zebrafish. *Cell* **129:** 69–82.

Houwing S, Berezikov E, Ketting RF. 2008. Zili is required for germ cell differentiation and meiosis in zebrafish. *EMBO J* **27:** 2702–2711.

Kaminker JS, Bergman CM, Kronmiller B, Carlson J, Svirskas R, Patel S, Frise E, Wheeler DA, Lewis SE, Rubin GM, et al. 2002. The transposable elements of the *Drosophila melanogaster* euchromatin: A genomics perspective. *Genome Biol* **3:** RESEARCH0084.

Kazazian HH. 2004. Mobile elements: Drivers of genome evolution. *Science* **303:** 1626–1632.

Kidwell MG. 1983. Evolution of hybrid dysgenesis determinants in *Drosophila melanogaster*. *Proc Natl Acad Sci* **80:** 1655–1659.

Kidwell MG, Kidwell JF, Sved JA. 1977. Hybrid dysgenesis in *Drosophila melanogaster*: A syndrome of aberrant traits including mutation, sterility and male recombination. *Genetics* **86:** 813–833.

Kim A, Terzian C, Santamaria P, Pélisson A, Prud'homme N, Bucheton A. 1994. Retroviruses in invertebrates: The gypsy retrotransposon is apparently an infectious retrovirus of *Drosophila melanogaster*. *Proc Natl Acad Sci* **91:** 1285–1289.

Klattenhoff C, Bratu DP, McGinnis-Schultz N, Koppetsch BS, Cook HA, Therkauf WE. 2007. *Drosophila* rasiRNA pathway mutations disrupt embryonic axis specification through activation of an ATR/Chk2 DNA damage response. *Dev Cell* **12:** 45–55.

Lau NC, Seto AG, Kim J, Kuramochi-Miyagawa S, Nakano T, Bartel DP, Kingston RE. 2006. Characterization of the piRNA complex from rat testes. *Science* **313:** 363–367.

Lau NC, Robine N, Martin R, Chung WJ, Niki Y, Berezikov E, Lai EC. 2009. Abundant primary piRNAs, endo-siRNAs, and microRNAs in a *Drosophila* ovary cell line. *Genome Res* **19:** 1776–1785.

Lécher P, Bucheton A, Pélisson A. 1997. Expression of the *Drosophila* retrovirus gypsy as ultrastructurally detectable particles in the ovaries of flies carrying a permissive *flamenco* allele. *J Gen Virol* **78:** 2379–2388.

Lee HY, Chou JY, Cheong L, Chang NH, Yang SY, Leu JY. 2008. Incompatibility of nuclear and mitochondrial genomes causes hybrid sterility between two yeast species. *Cell* **135:** 1065–1073.

Lepère G, Bétermier M, Meyer E, Duharcourt S. 2008. Maternal noncoding transcripts antagonize the targeting of DNA elimination by scanRNAs in *Paramecium tetraurelia*. *Genes Dev* **22:** 1501–1512.

Li C, Vagin VV, Lee S, Xu J, Ma S, Xi H, Seitz H, Horwich MD, Syrzycka M, Honda BM, et al. 2009. Collapse of germline piRNAs in the absence of Argonaute3 reveals somatic piRNAs in flies. *Cell* **137:** 509–521.

Malone CD, Hannon GJ. 2009. Small RNAs as guardians of the genome. *Cell* **136:** 656–668.

Malone CD, Anderson AM, Motl JA, Rexer CH, Chalker DL. 2005. Germ line transcripts are processed by a Dicer-like protein that is essential for developmentally programmed genome rearrangements of *Tetrahymena thermophila*. *Mol Cell Biol* **25:** 9151–9164.

Malone CD, Brennecke J, Dus M, Stark A, McCombie WR, Sachidanandam R, Hannon GJ. 2009. Specialized piRNA pathways act in germline and somatic tissues of the *Drosophila* ovary. *Cell* **137:** 522–535.

Marin L, Lehmann M, Nouaud D, Izaabel H, Anxolabéhère D, Ronsseray S. 2000. P-element repression in *Drosophila melanogaster* by a naturally occurring defective telomeric P copy. *Genetics* **155:** 1841–1854.

Maruyama K, Hartl DL. 1991. Evolution of the transposable element mariner in *Drosophila* species. *Genetics* **128:** 319–329.

McClintock B. 1950. The origin and behavior of mutable loci in maize. *Proc Natl Acad Sci* **36:** 344–355.

Mével-Ninio M, Pélisson A, Kinder J, Campos AR, Bucheton A. 2007. The *flamenco* locus controls the *gypsy* and *ZAM* retroviruses and is required for *Drosophila* oogenesis. *Genetics* **175:** 1615–1624.

Miller WJ, Hagemann S, Reiter E, Pinsker W. 1992. P-element homologous sequences are tandemly repeated in the genome of *Drosophila guanche*. *Proc Natl Acad Sci* **89:** 4018–4022.

Mochizuki K, Gorovsky MA. 2004. Small RNAs in genome rearrangement in *Tetrahymena*. *Curr Opin Genet Dev* **14:** 181–187.

Mochizuki K, Gorovsky MA. 2005. A Dicer-like protein in *Tetrahymena* has distinct functions in genome rearrangement, chromosome segregation, and meiotic prophase. *Genes Dev* **19**: 77–89.

Mochizuki K, Fine NA, Fujisawa T, Gorovsky MA. 2002. Analysis of a *piwi*-related gene implicates small RNAs in genome rearrangement in *Tetrahymena*. *Cell* **110**: 689–699.

Niki Y, Yamaguchi T, Mahowald AP. 2006. Establishment of stable cell lines of *Drosophila* germ-line stem cells. *Proc Natl Acad Sci* **103**: 16325–16330.

Pal-Bhadra M, Leibovitch BA, Gandhi SG, Rao M, Bhadra U, Birchler JA, Elgin SC. 2004. Heterochromatic silencing and HP1 localization in *Drosophila* are dependent on the RNAi machinery. *Science* **303**: 669–672.

Pélisson A. 1981. The I–R system of hybrid dysgenesis in *Drosophila melanogaster:* Are I factor insertions responsible for the mutator effect of the I–R interaction? *Mol Gen Genet* **183**: 123–129.

Pélisson A, Bregliano JC. 1987. Evidence for rapid limitation of the I element copy number in a genome submitted to several generations of I-R hybrid dysgenesis in *Drosophila melanogaster*. *Mol Gen Genet* **207**: 306–313.

Pélisson A, Song SU, Prud'homme N, Smith PA, Bucheton A, Corces VG. 1994. Gypsy transposition correlates with the production of a retroviral envelope-like protein under the tissue-specific control of the *Drosophila flamenco* gene. *EMBO J* **13**: 4401–4411.

Pélisson A, Payen-Groschêne G, Terzian C, Bucheton A. 2007. Restrictive *flamenco* alleles are maintained in *Drosophila melanogaster* population cages, despite the absence of their endogenous *gypsy* retroviral targets. *Mol Biol Evol* **24**: 498–504.

Picard G. 1976. Non-Mendelian female sterility in *Drosophila melanogaster:* Hereditary transmission of I factor. *Genetics* **83**: 107–123.

Picard G, L'Heritier P. 1971. A maternally inherited factor inducing sterility in *D. melanogaster*. Drosophila *Inf Serv* **46**: 54.

Prud'homme N, Gans M, Masson M, Terzian C, Bucheton A. 1995. *Flamenco*, a gene controlling the *gypsy* retrovirus of *Drosophila melanogaster*. *Genetics* **139**: 697–711.

Reinhart BJ, Bartel DP. 2002. Small RNAs correspond to centromere heterochromatic repeats. *Science* **297**: 1831.

Ronsseray S, Lehmann M, Anxolabéhère D. 1991. The maternally inherited regulation of P elements in *Drosophila melanogaster* can be elicited by two P copies at cytological site 1A on the X chromosome. *Genetics* **129**: 501–512.

Ronsseray S, Marin L, Lehmann M, Anxolabéhère D. 1998. Repression of hybrid dysgenesis in *Drosophila melanogaster* by combinations of telomeric P-element reporters and naturally occurring P elements. *Genetics* **149**: 1857–1866.

Rose MR, Doolittle WF. 1983. Molecular biological mechanisms of speciation. *Science* **220**: 157–162.

Saito K, Inagaki S, Mituyama T, Kawamura Y, Ono Y, Sakota E, Kotani H, Asai K, Siomi H, Siomi MC. 2009. A regulatory circuit for *piwi* by the large Maf gene *traffic jam* in *Drosophila*. *Nature* **461**: 1296–1299.

Simonelli M, Bazin C, Pélisson A, Bucheton A. 1988. Transposable and nontransposable elements similar to the I factor involved in inducer-reactive (IR) hybrid dysgenesis in *Drosophila melanogaster* coexist in various *Drosophila* species. *Proc Natl Acad Sci* **85**: 1141–1145.

Slotkin RK, Martienssen R. 2007. Transposable elements and the epigenetic regulation of the genome. *Nat Rev Genet* **8**: 272–285.

Slotkin RK, Vaughn M, Borges F, Tanurdzić, Becker JD, Feijó JA, Martienssen RA. 2009. Epigenetic reprogramming and small RNA silencing of transposable elements in pollen. *Cell* **136**: 461–472.

Song SU, Gerasimova T, Kurkulos M, Boeke JD, Corces VG. 1994. An env-like protein encoded by a *Drosophila* retroelement: Evidence that *gypsy* is an infectious retrovirus. *Genes Dev* **8**: 2046–2057.

Song SU, Kurkulos M, Boeke JD, Corces VG. 1997. Infection of the germ line by retroviral particles produced in the follicle cells: A possible mechanism for the mobilization of the *gypsy* retroelement of *Drosophila*. *Development* **124**: 2789–2798.

Symer DE, Connelly C, Szak ST, Caputo EM, Cost GJ, Parmigiani G, Boeke JD. 2002. Human l1 retrotransposition is associated with genetic instability in vivo. *Cell* **110**: 327–338.

Theurkauf WE, Klattenhoff C, Bratu DP, McGinnis-Schultz N, Koppetsch BS, Cook HA. 2006. rasiRNAs, DNA damage, and embryonic axis specification. *Cold Spring Harbor Symp Quant Biol* **71**: 171–180.

Vagin VV, Sigova A, Li C, Seitz H, Gvozdev V, Zamore PD. 2006. A distinct small RNA pathway silences selfish genetic elements in the germline. *Science* **313**: 320–324.

Van De Bor V, Hartswood E, Jones C, Finnegan D, Davis I. 2005. *gurken* and the *I* factor retrotransposon RNAs share common localization signals and machinery. *Dev Cell* **9**: 51–62.

Volpe TA, Kidner C, Hall IM, Teng G, Grewal SI, Martienssen RA. 2002. Regulation of heterochromatic silencing and histone H3 lysine-9 methylation by RNAi. *Science* **297**: 1833–1837.

White SE, Habera LF, Wessler SR. 1994. Retrotransposons in the flanking regions of normal plant genes: A role for copia-like elements in the evolution of gene structure and expression. *Proc Natl Acad Sci* **91**: 11792–11796.

Drosophila Brain Development: Closing the Gap between a Macroarchitectural and Microarchitectural Approach

A. Cardona,[1,2] S. Saalfeld,[3] P. Tomancak,[3] and V. Hartenstein[1]

[1]*Department of Molecular, Cell, and Developmental Biology, University of California at Los Angeles, Los Angeles, California 90095;* [2]*Institute of Neuroinformatics, Uni/ETH Zurich, CH-8057 Zurich, Switzerland;* [3]*Max-Planck Institute for Cell Biology and Genetics, Dresden, Germany*

Correspondence: volkerh@mcdb.ucla.edu

Neurobiologists address neural structure, development, and function at the level of "macrocircuits" (how different brain compartments are interconnected; what overall pattern of activity they produce) and at the level of "microcircuits" (how connectivity and physiology of individual neurons and their processes within a compartment determine the functional output of this compartment). Work in our lab aims at reconstructing the developing *Drosophila* brain at both levels. Macrocircuits can be approached conveniently by reconstructing the pattern of brain lineages, which form groups of neurons whose projections form cohesive fascicles interconnecting the compartments of the larval and adult brain. The reconstruction of microcircuits requires serial section electron microscopy, due to the small size of terminal neuronal processes and their synaptic contacts. Because of the amount of labor that traditionally comes with this approach, very little is known about microcircuitry in brains across the animal kingdom. Many of the problems of serial electron microscopy reconstruction are now solvable with digital image recording and specialized software for both image acquisition and postprocessing. In this chapter, we introduce our efforts to reconstruct the small *Drosophila* larval brain and discuss our results in light of the published data on neuropile ultrastructure in other animal taxa.

Studies of nervous system architecture and function typically approach the brain at two different levels of resolution: macrocircuitry and microcircuitry. The macrocircuitry-oriented approach asks these questions: What functions can be attributed to brain compartments such as the mammalian primary visual cortex or lateral geniculate nucleus and what is the connectivity between these and other brain compartments? In contrast, the study of microcircuitry zooms in on neurons, their dendrites, axons, and synapses. Thus, the way in which a given neuron is tuned to a specific input stimulus, or the pattern of activity triggered in this neuron when providing a specific input, depends on the distribution of excitatory and inhibitory synapses that connect the neuron with its neighbors (Douglas and Martin 1998; Silberberg et al. 2002; Toledo-Rodriguez et al. 2005; Silberberg 2008). The analysis of microcircuits is of great importance. All acts of fine motor control, memory formation, and cognition can only be understood if the microcircuitry within the brain compartments dealing with these functions is known. Likewise, the insight into psychiatric disease mechanisms and their pharmacological treatment requires that brain microcircuitry be known. For example, recent findings suggest that diseases such as schizophrenia can be understood in terms of abnormalities in the microcircuitry of the prefrontal cortex (Winterer and Weinberger 2004; Rolls et al. 2008).

This useful conceptualization of the nervous system as being composed of interconnected, structurally defined compartments, constituting microcircuits integrated into macrocircuits, also applies to the brain of invertebrates, such as the fruit fly *Drosophila*. The nervous system of *Drosophila* (and insects in general) is formed by a relatively small number of genetically and structurally defined modules, the neural lineages. The ventral nerve cord and subesophageal ganglion (containing the circuits controlling locomotion, flight, and feeding) are built of ~80 bilaterally symmetric pairs of lineages; the central brain, a mostly sensory and associative center, is formed by 100 paired lineages (Goodman and Doe 1993; Younossi-Hartenstein et al. 1996; Urbach and Technau 2003; Truman et al. 2004). Each lineage is derived from an asymmetrically dividing stem cell, called the neuroblast, that is born in the early embryo (Fig. 1A) (Hartenstein et al. 2008a,b). Neuroblasts and the lineages they produce represent genetic modules ("units of gene expression"). The expression pattern of more than 40 transcription factors in specific embryonic neuroblasts has been described previously (Urbach and Technau 2003). A given transcription factor becomes active in one, or a small number of, neuroblasts; a particular neuroblast thereby acquires a "genetic address," consisting of a specific set of transcription factors active in this cell. It is thought that this genetic address will essentially be involved in shaping the morphology and function of the lineage of neurons produced by the neuroblast.

Lineages also form structural modules. Thus, neurons that belong to one lineage remain together throughout development, forming compact clusters of cells (Fig.1B). More importantly, axons emitted by neurons of one lineage also form a coherent fascicle, the primary and secondary lineage axon tracts. This means that neurons of

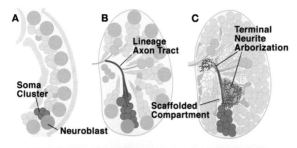

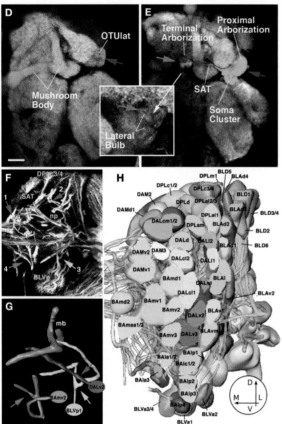

Figure 1. Developmental and structural characteristics of wild-type lineages. (*A–C*) Lineages as units of gene expression, projection, and connectivity. Stereotyped population of neuroblasts generates neurons in the embryo and larva (*A*). Neurons belonging to one lineage form a cohesive cluster and project their axons in one fascicle (*B*). Terminal branches of neurons of one lineage arborize in specific neuropile compartments (*C*). (*D,E*) Z-Projection of adult brain hemisphere labeled with anti-Bruchpilot (Nc82; Kittel et al. 2006) to visualize neuropile compartments (white). (*E*) One lineage, DALcl1, is labeled by expression of green fluorescent protein (GFP). Note dense proximal arborization restricted to lateral domain of optic tubercle (OTUlat; one of the optic foci); distal arborization is restricted to the lateral bulb, one of the input regions of the central complex. (*F*) Z-Projection of 10 successive 1-µm confocal cross sections at the level of the central neuropile. Secondary lineages, their axon tracts (SATs), and neuropile fascicles formed by convergence of SATs are labeled with antineurotactin antibody (white). Clusters of somata (so) belonging to lineages are located in the cortex; axon tracts project centripetally into the neuropile (np) Arrows point at lineages representing the types of SAT trajectories observed: SAT is unbranched and enters the neuropile in a straight course (*1*; DPMm lineage) or after a sharp turn at the cortex–neuropile boundary (*2*; DPLc3/4). (*3*) SAT bifurcates into two branches at cortex–neuropile boundary (BLVp1/2); (*4*) distal part of SAT bifurcates in neuropile (BAmv2). (*G*) Digital models of three representative lineage tracts illustrating typical branching behavior of SATs (DALv2: straight unbranched entry into neuropile; BLVp1: bifurcation at point of entry into neuropile [arrowhead]; BAmv2: bifurcation in distal leg of SAT). (*H*) Three-dimensional digital models of all clusters of neuronal somata representing all lineages of one brain hemisphere (anterior view). The polar region of the cortex was removed for a clearer view of lineages. Bar, 20 µm. (*F–H*, Modified from Fung et al. 2009 [©Elsevier].)

one lineage share their principal trajectory; they form a "unit of projection" (Fig.1C–G). Lineages thereby represent the most appropriate structural/developmental units of brain macrocircuitry. On the basis of their characteristic location and axon tract, we have generated an atlas of all lineages of the central brain for the larval stage (Fig. 1H) (Pereanu and Hartenstein 2006). Attempts are currently under way to link the larval lineage map with the adult stage, when each lineage has completed its terminal arborization (Pereanu et al. 2009).

We briefly summarize our recent findings pertaining to the structure and development of neural lineages of the *Drosophila* brain. We then outline our approach to reconstruct microcircuitry in the fly brain using computer-aided serial electron microscopy, an approach that is guided by the lineage-centered macroarchitectural map of the fly brain. Finally, we describe first results shedding light on *Drosophila* microcircuitry and discuss these data in the context of brain evolution.

LINEAGE-BASED ANALYSIS OF *DROSOPHILA* BRAIN STRUCTURE AND DEVELOPMENT

Using clonal marking techniques and specific Gal4 driver lines (Brand and Perrimon 1993), we have analyzed representative *Drosophila* brain lineages at all stages of development (Larsen et al. 2009). Most lineages have a number of important characteristics in common; our focus was on these generic lineage features. The early neurons of a lineage generated during embryogenesis (primary neurons; 15–20 neurons per neuroblast for the large majority of lineages) stay together as a coherent cluster (Fig. 2A,B). Likewise, axons of each lineage form a coherent bundle (primary axon tract [PAT]) that follows a stereotyped pathway in the neuropile (Younossi-Hartenstein et al. 2006). Apoptotic cell death removes an average of 30%–40% of neurons from the primary lineages around the time of hatching. Secondary neurons generated later, during the larval period, also form clusters that stay together all the

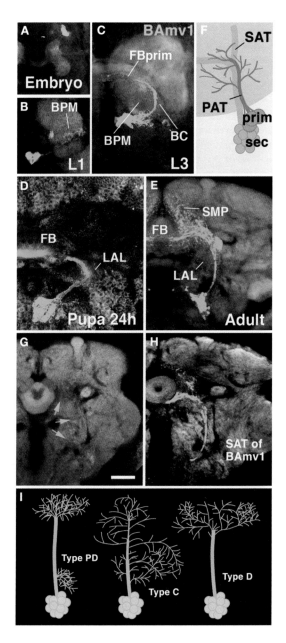

Figure 2. Morphogenesis of a brain lineage from embryo to adult. (*A–E*) Z-Projections of confocal sections of one brain hemisphere in which BAmv1 lineage is labeled by GFP driven by the line *per*-Gal4 (Kaneko and Hall 2000). BAmv1 has a conspicuous crescent-shaped tract, projecting first posteriorly, then dorsolaterally, and finally dorsomedially toward the primordium of the fan-shaped body (FBprim), which forms part of the CPM compartment of the larval brain. Arborizations of primary neurons occur in BC, BPM, and FBprim compartments (*B,C*). Secondary axons follow the same trajectory and branch in the lateral accessory lobe (LAL), fan-shaped body (FB), and superior medial protocerebrum (SMP; *D,E*). (*F*) Schematic illustrating that SAT typically fasciculates with, or at least grows close to, PAT of the corresponding lineage. (*G,H*) Secondary axon tracts develop into long fiber bundles of adult brain. (*G*) Frontal confocal section of adult brain hemisphere labeled with anti-Bruchpilot (Nc82; neuropile; white). (*H*) Secondary neurons of the BAmv1 lineage are labeled by GFP (driven by *per*-Gal4). Note that the coherent secondary axon tract of BAmv1 now forms a long fiber bundle that is visible as a Nc82-negative (i.e., synapse-free) "tunnel;" indicated by green arrows (*G*). (*I*) Schematic representation of different types of lineages encountered in *Drosophila* brain (PD, separate proximal and distal arborization; C, continuous arborization; D, distal arborization). Bar, (*G*) 20 μm (all photographic panels at same scale).

way to the adult stage and that form cohesive axon bundles (secondary axon tract [SAT]). SATs project into the neuropile compartment visited by the corresponding primary axon tract (Fig. 2C,F). SATs develop into the long fiber fascicles that interconnect the different compartments of the adult brain (Fig. 2D,E). These fascicles can be easily recognized by labeling brains with global markers. For example, labeling with synaptic markers such as the anti-DNcad or anti-Brp antibody fascicles (which only contain long fibers and lack synapses) stand out as signal negative channels (Fig. 2G,H).

With respect to overall arbor geometry, we distinguish between three types of lineages (Fig. 2I) (Larsen et al. 2009): Type-PD ("proximodistal") lineages are characterized by distinct, spatially separate, proximal (close to soma) and distal arborizations; type-C ("continuous") lineages have arborizations distributed more or less evenly along the entire length of their tract; and type-D lineages ("distal") lack proximal arborizations (their fiber tract extends for a considerable distance into the neuropile before branching into a more or less complex distal arbor).

Arborizations of lineages, in particular those of type PD, are restricted to distinct neuropile compartments; we propose that compartments are "scaffolded" by individual lineages or small groups thereof. Primary lineages set up the compartment map already in the embryo. Compartments then grow during the larval period simply by an increase in arbor volume of primary neurons. Arbors of secondary neurons form within the larval compartments, resulting in smaller compartment subdivisions and additional, adult-specific compartments. We proposed that each compartment has its own "scaffolding lineage" (or set of scaffolding lineages) that would be defined in the following manner. (1) During development, the outgrowth of neurites from a lineage S generates the compartment S′. If S is deleted, S′ also does not form; this has been shown for the calyx, the compartment scaffolded by the four MB (mushroom body) lineages (Ito et al. 1997), and for compartments of the central complex (JK Lovick and V Hartenstein, unpubl.). (2) The arborization of lineage S forms a dense matrix of terminal fibers on which synapses of S neurons themselves, as well as extrinsic neurons that enter compartment S′ from the outside, are made. Again, the calyx provides an example in this case: Electron microscopy (EM) investigations have shown that the majority of the postsynaptic terminal neurites belong to neurons of the MB lineages (Yasuyama et al. 2002). Our lab and others have found numerous other lineages whose proximal (or sometimes distal, axonal) terminal arborizations, in the adult brain, are highly focused in a small compartment. As an example in point, Figure 1, D and E, shows the labeled lineage DALcl1 with dense terminal arborizations in a small subcompartment, the lateral optic tubercle (OTUlat), and terminal arborizations confined even more narrowly to the lateral bulb, an input region of the central complex.

In summary, our data support the idea that lineages form the developmental/anatomical substrate of *Drosophila* brain macrocircuitry: Compartments and the long axon tracts interconnecting them can be assigned to specific (sets of) lineages. The reconstruction and digital representation of all lineages, including their axonal and dendritic arborization and interconnecting axon fascicle, will add up to a complete map of fly brain macrocircuitry.

ANALYSIS OF MICROCIRCUITRY

The anatomical reconstruction of microcircuitry requires the documentation of, at the level of single synapses and neuronal processes, how neurons in a given (small) volume of the brain are interconnected. Due to their small size, which lies in the range of 0.1–0.5 μm, fine neurites and their synaptic contacts can be conclusively shown only electron microscopically. However, the acquisition of complete series of EM sections and their photographic documentation and analysis requires a considerable effort, and therefore, studies of microcircuitry have traditionally been restricted to small parts of neurons or neuropile compartments in insects or other invertebrates (see, e.g., Watson and Burrows 1983; Meinertzhagen and O'Neil 1991; Yasuyama et al. 2002). The relatively complete reconstruction of the miniature *Caenorhabditis elegans* central nervous system, containing less than 400 unbranched neurons, represents a notable exception (White et al. 1986). The problem of image acquisition and reconstruction is now solvable with digital image recording and specialized software for image acquisition and postprocessing, and we and other groups have begun to generate stacks of digitized images from serial EM sections. These stacks can be segmented and analyzed in their entirety, which makes it possible to reconstruct the way a neuron (or neurite) is connected to its immediate neighbors.

It should be pointed out that even with the digital technology available today, digital serial EM-based analysis is still feasible only for relatively small objects in the millimeter range. That means that for a large vertebrate brain, only small "aliquots" of brain tissue can be processed. But this may be sufficient if one assumes (and evidence for this assumption is accumulating; for review, see Kozloski et al. 2001; Silberberg et al. 2002) that large brain areas such as the neocortex are essentially built in a stereotypical manner; this means that the microcircuitry characteristic of one cortical area is highly similar to that in a different area, and thus many of the principles of microcircuitry that emerge from the analysis of one tissue "aliquot" can be generalized to other cortical domains. In our view, this concept does not argue against location-specific properties. It merely suggests that it is practical to first undertake a comprehensive analysis of stereotyped aspects of microcircuitry; this then will make it easier to get at additional site-specific properties.

Insect brains offer the advantage of a much smaller size. The early first instar larval (L1) brain of *Drosophila*, formed by ~1500 differentiated and functional nerve cells (Larsen et al. 2009), has a diameter of ~50 μm. The neuropile that forms the center of a brain hemisphere measures less than 30 μm. Given these specifications, we are currently working with ~500 sections of 60-nm thickness; a more complete series of sections that includes both brain hemispheres and the ventral nerve cord is under way. We aimed at a resolution of ~2–3 nm per pixel; synaptic contacts and fine processes (diameter of ~100 nm), or even synaptic vesicles (20–30 nm), can be clearly resolved at this resolution. The size of the digitized image of one section contains ~15,000 x 15,000 pixels. We have packed the entire postprocessing pipeline in our own custom software TrakEM2, which is based on ImageJ, a National Institutes of Health (NIH)–sponsored image processing platform (www.ini.uzh.ch/~acardona/trakem2.html). The software allows us to stitch individual photographs covering one section into one seamless montage, to register montages of each section, and to navigate the resulting stack of sections efficiently (Saalfeld et al. 2009; Cardona et al., unpubl.).

MICROCIRCUITRY DATA ANALYSIS AND RECONSTRUCTION: MERGING EM AND LM DATA

When browsing EM sections, the staggering amount of data collected becomes very difficult to process without a priori "low-resolution knowledge" of the architecture of the object, i.e., the L1 *Drosophila* brain. The navigation of TEM (transmission electron microscopy) images must be guided by known, labeled, and registered confocal stacks that provide cues regarding the position of major "macroarchitectural" landmarks such as axon bundles or neuropile compartment boundaries, thus providing the necessary context for the analysis of microarchitectural components of the neuropile. In other words, if users want to zoom in on the pattern of connectivity in a small volume of neuropile, they need to know the compartment in which this volume is located in or the lineages and major tracts that connect the compartment to the rest of the brain. The TrakEM2 pipeline is designed to embed the analysis of the EM stack at the microcircuit level into the light-microscopically derived macroarchitectural framework.

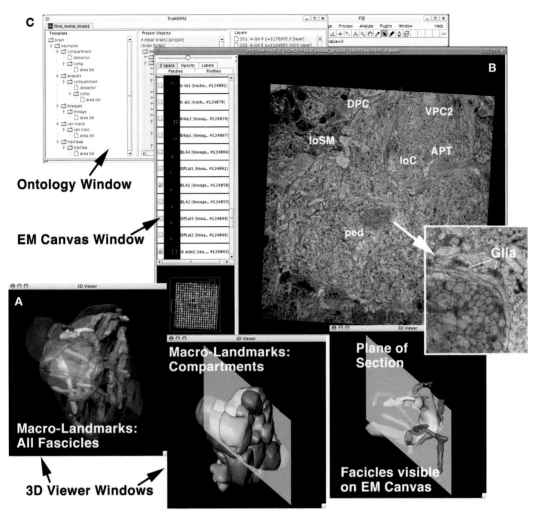

Figure 3. Graphical user interface of TrakEM2. The EM canvas presents the EM stack. The user can scroll and navigate through all sections in the "Google Map" style. Identified structures (e.g., compartments and lineage tracts) can be segmented. The ontology window displays all segmented objects as a hierarchically organized list. The interactive 3D viewer windows show selected objects segmented from the EM stack.

We have defined a set of lineages, compartments, and major axon tracts that can be recognized at all developmental stages in the adult and that serve as macroarchitectural landmarks. In our preliminary L1 EM stack, we could identify most of these landmarks, including compartments, neuropile fascicles, and individual lineage tracts (Fig. 3A). Thus, the lineage-associated primary axon tracts of the L1 brain contain 6–20 tightly bundled, straight axons that are visible in the EM stack if one knows where to look for them (Fig. 3B). Moreover, PATs of multiple lineages converge to form major axonal "thoroughfares" in the neuropile, such as the antennocerebral tract, peduncle, or longitudinal central tract. These tracts are all associated with high concentrations of glial processes that further facilitate their recognition in the EM stack. Glial densities also define many of the compartment boundaries (Younossi-Hartenstein et al. 2003). After identifying and segmenting the macroarchitectural landmarks in the EM stack, they become the objects of an intrinsic macromodel. Thus, with the help of the drawing tool, label fields are created on each section; the label fields stacked along the z axis form a given object for which a surface is generated and which can be displayed as an interactive digital three-dimensional (3D) model.

Figure 3 shows the current graphic user interface of TrakEM2. It is composed of three simultaneously active windows: the ontology window (Fig. 3C), EM (raw data) canvas (Fig. 3B), and 3D viewer windows (Fig. 3A). All objects of the model are displayed as a hierarchical list in the ontology window; here, individual objects can be activated or inactivated, and properties of the digital rendering of each object, such as color or transparency, can be changed. Activating an object in the ontology window will lead to its appearance in the EM canvas window, as well as in the 3D viewer. In both EM canvas and 3D viewer windows, one can further interact with the object in manifold ways. As a result, TrakEM2 allows us to focus on any part of the neuropile in the EM canvas and, at the same time, look up in the intrinsic macromodel (displayed in the 3D viewer) where we are and the lineages or tracts that are close by.

QUANTITATIVE ANALYSIS OF CONNECTIVITY IN "MICROVOLUMES"

Given that the problems of sectioning, EM imaging, and data handling of complete L1 brains are in principle solved, our ultimate goal is to reconstruct all neurons and their connections in their entirety. However, even a complete stack without any gaps will be difficult to register "perfectly" in toto, given the current technology. "Perfectly" would imply that, given the thinness of terminal processes, any pixel of the 15,000 × 15,000 pixel montage has to be <25 pixels ($\cong$100 nm) removed from its "true" x/y position. Considering the fact that sectioning, heating (electron beam), and handling add up to deformations of all sections (independent of one another), it will be understandable that it is currently impossible to register section montages with that accuracy. The strategy is therefore to break down the overall EM stack volume into smaller "microvolumes," in the range of 5 × 5 × 5 µm, and reconstruct these. TrakEM2 allows us to efficiently "cut out" microvolumes from the total stack and reregister them automatically, as well as manually, at an accuracy level high enough for segmenting even fine processes. Multiple microvolumes can be sampled from any desired position, guided by the macrolandmarks. It should be noted that, even irrespective of the technical factors that suggest the microvolume approach, this approach seems also to be conceptually well justified. Thus, as pointed out for the vertebrate brain above, it stands to reason that the *Drosophila* brain is structured in a stereotypical way: The biological "algorithm" underlying the neuronal wiring in a given microvolume may well resemble that of a neighboring volume. It seems appropriate to sample and compare microvolumes from different compartments, formed by different lineages and reached by different types of sensory inputs, before eventually reconstructing the entire neuropile in toto. The following fundamental parameters of microcircuitry can be gained from the microvolumes:

- Types of neurites (diameter, shape, branching behavior) that exist.
- Density/directionality of neurites (of different types) within a microvolume.
- Density of branchpoints and how branching of different neurites is coordinated.
- Density and distribution of input/output synapses, relative to neurite type, neurite directionality and branching behavior.

In the following section, we summarize our recent findings (Cardona et al., unpubl.) that have begun to provide information regarding these parameters.

DROSOPHILA BRAIN MICROCIRCUITRY: NEURITE PROFILES AND DISTRIBUTION OF SYNAPSES

The brain neuropile contains neurites connected by synapses, as well as glia. Recognizable by their lamellar shape and high electron density, glial processes form more or less continuous sheaths around the neuropile surface, as well as around many compartments and fascicles. Within neuropile compartments, fine glial processes are interspersed with terminal neurites and synapses. Neurites come in several different classes that relate in a systematic way to presynaptic and postsynaptic contact sites (Fig. 4). We distinguish among axiform, varicose, globular, and dendritiform neurites (Fig. 4A–C). These generic classes of neurites can be found in all compartments of the brain. Axiform neurites (axis = axle) are straight, unbranched processes of even diameter, ranging between 0.2 µm and 0.4 µm. These processes typically form bundles. The PAT emitted by the neurons belonging to one lineage consists of axiform neurites. Within the neuropile, thick bundles of axiform neurites (created by the coalescence of multiple PATs) form the long fiber tracts, such the antenno–protocerebral tract. Varicose neurites (varix = dilated vein) comprise most of the volume of the neuropile. They represent branched processes that vary in diameter and change direction. Along their length, thin segments (0.2–0.4 µm) alternate with swellings ("varicosities") that measure 0.5–1.5 µm. Globular neurites (globus = round body) contain one (terminal) or several large diameter (1.5–3 µm) round or irregularly shaped segments. These formations structurally and functionally resemble closely the end plates, or "boutons," of motor axons. Dendritiform neurites (dendron = tree) are thin, highly branched processes. They change direction frequently. Short, terminal branches of these "trees" are in the range of 0.1 µm.

Synapses are defined by the characteristic presynaptic site, an electron-dense patch of membrane bordered by the T bar, a cytoplasmic specialization involved in tethering and docking of synaptic vesicles (Fig. 4D) (Feeney et al. 1998; Kittel et al. 2006; Prokop and Meinertzhagen 2006). Synaptic vesicles can be observed at or near presynaptic sites. Presynaptic contact sites are relatively uniform in size, ranging from 0.1 µm to 0.3 µm, and are very predominantly found on large-diameter segments of neurites, i.e., the swellings of varicose and globular neurites (Fig. 4E); these neurites then constitute the output, or axonal, branches within the neuropile. Postsynaptic sites are less conspicuous membrane densities lacking T bars or synaptic vesicles; they are found almost exclusively on thin dendritiform neurites and, occasionally, thin side branches of varicose neurites. The discrepancy in diameter between presynaptic and postsynaptic neurites naturally creates the characteristic polyadic synapse: The relatively large size of the presynaptic element creates a "platform" that is in direct contact with multiple, thin, postsynaptic elements (Fig. 4D). It should be noted that for most synapses, it is difficult to infer exactly how many postsynaptic partners exist. Thus, postsynaptic membrane specializations are very subtle, and one can only base the assertion that a given profile represents a postsynaptic partner on whether its membrane is in direct contact with the presynaptic membrane (density). This, in numerous cases, is ambiguous (e.g., Fig. 4D).

In vertebrates, neurons show three different compartments: soma, dendrite, and axon (Fig. 4F). Dendrites arise from the soma and in terms of structure (e.g., microtubule

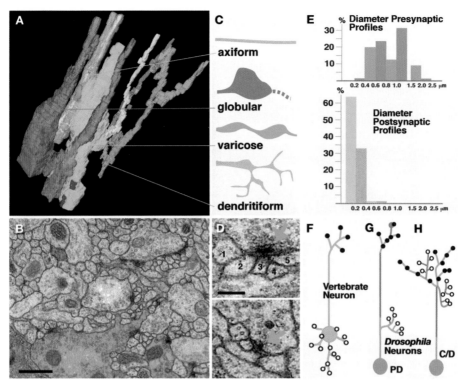

Figure 4. *Drosophila* neuropile ultrastructure. (*A–C*) Types of neurite profiles. (*A*) Three-dimensional digital model of several short neurites segmented from one microvolume. (*B*) Representative EM section of microvolume in which profiles of neurites modeled in *A* are shaded in the corresponding colors. (*C*) Schematic depiction of types of neurites. Axiform neurites (light blue) are straight unbranched processes of intermediate (0.2–0.4 µm) diameter. Globular neurites and varicose neurites (dark blue and green) have alternating segments of intermediate (0.2–0.4 µm) and large (0.5–1.5 µm for varicose; 1–3 µm for globular neurites) diameter. Dendritiform neurites (yellow, brown, orange) are highly branched and thin (<0.2 µm). (*D*) Section of two typical polyadic synapses. (Green arrow) Presynaptic specialization, consisting of the T bar and synaptic vesicles. Presynapses are contacted by multiple, thin branches of dendritiform processes. Numbers *1–5* in upper panel denote profiles of thin, postsynaptic profiles (dendritiform neurites); *2–5* have clear direct contact to presynaptic zone; *1* is located adjacent to the presynaptic site and represents a case that may represent a postsynaptic neurite of this synapse. (*E*) Correlation between frequency of presynaptic and postsynaptic sites and neurite diameter. Presynaptic sites (blue; *top*) are found predominantly on large-diameter profiles that correspond to thick segments of varicose and globular neurites. These neurites represent terminal axonal branches. Postsynaptic profiles (yellow, *bottom*) almost exclusively belong to thin dendritiform neurites; they represent terminal dendritic branches. (*F–H*) Distribution of axonal and dendritic branches. (*F*) Polarized vertebrate neuron, with dendrite/soma compartment carrying postsynaptic sites, and axon compartment with presynaptic sites. (*G*) *Drosophila* type-PD neuron that comes close to the vertebrate pattern, with postsynaptic dendritiform processes concentrated proximally and presynaptic varicose/globular processes distally. (*H*) In *Drosophila* type-C and -D neurons, terminal axons and dendrites are intermingled. Bars, (*B*) 0.5 µm; (*D*) 0.2 µm.

cytoskeleton) and function are similar to the soma and different from the axon (Peters et al. 1976; Baas and Yu 1996). This cell-biological distinction between dendrite/soma and axon goes hand in hand with a functional distinction: Dendrites and soma carry mostly postsynaptic membrane specializations; axons are specialized to conduct axon potentials and carry presynaptic sites at their terminals. The situation is different in insect central neurons. Here, the soma emits a single neurite that in the neuropile forms numerous branches that are dendritic (i.e., postsynaptic), axonal (i.e., presynaptic), or, frequently, mixed (both postsynaptic and presynaptic sites intermingled). For this reason, strictly speaking, one cannot refer to axons or dendrites, but just to neurites, when referring to neuronal processes in the *Drosophila* central nervous system. There clearly exist neurons (those that are part of PD lineages; see Figs. 1I and 4G) that resemble to some extent the typical vertebrate neuron, in the sense that they have postsynaptic (dendritic) branches proximally, close to the soma; a long, unbranched "axon" projects away from these dendrites and ends in multiple axonal terminal branches carrying presynaptic sites. The antennal projection neurons with dendrites in the antennal lobe (primary olfactory center; Stocker 1994; Lai et al. 2008) and axons in the calyx of the MB are an example in this case. But we estimate that the majority of brain lineages are of type C or D: How are presynaptic and postsynaptic sites arranged in neurons belonging to these lineages?

Axiform neurites in any of the microvolumes reconstructed are essentially devoid of synapses. They form the long axon fascicles interconnecting compartments ("macrocircuitry"). These processes then would most closely correspond to the long axons of vertebrates. The swellings of varicose and globular neurites contain almost all presynaptic sites; these then are the terminal axonal branches. The fine processes of dendritiform branches carry postsynaptic sites. Note that, at least at the level of several microns (which for the 30-μm L1 brain is a lot!), neurites are very predominantly either dendritic or varicose/globular (axonal). This does not mean that a given neuron could not have both types of branches close to each other; but at the level of a given branch, input and output is separated (Fig. 4H). This rule is only violated by the occasional thin "side projections" that occur on or near varicosities and that carry postsynaptic sites contacting nearby presynaptic neurites (see Fig. 6D) (Cardona et al., unpubl.).

MICROCIRCUITRY: DENSITY, BRANCHING, AND DIRECTIONALITY OF NEURITES

The above characterization of neurite ultrastructure applies to all regions sampled so far (six microvolumes: calyx, peduncle periphery, spur, CPL compartment, larval optic neuropile, ventral nerve cord), but there are significant quantitative differences. We summarize findings from three microvolumes: the calyx (input region of MB), spur (output region of mushroom body), and dorsolateral ventral nerve cord.

Afferent axons of the calyx are the antennal projection neurons receiving olfactory input in the antennal lobe (Stocker 1994; Ramaekers et al. 2005). We can identify in the calyx microvolume (Fig. 5A) the largely parallel array of varicose neurites, carrying presynaptic sites, as being derived from the incoming fiber bundle (antenno–protocerebral tract) carrying the antennal projection neurons. In overall volume, varicose/globular afferents comprise ~16% of the neuropile. At a medium diameter of 0.38 μm (reducing the shape of these neurites to smooth cylinders), this corresponds to a density of 4.2 varicose neurites terminating in or passing through a volume of 1 μm^3. Counting the number of branchpoints (for varicose neurites) in the microvolume yields a density of branches at one branch every 4 μm. Presynaptic sites are entirely restricted to the varicose/globular swellings of neurites. We find a density of about four presynaptic sites per 1 μm^3. Forming bundles between the varicose (and globular) presynaptic profiles are the fine terminal fibers of dendritiform neurites, mostly derived from MB neurons (Yasuyama et al. 2002). Synapses are mostly dyadic–tetradic, with two to four fine fibers "winding" around swellings of varicose/globular neurites and participating in several synapses. Connectivity is highly local: Individual varicose/globular neurites interact with multiple dendrites in their immediate vicinity.

Neuropile ultrastructure appears to be different in the spur, an output region of the MB (Fahrbach 2006), or the ventral nerve cord. Interestingly, the spur is characterized by the absence of intrinsic glia. Varicose neurites are aligned in bundles that travel along all three cardinal axes (Fig. 5B). Thin terminal filaments of dendritiform neurites are much less frequent than in the calyx (and most other compartments). The lateral connectivity is much more pronounced than in the calyx: If one visualizes all neurites connected to a given synapse, they reach throughout the entire microvolume. Intriguing are also the synaptic geometries: In most cases, two presynaptic sites and two to three postsynaptic sites are clustered together (not shown). We speculate that both wide-field connectivity and multi-input synapses are elements of a microcircuitry required for the specialized function (associative learning) of the MB.

In the dorsolateral domain of the ventral nerve cord (VNC), all neurites are predominantly arranged parallel to the longitudinal axis (Fig. 5C,D). Axonal cable length is 2.9 $\mu m/\mu m^3$ (axiform plus varicose/globular neurites). The branching density is low, with 0.18 branches/1 μm^3. The high quality of the VNC volume allowed us to reconstruct the pattern of thin dendritiform neurites with great confidence. These neurites have a much higher branch density than terminal axons (Fig. 5E,F). As a result, overall dendritic cable length exceeds that of axonal elements by a factor of almost 10. Dendritiform processes show an interesting convergence–divergence pattern (Fig. 5E–H). At a given level, dendritiform processes are not scattered evenly across the section, but they form several bundles of five to ten processes each; one such bundle is highlighted by arrows in Figure 5E,G. Bundles typically run between the loosely packed terminal, varicose axons, with which they form synapses. However, dendrites of a given bundle stay together only for a short interval (<1 μm); subsequently, they diverge and redistribute (compare G and H in Fig. 5), coming together with other dendritform processes in new combinations. This behavior is in contrast to that of axons: Axiform neurites (long axons) form tight bundles where neighborhood relationships among neurites are maintained over many micrometers; terminal varicose/globular axons are more loosely packed, but like axiform processes, they run largely parallel to one another and maintain their position relative to one another (Fig. 5D,G,H).

CONNECTIVITY AND NETWORK MOTIFS IN THE VNC MICROVOLUME

As a result of the high branching density of dendritiform processes, as well as the fact that each presynaptic site contacted multiple postsynaptic profiles, neurites within a microvolume are highly interconnected (as is shown in the proceeding section, this feature distinguishes the *Drosophila* neuropile from a vertebrate brain neuropile). The most common type of network is shown in Figure 6A–C. In the example shown here, one presynaptic element forms two adjacent presynaptic sites on one varicosity. Contacting these sites are six postsynaptic dendritiform processes. Quite frequently, as in this example, a given dendrite contacts its presynaptic partner multiple times at adjacent synapses. Each dendrite forms multiple branches that connect to other presynaptic partners; on average, a dendrite received input from 2.1 axons within the 85-μm^3 VNC microvolume. Figure 6B illustrates the presynaptic element, its associated six postsynaptic elements, and five additional presynaptic elements that are in contact with the same postsynaptic elements. This type of connectivity has been defined as a dense overlapping regulon motif (Reigl et al. 2004; Alon 2007): A given input element diverges onto multiple targets, and at the same time, each target element receives input from multiple presynaptic elements

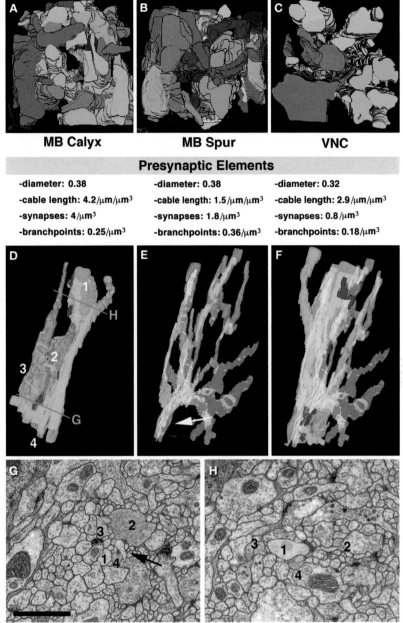

Figure 5. Structural network properties in *Drosophila* brain neuropile. (*A–C*) Three-dimensional digital models of presynaptic varicose/globular neurites from three different microvolumes. (*A,B*) Microvolumes from the calyx (*A*; input region) and spur (*B*; output region) of the MB. In both microvolumes, branches of terminal axonal neurites follow all directions. (*C*) Microvolume from dorsolateral neuropile of the VNC. All neurites are oriented predominantly along the longitudinal axis. Shown below each panel are some core parameters of neurite profiles seen in microvolumes. Note that the average diameter of the presynaptic elements (varicose and globular neurites) is very similar among the different regions. The density of presynaptic sites and branchpoints is significantly lower in the VNC compared to the MB. (*D–H*) Typical trajectories of presynaptic and postsynaptic terminal branches. (*D*) Three-dimensional digital models of four neighboring presynaptic neurites (*1–4*). Neurite *1* has a varicosity near the top of the panel; varicosities of the other neurites are more basal. (*E*) Bundle of dendritiform neurites extending in the vicinity of terminal axons shown in *D*. (*F*) Terminal axons and dendrites shown together. (*G,H*) EM sections close to top and bottom of VNC microvolume (levels of section shown in *D*). Profiles corresponding to the elements shown in models *D–F* are shaded in corresponding colors. As shown here, groups of dendritiform neurites (typically ranging between six and ten) form tight bundles in between adjacent preterminal axons (arrow in *E* and *G*). After forming synaptic contacts, dendritiform neurites typically splay apart (*E*) to then regroup with other dendrites in different configurations. Bar, (*G*), 1 μm.

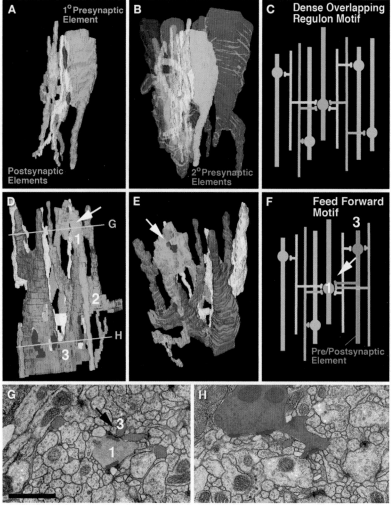

Figure 6. Network motifs that are most frequently encountered in microvolumes. (*A–C*) Dense overlapping regulon motif. (*A*) Segment of one "primary" presynaptic element (turquoise; varicose neurite) that contacts five postsynaptic elements (dendritiform neurites) at two synapses. (*B*) Same configuration of presynaptic and postsynaptic elements as in *A*; shown are several other "secondary" presynaptic elements (green) that form synapses with the same dendrites as the primary axon. (*C*) Schematic representation of this network motif. (*D–H*) Feed forward motif. (*D–F*) Three-dimensional digital models and schematic model of segments of three varicose neurites that form predominantly presynaptic contacts (terminal axons). The blue element has a thin branch that is postsynaptic to the light green element (white arrow in *D–F*). Shown are also the dendritiform postsynaptic neurites that are postsynaptic to both green and blue varicose neurites. (*G,H*) EM sections at levels shown by lines in *D*. *G* represents the top level and shows the synapse between the presynaptic/postsynaptic element (blue) and presynaptic element (green; arrow). Bar, (*G*), 1 µm.

(Fig. 6C). The large majority of presynaptic neurites and postsynaptic neurites within the VNC were engaged in dense overlapping regulon motifs.

Encountered less frequently is a type of connection that we labeled a feed forward motif. An example is shown in Figure 6D–H: Among the postsynaptic partners of one of the varicose neurite (1, light green) are thin branches of two other varicose neurites (2, dark green; 3, blue). In addition, axon A forms input onto several dendrites that at the same time receive input from B and C. In other words, an afferent neurite is at the same time presynaptic to one neurite and postsynaptic to another neighboring neurite; both elements are presynaptic to a common dendritic element. It will be informative to investigate how such network motifs are distributed throughout the brain and how they are correlated to different sensory modalities or types of output.

NEUROPILE ARCHITECTURE IN MAMMALIAN CORTEX AND FLY BRAIN: A FIRST COMPARISON

Quantitative statements about the architecture and connectivity of the neuropile of vertebrate brains, in particular mammalian cortex, were mainly based on statistical analyses of light microscopy preparations (e.g., Golgi-stained prepartions) and representative EM sections (see, e.g., Braitenberg and Schüz 1998). The first reconstruction of a microvolume of mammalian neuropile has recently become available (Mishchenko 2009). What can one learn from the comparison of these data with our data on *Drosophila* larval brain?

Figure 7 presents EM sections of mouse neocortex (A) and *Drosophila* brain (L1 larva, B; adult, C). Plotting the

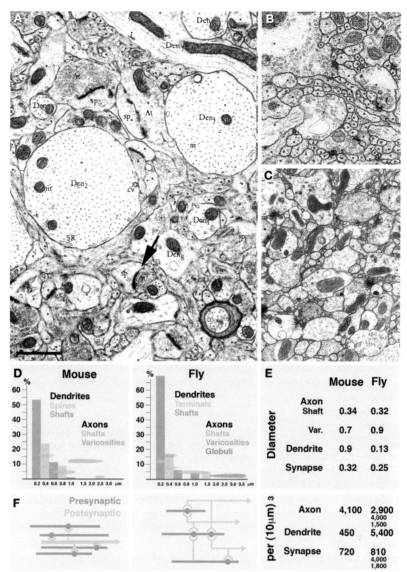

Figure 7. Parameters of microcircuitry in mammalian neocortex and *Drosophila* brain. (*A–C*) Representative EM sections of mouse neocortex (*A*), *Drosophila* larval brain (*B*), and *Drosophila* adult brain (*C*) shown at the same scale. Red dots in the *A* and *B* indicate profiles of individual sectioned neurites. (*D*) Frequency distribution of neurites with different diameters in the mammalian cortex and *Drosophila* brain. Indicated are also the range of diameters that correspond to different neuropile elements (yellow/orange: dendrites; green/blue: axons). (*E*) Comparison of several core parameters in the mouse and *Drosophila* neuropile. In both systems, terminal axons are varicose neurites that form presynaptic sites on their varicosities. Diameters of varicosities are in the range of 0.5–1.5 μm (average in mouse, 0.7 μm; in *Drosophila*; 0.9 μm); the thin segments of terminal axons have an average diameter of ~0.33 μm. The diameter of synapses (presynaptic sites) is also quite similar in both systems (0.32 μm in mammalian cortex, 0.25 μm in fly brain). The overall cable length of terminal axons per volume unit is also comparable: 4100 μm per 1000 μm^3 in mouse and between 1500 μm and 4,000 μm in different microvolumes of fly brain. The major difference between the mouse and fly neuropile lies in the size and branching density of dendrites. In *Drosophila*, dendrites are very thin (average diameter: 0.13 μm) and densely branched; in mammalian brain, dendrites are thick (average diameter: 0.9 μm; see even thicker examples of dendrites in *A*) and branches are much further apart. This is also reflected in the dendritic cable length, which is 450 mm/1000 μm^3 in mouse and more than 10-fold higher in *Drosophila*. Lower branch density as well as the absence of polyadic synapses in mouse cortex neuropile also results in a considerably less dense connectivity, schematically shown in *F*. Shown for *Drosophila* is the dense overlapping regulon motif, in which the large majority of neurite segments encountered in any microvolume of 100 μm^3 or more is engaged. In a mammalian cortical microvolume of that size, dendrite segments are unbranched; the only type of connectivity is convergence, whereby multiple terminal axons converge on a dendritic segment that happens to be within their range. Bar, (*A*), 1 μm.

frequency of profiles with different diameters shows quite similar distributions in mouse and fly (Fig. 7D). In particular, the average diameter of axon shafts and synapse-bearing varicosities, as well as presynaptic sites themselves, is quite conserved. The most conspicuous difference between the two species lies in the size of dendrites. In mouse cortex, dendrite shafts are large, measuring in average almost 1 µm (Fig. 7D,E) (Braitenberg and Schüz 1998). Many dendrites are considerably thicker (e.g., examples shown in Fig. 7A). Dendrites of pyramidal cells (and some other classes of neurons) bear spines, which are club-shaped processes of 1–2 µm in length with an average diameter of ~0.4 µm. Synapses are found on dendritic shafts and spines; synapses are typically monadic, which naturally follows from the fact that the presynaptic element is of equal size, or even smaller, than the postsynaptic element (arrow in Fig. 7A). The comparison between the mouse and fly in regard to axonal and dendritic cable length per volume unit is also informative (Fig. 7E). In the mouse, axonal cable adds up to ~4 mm/1000 µm^3; the length of dendrites is 0.5 mm. Axonal cable length in *Drosophila* lies in the same range; dendritic cable length, on the other hand, is 10 times higher, which is a reflection of the fact that dendritic processes are very thin and highly branched. Interestingly, judging from the inspection of EM photographs (see, e.g., Peters et al. 1976), processes in the 0.1-µm range seem to be quite numerous also in mammalian brain, but the nature of these thin neurites is unclear.

As described in the previous section, the overall synaptic density per volume unit seems to be quite different in different parts of the fly brain, ranging from ~1/µm^3 (VNC) to >4/µm^3 (input region of MB). In the mammalian cortex, synaptic density is toward the lower end of this range, at 0.72/µm^3 (Braitenberg and Schüz 1998). Approximately the same value was extracted from the microvolume of rat hippocampus (Chklovskii et al., unpubl.). Here, one axonal varicosity typically contained a single synapse; in *Drosophila*, most varicosities had between two and four synaptic sites. The higher density of synapses in *Drosophila* is accompanied by a higher number of branchpoints. In the rat hippocampus microvolume, measuring 8 x 8 x 8 µm, very vew axons or dendrites (out of hundreds) had any branches, not counting dendritic spines. Similarly, statistical analysis of Golgi preparations yielded typical distances of 10 µm and higher between branchpoints (Braitenberg and Schüz 1998). In contrast, in *Drosophila*, terminal axons had one branchpoint every 4 µm (MB calyx), 2.8 µm (MB, spur) or 7.5 µm (VNC), respectively. Dendritic branch density is even higher; in the VNC microvolume, dendritiform neurites had approximately twice as many branchpoints as varicose/globular neurites.

The fact that the spacing of branches and synapses is significantly higher in the *Drosophila* brain compared to the mammalian brain is reflected in the presence of a much higher connectedness of neurites in the former. As described in the previous section, the majority of axons are engaged in networks containing both dendrites and other axons within a microvolume of less than 100 µm^3. This is not the case in mammalian microvolumes of comparable, or even larger, size (Chklovskii et al., unpubl.). Here, the only type of connectivity is represented by a convergence of axonal segments onto isolated dendritic segments (Fig. 7H). However, unlike the fly brain, very few axonal segments will form input to more than one dendrite, so that network motifs such as the dense overlapping regulon motif or feed forward motif do not emerge. What this simply means is that the modules of mammalian brain that house microcircuits are considerably larger than those in *Drosophila*. In vertebrates, neurons reach much higher numbers than in the *Drosophila* brain. Furthermore, dendrites are larger in diameter, and polyadic synapses are largely absent. This results in connectivity occupying more space: in *Drosophila*, one presynaptic neurite reaches up to six postsynaptic elements in a single synapse. In mammalian brain, for the same purpose, six individual synapses, spaced apart by intervals of several micrometers, would have to be formed. One may speculate that, evolutionarily, each neuron grew in axonal and dendritic length to accommodate the higher number of synapses that had to form to connect a given neuron to a certain fraction of other neurons. As a result of this numerical increase in cell number, cable length, and synapse number, branches of dendrites and axons are spaced much further apart, such that in a volume of 8 x 8 x 8 µm almost no branchpoints (and consecutively no networks) occur.

A GLIMPSE AT THE PHYLOGENY OF NEUROPILE ARCHITECTURE

Systematic studies of neuropile architecture and connectivity from serial EM sections have not been performed for other animals, with the exception of the central nervous system of the nematode *C. elegans* (White et al. 1986). On the other hand, there is a rich literature documenting neurites and synapses in representative EM sections. These neuropile elements exist in all metazoans. Measurements of diameters of neurites and synapses, taken from representative photographs published in the literature (mollusk: Nagy and Elekes 2000; annelid: Riehl and Schlue 1998; vertebrate: Peters et al. 1976; nematode: White et al. 1986; acoela: our material) yielded figures similar to those for *Drosophila*. Presynaptic sites measured 0.2–0.35 µm in diameter. Neurite numbers, as in *Drosophila*, showed a peak in the range below 0.2 µm in diameter. In addition, a substantial number of neurites were in the range of 0.4–0.8 µm. Presynaptic sites are typically located on profiles with a diameter of 0.4–0.8 µm; these presynaptic sites contact a variable number of thin postsynaptic dendrites. Large profiles serving as postsynaptic sites (as in vertebrates) are rarely reported. This finding, which should be substantiated with further systematic serial reconstruction of microvolumes, suggests that the pattern of neurites shown here for *Drosophila*, with thin highly branched dendrites that engage with medium-sized varicose axons through polyadic synapses, may be primitive. The appearance of "gigantic" dendrites, accompanied by a significant increase in the number of neurons and overall cable length, appears to be an innovation of the vertebrate clade.

C. elegans with its miniaturized central nervous system, consisting of less than 400 neurons, presents an interesting case with respect to neuropile architecture. Thin profiles below 0.2 µm are almost entirely absent; the large majority of neurites have a diameter of 0.3–0.6 µm. This anomaly among invertebrate taxa is accompanied by the fact that neurons of *C. elegans* are almost all unbranched (White et al. 1986). Most somata project a single neurite that carries clusters of intermingled presynaptic and postsynaptic sites. In other words, thin terminal dendritiform branches as in other taxa do not exist in the worm.

OUTLOOK

We propose that the 3D reconstruction of neuropile architecture and connectivity will provide an important tool for the study of brain function and development. The technology we present here is well adapted to handle small volumes of neuropile, but a number of technical improvements are under way that will bring larger volumes (in the size range of adult fly brains) into the range of feasibility. The improvements are directed at three major bottlenecks of the procedure: better registration, automatic segmentation, and automatic feature extraction. Programs in the trial phase allow for automatic recognition of profile boundaries and subsequent segmentation (Chklosvskii et al., unpubl.). A step in microvolume analysis that is currently as labor-intensive as segmentation (which takes 5–10 working days per volume) is extracting relevant information from the segmented objects. For example, we currently manually input what synapse belongs to what neurite and manually generate lists that encapsulate the connectivity pattern. We are working on adding features to the EM canvas window that fully automate these tasks. Thus, after segmentation of neurite profiles, synapses can be assigned automatically to the underlying 3D reconstructed neuronal arborizations that host them. With not much extra effort than merely segmentation, a complete wiring diagram of the microvolume can be built. The next step would be to generate dynamic models, where lengths and diameters of neurites, intersynapse distances, and other parameters are taken into account to predict the temporal component of activity flow. We anticipate that these models of small volumes will provide suitable material for computational neuroscientists to build more accurate and heuristically valuable large-scale models of brain function.

ACKNOWLEDGMENTS

We thank many friends and collaborators, notably Dmitrii Chklovskii, Johannes Schindelin, James Truman, and Wayne Pereanu, for helpful discussions and comments. Parts of this work were supported by National Institutes of Health grant R01 NS054814 to V.H.

REFERENCES

Alon U. 2007. Network motifs: Theory and experimental approaches. *Nat Rev Genet* **8:** 450–461.
Baas PW, Yu W. 1996. A composite model for establishing the microtubule arrays of the neuron. *Mol Neurobiol* **12:** 145–161.
Braitenberg V, Schüz A. 1998. *Cortex: Statistics and geometry of neuronal connectivity,* 2nd ed. Springer, Berlin.
Brand AH, Perrimon N. 1993. Targeted gene expression as a means of altering cell fates and generating dominant phenotypes. *Development* **118:** 401–415.
Douglas R, Martin KAC. 1998. Neocortex. In *The synaptic organization of the brain* (ed. GM Shepherd), pp. 459–511. Oxford University Press, NY.
Fahrbach SE. 2006. Structure of the mushroom bodies of the insect brain. *Annu Rev Entomol* **51:** 209–232.
Feeney CJ, Karunanithi S, Pearce J, Govind CK, Atwood HL. 1998. Motor nerve terminals on abdominal muscles in larval flesh flies, *Sarcophaga bullata:* Comparisons with *Drosophila*. *J Comp Neurol* **402:** 197–209.
Fung S, Wang F, Spindler S, Hartenstein V. 2009. *Drosophila* E-cadherin and its binding partner Armadillo/β-catenin are required for axonal pathway choices in the developing larval brain. *Dev Biol* **332:** 371–382.
Goodman CS, Doe CQ. 1993. Embryonic development of the *Drosophila* central nervous system. In *The Development of Drosophila* (ed. M Bate and A Martinez-Arias), pp. 1131–1206. Cold Spring Harbor Laboratory Press, Cold Spring Harbor, NY.
Hartenstein V, Spindler S, Pereanu W, Fung S. 2008a. The development of the *Drosophila* larval brain. *Adv Exp Med Biol* **628:** 1–31.
Hartenstein V, Cardona A, Pereanu W, Younossi-Hartenstein A. 2008b. Modeling the developing *Drosophila* brain: Rationale, technique and application. *BioScience* **58:** 823–836.
Ito K, Awano W, Suzuki K, Hiromi Y, Yamamoto D. 1997. The *Drosophila* mushroom body is a quadruple structure of clonal units each of which contains a virtually identical set of neurones and glial cells. *Development* **124:** 761–771.
Kaneko M, Hall JC. 2000. Neuroanatomy of cells expressing clock genes in *Drosophila:* Transgenic manipulation of the *period* and *timeless* genes to mark the perikarya of circadian pacemaker neurons and their projections. *J Comp Neurol* **422:** 66–94.
Kittel RJ, Wichmann C, Rasse TM, Fouquet W, Schmidt M, Schmid A, Wagh DA, Pawlu C, Kellner RR, Willig KI, et al. 2006. Bruchpilot promotes active zone assembly, Ca^{2+} channel clustering, and vesicle release. *Science* **312:** 1051–1054.
Kozloski J, Hamzei-Sichani F, Yuste R. 2001. Stereotyped position of local synaptic targets in neocortex. *Science* **293:** 868–872.
Lai SL, Awasaki T, Ito K, Lee T. 2008. Clonal analysis of *Drosophila* antennal lobe neurons: Diverse neuronal architectures in the lateral neuroblast lineage. *Development* **135:** 2883–2893.
Larsen C, Shy D, Spindler S, Fung S, Younossi-Hartenstein A, Hartenstein V. 2009. Patterns of growth, axonal extension and axonal arborization of neuronal lineages in the developing *Drosophila* brain. *Dev Biol* **335:** 289–304.
Meinertzhagen IA, O'Neil SD. 1991. Synaptic organization of columnar elements in the lamina of the wild type in *Drosophila melanogaster*. *J Comp Neurol* **305:** 232–263.
Michchenko Y. 2009. Automation of 3D reconstruction of neural tissue from large volume of conventional serial section transmission electron micrographs. *J Neurosci Methods* **176:** 276–289.
Nagy T, Elekes K. 2000. Embryogenesis of the central nervous system of the pond snail *Lymnea stagnalis* L. An ultrastructural study. *J Neurocytol* **29:** 43–60.
Pereanu W, Hartenstein V. 2006. Neural lineages of the *Drosophila* brain: A 3D digital atlas of the pattern of lineage location and projection at the late larval stage. *J Neurosci* **26:** 5534–5553.
Pereanu W, Jennett A, Younossi-Hartenstein A, Hartenstein V. 2009. A development-based compartmentalization of the *Drosophila* central brain. *J. Comp. Neurol.* (in press).
Peters A, Palay S, Webster HE. 1976. *The fine structure of the nervous system*. W.B. Saunders, Philadelphia.

Prokop A, Meinertzhagen IA. 2006. Development and structure of synaptic contacts in *Drosophila*. *Semin Cell Dev Biol* **17:** 20–30.

Ramaekers A, Magnenat E, Marin EC, Gendre N, Jefferis GS, Luo L, Stocker RF. 2005. Glomerular maps without cellular redundancy at successive levels of the *Drosophila* larval olfactory circuit. *Curr Biol* **15:** 982–992.

Reigl M, Alon U, Chklovskii DB. 2004. Search for computational modules in the *C. elegans* brain. *BMC Biol* **2:** 2–25.

Riehl B, Schlue WR. 1998. Morphological organization of neuropile glial cells in the central nervous system of the medicinal leech (*Hirudo medicinalis*). *Tissue Cell* **30:** 177–186.

Rolls ET, Loh M, Deco G, Winterer G. 2008. Computational models of schizophrenia and dopamine modulation in the prefrontal cortex. *Nat Rev Neurosci* **9:** 696–709.

Saalfeld S, Cardona A, Hartenstein V, Tomancák P. 2009. CATMAID: Collaborative annotation toolkit for massive amounts of image data. *Bioinformatics* **25:** 1984–1986.

Silberberg G. 2008. Polysynaptic subcircuits in the neocortex: Spatial and temporal diversity. *Curr Opin Neurobiol* **18:** 332–337.

Silberberg G, Gupta A, Markram H. 2002. Stereotypy in neocortical microcircuits. *Trends Neurosci* **25:** 227–230.

Stocker RF. 1994. The organization of the chemosensory system in *Drosophila melanogaster:* A review. *Cell Tissue Res* **275:** 3–26.

Toledo-Rodriguez M, El Manira A, Wallén P, Svirskis G, Hounsgaard J. 2005. Cellular signalling properties in microcircuits. *Trends Neurosci* **28:** 534–540.

Truman JW, Schuppe H, Shepherd D, Williams DW. 2004. Developmental architecture of adult-specific lineages in the ventral CNS of *Drosophila*. *Development* **131:** 5167–5184.

Urbach R, Technau GM. 2003. Molecular markers for identified neuroblasts in the developing brain of *Drosophila*. *Development* **130:** 3621–3637.

Watson AH, Burrows M. 1983. The morphology, ultrastructure, and distribution of synapses on an intersegmental interneuron of the locust. *J Comp Neurol* **214:** 154–169.

White JG, Southgate E, Thomson JN, Brenner S. 1986. The structure of the nervous system of the nematode *Caenorhabditis elegans*. *Philos Trans R Soc Lond B* **314:** 1–340.

Winterer G, Weinberger DR. 2004. Genes, dopamine and cortical signal-to-noise ratio in schizophrenia. *Trends Neurosci* **27:** 683–690.

Yasuyama K, Meinertzhagen IA, Schürmann FW. 2002. Synaptic organization of the mushroom body calyx in *Drosophila melanogaster*. *J Comp Neurol* **445:** 211–226.

Younossi-Hartenstein A, Nassif C, Green P, Hartenstein V. 1996. Early neurogenesis of the *Drosophila* brain. *J Comp Neurol* **370:** 313–329.

Younossi-Hartenstein A, Salvaterra P, Hartenstein V. 2003. Early development of the *Drosophila* brain. IV. Larval neuropile compartments defined by glial septa. *J Comp Neurol* **455:** 435–450.

Younossi-Hartenstein A, Shy D, Hartenstein V. 2006. The embryonic formation of the *Drosophila* brain neuropile. *J Comp Neurol* **497:** 981–998.

A General Basis for Cognition in the Evolution of Synapse Signaling Complexes

S.G.N. Grant

Genes to Cognition Programme, Wellcome Trust Sanger Institute, Cambridge, United Kingdom

Correspondence: sg3@sanger.ac.uk

Beneath the complexity of the human brain are molecular principles shaped by evolution explaining the origins of the behavioral repertoire. The role of the nervous system is to provide a repertoire of behaviors allowing the animal to respond and adapt to changing environments during the course of its life. Multiprotein complexes in the postsynaptic terminal of synapses control adaptive and cognitive processes in metazoan nervous systems. These multiprotein complexes are organized into molecular networks that detect and respond to patterns of neural activity. Combinations of proteins are used to build different complexes and pathways producing great diversity. These complexes evolved from an ancestral core set of proteins controlling adaptive behaviors in unicellular organisms known as the protosynapse. Later expansion in numbers and interactions resulted in more complex synapses in invertebrates and vertebrates. The resultant combinatorial complexity has contributed to the neuroanatomical, neurophysiological, and behavioral diversity in these species. Mutations in genes encoding the complexes result in many human diseases of the nervous system. This general mechanism of cognition provides a useful template for studying evolution of behavior in all animals.

In his closing commentary on the Symposium, Dr. Brian Charlesworth (Edinburgh University) concluded that the two major evolutionary problems were the "origin of life" and the "evolution of human consciousness and cognition." Toward the latter issue, this chapter traces a path linking the molecular origins of adaptive behaviors in unicellular organisms to the synaptic mechanisms of human cognition and its medical disorders. This framework has emerged from research using proteomic approaches to synapses of the mouse brain, which exposed a hitherto unappreciated molecular complexity (Husi et al. 2000). Experiments probing this complexity using both single-molecule and large-scale methods, with a particular emphasis on integration of neurophysiological, neuroanatomical, and behavioral approaches, lead to a general molecular theory for the origins of cognition and how animals interact and adapt to their environments.

This chapter is structured along the following lines. First, cognition is defined, and an overview of behavioral adaptation to the environment is presented. Next, a description of the molecular organization of the mammalian postsynaptic proteome and its complexity is introduced as background to molecular evolutionary studies. The remaining sections address electrophysiological, anatomical, and behavioral implications of synapse evolution.

COGNITION: FROM UNICELLULAR ORGANISMS TO HUMANS

Cognition is defined as the biology of information processing (Wilson and Keil 1999). This definition is generally applied in psychology where it embraces the acquisition and processing of information obtained by the sensory organs of animals. Specific cognitive mechanisms underlie perception, attention, learning, strategy choice, and other processes, which are not dealt with in detail here.

Processing of sensory information allows animals to detect and adapt to their immediate environment and changes in that environment or movement between environments. Considering the range of species that show cognitive processes, the 19th century biologist C.L. Morgan (1890) noted that unicellular organisms shared mechanisms with humans:

> The primary end and object of the receptions of the influences (stimuli) of the external world or environment is to enable the organisms to answer or respond to these special modes of influence, or stimuli. In other words, their purpose is to set agoing certain activities. Now in the unicellular organisms, where both the reception and the response are effected by one and the same cell, the activities are for the most part simple, though even among these protozoa there are some which show no little complexity of response.

The importance of this observation arises when one considers the potential for conserved molecular mechanisms that underlie cognitive processes in animals, ranging from unicellular eukaryotic organisms to humans. Such core mechanisms, conserved across phyla, could form a basis around which species specificity and lineage-specific adaptations arose. In general, this aspect of neuroscience is neglected because most thinking is devoted to the function of neurons and their circuits, which are metazoan. Indeed most, if not all, models of cognition exclude unicellular organisms and are thus not general to all animals.

What form does cognitive processing take in unicellular organisms? To address this, we must consider how the information from the environment is converted into a meaningful adaptive response. Using yeast as an example of a unicellular organism that detects changes in nutrients, pH, stressors, or other stimuli in its environment, its response through activation of signal transduction pathways regulating the expression of many genes is well known (Gasch et al. 2000). The output of this response is an orchestrated change in expression of large sets of genes, some common to many types of environmental changes and some gene sets specific to certain stimuli. This ability to detect different features of the environment and respond using signal transduction machinery to orchestrate transcription of overlapping combinations of genes is similar to signaling at mammalian synapses where neurotransmitter receptors can be activated with different stimuli to drive the expression of various sets of genes (Coba et al. 2008; Alberini 2009). More generally, receptor-mediated signal transduction systems are found in all cells and represent elementary cognitive mechanisms.

Unicellular organisms exist in environmental niches where they are exposed to a range of stimuli external to the organism. In contrast, the environmental niche of the postsynaptic terminal, which receives the neurons input stimulus, is a highly specialized internal environment, built and maintained by the organism itself. We can envision the postsynaptic membrane as a specialized sensing point on the surface of neurons, receiving pulses of neurotransmitters released from the presynaptic terminal. Thus, it seems likely that synapses would share molecular features with unicellular organisms and have evolved molecular specializations for the purpose of responding to the information encoded within the patterns of neural activity.

That *patterns of neural activity* are a key aspect of the neurophysiological basis of cognition was discovered by E.D. Adrian (1928) in the 1920s. He found that nerve fibers from sensory end organs showed greater activity following greater stimulation, leading to the conclusion that information received by the brain from the environment is in the pattern or code of nerve action potentials. This concept of a "neural code" representing the environment has been extended from simple sensory inputs to abstract stimuli including perception of visual (Hubel and Wiesel 1959) and spatial information (O'Keefe and Nadel 1978), which is represented in the pattern of neural activity recorded in neurons in particular parts of the mammalian brain. It is therefore of paramount importance to understand how the postsynaptic terminal reads the neural code and ultimately to ask if these molecular mechanisms are relevant to the psychological representation of information in the brain.

Synapses have a major role in transmitting the neural code from one neuron to the next, as well as a role in "reading" or monitoring the information in the code. An action potential arriving at the presynaptic terminal elicits the release of neurotransmitter, which is then detected on the postsynaptic membrane by neurotransmitter receptors. The rapid (millisecond) role of these receptors is to change the electrical potential of the membrane through modulation of ion flux and thereby contribute to the generation of another action potential in the second neuron.

A slower (seconds to minutes) consequence of the activation of the neurotransmitter receptor is to activate enzymatic signal transduction pathways that can have prolonged effects on the function of the synapse or indeed the whole neuron. As noted earlier, these changes include transcriptional responses as well as local biochemical and structural changes in the synapse itself. It is this property of long-term change that E.R. Kandel has shown to have a role in memory processes (Goelet et al. 1986). Importantly, there is a direct link between the neural code and the induction of changes in the neuron. For example, simple neural codes (e.g., pairs of action potentials) have transient effects on synapses, and complex neural codes (e.g., a 5-Hz train of 30 sec) result in long-term changes. An extensive literature shows that various experimentally induced patterns and timings of synaptic activity induce changes in neuronal function, indicating that synapses are capable of detecting and discriminating many different neural codes and differentially activating signaling mechanisms.

This summary presents the view that mammalian synapses and unicellular organisms share a fundamental capacity for detecting and discriminating signals from their environments and that shared mechanisms exist at the level of signal transduction. The postsynaptic terminal of synapses detects a diverse range of neural codes and the unicellular organism must detect "environmental codes." This model by itself does not explain how complex nervous systems evolved or how the behavioral repertoire of animals arose. To address these issues, we need to consider the molecular composition of synapses, their signaling and information processing capacity, and their molecular evolution.

MOLECULAR COMPLEXITY AND ORGANIZATION OF THE POSTSYNAPTIC PROTEOME

The surprising complexity of the postsynaptic proteome was found using neuroproteomic methods (for review, see Bayes and Grant 2009). This complexity was first exposed when the excitatory neurotransmitter receptor known as the N-methyl-D-aspartate (NMDA) receptor was purified from mouse brain and found to have 77 proteins attached and named the NMDA receptor complex (NRC) (Husi et al. 2000). Since then, multiple affinity methods have been used to isolate NRCs, and the numbers of associated proteins have increased to ~100–200 (Collins et al. 2006; Fernandez et al. 2009). The NR2 subunit of the NMDA receptor binds to the Discs Large (Dlg) class of scaffold proteins (Kornau et al. 1995), also known as membrane-associated guanylate kinases (MAGUKs), of which mammalian PSD-95/Dlg4, SAP102/Dlg3, and PSD-93/Dlg2 are highly studied (Feng and Zhang 2009). Affinity isolation of PSD-95 allows MAGUK-associated signaling complexes (MASCs) to be isolated, which show considerable overlap in their composition with the NRC (Husi et al. 2000; Collins et al. 2006; Fernandez et al. 2009), and hereafter will be referred to as NRC/MASC. NRC/MASCs are core complexes within the postsynaptic terminal and contain the principle electrical machinery (e.g., glutamate receptors and voltage-gated potassium channels) necessary for the postsynaptic terminal to respond and transmit action potentials (Fernandez et al. 2009). The first direct evidence that NRC/MASC has a key role in cognition in animals was obtained using mice carrying mutations in PSD-95 (Migaud et al. 1998) or in the intracellular interaction domains of the NR2 subunits (Sprengel et al. 1998).

A larger set of proteins within the mammalian postsynaptic terminal of excitatory synapses can be isolated en masse using fractionation of a preparation known as the postsynaptic density (PSD). Approximately 1500 proteins were found in the rodent PSD, and the NRC/MASC represents ~10% of the PSD (Collins et al. 2006; Trinidad et al. 2008). Both the ionotropic (ion channel) and metabotropic (G-protein-coupled) glutamate receptors bind to scaffold proteins, which bind directly to signal transduction enzymes and structural proteins, forming multiprotein signaling complexes.

The concept of molecular networks complements the concept of multiprotein complexes and has proven to be useful in understanding the complexity imposed by the discovery of hundreds of postsynaptic proteins. For example, as shown in Figure 1, the proteins in NRC/MASC can be graphed as a molecular network of binary protein interactions (Pocklington et al. 2006). This reduces the complexity by grouping proteins into modules, and their relationship to one another can be examined. Specifically, the neurotransmitter receptors and their proximal interacting proteins form "upstream" modules that bind to "intermediate modules" of signaling proteins that in turn bind to "downstream modules" (Fig. 2). As discussed below, the molecular evolution of this hierarchy of modules shows preferential expansion of upstream components in organisms with complex nervous systems.

Beyond a static network, we must ask how dynamic pathways linking receptors to downstream proteins function. A pathway may be made from a subset of the proteins, comprising components from each level of the hierarchy (Fig. 3). In other words, the complexity provides a combinatorial system for building many different pathways. It is now known that different combinations of PSD proteins are activated by different neurotransmitter receptors, and the PSD networks integrate this signaling information in a combinatorial manner (Coba et al. 2009).

Direct evidence showing that neuronal activity regulates a subnetwork of hundreds of PSD proteins was found using large-scale phosphoproteomics (Coba et al. 2009). Stimulation of the NMDA receptor led to rapid changes in the phosphorylation of more than 100 proteins on more than 200 phosphorylation sites. The proteins that were modulated contained a wide range of functional classes including other neurotransmitter receptors and ion channels, structural and cytoskeletal proteins, and enzymes regulating protein translation and transcription. Stimulation of other neurotransmitter receptors led to changes in distinct and overlapping sets of phosphorylation sites on these substrates (Fig. 3). At the level of information processing, this shows that the information associated with activation of

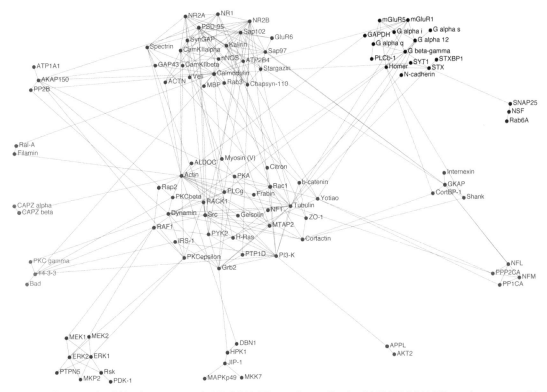

Figure 1. Protein network within the postsynaptic NRC/MASC complexes. Graph of 101 NRC/MASC proteins connected by 246 binary interactions and clustered into modules. The two uppermost modules contain the ionotropic (including NR2 and Dlg proteins) (pink) and metabotropic (brown) neurotransmitter receptors and thus "upstream modules." The large "intermediate module" is connected to "downstream modules" at the bottom of the figure. A detailed description can be found in Pocklington et al. (2006), from which this was adapted, and a further simplified view is shown in Figure 2.

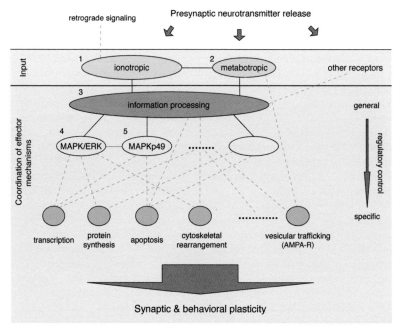

Figure 2. Modular structure and functional organization in the postsynaptic terminal. The synaptic cleft is at the top and neurotransmitters activate proteins in the input layer (purple top). The flow of information traverses the hierarchy of NRC/MASC modules through upstream (blue), intermediate (red), and downstream (yellow) modules connected by solid lines. These downstream modules regulate output effector mechanisms (green) that together produce the overall synaptic and neuronal changes. (Adapted from Pockington et al. 2006.)

different neurotransmitter receptors can recruit combinations of phosphorylation events in subsets of the many PSD proteins. Thus, the organization of the network and its differential activation provide a way to orchestrate the biological activity of hundreds of proteins.

The number of combinations of phosphorylation patterns induced by neurotransmitter receptor activation in the PSD is virtually limitless. Simple estimates indicate that 1000 proteins with an average of 10 phosphorylation sites can produce many more states ($2^{10,000}$) than there are synapses in the human brain. The full physiological range and diversity of PSD phosphorylation combinations have yet to be measured, but it is clear that differential phosphorylation patterns are functionally significant. For example, the phosphorylation of specific combinations of sites on the GluR1 subunit of the AMPA (α-amino-3-hydroxyl-5-methyl-4-isoxazole-propionate) receptor correlates with differential effects on synaptic plasticity (Delgado et al. 2007). In addition, the phosphorylation events regulate gene expression and differential activation of combinations of kinases, resulting in different patterns of gene expression (Coba et al. 2008). Thus, the picture is emerging that the molecular complexity of mammalian synapses provides a set of proteins from which combinations or subsets are used for processing of information.

MOLECULAR EVOLUTION AND DEVELOPMENT OF THE POSTSYNAPTIC PROTEOME

With the insight that synapses in the mammalian brain have evolved a signaling system capable of sensing many different types of stimuli and transducing these stimuli into a combinatorial multistate signaling output of modulated postsynaptic proteins, we must ask how this system originated and evolved. Through the analysis of species differences, this question can be answered in an outline form and summarized in the following six points.

Point 1. Examining the genomes of representative eukaryotes, metazoans, and chordates for orthologs of the mouse NRC/MASC and PSD genes showed that ~25% of the proteins were found in *Saccharomyces cerevisiae* and ~50% in invertebrate species (Fig. 4) (Emes et al. 2008;

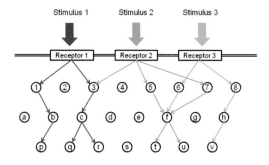

Figure 3. Combinations of postsynaptic proteins activated by different stimuli. Three stimuli activate three receptors that drive subsets of the cytoplasmic networks of proteins organized into a three-layer hierarchy (circles). Each stimulus modulates or uses a distinct subset of proteins shown as the postsynaptic code (*bottom*).

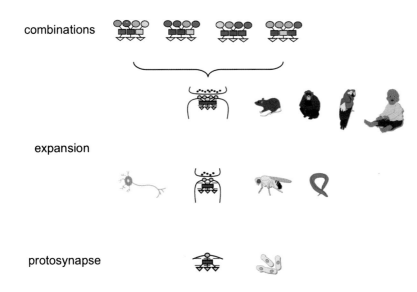

Figure 4. Evolution of postsynaptic complexes. Animals representing vertebrates (mouse, monkey, bird, and human), invertebrates (fly, worm), and unicellular eukaryotes (yeast) are shown adjacent to their complement of synaptic proteins represented using the modular representation of postsynaptic networks into upstream (blue circle), intermediate (red box), and downstream (yellow triangle). The protosynapse first found in unicellular organisms is expanded in invertebrates and vertebrates with preferential expansion of upstream (blue circles) proteins. A neuron is shown to the left of the invertebrate synapse, emphasizing its origins after the protosynapse. Bracketed above the vertebrate synapse is a set of four variants of NRC/MASCs, illustrating that combinations of proteins generate diversity.

Ryan et al. 2008; Ryan and Grant 2009). The core set of synapse proteins, conserved in all animals, is referred to as the *protosynapse*. In unicellular eukaryotes, protosynapse proteins are necessary for responding to changing environments and include genes (e.g., *NF1*) that in humans result in heritable forms of learning impairment, attesting to the conservation in cognitive processes. Molecular dating of eukaryotes places the origins of this protosynapse to ~1.4 billion years ago (Hedges and Kumar 2009; Ryan and Grant 2009), indicating that the origins of the brain are in deeply conserved adaptive machinery predating neurons.

Point 2. The eukaryote protosynapse set was expanded through addition of new proteins and functional domains. By 1 billion years ago, the protosynapse may have contained many of the essential signaling components of mammalian synapses as evidenced by the presence of homologous genes in the genomes of unicellular choanoflagellates (considered to be the last unicellular ancestor of animals) and the multicellular porifera (sponges), which lack neurons (King 2004; Emes et al. 2008; Kosik 2009; Ryan and Grant 2009). These include ion channels, neurotransmitter receptors, adhesion proteins, and their associated scaffold and signaling enzymes. Thus, the key functionality of the postsynaptic machinery arose in simpler animals before the origins of electrically active neurons, a model referred to as the "synapse first model" (Ryan and Grant 2009). With evolution of primitive metazoans, such as cnidarians, ionotropic glutamate receptors and other proteins were added, giving the invertebrate NRC/MASC the classes of receptors, scaffolds, and signaling enzymes that are commonly studied in mammalian synapses. Moreover, mutations in some of these genes in *Drosophila* impair learning, showing the functional conservation between invertebrates and vertebrates in this cognitive mechanism (Wu et al. 2007).

Point 3. The evolution of vertebrate synapses was associated with a major expansion of synaptic genes and proteins, shown by both proteomic and genomic methods (Emes et al. 2008). Vertebrates show a preferential expansion in the upstream proteins, rather than the downstream proteins and modules in the signaling networks (Fig. 4). The genomic mechanism for this expansion in gene families was gene and genome duplication. Two key examples that have been extensively investigated are the NMDA receptor *NR2* and *Dlg* genes that expanded from a single gene in invertebrates to four in vertebrates (Ryan et al. 2008).

Together, points 2 and 3 illustrate that major expansions in complexity occurred around the ancestral protosynapse proteins to give rise to an invertebrate synapse. The vertebrate postsynaptic proteome is largely characterized by a further major expansion in complexity in gene families, which means that more receptors and the upstream module components of the pathways that give specificity to signaling responses are available to synapses in these species (Fig. 4). Linking this increase in complexity to the earlier discussion of combinatorial signaling pathways found in phosphoproteomic studies, it is reasonable to expect that the vertebrate complexity provides a greater number of combinations of postsynaptic signaling subnetworks (Fig. 3) than those found in invertebrates.

Point 4. In addition to complexity in numbers, vertebrate molecular evolution in protein–protein interaction domains reveals that the *organization* of NRC/MASC complexes and networks has added to the complexity. Using the NMDA receptor NR2 example and considering its four vertebrate paralogs and comparing their structure with their single invertebrate homolog, it was found that the cytoplasmic carboxy-terminal domain (C-tail) in invertebrates lacks most of the enzyme interaction sites possessed by vertebrate NR2 subunits (Fig. 5) (Ryan et al. 2008).

Point 5. Expansion in complexity is amplified by the combinatorial nature of complexes. Again, NR2 and Dlg families have expanded from a single homolog in invertebrates to four in vertebrates. Because each NR2 binds a single Dlg to form a complex, invertebrates have only one complex, whereas vertebrates could generate tenfold

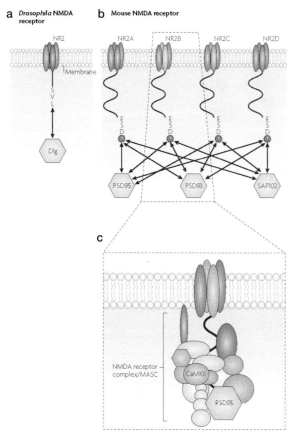

Figure 5. Evolution of NMDA interactions. Comparison of NMDA receptor complexes of *Drosophila melanogaster* containing a single NR2 subunit (*a*) with the mouse containing four NR2 subunits (*b*). PDZ-binding domains at the carboxyl termini of both *D. melanogaster* (NR2; SVL) and mouse NR2 (ESDV) are indicated. Note that the mouse NR2 intracellular domain is five times larger than that of *D. melanogaster*. The vertebrate NR2B carboxyl termini has numerous primary and secondary interacting proteins (colored shapes in *c*) and therefore a greater degree of NMDA-receptor-signaling complexity. Furthermore, the number of potential interactions of the NMDA receptor with Dlg proteins (yellow hexagons, Dlg, PSD95, SAP102, PSD93) is much greater for vertebrate synapses. (Reprinted, with permission, from Ryan and Grant 2009 [©Nature Publishing Group].)

more (Fig. 5). Considering the many other expanded families of proteins that also bind to NR2 and Dlg, one can readily appreciate the multiplicative increase in potential postsynaptic diversity.

Point 6. The organization of these complexes and networks of postsynaptic proteins requires them to be synthesized and transported to synapses during development of the nervous system. Genome-wide studies of messenger RNA (mRNA) (Valor et al. 2007) and microRNA (miRNA) (Manakov et al. 2009) show that the synthesis of PSD proteins proceeds in an orchestrated and orderly pattern of coregulation. As discussed in the next section, the coexpression of the sets of proteins is not identical in all neurons and all synapses. Variations in the relative levels of the proteins between neurons occurs, and this has a distinct evolutionary signature.

AN EVOLUTIONARY BASIS TO NEUROANATOMICAL MAPS OF SYNAPSE PROTEOME COMPOSITION

Vertebrates are characterized by having larger brains than invertebrates, and this difference is generally accepted as the explanation for their different cognitive abilities. This anatomical perspective also dominates the comparisons among vertebrates where brain size, encephalization quotients, and other measures of anatomical complexity are correlated with behavior. Not only is anatomical complexity a poor index of cognitive function and species-specific behaviors, but it lacks rigorous testing methods and is based almost exclusively on correlative approaches.

A large-scale study of the expression of NRC/MASC and PSD proteins in forebrain regions of the mouse found that more than 90% of the genes were coexpressed in neurons and different brain regions (Emes et al. 2008). Unsurprisingly, expression patterns of individual proteins were different, with some showing similar levels of expression in all regions and some showing very different levels in one or other region. At first glance, there was no prevailing logic to these patterns until the molecular phylogeny of the genes was considered. A clear picture emerged, in which the ancestral protosynapse set of proteins was expressed in all neurons and regions at equivalent levels, which was in contrast to more recently evolved genes that showed the most variety in expression patterns (Fig. 6). This pattern was functionally informative because the *upstream module* proteins, which contain neurotransmitter receptors, contributed more to anatomical diversification and regional specialization. Thus, the evolutionary expansions in postsynaptic complexity were used to generate diverse neurons and synaptic types (Fig. 4). These anatomical data again reflect on the combinatorial usage of postsynaptic proteins to generate diversity.

It is interesting to note that the ancestral chordate (~600 mya), which had the complex complement of synaptic proteins, evolved before species with large brains such as vertebrates. In other words, the complex synapse evolved before neuroanatomically complex brains.

NEUROPHYSIOLOGICAL CONSEQUENCES OF EXPANSIONS IN POSTSYNAPTIC COMPLEXITY

Returning to the fundamental neurophysiological basis of cognition—the encoding of information in patterns of action potentials—we can examine evidence supporting the following conclusion: Expanded molecular complexity conferred upon vertebrate synapses the ability to respond to a greater range of patterned stimuli. To specifically address this, different patterns of neural activity (e.g., trains of 1-Hz, 5-Hz or 100-Hz stimuli) are experimentally presented to the CA1 synapse in the hippocampus slice and their effects measured by examining the change in efficiency of synaptic transmission. These cellular phenomena (e.g., long-term potentiation [LTP], long-term depression, and spike-timing-dependent plasticity) share molecular mechanisms with behavioral cognitive processes. From

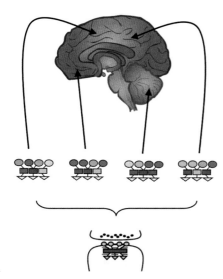

Figure 6. Combinatorial NRC/MASC diversity in brain regions. Modular representation of the NRC/MASC is shown in the synapse schematic (*bottom*). Bracketed above are four variants built from combinations of the evolutionarily expanded protein families, which are found in particular brain regions (arrows to brain).

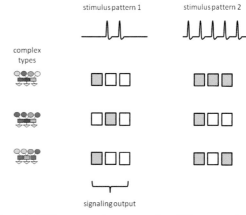

Figure 7. Different complexes respond to different neural codes. Patterns of action potentials shown as two distinct stimuli (*top*) and three types of NRC/MASC complexes composed of different proteins. The biochemical output is shown as three boxes and could represent transcription, receptor trafficking, and local protein synthesis or other downstream outputs of synaptic stimulation. The green shading indicates which of the three outputs is activated by the type of complex in response to the type of stimulus.

mice with deletions of PSD-95/dlg4, SAP102/dlg3, or PSD-93/dlg2, a distinct set of changes was observed in each mutant. For example, both Dlg4 and Dlg3 mutants showed enhanced LTP, whereas Dlg2 showed reduced LTP compared to normal mice (Migaud et al. 1998; Cuthbert et al. 2007; Carlisle et al. 2008). Further separating Dlg3 and Dlg4, some patterns of stimuli only showed abnormal effects in Dlg4. Other examples of paralog-specific effects were found in knockout mice of other protein families including the Src family (Grant et al. 1992) and NR2 subunits (Sprengel et al. 1998). Together, these data show that vertebrate expansion in synapse complexity has endowed the CA1 synapse with greater signaling specificity and a greater range of responsiveness to different neural codes (Fig. 7).

Because the NR2, Dlg, and Src family proteins interact in NRC/MASC complexes, these results indicate that complexes of different compositions could account for different synaptic responsiveness. Thus, the combinations of interacting postsynaptic proteins could generate an enormous diversity of *functional synapse types* in the mammalian brain. Because combinations of these proteins are characteristic of particular neurons and brain regions, it can then be expected that the mapping of synapse diversity at the molecular level will align with maps of synaptic electrophysiological diversity.

BEHAVIORAL ROLES FOR SYNAPSE EVOLUTION

The simplest evidence that evolutionary expansion in postsynaptic proteome complexity was important for behavior comes from vertebrate-specific genes, which if knocked out in mice or humans results in cognitive impairments (e.g., Dlg3/SAP102). A particularly important paradigm of how evolution of *complexity* and *organization* both influence behavior is provided by examples of the paralogs of the *NR2* and *Dlg* genes and their interactions via the NR2 cytoplasmic C-tail. The four paralogs of NR2 found in vertebrates have diverged at the level of protein sequence, particularly in the C-tail, whereas the structure of channel-forming and ligand-binding domains are highly conserved. This divergence in the C-tail sequence has been responsible for its selectivity in binding particular postsynaptic signaling proteins (Ryan et al. 2008) including PSD-95/Dlg4, SAP102/Dlg3, CamKII, and SynGAP, all of which when knocked out in mice result in cognitive deficits (Silva et al. 1992; Migaud et al. 1998; Komiyama et al. 2002; Cuthbert et al. 2007). In addition, the specific deletion of C-tails of the NR2A and NR2B subunits in mice results in distinct behavioral consequences (Sprengel et al. 1998).

More than 50 diseases in humans are caused by NRC/MASC gene mutations (Grant et al. 2005; Fernandez et al. 2009), reinforcing the observations in mice that different NRC/MASC and PSD proteins have common and distinct behavioral roles. Dlg proteins are again useful examples because Dlg2 mutations result in schizophrenia (Walsh et al. 2008), Dlg3 mutations in X-linked mental retardation (Tarpey et al. 2004), and the Neuroligins, which bind Dlg, underlie forms of autism (Jamain et al. 2003). Thus, different Dlgs have roles in different human cognitive disorders. In general, it appears that the evolution of increasing synaptic complexity has provided a larger set of molecular building blocks for the cognitive repertoire of animals.

IMPLICATIONS OF POSTSYNAPTIC COMPLEXITY AS A SOURCE OF SYNAPSE DIVERSITY

With the capacity to generate postsynaptic diversity from the combinatorial evolutionary molecular mechanisms described above, there could exist a vast or even limitless number of synapse types. If the number of synapse types is important for the overall function of the brain, it is then interesting to speculate that the traditional model of brain size as the index of cognitive capacity could be replaced by a model with a number of synapse types. Indeed, brain size may limit the number of synapse types. The full range of synapse diversity as measured by molecular markers or physiological properties is not known for any animal.

A general model of psychological function is that stimuli, ideation, action plans, and other constructs are *representations* of neural activity, as measured in functional brain imaging. Somehow, the neural activity is selectively activating particular brain regions. Considering that brain regions are characterized by combinations of PSD proteins and that these protein combinations lead to preferred responses to neural codes, synapse types in particular brain regions might "tune" that region to respond to the neural pattern of activity. Synapse diversity will also render different synapses susceptible to particular genetic disorders and influence the clinical manifestations including disease predilection to particular brain regions. At the cellular electrophysiological level, the importance of synapse diversity in connecting neural circuits is poorly understood.

CONCLUSIONS

The origins of human adaptive behavior and cognition from the ancestral protein machinery found in unicellular organisms was followed by the evolution of increasing complexity giving rise to synapses with extraordinary combinatorial computational signaling properties. The number and organization of these combinations were greatly enhanced during evolution, particularly with metazoans and chordates, and shown through genetic studies to directly contribute to fundamental cognitive properties at the synaptic level, including processing the patterns of neural activity and cognitive behaviors in the whole animal.

That the phylogeny of synapse complexity maps onto the anatomical organization of the brain, and that the global regulation of NRC/MASC and PSD development is coordinated during neural development, point to genomic mechanisms for generating and controlling the diversity in different neurons and synapses. This anatomical and functional diversity may be relevant to the formation of psychological representations or environment feature detection.

The differences and diversity of composition and organization in NRC/MASC and PSD proteins offer a molecular template for analyzing and understanding species diversity and behavioral specialization. Perhaps the increase in cognitive mechanisms that postsynaptic molecular evolutionary events provided contributed to the geographical spread and specialization of animals into novel environmental niches. This model helps to explain the evolutionary basis of the high degree of molecular complexity of mammalian synapses and its role in animal behavior. The data sets of proteins open many new approaches to the study of human brain disease, which must reflect the importance of these evolutionary mechanisms.

ACKNOWLEDGMENTS

Thanks to members of my lab in Edinburgh University, the Wellcom Trust Sanger Institute, and many collaborators for their contributions. Apologies to colleagues for the incomplete references imposed by space limitations. S.G.N.G. is supported by the Wellcome Trust.

REFERENCES

Adrian ED. 1928. *The basis of sensation: The action of the sense organs.* Norton, New York.

Alberini CM. 2009. Transcription factors in long-term memory and synaptic plasticity. *Physiol Rev* **89:** 121–145.

Bayes A, Grant SG. 2009. Neuroproteomics: Understanding the molecular organization and complexity of the brain. *Nat Rev* **10:** 635–646.

Carlisle HJ, Fink AE, Grant SG, O'Dell TJ. 2008. Opposing effects of PSD-93 and PSD-95 on long-term potentiation and spike timing-dependent plasticity. *J Physiol* **586:** 5885–5900.

Coba MP, Valor LM, Kopanitsa MV, Afinowi NO, Grant SG. 2008. Kinase networks integrate profiles of N-methyl-D-aspartate receptor-mediated gene expression in hippocampus. *J Biol Chem* **283:** 34101–34107.

Coba MP, Pocklington AJ, Collins MO, Kopanitsa MV, Uren RT, Swamy S, Croning MD, Choudhary JS, Grant SG. 2009. Neurotransmitters drive combinatorial multistate postsynaptic density networks. *Sci Signal* **2:** ra19.

Collins MO, Husi H, Yu L, Brandon JM, Anderson CN, Blackstock WP, Choudhary JS, Grant SG. 2006. Molecular characterization and comparison of the components and multiprotein complexes in the postsynaptic proteome. *J Neurochem* (suppl. 1) **9:** 16–23.

Cuthbert PC, Stanford LE, Coba MP, Ainge JA, Fink AE, Opazo P, Delgado JY, Komiyama NH, O'Dell TJ, Grant SG. 2007. Synapse-associated protein 102/dlgh3 couples the NMDA receptor to specific plasticity pathways and learning strategies. *J Neurosci* **27:** 2673–2682.

Delgado JY, Coba M, Anderson CN, Thompson KR, Gray EE, Heusner CL, Martin KC, Grant SG, O'Dell TJ. 2007. NMDA receptor activation dephosphorylates AMPA receptor glutamate receptor 1 subunits at threonine 840. *J Neurosci* **27:** 13210–13221.

Emes RD, Pocklington AJ, Anderson CN, Bayes A, Collins MO, Vickers CA, Croning MD, Malik BR, Choudhary JS, Armstrong JD, Grant SGN. 2008. Evolutionary expansion and anatomical specialization of synapse proteome complexity. *Nat Neurosci* **11:** 799–806.

Feng W, Zhang M. 2009. Organization and dynamics of PDZ-domain-related supramodules in the postsynaptic density. *Nat Rev* **10:** 87–99.

Fernandez E, Collins MO, Uren RT, Kopanitsa MV, Komiyama NH, Croning MD, Zografos L, Armstrong JD, Choudhary JS, Grant SG. 2009. Targeted tandem affinity purification of PSD-95 recovers core postsynaptic complexes and schizophrenia susceptibility proteins. *Mol Syst Biol* **5:** 269.

Gasch AP, Spellman PT, Kao CM, Carmel-Harel O, Eisen MB, Storz G, Botstein D, Brown PO. 2000. Genomic expression programs in the response of yeast cells to environmental changes. *Mol Biol Cell* **11:** 4241–4257.

Goelet P, Castellucci VF, Schacher S, Kandel ER. 1986. The long and the short of long-term memory—A molecular framework. *Nature* **322:** 419–422.

Grant SG, O'Dell TJ, Karl KA, Stein PL, Soriano P, Kandel ER. 1992. Impaired long-term potentiation, spatial learning, and hippocampal development in fyn mutant mice. *Science* **258:** 1903–1910.

Grant SG, Marshall MC, Page KL, Cumiskey MA, Armstrong JD. 2005. Synapse proteomics of multiprotein complexes: En route from genes to nervous system diseases. *Hum Mol Genet* (spec no. 2) **14:** R225–R234.

Hedges SB, Kumar S. 2009. *The timetree of life.* Oxford University Press, Oxford.

Hubel DH, Wiesel TN. 1959. Receptive fields of single neurones in the cat's striate cortex. *J Physiol* **14:** 574–591.

Husi H, Ward MA, Choudhary JS, Blackstock WP, Grant SG. 2000. Proteomic analysis of NMDA receptor-adhesion protein signaling complexes. *Nat Neurosci* **3:** 661–669.

Jamain S, Quach H, Betancur C, Rastam M, Colineaux C, Gillberg IC, Soderstrom H, Giros B, Leboyer M, Gillberg C, et al. 2003. Mutations of the X-linked genes encoding neuroligins NLGN3 and NLGN4 are associated with autism. *Nat Genet* **34:** 27–29.

King N. 2004. The unicellular ancestry of animal development. *Dev Cell* **7:** 313–325.

Komiyama NH, Watabe AM, Carlisle HJ, Porter K, Charlesworth P, Monti J, Strathdee DJ, O'Carroll CM, Martin SJ, Morris RG, O'Dell TJ, Grant SG. 2002. SynGAP regulates ERK/MAPK signaling, synaptic plasticity, and learning in the complex with postsynaptic density 95 and NMDA receptor. *J Neurosci* **22:** 9721–9732.

Kornau HC, Schenker LT, Kennedy MB, Seeburg PH. 1995. Domain interaction between NMDA receptor subunits and the postsynaptic density protein PSD-95. *Science* **269:** 1737–1740.

Kosik KS. 2009. Exploring the early origins of the synapse by comparative genomics. *Biol Lett* **5:** 108–111.

Manakov SA, Grant SG, Enright AJ. 2009. Reciprocal regulation of microRNA and mRNA profiles in neuronal development and synapse formation. *BMC Genomics* **10:** 419.

Migaud M, Charlesworth P, Dempster M, Webster LC, Watabe AM, Makhinson M, He Y, Ramsay MF, Morris RG, Morrison JH, et al. 1998. Enhanced long-term potentiation and impaired learning in mice with mutant postsynaptic density-95 protein. *Nature* **396:** 433–439.

Morgan CL. 1890. *Animal life and intelligence.* Arnold, London.

O'Keefe J, Nadel L. 1978. *The hippocampus as a cognitive map.* Clarendon/Oxford University Press, Oxford.

Pocklington AJ, Cumiskey M, Armstrong JD, Grant SG. 2006. The proteomes of neurotransmitter receptor complexes form modular networks with distributed functionality underlying plasticity and behaviour. *Mol Syst Biol* **2:** 2006.0023.

Ryan TJ, Grant SG. 2009. The origin and evolution of synapses. *Nat Rev Neurosci* **10:** 701–712.

Ryan TJ, Emes RD, Grant SG, Komiyama NH. 2008. Evolution of NMDA receptor cytoplasmic interaction domains: Implications for organisation of synaptic signalling complexes. *BMC Neurosci* **9:** 6.

Silva AJ, Paylor R, Wehner JM, Tonegawa S. 1992. Impaired spatial learning in α-calcium-calmodulin kinase II mutant mice. *Science* **257:** 206–211.

Sprengel R, Suchanek B, Amico C, Brusa R, Burnashev N, Rozov A, Hvalby O, Jensen V, Paulsen O, Andersen P, et al. 1998. Importance of the intracellular domain of NR2 subunits for NMDA receptor function in vivo. *Cell* **92:** 279–289.

Tarpey P, Parnau J, Blow M, Woffendin H, Bignell G, Cox C, Cox J, Davies H, Edkins S, Holden S, et al. 2004. Mutations in the *DLG3* gene cause nonsyndromic X-linked mental retardation. *Am J Hum Genet* **75:** 318–324.

Trinidad JC, Thalhammer A, Specht CG, Lynn AJ, Baker PR, Schoepfer R, Burlingame AL. 2008. Quantitative analysis of synaptic phosphorylation and protein expression. *Mol Cell Proteomics* **7:** 684–696.

Valor LM, Charlesworth P, Humphreys L, Anderson CN, Grant SG. 2007. Network activity-independent coordinated gene expression program for synapse assembly. *Proc Natl Acad Sci* **104:** 4658–4663.

Walsh T, McClellan JM, McCarthy SE, Addington AM, Pierce SB, Cooper GM, Nord AS, Kusenda M, Malhotra D, Bhandari A, et al. 2008. Rare structural variants disrupt multiple genes in neurodevelopmental pathways in schizophrenia. *Science* **320:** 539–543.

Wilson RA, Keil FC. 1999. *The MIT encyclopaedia of the cognitive sciences.* MIT Press, Cambridge, MA.

Wu CL, Xia S, Fu TF, Wang H, Chen YH, Leong D, Chiang AS, Tully T. 2007. Specific requirement of NMDA receptors for long-term memory consolidation in *Drosophila* ellipsoid body. *Nat Neurosci* **10:** 1578–1586.

Evolution in Reverse Gear: The Molecular Basis of Loss and Reversal

Q.C.B. CRONK

Centre for Biodiversity Research and Department of Botany, University of British Columbia, Vancouver V6T 1Z4, Canada

Correspondence: quentin.cronk@ubc.ca

Three types of regressive evolution are reviewed: loss, reversal, and regain after loss. Loss refers to the loss of a physical entity, either a structure or an organ, whereas reversals apply to character states returning to plesiomorphic from apomorphic conditions. The regain of characters after their loss represents a third type of evolutionary character change. The reconstruction of multiple losses and gains of characters by mapping on phylogenies is often problematic because of a lack of information about the relative likelihood of losses and gains. A developmental genetic approach using morphological, developmental, and molecular analysis is therefore an extremely important adjunct to phylogenetic approaches in interpreting losses, reversals, and regains. The molecular developmental basis of character loss and reversal is gradually becoming better understood. Loss of organs can occur by gain-of-function mutations (suppression) and loss-of-function mutations (that often leave a vestigial structure). The regain of characters after loss may occur by regulatory capture (a gain-of-function mutation) or by loss of function in suppressor genes. Reversals may occur by cryptic innovation (the formation of a new structure that mimics the old structure by gain-of-function mutations) or by loss of gene function associated with the apomorphic state (although this may have pleiotropic or neomorphic effects). The genetic landscape of reversal is illustrated by the reversal to polysymmetry from monosymmetry in flowers. The range of observed phenotypes, loss with vestige, cryptic innovation, and loss with neomorphism matches the range of changes predicted.

"In plants with separated sexes, the male flowers often have a rudiment of a pistil; and Kolreuter found that by crossing such male plants with an hermaphrodite species, the rudiment of the pistil in the hybrid offspring was much increased in size; and this shows that the rudiment and the perfect pistil are essentially alike in nature" (Darwin 1859).

IMPORTANCE OF REGRESSIVE EVOLUTION

Evolution does not run backward in any literal sense: The second law of thermodynamics determines that. However, as the quote above from Darwin indicates, regressive evolution, such as the loss of something once possessed, leaving only a vestige, is a very important facet of evolution and one of the strongest pieces of evidence for the evolutionary process. A famous example is the human vermiform appendix, which is hard to explain unless humans evolved from ancestors possessed of a leaf-eater's caecum. In the first edition of the *Origin* (1859), not only does Darwin have a section entitled "Rudimentary, atrophied, or aborted organs," but he also mentions rudiments or rudimentary organs more than 100 times throughout the work. Another section of the *Origin* is titled "Reversion to long lost characters," further indicating the importance of regressive evolution to the development of his theory.

In the age of molecular biology, we have the opportunity to explain these phenomena at the level of genes and molecules, an endeavor that both deepens the evolutionary synthesis and strengthens the evidence for evolution that Darwin realized comes from regressive evolution. In the quotation above, Darwin refers to the experiments by Koelreuter (1775) on crosses between the dioecious red campion (*Silene dioica*), with vestigial sex organs, and the hermaphrodite white sticky catchfly (*Silene viscosa*). Darwin realized that the intermediate size of the organs in the hybrids was excellent evidence that vestigial traces were indeed rudiments of the sexual organs and not some different class of structure, thus forestalling any arguments from his opponents on this score. This must be the first example of the use of genetics for homology determination, and Darwin's line of thought harmonizes well with modern evolutionary developmental genetics.

SOME TERMS: LOSS, REVERSAL, AND REGAIN AFTER LOSS

If we are to dissect regressive evolution at the molecular level, it is necessary to be very clear about the processes with which we are dealing. The terms "loss" and "reversal" are often used interchangeably for two rather different evolutionary events. First, there is the loss of a physical structure, such as legs or pappus scales, and second, there is the loss of a derived character state, e.g., the loss of red flower returning to the ancestral state of blue flowers. For this chapter, I propose to restrict the term "loss" to the first example and to restrict the term "reversal" to the second.

The first example, the loss of a physical entity, could be called a reversal in that the physical entity must have evolved in the first place, so loss of legs in snakes could be regarded as a reversal to the primitive state of leglessness as found in the primitive vertebrate *Amphioxus*. However, there are two objections to this:

1. The biological context of leglessness is so different in *Amphioxus* and snakes that it is not very useful to consider that a snake has reverted to an *Amphioxus*-like condition. The biological context of legs in reptiles is that they have evolved from fins and so a reversal would, more properly, be a reversal to fins.
2. If a loss is considered a reversal, this implies that the primitive and derived absences are similar states. However, shared absences are a very weak form of similarity. Bacteria and snakes are not similar because they both lack legs. They both lack an infinite number of real and potential structures, but that does not make them infinitely similar.

The second example, red petals reverting to the ancestral blue condition, is only a loss in the sense that any character state change is a loss—a loss of what went before it could equally well be considered a gain in that it is a secondary gain of blue flowers. It is not very useful to call a gain a loss, especially if it is only a loss in the sense that all change involves loss. It is, however, certainly a reversal, and this term describes it precisely.

Finally, it may be helpful in emphasizing the distinction between loss and reversal to analyze the terms in the context of phenotype ontology (Gkoutos et al 2004). In this, a lost organ is an *entity* (E), whereas a changed character state is a *quality* (Q), and philosophically, the difference is a fundamental one. Qualities are factors that are inherent to a particular instance of an entity. There cannot be an instance of an entity that does not exist. The absence of an organ is therefore not a feature of that entity but of a higher-level entity, such as the organism (E), which can have the quality "lacks part" (the lacking part specified by a second entity, E2). This may seem like an unnecessary precision of language, but it is important that genes and pathways are annotated to phenotypes in ways that are logically consistent (Mabee et al. 2007).

POTENTIAL PROBLEMS

Having made the case for separating the loss of an entity from the reversal of a character state, it must now be admitted that the two evolutionary events have many overlapping features. For instance, the reversal of a character state may often result from the loss of a gene (loss of a physical entity at the molecular level).

Furthermore, many morphological entities (usually minor ones) whose presence and absence are "flickering" (Marshall et al. 1994) on and off in evolution can plausibly be said to be reversing. An example is provided by the seed hairs of the Acanthaceae (Manktelow et al. 2001). Seed hairs appear to have evolved as a synapomorphy for the Whitfieldieae/Barlerieae clade and were subsequently lost in three component genera, an example of flicker over comparatively short evolutionary timescales. However, the multiple loss of pappus scales in European daisies (Bellideae) (Fiz et al. 2002) cannot be considered a reversal, because pappus scales evolved originally from sepals. This therefore represents a loss of the entity sepals/pappus. A reversal would be pappus scales evolving back into sepals.

The regain of an organ or entity after its loss is a special type of "reversal" that is worth considering separately because such changes break Dollo's law (discussed below). Purported examples include the regain of wings that have been lost during the evolution of flightlessness (Whiting et al. 2003). This is distinct from other types of reversals in that it involves the abolition of the quality "lacks part" in a group of organisms in which the presence of the part is plesiomorphic, so the gain is not an innovation but a reversal.

In conclusion, it therefore seems useful, for the reasons given above, to maintain a distinction among (1) loss of a physical entity (character state changes of a special type: present to absent), (2) reversal (character state change from apomorphic to plesiomorphic, but specifically excluding presence/absence cases), and (3) regain of a lost entity (i.e., abolition of a quality "lacks part" in an organism). Within the general concept of regressive evolution, three distinct types can therefore be recognized: *l*oss, *r*eversal and *r*egain (LRR) from loss. These may conveniently be referred to collectively as LRR events. Importantly, although these three types of evolutionary changes are related, they represent potentially distinct processes that may have different molecular mechanisms underlying them.

SINGLE AND MULTIPLE CHARACTER LOSSES AND REVERSALS

Phylogenetic versus Developmental Approaches to Loss and Reversal

The incongruence of molecular and morphological data under cladistic analysis has been remarked upon many times. The molecular evidence (particularly if it is based on multiple genes) is often regarded as giving the best estimate of the phylogeny. Incongruence of the tree with the evolution of morphological characters is then explained by parallel evolution (homoplasy). Homoplasy can result from multiple character state gains (parallel innovation), a single loss/reversal/regain (LRR) event, or multiple LRR events.

It can be problematic to reconstruct multiple LRR events by character mapping on phylogenies because these can also be modeled alternatively as multiple gain events. It is easy to find which is most parsimonious (or most likely under maximum likelihood) assuming symmetrical probabilities of losses and gains. However, if losses are thought to be more likely than gains, or vice versa, the situation becomes more complex. Most studies deal with this by determining the level of asymmetry of change probability at which the scenarios switch. Thus, Oakley and coworkers (Oakley and Cunningham 2002) noted that eyes in ostracods can be interpreted as a single parallel gain or as some 30 losses. As the authors note, both scenarios are somewhat improbable. It is problematic that something as complex as a compound eye should be able to evolve twice, but there is similarly no explanation for the alternative scenario of a massively repeated evolution of eyelessness. Similarly, a study of winged and wingless stick insects concluded that there were multiple gains of wings from a wingless ancestor, because under a multiple loss scenario, losses would have to be more than five times as likely as gains (Whiting et al. 2003). However, without evidence from developmental genetics, we have no means of knowing the relative likelihood of losses and gains, and such phylogenetics-only reconstructions will remain uncertain.

At this point, phylogenetic methods of homology assessment reach the limit of their usefulness, and character-based (developmental genetic) methods must take over (Cunningham 1999). In the example above, the single parallel gain hypothesis would imply that the ostrocod eye is not homologous to other compound eyes. In biology, the distinction between homology and analogy (coined by Richard Owen in 1843 [Boyden 1943]) is fundamental and determined by critical observation. If homology can be rejected at the level of similarity (whether molecular, developmental, or morphological), the two eyes are then (in Richard Owen's terminology) analogous, rather than homologous, and the parallel gain hypothesis is supported.

On the other hand, if molecular mechanisms can be discovered that support the relative ease of gains over losses (or vice versa), this is powerful evidence for constraining the asymmetry of gain/loss transition probabilities. It is one of the themes of this review that there has been an understandable overreliance on phylogenetic methods to study losses, reversals, and regains (driven by the ready availability of reliable phylogenies during the last two decades). However, the rise of molecular genetic and developmental understanding of character change permits a new, mechanism-based attack on the problems of losses and reversals.

Ecological and Developmental Factors Promoting Multiple Loss and Reversal

Multiple LRR events, like multiple convergent gains, are probably driven by strong selection in response to common ecological factors. Accessory olfactory bulbs in the brain, that function in responding to airborne pheromones, have been lost independently in the brains of sirenians, cetaceans, bats, and certain primates. The common ecological factor seems to be a departure from terrestrial habitats. In aquatic, aerial, or arboreal habitats, pheromones either are not dispersed or are too highly dispersed to function (Johnson et al. 1994).

In plants, the biochemical pathway of crassulacean acid metabolism (CAM) is an important mechanism for maintaining high rates of photosynthesis under drought conditions and is consequently commonly found in epiphytes. It appears to be a very labile character, gained and lost as plants adapt to shifting habitats. In Bromeliaceae, for instance, CAM has been gained and lost numerous times (Crayn et al. 2004). Breeding systems of plants are also labile traits because they are under very strong divergent selection for reproductive fitness (cross-pollination) or reproductive assurance (self-pollination) under different ecological conditions (Weller and Sakai 1999). A recent study of a small group of flax-flower plants, *Linanthus* sect. *Leptosiphon*, suggested four losses of self-incompatibility (SI) in response to conditions that favor partial selfing (Goodwillie 1999).

Finally, sexual selection functions as a very strong driver of reversals and losses (Omland and Lanyon 2000). Horn evolution in scarab beetles appears to have resulted in 25 separate gains and losses of five different horn types (Emlen et al. 2005). These labile and fast-evolving characters, under strong and opposing selectional forces, undergo what has been called evolutionary "flicker," i.e., relatively frequent gain and reversal or gain and loss in multiple lineages during a relatively short evolutionary timescale (Marshall et al. 1994).

In addition to ecological causes of multiple LRR events, it is possible to conceive that certain characters may be predisposed to multiple losses or reversals because of intrinsic developmental vulnerability. An example might be a character resulting from a gene network that is unstable for some reason. Conversely, other gene networks might be well buffered against change, thus preventing evolutionary loss or reversal. As the genetic basis of character change becomes increasingly well understood, it will be interesting to see whether examples of this sort emerge.

Multiple losses at the molecular level may also be due to mechanistic predisposition rather than ecological drivers. Intron losses provide a good example of basically stable characters that appear to be subject to repeated loss in some clades. Most flowering plants have an intron in their chloroplast *rpl2* gene, so this character is evolutionarily stable. However, in *Bauhinia*, a large genus of legumes, the intron is missing in many species and appears to have been lost independently in several different clades of *Bauhinia* (Lai et al. 1997). Similarly, there have been multiple losses of the rpoC1 intron from *Medicago* (Downie et al. 1998). In addition to molecular sequences that can undergo loss by deletion, secondary structures are another type of molecular feature that can undergo loss. A particular stem-loop structure in the mitochondrial genome has undergone as many as seven parallel losses (Macey et al. 2000).

IRREVERSIBILITY OF LOSS? DOLLO'S LAW AND GENE NETWORKS

In the introduction above, a distinction was made among (1) losses of organs, (2) reversals of character states, and (3) the regain of lost organs. The last of these is of particular evolutionary interest because it breaks Dollo's law. This law, in its modern form, states that a complex structure or organ, once lost, will not be regained. The Belgian palaeontologist Louis Dollo (1857–1931) articulated this law in 1893 as a result of his studies of fossil vertebrates. Stephen Jay Gould gave added prominence to the law in an influential essay (Gould 1970) and illustrated it with examples of the irreversibility of the loss of shell-coiling in the mollusca.

It has been suggested that there is molecular and developmental logic to Dollo's law, which is that the gene network underpinning a structure will rapidly decay when the structure is no longer phenotypically expressed and the underlying network is not subject to purifying selection. Marshall et al. (1994) suggested that silenced genes or lost developmental programs may be reactivated as long as this happens in under 6 million years. After 10 million years, they suggested that the chances are unrealistic of vestigial gene networks still being intact enough for reactivation. A good example of this, at a biochemical level, is provided by the loss of the cyanidin branch of the anthocyanin biosynthetic pathway in *Ipomoea* (Zufall and Rausher 2004), which is a key step in the production of blue pigment. The block of the cyanidin branch is therefore important in the evolution of red flowers in species such as *Ipomoea quamoclit* from blue-flowered ancestors. There is evidence that the relaxation of selection that occurred after the initial blockage of the pathway will now make it very unlikely that blue flowers will reevolve in that lineage.

On the other hand, mechanisms have been proposed to explain the conservation of genetic pathways after loss. Conservation may occur if the network has another function and was merely coopted to create the character that was subsequently lost. There is then no reason that the network should not be coopted again to remake the original character. The reversal of loss is therefore akin to a multiple gain. An example of how this might work is provided by the wing spots in *Drosophila* (Gompel et al. 2005). *D. biarmipes* has evolved a prominent wing spot caused by novel expression of the pigment gene, *yellow*. The *yellow* gene has become wing-expressed in a highly specific pattern because it has come under the regulation of the conserved *Engrailed* (*En*) gene regulatory network, by gaining (as a result of random mutation) a *cis* element that interacts with *En*. It is easy to see that the wing spot could be lost by a loss of the *cis* element. However, neither the *En* network nor the *yellow* gene would be lost because they have other important functions in *Drosophila*. If the wing spot were to be lost, it could be regained by the same means. It would, however, represent an independent gain, and the *cis* element might be in a slightly different location.

A similar example appears to be provided by the developmental genes *Notch* (*N*), *Distalless* (*Dll*), *En*, and *Spalt* (*Sal*) in butterflies, which are responsible for body patterning, are expressed at the eyespot, and may be coopted as part of the eyespot regulatory pathway (Beldade et al. 2005). An eyespot loss mutant has been shown to have reduced *N* and *Dll* expression (Reed and Serfas 2004). However, if eyespots are lost, the underlying network will not be because it is of vital importance to the whole organism, and eyespots can potentially be regained. Regulatory capture of this sort may prove to be common in evolution.

In the example above, loss of *cis* elements causes loss of gene expression and thus loss of a feature. This is a loss-of-function mutation, which would be genetically recessive. Alternatively, the loss of an organ or feature may be caused by active suppression as a result of a gain-of-function mutation in a suppressor gene. Such a mutation would be genetically dominant. An example of this appears to be the loss of bracts in the Brassicaceae, including *Arabidopsis*. In most eudicots, the inflorescence branches are subtended by bracts. However, bracts are usually missing from the inflorescences of the Brassicaceae, although there is vestigial expression of bract-specific genes where bracts would be expected. These vestiges have been called "cryptic bracts." The leaf developmental gene *JAGGED* is required to form bracts (Dinnney et al. 2004), and this is excluded from the inflorescence of *Arabidopsis* by the action of the *BLADE ON PETIOLE* (*BOP*) genes acting in concert with the developmental gene *LEAFY* (Norberg et al. 2005). Thus, bracts are actively suppressed in *Arabidopsis*. Loss-of-function mutations in the suppression gene network allows bracts to form, after having been evolutionarily lost, by desuppression. Because the suppressed gene has an important role in leaf development elsewhere in the plant, there is no likelihood that the *JAGGED* gene will degenerate, making the reoccurrence of bracts impossible.

The key feature of both loss by suppression and regulatory loss is that the gene that is no longer expressed in a particular place is expressed elsewhere in the organism. What is therefore important is a reduction in expression domain of a gene that has pleiotropic effects and not an elimination of expression. Such a gene will still be exposed to purifying selection, and the original expression pattern could be readily regained by either desuppression or regulatory capture, thus breaking Dollo's law. In such cases, the main factors preventing evolutionary recall may be ecological conditions that continue to select for loss.

Breaking Dollo's Law by Heterochrony

The reduction in expression domain may be in space (heterotopy) or time (heterochrony). Gastropods have lost shell-coiling multiple times, but two purported occurrences of the reevolution of shell-coiling have also been identified (Collin and Cipriani 2003). Because many gastropods without coiling in the adult have planktonic larvae with coiled shells, it is probable that the genes for shell-coiling continued to be exposed to purifying selection in juveniles, having been eliminated from adults by suppression or regulatory loss. It is therefore unsurprising that these genes could be reexpressed in the adult by either desuppression or regulatory capture, in which they would come under the control of adult developmental gene networks. Ironically, this study of the reversibility of the loss of shell-coiling is in direct contrast to Gould's earlier use of shell-coiling to illustrate Dollo's law (Gould 1970).

EVOLUTIONARY–DEVELOPMENTAL MECHANISMS OF LOSS AND REVERSAL

A Genetic Landscape of Loss and Reversal

In the introduction, a distinction was made among character loss, character state reversal, and regain after loss. These different evolutionary phenomena may have different processes underlying them. Another axis that is relevant to dissecting the evolutionary developmental mechanisms underlying loss and reversal is the gain-/loss-of-function axis. These two axes are used to produce a landscape of loss and reversal in 12 categories as shown in Figure 1.

A loss may be due to a gain-of-function (dominant) mutation in a gene, or genes, that causes active suppression of a trait. Similarly, a reversal may result from a "cryptic innovation" that produces what appears to be a reversal but is in fact an innovation caused by a gain of gene function.

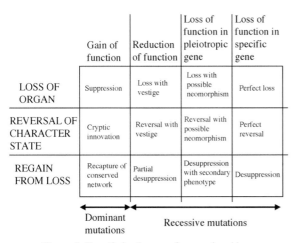

Figure 1. Genetic landscape of reversal and loss.

Another means of generating losses and reversals is through recessive loss-of-function mutations. If the loss of function is not complete, as it would be if a protein were to be merely down-regulated (and not completely eliminated as by a knockout mutation), then vestigial traces of a character are to be expected. This is the phenomenon of loss-with-vestige. Vestigial characters may be reduced forms of the original organs or more subtle traces. A common example of reduced organs is found in staminodes, which are stamens whose development was arrested very early.

Vestigial structures are very common mainly because when a structure becomes minute, the selective pressure to eliminate it further drops: At a certain size, it has not selective consequences. However, many structures are eliminated to a much greater degree and are only visible as vestiges by microscopic investigation very early in development. *Arabidopsis* lacks bracts in the inflorescence but "cryptic bracts" are distinguishable as patches of gene expression where bracts would be expected.

If the reversal results from the knockout of a gene that has other pleiotropic effects (as is the case in many artificially induced mutants), neomorphic characters are likely. Neomorphisms are new phenotypes produced as by-products of gene mutations, in addition to the main phenotypic effects. A good example of a neomorphism is hexamerous flowers commonly produced as a result of knockout of the floral symmetry gene *CYCLOIDEA* (*CYC*).

If there are no vestiges or neomorphic side effects, it may then be possible to get a "perfect" loss or reversal, although this appears to be rare. An example of loss-without-vestige (Cronk 2001) is found in the stamens of the Detarieae (Tucker 2001), in which, rather than being reduced to staminodes, certain stamens are lost without any ontogenetic trace. Such a phenomenon may result from the clean deletion of an ontogenetic pathway or from very complete but progressive reduction.

Reversal from loss, the regain of a structure or organ once lost, is a particularly interesting type of reversal because it is not degenerative but involves a new innovation. If the original loss was by a gain-of-function or suppressive mutation, regain is then easy—it just requires a desuppression by a loss-of-function mutation of the suppressing genes. However, if the original loss was a loss of function, then a gain of function will be needed to reinnovate the structure. This is more problematic because it involves reevolution of a character. In the case of complex structures such as eyes and wings, this is intuitively implausible. However, if the evolution of complex characters can occur by the capture of regulators by a control gene that is then expressed ectopically to produce the structure anew, the process may be more plausible. An example of this mechanism of evolution is found in the wing spots of insects as discussed in the section above, under Dollo's law.

Soft- versus Hard-wiring

Endress (2001) proposed that early in the history of an innovation, reversals are common, but over time, the innovations become "deeply rooted genetically in the organization" and so are less prone to reversal. In support of this interesting idea, similar to the concept of canalization, Endress cites the distribution of flowering plant characters on plant phylogenies. He tentatively suggests that character innovations in flowering plants such as sympetaly and tenuinucellate/unitegmic ovules may have begun as minor changes with, at first, frequent reversals. If this is correct, it implies that the gene networks underlying character innovations are at first "soft-wired," i.e., there is little developmental canalization (Waddington 1957). However, as the networks develop, they become buffered against reversals, and thus development is more canalized (hard-wired).

Figure 2 shows a simple graphic representation of this as a genotype of a lineage changing over time, across a critical threshold from phenotype change. At the start, away from the critical threshold, the phenotype is stable at A. Near the critical threshold, the phenotype can change rapidly as the genotypic "walk" crosses and recrosses the critical threshold. Later, when the walk has moved away from the critical threshold, the phenotype is again stable but at state B. This might be a useful analogy for Endressian soft-

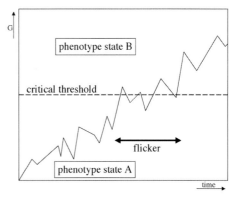

Figure 2. Graphic illustration of the concept of "flicker." Phenotypic traits when they first appear may not be stable, but they may acquire evolutionarily stable expression in later diverging lineages. In this diagram, the evolution of a lineage through time (*x* axis) is plotted with the *y* axis representing change in some feature or features of the genome.

wiring if there is a critical threshold in the evolution of a particular developmental gene network.

Endress' idea resonates with the idea of "flicker" put forward by Marshall et al. (1994), interpreting Shaffer's study of Mexican ambystomatid salamanders. In this study, metamorphosing salamanders, neotenic salamanders, and facultatively neotenic salamanders were found to exist in a complex phylogenetic pattern, apparently indicating frequent evolutionary change and reversals. Similar reversals have also been postulated in plethodontid salamanders. These have evolved from having a larval life stage to direct development, and in some of the direct-developing lineages, larvae have apparently reevolved (Mueller et al. 2004).

Under this interpretation of a labile character undergoing change and reversal, in appears that rapidly alternating character state changes can occur over relatively short evolutionary time periods. However, it has been argued that over longer periods, one of the character states will be irretrievably lost by mutational decay of the relevant gene networks and the time course of mutational decay of gene function has been modeled (Marshall et al. 1994).

The test of these ideas requires a good understanding of the developmental networks underpinning traits so that their evolution can be reconstructed and the likelihood of response to perturbation modeled.

AN EXAMPLE: REVERSALS FROM MONOSYMMETRY TO POLYSYMMETRY IN FLOWERS

It is thought that angiosperms ancestrally had spirally arranged floral parts. However, in the eudicots, the petals are typically inserted at the same level and are definite in number (typically five) arranged radially. These radial whorls are therefore "polysymmetric," having rotational symmetry and five planes of reflectional symmetry. In some clades of eudicots, however, asymmetry has developed along the adaxial–abaxial axis, with the adaxially oriented petals assuming a different form from the abaxially oriented petals. This results in the petal whorl being monosymmetric, having no rotational symmetry and only one plane of reflectional symmetry (Cubas 2004). This change of symmetry is analogous to the change from typical polysymmetric (pentamerous) sea urchins to monosymmetric heart urchins. Polysymmetric and monosymmetric petal whorls are illustrated in Figure 3.

The transition to monosymmetry from polysymmetry is well studied in the snapdragon (*Antirrhinum*). It results from the adaxial expression of *CYCLOIDEA*-like genes (*CYC* and its close paralog *DICHOTOMA* [*DICH*]) (Luo et al. 1996), which not only directly affect the morphology of the domain in which they are expressed, but also activate a second gene *RADIALIS* (*RAD*) that has a greater effect (Corley et al. 2005). In *Antirrhinum*, the expression domain of *RAD* is somewhat greater than the expression domain of *CYC* probably due to movement of the *RAD* transcript (Corley et al. 2005). The asymmetric expression of *CYC* and *RAD* in the adaxial domain leads to asymmetric petal morphology and hence the transition from polysymmetry to monosymmetry.

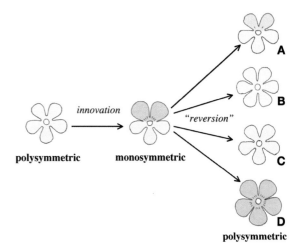

Figure 3. Different types of reversals from monosymmetric to polysymmetric flowers. (*A*) Reversal with vestige, (*B*) reversal with neomorphism (hexamery), (*C*) reversal without vestige, (*D*) cryptic innovation.

Many clades of flowering plants are characteristically monosymmetric and reversals seem to be rare. Nevertheless they do occur, both in nature and as horticulturally important mutants. Thus, floral symmetry is potentially a useful model for studying character reversal in a trait (reversal of monosymmetry back to polysymmetry) at the molecular level. Examples occur in the pea family (Leguminosae), for instance, all species of the peculiar genus *Cadia* (discussed further below), and horticultural varieties of the butterfly pea (*Clitoria ternatea*). Other examples occur in the snapdragon family (Scrophulariaceae), as in the so-called "peloric" mutants of snapdragon and toadflax (*Linaria*) and in the related African violet family Gesneriaceae, for instance, all members of the genus *Ramonda* and horticultural varieties of African violet (*Saintpaulia*) and gloxinia (*Sinningia*).

Not all reversals to polysymmetry are morphologically equivalent, and subtle differences may be found (see Fig. 3). Some polysymmetric revertants are hexamerous, as opposed to the original pentamerous state. This condition is found in the peloric snapdragon and in the horticultural polysymmetric *Sinningia*. Hexamery here is a "neomorphism" that results as a by-product of the reversion to polysymmetry (Fig. 3C). In the snapdragon, such mutants result from the elimination of the expression of *CYC* and *DICH*. This eliminates adaxial identity. Thus, all corolla lobes have default abaxial identity and are identical, leading to a polysymmetric flower. However, the elimination of *CYC/DICH* expression leads to the neomorphism of six corolla lobes. This implies that these genes have an additional, pleiotropic, role (that of meristic control) in addition to their role as identity genes. Because this mutation involves a loss of function, it is recessive.

Other polysymmetric revertants have barely noticeable vestigial monosymmetry. This is true of polysymmetric *Saintpaulia*, in which the mature flower appears nearly perfectly polysymmetric, but the early stages of stamen development are asymmetric. In also occurs in *Ramonda*, in which the petal shape is polysymmetric, but small differences in pigmentation at the base of the corolla lobes are asymmetric. Such reversion might be caused by partial loss of function of *CYCLOIDEA*, or loss of function of downstream genes in the *CYC* pathway, so that the direct effects of *CYC* on floral zygomorphy are still expressed, giving a vestigial phenotype. Both scenarios result from recessive (loss-of-function) mutations, although neither is as extreme as the complete *CYC* loss-of-function mutation detailed above.

Cadia and Cryptic Innovation

The original innovation to give monosymmetry involved a specific novel expression of a pathway in the adaxial (dorsal) part of the flower. It is this difference between the dorsal and ventral part of the flower that has to be abolished to return to polysymmetry. There are therefore two mechanisms to revert to polysymmetry. The polysymmetric petals may be equivalent either to what were formerly the abaxial petals or to what were formerly the adaxial petals.

In *Cadia* and *Clitoria*, the five rather large petals in the polysymmetric petal whorl appear to be equivalent to the adaxial or flag petal. This is not, strictly speaking, a reversal because the adaxial petal identity is an innovation in the origin of monosymmetry, so a true reversal would be a return to abaxial identity for all petals. A return to polysymmetry by a spread of adaxial identity is therefore a "cryptic innovation" rather than a reversal. Such cryptic innovation can result from gain-of-function mutations in *CYC*, for instance, in *cis*-regulatory elements, to expand the expression domain from adaxial regions to throughout the whole flower so that every petal takes on an adaxial identity. This is what has happened in *Cadia* (Citerne et al. 2006). Such mutations would be expected to be dominant and would be immediately exposed to selection. They would therefore only persist if a preexisting floral biology niche existed for such mutants to occupy. In the case of *Cadia*, a shift to bird pollination from bee pollination seems to have been involved.

It is conceivable that a perfect return to polysymmetric petals is possible, all with abaxial identity and no vestige, neomorphism, or "cryptic innovation." However, such perfect reversals appear to be nonexistent, or at least very rare. It would require partial inactivation of the genes responsible for monosymmetry, eliminating their promotion of adaxial identity and leaving untouched their other pleiotropic functions. It would thus require very specific and targeted loss of gene function. This might, however, be possible through natural selection over long periods of time, progressively eliminating all vestigial traces of monosymmetry. However, selection coefficients against vestigial characters is generally thought to be low, because such characters are likely to have negligible adaptive significance. The four types of polysymmetric revertants—reversal with vestige, reversal with neomorphism, cryptic innovation, and perfect reversal—are shown diagrammatically in Figure 3. These also correspond to the four types of reversals given in Figure 1.

DISCUSSION: REASSESSMENT OF LOSSES AND REVERSALS AT THE MORPHOLOGICAL AND MOLECULAR DEVELOPMENTAL LEVEL

The subject of evolutionary developmental genetics often, and rightly, focuses on the origin of innovations. However, every loss and reversal represents a challenge for the science of evo-devo. More specifically, the limitations of the phylogenetic pattern-based approach to distinguishing multiple gains from multiple losses means that a developmental, or process-based, examination of losses and gains is needed.

The phylogenetic method seeks to study character state changes by reconstructing ancestral states, generally using maximum likelihood or parsimony, of characters mapped on trees. However, such phylogenetic methods are prone to problems of interpretation, given that the asymmetry in the probabilities of character gain and loss are not known a priori. Ultimately, information about the relative ease of different transitions is needed, information that can only come from knowledge of the mechanisms underlying the processes. Thus, developmental genetic methods involving morphology, development, and molecular mechanism are required.

The first step in a developmental genetic study of loss is first-rate morphological and developmental descriptions. If multiple gains are postulated over multiple losses, differences in phenotype and development should be evident from minute morphological inspection, which should be confirmable by examining similarities and differences at the level of molecular process. Furthermore, much can be inferred about the molecular mechanism from morphological clues. Of particular importance are neomorphisms resulting from gene knockout and various vestiges left from incomplete loss of function. If what are apparently reversals are cryptic innovations, this too may be evident from careful morphological examination.

Finally, detailed understanding of the molecular mechanisms involved in loss, reversal, and regain after loss should be the most telling. It will be a challenge, although not impossible, to place probabilities on the occurrence of particular molecular events, but it is easy to distinguish between those events that are likely to be exceedingly rare and those events that are likely to happen repeatedly in evolutionary time.

ACKNOWLEDGMENTS

I thank Cliff Cunningham (Duke), Owen McMillan (NCSU), and Derrick Zwickl (Nescent) for helpful discussion. Research in the Cronk Lab is funded by the Natural Sciences and Engineering Research Council of Canada.

REFERENCES

Beldade P, Brakefield PM, Long AD. 2005. Generating phenotypic variation: Prospects From "evo-devo" research on *Bicyclus anyana* wing patterns. *Evol Dev* **7:** 101–107.

Boyden A. 1943. homology and analogy a century after the definitions of "homologue" and "analogue" of Richard Owen. *Q Rev Biol* **18:** 228–241.

Citerne HL, Pennington RT, Cronk QC. 2006. An apparent reversal in floral symmetry in the legume *Cadia* is a homeotic transformation. *Proc Natl Acad Sci* **103**: 12017–12020.

Collin R, Cipriani R. 2003. Dollo's law and the re-evolution of shell coiling. *Proc R Soc Lond B Biol Sci* **270**: 2551–2555.

Corley SB, Carpenter R, Copsey L, Coen E. 2005. Floral asymmetry involves an interplay between TCP and MYB transcription factors in *Antirrhinum*. *Proc Natl Acad Sci* **102**: 5068–5073.

Crayn DM, Winter K, and Smith JAC. 2004. Multiple origins of crassulacean acid metabolism and the epiphytic habit in the neotropical family Bromeliaceae. *Proc Natl Acad Sci* **101**: 3703–3708.

Cronk QC. 2001. Plant evolution and development in a postgenomic context. *Nat Rev Genet* **2**: 607–619.

Cubas P. 2004. Floral zygomorphy, the recurring evolution of a successful trait. *Bioessays* **26**: 1175–1184.

Cunningham CW. 1999. Some limitations of ancestral character state reconstruction when testing evolutionary hypotheses. *Syst Biol* **48**: 665–674.

Darwin C. 1859. *On the origin of species by means of natural selection,* 1st ed. Murray, London.

Dinneny JR, Yadegari R, Fischer RL, Yanofsky MF, Weigel D. 2004. The role of *JAGGED* in shaping lateral organs. *Development* **131**: 1101–1110.

Downie SR, Katz-Downie DS, Rogers EJ, Zujewski HL, Small E. 1998. Multiple independent losses of the plastid *rpo*C1 intron in *Medicago* (Fabaceae) as inferred from phylogenetic analyses of nuclear ribosomal DNA internal transcribed spacer sequences. *Can J Bot* **76**: 791–803.

Emlen DJ, Marangelo J, Ball B, Cunningham CW. 2005. Diversity in the weapons of sexual selection: Horn evolution in the beetle genus *Onthophagus* (Coleoptera: Scarabaeidae). *Evolution* **59**: 1060 1084.

Endress PK. 2001. Origins of flower morphology. *J Exp Zool* **291**: 105–115.

Fiz O, Valcarcel V, Vargas P. 2002. Phylogenetic position of Mediterranean Astereae and character evolution of daisies (*Bellis*, Asteraceae) inferred from nrDNA ITS sequences. *Mol Phylogenet Evol* **25**: 157–171.

Gkoutos GV, Green EC, Mallon A-M, Blake A, Greenaway S, Hancock JM, Davidson D. 2004. Ontologies for the description of mouse phenotypes. *Comp Funct Genomics* **5**: 545–551.

Gompel N, Prud'homme B, Wittkopp PJ, Kassner VA, Carroll SB. 2005. Chance caught on the wing: *cis*-regulatory evolution and the origin of pigment patterns in *Drosophila*. *Nature* **433**: 481–487.

Goodwillie C. 1999. Multiple origins of self-compatibility in *Linanthus* section Leptosiphon (Polemoniaceae): Phylogenetic evidence from internal-transcribed-spacer sequence data. *Evolution* **53**: 1387–1395.

Gould SJ. 1970. Dollo on Dollo's law: Irreversibility and the status of evolutionary laws. *J Hist Biol* **3**: 189–212.

Johnson JI, Kirsch JA, Reep RL, Switzer RC 3rd. 1994. Phylogeny through brain traits: More characters for the analysis of mammalian evolution. *Brain Behav Evol* **43**: 319–347.

Koelreuter JT. 1775. Lychni-cucubalus: Novum plantae hybridae genus. Novi comment. *Acad Sci Imp Petropol* **20**: 431–448.

Lai M, Sceppa J, Ballenger JA, Doyle JJ, Wunderlin RP. 1997. Polymorphism for the presence of the *rpL2* intron in chloroplast genomes of *Bauhinia* (Leguminosae). *Syst Bot* **22**: 519–528.

Luo D, Carpenter R, Vincent C, Copsey L, Coen E. 1996. Origin of floral asymmetry in *Antirrhinum*. *Nature* **383**: 794–799.

Mabee PM, Ashburner M, Cronk QCB, Gkoutos GV, Haendel M, Segerdell E, Mungall C, Westerfield M. 2007. Phenotype ontologies: The bridge between genomics and evolution. *Trends Ecol Evol* **22**: 345–350.

Macey JR, Schulte JA, Larson A. 2000. Evolution and phylogenetic information content of mitochondrial genomic structural features illustrated with acrodont lizards. *Syst Biol* **49**: 257–277.

Manktelow M, McDade LA, Oxelman B, Furness CA, Balkwill MJ. 2001. The enigmatic tribe whitfieldieae (Acanthaceae): Delimitation and phylogenetic relationships based on molecular and morphological data. *Syst Bot* **26**: 104–119.

Marshall CR, Raff EC, Raff RA. 1994. Dollo's law and the death and resurrection of genes. *Proc Natl Acad Sci* **91**: 12283–12287.

Mueller RL, Macey JR, Jaekel M, Wake DB, Boore JL. 2004. Morphological homoplasy, life history evolution, and historical biogeography of plethodontid salamanders inferred from complete mitochondrial genomes. *Proc Natl Acad Sci* **101**: 13820–13825.

Norberg M, Holmlund M, Nilsson O. 2005. The *BLADE ON PETIOLE* genes act redundantly to control the growth and development of lateral organs. *Development* **132**: 2203–2213.

Oakley TH, Cunningham CW. 2002. Molecular phylogenetic evidence for the independent evolutionary origin of an arthropod compound eye. *Proc Natl Acad Sci* **99**: 1426–1430.

Omland KE, Lanyon SM. 2000. Reconstructing plumage evolution in orioles (*Icterus*): Repeated convergence and reversal in patterns. *Evolution* **54**: 2119–2133.

Reed RD, Serfas MS. 2004. Butterfly wing pattern evolution is associated with changes in a Notch/Distal-less temporal pattern formation process. *Curr Biol* **14**: 1159–1166.

Tucker SC. 2001. The ontogenetic basis for missing petals in *Crudia* (Leguminosae: Caesalpinioideae: Detarieae). *Int J Plant Sci* **162**: 83–89.

Waddington CH. 1957. *The strategy of the genes*. Allen and Unwin, London.

Weller SG, Sakai AK. 1999. Using phylogenetic approaches for the analysis of plant breeding system evolution. *Annu Rev Ecol Syst* **30**: 167–199.

Whiting MF, Bradler S, Maxwell T. 2003. Loss and recovery of wings in stick insects. *Nature* **421**: 264–267.

Zufall RA, Rausher MD. 2004. Genetic changes associated with floral adaptation restrict future evolutionary potential. *Nature* **428**: 847–850.

Darwin's "Abominable Mystery": The Role of RNA Interference in the Evolution of Flowering Plants

A. CIBRIÁN-JARAMILLO[1,2,4] AND R.A. MARTIENSSEN[3,4]

[1]Sackler Institute for Comparative Genomics, American Museum of Natural History, New York, New York 10024; [2]The New York Botanical Garden, Bronx, New York 10458; [3]Cold Spring Harbor Laboratory, Cold Spring Harbor, New York 11724; [4]The New York Plant Genomics Consortium, Center for Genomics and Systems Biology, Department of Biology, New York University, New York, New York 10003*

Correspondence: martiens@cshl.edu

Darwin was famously concerned that the sudden appearance and rapid diversification of flowering plants in the mid-Cretaceous could not have occurred by gradual change. Here, we review our attempts to resolve the relationships among the major seed plant groups, i.e., cycads, ginkgo, conifers, gnetophytes, and flowering plants, and to provide a pipeline in which these relationships can be used as a platform for identifying genes of functional importance in plant diversification. Using complete gene sets and unigenes from 16 plant species, genes with positive partitioned Bremer support at major nodes were used to identify overrepresented gene ontology (GO) terms. Posttranscriptional silencing via RNA interference (RNAi) was overrepresented at several major nodes, including between monocots and dicots during early angiosperm divergence. One of these genes, *RNA-dependent RNA polymerase 6*, is required for the biogenesis of *trans*-acting small interfering RNA (tasiRNA), confers heteroblasty and organ polarity, and restricts maternal specification of the germline. Processing of small RNA and transfer between neighboring cells underlies these roles and may have contributed to distinct mutant phenotypes in plants, and in particular in the early split of the monocots and eudicots.

In an 1879 letter to J.D. Hooker, Darwin described the sudden appearance and rapid diversification of flowering plants in the mid Cretaceous as an "abominable mystery." Indeed, angiosperm diversification patterns presented an exception to his notion that nature evolves gradually *natura non facit saltum* (Friedman 2009). The key evolutionary processes that enable plants to adapt and diversify are only partially understood and mostly at the level of species, not at the level of major nodes, and less at the level of the evolving genome. On the 150th anniversary of *On the Origin of Species* (Darwin 1859), Darwin's mystery remains a fundamental issue in plant evolutionary biology not only within the angiosperms, but also within the gymnosperms. Namely, what are the fundamental evolutionary processes that enable plant species to adapt and diversify? One approach to understanding the origin and diversification of angiosperms and other seed plants is to identify genes or sets of genes that were critical in the divergence of key branches in plant evolution. By knowing gene function in at least some extant species, it should then be possible to correlate functional processes of interest with key steps such as the transition of plants from water to land or the evolution of the seed.

In principle, genes that were functionally important for branch divergence and plant diversification will have a phylogenetic signal that we can measure when we reconstruct phylogenies, through analysis of their effect on tree topology and branch support. Phylogenetic incongruence between a partitioned functional class of genes and the organismal phylogeny would suggest that the partition has experienced a unique evolutionary history relative to the organisms involved. In this way, incongruence of a particular class of genes in a partitioned analysis allows us to establish hypotheses about the evolution and potential function of these gene classes. Here, we use congruence measures of character evolution to mine genomes for patterns of protein function. In particular, we use modified elaborations of Bremer support—partitioned branch support (PBS) (Baker and DeSalle 1997), and partitioned hidden branch support (PHBS) (Gatesy et al. 1999). These measures can be used to evaluate the overall contribution (positive, negative, or neutral) of a particular gene to the various nodes or branches in a phylogenetic hypothesis. If one assumes that the tree obtained best represents the evolutionary history of the taxa involved, partitions that are in agreement or in conflict with the overall evolutionary history of the groups in the analysis can be detected and used to explain some of the more interesting organismal differences among taxa. This phylogenomic framework is powerful because it integrates experimental and genomic data to enable predictions of gene function, allowing us to tease apart the role of evolutionary change in protein function (Eisen 1998; Eisen and Fraser 2003; Sjölander 2004; Brown and Sjölander 2006).

---

*Members of the Consortium are Gloria Coruzzi (New York University), Rob DeSalle (American Museum of Natural History), Dennis Stevenson (The New York Botanical Garden), W. Richard McCombie (Cold Spring Harbor Laboratory), Ernest Lee (American Museum of Natural History), Sergios-Orestis Kolokotronis (American Museum of Natural History), Manpreet Katari (New York University), A.C.J. (American Museum of Natural History/The New York Botanical Garden), A.C.-J. (The New York Botanical Gardens), and R.M. (Cold Spring Harbor Laboratory).

Plant phylogenies to date are based on a few nuclear and plastid markers (for review, see APGIII 2009; Mathews 2009). By making use of the increasingly available genomic and expressed sequence tag (EST) data, as well as in-house data provided by The New York Plant Genomics Consortium, it was possible to construct a phylogeny of protein sequences from 2557 orthologous genes spanning 16 plant species. Species were chosen as representatives of major groups of angiosperms, gymnosperms, and nonseed plants, and measures of support were used to identify proteins and characters that may have functional significance across those 16 taxa (Cibrián-Jaramillo et al. 2010).

Our phylogenomic approach allows all character information to interact freely and reveal a more accurate description of species relationships, and at the same time, it makes it possible to observe snapshots of how genes or groups of genes may have evolved in the context of the overall phylogeny. These sets of genes themselves are a hypothesis, and their relevance to that node can be tested further based on measures of selection and explicit experimental analyses. It is clear that genome-level sequencing and large EST studies are rapidly growing, expanding the number of gene partitions and ways of partitioning phylogenetic information that are available. Our platform can easily incorporate the information simultaneously, contributing to the efficient integration of genomic and experimental data and enlightening the evolutionary processes driving plant diversification (Chiu et al. 2006; Cibrián-Jaramillo et al. 2010).

METHODS

Expanding on the analysis of De la Torre-Bárcena et al. (2009), we assembled a matrix of all available genomic and EST data to date for 16 plant species including 11 seed plants—five angiosperms (*Amborella*, rice, *Arabidopsis*, poplar, and grape) and six gymnosperms (*Cryptomeria*, pine, two cycads, gingko, *Gnetum*, and *Welwitschia*)—and four seed-free plants—Filicalian fern (*Adiantum*), a thalloid liverwort (*Marchantia*), a moss (*Physcomitrella*), and a Lycophyte (*Selaginella*) (Table 1).

Orthology of genes was established using OrthologID (Chiu et al. 2006), http://nypg.bio.nyu.edu/orthologid. OrthologID is an automated approach to sort query sequences into gene family membership and determine sets of orthologs from the gene trees. All ortholog groups reflecting coding genes are then assembled into a concatenated matrix of 1,062,841 amino acids representing 2557 proteins (genes), with delineated data partitions for each gene (for other methodological details, see Cibrián-Jaramillo et al. 2010). A maximum parsimony tree was generated using all concatenated genes in a simultaneous analysis (SA) and individually (partitioned data). Parsimony analysis was performed in PAUP* 4b10 using equal weights (Swofford 2003). Branch support was evaluated using the nonparametric bootstrap (2000 replicates) and jackknife (50% and 30% removal) methods in PAUP (Felsenstein 1985; Farris et al. 1996).

Once the most parsimonious tree is identified through character congruence, we can examine the partitions to say

Table 1. List of Species and Genomic Sources

Species	Genomic database
Adiantum capillus-veneris	TIGR PlantTA
Amborella trichopoda	TIGR PlantTA
Arabidopsis thaliana[a]	TAIR
Cryptomeria japonica	TIGR PlantTA
Cycas rumphii	CSHL/TIGR PlantTA
Ginkgo biloba	CSHL/TIGR PlantTA
Gnetum gnemon	CSHL/TIGR PlantTA
Marchantia polymorpha	JCVI
Oryza sativa[a]	JGI
Pinus taeda	TIGR PlantTA
Populus trichocarpa[a]	JGI
Selaginella moellendorffii	TIGR PlantTA
Vitis vinifera[a]	Genoscope
Welwitschia mirabilis	TIGR PlantTA
Zamia fischeri	CSHL/TIGR PlantTA

[a]Complete genomes: (TIGR) http://compbio.dfci.harvard.edu/tgi/plant.html; (PlantTA) http://plantta.jcvi.org; (CSHL) http://www.cshl.edu; (JCVI) http://www.jcvi.org; (JGI) http://www.jgi.doe.gov; (Genoscope) http://www.genoscope.cns.fr/spip.

something about their function. The delineation of data partitions allows the contribution of a gene (partition) to a branch to be assessed using congruence measures of support. We used a customized Perl script to calculate individual tree statistics including PBS and PHBS (Cibrián-Jaramillo et al. 2010). By definition, for a particular combined data set, a particular node (branch), and a particular data partition, PBS is the minimum number of character steps for that partition on the shortest topologies for the combined data set which do not contain that node, minus the minimum number of character steps for that partition on the shortest topologies for the combined data set that do contain that node (Baker and DeSalle 1997). PHBS is the difference between PBS for that data partition and the Bremer support value (Bremer 1988, 1994) for that node for that data partition (Gatesy et al. 1999). Values for these metrics can be positive, zero, or negative, and the value can indicate the direction of support for the overall concatenated hypothesis: Positive lends support, zero is neutral, and negative gives conflicting support (Gatesy et al. 1999).

A gene ontology (GO) term was established for each gene based on orthology with an *Arabidopsis* gene ID number using the current TAIR v8 database (http://www.arabidopsis.org). To determine which branch contains enrichment of a certain molecular or biological function, statistically overrepresented GO categories at each partition were compared to the distribution of that GO term in the *Arabidopsis* genome (considered a "baseline distribution"; Cibrián-Jaramillo et al. 2010). Because each branch is composed of partitions which represent genes that provide positive, negative, or neutral support, genes were first grouped into four sets: (1) genes that had a positive value for PBS (apparent), (2) genes that had a positive value for PHBS (hidden) support, (3) genes with neutral PBS, and (4) genes with neutral (zero) PHBS (no evolutionary signature for each branch).

Sungear (Poultney et al. 2007) implemented in Virtual Plant (http://www.virtualplant.org) was used to compare different sets of gene lists against *Arabidopsis*. GO term overrepresentation is measured by a z-score representing the number of standard deviations by which a particular observation (i.e., number of genes) is above or below the mean (Dudoit et al. 2004; Gutiérrez et al. 2007). Partitions with overrepresented GO terms and positive PBS within the angiosperms, nodes 4 through 7, were further investigated using Biomaps (Wang et al. 2004) as implemented in VirtualPlant. This tool provides a different measure of overrepresentation by using a hypergeometric distribution and significance based on a p-value ($p < 0.05$). Biomaps was used to compare the observed distribution of genes at each branch to the distribution of those GOs terms associated with *Arabidopsis* genes found in the matrix.

RESULTS

The resulting tree is identical in topology to a tree previously obtained with maximum parsimony (MP) and maximum likelihood (ML), although with fewer partitions (1200) and various combinations of ingroup and outgroup taxa (De la Torre-Bárcena et al. 2009) (Fig. 1). Other tree manipulations and details regarding phylogenetic analyses are summarized in De la Torre-Bárcena et al. (2009) and Cibrián-Jaramillo et al. (2010). A subset of *Arabidopsis* orthologs 1503 (58.7%) had at least one functional GO category (the total number of GO categories matched is 1872). The overall GO term distribution of the matrix had no significant biases (that would suggest a methodological bias) compared to *Arabidopsis* (Cibrián-Jaramillo et al. 2010).

A number of genes were found belonging to GO categories with very low probabilities of occurring by chance at the observed frequencies (z-scores) for both positive PBS and PHBS, with no significant outliers with neutral PHBS or PBS. Positive PHBS genes provide additional support at a particular node in the simultaneous analysis of all data partitions. Figure 2 illustrates the distribution of sets of overrepresented genes (represented by vessels) for PHBS values at each node in the tree. Node 6 (*Arabidopsis*, *Populus*, *Vitis*) and node 7 (*Populus*, *Vitis*) have the largest outlier vessels.

Within the angiosperms (Biomaps), overrepresented GO terms with both PBS and PHBS support included photosynthesis, development, and hormone-related functional categories (Figs. 1 and 2). A functional group was of exceptional interest: genes involved in posttranscriptional gene silencing, in particular *AGO1* and *RDR6* within the rosids (*Arabidopsis*, *Populus*, *Vitis*). Notably, character comparison for *AGO1* (not shown) and *RDR6* (Fig. 3) revealed a number of amino acid substitutions at regions in proteins with known functional importance (Marchler-Bauer et al. 2007). For *RDR6*, the *SHOOT LESS2* gene (*SHL2*) is the rice ortholog of *RDR6* in *Arabidopsis*. The *shl2-10* allele, *shl2*, has a G614D mutation, responsible for that mutant phenotype (Nagasaki et al. 2007). This specific site is one of those supporting cladogenetic variations in our matrix (Fig. 3), providing positive branch support as apomorphic for monocots (Cibrián-Jaramillo et al. 2010).

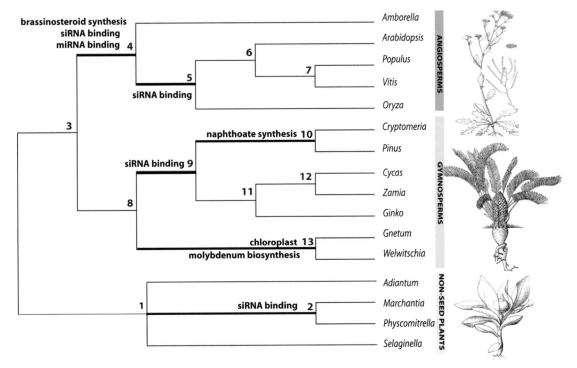

Figure 1. Phylogenetic relationships of seed plants using 2557 proteins inferred with maximum parsimony. All nodes showed bootstrap (2000 replicates) and jackknife (1000 replicates) support values above 99%. Overrepresented GOs with the most important functional categories are shown at the base of the nodes. (Modified, with permission, from Cibrián-Jaramillo et al. 2010 [© Oxford University Press].)

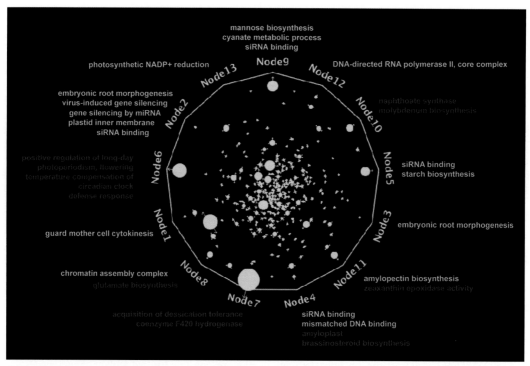

Figure 2. Distribution of genes across nodes. Sungear allows for the visual and statistical analysis of overlapping relationships among different lists of data and Boolean combinations. Each polygon corresponds to a particular node in the phylogeny. The circles with arrows within the polygon are called vessels, which represent genes with a positive z-score, from the set of categories with positive PHBS values. The position and the arrows of the vessels identify which node the genes are from, and the size of the vessel is relative to the number of genes in that vessel. The most interesting gene categories are written on each node (red categories are only found in that node, whereas yellow categories are shared across various nodes).

DISCUSSION

The topology described here recovers major groups of seed plants as all previous morphological analyses and most molecular analyses with monophyletic seed plants have done: the cycads, the conifers, the gnetophytes, and the angiosperms. We support the gnetophytes as the sister group to all other gymnosperms, congruent with phylogenetic studies using phytochrome genes (Mathews and Donoghue 2000; Schmidt and Schneider-Poetsch 2002; Mathews 2009), *AGAMOUS*-like genes (Winter et al. 1999; Becker et al. 2003), and *FLORICAULA/LEAFY* (Frohlich and Parker 2000). These results are to a great extent congruent with the angiosperm phylogeny group (APG III) (APGIII 2009). A more detailed explanation of the importance of a simultaneous analysis, as well as of support dynamics, the role of outgroup choice, taxon sampling, and missing data is presented elsewhere (De la Torre-Bárcena et al. 2009; Cibrián-Jaramillo et al. 2010).

Genes with Evolutionary and Functional Relevance

Overrepresented functional categories that are common throughout nodes are largely metabolic processes, such as photosynthesis. This distribution is concordant with their importance in key biochemical pathways that plants have developed in response to major environmental stress. For example, changes in photosynthetic chemical pathways are used not only to adapt to novel light conditions, but also to reduce evaporative water loss that was probably required from the transition from water to land and when plants colonized new environments (Bohnert et al. 1988). Interestingly, the Gnetophyta (node 13) had the highest number of overrepresented photosynthetic genes (Cibrián-Jaramillo et al. 2010). The gnetophyte *Welwitschia mirabilis* has a crassulacean acid metabolism (CAM) photosynthetic pathway in which stomata are open at night, avoiding water diffusion during the day (von Willert et al. 2005). Another member, *Ephedra*, is found in semiarid to desert conditions exposed to water stress during part of the year. Most *Gnetum* species are distributed in lowland tropical rainforests and are uniquely characterized by a relatively lower photosynthetic capacity as well as reduced capacity for stem water transport (Feild and Balun 2008).

At the other end of the spectrum are overrepresented gene categories directly related to specific traits or phenotype characteristics of that clade. For example, overrepresented amylopectin genes at the conifers (node 9) and mannose biosynthesis genes at the cycad node (node 11) may have a direct association to their morphology (Fig. 2). Amylopectin is fundamental to the manoxylic wood in cycads, and it differs from the pycnoxylic wood in conifers and the Gnetales, in which mannose is an important component (Greguss 1955).

Within the angiosperms, plant hormones and genes involved in circadian clock and photoperiodism were among the most interesting overrepresented partitions.

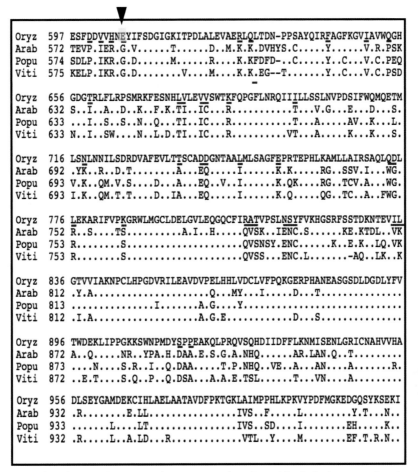

Figure 3. Character comparison for *RDR6* among angiosperms reveals amino acid substitutions at regions in proteins with known functional importance. Shown is part of the alignment in our matrix that corresponds to the *RDR6* domain. The *SHOOTLESS2* gene (*SHL2*) is the rice ortholog of *RDR6* (from *Arabidopsis*). In the *shl2-10* allele, *shl2* has a G614D mutation, responsible for the mutant phenotype, i.e., functionally important site (Nagasaki et al. 2007). This site is one of those supporting cladogenetic variations in our matrix, i.e., providing positive branch support for the split between monocots and the rest of the angiosperms. Substitutions unique to rice throughout the domain are underlined in red. Approximations of domain span are based on Marchler-Bauer et al. (2007). (Modified, with permission, from Cibrián-Jaramillo et al. 2010 [© Oxford University Press].)

Brassinosteroid genes were found to be uniquely overrepresented in the angiosperm clade (node 4) (Fig. 2). Brassinosteroid hormones differ in their signaling from other hormones, with a relatively longer pathway than either auxin or gibberellin (Bajguz and Tretyn 2003). Carotenoid biosynthesis factors, involved in shoot-branching and long-range signaling (Mouchel and Leyser 2007), were identified in the same node. Genes that are involved in the regulation of the circadian clock, photoperiodism, and growth habit (Balasubramanian et al. 2006) are overrepresented in the rosids (node 6). Patterns of hormone expression and regulation of circadian clock, and their specific function in the rest of the angiosperms, must be tested in future molecular and developmental studies, but their relevance is highlighted here.

RNAi in Plant Evolution

Genes involved in posttranscriptional regulation by small RNAs are highly overrepresented functional categories in both gymnosperms and angiosperms (Fig. 1). They have among the highest significance values for overrepresented genes in the split between *Amborella* and the rest of the angiosperms (node 4) and in the split between monocots and eudicots (node 5). microRNAs (miRNAs) and small interfering RNAs (siRNAs) are important for developmental phenotypes (Willman and Poethig 2005; Sunkar and Zhu 2007; Wang et al. 2007), and some are highly conserved.

Two genes in particular, *Argonaute* (*AGO1*) and *RNA-dependent RNA polymerase 6* (*RDR6*), are critical in developmental aspects of RNAi. They have roles in various stages of embryo and leaf development, polarity, and shape through *trans*-acting siRNA and miRNA pathways (Kidner and Martienssen 2004, 2005; Peragine et al. 2004). *AGO1* provides positive phylogenetic support for the angiosperm clade (node 4), for the split of *Amborella* and the rest of the angiosperms (node 5), and for the eudicot clade only (node 6). *RDR6* provides support for the split of the eudicots from *Amborella* and rice (node 6).

Their overrepresentation and phylogenetic contribution are highly relevant given character analysis in our phylogeny. In particular, we found a number of amino acid substitutions among species in the clades with high support, at regions in proteins with known functional importance. For *AGO1*, mutations unique to rice were found in the PAZ nucleic-acid-binding interface and in regions that correspond to the 5′ guide strand anchoring domain and the PIWI catalytically active domain. Interestingly, mutants in *RDR6*, which support the dicot clade, have much milder phenotypes in *Arabidopsis* (Adenot et al. 2006; Fahlgren et al. 2006; Garcia et al. 2006) than in the monocot rice (Nagasaki et al. 2007). Asymmetry and shoot meristem organization in the monocotyledonous embryo of rice are strongly affected, whereas the dicotyledonous embryo of *Arabidopsis* remains radically symmetric and germinates normally. In terms of target gene expression, rice *rdr6* mutant embryos lose some of their asymmetry, resembling dicotyledonous embryos in this respect, although profound differences remain. Unique changes to rice are sites of potentially important mutants. Overall, *AGO1* and *RDR6* (and then, the processing or transport of *trans*-acting siRNA) are implicated in this defining feature of the angiosperm seed. Recently, *RDR6* has also been implicated in small-RNA-mediated suppression of gamete formation in the *Arabidopsis* ovule, a phenotype related to asexual seed formation, or apomixis (Olmedo-Monfil et al. 2010). Expression analysis using these sequence variants will confirm a role for these amino acid residues in determining significant phenotypic effects of ecological and evolutionary importance.

CONCLUSIONS

We demonstrate a novel method using phylogenomic tools to postulate hypotheses of gene function in the evolution of major plant groups. Functional hypotheses can be further coupled with expression and genetic data to arrive at better gene annotations and functional analyses for genome-level studies. Our findings help to guide plant ecological genomics studies and enlighten the precise evolutionary mechanisms driving the diversification of plant species, helping to gradually unravel Darwin's abiding and perplexing mystery.

ACKNOWLEDGMENTS

We are grateful to our colleagues in the New York Plant Genomics (NYPG) Consortium whose work we review here, especially Ernest Lee, Sergios Kolokotronis, and Dennis Stevenson. A.C.J. is funded by the Lewis B. and Dorothy Cullman Fellowship at the American Museum of Natural History and The New York Botanical Garden. The NYPG Consortium is supported by National Science Foundation grant 0421604 "Genomics of Comparative Seed Evolution" to Gloria Coruzzi (New York University), Rob DeSalle (American Museum of Natural History), Dennis Stevenson (The New York Botanical Garden), Dick McCombie (Cold Spring Harbor Laboratory) and R.M. (Cold Spring Harbor Laboratory).

REFERENCES

Adenot X, Elmayan T, Lauressergues D, Boutet S, Bouché N, Gasciolli V, Vaucheret H. 2006. DRB4-dependent *TAS3* trans-acting siRNAs control leaf morphology through AGO7. *Curr Biol* **16:** 927–932.

APGIII. 2009. An update of the Angiosperm Phylogeny Group classification for the orders and families of flowering plants: APG III. *Bot J Linn Soc* **161:** 105–121.

Bajguz A, Tretyn A. 2003. The chemical characteristic and distribution of brassinosteroids in plants. *Phytochemistry* **62:** 1027–1046.

Baker RH, DeSalle R. 1997. Multiple sources of character information and the phylogeny of Hawaiian drosophilids. *Syst Biol* **46:** 654–673.

Balasubramanian S, Sureshkumar S, Agrawal M, Michael TP, Wessinger C, Maloof JN, Clark R, Warthmann JC, Weigel D. 2006. The PHYTOCHROME C photoreceptor gene mediates natural variation in flowering and growth responses of *Arabidopsis thaliana*. *Nat Genet* **38:** 711–715.

Becker A, Saedler H, Theissen G. 2003. Distinct MADS-box gene expression patterns in the reproductive cones of the gymnosperm *Gnetum gnemon*. *Dev Genes Evol* **213:** 567–572.

Bohnert HJ, Ostrem JA, Cushman JC, Michalowski CB, Rickers J, Meyer G, deRocher EJ, Vernon DM, Krueger M, Vazquez-Moreno L, et al. 1988. *Mesembryanthemum crystallinum*, a higher plant model for the study of environmentally induced changes in gene expression. *Plant Mol Biol Rep* **6:** 10–28.

Bremer K. 1988. The limits of amino acid sequence data in angiosperm phylogenetic reconstruction. *Evolution* **42:** 795–803.

Bremer K. 1994. Branch support and tree stability. *Cladistics* **10:** 295–304.

Brown D, Sjölander K. 2006. Functional classification using phylogenomic inference. *PLoS Comput Biol* **2:** e77.

Chiu JC, Lee EK, Egan MG, Sarkar IN, Coruzzi GM, DeSalle R. 2006. OrthologID: Automation of genome-scale ortholog identification within a parsimony framework. *Bioinformatics* **22:** 699–707.

Cibrián-Jaramillo A, De la Torre-Bárcena JE, Lee KE, Katari MS, Little DP, Stevenson DW, Martienssen R, Coruzzi G, DeSalle R. 2010. Using phylogenomic patterns and gene ontology to identify proteins of importance in plant evolution. *Genome Biol Evol* (in press).

Darwin C. 1859. *On the origin of species by means of natural selection*, 1st ed. Murray, London.

De la Torre-Bárcena JE, Kolokotronis SO, Lee EK, Stevenson DW, Coruzzi GM, DeSalle R. 2009. The impact of outgroup choice and missing data on major seed plant phylogenetics using genome-wide EST data. *PLoS ONE* **4:** e5764.

Dudoit S, van der Laan MJ, Pollard KS. 2004. Multiple testing. I. Single-step procedures for control of general type I error rates. *Stat Appl Genet Mol Biol* **3:** 1–69.

Eisen JA. 1998. Phylogenomics: Improving functional predictions for uncharacterized genes by evolutionary analysis. *Genome Res* **8:** 163–167.

Eisen JA, Fraser CM. 2003. Phylogenomics: Intersection of evolution and genomics. *Science* **300:** 1706–1707.

Fahlgren N, Montgomery TA, Howell MD, Allen E, Dvorak SK, Alexander AL, Carrington J. 2006. Regulation of AUXIN RESPONSE FACTOR3 by TAS3 ta-siRNA affects developmental timing and patterning in *Arabidopsis*. *Curr Biol* **9:** 939–944.

Farris J, Albert V, Källersjö M, Lipscomb D, Kluge A. 1996. Parsimony jackknifing outperforms neighbor-joining. *Cladistics* **12:** 99–124.

Feild TS, Balun L. 2008. Xylem hydraulic and photosynthetic function of *Gnetum* (Gnetales) species from Papua New Guinea. *New Phytol* **177:** 665–675.

Felsenstein J. 1985. Confidence limits on phylogenies: An approach using the bootstrap. *Evolution* **39:** 783–791.

Friedman WE. 2009. The meaning of Darwin's 'abominable mystery.' *Am J Bot* **96:** 5.

Frohlich MW, Parker DS. 2000. The mostly male theory of flower

evolutionary origins: From genes to fossils. *Syst Bot* **25:** 155–170.
Garcia D, Collier SA, Byrne ME, Martienssen RA. 2006. Specification of leaf polarity in *Arabidopsis* via the *trans*-acting siRNA pathway. *Curr Biol* **16:** 933–938.
Gatesy J, O'Grady P, Baker RH. 1999. Corroboration among data sets in simultaneous analysis: Hidden support for phylogenetic relationships among higher level artiodactyl taxa. *Cladistics* **15:** 271–313.
Greguss P. 1955. *Identification of living gymnosperms on the basis of xylotomy* (transl. L Jocsik). Akademiai Kiado, Budapest.
Gutiérrez RA, Gifford ML, Poultney C, Wang R, Shasha DE, Coruzzi GM, Crawford NM. 2007. Insights into the genomic nitrate response using genetics and the Sungear Software System. *J Exp Bot* **58:** 2359–2367.
Kidner CA, Martienssen RA. 2004. Spatially restricted microRNA directs leaf polarity through ARGONAUTE1. *Nature* **428:** 81–84.
Kidner CA, Martienssen RA. 2005. The role of ARGONAUTE1 (AGO1) in meristem formation and identity. *Dev Biol* **280:** 504–517.
Marchler-Bauer A, Anderson JB, Derbyshire MK, DeWeese-Scott C, Gonzales NR, Gwadz M, Hao L, He S, Hurwitz DI, Jackson JD, et al. 2007. CDD: A conserved domain database for interactive domain family analysis. *Nucleic Acids Res* **35:** D237–D240.
Mathews S. 2009. Phylogenetic relationships among seed plants: Persistent questions and the limits of molecular data. *Am J Bot* **96:** 228–236.
Mathews S, Donoghue MJ. 2000. Basal angiosperm phylogeny inferred from duplicate phytochromes A and C. *Int J Plant Sci* **161:** 41–55.
Mouchel CF, Leyser O. 2007. Novel phytohormones involved in long-range signaling. *Curr Opin Plant Biol* **10:** 473–476.
Nagasaki H, Itoh J, Hayashi K, Hibara K, Satoh-Nagasawa N, Nosaka M, Mukouhata M, Ashikari M, Kitano H, Matsuoka M, et al. 2007. The small interfering RNA production pathway is required for shoot meristem initiation in rice. *Proc Natl Acad Sci* **104:** 14867–14871.
Olmedo-Monfil V, Durán-Figueroa N, Arteaga-Vázquez M, Demesa-Arévalo E, Autran D, Grimanelli D, Slotkin RK, Martienssen RA, Vielle-Calzada JP. 2010. Control of female gamete formation by a small RNA pathway in *Arabidopsis*. *Nature* **464:** 628–632.
Peragine A, Yoshikawa M, Wu G, Albrecht HL, Poethig RS. 2004. *SGS3* and *SGS2/SDE1/RDR6* are required for juvenile development and the production of *trans*-acting siRNAs in *Arabidopsis*. *Genes Dev* **18:** 2368–2379.
Poultney CS, Gutiérrez RA, Katari MS, Gifford ML, Palen WB, Coruzzi GM, Shasha DE. 2007. Sungear: Interactive visualization and functional analysis of genomic datasets. *Bioinformatics* **23:** 259–261.
Schmidt M, Schneider-Poetsch HA. 2002. The evolution of gymnosperms redrawn by phytochrome genes: The Gnetatae appear at the base of the gymnosperms. *J Mol Evol* **54:** 715–724.
Sjölander K. 2004. Phylogenomic inference of protein molecular function: Advances and challenges. *Bioinformatics* **20:** 170–179.
Sunkar R, Zhu JK. 2007. Micro RNAs and short-interfering RNAs in plants. *J Integr Plant Biol* **49:** 817–826.
Swofford D. 2003. *PAUP*: Phylogenetic analysis using parsimony (*and other methods)*. Sinauer, Sunderland, MA.
von Willert DJ, Armbruster N, Drees T, Zaborowski M. 2005. *Welwitschia mirabilis:* CAM or not CAM—What is the answer? *Funct Plant Biol* **32:** 389–395.
Wang R, Tischner R, Gutiérrez RA, Hoffman M, Xing X, Chen M, Coruzzi G, Crawford NM. 2004. Genomic analysis of the nitrate response using a nitrate reductase-null mutant of *Arabidopsis*. *Plant Physiol* **136:** 2512–2522.
Wang Y, Stricker HM, Gou D, Liu L. 2007. MicroRNA: Past and present. *Front Biosci* **12:** 2316–2329.
Willman MR, Poethig RS. 2005. Time to grow up: The temporal role of small RNAs in plants. *Curr Opin Plant Biol* **8:** 548–552.
Winter KU, Becker A, Munster T, Kim JT, Saedler H, Theissen G. 1999. MADS-box genes reveal that gnetophytes are more closely related to conifers than to flowering plants. *Proc Natl Acad Sci* **96:** 7342–7347.

Evolution of Insect Dorsoventral Patterning Mechanisms

M.W. Perry,[1] J.D. Cande,[2,4] A.N. Boettiger,[3] and M. Levine[2]

[1]Department of Integrative Biology, University of California, Berkeley, California 94720-3140;
[2]Department of Molecular and Cell Biology, Division of Genetics, Genomics and Development,
Center for Integrative Genomics, University of California, Berkeley, California 94720-3200;
[3]Biophysics Program, University of California, Berkeley, California 94720-3200

Correspondence: mlevine@berkeley.edu

The dorsoventral (DV) patterning of the early *Drosophila* embryo depends on Dorsal, a maternal sequence-specific transcription factor related to mammalian NF-κB. Dorsal controls DV patterning through the differential regulation of ~50 target genes in a concentration-dependent manner. Whole-genome methods, including ChIP-chip and ChIP-seq assays, have identified ~100 Dorsal target enhancers, and more than one-third of these have been experimentally confirmed via transgenic embryo assays. Despite differences in DV patterning among divergent insects, a number of the Dorsal target enhancers are located in conserved positions relative to the associated transcription units. Thus, the evolution of novel patterns of gene expression might depend on the modification of old enhancers, rather than the invention of new ones. As many as half of all Dorsal target genes appear to contain "shadow" enhancers: a second enhancer that directs the same or similar expression pattern as the primary enhancer. Preliminary studies suggest that shadow enhancers might help to ensure resilience of gene expression in response to environmental and genetic perturbations. Finally, most Dorsal target genes appear to contain RNA polymerase II (pol II) prior to their activation. Stalled pol II fosters synchronous patterns of gene activation in the early embryo. In contrast, DV patterning genes lacking stalled pol II are initially activated in an erratic or stochastic fashion. It is possible that stalled pol II confers fitness to a population by ensuring coordinate deployment of the gene networks controlling embryogenesis.

DV patterning of the *Drosophila* embryo is controlled by Dorsal, a sequence-specific transcription factor related to mammalian nuclear factor κB (NF-κB) (Roth et al. 1989; Rushlow et al. 1989; Ip et al. 1991). The Dorsal protein is distributed in a broad nuclear gradient, with peak levels present in ventral nuclei and progressively lower levels in lateral and dorsal regions (Roth et al. 1989; Rushlow et al. 1989; Steward 1989). This Dorsal nuclear gradient initiates DV patterning by regulating 50–60 target genes in a concentration-dependent fashion (Stathopoulous et al. 2002; Zeitlinger et al. 2007a).

Whole-genome chromatin immunoprecipitation (ChIP)-chip assays (see below) identified ~100 potential Dorsal target enhancers, and more than 30 of these have been directly tested in transgenic embryos (see, e.g., Zeitlinger et al. 2007a; Hong et al. 2008a). Altogether, these enhancers direct six distinct patterns of gene expression across the DV axis of precellular embryos. Dorsal works in a highly combinatorial manner to generate these diverse patterns (for review, see Hong et al. 2008b). For example, Dorsal and SuH, a transcriptional effector of Notch signaling, activate *single-minded* (*sim*) expression in a single line of cells (central nervous system [CNS] ventral midline) on either side of the mesoderm (Cowden and Levine 2002; Morel et al. 2003). In contrast, Dorsal works together with a different sequence-specific transcription factor, Pointed (an effector of epidermal growth factor [EGF] signaling), to activate gene expression within lateral stripes in intermediate regions of the future ventral nerve cord (Gabay et al. 1996).

ENHANCER EVOLUTION

In principle, substitutions of "coactivator" binding sites within Dorsal target enhancers can alter the DV limits of gene expression. For example, replacing SuH-binding sites with Twist sites results in expanded expression of the modified enhancer within the presumptive neurogenic ectoderm (Gray and Levine 1996; Zinzen et al. 2006). Analysis of Dorsal target enhancers in divergent insects, including mosquitoes (*Anopheles gambiae*), flour beetles (*Tribolium castaneum*), and honeybees (*Apis mellifera*), suggests that such changes might occur during evolution to produce distinctive DV patterning mechanisms (Zinzen et al. 2006).

One such example is seen for the ventral midline of *A. mellifera*. In *Drosophila*, the ventral midline is just two cells in width and arises from two lines of *sim*-expressing cells that straddle the mesoderm before gastrulation (Fig. 1). In contrast, the ventral midline of the *A. mellifera* CNS is considerably wider, encompassing about five to six cells. An expanded ventral midline is also seen in *T. castaneum*, suggesting that the broad pattern is ancestral, and the narrow midline of *Drosophila* (and *A. gambiae*) is a derived feature of the dipteran CNS (Zinzen et al. 2006).

Expansion of the *sim* expression pattern is sufficient to account for the broad ventral midlines of the *A. mellifera* and *T. castaneum* CNS. In *Drosophila*, ectopic activation of *sim* expression using the *eve* stripe-2 enhancer results in

---

[4]Present address: Developmental Biology Institute of Marseilles (IBDML), Marseille, France.

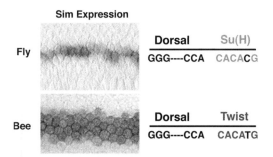

Figure 1. *sim* exhibits a broader pattern of expression in the honeybee CNS as compared with *Drosophila*. This expansion appears to result from the replacement of Suppressor of Hairless sites (Notch signaling) with Twist sites in the respective 5′ *sim* enhancers.

the formation of an ectopic ventral midline throughout the neurogenic ectoderm of transgenic embryos (Zinzen et al. 2006). The *sim* regulatory region contains two distinct enhancers: One mediates activation by Dorsal and Notch signaling (establishment enhancer), and the other mediates positive autofeedback through direct binding of the Sim transcription factor to the autoregulatory enhancer (Kasai et al. 1992). Once Sim is misexpressed, the expanded pattern is maintained by autofeedback.

Sim establishment enhancers were identified in the 5′-flanking regions of the *sim* loci in *A. gambiae*, *T. castaneum*, and *A. mellifera*. The *sim* enhancer from *A. gambiae* directs sharp lateral lines when expressed in transgenic *Drosophila* embryos. In contrast, the enhancers obtained from the *sim* loci of *T. castaneum* and *A. mellifera* produce broader expression patterns. The *A. gambiae* enhancer resembles the *Drosophila* enhancer in that it contains a series of Dorsal- and SuH-binding sites. However, the *T. castaneum* and *A. mellifera* enhancers contain Twist sites rather than SuH sites, and consequently, they direct broader patterns of gene expression (Zinzen et al. 2006; Cande 2009).

CONSTANCY OF ENHANCER LOCATION

The *sim* enhancers of flies, mosquitoes, flour beetles, and bees lack simple sequence similarity. Despite this extensive sequence divergence, comparable enhancers are located in the same relative positions: in the immediate 5′-flanking regions of the respective *sim* loci (e.g., Fig. 2).

Because this is a relatively common location for developmental enhancers, additional studies were done to determine whether enhancer locations are conserved for other critical DV patterning genes (Cande et al. 2009). These studies identified enhancers for five additional genes: *cactus*, *sog*, *twist*, *brinker*, and *vnd*. *cactus* is a key component of the Toll signaling pathway that regulates Dorsal nuclear transport (Roth et al. 1991; Stein and

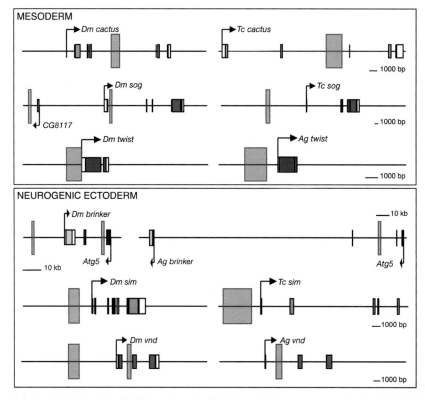

Figure 2. Conservation of enhancer location in divergent insects. (Pink boxes) Enhancers regulating the associated transcription units, (colored rectangles) coding exons. Note the conservation of a *brinker* enhancer within the intron of the neighboring *Atg5* loci of flies and mosquitoes. (Ag) *Anopheles gambiae*, (Dm) *Drosophila melanogaster*, (Tc) *Tribolium castaneum*, (*sim*) *single minded*, (*sog*) *short gastrulation*, (*vnd*) *ventral nervous system defective*. (Reprinted, with permission, from Cande et al. 2009 [© National Academy of Sciences].)

Nüsslein-Volhard 1992). It is activated by high levels of the Dorsal gradient in the presumptive mesoderm of both *Drosophila* and *T. castaneum* embryos (Maxton-Kuchenmeister et al. 1999; Nunes da Fonseca et al. 2008). The enhancers that are responsible for these expression patterns are located in 3′ introns of the respective *cactus* transcription units (Cande et al. 2009).

Enhancer conservation at the *brinker* (*brk*) locus is even more dramatic. *brk* encodes a sequence-specific transcriptional repressor that helps to restrict Dpp (bone morphogenetic protein [BMP]) signaling to the dorsal ectoderm (Jaźwińska et al. 1999). In *Drosophila*, two separate enhancers regulate *brk* expression in the presumptive neurogenic ectoderm of pregastrular embryos (Hong et al. 2008a). One of the enhancers is located ~10 kb 5′ of the *brk* transcription start site. The other is located 13 kb downstream from the start site, within the intron of a neighboring gene, *Atg5*. The major enhancer regulating *brk* expression in the *A. gambiae* embryo is located within the *Atg5* gene, even though the *brk* transcription unit is inverted relative to its orientation in *Drosophila* and *Atg5* is located quite far, ~100 kb, from *brk* in the mosquito genome (Fig. 2) (Cande et al. 2009).

Binding-site turnover has been well documented in insect enhancers (Moses et al. 2006; for review, see Ludwig 2002). Despite this turnover within existing enhancers, there might be constraints on the de novo evolution of developmental enhancers. We suggest that the evolution of novel patterns of gene expression depends primarily on the modification of ancestral enhancers, rather than the invention of new ones.

SHADOW ENHANCERS

ChIP-chip assays led to the comprehensive identification of Dorsal target enhancers in the *Drosophila* genome (Zeitlinger et al. 2007a). These studies identified multiple enhancers at more than one-third of the target genes that are directly regulated by the Dorsal gradient. For example, the *vnd* gene encodes a sequence-specific transcription factor that specifies the ventral-most neuronal cell identities of the ventral nerve cord (see, e.g., Weiss et al. 1998). It is activated by enhancers located in both the 5′-flanking region and within the first intron of the transcription unit (Shao et al. 2002; Stathopoulous et al. 2002; Zeitlinger et al. 2007a). Similarly, *sog* is regulated by both a 5′ enhancer and an intronic enhancer (Fig. 3), and as discussed above, *brk* is activated by enhancers located in both 5′- and 3′-flanking regions (Zeitlinger et al. 2007a, Hong et al. 2008a).

We refer to the secondary enhancers located in remote 5′ or 3′ positions as shadow enhancers (Hong et al. 2008a). Preliminary studies suggest that they might help to confer resilience in gene expression in response to genetic and environmental perturbations. For example, *vnd* and *sog* exhibit normal patterns of transcriptional activation in embryos derived from dl/+ heterozygotes (half of the normal dose of the Dorsal gradient), whereas *Neu3* and *rho* display erratic patterns of activation (Fig. 4) (Boettiger and Levine 2009). *vnd* and *sog* contain shadow enhancers, whereas *Neu3* and *rho* do not. It is possible that dual enhancers for a common expression pattern ensure accurate and reproducible activation in large populations of embryos subject to environmental fluctuations.

It is possible that shadow enhancers arise from "cryptic" duplication events. Of course, other scenarios can be envisioned, but regardless of mechanism, once they arise, shadow enhancers might confer an adaptive advantage to a population by ensuring accurate activation of critical developmental control genes. Shadow enhancers offer an opportunity for producing novel patterns of gene expression without disrupting the core function of the primary enhancer and associated gene. According to this view, the evolution of shadow enhancers might come at a cost to the fitness of a population, but this cost could be compensated by the advantages conferred by the novel mode of gene expression.

TRANSCRIPTIONAL SYNCHRONY

Recent studies with mammalian progenitor cells, including stem cells, suggest that many critical developmental control genes (e.g., Hox genes) are repressed but poised for

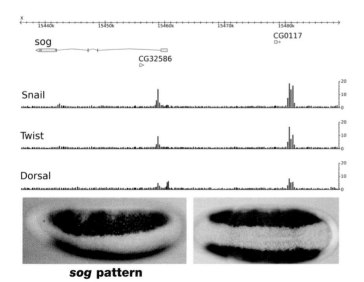

Figure 3. ChIP-chip assays identified two enhancers for the early *sog* expression pattern. (Light yellow) *sog* transcription unit. The locations of Dorsal-, Twist-, and Snail-binding sites are indicated below. There are two clusters of binding sites: in the first intron and more than 20 kb 5′ of the start site. The intronic cluster was previously shown to function as an enhancer for the *sog* expression pattern (*left*, embryo stained to show the endogenous *sog* expression pattern). The distal cluster generates a similar pattern of expression when attached to a *lacZ* reporter gene and expressed in transgenic embryos (*right*). (Modified, with permission, from Hong et al. 2008a [© AAAS].)

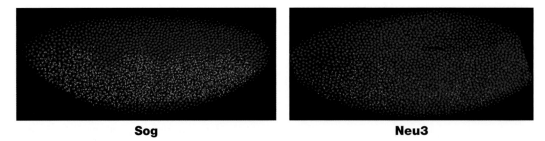

Figure 4. Onset of *sog* and *Neu3* expression in precellular embryos at the early phases of nuclear cleavage cycle 14. The embryos were collected from *dorsal*/+ females and therefore contain half of the normal levels of the Dorsal nuclear gradient. The *sog* pattern is normal, but *Neu3* displays erratic activation. *sog* contains a shadow enhancer, whereas *Neu3* does not. (Modified, with permission, from Boettiger and Levine 2009 [© AAAS].)

rapid induction (Guenther et al. 2007). Many such genes contain bivalent histone marks, H3K4 trimethylation and H3K27 methylation, which are indicative of genes that are active and repressed, respectively (Bernstein et al. 2006).

ChIP-chip assays were done in *Drosophila* using a mixture of antibodies directed against pol II (Muse et al. 2007; Zeitlinger et al. 2007b). These studies suggest that most DV patterning genes contain stalled pol II at the core promoter before their activation in response to the Dorsal gradient. Like the bivalent marks seen in mammalian progenitor cells, stalled pol II is likely to render the associated genes repressed but poised for rapid activation.

Classical studies on *Drosophila* heat shock genes have documented that stalled, or paused, pol II accelerates their activation in response to stress as compared with comparable promoters lacking paused pol II (Lis and Wu 1993; Conaway et al. 2000; Saunders et al. 2006). This paradigm of gene expression was seen as a specialized stress response. However, the finding that many developmental control genes contain stalled pol II in the early *Drosophila* embryo raises the possibility that the control of transcription elongation is an important strategy for differential gene regulation during development (Lis 2007; Zeitlinger et al. 2007b; Hendrix et al. 2008; Chopra et al. 2009).

It is possible that stalled pol II suppresses transcriptional noise during development. In principle, a major source of such noise is variability in pol II recruitment and promoter escape (Raser and O'Shea 2004, 2005; Raj et al. 2006; Darzacq et al. 2007; Raj and van Oudenaarden 2008). For example, just a fraction of the pol II that interacts with a promoter succeeds in melting the double-stranded DNA and launching transcription. In contrast, stalled pol II has already passed the "checkpoint" (promoter escape) and is more likely to succeed in transcribing the associated gene as compared with a naïve pol II complex that is newly recruited to the DNA template.

These considerations raise the possibility that genes containing stalled pol II might be activated in a synchronous fashion due to diminished nucleus-to-nucleus variation in de novo transcription upon induction. To test this possibility, a quantitative in situ hybridization method was developed to identify the initial de novo transcripts in all of the nuclei of a large number of embryos (Boettiger and Levine 2009). These studies suggest that genes containing stalled pol II are activated in a coordinated fashion throughout the field of nuclei where the gene is expressed (Fig. 5). In contrast, genes lacking stalled pol II are activated in an erratic fashion, whereby nuclei displaying de novo transcripts are surrounded by those lacking expression (Boettiger and Levine 2009).

It is possible that transcriptional synchrony is a manifestation of metazoan development, whereby groups of cells function in a highly coordinated fashion. Stalled pol II and transcriptional synchrony might help to foster such coordinate behavior. We propose that stalled pol II contributes to population fitness, in that it helps to ensure the accurate and reproducible regulation of key developmental control genes.

Figure 5. High-magnification view of the presumptive mesoderm of a precellular embryo at the early phase of nuclear cleavage cycle 14. Intronic probes were used to visualize nascent transcripts from the *Mes2* (green) and *Mes4* (red) genes. *Mes2* displays expression in most of the mesodermal nuclei, whereas *Mes4* is expressed in less than half of the nuclei. *Mes2* contains stalled pol II, whereas *Mes4* does not. (Modified, with permission, from Boettiger and Levine 2009 [© AAAS].)

ACKNOWLEDGMENTS

This work was supported in part by the National Institutes of Health grants GM46638 and GM34431. A.N.B. and M.W.P. are supported by National Science Foundation graduate research fellowships and J.D.C. is supported by a Chang Ling Tien graduate fellowship.

REFERENCES

Bernstein BE, Mikkelsen TS, Xie X, Kamal M, Huebert DJ, Cuff J, Fry B, Meissner A, Wernig M, Plath K, et al. 2006. A bivalent chromatin structure marks key developmental genes in embryonic stem cells. *Cell* **125:** 315–326.

Boettiger AN, Levine M. 2009. Synchronous and stochastic patterns of gene activation in the *Drosophila* embryo. *Science* **325:** 471–473.

Cande J, Goltsev Y, Levine M. 2009. Conservation of enhancer positions in divergent insects. *Proc Natl Acad Sci* **106:** 14414–14419.

Chopra VS, Hong JW, Levine M. 2009. Regulation of Hox gene activity by transcriptional elongation in *Drosophila*. *Curr Biol* **19:** 688–693.

Conaway JW, Shilatifard A, Dvir A, Conaway RC. 2000. Control of elongation by RNA polymerase II. *Trends Biochem Sci* **25:** 375–380.

Cowden J, Levine M. 2002. The Snail repressor positions Notch signaling in the *Drosophila* embryo. *Development* **129:** 1785–1793.

Darzacq X, Shav-Tal Y, de Turris V, Brody Y, Shenoy SM, Phair RD, Singer RH. 2007. In vivo dynamics of RNA polymerase II transcription. *Nat Struct Mol Biol* **14:** 796–806.

Gabay L, Scholz H, Golembo M, Klaes A, Shilo BZ, Klambt C. 1996. EGF receptor signaling induces pointed P1 transcription and inactivates Yan protein in the *Drosophila* embryonic ventral ectoderm. *Development* **122:** 3355–3362.

Gray S, Levine M. 1996. Short-range transcriptional repressors mediate both quenching and direct repression within complex loci in *Drosophila*. *Genes Dev* **10:** 700–710.

Guenther MG, Levine SS, Boyer LA, Jaenisch R, Young RA. 2007. A chromatin landmark and transcription initiation at most promoters in human cells. *Cell* **130:** 77–88.

Hendrix DA, Hong JW, Zeitlinger J, Rokhsar DS, Levine MS. 2008. Promoter elements associated with RNA Pol II stalling in the *Drosophila* embryo. *Proc Natl Acad Sci* **105:** 7762–7767.

Hong JW, Hendrix DA, Levine MS. 2008a. Shadow enhancers as a source of evolutionary novelty. *Science* **321:** 1314.

Hong JW, Hendrix DA, Papatsenko D, Levine MS. 2008b. How the Dorsal gradient works: Insights from postgenome technologies. *Proc Natl Acad Sci* **105:** 20072–20076.

Ip YT, Kraut R, Levine M, Rushlow CA. 1991. The dorsal morphogen is a sequence-specific DNA-binding protein that interacts with a long-range repression element in *Drosophila*. *Cell* **64:** 439–446.

Jaźwińska A, Kirov N, Wieschaus E, Roth S, Rushlow C. 1999. The *Drosophila* gene *brinker* reveals a novel mechanism of Dpp target gene regulation. *Cell* **96:** 563–573.

Kasai Y, Nambu JR, Lieberman PM, Crews ST. 1992. Dorsal-ventral patterning in *Drosophila*: DNA binding of snail protein to the *single-minded* gene. *Proc Natl Acad Sci* **89:** 3414–3418.

Lis JT. 2007. Imaging *Drosophila* gene activation and polymerase pausing in vivo. *Nature* **450:** 198–202.

Lis J, Wu C. 1993. Protein traffic on the heat shock promoter: Parking, stalling, and trucking along. *Cell* **74:** 1–4.

Ludwig MZ. 2002. Functional evolution of noncoding DNA. *Curr Opin Genet Dev* **12:** 634–639.

Maxton-Küchenmeister J, Handel K, Schmidt-Ott U, Roth S, Jäckle H. 1999. Toll homologue expression in the beetle *Tribolium* suggests a different mode of dorsoventral patterning than in *Drosophila* embryos. *Mech Dev* **83:** 107–114.

Morel V, Le Borgne R, Schweisguth F. 2003. Snail is required for Delta endocytosis and Notch-dependent activation of *single-minded* expression. *Dev Genes Evol* **213:** 65–72.

Moses AM, Pollard DA, Nix DA, Iyer VN, Li XY, Biggin MD, Eisen MB. 2006. Large-scale turnover of functional transcription factor binding sites in *Drosophila*. *PLoS Comput Biol* **2:** e130.

Muse GW, Gilchrist DA, Nechaev S, Shah R, Parker JS, Grissom SF, Zeitlinger J, Adelman K. 2007. RNA polymerase is poised for activation across the genome. *Nat Genet* **39:** 1507–1511.

Nunes da Fonseca R, von Levetzow C, Kalscheuer P, Basal A, van der Zee M, Roth S. 2008. Self-regulatory circuits in dorsoventral axis formation of the short-germ beetle *Tribolium castaneum*. *Dev Cell* **14:** 605–615.

Raj A, van Oudenaarden A. 2008. Nature, nurture, or chance: Stochastic gene expression and its consequences. *Cell* **135:** 216–226.

Raj A, Peskin CS, Tranchina D, Vargas DY, Tyagi S. 2006. Stochastic mRNA synthesis in mammalian cells. *PLoS Biol* **4:** e309.

Raser JM, O'Shea EK. 2004. Control of stochasticity in eukaryotic gene expression. *Science* **304:** 1811–1814.

Raser JM, O'Shea EK. 2005. Noise in gene expression: Origins, consequences, and control. *Science* **309:** 2010–2013.

Roth S, Stein D, Nüsslein-Volhard C. 1989. A gradient of nuclear localization of the dorsal protein determines dorsoventral pattern in the *Drosophila* embryo. *Cell* **59:** 1189–1202.

Roth S, Hiromi Y, Godt D, Nüsslein-Volhard C. 1991. *cactus*, a maternal gene required for proper formation of the dorsoventral morphogen gradient in *Drosophila* embryos. *Development* **112:** 371–388.

Rushlow CA, Han K, Manley JL, Levine M. 1989. The graded distribution of the dorsal morphogen is initiated by selective nuclear transport in *Drosophila*. *Cell* **59:** 1165–1177.

Saunders A, Core LJ, Lis JT. 2006. Breaking barriers to transcription elongation. *Nat Rev Mol Cell Biol* **7:** 557–567.

Shao X, Koizumi K, Nosworthy N, Tan DP, Odenwald W, Nirenberg M. 2002. Regulatory DNA required for vnd/NK-2 homeobox gene expression pattern in neuroblasts. *Proc Natl Acad Sci* **99:** 113–137.

Stathopoulos A, Van Drenth M, Erives A, Markstein M, Levine M. 2002. Whole-genome analysis of dorsal-ventral patterning in the *Drosophila* embryo. *Cell* **111:** 687–701.

Stein D, Nüsslein-Volhard C. 1992. Multiple extracellular activities in *Drosophila* egg perivitelline fluid are required for establishment of embryonic dorsal-ventral polarity. *Cell* **68:** 429–440.

Steward R. 1989. Relocalization of the dorsal protein from the cytoplasm to the nucleus correlates with its function. *Cell* **59:** 1179–1188.

Weiss JB, Von Ohlen T, Mellerick DM, Dressler G, Doe CQ, Scott MP. 1998. Dorsoventral patterning in the *Drosophila* central nervous system: The intermediate neuroblasts defective homeobox gene specifies intermediate column identity. *Genes Dev* **12:** 3591–3602.

Zeitlinger J, Zinzen RP, Stark A, Kellis M, Zhang H, Young RA, Levine M. 2007a. Whole-genome ChIP-chip analysis of Dorsal, Twist, and Snail suggests integration of diverse patterning processes in the *Drosophila* embryo. *Genes Dev* **21:** 385–390.

Zeitlinger J, Stark A, Kellis M, Hong JW, Nechaev S, Adelman K, Levine M, Young RA. 2007b. RNA polymerase stalling at developmental control genes in the *Drosophila melanogaster* embryo. *Nat Genet* **39:** 1512–1516.

Zinzen R, Cande JD, Papatsenko D, Levine M. 2006. Evolution of the ventral midline in insect embryos. *Dev Cell* **11:** 895–902.

Lophotrochozoa Get into the Game: The Nodal Pathway and Left/Right Asymmetry in Bilateria

C. GRANDE[1] AND N.H. PATEL[2]

[1]*Centro de Biología Molecular Severo Ochoa. Universidad Autónoma de Madrid. 28049, Madrid, Spain;* [2]*Departments of Molecular and Cell Biology and of Integrative Biology, University of California, Berkeley, California 94720-3200*

Correspondence: nipam@uclink.berkeley.edu

Animals as diverse as humans, flies, crabs, and snails show overall bilateral symmetry, but each species has specific structures and organs that display left/right asymmetry, and the presence of these asymmetries is vital to the organism. Here, we review recent results showing that part of the molecular pathway that sets left/right asymmetry in vertebrates is also conserved in snails, suggesting that left/right asymmetry was present in the common ancestor of all bilaterians. More specifically, we can now predict that the signaling molecule Nodal and the transcription factor Pitx were expressed on the right side of the bilaterian ancestor. These results also allow us to understand how the direction of shell coiling (chirality) is regulated in snails and provides interesting insights into the possible inversion of the dorsoventral axis in the lineage leading to chordates.

Despite great variation in form, most animals are classified as bilaterians because they possess clear anteroposterior (AP) and dorsoventral (DV) axes, which then together position the axis of bilateral symmetry separating the left and right sides of the body. However, although bilaterians show a general symmetry between the left and right sides of the body, most deviate from perfect symmetry in some reproducible way. This can be as subtle as the direction of looping of an internal organ in a fly or as obvious as the asymmetric specialization of claws in certain crab species. Other examples of left/right (LR) asymmetry include the positioning of organs, such as the heart and liver in humans, and the LR asymmetric growth pattern of snail shells that leads to their coiling in either clockwise or anticlockwise directions. Some of these asymmetries are essential to properly pack and connect many internal organs. Individuals that are perfect mirror images of normal are generally viable, whereas individuals in which there are no asymmetries, or the pattern of asymmetry is random between organs, are often severely compromised.

Phylogenetic studies agree on the existence of three main groups within the Bilateria (Aquinaldo et al. 1997): Deuterostomia (including echinoderms, hemichordates, and chordates), Ecdysozoa (including arthropods and nematodes, among others), and Lophotrochozoa (including, for example, snails and annelids); the Ecdysozoa and Lophotrochozoa are closer to each other than either is to Deuterostomia. Although examples of LR asymmetry can be found in representatives of these three main groups, the molecular mechanisms controlling the development of this asymmetry have been studied in detail in just a few model systems in Deuterostomia and Ecdysozoa. The initial identification of several genes that are asymmetrically expressed with respect to the midline in vertebrates (Levin et al. 1995), combined with subsequent functional studies, led to rapid progress in understanding the molecular and genetic mechanisms underlying LR asymmetry in the deuterostomes. These studies led to the identification of Nodal, a transforming growth factor-β (TGF-β) ligand, as a signaling molecule that is asymmetrically expressed on the LR axis and functionally critical for establishing LR asymmetry in all deuterostomes studied so far (Hamada et al. 2002; Morokuma et al. 2002; Yu et al. 2002; Duboc and Lepage 2008). Another TGF-β family member, Lefty, and the homeobox gene *Pitx2*, are also key components of the LR pathway (Boorman and Shimeld 2002; Hamada et al. 2002; Duboc et al. 2005).

Like *nodal, lefty* is also asymmetrically expressed transiently, and both genes positively regulate *Pitx,* which is critical in cellular processes of differential migration, proliferation, adhesion, and asymmetric morphogenesis in deuterostomes. Loss of expression, as well as inappropriate spatial misexpression of these genes, leads to heterotaxia, i.e., the abnormal LR arrangement of organs (Levin et al. 1997; Sampath et al. 1997). Comparative studies in deuterostomes (echinoderms, ascidians, *Amphioxus*, and vertebrates) have revealed that *nodal, lefty*, and *Pitx* have evolutionarily conserved roles in development and mechanisms of regulation (Hamada et al. 2002; Duboc et al. 2005; Shimeld and Levin 2006).

In contrast, this particular cascade of genes that regulates LR asymmetries in all deuterostomes is not evolutionarily conserved in at least some, if not all, Ecdysozoa. To date, no *nodal* or *lefty* orthologs have been identified in the genomes of model ecdysozoans, and although an ortholog of *Pitx* is present in *Drosophila*, this gene shows a bilaterally symmetric expression pattern, and no LR defects are seen in *Drosophila Pitx* mutants (Vorbrüggen et al. 1997), suggesting that some other mechanisms must be involved in the specification of LR asymmetry in *Drosophila* and possibly other Ecdysozoa. The identification of some

Drosophila mutants with altered LR asymmetry patterns (*fas2*, *sim*, *puc*, *hkb*, *byn*, *ptc*, and *myo31DF* among others; for review, see Coutelis et al. 2008; Okumura et al. 2008) has provided some clues about the mechanisms that establish LR asymmetry in this group, but further studies are needed to understand the potential links between LR asymmetry in deuterostomes and ecdysozoans.

Initial data on the phylogenetic distribution of *nodal* orthologs (present in deuterostomes but absent in ecdysozoans and cnidarians) suggested that the Nodal pathway first originated in the common ancestor of deuterostomes (Fig. 1). However, the recent discovery of a *nodal* ortholog in snails and annelids (Grande and Patel 2009), both members of the Lophotrochozoa, has established a new hypothesis in which the Nodal pathway evolved before the bilaterians split off into the Ecdysozoa, the Lophotrochozoa, and the Deuterostomia (Fig. 1). Below, we describe the implications of these striking new data on the origin and evolution of the Nodal pathway and LR asymmetry within an evolutionary perspective.

LR ASYMMETRY IN SNAILS

On the basis of direction of shell coiling and body organization, snails can be left-handed (sinistral, shell coiled in an anticlockwise direction) or right-handed (dextral, shell coiled in a clockwise direction), with dextral coiling as the ancestral configuration as suggested by the fossil record (Ponder and Lindberg 1997). These two mirror-imaged configurations are a manifestation of LR asymmetry in snails. Most snail species are dextral, but a few are sinistral. In a small number of cases, both types can exist within a population of a single species. Classical genetic studies showed that the handedness of snails is genetically determined by a maternal effect locus (Boycott and Diver 1923; Sturtevant 1923; Boycott et al. 1930; Freeman and Lundelius 1982), but the molecular nature of the gene involved remains unknown. The handedness of a snail can be detected during the third cell division of the embryo when the first spiral cleavage occurs (Meshcheryakov and Beloussov 1975). The spiral cleavage is characterized by the oblique angle of the early cleavage planes and the alternating directions of successive divisions. For dextral individuals, the third cleavage generates a set of four micromeres that are oriented with a clockwise spiral relative to their underlying macromere sisters, whereas a sinistral individual displays the opposite rotation pattern.

Recent work on dextral and sinistral forms of the snail *Lymnaea stagnalis* has shown that cytoskeletal dynamics have a crucial role in determination of body handedness (Shibazaki et al. 2004). In this case, helical spindle inclination and clockwise spiral micromere formation are observed in the dominant dextral embryos at the meta-

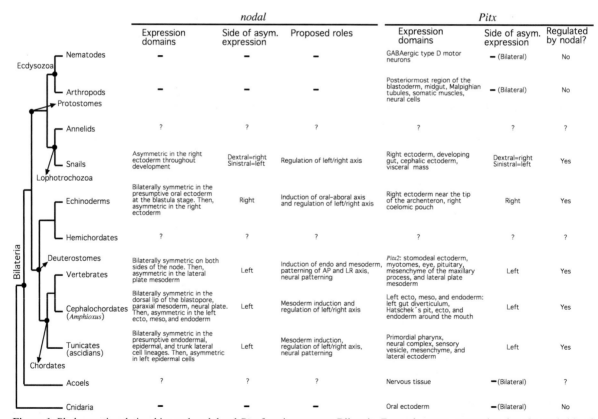

Figure 1. Phylogenetic relationships and *nodal* and *Pitx* functions among Bilateria. For each group, expression domains and side of the body that expresses *nodal* and *Pitx* are shown. Proposed roles for *nodal* and the relationship between *nodal* and *Pitx* are also given. Question marks indicate that no information is available. Boldface dashes indicate that the gene has not been reported in this group. The same information is also provided for Acoels and Cnidaria.

phase–anaphase transition initiating the third division, whereas in the recessive sinistral embryos, the third cleavage at first appears to be radially oriented, and helicity first emerges during furrow formation. The actual importance of this cleavage pattern to overall chirality has been tested by mechanically manipulating the orientation of the third cleavage (Kuroda et al. 2009). If an embryo that would have developed as a dextral animal is mechanically manipulated so that the micromeres are forced to rotate in an anticlockwise spiral (opposite to what would have normally occurred), the subsequently formed individual is instead sinistral. Likewise, if an embryo that would have developed as a sinistral animal is mechanically manipulated so that the micromeres are forced to rotate in a clockwise spiral (opposite to what would have normally occurred), the subsequently formed individual is instead dextral.

Interestingly, correct spindle orientation by a G protein at the third division has been also shown to be critical for LR asymmetry in the nematode *Caenorhabditis elegans* (Wood 1991; Bergmann et al. 2003). Although G proteins have been also shown to be involved in cytoskeletal spindle alignment in *Drosophila* (Ahringer 2003), it is unknown if this has any role in LR asymmetry for *Drosophila*. Nevertheless, actin and associated molecular motors such as myosins have been shown to be involved in LR determination in *Drosophila* (Speder et al. 2006) and vertebrates (Danilchik et al. 2006). These results support the role of the actin cytoskeleton in defining early LR asymmetry in Bilateria, although further molecular and cellular work are needed to understand the actual mechanisms behind these observations. An interesting hypothesis proposes the cytoskeleton as a vector for orienting key polarized molecules within cells (Aw and Levin 2009), but further studies are needed to test this hypothesis.

A widely accepted mechanism of symmetry breaking in vertebrates involves the rotation of cilia and the generation of a leftward flow in or close to the embryonic node in several vertebrates (Nonaka et al. 2002; Kramer-Zucker et al. 2005; Schweickert et al. 2007). One model is that *nodal* expression is initially symmetric, but the leftward flow at the node causes a leftward bias in the concentration of Nodal protein, and through its own autoregulation, *nodal* then becomes asymmetrically expressed on the left side. Supporting this model are the observations that mutations affecting inversin and motor proteins that modify nodal flow cause LR defects in several vertebrates (Afzelius 1976; Supp et al. 1997). Recent studies in zebrafish and frogs have shown a role for fibroblast growth factor signaling in LR regulation through ciliogenesis (Hong and Dawid 2009; Neugebauer et al. 2009). These results support the key role of cilia in LR asymmetry in vertebrates. However, the observation of an early leftward movement of cells around the node in chicks before the appearance of cilia suggests that alternative modes of symmetry breaking are possible within vertebrates (Gros et al. 2009). Furthermore, the fact that modification of ciliogenesis in snails (Shibazaki et al. 2004) and flies (Speder et al. 2006) does not lead to LR defects, together with the fact that asymmetrical *nodal* expression in snails starts long before the appearance of motile cilia (Grande and Patel 2009; see below), suggests that cilia may not be involved in LR determination in snails and flies.

Ion flow has also been described as a mechanism for early symmetry breaking in deuterostomes (excluding mammals) (Levin 2005) and nematodes (Bauer Huang et al. 2007). The ion flow could actively transport small LR determinants through gap junctions to one side, where they would initiate asymmetric regulation. However, no information regarding the potential role of ion flow in Lophotrochozoa is currently available and thus should be addressed in future studies to explore the possibility that this is an evolutionary ancient mechanism for symmetry breaking.

In summary, although further studies are required, current data support the existence of different asymmetry-breaking mechanisms among the Bilateria. However, the conserved subsequent asymmetric activation of the Nodal pathway in Deuterostomia and Lophotrochozoa (see below) argues that, despite variation in the mechanisms of the initial symmetry-breaking event, the LR asymmetry pathways of bilaterians converge on generating the asymmetric expression of *nodal* and traces this asymmetric expression of *nodal* back to the ancestor of all bilaterians.

THE *NODAL* PATHWAY IN BILATERIA: AN EVOLUTIONARY PERSPECTIVE

Comparative molecular analysis of *nodal* in deuterostomes and lophotrochozoans has turned out to be useful in determining the extent to which the LR asymmetry pathway is conserved in evolution and in inferring the ancestral features of the Nodal pathway in Bilateria. As described above, *nodal* was initially believed to be an innovation of the deuterostomes, but the discovery of a *nodal* ortholog in Lophotrochozoa (Grande and Patel 2009) suggested that Nodal predated the separation of the three main bilaterian lineages (see Fig. 1). Furthermore, the finding that *nodal* is asymmetrically expressed along the LR axis and functions in LR asymmetry in snails suggests that the role of the Nodal signaling pathway in LR asymmetry also predates the split of the Bilateria.

Several evolutionary relevant issues can be highlighted here. First, as described previously, *nodal* expression in vertebrates is initially symmetrical and depends on its own regulation in a positive and negative feedback loop to achieve an asymmetric pattern (Hamada et al. 2002). As in vertebrates, *nodal* expression is also first symmetric and then later asymmetric in nonvertebrate deuterostomes (Morokuma et al. 2002; Yu et al. 2002; Duboc et al. 2005). But in contrast, *nodal* expression in snails is asymmetric from the onset (early cleavage stages), and it does not seem to be regulated by its own expression (Grande and Patel 2009). The previously described differences in symmetry-breaking mechanisms observed in different bilaterians might be responsible for these differences in the onset of *nodal* expression.

Second, *nodal* expression (as it relates to LR asymmetry) occurs in different germ layers depending on the group considered: the ectoderm in echinoderms, the three germ layers in *Amphioxus* and probably in ascidians, and the mesoderm in vertebrates (Hamada et al. 2002; Morokuma et al. 2002; Yu et al. 2002; Duboc et al. 2005). The restriction of *nodal* expression exclusively to the ectoderm in snails and echinoderms suggests that the ancestral state is ectodermal and that it was expanded to the three germ layers in the ancestor of chordates and secondarily restricted to the mesoderm in vertebrates (Chea et al. 2005).

Third, the functional significance of the Nodal pathway outside of LR patterning (e.g., in gastrulation) also seems to vary among different bilaterians. The Nodal pathway is essential for endoderm and mesoderm formation in vertebrates (Shen 2007) and mesoderm patterning in *Amphioxus* (Yu et al. 2002). In ascidians, however, *nodal* signaling is not critical for endoderm and mesoderm specification because only secondary muscle induction seems to require *nodal* signaling (Hudson and Yasuo 2005). In addition, echinoderms specify germ layers independently of *nodal* activity (Duboc and Lepage 2008). On the other hand, modification of the Nodal pathway in snails produces embryos that fail to gastrulate (Grande and Patel 2009), suggesting that germ-layer specification may require the Nodal pathway, although more thorough studies are required to confirm the generality of this finding.

The Nodal pathway also has a role in the regulation of the AP and DV axes in vertebrates (Hamada et al. 2002) as well as the oral–aboral axis in echinoderms (Duboc et al. 2004). Preliminary data on the disruption of the Nodal pathway in snails show modifications in LR morphology, although neither the AP nor DV axis seems to be modified (Grande and Patel 2009). Hence, currently available data support the ancestral function of Nodal signaling in LR asymmetry in Bilateria. Functional manipulation of *nodal* orthologs in *Amphioxus* and ascidians, as well as additional lophotrochozoans, will clarify the evolutionary history of the role of *nodal* in DV and AP axis regulation.

Besides germ-layer induction and axis specification, Nodal signaling is also required in neural patterning in vertebrates (Shen 2007) and ascidians (Hudson and Yasuo 2005). In vertebrates, generation of anterior neuronal tissue requires inhibition of signaling, although high levels of *nodal* activity in the posterior epiblast are necessary for ventral patterning of the neural tubes and maintenance of anterior forebrain territories (Shen 2007). In contrast, Nodal signaling is required for lateral fate in the neural tube of ascidians (Hudson and Yasuo 2005). In snails, *nodal* transcription is seen in the cephalic regions but in an asymmetric pattern. Whether this expression correlates to LR symmetries in the nervous system in snails is still unknown. If true, however, this would support the hypothesis that *nodal* has an ancestral role in regulating brain asymmetry (Palmer 2004).

The Nodal signaling pathway acts on the left side in vertebrates, and it is a determinant for leftness in this group because inappropriate induction of *nodal* activity on the right side produces left-side morphologies (Hamada et al. 2002). All other deuterostomes conserve this pattern of expression except for echinoderms, which show Nodal signaling on the right side (Duboc et al. 2005). Unlike vertebrates, Nodal signaling is a determinant for rightness in echinoderms because inappropriate induction of *nodal* activity on the left side produces right-side morphologies and prevents the formation of the rudiment (an imaginal structure that gives rise to most of the adult organs and is normally present on just the left side). This suggests that the rudiment normally forms on the left side by default (Duboc et al. 2005). Thus, in echinoderms, the presence of *nodal* on the right modifies the "default" condition. In snails, the side that expresses *nodal* is correlated with chirality (i.e., the right side in dextral snails and the left in sinistral snails) (Grande and Patel 2009). The first indication of morphological asymmetry in snails is given by a displacement of the shell gland to the left in dextral species and to the right in sinistral species due to a differential formation of shell-producing cells on one side of the embryo. Inhibition of the Nodal pathway in sinistral snails produces noncoiled shells (Grande and Patel 2009), which suggests equal formation of shell-producing cells on both sides of the embryo or, in other words, that the default condition is established on both sides of the embryo when *nodal* function is eliminated. In summary, these results suggest that the side of the embryo that expresses *nodal* is the one that acquires new features and defers from the default condition.

Finally, several lines of evidence indicate that the last common ancestor of snails had a dextral body (Ponder and Lindberg 1997) and therefore expressed *nodal* on the right side. This, together with the fact that *nodal* is also expressed on the right side in echinoderms, raises the hypothesis that the bilaterian ancestor expressed *nodal* on the right side as well.

CONSERVED GENE CASCADE *NODAL-PITX* IN BILATERIA

Pitx is a bicoid-type homeobox transcription factor that shows multiple sites of expression during deuterostome development. Two roles have been described for *Pitx* genes in chordates: regulation of the pituitary and its homologs and control of directional asymmetry in visceral organ organization (Boorman and Shimeld 2002). Data from echinoderms, ascidians, *Amphioxus*, and vertebrates show that this gene is important in asymmetric morphogenesis and that it is up-regulated by *nodal* (Boorman and Shimeld 2002; Hamada et al. 2002; Yoshida and Saiga 2008). Following the asymmetric expression of *nodal*, *Pitx* is also asymmetrically expressed, persists longer than *nodal*, and is localized in the primordia of most asymmetric organs. Although *Pitx* orthologs have been described for members of the Ecdysozoa (*Drosophila* and *C. elegans*) (McIntire et al. 1993; Vorbrüggen et al. 1997), asymmetrical expression patterns have not been reported in this group, and functional experiments suggest that *Pitx* is not involved in pattern formation during development in Ecdysozoa. Interestingly, a *Pitx* ortholog has been recently identified in snails, and, like deuterostomes and unlike ecdysozoans, *Pitx* is asymmetrically expressed and regulated by *nodal* (Grande and Patel 2009). Asymmetric expression of *Pitx* in snails starts right after the activation of *nodal* and also correlates with body handedness (it is expressed on the right side in dextral snails and on the left in sinistral snails) (Grande and Patel 2009). Hence, these new data suggest that the ancestor of bilaterians already had the gene cascade *nodal-Pitx* controlling LR asymmetry (see Fig. 1).

In addition to asymmetric expression, *Pitx* is a stomodeal (mouth) marker for both chordates and echinoderms (Christiaen et al. 2007). However, in another group of deuterostomes, hemichordates, this gene is expressed at the opposite end of the site of mouth formation (Lowe et al. 2006). This result, together with the absence of stomodeal expression of the *Pitx* ortholog in ecdysozoans (McIntire et al. 1993; Vorbrüggen et al. 1997), initially argued against stomadeal expression as an ancestral state in deuterostomes. However, new data from snails show that *Pitx* is also expressed in the stomodeum (Grande and Patel 2009) and therefore supports the hypothesis that *Pitx* was expressed not only in the stomodeum of the ancestor of deuterostomes, but also in the stomodeum of the ancestor of bilaterians.

Finally, the *Pitx* transcript in snails is localized simultaneously in different asymmetric and symmetric domains (Grande and Patel 2009). Preliminary data on the inhibition of the Nodal pathway in snails show that *nodal* regulates exclusively the asymmetric domain of *Pitx* expression, suggesting different independent enhancers for *Pitx* in snails, as described for vertebrates (Christiaen et al. 2005, 2007).

THE DV INVERSION HYPOTHESIS AND THE LR AXIS

In the 19th century, Geoffroy Saint-Hilaire (1822) proposed that the DV axis of deuterostomes is inverted with respect to that observed in protostomes (Lophotrochozoa plus Ecdysozoa). He based this idea on the relative placement of the nerve cord (ventral in protostomes and dorsal in deuterostomes) and the heart and circulatory vessels (dorsal in protostomes and ventral in deuterostomes). Although this hypothesis was eventually dismissed, recent molecular data on developmental expression and function of homologous genes in distantly related taxa have been interpreted as new evidence for the inversion of the axis (Arendt and Nübler-Jung 1994; Holley et al. 1995; DeRobertis and Sasai 1996; Gerhart 2000). Several orthologous genes of chordates and protostomes show similar but inverted expression domains. This is especially striking in those related with DV patterning, such as certain TGF-β ligands (Bmp4 in chordates and its ortholog Dpp in *Drosophila*) and TGF-β inhibitors (Chordin in chordates and its ortholog Sog in *Drosophila*). *bmp4* is expressed in ventral domains and has ventralizing activity in chordates, whereas *chordin* is expressed dorsally and has an antagonistic role (Holley et al. 1995; DeRobertis and Sasai 1996). On the other hand, *dpp* is expressed in dorsal domains and has dorsalizing activity in *Drosophila*, whereas *sog* is expressed in ventral domains and has an antagonistic role (Holley et al. 1995; DeRobertis and Sasai 1996). However, both *chordin* and *sog* promote the development of the central nervous system wherever they are expressed, suggesting that the ventral nerve cord of insects and the dorsal nerve cord of chordates are built by the same genetic machinery and may have not evolved independently. In addition—*hedgehog*, an important gene in dorsal midline formation, differentiation of neural-tube structures,

and DV patterning of adjacent tissues in chordates (Sasai and DeRobertis 1997)—is expressed in the ventral midline in Lophotrochozoa (Nederbragt et al. 2002), suggesting that similar mechanisms are involved in the development of the midline in both groups and adding new evidence for DV axis inversion. Furthermore, several other genes involved in heart and nervous system development, such as *Nkx2-5*, *netrin*, *vnd/Nk2*, *ind/Gsh1,2* and *Msh/Msx1,3*, are expressed in opposite domains in chordates and protostomes (Weiss et al. 1998; Tanaka et al. 1999). Thus, on the basis of both morphological and molecular data, the DV inversion hypothesis postulates that the common ancestor of deuterostomes and protostomes already had a nerve cord, a heart, and circulatory vessels, as well as the genetic modules to build them, and that the DV axis was inverted throughout the course of evolution in one of the two lineages.

The LR axis is always specified with reference to AP and DV axes. Experimental manipulations have shown that the LR axis can be reversed by reversing the AP axis or the DV axis (Ligoxygakis et al. 2001). Thus, according to the DV inversion hypothesis, genes that are expressed on one side of the body (left versus right) in one of the two groups (either protostomes or deuterostomes) are expected to be expressed on the opposite side in the other group. This is exactly the case for the *nodal-Pitx* gene cascade, which is expressed on the right side in dextral snails (representing the ancestral state) (Grande and Patel 2009) and on the left side in chordates (Hamada et al. 2002). The fact that *nodal-Pitx* is also expressed on the right side in echinoderms (Duboc et al. 2005) suggests that the inversion must have occurred in the chordate lineage.

Some alternatives to the DV inversion hypothesis have been proposed. Some investigators have suggested that the differences in morphogenetic movements during gastrulation of protostomes versus deuterostomes result in inverted expression patterns of developmental genes (van den Biggelaar et al. 2002). In chordates, the AP axis is extended evenly during gastrulation, whereas in protostomes, the growth along the AP axis is uneven, almost exclusively limited to the dorsal side. Thus, originally, dorsal cells will be located in a ventral position in protostomes after gastrulation, explaining the ventral expression patterns of certain genes with respect to chordates (van den Biggelaar et al. 2002). However, these differences in migration of cells and growth along the AP axis would not explain left versus right differences in expression patterns of the *nodal-Pitx* cascade described above. In addition, it is important to note that *nodal* initiates its expression before gastrulation and hence before those morphogenetic movements and uneven growth. Furthermore, the onset of the expression of the ortholog of *dpp* in snails is also detected in the dorsal domain of the late blastula (C Grande and NH Patel, unpubl.), whereas it is ventral in the early gastrula of vertebrates. Hence, the shift of initially dorsally specified cells to a secondarily ventral position does not fully explain the inverted gene expression domains described above.

Other investigators have suggested that the DV axis of the last common ancestor of protostomes and deuterostomes was relatively undifferentiated, and subsequently both lineages organized the nerve cord and organogenesis independently in opposite locations, and thus inversion never occurred (Gerhart 2000). This hypothesis was traditionally based on the presence of a diffuse nerve net in some hemichordates, which supports an undefined central nervous system for the ancestor of deuterostomes. However, recent studies have shown that the nervous system of hemichordates is much more localized than previously thought and therefore that the centralization of the deuterostome nervous system predates the origin of chordates (Nomaksteinsky et al. 2009). Whether nervous system centralization does or does not predate the spilt between deuterostomes and protostomes, the observations on *nodal* expression are still most consistent with a scenario of some type of DV axis inversion. Further studies of cell and morphological homologies, as well as the gene modules that regulate body organization in a comparative phylogenetic framework, will help to elucidate the evolution of the DV axis and hence of the LR axis as well.

CONCLUSIONS

A central focus in evolutionary developmental biology is to understand the evolution of the mechanisms that establish and elaborate the body axes of the developing embryo. Recent studies have suggested that DV axis specification is controlled by ancient molecular patterning mechanisms, but whether conserved mechanisms regulate LR asymmetries constitutes an open question for which initial answers have been only recently proposed. Striking support for an ancient origin of LR axis patterning is provided by data on the Nodal pathway in Lophotrochozoa and its function in LR regulation. These results provide evidence of the ancient role of this pathway in LR patterning in Bilateria and sheds light on the evolutionary events that have led to the current diversity in body plans. Finally, new avenues of research are opening up novel approaches to understanding symmetry-breaking events in bilaterians, the regulatory mechanisms involved in LR determination, and the evolutionary implications of variation in these pathways for generating morphological diversity.

ACKNOWLEDGMENTS

C.G. is currently a "Ramon y Cajal" postdoctoral fellow supported by the Spanish Ministerio de Ciencia e Innovacion and the Universidad Autonoma de Madrid.

REFERENCES

Afzelius BA. 1976. Human syndrome caused by immotile cilia. *Science* **193:** 317–319.

Aguinaldo AM, Turbeville JM, Linford LS, Rivera MC, Garey JR, Raff RA, Lake JA. 1997. Evidence for a clade of nematodes, arthropods and other moulting animals. *Nature* **387:** 489–493.

Ahringer J. 2003. Control of cell polarity and mitotic spindle positioning in animal cells. *Curr Opin Cell Biol* **15:** 73–81.

Arendt D, Nübler-Jung K. 1994. Inversion of dorsoventral axis? *Nature* **371:** 26.

Aw S, Levin M. 2009. Is left-right asymmetry a form of planar cell polarity? *Development* **136:** 355–366.

Bauer Huang SL, Saheki Y, VanHoven MK, Torayama I, Ishihara T, Katsura I, van der Linden A, Sengupta P, Bargmann CI. 2007. Left-right olfactory asymmetry results from antagonistic functions of voltage-activated calcium channels and the Raw repeat protein OLRN-1 in *C. elegans*. *Neural Dev* **2:** 24.

Bergmann DC, Lee M, Robertson B, Tsou MF, Rose LS, Wood WB. 2003. Embryonic handedness choice in *C. elegans* involves the Gα protein GPA-16. *Development* **130:** 5731–5740.

Boorman CJ, Shimeld SM. 2002. *Pitx* homeobox genes in *Ciona* and *Amphioxus* show left-right asymmetry is a conserved chordate character and define the ascidian adenohypophysis. *Evol Dev* **4:** 354–365.

Boycott AE, Diver C. 1923. On the inheritance of sinistrality in *Limnaea peregra*. *Proc R Soc Lond B Biol Sci* **95:** 207–213.

Boycott AE, Diver C, Garstang SL, Hardy MAC, Turner FM. 1930. The inheritance of sinistrality in *Limnaea peregra*. *Philos Trans R Soc Lond B Biol Sci* **219:** 51–130.

Chea HK, Wright CV, Swalla BJ. 2005. Nodal signaling and the evolution of deuterostome gastrulation. *Dev Dyn* **234:** 269–278.

Christiaen L, Bourrat F, Joly JS. 2005. A modular *cis*-regulatory system controls isoform-specific *pitx* expression in ascidian stomodaeum. *Dev Biol* **277:** 557–566.

Christiaen L, Jaszczyszyn Y, Kerfant M, Kano S, Thermes V, Joly JS. 2007. Evolutionary modification of mouth position in deuterostomes. *Semin Cell Dev Biol* **18:** 502–511.

Coutelis JB, Petzoldt AG, Speder P, Suzanne M, Noselli S. 2008. Left-right asymmetry in *Drosophila*. *Semin Cell Dev Biol* **19:** 252–262.

Danilchik MV, Brown EE, Riegert K. 2006. Intrinsic chiral properties of the *Xenopus* egg cortex: An early indicator of left-right asymmetry? *Development* **133:** 4517–4526.

DeRobertis EM, Sasai Y. 1996. A common plan for dorsoventral patterning in Bilateria. *Nature* **380:** 37–40.

Duboc V, Lepage T. 2008. A conserved role for the nodal signaling pathway in the establishment of dorso-ventral and left-right axes in deuterostomes. *J Exp Zool B Mol Dev Evol* **310:** 41–53.

Duboc V, Rottinger E, Besnardeau L, Lepage T. 2004. *Nodal* and *BMP2/4* signaling organizes the oral-aboral axis of the sea urchin embryo. *Dev Cell* **6:** 397–410.

Duboc V, Rottinger E, Lapraz F, Besnardeau L, Lepage T. 2005. Left-right asymmetry in the sea urchin embryo is regulated by nodal signaling on the right side. *Dev Cell* **9:** 147–158.

Freeman G, Lundelius J. 1982. The developmental genetics of dextrality and sinistrality in the gastropod *Lymnaea peregra*. *Wilhelm Roux's Arch Dev Biol* **191:** 69–83.

Geoffroy St. Hillaire E. 1822. Considerations generales sur la vertebre. *Mem Mus Hist Nat* **9:** 89–119.

Gerhart J. 2000. Inversion of the chordate body axis: Are there alternatives? *Proc Natl Acad Sci* **97:** 4445–4448.

Grande C, Patel NH. 2009. Nodal signalling is involved in left-right asymmetry in snails. *Nature* **457:** 1007–1011.

Gros J, Feistel K, Viebahn C, Blum M, Tabin C. 2009. Cell movements at Hensen's node establish left/right asymmetric gene expression in the chick. *Science* **324:** 941–944.

Hamada H, Meno C, Watanabe D, Saijoh Y. 2002. Establishment of vertebrate left-right asymmetry. *Nat Rev Genet* **2:** 103–113.

Holley SA, Jackson PD, Sasai Y, Lu B, DeRobertis EM, Hoffmann FM, Ferguson EL. 1995. A conserved system for dorsal-ventral patterning in insects and vertebrates involving *sog* and *chordin*. *Nature* **376:** 249–253.

Hong SK, Dawid IB. 2009. FGF-dependent left-right asymmetry patterning in zebrafish is mediated by Ier2 and Fibp1. *Proc Natl Acad Sci* **106:** 2230–2235.

Hudson C, Yasuo H. 2005. Patterning across the ascidian neural plate by lateral Nodal signalling sources. *Development* **132:** 1199–1210.

Kramer-Zucker AG, Olale F, Haycraft CJ, Yoder BK, Schier AF, Drummond IA. 2005. Cilia-driven fluid flow in the zebrafish pronephros, brain and Kupffer's vesicle is required for normal organogenesis. *Development* **132:** 1907–1921.

Kuroda R, Endo B, Masanri A, Shimuzu M. 2009. Chiral blastomere arrangement dictates zygotic left-right asymmetry pathway in snails. *Nature* **462:** 790–794.

Levin M. 2005. Left-right asymmetry in embryonic development: A comprehensive review. *Mech Dev* **122:** 3–25.

Levin M, Johnson RL, Stern CD, Kuehn M, Tabin C. 1995. A molecular pathway determining left-right asymmetry in chick embryogenesis. *Cell* **82:** 803–814.

Levin M, Pagan S, Roberts DJ, Cooke J, Kuehn MR, Tabin CJ. 1997. Left/right patterning signals and the independent regulation of different aspects of situs in the chick embryo. *Dev Biol* **189:** 57–67.

Ligoxygakis P, Strigini M, Averof M. 2001. Specification of left-right asymmetry in the embryonic gut of *Drosophila*. *Development* **128:** 1171–1174.

Lowe CJ, Terasaki M, Wu M, Freeman RM, Runft L, Kwan K, Haigo S, Aronowicz J, Lander E, Gruber C, et al. 2006. Dorsoventral patterning in hemichordates: Insights into early chordate evolution. *PLoS Biol* **4:** 1603–1619.

McIntire SL, Jorgensen E, Horvitz HR. 1993. Genes required for Gaba function in *Caenorhabditis elegans*. *Nature* **364:** 334–337.

Meshcheryakov VN, Beloussov LV. 1975. Asymmetrical rotations of blastomeres in early cleavage of gastropoda. *Wilhelm Roux's Arch Dev Biol* **177:** 193–203.

Morokuma J, Ueno M, Kawanishi H, Saiga H, Nishida H. 2002. *HrNodal*, the ascidian nodal-related gene, is expressed in the left side of the epidermis, and lies upstream of *HrPitx*. *Dev Genes Evol* **212:** 439–446.

Nederbragt AJ, van Loon AE, Dictus WJ. 2002. Evolutionary biology: Hedgehog crosses the snail's midline. *Nature* **417:** 811–812.

Neugebauer JM, Amack JD, Peterson AG, Bisgrove BW, Yost HJ. 2009. FGF signalling during embryo development regulates cilia length in diverse epithelia. *Nature* **458:** 651–654.

Nomaksteinsky M, Rottinger E, Dufour HD, Chettouh Z, Lowe CJ, Martindale MQ, Brunet JF. 2009. Centralization of the deuterostome nervous system predates chordates. *Curr Biol* **19:** 1264–1269.

Nonaka S, Shiratori H, Saijoh Y, Hamada H. 2002. Determination of left-right patterning of the mouse embryo by artificial nodal flow. *Nature* **418:** 96–99.

Okumura T, Utsuno H, Kuroda J, Gittenberger E, Asami T, Matsuno K. 2008. The development and evolution of left-right asymmetry in invertebrates: Lessons from *Drosophila* and snails. *Dev Dyn* **237:** 3497–3515.

Palmer AR. 2004. Symmetry breaking and the evolution of development. *Science* **306:** 828–833.

Ponder WF, Lindberg DR. 1997. Towards a phylogeny of gastropod molluscs: An analysis using morphological characters. *Zool J Linn Soc* **119:** 83–265.

Sampath K, Cheng AM, Frisch A, Wright CV. 1997. Functional differences among *Xenopus* nodal-related genes in left-right axis determination. *Development* **124:** 3293–3302.

Sasai Y, DeRobertis EM. 1997. Ectodermal patterning in vertebrate embryos. *Dev Biol* **182:** 5–20.

Schweickert A, Weber T, Beyer T, Vick P, Bogusch S, Feistel K, Blum M. 2007. Cilia-driven leftward flow determines laterality in *Xenopus*. *Curr Biol* **17:** 60–66.

Shen M. 2007. Nodal signaling: Developmental roles and regulation. *Development* **134:** 1023–1034.

Shibazaki Y, Shimizu M, Kuroda R. 2004. Body handedness is directed by genetically determined cytoskeletal dynamics in the early embryo. *Curr Biol* **14:** 1462–1467.

Shimeld SM, Levin M. 2006. Evidence for the regulation of left-right asymmetry in *Ciona intestinalis* by ion flux. *Dev Dyn* **235:** 1543–1553.

Speder P, Adam G, Noselli S. 2006. Type ID unconventional myosin controls left-right asymmetry in *Drosophila*. *Nature* **440:** 803–807.

Sturtevant AH. 1923. Inheritance of direction of coilling in *Lymnaea*. *Science* **58:** 269–207.

Supp DM, Witte DP, Potter SS, Brueckner M. 1997. Mutation of an axonemal dynein affects left-right asymmetry in inversus viscerum mice. *Nature* **389:** 963–966.

Tanaka M, Wechsler SB, Lee IW, Yamasaki N, Lawitts JA, Izumo S. 1999. Complex modular *cis*-acting elements regulate expression of the cardiac specifying homeobox gene *Csx/Nkx2.5*. *Development* **126:** 1439–1450.

van den Biggelaar JAM, Edsinger-Gonzales E, Schram FR. 2002. The improbability of dorso-ventral axis inversion during animal evolution, as presumed by Geoffroy Saint Hilaire. *Contrib Zool* **71:** 29–36.

Vorbrüggen G, Constien R, Zilian O, Wimmer EA, Dowe G, Taubert H, Noll M, Jäckle H. 1997. Embryonic expression and characterization of a *Ptx1* homolog in *Drosophila*. *Mech Dev* **68:** 139–147.

Weiss JB, Von Ohlen T, Mellerick DM, Dressler G, Doe CQ, Scott MP. 1998. Dorsoventral patterning in the *Drosophila* central nervous system: The intermediate neuroblasts defective homeobox gene specifies intermediate column identity. *Genes Dev* **12:** 3591–3602.

Wood WB. 1991. Evidence from reversal of handedness in *C. elegans* embryos for early cell-interactions determining cell fates. *Nature* **349:** 536–538.

Yoshida K, Saiga H. 2008. Left-right asymmetric expression of Pitx is regulated by the asymmetric Nodal signaling through an intronic enhancer in *Ciona intestinalis*. *Dev Genes Evol* **218:** 353–360.

Yu JK, Holland LZ, Holland ND. 2002. An amphioxus *nodal* gene (*AmphiNodal*) with early symmetrical expression in the organizer and mesoderm and later asymmetrical expression associated with left-right axis formation. *Evol Dev* **4:** 418–425.

On the Origins of Novelty and Diversity in Development and Evolution: A Case Study on Beetle Horns

A.P. MOCZEK

Department of Biology, Indiana University, Bloomington, Indiana 47405
Correspondence: armin@indiana.edu

The origin of novel features continues to represent a major frontier in evolutionary biology. What are the genetic, developmental, and ecological processes that mediate not just the modification of preexisting traits, but the origin of novel traits that lack obvious homology with other structures? In this chapter, I highlight a class of traits and organisms that are emerging as new models for exploring the mechanisms of innovation and diversification in nature: beetle horns and horned beetles. Here, I review recent significant findings and their contributions to current frontiers in evolutionary developmental biology.

The process of innovation in evolution has captivated evolutionary biologists ever since the inception of the discipline (Raff 1996; West-Eberhard 2003). What does it take genetically, developmentally, or ecologically for novel traits to arise and diversify in nature? Is the origin of novel traits underlain by processes separate from those that govern quantitative changes in preexisting traits or is *innovation* merely an extrapolation of *diversification* over time (Erwin 2000; Davidson and Erwin 2006; Moczek 2008)? In this chapter, I propose that beetle horns and horned beetles offer unusual opportunities to integrate genetic, developmental, and ecological mechanisms into a holistic understanding of how novel complex traits originate and diversify during both development and evolution. Specifically, I highlight and synthesize recent advances in our understanding of the genetic, developmental, and ecological origins of horns and horn diversity, as well as their consequences for diversification and radiation of horned beetles. I begin with a brief review of the basic biology and natural history of horned beetles.

A BRIEF NATURAL HISTORY OF BEETLE HORNS AND HORNED BEETLES

Beetle horns unite several characteristics that make them outstanding models for exploring the origin and diversification of novel traits (Moczek 2005). First, beetle horns are massive, solid, three-dimensional outgrowths that often severely transform the shape of whoever bears them (Fig. 1). Second, beetle horns function as weapons in male competition over breeding opportunities, thereby defining the behavioral ecology of individuals and populations. Third, beetle horns are inordinately diverse (Figs. 1 and 2A). Horns differ in size, shape, number, and location of expression, and much of this variation can be found not only among species, but also between, and oftentimes within, sexes. Finally, and most importantly, beetle horns are *unique* structures in the sense that they lack clear homology with other traits in insects. Horns are not modified mouthparts or legs; instead, they exist *alongside* these structures in body regions in which insects normally do not produce any outgrowths (Moczek 2005, 2006a). Hence, we can look at horns as an example of an evolutionary novelty that horned beetles invented at some point during their history and which has since fueled one of the most impressive radiations of secondary sexual traits in the animal kingdom (Arrow 1951; Balthasar 1963; Emlen et al. 2007). Here, I explore the mechanisms that have mediated the initial origin and subsequent diversification of horns. Specifically, I argue that the origin and diversification of horns was made possible through (1) widespread cooption of preexisting developmental mechanisms into new developmental contexts, (2) exaptation of preexisting structures that originally performed unrelated functions, and (3) trade-offs arising during development that bias, and possibly accelerate, patterns of diversification. However, before introducing any of these mechanisms, we must first briefly review the basic biology of horn development and formation.

THE ONTOGENY OF HORNS

The horns of beetles first become discernible during the last larval stage as the animal nears the larval-to-pupal molt (Fig. 2B) (for review, see Moczek 2006a; Moczek and Rose 2009). At this stage, selected regions of the epidermis detach from the larval cuticle and proliferate underneath. The resulting tissue is thrown into folds as it is trapped underneath the larval cuticle, but it expands once the animal molts to the pupal stage. It is at this stage that horns become visible externally for the first time. This period of *prepupal horn growth* is then followed by a period of *pupal remodeling* of horn primordia. During the pupal stage, pupal horns undergo at times substantial remodeling in both size and shape, including, in some cases, the complete resorption of horns before the adult molt. After the competition of the pupal-to-adult molt,

Figure 1. Examples of horned beetles illustrating diversity and magnitude of horn expression in adult beetles. (*Clockwise from top left*) *Phanaeus imperator*, *Eupatorus gracilicornis*, *Onthophagus watanabei*, *Golofa claviger*, and *Trypoxylus (Allomyrina) dichotoma*.

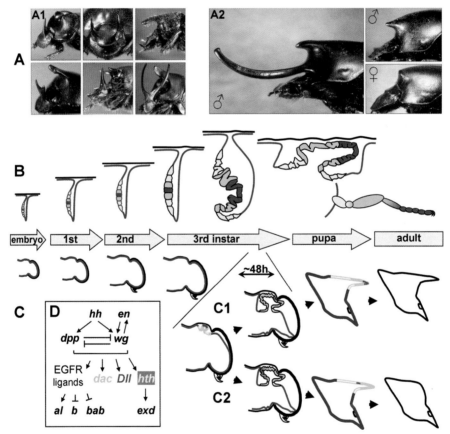

Figure 2. Diversity and development of beetle horns. (*A*) Diversity in number, size, location, and shape of horn expression between (*A1*) and within (*A2*) species of *Onthophagus*. (*B*) *Drosophila* model of limb formation compared to (*C*) the development of a thoracic beetle horn from embryo to adult. (Black) Cuticle, (blue) epidermis, including schematic expression domains of the proximo/distal patterning genes *homothorax* (*hth*, yellow), *dachshund* (*dac*, green), and *Distal-less* (*Dll*, red). *Drosophila* legs develop from imaginal discs, epidermal invaginations specified during embryonic development, which grow throughout larval development. Patterning takes place while the disc is a two-dimensional sheet of tissue, and all disc growth occurs while the disc is invaginated into the body interior. In contrast, beetle horns appear not to be specified during embryonic development. Instead, horns grow from the start as three-dimensional epidermal outbuddings, and all growth is confined to the relatively brief prepupal stage and takes place while the primordium is evaginated into the space between epidermis and larval cuticle. In addition to a rapid prepupal growth phase, horn expression is also affected by an at times drastic pupal remodeling phase (*C1, C2*) during the early pupal stage. During this stage, pupal horn primordia are either converted into a future adult structure (*C1*) or resorbed (*C2*) via programmed cell death (PCD). In the latter case, expression of *Dll*, but not *hth* or *dac*, is shifted more posteriorly. (*D*) Position of *dac*, *hth*, and *Dll* within the basic *Drosophila* limb patterning network (*hh*, hedgehog; *en*, engrailed; *dpp*, decapentaplegic; *wg*, wingless; *EGFR*, epidermal growth factor receptor; *al*, aristaless; *b*, bar; *bab*, bric a brac; *exd*, extradenticle). (Modified from Moczek and Rose 2009.)

horns have then attained their final adult size and shape. The horns of adult beetles are thus the product of (1) a prepupal growth phase followed by a (2) pupal remodeling phase, and therefore develop in many ways similar to traditional appendages in other holometabolous insects (Svácha 1992). However, more substantial differences exist when compared to appendage formation in the best studied insect: *Drosophila*. Here, appendages develop from imaginal discs, which represents a highly derived mode of appendage formation absent in the majority of insect orders (Kojima 2004). Imaginal discs are epidermal invaginations specified during embryonic development that grow throughout larval development. Moreover, many important patterning steps take place while the disc is a two-dimensional sheet of tissue, and most disc growth occurs while the disc is *in*vaginated into the body interior (Fig. 3A). Beetle horns differ in that they appear not to be specified during embryonic development, grow right away as three-dimensional epidermal outbuddings, have their growth confined to the relatively brief period just preceding the larval-to-pupal molt, and as they grow, *e*vaginate into the space between epidermis and larval cuticle (see Fig. 2) (Moczek 2006a). Unfortunately, most of our understanding of insect appendage formation comes from studies of imaginal disc development in *Drosophila*, and as such, the *Drosophila* model of limb development represents our best starting point to begin exploring the regulation of horn growth and differentiation.

THE ORIGIN OF *NEW* THROUGH COOPTION OF THE *OLD*. I: THE REGULATION OF HORN GROWTH

Beetle horns can be thought of, at least in some ways, as simplified appendages. Although they lack muscles, nerves, or joints, they are three-dimensional outgrowths of epidermal origin with clearly defined proximodistal, mediolateral, and anteroposterior axes. Recent studies show that the similarities do not end on the surface but that much of the underlying developmental machinery used in the making of traditional appendages such as legs and antennae has been redeployed in the development and evolution of horns.

The first clues came from a series of expression studies which showed that several cardinal appendage patterning genes known to have important roles in establishing the proximodistal axis of insect appendages (*Distal-less* [*Dll*], *aristaless* [*al*], *dachshund* [*dac*], *homothorax* [*hth*], *extradenticle* [*exd*]) were expressed during horn formation (Moczek and Nagy 2005; Moczek et al. 2006). Recent gene function analyses using RNA interference (RNAi)-

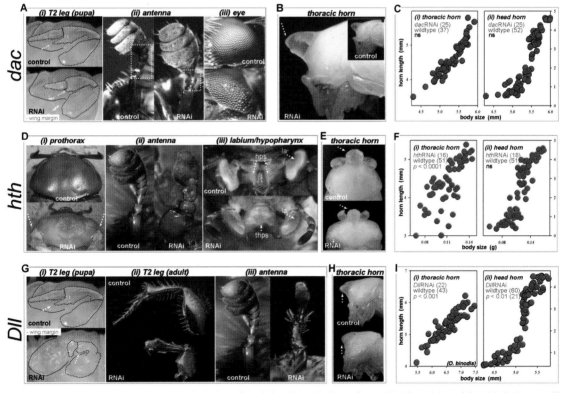

Figure 3. Larval RNAi-mediated transcript depletion of *dachshund* (*A–C*), *homothorax* (*D–F*), and *Distal-less* (*G–I*). Images illustrate typical phenotypes observed in each experiment compared to wild-type phenotypes. Graphs depict scaling relationships between pupal body size and horn length for thoracic horns (i) and head horns (ii). Pupal body size was measured as thorax width for *dac* and *Dll*. *hth*RNAi affected thorax shape and we therefore used pupal mass as an estimator of body size. (Blue) Wild type, (red) RNAi-treated individuals. All data are from male *O. taurus* except I(i), which were collected from female *O. binodis*. Sample sizes are given in parentheses. (Modified from Moczek and Rose 2009.)

mediated transcript depletion further refined our insights into the roles of a subset of these genes. Specifically, Moczek and Rose (2009) examined the function of *dac*, *hth*, and *Dll* during horn development in *Onthophagus* beetles, a focal genus of horned beetles for evo-devo studies. Independent of any involvement in horn development, larval RNAi-mediated transcript depletion of all three patterning genes generated phenotypic effects identical or similar to those documented by previous studies in other taxa, such as loss or fusion of distal and medial leg and antennal regions in the case of *Dll* (Prpic et al. 2001; Angelini and Kaufman 2004; Koijma 2004) or accelerated eye differentiation and ectopic wing tissue formation on the first thoracic segment in the case of *hth* (Ryoo et al. 1999; Yao et al. 1999; Bessa et al. 2002). These results documented that all three patterning genes exhibited conservation of function with respect to the patterning of traditional appendages as well as the general feasibility of larval RNAi in *Onthophagus* beetles. In addition, however, this study yielded many surprising insights into the functional regulation of horn development, including unexpected twists that would have been undetectable using a traditional gene expression approach. For instance, despite being widely expressed throughout prepupal horn primordia in *Onthophagus* (Moczek et al. 2006), *dac* did not appear to have any obvious role in the regulation of size, shape, or identity of horns. Instead, *dac*RNAi individuals expressed thoracic and head horns of precisely the same size and overall shape as control animals despite severe *dac* knockdown phenotypes elsewhere in the same individuals. In contrast, *hth*RNAi had a dramatic effect on horn expression, but only affected horns produced by the prothorax. Head horns expressed by the *same* individuals were completely unaffected, despite severe effects on other head appendages such as antennae. The results of *Dll*RNAi only added to the complexity. Unlike *hth*, *Dll*RNAi affected the expression of both head and thoracic horns but not in the same individuals or even species. In *O. taurus*, head horn expression was only affected in large males otherwise determined to express a full set of head horns, whereas horn expression in small- and medium-sized males was unaffected. Similarly unaffected was the expression of pupal thoracic horns in both males and females regardless of body size. In the congener *O. binodis*, however, *Dll*RNAi affected the expression of thoracic horns in both males and females, although the effect was strongest in large individuals. Combined, these results illustrate that *Onthophagus Dll* and *hth*, but not *dac*, alter horn expression in a sex-, body-region-, and body-size-specific manner and that even closely related species can diverge rather substantially in aspects of this regulation.

More generally, these results suggest that horn development evolved via differential recruitment of at least some proximodistal axis patterning genes normally involved in the formation of traditional appendages. On one hand, these results are not surprising because they confirm a general theme in the evolution of novel traits: New morphologies do not require new genes or developmental pathways and instead may arise by recruiting existing developmental mechanisms into new contexts (Shubin et al. 2009). On the other hand, these results also highlighted an unexpected degree of evolutionary lability, ranging from the absence of patterning function (*dac*) to patterning function in selected horn types only (*hth*, *Dll*) to function in one size class, sex, or species but not another (*Dll*). Combined, these data suggest that different horn types, and even the same horn type in different species, may be regulated at least in part by different pathways and that different horn types may therefore have experienced distinct, and possibly independent, evolutionary histories. Similar conclusions emerge when we take a closer look at the regulation of the second developmental period relevant to adult horn expression: pupal remodeling.

THE ORIGIN OF *NEW* THROUGH COOPTION OF THE *OLD*. II: THE REGULATION OF PUPAL REMODELING

During the pupal stage, horns are sculpted into their final adult shape. As such, pupal remodeling of horns is not unusual; instead, all pupal appendages and body regions of holometabolous insects undergo at least some degree of sculpting during the pupal stage (Cullen and McCall 2004). What is unusual, however, is the often extreme nature of pupal horn remodeling, especially with respect to horns emanating from the thorax (Moczek 2006b). Here, horn remodeling is so extreme that it often results in the complete resorption of pupal horn primordia, causing fully horned pupae to molt into thorax hornless adults lacking any signs of the previous existence of a thoracic horn primordium (Fig. 4). Moreover, pupal horn resorption may or may not affect both sexes equally. For instance, of 19 *Onthophagus* species studied thus far, three species used female-specific resorption of thoracic horn tissue to generate sexual dimorphism (only males remained horned) and one species used male-specific resorption of thoracic horn tissue, leaving females as the only sex with thoracic horns. The remaining 15 species used the same process to remove thoracic horn primordia in *both* sexes. In at least one of these, *O. taurus*, pupal thoracic horn resorption actually *eliminated* a pronounced sexual dimorphism in thoracic horns evident in pupae but not in the resulting adults (Moczek et al. 2006). Combined, these data suggest that pupal horn resorption is incredibly widespread at least in *Onthophagus*, but at the same time, it is evolutionarily labile with respect to the sex in which it operates (Fig. 4). Recent work now strongly implicates programmed cell death (PCD) in the resorption of horn primordial tissue.

PCD mediates the coordinated destruction of cells and their content (Potten and Wilson 2004). As such, PCD uses a complex cascade of developmental and cellular processes. Despite this apparent complexity, PCD is an ancient physiological process used by all metazoan organisms to eliminate cells during development (Potten and Wilson 2004). Recent work has now shown that primordial horn epidermis fated to be resorbed undergoes premature PCD during the first 48 hours of the pupal stage (Kijimoto et al. 2009). Relying on two different biochemical assays, the same study then showed that PCD is considerably more frequent among

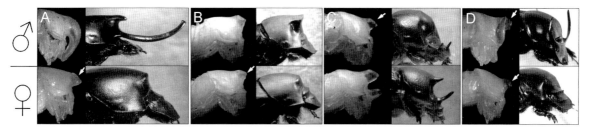

Figure 4. Pupal horn remodeling is common in the genus *Onthophagus* yet variable among species. Males (*top*) and females (*bottom*) of four *Onthophagus* species. Pupae are shown on the left and corresponding adults on the right. (*A*) *O. nigriventris*, (*B*) *O. binodis*, (*C*) *O. sagittarius*, (*D*) *O. taurus*. Arrows highlight cases of pupal horn resorption. (Modified from Moczek 2009.)

horn primordial cells of transient horns compared to individuals whose pupal horns persist and become converted into an adult structure. Comparisons across species suggest that the exact position and timing of PCD-mediated horn remodeling can differ remarkably from one species to the next. Combined, the regulation of pupal remodeling therefore reveals many of the same features highlighted above for the regulation of horn growth. On one hand, pupal remodeling and resorption of horns appears to rely on a preexisting developmental machinery that became recruited into a new developmental context. On the other hand, remarkable variation exists within and among species regarding when, where, and how much remodeling and resorption of horns occurs. By extension, this variation suggests the existence of rapidly evolving modifier mechanisms that regulate species-, sex-, and body-region-specific resorption of horns. The identity and nature of these modifier mechanisms are currently being investigated.

It is important to realize that this is clearly just the beginning of a more-detailed analysis of the developmental regulation and diversification of beetle horns. The recent development of *Onthophagus* expressed sequence tag (EST) libraries, microarrays, and transcriptome profiles has now rapidly increased the number and diversity of genes and pathways implicated in the regulation of horns and horn diversity (Kijimoto et al. 2009). Although much work clearly lies ahead before we achieve a satisfactory understanding of the origins of horn development, it appears that the most critical resources are now available to place this goal within reach.

EXAPTATION AND THE ORIGIN OF ADULT THORACIC HORNS

PCD-mediated resorption of pupal thoracic horn tissue is ubiquitous among *Onthophagus* species, raising the question as to the adaptive significance, if any, of such transient horn expression. Why expend all this energy to build a conspicuous pupal outgrowth if seemingly all that happens to it is its subsequent removal through PCD? Experimental approaches have now revealed that pupal horns, regardless of whether they are resorbed or converted into an adult structure, have a crucial role during the larval-to-pupal molt and especially the removal of the larval head capsule (Moczek et al. 2006). Unlike in larval-to-larval and pupal-to-adult molts, larvae that molt into pupae have very little muscle tissue left that could aid in the shedding of the larval cuticle. Instead, animals rely on peristaltic contractions to pump hemolymph and the swallowing of air to inflate selected body regions and to force old cuticle to rupture. This suffices for the removal of the thoracic and abdominal cuticles of larval scarab beetles that are highly membranous and paper-thin. However, shedding the larval head capsule poses much greater challenges because it is composed of extremely thick cuticle. During larval life, this robust cuticle provides important attachment points for the powerful jaw muscles of fiber-feeding scarab larvae, such as *Onthophagus*. Histological studies have now shown that during *Onthophagus*' prepupal stage, thoracic horn primordia force themselves into the space vacated between the larval head capsule and corresponding epidermis, fill with hemolymph, and expand. This expansion forces the larval head capsule to fracture along preexisting lines of weakness. As a consequence, as the larval head molts into a pupal head, the first pupal structure visible from the outside is not a part of the head, but the thoracic horn primordium as it bursts through the head capsule. When the precursor cells that would normally give rise to thoracic horn primordia are removed before the prepupal stage, the resulting pupae not only lacked a thoracic horn, but also failed to shed their larval head capsule (Moczek et al. 2006). Replicating this approach across *Onthophagus* species as well as outside the genus showed that this putative dual function of thoracic horn primordia appears to be unique to onthophagine beetles. Phylogenetic analyses suggested that the pupal molting function of horns may have preceded the horns-as-a-weapon function of the adult counterparts and that ancestrally, pupal horns may have always been resorbed before the adult molt (Moczek et al. 2006). If correct, this would explain why prepupal thoracic horn growth has been maintained in so many *Onthophagus* species even though the resulting pupal horns are not used to form a functional structure in the adult.

These results also raise the possibility that the first origin of *adult* horns could have involved a simple failure to remove otherwise pupal-specific projections via PCD. Anecdotal evidence suggests that such events occur in natural populations frequently enough to be detected by entomologists (see, e.g., Paulian 1945; Ziani 1994; Ballerio 1999). Although such a failure would have resulted in an adult outgrowth that at first would have been rather small, behavioral studies have shown that even very small increases in horn length are sufficient to bring about significant increases in fighting success and fitness (Emlen 1997; Moczek and Emlen 2000). Behavioral studies have also shown that fighting behavior is widespread among beetles, occurring well outside horned taxa, and that possession of horns is not a prerequisite for fighting. Combined, this may have created a selective environment in which the first pupal horn that failed to be removed before the pupal-to-adult molt could have provided an immediate fitness advantage. Thoracic beetle horns may thus be a good example of a novelty that arose via exaptation from traits originally selected for providing a very different function at a different stage of development. However, it is important to keep in mind that none of these arguments holds up for other horn types such as head horns. Head horns, at least in *Onthophagus*, undergo little to no remodeling, and morphological differences among adults are already largely established in the preceding pupal stage (Moczek 2007). These basic differences further underscore the likely evolutionary and developmental independence already highlighted above that characterizes different types of horns and most likely different lineages of horned beetles (Emlen et al. 2007). At the same time, the presumed origin of adult thoracic horns from ancestral molting devices vividly illustrates the crooked routes that developmental evolution can take as it generates what in the end may be perceived as an evolutionary novelty. The same complexity in the interactions among development, morphology, and ecology emerges when our emphasis is shifted away from the origin and diversification of *horns* and toward the diversification of *horned beetles*, as the last section of this chapter hopes to illustrate.

TRADE-OFFS DURING DEVELOPMENT AND EVOLUTION OF HORNED BEETLES

In an important study, Kazuo Kawano (2002) showed that two species of giant southeast Asian rhinoceros beetles (genus *Chalcosoma*) had diverged in both relative horn sizes and copulatory organ sizes and that this divergence was more pronounced among sympatric (overlapping) populations than among allopatric (separated) populations. His findings were fully consistent with reproductive character displacement reinforced in sympatry but not allopatry. What was intriguing, however, was the observation that the species which had evolved relatively longer horns had also evolved relatively shorter copulatory organs, i.e., male horn copulatory organ sizes had coevolved in opposite directions or *antagonistically*. Later experimental work on *O. taurus* (Moczek and Nijhout 2004) suggested that this antagonistic coevolution may not have been a coincidence. In this study, surgical removal of the genital primordia during larval development resulted in males lacking a copulatory organ but they had disproportionately longer horns, suggesting that there may indeed be a connection between how horns and copulatory organs developed, and therefore possibly how they evolved. This was particularly intriguing because changes in male copulatory organs are thought to be closely associated with the origin of reproductive isolation and thus speciation (Eberhard 1985). In fact, copulatory organ morphology is often the only way to distinguish cryptic species, suggesting that whatever mechanism is able to influence how copulatory organs develop in a population may have immediate repercussions for that population's ability to interbreed with others.

A recent study examining both within and between species covariation in horn versus copulatory organ investment provides the strongest evidence to date suggesting exactly that kind of interaction between horn and copulatory organ evolution (Parzer and Moczek 2008). Specifically, this study focused first on three rapidly diverging exotic *O. taurus* populations. These populations were introduced from their native Mediterranean range to the eastern United States as well as to eastern and western Australia <50 years ago and since then had evolved significant differences in male horn investment. Ecological studies are consistent with the hypothesis that this diversification was driven by selection acting directly on horn expression, rather than other traits. As a result, present-day western Australian males grow the relatively shortest horns of any population, whereas eastern United States males grow the relatively longest, with the other two populations (eastern Australia and Mediterranean) being intermediate (Moczek 2003; Moczek and Nijhout 2003). The study showed that across these four populations, there was a perfect negative correlation between relative investment into horns and copulatory organ size (Fig. 5a,b). Western Australian males invested the least in horns but by far the most in copulatory organ size, whereas the relationship was reversed for eastern United States males, and intermediate for males from the other two populations.

As a second step, the study applied the same approach to nine different *Onthophagus* species, and the same highly significant negative correlation between relative investment into horns and copulatory organ size emerged (Fig. 5c). Importantly, the greatest differences observed between *populations* were similar in nature and magnitude to some of the differences detected between *species*. Combined, these results had three important implications. First, they suggest that copulatory organ size may, under certain circumstances, diverge as a by-product of evolutionary changes occurring in horns. Second, the resulting signatures of antagonistic coevolution detectable during both very recent divergences between populations and macroevolutionary divergence between species suggested that this tradeoff can bias evolutionary trajectories over a range of phylogenetic distances. Third, and most remarkable, given the significance of copulatory organ morphology for the evolution of reproductive isolation, these findings raise the possibility that horn diversification may promote speciation as a by-product. If correct, this might help to explain how the genus *Onthophagus*, famous for its dramatic diversity in patterns of horn expression, was able to radiate into more than 2400 extant species, making it the most speciose genus in the animal kingdom (Arrow 1951).

CONCLUSIONS

I have argued here that the origin and diversification of horns were mediated by widespread cooption of preexist-

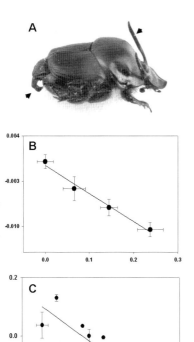

Figure 5. Trade-offs between primary and secondary sexual characters in populations and species of *Onthophagus* beetles. (*A*) Horned male *O. taurus*. Arrows highlight horns and copulatory organs. (*B*) Relative investment into copulatory organ size as a function of relative investment into horn size in four different populations of *O. taurus*. Error bars represent one standard error. (*C*) Relative investment into copulatory organ size as a function of relative investment into horn size in nine different *Onthophagus* species. Data are corrected for differences in body size. (Modified from Parzer and Moczek 2008.)

ing developmental mechanisms into a new spatiotemporal context, exaptation of preexisting structures that originally performed unrelated functions, and trade-offs arising during development that bias, and possibly accelerate, patterns of diversification. Beetle horns and horned beetles emerge as a rich microcosm within which to explore the mechanisms underlying organismal innovation and diversification. Combining ecological richness and morphological exuberance with increasing accessibility to genetic and developmental manipulation alongside the development of genomic resources, horned beetles emerge as an increasingly powerful model system in evo-devo and eco-devo research.

ACKNOWLEDGMENTS

I thank Cold Spring Harbor Laboratories for organizing a fantastic and memorable conference. Research presented here was supported by National Science Foundation grants IOS 0445661 and IOS 0718522 to A.P.M.

REFERENCES

Angelini D, Kaufman TC. 2004. Functional analyses in the hemipteran *Oncopeltus fasciatus* reveal conserved and derived aspects of appendage patterning in insects. *Dev Biol* **271:** 306–321.

Arrow GH. 1951. *Horned beetles*. W. Junk, The Hague, The Netherlands.

Ballerio A. 1999. Revision of the genus *Pterorthochroaetes* first contribution (Coleoptera: Scarabaeoidea: Ceratocanthidae). *Folia Heyrovskyana* **7:** 221–228.

Balthasar V. 1963. *Monographie der Scarabaeidae und Aphodiidae der palaearktischen und orientalischen Region (Coleoptera: Lamellicornia). Band 2, Coprinae*. Verlag der tschechoslowakischen Akademie der Wissenschaften, Prague.

Bessa J, Gebelein B, Pichaud F, Casares F, Mann RS. 2002. Combinatorial control of *Drosophila* eye development by Eyeless, Homothorax, and Teashirt. *Genes Dev* **16:** 2415–2427.

Cullen K, McCall K. 2004. Role of programmed cell death in patterning the *Drosophila* antennal arista. *Dev Biol* **275:** 82–92.

Davidson EH, Erwin DH. 2006. Gene regulatory networks and the evolution of animal body plans. *Science* **311:** 796–800.

Eberhard WG. 1985. *Sexual selection and animal genitalia*. Harvard University Press, Cambridge, MA.

Emlen DJ. 1997. Alternative reproductive tactics and male dimorphism in the horned beetle *Onthophagus acuminatus*. *Behav Ecol Sociobiol* **41:** 335–341.

Emlen DJ, Corley Lavine L, Ewen-Campen B. 2007. On the origin and evolutionary diversification of beetle horns. *Proc Natl Acad Sci* **104:** 8661–8668.

Erwin DH. 2000. Macroevolution is more than repeated rounds of microevolution. *Evol Dev* **2:** 78–84.

Kawano K. 2002. Character displacement in giant rhinoceros beetles. *Am Nat* **159:** 255–271.

Kijimoto T, Costello J, Tang Z, Moczek AP, Andrews J. 2009. Candidate genes for the development and evolution of beetle horns. *BMC Genomics* (in press).

Kojima T. 2004. The mechanism of *Drosophila* leg development along the proximodistal axis. *Dev Growth Differ* **46:** 115–129.

Moczek AP. 2003. The behavioral ecology of threshold evolution in a polyphenic beetle. *Behav Ecol* **14:** 831–854.

Moczek AP. 2005. The evolution and development of novel traits, or how beetles got their horns. *BioScience* **55:** 937–951.

Moczek AP. 2006a. Integrating micro- and macroevolution of development through the study of horned beetles. *Heredity* **97:** 168–178.

Moczek AP. 2006b. Pupal remodeling and the development and evolution of sexual dimorphism in horned beetles. *Am Nat* **168:** 711–729.

Moczek AP. 2007. Pupal remodeling and the evolution and development of alternative male morphologies in horned beetles. *BMC Evol Biol* **7:** 151.

Moczek AP. 2008. On the origin of novelty in development and evolution. *BioEssays* **5:** 432–447.

Moczek AP. 2009. The origin and diversification of complex traits through micro- and macro-evolution of development: Insights from horned beetles. *Curr Top Dev Biol* **86:** 135–162.

Moczek AP, Emlen DJ. 2000. Male horn dimorphism in the scarab beetle *Onthophagus taurus:* Do alternative reproductive tactics favor alternative phenotypes? *Anim Behav* **59:** 459–466.

Moczek AP, Nagy LM. 2005. Diverse developmental mechanisms contribute to different levels of diversity in horned beetles. *Evol Dev* **7:** 175–185.

Moczek AP, Nijhout HF. 2003. Rapid evolution of a polyphenic threshold. *Evol Dev* **5:** 259–268.

Moczek AP, Nijhout HF. 2004. Trade-offs during the development of primary and secondary sexual traits in a horned beetle. *Am Nat* **163:** 184–191.

Moczek AP, Rose DJ. 2009. Differential recruitment of limb patterning genes during development and diversification of beetle horns. *Proc Natl Acad Sci* **106:** 8992–8997.

Moczek AP, Rose D, Sewell W, Kesselring BR. 2006. Conservation, innovation, and the evolution of horned beetle diversity. *Dev Genes Evol* **216:** 655–665.

Parzer HF, Moczek AP. 2008. Rapid antagonistic coevolution between primary and secondary sexual characters in horned beetles. *Evolution* **62:** 2423–2428.

Paulian R. 1945. *Coléoptère Scarabéides de l'Indochine,* première partie. Faune de l'Empire Français III. Librairie Larose, Paris.

Potten C, Wilson J. 2004. *Apoptosis: The life and death of cells.* Cambridge University Press, Cambridge.

Prpic NM, Wigand B, Damen WG, Klingler M. 2001. Expression of *dachshund* in wild-type and *Distal-less* mutant *Tribolium* corroborates serial homologies in insect appendages. *Dev Genes Evol* **211:** 467–477.

Raff R. 1996. *The shape of life: Genes, development, and the evolution of animal form.* University of Chicago Press, Chicago.

Ryoo HD, Marty T, Casares F, Affolter M, Mann RS. 1999. Regulation of Hox target genes by a DNA bound *Homothorax/Hox/Extradenticle* complex. *Development* **126:** 5137–5148.

Shubin N, Tabin C, Carroll S. 2009. Deep homology and the origins of evolutionary novelty. *Nature* **457:** 818–823.

Svácha P. 1992. What are and what are not imaginal discs: Reevaluation of some basic concepts (Insecta, Holometabola). *Dev Biol* **154:** 101–117.

West-Eberhard MJ. 2003. *Developmental plasticity and evolution.* Oxford University Press, New York.

Yao LC, Liaw GJ, Pai CY, Sun YH. 1999. A common mechanism for antenna-to-Leg transformation in *Drosophila:* Suppression of *homothorax* transcription by four HOM-C genes. *Dev Biol* **211:** 268–276.

Ziani S. 1994. Un interessante caso di teraologia simmetrica in *Onthophagus* (*Paleonthophagus*) *fracticornis* (Coleoptera, Scarabaeidae). *Boll Assoc Rom Entomol* **49:** 165–167.

Genetic Regulation of Mammalian Diversity

R.R. Behringer,[1] J.J. Rasweiler IV,[2] C.-H. Chen,[1] and C.J. Cretekos[3]

[1]*Department of Genetics, University of Texas M.D. Anderson Cancer Center, Houston, Texas 77030;*
[2]*Department of Obstetrics and Gynecology, State University of New York Downstate Medical Center, Brooklyn, New York 11203;* [3]*Department of Biological Sciences, Idaho State University, Pocatello, Idaho 83209*

Correspondence: rrb@mdanderson.org

Mammals have evolved a variety of morphological adaptations that have allowed them to compete in their natural environments. The developmental genetic basis of this morphological diversity remains largely unknown. Bats are mammals that have the unique ability of powered flight. We have examined the molecular embryology of bats and investigated the developmental genetic basis for their highly derived limbs used for flight. Initially, we developed an embryo staging system for a model chiropteran, *Carollia perspicillata*, the short-tailed fruit bat that has subsequently been used for staging other bat species. Expression studies focusing on genes that regulate limb development indicate that there are similarities and differences between bats and mice. To determine the consequences of these expression differences, we have conducted an enhancer switch assay by gene targeting in mouse embryonic stem cells to create mice whose genes are regulated by bat sequences. Our studies indicate that *cis*-regulatory elements contribute to the morphological differences that have evolved among mammalian species.

The approximately 4800 extant species of the class Mammalia exhibit remarkable morphological and physiological diversity (Behringer et al. 2006). Most of these live on land, whereas others are born and live their entire lives swimming in aquatic environments or are capable of powered flight exploiting aerial niches. We are interested in the developmental genetic basis of the morphological and physiological variation observed among mammalian species.

Mammals are divided into three subclasses: the oviparous (egg-laying) monotremes, the viviparous (live-bearing) marsupials, and the eutherians. Most mammalian species are contained in the eutherian subclass, with a majority belonging to two orders: the Rodentia (~2000 species) and the Chiroptera or bats (~1200 species). We have focused our developmental studies on one of the bats, while also exploiting some of the numerous advantages of the laboratory mouse model.

The chiropterans have traditionally been divided into the megachiroptera and microchiroptera, i.e., simplistically, big bats and small bats. However, the biggest microchiropteran is larger than the smallest megachiropteran. Bats are widely dispersed throughout the world except at the poles. They feed on a wide variety of food sources, including pollen, nectar, fruits, other plant parts (flowers, leaves), insects, spiders, blood, and vertebrates. Many species have the ability of echolocation, using sound to "visualize" their environments and locate food. Among the mammals, bats have evolved the unique ability of powered flight that relies on highly modified limbs. This ability is distinct from other mammals such as flying squirrels (subfamily Petauristinae) and flying lemurs (order Dermoptera) that glide rather than actively sustain flight. The evolution of flight by bats allowed them to exploit a unique niche, the night sky.

The primary model we have chosen for most of our studies is a microchiropteran, *Carollia perspillata* (the short-tailed fruit bat). *Carollia* possesses many useful attributes as a model organism for studies of the Chiroptera. It is probably the most abundant mammal inhabiting forested areas in the lowland tropics of the New World and frequently establishes colonies in a variety of man-made structures, from which animals may be easily collected with hand nets. *Carollia* also readily adapts to captivity in a research setting. Indeed, it is feasible to accomplish this with no or negligible mortalities from the time of initial capture. In captivity, these animals are maintained in modest-sized cages that permit flight, but still facilitate efficient and noninjurious capture for experimental purposes. They are fed a fruit-based diet readily prepared from inexpensive canned and powdered components (Rasweiler et al. 2009). Because *Carollia* breeds very successfully in captivity with this husbandry program, self-sustaining colonies can be easily established. Finally, pregnancies can be conveniently timed from the first appearance of spermatozoa in daily vaginal smears. This has permitted the accumulation of substantial information on the timing of reproductive and developmental events in captive-maintained animals (Badwaik et al. 1997; Rasweiler et al. 2000; Cretekos et al. 2005). Using embryos derived from timed matings from our laboratory colony, we generated a standardized embryo staging system for *Carollia* (Cretekos et al. 2005), which has subsequently been modified for use with other species of bats (Giannini et al. 2006; Tokita 2006; Hockman et al. 2009; Nolte et al. 2009). These various factors have enabled our studies on the genetic basis of limb diversity among mammals.

LIMB DIVERSITY AND DEVELOPMENT BETWEEN MICE AND BATS

Comparisons of the forelimbs of adult mice and bats show significant differences in morphology (Fig. 1). The upper and lower arms of the mouse are approximately

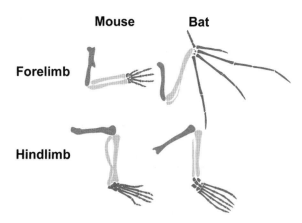

FIGURE 1. Limb comparisons between mouse and bat. Diagram of the limb skeletal elements of the mouse and bat.

equal in length, whereas in the bat the lower arm is about twice as long as the upper arm. The mouse has digits that are separated due to programmed cell death of interdigital tissue during embryogenesis. In the bat, although the first digit of the forelimb is free of interdigital tissue, the four posterior digits (II–V) are connected by the wing membrane. In addition, digits II–V are highly elongated. In contrast, the relative lengths of the segments of the hindlimbs of the mouse and bat are similar. Another bat-specific feature includes a significant reduction of the ulna relative to that of the mouse. These morphological adaptations promote the quadrapedal and volant lifestyles of the mouse and bat, respectively.

Perhaps not surprisingly, the embryonic staging systems for all bat species examined so far are very similar in terms of the relative timing, pattern, and initial appearance of major anatomical features (Giannini et al. 2006; Tokita 2006; Hockman et al. 2009; Nolte et al. 2009). Species-specific differences in the shapes and sizes of some structures, such as craniofacial structures, limbs, and tail, appear after initial morphogenesis. In particular, limb outgrowth and early morphogenesis appear to be highly conserved among all bat species examined. Differences in wing proportions corresponding to different flight behaviors of the various species appear during the last third of development (Giannini et al. 2006; Tokita 2006; Hockman et al. 2009; Nolte et al. 2009). For example, digits II–V are spaced widely in the anteroposterior axis, with abundant wing membrane between them in *C. perspicillata* and *Pipistrellus abramus* (Cretekos et al. 2005; Tokita 2006), whereas digits II–IV are relatively closer together, with less interdigital tissue in *Molossus rufus* forelimbs, reflecting the relatively broad wing shape in the former species and the narrower wing shape of *M. rufus* (Nolte et al. 2009).

When limb morphogenesis is compared between bats and mice, a similar progression from early similarity to later differences is observed (Fig. 2). Limb buds form in both mice and bats at a similar stage of development and are morphologically similar in appearance during the early bud stage. Shortly after the formation of the forelimb handplate, the first overt differences appear, with the bat developing a proportionally larger handplate that is posteriorly expanded relative to the symmetric mouse handplate. During subsequent stages, programmed cell death occurs among all of the digits in the mouse, whereas only the interdigital tissue between digits I and II is lost in bat. The remaining interdigital tissue persists and later develops into the chiropatagium or "handwing" of the bat. Finally, during later stages of development, the bat forelimb shows a distinct proximal-to-distal bias in skeletal elongation, whereas the mouse limb skeleton elongates in a relatively uniform proximodistal pattern.

Formation of the axial skeleton begins during *Carollia* stage (CS) 14, with alcian-blue-positive cartilaginous condensations forming in the midline just ventral to the

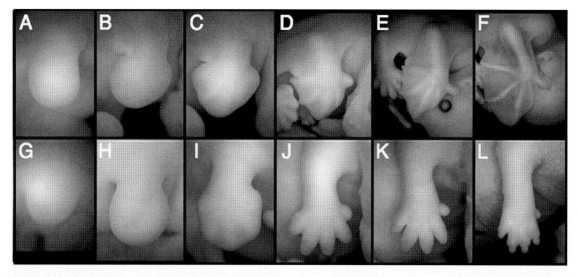

FIGURE 2. Limb development in the bat and mouse. Whole-mount images of forelimbs from bat (*A–F*) and mouse (*G–L*). *Carollia* stages 14 (*A*), 15E (*B*), 15 (*C*), 16 (*D*), 16L (*E*), 19 (*F*); mouse stages 10 dpc (*G*), 11 dpc (*H*), 12 dpc (*I*), 13 dpc (*J*), 14 dpc (*K*), 16 dpc (*L*). Anterior is to the right. (Reprinted, with permission, from Cretekos et al. 2008.)

spinal cord (Fig. 3). During subsequent stages, the axial skeleton elaborates, with neural arches extending dorsally from the vertebral bodies by late-stage CS 15 and closing dorsally over the spinal cord by early-stage CS 17 and ribs extending ventrally beginning during CS 16. The first indication of the appendicular skeleton appears as an alcian-blue-positive condensation in the proximal forelimb during CS 14 (Fig. 3). During subsequent stages, the cartilaginous anlagen of each of the adult appendicular skeletal elements condense sequentially from proximal to distal (Hockman et al. 2008). In the forelimb, the scapula and humerus precursors appear during late-stage CS 14, followed by radius and ulna during early-stage CS 15, carpals and metacarpals by late-stage CS 15, and phalanges during CS 16. During subsequent stages, the alcian-blue-staining cartilages are gradually replaced by ossified bone, again with a proximal-to-distal temporal bias, starting in the central region of each element and spreading with time toward the ends. Hindlimb skeletogenesis, consistent with the timing of limb bud formation, begins about one stage later than that of the forelimb, but catches up by CS 17 (Fig. 3).

Analysis of the cartilaginous skeleton of a stage-17 *Carollia* embryo reveals that the forming skeletal elements are relatively similar to the mouse, i.e., the upper and lower arm bones are of approximately equal length and each of the five digits are also approximately equal in length (Fig. 4A). However, by the time of birth, the forelimb skeletal elements have taken on the bat-specific morphological characters of the adult, including bone lengthening and ulna reduction (Fig. 4B). These findings suggest that limb form initially follows a general vertebrate pattern of development but, at later stages of gestation, species-specific characters emerge.

GENE EXPRESSION DURING BAT LIMB DEVELOPMENT

Many genes have been identified, predominantly in the chick and mouse systems, that are expressed in unique temporal and spatial patterns during vertebrate limb development. Their functions in limb patterning and morphogenesis have been determined for a subset of these genes. The expression of some of these genes have also been determined during bat limb development, including *Hoxd13* (Chen et al. 2005; Ray and Capecchi 2008), *Fgf8* (Weatherbee et al. 2006; Cretekos et al. 2007), *Bmp2*, *Bmp4*, *Bmp7*, *Msx2*, *Gremlin*, and *Spry2* (Weatherbee et al. 2006), phospho-Smad1/5/8 (Sears et al. 2006), *Prx1* (Cretekos et al. 2008), and *Shh* and *Ptc1* (Hockman et al. 2008). The general findings from these studies are that the expression patterns of orthologous genes in the mouse

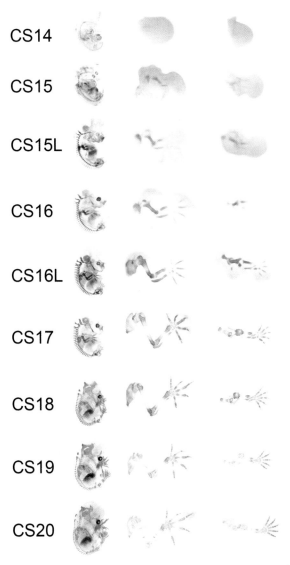

FIGURE 3. Development of the cartilaginous skeleton of *Carollia perspicillatta*. Whole-mount alcian blue staining of bat embryos from CS 14 through CS 20. The first column shows lateral views with anterior to the top and dorsal to the left. The second column shows high-magnification views of dissected forelimbs shown with anterior to the top and proximal to the left. The third column shows high-magnification views of dissected hindlimbs shown with anterior to the top and proximal to the left. Embryos are staged according to Cretekos et al. (2005) and indicated for each row. Views are not to scale.

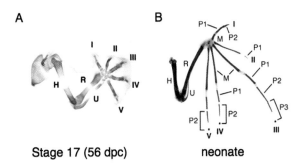

FIGURE 4. Bat-specific characters develop late in gestation. *Carollia* forelimb skeletons at CS 17 (*A*) and newborn (*B*). (*A*) Alcian blue preparation, (*B*) alcian blue/alizarin red preparation. Digits are numbered I–V. (dpc) Days postcoitus, (H) humerus, (M) metacarpel, (P) phalange, (R) radius, (U) ulna. (*B*, Reprinted, with permission, from Chen et al. 2005 [© Wiley].)

and bat are generally similar, but many times, species-specific differences occur notably at later stages of embryogenesis. For example, *Hoxd13* is expressed in the posterior region of the developing mouse limb bud. Although *Hoxd13* expression in the bat forelimb is initially similar to that of the mouse, at later stages the anterior boundary of expression is shifted to a more posterior position relative to mouse *Hoxd13* (Chen et al. 2005). Likewise, *Fgf8* expression is detected in the apical ectodermal ridge (AER) of the forming mouse and bat limb buds. In the bat, the width of *Fgf8* AER expression is significantly greater than that of the mouse (Cretekos et al. 2007). These expression studies provided a foundation for assaying the function of bat genes in the mouse.

PRX1

Prx1 (also called *MHox* or *Prrx1*) encodes a paired-class homeodomain transcription factor. The mouse locus produces two isoforms by alternative splicing, encoding proteins that differ only in the carboxy-terminal region. Alignment of the deduced amino acid sequences of *Carollia* PRX1 isoforms shows more than 99% identity with the orthologous isoforms of mouse PRX1 (Cretekos et al. 2008). Only two amino acid differences are found between bat and mouse PRX1 proteins, and these occur outside of the conserved homeodomain, PRX and OAR motifs, suggesting that the orthologs have identical biochemical activities.

Prx1 is expressed in a complex and dynamic pattern in the limb from the early undifferentiated bud stage through outgrowth and morphogenesis, suggesting roles in multiple steps of limb development (Fig. 5A) (Cserjesi et al. 1992; Martin et al. 1995). Expression of *Prx1* in the developing bat limb shows a striking up-regulation in the distal limb compared to mouse (Cretekos et al. 2008). Analysis of the mouse *Prx1* knockout (null) phenotype demonstrates an essential role in the morphogenesis of limb and craniofacial tissues. The limbs of mice with *Prx1* function eliminated display shortened zeugopod (forearm and lower leg) elements and numerous autopod (hand and foot) element reductions and malformations (Fig. 5B) (Martin et al. 1995; ten Berge et al. 1998; Lu et al. 1999).

A region in which sequences capable of regulating expression in the limb reside, just upstream of the start of transcription, had been previously identified (Martin and Olson 2000). We identified the bat ortholog of the *Prx1* limb-specific enhancer by virtue of its conserved sequence and genomic position relative to mouse and human orthologs, and characterized the activity of this enhancer by combining it with a basal promoter and *lacZ* reporter in transgenic mice. The *Carollia Prx1* limb enhancer shares ~80% sequence identity with that of mouse and directs the expression of the reporter in a pattern similar to that of the orthologous mouse enhancer in the limbs of transgenic mice (Cretekos et al. 2008).

ENHANCER SWITCH AND DELETION BY GENE TARGETING

The challenge in studying the evolution of gene regulation is to distinguish between unselected genetic drift and significant functional divergence in noncoding *cis*-regulatory sequences (Cretekos et al. 2001). To address this challenge, we replaced the mouse *Prx1* limb enhancer with the bat limb enhancer within the endogenous locus by gene targeting in mouse embryonic stem (ES) cells and generated mice from these ES cells (*Prx1^{BatE}*) (Cretekos et al. 2008). We replaced the limb enhancer sequences without altering the coding sequence; thus, any change in *Prx* function in the *Prx1^{BatE}* mice is directly attributable to the bat regulatory sequences. Heterozygous and homozygous mice for the bat *Prx1* limb enhancer allele are viable, fertile, segregate in the expected Mendelian ratio, and show no readily apparent limb or craniofacial phenotypes. However, skeletal analysis of *Prx1$^{BatE/BatE}$* mice reveals that the forelimbs of these animals are about 6% longer than their wild-type littermates by the end of gestation, and this difference is statistically significant. Histological and molecular analyses indicate that bat enhancer-driven expression of *Prx1* results in accelerated endochondral bone growth in the mutant forelimbs, caused at least in part by increased proliferation of chondrocyte precursors at the ends of the long bones. Quantitative reverse transcriptase–polymerase chain reaction (Q-RT-PCR) analysis of 17.5-day postcoitus (dpc) humerus from *Prx1^{BatE}* homozygotes showed ~1.7-fold higher levels of *Prx1* transcripts, suggesting that *Prx1^{BatE}* is a hypermorphic allele.

To test whether this *Prx1* limb enhancer is essential for limb-specific transcription, we also generated mice lacking it using a similar gene targeting strategy. To our surprise, mice homozygous for the *Prx1* enhancer deletion allele are viable, fertile, and do not exhibit the limb defects of *Prx1* null mutants. Q-RT-PCR analysis of 17.5-dpc humerus from enhancer deletion homozygotes showed no significant difference in *Prx1* RNA levels in comparison to controls. Thus, although the mouse limb enhancer is

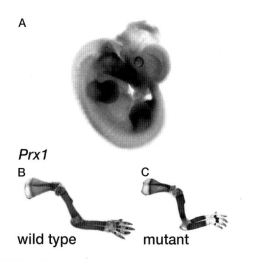

FIGURE 5. *Prx1* expression and mutant forelimb phenotype in the mouse. (*A*) *Prx1* whole-mount in situ hybridization of 11.5-dpc mouse embryo. Expression is detected predominantly in the craniofacial region and limbs. (*B,C*) Alcian blue/alizarin red forelimb skeleton preparations from wild-type (*B*) and *Prx1$^{-/-}$* (*C*) newborn mice.

sufficient to direct limb-specific expression of heterologous genes, it is not required for limb transcription.

These results show that divergence of a single nonessential *cis*-regulatory region of one gene is by itself sufficient to elicit a discrete and measurable change in limb morphogenesis. They also suggest that regulatory redundancy may be part of a general mechanism for gain of novel expression patterns among conserved essential functions. In addition to *Prx1*, the mouse genome contains a *Prx2* gene, most likely a result of gene duplication. *Prx2* is also expressed in the mouse limb. *Prx2* homozygous mutant mice are viable with normal limbs (ten Berge et al. 1998). However, *Prx1;Prx2* double-mutant mice have severely shortened limbs, suggesting that *Prx1* and *Prx2* interact to regulate limb length (ten Berge et al. 1998). In our $Prx1^{BatE}$ mutants, it is possible that PRX2 buffers the actions of bat-regulated PRX1 for limb development. These findings suggest that there are multiple levels of redundancy for buffering the effects of mutations on genes that regulate developmental processes.

CONCLUSIONS

The bat provides an excellent mammalian model system to explore the developmental genetics of animal diversity. The development of *C. perspicillata* as a model system has greatly facilitated research in the reproductive biology and embryology of bats. There are many morphological and physiological characteristics of developing bat embryos that show diversity relative to primary model organisms, such as the chick and mouse, beyond the limbs. For example, studying their craniofacial development would be a very exciting avenue for future research. Gene expression studies of developing bat embryos are still very limited. It would be useful to have a comprehensive analysis of gene expression in the developing limbs of bats. Such studies would likely generate candidate genes that are responsible for the generation of a wing. The reproductive biology of most bats, i.e., a relatively long gestation resulting in a single offspring and length to sexual maturity, makes the possibility of genetic studies, at best, unlikely. Thus, assaying the activity of bat sequences in the mouse with its tremendous genetic advantages can be a productive way to identify factors that lead to divergent morphologies among species.

ACKNOWLEDGMENTS

We are grateful for the kindness and assistance given to us by the many people of Trinidad who have helped us to find bat roosts, loaned us ladders, fed us wonderful food, and educated us about West Indian natural history and culture. This work was supported by National Science Foundation grant IBN 0220458 and the Ben F. Love Endowment to R.R.B.

REFERENCES

Badwaik NK, Rasweiler JJ IV, Oliveira SF. 1997. Formation of reticulated endoderm, Reichert's membrane, and amniogenesis in blastocysts of captive-bred, short-tailed fruit bats, *Carollia perspicillata*. *Anat Rec* **247**: 85–101.

Behringer RR, Eakin GS, Renfree MB. 2006. Mammalian diversity: Gametes, embryos, and reproduction. *Mol Reprod Dev* **18**: 99–107.

Chen C-H, Cretekos CJ, Rasweiler JJ IV, Behringer RR. 2005. *Hoxd13* expression in the developing limbs of the short-tailed fruit bat, *Carollia perspicillata*. *Evol Dev* **7**: 130–141.

Cretekos CJ, Rasweiler JJ IV, Behringer RR. 2001. Comparative studies on limb morphogenesis in mice and bats, a functional genetic approach towards a molecular understanding of diversity in organ formation. *Reprod Fertil Dev* **13**: 691–695.

Cretekos CJ, Weatherbee SD, Chen C-H, Badwaik NK, Niswander L, Behringer RR, and Rasweiler JJ IV. 2005. Embryonic staging system for the short-tailed fruit bat, *Carollia perspicillata*, a model organism for the mammalian order *Chiroptera*, based upon timed pregnancies in captive-bred animals. *Dev Dyn* **233**: 721–738.

Cretekos CJ, Deng JM, Green ED, NISC Comparative Sequencing Program, Rasweiler JJ IV, Behringer RR. 2007. Isolation, genomic structure and developmental expression of Fgf8 in the short-tailed fruit bat, *Carollia perspicillata*. *Int J Dev Biol* **51**: 333–338.

Cretekos CJ., Wang Y, Green ED, NISC Comparative Sequencing Program, Martin JF, Rasweiler JJ IV, Behringer RR. 2008. Regulatory divergence modifies forelimb length in mammals. *Genes Dev* **22**: 141–151.

Cserjesi P, Lilly B, Bryson L, Wang Y, Sassoon DA, Olson EN. 1992. MHox: A mesodermally restricted homeodomain protein that binds an essential site in the muscle creatine kinase enhancer. *Development* **115**: 1087–1101.

Giannini N, Goswami A, Sanchez-Villagra MR. 2006. Development of integumentary structures in *Rousettus amplexicaudatus* (Mammalia: Chiroptera: Pteropodidae) during late-embryonic and fetal stages. *J Mammalogy* **87**: 993–1001.

Hockman D, Mason MK, Cretekos CJ, Behringer RR, Jacobs DS, Illing N. 2008. A second wave of *Sonic Hedgehog* expression during the development of the bat limb. *Proc Natl Acad Sci* **105**: 16982–16987.

Hockman D, Mason MK, Jacobs DS, Illing N. 2009. The role of early development in mammalian limb diversification: A descriptive comparison of early limb development between the Natal long-fingered bat (*Miniopterus natalensis*) and the mouse (*Mus musculus*). *Dev Dyn* **238**: 965–979.

Lu MF, Cheng HT, Lacy AR, Kern MJ, Argao EA, Potter SS, Olson EN, Martin JF. 1999. Paired-related homeobox genes cooperate in handplate and hindlimb zeugopod morphogenesis. *Dev Biol* **205**: 145–157.

Martin JF, Olson EN. 2000. Identification of a *prx1* limb enhancer. *Genesis* **2**: 225–229.

Martin JF, Bradley A, Olson EN. 1995. The *paired*-like homeo box gene *MHox* is required for early events of skeletogenesis in multiple lineages. *Genes Dev* **9**: 1237–1249.

Nolte MJ, Hockman D, Cretekos CJ, Behringer RR, Rasweiler JJ IV. 2009. Embryonic staging system for the black mastiff bat, *Molossus rufus* (Molossidae), correlated with structure-function relationships in the adult. *Anat Rec* **292**: 155–168.

Rasweiler JJ IV, Oliveira SF, Badwaik S.F. 2000. An ultrastructural study of interstitial implantation in captive-bred, short-tailed fruit bats, *Carollia perspicillata:* Trophoblastic adhesion and penetration of the uterine epithelium. *Anat Embryol* **205**: 371–391.

Rasweiler JJ IV, Cretekos CJ, Behringer RR. 2009. The short-tailed fruit bat *Carollia perspicillata*: A model for studies in reproduction and development. In *Emerging model organisms: A laboratory manual*, vol. 1, pp. 519–555. Cold Spring Harbor Laboratory Press, Cold Spring Harbor, NY.

Ray R, Capecchi M. 2008. An examination of the chiropteran HoxD locus from an evolutionary perspective. *Evol Dev* **10**: 657–670.

Sears KE, Behringer RR, Rasweiler JJ IV, Niswander LA. 2006. The development of bat flight, morphological and molecular evolution of bat wing digits. *Proc Natl Acad Sci* **103**: 6581–6586.

ten Berge D, Brouwer A, Korving J, Martin JF, Meijlink F. 1998.

Prx1 and *Prx2* in skeletogenesis: Roles in the craniofacial region, inner ear and limbs. *Development* **125**: 3831–3842.

Tokita M. 2006. Normal embryonic development of the Japanese pipistrelle, *Pipistrellus abramus*. *Zoology* **109**: 137–147.

Weatherbee SD, Behringer RR, Rasweiler JJ IV, Niswander LA. 2006. Interdigital webbing retention in bat wings illustrates genetic changes underlying amniote limb diversification. *Proc Natl Acad Sci* **103**: 15103–15107.

Genomics, Domestication, and Evolution of Forest Trees

R. SEDEROFF,[1] A. MYBURG,[2] AND M. KIRST[3]

[1]*Forest Biotechnology Group, North Carolina State University, Raleigh, North Carolina 27695;* [2]*Department of Genetics, Forestry and Agricultural Biotechnology Institute, University of Pretoria, Pretoria, 0002, South Africa;* [3]*School of Forest Resources and Conservation, University of Florida, Gainesville, Florida 32611*

Correspondence: ron_sederoff@ncsu.edu

The forests of the world continue to be threatened by climate change, population growth, and loss to agriculture. Our ability to conserve natural forests and to meet the increasing demand for fuel, biomass, wood, and paper depends on our fundamental understanding of tree growth and adaptation (FAO 2001; Fenning and Gershenzon 2002; Campbell et al. 2003; Gray et al. 2006). Our knowledge of the unique biology of trees will be greatly advanced through the application of genomics. The purpose of this chapter is to describe this emergent genomic paradigm as it is being applied to trees.

Genomics provides a platform to learn the relationships of genes and phenotypes and to integrate molecular and quantitative genetics for the analysis of adaptive traits (Frewen et al. 2000; Wullschleger et al. 2002; Brunner et al. 2004). Genomics provides answers to fundamental questions of tree biology and addresses practical problems in tree plantations, forest management, and conservation. Genomic association of natural allelic variation with improved growth and wood properties is being incorporated into tree-breeding programs through allele-specific evaluation and selection (Neale 2007; Grattapaglia et al. 2009). The following are four main areas of genome-related investigation in forest trees.

1. **Fundamental biology of growth, metabolism, and development.** Genomic analysis elucidates the mechanisms regulating the vascular cambium, the lateral meristem (stem cell layer) that produces wood. There is also considerable interest in the control of flowering and dormancy, because of the adaptive significance of these traits. Abundant flowering and short growing seasons reduce wood yield and biomass. The ability to induce flowering can accelerate breeding.

2. *Abiotic stress.* Our understanding of the mechanisms of tree responses to abiotic stress can be expanded by genomic analysis. Stresses may be due to water and temperature or the effects of environmental chemistry or environmental pollution. Abiotic stress is a major consequence of climate change, and the understanding of the genetic architecture of stress responses is essential for predicting adaptation in forests.

3. *Pests and pathogens.* Understanding the interactions of trees with their symbionts, pathogens, and pests is important for predicting the biological consequences of climate change and for developing more tolerant plantation trees. Trees, pathogens, and pests are co-evolving systems that often determine the fates of their ecosystems. Pest and pathogen interactions can now be explored at a genomic level to ameliorate ecosystem threats (Steiner and Carlson 2004; Ralph et al. 2006a,b; Whitham et al. 2006).

4. *Adaptation and evolution.* Genomics provides valuable information about genetic diversity of trees, its distribution, and effects on adaptation and evolution (Neale and Ingvarsson 2008). This knowledge is being applied to forest management and conservation in the context of past history and predicted change in climate.

Molecular genetic studies of forest trees have followed three main approaches: (1) Comparative analysis of trees with *Arabidopsis* and model crops has been used to identify functional genetic conservation and to expand our understanding of the similarities and differences between woody and herbaceous plants. (2) Transgenic technology in trees has allowed a reverse genetics approach, using activation or suppression of gene expression to investigate function. (3) Forward genetics approaches, based on quantitative and qualitative analysis of natural variation, have been used to understand the genetic architecture and molecular basis of specific traits.

FOREST TREE DIVERSITY AND DOMESTICATION

In contrast to many of our herbaceous crops that have been bred and selected for thousands of generations, forest trees are essentially undomesticated. They are also characterized by high genetic diversity and high genetic load (Hamrick et al. 1992; Remington and O'Malley 2000). An exception, red pine, exhibits low genetic diversity, attributed to a bottleneck in its recent evolution (DeVerno and Mosseler 1997). Clones of quaking aspen (*Populus tremuloides*) are found in massive monoclonal stands, consisting of tens of thousands of genetically identical trees, all connected through a massive underground root system. One such clone has been described as the largest single plant and potentially the largest organism on earth (Mitton and Grant 1996).

The high levels of genetic diversity can be exploited for tree breeding (Zobel and Talbert 1984). Trees used for plantation forestry are very much like their natural counterparts. Early domestication in annual crops such as maize is thought to have involved relatively small numbers of genetic events (Doebley et al. 1997; Doebley and Lukens 1998), followed by rapid changes in allele frequencies at loci that affect plant architecture and biochemistry (Lev-Yadun et al. 2002; Jaenicke-Després et al. 2003; J. Doebley, pers. comm.). Major genetic changes of a kind that have accompanied domestication of cereals, for example, have not yet occurred in forest trees. It is not yet clear what kind of genetic changes might become associated with tree domestication (Boerjan 2005).

WOOD DEVELOPMENT AND EVOLUTION

Wood provides the basis for the mechanical support and water transport that allow trees to attain great age and size. Wood formation (xylogenesis) takes place through cell division and differentiation of the vascular cambium, a secondary lateral meristem in the stems, branches, and roots of woody plants. Cambial initials divide and differentiate into a vertical support and water transport system consisting of tracheids in gymnosperms or fibers and vessels in angiosperms. A lateral transport system is predominantly composed of rays (Larson 1994). Growth rate slows as trees age, thought to be a consequence of the physiological burden of increasing size (Mencuccini et al. 2007). With abundant water and nutrients, absence of biotic and abiotic stresses, and competition for light, trees may have the potential to grow as high as 120 m (Koch et al. 2004). The current tallest tree—a costal redwood— (Hyperion) is 115 m. Records of *Eucalyptus regnans* from the 19th century exceed this height by more than 30 m.

Vascular plants are derived from a rootless leafless freshwater filamentous alga in the genus Cooksonia, ~400 million years ago (mya) (Chaloner and Macdonald 1980). Five extant plant lines derived from this early ancestor are separated into two major groups, the angiosperms and the gymnosperms (Fig. 1). The five lines are the Gnetales, cycads, ginkgos, conifers (all gymnosperms), and angiosperms (Qui et al. 2007). Angiosperms and gymnosperms last shared a common ancestor ~300 mya (Bowe et al. 2000). The angiosperms include the largest number (250,000+) and the most diverse number of species, with both herbaceous and woody plants, and include all the major food crops. The gymnosperms include only ~1000 species, but a small number of those dominate many of the Northern hemisphere's temperate forests. All gymnosperms are woody plants.

Small numbers of genetic changes may be involved in the transition between herbaceous and woody plants. DNA sequence relationships support the relatively recent evolution of woody plants found on islands descending from continental herbaceous founders (Böhle et al. 1996; Helfgott et al. 2000). *Arabidopsis thaliana*, typically small, fast growing, and herbaceous, can be induced to form secondary xylem (wood) by delaying flowering and senescence (Lev-Yadun 1994; Dolan and Roberts 1995; Chaffey et al. 2002). The genes that determine the woody growth habit appear not to be unique to woody plants but are likely to be common genes, regulated in different ways (Groover 2005). A suite of transcription factors that regulate secondary cell wall formation in xylem vessel and fiber cells have recently been characterized in *Arabidopsis* (Zhong et al. 2008) and may perform similar functions in woody plants.

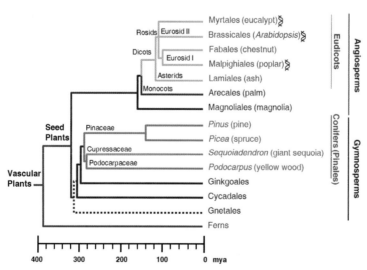

Figure 1. Simplified phylogeny of the major groups of woody plants. Species for which genome sequencing has been completed, or is under way, are indicated by DNA helices. The position of the Gnetales in the seed plant phylogeny has not been resolved. Some studies place Gnetales as a sister group to the Coniferales or the Pinaceae, but a phytochrome phylogeny placed it at the base of the gymnosperms (Chaw et al. 2000; Magallón and Sanderson 2002; Schmidt and Schneider-Poetsch 2002; Burleigh and Mathews 2004; de la Torre et al. 2006).

APPLICATION OF GENOMICS IN FOREST TREES

Although they are large, unwieldy, and have long generation times, many features of trees make them suitable for genomic studies. Trees provide excellent material for genome-wide studies of adaptation and evolution in natural populations. Genomics has also provided insight into the molecular basis of tree growth (Kirst et al. 2004a; Poke et al. 2005; Grattapaglia et al. 2009), development (Hertzberg et al. 2001; Schrader et al. 2004; Ko et al. 2006), adaptation (Yang and Loopstra 2005), and the evolution of herbaceous and woody growth habits (Kirst et al. 2003; Tuskan et al. 2006). The molecular basis of wood formation, the chemical diversity of woody plants, and the basis for extensive longevity in forest trees are topics being addressed. Genomics has also provided a paradigm for the investigation of the molecular basis of complex traits in forest trees, by associating genotype with phenotype. A more complete understanding of this relationship leads to predictive models of growth and adaptation of individual organisms and, perhaps, of species interactions and ecosystems.

Traditionally, forest genetics was built on three approaches: (1) studies of natural variation, particularly, for population and quantitative genetics, (2) micropropagation directed to plantation forestry, and (3) traditional biochemistry for studies of enzymes and metabolites, particularly for secondary metabolism. Early genetic studies of trees focused on quantitative trait variation in populations, using long-term crosses for the purpose of improving breeding material (Zobel and Talbert 1984). Genetic studies of trees focused on isozymes and population genetics (Adams et al. 1992; White et al. 2007).

The advent of DNA-based molecular markers led to the first genomic maps of forest trees (Bradshaw et al. 1994; Devey et al. 1994). Genetic analysis was greatly facilitated by the simplified segregation in the haploid megagametophyte of conifers and by the availability of polymerase chain reaction (PCR)–based molecular markers. This led to single-tree genetic maps (Tulsieram et al. 1992; Grattapaglia and Sederoff 1994; O'Malley et al. 1996) and quantitative trait mapping, e.g., to identify loci that control host resistance to fusiform rust disease in pines (Wilcox et al. 1996). Dozens of genetic maps have since been made for forest trees (Kirst et al. 2004b; http://dendrome.ucdavis.edu).

The next major advance in forest genomics came with the advent of automated DNA sequencing, which enabled high-throughput expressed sequence tag (EST) analysis in trees (Allona et al. 1998; Sterky et al. 1998, 2004; Kirst et al. 2003). A large number of ESTs are now available for forest tree species (Table 1). Microarrays developed from ESTs have been used for functional genomic studies for many of these species.

Gene transfer technology has been developed for many broadleaf species and for some conifers, particularly pine and spruce. The best-developed transformation methods are for poplars. Gene transfer experiments have shown that specific genes can have major effects on wood properties and control of flowering (Strauss et al. 1995; Chiang 2006). Many genes encoding enzymes of the monolignol biosynthetic pathway affect the chemical composition of wood (Chiang 2006).

Table 1. Some Tree Genomes Being Studied

Species	Common name	Genome size (Mbp)[a]	ESTs in NCBI[b] (GenBank)	Physical map (Mbp)	Genome sequence	Properties of interest
Castanea mollissima	Chinese chestnut	800	none	in progress	none	disease resistance
Cryptomeria japonica	Sugi	10,824	19,605	none	none	timber
Eucalyptus camaldulensis	River red gum		none	none	in progress	biomass and pulp
Eucalyptus grandis	Flooded gum	640	1,574	in progress	in progress	biomass and pulp
Eucalyptus globulus (incl. subspecies)	Blue gum	564	3,951	none	none	biomass and pulp
Ginkgo biloba	Ginko	19,462	6,432	none	none	plant evolution and medicine
Gnetum gnemon	Melindjo	3,793	4,323	none	none	plant evolution
Picea abies	Norway spruce	18,228	10,217	none	none	timber
Picea glauca	White spruce	19,796	132,624	none	none	biomass and adaptation
Picea sitchensis	Sitka spruce		139,570	none	none	biomass, adaptation and disease resistance
Pinus elliottii	Slash pine	22,834	150	none	none	timber and pulp
Pinus taeda	Loblolly pine	21,658	329,469	in progress	none	biomass and adaptation
Pinus pinaster	Maritime pine	23,863	27,288	none	none	timber and pulp
Pinus radiata	Monterey pine	21,560	151	none	none	timber and pulp
Pinus sylvestris	Scots pine	27,244	1,689	none	none	timber and pulp
Populus trichocarpa	Black cottonwood	485	89,943	410	draft sequence	model tree
Pseudotsuga menziesii	Douglas fir	18,669	18,142	none	none	adaptation
Quercus robur	Oak	907	1,439	none	none	environmental adaptation
Robinia pseudoacacia	Black locust	2,933	637	none	none	heartwood formation

[a]Genome sizes are only listed where published estimates exist.
[b]Numbers of ESTs listed in NCBI for each species in September 2007.

Comparative genetics has been a powerful tool for forest trees by the identification of tree candidate genes affecting related traits in the better-characterized model systems. This is possible because of the substantial sequence conservation (Kirst et al. 2003) and functional homology (Böhlenius et al. 2006) among genes of vascular plants. Genomic approaches can identify important candidate genes for molecular breeding or genetic engineering. Genes associated with variation in wood chemistry (lignin), wood density, and growth rate have been identified using such approaches (Kirst et al. 2004a; Gonzales-Martinez et al. 2006a; Li et al. 2006).

The most intensively studied forest trees are pines, spruces, poplars, and eucalypts (Table 1; Fig. 1). Within each of these genera, several species and hybrid combinations are widely planted because of their fast growth, broad adaptation to site, and superior wood properties. Many are grown as exotics in different parts of the world. Poplar has emerged as the model system for molecular studies in trees (Bradshaw et al. 2000; Taylor 2002) because of its rapid growth rate, facile propagation from cuttings, well-developed cell and tissue culture, transformability, small genome size, saturated genetic maps, and genome sequence (Tuskan et al. 2006). A chestnut genome is being characterized to advance restoration of American chestnut (*Castanea dentata*), a species highly valued in the 19th century that was essentially lost due to a devastating fungal blight (Steiner and Carlson 2004).

Poplar is a good model tree, but it is not widely planted compared to the pines and the eucalypts. Eucalyptus species and their hybrids are the most widely planted hardwoods. Draft sequencing of the genome of *Eucalyptus camaldulensis* is in progress at the Kazusa DNA Research Institute (http://www.kazusa.or.jp/e/index.html), and the Department of Energy (DOE) Joint Genome Institute expects to complete a draft of the genome sequence of *Eucalyptus grandis* (the rose gum) in 2010 (Myburg et al. 2008). Genomic resources developed for eucalypts include large-insert bacterial artificial chromosome (BAC) genomic libraries, ESTs, and high-resolution genetic maps (http://www.eucagen.org). Similar genomic resources are being developed for other species with the goal of genome sequencing, for example, pine, spruce, and chestnut (Table 1).

Pines and spruces are the most widely planted gymnosperms. Full-genome sequencing is a challenge in conifers due to their large genome sizes and the high proportion of repeated sequences (Kamm et al. 1996; Kossack and Kinlaw 1999; Schmidt et al. 2000; Ahuja and Neale 2005). Genome sizes of conifers range from 6500 Mb to ~37,000 Mb. The loblolly pine genome, estimated at 21,500 Mb, is about seven times larger than the human genome. A BAC library (6x) of loblolly pine has been constructed (Plomion et al. 2007), and pilot scale BAC sequencing projects are under way in the United States and Canada. With new high-throughput sequencing methods (Margulies et al. 2005; Wicker et al. 2006), draft sequences of a pine and a spruce genome are likely to be initiated soon.

INFERRING FUNCTIONAL ROLES FOR GENES IN TREES

Forward Genetics

In model plant systems such as *Arabidopsis*, it is possible to induce and identify mutations in most genes and to infer phenotypic and functional relationships with genes in other plants (Alonso and Ecker 2006). In most forest tree populations, the frequency of recessive mutations is high, but analysis of mutations is limited by lethality, redundancy, and difficulty in the detection of quantitative phenotypes. Due to limited selfing and long juvenility, it is difficult to make heterozygous mutations homozygous in forest trees. Some mutations can be detected in the haploid megagametophytes that surround the embryo in the seed of gymnosperms. For example, a null mutation in the monolignol biosynthetic gene *Cad* was discovered in megagametophytes and was subsequently mapped, sequenced, and characterized (MacKay et al. 1997; Ralph et al. 1997; Gill et al. 2003).

Forward genetics approaches such as population-wide association genetics are feasible in trees because of the high levels of genetic diversity in natural populations and breeding populations (Neale and Savolainen 2004). Experimental populations derived from interspecific hybrids have even higher levels of variation and high levels of linkage disequilibrium (Grattapaglia et al. 1996).

Another forward genetics approach that has been successful in trees is activation tagging (Busov et al. 2003), i.e., using random transgenic insertions of strong enhancers to create dominant hypermorphic phenotypes. When an enhancer is inserted near the transcriptional start site of a gene, ectopic expression of the gene may give rise to an unusual phenotype (Fig. 2). Activation-tagged trees can be screened in the field, and altered phenotypes can be detected by inspection or aerial photography. In hybrid poplar, dominant variants occurred at a frequency of >1% and included a dwarf caused by overexpression of GA 2-oxidase, the major gibberellin catabolic enzyme (Busov et al. 2003).

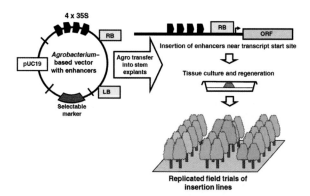

Figure 2. Activation tagging in trees. Explants of leaves or stems are cocultivated with *Agrobacterium* containing a pUC19-based vector carrying multiple (4x) (cauliflower mosaic virus) CaMV 35S enhancer sequences. Ectopic activation of genes is detected in replicated field trials showing altered growth and developmental traits in some insertion lines.

Reverse Genetics

For more than a decade, where candidate genes have been identified, specific transgene expression has been modified in forest trees using antisense or cosuppression (Hu et al. 1999; Li et al. 2003). More recently, suppression by RNA interference (RNAi) (Wang and Waterhouse 2002) has been used in forest trees to perform functional tests. RNAi suppression has been used to test for stability of transgene suppression in trees in the field over time (Meilan et al. 2002; Li 2006). RNAi was used in grey poplar (*Populus canescens*) to suppress isoprene synthase and demonstrate that isoprene protects leaves from thermal and oxidative stress (Behnke et al. 2007). Coleman et al. (2008a) used RNAi suppression of *p*-coumaroyl–CoA 3′ hydrolase (C3H) in hybrid poplar to verify the role of C3H in lignin biosynthesis. The perturbation of lignification reduced tree growth and compromised vascular integrity (Coleman et al. 2008b).

A Novel Method to Test Candidate Genes in Wood Formation: Somatic Sector Analysis in Developing Stem Tissues

The lateral secondary meristem of trees is abundant and accessible and has been used to test gene function by direct transformation of cambial cells in planta (Van Beveren et al. 2006). A segment of bark is carefully removed to provide access to cambial cells that are then inoculated with *Agrobacterium* containing the GUS marker and a candidate gene. The bark is replaced and cambial growth resumes. Subsequently, stably transformed cambial initials differentiate, giving rise to transformed xylem and/or phloem sectors identified by GUS staining. In this way, the role of a β-tubulin gene (*EgrTUB1*) in microfibril angle (MFA) formation was demonstrated in *Eucalyptus* (Spokevicius et al. 2007). The transformed wood segments were macerated, and MFA was determined microscopically from the angle of the pit apertures in the secondary cell walls relative to the long axis of the fibers. MFA is a function of the orientation of cellulose microfibrils in the secondary walls of wood fiber cells and is a widely used indicator of the strength and dimensional stability of wood.

Which Tree Genomes Are Being Studied?

Many of the traits studied in forest trees are quantitative and are affected by multiple genes of varying degrees of effect. The genetic dissection of complex traits into defined components has been advanced greatly by the development of genetic mapping in trees. The first step in the analysis of the genetic basis of complex traits has been the localization of regions of the genome, quantitative trait loci (QTLs) that harbor factors with significant effects on the phenotype. The second step is the identification of one or more specific genes or regulatory sequences in each QTL interval responsible for the effect. Many kinds of genomic data can be analyzed as quantitative molecular phenotypes (Brem et al. 2002; Schadt et al. 2003). Transcript abundance, protein abundance, enzyme activity, and metabolite concentrations vary quantitatively in genetically diverse individuals or populations and may be regulated by alleles in one or more genes.

QTL mapping was initially considered likely to fail in forest trees, due to the complexity of growth and adaptation associated with long lifetimes (Strauss et al. 1992). However, tree growth (Bradshaw and Stettler 1995; Grattapaglia et al. 1996, 2009; Plomion et al. 1996; Verhaegen et al. 1997; Wu 1998; Kaya et al. 1999) and wood properties (Grattapaglia et al. 1996; Sewell et al. 2000, 2002; Ukrainetz et al. 2007) have been successfully studied using QTL mapping approaches. A great many quantitative traits in trees appear to be oligogenic with substantial proportions of the phenotypic variance attributable to a few genetic loci (Table 2). QTL mapping also provides a way to integrate various forms of genomic data to make associations of alleles and/or transcripts, with other quantitative traits or phenotypes, provided there is sufficient statistical power to validate such relationships (Beavis 1998; Yvert et al. 2003; Kirst et al. 2004b).

Growth rate has been the single most important trait in tree breeding and plantation forestry. Multiple QTLs for growth have been identified within species of *Populus*, *Eucalyptus*, and *Pinus* (Table 2) (Grattapaglia et al. 2009). Genetic differences are more readily detected in interspecific hybrids. In F_1 progeny, it is possible to investigate variation within each pure-species parent, whereas an F_2 cross provides detection of interspecific polymorphisms segregating in the F_1 parent. Maturation may induce a progressive shift in the genetic control of height growth, providing an explanation for the low juvenile/mature correlation for height growth in pines.

Eucalyptus is grown widely by vegetative propagation leading to deployment of specific clones for high-quality pulp properties. QTL analysis was performed in an *E. grandis* × *urophylla* F_1 hybrid progeny set for traits affecting vegetative propagation (Grattapaglia et al. 1995). QTLs were detected for rooting ability, stump sprouting, and shoot fresh weight. Similarly, genetic dissection of vegetative propagation traits was performed for *E. tereticornis* and *E. globulus* (Marques et al. 1999).

The genetic architecture of chemical and physical properties of wood also shows oligogenic structure. In an *E. grandis* study (Grattapaglia et al. 1996), QTLs for wood-specific gravity were identified, and in loblolly pine, QTLs were identified for MFA and wood-specific gravity (Sewell et al. 2000). QTLs for chemical properties of wood have been described in *Eucalyptus* (Kirst et al. 2004a) and in conifers (Sewell et al. 2002; Ukrainetz et al. 2007). QTLs have been identified for lignin content and composition and hemicellulose composition.

It had been assumed for many years that host resistance to disease in trees should be controlled by many genes of small effect because of the long lifetime of the host and the far shorter generation time of the fungal pathogens. White pine blister rust resistance in sugar pine, controlled by a single locus (Kinloch et al. 1970: Devey et al. 1995), had been viewed as an exception. The resistance phenotype of seedlings could be genetically mapped using the haploid maternal genetics of the megagametophytes (Devey et al. 1995).

Table 2. Diversity of Traits for Which QTLs Have Been Mapped in Forest Tree Genomes (Some Examples)

Species	Type of cross and size of population[a]	Trait	Number of QTLs[b]	Proportion phenotypic variance explained[c]	Reference
Populus trichocarpa × P. deltoides	interspecific F_2 ($n = 55$)	stem volume (age 2 yr)	2	44.7%	1
		spring bud flush	5	84.7%	1
Eucalyptus grandis	half sib ($n = 300$)	volume growth	3	13.7%	2
		wood-specific gravity	5	25.7%	2
Eucalyptus urophylla × E. grandis	interspecific F_1 ($n = 200$)	wood density at 38 mo	4	26.4%	3
		vigor at 38 mo	6	39.8%	3
		height:diameter ratio at 38 mo	4	31.2%	3
Pinus taeda	full sib ($n = 172$)	percentage latewood	5	28.6%[d]	4
		microfibril angle	5	44.8%[d]	4
		wood-specific gravity	9	56.0%[d]	4
Pinus pinaster	full sib ($n = 202$)	water use efficiency ($\delta^{13}C$)	4	51.4%	5
		diameter growth (ring width)	2	42.9%	5
Pinus sylvestris	full sib ($n = 94$)	height, frost hardiness	12	9%–23%	6
Pinus taeda	full sib ($n = 172$)	PLS of cell wall chemistry (e.g., α-cellulose, galactan, lignin)[e]	8	52.7%	7
Populus trichocarpa × P. deltoides	interspecific F_2 ($n = 167$)	response to drought (Abscission)	5	43.0%	8
Populus deltoides × P. nigra and P. deltoides × P. trichocarpa	interspecific F_1 ($n = 152$ and $n = 202$)	metabolite concentration (15 flavonoid compounds)	4	24%–44%	9
Quercus robur	full sib ($n = 120$)	water logging tolerance traits (e.g., hypertrophied lenticels, leaf epinasty)	5	56.0%[d]	10
Populus trichocarpa × P. deltoides	interspecific F_2 ($n = 285$)	stem height under elevated CO_2	2	12.2%[d]	11
		stem diameter under elevated CO_2	4	18.1%[d]	11
Eucalyptus nitens	full sib ($n = 296$)	foliar terpene (α-pinene)	3	33.3%[d]	12
		foliar terpene (trans-pinocarveol)	4	30.6%[d]	12
Eucalyptus globulus	full sib ($n = 112$)	leaf damage (Mycosphaerella leaf disease)	2	52.0%	13

[a]Half-sib families were typically from two-generation pedigrees (seed parent and progeny, grandparents unknown), whereas full-sib families were typically from three-generation pedigrees (at least one grandparent known).
[b]Number of significant QTLs detected for each trait.
[c]Total proportion of the phenotypic variance explained by the QTLs.
[d]Sum of the phenotypic variance explained by individual QTLs is given where multipoint estimates were not reported. These totals may be inflated relative to multipoint estimates of the joint effects of the same QTLs.
[e]PLS (Projection to latent structures) was used to identify shared variance components of several cell wall chemistry traits.
References: (1) Bradshaw and Stettler 1995; (2) Grattapaglia et al. 1996; (3) Verhaegen et al. 1997; (4) Sewell et al. 2000; (5) Brendel et al. 2002; (6) Lerceteau et al. 2000; (7) Sewell et al. 2002; (8) Street et al. 2006; (9) Morreel et al. 2006; (10) Parelle et al. 2007; (11) Rae et al. 2007; (12) Henery et al. 2007; (13) Freeman et al. 2008.

Host resistance to fusiform rust disease in loblolly pine, long thought to be a complex quantitative trait, was shown to be a simple gene-for-gene pathosystem once genome mapping was applied (Wilcox et al. 1996). Similarly, in poplar, host resistance to *Melamspora medusae* leaf rust was mapped to a single locus in a hybrid poplar F_2 pedigree (Newcombe et al. 1996). The severity of rust in the field and in growth-room experiments was associated with Mmd1, a single factor with a major role in resistance. More recently, two QTLs were found to explain 52% of the phenotypic variation in leaf damage of *E. globulus*, by the fungal pathogen *Mycosphaerella cryptica* (Freeman et al. 2008). Therefore, the genetic basis of host disease resistance in forest trees appears to be similar to that of annual crops.

Even when genome sequence is available, as it is for poplar, QTL regions often include hundreds of genes, and the identification of a specific candidate is not obvious. Associations can be made, however, using functional data, such as transcript abundance in combination with QTL data, or from direct allele–trait associations in large populations in high linkage equilibrium.

Transcriptomics

Microarrays have been used to study transcriptome changes in trees during adaptation to biotic and abiotic stress, development, wood formation, maturation, and dormancy. Even relatively small numbers of ESTs in libraries from pines, spruces, eucalypts, and other species have permitted a broad scale of gene expression studies defining specificity and coordinated control. Microarrays have also been used in combination with genetic markers to map the controlling factors that regulate variation in expression of specific genes, called expression QTLs (eQTLs) (Kirst et al. 2004a, 2005).

Adaptive stress responses. Expression profiles of loblolly pine seedlings subjected to drought reflected photosynthetic acclimation to mild stress and photosynthetic failure during severe stress (Watkinson et al. 2003). Distinctive responses to acclimation were found for transcripts encoding heat shock proteins (e.g., HSP70 and HSP90), late embryo-abundant proteins (e.g., LEA g3), as well as enzymes of the aromatic acid and flavonoid biosynthetic pathways. Many of the responses have clear functional relationships to drought.

The response of Scots pine (*Pinus sylvestris*) root tissue to invasion by *Heterobasidium annosum*, the root and butt rot pathogen of forest trees, has been investigated using transcript profiling (Adomas et al. 2007). In general, the total number of expressed genes decreases over time of infection, but genes were up-regulated for the phenylpropanoid pathway (shikimate, lignin, and flavonoid pathways) and for oxidases (peroxidases and laccases). In a related study (Adomas et al. 2008), Scots pine transcript abundances were compared after infection with fungi of different trophic strategies, the saprotrophic fungus *Trichoderma aurioviride*, the pathogen *Heterobasidium annosum*, and the mutualistic ectomycorrhizal symbiont *Laccaria bicolor*. The diverse fungi elicited different transcriptional responses. The results of challenge by *L. bicolor* indicate substantial similarities among infection by ectomycorrhiza in angiosperms, with some differences in timing and regulation (Heller et al. 2008). The similarity of expression profiles to the responses known in agricultural crops suggests that angiosperms and gymnosperms may have similar responses to invasive pathogens.

Barakat et al. (2009) used high-throughput sequencing to compare transcripts of susceptible American chestnut (*Castanea dentata*) and resistant Chinese chestnut (*Castanea mollissima*) induced in response to chestnut blight (*Cryphonectria parasitica*) infection. A large number of genes associated with resistance to stresses were expressed in canker tissues, some differentially expressed in canker tissues of the two species.

cDNA microarrays were also used to study genome-scale changes in transcript abundance in hybrid poplar leaves following herbivory by forest tent caterpillars (Ralph et al. 2006a). The up-regulated genes included many involved in plant defense (e.g., endochitinases), octadecanoid and ethylene signaling (e.g., lipoxygenase), secondary metabolism, and transcriptional regulation. Similarly, large-scale changes in the transcriptome were observed in Sitka spruce (*Picea sitchensis*) following either budworm or weevil attack (Ralph et al. 2006b).

Transcript changes in development. Variation in transcript abundance was studied during somatic embryogenesis of Norway spruce (*Picea abies*) to investigate early differentiation in gymnosperms (van Zyl et al. 2002; Stasolla et al. 2004). Early phases of normal embryo development are characterized by a precise pattern of induction, repression, followed by induction (up, down, and up again) as cells differentiate from proembryos to early embryos (repression), and then to late embryos (induction).

Transcriptome changes during adventitious root development (induced by an auxin pulse) in lodgepole pine (*P. contorta*) were monitored during initiation, meristem formation, and root elongation (Brinker et al. 2004). During initiation, transcript abundance reflected cell wall weakening and cell replication. During root initiation, a PINHEAD/ZWILLE-like protein (with a potential role in meristem formation) was up-regulated, whereas auxin related genes were down-regulated. During meristem formation and root differentiation, transcript changes indicated that the roots had become functional in water transport and auxin transport. Expression patterns therefore provided information on general physiological responses and specific genes that may regulate those processes.

Wood formation. Microarrays have been used to investigate the variation in transcript abundance associated with wood formation (Hertzberg et al. 2001; Yang et al. 2003; Geisler-Lee et al. 2006). Wood-forming tissue has a complex, highly regulated transcriptome, with abundant expression of genes dedicated to secondary cell wall biosynthesis (Hertzberg et al. 2001; Kirst et al. 2003, 2004a; Yang et al. 2003; Paux et al. 2005; Lu et al. 2005; Geisler-Lee et al. 2006). A microarray study in *Eucalyptus* suggested a negative correlation of tree growth and lignification at the transcript level (Kirst et al. 2004a). Transcript abundance of lignin biosynthetic genes was highest in slow-growing trees, along with increased lignin content and abundance of syringyl lignin subunits. The most elegant use of microarray technology combined transcriptome analysis with tangential cryosectioning of the developmental gradient of differentiating xylem in poplar (Hertzberg et al. 2001) to identify stage-specific gene expression patterns. Transcripts for biosynthetic enzymes of lignin and cellulose as well as potential regulators of wood formation were found to be under stage-specific transcriptional control.

An analysis of homology and tissue specificity of transcripts for carbohydrate-active enzymes (CAZymes) in poplar and *Arabidopsis* (Geisler-Lee et al. 2006) compared glycosyl transferases, glycoside hydrolases, carbohydrate esterases, polysaccharidelyases, and expansins in both species. *Populus* has ~1600 CAZyme genes compared to ~1000 for *Arabidopsis*. This difference may be the result of a recent genome-wide duplication event in the poplar genome (Tuskan et al. 2006) and greater emphasis on carbohydrate biosynthesis in woody tissues of poplar. Woody tissues have abundant CAZyme transcripts related to cell wall formation but low levels associated with starch formation.

Variation in gene expression has been studied in pines and poplars during growth seasons and over winter dormancy. In European aspen (*P. tremula*), seasonal growth and cambial dormancy involve extensive remodeling of the transcriptome (Schrader et al. 2004). The complexity of transcripts is greatly reduced in the dormant state; however, a minimal level of transcripts for some cell cycle regulators is maintained. Transcript abundance for starch enzymes increases during dormancy. Less-extensive studies in pine showed changes in abundance of transcripts for genes of cell wall biosynthesis associated with earlywood and latewood (Egertsdotter et al. 2004; Yang and Loopstra 2005).

The trunk wood of trees is differentiated into sapwood (involved in vertical transport) and heartwood, which is rich in chemical extractives, composed of waxes, fats, and resins. RNA obtained from the sapwood/heartwood transition zone shows transcript profiles that are more abundant for genes in flavonoid biosynthesis and many genes of unknown function (Yang et al. 2003).

Integration of quantitative and molecular genomics.
Mapping QTLs provides valuable information regarding the genetic structure of complex traits and identifies genomic regions that contain genes with polymorphisms affecting traits of interest in a particular cross. This kind of information has led to the identification of positional candidate genes underlying quantitative traits in plants (Morgante and Salamini 2003); however, few genes have been identified in this way. Association of genes and QTLs requires high-resolution genetic maps, identification of genes by sequence and location, and verification by direct gene transfer. Such experiments are difficult and slow even in the most tractable forest tree species. Attempts to clone a disease resistance gene from *Populus* are hindered by low recombination around the resistance locus (Stirling et al. 2001). The large amount of DNA per recombination unit in conifer genomes (e.g., 12 Mb/cM in pines) makes the identification of genes by positional cloning appear to be infeasible. However, the recombination rate per chromosome in conifers is similar to that in *Arabidopsis,* despite a difference in genome DNA content of nearly two orders of magnitude. The density and spacing of genes per recombination unit will ultimately be a critical factor determining the success of map-based cloning. Information on gene spacing and crossover distribution has not yet been obtained for any gymnosperm, although exploratory BAC sequencing is under way for spruce and pine.

Association studies in tree populations.
In many organisms, including humans, association genetics has become an established approach to detect genetic variants that may affect phenotypic variation. Large populations are screened for associations of specific markers and phenotypes. Single-nucleotide polymorphisms (SNPs) are abundant in plant genomes and have been used to find associations (Rafalski 2002; Gupta et al. 2005; Buckler et al. 2009).

Associating genes and traits in trees is challenging because of the high levels of genetic diversity, large genomes, substantial effects of environment on phenotypes, and the life history of trees (Table 3) (Neale and Savolainen 2004). Only a few specific alleles in an entire genome may be expected to have detectable quantitative effects for any given phenotype. Association genetics has been applied to wood property traits (Thumma et al. 2005; Gonzalez-Martinez et al. 2006a) and the drought stress-response (Gonzalez-Martinez et al. 2006b). In loblolly pine, a study of 58 SNPs in 20 wood and drought-related candidate genes found association of allelic variation in α-tubulin with earlywood microfibril angle (Gonzalez-Martinez et al. 2006a). Association was also found for SNPs in two genes in the lignin biosynthetic pathway. Cinnamyl alcohol dehydrogenase (CAD) alleles were associated with earlywood-specific gravity, and specific alleles of the 4-coumarate CoA-ligase (*4CL*) gene were associated with percentage of latewood. In *Eucalyptus*, allelic variants of cinnamoyl CoA reductase (CCR) were associated with variation in microfibril angle (Thumma et al. 2005). Detection of gene effects at the population level in trees therefore appears to be feasible, but much larger numbers of genes and alleles need to be evaluated.

Association genetics has strengths and limitations. The power of the method depends on the size of the experimental population, heritability of the trait, number and effects of contributing genes, and allelic frequencies for these genes. Also critical for association genetics is the amount and distribution of linkage disequilibrium (LD), which determines the number of markers and the experimental approach (genome-wide scans vs. gene-based assays) used to find marker–trait associations (Gratta-paglia et al. 2009). LD decays rapidly in pines with average r^2 values of

Table 3. Genes and Markers Identified in Association with Quantitative Traits in Forest Tree Species

Gene/allele or marker	Location (genic/intergenic)	Trait	Species	Detection approach	Population	Reference
cad/cad-n1	genic (exon)	wood density	loblolly pine (*Pinus taeda*)	n.a.	n.a.	Wu et al. (1999)
RCI2/marker unknown	unknown	wood density	*E. grandis* and *E. globulus*	QTL/genetical genomics	pseudo-backcross hybrid population	M. Kirst et al. (unpubl.)
ccr/SNP20 and SNP21	genic (intron)	microfibril angle	*E. globulus*	LD mappin	unrelated gentoypes ($n = 290$)	Thumma et al. (2005)
sams-2/M44	genic (intron)	earlywood-specific gravity	loblolly pine (*Pinus taeda*)	LD mapping	unrelated gentoypes ($n = 435$)	Gonzales-Martinez et al. (2006a)
cad/M28	genic (exon)	earlywood-specific gravity	loblolly pine (*Pinus taeda*)	LD mapping	unrelated gentoypes ($n = 435$)	Gonzales-Martinez et al. (2006a)
lp3-1/Q5	genic (UTR)	percentage of latewood	loblolly pine (*Pinus taeda*)	LD mapping	unrelated gentoypes ($n = 435$)	Gonzales-Martinez et al. (2006a)
α-*tubulin*/M10	genic (intron)	earlywood microfibril angle	loblolly pine (*Pinus taeda*)	LD mapping	unrelated gentoypes ($n = 422$)	Gonzales-Martinez et al. (2006a)

Abbreviations: (*cad*) Cinnamyl alcohol dehydrogenase; (*ccr-1*), cinnamoyl CoA reductase 1; (*sams-2*) S-adenosylmethioninesynthetase 2; (*lp3-1*) water-stress-inducible protein 1; (n.a.) not applicable.

0.2 for sites within 2 kb of each other (Brown et al. 2004). Similar rates of decay have been reported in *Eucalyptus* (Thumma et al. 2005) and *Populus* (Ingvarsson 2005). At such low levels of LD, there should be little correlation between adjacent genes, and an association of a gene and a trait implies a direct effect. However, to achieve such resolution, multiple, well-spaced markers are needed, preferably for all genes. New ultra-high-throughput DNA sequencing technologies (Wicker et al. 2006) enable the discovery of many SNPs in large numbers of tree genes, paving the way for genome-wide gene-based association genetics in trees. A recent study in *Eucalyptus* (Novaes et al. 2008) shows that very large numbers of informative SNPs can be readily obtained using this technology.

Genetical genomics in forest trees. Quantitative genetic analysis of transcript abundance and trait variation in segregating populations can identify functional associations of genes and traits, providing a powerful complement to association genetic studies. This approach, called genetical genomics (Jansen and Nap 2001), is based on the expectation that transcript level variation should be correlated with trait variation where allelic polymorphisms affect gene expression. Many genes show significant variation in transcript abundance associated with segregating genotypes, which can be considered quantitative traits and therefore be mapped as eQTLs. Studies in yeast, *Drosophila*, mice, maize, humans, and forest trees (Brem et al. 2002; Wayne and McIntyre 2002; Schadt et al. 2003; Yvert et al. 2003; Kirst et al. 2004a; Bystrykh et al. 2005; Hubner et al. 2005) revealed large numbers of specific genes with heritable variation in transcript abundance. Transcript abundance is measured in mapping pedigrees by microarray analysis, and gene–trait associations are inferred by the colocation of the resulting eQTLs and trait QTLs.

eQTLs provide information about gene regulation in two ways. If a site regulating transcript variation is coincident with the location of the gene encoding the transcript, then variation of transcript abundance is inferred to be *cis* regulated, (i.e., the controlling element is within or very near to the gene itself). If the eQTL is located elsewhere in the genome, gene expression is inferred to be regulated in *trans*, by a polymorphism in another molecule, presumably RNA or protein. A *trans*-acting eQTL shared by several genes (giving rise to an "eQTL hot spot") may indicate polymorphism in a transcription factor that regulates many genes coordinately.

Besides transcript profiling tools, genetical genomics in trees requires large-scale determination of phenotypes such as wood chemistry, tree physiology and disease resistance. High-throughput nondestructive measurement systems of wood chemistry by near-infrared (NIR) spectroscopy (Schimleck et al. 2000; Yeh et al. 2005) have been developed that allow genetic dissection of wood properties in large populations. Gene expression arrays are now available for several tree species (Hertzberg et al. 2001; Yang et al. 2003; Ralph et al. 2006a; EUCAGEN [www.eucagen.org]; NSF-GLPEP 2007 [http://compbio.dfci.harvard.edu/cgi-bin/tgi/gimain.pl?gudb=pine]). The costs of phenotyping and microarray analysis for large populations and the establishment of new large experimental populations may be the most serious limitations for application of genetical genomics.

We have studied a wide interspecific cross of *E. grandis* × *globulus* species that are planted for rapid growth (*E. grandis*) and high wood density (*E. globulus*) (Myburg 2001; Myburg et al. 2003, 2004). QTLs were identified for many traits, including wood density, growth rate, and wood properties (Myburg 2001; Kirst et al. 2004a). At least two chromosomal regions showed a significant correlation with lignin content (Fig. 3). A segment from the *E. globulus* parent on linkage group 3 (LG3) was associated with decreased lignin relative to the *E. grandis* allele, and a site on LG7 associated increased lignin and a segment from *E. globulus*. Segregating eQTLs were detected for large numbers of genes (Kirst et al. 2005). eQTLs were detected for 41% of 2304 genes represented on a microarray, of which 77% had a single eQTL. For the remaining 23%, more than one significant eQTL was detected.

eQTL analysis has also allowed us to identify a candidate gene for wood density in this *Eucalyptus* hybrid population (Myburg 2001; Kirst et al. 2004a). The abundance of one gene transcript, a *Eucalyptus* cDNA that has homology with the *RCI-2* gene of *Arabidopsis* (Jarillo et al. 1994), was strongly associated with wood density (Fig. 4, left). Several other transcripts were associated with diameter growth rate and lignin properties and appeared

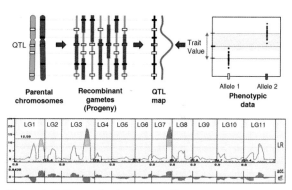

Figure 3. QTL mapping in progeny of heterozygous trees. (*Top*) Schematic representation of the use of dominant molecular markers to track the inheritance of parental chromosome segments in segregating progeny and associate the presence of alternative alleles with differences in trait values. The strength of the association is represented by the QTL graph (thick red line), which is a likelihood ratio (LR) test statistic for the difference in trait means of progeny with alternative marker alleles. QTL allele effects (relative to the trait mean) are indicated by red arrows. (*Bottom*) Example of a QTL scan for lignin content in the genome of a F_1 hybrid of *E. grandis* × *globulus* backcrossed to a unrelated *E. grandis* individual (i.e., trait variation in F_2 pseudo-backcross progeny (Myburg 2001; Kirst et al. 2004a). Significant and suggestive composite interval mapping LR thresholds (solid and broken lines) correspond to approximate experiment-wise $a = 0.05$ and $a = 0.3$, respectively. The additive effect of putative QTLs is indicated below the LR plot and shows the magnitude and direction of the effect of the *E. globulus* allele relative to that of the *E. grandis* allele in the F_1 hybrid. The 11 linkage groups are shown end to end, and positions of framework markers are indicated by triangles (Δ) along the *x* axis.

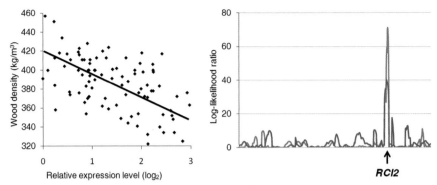

Figure 4. Association of variation in transcript abundance of the *RCI-2* homolog of *Eucalyptus* with wood density in an *E. grandis* × *globulus* hybrid F_2 progeny. (*Left*) Correlation analysis identified a significant association between *RCI-2* relative expression level and wood density measured in the F_2 population ($r^2 = 0.28$, $p < 0.0001$). (*Right*) Colocalization of the *RCI-2* gene locus, *RCI-2* eQTL (black line), and wood density QTL (gray line) on the hybrid parent genetic map further defined *RCI-2* as a strong candidate for regulation of the trait variation. The arrow indicates the genetic location of *RCI-2*.

to be coordinately controlled (Kirst et al. 2004a). Among the genes most strongly and negatively associated with growth were several genes encoding enzymes for monolignol biosynthesis. The strongest correlation with growth was found for Cald5H ($r^2 = 0.38$), known to regulate the angiosperm-specific branch of monolignol biosynthesis forming sinapyl alcohol (Chiang 2006).

The association of transcript variation and trait variation can be confirmed by evaluating the colocation of the genes, the eQTL regulating transcript level, and the QTL(s) for the trait of interest. In the case of the *RCI-2* transcript, all three (the gene, eQTL, and wood density QTL) were colocalized (Fig. 4, right), suggesting that the RCI-2eQTL affects *cis* regulation of the gene itself. In the case of the monolignol genes (Kirst et al. 2004a), the polymorphisms affecting their transcript levels (i.e., the eQTLs) did not colocalize with the genes, but they appeared to exert coordinate control. The shared eQTLs (eQTL hot spots) may be *trans*-acting regulators of the lignin biosynthetic pathway.

Under appropriate conditions, eQTLs can explain a larger component of the total phenotypic variance than traditional marker-based trait QTLs. In the case of *RCI-2*, the eQTL based on transcript level variation accounted for 28% of the total phenotypic variance, whereas the marker-based QTL in the same region accounted for only 11% of the variance. Phenotypic traits are subject to additional developmental and environmental variation that affects their transcript levels. Transcript level measurements can therefore have greater power to detect genetic associations than markers alone.

It has not yet been possible to determine the nature of the eQTLs associated with lignin biosynthesis and growth in the *E. grandis* × *globulus* backcross pedigree in the absence of genome sequence. Preliminary genetical genomics studies in F_2 hybrid progeny of *P. trichocarpa* and *P. deltoides* succeeded in identifying candidate genes that colocate with drought resistance QTLs (Street et al. 2006).

In the near future, we may expect genetical genomics and association genetics to uncover many associations of specific genes with disease resistance, insect resistance, climatic adaptation, wood properties, and carbon allocation in both hardwoods and conifers.

CONCLUSIONS

Genome Content and Evolution

Research should focus on obtaining fundamental information on gene content, gene location, and gene function in key species through comparative genomics. A pressing goal will be the ability to infer functional gene orthology across the major taxa of higher plants. Comparison of the genomes and transcriptomes of herbaceous and woody plants will help to elucidate the genetic basis of the woody growth form and perennial plant life. A major objective should be the de novo sequencing of genomes within the gymnosperms and a great many of angiosperm tree genomes to gain a detailed understanding of genome evolution in trees.

Population Genomics

Further integration of molecular and quantitative research tools will come from studies of quantitative variation in populations and, in particular, where associations can be made directly between alleles and phenotypes on a genomic scale. New methods for high-throughput phenotyping of traits and the physical and chemical properties of cells and tissues will make more associations possible. Applications may be expected in both plantations and natural populations. Comparisons of populations within and between related species will provide new insights into the evolution of many woody plant genera.

Metabolomics

The characterization of large numbers of metabolites from cells and tissues (Fiehn et al. 2000) would be of great value for forest trees to study chemical diversity and to associate metabolites with genes. Such associations could aid in assigning function to unknown genes, many of which may be involved in secondary metabolic pathways that may be less readily studied in model plants. Specific applications of global metabolite profiling will be particularly important for understanding interactions of insect pests and the chemical defenses of tree hosts (Henery et al. 2007).

Environmental Genomics

Environmental genomics (also called Ecogenomics) is an emerging area of tree biology (Whitham et al. 2006). The application of genomics to the breeding or engineering of a blight-resistant American chestnut (Steiner and Carlson 2004; Wheeler and Sederoff 2009) aimed at species restoration is a model for many endangered tree species where both conservation and genetic intervention may be needed. Such species include fraser fir, lodgepole pine, American beech, sugar pine, and American elm. Trees are desirable for bioremediation with the added advantage of carbon sequestration. Recently, transgenic poplar plants expressing a cytochrome P450 were shown to increase metabolism and remove halogenated hydrocarbons from hydroponic solution (Doty et al. 2007).

Woody Biomass for Energy

Large efforts in tree genomics will be directed to the improved use of woody biomass for fermentation to ethanol or for thermal conversion to electrical energy. The successful use of trees as bioenergy crops will require increased forest tree productivity and changes in wood biochemistry that will make cellulosics from wood more accessible for saccharification and fermentation. Alternatively, wood with high lignin content is a superior source of thermal energy, and efforts may also be directed to produce fast-growing high-content lignin trees. A substantial gap exists between the current information and what is needed to describe the relationship between genotype and phenotype by predictive models for important traits such as biomass production in forest trees.

ACKNOWLEDGMENTS

This work was supported in part by grants from the National Science Foundation (DBI-0605135) and the Consortium for Plant Biotechnology Research (GO-12026-225).

REFERENCES

Adams WT, Strauss SH, Copes DL, Griffin AR. 1992. *Population genetics of forest trees*. Kluwer, Dordrecht, The Netherlands.
Adomas A, Heller G, Li G, Olson A, Chu T-M, Osborne J, Craig D, Van Zyl L, Wolfinger R, Sederoff R, et al. 2007. Transcript profiling of conifer pathosystem: Response of *Pinus sylvestris* root tissues to pathogen (*Heterobasidion annosum*) invasion. *Tree Physiol* **27**: 1441–1458.
Adomas A, Heller G, Olson A, Osbourne J, Karlsson M, Nahalkova J, van Zyl L, Sederoff R, Stenlid J, Finlay R, Asiegbu FO. 2008. Comparative analysis of transcript abundance in *Pinus sylvestris* after challenge with a saprophytic, pathogenic or mutualistic fungus. *Tree Physiol* **28**: 885–897.
Ahuja MR, Neale DB. 2005. Evolution of genome size in conifers. *Silvae Genet* **54**: 126–137.
Allona I, Quinn M, Shoop E, Swope K, St. Cyr S, Carlis J, Riedl J, Retzel E, Campbell MM, Sederoff RR, et al. 1998. Analysis of xylem formation in pine by cDNA sequencing. *Proc Natl Acad Sci* **95**: 9693–9698.
Alonso JM, Ecker JR. 2006. Moving forward in reverse: Genetic technologies to enable genome-wide phenomic screens in *Arabidopsis*. *Nat Rev Genet* **7**: 524–536.
Barakat A, DiLoreto DS, Zhang Y, Smith C, Baier K, Powell WA, Wheeler N, Sederoff R, Carlson JE. 2009. Comparison of the transcriptomes of American chestnut (*Castanea dentata*) and Chinese chestnut (*Castanea mollissima*) in response to the chestnut blight infection. *BMC Plant Biol* **9**: 51.
Beavis WD. 1998. QTL analysis: Power, precision, and accuracy. In *Molecular dissection of complex traits* (ed. HA Patterson), pp. 145–162. CRC, Boca Raton, FL.
Behnke K, Ehlting B, Teuber M, Bauerfeind M, Louis S, Hansch R, Polle A, Bohlmann J, Schnitzler J-P. 2007. Transgenic, non-isoprene emitting poplars don't like it hot. *Plant J* **51**: 485–499.
Boerjan W. 2005. Biotechnology and the domestication of forest trees. *Curr Opin Biotechnol* **16**: 159–166.
Böhle U, Hilger H, and Martin WF. 1996. Island colonization and evolution of the insular woody habit in *Echium* L. (Boraginaceae). *Proc Natl Acad Sci* **93**: 11740–11745.
Böhlenius H, Huang T, Charbonnel-Campaa L, Brunner AM, Jansson S, Strauss SH, Nilsson O. 2006. Co/FT regulatory module controls timing of flowering and seasonal growth cessation in trees. *Science* **312**: 1040–1043.
Bowe LM, Coat G, de Pamphilis C. 2000. Phylogeny of seed plants based on all three genomic compartments: Extant gymnosperms are monophyletic and Gnetales' closest relatives are conifers. *Proc Natl Acad Sci* **97**: 4092–4097.
Bradshaw HD Jr, Stettler RF. 1995. Molecular genetics of growth and development in *Populus*. 4. Mapping QTLs with large effects on growth, form, and phenology traits in a forest tree. *Genetics* **139**: 963–973.
Bradshaw HD Jr, Villar M, Watson BD, Otto KG, Stewart S, Stettler RF. 1994. Molecular genetics of growth and development in *Populus*. III. A genetic linkage map of a hybrid poplar composed of RFLP, STS, and RAPD markers. *Theor Appl Genet* **89**: 167–178.
Bradshaw HD Jr, Ceulemans R, Davis J, Stettler R. 2000. Emerging model systems in plant biology: Poplar (*Populus*) as a model forest tree. *J Plant Growth Regul* **19**: 306–313.
Brem RB, Yvert G, Clinton R, Kruglyak L. 2002. Genetic dissection of transcriptional regulation in budding yeast. *Science* **296**: 752–755.
Brendel O, Pot D, Plomion C, Rozenberg P, Guehl JM. 2002. Genetic parameters and QTL analysis of d^{13}C and ring width in maritime pine. *Plant Cell Environ* **25**: 945–953.
Brinker M, van Zyl L, Liu W, Craig D, Sederoff R, Clapham D, von Arnold S. 2004. Microarray analyses of gene expression during adventitious root development in *Pinus contorta*. *Plant Physiol* **135**: 1526–1539.
Brown GR, Gill GP, Kuntz RJ, Langley CH, Neale DB. 2004. Nucleotide diversity and linkage disequilibrium in loblolly pine. *Proc Natl Acad Sci* **10**: 15255–15260.
Brunner AM, Busov VB, Strauss SH. 2004. Poplar genome sequence: Functional genomics in an ecologically dominant plant species. *Trends Plant Sci* **9**: 49–56.
Buckler ES, Holland JB, Bradbury PJ, Acharya CB, Brown PJ, Browne C, Ersoz E, Flint-Garcia S, Garcia A, Glaubitz JC, et al. 2009. The genetic architecture of maize flowering time. *Science* **235**: 714–718.
Burleigh JG, Mathews S. 2004. Phylogenetic signal in nucleotide data from seed plants: Implications for resolving the seed plant tree of life. *Am J Bot* **91**: 1599–1613.
Busov VB, Meilan R, Pearce DW, Ma C, Rood SB, Strauss SH. 2003. Activation tagging of a dominant gibberellin catabo-

lism gene (*GA 2-oxidase*) from poplar that regulates tree stature. *Plant Physiol* **132:** 1283–1291.

Bystrykh L, Weersing E, Dontje B, Sutton S, Pletcher MT, WiltshireT, Su AI, Vellenga E, Wang J, Manly KF, et al. 2005. Uncovering regulatory pathways that affect hematopoietic stem cell function using "genetical genomics." *Nat Genet* **37:** 225–232.

Campbell MM, Brunner AM, Jones HM, Strauss SH. 2003. Forestry's fertile crescent: The application of biotechnology to forest trees. *Plant Biotechnol* **1:** 141–154.

Chaffey N, Cholewa E, Regan S, Sundberg B. 2002. Secondary xylem development in *Arabidopsis:* A model for wood formation. *Physiol Plant* **114:** 594–600.

Chaloner WG, Macdonald P. 1980. *Plants invade the land*. Royal Scottish Museum, Edinburgh.

Chaw SM, Parkinson CL, Cheng Y, Vincent TM, Palmer JD. 2000. Seed plant phylogeny inferred from all three plant genomes: Monophyly of extant gymnosperms and origin of Gnetales from conifers. *Proc Natl Acad Sci* **97:** 4086–4091.

Chiang VL. 2006. Monolignol biosynthesis and genetic engineering of lignin in trees, a review. *Environ Chem Lett* **4:** 143–146.

Coleman HD, Park J-Y, Nair R, Chapple C, Mansfield SD. 2008a. RNAi-mediated suppression of *p*-coumaroyl-CoA 3′-hydrolase in hybrid poplar impacts lignin deposition and soluble secondary metabolism. *Proc Natl Acad Sci* **105:** 4501–4506.

Coleman HD, Samuels AL, Guy RD, Mansfield SD. 2008b. Perturbed lignification impacts tree growth in hybrid poplar: A function of sink strength, vascular integrity and photosynthetic assimilation. *Plant Physiol* **148:** 1229–1237.

de la Torre JEB, Egan MG, Katari MS, Brenner ED, Stevenson DW, Coruzzi GM, DeSalle R. 2006. ESTimating plant phylogeny: Lessons from partitioning. *BMC Evol Biol* **6:** 48.

DeVerno LL, Mosseler A. 1997. Genetic variation in red pine (*Pinus resinosa*) revealed by RAPD and RAPD-RFLP analysis. *Can J For Res* **27:** 1316–1320.

Devey ME, Fiddler TA, Liu BH, Knapp SJ, Neale DB. 1994. An RFLP linkage map for loblolly pine based on a 3-generation outbred pedigree. *Theor Appl Genet* **88:** 273–278.

Devey ME, Delfinomix A, Kinloch BB, Neale DB. 1995. Random amplified polymorphic DNA markers tightly linked to a gene for resistance to white-pine blister rust in sugar pine. *Proc Natl Acad Sci* **92:** 2066–2070.

Doebley J, Lukens L. 1998. Transcriptional regulators and the evolution of plant form. *Plant Cell* **10:** 1075–1082.

Doebley J, Stec A, Hubbard L. 1997. The evolution of apical dominance in maize. *Nature* **386:** 485–488.

Dolan L, Roberts K. 1995. Secondary thickening in roots of *Arabidopsis thaliana:* Anatomy and cell-surface changes. *New Phytol* **131:** 121–128.

Doty SL, James CA, Moore AL, Vaizovic A, Singleton GL, Ma C, Kahn Z, Xin G, Kang JW, Park JY, et al. 2007. Enhanced phytoremediation of volatile environmental pollutants with transgenic trees. *Proc Natl Acad Sci* **104:** 16816–16821.

Egertsdotter U, van Zyl LM, MacKay J, Peter G, Kirst M, Clark C, Whetten R, Sederoff R. 2004. Gene expression during formation of earlywood and latewood in loblolly pine: Expression profiles of 350 genes. *Plant Biol* **6:** 654–663.

EUCAGEN. The Eucalyptus Genome Network (http://www.eucagen.org).

FAO (Food and Agriculture Organization). 2001. Global forest resources assessment 2000: Main report. *Food and Agriculture Organization (FAO) Forestry Paper* 140 (Rome).

Fenning TM, Gershenzon J. 2002. Where will the wood come from? Plantation forests and the role of biotechnology. *Trends Biotechnol* **20:** 291–296.

Fiehn O, Kopka J, Dörmann P, Altmann T, Trethewey RN, Willmitzer L. 2000. Metabolite profiling for plant functional genomics. *Nat Biotechnol* **18:** 1157–1161.

Freeman JS, Potts BM, Vaillancourt RE. 2008. Few Mendelian genes underlie the quantitative response of a forest tree, *Eucalyptus globulus*, to a natural fungal epidemic. *Genetics* **178:** 563–571.

Frewen BE, Chen THH, Howe GT, Davis J, Rohde A, Boerjan W, Bradshaw HD Jr. 2000. Quantitative trait loci and candidate gene mapping of bud set and bud flush in *Populus*. *Genetics* **154:** 837–845.

Geisler-Lee J, Geisler M, Coutinho P, Segerman B, Nishikubo N, Takahashi J, Aspenborg H, Djerbi S, Masters E, Andersson-Gunner S, et al. 2006. Poplar carbohydrate-active enzymes (CAZymes). Gene identification and expression analysis. *Plant Physiol* **140:** 946–962.

Gill GP, Brown GR, Neale DB. 2003. A sequence mutation in the cinnamyl alcohol dehydrogenase gene associated with altered lignification in loblolly pine. *Plant Biotechnol* **1:** 253–258.

Gonzalez-Martinez SC, Wheeler NC, Ersoz E, Nelson DR, Neale DB. 2006a. Association genetics in *Pinus taeda* L. I. Wood property traits. *Genetics* **175:** 399–409.

Gonzalez-Martinez SC, Ersoz E, Brown GR, Wheeler NC, Neale DB. 2006b. DNA sequence variation and selection of tag SNPs at candidate genes for drought-stress response in *Pinus taeda* L. *Genetics* **172:** 1915–1926.

Grattapaglia D, Sederoff R. 1994. Genetic linkage maps of *Eucalyptus grandis* and *Eucalyptus urophylla* using a pseudo-testcross mapping strategy and RAPD markers. *Genetics* **137:** 1121–1137.

Grattapaglia D, Bertolucci FL, Sederoff RR. 1995. Genetic mapping of QTLs controlling vegetative propagation in *Eucalyptus grandis* and *E. urophylla* using a pseudo-testcross strategy and RAPD markers. *Theor Appl Genet* **90:** 933–947.

Grattapaglia D, Bertolucci FL, Penchel R, Sederoff RR. 1996. Genetic mapping of quantitative trait loci controlling growth and wood quality traits in *Eucalyptus grandis* using a maternal half-sib family and RAPD markers. *Genetics* **144:** 1205–1214.

Grattapaglia D, Plomion C, Kirst M, Sederoff RR. 2009. Genomics of growth traits in forest trees. *Curr Opin Plant Biol* **12:** 148–156.

Gray KA, Zhao L, Emptage M. 2006. Bioethanol. *Curr Opin Chem Biol* **10:** 141–146.

Groover AT. 2005. What genes make a tree a tree? *Trends Plant Sci* **10:** 210–214.

Gupta PK, Rustgi S, Kulwal PL. 2005. Linkage disequilibrium and association studies in higher plants: Present status and future prospects. *Plant Mol Biol* **57:** 461–485.

Hamrick JL, Godt MJW, Sherman-Broyles SL. 1992. Factors influencing levels of genetic diversity in woody plant species. *New Forests* **6:** 95–124.

Helfgott DM, Fransisco-Ortega J, Santos-Guerra A, Jansen RK, Simpson BB. 2000. Biogeography and breeding system evolution of the woody *Bencomia* alliance (Rosaceae) in Macronesia based on ITS sequence data. *Syst Bot* **25:** 82–97.

Heller G, Adomas A, Li G, Osbourne J, van Zyl L, Sederoff R, Finlay R, Stenlid J, Asiegbu F. 2008. Transcriptional analysis of *Pinus sylvestris* roots challenged with the ectomycorrhizal fungus *Lacciaria bicolor*. *BMC Plant Biol* **8:** 19.

Henery ML, Moran GF, Wallis IR, Foley WJ. 2007. Identification of quantitative trait loci influencing foliar concentrations of terpenes and formylated phloroglucinol compounds in *Eucalyptus nitens*. *New Phytol* **176:** 82–95.

Hertzberg M, Aspeborg H, Schrader J, Andersson A, Erlandsson R, Blomqvist K., Bhalero R, Uhlen M, Teeri TT, Lundeberg J, et al. 2001. A transcriptional roadmap to wood formation. *Proc Natl Acad Sci* **98:** 14732–14737.

Hu W-J, Harding SA, Lung J, Popko JL, Ralph J, Stokke DD, Tsai C-J, Chiang VL. 1999. Repression of lignin biosynthesis promotes cellulose accumulation and growth in transgenic trees. *Nat Biotechnol* **17:** 808–812.

Hubner N, Wallace CA, Zimdahl H, Petretto E, Schulz H, Maciver F, Mueller M, Hummel O, Monti J, Zidek V, et al. 2005. Integrated transcriptional profiling and linkage analysis for identification of genes underlying disease. *Nat Genet* **37:** 243–253.

Ingvarsson PK. 2005. Nucleotide polymorphism and linkage disequilibrium within and among natural populations of European aspen (*Populus tremula* L., Salicaceae). *Genetics* **169:** 945–953.

Jaenicke-Després V, Buckler ES, Smith BD, Gilbert MTP, Cooper A, Doebley J, Pääbo S. 2003. Early allelic selection in

maize as revealed by ancient DNA. *Science* **302**: 1206–1208.
Jansen RC, Nap JP. 2001. Genetical genomics: The added value from segregation. *Trends Genet* **17**: 388–391.
Jarillo JA, Capel J, Leyva A, Martinez-Zapater JM, Salinas J. 1994. Two related low-temperature-inducible genes of *Arabidopsis* encode proteins showing high homology to 14-3-3 proteins, a family of putative kinase regulators. *Plant Mol Biol* **25**: 693–704.
Kamm A, Doudrick RL, Heslop-Harrison JS, Schmidt T. 1996. The genomic and physical organization of Ty1-copia-like sequences as a component of large genomes in *Pinus elliottii* var. *elliottii* and other gymnosperms. *Proc Natl Acad Sci* **93**: 2708–2713.
Kaya Z, Sewell MM, Neale DB. 1999. Identification of quantitative trait loci influencing annual height- and diameter-increment growth in loblolly pine (*Pinus taeda* L.). *Theor Appl Genet* **98**: 586–592.
Kinloch BB Jr, Parks GK, Fowler CW. 1970. White pine blister rust: Simply inherited resistance in sugar pine. *Science* **167**: 193–195.
Kirst M, Johnson A, Baucom C, Ulrich E, Hubbard C, Staggs R, Paule C, Retzel E, Whetten R, Sederoff R. 2003. Apparent homology of expressed genes from wood-forming tissues of loblolly pine (*Pinus taeda* L.) with *Arabidopsis thaliana*. *Proc Natl Acad Sci* **100**: 7383–7388.
Kirst M, Myburg AA, Kirst ME, Scott J, Sederoff RR. 2004a. Coordinated genetic regulation of growth and lignin content revealed by QTL analysis of cDNA microarray data in an interspecific cross of *Eucalyptus*. *Plant Physiol* **135**: 2368–2378.
Kirst M, Myburg A, Sederoff R. 2004b. Genetic mapping in forest trees: Markers linkage analysis and genomics. *Genet Eng* **26**: 106–141.
Kirst M, Basten CJ, Myburg AA, Zeng ZB, Sederoff RR. 2005. Genetic architecture of transcript-level variation in differentiating xylem of a eucalyptus hybrid. *Genetics* **169**: 2295–2303.
Ko JH, Prassinos C, Han KH. 2006. Developmental and seasonal expression of *PtaHB1*, a *Populus* gene encoding a class III HD-Zip protein, is closely associated with secondary growth and inversely correlated with the level of microRNA (*miR166*). *New Phytol* **169**: 469–478.
Koch GW, Sillett SC, Jennings GM, Davis SD. 2004. The limits to tree height. *Nature* **428**: 851–854.
Kossack D, Kinlaw C. 1999. IFG, a gypsy-like retrotransposon in *Pinus* (Pinaceae) has an extensive history in pines. *Plant Mol Biol* **39**: 417–426.
Larson PR. 1994. *The vascular cambium: Development and structure*. Springer, Berlin.
Lerceteau E, Plomion C, Andersson B. 2000. AFLP mapping and detection of quantitative trait loci (QTLs) for economically important traits in *Pinus sylvestris*: A preliminary study. *Mol Breed* **6**: 451–458.
Lev-Yadun S. 1994. Induction of sclereid differentiation in the pith of *Arabidopsis thaliana* (L.) Heynh. *J Exp Bot* **45**: 1845–1849.
Lev-Yadun S, Abbo S, Doebley J. 2002. Wheat, rye, and barley on the cob? *Nat Biotechnol* **20**: 337–338.
Li J. 2006. "Stability of reporter gene expression and RNAi in transgenic poplars over multiple years in the field under vegetative propogation." PhD thesis, Oregon State University, Portland.
Li L, Zhou Y, Cheng XF, Sun J, Marita J, Ralph J, Chiang VL. 2003. Combinatorial genetic reduction of lignin quantity and augmentation of lignin syringyl constituents in trees through multigene co-transformation. *Proc Natl Acad Sci* **100**: 4939–4944.
Li L, Lu S, Chiang V. 2006. A genomic and molecular view of wood formation. *Crit Rev Plant Sci* **25**: 215–233.
Lu S, Sun Y-H, Shi R, Clark C, Li L, Chiang VL. 2005. Novel and mechanical stress-responsive microRNAs in *Populus trichocarpa* that are absent from *Arabidopsis*. *Plant Cell* **17**: 2186–2203.
MacKay J, O'Malley DM, Presnell T, Booker FL, Campbell MM, Whetten RW, Sederoff RR. 1997. Inheritance, gene expression, and lignin characterization in a mutant pine deficient in cinnamyl alcohol dehydrogenase. *Proc Natl Acad Sci* **94**: 8255–8260.

Magallón S, Sanderson MJ. 2002. Relationships among seed plants according to highly conserved genes: Sorting conflicting phylogenetic signals among ancient lineages. *Am J Bot* **89**: 1991–2006.
Margulies M, Egholm M, Altman WE, Attiya S, Bader JS, Bemben LA, Berka J, Braverman MS, Chen YJ, Chen Z, et al. 2005. Genome sequencing in microfabricated high-density picolitre reactors. *Nature* **437**: 376–380.
Marques CM, Vasquez-Kool J, Carocha VJ, Ferreira JG, O'Malley DM, Liu B-H, Sederoff R. 1999. Genetic dissection of vegetative propagation traits in *Eucalyptus tereticornis* and *E. globulus*. *Theor Appl Genet* **99**: 936–946.
Meilan R, Auerbach DJ, Ma C, DiFazio SP, Strauss SH. 2002. Stability of herbicide resistance and GUS expression in transgenic hybrid poplars (*Populus* sp.) during several years of field trials and vegetative propagation. *Hortscience* **37**: 277–280.
Mencuccini M, Martinez-Vilalta J, Hamid HA, Korakaki E, Vanderklein D. 2007. Evidence for age- and size-mediated controls of tree growth from grafting studies. *Tree Physiol* **27**: 463–473.
Mitton JB, Grant MC. 1996. Genetic variation and the natural history of quaking aspen. *Bioscience* **46**: 25–31.
Morgante M, Salamini F. 2003. From plant genomics to breeding practice. *Curr Opin Biotechnol* **14**: 214–219.
Morreel K, Goeminne G, Storme V, Sterck L, Ralph J, Coppieters W, Breyne P, Steenackers M, Georges M, Messens E, et al. 2006. Genetical metabolomics of flavonoid biosynthesis in *Populus*: A case study. *Plant J* **47**: 224–237.
Myburg AA. 2001. "Genetic architecture of hybrid fitness and wood quality traits in a wide interspecific cross of *Eucalyptus* tree species." PhD thesis, North Carolina State University, Raleigh.
Myburg AA, Griffin AR, Sederoff RR, Whetten RW. 2003. Comparative genetic linkage maps of *Eucalyptus grandis*, *Eucalyptus globulus* and their F_1 hybrid based on a double pseudo-backcross mapping approach. *Theor Appl Genet* **107**: 1028–1042.
Myburg AA, Vogl C, Griffin AR, Sederoff RR, Whetten RW. 2004. Genetics of postzygotic isolation in eucalyptus: Whole-genome analysis of barriers to introgression in a wide interspecific cross of *Eucalyptus grandis* and *E. globulus*. *Genetics* **166**: 1405–1418.
Myburg AA, Grattapaglia D, Tuskan GA, Schmutz J, Barry K, Bristow J. 2008. The *Eucalyptus* genome network. *Sequencing the Eucalyptus genome: Genomic resources for renewable energy and fiber production* (Plant and Animal Genome XVI Conference W195, January 12–16, San Diego).
Neale DB. 2007. Genomics to tree breeding and forest health. *Curr Opin Genet Dev* **17**: 539–544.
Neale DB, Ingvarsson PK. 2008. Population, quantitative and comparative genomics of adaptation in forest trees. *Curr Opin Plant Biol* **11**: 149–155.
Neale DB, Savolainen O. 2004. Association genetics of complex traits in conifers. *Trends Plant Sci* **9**: 325–330.
Newcombe G, Bradshaw HD Jr, Chastagner GA, Stettler RF. 1996. A major gene for resistance to *Melampsora medusae* f.sp. *deltoidae* in a hybrid poplar pedigree. *Phytopathology* **86**: 87–94.
Novaes E, Drost DR, Farmerie WG, Pappas GJ Jr, Grattapaglia D, Sederoff R, Kirst M. 2008. High-throughput gene and SNP discovery in *Eucalyptus grandis*, an uncharacterized genome. *BMC Genomics* **9**: 312.
NSF-GLPEP. 2007. NSF Genomics of loblolly pine embryogenesis project (http://compbio.dfci.harvard.edu/cgi-bin/tgi/gimain.pl?gudb=pine).
O'Malley DM, Grattapaglia D, Chaparro JX, Wilcox PL, Amerson HV, Liu B-H, Whetten R, McKeand S, Kuhlman EG, McCord S, et al. 1996. Molecular markers, forest genetics and tree breeding. In *Genomes of plants and animals: 21st Genetics Symposium* (ed. JP Gustafson and RB Flavell), pp 87–102. Plenum, New York.
Parelle J, Zapater M, Scotti-Saintagne C, Kremer A, Jolivet Y, Dreyer E, Brendel O. 2007. Quantitative trait loci of tolerance to waterlogging in a European oak (*Quercus robur* L.):

Physiological relevance and temporal effect patterns. *Plant Cell Environ* **30**: 422–434.

Paux E, Carocha V, Marques C, Mendes de Sousa A, Borralho N, Sivadon P, Grima-Pettenati J. 2005. Transcript profiling of *Eucalyptus* xylem genes during tension wood formation. *New Phytol* **167**: 89–100.

Plomion C, Durel C-E, O'Malley DM. 1996. Genetic dissection of height in maritime pine seedlings raised under accelerated growth conditions. *Theor Appl Genet* **93**: 849–858.

Plomion C, Chagne D, Pot D, Kumar S, Wilcox P, Burdon R, Prat D, Peterson DG, Pavia J, Chaumeil P, et al. 2007. The pines. In *Genome mapping and molecular breeding in plants: Forest trees* (ed. CR Kole), Vol. 7, pp. 29–78. Springer, Heidelberg.

Poke FS, Vaillancourt RE, Potts BM, Reid JB. 2005. Genomic research in *Eucalyptus*. *Genetica* **125**: 79–101.

Qui Y-L, Li L, Wang B, Chen Z, Dombrovska O, Lee J, Kent L, Li R, Jobson RW, Hendry TA, et al. 2007. A nonflowering land plant phylogeny inferred from nucleotide sequences of seven chloroplast mitochondrial and nuclear genes. *Int J Plant Sci* **168**: 691–708.

Rae AM, Tricker PJ, Bunn SM, Taylor G. 2007. Adaptation of tree growth to elevated CO_2: Quantitative trait loci for biomass in *Populus*. *New Phytol* **175**: 59–69.

Rafalski A. 2002. Applications of single nucleotide polymorphisms in crop genetics. *Curr Opin Plant Biol* **5**: 94–100.

Ralph J, MacKay J, Hatfield R, Whetten RW, O'Malley DM, Sederoff RR. 1997 Abnormal lignin in a loblolly pine mutant. *Science* **277**: 235–239.

Ralph S, Oddy C, Cooper D, Yueh H, Jancsik S, Kolosova N, Phillipe RN, Aeschliman D, White R, Huber D, et al. 2006a. Genomics of hybrid poplar (*Populus trichocarpa* × *deltoides*) interacting with forest tent caterpillars (*Malacosoma disstria*): Normalized and full-length cDNA libraries, expressed sequence tags, and a cDNA microarray for the study of insect-induced defences in poplar. *Mol Ecol* **15**: 1275–1297.

Ralph SG, Yueh H, Friedmann M, Aeschliman D, Zeznik JA, Nelson CC, Butterfield YSN, Kirkpatrick R, Liu J, Jones SJ, et al. 2006b. Conifer defence against insects: Microarray gene expression profiling of Sitka spruce (*Picea sitchensis*) induced by mechanical wounding or feeding by spruce budworms (*Choristoneura occidentalis*) or white pine weevils (*Pissodes strobi*) reveals large-scale changes of the host transcriptome. *Plant Cell Environ* **29**: 1545–1570.

Remington DL, O'Malley DM. 2000. Whole-genome characterization of embryonic stage inbreeding depression in a selfed loblolly pine family. *Genetics* **155**: 337–348.

Schadt EE, Monks SA, Drake TA, Lusis AJ, Che N, Colinayo V, Ruff TG, Milligan SB, Lamb JR, Cavet G, et al. 2003. Genetics of gene expression surveyed in maize, mouse and man. *Nature* **422**: 297–302.

Schimleck LR, Raymond CA, Beadle CL, Downes GM, Kube PD, French J. 2000. Application of NIR spectroscopy to forest research. *Appita J* **53**: 458–464.

Schmidt M, Schneider-Poetsch HA. 2002. The evolution of gymnosperms redrawn by phytochrome genes: The Gnetales appear at the base of the gymnosperms. *J Mol Evol* **54**: 715–724.

Schmidt A, Doudrick RL, Heslop-Harrison JS, Schmidt T. 2000. The contribution of short repeats of low sequence complexity to large conifer genomes *Theor Appl Genet* **101**: 7–14.

Schrader J, Moyle R, Bhalerao R, Hertzberg M, Lundeberg J, Nilsson P, Bhalerao RP. 2004. Cambial meristem dormancy in trees involves extensive remodeling of the transcriptome. *Plant J* **40**: 173–187.

Sewell MM, Bassoni DL, Megraw RA, Wheeler NC, Neale DB. 2000. Identification of QTLs influencing wood property traits in loblolly pine (*Pinus taeda* L.). I. Physical wood properties. *Theor Appl Genet* **101**: 1273–1281.

Sewell MM, Davis MF, Tuskan GA, Wheeler NC, Elam CC, Bassoni DL, Neale DB. 2002. Identification of QTLs influencing wood property traits in loblolly pine (*Pinus taeda* L.). II. Chemical wood properties. *Theor Appl Genet* **104**: 214–222.

Spokevicius AV, Southerton SG, Colleen P, MacMillan CP, Qiu D, Gan S, Josquin FG, Tibbits JFG, Moran GF, Bossinger G. 2007. β-tubulin affects cellulose microfibril orientation in plant secondary fibre cell walls. *Plant J* **51**: 717–726.

Stasolla C, Bozhkov PV, Chu T-M, van Zyl L, Egertsdotter U, Suarez MF, Craig D, Wolfinger RD, von Arnold S, Sederoff RR. 2004. A transcriptional pathway during somatic embryogenesis in gymnosperms. *Tree Physiol* **24**: 1073–1085.

Steiner KC, Carlson JE, Eds. 2004. *Restoration of American chestnut to forest lands.* Pennsylvania State University School of Forest Resources, University Park, PA.

Sterky F, Regan S, Karlsson J, Hertzberg M, Rohde A, Holmberg A, Amini B, Bhalerao R, Larsson M, Villarroel R, et al. 1998. Gene discovery in the wood-forming tissues of poplar: Analysis of 5,692 expressed sequence tags. *Proc Natl Acad Sci* **95**: 13330–13335.

Sterky F, Bhalerao RR, Unneberg P, Segerman B, Nilsson P, Brunner AM, Charbonnel-Campaa L, Lindvall JJ, Tandre K, Strauss SH, et al. 2004. A *Populus* EST resource for plant functional genomics. *Proc Natl Acad Sci* **101**: 13951–13956.

Stirling B, Newcombe G, Vrebalov J, Bosdet I, Bradshaw HD. 2001. Suppressed recombination around the *MXC3* locus, a major gene for resistance to poplar leaf rust. *Theor Appl Genet* **103**: 1129–1137.

Strauss SH, Lande R, Namkoong G. 1992. Limitations of molecular marker-aided selection in forest tree breeding. *Can J Forest Res* **22**: 1050–1061.

Strauss SH, Rottman WH, Brunner AM, Shephard L. 1995. Genetic engineering of reproductive sterility in forest trees. *Mol Breed* **1**: 5–26.

Street NR, Skogstrom O, Sjodin A, Tucker J, Rodriguez-Acosta M, Nilsson P, Jansson S, Taylor G. 2006. The genetics and genomics of the drought response in *Populus*. *Plant J* **48**: 321–341.

Taylor G. 2002. *Populus*: *Arabidopsis* for forestry. Do we need a model tree? *Ann Bot* **90**: 681–689.

Thumma BR, Nolan MR, Evans R, Moran GF. 2005. Polymorphisms in cinnamoyl CoA reductase (CCR) are associated with variation in microfibril angle in *Eucalyptus* spp. *Genetics* **171**: 1257–1265.

Tulsieram LK, Glaubitz JC, Kiss G, Carlson JE. 1992. Single tree genetic linkage mapping in conifers using haploid DNA from megagametophytes. *Bio-Technology* **10**: 686–690.

Tuskan GA, DiFazio S, Jansson S, Bohlmann J, Grigoriev I, Hellsten U, Putnam N, Ralph S, Rombauts S, Salamov A, et al. 2006. The genome of black cottonwood, *Populus trichocarpa* (Torr. and Gray). *Science* **313**: 1596–1604.

Ukrainetz NK, Ritland K, Mansfield SD. 2007. Identification of quantitative trait loci for wood quality and growth across eight full-sib coastal Douglas-fir families. *Tree Genet Genomes* **4**: 159–170.

Van Beveren KS, Spokevicius AV, Tibbits J, Qing W, Bossinger G. 2006. Transformation of cambial tissue in vivo provides an efficient means for induced somatic sector analysis and gene testing in stems of woody plant species. *Funct Plant Biol* **33**: 629–638.

Van Zyl LM, Bozhkov P, Clapham D, Sederoff RR, von Arnold S. 2002. Up, down and up again is a signature global gene expression pattern at the beginning of gymnosperm embryogenesis. *Gene Expr Patterns* **3**: 83–91.

Verhaegen D, Plomion C, Gion J-M, Poitel M, Costa P, Kremer A. 1997. Quantitative trait dissection analysis in *Eucalyptus* using RAPD markers. 1. Detection of QTL in interspecific hybrid progeny, stability of QTL expression across different ages. *Theor Appl Genet* **95**: 597–608.

Wang MB, Waterhouse PM. 2002. Application of gene silencing in plants. *Curr Opin Plant Biol* **5**: 146–150.

Watkinson JI, Sioson AA, Vasquez-Robinet C, Shukla M, Ellis M, Heath LS, Ramakrishnan N, Chevone B, Watson LT, Egertsdotter U, et al. 2003. Photosynthetic acclimation is reflected in specific patterns of gene expression in drought-stressed loblolly pine. *Plant Physiol* **133**: 1702–1716.

Wayne ML, McIntyre LM. 2002. Combining mapping and arraying: An approach to candidate gene identification. *Proc Natl Acad Sci* **99**: 14903–14906.

Wheeler N, Sederoff R. 2009. Role of genomics in the restoration of the American chestnut. *Tree Genet Genomes* **5:** 181–187.

White TL, Adams WT, Neale DB. 2007. *Forest genetics*. CAB International, Cambridge, MA.

Whitham TG, Bailey JK, Schweitzer JA, Shuster SM, Bangert RK, LeRoy CJ, Lonsdorf EV, Allan GJ, DiFazio SP, Potts BM, et al. 2006. A framework for community and ecosystem genetics: From genes to ecosystems. *Nat Rev Genet* **7:** 510–523.

Wicker T, Schlagenhauf E, Graner A, Close TJ, Keller B, Stein N. 2006. 454 sequencing put to the test using the complex genome of barley. *BMC Genomics* **7:** 275.

Wilcox PL, Amerson HV, Kuhlman EG, Liu B-H, O'Malley DM, Sederoff RR. 1966. Detection of a major gene for resistance to fusiform rust disease in loblolly pine by genomic mapping. *Proc Natl Acad Sci* **93:** 3859–3864.

Wu RL. 1998. Genetic mapping of QTLs affecting tree growth and architecture in *Populus:* Implication for ideotype breeding. *Theor Appl Genet* **96:** 447–457.

Wu RL, Remington DL, MacKay JJ, McKeand SE, O'Malley DM. 1999. Average effect of a mutation in lignin biosynthesis in loblolly pine. *Theor Appl Genet* **99:** 705–710.

Wullschleger SD, Tuskan GA, DiFazio SP. 2002. Genomics and the tree physiologist. *Tree Physiol* **22:** 1273–1276.

Yang SH, Loopstra CA. 2005. Seasonal variation in gene expression for loblolly pines (*Pinus taeda*) from different geographical regions. *Tree Physiol* **25:** 1063–1073.

Yang J, Park S, Kamdem DP, Keathley DE, Retzel E, Paule C, Kapur V, Han K-H. 2003. Novel gene expression profiles define the metabolic and physiological processes characteristic of wood and its extractive formation in a hardwood tree species, *Robinia pseudoacacia. Plant Mol Biol* **52:** 935–956.

Yeh TF, Yamada T, Capanema E, Chang H-M, Chiang V, Kadla JF. 2005. Rapid screening of wood chemical component variations using transmittance near-infrared spectroscopy. *J Agric Food Chem* **53:** 3328–3332.

Yvert G, Brem RB, Whittle J, Akey JM, Foss E, Smith EN, Mackelprang R, Kruglyak L. 2003. *Trans*-acting regulatory variation in *Saccharomyces cerevisiae* and the role of transcription factors. *Nat Genet* **35:** 57–64.

Zhong R, Lee C, Zhou J, McCarthy RL, Ye Z-H. 2008. A battery of transcription factors involved in the regulation of secondary cell wall biosynthesis in *Arabidopsis. Plant Cell* **20:** 2763–2782.

Zobel BJ, Talbert J. 1984. *Applied forest tree improvement*. Wiley, New York.

Studying Phenotypic Evolution in Domestic Animals: A Walk in the Footsteps of Charles Darwin

L. ANDERSSON

Department of Medical Biochemistry and Microbiology, Uppsala University, SE-75124 Uppsala, Sweden, and Department of Animal Breeding and Genetics, Swedish University of Agricultural Sciences, SE-75123 Uppsala, Sweden

Correspondence: leif.andersson@imbim.uu.se

Charles Darwin used domesticated plants and animals as proof of principle for his theory on phenotypic evolution by means of natural selection. Inspired by Darwin's work, we developed an intercross between the wild boar and domestic pigs to study the genetic basis for phenotypic changes during domestication. The difference in coat color is controlled by two major loci. Dominant white color is due to two consecutive mutations in the *KIT* gene: a 450-kb duplication and a splice mutation. Black spotting is caused by the combined effect of two mutations in *MC1R*: a missense mutation for dominant black color and a 2-bp insertion leading to a frameshift. A major discovery made using this pedigree is the identification of a single-nucleotide substitution in intron 3 of the gene for insulin-like growth factor 2 (*IGF2*) that is underlying a quantitative trait locus affecting muscle growth, size of the heart, and fat deposition. The mutation disrupts the interaction with a repressor and leads to threefold increased *IGF2* expression in postnatal muscle. In a recent study, we have identified the *IGF2* repressor, and this previously unknown protein, named ZBED6, is specific for placental mammals and derived from a domesticated DNA transposon.

Charles Darwin was the first to realize that phenotypic changes in domesticated animals and plants, caused by selective breeding, mimic the process of phenotypic evolution in natural populations (Darwin 1859). In fact, Darwin himself became an animal breeder and performed breeding experiments with pigeons to prove that the phenotype could be altered by selection. Nine years after the publication of *On the origins of species*, Darwin summarized his studies of domesticated species in the book *The Variation of Animals and Plants Under Domestication* (Darwin 1868). The powerful genomic tools now available allow us to further use domestic animals as models by revealing the genes and mutations that have contributed to their phenotypic evolution (Andersson 2001). Some of these mutations predate domestication (standing genetic variation), whereas others have occurred subsequent to domestication and were picked up by human selection. This screen for mutations with phenotypic effects is extremely deep; it has been going on worldwide for thousands of years and has led to genetic adaptations to various climates and production systems, as we have taken our domestic animals around the globe. These genetic changes are the reason why domesticated plants and animals are outstanding models for phenotypic evolution by means of natural selection.

In 1989, inspired by Charles Darwin's studies of phenotypic differences between wild and domesticated species, we generated an intercross between the European wild boar and domestic pigs, with the objective of using the emerging genomic tools in an attempt to reveal some of the genes that have been under selection during pig domestication. The aim of this chapter is to summarize some of the highlights from this project and their implications for evolutionary biology.

We crossed two European wild boars with eight Swedish Yorkshire (Large White) sows and generated 200 F_2 progeny. The pedigree was established with two major aims: as a resource population for developing a linkage map for the pig and for mapping trait loci of biological interest. When the project was initiated, the porcine linkage map was restricted to ~20 loci, but it grew rapidly as microsatellite markers became available for the pig. This pedigree contributed to the development of the first-generation genome-wide linkage map for the pig (Ellegren et al. 1994). We recorded a large number of phenotypes that differed markedly between wild boars and domestic pigs including coat color, growth rate, body composition, immunological traits, and some skeletal measures. Most of these traits have a complex genetic background, and we considered this a pilot experiment because it was not known whether it would be possible to detect any convincing quantitative trait loci (QTL) using a sample size of only 200 F_2 animals derived from an intercross between outbred populations.

In one of the volumes of *The Variation of Animals and Plants Under Domestication* (Darwin 1868) is an illustration of a wild boar in comparison with a pig of the Large White breed, i.e., the same breed group as used in our experiment (Fig. 1). The Large White pig was white at that time and it still is today, but there is a very striking difference in body composition because the 1850 version is fat, whereas the modern Large White pig is very lean (high muscle content, low fat content). During the 19th century, the breeding goal was to produce fat pigs because there was a high consumer demand for energy-rich food. Due to the change in our life style, the breeding goal has changed, and during the last century, there has been an intense selection to increase muscle growth and reduce

Figure 1. Illustration of a wild boar (*top*) and a Yorkshire Large White pig from Charles Darwin's book *The Variation of Animals and Plants Under Domestication* (Darwin 1868).

fat deposition, because consumers pay much more for meat than for lard. Our wild boar project has led to the identification of the mutations that cause the white color, a selection that primarily took place before 1850, and a mutation that has contributed very significantly to the transformation of a rather fat pig to a lean one.

SELECTION AGAINST CAMOUFLAGE ON THE FARM

There is a striking difference in coat color between the wild boar and Large White pigs and we observed a considerable amount of coat color diversity in the F_2 generation (Fig. 2). The analysis of the segregation data indicated that this diversity was controlled by two major loci: the *Dominant white* (*I*) and the *Extension* (*E*). Large White pigs were assumed to be homozygous for the dominant allele (*I*) for white color and the recessive allele (*Ep*) for black spotting, whereas wild boar were homozygous for the recessive *i* allele and the dominant *E+* allele. The segregation data for the F_2 generation did not deviate significantly from the expected 12:3:1 ratio of white (*I/–*), wild-type (*i/i, E+/–*), and black-spotted (*i/i, Ep/Ep*) pigs (Johansson et al. 1992). However, a surprising finding was the observation of a partially white phenotype that we denoted Patch and that was found to be controlled by a third allele at the *Dominant white* locus, *Ip* (Fig. 2). We mapped the *Dominant white* locus to pig chromosome 8 and identified *KIT* as a strong positional candidate gene (Johansson et al. 1992). Previous studies had demonstrated that *KIT* encodes a tyrosine kinase receptor with an essential function for the migration and survival of melanoblasts and that *KIT* mutations cause pigmentation disorders in both mice (Dominant white spotting) and humans (Piebald) (Spritz 1994).

Further characterization of the *Dominant white* locus over several years revealed that the *Dominant white* allele is caused by the combined effect of two independent mutations: (1) a ~450-kb tandem duplication that encompasses the entire coding sequence of *KIT* and ~150 kb upstream of exon 1 and (2) a splice mutation at the first nucleotide of intron 17 in one of the *KIT* copies that leads to exon skipping (Johansson Moller et al. 1996; Marklund et al. 1998; Giuffra et al. 2002). The *Patch* allele possesses the duplication but not the splice mutation. The data imply an evolutionary scenario whereby the duplication first occurred and resulted in a white-spotted phenotype that was selected by humans. The splice mutation occurred subsequently and resulted in a completely white phenotype but with normally colored eyes. The explanation of why *KIT* mutants have normally colored eyes is that *KIT* has an essential function for the development of neural-crest-derived melanocytes found in hair, skin, and ear but not for the melanocytes forming the *retinal pigment epithelium* (*RPE*). Our hypothesis is that the duplication is causing white spotting because the duplicated *KIT* copy has been separated from regulatory elements located far upstream of *KIT* (Giuffra et al. 2002), and the splice mutation enhances the defect in KIT signaling. The skipping of exon 17 in the mature transcript removes a crucial part of the tyrosine kinase domain, and thus, one of the *KIT* copies is expected to code for a receptor with normal ligand binding but defective kinase signaling. An allele carrying only the splice mutation but not the duplication is expected to be homozygous lethal because KIT signaling is absolutely required for hematopoiesis. The presence of one normal *KIT* copy ensures that white pigs have a sufficient amount of KIT signaling to avoid severe pleiotropic effects on hematopoiesis and germ-cell development. Thus, by combining the effect of two mutations—the duplication and the splice mutation—a *KIT* allele has been "created" with more severe effect on pigmentation than any known mouse mutant, despite the fact that the allele is fully viable in the homozygous condition, whereas *KIT*-null alleles are homozygous lethal.

In his work, Charles Darwin discussed the observation that domestic animals may revert toward a wild-type phenotype "when (they) run wild" (Darwin 1859). One important explanation for this phenomenon is the altered selection pressure animals experience in a natural environment. However, the *Dominant white* allele may revert to the wild-type allele due to unequal crossing-over between the two copies of the *KIT* duplication. We revealed extensive genetic diversity at the *KIT* locus in populations of white pigs, where the copy number varied from one to three, and one or two of the copies carried the splice mutation (Pielberg et al. 2002, 2003). This genetic instability explains why breeders have never been able to

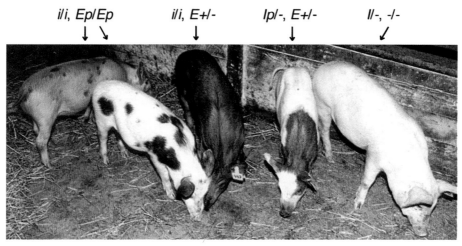

Figure 2. Segregation of coat color among F_2 animals from an intercross between European wild boar and Large White domestic pigs. Two alleles are segregating at the *Extension/MC1R* locus: the wild-type allele ($E+$) and the recessive allele for black spotting (Ep). Three alleles are segregating at the *Dominant White/KIT* locus: the recessive wild-type allele and the alleles for Patch (Ip) and Dominant white color (I). Photo by Mats Gerentz, Swedish University of Agricultural Sciences.

fix the dominant white phenotype despite it being a monogenic trait, and they have been selecting for white color for hundreds of years. Furthermore, the genetic instability implies that if white pigs were released in the wild and became feral pigs, they would soon revert to wild type due to unequal crossing-over combined with selection for sun protection and camouflage in the wild.

The second major locus explaining the difference in coat color between wild and Large White domestic pigs is the *Extension* (*E*) locus (Fig. 2). The *E* locus is one of the classical coat color loci in mammals, and alleles at this locus determine the relative distribution of red pheomelanin and black eumelanin. The *E* locus encodes the melanocortin 1 receptor (MC1R) (Robbins et al. 1993). The wild boar carries the wild-type allele *E+*, and some hairs show the classical agouti pattern with alternating black and red stripes created by the interaction among MC1R, its ligand melanocyte-stimulating hormone (MSH), and the antagonist agouti-signaling protein (ASIP). The *Ep* allele for black spotting involves two causative mutations that must have occurred consecutively (Kijas et al. 1998, 2001). First, there is a missense mutation D124N causing dominant black color found, for instance, in Hampshire pigs. This mutation is assumed to cause a constitutively active receptor that leads to the production of black pigment. Second, the *Ep* allele also carries a 2-bp (CC) insertion at codon 22 that expands a stretch of six cytosine nucleotides to eight. This causes a frameshift and an expected complete loss of function. The predicted phenotype of an *MC1R*-null animal is a uniform red coat color. Some *Ep/Ep* homozygotes are uniformly red, but most of them are red with black spots or white with black spots (Fig. 2). So why do most of these pigs exhibit black spots that is inconsistent with a complete absence of MC1R signaling? The explanation is that the stretch of eight C nucleotides behaves as a small microsatellite that may be affected by slippage during DNA replication in somatic cells and possibly also in germ cells. Reverse transcriptase–polymerase chain reaction (RT-PCR) analysis using skin from black spots revealed that one copy of *MC1R* had reverted to six Cs and thereby restored the reading frame (Kijas et al. 2001). The consequence is that the missense mutation D124N becomes activated and the pigment cells produce black eumelanin.

We have not yet been able to find the genetic basis of why some *Ep/Ep* pigs are red with small black spots and others are white with large black spots (Fig. 2). The numbers of these two classes of pigs did not fit any Mendelian ratio, suggesting that one of the founder populations (most likely the domestic pigs) are not fixed at this locus or that the phenotype has a more complex genetic background. Interestingly, there is a very clear trend that the black spots are larger on a white background than on a red background (Fig. 2).

After describing the alleles for dominant black color (*ED*), black spotting (*Ep*), and recessive red (*e*) color (Kijas et al. 1998, 2001), we decided to screen for *MC1R* diversity among pigs representing 51 different Asian and European breeds of domestic pigs, as well as populations of wild boars from both Europe and China (Fang et al. 2009). The results revealed contrasting modes of evolution at the *MC1R* locus in wild and domestic pigs (Fig. 3). We have previously estimated that Asian and European wild boars diverged from a common ancestor more than 500,000 years before present on the basis of the sequence divergence of mitochondrial DNA (Kijas and Andersson 2001). We found a total of seven nucleotide substitutions in *MC1R* among European and Asian wild boars, and they were all synonymous, providing evidence for purifying selection maintaining normal MC1R function (Fig. 3). The most likely selection pressure is to maintain a camouflage color. In sharp contrast, we discovered a total of nine genetic changes unique to domestic pigs, and they all changed the coding sequence; eight were nonsynonymous substitutions and one was the frameshift mutation

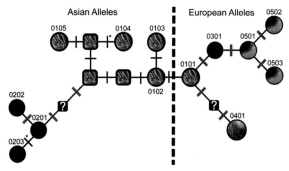

Figure 3. Median-joining network tree of *MC1R* alleles in wild and domestic pigs from Europe and Asia. All known alleles are represented by circles and a four-digit allele designation, whereas squares represent predicted intermediate forms that have not yet been found. Thin black lines perpendicular to lines connecting alleles represent synonymous changes, whereas thicker red lines represent nonsynonymous change. Colors inside circles and squares represent observed and predicted phenotypes, respectively; question marks inside two of the squares indicate that the associated phenotype cannot be predicted because they are intermediate forms between alleles differing by two nonsynonymous substitutions. The asterisk associated with the synonymous substitutions leading to alleles *0203* and *0104* indicates the only instance of an identical mutation at different locations on the tree. (Reprinted, from Fang et al. 2009.)

described above. This highly significant excess of substitutions changing the coding sequence demonstrates that the change in coat color in domestic pigs is caused by a direct selection to alter coat color, rather than relaxed purifying selection. This provides evidence for selection against camouflage in domestic pigs, and this selection on color must have been initiated early during domestication. We have hypothesized that humans have selected on color because it distinguished the early domestic forms from their wild ancestors. Alternatively, it facilitated animal husbandry, because it is easier to keep track of a colorful pig than a camouflage-colored pig and/or simply because we were attracted by new color variants as they arose due to spontaneous mutations. Selection for black color in pigs happened independently in Asia and Europe (Fig. 3) and involved two different missense mutations (L102P and D124N) with very similar phenotypic effects (Kijas et al. 1998). The *MC1R* locus is a beautiful illustration of why domestic animals are such good models for phenotypic evolution; humans have transformed their phenotype during the last 10,000 years by cherry-picking mutations with phenotypic effects.

Both *KIT* and *MC1R* in pigs illustrate that our domestic animals have a sufficiently old history that allows the evolution of alleles differing by multiple causative mutations. We have described several additional such examples. For instance, the *Smoky* allele affecting plumage color in the chicken involves both a 9-bp insertion in the *PMEL17* gene, causing *Dominant white* color, and a 12-bp deletion in the same gene that partially restores pigmentation (Kerje et al. 2004). Similarly, alleles causing different degrees of white spotting in dogs (*Irish*, *Piebald*, and *Extreme white*) are caused by different combinations of regulatory mutations at the *MITF* locus (Karlsson et al. 2007). This is an important lesson for studies of genotype/phenotype relationships in natural populations, including humans, because it suggests that we may often find alleles that differ by multiple functionally important mutations, in contrast to mouse mutants or monogenic disorders in humans where the paradigm has been that a phenotype or disorder almost always is caused by a single causative mutation.

GENETIC DISSECTION OF A MAJOR LOCUS CONTROLLING MUSCLE GROWTH, HEART SIZE, AND FAT DEPOSITION

The intercross between the European wild boar and Large White domestic pigs was designed to allow for the mapping of QTL controlling some of the multifactorial traits that show striking differences between the founder populations. Domestic pigs are today very lean due to the strong selection for high muscle growth and reduced fat deposition. A striking observation was that the F_1 and F_2 hybrid progeny on average became more obese than purebred domestic pigs because the wild boar founders transmitted alleles for high fat deposition that most certainly are adaptive for the survival of wild boars during periods of sparse food resources under natural conditions. This is in line with the "thrifty gene hypothesis," which implies that alleles associated with an increased risk to develop obesity and other metabolic disorders in humans have a relatively high frequency in many populations because they were advantageous during periods of starvation (Neel 1962).

In 1994, we published the first QTL mapping paper based on the wild boar/Large White intercross (Andersson et al. 1994). This study was the first genome-wide QTL mapping study in an outbred organism, because previously published QTL mapping studies were based on intercrosses among inbred lines (see, e.g., Paterson et al. 1988). The study revealed several QTL reaching genome-wide significance, and the most prominent one was detected on chromosome 4. For each trait, the estimated QTL effects were in the expected direction, so that the QTL allele from the domestic pig increased growth, reduced fat deposition, and increased the length of the small intestine (Andersson et al. 1994). Interestingly, Charles Darwin in his book *The Variation of Animals and Plants Under Domestication* noted that one of the phenotypic changes that happened during pig domestication is an increased length of the small intestine (Darwin 1868). The assumption is that this phenotypic change is related to selection for higher feed efficiency. Subsequent studies have shown that the major QTL on pig chromosome 4 affecting multiple traits represents multiple linked loci (Berg et al. 2006).

In 1999, we (Jeon et al. 1999) and Michel Georges' group (Nezer et al. 1999) identified a major QTL located at the distal tip of pig chromosome 2. The QTL was shown to have major effects on muscle growth, heart size, and subcutaneous fat deposition but had no significant effect on birth weight, adult body weight, or abdominal fat deposition. The QTL allele derived from the domestic Large White pig in our intercross made the pig leaner (more muscle and less fat), and this locus alone controlled ~30% of the residual phenotypic variance for muscle traits among the F_2 animals (Jeon et al. 1999). This QTL showed paternal expression, i.e., the phenotypic effects observed in the F_2 animals were determined by the paternal allele they had received from their F_1 sire. This immediately revealed the gene coding for insulin-like growth factor 2 (*IGF2*) as the prime positional candidate gene, because this is one of the very few paternally expressed genes located in this chromosomal region and it is a well-known growth factor. We could immediately conclude that if *IGF2* was the gene, then the causative mutation must be a regulatory one because there was no difference in IGF2 protein sequence between wild and domestic pigs. Without access to a genome assembly, we decided to sequence a porcine bacterial artificial chromosome (BAC) comprising the *IGF2* locus, and the results revealed an extensive number of evolutionarily conserved noncoding sequences consistent with the fact that *IGF2* has four different promoters and a complex regulation (Amarger et al. 2002).

In collaboration with Michel Georges' group in Liege, we collected a set of chromosomes for which we could establish the genotype at the *IGF2* QTL with confidence based on family segregation data. We named the allele associated with high muscle growth Q and the wild-type allele q. We then sequenced 28.6 kb of genomic DNA from five different Q haplotypes representing four different breeds and 10 q haplotypes derived from wild boars and domestic pigs. The region comprised the insulin (*INS*) and *IGF2* genes and their upstream regions. The resequencing data revealed that all five Q haplotypes were identical for a 20-kb region spanning from intron 1 to the 3' untranslated region (3'UTR) of *IGF2*, whereas extensive sequence diversity existed in this region between Q and q haplotypes as well as among q haplotypes (Van Laere et al. 2003). This implied that the five Q haplotypes are identical by descent (IBD) for this region and that the causative mutation(s) should be located within the 20-kb region because the sequence identity among Q chromosomes broke up outside the region.

The problem we were facing at this stage was the extensive sequence diversity, approaching 1% (!), between Q and q chromosomes, which meant that although we had achieved an exceedingly high resolution in the QTL mapping, ~150 sequence polymorphisms were still showing complete linkage disequilibrium with the two QTL alleles. At this stage, we considered the possibility that the Q haplotype had an Asian origin, which would explain the large sequence diversity to the q chromosomes. In an earlier study, we had shown that Asian pigs were crossed with European pigs primarily during the 18th and 19th centuries and that many breeds of European pigs have a hybrid origin (Giuffra et al. 2000). For this reason, we incorporated into the sequence analysis a Chinese Meishan chromosome carried by one of the founder animals of a Meishan/Large White intercross developed by Alan Archibald and Chris Haley at the Roslin Institute. Segregation data proved that this chromosome should be classified as a q chromosome. Resequencing the Meishan chromosome revealed that it was identical to the Q chromosomes in the critical 20-kb region, but with one important exception: At nucleotide 3072 in *IGF2* intron 3, the Q chromosomes had an A nucleotide, whereas the Meishan chromosome as well as all other q chromosomes had a G nucleotide (Van Laere et al. 2003). A further examination showed that the G3072A substitution occurs in a CpG island that is evolutionarily conserved among placental mammals, and seven other mammals, including humans, have a G at the mutated site (Fig. 4). We concluded that this nucleotide substitution must be the quantitative trait nucleotide (QTN) causing the *IGF2* QTL in pigs (Van Laere et al. 2003).

The functional characterization of the QTN showed that it constitutes a *cis*-acting regulatory mutation that up-regulates postnatal *IGF2* expression in skeletal and cardiac muscle but not in liver (Van Laere et al. 2003). First, we showed by bisulfide sequencing that the mutation does not alter the DNA methylation pattern. Second, we used an electrophoretic mobility-shift assay (EMSA) to show that the mutation disrupts the interaction with an unknown nuclear factor and that this protein only binds the DNA sequence when it is unmethylated. Third, we used northern blot and real-time PCR analysis to show tissue-specific up-regulation of *IGF2* mRNA expression in postnatal muscle. Interestingly, the mutation had no significant effect on *IGF2* expression in prenatal muscle or postnatal liver. The latter finding was in perfect agreement with the observation that the serum IGF2 level was unaltered between genotypes, because liver is the major source of circulating IGF2. Finally, transfection of a luciferase construct including the porcine QTN region and the porcine P3 promoter into mouse C2C12 myoblasts demonstrated that the unknown factor acts as a repressor at the *IGF2* locus.

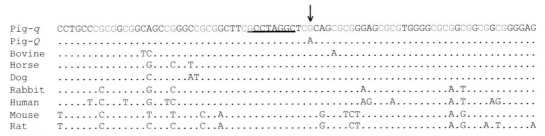

Figure 4. Alignment of the nucleotide sequence for the *IGF2* QTN region in pig intron 3 with the corresponding sequences from seven other placental mammals. The pig wild-type alleles (Pig-*q*) are used as the master sequence; a dot represents sequence identity to the master sequence. Pig-*Q* represents the mutant sequence. Red letters highlight the CG dinucleotides. A short palindrome in the near vicinity of the QTN is underlined. (Modified from Van Laere et al. 2003 [©Nature Publishing Group].)

The initial characterization of the *IGF2* QTN showed that this mutation affects the expression from the *IGF2* P2, P3, and P4 promoters (Van Laere et al. 2003). A subsequent study revealed that the mutation has a similar effect on the *IGF2* antisense transcript (Braunschweig et al. 2004). Thus, the binding of the repressor affects transcription from four promoters spread over 4 kb in the porcine genome. The highly significant effect of the *IGF2* QTN on several phenotypic traits has been confirmed in several subsequent studies (Jungerius et al. 2004; Estellé et al. 2005; Oczkowicz et al. 2009).

The very specific interaction between the wild-type sequence at the *IGF2* QTN and the unknown nuclear factor implied that it should be possible to fish out the factor using a biotin-labeled nucleotide and then determine peptide sequences by mass spectrometry. We have recently achieved this by labeling nuclear proteins from mouse C2C12 cells using the sensitive SILAC (stable isotope labeling of amino acids in culture) method (Markljung et al. 2009). Mass spectrometry analysis combined with database searches against the mouse proteome revealed that the peptides enriched using the wild-type q oligonucleotide but not with the mutant Q oligonucleotide were encoded by an open reading frame (ORF) located in intron 1 of the *Zc3h11a* gene that encodes a poorly characterized zinc finger transcription factor. Further bioinformatic analysis revealed that this ORF encodes a distinct protein comprising more than 900 amino acids with no sequence similarity to ZC3H11A. We have named this protein ZBED6 because it is the sixth mammalian protein containing the BED zinc finger domain. Furthermore, the *ZBED6* coding sequence represents a domesticated DNA transposon belonging to the hAC family that contains many active DNA transposons in fruit flies, maize, and the house fly. *ZBED6* is unique to placental mammals, but remnants of this gene are also found in platypus and opossum; however, the coding sequence is disrupted and not conserved. In contrast, the *ZBED6* coding sequence is highly conserved among all placental mammals sequenced to date. This implies that the integration occurred in a primitive mammal before the divergence of monotremes and other mammals and that it evolved an essential function in a primordial placental mammal after the split between marsupials and placental mammals but before the radiation of different families of placental mammals. Thus, *ZBED6* function is an invention shared by all placental mammals, and it may have contributed to their evolution.

The functional characterization of ZBED6 demonstrated that it is the bona fide repressor binding the QTN site in pig *IGF2* intron 3 (Markljung et al. 2009). An EMSA with recombinant ZBED6 protein confirmed the specific binding to the wild-type but not the mutant oligonucleotide, and a specific supershift of the EMSA complex was observed using an anti-ZBED6 antibody and nuclear extracts from mouse C2C12 cells. Expression analysis in mouse using northern blot analysis, real-time PCR analysis, and immunohistochemistry revealed that ZBED6 has a broad tissue distribution, both in adult animals and during development, indicating that it has a much wider function than regulating muscle growth. This was further supported by the results of chromatin immunoprecipitation using our ZBED6 antibody and mouse C2C12 cells followed by new-generation sequencing (chromatin immunoprecipitation high-throughput sequencing [ChIP-seq]). This experiment revealed 2499 putative ZBED6 binding sites, with a minimum of 15 overlapping reads, which were considered statistically significant. The region corresponding to the pig QTN site in the mouse *Igf2* gene was one of the most highly enriched regions. A consensus binding motif of 5'-GCTCGC-3' was established based on all hits excluding *Igf2*, and this sequence is a perfect match to the pig wild-type sequence at the *IGF2* QTN, whereas the mutant pig sequence is 5'-GCTCAC-3' (Fig. 4). Thus, the results of the ChIP-seq analysis provided further support for ZBED6 being the repressor binding the pig QTN.

As many as 1200 genes in the mouse genome were associated with one or more putative ZBED6-binding sites located within 5 kb of the gene. This gene list was used to search for an enrichment of specific gene ontology (GO) classifications. This analysis revealed that the list of putative ZBED6 downstream targets was a highly nonrandom collection. Genes associated with development, regulation of biological processes, transcriptional regulation, and cell differentiation were highly enriched. More than 20% of the putative ZBED6 targets were other transcription factors. The results suggest that ZBED6 could be a master regulator of transcription in placental mammals.

The pig mutation at the QTN site in intron 3 of *IGF2* has opened up a door to start exploring the biological significance of ZBED6, and further research on this interesting transcription factor may lead to new knowledge concerning the evolution and development of placental mammals. Charles Darwin used domestic animals as a proof of principle for his theory on phenotypic evolution by means of natural selection. The IGF2/ZBED6 story now provides a proof of principle for how research on domestic animals can provide novel insight into basic biology.

ACKNOWLEDGMENTS

Sincere thanks are due to Jennifer Meadows for comments on the manuscript and for assistance in preparing the illustrations. The work summarized in this chapter has been supported by the Swedish Foundation for Strategic Research, the Knut och Alice Wallenbergs Foundation, and The Swedish Research Council for Environment, Agricultural Sciences, and Spatial Planning.

REFERENCES

Amarger V, Nguyen M, Van Laere AS, Braunschweig M, Nezer C, Georges M, Andersson L. 2002. Comparative sequence analysis of the *Insulin-IGF2-H19* gene cluster in pigs. *Mamm Genome* **13:** 388–398.

Andersson L. 2001. Genetic dissection of phenotypic diversity in farm animals. *Nat Rev Genet* **2:** 130–138.

Andersson L, Haley CS, Ellegren H, Knott SA, Johansson M, Andersson K, Andersson-Eklund L, Edfors-Lilja I, Fredholm M, Hansson I, et al. 1994. Genetic mapping of quantitative trait loci for growth and fatness in pigs. *Science* **263:** 1771–1774.

Berg F, Stern S, Andersson K, Andersson L, Moller M. 2006.

Refined localization of the FAT1 quantitative trait locus on pig chromosome 4 by marker-assisted backcrossing. *BMC Genet* **7:** 17.

Braunschweig MH, Van Laere A-S, Buys N, Andersson L, Andersson G. 2004. IGF2 antisense transcript expression in porcine postnatal muscle is affected by a quantitative trait nucleotide in intron 3. *Genomics* **84:** 1021–1029.

Darwin C. 1859. *On the origins of species by means of natural selection or the preservation of favoured races in the struggle for life.* Murray, London.

Darwin C. 1868. *The variation of animals and plants under domestication.* Murray, London.

Ellegren H, Chowdhary BP, Johansson M, Marklund L, Fredholm M, Gustavsson I, Andersson L. 1994. A primary linkage map of the porcine genome reveals a low rate of genetic recombination. *Genetics* **137:** 1089–1100.

Estellé J, Mercadé A, Noguera JL, Pérez-Enciso M, Ovilo C, Sánchez A, Folch JM. 2005. Effect of the porcine IGF2-intron3-G3072A substitution in an outbred Large White population and in an Iberian × Landrace cross. *J Anim Sci* **83:** 2723–2728.

Fang M, Larson G, Soares Ribeiro H, Li N, Andersson L. 2009. Contrasting mode of evolution at a coat color locus in wild and domestic pigs. *PLoS Genet* **5:** e1000341.

Giuffra E, Kijas JMH, Amarger V, Carlborg Ö, Jeon J-T, Andersson L. 2000. The origin of the domestic pig: Independent domestication and subsequent introgression. *Genetics* **154:** 1785–1791.

Giuffra E, Törnsten A, Marklund S, Bongcam-Rudloff E, Chardon P, Kijas JMH, Anderson SI, Archibald AL, Andersson L. 2002. A large duplication associated with Dominant White color in pigs originated by homologous recombination between LINE elements flanking *KIT*. *Mamm Genome* **13:** 569–577.

Jeon J-T, Carlborg Ö, Törnsten A, Giuffra E, Amarger V, Chardon P, Andersson-Eklund L, Andersson K, Hansson I, Lundström K, Andersson L. 1999. A paternally expressed QTL affecting skeletal and cardiac muscle mass in pigs maps to the *IGF2* locus. *Nat Genet* **21:** 157–158.

Johansson M, Ellegren H, Marklund L, Gustavsson U, Ringmar-Cederberg E, Andersson K, Edfors-Lilja I, Andersson L. 1992. The gene for dominant white color in the pig is closely linked to *ALB* and *PDGRFRA* on chromosome 8. *Genomics* **14:** 965–969.

Johansson Moller M, Chaudhary R, Hellmen E, Hoyheim B, Chowdhary B, Andersson L. 1996. Pigs with the dominant white coat color phenotype carry a duplication of the *KIT* gene encoding the mast/stem cell growth factor receptor. *Mamm Genome* **7:** 822–830.

Jungerius BJ, Van Laere A-S, te Pas MFW, van Oost BA, Andersson L, Groenen MAM. 2004. The IGF2-intron3-G3072A substitution explains a major imprinted QTL effect on backfat thickness in a Meishan × European white pig intercross. *Genet Res* **84:** 95–101.

Karlsson EK, Baranowska I, Wade CM, Salmon Hillbertz NH, Zody MC, Andersson N, Biagi TM, Patterson N, Pielberg GR, Kulbokas EJ III, et al. 2007. Efficient mapping of Mendelian traits in dogs through genome-wide association analysis. *Nat Genet* **39:** 1321–1328.

Kerje S, Sharma P, Gunnarsson U, Kim H, Bagchi S, Fredriksson R, Schütz K, Jensen P, von Heijne G, Okimoto R, Andersson L. 2004. The *Dominant white*, *Dun* and *Smoky* color variants in chicken are associated with insertion/deletion polymorphisms in the *PMEL17* gene. *Genetics* **168:** 1507–1518.

Kijas JMH, Andersson L. 2001. A phylogenetic study of the origin of the domestic pig estimated from the near complete mtDNA genome. *J Mol Evol* **52:** 302–308.

Kijas JMH, Wales R, Törnsten A, Chardon P, Moller M, Andersson L. 1998. Melanocortin receptor 1 (*MC1R*) mutations and coat color in pigs. *Genetics* **150:** 1177–1185.

Kijas JMH, Moller M, Plastow G, Andersson L. 2001. A frameshift mutation in *MC1R* and a high frequency of somatic reversions cause black spotting in pigs. *Genetics* **158:** 779–785.

Marklund S, Kijas J, Rodriguez-Martinez H, Ronnstrand L, Funa K, Moller M, Lange D, Edfors-Lilja I, Andersson L. 1998. Molecular basis for the dominant white phenotype in the domestic pig. *Genome Res* **8:** 826–833.

Markljung E, Jiang L, Jaffe JD, Mikkelsen TS, Wallerman O, Larhammar M, Zhang X, Wang L, Saenz-Vash V, Gnirke A, et al. 2009. ZBED6, a novel transcription factor derived from a domesticated DNA transposon regulates IGF2 expression and muscle growth. *PLoS Biol* **7:** e1000256.

Neel JV. 1962. Diabetes mellitus: A "thrifty" genotype rendered detrimental by "progress"? *Am J Hum Genet* **14:** 353–362.

Nezer C, Moreau L, Brouwers B, Coppieters W, Detilleux J, Hanset R, Karim L, Kvasz A, Leroy P, Georges M. 1999. An imprinted QTL with major effect on muscle mass and fat deposition maps to the IGF2 locus in pigs. *Nat Genet* **21:** 155–156.

Oczkowicz M, Tyra M, Walinowicz K, Rózycki M, Rejduch B. 2009. Known mutation (A3072G) in intron 3 of the *IGF2* gene is associated with growth and carcass composition in Polish pig breeds. *J Appl Genet* **50:** 257–259.

Paterson AH, Lander ES, Hewitt JD, Peterson S, Lincoln SE, Tanksley SD. 1988. Resolution of quantitative traits into Mendelian factors by using a complete linkage map of restriction fragment length polymorphisms. *Nature* **335:** 721–726.

Pielberg G, Olsson C, Syvänen A-C, Andersson L. 2002. Unexpectedly high allelic diversity at the *KIT* locus causing dominant white color in the domestic pig. *Genetics* **160:** 305–311.

Pielberg G, Day AE, Plastow GS, Andersson L. 2003. A sensitive method for detecting variation in copy numbers of duplicated genes. *Genome Res* **13:** 2171–2177.

Robbins LS, Nadeau JH, Johnson KR, Kelly MA, Roselli-Rehfuss L, Baack E, Mountjoy KG, Cone RD. 1993. Pigmentation phenotypes of variant extension locus alleles result from point mutations that alter MSH receptor function. *Cell* **72:** 827–834.

Spritz RA. 1994. Molecular basis of human piebaldism. *J Invest Dermatol* **103:** 137S–140S.

Van Laere AS, Nguyen M, Braunschweig M, Nezer C, Collette C, Moreau L, Archibald AL, Haley CS, Buys N, Andersson G, et al. 2003. A regulatory mutation in *IGF2* causes a major QTL effect on muscle growth in the pig. *Nature* **425:** 832–836.

Fine Mapping a Locus Controlling Leg Morphology in the Domestic Dog

P. QUIGNON,[1] J.J. SCHOENEBECK,[1] K. CHASE,[2] H.G. PARKER,[1] D.S. MOSHER,[1] G.S. JOHNSON,[3] K.G. LARK,[2] AND E.A. OSTRANDER[1]

[1]Cancer Genetics Branch, National Human Genome Research Institute, National Institutes of Health, Bethesda, Maryland 20892; [2]Department of Biology, University of Utah, Salt Lake City, Utah 84112-0840; [3]Department of Veterinary Pathobiology, University of Missouri, Columbia, Missouri 65211

Correspondence: eostrand@mail.nih.gov

The domestic dog offers a remarkable opportunity to disentangle the genetics of complex phenotypes. Here, we explore a locus, previously identified in the Portuguese water dog (PWD), associated with PC2, a morphological principal component characterized as leg width versus leg length. The locus was initially mapped to a region of 26 Mb on canine chromosome 12 (CFA12) following a genome-wide scan. Subsequent and extensive genotyping of single-nucleotide polymorphisms (SNPs) and haplotype analysis in both the PWD and selected breeds representing phenotypic extremes of PC2 reduced the region from 26 Mb to 500 kb. The proximity of the critical interval to two collagen genes suggests that the phenotype may be controlled by *cis*-acting mechanisms.

Identification of the molecular underpinnings that give rise to morphological variation in mammals represents a formidable challenge (Chase et al. 1999; Lark et al. 2005; Wayne and Ostrander 2007). The genetic basis of many morphological traits is presumed to be polygenic, as evidenced by deviation from simple Mendelian inheritance. Individually, each genetic variant may contribute very little to altering the basic body plan of an organism (Chase et al. 1999; Lark et al. 2005). Collectively, these variants generate a quantitative effect whose sum imparts the basis of individuality (Chase et al. 2002; Lark et al. 2005). Further complicating their identification is the fact that many genetic variants responsible for complex traits are rare and tend to segregate within subpopulations (Ostrander and Kruglyak 2000). Thus, it is impractical to map complex traits using family-based approaches, which are typically underpowered.

The canine model system provides fertile ground for identifying the genetic basis that underlies complex, quantitative traits such as morphology (Lark et al. 2005; Ostrander and Wayne 2005; Wayne and Ostrander 2007). Modern day, domestic, purebred dogs display morphological diversity unparalleled by any other terrestrial species: Skeletal size, weight, and height vary tremendously from breed to breed and can be readily quantified with minimal invasiveness (Wayne 1986a,b; Wayne and Ostrander 2007). Belying their physical diversity, the underlying genetic complexity of purebred dogs is comparatively simple relative to humans. Population bottlenecks, closed breeding practices, and the use of popular sires have gradually reduced genetic diversity within breeds while exaggerating diversity between breeds (Ostrander et al. 2000; Parker et al. 2007). As a direct consequence, intrabreed linkage disequilibrium (LD) extends often on the order of megabases, at times nearly 10-fold that observed in humans (Sutter et al. 2004; Lindblad-Toh et al. 2005), whereas between breeds, LD rapidly decays (Lindblad-Toh et al. 2005). In practice, the long LD encountered in pedigree analyses is useful for coarse mapping, such as pinning quantitative trait loci (QTLs) to broad, subchromosomal regions, and multibreed comparisons facilitate fine mapping and haplotype identification, as evidenced by the recent identification of genes for a variety of diseases and morphologic traits (Zangerl et al. 2006; Parker et al. 2007, 2009; Sutter et al. 2007; Karlsson and Lindblad-Toh 2008). Coupled with a 7.5x, high-quality genomic assembly and a dense set of identified single-nucleotide polymorphisms (SNPs) (Kirkness et al. 2003; Lindblad-Toh et al. 2005), the dog model offers the potential to map QTLs down to their causal variants.

Previously, pedigree studies using microsatellite markers were used to demonstrate the feasibility of mapping quantitative traits in PWDs, which display skeletal heterogeneity despite originating from only 30 founders (Chase et al. 1999, 2002). Using principal components analysis (PCA) to categorize radiograph measurements from 463 dogs, we previously mapped QTLs for several morphological traits. Principal component 1 (PC1) accounted for 54% of the skeletal variation observed in the PWDs and its loadings described overall skeletal size. The best marker association for PC1 was for a QTL located on chromosome 15 (Chase et al. 2002), and fine mapping at this locus revealed a single haplotype spanning the insulin growth factor 1 (*IGF1*) gene that cosegregated with small size in PWDs and approached fixation among many additional small-breed dogs (Sutter et al. 2007). These studies demonstrated that a genetic variant responsible for intrabreed morphological traits can also segregate among breeds that represent phenotypic extremes for the same trait. Furthermore, because LD varies among breeds, cross-breed comparisons proved integral to resolving a minimal haplotype (Sutter et al. 2007).

The initial mapping studies of Chase et al. (2002) identified PC2 as an inverse correlation between leg length and width. A QTL for PC2 mapped to chromosome 12 (CFA12) (Lark et al. 2006). Our analysis of SNPs segregating in PWDs has confirmed that association and prompted our fine mapping of the locus, which is summarized below.

MATERIALS AND METHODS

Sample collection and DNA isolation. DNA was isolated from blood samples collected from registered dogs of established breeds with at least three generations of available pedigree data. Individuals were recruited through American Kennel Club (AKC)-sanctioned dog shows, specialty events, breed clubs, and veterinary clinics. Samples were collected by licensed veterinarians or trained veterinary technicians through venipuncture of the cephalic vein using standard protocols approved by the national Human Genome Research Institute (NHGRI) Animal Care and Use Committee. In all cases, blood samples were collected into ACD or EDTA anticoagulant and shipped at room temperature to the Ostrander lab. Samples were stored at 4°C before DNA extraction. DNA isolation was performed by HealthGene, Inc. (Ontario, Canada) using standard proteinase K/phenol:chloroform isolation procedures. Quantitated DNA samples were suspended in 10 mM Tris base, 0.1 mM EDTA, aliquoted, and stored at –80°C.

Marker discovery and selection. All SNP positions given correspond to CanFam2. A total of 354 SNPs were selected for genotyping based on known markers included in the dog assembly (Lindblad-Toh et al. 2005) (http://www.broad.mit.edu/node/459, http://www.ncbi.nlm.nih.gov/projects/SNP/) or those discovered during Sanger sequencing of amplicons. Primers were designed with Primer3 (Rozen and Skaletsky 2000) using standard parameters, T_m = 60°C, and an optimal amplicon length of 500–700 bp. Segments were amplified using standard polymerase chain reaction (PCR) protocols and a 40-cycle touch-down thermocycler program at 61°C–51°C. Sanger sequencing was done using BigDye terminator sequencing kits and standard protocols (Applied Biosystems, Santa Clara, California). Traces were analyzed using PhredPhrap and Consed (Ewing and Green 1998; Ewing et al. 1998; Gordon et al. 1998). For fine mapping in the 500-kb region, 118 amplicons were designed and a total of 65,938 bp out of 416,264 bp were entirely sequenced, i.e., 16% of the region.

Phenotypic assessment of breeds. Phenotype assessment was based on tape measurement data (Sutter et al. 2008) collected from individual dogs of distinct breeds. For symmetrical left and right leg measurements, data were averaged. In total, 19 measurements (Table 1) were included for PCA from only those breeds with at least three representative members (707 dogs from 66 breeds). Analysis of measurement data was done using the nipalsPca method, part of the pcaMethods package implementable for R statistical software.

Table 1. Factor Loadings of PC2 in a Set of 19 Measurements

Measurements	PC2 loadings
Width/circumference	
chest	–0.375
head	–0.374
forefoot	–0.333
neck girth	–0.311
eye width	–0.305
hindfoot	–0.275
abdominal girth	–0.070
Length	
head	–0.086
body	0.033
forefoot	0.071
upper hind leg	0.097
upper fore leg	0.150
hindfoot	0.150
neck	0.173
snout	0.185
height at base of tail	0.196
height at withers	0.222
lower fore leg	0.243
lower hind leg	0.244

A total of 707 dogs from 66 breeds were used for PCA. Factor loadings for PC2 indicate that leg circumference measurements are inversely correlated with length measurements.

Genotyping and association analyses. For the Illumina Golden Gate assay (Fan et al. 2006), 123 PWDs were genotyped according to the manufacturer's protocol (Illumina Inc., San Diego, California). Genotypes were called using Beadstudio data analysis software v3.1 (Illumina Inc.). Association analysis was done using the Wald test implemented in PLINK (Purcell et al. 2007). Additional SNP genotyping was performed on 296 dogs using the SNPlex genotyping system and an ABI 3730XL genome analyzer with standard protocols (Applied Biosystems, Santa Clara, California). Genotype calls were generated using GeneMapper software v4.0 (Applied Biosystems). Of the 354 SNPs assayed, those with >20% missing data or a minor allele frequency of <1% were discarded, leaving 234 SNPs for further analysis. Individuals with >40% missing genotypes were also discarded from analysis, leaving 243 dogs for analyses (35 ThL dogs and 208 TiS dogs). Single-marker χ^2 association was performed using PLINK (Purcell et al. 2007). Association testing of predicted haplotypes was done using Haploview (Barrett et al. 2005). The 1-Mb region defined by Haploview contains 90 SNPs. Those 90 SNPs were used to infer haplotypes using PHASE v2.1 (Stephens et al. 2001) for comparison between breeds.

RESULTS

A whole-genome scan of 463 PWDs using microsatellites identified a locus on CFA12 that controls PC2, the ratio of leg length to width (Fig. 1; Table 2) (Lark et al. 2006). The best-associated microsatellite, FH3585, was located at position 39,124,161 and had a p-value of 1.11×10^{-16}. To confirm this association, an Illumina Golden Gate custom assay with 384 SNPs spanning the 26 Mb from base pairs 23,066,909 to 49,074,637 was designed; 123 PWDs for which PC2 values were available were geno-

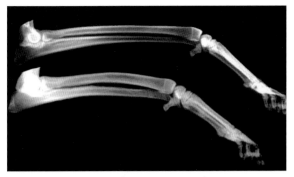

Figure 1. Radiographs of radius and ulna forelimb bones from PWDs with extreme PC2 values. Compare thin, longer limb bones (*top*) to thick, shorter limb bones (*bottom*).

Table 2. Effects of the D Allele of FH3585 on Limb Length and Width in PWDs

Trait[a]	tstat[b]	p-value[c]
Length		
radius	−12.5	$<10^{-10}$
tibia	−11.3	$<10^{-10}$
humerus	−9.8	$<10^{-10}$
femur	−6.2	8.22×10^{-10}
metatarsal	−5.6	2.14×10^{-08}
metacarpal	−3.8	9.93×10^{-05}
foot	−3.6	1.92×10^{-04}
Width		
femur.I.D.	2.7	3.15×10^{-03}
humerus.I.D.	3.0	1.45×10^{-03}
tibia.I.D.	3.2	7.94×10^{-04}
tibia.O.D.	3.3	4.70×10^{-04}
humerus.O.D.	3.4	3.54×10^{-04}
femur.O.D.	4.7	2.15×10^{-06}
radius.O.D.	5.6	1.62×10^{-08}
radius.I.D.	5.9	3.04×10^{-09}

Allele D effects on length and width were estimated using a mixed model with consanguinity between pairs of dogs as the random effect (pedigree effect) and allele count (0,1,2) as the fixed effect.
[a]Values are residuals after removing the effect of size (PC1).
[b]Represents the magnitude (number of standard errors) and direction of the allele effect with respect to the population as a whole.
[c]Indicates the significance of the effect.

typed. The best-associated SNP had a *p*-value of 2.50×10^{-08} and was located at position 35,077,940 bp. Seven additional contiguous SNPs located from base pairs 37,890,443 to 38,165,577 yielded significant results with *p*-values ranging from 5.55×10^{-05} to 6.14×10^{-07} (Fig. 2, top). Analysis of assembled haplotypes revealed that PWDs with positive PC2 scores were more homozygous for one haplotype in the 275,134-bp region spanning base pairs

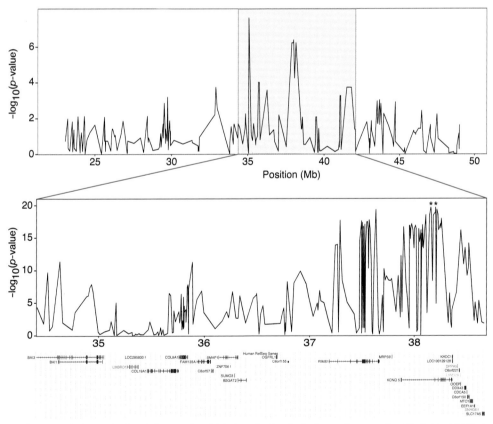

Figure 2. Association analysis results from PWDs and multiple breeds. Graphs indicate χ^2 test results of individual SNPs (*top and bottom*). (*Top*) Association analysis of 384 SNPs spanning 26 Mb in 123 PWDs. Positional overlap between PWDs and multiple breed analyses is indicated by the pink box. (*Bottom*) Expanded view of the multiple breed association analysis, which includes 234 SNPs from 243 dogs. The two best associated SNPs are indicated by asterisks. University of California at Santa Cruz browser view of the 4-Mb region genotyped in the multibreed fine mapping depicts human orthologs found within the CFA12 locus.

37,890,443 to 38,165,577 than were PWDs with negative PC2 scores (25/31 compared to 7/30, respectively, p-value = 7.43×10^{-6}). However, LD was, as expected, extensive in the breed due to the small number of founders.

To reduce LD and thereby eliminate regions not associated with PC2, we genotyped dogs from several unrelated breeds. To assess breed average leg length/thickness variation, we used tape measurement data from 707 dogs of 66 distinct breeds to run PCA (Sutter et al. 2008). Although our measurements did not include leg bone diameter, they did include leg circumference, which we assumed would serve as a surrogate for leg bone diameter. As expected, factor loadings for the multibreed PC2 indicated that leg length was inversely correlated to leg circumference (Table 1). Mean PC2 values were calculated for each of the 66 breeds and those with extreme PC2 mean values were selected for fine-mapping studies (Fig. 3). For simplicity, we refer to individuals at the ends of the PC2 continuum as thin, long-legged (ThL) or thick, short-legged (TiS) dogs.

The PWD results indicated that the region of strongest interest was between base pairs 35,077,940 and 38,165,577. We selected 354 putative SNPs listed in the CanFam assembly (Lindblad-Toh et al. 2005) as well as SNPs discovered by sequencing in the region spanning base pairs 34,414,776 to 38,676,086 (i.e., >4 Mb) and containing 30 genes (Fig. 2, bottom). In total, we genotyped 296 dogs from 32 breeds (60 dogs from 8 ThL breeds and 236 dogs from 24 TiS breeds) using the expanded marker set. Testing for association using χ^2 analysis of individual SNPs revealed two SNPs at positions 38,159,975 and 38,207,563, which demonstrated highly significant associations (p-value = 2.24×10^{-20}; Fig. 2, bottom).

Using the program Haploview (Barrett et al. 2005), we identified associated haplotypes across the region. This analysis showed that seven contiguous LD blocks between base pairs 37,241,446 and 38,263,102 had significant p-values <10^{-13}, thus reducing our region of interest to ~1 Mb. This region includes only three genes: *RIMS1*, *MRPS9*, and *KCNQ5*. Haplotype sharing among the different breeds was evaluated using PHASE inference (Stephens et al. 2001) and the 90 SNPs contained in the 1-Mb region. The most common haplotypes in the region were carried by 8 of 10 "chondrodysplastic" breeds, as well as the American cocker spaniel, all of whom are considered to be TiS-type breeds. Chondrodysplastic breeds are those that display fore-shortened limbs resembling disproportional dwarfism (American

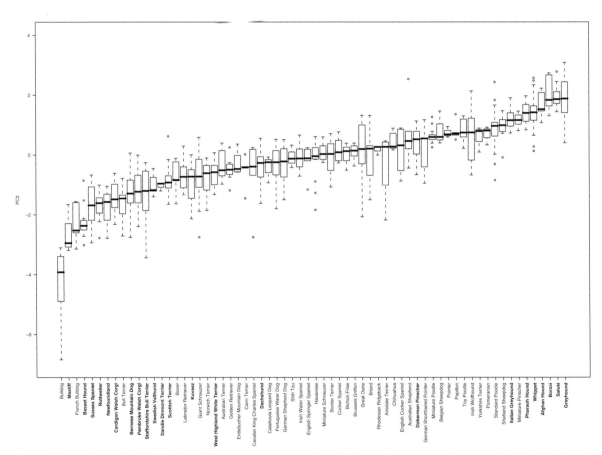

Figure 3. Box whisker plot of PC2 by breed. Nineteen measurements from 707 dogs of various breeds were analyzed by PCA. In each box, the thick black line indicates the median PC2 value, whereas the upper and lower limits of each box represent the first and third quartiles, respectively. Dashed lines extend for data points within 1.5× the interquartile range. Open circles indicate outliers. A continuum of leg length/width ratio is observed among dogs of various breeds. Names in bold indicate breeds that were chosen for the multibreed fine mapping of the locus.

Kennel Club 1998; Parker et al. 2009). We also generated PHASE-inferred haplotypes using 20 SNP sliding windows at 10 SNP intervals for a total of eight windows. Only the first two windows contributed to the PHASE result that was observed with the entire region; subsequent windows indicated that an Italian greyhound, a ThL breed, shared the haplotype carried by the majority of TiS breeds. Thus, this last analysis allowed us to narrow the region of interest to between base pairs 37,241,446 and 37,657,100, which is less than 500 kb.

To further refine this 500-kb region, amplicons were designed and sequenced to discover new SNPs. These SNPs were then genotyped on a panel of 55 dogs: 22 ThL, 26 chondrodysplastic TiS, and 7 TiS nonchondrodysplastic. In total, 119 SNPs and indels were genotyped. Following quality filtering, 85 SNPs from 55 dogs were kept for association analysis using PLINK (Purcell et al. 2007). This analysis revealed 14 SNPs and one indel with p-values $<10^{-9}$. Six of these SNPs were homozygous among ThL dogs (the exception being one heterozygous Italian greyhound), whereas thick-legged dogs carried all three genotypes. Those six SNPs are localized between base pairs 37,577,209 and 37,581,015, in a region where 25,884 bp was entirely sequenced (between base pairs 37,560,863 and 37,586,747) (Fig. 4). We also ran PHASE to infer haplotypes within the 26-kb region that was sequenced (Fig. 5).

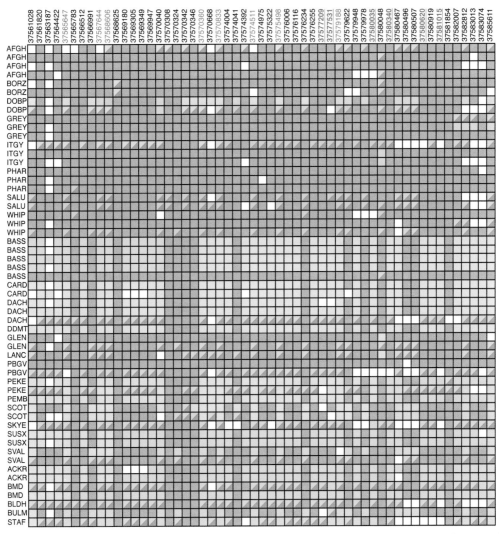

Figure 4. Genotyping results of 26 kb using 55 dogs. Each line represents a dog and each column represents a SNP or an indel. The first 22 dogs are ThL dogs, and the remaining 33 dogs are TiS dogs (26 "chondrodysplastic" and 7 nonchondrodysplastic). (Blue) Homozygous genotypes of the major allele among ThL dogs, (yellow) homozygous genotypes of the minor allele, (double-colored) heterozygous genotypes, (white spaces) missing genotypes. SNPs with p-values $<10^{-9}$ are indicated in red, and those in homozygous regions in ThL breeds are boldfaced and underlined. Breed abbreviations: Afghan hound (AFGH), borzoi (BORZ), doberman pinscher (DOBP), greyhound (GREY), Italian greyhound (ITGY), Pharaoh hound (PHAR), saluki (SALU), whippet (WHIP), basset hound (BASS), Cardigan Welsh corgi (CARD), dachshund (DACH), dandie dinmont terrier (DDMT), Glen of Imaal terrier (GLEN), Lancashire terrier (LANC), petit basset griffon vendéen (PBGV), pekingese (PEKE), Pembroke Welsh corgi (PEMB), Scottish terrier (SCOT), Skye terrier (SKYE), Sussex spaniel (SUSX), Swedish Vallhund (SVAL), American cocker spaniel (ACKR), Bernese mountain dog (BMD), bloodhound (BLDH), bullmastiff (BULM), Staffordshire bull terrier (STAF).

37561028	37561820	37564422	37565647	37566512	37566991	37567644	37568606	37569180	37569305	37569349	37569947	37570040	37570324	37570342	37570346	37570360	37570668	37570833	37574004	37574041	37574392	37574511	37574975	37575322	37575498	37576006	37576116	37576234	37576255	37577209	37577531	37579188	37579622	37579948	37579978	37580035	37580048	37580340	37580467	37580496	37580507	37580620	37580919	37581015	37582007	37582812	37583013	37585611	% ThL breeds	% TiS breeds
A	C	A	C	C	T	G	C	A	C	A	D	G	G	A	I	T	C	D	A	C	C	D	G	T	D	C	A	A	G	G	C	T	G	G	A	A	I	T	G	C	T	G	D	A	T	G	A	C	0	20.1
G	G	C	A	T	T	A	T	G	T	G	I	A	G	A	I	C	T	I	G	C	T	I	A	C	I	T	G	A	A	T	T	C	G	A	A	G	I	C	C	C	T	A	I	G	G	A	C	T	43.6	2.9
G	G	C	A	T	T	A	T	G	T	G	I	A	G	A	I	C	T	I	G	C	T	I	A	C	I	T	G	A	A	T	T	C	G	A	A	G	D	C	C	C	T	A	I	G	G	A	C	T	27.2	2.9

Figure 5. PHASE-inferred haplotypes for 57 dogs. PHASE inference was used to generate haplotypes from polymorphisms genotyped within the 26-kb region that was sequenced. Alleles of the three most common haplotypes (>5 chromosomes) are depicted according to position; the major allele among ThL dogs are colored blue and the minor allele yellow. Percentages indicate the number of haplotypes counted, divided by the total number of chromosomes for ThL or TiS dogs.

Two closely related haplotypes were found among 70.8% of ThL chromosomes, whereas among TiS breeds, the most common haplotype accounted for only 20% of chromosomes. These results suggest that the CFA12 genetic variant was selected to produce animals with long, thin legs.

DISCUSSION

A pedigree-based, genome-wide microsatellite scan using PWDs indicated that a QTL on CFA12 is a major determinant of skeletal PC2 (leg length vs. leg width; Lark et al. 2006). By using a multibreed approach for fine mapping, the region of interest was initially reduced from 26 Mb to just over 3 Mb, and then finally to ~500 kb. The fine mapping results define a broad region of homozygosity found only among ThL dogs, suggesting that the CFA12 variant mapped here may function specifically to increase the ratio of leg length versus width.

Additional evidence of a leg morphology determinant on CFA12 comes from a separate study investigating the genetics of chondrodysplasia in domestic dogs. In this multibreed GWAS (genome-wide association study), a highly significant association was found on chromosome 18, corresponding to a *FGF4* retrogene insertion (Parker et al. 2009). Other loci with less significant associations were identified in that study, one of them located on CFA12 (p-value = 5×10^{-33}) in a region corresponding to the CFA12 locus described here. Given the aforementioned homozygosity found among ThL breeds, we speculate that this genetic variant is incompatible with the *FGF4* retrogene, because most chondrodysplastic dogs display the retrogene insertion and, simultaneously, rarely carry the ThL haplotype.

According to the CanFam2 assembly (Lindblad-Toh et al. 2005), the QTL defined by the initial PWD analyses contain at least 30 genes. However, two exceptional candidates, *COL19A1* and *COL9A1*, located within this 3-Mb region, stand out: Both encode collagen genes, whose function in bone matrix formation makes them appealing candidates for regulating limb bone morphology. *Col9a1*-null mice display a reduction in long-bone growth that coincides with bone thickening (Dreier et al. 2008). Furthermore, *TSP3/TSP5/Col9a1* combinatorial knockout mice display skeletal defects including a reduction in limb length, a phenotype attributed to disorganization of the growth plates (Posey et al. 2008). Despite reducing the critical interval nearly sixfold, we were unable to determine the actual causal variant using our multibreed approach. In fact, the strongest associations and best haplotype from our multibreed analyses indicate that the relevant genetic variant is located upstream of both collagen genes. The 500-kb region identified by our fine mapping is devoid of genes with known osteogenic function, and sequencing of the *RIMS1* exons located in this region failed to reveal SNPs in strong association with PC2. On the basis of these results, it is likely that the causal mutation is hidden within intergenic sequences, perhaps affecting a cryptic distant enhancer of a neighboring gene, or the mutation could occur as a novel gain-of-function insertion, or rearrangement, as recently demonstrated by Parker et al. (2009). Ultimately then, discovery of the causal mutation will likely require sequencing of the entire 500-kb region in a panel of TiS and ThL dog breeds. Although such an endeavor would have previously been impractical, targeted sequence capture technologies coupled with next-generation sequencing should allow deep sequencing across the 500-kb region in the very near future.

PC2 highlights a trade-off in functional morphology between energy-efficient speed (long, thin [light] bones, e.g., the greyhound) and the generation of force (short, thick bones, e.g., the pit bull) (Chase et al. 2002). This trade-off appears to be ancient and is found in the red fox (*Vulpes vulpes*) (Kharlamova et al. 2007), a lineage that represents the outgroup of modern canids separated from *Canis familiaris* by 10 million years (Wayne 1993; Vila et al. 1997). This indicates an evolutionarily conserved coordinate network of growth regulation. The locus on CFA12 may be modifying part of this network along a functional continuum, regulating elongation—the endochondrial growth that takes place at the epiphysial plate—versus periosteal growth that increases bone width. It remains unclear whether one gene controls both growth zones or whether there are two closely linked loci within the large LD spanned by the 500-kb interval. Future studies using the approach exemplified here should discriminate between these two possibilities. However, the fact that one linkage group can regulate such a trade-off explains why it has been possible for breeders to rapidly change the functional morphology of dogs encompassing a range that includes so many different morphological types.

ACKNOWLEDGMENTS

We gratefully acknowledge the National Institutes of Health grant GM063056 (K.G.L. and K.C.), the AKC Canine Health Foundation, and the Intramural Program of the National Human Genome Research Institute. Radiographs of PWDs were obtained through the Georgie Project, Karen Miller, Director. K.G.L. acknowledges gifts from the Judith Chiara Charitable Trust and the Nestle Purina Company. We thank Dr. Robert K. Wayne and members of our laboratories for helpful comments. Finally, we thank also the many dog owners who generously provided us with samples from their pets.

REFERENCES

American Kennel Club. 1998. *The complete dog book*. Howell, New York.

Barrett JC, Fry B, Maller J, Daly MJ. 2005. Haploview: Analysis and visualization of LD and haplotype maps. *Bioinformatics* **21**: 263–265.

Chase K, Adler FR, Miller-Stebbings K, Lark KG. 1999. Teaching a new dog old tricks: Identifying quantitative trait loci using lessons from plants. *J Hered* **90**: 43–51.

Chase K, Carrier DR, Adler FR, Jarvik T, Ostrander EA, Lorentzen TD, Lark KG. 2002. Genetic basis for systems of skeletal quantitative traits: Principal component analysis of the canid skeleton. *Proc Natl Acad Sci* **99**: 9930–9935.

Dreier R, Opolka A, Grifka J, Bruckner P, Grossel S. 2008. Collagen IX-deficiency seriously compromises growth cartilage development in mice. *Matrix Biol* **27**: 319–329.

Ewing B, Green P. 1998. Base-calling of automated sequencer traces using phred. II. Error probabilities. *Genome Res* **8**: 186–194.

Ewing B, Hillier L, Wendl MC, Green P. 1998. Base-calling of automated sequencer traces using phred. I. Accuracy assessment. *Genome Res* **8**: 175–185.

Fan JB, Chee MS, Gunderson KL. 2006. Highly parallel genomic assays. *Nat Rev Genet* **7**: 632–644.

Gordon D, Abajian C, Green P. 1998. Consed: A graphical tool for sequence finishing. *Genome Res* **8**: 195–202.

Karlsson EK, Lindblad-Toh K. 2008. Leader of the pack: Gene mapping in dogs and other model organisms. *Nat Rev Genet* **9**: 713–725.

Kharlamova AV, Trut LN, Carrier DR, Chase K, Lark KG. 2007. Genetic regulation of canine skeletal traits: Trade-offs between the hind limbs and forelimbs in the fox and dog. *Integr Comp Biol* **47**: 373–381.

Kirkness EF, Bafna V, Halpern AL, Levy S, Remington K, Rusch DB, Delcher AL, Pop M, Wang W, Fraser CM, Venter JC. 2003. The dog genome: Survey sequencing and comparative analysis. *Science* **301**: 1898–1903.

Lark KG, Chase K, Sutter NB. 2006. Genetic architecture of the dog: Sexual size dimorphism and functional morphology. *Trends Genet* **22**: 537–544.

Lark KG, Chase K, Carrier DR, Adler FR. 2005. Genetic analysis of the canid skeleton: Morphological loci in the Portuguese water dog population. In *The dog and its genome*, (ed. EA Ostrander et al.), pp. 67–80. Cold Spring Harbor Laboratory Press, Cold Spring Harbor, New York.

Lindblad-Toh K, Wade CM, Mikkelsen TS, Karlsson EK, Jaffe DB, Kamal M, Clamp M, Chang JL, Kulbokas EJ, Zody MC, et al. 2005. Genome sequence, comparative analysis and haplotype structure of the domestic dog. *Nature* **438**: 803–819.

Ostrander EA, Kruglyak L. 2000. Unleashing the canine genome. *Genome Res* **10**: 1271–1274.

Ostrander EA, Wayne RK. 2005. The canine genome. *Genome Res* **15**: 1706–1716.

Ostrander EA, Galibert F, Patterson DF. 2000. Canine genetics comes of age. *Trends Genet* **16**: 117–124.

Parker H, Kukekova A, Akey D, Goldstein O, Kirkness EF, Baysac K, Mosher DS, Aguirre G, Acland GM, Ostrander EA. 2007. Breed relationships facilitate fine mapping studies: A 7.8 Kb deletion cosegregates with collie eye anomaly across multiple dog breeds. *Genome Res* **17**: 1562–1571.

Parker HG, VonHoldt BM, Quignon P, Margulies EH, Shao S, Mosher DS, Spady TC, Elkahloun A, Cargill M, Jones PG, et al. 2009. An expressed fgf4 retrogene is associated with breed-defining chondrodysplasia in domestic dogs. *Science* (in press).

Posey KL, Hankenson K, Veerisetty AC, Bornstein P, Lawler J, Hecht JT. 2008. Skeletal abnormalities in mice lacking extracellular matrix proteins, thrombospondin-1, thrombospondin-3, thrombospondin-5, and type IX collagen. *Am J Pathol* **172**: 1664–1674.

Purcell S, Neale B, Todd-Brown K, Thomas L, Ferreira MA, Bender D, Maller J, Sklar P, de Bakker PI, Daly MJ, Sham PC. 2007. PLINK: A tool set for whole-genome association and population-based linkage analyses. *Am J Hum Genet* **81**: 559–575.

Rozen S, Skaletsky H. 2000. Primer3 on the WWW for general users and for biologist programmers. *Methods Mol Biol* **132**: 365–386.

Stephens M, Smith NJ, Donnelly P. 2001. A new statistical method for haplotype reconstruction from population data. *Am J Hum Genet* **68**: 978–989.

Sutter NB, Mosher DS, Gray MM, Ostrander EA. 2008. Morphometrics within dog breeds are highly reproducible and dispute Rensch's rule. *Mamm Genome* **19**: 713–723.

Sutter NB, Eberle MA, Parker HG, Pullar BJ, Kirkness EF, Kruglyak L, Ostrander EA. 2004. Extensive and breed-specific linkage disequilibrium in *Canis familiaris*. *Genome Res* **14**: 2388–2396.

Sutter NB, Bustamante CD, Chase K, Gray MM, Zhao K, Zhu L, Padhukasahasram B, Karlins E, Davis S, Jones PG, et al. 2007. A single *IGF1* allele is a major determinant of small size in dogs. *Science* **316**: 112–115.

Vila C, Savolainen P, Maldonado JE, Amorim IR, Rice JE, Honeycutt RL, Crandall KA, Lundeberg J, Wayne RK. 1997. Multiple and ancient origins of the domestic dog. *Science* **276**: 1687–1689.

Wayne RK. 1986a. Cranial morphology of domestic and wild canids: The influence of development on morphological change. *Evolution* **40**: 243–261.

Wayne RK. 1986b. Limb morphology of domestic and wild canids: The influence of development on morphologic change. *J Morphol* **187**: 301–319.

Wayne RK. 1993. Molecular evolution of the dog family. *Trends Genet* **9**: 218–224.

Wayne RK, Ostrander EA. 2007. Lessons learned from the dog genome. *Trends Genet* **23**: 557–567.

Zangerl B, Goldstein O, Philp AR, Lindauer SJ, Pearce-Kelling SE, Mullins RF, Graphodatsky AS, Ripoll D, Felix JS, Stone EM, et al. 2006. Identical mutation in a novel retinal gene causes progressive rod-cone degeneration in dogs and retinitis pigmentosa in humans. *Genomics* **88**: 551–563.

Human Origins and Evolution: Cold Spring Harbor, Déjà Vu

T.D. WHITE

Department of Integrative Biology, and Human Evolution Research Center,
University of California, Berkeley, California 94720
Correspondence: timwhite@berkeley.edu

The Cold Spring Harbor Symposia of the 1950s were key to integrating human evolutionary studies into biology. That integration provided a solid foundation for systematic and functional interpretations of an expanding base of fossil and molecular evidence during the latter half of the 20th century. Today, the paleontological record of human evolution amassed during the last 150 years illuminates the human clade on life's tree. However, the rise of Hennegian parsimony cladistics and punctuationalism during the end of the last century witnessed the partial abandonment of classificatory conventions cemented by Mayr, Simpson, Dobzhansky, and others at Cold Spring Harbor. This has led to an artificial, postmillennial amplification of apparent species diversity in the hominid clade. Work on a stratigraphically thick and temporally deep sedimentary sequence in the Middle Awash study area of Ethiopia's Afar Depression reveals an assembly order of hominid anatomies and behaviors that was impossible for Darwin to discern. Large parts of that record appear to reflect phyletic evolution, consistent with the lessons and expectations of Cold Spring Harbor in 1950. Molecular biology cannot reveal the assembly sequences or contexts of human origins and evolution without reference to adequate geological, geochronological, paleobiological, and archaeological records. Today's consilience of these disparate data sets would have impressed Charles Darwin.

Darwin would have been astonished and delighted to witness the 2009 Cold Spring Harbor (CSH) Laboratory's Symposium on Quantitative Biology anniversary celebrations of his birth and book. He would have recognized the many persistent themes we discussed, taken satisfaction in the hundreds of mechanisms revealed, and been amazed by the broad advancing front of modern evolutionary biology. From "shadow" enhancers (Hong et al. 2008) to segmental duplications (Marques-Bonet et al. 2009) and from ancient fossils to the "cognitive niche" (Pinker 2003), ours is a world full of insights unavailable to Darwin in 1859.

In his historical scientist mode, Darwin was directly concerned with the paleontological, neontological, and contextual data resulting from the natural, one-time, uncontrolled experiment of life on earth. Darwin clearly understood how the rich data sources of the neontological realm were living products of that vast experiment. And the phylogenetic and functional elucidation of how extant diversity has arisen—now provided by the modern landscape of molecular biology—is truly astonishing, even in the hindsight of a single decade. These revelations make it too easy to forget what Darwin clearly appreciated—that the historical record of fossils, artifacts, and contexts is crucial to the fullest understanding of our evolution. The organizers of this Symposium deserve congratulations for recognizing this fact and for welcoming a broad community to join the celebration.

I am not a historian of science, but the opportunity to participate in the Symposium provides an opportunity to highlight the important historical role that earlier gatherings at the Cold Spring Harbor Laboratory had in the development of human evolutionary studies. Celebrations here in 1950 and 1959 cemented major elements of the modern synthesis and also served to catalyze the integration of human evolutionary studies into modern biology. The integration came with important lessons sometimes lost on modern practitioners.

Human evolution was touched upon ever so lightly in Darwin's 1859 *On the Origin of Species* that we celebrate this year. There, Darwin devoted detailed attention to "Imperfections in the Geological Record," perhaps because he saw such gaps as rendering his theory vulnerable to critics (Sepkoski and Ruse 2009). His 1871 treatise on human evolution pondered what was then one of the largest imperfections of earth's historical record—the paucity of truly early hominid fossil remains (family Hominidae bounds genera in the human clade after the last common ancestor we shared with chimpanzees). We have come a long way. I review this progress, discuss the impacts of the 1950 and 1959 CSH Symposia, and illustrate current investigations of hominid evolution by using examples from earth's most important repository of data on human origins and evolution, Ethiopia's Afar Depression.

DARWIN ON HOMINIDS

Living humans are obviously anatomically, physiologically, and behaviorally uniquely different from our closest living relatives, the African apes. What was the sequence by which natural selection assembled our obvious derivations of brain expansion, canine reduction, technology, and bipedality?

Darwin famously avoided these topics in 1859, but despite this, *Origin*'s implications for human evolution could scarcely be concealed. Indeed, they generated even more immediate discussion and debate than did his later (1871) treatise on humans (Browne 2002). When Huxley wrote on the subject in 1863—followed by Darwin in

1871—the poverty of the human paleontological record was overwhelming. Darwinian scholars had only a small, mostly European paleontological record (extracted primarily from archaeological contexts) with which to address the question of human origins and evolution. A handful of western European Neanderthals had been labeled everything from ancestral to pathological. Even the extant great apes were barely known. So Darwin and Huxley turned to the extant hominoid primates to serve as their "outgroup" for humans and as proxies for the common ancestors we once shared with these now relict forms.

The late Stephen J. Gould famously characterized hominid paleontology as follows: "…no true consensus exists in this most contentious of all scientific professions…a field that features more minds at work than bones to study" (Gould 2002, p. 910). Hominoid primates are, in general, highly variable as judged by any of their living representatives. All workers agree that there is rampant homoplasy within the clade. Hominids have always lived fairly high on the food chain. Relative to many other mammals, they are K-selected, and therefore rare as fossils. These factors all contribute to make the delineation of hominid species lineages very difficult…and contentious. Contention is difficult to quantify, but given the literally thousands of hominid fossils—and the relatively few professionals who work to interpret them—Gould's characterization has surely been invalid since early in the 20th century. The fossil samples are today relatively large, even though the hominid clade's record is terrestrial and therefore still full of imperfections.

HOMINIDS APLENTY: 1900–1950

The hominid fossil record was pushed deep into the Pleistocene by Eugene Dubois' discovery of *Pithecanthropus* (now *Homo*) *erectus* on Java near the end of the 18th century. In less than a decade, recovery of a large sample of Neanderthals in Croatia would barely precede the extraction of even more complete remains from France and elsewhere. By 1925, Raymond Dart had recognized *Australopithecus* from Pliocene antiquity in South Africa. By 1938, Robert Broom had recognized what he called *Paranthropus* (now *Australopithecus*) *robustus*.

Decades would pass before *Australopithecus* was afforded hominid status. Until the advent of radioisotopic dating decades later, the age of all of these discoveries remained largely in the realm of informed speculation. As time's veil was lifted progressively from younger to older, arguments about the place of these various fossils on the family tree ran the gamut from unilineal to speciose. There was even a joker in the deck, in the form of the Piltdown forgery. Different schools of thought arose, practitioners of human evolutionary studies came increasingly from anthropology rather than anatomy (Delisle 2006; White 2009), but the popular appeal of the quest to understand human evolution persisted.

HOMINIDS IN THE MODERN SYNTHESIS

The integration of 19th century selection theory with Mendelian and population genetics was labeled the "modern synthesis" by Julian Huxley (Huxley 1942). Key among its architects were Mayr, Simpson, and Dobzhansky. All of them grappled with the hominid fossil record and all were present at the 1950 CSH Symposium. Dobzhansky had turned his attention to human evolution in 1944. In tune with Simpson, he recognized both "horizontal" and "vertical" species (the latter are "chronospecies" that constitute arbitrary divisions of species lineages): "In practice, the incompleteness of the geological record is made use of in making "vertical" classifications: the gaps in the fossil series subdivide the continuous succession of forms into discrete sections" (Dobzhansky 1944, p. 256). Gaps became an ally instead of a problem. Dobzhansky even diagnosed Piltdown correctly: "…it now seems fairly clear that these remains are a mixture of ape and human bone fragments" (1944, p. 257). Hominid classification did not escape his attention: "The abuse of generic and specific names by students of the hominid evolution is notorious; it is making this fascinating field rather bewildering to other biologists" (1944, p. 257). And on the matter of phylogeny, Dobzhansky outlined the null hypothesis (following earlier writers such as Schwalbe and Weidenreich in this regard) as follows: "…as far as known, no more than a single hominid species existed at any one time level." (1944, p. 261).

THE 1950 CSH SYMPOSIUM

In June of 1950, the 9-day symposium "Origin and Evolution of Man" was held at the Biological Laboratory, Cold Spring Harbor. The program was worked out in cooperation between Dobzhansky and anthropologist Sherwood Washburn. The 129 registered participants included prominent anthropologists (A. Kroeber, E. Hooton, W. Howells, C. Coon, W. Pollitzer, A. Schultz, T. Stewart) and geneticists (B. Glass, E. Hunt, J. Neel, J. Spuhler, C. Stern), as well as E. Mayr and G. Simpson (Fig. 1).

As Browne notes, by this time in the mid 20th century, the modern synthesis "…had almost taken the form of a political treaty" (Browne 2008, p. 324). The CSH Symposium created a historically pivotal intersection for the architects of that synthesis and the anthropological community, whose practitioners had been slow on the uptake, but who would (mostly) promptly assimilate two key concepts that Dobzhansky had already chastised them for ignoring in his 1944 paper: populational thinking within an evolutionary species concept, and the classification that went along with it. As Mayr would later write of the CSH 1950 gathering, "It was on that occasion that the study of fossil man was integrated into the evolutionary snythesis [sic]" (Mayr 1982, p. 231).

In his 1944 paper, Dobzhansky had written, "The time is not far past when many systematists designated as species any two populations which they could (or thought that they could) distinguish by examining the morphology of a single specimen…. Lately such extravagances are becoming rare, but conservative systematists still cling to a purely morphological species concept." Adolph Schultz was an anatomist not widely associated with the modern synthesis, but his contribution to the 1950 CSH

Figure 1. Clash of academic cultures. The CSH Symposium of 1950 brought the architects of the modern synthesis face-to-face with the anthropological community. It was this Symposium that moved the latter to adopt modern systematic practices and a populational view of human evolution. (*Left to right*) Stanley Garn (Michigan physical anthropologist), Ernst Mayr; George Gaylord Simpson, Mayr, Ted McCown (Berkeley physical anthropologist). Courtesy of Cold Spring Harbor Laboratory.

Symposium was crucial for hominid systematics. His studies had established high levels of within-species variation based on geographic, idiosyncratic, sexual, and ontogenetic factors in extant higher-primate species. In the transcribed discussions about his paper at the CSH Symposium, he stated, "It is the intra-populational variability of the anthropoids, however, which appears so impressively great to all students with access to large series of these primates" (Schultz 1951, p. 53). Dobzhansky commented in Schultz (1951, p. 52), "The finding by Professor Schultz of a great variability in some species of anthropoids is very important" (see Fig. 2).

By 1950, nearly 30 generic names and more than 100 specific names had been applied to hominid fossils. Washburn's CSH paper characterized hominid classification as being "in a complete state of confusion" (Washburn 1951, p. 67). Extant chimpanzees, by Schultz's count, had been bestowed 21 different generic names and 73 specific names.

As an antidote for such typological thinking about hominoid primates, at the CSH Symposium, Mayr collapsed the entire hominid fossil record into a single evolving lineage of genus *Homo*: "...all the now available evidence can be interpreted as indicating that, in spite of

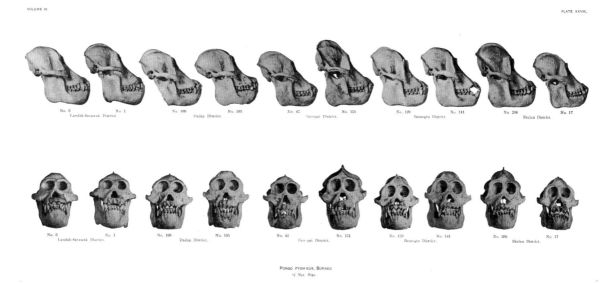

Figure 2. A key participant of the 1950 CSH Symposium was Adolph Schultz, who forcefully drew attention to normal natural intraspecific variation seen in modern hominoid primates. As ironically illustrated by this plate from Eliot's American Museum of Natural History three-volume set (1912), such variation can be considerable, as reflected in views of these skulls. All are modern orangutans, from one island (Borneo), adult, and male. Variation would only increase were geographic, ontogenetic, and sexual sources of variation to be included. Shape differences far less than those apparent here are currently being used to "diagnose" different "species" of fossil hominids, where yet another source of variation—temporal variation—is expected to increase variation even more than seen in this neontological series. If this osteological series were to be assessed by some modern techniques/practitioners blind to the species identity of the remains, the result would likely be a false inflation of actual biological species diversity.

much geographical variation never more than one species of man existed on earth at any one time" (Mayr 1951, p. 112). He based his interpretation, at least in part, on ecological principles: "What is the cause for this puzzling trait of the hominid stock to stop speciating in spite of its eminent evolutionary success? It seems to me that the reason is man's great ecological diversity. Man has, so to speak, specialized into despecialization.... The ecological diversity of man and his slowness in acquiring reproductive isolating mechanisms have prevented the breaking up of *Homo* into several species" (Mayr 1951, pp. 116–117).

For his part, Simpson saw his CSH mission as teaching the principles of modern systematics (phylogenetics, classification) to an audience specialized on human evolution. He criticized the anthropologists for using the circular logic of "morphological dating," and was "...appalled at the extent of restoration indulged in by the anthropologists" (Simpson 1951, p. 57). Simpson joined Mayr and Dobzhansky in urging anthropologists and anatomists to use a modern biological species concept and the classificatory nomenclature consistent with it, instead of continuing the widespread practice of using genus and species nomina as convenient labels for fossil fragments.

The anthropologists and anatomists involved in the study of human evolution were thus pulled under the umbrella of the modern synthesis by its architects at the 1950 CSH Symposium. The gathering also witnessed full agreement on the importance of the fossil record for revealing human origins and evolution. A century earlier, Darwin had made important predictions about the historical data he predicted would one day be found in Africa. He had made extensive use of extant African apes to triangulate on still-missing human ancestors. This was appreciated by many participants at the 1950 CSH Symposium. W.W. Howells wrote, "Darwin and Huxley proceeded of course from comparative anatomy.... Their allying of man and the apes was a great victory for the day. At the same time, however, men were men and apes were apes. The two could meet, all right, at a hypothetical crotch where their branches came together in the past. This is the diagram that has fascinated us, and plagued us, ever since. Added to this, the apes outnumbered us four to one which, with other facts, made us look like the aberrant animal, and the apes like the more natural, conservative primates" (Howells 1951, p. 80).

By 1950, more than a century of fossil and primate behavioral discoveries had armed the architects of the modern synthesis with historical and neontological information unavailable to Darwin and Huxley, and had reinforced the mistaken notion that modern apes were suitable proxies for early hominids. Simpson, the paleontologist, was wary of this, noting that "non-historical seriation" of living forms "...may approximate, although it can never equal, a historical sequence" (Simpson 1951, p. 56). Co-organizer Sherwood Washburn concluded, "Without fossils, ancestors can be reconstructed only by what has been called mental triangulation.... The actual course of evolution can be determined only from fossils..." (Washburn 1951, p. 76).

Nine years later, the CSH Symposium centenary celebration of Darwin's *Origin* took a broader look at evolution. Not focused on humans as the 1950 gathering had been, it covered genetics, race, ecology, speciation, the fossil record, and evolutionary trends. Mayr's paper cited the "almost universal acceptance of the synthetic theory of evolution" (Mayr 1960). Simpson reviewed the entire Mammalia, but about hominids he said little: "Finally, from some lineage that probably separated in the early part of the pongid radiation developed a poorly known and apparently quite restricted radiation, probably mainly in the Pliocene, characterized especially by upright posture with bipedal terrestrial locomotion and later by exaggerated development of the brain. So few lineages are known that this may not have been a radiation, strictly speaking, but rather only progressive advance in a rather unified group. The surviving product is *Homo* and the grade is hominid" (Simpson 1960, p. 270).

IN THE WAKE OF THE SYNTHESIS: 1950–1975

Paleoanthropology witnessed relative systematic stability in the face of an expanding fossil record in the 30 years following the 1950 and 1959 CSH Symposia. Piltdown was exposed as a fraud in 1953. With it went most remaining objections to *Australopithecus* as a hominid. Washburn would convene the next conference on hominid systematics 12 years after the 1950 CSH gathering. By then, Mayr had abandoned the null hypothesis of a single evolving hominid lineage, recognizing the "robust" *Australopithecus* from South Africa and the Leakeys' "nutcracker" cranium from Olduvai Gorge as a second hominid lineage. But that was as far as he would go: "When one reads the older anthropological literature with its rich proliferation of generic names, one has the impression of large numbers of species of fossil man and other hominids coexisting with each other. When these finds are properly placed into a multi-dimensional framework of space and time, the extreme rarity of the coexistence of two hominids became at once apparent" (Mayr 1963, p. 339).

During the 1960s and early 1970s, the evolutionary species concept was widely used within and beyond anthropology. As Tattersall and Schwartz (2009, p. 69) point out, anthropologists became "name shy." This stability was upset by the rise of parsimony cladistics and punctuationism (White 2009). The adoption of these by many practitioners of paleoanthropology represented an abandonment of the synthesis, which Tattersall has repeatedly and emphatically applauded.

The revealing of DNA's structure and the rise of molecular biology impacted human paleontologists, who first poorly received the immunologically based temporal estimates of chimp–human divergence of Sarich and Wilson (1967). Subsequent events have even emboldened some molecular biologists to have a hand at classification (Wildman et al. 2003).

THE RISE OF HOMINID DIVERSITY SYSTEMATICS: 1975–2009

In his 1982 history of human paleontology, even as Mayr took another swipe at human paleontologists for being typological, he underestimated the degree to which the modern synthesis had been perturbed by the rise of cladistics and punctuated equilibria (White 2009). By this time, the evidentiary record of early hominids had been pushed more than a million years (Ma) deeper by the discovery of *Au. afarensis* in Ethiopia and Tanzania during the 1970s. Mayr assimilated this taxon as an earlier chronospecies of *Au. africanus*.

The significance of *Au. afarensis* was the extension of the main attributes of its genus far deeper into the Pliocene. Evidence such as the 3.2-million-year-old A.L. 288-1 ("Lucy") partial skeleton allowed human paleontologists to go beyond "mental triangulation" and to demonstrate that Darwin's inferences about the assembly sequence of human evolution (drawn on the basis of such triangulation, forced by the lack of a fossil record) had been wrong; bipedality and canine reduction had actually preceded lithic technology and brain expansion by millions of years.

By the mid 1970s, Stephen J. Gould's *Natural History* essays were widely read by the public and scholars alike. His 1976 article "Ladders and Bushes in Human Evolution" was inspired by *Nature*'s publication of the *Au. afarensis* fossils at Laetoli. Gould asserted, "We are merely the surviving branch of a once luxuriant bush" (Gould 1976, p. 31). He even took the liberty of predicting when and what paleoanthropologists would find next: "We know about three coexisting branches of the human bush. I will be surprised if twice as many more are not discovered before the end of the century" (Gould 1976, p. 31).

Gould's 1976 prediction of bushiness became a paleoanthropological obsession in the years that followed, and modern practitioners and observers today have widely adopted an increasingly speciose, or "bushy," view of hominid phylogeny. The campaign to effectuate this was led by Gould, and Niles Eldredge and Ian Tattersall of the American Museum of Natural History. Quick to follow Gould's lead, they asserted in their 1982 *The Myths of Human Evolution*, "We have debunked the myth that evolutionary change is gradual and progressive" (Eldredge and Tattersall 1982, p. 175). This was a dramatic claim for any terrestrial vertebrate clade, let alone hominids. Tattersall and Eldredge sustained a persistent campaign of popular and professional writing. A recurrent theme was the heaping of abuse on their straw men of the modern synthesis, particularly its architects who dared to entertain the notion that phyletic (nonbranching) evolution might have characterized much of hominid evolution.

What motivated the campaign for what Eldredge has termed the "taxic" (as opposed to the transformational) approach to hominid paleobiology? Hominids have been used as exemplars ever since Darwin, so it is no wonder that the advocates of punctuationism and cladistics strove to accommodate this tiny but highly visible clade. And Tattersall and Eldredge were not the only front in the sustained war on the modern synthesis and its architects. As Cain (2009) contends, the breadth and persistence of Gould's attacks on the paleontologist Simpson may well constitute a case of "ritual patricide."

Possible motives and connections aside, hominid diversity advocates have been very successful. With the deliberate maligning of the synthesis, a radically different environment for hominid paleobiology was structured. As diversity politics of the academy, mediaphilic journals, biology envy, new fossils from previously unplumbed periods, and the inevitable public interest inherent in hominid paleobiology were synthesized in the 1990s, it became evident that the simple 1960s would never return (White 2009). After a respite of nearly half a century since Cold Spring Harbor, it had become permissible, indeed fashionable, to split hominid fossils into multiple contemporary branches on a relatively bushy tree.

The regular pronouncements in *Nature* of correspondents and editors alike are but one measure of this diversity mania: "It thus appears that the phylogeny of hominids, like that of many other mammalian groups, is very bushy at its base" (see, e.g., Kappelman and Fleagle 1995, p. 559). In public displays such as the prominent and permanent new Hall of Human Origins at New York's American Museum of Natural History, the public views a bushy hominid tree with 23 named species and more than 10 clades. A recent book entitled *The Last Human: A Guide to Twenty-two Species of Extinct Humans* (Sawyer and Deak 2007) is available to navigate this maze.

In 2001, the cover of *Nature* featured a single cranium, and a new genus was created. *Nature*'s "News and Views" author wrote, "We can now say with confidence that hominin evolution, like that of many other mammalian groups, occurred through a series of complex radiations, in which many new species evolve and diversify rapidly" (Lieberman 2001, p. 420). Gould's 2002 opus would use the same fossil as follows: "Multiple events of speciation now seem to operate as the primary drivers of human phylogeny" (Gould 2002, p. 909). These views of early hominid diversity have now been adopted so thoroughly that in the case of one prominent recent fossil discovery from Chad (Brunet et al. 2002), a reviewer attempted to rewrite the paper to accommodate this single cranium from the previously unsampled, 6 Ma time horizon to his diversity viewpoint, because, after all, there must be other species "out there" waiting to be discovered. This is "X-files paleontology" (White 2000, 2009).

Tattersall heralded the discovery of this single Chadian cranium as follows: "From the beginning, the hominid pattern had been to diversify, and for multiple hominid species to be in existence at any one time" (see Tattersall and Schwartz 2009, p. 85 and reference to Tattersall 2000 cited therein). The biological reality of such assertions is questionable (White 2003, p. 1996).

Secondary sources, from prominent newspapers to the major scientific journals, have uncritically reported each new taxonomic pronouncement as reflecting ever more diversity. A recent review stated about the hominid clade, "Its evolution has been bushy or tree-like, not a progressive line leading inevitably to us. More than 20 species…have been identified from fossils…up to five different hominin species have coexisted" (Pagel 2009, p. 809). It also cites

Tattersall. Today, Tattersall and his colleague Schwartz continue to promote hominid species diversity in popular books and scientific articles: "Yet the inescapable reality is that, with almost every new discovery, the rapidly expanding hominid fossil record amplifies the signal of past hominid diversity. As a result, pulled in opposite directions by received wisdom and by the accumulating morphological evidence, paleoanthropology is currently in a state of flux" (Tattersall and Schwartz 2009, p. 69). The reality may seem inescapable—to them—but a serious reality check seems to be in order.

THE MIDDLE AWASH

Paleoanthropology is often thought of as being driven by fossil data. As Mayr wrote in 1982 (p. 232), "Nothing, of course, has shed as much light on the history of the hominids as new fossil discoveries." Such discoveries—while conditioned by theory and interpretation—do have evidentiary value independent of prevailing academic fashions.

Ethiopia's Afar Depression is a vast desert region at the junction of the Red Sea, Gulf of Aden, and continental African rifts. Had the H.M.S. Beagle sailed to explore this region on its return journey in 1836, Darwin's impact on biology might never have been felt. The region remained geographically unexplored by Europeans until well into the last century. Many of the earliest explorers to the region succumbed to intense temperatures and hostile inhabitants. Even today, the region is very difficult to investigate.

Late in the 1960s, exploration of the region by the French geologist Maurice Taieb revealed its rich geological, paleontological, and archaeological resources. Discovery of a partial skeleton at the Hadar site, now dated to 3.2 Ma, focused the world's attention here beginning in 1974.

The Middle Awash study area lies to the south of Hadar, is bounded by an escarpment at the base of the Ethiopian plateau to the west, and is today bisected by the modern Awash river (Gilbert and Asfaw 2008; Haile-Selassie and WoldeGabriel 2009). Our team has worked here since 1981, with the twin goals of conducting research and building local capacity in paleoanthropology. More than 70 Ph.D.-level scholars, spanning archaeology to zoology, have contributed. Nineteen nations have been represented, and more than 19,000 vertebrate fossils, nearly 2000 geological samples, and thousands of artifacts have been collected from a combined stratigraphic thickness of ~1 km that spans the last 6 million years (Table 1). Figure 3 shows the geography and content of the study area and illustrates how the imperfections of the geological record continue to hamper our understanding of human evolution.

Discovery and analysis of *Ardipitheus ramidus* in the Middle Awash (White et al. 1994) and the geologically younger *Australopithecus anamensis* in Kenya (Leakey et al. 1995) during the 1990s established these as cladistic sister taxa. Their ages opened two possibilities. The null hypothesis is that these are chronospecies along one lineage (and in that sense, also "chronogenera"). Alternatively, the earlier taxon might be the dead-end, relict mother species of the younger genus. Only the discovery of more fossils would resolve this key problem in understanding mode and tempo in early hominid evolution. Unfortunately, the necessary temporal interval of the Middle Awash stratigraphic succession is marked by the incursion of a lake, in which fine fossils of a new fish species have been found (Murray and Stewart 1999), but no primates. One paleontologist's imperfection is another's bonanza, as we all celebrate the ever-narrowing gaps in the fossil record.

This example emphasizes that the mostly terrestrial, continental nature of the Middle Awash succession is better viewed as a series of sporadic snapshots than a continuous videotape of species through time (particularly hominids). But seen from a distance, here in this single Ethiopian valley, more than a dozen time-successive strata have yielded hominid fossils (along with stone tools after 2.5 Ma). Nearly 300 hominid individuals have been sampled across 6 million years. If hominid biodiversity was so high, it seems fair to ask whether there is evidence of it here. The answer is "no, not so far." Unlike many Plio-Pleistocene African assemblages in which two or sometimes (arguably) three hominid taxa are present, not one Middle Awash case demonstrates the presence of contemporary—let alone

Table 1. Main Discoveries in the Middle Awash Study Area, Afar Rift, Ethiopia

Age (Ma)	Localities	Faunal NISP	Archaeology	Hominid NISP	Hominid publication
0.08	Aduma	21	MSA	3	*AJPA* 03
0.10	Halibee	2646	MSA	16	
0.16	Herto	104	MSA/late Acheulean	12	*Nature* 03
0.2	Talalak	575	MSA/late Acheulean	2	
0.5	Bodo	25	typical Acheulean	3	*Nature* 84/*Science* 04
1.0	Bouri Daka	753	early Acheulean	11	*Nature* 02
2.0	Guneta	753	Olduwan+	2	
2.5	Bouri Hata	576	Olduwan	11	*Nature* 99
3.5	Maka	171		12	*Nature* 84, 93
3.9	Belohdelie	54		1	*Nature* 84
4.1	Asa Issie	605		37	*Nature* 06
4.4	Aramis	6432		114	*Nature* 94
5.3	Amba	550		1	*Nature* 01
5.8	Western Margin	2230		19	*Nature* 01/*Science* 04

NISP indicates numbers of identified specimens. Only major hominid announcements listed in publications. For more details, see text, Gilbert and Asfaw (2008), and Haile-Selassie and WoldeGabriel (2009).

Figure 3. The Middle Awash paleoanthropological study area, Afar Rift, Ethiopia. View is to the north. A composite stratigraphic thickness of more than 1 km of predominantly fluviatile, lacustrine, and volcaniclastic sediment has sampled 14 different time horizons containing hominid fossils. Small red dots in the upper frame are geological sample points, usually indicating rocks extracted for radioisotopic, geomagnetic, isotopic, and sedimentological studies. Yellow triangles show the location of major hominid-bearing localities. Inset photographs show an antelope molar from the terrestrial 4.4 Ma *Ardipithecus ramidus* horizon, a new species of fish from the volcaniclastic lacustrine Beidareem horizon, and a molar attributed to *Australopithecus anamensis* from the overlying Adgantole Member. Even with a very thick (>1 km) succession, temporal, depositional, and facies-related gaps continue to make stratophenetics difficult for the many organisms fossilized in these rocks. See text for details.

sympatric—hominid species. The ghosts of CSH 1950 would not be surprised by these data.

THE PRESENT THROUGH THE LENS OF CSH 1950

Most current inventories tally around 25 separate named hominid species (Wood and Richmond 2000; Sawyer and Deak 2007). Figure 4 lists these and plots them chronologically. Since CSH 1950, it has been abundantly clear that there are several major sources of hominid "species diversity inflation." This inflation has been driven by Hennigian cladistics and the punctuationist advocacy of Gould and followers. Taxonomic inflation has been arrived at via the following practices, all recognized as problematic during the 1950 CSH Symposium:

- **The bestowal of invalid names (invalid subjective synonyms) to contemporary specimens from single** species. This practice has created a long species list for Hominidae. However, these biologically superfluous names only indicate improper systematic practice, often based on the failure to appreciate the extensive within-species variation that characterizes hominoid primates. Invalid names do not constitute evidence for branching speciation.

- **The uncritical acceptance of arbitrary names that merely designate chronospecies (arbitrary segments) of species lineages.** This practice also artificially lengthens the species list, but this is not evidence for biological species diversity. Chronospecies-based diversity inflation is not an issue for some phylogeneticists who refuse to recognize such parsing of apparently phyletic evolution: "No presumed separate, single, evolutionary lineage may be subdivided into a series of ancestral and descendant 'species'" (Wiley 1978, p. 21). It is ironic that if Wiley's admonition is applied to contemporary hominid trees, their diversity is considerably pruned (Fig. 4) (White 2009).

- **The chronological exaggeration of species extinctions and/or first appearances.** This practice has resulted in apparent temporal overlap among taxa that are actually time-successive (and therefore possibly sampling single evolving lineages). This erroneous time range extension should not be confused for actual biological species diversity. The proper measure of hominid species diversity through the Neogene should involve only the accurate delineation of species lineages at any given time horizon.

Once the 25+ modern hominid "species" are organized in realistic chronological and taxonomic space, an obvious pattern emerges (Fig. 4). It is remarkable how it conforms to predictions made at the 1950 CSH Symposium. This pattern continues to evoke the biological issue recognized by Dobzhansky so long ago, but since forgotten during the diversity mania described above. The question is not why there was so *much* hominid biological species diversity, but why—compared to other mammals—so *little* evolved within the hominid clade.

It is ironic, nearly 30 years after Eldredge and Tattersall's book (1982), and after such widespread acceptance of their "taxic" model of hominid evolution, that even today, any slice of geological time across Figure 4 intersects only a maximum or three or four hominid species lineages, and these are often geographically widely separated.

As Mayr, Simpson, Dobzhansky, and Schultz recognized in 1950, hominids were large terrestrial generalists. At least two species lineages specialized in intelligence and culture. Intelligence and culture increased the breadth of the niches occupied by these hominid species, further reducing opportunities for sympatry.

Gould's prediction about doubling the number of hominid lineages in a hominid "bush" was a bold one, but more than 30 years later, we still await evidence of a fourth (let alone a sixth) contemporary hominid lineage. So convinced was Gould by the taxic interpretations he had inspired, that by the time he published *The Structure of Evolutionary Theory* in 2002, the study of human evolution had, for him, been "recast" in "speciational terms" (Gould 2002, p. 910).

But how has Gould's 1976 prediction really done in the case of fossil hominids? If one measures species richness by the number of named taxa, the prediction has fared very well. But if one measures actual species diversity at any time slice, the prediction has not fared well, despite a great increase in discovered fossil specimens. Today, the record before us (Fig. 4) hardly constitutes a signal of adaptive radiation in the sense of contemporary biological science (Jolly 2001; Gavrilets and Losos 2009; Reznick and Ricklefs 2009; Schluter 2009).

Eldredge and Gould's work (1972 and thereafter) generated a large body of empirical evidence concerning what Simpson had called "quantum evolution" (punctuated equilibrium; see Cain 2009). Much of this has shown

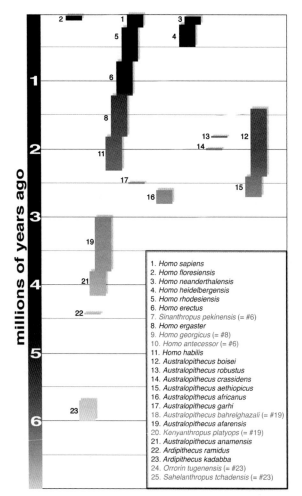

Figure 4. The synonomy, inferred phylogenetic relationships, and temporal distribution of 25 hominid species currently in wide use. Names in gray are considered to be junior subjective synonyms of previously established taxa. The first and last reliable appearance dates for the remaining taxa are plotted. Phyletic evolution between chronospecies is indicated by the abutting adjacent boxes. Note that the maximum number of separate species at any one time horizon is small, sampling only three or four lineages at ~2 Ma.

that rectangular evolution was more frequent in the record of life than had previously been recognized. In a long argument, Gould, Eldredge, and Tattersall have repeatedly deployed fossil hominid evidence in their attacks on the modern synthesis. They have persistently claimed that punctuational, cladogenetic species change has dominated in hominid evolution, as opposed to phyletic evolution. Ironically, the paucity of hominid lineages (as opposed to specimens) through geological time may actually turn out to be the exception that proves the rule among mammals. And ironically, the persistence of the chronospecies concept to describe what appears to be widespread phyletic evolution among hominids may even itself be evidence for the latter.

As Dobzhansky appreciated, there are good ecological and theoretical reasons to expect that all of this would be true. Indeed, studies of Neogene mammals have shown that generalists are predictably less speciose and have longer species durations (unless divided into chronospecies or invalid taxa) than specialists (Vrba 1993; Benton 2009). Culture-bearing hominids are consummate generalists, and the minimal hominid species diversity seen in the Pleistocene (notably the European Neanderthal clade) appears largely tied to the wide geographic spread of *Homo erectus*. The recent discovery of what may be a case of island dwarfing in Indonesia (Brown et al. 2004) may also be informative in this regard. If *Homo floresiensis* turns out (with the discovery of critically needed additional specimens) to be a separate hominid species, it would represent another exception to prove another rule, namely that hominids only speciate under very rare conditions (for a review of speciation, see Hendry [2009] and Schluter [2009]).

CONCLUSIONS: CONSILIENCE AND FUTURE DIRECTIONS

We all work on the neontological and paleontological results pertaining to a single tree of life whose primate species lineages—as Darwin surely recognized—virtually never anastomosed. Today, hominid paleobiology is probing at the root and branches of the hominid clade with a wide assortment of tools ranging from DNA sequencers to micro-CT scanners. Our interpretations must synthesize the data from those disparate activities. Major progress has recently been achieved. The 150-year-old debate about the place of Neanderthals and the origins of *Homo sapiens* has only recently abated in the face of breakthroughs in the paleontological (White et al. 2003) and neontological (biomolecular; Noonan et al. 2006; Briggs et al. 2009) realms.

Such consilience (Wilson 1998) shows the combinatorial power of the multidisciplinary approach. However, from Neanderthals on back, the prospects of similar recovery and analysis of ancient biomolecules are not good. Beyond the special case of a few Neanderthals, the DNA that created early hominids has all but disappeared. Darwin could never have predicted *Australopithecus* by triangulating between living chimpanzee and modern human anatomies and behaviors. Nor will it be possible to reveal the anatomy, habitats, or lifeways of our ancient ancestors by genomic triangulation or by more study of our relict ape relatives who have been independently evolving their own specializations for more than 6 million years.

The global experiment of human evolution cannot be repeated in a laboratory. We must infer what happened from the one-time experimental results, fragmentary and scattered as they may be. The good news about understanding our behavioral evolution is that there is a 2.5 Ma archaeological record. The good news about understanding our anatomical evolution is that some of the tissues shaped by that disappeared DNA can still be recovered from unique paleontological records derived from ancient landscapes. The order in which our unique human characteristics have been assembled via evolution is susceptible to investigation, and the temporal and anatomical perspective of the fossil record will continue to be key to its success. Crucial in that investigation will be the understanding of how the hard tissues we recover as fossils were formed via development (see, e.g., Prabhakar et al. 2008). Integration will continue to be the key to better understanding human origins and evolution, just as it was during the CSH Symposium in 1950.

Huxley's *Nature* 1882 obituary said it well on the occasion of Charles Darwin's death: "He found a great truth, trodden under foot… ." A century and a half ago, Charles Darwin wrote in *Origin* (1959) only that "…light will be thrown on the origin of man and his history." He could not have imagined the illumination already thrown on our ancestry through the integration of ever-expanding constellations of evidence about our evolution. What wonders await our intellectual descendants at the Cold Spring Harbor Laboratory's bicentennial celebration of Darwin's great book?

ACKNOWLEDGMENTS

Thanks to the organizers of the 2009 CSH Symposium for a thoroughly stimulating, memorable, and enjoyable event. Thanks to all members of the Middle Awash research project for the field and laboratory work behind the summary presented here and to the Ethiopian and Afar Regional governments. Thanks to Josh Carlson, Kyle Brudvik, and Henry Gilbert for illustrations, editorial, and bibliographic work. Thanks to all others for their patience.

REFERENCES

Benton MJ. 2009. The red queen and the court jester: Species diversity and the role of biotic and abiotic factors through time. *Science* **323:** 728–732.

Briggs A, Good JM, Green RE, Krause J, Maricic T, Stenzel U, Lalueza-Fox C, Rudan P, Brajković D, Kućan Ž, et al. 2009. Targeted retrieval and analysis of five Neandertal mtDNA genomes. *Science* **325:** 318–321.

Brown P, Sutikna T, Morwood MJ, Soejono RP, Jatmiko, Saptomo EW, Due RA. 2004. A new small-bodied hominin from the Late Pleistocene of Flores, Indonesia. *Nature* **431:** 1055–1061.

Browne J. 2002. *Charles Darwin: A biography*, Vol. 2. *The Power of place*. Knopf, New York.

Browne J. 2008. Essay: Birthdays to remember. *Nature* **456:** 324–325.

Brunet M, Guy F, Pilbeam D, Mackaye HT, Likius A, Ahounta D, Beauvilain A, Blondel C, Bocherens H, Boisserie J-R, et al. 2002. A new hominid from the Upper Miocene of Chad, Central Africa. *Nature* **418**: 145–151.

Cain J. 2009. Ritual patricide: Why Stephen Jay Gould assassinated George Gaylord Simpson. In *The palobiological revolution: Essays on the growth of modern paleontology* (ed. D Sepkoski and M Ruse), pp. 346–363. University of Chicago Press, Chicago.

Darwin C. 1859. *On the origin of species by means of natural selection*, 1st ed. Murray, London.

Darwin C. 1871. *The descent of man, and selection in relation to sex*. Murray, London.

Delisle RG. 2006. *Debating humankind's place in nature, 1860–2000*. Pearson/Prentice Hall, Upper Saddle River, NJ.

Dobzhansky T. 1944. On species and races of living and fossil man. *Am J Phys Anthropol* **2**: 251–265.

Eldredge N, Gould SJ. 1972. Punctuated equilibria: An alternative to phyletic gradualism. In *Models in paleobiology* (ed. TJM Schopf), pp. 82–115. Freeman, Cooper, San Francisco.

Eldredge N, Tattersall I. 1982. *The myths of human evolution*. Columbia University Press, New York.

Elliot DG. 1912. *A review of the Primates: Anthropoidea,* Vol. 3. American Museum of Natural History, New York.

Gavrilets S, Losos JB. 2009. Adaptive radiation: Contrasting theory with data. *Science* **323**: 732–737.

Gilbert WH, Asfaw B, Eds. 2008. *Homo erectus: Pleistocene evidence from the Middle Awash, Ethiopia*. University of California Press, Berkeley.

Gould SJ. 1976. Ladders, bushes, and human evolution. *Nat Hist Mag* **85**: 24–31.

Gould SJ. 2002. *The structure of evolutionary theory*. Belknap, Cambridge, MA.

Haile-Selassie Y, WoldeGabriel G, Eds. 2009. *Ardipithecus kadabba: Late Miocene evidence from the Middle Awash, Ethiopia*. University of California Press, Berkeley.

Hendry AP. 2009. Speciation. *Nature* **458**: 162–164.

Hong J-W, Hendrix DA, Levine MS. 2008. Shadow enhancers as a source of evolutionary novelty. *Science* **321**: 1314.

Howells WW. 1951. Origin of the human stock: Concluding remarks of the chairman. *Cold Spring Harbor Symp Quant Biol* **15**: 79–86.

Huxley TH. 1863. *Man's place in nature*. Williams and Norgate, London.

Huxley TH. 1882. Charles Darwin. *Nature* **25**: 597.

Huxley J. 1942. *Evolution: The modern synthesis* (1964 edition). Wiley, New York.

Jolly CJ. 2001. A proper study for mankind: Analogies from the papionin monkeys and their implications for human evolution. *Yearb Phys Anthropol* **44**: 177–204.

Kappelman J, Fleagle JG. 1995. Age of early hominids. *Nature* **376**: 558–559.

Leakey MG, Feibel CS, McDougall I, Walker A. 1995. New four-million-year-old hominid species from Kanapoi and Allia Bay, Kenya. *Nature* **376**: 565–571.

Leakey MG, Spoor F, Brown FH, Gathogo PN, Kiarie C, Leakey LN, McDougall I. 2001. New hominin genus from eastern Africa shows diverse middle Pliocene lineages. *Nature* **410**: 433–440.

Lieberman DE. 2001. Another face in our family tree. *Nature* **410**: 419–420.

Marques-Bonet T, Kidd JM, Ventura M, Graves TA, Cheung Z, Hiller LW, Jiang Z, Baker C, Malfavon-Borja R, Fulton LA, et al. 2009. A burst of segmental duplications in the genome of the African great ape ancestor. *Nature* **457**: 877–881.

Mayr E. 1951. Taxonomic categories in fossil hominids. *Cold Spring Harbor Symp Quant Biol* **15**: 109–118.

Mayr E. 1960. Where are we? *Cold Spring Harbor Symp Quant Biol* **24**: 409–440.

Mayr E. 1963. The taxonomic evaluation of fossil hominids. Classification and human evolution. *Viking Fund Publ Anthropol* **37**: 332–345.

Mayr E. 1982. Reflections on human paleontology. In *A history of American physical anthropology 1930–1982* (ed. F Spencer), pp. 231–237. Academic, New York.

Murray AM, Stewart KM. 1999. A new species of tilapiine cichlid from the Pliocene, Middle Awash, Ethiopia. *J Vertebr Paleontol* **19**: 293–301.

Noonan JP, Coop G, Kudaravalli S, Smith D, Krause J, Alessi J, Chen F, Platt D, Pääbo A, Pritchard JK, Rubin E. 2006. Sequencing and analysis of Neanderthal genomic mtDNA. *Science* **314**: 1113–1118.

Pagel M. 2009. Natural selection 150 years on. *Nature* **457**: 808–811.

Pinker S. 2003. Language as an adaptation to the cognitive niche. In *Language evolution: The states of the art* (ed. MH Christiansen and S Kirby), pp. 16–37. Oxford University Press, Oxford.

Prabhakar S, Visel A, Akiyama JA, Shoukry M, Lewis KD, Holt A, Plajzer-Frick I, Morrison H, FitzPatrick DR, Afzal V, et al. 2008. Human-specific gain of function in a developmental enhancer. *Science* **321**: 1346–1349.

Reznick DN, Ricklefs RE. 2009. Darwin's bridge between microevolution and macroevolution. *Nature* **457**: 837–842.

Sarich VM, Wilson AC. 1967. Immunological time scale for hominid evolution. *Science* **158**: 1200–1203.

Sawyer GJ, Deak V. 2007. *The last human: A guide to twenty-two species of extinct humans*. Nèvraumont, New York.

Schluter D. 2009. Evidence for ecological speciation and its alternative. *Science* **323**: 737–741.

Schultz, A. 1951. The specializations of man and his place among the catarrhine primates. *Cold Spring Harbor Symp Quant Biol* **15**: 37–53.

Sepkoski D, Ruse M, Eds. 2009. Introduction: Paleontology at the high table. In *The palobiological revolution: Essays on the growth of modern paleontology*, pp. 1–13. University of Chicago Press, Chicago.

Simpson GG. 1951. Some principles of historical biology bearing on human origins. *Cold Spring Harbor Symp Quant Biol* **15**: 55–66.

Simpson GG. 1960. The nature and origin of supraspecific taxa. *Cold Spring Harbor Symp Quant Biol* **24**: 255–271.

Tattersall I. 2000. Once we were not alone. *Sci Am* **282**: 56–62.

Tattersall I, Schwartz J. 2009. Evolution of the genus *Homo*. *Annu Rev Earth Planetary Sci* **37**: 67–92.

Vrba ES. 1993. Turnover-pulses, the Red Queen, and related topics. *Am J Sci* **293**: 418–452.

Washburn S. 1951. The analysis of primate evolution with particular reference to the origin of man. *Cold Spring Harbor Symp Quant Biol* **15**: 67–78.

White TD. 2000. A view on the science: Physical anthropology at the millennium. *Am J Phys Anthropol* **113**: 287–292.

White TD. 2003. Early hominids: Diversity or distortion? *Science* **299**: 1994–1996.

White TD. 2009. Ladders, bushes, punctuations, and clades: Hominid paleobiology in the late twentieth century. In *The Palobiological revolution: Essays on the growth of modern paleontology* (ed. D Sepkoski and M Ruse), pp. 122–148, University of Chicago Press, Chicago.

White TD, Suwa G, Asfaw B. 1994. *Australopithecus ramidus*, a new species of early hominid from Aramis, Ethiopia. *Nature* **371**: 306–312.

White TD, Asfaw B, DeGusta D, Gilbert H, Richards GD, Suwa G, Howell FC. 2003. Pleistocene *Homo sapiens* from Middle Awash, Ethiopia. *Nature* **423**: 742–747.

Wildman DE, Uddin M, Liu G, Grossman LI, Goodman M. 2003. Implications of natural selection in shaping 99.4% nonsynonymous DNA identity between humans and chimpanzees: Enlarging genus *Homo*. *Proc Natl Acad Sci* **100**: 7181–7188.

Wiley EO. 1978. The evolutionary species concept reconsidered. *Syst Zool* **27**: 17–26.

Wilson EO. 1998. *Consilience: The unity of knowledge*. Knopf, New York.

Wood BA, Richmond BG. 2000. Human evolution: Taxonomy and paleobiology. *J Anat* **196**: 19–60.

Reconstructing the Evolution of Vertebrate Sex Chromosomes

D.W. BELLOTT AND D.C. PAGE

*Howard Hughes Medical Institute, Whitehead Institute, and Department of Biology,
Massachusetts Institute of Technology, Cambridge, Massachusetts 02142*
Correspondence: dcpage@wi.mit.edu

Sex chromosomes and their evolution have captivated researchers since their discovery. For more than 100 years, the dominant model of sex chromosome evolution has held that differentiated sex chromosomes, such as the X and Y chromosomes of mammals or the Z and W chromosomes of birds, evolved from ordinary autosomes, primarily through the degeneration of the sex-specific Y or W chromosome. At the same time, the sex chromosomes shared between sexes, the X and Z chromosomes, are expected to remain essentially untouched. This model was based on limited cytogenetic and genetic data. Only in the last decade, with the advent of genomics, has the complete sequence of any sex chromosome pair become available. High-quality finished sequences of the human and chimpanzee Y chromosomes, as well as the human X chromosome, have revealed sequence features unanticipated by the traditional model of sex chromosome evolution. Large, highly identical, tandem and inverted arrays of testis-expressed genes are major sources of innovation in gene content on sex-specific chromosomes as well as sex-shared chromosomes. Accounting for the emergence of these ampliconic structures presents a challenge for future studies of sex chromosome evolution.

Since the discovery of sex chromosomes, researchers have sought to explain the evolutionary forces that could produce a pair of chromosomes that differed between the sexes. During the 20th century, the fields of classical genetics, evolutionary and population genetics, and cytology converged on a single explanation for the evolution of heteromorphic sex chromosomes: Sex chromosomes evolved from autosomes primarily through the degeneration of the sex-specific Y or W chromosome, whereas the X or Z chromosome faithfully preserved the gene content of the ancestral autosome pair. X and Z chromosomes were museums; Y and W chromosomes were ruins, destined to be lost to the sands of time.

In the last 10 years, genomics has revolutionized the study of evolution. Evolution changes the sequence of DNA molecules, and comparing DNA sequences allows us to reconstruct evolutionary events from the past. The availability of DNA sequences from multiple vertebrates has confirmed that the process of sex chromosome evolution envisioned by theorists has played out multiple times in the evolution of vertebrate sex chromosomes. However, complete high-quality sequences of sex chromosomes have led to discoveries that were unanticipated by existing theory. Sex-specific chromosomes are not doomed to decay, but selection can act to preserve their gene content over long timescales. Amplicons, massive and highly identical arrays of duplicated genes, are sources of innovation in gene content on sex-specific as well as sex-shared chromosomes. These arrays consist of genes expressed exclusively or predominantly in the testis.

The unexpected results of genomic analyses have challenged long-standing assumptions about the evolution of sex chromosomes. It is now clear that sex chromosomes are subject to constant remodeling; they resemble Theseus' ship rather than museums or ruins. The dramatic nature of innovation in gene content on sex chromosomes presents major theoretical challenges for the field of sex chromosome evolution. What selective forces can generate ampliconic structures? What is the relationship between ampliconic genes and male reproduction? As more sex chromosome sequences become available, including those of multiple mammals as well as the Z and W chromosomes of birds, they will enhance our ability to address these questions.

THEORETICAL MODELS OF SEX CHROMOSOME EVOLUTION

The study of sex chromosome evolution shares its origin with that of genetics, in Thomas Hunt Morgan's fly room at Columbia University. In 1913, Alfred Sturtevant produced the first genetic map, consisting of six sex-linked genes (Sturtevant 1913). The following year, his colleague, Calvin Bridges, combined Sturtevant's linear map of sex-linked genes with his own work on nondisjunction of sex chromosomes to demonstrate that Sturtevant's map was that of the X chromosome, and the chromosomes were the material of heredity (Bridges 1914). This suggested that sex chromosomes were not merely a sign, but instead the root cause of sexual dimorphism. The following year, a third member of Morgan's lab, Hermann Muller, established the linkage of a gene with the fourth chromosome, the smallest *Drosophila* autosome (Muller 1914). With Muller's publication, all *Drosophila* chromosomes, with the exception of the Y chromosome, had at least one known gene. This fact troubled Muller, who explained it with the first theory of sex chromosome evolution: The X and Y chromosomes evolved from an ordinary pair of autosomes, but the Y chromosome, unable to recombine in males, had accumulated deleterious mutations, eliminating all of its genes.

This simple theory, that heteromorphic sex chromosomes evolve from autosomes through the decay of the sex-specific chromosome, has been fundamental to the study of sex chromosome evolution for nearly 100 years.

Muller's theory that heteromorphic sex chromosomes were the result of degradation of the sex-specific chromosome was corroborated by the lack of credible Y-linked phenotypes in humans. As was the case in *Drosophila*, the first traits mapped to a human chromosome were mapped to the X chromosome (Morgan 1911a,b; Wilson 1911). By the middle of the century, X-linked inheritance had been reported for dozens of traits, whereas only a handful of traits had been mapped to the Y chromosome (Stern 1957; McKusick 1962). In 1957, Curt Stern, another former student of Morgan's and the President of the American Society of Human Genetics, addressed the society's annual meeting (Stern 1957). Stern used his address to systematically debunk every reported case of Y linkage in humans. Stern noted that no Y-linked trait had been discovered in experimental mammals, but he cautioned investigators not to give up the search for Y-linked traits. Two years later, it was discovered that the human and mouse Y chromosomes contained the male sex-determining gene (Ford et al. 1959; Jacobs and Strong 1959; Welshons and Russell 1959), but the reputation of the Y chromosome had been irreparably damaged. Apart from sex determination, geneticists viewed the sex-specific Y chromosome as a "dud" (McKusick 1962).

Not only was the idea of the sex-specific chromosome as a degenerate autosome in accord with the genetic data from flies and mammals, but it could also account for the diverse sex-determining mechanisms of vertebrates. Many vertebrate species have no sex chromosomes; in these species, sex is determined by an environmental cue such as temperature. Some species have homomorphic sex chromosomes. Homomorphic sex chromosomes are not cytologically distinguishable, but they can be revealed by experiments with artificially sex-reversed animals. Heteromorphic sex chromosomes of the type seen in *Drosophila* predominate in three vertebrate lineages: mammals, birds, and snakes. Susumu Ohno (1967) argued that these three states—the absence of sex chromosomes, homomorphic sex chromosomes, and heteromorphic sex chromosomes—represented a continuum that revealed the evolutionary trajectory of the heteromorphic vertebrate sex chromosomes. Ohno conjectured that the common ancestor of vertebrates possessed no sex chromosomes, but in some lineages, a mutation had arisen that caused an ordinary pair of autosomes to behave as homomorphic sex chromosomes, and after this event, the sex-specific chromosome decayed, producing heteromorphic sex chromosomes like those of mammals, birds, and snakes.

Ohno also modified Muller's theory to account for differences in recombination between *Drosophila* and vertebrates. Muller's theory relied on the absence of crossing-over between homologous chromosomes in *Drosophila* males to automatically isolate any Y chromosome from crossing-over, but because recombination occurs in both sexes in vertebrates, the sex-specific chromosomes of vertebrates would not spontaneously begin to degenerate. After the emergence of a new sex-determining gene in a vertebrate, a second event is required to suppress crossing-over. Ohno proposed that a pericentric inversion on the sex-specific chromosome that encompassed the region of the sex-determining gene could suppress crossing-over between sex chromosomes in the heterogametic sex (Ohno 1967). If crossing-over occurs within the boundaries of a pericentric inversion, the recombinant chromosomes will be duplicated in part of the inversion and deficient in the other; if essential genes fall within the boundaries of the inversion, recombinant progeny will die and only those whose sex chromosomes did not recombine will survive. Once the sex-specific Y or W chromosome was isolated, it would begin to diverge from the shared X or Z chromosome by losing its gene content as Muller had predicted.

As the study of population genetics emerged, it became clear that Muller's explanation for the degeneration of the sex-specific chromosome was inadequate. Inspired by his work on chromosomes carrying balanced lethal mutations, Muller initially proposed that a lack of crossing-over was sufficient to lead to genetic decay. Each chromosome in a pair carrying balanced lethal mutations exists only in the heterozygous state; recessive mutations on one chromosome are not exposed to selection as long as the other chromosome maintains the ancestral allele. Thus, both chromosomes can accumulate complementary recessive mutations. Muller believed that Y and W chromosomes, held in a heterozygous state by linkage to the sex-determining locus, would be sheltered from selection by their partner, whereas X and Z chromosomes would be exposed to selection against recessive mutations in the homogametic sex (Muller 1918). Fisher demonstrated that this explanation could not account for the degeneration of the sex-specific chromosome, because mutation must affect incipient sex chromosomes equally (Fisher 1935). If an X-linked or Z-linked gene suffered a loss of function, the result would be selection against a parallel loss of function in the Y-linked or W-linked counterpart. Fisher showed that for an infinite population, degeneration of the type Muller described could only occur if the mutation rate was much higher on the sex-specific chromosome than in the rest of the genome. In light of this difficulty, it was necessary to modify Muller's theory to explain why only the sex-specific chromosome was subject to degeneration.

Although Muller's initial explanation for the degeneration of the sex-specific chromosome proved to be inadequate, population genetic theories designed to explain the benefits of sex and recombination became the source of alternative models that could account for the degeneration of a nonrecombining chromosome. Muller proposed that genetic drift could account for the degeneration of nonrecombining chromosomes through a mechanism that is now known as "Muller's ratchet" (Muller 1964; Felsenstein 1974). Muller's ratchet is the idea that, in the absence of crossing-over, a population cannot generate chromosomes with a smaller mutational load than those that currently exist within the population. If the least-mutated class of chromosomes is lost to drift, it is replaced by one that carries more mutations, and the "ratchet" has clicked irreversibly toward the decay of the nonrecombining chromosome.

Alternative models of degeneration rely on the absolute linkage between all of the sites on a nonrecombining chromosome. Selection at one site interferes with selection at linked sites, preventing the efficient elimination of deleterious mutations and slowing the spread of beneficial mutations (Felsenstein 1974). Strongly beneficial mutations can sweep through a population, dragging many weakly deleterious mutations along with them (genetic hitchhiking) (Maynard Smith and Haigh 1974; Rice 1987); chromosomes with strongly deleterious alleles will be lost from the population before they can spread, increasing the chances that weakly deleterious alleles will become fixed by drift (background selection) (Charlesworth et al. 1993; Charlesworth 1994). Both of these models predict reductions in the effective population size of a nonrecombining chromosome, increasing the effects of genetic drift (Charlesworth 1978). Thus, both genetic hitchhiking and background selection should act synergistically with Muller's ratchet to hasten the degeneration of a nonrecombining chromosome (Charlesworth 1978; Bachtrog 2008).

Theoretical models of sex chromosome evolution based on population genetics implicitly assumed that the sex-shared X and Z chromosomes were unchanging; Ohno (1967) codified this as an explicit prediction. Ohno predicted that the X and Z chromosomes should preserve the gene content of the ancestral autosome pair from which they evolved. As a corollary, the sex chromosomes of species that share a common origin are expected to share the same ancestral gene content. This concept is now most familiar as "Ohno's Law," i.e., genes that are X linked in one mammal should be X linked in all others, but Ohno applied his predictions equally to the Z chromosomes of birds and snakes. Ohno and other investigators reasoned that the degeneration of the sex-specific chromosome would result in the evolution of dosage compensation on the sex chromosome shared between the sexes (Ohno 1967; Charlesworth 1978; Jegalian and Page 1998). Once genes were lost from the sex-specific chromosome, the heterogametic sex would only have half of the original dose of X-linked genes (Ohno 1967; Charlesworth 1978; Jegalian and Page 1998). A system of dosage compensation would evolve to provide males with the correct expression level for X-linked genes (Ohno 1967; Charlesworth 1978; Jegalian and Page 1998). Ohno argued that autosomal genes could not be added to the X chromosome because they would be expressed at too low a level in males, and X-linked genes could not move to autosomes because they were dependent on the dosage compensation mechanism for proper expression (Ohno 1967). Thus, although Y and W chromosomes were subject to drastic changes in gene content, X and Z chromosomes were locked into stably retaining their ancestral genes.

EVOLUTIONARY STRATA: RECONSTRUCTING THE DEGENERATION OF SEX-SPECIFIC CHROMOSOMES

As DNA sequences from vertebrate sex chromosomes became available, researchers interpreted them in the context of the theories built on Muller's ideas. Pairing and crossing-over between the human X and Y chromosomes at meiosis implied that some vestige of the original autosomal homology between them remained (Solari and Tres 1970; Rasmussen and Holm 1978). This suspicion was confirmed by the discovery of pseudoautosomal genes on the mammalian X and Y chromosomes (Cooke et al. 1985; Simmler et al. 1985; Goodfellow et al. 1986). The first sequence map of the Y chromosome showed that even outside of the pseudoautosomal region, the human X and Y chromosomes carried homologous genes (Foote et al. 1992; Vollrath et al. 1992). The sequence of these Y-linked genes, when compared to the sequence of their X-linked homologs, revealed a pattern that suggested a pathway for X-Y evolution (Lahn and Page 1999b). Nucleotide divergence between X-linked and Y-linked gene copies was strongly correlated with the position of the X-linked gene copy, such that X-Y pairs formed several groups of increasing divergence from the short arm to the long arm of the X chromosome. Bruce Lahn likened the surviving gene pairs to fossils preserved in layers of stone from different periods in the past, and he christened these groups "evolutionary strata." Each stratum contains genes isolated from recombination by the same event; thus, the genes share similar levels of divergence. Lahn postulated at least four inversion events on the Y chromosome to account for his observations, in accordance with Ohno's prediction that inversion events would initiate Y-chromosome divergence and that the X chromosome would remain untouched.

Subsequent work on the human X chromosome and the chicken Z and W chromosomes provided further evidence for the degeneration of the sex-specific chromosome. The finished sequence of the human X chromosome was presented as a foil for the Y chromosome, revealing further details of Y-chromosome degeneration (Ross et al. 2005). Ross and colleagues confirmed the existence of the strata identified by Lahn and identified an additional, more recent stratum. As was the case for the X and Y chromosomes of mammals, the first sequence data from the chicken sex chromosomes showed that the Z and W chromosomes shared genes, suggesting that they too had evolved from a homologous pair of autosomes (Fridolfsson et al. 1998). As more W-linked genes were identified, Handley et al. (2004) compared them to their Z-linked homologs and identified strata. The sex-specific Y and W chromosomes evolved from autosomes along the same pathway of progressive isolation from recombination followed by degeneration.

CONSERVATION, RECOMBINATION, AND INNOVATION ON THE Y CHROMOSOME

The finished sequence of the human Y chromosome, published almost 90 years after Muller's original paper anticipating the degeneration of the nonrecombining sex chromosome, represented the first sequence of any sex-specific chromosome (Skaletsky et al. 2003). The human Y-chromosome sequence was assembled from individual BAC (bacterial artificial chromosome) clones from a single man's Y chromosome, allowing a greater degree of completeness in repetitive regions than had been achieved for other human chromosomes (Skaletsky et al. 2003). This effort enabled genomic comparisons that could, for the first time, rigorously test theoretical predictions of the course of sex chromosome evolution. Although it was clear that the human X and Y chromosomes had evolved from autosomes, unanticipated findings called into question some of the core assumptions of sex chromosome evolutionary theory. The human Y chromosome appeared to be a mosaic of different sequence classes that had different evolutionary trajectories (Skaletsky et al. 2003). The divergence evident in X-degenerate sequences had defined the evolutionary strata, but subsequent work would show that selection was more effective at preserving the surviving genes from degeneration than had been anticipated. The Y chromosome also gained genes in X-transposed and ampliconic sequences; these sequences demonstrated that Y chromosomes evolved not only by degeneration, but also by growth and elaboration.

Nearly half of the human Y chromosome is composed of X-degenerate sequences that contain genes that have survived the stepwise process of Y degeneration from the ancestral autosome pair that gave rise to the X and Y chromosomes (Skaletsky et al. 2003). The X-degenerate portion of the Y chromosome has unquestionably lost most genes that were present on the ancestral autosome pair; only 16 single-copy genes have survived out of the hundreds that are inferred to have been present on the ancestor of the X and Y chromosomes (Skaletsky et al. 2003). This has led to prominent claims that the Y chromosome is decaying at such a rapid pace that it will be devoid of genes in 10 million years (Aitken and Graves 2002). However, there is abundant evidence that the Y chromosome will not "self-destruct" any time soon. Rozen et al. (2009) examined variation in these surviving genes across a panel of 105 men representing worldwide Y-chromosome diversity. They discovered that there is remarkably little variation in X-degenerate protein-coding sequences; on average, two randomly chosen Y chromosomes differ by only a single-amino-acid change (Rozen et al. 2009). They found that both nucleotide diversity and the proportion of variant sites are higher for silent substitutions than for substitutions that would lead to amino acid changes, implying that natural selection has operated effectively to preserve the coding sequences of the X-degenerate genes during human history (Rozen et al. 2009). Nonrecombining sequences can be stable over even longer timescales. Hughes et al. (2005) systematically compared the human X-degenerate genes to those of the chimpanzee. They found that the human Y has preserved all X-degenerate genes that were present in the common ancestor of humans and chimps. Thus, the X-degenerate sequences of the human Y chromosome have been stable for at least the past 6 million years.

The sequence of the human Y chromosome showed not only that the human Y has avoided destruction, but that it is also undergoing growth and innovation in gene content. The rest of the human Y chromosome is composed of two sequence classes, X-transposed and ampliconic, many of whose genes have been added to the Y chromosome since it began to diverge from the X (Skaletsky et al. 2003). After the divergence of humans and chimpanzees, a transposition event restored a block of two-single-copy X-transposed genes to the human Y chromosome (Skaletsky et al. 2003). Ampliconic sequences form highly identical (>99.9% nucleotide identity) tandem arrays and inverted repeats that could only be resolved by BAC-based finishing strategies. The largest was a nearly perfect palindrome almost 3 Mb across (Kuroda-Kawaguchi et al. 2001; Skaletsky et al. 2003). The ampliconic portion of the Y chromosome contains nine multicopy gene families, totaling ~60 transcription units (Skaletsky et al. 2003). Two gene families are survivors of Y-chromosome decay that have become amplified, whereas others appear to have moved to the Y chromosome from autosomes (Saxena et al. 1996; Lahn and Page 1999a; Skaletsky et al. 2003). All of these genes are expressed in the testis (Skaletsky et al. 2003), and deletions in these sequences are the most common known genetic cause of spermatogenic failure in humans (Kuroda-Kawaguchi et al. 2001; Repping et al. 2002, 2003). Muller's theory did not predict the existence of this crucial part of the Y chromosome.

Further characterization of mammalian Y chromosomes demonstrated that ampliconic sequences represent a major exception to Muller's theory. The high nucleotide identity between the genes in palindromes on the human Y chromosome could be interpreted as evidence that the ampliconic sequences evolved relatively recently in human evolution, within the last 100,000 years. However, Rozen et al. (2003) used comparative sequencing in great apes to show that at least six of the eight human Y-chromosome palindromes predate the divergence of chimpanzees and humans more than 6 million years ago. To explain this result, they hypothesized that the arms of these palindromes must engage in gene conversion, driving the paired arms to evolve in concert. They confirmed this by surveying the diversity of human Y chromosomes to capture instances of gene conversion within the human lineage (Rozen et al. 2003). Muller and other investigators had assumed that the Y chromosome could not engage in recombination and would inevitably decay, but gene conversion allows for productive recombination between palindrome arms as though they were two alleles on homologous autosomes (Rozen et al. 2003; Skaletsky et al. 2003). This has allowed the ampliconic genes of the Y chromosomes to survive and expand during primate evolution while many single-copy genes have decayed.

Not only are ampliconic regions capable of recombination, but this recombination results in the continual remodeling of Y-chromosome sequence. Because ampliconic regions are, by definition, highly identical sequences in tandem or inverted repeats, they are prone to rearrangements that lead to variations in copy number as well as inversions. Repping et al. (2006) surveyed a panel of diverse Y chromosomes and observed extensive structural variation among human Y chromosomes. Using the phylogentic tree of human Y chromosomes, they were able to place a lower bound on the rate of rearrangements; most rearrangements occur on the order of 10^{-4} events per father-to-son transmission (Repping et al. 2006). This high rate of rearrangement causes the structure of ampliconic sequences to evolve much more rapidly than X-degenerate sequences. Hughes et al. (2010) found that although all ampliconic gene families are conserved between humans and chimpanzees, the chimpanzee ampliconic sequences have experienced many more rearrangements than the X-degenerate sequences, producing a completely different structure. Unlike the X-degenerate regions of the Y, the ampliconic regions are a source of continual growth and change.

INNOVATION ON THE X CHROMOSOME

Although the finished sequence of the human Y chromosome led to discoveries that challenged the traditional model of the Y chromosome as a rotting autosome by showing growth and change on the Y chromosome, it also reinforced the view of the X chromosome as unchanging. Muller's theory predicts that the decay of genes on Y and W chromosomes constrains X and Z chromosomes to stably maintain the gene content of the autosomes from which they evolved. In formulating Ohno's Law, Ohno (1967) reasoned that an elaborate chromosome-wide mechanism of dosage compensation would also stabilize the gene content of X and Z chromosomes, because genes that translocated to or from an X or Z chromosome would become misregulated. As a result, most genomic studies have treated the X chromosome as a control to show the dramatic changes on the Y chromosome, leaving the question of changes in X-chromosome gene content unexamined. Only comparisons among X chromosomes or between X chromosomes and the autosomes of other species can test whether the gene content of the X chromosome has changed through the course of X-chromosome evolution.

Initial comparisons of X and Z chromosomes among species have generally supported Muller and Ohno's predictions of conservation. Comparative mapping experiments have repeatedly shown that the genes of the X chromosome are well conserved among placental mammals (O'Brien et al. 1993; Carver and Stubbs 1997; Chowdhary et al. 1998; Ross et al. 2005). Although mammalian X chromosomes have experienced a number of rearrangements, particularly in the rodent lineage, over the course of mammalian evolution, they have sustained fewer interchromosomal translocations than mammalian autosomes (Carver and Stubbs 1997). Outside of mammals, comparative mapping of Z-linked genes in birds by FISH (fluorescence in situ hydridization) has indicated that the Z chromosome is conserved among avian species (Nanda et al. 2008). Similar results have been reported in comparisons of several snake species (Matsubara et al. 2006). Because comparative mapping experiments are designed to locate the orthologs of the genes from one species on the chromosomes of another, the results of these experiments are biased toward finding conservation rather than novelty.

In line with the predictions of Ohno's Law, PARs (pseudoautosomal regions) have not been as well conserved as the rest of the X chromosome. Several genes in the mammalian pseudoautosomal region have moved from the PAR to autosomes in mice (Palmer et al. 1995; Carver and Stubbs 1997). Wilcox et al. (1996) examined the locations of human X-linked genes in marsupials and monotremes. They discovered that the genes composing the short arm of the human X were present on the autosomes of monotremes and marsupials (Wilcox et al. 1996). This gene traffic to and from the mammalian X chromosome seems like a violation of Ohno's Law, but it is actually in accord with Ohno's predictions. The region added to the X in eutherian mammals falls into the three most recent strata of the human sex chromosomes; when it translocated to the ancestral eutherian X chromosome, it was added to the PAR and shared with the Y chromosome. Because PARs still participate in crossing-over, Y-linked gene copies do not decay and the X-linked copies are not subject to dosage compensation. The genes in the PAR are free to move between autosomes and the sex chromosomes until they are locked in by an event that expands the region of suppressed recombination between the sex chromosomes.

Even outside the PARs, the gene content of the mammalian X chromosome is not completely stable. Genomic data from humans and mice have allowed researchers to systematically identify gene movement to and from the mammalian X chromosome. Emerson et al. (2004) found that the mouse and human X chromosomes have both generated and received an excess of genes through retrotransposition. By comparing the human and mouse X chromosomes, they found that this process began before humans and mice diverged and has continued after that divergence in both lineages. Mammalian X chromosomes have also gained genes through the duplication of existing X-linked genes. Warburton et al. (2004) found that the human X chromosome is enriched for amplicons that contain testis-expressed genes. These X-chromosome amplicons primarily contain the cancer-testis antigen (CTA) genes. Comparative studies have shown that several CTA gene families expanded in the primate lineage (De Backer et al. 1999; Aradhya et al. 2001; Kouprina et al. 2004). Other CTA gene families, including the MAGE genes, the most abundant gene family on the human X chromosome, have independently expanded in both rodent and primate lineages (Chomez et al. 2001; Chen et al. 2003; Birtle et al. 2005; Ross et al. 2005). Mueller et al. (2008) found that the mouse X chromosome contained 33 multicopy gene families, which, like human CTA genes, are expressed in the testis. These multicopy families were arranged in elaborate ampliconic structures covering 19 Mb of the mouse X chromosome (Mueller et al. 2008). Just as ampliconic gene families are a source of unexpected novel gene content on mammalian Y chromosomes, they are also a source of innovation on X chromosomes as well.

Contrary to the expectations of Muller's theory and Ohno's Law, recent research has shown that the gene content of X chromosomes is not static. On the one hand, conservation of gene content is observed throughout the majority of the mammalian X chromosome, where gene loss from the Y chromosome and the subsequent evolution of dosage compensation restrict the flow of genes off of and onto the X. On the other hand, PARs have been sites of gene movement to and from the X chromosome, the most dramatic being the X added region of placental mammals, which accounts for nearly the entire short arm of the human X chromosome. Even outside of PARs, retrotransposition and gene duplication have reshaped the gene content of mammalian X chromosomes, creating amplicons of testis-expressed genes parallel to those observed on mammalian Y chromosomes. The changes to X chromosomes are as impressive as their conservation.

CURRENT CHALLENGES AND FUTURE DIRECTIONS

For nearly 100 years, the evolution of sex chromosomes has been described in the context of Muller's theory that sex chromosomes evolve from autosomes through the degeneration of the sex-specific chromosome. This hypothesis accounts for nearly all of the data that were available before the sequences of sex chromosomes were completed. However, Muller's theory does not account for the degree to which gene movement and duplication have shaped the evolution of sex chromosomes. The ampliconic sequences of the human Y chromosome are essential for male fertility and therefore for the continued survival of the Y chromosome, but they were unanticipated in Muller's theory. Amplicons on X chromosomes represent unexpected innovations in gene content on what was presumed to be an unchanging chromosome. In the same way that the development of population genetics reshaped the description of Y degeneration under Muller's theory, it is necessary to amend Muller's hypothesis in light of genomic data.

A greater understanding of the forces that generate amplicons will result from a more complete description of their function. One possibility is that the high copy number of ampliconic genes reflects selection for increased expression. Ampliconic genes might be duplicated to facilitate high levels of transcription, as has been proposed for ribosomal RNAs, transfer RNAs, and histone genes (Finnegan et al. 1978; Kedes 1979; Long and Dawid 1980). The high frequency of transcription of mouse X ampliconic genes despite the general postmeiotic silencing of single-copy genes on the X chromosome would be consistent with this hypothesis. The universal expression of ampliconic genes in the testis provides a second possible explanation: Repetitive DNA structures provide a chromatin environment that is permissive for gene expression in germ cells. As an alternative to hypotheses based on gene expression, amplicons may have a role in preserving functional gene copies in regions where crossing-over with a homologous chromosome rarely, if ever, occurs. The amplicons on the Y chromosome of primates engage in gene conversion, providing a mechanism to preserve the function of genes in the face of chromosome-wide degradation. Ideally, a unified theory would explain why amplicons are more prevalent on sex chromosomes than in the rest of the genome, but it is possible that amplicons are present on different sex chromosomes for different reasons.

Escape from postmeiotic silencing on sex chromosomes could serve as a compelling explanation for the location of amplicons in mammals, but silencing of sex chromosomes is far from universal. Unlike XY male mammals, ZW female birds do not appear to silence unpaired chromosomes during meiosis (Solari 1977).

During the diplotene stage of female meiosis, the Z and W chromosomes of chickens are highly transcriptionally active, forming lampbrush chromosomes (Hutchison 1987). If ampliconic sequences exist in birds, they will require an alternative explanation.

An alternative to the avoidance of meiotic silencing is that sex-linked amplicons are the result of sexually antagonistic selection. Sexually antagonistic genes are those that produce a phenotype that benefits one sex more than the other. These traits are more likely to become fixed on sex chromosomes than on autosomes because the sex chromosomes are not evenly exposed to selection in both sexes (Rice 1984). Male-benefit genes should accumulate on Y chromosomes, and female-benefit genes should accumulate on W chromosomes. The case for X and Z chromosomes is more complex. Dominant traits that benefit the homogametic sex should accumulate because they are exposed to selection twice as often in the homogametic sex. Recessive traits that benefit the heterogametic sex should accumulate because they are always exposed to stronger selection in the heterogametic sex than in the homogametic sex, where they can be masked by other alleles. Eventually, sexually antagonistic genes are expected to evolve sex-limited expression to avoid costs to the sex where they are not beneficial (Rice 1984). As a result, one would expect to find that sex chromosomes would become enriched for genes expressed only in one sex.

Sexually antagonistic selection is an attractive explanation for the enrichment of amplicons on the sex chromosomes, but there are incongruities with the existing data. There do not appear to be any female-benefit amplicons on X chromosomes, where they might be expected to arise because the X chromosome is exposed to more frequent selection in females than in males. All known ampliconic sequences, including those on X chromosomes, are expressed in the testis. The presence of testis-expressed amplicons on X chromosomes is striking because gene duplication was classically imagined as a dominant gain-of-function mutation (Muller 1932), but the theory of sexually antagonistic selection predicts that only recessive male-benefit alleles should accumulate on X chromosomes. If sexually antagonistic selection is responsible for the generation of testis-expressed amplicons, gene duplication on the X chromosome may be preceded by the evolution of male-limited expression, so that duplications are only subjected to selection in males.

Amplicons could also be involved in intragenomic conflict through segregation distortion in the germline. Autosomal segregation distortion due to the t haplotype of chromosome 17 in mice is well known (Silver 1993). On the sex chromosomes, a segregation-distorting locus could function as a sex ratio distorter. Because most organisms are constrained to a 1:1 sex ratio, any sex ratio distorter that meets with success immediately increases the selective advantage for a second distorter to restore the sex ratio to equilibrium (Fisher 1930; Nur 1974). This could lead to an evolutionary arms race between sex chromosomes. There are indications that the mouse X and Y chromosomes are involved in segregation distortion; deletions on the long arm of the mouse Y chromosome lead to an excess of female offspring, suggesting that the multicopy genes on the mouse Y chromosome may suppress X-chromosome segregation distortion (Conway et al. 1994). If amplicons are primarily generated as a result of intragenomic conflict between the sex chromosomes, birds and snakes would be expected to accumulate genes that are expressed during female meiosis to influence the partition of the Z and W chromosomes between the oocyte and the first polar body (Rutkowska and Badyaev 2008).

In the past 10 years, genomic data from vertebrate sex chromosomes have allowed reconstructions of the process of sex chromosome evolution, and these reconstructions have revealed surprising exceptions to Muller's theory. We can look forward to the availability of additional sex chromosome sequences that will enable us to extend our analyses of sex chromosomes. Sequencing efforts for several mammalian Y chromosomes are under way. These will allow us to extend our comparisons of Y chromosomes from the divergence of human populations, through primate evolution, to the very base of the mammalian tree. The sequences of the chicken sex chromosomes will allow us to extend our evolutionary comparisons even further. The chicken sex chromosomes have evolved independently of mammalian sex chromosomes for more than 300 million years. As a result, the chicken sex chromosomes and the human sex chromosomes represent the outcome of two parallel experiments of nature. Reciprocal comparisons of the finished sequences of the chicken Z and human X chromosomes to the orthologous autosomal regions in the other species will enable us to trace changes that occurred on the Z and X chromosomes during the course of sex chromosome evolution. Intraspecific comparisons between the finished sequences of the Z and W chromosomes will reveal whether the course of W evolution has been parallel to that of the degeneration and elaboration of the human Y chromosome. The description of ampliconic sequences on the W chromosome is also likely to be revealing. There are at least two multicopy gene families on the W chromosome, but they are ubiquitously expressed and their genomic structure is unknown. W amplicons, if they exist, may show a functional coherence like that of the human Y, revealing genes that are essential for female fertility.

Additional insights on par with those obtained from the sequence of the human X and Y chromosomes can only come with additional high-quality finished sequencing efforts. Ampliconic sequences could not have been described without the BAC-based "clone-by-clone" methods used to determine the sequence of the human sex chromosomes. Shotgun sequencing technologies collapse highly identical repeats into single contigs, obscuring rather than revealing their structure and organization. This deficiency of shotgun methods only worsens with shorter read lengths. Only BAC-based sequencing provides the positional information needed to disentangle long repeats. Although these BAC-based sequencing technologies are slower and more expensive than their whole-genome shotgun counterparts, they have resulted in insights that would have been impossible to obtain in any other way and which were unanticipated by a century of theory.

ACKNOWLEDGMENTS

We thank H. Skaletsky, J. Hughes, J. Mueller, A. Larracuente, S. Soh, and K. Romer for comments on the manuscript. Our work is supported by the National Institutes of Health and the Howard Hughes Medical Institute.

REFERENCES

Aitken RJ, Graves JAM. 2002. Human spermatozoa: The future of sex. *Nature* **415:** 963–964.

Aradhya S, Bardaro T, Galgoczy P, Yamagata T, Esposito T, Patlan H, Ciccodicola A, Munnich A, Kenwrick S, Platzer M. 2001. Multiple pathogenic and benign genomic rearrangements occur at a 35 kb duplication involving the *NEMO* and *LAGE2* genes. *Hum Mol Genet* **10:** 2557–2567.

Bachtrog D. 2008. The temporal dynamics of processes underlying Y chromosome degeneration. *Genetics* **179:** 1513–1525.

Birtle Z, Goodstadt L, Ponting C. 2005. Duplication and positive selection among hominin-specific *PRAME* genes. *BMC Genomics* **6:** 120–138.

Bridges CB. 1914. Direct proof through non-disjunction that the sex-linked genes of *Drosophila* are borne by the X-chromosome. *Science* **40:** 107–109.

Carver EA, Stubbs L. 1997. Zooming in on the human-mouse comparative map: Genome conservation re-examined on a high-resolution scale. *Genome Res* **7:** 1123–1137.

Charlesworth B. 1978. Model for evolution of Y chromosomes and dosage compensation. *Proc Natl Acad Sci* **75:** 5618–5622.

Charlesworth B. 1994. The effect of background selection against deleterious mutations on weakly selected, linked variants. *Genet Res* **63:** 213–227.

Charlesworth B, Morgan MT, Charlesworth D. 1993. The effect of deleterious mutations on neutral molecular variation. *Genetics* **134:** 1289–1303.

Chen YT, Alpen B, Ono T, Gure AO, Scanlan MA, Biggs WH, Arden K, Nakayama E, Old LJ. 2003. Identification and characterization of mouse *SSX* genes: A multigene family on the X chromosome with restricted cancer/testis expression. *Genomics* **82:** 628–636.

Chomez P, De Backer O, Bertrand M, De Plaen E, Boon T, Lucas S. 2001. An overview of the MAGE gene family with the identification of all human members of the family. *Cancer Res* **61:** 5544–5551.

Chowdhary BP, Raudsepp T, Frönicke L, Scherthan H. 1998. Emerging patterns of comparative genome organization in some mammalian species as revealed by Zoo-FISH. *Genome Res* **8:** 577–589.

Conway S, Mahadevaiah S, Darling S, Capel B, Rattigan A, Burgoyne P. 1994. *Y353/B*: A candidate multiple-copy spermiogenesis gene on the mouse Y chromosome. *Mamm Genome* **5:** 203–210.

Cooke H, Brown W, Rappold G. 1985. Hypervariable telomeric sequences from the human sex chromosomes are pseudoautosomal. *Nature* **317:** 687–692.

De Backer O, Arden KC, Boretti M, Vantomme V, De Smet C, Czekay S, Viars CS, De Plaen E, Brasseur F, Chomez P. 1999. Characterization of the *GAGE* genes that are expressed in various human cancers and in normal testis. *Cancer Res* **59:** 3157–3165.

Emerson JJ, Kaessmann H, Betran E, Long M. 2004. Extensive gene traffic on the mammalian X chromosome. *Science* **303:** 537–540.

Felsenstein J. 1974. The evolutionary advantage of recombination. *Genetics* **78:** 737–756.

Finnegan D, Rubin G, Young M, Hogness D. 1978. Repeated gene families in *Drosophila melanogaster*. *Cold Spring Harbor Symp Quant Biol* **42:** 1053–1063.

Fisher R. 1930. *The genetical theory of natural selection*. Dover, New York.

Fisher R. 1935. The sheltering of lethals. *Am Nat* **69:** 446–455.
Foote S, Vollrath D, Hilton A, Page DC. 1992. The human Y chromosome: Overlapping DNA clones spanning the euchromatic region. *Science* **258:** 60–66.
Ford CE, Jones KW, Polani PE, De Almeida JC, Briggs JH. 1959. A sex-chromosome anomaly in a case of gonadal dysgenesis (Turner's syndrome). *Lancet* **1:** 711–713.
Fridolfsson AK, Cheng H, Copeland NG, Jenkins NA, Liu HC, Raudsepp T, Woodage T, Chowdhary B, Halverson J, Ellegren H. 1998. Evolution of the avian sex chromosomes from an ancestral pair of autosomes. *Proc Natl Acad Sci* **95:** 8147–8152.
Goodfellow PJ, Darling SM, Thomas NS, Goodfellow PN. 1986. A pseudoautosomal gene in man. *Science* **234:** 740–743.
Handley LJ, Ceplitis H, Ellegren H. 2004. Evolutionary strata on the chicken Z chromosome: Implications for sex chromosome evolution. *Genetics* **167:** 367–376.
Hughes JF, Skaletsky H, Pyntikova T, Minx PJ, Graves T, Rozen S, Wilson RK, Page DC. 2005. Conservation of Y-linked genes during human evolution revealed by comparative sequencing in chimpanzee. *Nature* **437:** 100–103.
Hughes JF, Skaletsky H, Pyntikova T, Graves TA, van Daalen SK, Minx PJ, Fulton RS, McGrath SD, Locke DP, Friedman C, et al. 2010. Chimpanzee and human Y chromosomes are remarkably divergent in structure and gene content. *Nature* **463:** 536–539.
Hutchison N. 1987. Lampbrush chromosomes of the chicken, *Gallus domesticus*. *J Cell Biol* **105:** 1493–1500.
Jacobs PA, Strong JA. 1959. A case of human intersexuality having a possible XXY sex-determining mechanism. *Nature* **183:** 302.
Jegalian K, Page DC. 1998. A proposed path by which genes common to mammalian X and Y chromosomes evolve to become X inactivated. *Nature* **394:** 776–780.
Kedes L. 1979. Histone genes and histone messengers. *Annu Rev Biochem* **48:** 837–870.
Kouprina N, Mullokandov M, Rogozin IB, Collins NK, Solomon G, Otstot J, Risinger JI, Koonin EV, Barrett JC, Larionov V. 2004. The *SPANX* gene family of cancer/testis-specific antigens: Rapid evolution and amplification in African great apes and hominids. *Proc Natl Acad Sci* **101:** 3077–3082.
Kuroda-Kawaguchi T, Skaletsky H, Brown LG, Minx PJ, Cordum HS, Waterston RH, Wilson RK, Silber S, Oates R, Rozen S, et al. 2001. The AZFc region of the Y chromosome features massive palindromes and uniform recurrent deletions in infertile men. *Nat Genet* **29:** 279–286.
Lahn B, Page D. 1999a. Retroposition of autosomal mRNA yielded testis-specific gene family on human Y chromosome. *Nat Genet* **21:** 429–433.
Lahn BT, Page DC. 1999b. Four evolutionary strata on the human X chromosome. *Science* **286:** 964–967.
Long E, Dawid I. 1980. Repeated genes in eukaryotes. *Annu Rev Biochem* **49:** 727–764.
Matsubara K, Tarui H, Toriba M, Yamada K, Nishida-Umehara C, Agata K, Matsuda Y. 2006. Evidence for different origin of sex chromosomes in snakes, birds, and mammals and step-wise differentiation of snake sex chromosomes. *Proc Natl Acad Sci* **103:** 18190–18195.
Maynard Smith J, Haigh J. 1974. The hitchhiking effect of a favorable gene. *Genet Res* **23:** 23–35.
McKusick VA. 1962. On the X chromosome of man. *Q Rev Biol* **37:** 69–175.
Morgan TH. 1911a. An attempt to analyze the constitution of the chromosomes on the basis of sex-limited inheritance in *Drosophila*. *J Exp Zool* **11:** 365–414.
Morgan TH. 1911b. The application of the conception of pure lines to sex-limited inheritance and to sexual dimorphism. *Am Nat* **45:** 65–78.
Mueller JL, Mahadevaiah SK, Park PJ, Warburton PE, Page DC, Turner JM. 2008. The mouse X chromosome is enriched for multicopy testis genes showing postmeiotic expression. *Nat Genet* **40:** 794–799.
Muller HJ. 1914. A gene for the fourth chromosome of *Drosophila*. *J Exp Zool* **17:** 325–336.
Muller HJ. 1918. Genetic variability, twin hybrids and constant hybrids, in a case of balanced lethal factors. *Genetics* **3:** 422–499.
Muller HJ. 1932. Further studies on the nature and causes of gene mutations. In *Proceedings of the 6th International Congress of Genetics,* pp. 213–255.
Muller HJ. 1964. The relation of recombination to mutational advance. *Mutat Res* **106:** 2–9.
Nanda I, Schlegelmilch K, Haaf T, Schartl M, Schmid M. 2008. Synteny conservation of the Z chromosome in 14 avian species (11 families) supports a role for Z dosage in avian sex determination. *Cytogenet Genome Res* **122:** 150–156.
Nur U. 1974. The expected changes in the frequency of alleles affecting the sex ratio. *Theor Popul Biol* **5:** 143–147.
O'Brien SJ, Womack JE, Lyons LA, Moore KJ, Jenkins NA, Copeland NG. 1993. Anchored reference loci for comparative genome mapping in mammals. *Nat Genet* **3:** 103–112.
Ohno S. 1967. *Sex chromosomes and sex-linked genes*. Springer-Verlag, New York.
Palmer S, Perry J, Ashworth A. 1995. A contravention of Ohno's law in mice. *Nat Genet* **10:** 472–476.
Rasmussen SW, Holm PB. 1978. Human meiosis. II. Chromosome pairing and recombination nodules in human spermatocytes. *Carlsberg Res Commun* **43:** 275–327.
Repping S, Skaletsky H, Lange J, Silber S, Van Der Veen F, Oates RD, Page DC, Rozen S. 2002. Recombination between palindromes P5 and P1 on the human Y chromosome causes massive deletions and spermatogenic failure. *Am J Hum Genet* **71:** 906–922.
Repping S, Skaletsky H, Brown L, van Daalen SK, Korver CM, Pyntikova T, Kuroda-Kawaguchi T, de Vries JW, Oates RD, Silber S, et al. 2003. Polymorphism for a 1.6-Mb deletion of the human Y chromosome persists through balance between recurrent mutation and haploid selection. *Nat Genet* **35:** 247–251.
Repping S, van Daalen SK, Brown LG, Korver CM, Lange J, Marszalek JD, Pyntikova T, van der Veen F, Skaletsky H, Page DC, et al. 2006. High mutation rates have driven extensive structural polymorphism among human Y chromosomes. *Nat Genet* **38:** 463–467.
Rice WR. 1984. Sex chromosomes and the evolution of sexual dimorphism. *Evolution* **38:** 735–742.
Rice WR. 1987. Genetic hitchhiking and the evolution of reduced genetic activity of the Y sex chromosome. *Genetics* **116:** 161–167.
Ross MT, Grafham DV, Coffey AJ, Scherer S, McLay K, Muzny D, Platzer M, Howell GR, Burrows C, Bird CP, et al. 2005. The DNA sequence of the human X chromosome. *Nature* **434:** 325–337.
Rozen S, Skaletsky H, Marszalek JD, Minx PJ, Cordum HS, Waterston RH, Wilson RK, Page DC. 2003. Abundant gene conversion between arms of palindromes in human and ape Y chromosomes. *Nature* **423:** 873–876.
Rozen S, Marszalek JD, Alagappan RK, Skaletsky H, Page DC. 2009. Remarkably little variation in proteins encoded by the Y chromosome's single-copy genes, implying effective purifying selection. *Am J Hum Genet* **85:** 923–928.
Rutkowska J, Badyaev A. 2008. Meiotic drive and sex determination: Molecular and cytological mechanisms of sex ratio adjustment in birds. *Philos Trans R Soc Lond B Biol Sci* **363:** 1675–1686.
Saxena R, Brown LG, Hawkins T, Alagappan RK, Skaletsky H, Reeve MP, Reijo R, Rozen S, Dinulos MB, Disteche CM, et al. 1996. The *DAZ* gene cluster on the human Y chromosome arose from an autosomal gene that was transposed, repeatedly amplified and pruned. *Nat Genet* **14:** 292–299.
Silver LM. 1993. The peculiar journey of a selfish chromosome: Mouse t haplotypes and meiotic drive. *Trends Genet* **9:** 250–254.
Simmler M, Rouyer F, Vergnaud G, Nyström-Lahti M, Ngo K, de La Chapelle A, Weissenbach J. 1985. Pseudoautosomal DNA sequences in the pairing region of the human sex chromosomes. *Nature* **317:** 692–697.
Skaletsky H, Kuroda-Kawaguchi T, Minx PJ, Cordum HS, Hillier

L, Brown LG, Repping S, Pyntikova T, Ali J, Bieri T, et al. 2003. The male-specific region of the human Y chromosome is a mosaic of discrete sequence classes. *Nature* **423**: 825–837.

Solari AJ. 1977. Ultrastructure of the synaptic autosomes and the ZW bivalent in chicken oocytes. *Chromosoma* **64**: 155–165.

Solari AJ, Tres LL. 1970. The three-dimensional reconstruction of the XY chromosomal pair in human spermatocytes. *J Cell Biol* **45**: 43–53.

Stern C. 1957. The problem of complete Y-linkage in man. *Am J Hum Genet* **9**: 147–166.

Sturtevant AH. 1913. The linear arrangement of six sex-linked factors in *Drosophila*, as shown by their mode of association. *J Exp Zool* **14**: 43–59.

Vollrath D, Foote S, Hilton A, Brown LG, Beer-Romero P, Bogan JS, Page DC. 1992. The human Y chromosome: A 43-interval map based on naturally occurring deletions. *Science* **258**: 52–59.

Warburton PE, Giordano J, Cheung F, Gelfand Y, Benson G. 2004. Inverted repeat structure of the human genome: The X-chromosome contains a preponderance of large, highly homologous inverted repeats that contain testes genes. *Genome Res* **14**: 1861–1869.

Welshons WJ, Russell LB. 1959. The Y-chromosome as the bearer of male determining factors in the mouse. *Proc Natl Acad Sci* **45**: 560–566.

Wilcox SA, Watson JM, Spencer JA, Graves JAM. 1996. Comparative mapping identifies the fusion point of an ancient mammalian X-autosomal rearrangement. *Genomics* **35**: 66–70.

Wilson EB. 1911. The sex chromosomes. *Arch Mikrosk Anat* **77**: 249–271.

The Evolution of Human Segmental Duplications and the Core Duplicon Hypothesis

T. MARQUES-BONET[1,2] AND E.E. EICHLER[1,3]

[1]*Department of Genome Sciences, University of Washington, Seattle, Washington 98195;*
[2]*Institut de Biologia Evolutiva (UPF-CSIC), 08003 Barcelona, Catalonia, Spain;*
[3]*Howard Hughes Medical Institute, Seattle, Washington 98195*

Correspondence: eee@gs.washington.edu

Duplicated sequences are important sources of genetic instability and in the evolution of new gene function within species. Hominids have a preponderance of intrachromosomal duplications organized in an interspersed fashion, as opposed to tandem duplications, which are common in other mammalian genomes such as mouse, dog, and cow. Multiple lines of evidence, including sequence divergence, comparative primate genomes, and fluorescence in situ hybridization (FISH) analyses, point to an excess of segmental duplications in the common ancestor of humans and African great apes. We find that much of the interspersed human duplication architecture within chromosomes is focused around common sequence elements referred to as "core duplicons." These cores correspond to the expansion of gene families, some of which show signatures of positive selection and lack orthologs present in other mammalian species. This genomic architecture predisposes apes and humans not only to extensive genetic diversity, but also to large-scale structural diversity mediated by nonallelic homologous recombination. In humans, many de novo large-scale genomic changes mediated by these duplications are associated with neuropsychiatric and neurodevelopmental disease. We propose that the disadvantage of a high rate of new mutations is offset by the selective advantage of newly minted genes within the cores.

Geneticists have long appreciated the dual nature of duplicated sequences as sources of evolutionary innovation and regions of genomic instability. Muller et al. (1936), Bridges (1936), and Sturtevant (1925) were among the first to recognize the role duplicated sequences have in both phenotype and genetic instability by their association of unequal crossing-over of the *Bar* locus in *Drosophila* and the eye-reduction phenotype. The frequency and phenotypic consequences of new mutations among tandem duplicates were noted by Bridges in 1936 when he commented, "The production of *Bar*-double and of *Bar*-reverted is seen to be the insertion of this extra section twice, or conversely, its total loss—both presumably by a process of unequal crossing-over. A remarkable peculiarity of the mutant is that occasionally the homozygous stock gives rise to a fly indistinguishable in appearance and genetic behavior from wild-type." Ohno highlighted the importance of duplication in the "birth" of new genes during evolution. To Ohno, the process of duplication liberated genes from the constraint of ancestral function, allowing new mutations to give rise to modified or novel function. This was an extension of Muller's dictum "all life from pre-existing life…and every gene from a pre-existing gene" (Muller et al. 1936). Ohno posited that the origin of vertebrate complexity lies in the large whole-genome duplications providing a burst of functional redundancy and subsequent specialization (Ohno et al. 1968).

It follows that if one is interested in areas of rapid evolutionary change and the discovery of genes important in the specification of the human condition, then the recently duplicated regions of our genome represent fertile areas of investigation (Eichler 2001). The study of these regions has revealed unexpected complexities in the evolution of our genome, led to the identification of novel human/great ape genes, and provided a road map for the discovery of new mutations associated with a wide range of pediatric and adult-onset disease. Although the sequencing of entire genomes has accelerated at a breakneck pace, sequencing of recently duplicated regions of the genome has proved more challenging and proceeded much more slowly. By dint of their high sequence identity and their large size (frequently >100 kbp in length) (She et al. 2004), sequence assemblies based strictly on short whole-genome shotgun sequences (<600 bp) have often failed to resolve these aspects of genome organization. Among mammals, only two genomes—mouse and human—have been sequenced to the level of rigor required to accurately infer the structure and organization from the assembled genome sequence.

HUMAN VERSUS MOUSE SEGMENTAL DUPLICATION PROPERTIES

The most recent comparisons of the mouse and human finished genomes (Collins et al. 2004; Church et al. 2009) show that the two species are comparable in terms of the number of base pairs mapping to high-identity (>90%) duplications. However, there are three notable differences. Almost all large segmental duplications (SDs) in the mouse lineage are tandemly organized, whereas >59% of the duplications in humans are interspersed—being separated from their nearest paralog by more than 1 Mbp or mapping to a nonhomologous chromosome (She et al. 2008). Experimental and computational analyses of other genomes, such as the dog, rat, and cow, suggest that the tandem configuration likely represents the mammalian archetype (Tuzun et al. 2004; Elsik et al. 2009; Nicholas et al. 2009). Second, human duplications tend to be significantly enriched in spliced transcripts when compared to mouse, which appear to be more deficient in transcripts and, possibly, genes (She et al. 2008). Third, within the human genome, there is a skew toward higher sequence identity duplications, which suggests a potential excess of evolutionarily young SDs (Fig. 1). The presence of large, high-identity duplications at more locations has sensitized more of the human genome to the dosage and potential position effects as a result of unequal crossing-over.

PRIMATE COMPARISONS

Despite the working draft nature of other nonhuman primate genome assemblies, the random nature of the underlying whole-genome shotgun (WGS) sequence data provides a means to detect duplications in the absence of an assembly. By mapping regions of excess WGS read-depth against the finished human reference sequence, we can predict the content of duplication in closely related primates such as chimpanzee, orangutan, and macaque. We can, then, parsimoniously infer the age of human duplications based on their shared or lineage-specific nature within the context of the generally accepted primate phylogeny. The analysis shows that the proportion of lineage-specific duplications in the chimpanzee and human lineages is approximately equal (Cheng et al. 2005; Marques-Bonet et al. 2009). We, however, predict a two to fourfold excess of new SDs in the common ancestor of humans and African great apes when compared to Asian apes (orangutan) and Old World monkey lineages (represented by macaque) (Fig. 2). The effect is most pronounced for intrachromosomal SDs. These findings are consistent with the excess of high-identity (>97%) pairwise alignments noted within the human genome assembly for intrachromosomal duplications (Fig. 1) and studies of gene duplication (Fortna et al. 2004; Dumas et al. 2007; Hahn et al. 2007) that suggest a burst of duplication activity during primate evolution. Notably, this duplication acceleration occurs at a period of time when most other mutational processes, including point mutation and retrotransposon activity, were slowing down (Wu and Li 1985; Li and Tanimura 1987; Waterston et al. 2002; Consortium 2005).

DUPLICATION ORGANIZATION AND CORE DUPLICONS

Within the human genome, ancestral duplications (termed duplicons) of diverse interspersed origin juxtapose one another, forming complex mosaic duplication blocks that are hundreds of kilobase pairs in length (Rouquier et al. 1998; Johnson et al. 2006). This is in contrast to the organization in the mouse where most duplication blocks consist of tandemly organized SDs. Using a modified de Brujin graph theory approach along with comparative sequence data, we identified the ancestral origin of 4692 human duplication loci and deconvoluted the architecture of 437 duplication blocks in the human genome (Jiang et al. 2007). A complex pattern of duplication within duplications emerges, confirming the stepwise accretion of SDs on a genome-wide scale during hominid evolution (Eichler et al. 1997; Horvath et al. 2000; Courseaux et al. 2003; Stankiewicz et al. 2004; Johnson et al. 2006). Hierarchical clustering of these duplication blocks based on shared duplicon content organizes duplication blocks into 24 distinct groups (Fig. 3). Two distinct types of duplication blocks are distinguished: those in which the evolutionary flow of genetic information has occurred between nonhomologous chromosomes ($n = 10$) and those where the mosaic architectures have largely formed within a specific chromosome ($n = 14$). The former consists mainly of subtelomeric and

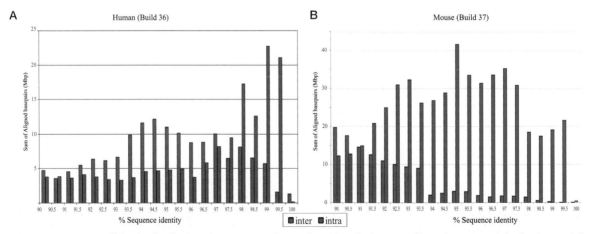

Figure 1. Percentage of identity distribution of mouse versus human SDs. Note the increase of interchromosomal duplications and the higher proportion of recent SDs in humans and the excess of intrachromosomal (tandem) duplications in mouse.

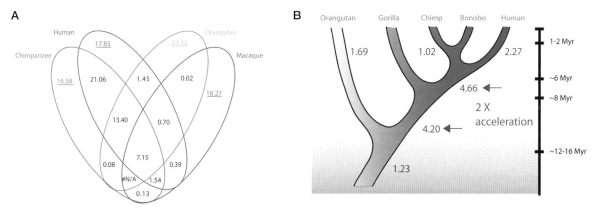

Figure 2. (*A*) Venn diagram showing shared and lineage-specific duplications among four primate genomes. Estimates were based on identifying regions of excess read-depth to the human assembly genome. Numbers underlined are copy-number corrected to avoid the bias of nonhuman-specific SDs. (*B*) Assignment of duplications and rate estimation of Mbp/Myr for each branch. Note the excess of duplication rate in the branch leading to the common ancestor of human and chimpanzee (Marques-Bonet et al. 2009).

pericentromeric duplications, and the latter corresponds almost exclusively to the intrachromosomal burst of SDs discussed above.

The hierarchical clustering suggests that the duplication blocks have been formed around a core or seed duplicon (defined as an ancestral duplicon that populates >67% of all duplication blocks within a group). These core sequences are among the most abundant and most ancient; they are particularly enriched for RefSeq genes and spliced expressed sequence tags (ESTs) when compared to flanking duplicons, and a few have been subjected to independent and recurrent duplications in different primate lineages (Johnson et al. 2006). Several of the corresponding genes and gene families encoded by these core duplicons lack orthologs in other mammalian species and have been highlighted as human–great ape

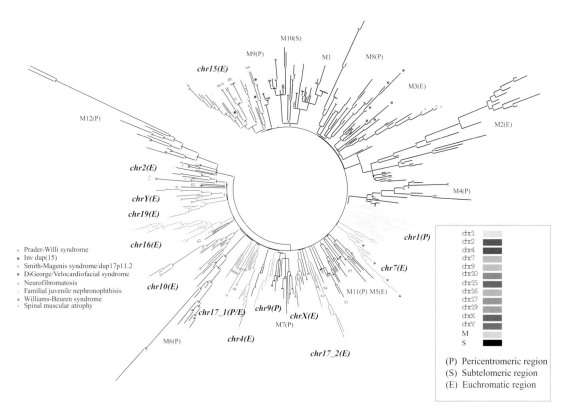

Figure 3. Hierarchical clustering of human duplication blocks based on ancestral duplicon content. The termini of each branch represent one of 437 duplication blocks, which cluster into 24 distinct groups, 14 of which are restricted to a specific chromosome and 10 of which are mixed (M) among chromosomes mapping largely to subtelomeric (S) or pericentromeric (P) regions of the genome. An expanded view of chromosome 16 is shown (Fig. 5) (Jiang et al. 2007).

gene family innovations (Johnson et al. 2001; Paulding et al. 2003; Ciccarelli et al. 2005). The *TRE2* oncogene, for example, is a fusion of a USP32 protease and a TBC1D3 core duplicon. The resulting fusion gene is expressed solely in humans and African great apes (Paulding et al. 2003). The *RANBP2*, morpheus (*NPIP*), and *NBPF11* (also known by its protein domain DUF1220) gene families show evidence of positive selection. Data from numerous copy-number variation studies (Sharp et al. 2005; Redon et al. 2006) suggest that these gene families are copy-number polymorphic in the human population. The functional significance of most of these genes is unknown. Functional characterization of the TBC1D3 core suggests that it may be important in modulating signaling of growth factors during development (Hodzic et al. 2006; Wainszelbaum et al. 2008). It is interesting that the copy-number polymorphism of one of these genes (*NPBF23*) has recently been implicated in pediatric neuroblastoma, with certain gene family members showing preferential expression in fetal brain and fetal sympathetic nervous tissue (Diskin et al. 2009).

PRIMATE SEQUENCE CHARACTERIZATION OF LCR16A

Detailed comparative primate sequencing of one of the core duplicons (LCR16a—seat of the *NPIP*/morpheus gene family expansion) is illustrative of the evolutionary dynamism that occurred during the human–great ape evolution. In the human genome reference sequence, there are 23 copies of the LCR16a sequence distributed among 17 complex duplication blocks ranging in size from ~40 to 609 kbp (Figs. 4 and 5). In addition to LCR16, 11 additional SDs of distinct evolutionary origin populate the duplication blocks on chromosome 16. Although the 20-kbp LCR16a occasionally occurs as a solitary duplicon (i.e., without flanking duplicons), almost all other LCR16 elements occur in association with the LCR16a core duplicon. Phylogenetic reconstruction indicates that the flanking duplicons duplicated more recently have accumulated at the periphery of LCR16a duplications, leading to the formation of the complicated duplication blocks now observed in the human genome. Comparative sequence analysis in macaque and baboon (Old World outgroup species) reveals that each of the SDs originated as a single-copy sequence on chromosome 16 (Fig. 4). Remarkably, bacterial artificial chromosome (BAC)-based sequencing of LCR16a elements in the orangutan shows that the LCR16a core duplicon has duplicated independently and to nonorthologous locations when compared to human and African great apes. Moreover, the LCR16a has colonized chromosome 13 in the orangutan and has accumulated its own set of orangutan-specific flanking SDs on the periphery. Most of these flanking duplicons are single copy in humans and African great ape genomes. These data suggest that the LCR16a core duplicon has an inherent proclivity to duplicate and has served to prime lineage-specific duplications contributing to the emergence of large duplication blocks in both lineages. Thus, two independent bursts of the LCR16a have occurred in the last 12 million years in two different ape lineages.

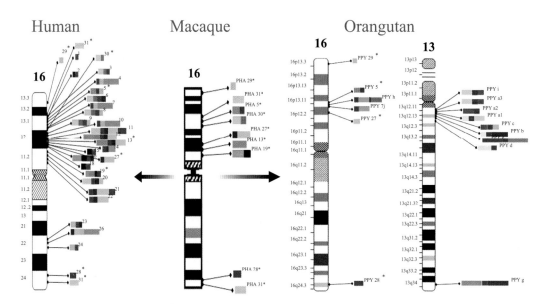

∗ Location of ancestral loci

Figure 4. Comparative schematic showing the distribution of LCR16 duplications. Color bars shows LCR16 duplicons. In human, the LCR16a core duplicon (red) is present within most duplication blocks on chromosome 16; all corresponding duplications are single copy in baboon, but in orangutan, LCR16a exists at nonorthologous locations and on different chromosomes (chromosome 13) in association with a new suite of orangutan-specific duplications at the periphery. Map locations are numbered according to the human reference with ancestral locations flagged by an asterisk.

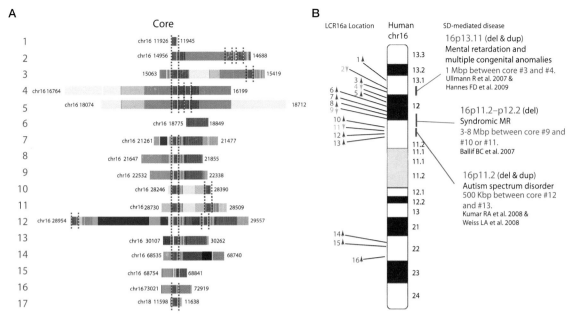

Figure 5. SDs and disease. Detailed duplicon composition of duplication blocks are shown along an ideogram of human chromosome 16. Duplications mediating recurrent deletions and duplications associated with disease are indicated (Ballif et al. 2007; Jiang et al. 2007; Ullmann et al. 2007; Kumar et al. 2008; Weiss et al. 2008; Hannes et al. 2009).

DISEASE CONSEQUENCES AND COPY-NUMBER VARIATION

Similar to Bridges and Muller's *Bar* locus, the presence of these large, high-identity duplications predisposes to recurrent deletions and duplications as a result of unequal crossing-over events during meiosis and/or mitosis. Not surprisingly, SDs are significantly enriched for copy-number polymorphisms (Iafrate et al. 2004; Sharp et al. 2005; Redon et al. 2006) with most of the genic copy-number polymorphisms mapping to these regions of the genome (Cooper et al. 2007; Bailey et al. 2008). The fact that so many of these duplications are interspersed, however, is double jeopardy for humans and its most closely related ape species. An unequal crossover event between two directly oriented duplications separated by a unique gene-rich region of the genome means that both the duplicated sequence and the unique sequence are subjected to copy-number variation (Lupski 1998). Nearly 10% of human euchromatin maps to ~110 such hot-spot regions of the genome, which is now sensitized to recurrent copy-number changes due to the evolution of this genomic architecture. More than 30 of these regions have been associated with both syndromic and complex diseases (Stankiewicz et al. 2004; Lupski 2007; Mefford and Eichler 2009). Interestingly, the majority of the pathogenic rearrangements involve neurocognitive and neurobehavioral diseases including intellectual disability, developmental delay, autism, schizophrenia, and epilepsy. Ironically, the breakpoints of many of these disease-causing rearrangements map to the same duplication blocks carrying core duplicons that emerged specifically within the human–great ape lineage (Tables 1 and 2). Although most of these large-scale copy-number changes appear to be under strong negative selection (Itsara et al. 2009), there is also evidence that SD-mediated rearrangements, such as the inversion on 17q21.31, may be positively selected, resulting in increased fecundity in specific human populations (Stefansson et al. 2005).

CONCLUSIONS

Both experimental and computational data support an acceleration of SDs in the common ancestor of humans and African great apes. This apparent burst in mutational process occurred at a time when most other mutational processes such as single base pair substitutions experienced a slowdown. At a base per base level, SDs contribute

Table 1. Core Duplicons and Disease-causing Rearrangements

Core	Locus	Phenotype[a]
NPIP	16p11.2	autism (1%), ID (0.6%)
NPIP	16p13.1	nonsyndromic ID (1%)
GLP/GOLGA -like protein	15q11.2	PW/AS, autism (1%)
GLP/GOLGA -like protein	15q13.3	epilepsy (1%), autism/ID (0.3%), schizophrenia (0.2%)
GLP/GOLGA -like protein	15q24	rare autism spectrum disorder
LRRC37	17q21.31	0.5% European ID syndrome
TBC1D3	17q12	renal cyst and diabetes (RCAD)
TBC1D3	17p11.2	Smith Magenis syndrome
NPBF	1q21.1	ID (0.5%), schizophrenia (0.3%), congenital heart defects

[a]ID indicates intellectual disability and developmental delay.

Table 2. Examples of Genes/Gene Families Mapped to the Most Representative Core Duplication

Duplicon clade	RefSeq gene	Gene name	Significant expression	Subcellular localization	Description	Possible function	Cancer association	Refs.
chr1	NM_183372	*NBPF11*	soft tissue	cytoplasm	neuroblastoma breakpoint gene family, DUF1220	unknown	neuroblastoma	1,2
chr2	NM_005054	*RANBP2*	testis	nuclear pore	RANBP2-like and GRIP domain-containing 5 isoform	Ran GTPase binding	highly expressed in leukemia	3
chr7_2	NM_174930	*PMS2L5*	ubiquitous	nuclear	postmeiotic segregation increased 2-like 5	DNA mismatch repair	unknown	4
chr15	NM_001012423	*GLP*	exclusively	unknown in testis	golgin-like protein, golgi autoantigen, golgin subfamily a, 8E (GOLGA8E)	DNA binding	unknown	5,6
chr16	NM_006985	*NPIP*	ubiquitous	nuclear membrane	nuclear pore complex interacting protein, morpheus gene family	Nuclear-pore-associated	highly expressed	7
chr17_1	NM_001006607	*LRRC37B*	ubiquitous	unknown	leucine-rich repeat, c114 SLIT-like testicular protein	ATP-dependent peptidase activity	unknown	8
chr17_2	NM_001001418	*TBC1/USP6*	testis	unknown	TBC1 domain family member 3C	GTPase activator activity	highly expressed in lymphoma	9

(1) Laureys et al. (1990), (2) Popescu et al. (2006), (3) Ciccarelli et al. (2005), (4) Horii et al. (1994), (5) Pujana et al. (2002), (6) Gilles et al. (2000), (7) Johnson et al. (2001), (8) Ota et al. (2004), (9) Paulding et al. (2003).

to more genetic variation than single base pair changes. SDs have restructured great ape and human chromosomes, creating complex lineage-specific duplication blocks distributed throughout specific chromosomes where novel gene structures have been formed by shuffling and juxtaposition of different exon cassettes. Much of the intrachromosomal duplication acceleration is centered around core duplicons that are also the seats of rapidly evolving genes that have expanded in the human and African great ape lineage. The concomitant large blocks of SDs are now predisposing to recurrent rearrangements that are associated with intellectual disability, autism, and schizophrenia. We hypothesize that the negative selection of disease-causing microdeletions and microduplications is balanced by positive selection of newly minted gene families embedded in cores and distributed to new locations. Elucidating the function of the genes embedded within the core duplicons remains an unmet challenge of human genetics and evolutionary biology.

ACKNOWLEDGMENTS

We thank Lin Chen, Ze Cheng, Santhosh Girirajan, and Tonia Brown for valuable comments and help in the preparation of this manuscript. This work was supported, in part, by National Institutes of Health grants GM-058815 and HG002385 to E.E.E. T.M.-B. is supported by a Marie Curie fellowship. E.E.E. is an investigator of the Howard Hughes Medical Institute.

REFERENCES

Bailey JA, Kidd JM, Eichler EE. 2008. Human copy number polymorphic genes. *Cytogenet Genome Res* **123:** 234–243.

Ballif BC, Hornor SA, Jenkins E, Madan-Khetarpal S, Surti U, Jackson KE, Asamoah A, Brock PL, Gowans GC, Conway RL, et al. 2007. Discovery of a previously unrecognized microdeletion syndrome of 16p11.2–p12.2. *Nat Genet* **39:** 1071–1073.

Bridges CB. 1936. The Bar "gene"—A duplication. *Science* **83:** 210–211.

Cheng Z, Ventura M, She X, Khaitovich P, Graves T, Osoegawa K, Church D, DeJong P, Wilson RK, Paabo S, et al. 2005. A genome-wide comparison of recent chimpanzee and human segmental duplications. *Nature* **437:** 88–93.

Church DM, Goodstadt L, Hillier LW, Zody MC, Goldstein S, She X, Bult CJ, Agarwala R, Cherry JL, DiCuccio M, et al. 2009. Lineage-specific biology revealed by a finished genome assembly of the mouse. *PLoS Biol* **7:** e1000112.

Ciccarelli FD, von Mering C, Suyama M, Harrington ED, Izaurralde E, Bork P. 2005. Complex genomic rearrangements lead to novel primate gene function. *Genome Res* **15:** 343–351.

Collins FS, Lander ES, Rogers J, Waterston RH, et al. (International Human Genome Sequencing Consortium). 2004. Finishing the euchromatic sequence of the human genome. *Nature* **431:** 931–945.

Consortium (Chimpanzee Sequencing and Analysis Consortium). 2005. Initial sequence of the chimpanzee genome and comparison with the human genome. *Nature* **437:** 69–87.

Cooper GM, Nickerson DA, Eichler EE. 2007. Mutational and selective effects on copy-number variants in the human genome. *Nat Genet* **39:** S22–S29.

Courseaux A, Richard F, Grosgeorge J, Ortola C, Viale A, Turc-Carel C, Dutrillaux B, Gaudray P, Nahon JL. 2003. Segmental duplications in euchromatic regions of human chromosome 5: A source of evolutionary instability and transcriptional innovation. *Genome Res* **13:** 369–381.

Diskin SJ, Hou CP, Glessner JT, Attiyeh EF, Laudenslager M, Bosse K, Cole K, Mosse YP, Wood A, Lynch JE, et al. 2009. Copy number variation at 1q21.1 associated with neuroblastoma. *Nature* **459:** 987–991.

Dumas L, Kim YH, Karimpour-Fard A, Cox M, Hopkins J, Pollack JR, Sikela JM. 2007. Gene copy number variation spanning 60 million years of human and primate evolution. *Genome Res* **17:** 1266–1277.

Eichler EE. 2001. Segmental duplications: What's missing, misassigned, and misassembled—and should we care? *Genome Res* **11:** 653–656.

Eichler EE, Budarf ML, Rocchi M, Deaven LL, Doggett NA, Baldini A, Nelson DL, Mohrenweiser HW. 1997. Interchromosomal duplications of the adrenoleukodystrophy locus: A phenomenon of pericentromeric plasticity. *Hum Mol Genet* **6:** 991–1002.

Elsik CG, Tellam RL, Worley KC, Gibbs RA, Muzny DM, Weinstock GM, Adelson DL, Eichler EE, Elnitski L, Guigo R, et al. 2009. The genome sequence of taurine cattle: A window to ruminant biology and evolution. *Science* **324:** 522–528.

Fortna A, Kim Y, MacLaren E, Marshall K, Hahn G, Meltesen L, Brenton M, Hink R, Burgers S, Hernandez-Boussard T, et al. 2004. Lineage-specific gene duplication and loss in human and great ape evolution. *PLoS Biol* **2:** E207.

Gilles F, Goy A, Remache Y, Manova K, Zelenetz AD. 2000. Cloning and characterization of a Golgin-related gene from the large-scale polymorphism linked to the PML gene. *Genomics* **70:** 364–374.

Hahn MW, Demuth JP, Han SG. 2007. Accelerated rate of gene gain and loss in primates. *Genetics* **177:** 1941–1949.

Hannes FD, Sharp AJ, Mefford HC, de Ravel T, Ruivenkamp CA, Breuning MH, Fryns JP, Devriendt K, Van Buggenhout G, Vogels A, et al. 2009. Recurrent reciprocal deletions and duplications of 16p13.11: The deletion is a risk factor for MR/MCA while the duplication may be a rare benign variant. *J Med Genet* **46:** 223–232.

Hodzic D, Kong C, Wainszelbaum MJ, Charron AJ, Su XO, Stahl PD. 2006. TBC1D3, a hominoid oncoprotein, is encoded by a cluster of paralogues located on chromosome 17q12. *Genomics* **88:** 731–736.

Horii A, Han HJ, Sasaki S, Shimada M, Nakamura Y. 1994. Cloning, characterization and chromosomal assignment of the human genes homologous to yeast *PMS1*, a member of mismatch repair genes. *Biochem Biophys Res Commun* **204:** 1257–1264.

Horvath JE, Schwartz S, Eichler EE. 2000. The mosaic structure of human pericentromeric DNA: A strategy for characterizing complex regions of the human genome. *Genome Res* **10:** 839–852.

Iafrate AJ, Feuk L, Rivera MN, Listewnik ML, Donahoe PK, Qi Y, Scherer SW, Lee C. 2004. Detection of large-scale variation in the human genome. *Nat Genet* **36:** 949–951.

Itsara A, Cooper GM, Baker C, Girirajan S, Li J, Absher D, Krauss RM, Myers RM, Ridker PM, Chasman DI, et al. 2009. Population analysis of large copy number variants and hotspots of human genetic disease. *Am J Hum Genet* **84:** 148–161.

Jiang Z, Tang H, Ventura M, Cardone MF, Marques-Bonet T, She X, Pevzner PA, Eichler EE. 2007. Ancestral reconstruction of segmental duplications reveals punctuated cores of human genome evolution. *Nat Genet* **39:** 1361–1368.

Johnson ME, Viggiano L., Bailey JA, Abdul-Rauf M, Goodwin G, Rocchi M, Eichler EE. 2001. Positive selection of a gene family during the emergence of humans and African apes. *Nature* **413:** 514–519.

Johnson ME, Cheng Z, Morrison VA, Scherer S, Ventura M, Gibbs RA, Green ED, Eichler EE. 2006. Recurrent duplication-driven transposition of DNA during hominoid evolution. *Proc Natl Acad Sci* **103:** 17626–17631.

Kumar RA, KaraMohamed S, Sudi J, Conrad DF, Brune C, Badner JA, Gilliam TC, Nowak NJ, Cook EH, Dobyns WB, et al. 2008. Recurrent 16p11.2 microdeletions in autism. *Hum Mol Genet* **17:** 628–638.

Laureys G, Speleman F, Opdenakker G, Benoit Y, Leroy J. 1990. Constitutional translocation t(1;17)(P36;Q12-21) in a patient with neuroblastoma. *Genes Chromosome Cancer* **2:** 252–254.

Li WH, Tanimura M. 1987. The molecular clock runs more slowly in man than in apes and monkeys. *Nature* **326:** 93–96.

Lupski JR. 1998. Genomic disorders: Structural features of the genome can lead to DNA rearrangements and human disease traits. *Trends Genet* **14:** 417–422.

Lupski JR. 2007. Genomic rearrangements and sporadic disease. *Nat Genet* **39:** S43–S47.

Marques-Bonet T, Kidd JM, Ventura M, Graves TA, Cheng Z, Hillier LW, Jiang ZS, Baker C, Malfavon-Borja R, Fulton LA, et al. 2009. A burst of segmental duplications in the genome of the African great ape ancestor. *Nature* **457:** 877–881.

Mefford HC, Eichler EE. 2009. Duplication hotspots, rare genomic disorders, and common disease. *Curr Opin Genet Dev* **19:** 196–204.

Muller HJ, Prokofjeva-Belgovskaja AA, Kossikov KV. 1936. Unequal crossing-over in the bar mutant as a result of duplication of a minute chromosome section. *C R Acad Sci USSR* **2:** 87–88.

Nicholas TJ, Cheng Z, Ventura M, Mealey K, Eichler EE, Akey JM. 2009. The genomic architecture of segmental duplications and associated copy number variants in dogs. *Genome Res* **19:** 491–499.

Ohno S, Wolf U, Atkin NB. 1968. Evolution from fish to mammals by gene duplication. *Hereditas* **59:** 169–187.

Ota T, Suzuki Y, Nishikawa T, Otsuki T, Sugiyama T, Irie R, Wakamatsu A, Hayashi K, Sato H, Nagai K, et al. 2004. Complete sequencing and characterization of 21,243 full-length human cDNAs. *Nat Genet* **36:** 40–45.

Paulding CA, Ruvolo M, Haber DA. 2003. The *Tre2* (*USP6*) oncogene is a hominoid-specific gene. *Proc Natl Acad Sci* **100:** 2507–2511.

Popesco MC, Maclaren EJ, Hopkins J, Dumas L, Cox M, Meltesen L, McGavran L, Wyckoff GJ, Sikela JM. 2006. Human lineage-specific amplification, selection, and neuronal expression of DUF1220 domains. *Science* **313:** 1304–1307.

Pujana MA, Nadal M, Guitart M, Armengol L, Gratacos M, Estivill X. 2002. Human chromosome 15q11-q14 regions of rearrangements contain clusters of LCR15 duplicons. *Eur J Hum Genet* **10:** 26–35.

Redon R, Ishikawa S, Fitch KR, Feuk L, Perry GH, Andrews TD, Fiegler H, Shapero MH, Carson AR, Chen W, et al. 2006. Global variation in copy number in the human genome. *Nature* **444:** 444–454.

Rouquier S, Taviaux S, Trask BJ, Brand-Arpon V, van den Engh G, Demaille J, Giorgi D. 1998. Distribution of olfactory receptor genes in the human genome. *Nat Genet* **18:** 243–250.

Sharp AJ, Locke DP, McGrath SD, Cheng Z, Bailey JA, Vallente RU, Pertz LM, Clark RA, Schwartz S, Segraves R, et al. 2005. Segmental duplications and copy-number variation in the human genome. *Am J Hum Genet* **77:** 78–88.

She XW, Jiang ZX, Clark RL, Liu G, Cheng Z, Tuzun E, Church DM, Sutton G, Halpern AL, Eichler EE. 2004. Shotgun sequence assembly and recent segmental duplications within the human genome. *Nature* **431:** 927–930.

She X, Cheng Z, Zöllner S, Church DM, Eichler EE. 2008. Mouse segmental duplication and copy number variation. *Nat Genet.* **40:** 909–914.

Stankiewicz P, Shaw CJ, Withers M, Inoue K, Lupski JR. 2004. Serial segmental duplications during primate evolution result in complex human genome architecture. *Genome Res* **14:** 2209–2220.

Stefansson H, Helgason A, Thorleifsson G, Steinthorsdottir V, Masson G, Barnard J, Baker A, Jonasdottir A, Ingason A, Gudnadottir VG, et al. 2005. A common inversion under selection in Europeans. *Nat Genet* **37:** 129–137.

Sturtevant AH. 1925. The effects of unequal crossing over at the bar locus in *Drosophila*. *Genetics* **10:** 117–147.

Tuzun E, Bailey JA, Eichler EE. 2004. Recent segmental duplications in the working draft assembly of the brown Norway rat. *Genome Res* **14:** 493–506.

Ullmann R, Turner G, Kirchhoff M, Chen W, Tonge B,

Rosenberg C, Field M, Vianna-Morgante AM, Christie L, Krepischi-Santos AC, et al. 2007. Array CGH identifies reciprocal 16p13.1 duplications and deletions that predispose to autism and/or mental retardation. *Hum Mutat* **28:** 674–682.

Wainszelbaum MJ, Charron AJ, Kong C, Kirkpatrick DS, Srikanth P, Barbieri MA, Gygi SP, Stahl PD. 2008. The hominoid-specific oncogene TBC1D3 activates *ras* and modulates epidermal growth factor receptor signaling and trafficking. *J Biol Chem* **283:** 13233–13242.

Waterston RH, Lindblad-Toh K, Birney E, Rogers J, Abril JF, Agarwal P, Agarwala R, Ainscough R, Alexandersson M, An P, et al. 2002. Initial sequencing and comparative analysis of the mouse genome. *Nature* **420:** 520–562.

Weiss LA, Shen YP, Korn JM, Arking DE, Miller DT, Fossdal R, Saemundsen E, Stefansson H, Ferreira MAR, Green T, et al. 2008. Association between microdeletion and microduplication at 16p11.2 and autism. *N Engl J Med* **358:** 667–675.

Wu CI, Li WH. 1985. Evidence for higher rates of nucleotide substitution in rodents than in man. *Proc Natl Acad Sci* **82:** 1741–1745.

snaR Genes: Recent Descendants of *Alu* Involved in the Evolution of Chorionic Gonadotropins

A.M. Parrott and M.B. Mathews

Department of Biochemistry and Molecular Biology, New Jersey Medical School, UMDNJ, Newark, New Jersey 07101-1709

Correspondence: mathews@umdnj.edu

We identified a novel family of human noncoding RNAs by in vivo cross-linking to the nuclear factor 90 (NF90) protein. These small NF90-associated RNAs (snaRs) are transcribed by RNA polymerase III and display restricted tissue distribution, with high expression in testis and discrete areas of the brain. The most abundant human transcript, snaR-A, interacts with the cell's transcription and translation systems. *snaR* genes have evolved in African Great Apes (human, chimpanzee, and gorilla) and some are unique to humans. We traced their ancestry to the *Alu* SINE (short interspersed nucleotide element) family, via two hitherto unreported sets of short genetic elements termed ASR (*Alu/snaR*-related) and CAS (Catarrhine ancestor of *snaR*). This derivation entails a series of internal deletions followed by expansions. The evolution of these genes coincides with major primate speciation events: ASR elements are found in all monkeys and apes, whereas CAS elements are limited to Old World monkeys and apes. In contrast to ASR and CAS elements, which are retrotransposons, human *snaR* genes are predominantly located in three clusters on chromosome 19 and have been duplicated as part of a larger genetic element. Insertion of the element containing snaR-G into a gene encoding a chorionic gonadotropin β subunit generated new hormone genes in African Great Apes.

Noncoding RNAs are essential to many biological processes. They serve as structural and catalytic components of cellular machines, such as the ribosome and spliceosome, and can exert exquisite gene regulation through processes such as RNA silencing. Recent work has revealed that the transcribed portions of higher eukaryotic genomes are larger and more complex than previously thought, with the majority of transcripts being noncoding (Kapranov et al. 2007; Wilhelm et al. 2008). In one view, the number of such transcripts scales with evolutionary complexity as the need for sophisticated control circuitry expands (Mattick and Makunin 2005). Undoubtedly, large numbers of noncoding RNAs remain to be discovered; of those that are known, many are understudied and functions have been ascribed to only a minority (Mattick and Makunin 2005; Prasanth and Spector 2007; Wilusz et al. 2009). Here, we describe the properties and ancestry of the snaRs, a rapidly evolved family of small noncoding RNAs, and discuss their possible functions. We begin with the historical context in which the RNAs were discovered and conclude with the ramifications of their mode of duplication, including the genesis of new human chorionic gonadotropin (hCG) genes.

NF90: A MULTIFUNCTIONAL PROTEIN

Human adenoviruses transcribe two short (~160 nucleotides) noncoding RNAs, VA RNA$_I$ and VA RNA$_{II}$, that accumulate to high cytoplasmic concentration in the late phase of viral infection (Mathews and Shenk 1991). Although VA RNA$_I$ is well characterized as a kinase inhibitor, counteracting the interferon-induced shut off of host cell protein synthesis, the function of VA RNA$_{II}$ remains obscure. To approach the role of VA RNA$_{II}$, our laboratory set out to characterize its cellular binding partners. Two RNA-binding proteins were identified: RNA helicase A and nuclear factor 90 (NF90), together with nuclear factor 45 (NF45) (Liao et al. 1998).

NF90 is the founding member of a family of RNA-binding proteins transcribed from the vertebrate-specific interleukin enhancer binding factor 3 (*ILF3*) gene (Duchange et al. 2000). Alternative splicing of *ILF3* transcripts gives rise to two major protein isoforms: NF90 and NF110 (Fig. 1). These proteins carry two canonical double-stranded RNA-binding motifs (dsRBMs), a single-stranded nucleic-acid-binding RGG motif, and a DZF (dsRBM and zinc-finger-associated) motif that is important in fly and mouse development (Meagher et al. 1999). NF90 and NF110 have distinct carboxyl termini that, in the case of NF110, are extended and contain a GQSY RNA-binding domain (Fig. 1). Both NF90 and NF110 have homology with the NF45 protein (Fig. 1), with which they exist in mutually stabilizing 1:1 heterodimeric complexes (Guan et al. 2008). As their names suggest, both NF90 and NF110 are present in the nucleus, but they carry a nuclear export signal (NES) as well as a nuclear localization signal (NLS), and a significant proportion (~20%) of NF90 is present in the cytoplasm (Parrott et al. 2005).

NF90 and NF110 bind numerous viral and cellular RNA and protein partners, and they have been associated with several biological processes. First recognized by binding to a promoter element (Corthesy and Kao 1994) and to synthetic duplex RNA (Bass et al. 1994), they have been assigned a number of functions, many of which await full investigation. Perhaps the best-characterized role is in the rapid induction of the interleukin-2 (*IL-2*) gene upon T-cell

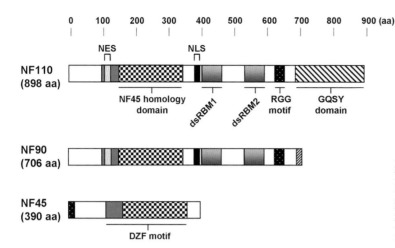

Figure 1. Schematic of NF90, NF110, and NF45 proteins. Motifs and domains are highlighted. Differential cross-hatching emphasizes the distinct carboxyl termini of NF90 and NF110. Illustrated are the "b" forms of NF90/110, which have a four-amino-acid insert that is lacking in the "a" forms.

activation. NF90 facilitates *IL-2* transcription (Shi et al. 2007), is specifically phosphorylated, and translocates to the cytoplasm where it stabilizes *IL-2* mRNA (Shim et al. 2002; Pei et al. 2008). Increasingly, NF90 has been implicated in aspects of gene regulation, for example, through control of nuclear export (Pfeifer et al. 2008), mRNA stability (Pullmann et al. 2007), and microRNA (miRNA) processing (Sakamoto et al. 2009), making it a likely target for viral manipulation. Interestingly, NF90 binds VA RNA$_{II}$ considerably more efficiently than it binds VA RNA$_I$, despite similarities between these two highly structured single-stranded viral RNAs (Liao et al. 1998). Its high affinity for NF90 suggests that VA RNA$_{II}$ may have evolved to displace cellular RNA from NF90, bringing some as yet undefined benefit to the virus.

NF90 BINDS TO A NOVEL CELLULAR RNA FAMILY

The dsRBMs of NF90 are largely occupied by cellular RNA throughout the cell cycle (Parrott et al. 2005). To verify known NF90-binding partners and to identify new ones, we isolated ribonucleoprotein (RNP) complexes after in vivo cross-linking. This technique stabilizes complexes in live cells and obviates partner reassortment that can occur during cell rupture and fractionation. As an additional advantage, it is not focused on detecting a particular RNA of interest. We used a human embryonic kidney 293 (HEK-293) stable cell line that expresses a tagged form of NF90 at moderate levels, comparable to endogenous NF90. RNP complexes containing NF90 were isolated by immunoprecipitation, and RNA ligands were extracted after reversal of the cross-links (Parrott et al. 2007). A heterogeneous mix of transcripts generated by RNA polymerase (pol) I, II, and III were found cross-linked to NF90, and all proved to be previously unidentified partners. The most intense band, visualized by autoradiography of 3′-end-labeled RNA and confirmed by oligonucleotide-directed RNase H digestion, was 5.8S rRNA. Other NF90-associated RNAs were cloned using 5′-rapid amplification of cDNA ends (5′-RACE) reverse transcriptase–polymerase chain reaction (RT-PCR) schemes (Parrott et al. 2007). The most abundant species identified by this means were members of a novel family of short RNAs that we termed snaR (small NF90 associated). Dimeric *Alu* transcripts, which are relatively rare in cells (see below), comprised the second most abundant group.

THE snaR FAMILY OF NONCODING RNAs

Two subsets of the snaR family of noncoding RNAs, snaR-A and snaR-B, were initially isolated with NF90. Search of the human genome identified genes encoding a further subset, snaR-C, and several outliers, mainly on chromosome 19 (Table 1 and Fig. 2) (Parrott and Mathews 2007). As shown experimentally for snaR-A (Parrott and Mathews 2007), *snaR* genes are predicted to be transcribed by pol III into highly structured RNAs of ~117 nucleotides that bind to the dsRBMs of NF90. The dominant snaR species, in terms of gene multiplicity (14 of 30 genes) as well as transcript abundance (~70,000 copies per HEK-293 cell), is snaR-A. Of the remaining genes, about half (seven genes) encode snaR-B and snaR-C, which are closely related to one another (Fig. 3A). snaR-A is found in a wide array of immortal cells, but it exhibits a sharply restricted distribution in human tissue, being expressed at a high level in testis (Parrott and Mathews 2007) and to a lesser extent in discrete parts of the brain (AM Parrott and MB Mathews, in prep.). Expression in testis and differential expression in the brain have also been observed for other snaRs. The *snaR-A* genes, together with those of *snaR-B*, *-C* and *-D*, are arrayed in two large inverted regions of tandem repeats on the q-arm of chromosome 19 (Fig. 2). The 2-Mb region separating the two clusters hosts more than 100 protein-coding genes including the luteinizing hormone (LH)/ chorionic gonadotropin (CG) β subunit (*LHβ/CGβ*) gene cluster and a number of genes essential to sperm biology. In this region are two additional *snaR* genes, encoding snaR-G1 and snaR-G2. They are situated within the proximal promoters of the two most recent additions to the CGβ family, *CGβ1* and *CGβ2*, respectively (Fig. 2), as discussed below.

Table 1. Subsets of African Great Ape *snaR* Genes

snaR[a]	NCBI gene ID	Chromosome[b]	Transcribed[c]	Human[d]	Pan[d,e]	Gorilla[e]
A1-14	100126798-99, 100169951-59, 100170216, 100191063	19q13 (53.102-53.140, 55.287-55.324)	+	14		+
B1-2	100170217, 100170224	19q13 (55.328-55.334)	+	2		
C1-5	100170218-19, 100170223, 100170225-26	19q13 (53.107-53.152)	+	5		
D	100170227	19q13 (55.335)	n.d.[f]	1		
E	100170220	19q13 (52.026)	n.d.[f]	1		
F	100126781	19q13 (55.800)	n.d.	1		
G1	100126780	19q13 (54.232)	–	1	2, +	+
G2	100170228	19q13 (54.227)	+	1		+
H (2)	100170221	2p12 (78.036)	n.d.	1	1	
I (3)	100170222	3q28 (192.078)	n.d.	1	7, +	
12		12p12 (21.158)	n.d.	1	1	
21		21q21 (25.032)	n.d.	1	1	

[a]*snaR* nomenclature as defined by the HUGO Gene Nomenclature Committee. Former name is in parentheses. snaR-12 and -21 are not yet cataloged.
[b]Human chromosome location (in Mb).
[c]*snaR* expression as confirmed by northern blot and/or RT-PCR of human tissue. n.d. indicates not determined.
[d]Number of gene copies in human and chimpanzee genomes.
[e]+ denotes *snaR* subset presence in bonobo and gorilla as confirmed by genomic PCR.
[f]*snaR-D* and *-E* are pseudogenes of *snaR-A* and *-B*, respectively.

STEPWISE EVOLUTION OF *snaR* COINCIDES WITH PRIMATE SPECIATION

Examination of available genome sequences revealed *snaR* genes in the chimpanzee (Table 1), but not in other mammals such as rhesus macaque and rodents. To trace the origin of this recently evolved gene family, we exploited a notable feature of the *snaR* genes, namely, that their flanking sequences are often more highly conserved than the genes themselves (Parrott and Mathews 2007). Analysis of great ape genomic DNA by PCR amplification using primers designed against these *snaR* flanking sequences, supplemented by database searches, revealed *snaR* genes in all of the African Great Apes (human, chimpanzee, bonobo, and gorilla) (Table 1) (AM Parrott et al., in prep.). In the Sumatran orangutan (an Asian Great Ape), however, the sequence spanned by the flanking homology exhibited only limited identity with *snaR* genes. Searches of the orangutan draft assembly failed to disclose authentic *snaR* sequences. We conclude that authentic *snaR* genes evolved in the African Great Apes.

The amplified orangutan sequence originates from chromosome 19 and is orthologous to a chromosome 19 region in macaque (Fig. 3B). In the macaque, the orthologous region is embedded in a triplet of adjacent 1.9-kb repeats that are demarcated by *Alu* sequences (Fig. 3B). This triplet repeat pattern is conserved in humans but fragmented in the orangutan, possibly due to draft assembly errors. *snaR-F* is present in the first repeat in human; elements resembling truncated *snaR* genes are present at the same position in both the first (Mm19) and the second (Mm19i) repeat in the macaque and in the first partial repeat in the orangutan (Pa19). Further searches disclosed that this class of truncated elements is restricted to Old World monkeys and apes (Catarrhines), and they were therefore termed CAS (Catarrhine ancestor of snaR) elements.

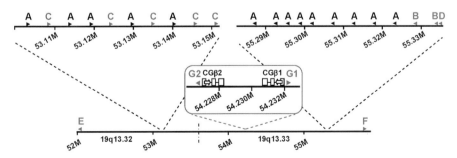

Figure 2. *snaR* gene clusters. Shown is the region of human chromosome 19q13.32-33 containing *snaR* genes. The locations and direction of transcription of *snaR-A* genes (black) and *snaR-B*, *-C*, *-D*, *-E*, *-F*, *-G1*, and *-G2* genes (gray) are denoted by arrowheads. *CGβ1* and *CGβ2* exons are shown as open boxes and their direction of transcription by open arrows. Distances are in megabases (M).

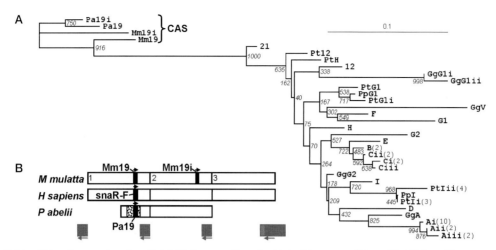

Figure 3. Identifying the ancestor of *snaR*. (*A*) Phylogram of bonobo (Pp), chimpanzee (Pt), gorilla (Gg), and human *snaR* genes (Ai, Aii, B, etc.) and macaque CAS elements (Mm19 and Mm19i), rooted to orangutan CAS elements (Pa19 and Pa19i). For clarity and to reduce bias, multiple identical alleles were represented once in the alignment; their multiplicity is shown in parentheses. The scale bar represents 0.1 nucleotide substitutions per site and "bootstrap" node values are given. (*B*) Schematic of syntenic regions of chromosome 19 in macaque (56.819–56.825 Mb), human (55.798–55.804 Mb), and orangutan (52.171–52.174 Mb), containing ~1.9-kb repeats (numbered open boxes). CAS elements and snaR-F are indicated in black and *Alu*Sx in gray, with direction of transcription indicated by arrows. The amplified orangutan sequence is stippled.

The synteny evident in Figure 3B suggested that *snaR* genes evolved from CAS elements. Alignment of all *snaR*s with CAS elements shows how this could have taken place at the molecular level (Fig. 4A). Compared to CAS elements, *snaR* genes have two internal 8-nucleotide expansions between pol III A- and B-box transcription elements. The first expansion (ϵ1), but not the second (ϵ2), is preserved in snaR-12, a putative "fossil" species that hints at the sequential nature of this evolutionary transition. A phylogram generated from this alignment emphasizes the distinction between the *snaR* family and CAS element sequences (Fig. 3A).

Searches for the triplet syntenic repeat in New World monkeys (Platyrrhines) disclosed a slightly longer element occupying the position equivalent to that of CAS in Catarrhine species. This longer element has homology with FLAM-C (free left *Alu* monomer of the C subtype), a forerunner of the left monomer of *Alu* (sc*Alu*) (Quentin 1992), both of which are members of the most populous, short interspersed nucleotide element (SINE) family in primates. Because of its descent from *Alu* and ancestral relationship to CAS and thence *snaR*, the element is named ASR (*Alu/snaR*-related). ASR elements are present in all Simiiformes (monkeys and apes) tested, but not in prosimians such as tarsiers, galagos, and lemurs. ASR lacks a 19-nucleotide section (δ1) between the A and B boxes of FLAM-C, and CAS appears to have arisen from ASR through a further 13-nucleotide 3′ deletion (δ2; Fig. 4B). In summary, we deduce that *snaR* has evolved from a monomeric *Alu* element through a series of deletions followed by expansions. It is likely that these major molecular rearrangements occurred within a single locus on the q-arm of chromosome 19 and seemingly only after major primate speciation events (Fig. 4B). The Sumatran orangutan CAS element Pa19 carries a unique expansion and provides further demonstration of the innate instability of this locus and its propensity for sequence rearrangement (Fig. 4). Nucleotide substitutions also separate *snaR* from its ancestors. Most of the substitutions (~11 of 14) between CAS and *snaR* sequences are present in snaR-21, another probable "fossil" species that lacks both internal expansions ϵ1 and ϵ2, indicating that the majority of the substitutions preceded the internal expansions (Fig. 4A).

Alu SINE: PRESUMPTIVE ANCESTOR OF snaR

Dimeric *Alu* transcripts are ~300 nucleotides long and typically have a poly(A) tail and an internal AU-rich region separating two bulged hairpin domains, known as the sc*Alu* (left) and *Alu*-RA (right) monomers (Fig. 5A). The left monomer region carries the pol III basal transcription promoter and enhancer. *Alu*'s dimeric form arose in primates through the fusion of a FLAM-C and a free right *Alu* monomer (FRAM) (Quentin 1992). Both monomers are independently descended from the highly conserved 7SL RNA that forms the scaffold of the signal recognition particle (SRP), which has a key role in the synthesis of secreted proteins (Kriegs et al. 2007). *Alu* genes underwent a series of three distinct expansion events through the self-replication mechanism of retrotransposition, with the most extensive undertaken by the *Alu*-S subfamily ~35–40 million years ago (mya) (Bailey et al. 2003). *Alu* sequences now constitute more than 10% of the human genome, but most copies are transcriptionally silent because of mutations during retrotransposition and through neutral drift (Lander et al. 2001; Bennett et al. 2008). Consequently, pol-III-transcribed *Alu* RNAs are relatively rare, although they are rapidly induced up to 20-fold by cellular insults such as adenovirus infection and heat shock (Liu et al. 1995; Chang et al. 1996).

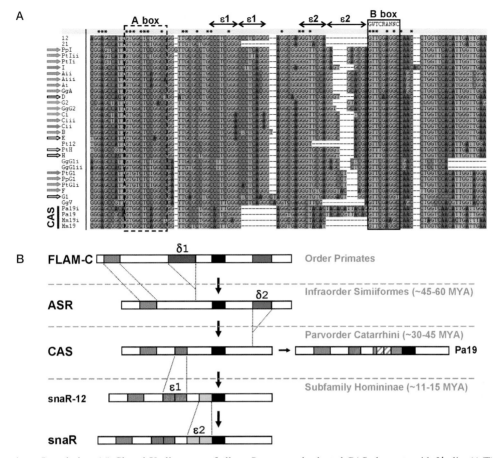

Figure 4. *snaR* evolution. (*A*) Clustal-X alignment of all *snaR* genes and selected CAS elements with 3′-oligo(A/T) tracts omitted. Full-length *snaR*s are marked with arrows: (green arrows) *snaR*s predicted to be transcriptionally active, (yellow arrows) *snaR*s with mutations in their pol III B box (solid box) or a truncated 3′-oligo(T) tract. Expansions ε1 and ε2 and the putative pol III A box are indicated. Asterisks show conserved nucleotides. This alignment generated the phylogram of Fig. 3A. (*B*) Schematic of the major deletions (red) and expansions (green) relating FLAM-C, ASR, CAS, and *snaR* genes. These molecular events apparently occurred contemporaneously with major events in primate evolution (dashed gray lines), allowing their timing to be estimated. pol III A and B boxes are represented by gray and black boxes, respectively. Two unique overlapping expansions in the Sumatran orangutan Pa19 CAS element are shown by orange and orange hatching. snaR-12 appears to be an intermediate species between CAS and *snaR*.

Alu retrotransposition is dependent on the reverse transcriptase and integrase/endonuclease genes encoded by long interspersed nucleotide elements (LINEs). As such, *Alu* has been regarded as a parasite's parasite and a burden on the host genome. However, it is now realized that *Alu* has had a major role in shaping primate genomes, and it is a source of novel noncoding RNA species including *snaR*. *Alu* insertion into protein-coding genes and their promoters can lead to alternative splice site selection or transcriptional regulation (Jurka 2004; Hasler and Strub 2006a). *Alu* sequences also promote genomic remodeling through nonallelic homologous recombination, resulting in localized gene disruption, duplication, deletion, or larger segmental duplications (Bailey et al. 2003). Segmental duplications are now considered to be critical in the evolution of the primate lineage; as a source of novel genes, they can facilitate both phenotypic variation and susceptibility to disease (Bailey and Eichler 2006).

snaR GENES DISSEMINATED BY SEGMENTAL DUPLICATION

Like *Alu*, ASR and CAS elements are scattered throughout primate genomes and the majority are flanked by short direct repeats. These are hallmarks of retrotransposon activity (Grindley 1978; Maraia et al. 1992). Thus, it appears that ASR and CAS elements are novel SINEs and are (or were) genes. With a single exception in chimpanzee, all human and chimpanzee (Tribe Hominini) CAS elements occupy syntenic positions. We conclude that almost all CAS elements have become transcriptionally silent in Hominini since divergence from the common ancestor. Thus, only five discrete CAS clones have been reported in expressed sequence tag (EST) databases (BU567151, BG526971, DW458066, DW458333, and BU56539), and they all originate from the same human locus. In contrast, at least half of the CAS elements in macaque and orangutan occupy unique loci, suggesting that CAS was an active retrotrans-

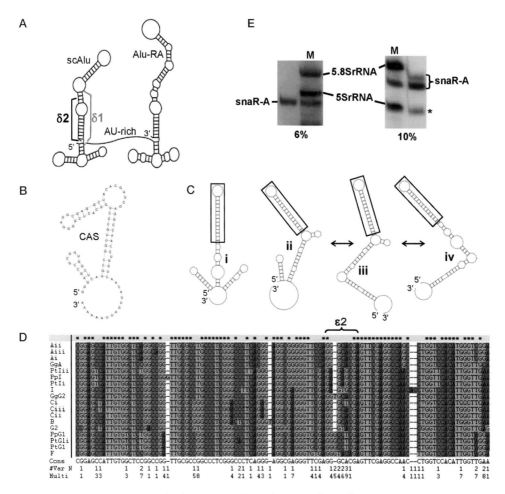

Figure 5. *snaR*s appear to have structural plasticity. (*A*) Positions of deletions δ1 and δ2 superimposed on the *Alu* secondary structure model. (*B*) Predicted secondary structure of CAS RNA. A consensus sequence, derived from a multiple alignment of 120 primate CAS elements (after omitting 26 elements with single insertions/deletions) using WebLogo (Crooks et al. 2004), was folded using MFOLD (Zuker 2003). (*C*) Alternative secondary structures predicted for 23 of 29 *snaR*s. Six *snaR*s cannot adopt these structures due to internal deletions. The conserved apical stem-loop is boxed. (*D*) Clustal-X alignment of transcriptionally active (as defined in Fig. 4A) *snaR* genes with 3′-oligo(A/T) tracts omitted. (Cons) Consensus sequence, (#Var N) number of variable nucleotides at each position, (Multi) multiplicity of variable nucleotides at each position. The most recent *snaR* expansion, ε2, is indicated. (Asterisks) Conserved nucleotides. (*E*) Northern blots of cytoplasmic RNA from 293 cells resolved in 6% and 10% acrylamide/7 M urea gels and probed for snaR-A. 3′-pCp–labeled RNA served as a marker (M). (Asterisk) Band of unknown origin. (*A*, Adapted, with permission, from Mariner et al. 2008 [© Elsevier].)

poson until quite recently in Catarrhines. These observations reflect a general decline in retrotransposon activity along the Hominoid lineage (apes) (Liu et al. 2009).

Unlike their ancestors, most *snaR* genes seem to have disseminated through segmental duplication of an encompassing locus, possibly as a result of nonallelic homologous recombination of flanking *Alu* sequences. Exceptionally, snaR-D and -E appear to represent cases of *snaR* retrotransposition (see Table 1) (Parrott and Mathews 2007). Strikingly, *snaR* gene radiation has coincided with a genome-wide burst of segmental duplication within the African Great Ape ancestor (Marques-Bonet et al. 2009). *snaR*'s mode of dissemination, which can result in gene duplication as well as deletion, has seemingly resulted in dramatic copy-number variation and *snaR* subset diversity within the African Great Apes (Table 1). For example, goril-

las and humans have *snaR-A* genes, whereas chimpanzees do not; a single *snaR-I* gene is present in humans, whereas the chimpanzee has seven copies, and perhaps most interestingly, *snaR* gene duplication has contributed to the expansion of the well-studied *LHβ/CGβ* gene cluster (see below).

PREDICTED *snaR* STRUCTURES

The radical sequence changes, deletions, and expansions that took place during *snaR* evolution from a left monomeric *Alu* element (Fig. 4B) undoubtedly impacted the secondary structures of the RNA species in profound ways. If the successive deletions leading to CAS are mapped onto the experimentally determined model of *Alu* secondary structure (Sinnett et al. 1991), they constitute the sequential loss of a middle stem of the left monomer of *Alu* (sc*Alu*; Fig. 5A). The second deletion could be a consequence of the first deletion, perhaps resulting in stabilization of the RNA structure. However, considering that numerous examples of the intermediate deletion (ASR) exist, some structural reorganization is likely. The predicted CAS RNA secondary structure (Fig. 5B) is stable, with a ΔG of folding of –31.8 kcal/mol, and distinct from that of *Alu*.

Four predicted secondary structures can describe the lowest energy structure of most *snaR*s (Fig. 5C). All four structures share a common apical stem-loop. Primer-extension analysis of snaR-A uncovered a "strong-stop" at the predicted start of its apical stem-loop, supporting the existence of this conserved motif (data not shown). The apical stem-loop should make an ideal NF90 dsRBM-binding site (Parrott and Mathews 2007), but further work is needed to establish whether this motif is the point of contact between the two molecules because dsRBMs bind to imperfect as well as canonical dsRNAs (Bevilacqua et al. 1998; Tian et al. 2004). Three of the predicted secondary structures (ii–iv, Fig. 5C) have similar minimal folding energies (within 6% of the optimal folding energy). We attempted to deduce the preferred secondary structure by folding a *snaR* consensus sequence. This was derived from an alignment of 18 *snaR* sequences with an intact B box and 3′-oligo (T) tract (Fig. 5D), considering species that lack these transcriptional requirements and/or harbor internal deletions (Fig. 4A) to be potential pseudogenes. The most variable region of the alignment, which is encompassed by the most recent expansion (ε2), dictates the length of the conserved apical stem-loop (Fig. 5D). No satisfactory unifying secondary structure model could be derived for the consensus sequence, however, because all predictions placed variant nucleotides in stabilizing G:C base pairs. One interpretation of this finding is that the *snaR* subsets adopt different secondary structures while retaining a common apical stem-loop.

RAPIDLY EVOLVED snaRs MAY BE PARALOGS

That *snaR* genes are generally more divergent in sequence than their flanking regions hints at adaptation for distinct functions or maybe refinement of one function. Human accelerated region 1 (HAR1), a small noncoding transcript that is considered to be one of the most rapidly evolved RNAs in humans, exhibits 18 base changes in 118 nucleotides (15.3% difference) between its human and chimpanzee forms; consequently, it adopts a cloverleaf structure in humans but a hairpin structure in chimpanzees (Beniaminov et al. 2008). The function of HAR1 is not known, although a role in human neurodevelopment has been posited (Pollard et al. 2006), and it is surmised that the cloverleaf structure of the human ortholog adapts it for this role (Beniaminov et al. 2008).

The most abundant human *snaR*s form two distinct subsets, snaR-A and snaR-B/C, that have undergone rapid divergence from other *snaR* genes and from each other (see Fig. 3A). snaR-A is present in humans and gorillas, arguing for an origin in the common ancestor of African Great Apes followed by loss in *Pan*, whereas snaR-B and -C are unique to humans (Fig. 3A and Table 1) and therefore probably evolved rapidly after the *Homo-Pan* species divergence. snaR-A underwent copy-number expansion, leading to 14 highly homologous alleles in humans and an unknown number in gorilla. Intriguingly, *snaR-B* and *-C* genes are interspersed with *snaR-A* genes in the two *snaR* clusters in a tandem repeat pattern, suggesting that *snaR-B/C* evolved from redundant *snaR-A* alleles (see Fig. 2).

Conservatively, there are 12 base changes in 103 nucleotides (excluding the oligo[A/T] tract: 11.6% difference) between snaR-A and snaR-C species (Fig. 5D), and these two subsets are predicted to fold into distinctly different structures (Parrott and Mathews 2007). snaR-B and -C are predicted to form a unique and very stable (–58.3 kcal/mol) secondary structure (structure i), reminiscent of that of sc*Alu*, whereas snaR-A adopts an equally robust linear structure (structure iv, Fig. 5C). This species exhibits mobility differences in urea gels depending on the concentration of acrylamide, consistent with the existence of a structured region that persists even under denaturing conditions (Fig. 5E). snaR-A is expressed in the pituitary gland, whereas snaR-B/C is not (AM Parrott and MB Mathews, in prep.), so we speculate that nucleotide substitutions have allowed these two subsets to adopt different secondary structures for different functions—in other words, they are paralogs.

snaR BIOLOGY AND POSSIBLE FUNCTIONS

Despite its abundance, snaR-A has a short half-life (only ~20 min) in HeLa cells, implying a rapid rate of transcription (Parrott and Mathews 2007) and suggestive of a regulatory role. snaR-A is distributed between the nucleus and cytoplasm of immortal cells and appears to undergo processing to a slightly shortened form in the cytoplasm (Fig. 6A). NF90 is the only protein yet identified to bind *snaR*, but its role in *snaR* biology is undefined. Knockdown of NF90 by RNA interference (RNAi) resulted in reduction of snaR-A levels in both the nucleus and cytoplasm (Fig. 6B). Hence, NF90 protein is necessary for the synthesis or stability of snaR-A within the cell but possibly not for its maturation or nucleocytoplasmic shuttling. Knockdown of NF45 by RNAi did not impact snaR-A levels, however, suggesting that this NF90 function may not depend on formation of a heterodimer with NF45.

The presence of *snaR* RNA in the nucleus and cytoplasm is consistent with functions in either or both cellular compartments. *Alu* RNAs offer a precedent: They have a role in translation, as discussed in the next section, and in transcription. As elucidated by Goodrich and colleagues, dimeric *Alu* can bind tightly to pol II and inhibits its function (Mariner et al. 2008). The left and right monomers bind independently to pol II, but only *Alu*-RA (Fig. 5A) is able to inhibit polymerase function in vitro. When tested in the same way, snaR-A was found to mimic its sc*Alu* ancestor; it bound tightly to pol II but did not inhibit its function (J Goodrich, pers. comm.). These findings are compatible with the evolutionary origin of *snaR*

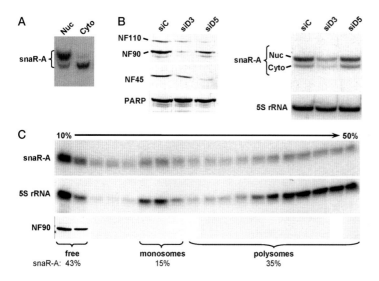

Figure 6. Intracellular distribution of snaR-A. (*A*) Subcellular fractionation. Nuclear (Nuc) and cytoplasmic (Cyto) RNA from 293 cells was resolved in a northern blot and probed for snaR-A. (*B*) snaR-A is depleted by NF90 knockdown. HeLa cells were transfected with control siRNA (siC), or siRNA directed against NF90 (siD3) or NF45 (siD5); extracts were prepared 48 h later for analysis by western blot (*left panel*) or northern blot (*right panel*). Nuclear (Nuc) and cytoplasmic (Cyto) forms of snaR-A are marked. (*C*) snaR-A cosediments with ribosomes. HeLa cell cytoplasmic extract was centrifuged through a 10%–50% sucrose gradient. Fractions were analyzed by northern blot for snaR-A (*top panel*) and 5S rRNA (*middle panel*) and by western blot for NF90 (*bottom panel*). The distribution of snaR-A in three regions of the gradient was determined by phosphorimager quantitation of the blot.

genes discussed above, but they shed little light on a putative nuclear function at present because the significance of sc*Alu* binding to pol II is still obscure.

Sucrose gradient sedimentation analysis revealed that a substantial fraction of cytoplasmic snaR-A is associated with ribosomes, both with monosomes and polysomes (Fig. 6C). This suggests that *snaR*, like 7SL RNA and some of its relatives, has a role in translation. *Alu* RNA has retained sufficient similarity to its 7SL ancestor to form a complex with the SRP9/14 heterodimer (Chang et al. 1996; Hasler and Strub 2006b). The SRP9/14 heterodimer positions the *Alu* domain of 7SL in the elongation-factor-binding site of the large 60S ribosomal subunit, such that the SRP can arrest translation (Halic et al. 2004). Interestingly, the *Alu*-SRP9/14 RNP can also inhibit translation in vitro (Hasler and Strub 2006b). BC200, a 200-nucleotide-long noncoding RNA derived from the left monomer of *Alu*/FLAM, also associates with the SRP9/14 heterodimer (Martignetti and Brosius 1993; Kremerskothen et al. 1998). BC200 is specific to Simiiformes and is expressed in the cytoplasm of dendritic cells, especially at the synapse (Tiedge et al. 1993; Skryabin et al. 1998). Like its murine functional analog BC1 (which is unrelated to *Alu*), BC200 binds to the eukaryotic initiation factor 4A (eIF4A) helicase and uncouples ATPase hydrolysis from the factor's RNA unwinding ability, thereby repressing the translation of mRNAs dependent on eIF4A for unwinding of their structured 5'UTRs (untranslated regions) (Lin et al. 2008). Thus, two RNAs descended from 7SL retain the ability to bind protein components of the SRP, associate with ribosomes, and perform translational control functions.

It is not known whether SRP proteins can bind *snaR* RNA, but NF90 binds directly to dimeric *Alu* and 5.8S rRNA (Parrott et al. 2007). The latter is a structural component of the 60S ribosomal subunit, present at its surface and critical to ribosome translocation (Abou Elela and Nazar 1997; Graifer et al. 2005). NF90 is also present in the ribosomal salt wash fraction (Langland et al. 1999), possibly as a result of its association with 5.8S rRNA, but is dissociated from ribosomes in a sucrose gradient (Fig. 6C). Hence, *snaR*'s defined protein partner may transiently associate with ribosomes and perhaps transport *snaR* to them. The restricted tissue distribution of *snaR*s raises the possibility that they function as tissue-specific gene regulators acting at the level of translation.

NEW CHORIONIC GONADOTROPIN GENES

Chorionic gonadotropin (CG) belongs to the same glycoprotein hormone family as luteinizing hormone, follicle-stimulating hormone, and thyroid-stimulating hormone, each composed of a common α subunit and a distinct β subunit. The *CGβ* subunit genes are located in a cluster on chromosome 19q13.33 together with *LHβ*, from which they evolved (Fig. 7, bottom). *CGβ* subunit genes have undergone several duplications during primate speciation (Maston and Ruvolo 2002; Hallast et al. 2008). Humans possess six CG genes: *hCGβ3*, *5*, *7*, and *8*, which share 96% identity with *LHβ*, and *hCGβ1* and *hCGβ2*, which have a unique 5' region (Hallast et al. 2007). It is likely that *CGβ1* and *CGβ2* evolved stepwise in the African Great Apes (Fig. 7). In the first step, the 5' region of an ancestral *CGβ* gene was substituted by a segmental duplication encompassing a *snaR-G* gene, giving rise to *CGβ1* (AM Parrott et al., in prep.). The gene sequence that was replaced contained the proximal promoter, 5'UTR, and first exon of the ancestral *CGβ* gene (Bo and Boime 1992). In the second step, *CGβ1* was duplicated together with flanking sequences that included *CGβ5*, to generate *CGβ2* and *CGβ3* (Hallast et al. 2008). These events introduced *snaR-G1* and *snaR-G2* into the proximal promoters of *CGβ1* and *CGβ2*, respectively.

Both gorillas and humans have *snaR-G1* and *snaR-G2* genes, whereas the genus *Pan* has two copies of *snaR-G1* but no *snaR-G2* (Table 1). This observation is consistent with recent sequencing of Hominini genomes, which found chimpanzees to possess two copies of *CGβ1* but to lack *CGβ2* and *CGβ3*, indicating that *Pan*-specific duplication of *CGβ1* occurred after the *Homo-Pan* divergence (Hallast et al. 2008). Considering that the gorilla possesses

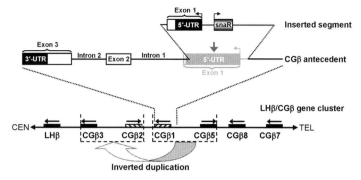

Figure 7. New hormone genes arise by consecutive segmental duplications. (*Top*) Schematic of *CGβ1* and *CGβ2* gene structure illustrating the substitution of sequence common to other *CGβ* genes (gray) with a segment containing a *snaR-G* gene. (Black/gray boxes) Untranslated regions (UTRs), (open boxes) open reading frame, (solid lines) introns or promoter regions, (crooked arrows) direction of transcription. (*Bottom*) Diagram of the human *LHβ/CGβ* gene cluster at chromosome 19q13.33, oriented with respect to centromere (CEN) and telomere (TEL). Straight arrows indicate direction of transcription. (Large curved arrow) Inverted segmental duplication of the original *CGβ1* gene formed by *snaR* substitution.

CGβ2 (Hallast et al. 2007; AM Parrott et al., in prep.), it is likely that the common ancestor of African Great Apes possessed *CGβ2* and *CGβ3* and that these genes were deleted in the *Pan*-specific *CGβ1* duplication event.

The *snaR-G* genes are transcribed in the opposite direction to their host *hCGβ1/2* genes (Fig. 7), and the 5′ ends of the transcripts are separated by only 86 nucleotides. Thus, the *snaR-G* genes are well positioned to influence *hCGβ1/2* gene transcription, either by virtue of their transcription or because of the novel transcription-factor-binding sites that they have introduced. The first mechanism envisages that pol III transcription opens chromatin, thereby allowing pol II access; retention of pol III transcription signals would be required for this action. The second mechanism would necessitate preservation of discrete sequences within the *snaR-G* gene, although pol III transcription signals could be lost and the integrity of the pol III gene could be compromised. Interestingly, *snaR-G1* orthology appears to have diverged more than that of *snaR-G2*, with multiple substitutions in human *snaR-G1* including in its B box and a 3′ deletion in gorilla *snaR-G1* (see Figs. 3A and 4A). Correspondingly, we have detected the *snaR-G2*, but not *snaR-G1*, transcript in human tissues (AM Parrott and MB Mathews, in prep.). Moreover, *snaR-G1* is predicted to contain a number of proximal transcription-factor-binding sites that are not in *snaR-G2* (Hallast et al. 2007). Conceivably, *snaR-G1* provides binding sites for transcription factors, whereas *snaR-G2* acts in *cis* through its continued transcription or in *trans* via its expressed transcript.

FUNCTION AND REGULATION OF hCG

CG is essential for the successful onset and progression of pregnancy in primates. Its classical function is to maintain the corpus luteum by sustaining progesterone production (Jameson and Hollenberg 1993). In humans, hCG has additional roles in angiogenesis, blastocyst implantation, and nourishment of the growing fetus (Lei et al. 1992; Zygmunt et al. 2002; Cole et al. 2006; Handschuh et al. 2007), as well as development of tolerance in the maternal immune system (Wan et al. 2008; Schumacher et al. 2009). Whether *CGβ1* and *CGβ2* contribute to these roles is an open question, and even their coding capacity has been held suspect.

Substitution of the first CG exon by the *snaR*-containing element introduced two new potential start codons in *CGβ1/2*, presumptively leading to alternate reading frame usage and the synthesis of novel proteins lacking recognizable functional motifs (Bo and Boime 1992; Hallast et al. 2007). However, an experimental analysis of start codon selection demonstrated that these genes indeed give rise to authentic *hCGβ* (AM Parrott et al., in prep.). On the other hand, they display a distinct pattern of tissue expression, probably by virtue of their altered proximal promoter. Unlike other *hCGβ* genes, *hCGβ1/2* are expressed weakly in placenta but at relatively high levels in testis, to the extent that they account for approximately one-third of the total *hCGβ* mRNA in this tissue (Rull et al. 2008). Substantial concentrations of *hCGβ* peptide have been detected in human semen and in fetal testis, where it is thought to act as the primary stimulus of fetal Leydig cells, resulting in early testosterone secretion and masculine differentiation of the male fetus (Clements et al. 1976; Saito et al. 1988; Brotherton 1989). Thus, *hCGβ1/2* may have evolved to generate testicular *hCGβ*, functioning as a paracrine alternative to LHβ endocrine stimulation of Leydig cells.

CONCLUSION

Noncoding RNAs are the most common transcripts in higher eukaryotes and a likely source of genes that propel speciation events. We have traced the ancestry of the *snaR* family of small noncoding RNA to *Alu* elements, the most populous SINE family in primates. The *snaR* genes evolved recently in the line leading to the African Great Apes, via two retrotransposon intermediates (ASR and CAS). Members of the *snaR* family display rapid sequence divergence, suggesting their accelerated evolution in the

African Great Apes toward new functions. *snaR-A* stably associates with pol II and ribosomes, possibly linking this descendant of the *Alu* family to control of gene expression. *snaR* genes display species-specific amplification and loss by recombination. Such events led to the generation of new hormone genes in the African Great Apes, with expression that implies CGβ function in a new target organ.

ACKNOWLEDGMENTS

We thank our collaborators Dr. James Goodrich and Linda Drullinger for their work on pol II binding, and Dr. Tsafi Pe'ery for invaluable discussion. Funding of this work is provided by National Institutes of Health (grant R01 A1034552).

REFERENCES

Abou Elela S, Nazar RN. 1997. Role of the 5.8S rRNA in ribosome translocation. *Nucleic Acids Res* **25:** 1788–1794.

Bailey JA, Eichler EE. 2006. Primate segmental duplications: Crucibles of evolution, diversity and disease. *Nat Rev* **7:** 552–564.

Bailey JA, Liu G, Eichler EE. 2003. An *Alu* transposition model for the origin and expansion of human segmental duplications. *Am J Hum Genet* **73:** 823–834.

Bass BL, Hurst SR, Singer JD. 1994. Binding properties of newly identified *Xenopus* proteins containing dsRNA-binding motifs. *Curr Biol* **4:** 301 314.

Beniaminov A, Westhof E, Krol A. 2008. Distinctive structures between chimpanzee and human in a brain noncoding RNA. *RNA* **14:** 1270–1275.

Bennett EA, Keller H, Mills RE, Schmidt S, Moran JV, Weichenrieder O, Devine SE. 2008. Active *Alu* retrotransposons in the human genome. *Genome Res* **18:** 1875–1883.

Bevilacqua PC, George CX, Samuel CE, Cech TR. 1998. Binding of the protein kinase PKR to RNAs with secondary structure defects: Role of the tandem A-G mismatch and noncontiguous helixes. *Biochemistry* **37:** 6303–6316.

Bo M, Boime I. 1992. Identification of the transcriptionally active genes of the chorionic gonadotropin β gene cluster in vivo. *J Biol Chem* **267:** 3179–3184.

Brotherton J. 1989. Human chorionic gonadotrophin in human seminal plasma as shown with assays using monoclonal antibodies. *Andrologia* **21:** 407–415.

Chang DY, Hsu K, Maraia RJ. 1996. Monomeric scAlu and nascent dimeric Alu RNAs induced by adenovirus are assembled into SRP9/14-containing RNPs in HeLa cells. *Nucleic Acids Res* **24:** 4165–4170.

Clements JA, Reyes FI, Winter JS, Faiman C. 1976. Studies on human sexual development. III. Fetal pituitary and serum, and amniotic fluid concentrations of LH, CG, and FSH. *J Clin Endocrinol Metab* **42:** 9–19.

Cole LA, Khanlian SA, Riley JM, Butler SA. 2006. Hyperglycosylated hCG in gestational implantation and in choriocarcinoma and testicular germ cell malignancy tumorigenesis. *J Reprod Med* **51:** 919–929.

Corthesy B, Kao PN. 1994. Purification by DNA affinity chromatography of two polypeptides that contact the NF-AT DNA binding site in the interleukin 2 promoter. *J Biol Chem* **269:** 20682–20690.

Crooks GE, Hon G, Chandonia JM, Brenner SE. 2004. WebLogo: A sequence logo generator. *Genome Res* **14:** 1188–1190.

Duchange N, Pidoux J, Camus E, Sauvaget D. 2000. Alternative splicing in the human interleukin enhancer binding factor 3 (*ILF3*) gene. *Gene* **261:** 345–353.

Graifer D, Molotkov M, Eremina A, Ven'yaminova A, Repkova M, Karpova G. 2005. The central part of the 5.8 S rRNA is differently arranged in programmed and free human ribosomes. *Biochem J* **387:** 139–145.

Grindley ND. 1978. IS1 insertion generates duplication of a nine base pair sequence at its target site. *Cell* **13:** 419–426.

Guan D, Altan-Bonnet N, Parrott AM, Arrigo CJ, Li Q, Khaleduzzaman M, Li H, Lee CG, Pe'ery T, Mathews MB. 2008. Nuclear factor 45 (NF45) is a regulatory subunit of complexes with NF90/110 involved in mitotic control. *Mol Cell Biol* **28:** 4629–4641.

Halic M, Becker T, Pool MR, Spahn CM, Grassucci RA, Frank J, Beckmann R. 2004. Structure of the signal recognition particle interacting with the elongation-arrested ribosome. *Nature* **427:** 808–814.

Hallast P, Rull K, Laan M. 2007. The evolution and genomic landscape of *CGB1* and *CGB2* genes. *Mol Cell Endocrinol* **260–262:** 2–11.

Hallast P, Saarela J, Palotie A, Laan M. 2008. High divergence in primate-specific duplicated regions: Human and chimpanzee *Chorionic Gonadotropin Beta* genes. *BMC Evol Biol* **8:** 195.

Handschuh K, Guibourdenche J, Tsatsaris V, Guesnon M, Laurendeau I, Evain-Brion D, Fournier T. 2007. Human chorionic gonadotropin produced by the invasive trophoblast but not the villous trophoblast promotes cell invasion and is down-regulated by peroxisome proliferator-activated receptor-γ. *Endocrinology* **148:** 5011–5019.

Hasler J, Strub K. 2006a. *Alu* elements as regulators of gene expression. *Nucleic Acids Res* **34:** 5491–5497.

Hasler J, Strub K. 2006b. *Alu* RNP and *Alu* RNA regulate translation initiation in vitro. *Nucleic Acids Res* **34:** 2374–2385.

Jameson JL, Hollenberg AN. 1993. Regulation of chorionic gonadotropin gene expression. *Endocr Rev* **14:** 203–221.

Jurka J. 2004. Evolutionary impact of human *Alu* repetitive elements. *Curr Opin Genet Dev* **14:** 603–608.

Kapranov P, Cheng J, Dike S, Nix DA, Duttagupta R, Willingham AT, Stadler PF, Hertel J, Hackermuller J, Hofacker IL, et al. 2007. RNA maps reveal new RNA classes and a possible function for pervasive transcription. *Science* **316:** 1484–1488.

Kremerskothen J, Zopf D, Walter P, Cheng JG, Nettermann M, Niewerth U, Maraia RJ, Brosius J. 1998. Heterodimer SRP9/14 is an integral part of the neural BC200 RNP in primate brain. *Neurosci Lett* **245:** 123–126.

Kriegs JO, Churakov G, Jurka J, Brosius J, Schmitz J. 2007. Evolutionary history of 7SL RNA-derived SINEs in Supraprimates. *Trends Genet* **23:** 158–161.

Lander ES, Linton LM, Birren B, Nusbaum C, Zody MC, Baldwin J, Devon K, Dewar K, Doyle M, FitzHugh W, et al. (International Human Genome Sequencing Consortium). 2001. Initial sequencing and analysis of the human genome. *Nature* **409:** 860–921.

Langland JO, Kao PN, Jacobs BL. 1999. Nuclear factor-90 of activated T-cells: A double-stranded RNA-binding protein and substrate for the double-stranded RNA-dependent protein kinase, PKR. *Biochemistry* **38:** 6361–6368.

Lei ZM, Reshef E, Rao V. 1992. The expression of human chorionic gonadotropin/luteinizing hormone receptors in human endometrial and myometrial blood vessels. *J Clin Endocrinol Metab* **75:** 651–659.

Liao HJ, Kobayashi R, Mathews MB. 1998. Activities of adenovirus virus-associated RNAs: Purification and characterization of RNA binding proteins. *Proc Natl Acad Sci* **95:** 8514–8519.

Lin D, Pestova TV, Hellen CU, Tiedge H. 2008. Translational control by a small RNA: Dendritic BC1 RNA targets the eukaryotic initiation factor 4A helicase mechanism. *Mol Cell Biol* **28:** 3008–3019.

Liu WM, Chu WM, Choudary PV, Schmid CW. 1995. Cell stress and translational inhibitors transiently increase the abundance of mammalian SINE transcripts. *Nucleic Acids Res* **23:** 1758–1765.

Liu GE, Alkan C, Jiang L, Zhao S, Eichler EE. 2009. Comparative analysis of *Alu* repeats in primate genomes. *Genome Res* **19:** 876–885.

Maraia RJ, Chang DY, Wolffe AP, Vorce RL, Hsu K. 1992. The RNA polymerase III terminator used by a B1-*Alu* element can modulate 3′ processing of the intermediate RNA product. *Mol*

Cell Biol 12: 1500–1506.

Mariner PD, Walters RD, Espinoza CA, Drullinger LF, Wagner SD, Kugel JF, Goodrich JA. 2008. Human Alu RNA is a modular transacting repressor of mRNA transcription during heat shock. *Mol Cell* **29:** 499–509.

Marques-Bonet T, Kidd JM, Ventura M, Graves TA, Cheng Z, Hillier LW, Jiang Z, Baker C, Malfavon-Borja R, Fulton LA, et al. 2009. A burst of segmental duplications in the genome of the African Great Ape ancestor. *Nature* **457:** 877–881.

Martignetti JA, Brosius J. 1993. BC200 RNA: A neural RNA polymerase III product encoded by a monomeric *Alu* element. *Proc Natl Acad Sci* **90:** 11563–11567.

Maston GA, Ruvolo M. 2002. Chorionic gonadotropin has a recent origin within primates and an evolutionary history of selection. *Mol Biol Evol* **19:** 320–335.

Mathews MB, Shenk T. 1991. Adenovirus virus-associated RNA and translation control. *J Virol* **65:** 5657–5662.

Mattick JS, Makunin IV. 2005. Small regulatory RNAs in mammals. *Hum Mol Genet* (spec no 1) **14:** R121–R132.

Meagher MJ, Schumacher JM, Lee K, Holdcraft RW, Edelhoff S, Disteche C, Braun RE. 1999. Identification of ZFR, an ancient and highly conserved murine chromosome-associated zinc finger protein. *Gene* **228:** 197–211.

Parrott AM, Mathews MB. 2007. Novel rapidly evolving hominid RNAs bind nuclear factor 90 and display tissue-restricted distribution. *Nucleic Acids Res* **35:** 6249–6258.

Parrott AM, Walsh MR, Reichman TW, Mathews MB. 2005. RNA binding and phosphorylation determine the intracellular distribution of nuclear factors 90 and 110. *J Mol Biol* **348:** 281–293.

Parrott AM, Walsh MR, Mathews MB. 2007. Analysis of RNA: protein interactions in vivo: Identification of RNA-binding partners of nuclear factor 90. *Methods Enzymol* **429:** 243–260.

Pei Y, Zhu P, Dang Y, Wu J, Yang X, Wan B, Liu JO, Yi Q, Yu L. 2008. Nuclear export of NF90 to stabilize IL-2 mRNA is mediated by AKT-dependent phosphorylation at Ser^{647} in response to CD28 costimulation. *J Immunol* **180:** 222–229.

Pfeifer I, Elsby R, Fernandez M, Faria PA, Nussenzveig DR, Lossos IS, Fontoura BM, Martin WD, Barber GN. 2008. NFAR-1 and -2 modulate translation and are required for efficient host defense. *Proc Natl Acad Sci* **105:** 4173–4178.

Pollard KS, Salama SR, Lambert N, Lambot MA, Coppens S, Pedersen JS, Katzman S, King B, Onodera C, Siepel A, et al. 2006. An RNA gene expressed during cortical development evolved rapidly in humans. *Nature* **443:** 167–172.

Prasanth KV, Spector DL. 2007. Eukaryotic regulatory RNAs: An answer to the 'genome complexity' conundrum. *Genes Dev* **21:** 11–42.

Pullmann R Jr, Kim HH, Abdelmohsen K, Lal A, Martindale JL, Yang X, Gorospe M. 2007. Analysis of turnover and translation regulatory RNA-binding protein expression through binding to cognate mRNAs. *Mol Cell Biol* **27:** 6265–6278.

Quentin Y. 1992. Fusion of a free left Alu monomer and a free right Alu monomer at the origin of the Alu family in the primate genomes. *Nucleic Acids Res* **20:** 487–493.

Rull K, Hallast P, Uuskula L, Jackson J, Punab M, Salumets A, Campbell RK, Laan M. 2008. Fine-scale quantification of HCG β gene transcription in human trophoblastic and non-malignant non-trophoblastic tissues. *Mol Hum Reprod* **14:** 23–31.

Saito S, Kumamoto Y, Ito N, Kurohata T. 1988. Human chorionic gonadotropin β-subunit in human semen. *Arch Androl* **20:** 87–99.

Sakamoto S, Aoki K, Higuchi T, Todaka H, Morisawa K, Tamaki N, Hatano E, Fukushima A, Taniguchi T, Agata Y. 2009. The NF90-NF45 complex functions as a negative regulator in the microRNA processing pathway. *Mol Cell Biol.* **29:** 3754–3769.

Schumacher A, Brachwitz N, Sohr S, Engeland K, Langwisch S, Dolaptchieva M, Alexander T, Taran A, Malfertheiner SF, Costa SD, et al. 2009. Human chorionic gonadotropin attracts regulatory T cells into the fetal-maternal interface during early human pregnancy. *J Immunol* **182:** 5488–5497.

Shi L, Godfrey WR, Lin J, Zhao G, Kao PN. 2007. NF90 regulates inducible IL-2 gene expression in T cells. *J Exp Med* **204:** 971–977.

Shim J, Lim H, Yates JR, Karin M. 2002. Nuclear export of NF90 is required for interleukin-2 mRNA stabilization. *Mol Cell* **10:** 1331–1344.

Sinnett D, Richer C, Deragon JM, Labuda D. 1991. *Alu* RNA secondary structure consists of two independent 7 SL RNA-like folding units. *J Biol Chem* **266:** 8675–8678.

Skryabin BV, Kremerskothen J, Vassilacopoulou D, Disotell TR, Kapitonov VV, Jurka J, Brosius J. 1998. The BC200 RNA gene and its neural expression are conserved in Anthropoidea (Primates). *J Mol Evol* **47:** 677–685.

Tian B, Bevilacqua PC, Diegelman-Parente A, Mathews MB. 2004. The double-stranded-RNA-binding motif: Interference and much more. *Nat Rev Mol Cell Biol* **5:** 1013–1023.

Tiedge H, Chen W, Brosius J. 1993. Primary structure, neural-specific expression, and dendritic location of human BC200 RNA. *J Neurosci* **13:** 2382–2390.

Wan H, Versnel MA, Leijten LM, van Helden-Meeuwsen CG, Fekkes D, Leenen PJ, Khan NA, Benner R, Kiekens RC. 2008. Chorionic gonadotropin induces dendritic cells to express a tolerogenic phenotype. *J Leukoc Biol* **83:** 894–901.

Wilhelm BT, Marguerat S, Watt S, Schubert F, Wood V, Goodhead I, Penkett CJ, Rogers J, Bahler J. 2008. Dynamic repertoire of a eukaryotic transcriptome surveyed at single-nucleotide resolution. *Nature* **453:** 1239–1243.

Wilusz JE, Sunwoo H, Spector DL. 2009. Long noncoding RNAs: Functional surprises from the RNA world. *Genes Dev* **23:** 1494–1504.

Zuker M. 2003. Mfold web server for nucleic acid folding and hybridization prediction. *Nucleic Acids Res* **31:** 3406–3415.

Zygmunt M, Herr F, Keller-Schoenwetter S, Kunzi-Rapp K, Munstedt K, Rao CV, Lang U, Preissner KT. 2002. Characterization of human chorionic gonadotropin as a novel angiogenic factor. *J Clin Endocrinol Metab* **87:** 5290–5296.

DUF1220 Domains, Cognitive Disease, and Human Brain Evolution

L. DUMAS AND J.M. SIKELA

University of Colorado Denver School of Medicine, Aurora, Colorado 80045
Correspondence: james.sikela@ucdenver.edu

We have established that human genome sequences encoding a novel protein domain, DUF1220, show a dramatically elevated copy number in the human lineage (>200 copies in humans vs. 1 in mouse/rat) and may be important to human evolutionary adaptation. Copy-number variations (CNVs) in the 1q21.1 region, where most DUF1220 sequences map, have now been implicated in numerous diseases associated with cognitive dysfunction, including autism, autism spectrum disorder, mental retardation, schizophrenia, microcephaly, and macrocephaly.

We report here that these disease-related 1q21.1 CNVs either encompass or are directly flanked by DUF1220 sequences and exhibit a dosage-related correlation with human brain size. Microcephaly-producing 1q21.1 CNVs are deletions, whereas macrocephaly-producing 1q21.1 CNVs are duplications. Similarly, 1q21.1 deletions and smaller brain size are linked with schizophrenia, whereas 1q21.1 duplications and larger brain size are associated with autism. Interestingly, these two diseases are thought to be phenotypic opposites. These data suggest a model which proposes that (1) DUF1220 domain copy number may be involved in influencing human brain size and (2) the evolutionary advantage of rapidly increasing DUF1220 copy number in the human lineage has resulted in favoring retention of the high genomic instability of the 1q21.1 region, which, in turn, has precipitated a spectrum of recurrent human brain and developmental disorders.

GENOME-WIDE SURVEY OF LINEAGE-SPECIFIC GENE COPY-NUMBER GAIN AND LOSS IN HUMAN AND GREAT APE EVOLUTION

At a fundamental level, evolution has been characterized as a change in the allele frequency of a gene. More precisely, it is an alteration in the frequency of a genome sequence and may or may not involve a gene. It has been proposed that the primary types of genome alterations that underlie evolutionary change are single-nucleotide substitution, chromosomal rearrangement, and gene duplication. In 1970, Ohno (1970) put forth the argument that gene duplication is a primary mechanism of evolutionary change, due to the relaxation of selection and increase in variation afforded by its built-in redundancy. This view has also been expressed by W.H. Li (1997): "There is now ample evidence that gene duplication is the most important mechanism for generating new genes and new biochemical processes that have facilitated the evolution of complex organisms from primitive ones." More recently, a similar view has been expressed, albeit more succinctly, by E.E. Eichler (2001): "Exceptional duplicated regions underlie exceptional biology."

Given the importance of gene duplication to evolutionary change, we initiated a collaboration with Jonathan Pollack at Stanford to generate the first genome-wide and first gene-based application of array comparative genomic hybridization (arrayCGH) across human and nonhuman primate lineages (Fortna et al. 2004). The approach surveyed humans and four great ape species (bonobo, chimpanzee, gorilla, and orangutan) using multiple individuals from each. The arrays were generated from >41,000 human cDNA clones (~24,000 genes), from which full inserts were amplified by polymerase chain reaction (PCR) and spotted. Each arrayCGH experiment represented a pairwise comparison of a reference DNA sample (always human) labeled with a green fluorescent dye and a test DNA sample (either human or a sample from one of the four great apes) labeled with a red fluorescent dye. Comparison of resulting array signals indicated that at least 1004 genes could be identified that gave changes in hybridization intensity consistent with lineage-specific gains or losses in gene copy number (Fig. 1). Several sets of control experiments were used to verify that the data were lineage-specific the result of copy-number changes, rather than sequence divergence.

Analysis of resulting arrayCGH data not only identified numerous candidate genes for a wide range of lineage-specific traits, but, by parsimony, also allowed one to determine when, in recent primate evolution, each of these events was likely to have occurred. In this manner, a genome-wide evolutionary history of lineage-specific gene-copy-number gain (duplication) and loss could be reconstructed covering the past 16 million years, extending from the last common ancestor (LCA) of the human and great ape lineages to the present (Fortna et al. 2004).

In addition, the data revealed that gene-copy-number changes showed a strong positional bias with respect to genome location and, in many cases, corresponded to regions known to be evolutionarily active. Among the most prominent of these were 1q21.1 (where the majority of DUF1220 domain sequences map; Fig. 2), the pericentromeric region of chromosome 9, and the fusion site on chromosome 2, which is the site at which two ancestral ape

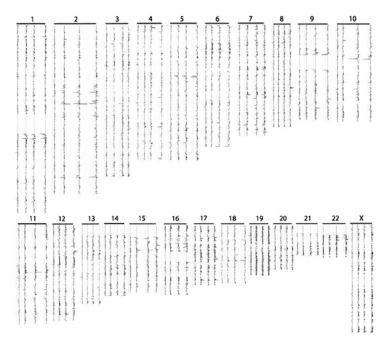

Figure 1. Caryoscope view of genome-wide interhominoid arrayCGH analysis. cDNA arrays were used to perform pairwise comparisons of a reference genomic DNA (always human) labeled green and a test genomic DNA (either human, chimp, gorilla, or orangutan) labeled red. Caryoscope data are plotted as results of a five cDNA-sliding window with results from human, chimp, gorilla, and orangutan (reading right to left) for each human chromosome (Fortna et al. 2004).

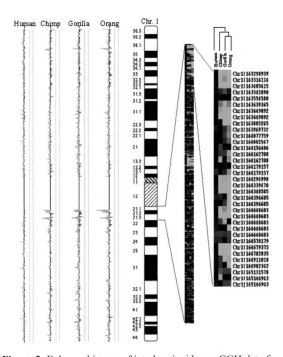

Figure 2. Enhanced image of interhominoid arrayCGH data for chromosome 1 and the 1q21.1 region. (*Left*) Caryoscope image of arrayCGH data from human, chimp, gorilla, and orangutan, as described in Fig. 1. (*Right*) TreeView enlargement of the same data for the 1q21.1 region. In this view, each vertical column shows data from an arrayCGH experiment, whereas each horizontal row represents data for a single cDNA, organized in the order in which the genes occur in the genome (Fortna et al. 2004).

chromosomes fused to generate human Chr 2. Whereas 1q21.1 and 9p13/9q13 showed the highest concentration of genes giving human lineage-specific duplications, the fusion region on chromosome 2 contained a number of virtually contiguous genes that exhibited dramatic copy-number changes specific for several different lineages, all within a narrow (400 kb) genomic interval (Fortna et al. 2004).

There was also a pronounced enrichment of genes that showed increases in copy number specifically in the African Great Apes, compared to humans and orangutans. This finding has been reinforced by a recent report (Marques-Bonet et al. 2009) indicating that a burst of gene duplication occurred in the African Great Ape lineages. Interestingly, at least some of these African Great Ape–specific gene-copy-number increases may represent independent expansions within the African Great Ape clade; e.g., the gene was duplicated in the gorilla independent of its duplication in the *Pan* (chimp/bonobo) lineages (Fortna et al. 2004; Marques-Bonet et al. 2009).

LINEAGE-SPECIFIC GENE-COPY-NUMBER VARIATION SPANNING 60 MILLION YEARS OF HUMAN AND PRIMATE EVOLUTION

Because of the highly informative results that we obtained from a comparison of the human and great ape genomes, we extended the application of arrayCGH to five evolutionarily more distant primate lineages (Dumas et al. 2007). By using human cDNA arrays and retaining humans as the reference DNA in all arrayCGH experiments, we were able to interrelate data from all 10 primate species tested: human, bonobo, chimp, gorilla, orangutan, gibbon, macaque, baboon, marmoset, and lemur. This series of experiments further extended the period of primate evolutionary history that we could survey. The most distant comparison was between lemur and human, the LCA of which is estimated to have occurred ~60 million years ago. Although we were aware that sequence divergence may contribute to arrayCGH signals and that this would be more pronounced as more and more distantly related species were compared, these data indicated that valid copy-number changes could be detected even among the most distant primate species. For example, the many genes giving strong red signals in the lemur to human comparison (indicative of copy-number gains in lemur relative to humans) indicated that the method could reliably pick up copy-number gains between these highly divergent species and that the gene-copy-number expansions which were giving hybridization signals indicative of lemur-specific copy-number increases must be dramatic enough to override any contribution that sequence divergence was making to the arrayCGH signals.

This study identified 4159 genes that showed lineage-specific copy-number changes among these 10 primate species, including many that were specific to multiple lineages (Fig. 3). Among these were genes that were (1) increased in human and ape lineages relative to monkey lineages, (2) increased only in Old World monkeys relative to the other primates tested, and (3) increased in African Great Apes relative to the other species tested. If one applies parsimony to the results, the copy-number changes can be positioned at specific points in primate evolutionary time.

The number of genes showing lineage-specific copy-number changes for each lineage showed a general correlation with the age of the lineage (the age of the LCA with humans). Although the gibbon was an exception, showing an increased number of lineage-specific changes relative to its evolutionary age, this was consistent with the unusually high degree of chromosomal rearrangement that has been found within this species. In agreement with the overall trend, application of a tree-building program to this extended arrayCGH data set recapitulated the established evolutionary relationship of the species to a very high approximation (Fig. 4) (Dumas et al. 2007).

HUMAN LINEAGE-SPECIFIC AMPLIFICATION OF DUF1220 DOMAINS

Among the genes identified in the human and great ape arrayCGH study were 134 that exhibited hybridization signals consistent with human lineage-specific increases in copy number. To characterize these more completely, the cDNA inserts of each clone were sequenced and used as BLAT queries against available primate genome sequences (Popesco et al. 2006). The most striking finding was that one gene, *MGC8902*, encoded by cDNA IMAGE:843276, gave more than 49 hits in the human genome but only 10 and 4 in chimp and macaque genomes, respectively. On the

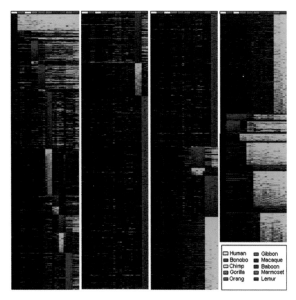

Figure 3. TreeView image of genes showing lineage-specific (LS) copy-number changes among 10 primate species. TreeView image of 7318 genes giving LS arrayCGH signatures are shown for each of 10 lineages, including human (blue-gray), bonobo (rust), chimp (yellow), gorilla (orange), orangutan (purple), gibbon (green), macaque (blue), baboon (brown), marmoset (magenta), and lemur (light purple), as well as Old World monkeys (OWMs), marmoset and lemur, African Great Apes, *Pan* lineage (bonobo and chimp together), and combined arrayCGH-predicted changes relative to the remaining extended primates for the following groups: human and *Pan* lineage, human and African Great Apes, human and great apes, and human and all apes (great and lesser). The LS signals are grouped according to lineage and within each lineage are ordered, highest to lowest, according to the $\log_2$ fluorescence ratio of the signal intensity of test sample to reference sample. Colors are displayed using a pseudocolor scale. Green and red signals correspond to LS decreases and increases, respectively, with respect to humans. (Modified, with permission, from Dumas et al. 2007.)

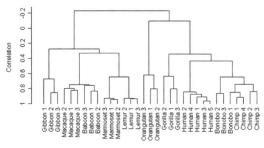

Figure 4. Correlation of arrayCGH data with established primate phylogeny. An unsupervised clustering of the raw data (arrayCGH $\log_2$ ratios) was conducted to determine whether the individuals within each lineage cluster together as expected from the established phylogeny of these species. The complete raw data set, with each sample numbered as indicated, was used for the unbiased cluster analysis using BRB Array Tools (http://linus.nci.nih.gov/BRB-ArrayTools.html). Hierarchical clustering, using default settings for filtering, was conducted by array and included all 33 arrays from the study. The scale on the left represents the centered correlation (Pearson's coefficient) using average linkage analysis.

basis of BLAT hit span size, all human hits were shown to contain both exonic and intronic sequences and thus were not the products of retrotransposition. Further investigation of the IMAGE:843276 cDNA insert sequence indicated that it encoded six closely spaced copies of a predicted PFAM protein domain of unknown function (DUF1220). Follow-up analysis of DUF1220 domains indicated that they are ~65 amino acids in length and composed of a two-exon doublet (Fig. 5).

Sequences encoding DUF1220 domains continued to show a human lineage-specific increase in copy number when the five additional primate lineages were compared. PFAM has divided DUF1220 sequences into 11 seed domains, only one of which (O75042) is found outside primates, and there only as a single copy. All of the remaining 10 seed domains are primate specific, and, of these, Q8IX162 gave the highest number of BLAT hits in the human genome. Remarkably, 37 of the 90 hits produced by Q8IX162 in the human genome gave a perfect sequence match. In humans, the DUF1220 family is found on 20–34 genes (also called the NBPF family; Vandepoele et al. 2005; Popesco et al. 2006) where they are present in from 1 to 50+ copies. Analysis of predicted DUF1220-encoding proteins indicates that the genes appear to be virtually devoid of other domains or functional signatures (Fig. 6), suggesting that the primary focus of selection was on increasing the number of DUF1220 domain copies in the genome, rather than integration of DUF1220 copies into known genes.

When all 11 DUF1220 seed domains are used for BLAT analysis and redundant hits are removed, we estimate that the human genome encodes 212 DUF1220 domains, whereas chimp, macaque, and mouse genomes encode 34, 30, and 1, respectively (Popesco et al. 2006). It should be noted that, with the exception of humans, these estimates are based on draft genome sequences, and these numbers are likely to vary when more recent genome assemblies are surveyed.

DUF1220 domains are thought to have undergone recent positive selection based on elevated K_a/K_s ratios (Popesco et al. 2006). Western analysis with antibodies directed against a human DUF1220 peptide show DUF1220-positive signals in several human tissues, including the brain. Immunocytochemistry of postmortem brain indicates that DUF1220 proteins are expressed exclusively in neurons, and they seem to be enriched in cell bodies and dendrites (Popesco et al. 2006). Although DUF1220 sequences are found at three cytogenetic locations on chromosome 1, the great majority map to 1q21.1, which, along with the pericentromeric region of chromosome 9, are the two genomic regions that show the highest concentration of genes that exhibit human lineage-specific increases in copy number (Fig. 7).

LINKING DUF1220 DOMAINS, BRAIN SIZE, AND COGNITIVE DISEASE

During the last few years, CNVs in the 1q21.1 region have been implicated in an increasingly higher number of human diseases, including idiopathic mental retardation (de Vries et al. 2005; Sharp et al. 2006), autism/autism spectrum disorder (Sharp et al. 2006; Autism Genome Project Consortium 2007; Mefford et al. 2008), congenital heart disease (Christiansen et al. 2004), schizophrenia (ISC 2008; Stefansson et al. 2008; Walsh et al. 2008; Need et al. 2009), microcephaly/macrocephaly (Brunetti-Pierri et al. 2008; Mefford et al. 2008), and neuroblastoma (Vandepoele et al. 2005; Diskin et al. 2009). These results have provided a number of provocative hints as to the possible role DUF1220 amplification may have in human evolution and human disease. For example, the evolutionary gene-copy-number studies that we have reported show that DUF1220 copy-number increases roughly parallel the increase in relative brain size, and/or neocortex expansion, that has occurred over recent primate and especially human evolution (Popesco et al. 2006; Dumas et al. 2007). Given this observation, it is intriguing that two reports that studied 1q21.1 CNVs and brain size found that deletions were associated with microcephaly, and duplications were associated with macrocephaly (Brunetti-Pierri et al. 2008; Mefford et al. 2008). Although no mention was made in these reports that DUF1220 copy number might underlie or be related to these phenotypes, we note here that these CNVs either encompassed or flanked DUF1220 domains (which, because of their highly duplicated nature, were not directly interrogated in these studies), raising the prospect that DUF1220 domain copy number (i.e., DUF1220 dosage) may be causally related to the observed differences in human brain size.

We point out here that a similar argument can be made regarding the 1q21.1 CNVs that have been reported to be involved in autism and schizophrenia. These diseases, as well as a number of pairs of other disorders, have been labeled genomic "sister disorders" by virtue of their tendency to exhibit diametrically opposite phenotypes and be caused by duplications versus deletions of the same genomic sequences (Crespi et al. 2009). Among the opposing phenotypes that distinguish autism and schizophrenia are a larger and smaller brain size, respectively (Crespi and Badcock 2008). It is therefore intriguing that the 1q21.1 CNVs that have been reported to underlie autism and schizophrenia tend to be duplications and deletions, respectively. In addition, these 1q21.1 CNVs, as with those

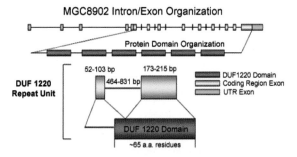

Figure 5. Genomic intron/exon organization of the DUF1220-encoding *MGC8902* gene and the predicted domain structure of the translated protein. A representative DUF1220 repeat unit is also shown. (Modified, with permission, from Popesco et al. 2006 [© AAAS].)

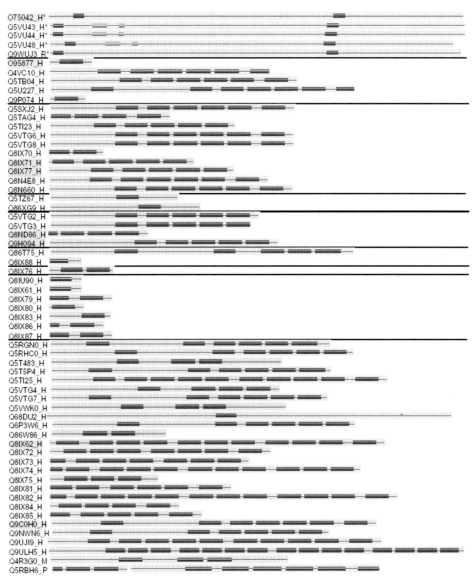

Figure 6. All InterPro proteins with DUF1220 domains. Schematic representations of all proteins in the InterPro database (www.ebi.ac.uk/interpro) that are predicted to have DUF1220 domains are shown. Protein identifications are shown on the right with accession numbers and letters. _H, _R, _M, and _P refer to proteins derived from human, rat, macaque, and orangutan, respectively. (Yellow) Accession numbers of proteins from which the 11 seed domains were derived. All protein schematics are to scale, but those marked with an asterisk are half-scale compared to the others. Horizontal lines separate groups of proteins derived from the same gene with the exception of the last group, which includes human proteins that have not been assigned to a gene. Red, green, yellow, and blue boxes are indicative of predicted DUF1220, ribosomal protein L29, prefoldin, and spindle-associated domains, respectively. (Modified, with permission, from Popesco et al. 2006 [© AAAS].)

underlying microcephaly and macrocephaly, encompass or are flanked by DUF1220 domains and span the same genomic interval. Taken together, these observations support the involvement of DUF1220 domains in the brain size differences that are known to exist between autistic and schizophrenic populations and between microcephalic and macrocephalic populations, and they suggest that a mechanistic link exists among 1q21.1 instability, the evolutionarily rapid DUF1220 copy-number amplification that has occurred in the human lineage, and the high prevalence of these diseases in human populations.

During the past several years, a number of microcephaly disease genes have been identified and these provide another link between DUF1220 and brain size. Specifically, it has been shown that of the half-dozen or so genes identified that cause microcephaly, a majority appear to encode proteins that are associated with the centrosome and control of the cell cycle (Bond and Woods 2006). Cell cycle control and the timing of when and where cells switch from symmetric to asymmetric cell division have been postulated to be key factors in changes in neuron number and brain size that have characterized mammalian and primate brain evo-

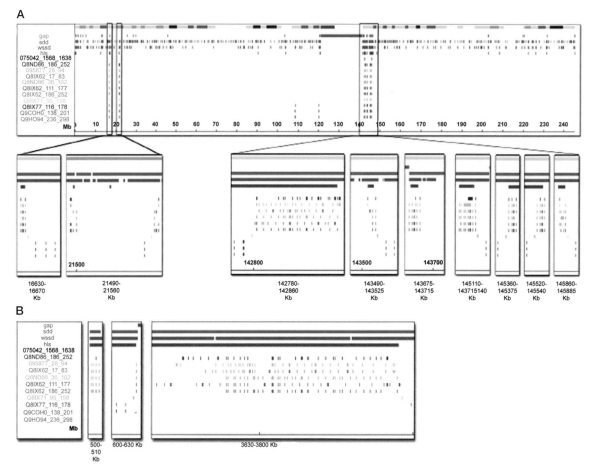

Figure 7. Human genomic locations of DUF1220-encoding sequences. The positions of genomic BLAT alignments to the 11 DUF1220 PFAM-predicted seed domains on chromosome 1 (*A*) and unordered chromosome 1 fragments (*B*) are noted along with the positions of gaps in the genome sequence. Notably, DUF1220 sequences are clustered around the pericentromeric C-band region on chromosome 1 and also on the unordered chromosome 1 sequences. Regions of particular interest are enlarged to show greater detail. The G-banding pattern is also shown, as well as positions of the centromere, known segmental duplications (sdd, wssd), and the BLAT hit positions of genes showing human lineage-specific increases in copy number.

lution (Rakic 1995). Given these observations, it is noteworthy that the ancestral DUF1220 domain (found once in mouse/rat and other nonprimate mammals) is found on myomegalin (PDE4DIP), a centrosomal protein whose gene is a homolog of *CDK5RAP2*, a non-DUF1220 microcephaly disease gene that also encodes a centrosomal protein (Bond et al. 2005; Dumas et al. 2007).

Recently, it has been reported that DUF1220 domains are one of only a small number of core duplicons that exist in the human genome (Jiang et al. 2007). These highly duplicated sequences have been the focal point for large-copy-number expansions in the human lineage, and in many cases, each core duplicon has been instrumental in recent chromosome-specific copy-number expansions. Typically, core duplicons are flanked by a mosaic of other duplicated sequences that appear to have been disseminated by being carried along during core duplicon transpositions. The DUF1220 domain duplicon appears to be responsible for much of the duplicated sequences found in the pericentromeric region of chromosome 1 (Jiang et al. 2007). Core duplicon sequences are often interspersed and separated by single- or low-copy-number genes and often promote nonallelic homologous recombination (NAHR) events. Such a duplicon-rich genome architecture can often lead to NAHR between distantly spaced highly similar sequences (e.g., those of a core duplicon) that simultaneously produce duplications or deletions of the intervening sequences (segmental aneusomy). These intervening sequences often contain dosage-sensitive genes that are carried along with core duplicon duplications and deletions, resulting in disease-causing CNVs. The fact that DUF1220 sequences are associated with the numerous disease-related 1q21.1 CNVs mentioned above is fully consistent with such predictions of core duplicon behavior.

On the basis of these observations, we propose a model that links DUF1220 sequences to both human disease and human brain evolution. The model proposes the following specific testable hypotheses: (1) Increasing DUF1220 copy number may be related to increases in brain size; this is based both on a general correlation of DUF1220 copy number with brain size between species (Fortna et al. 2004; Popesco et al. 2006; Dumas et al. 2007), as well as within the human population (Brunetti-Pierri et al. 2008; Mefford et al. 2008), (2) 1q21.1 instability promoted the rapid evolutionary copy-number increase of DUF1220 sequences, and (3) the evolutionary advantage that increased DUF1220 copy number conferred favored retention of high 1q21.1 instability, which, in turn, has resulted in many recurrent deleterious duplications (macrocephaly, autism) and deletions (microcephaly, schizophrenia) of dosage-sensitive 1q21.1 genes.

It is also noteworthy that the extreme instability of the 1q21.1 region increases the chances that somatic or de novo germ-line CNVs will occur, an observation that provides a potential explanation for the frequent examples of non-Mendelian inheritance that have been found to be associated with schizophrenia and autism. Finally, this mechanistic linking of 1q21.1 genomic instability with an evolutionarily favored process (i.e., DUF1220 copy-number increase) provides a means of reconciling the central paradox associated with schizophrenia and autism. Namely, why do diseases that are clearly maladaptive and have a significant genetic component persist at unusually high frequency throughout human populations?

FINISHING THE 1q21.1 GENOME SEQUENCE

The 1q21.1 region is highly complex and shows a disproportionately high number of segmental duplications and sequence gaps (18), making precise assessment of its sequence content and organization difficult. As a result, caution should be exercised when relying on the current human and nonhuman primate genome sequences for 1q21.1. Before the complete evolutionary ontogeny of DUF1220 sequences can be reconstructed, and before direct comprehensive comparisons of the 1q21.1 region can be made among different individuals and among different species, a much more complete genome assembly of the region will be required. One potentially useful strategy will be the use of a haploid genomic resource: a human hydatiform mole bacterial artificial chromosome (BAC) library that has been constructed with this type of application in mind. Because of the repeat-rich nature of 1q21.1 and other genomic regions, the development of longer read sequencing technology would also provide a valuable tool for accurate genome finishing (Eid et al. 2009). Current available sequencing platforms have relatively short read lengths, a limitation that prevents accurate assembly of many of these repeat-rich regions. Such regions are increasingly being implicated in human disease, yet, due to this sequencing limitation, they continue to remain largely unexamined. Finishing 1q21.1 will provide a framework both for correctly annotating the DUF1220 domain content and organization in the human genome, and for precisely defining breakpoints and content of the many disease-related CNVs that are being identified in this evolutionarily important genomic region.

CONCLUSIONS

Presently available data suggest that the striking increase in DUF1220 domain copy number in recent evolutionary time has occurred because there is a clear adaptive advantage to greater numbers of DUF1220 domains and that this advantage and selection process has persisted throughout primate, and especially human, evolution. Indeed, as we have previously reported, sequences encoding DUF1220 domains are virtually all primate specific, show signs of strong positive selection, and are increasingly amplified generally as a function of a species' evolutionary proximity to humans, where the greatest number of copies (212) is found (Popesco et al. 2006). On the basis of the genome organization of DUF1220 sequences, the large number of human copies is not likely to have arisen by a single event, but rather by a series of small and large incremental increases, each of which conferred an adaptive advantage. This series of increases (involving multiple rounds of both gene and domain duplications) is likely to have been promoted by the high degree of genomic instability associated with the human 1q21.1 region. Such instability would have more frequently produced duplications and deletions in the region, with the DUF1220 duplications conferring a selective advantage, which in turn would result in retention of the instability in those who had more DUF1220 copies. This process could be viewed as a recurring cycle in which the increased 1q21.1 instability facilitated DUF1220 copy-number increases in certain individuals, followed by selection of individuals who exhibited the increased DUF1220 copy number, resulting in retention of the 1q21.1 instability in these individuals. In this manner, high 1q21.1 instability may have been selected for and retained in the human lineage because it more rapidly produced increases in DUF1220 copy number.

The 1q21.1 instability is a product of the genome architecture of the region and, because it works by promoting rearrangements in sequence organization and copy number, many deleterious events can be expected to occur along with the beneficial DUF1220 duplications. In this regard, the large number of recent reports implicating 1q21.1 CNVs in multiple human diseases should not be unexpected and should be viewed as a natural outcome of the genomic instability that is being favored because of the adaptive value that more DUF1220 copies may be conferring. In summary, the high number of 1q21.1 CNVs that are disease-causing may be the price that our species paid, and continues to pay, for the adaptive benefit of large numbers of DUF1220 domains.

ACKNOWLEDGMENTS

We thank the individuals who have contributed to the data presented here, including J. Pollack, A. Fortna, M. Popesco, E. MacLaren, Y. Kim, M. O'Bleness, J. Keeney, J. Hopkins, A. Karimpour-Fard, M. Cox, R. Berry, L. Meltesen, L. McGavran, G. Wyckoff, and L. Jorde. J.M.S. is supported by National Institute of Mental Health grant R01 MH81203, NIAAA grant 2 R01 AA11853, and a Butcher Foundation grant.

REFERENCES

Autism Genome Project Consortium. 2007. Mapping autism risk loci using genetic linkage and chromosomal rearrangements. *Nat Genet* **39:** 319–328.

Bond J, Woods CG. 2006. Cytoskeletal genes regulating brain size. *Curr Opin Cell Biol* **18:** 95–101.

Bond J, Roberts E, Springell K, Lizarraga SB, Scott S, Higgins J, Hampshire DJ, Morrison EE, Leal GF, Silva EO, et al. 2005. A centrosomal mechanism involving CDK5RAP2 and CENPJ controls brain size. *Nat Genet* **37:** 353–355.

Brunetti-Pierri N, Berg JS, Scaglia F, Belmont J, Bacino CA, Sahoo T, Lalani SR, Graham B, Lee B, Shinawi M, et al. 2008. Recurrent reciprocal 1q21.1 deletions and duplications associated with microcephaly or macrocephaly and developmental and behavioral abnormalities. *Nat Genet* **40:** 1466–1471.

Christiansen J, Dyck JD., Elyas BG, Lilley M, Bamforth JS, Hicks M, Sprysak KA, Tomaszewski R, Haase SM, Vicen-Wyhony LM, Somerville MJ. 2004. Chromosome 1q21.1 contiguous gene deletion is associated with congenital heart disease. *Circ Res* **94:** 1429–1435.

Crespi B, Badcock C. 2008. Psychosis and autism as diametric disorders of the social brain. *Behav Brain Sci* **31:** 241–320.

Crespi B, Summers K, Dorus S. 2009. Genomic sister-disorders of neurodevelopment: An evolutionary approach. *Evol Appl* **2:** 81–100.

de Vries BB, Pfundt R, Leisink M, Koolen DA, Vissers LE, Janssen IM, Reijmersdal S, Nillesen WM, Huys EH, Leeuw N, et al. 2005. Diagnostic genome profiling in mental retardation. *Am J Hum Genet* **77:** 606–616.

Diskin SJ, Hou C, Glessner JT, Attiyeh EF, Laudenslager M, Bosse K, Cole K, Mosse YP, Wood A, Lynch JE, et al. 2009. Copy number variation at 1q21.1 associated with neuroblastoma. *Nature* **459:** 987–991.

Dumas L, Kim YH, Karimpour-Fard A, Cox M, Hopkins J, Pollack JR, Sikela JM. 2007. Gene copy number variation spanning 60 million years of human and primate evolution. *Genome Res* **17:** 1266–1277.

Eichler EE. 2001. Segmental duplications: What's missing, misassigned, and misassembled—And should we care? *Genome Res* **11:** 653–656.

Eid J, Fehr A, Gray J, Luong K, Lyle J, Otto G, Peluso P, Rank D, Baybayan P, Bettman B, et al. 2009. Real-time DNA sequencing from single polymerase molecules. *Science* **323:** 133–138.

Fortna A, Kim Y, MacLaren E, Marshall K, Hahn G, Meltesen L, Brenton M, Hink R, Burgers S, Hernandez-Boussard T, et al. 2004. Lineage-specific gene duplication and loss in human and great ape evolution. *PLoS Biol* **2:** 937–954.

International Schizophrenia Consortium (ISC). 2008. Rare chromosomal deletions and duplications increase risk of schizophrenia. *Nature* **455:** 237–241.

Jiang Z, Tang H, Ventura M, Cardone MF, Marques-Bonet T, She X, Pevzner PA, Eichler EE. 2007. Ancestral reconstruction of segmental duplications reveals punctuated cores of human genome evolution. *Nat Genet* **39:** 1361–1368.

Li WH. 1997. *Molecular evolution*. Sinauer, Sunderland, MA.

Marques-Bonet T, Kidd JM, Ventura M, Graves TA, Cheng Z, Hillier LW, Jiang Z, Baker C, Malfavon-Borja R, Fulton LA, et al. 2009. A burst of segmental duplications in the genome of the African great ape ancestor. *Nature* **457:** 877–881.

Mefford HC, Sharp AJ, Baker C, Itsara A, Jiang Z, Buysse K, Huang S, Maloney VK, Crolla JA, Baralle D, et al. 2008. Recurrent rearrangements of chromosome 1q21.1 and variable pediatric phenotypes. *N Engl J Med* **359:** 1685–1699.

Need AC, Ge D, Weale ME, Maia J, Feng S, Heinzen EL, Shianna KV, Yoon W, Kasperaviciute D, Gennarelli M, et al. 2009. A genome-wide investigation of SNPs and CNVs in schizophrenia. *PLoS. Genet* **5:** e1000373.

Ohno S. 1970. *Evolution by gene and genome duplication*. Springer, Berlin.

Popesco MC, Maclaren EJ, Hopkins J, Dumas L, Cox M, Meltesen L, McGavran L, Wyckoff GJ, Sikela JM. 2006. Human lineage-specific amplification, selection, and neuronal expression of DUF1220 domains. *Science* **313:** 1304–1307.

Rakic P. 1995. A small step for the cell, a giant leap for mankind: A hypothesis of neocortical expansion during evolution. *Trends Neurosci* **18:** 383–388.

Sharp AJ, Hansen S, Selzer RR, Cheng Z, Regan R, Hurst JA, Stewart H, Price SM, Blair E, Hennekam RC, et al. 2006. Discovery of previously unidentified genomic disorders from the duplication architecture of the human genome. *Nat Genet* **38:** 1038–1042.

Stefansson H, Rujescu D, Cichon S, Pietilainen OPH, Ingason A, Steinberg S, Fossdal R, Sigurdsson E, Sigmundsson T, Buizer-Voskamp JE, et al. 2008. Large recurrent microdeletions associated with schizophrenia. *Nature* **455:** 232–236.

Vandepoele K, Van Roy N, Staes K, Speleman F, van Roy F. 2005. A novel gene family NBPF: Intricate structure generated by gene duplications during primate evolution. *Mol Biol Evol* **22:** 2265–2274.

Walsh T, McClellan JM, McCarthy SE, Addington AM, Pierce SB, Cooper GM, Nord AS, Kusenda M, Malhotra D, Bhandari A, et al. 2008. Rare structural variants disrupt multiple genes in neurodevelopmental pathways in schizophrenia. *Science* **320:** 539–543.

Mitochondria, Bioenergetics, and the Epigenome in Eukaryotic and Human Evolution

D.C. WALLACE

*ORU for Molecular and Mitochondrial Medicine and Genetics,
University of California, Irvine, California 92697-3940*

Correspondence: dwallace@uci.edu

Studies on the origin of species have focused largely on anatomy, yet animal populations are generally limited by energy. Animals can adapt to available energy resources at three levels: (1) evolution of different anatomical forms between groups of animals through nuclear DNA (nDNA) mutations, permitting exploitation of alternative energy reservoirs and resulting in new species with novel niches, (2) evolution of different physiologies within intraspecific populations through mutations in mitochondrial DNA (mtDNA) and nDNA bioenergetic genes, permitting adjustment to energetic variation within a species' niche, and (3) epigenomic regulation of dispersed bioenergetic genes within an individual via mitochondrially generated high-energy intermediates, permitting individual adjustment to environmental fluctuations. Because medicine focuses on changes within our species, clinically relevant variation is more likely to involve changes in bioenergetics than anatomy. This may explain why mitochondrial diseases and epigenomic diseases frequently have similar phenotypes and why epigenomic diseases are being found to involve mitochondrial dysfunction. Therefore, common complex diseases may be the result of changes in any of a large number of mtDNA and nDNA bioenergetic genes or to altered epigenomic regulation of these bioenergetic genes. All of these changes result in similar bioenergetic failure and consequently related phenotypes.

BIOENERGETICS AND THE ORIGIN OF SPECIES

Darwin and Wallace (1858; Darwin 1859) proposed that natural selection acts on random variation in plants and animals to shape new species. Natural selection encompasses the environmental constraints acting on the organism, one of the most important being energy availability.

Life exists in a nonequilibrium thermodynamic state requiring the constant flow of energy to sustain its complex structures and to permit the accumulation and transmission of biological information. In the absence of energy flow, complex systems decay. Therefore, life exists through the interplay among structure, energy, and information.

The source of most biological energy is the Sun. The high-energy photons collected by photosynthetic cyanobacteria and their chloroplast descendants are used to split water into hydrogen and oxygen. The oxygen is released into the atmosphere and the hydrogen is used to reduce CO_2 to generate glucose. Plant glucose is consumed by herbivores and the energy therein sequentially passed through the animal and fungal food chains. Ultimately, the degraded energy is dissipated as infrared radiation into space.

Animal populations grow and multiply until energy becomes limiting. Animals can adapt to available energy resources at three levels: interspecific exploitation of different energy reservoirs that define the species' niche, intraspecific exploitation of differences in regional energy resources, and individual responses to oscillating environmental energy resources (Fig. 1).

Interspecific energy resource adaptation involves exploitation of distinct energy reservoirs through anatomical variation. This helps to delineate species and genera. Energy reservoirs are the distinct source of calories used by a species throughout its existence, perhaps over hundreds of thousands of years. Factors that can define an energy reservoir include being an herbivore versus a carnivore. For herbivores, these could involve eating fruits versus nuts or leaves versus sap. For carnivores, they could encompass being a predator versus a scavenger or consuming blood versus bugs, etc. To exploit alternative energy reservoirs, selection acts on nDNA genetic variation to alter animal anatomical traits, permitting access to novel energy reservoirs. This interaction between the species' anatomy and the energy reservoir helps to define the species' niche (Fig. 1).

Intraspecific energy resource adaptation involves the exploitation of regional energy environments through changes in physiology. Within a species' niche, regional energy resources can vary in energy type (carbohydrate, fat, protein), amount, and demand (activity level, stress, etc.). These regional differences can be stable over thousands of years and lead to development of distinct geographically constrained populations within the species. To exploit different regional energy environments, genetic variants in mtDNA and nDNA bioenergetic genes are selected that adjust the bioenergetic metabolism to be in balance with local energy resources (Fig. 2).

Individual energy resource adaptation involves adjustments within the individual to short-term oscillations in energy availability and demand. These changes can occur within days to years and include seasonal changes and

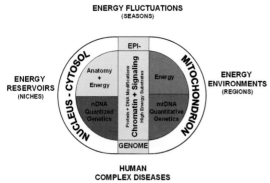

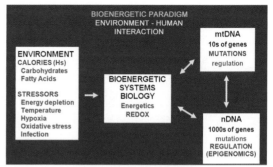

Figure 1. Three hypothesized levels of animal eukaryotic cell adaptation to varying energy resources. The original eukaryotic symbiosis brought together the glycolytic nucleus-cytosol with the oxidative mitochondrion. Most of the mitochondrial genome was then transferred to the nDNA, such that the current animal cell nucleus encodes the genes for determining cellular and organismal structure plus the genes for glycolysis and most of the genes for oxidative metabolism, all inherited according to the laws of Mendel. Maternally inherited mtDNA retains the core genes for generating, maintaining, and using the mitochondrial inner membrane potential, ΔP, which links the calories metabolized with cellular energy metabolism. The epigenome evolved to coordinate nDNA gene expression in relation to the availability of environmental calories. This is mediated by the modification of proteins and DNA elements of the epigenome via intracellular levels of the high-energy intermediates: ATP, acetyl-CoA, SAM, plus the redox status of the cell. When calories are abundant, the bioenergetic intermediates increase, chromatin is modified and decondensed, gene expression increases, and growth and reproduction are stimulated. When calories are limited, the reverse is true. Between species and higher taxa, mutations in the nDNA developmental genes change anatomy and permit the exploitation of different energy reservoirs. This creates species and defines niches. Within a species, mutations in the mtDNA change the cellular physiology to permit adaptation of regional populations to consistent regional differences in energy resources. Frequent fluctuations in energy resources of a species are addressed by changes in the epigenome that modulate the coordinate expression of *cis*- and *trans*-distributed nDNA bioenergetic genes. De novo mutations in nDNA bioenergetic genes, mutations and polymorphisms in mtDNA bioenergetic genes, and mutational or environmentally induced alterations in the epigenomic regulation of bioenergetic genes can all perturb bioenergetic homeostasis and contribute to the pathophysiology of common diseases, cancer, and aging.

Figure 2. Bioenergetic interface with the environment explains the importance of mtDNA and epigenomic variation of intraspecific animal adaptation. Energy availability and demand are the central factors in an animal's environment, the energy environment. Primary sources of available energy for omnivores, such as humans, are dietary calories generally obtained as carbohydrates and fats. Demands for calories include physical activity, thermal stress, hypoxia, oxidative stress, infection, body maintenance, and reproduction. Available calories are processed through cellular and mitochondrial bioenergetic pathways. The bioenergetic system is assembled from both mtDNA and nDNA genes. The mtDNA encodes core genes of OXPHOS. It has a very high mutation rate, resulting in the continual generation of functional variants, thus providing the genetic variation to permit animals to adapt to regional variation in the energetic environment. The nDNA encodes all of the genes for glycolysis, most of the genes for mitochondrial biogenesis and energy production, and all of the genes for the energetic- and redox-regulated signal transduction systems. These nDNA genes have a low mutation rate, within the time range for speciation. However, expression of the ~2000 nDNA-encoded energy genes is regulated by the production of high-energy intermediates by glycolysis and OXPHOS including ATP, acetyl-CoA, and SAM. These cellular bioenergetic substrates then drive the modification of the chromatin by phosphorylation, acetylation, and methylation, thus coordinating gene expression in relation to short-term fluctuations in the individual's energetic environment.

inflammation. Adjustment to these changes is accomplished through epigenomic modulation of nDNA and mtDNA bioenergetic gene expression (Fig. 2).

These three levels of energy adaptation strategy result from the fact that nDNA has a low mutation rate and encodes all of the genes for determining anatomy. Hence, changes in nDNA genes will define the species' structure and thus the energy reservoir that it can exploit. In contrast, the mtDNA has a very high mutation rate and encodes the central mitochondrial bioenergetic genes, augmented by hundreds of nDNA bioenergetic genes. Hence, mtDNA plus nDNA variation in bioenergetic genes has a dominant role in adapting to regional energy environments. Finally, the bioenergetic genes of mtDNA and nDNA must be coordinately regulated in response to environmental energetic oscillations. This is accomplished by modification of the epigenome, mediated by high-energy intermediates generated by the flow of calories through cellular bioenergetic systems (Figs. 1 and 2).

Interspecific variation leading to species, genera, families, orders, and classes is therefore the product of genetic variation in anatomical genes of nDNA. Alternatively, intraspecific variation resulting in population and individual adaptation to energy environments is dominated by mutations in bioenergetic genes, particularly those of the mtDNA, plus alterations in bioenergetic gene regulation through modifications of the epigenome (Fig. 1).

Since Darwin and Wallace proposed natural selection, considerable attention has been focused on changes in the genes and processes that dictate animal structures. However, little consideration has been given to the role of energy flow through the biosphere. Yet it is the flow of energy that drives the generation of complexity, the accumulation of information, and animal radiation. Indeed, it is precisely the intraspecific adaptation to regional energy resources that determines whether a population will survive, reproduce, and radiate.

The role of intraspecific energy adaptation is particularly pertinent for medicine, which is concerned with the health and well being of a single species, *Homo sapiens*. Therefore, genetic and epigenetic adaptations affecting energy metabolism are more likely to mediate environmental interactions than are changes in anatomy. Hence, the strong anatomical and nDNA focus of Western medicine may account for the reason that it has been so difficult to understand genetics and environmental interactions associated with common "complex diseases" that have a bioenergetic component to their etiology (Fig. 2).

BIOENERGETICS IN CELLULAR EVOLUTION

The duality of structure and energy for our cells became delineated with the symbiosis that created the original eukaryotic cell ~2 × 10⁹ years ago. This symbiosis is thought to have combined a glycolytic motile microorganism, the protonucleus-cytosol, and an oxidative α-protobacterium, the protomitochondrion. One impetus for this association was the increase in atmospheric oxygen generated by cyanobacteria. Oxygen provides the terminal electron acceptor for the efficient burning of the reducing equivalents (hydrogens) from calories consumed by animals.

For the first 1.2 × 10⁹ years after symbiosis, the two symbiotic organisms consolidated their metabolic pathways and exchanged genes, with natural selection enriching for more efficient forms. Ultimately, one genetic and metabolic combination proved to be sufficiently efficient to permit the advent of multicellularity. During the ensuing intersymbiont reorganization, most of the genes of the mitochondrial genome were transferred sequentially into nDNA. As a result, mitochondrial genes became randomly dispersed across all of the nuclear chromosomes (Wallace 2007).

In the eukaryotic cell that gave rise to multicellular organisms, all of the polypeptide genes for mitochondrial growth, reproduction, metabolism, and energy production came to reside in nDNA. These include genes for mitochondrial biogenesis such as the mtDNA polymerase γ, mtDNA RNA polymerase, the Twinkle helicase, ribosomal proteins, elongation factors, and tRNA synthetases; intermediate metabolism proteins including those for the tricarboxylic acid cycle (TCA), amino acid metabolism, folate metabolism, and nucleotide biogenesis; structural and assembly proteins including import complex proteins, chaperones, proteases; and ~80 mitochondrial OXPHOS proteins including all four of the complex II polypeptides and polypeptides for other dehydrogenases that feed electrons into the electron transport chain (ETC) (Wallace et al. 2007).

It has been estimated that mammalian nDNA encompasses ~1500 genes of the mitochondrial genome. These mitochondrial genes were added to the already existing nDNA anaerobic energy metabolism genes. Therefore, the genes involved in energy metabolism encoded by nDNA must number in the thousands.

By the time that the fungal–animal lineage was established, mtDNA retained only 13 polypeptide genes. Although small in number, these genes are by no means inconsequential because these are core polypeptides for the mitochondrial energy–generating system, oxidative phosphorylation (OXPHOS). mtDNA polypeptides include seven (ND1, 2, 3, 4L, 4, 5, 6) of the 45 subunits of complex I, one (cytochrome b, cytb) of the 11 subunits of complex III, three (COI, II, III) of the 13 subunits of complex IV, and two (ATP 6 and 8) of the ~17 subunits of complex V. Because these mtDNA polypeptides must be translated within the mitochondrion, the animal mtDNA also retains rRNAs and tRNA genes for mitochondrial protein synthesis (Wallace 2007).

The retention of these polypeptide genes by mtDNA is likely the product of their being central electron- and/or proton-carrying polypeptides for complexes I, III, IV, and V. These are the only OXPHOS complexes that transport protons. Consequently, they are critical for the generation, maintenance, and use of the mitochondrial inner membrane electrochemical gradient ($\Delta P = \Delta\psi + \Delta\mu^{H+}$). ΔP, in turn, is central to the conversion of dietary calories into ATP to perform work or generate heat to maintain body temperature (Wallace 2007).

ΔP is generated by the ETC through the oxidation of reducing equivalents recovered from carbohydrates and fats, the electrons passing sequentially through complexes I, III, and IV to reduce ½ O_2 into H_2O. The energy that is released is used by these three complexes to pump protons out of the mitochondrial matrix into the intermembrane space, generating ΔP. The energy stored in ΔP is then used by complex V to condense ADP + Pi to ATP, which is exported to the cytosol by adenine nucleotide translocators (ANTs). ΔP can also be used to import molecules and ions including Ca^{2+} into the mitochondrial matrix. If ΔP increases to its maximum, the ETC stalls and the electron carriers become saturated with electrons. In the presence of oxygen, electrons in complexes I and III can be donated directly to O_2 to give superoxide anion, the first of the reactive oxygen species (ROS). Mitochondrial ROS provides an important signal transduction system from the mitochondrion to the nucleus cytosol. However, excessive ROS can damage mitochondrial and cellular lipids, proteins, and DNA, ultimately resulting in cell death (Wallace et al. 2009).

The efficiency of complexes I, III, and IV at producing ΔP is determined by the ratio of the number of protons pumped out of the mitochondrial matrix relative to the number of electrons that move down the ETC. The efficiency of complex V, the ATP synthase, is defined by the number of protons that pass back into the matrix relative to ATP generated. Taken together, these two parameters make up the OXPHOS coupling efficiency. Tightly coupled OXPHOS will generate the maximum ATP per calorie burned, whereas a more loosely coupled mitochondria will require the oxidation of more calories to generate the same amount of ATP, the differential energy being dissipated as increased heat production. Therefore, the mitochondrial coupling efficiency determines the relative allocation of the calories consumed to generate ATP versus heat produced.

Because all mitochondrial energy production is contingent on ΔP, proton permeability of complexes I, III, IV, and V must be balanced to ensure that one complex is not more permeable to protons than the others. Otherwise, the leaky complex will short circuit the capacitor (ΔP) and negate the coupling efficiency of the other complexes. Having the core proton and electron carrier protein genes linked together in the mtDNA ensures that the genes of the complexes will be selected as a unit and thus work optimally with one another (Wallace 2007).

Assurance that complex I, III, IV, and V mtDNA polypeptides will coevolve is achieved by maternal inheritance of the mtDNA. Exclusive maternal inheritance prohibits mixing within the same cell of the mtDNAs from two individuals with different coupling efficiencies, thus blocking inter-mtDNA recombination. Therefore, the only way that mtDNA genes can change is by the sequential accumulation of mutations along radiating maternal lineages. Each new mutation is then tested by selection against the background of the previously existing mtDNA genetic variants (Wallace 2007).

Because each cell has hundreds to thousands of copies of mtDNA, when a new mtDNA mutation arises within an oocyte or cell, it generates an intracellular mixture of mutant and normal mtDNAs, a state known as heteroplasmy. Because mtDNAs are distributed randomly between daughter cells during mitotic division, the percentage of mutant and normal mtDNAs can drift during mitosis and meiosis, the process of replicative segregation. Thus, neutral, beneficial, or mildly deleterious mtDNA mutations can become fixed within a maternal lineage through intracellular genetic drift. However, as the percentage of deleterious mutations increases, the energy output of the cell declines until it drops below the minimum energy output required for that cell type to function, the bioenergetic threshold, and symptoms ensue (Wallace 2005, 2007).

BIOENERGETICS IN HUMAN ORIGINS AND DISEASE

The most important functions of the mitochondrion for animal cellular and tissue physiology and thus environmental adaptation and health are its (1) production of most of the cellular energy in the form of ATP and heat, (2) generation of much of the endogenous ROS, (3) uptake of cytosolic Ca^{2+} thus maintaining Ca^{2+} homeostasis, and (4) regulation of cell death through activation of the mitochondrial permeability transition pore (mtPTP). The mtPTP monitors mitochondrial ΔP, adenine nucleotides, ROS, and Ca^{2+} levels, and when energy production is too low or ROS production and matrix Ca^{2+} levels are too high, mtPTP opens a channel through the mitochondrial membrane, shorting ΔP. This causes the release of proapoptotic proteins from the mitochondrial intermembrane space into the cytosol, initiating cell death. Thus, mitochondrially mediated apoptosis removes energetically impaired cells from the tissue, eliminating their disruption of tissue function (Wallace 2005).

All four of these critical physiological functions of the mitochondrion can be modulated by variation in mitochondrial genes, both nDNA and mtDNA (Fig. 3). Phenotypi-

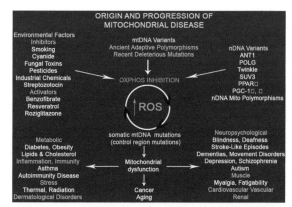

Figure 3. Classes of human mitochondrial gene mutations in the origin of metabolic and degenerative diseases, cancer, and aging. The "mitochondrial genome" encompasses ~1500 nDNA genes dispersed across the chromosomes plus 37 critical energetic genes within mtDNA. Genetic variation in any of these mitochondrial genes may perturb the mitochondrial OXPHOS. An array of common environmental agents and pharmacological agents can also modulate mitochondrial bioenergetics and/or biogenesis. Inhibition of OXPHOS can increase mitochondrial ROS production, which will damage mtDNA, gradually erode the cellular capacity to generate energy, and create the clock central to aging and adult cancers. OXPHOS dysfunction will have the greatest effect on tissues having the highest energy demand (brain, heart, skeletal muscle, kidney, endocrine system) to cause degenerative diseases. Altered mitochondrial energy production will also perturb caloric sensing and use, resulting in common metabolic diseases such as diabetes and obesity. Finally, altered mitochondrial ROS production and redox biology will precipitate inflammatory disease and change mitochondrial coupling efficiency to affect thermal modulation and sensitivity to radiation-induced cellular toxicity.

cally relevant nDNA-encoded mitochondrial gene variants alter the structural and assembly genes for building functional mitochondrial OXPHOS complexes. Mutations in the heart muscle ANT (ANT1) result in myopathy and cardiomyopathy (Palmieri et al. 2005; Wallace et al. 2007). Functional polymorphisms in nDNA genes can alter transcriptional and translational systems of nDNA or mtDNA mitochondrial biogenesis. For example, polymorphisms in peroxisome proliferation-activated receptor γ (*PPARγ*) (Altshuler et al. 2000) and *PGC-1α* (Ek et al. 2001; Muller et al. 2003) genes have been associated with regional predilection to diabetes. Finally, mutations in nDNA-encoded genes involved in mitochondrial biogenesis, such as mtDNA polymerase γ (POLG) and the Twinkle helicase, can inhibit mtDNA replication resulting in multiple deletions and/or depletion of mtDNA (Wallace et al. 2007).

Phenotypically relevant inherited mtDNA sequence variants include *recent deleterious mutations* and *ancient adaptive polymorphisms*. Deleterious mtDNA mutations can cause an array of human degenerative and metabolic diseases, primarily affecting the more oxidative tissues: brain, heart, muscle, kidney, and endocrine system. Adaptive mutations permitted ancestral human populations to adapt to different energetic environments as they migrated throughout the world. Although an adaptive mutation can be beneficial in one energetic environment, it can be deleterious in another. Hence, population-specific variants have been found to be risk factors for a variety of clinical problems (Khusnutdinova et al. 2008; Wallace 2008). The mitochondrial energy production system is also specifically inhibited by a wide range of environmental agents including cyanide, MPTP, and rotenone. Therefore, the mitochondria provide a direct link among the environment, cellular physiology, and genes (Fig. 3).

Because mtDNAs within the cell are continuously turning over, mtDNA mutations accumulate in tissue cells. An important source of mtDNA mutations is ROS damage. The resulting *somatic mtDNA mutations* accumulate with age in postmitotic tissues, eroding cellular energetics, exacerbating inherited mitochondrial defects, and producing the "aging clock." The accumulation of these somatic mutations may explain the delayed onset and progressive course of multiple age-related diseases and certain forms of cancer (Fig. 3) (Wallace 2005).

Because mitochondria provide a direct link between the energetic environment and the bioenergetics of an individual's cells and tissues, they must also be central to diseases influenced by environmental variation. Accordingly, mitochondrial dysfunction has been associated with a wide range of common metabolic and generative disease symptoms (Wallace 2005; Wallace et al. 2007).

That mitochondrial dysfunction is sufficient to cause disease has been demonstrated by the introduction of mtDNA mutations into the mouse. Mice harboring mutant mtDNAs developed the same symptoms as observed in common age-related diseases (Fan et al. 2008; Wallace and Fan 2009b). Thus, mitochondrial dysfunction must have a central role in the pathophysiology of metabolic and degenerative diseases, cancer, and aging (Wallace 2005).

The discovery that human mtDNAs can harbor adaptive mutations offers a novel perspective on how species adapt to the regional energetic environment. Unlike nDNA variation, mtDNA variation correlates strongly with indigenous populations and their geographic location. This is the result of both maternal inheritance and energetic environmental selection. Mutations in mtDNA accumulated sequentially on radiating maternal lineages as women migrated out of Africa to colonize Eurasia and the Americas. Those mutations that resulted in an energy metabolism better suited for the new energy environment became enriched, creating region-specific branches of the mtDNA tree. These branches represent clusters of related mtDNA haplotypes constituting a haplogroup. Haplogroups are frequently founded by one or more adaptive mutations. Moreover, the same adaptive mutation can be observed on different mtDNA backgrounds, indicating that the mutation has arisen several independent times and in each case has been enriched in the population by natural selection. This convergent evolution provides direct evidence that mtDNA mutations can be adaptive.

The greatest degree of mtDNA diversity resides in Africa and constitutes macrohaplogroup L. This high African mtDNA diversity demonstrates an African origin for the mtDNA tree (Johnson et al. 1983; Cann et al. 1987; Merriwether et al. 1991). However, of all African mtDNA variation, only two mtDNAs successfully left Africa and colonized all of Eurasia, founding macrohaplogroups M and N (Wallace et al. 1999; Mishmar et al. 2003; Ruiz-Pesini et al. 2004, 2007). The mtDNA that founded the macrohaplogroup N lineage harbored two missense mutations, ND3 10398 A114T and ATP6 8701 A59T, which have been shown to alter several mitochondrial physiological parameters (Kazuno et al. 2006). From Africa, the N lineage radiated into Europe, generating the European-specific lineages H, I, J, Uk, T, U, V, W, and X. Both the M and N mtDNA founders radiated into Asia, generating a plethora of mtDNA lineages. Of these, A, C, and D became enriched in northeastern Siberia and were in position to cross the Bering land bridge to colonize the Americas (Wallace et al. 1999; Ruiz-Pesini et al. 2007).

The ROS-induced high mtDNA mutation rate rapidly produces new variants within a population that can be acted on by selection. Alterations in mtDNA genes can affect OXPHOS coupling efficiency, changing the allocation of calories between ATP versus heat, thus optimizing energy metabolism to warmer versus colder climates. Other mtDNA alterations can increase or decrease mitochondrial ROS production at particular caloric loads, modulate Ca^{2+} buffering, or regulate the rate of apoptosis (Wallace 1994; Mishmar et al. 2003; Ruiz-Pesini et al. 2004; Ruiz-Pesini and Wallace 2006). The contribution of mtDNA variation to climate adaptation has been supported by demonstrating that mtDNA variation but not nDNA variation correlates with regional temperature extremes (Balloux et al. 2009). Therefore, mtDNA variation is of central importance in animal and human adaptation to the changing energy environments.

The mtDNA functional mutations that have permitted human populations to adapt to regional environmental differences have accumulated over thousands of years. These functional changes in mtDNA polypeptide and structural RNA genes impart a shift in the range of environmental conditions that the individual can tolerate, in effect defining the outer limits of individual environmental tolerance. The changes in mtDNA structural genes can be further modulated by mutations in the mtDNA regulatory control region, which can alter both mtDNA transcription levels and copy number (Suissa et al. 2009). Thus, control region variation may permit the fine-tuning of an individual's mitochondrial energetics (Fig. 3).

BIOENERGETIC ORIGIN OF THE EPIGENOME

Because nDNA has a much lower mutation rate than mtDNA, the thousands of nDNA energy genes must adjust to changes in the energetic environment by alterations in gene expression, not mutation (Fig. 2).

Calories are the limiting factor for cellular and organismal growth and reproduction, and thus the expression of nDNA genes for energy metabolism and cell growth must be modulated according to the availability of calories. For unicellular organisms, a direct link is required among calories, growth, and gene expression. For multicellular organisms, with specialized organs and tissues, this more primitive connection between calories and gene expression can be overlaid with other genetic programs that regulate gene expression and growth within the context of specific tissue requirements (Wallace and Fan 2009a).

In humans, calories can come from either carbohydrates or fatty acids and ketone bodies (acetoacetate and β-hydroxybutyrate). Carbohydrates such as glucose are initially processed through cytosolic glycolysis to generate pyruvate, ATP, and reduced NAD^+, $NADH + H^+$. The pyruvate then enters the mitochondrion where it is converted to acetyl-CoA. Acetyl-CoA condenses with oxaloacetate (OAA) to give citrate that fuels the tricarboxylic acid cycle (TCA). In contrast, fatty acids and ketone bodies are transferred directly into the mitochondrion where they are converted to acetyl-CoA and intramitochondrial reducing equivalents in the form of $NADH + H^+$ and $FADH_2$ (Fig. 4) (Wallace and Fan 2009a; Wallace et al. 2009).

At the level of the unicellular eukaryote, the connection between calorie availability and gene expression is mediated through high-energy intermediates of glycolysis and OXPHOS, because these intermediates can reflect caloric availability for the cell. The high-energy intermediates used are ATP, acetyl-CoA, and S-adenosylmethionine (SAM). These three bioenergetic intermediates are the primary substrates for modifying and modulating the epigenome (Wallace and Fan 2009a).

nDNA is packaged in nucleosomes that encompass ~146–147 nucleotides of DNA wrapped around two molecules each of histones H2A, H2B, H3, and H4. Each of these histones has an amino-terminal tail that is basic and thus has high affinity for the negatively charged sugar-phosphate backbone of DNA. When these amino-terminal tails are unmodified, they bind to the acidic DNA, stabilize its packaging, and shut down transcription. However, when the tails are modified by phosphorylation via kinases using ATP, acetylation via histone acetylases (HATs) using acetyl-CoA, and methylation via methyltransferases using SAM, these modifications increase the negative charge, decrease the positive charge, and alter the molecular interactions of histone tails with the negatively charged DNA, thus opening the chromatin and increasing transcription (Figs. 3 and 4) (Wallace and Fan 2009a).

ATP can be generated by both glycolysis and OXPHOS. However, acetyl-CoA is generated predominantly within

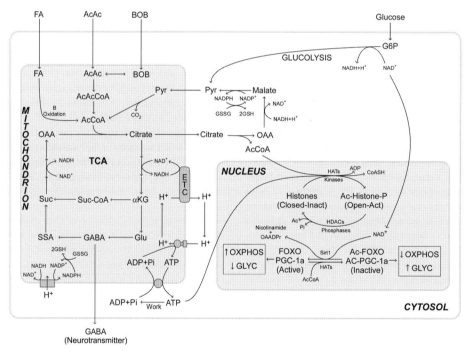

Figure 4. Mitochondrial bioenergetic coupling with the epigenome. The flow of reducing equivalents (calories) through glycolysis and mitochondrial oxidation regulates levels of cellular ATP and acetyl-CoA. High-calorie intake increases ATP and acetyl-CoA levels, thereby increasing histone phosphorylation and acetylation. This opens the chromatin to stimulate transcription, growth, and replication. Acetylation also regulates major signal transduction pathways, including the insulin signaling pathway. Forkhead box class O (FOXO) transcription factors that regulate expression of PGC-1α, the key transcription factor to regulate mitochondrial OXPHOS and biogenesis, can be acetylated and inactivated. During carbohydrate metabolism by glycolysis (Glyc), the cytosolic $NADH/NAD^+$ ratio increases, thus limiting NAD^+ availability for Sirt1-mediated FOXO and PGC-1α deacetylation. This inhibits FOXOs and PGC-1α, which down-regulate mitochondrial OXPHOS and biogenesis. Fatty acid and ketone body oxidation within the mitochondrion leaves the cytosolic $NADH/NAD^+$ more oxidized, thereby stimulating Sirt1-mediated deacetylation and activation of FOXOs and PGC-1α, causing up-regulation of mitochondrial OXPHOS. Thus, the availability and nature of calories directly regulates the epigenome and modulates the bioenergetic pathways required for optimal caloric exploitation. (AcCoA) Acetyl-CoA, (AcAc) acetylaldehyde, (BOB) β-hydroxybutyrate, (OAA) oxaloacetate, (Pyr) pyruvate, (suc) succinate, (Suc-CoA) succinyl-CoA, (SSA) succinate semialdehyde, (GABA) γ-aminobutyric acid, (αKG) α-ketoglutarate, (G6P) glucose-6-phosphate, (HATs) histoacetyltransferases, (HDAC) histone deacetylase, (GSSG and GSH) oxidized and reduced glutathione.

the mitochondrion. After condensation with OAA, it is exported to the cytosol within citrate. Once in the cytosol, citrate can be cleaved back to acetyl-CoA and OAA by ATP-citrate lyase (Wallace and Fan 2009a).

SAM is generated from methyl groups recovered from the export of serine from the mitochondrion, the methyl group being transferred to folate by serine hydroxymethyltransferase and then transferred to homocysteine to generate methionine. This methionine is subsequently converted to SAM by condensation with ATP. In addition to modifying histones, SAM is also used to add methyl groups to the cytosines of CpG dinucleotides, modulating DNA–protein interactions (Wallace and Fan 2009a).

When calories are abundant, cytosolic and nuclear ATP, acetyl-CoA, and SAM therefore increase. This drives the modification of the histone tails and opens the chromatin, thus turning on transcription and stimulating growth and reproduction. When calories become limiting, ATP, acetyl-CoA, and SAM decline and phosphatases remove the phosphate groups, histone deacetylases (HDACs) remove the acetyl groups, and demethylases remove the methyl groups. The chromatin then condenses, shutting down transcription until calories are once again available. Thus, the major factor in the evolution of the epigenome must have been the need to provide a global mechanism for coordinating nDNA gene expression in response to energy availability, mediated through cellular energy metabolism (Wallace and Fan 2009a).

Pangenetic transcriptional regulation is particularly pertinent for the regulation of the thousands of dispersed nDNA energy-generating genes. These genes need to be up-regulated when calories are plentiful and down-regulated when calories are limited. However, their scattered distribution would require both regulation in *cis* for genes relatively close to one another on the same chromosome and in *trans* for genes that are dispersed on different chromosomes. The former could have provided the impetus for the formation of transcriptional loops and the latter for the development of nDNA transcriptional islands, where genes from multiple chromosomes come together for coordinate gene transcription (Wallace and Fan 2009a). Once established in unicellular eukaryotes, these systems could be generalized for the coordinate regulation of other interrelated genes.

Although changes in chromatin could provide an effective means of linking caloric availability to gene expression and growth, this approach does not discriminate between type of calorie, carbohydrate versus fat, and ketone bodies. In multicellular animals, switching reliance between glycolysis and OXPHOS must be controlled both systemically and intracellularly.

Systemic control is achieved through hormonal secretion by pancreatic α and β cells. When carbohydrates are abundant, insulin is released from mammalian β cells. Insulin binds to insulin receptors on target cells, activating the phosphoinositol-3 kinase (PI3K) pathway. This activates Akt/PKB (protein kinase B) to phosphorylate and inactivate FOXO (Forkhead box class O) transcription factors. When active, FOXOs bind to insulin response elements (IREs) of DNA, some of which are upstream of the transcriptional coactivator PPARγ–coactivator-1α (PGC-1α). PGC-1α, in turn, binds to a range of transcription factors in various tissues including PPARγ and up-regulates mitochondrial biogenesis and OXPHOS. Therefore, when carbohydrates are abundant, FOXOs are inactive, PGC-1α is down-regulated, OXPHOS declines, and the animal relies increasingly on glycolysis for ATP production. The excess dietary calories are then stored as fat (Wallace 2005, 2007).

During starvation, when carbohydrates are limited, the animal mobilizes its stored fat and up-regulates mitochondrial OXPHOS, because only OXPHOS can burn fat to generate ATP. In the absence of carbohydrates, insulin production declines, inactivating Akt/PKB, and resulting in the dephosphorylation of FOXOs. Activated FOXOs then induce PGC-1α and this up-regulates OXPHOS. Concurrently, the mammalian pancreatic α cells sense the low glucose and respond by secreting glycagon. This hormone binds to the glucagon receptors of target cells, activating adenylylcyclase. The increased cAMP activates PKA (protein kinase A) to phosphorylate and activates CREB (cAMP response element binding). Phospho-CREB then enters the nucleus and binds to cAMP response elements (CRE), one of which is also upstream of PGC-1α. Therefore, during fasting and starvation, PGC-1α and thus OXPHOS is doubly induced by dephosphorylated FOXOs and phosphorylation of CREB, thus shifting metabolism toward OXPHOS to burn stored fat (Wallace 2005, 2007).

This systemic regulation is complemented and extended by regulation of the activity of FOXOs and PGC-1α transcription factors. FOXOs and PGC-1α can be acetylated by HATs using acetyl-CoA and deacetylated by the class III deacetylase Sirt1. Sirt1 requires NAD^+ as a coreactant for deacetylation; it cannot use NADH. When carbohydrate or fat calories are abundant, acetyl-CoA is abundant and FOXOs and PGC-1α are acetylated. When carbohydrates are metabolized by glycolysis, cytosolic NAD^+ is reduced to NADH + H^+, rendering Sirt1 inactive. As a result, FOXOs and PGC-1α stay acetylated, the transcription and activity of PGC-1α remains suppressed, OXPHOS is inhibited, and glycolysis is favored. In contrast, when fatty acids or ketones are metabolized, cytosolic NAD^+ remains oxidized, Sirt1 is activated to deacetylate FOXOs and PGC-1α, OXPHOS is up-regulated, and fats and ketones are oxidized (Fig. 4) (Wallace and Fan 2009a; Wallace et al. 2009).

BIOENERGETICS AND EPIGENOMIC DISEASE

Although mtDNA variation determines the outer limits of the environmental challenge that an individual can tolerate, animals are also subjected to daily, monthly, and annual changes in their energetic environment (e.g., available calories, temperature, and activity). These short-term changes in energy availability and demand must be accommodated by changes in the expression of nDNA-encoded energetic genes (Fig. 2).

It has been proposed that the epigenome provides the interface between the environment and nDNA gene expression (Feinberg 2007, 2008). Yet the mechanism by which the epigenome perceives the environment has not been specified. The fact that the epigenome is modulated by high-energy intermediates (ATP, acetyl-CoA, and SAM) immediately suggests that the primary interface between the environment and the epigenome is the cellular bioenergetic system (Fig. 2). Not only would the availability and demand for calories require coordinate changes in the nDNA genes for energy metabolism, but changes in the energetic status of the cell and individual would also determine which developmental pathways would be optimal for growth, differentiation, and death. Therefore, the primary interface between the environment and the individual must be the cytosolic and mitochondrial bioenergetic systems (Fig. 2) (Wallace and Fan 2009a; Wallace et al. 2009).

The role of the epigenome in regulating bioenergetics is supported by the observations that (1) pathogenic mutations in the mtDNA result in symptoms that overlap with those attributed to the epigenome disease and (2) several epigenomic diseases have been found to be associated with mitochondrial dysfunction. This leads to a bioenergetic hypothesis for common diseases, which is based on six key tenets:

1. Energy is derived from the calories in our diet.
2. Energy is required by every organ, but in different ways and to different extents.
3. Hundreds of critical bioenergetic genes are dispersed throughout mtDNA and nDNA.
4. The expression of bioenergetic genes is coordinated in relation to calories via modification of the epigenome.
5. High-energy intermediates generated by cellular bioenergetic systems mediate changes in the epigenome.
6. Defects in either bioenergetic genes or the epigenome can disrupt energy flow and cause multisystem disease.

"Epigenomic diseases" affect imprinting, methylation, and chromatin organization (Feinberg 2007, 2008) and alter transcription regulatory proteins such as PPARγ and PGC-1α (Fig. 5). Thus, the epigenome must be regulating bioenergetic gene expression in the *cis* configuration for adjacent bioenergetic genes within individual chromatin loop domains and in the *trans* configuration for genetic elements on multiple chromosomes, possibly through transcriptional islands.

Imprinting diseases are examples of *cis*-acting diseases affecting transcriptional chromatin domains. Alterations in DNA methylation and methylation-binding proteins or in overall chromatin organization are examples of *trans*-acting factors (Fig. 5).

The perturbation of imprinting seen in Angelman and Prader-Willi syndromes involves inactivation of the active allele on chromosome 15q11-13 in the context of an imprinted and inactive maternal or paternal allele. In a mouse model of Angelman syndrome, hippocampal neurons were found to have reduced synaptic vesicle density and shrunken mitochondria in association with a reduction of brain mitochondrial complex II + III (Su et al. 2009). In Beckwith-Wiedemann syndrome and Wilm's tumor, there

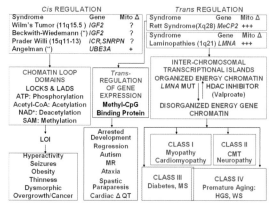

Figure 5. Epigenomic disease association with mitochondrial dysfunction. (*Left side*) Diseases associated with loss of imprinting (LOI) potentially alter chromatin loop domain structure and cause aberrant *cis* interactions of bioenergetic genes. Chromatin domains affected by LOI can correlate with LOCHs (large organized chromatin K9 modifications blocks [Wen et al. 2009]) and LADs (lamina-associated domains [Guelen et al. 2008]). (*Right side*) Diseases associated with aberrations in the interaction of chromatin domains on different chromosomes or chromosomal domains resulting in aberrant *trans* interactions on bioenergetic genes. (Mito Δ) Extent of evidence for mitochondrial dysfunction associated with epigenetic disease. (IGF2) insulin-like growth factor 2 gene, (SNRPN) small nuclear ribonuclear polypeptide N, (ICR) imprinting control region, (UBE3A) ubiquitin-protein ligase E3A, (MeCP2) methyl C binding protein 2, (LMNA) lamin A gene, (MUT) mutant, (HDAC) histone deacetylase.

is loss of imprinting (LOI) on chromosome 11q15.5 within a chromatin loop domain encompassing the insulin-like growth factor 2 (*IGF2*) gene (Bjornsson et al. 2007; Feinberg 2007, 2008). *IGF2* could act through the PI3K-Akt-FOXO pathway to modulate energy metabolism (Kaneda et al. 2007; Wallace and Fan 2009a).

Alterations in *trans*-acting transcription regulatory factors affecting mitochondrial function are seen in Rett syndrome and the laminopathies. Rett syndrome is caused by mutations in methyl-CpG-binding protein 2 (*MeCP2*) (Loat et al. 2008). Yet multiple studies have reported abnormal mitochondria and mitochondrial function in Rett patients (Eeg-Olofsson et al. 1989; Heilstedt et al. 2002; Kriaucionis et al. 2006; Wallace and Fan 2009a). The laminopathies are caused by mutations in the laminin A/C gene (*LMNA*), with different mutations encompassing most of the clinical phenotypes observed in mitochondrial diseases (Stewart et al. 2007; Liu and Zhou 2008). Class I laminopathies result in myopathy and cardiomyopathy. Class II laminopathies result in the peripheral neuropathy Charcot-Marie-Tooth. Class III laminopathies result in diabetes and metabolic syndrome, and Class IV laminopathies result in premature aging syndromes including Hutchinson-Guilford progeria and Werner syndrome. That these phenotypes are generated by bioenergetic dysfunction is supported by the report that cells harboring *LMNA* mutations exhibit mitochondrial biochemical defects (Caron et al. 2007; Wallace and Fan 2009a).

Data on "epigenomic diseases," although limited, therefore support their association with mitochondrial dysfunction. If confirmed, then epigenomic and mitochondrial diseases share a common pathophysiology, explaining their similar clinical phenotypes. This would establish the central role of energy metabolism in modulating the epigenome, thus providing the direct link among the nucleus-cytosol, mitochondrion, and environment (Wallace and Fan 2009a).

BIOENERGETIC PARADIGM IN ANIMAL EVOLUTION AND HUMAN DISEASE

The bioenergetic paradigm now provides a framework within which to link animal evolution and human disease (see Figs. 1 and 2). Nuclear gene mutations, both anatomical and bioenergetic, permit speciation. The mutation rate of nDNA is low. Hence, it may take hundreds of thousands of years for sufficient mild nDNA mutations to accumulate in anatomical genes to result in a shift in animal anatomy to permit exploitation of a new energy reservoir and, thus, speciation. For species in established niches, more severe anatomical mutations will render the individual incompatible with its niche, resulting in elimination by selection. In humans, such mutations cause classical Mendelian diseases such as osteogenesis imperfecta. Relatively mild nDNA mutations in bioenergetic genes can be compatible with survival within a niche because there is some latitude for differences in energy efficiency within a species. Functional variants in important energy regulatory genes such as *PPARγ* and *PGC-1α* can reach polymorphic frequencies in regional populations. These variants may be beneficial in one local energy environment. However, when found outside the appropriate environment, they become risk factors for disease. Thus, PPARγ (Altshuler et al. 2000) and PGC-1α (Ek et al. 2001; Muller et al. 2003) polypeptide polymorphisms show very high correlations with diabetes in certain regional populations, but no association with that phenotype in large interpopulation studies (Saxena et al. 2007; Scott et al. 2007; Sladek et al. 2007; Zeggini et al. 2007). More significant bioenergetic variants might also permit exploitation of adjacent energy reservoirs, ultimately leading to speciation. For example, changes in certain nDNA-encoded complex I genes have been associated with branches in primate radiation (Mishmar et al. 2006).

In contrast to nDNA, mtDNA mutations occur at a very high rate. Because mtDNA encodes only genes related to OXPHOS, mtDNA mutations primarily alter bioenergetics. Because mtDNA genes are vital and highly conversed, most functional mtDNA mutations are deleterious. However, the most severe mtDNA mutations are eliminated by intraovarian selection before ovulation and thus do not contribute to the genetic load of the population (Fan et al. 2008; Stewart et al. 2008). This explains why there are relatively few common pathogenic mtDNA mutations (Wallace et al. 2007). As a consequence of intraovarian selection, most mtDNA mutations are mildly deleterious, neutral, or beneficial variants. Still, the ovarian threshold for functional defects is relatively high, ensuring that enough mildly deleterious mutations are introduced into the population that at least some variants will be available to permit adaptation to sustained energy environmental change.

Because of the lack of recombination, beneficial mtDNA mutations must be continually regenerated through de novo mutation. There are a finite number of physiologically beneficial mtDNA-coding gene changes, and thus the same mutations are observed repeatedly in different regional populations on different mtDNA backgrounds. This convergent evolution confirms the beneficial nature of these variants (Wallace et al. 2003; Ruiz-Pesini et al. 2004; Ruiz-Pesini and Wallace 2006). Because beneficial coding region mutations are relatively rare, mtDNA bioenergetic flexibility is expanded through mtDNA control region mutations. Because control region mutations can alter mtDNA transcription and copy number, they can have a significant effect on mitochondrial function without disrupting bioenergetic pathways. Consequently, polymorphic mtDNA control region variants are much more common than coding region variants (Suissa et al. 2009). Therefore, the mtDNA provides a powerful system for bioenergetic adaptation to environmental change. However, the negative consequence of this system is that both recent deleterious and ancient polymorphic mtDNA variants contribute substantially to the etiology of common metabolic and degenerative diseases.

Changes in the epigenome permit modulation of bioenergetics in response to short-term fluctuations in the energy environment. This results from alterations in gene expression mediated by the modification of histones, signal transduction proteins, and DNA, driven by the availability of high-energy intermediates (Wallace and Fan 2009a) and the cellular redox potential (Wallace et al. 2009). These cellular metabolic intermediates and parameters, in turn, reflect the flux of reducing equivalents through mitochondrial and glycolytic bioenergetic pathways of the cell. Thus, environmental conditions are linked to coordinate gene expression through bioenergetics. Daily and weekly changes can be accommodated by activation of signal transduction systems (e.g., insulin and glucagon) or the expression and activity of transcription factors that regulate bioenergetics (e.g., FOXOs and PGC-1α) (Wallace 2005, 2007). More long-term regulation can occur through *cis*-acting processes, as is seen for imprinted loci associated with chromatic loop domains (e.g., Angelman, Prader-Willi, and Beckwith-Weidemann syndromes). Longer-term regulation can also be mediated by *trans*-acting factors, as is observed for MeCP2 protein and perhaps lamin A. DNA methylation is another way that bioenergetic gene expression could be modulated over prolonged periods, and DNA methylation patterns can be altered in respiratory-deficient cancer cells (Feinberg 2007, 2008; Smiraglia et al. 2008). Genetic diseases affecting *cis*-acting imprinted loci can involve either inactivation or activation of a functional chromosomal allele and have been documented to be associated with mitochondrial dysfunction (Wallace and Fan 2009a). Similarly, genetic diseases affecting *trans*-acting systems involving DNA-binding factors or chromatin structural elements have also been associated with mitochondrial dysfunction.

Perturbation of cellular bioenergetics is therefore a common feature of metabolic and degenerative diseases. Bioenergetic defects can be caused by an array of genetic and epigenetic perturbations, including mutations in nDNA and mtDNA-encoded bioenergetic genes and in the epigenetic regulation of bioenergetic genes. Consequently, the complexity of "complex diseases" is the product of the large number of ways that bioenergetic genes can be perturbed. However, the pathophysiology of common diseases is the same, bioenergetic dysfunction, which explains why all of these diseases affect the more metabolically active tissues: brain, heart, muscle, renal, endocrine, and hepatic systems. Although this analysis confirms the complexity of the factors that can cause common diseases, it implies that the clinical symptoms of a wide range of common diseases might be treated by addressing the same target, the bioenergetic systems of the cell.

ACKNOWLEDGMENTS

The author thanks Dr. Weiwei Fan, Ms. Marie T. Lott, and Kate Hartshorn for their assistance in assembling this document. The work has been supported by National Institutes of Health grants NS21328, AG24373, DK73691, AG13154, AG16573, a CIRM comprehensive grant RC1-00353-1, and a Doris Duke Clinical Interfaces award 2005 to D.C.W.

REFERENCES

Altshuler D, Hirschhorn JN, Klannemark M, Lindgren CM, Vohl MC, Nemesh J, Lane CR, Schaffner SF, Bolk S, Brewer C, et al. 2000. The common PPARγ Pro12Ala polymorphism is associated with decreased risk of type 2 diabetes. *Nat Genet* **26:** 76–80.

Balloux F, Handley LJ, Jombart T, Liu H, Manica A. 2009. Climate shaped the worldwide distribution of human mitochondrial DNA sequence variation. *Proc Biol Sci* **276:** 3447–3455.

Bjornsson HT, Brown LJ, Fallin MD, Rongione MA, Bibikova M, Wickham E, Fan JB, Feinberg AP. 2007. Epigenetic specificity of loss of imprinting of the *IGF2* gene in Wilms tumors. *J Natl Cancer Inst* **99:** 1270–1273.

Cann RL, Stoneking M, Wilson AC. 1987. Mitochondrial DNA and human evolution. *Nature* **325:** 31–36.

Caron M, Auclair M, Donadille B, Bereziat V, Guerci B, Laville M, Narbonne H, Bodemer C, Lascols O, Capeau J, Vigouroux C. 2007. Human lipodystrophies linked to mutations in A-type lamins and to HIV protease inhibitor therapy are both associated with prelamin A accumulation, oxidative stress and premature cellular senescence. *Cell Death Differ* **14:** 1759–1767.

Darwin C. 1859. *On the origin of species by means of natural selection (or the preservation of favoured races in the struggle for life)*. Murray, London.

Darwin CR, Wallace AR. 1858. On the tendency of species to form varieties; and on the perpetuation of varieties and species by natural means of selection. *J Proc Linn Soc Lond Zool* **3:** 46–50.

Eeg-Olofsson O, al-Zuhair AG, Teebi AS, al-Essa MM. 1989. Rett syndrome: Genetic clues based on mitochondrial changes in muscle. *Am J Med Genet* **32:** 142–144.

Ek J, Andersen G, Urhammer SA, Gaede PH, Drivsholm T, Borch-Johnsen K, Hansen T, Pedersen O. 2001. Mutation analysis of peroxisome proliferator-activated receptor-γ coactivator-1 (PGC-1) and relationships of identified amino acid polymorphisms to type II diabetes mellitus. *Diabetologia* **44:** 2220–2226.

Fan W, Waymire K, Narula N, Li P, Rocher C, Coskun PE, Vannan MA, Narula J, MacGregor GR, Wallace DC. 2008. A mouse model of mitochondrial disease reveals germline selection against severe mtDNA mutations. *Science* **319:** 958–962.

Feinberg AP. 2007. Phenotypic plasticity and the epigenetics of human disease. *Nature* **447:** 433–440.

Feinberg AP. 2008. Epigenetics at the epicenter of modern medicine. *J Am Med Assoc* **299:** 1345–1350.

Guelen L, Pagie L, Brasset E, Meuleman W, Faza MB, Talhout W, Eussen BH, de Klein A, Wessels L, de Laat W, van Steensel B. 2008. Domain organization of human chromosomes revealed by mapping of nuclear lamina interactions. *Nature* **453:** 948–951.

Heilstedt HA, Shahbazian MD, Lee B. 2002. Infantile hypotonia as a presentation of Rett syndrome. *Am J Med Genet* **111:** 238–242.

Johnson MJ, Wallace DC, Ferris SD, Rattazzi MC, Cavalli-Sforza LL. 1983. Radiation of human mitochondria DNA types analyzed by restriction endonuclease cleavage patterns. *J Mol Evol* **19:** 255–271.

Kaneda A, Wang CJ, Cheong R, Timp W, Onyango P, Wen B, Iacobuzio-Donahue CA, Ohlsson R, Andraos R, Pearson MA, et al. 2007. Enhanced sensitivity to IGF-II signaling links loss of imprinting of IGF2 to increased cell proliferation and tumor risk. *Proc Natl Acad Sci* **104:** 20926–20931.

Kazuno AA, Munakata K, Nagai T, Shimozono S, Tanaka M, Yoneda M, Kato N, Miyawaki A, Kato T. 2006. Identification of mitochondrial DNA polymorphisms that alter mitochondrial matrix pH and intracellular calcium dynamics. *PLoS Genet* **2:** e128.

Khusnutdinova E, Gilyazova I, Ruiz-Pesini E, Derbeneva O, Khusainova R, Khidiyatova I, Magzhanov R, Wallace DC. 2008. A mitochondrial etiology of neurodegenerative diseases: Evidence from Parkinson's disease. *Ann NY Acad Sci* **1147:** 1–20.

Kriaucionis S, Paterson A, Curtis J, Guy J, Macleod N, Bird A. 2006. Gene expression analysis exposes mitochondrial abnormalities in a mouse model of Rett syndrome. *Mol Cell Biol* **26:** 5033–5042.

Liu B, Zhou Z. 2008. Lamin A/C, laminopathies and premature ageing. *Histol Histopathol* **23:** 747–763.

Loat CS, Curran S, Lewis CM, Duvall J, Geschwind D, Bolton P, Craig IW. 2008. Methyl-CpG-binding protein 2 polymorphisms and vulnerability to autism. *Genes Brain Behav* **7:** 754–760.

Merriwether DA, Clark AG, Ballinger SW, Schurr TG, Soodyall H, Jenkins T, Sherry ST, Wallace DC. 1991. The structure of human mitochondrial DNA variation. *J Mol Evol* **33:** 543–555.

Mishmar D, Ruiz-Pesini EE, Golik P, Macaulay V, Clark AG, Hosseini S, Brandon M, Easley K, Chen E, Brown MD, et al. 2003. Natural selection shaped regional mtDNA variation in humans. *Proc Natl Acad Sci* **100:** 171–176.

Mishmar D, Ruiz-Pesini E, Mondragon-Palomino M, Procaccio V, Gaut B, Wallace DC. 2006. Adaptive selection of mitochondrial complex I subunits during primate radiation. *Gene* **378:** 11–18.

Muller YL, Bogardus C, Pedersen O, Baier L. 2003. A Gly482Ser missense mutation in the peroxisome proliferator-activated receptor γ coactivator-1 is associated with altered lipid oxidation and early insulin secretion in Pima Indians. *Diabetes* **52:** 895–898.

Palmieri L, Alberio S, Pisano I, Lodi T, Meznaric-Petrusa M, Zidar J, Santoro A, Scarcia P, Fontanesi F, Lamantea E, Ferrero I, Zeviani M. 2005. Complete loss-of-function of the heart/muscle-specific adenine nucleotide translocator is associated with mitochondrial myopathy and cardiomyopathy. *Hum Mol Genet* **14:** 3079–3088.

Ruiz-Pesini E, Wallace DC. 2006. Evidence for adaptive selection acting on the tRNA and rRNA genes of the human mitochondrial DNA. *Hum Mutat* **27:** 1072–1081.

Ruiz-Pesini E, Mishmar D, Brandon M, Procaccio V, Wallace DC. 2004. Effects of purifying and adaptive selection on regional variation in human mtDNA. *Science* **303:** 223–226.

Ruiz-Pesini E, Lott MT, Procaccio V, Poole J, Brandon MC, Mishmar D, Yi C, Kreuziger J, Baldi P, Wallace DC. 2007. An enhanced MITOMAP with a global mtDNA mutational phylogeny. *Nucleic Acids Res* (database issue) **35:** D823–D828.

Saxena R, Voight BF, Lyssenko V, Burtt NP, de Bakker PI, Chen H, Roix JJ, Kathiresan S, Hirschhorn JN, Daly MJ, et al. 2007.

Genome-wide association analysis identifies loci for type 2 diabetes and triglyceride levels. *Science* **316:** 1331–1336.

Scott LJ, Mohlke KL, Bonnycastle LL, Willer CJ, Li Y, Duren WL, Erdos MR, Stringham HM, Chines PS, Jackson AU, et al. 2007. A genome-wide association study of type 2 diabetes in Finns detects multiple susceptibility variants. *Science* **316:** 1341–1345.

Sladek R, Rocheleau G, Rung J, Dina C, Shen L, Serre D, Boutin P, Vincent D, Belisle A, Hadjadj S, et al. 2007. A genome-wide association study identifies novel risk loci for type 2 diabetes. *Nature* **445:** 881–885.

Smiraglia DJ, Kulawiec M, Bistulfi GL, Gupta SG, Singh KK. 2008. A novel role for mitochondria in regulating epigenetic modification in the nucleus. *Cancer Biol Ther* **7:** 1182–1190.

Stewart CL, Kozlov S, Fong LG, Young SG. 2007. Mouse models of the laminopathies. *Exp Cell Res* **313:** 2144–2156.

Stewart JB, Freyer C, Elson JL, Wredenberg A, Cansu Z, Trifunovic A, Larsson NG. 2008. Strong purifying selection in transmission of mammalian mitochondrial DNA. *PLoS Biol* **6:** e10.

Su H, Fan W, Coskun PE, Vesa J, Gold JA, Jiang YH, Potluri P, Procaccio V, Acab A, Weiss JH, Wallace DC, Kimonis VE. 2009. Mitochondrial dysfunction in CA1 hippocampal neurons of the *UBE3A* deficient mouse model for Angelman syndrome. *Neurosci Lett* (in press).

Suissa S, Wang Z, Poole J, Wittkopp S, Feder J, Shutt TE, Wallace DC, Shadel GS, Mishmar D. 2009. Ancient mtDNA genetic variants modulate mtDNA transcription and replication. *PLoS Genet* **5:** e1000474.

Wallace DC. 1994. Mitochondrial DNA sequence variation in human evolution and disease. *Proc Natl Acad Sci* **91:** 8739–8746.

Wallace DC. 2005. A mitochondrial paradigm of metabolic and degenerative diseases, aging, and cancer: A dawn for evolutionary medicine. *Annu Rev Genet* **39:** 359–407.

Wallace DC. 2007. Why do we have a maternally inherited mitochondrial DNA? Insights from evolutionary medicine. *Annu Rev Biochem* **76:** 781–821.

Wallace DC. 2008. Mitochondria as chi. *Genetics* **179:** 727–735.

Wallace DC, Fan W. 2009a. Energetics, epigenetics, mitochondrial genetics. *Mitochondrion* (in press).

Wallace DC, Fan W. 2009b. The pathophysiology of mitochondrial disease as modeled in the mouse. *Genes Dev* **23:** 1714–1736.

Wallace DC, Brown MD, Lott MT. 1999. Mitochondrial DNA variation in human evolution and disease. *Gene* **238:** 211–230.

Wallace DC, Ruiz-Pesini E, Mishmar D. 2003. mtDNA variation, climatic adaptation, degenerative diseases, and longevity. *Cold Spring Harbor Symp Quant Biol* **68:** 479–486.

Wallace DC, Lott MT, Procaccio V. 2007. Mitochondrial genes in degenerative diseases, cancer and aging. In *Emery and Rimoin's principles and practice of medical genetics,* 5th ed. (ed. DL Rimoin et al.), pp. 194–298. Churchill Livingstone Elsevier, Philadelphia.

Wallace DC, Fan W, Procaccio V. 2010. Mitochondrial energetics and therapeutics. *Annu Rev Pathol Mech Dis* **5:** 297–348.

Wen B, Wu H, Shinkai Y, Irizarry RA, Feinberg AP. 2009. Large histone H3 lysine 9 dimethylated chromatin blocks distinguish differentiated from embryonic stem cells. *Nat Genet* **49:** 246–250.

Zeggini E, Weedon MN, Lindgren CM, Frayling TM, Elliott KS, Lango H, Timpson NJ, Perry JR, Rayner NW, Freathy RM, et al. 2007. Replication of genome-wide association signals in UK samples reveals risk loci for type 2 diabetes. *Science* **316:** 1336–1341.

Genetic Structure in African Populations: Implications for Human Demographic History

C.A. LAMBERT[1] AND S.A. TISHKOFF[1,2]

[1]*Department of Genetics and* [2]*Department of Biology,*
University of Pennsylvania, Philadelphia, Pennsylvania 19104
Correspondence: tishkoff@mail.med.upenn.edu

The continent of Africa is the source of all anatomically modern humans that dispersed across the planet during the past 100,000 years. As such, African populations are characterized by high genetic diversity and low levels of linkage disequilibrium (LD) among loci, as compared to populations from other continents. African populations also possess a number of genetic adaptations that have evolved in response to the diverse climates, diets, geographic environments, and infectious agents that characterize the African continent. Recently, Tishkoff et al. (2009) performed a genome-wide analysis of substructure based on DNA from 2432 Africans from 121 geographically diverse populations. The authors analyzed patterns of variation at 1327 nuclear microsatellite and insertion/deletion markers and identified 14 ancestral population clusters that correlate well with self-described ethnicity and shared cultural or linguistic properties. The results suggest that African populations may have maintained a large and subdivided population structure throughout much of their evolutionary history. In this chapter, we synthesize recent work documenting evidence of African population structure and discuss the implications for inferences about evolutionary history in both African populations and anatomically modern humans as a whole.

Africa is a continent of considerable genetic, linguistic, cultural, and phenotypic diversity. It contains more than 2000 distinct ethno-linguistic groups, speaking languages that constitute nearly a third of the world's languages (http://www.ethnologue.com/). The populations within Africa practice a wide range of subsistence patterns, including various modes of agriculture, pastoralism, and hunting and gathering. Africans also live in climates that range from the world's largest desert (the Sahara) and second largest tropical rainforest (the Congo Basin) to savanna, swamps, and mountain highlands. This dramatic range in culture, geography, and diet has given rise to a complex history across the African continent, characterized by high levels of both genetic and phenotypic variation.

Africa is also the source of all modern humans, making its populations the oldest and most genetically diverse among the world's human populations. According to the Recent African Origin (RAO) model, anatomically modern humans originated in Africa and then migrated to all other regions of the globe within the past ~100,000 years (Tishkoff and Verrelli 2003). The transition to modern humans within Africa was not sudden. Rather, the paleobiological record indicates an irregular mosaic of modern, archaic, and regional traits occurring over a substantial period of time and across a broad geographic range (McBrearty and Brooks 2000). The earliest known suite of morphological traits associated with modern humans appears in fossil remains from Ethiopia that are dated to ~150–190 thousand years ago (kya) (White et al. 2003; McDougall et al. 2005). However, this finding does not preclude the existence of modern morphological traits in other African regions before 100 kya; paleobiological specimens from other regions may be less preserved and thus less informative than those discovered in the arid climate of Ethiopia, and extensive archaeological investigations have yet to be conducted across all of Africa (Reed and Tishkoff 2006). A more modern suite of traits appears in East Africa and Southwest Asia ~90 kya, followed by a rapid spread of modern humans throughout the rest of Africa and Eurasia within the past 40,000–80,000 years (Macaulay et al. 2005).

Patterns of genetic variation in modern African populations are shaped by demographic forces that influence variation on a genome-wide scale, such as ancient migration events and fluctuations in population size, and by evolutionary forces that influence individual loci, such as natural selection and mutation. The Bantu expansion is one example that dramatically illustrates the impact of migration on extant patterns of African genetic variation. Within the past ~4000 years, Bantu speakers from West Africa practicing agricultural subsistence migrated throughout sub-Saharan Africa and subsequently admixed with indigenous populations (Ehret 1998; Tishkoff et al. 2009). This expansion greatly influenced genome-wide patterns of genetic variation in modern African populations, an impact that can be observed readily in studies of mitochondrial loci (Soodyall et al. 1996; Behar et al. 2008; Quintana-Murci et al. 2008; Castri et al. 2009), Y-chromosome DNA (Poloni et al. 1997; Hammer et al. 2001), or both (Passarino et al. 1998; Wood et al. 2005; Tishkoff et al. 2007, 2009; Pilkington et al. 2008; Coelho et al. 2009; de Filippo et al. 2009). A classic example of positive natural selection in African populations involves a single-base mutation upstream of *FY*, the gene encoding the Duffy blood group system. Individuals homozygous for the Duffy-O mutation do not express *FY* in their bone marrow and are resistant to malaria caused by the parasite *Plasmodium vivax* (Miller et al. 1978; Barnwell et al. 1989). The mutation exists at high frequencies in populations from sub-Saharan Africa but is virtually nonexistent elsewhere (Mourant 1976). Because of this unusual geographic distribution, it has long been hypothesized that Duffy-O was subject to selection pressure caused by the presence of either *P. vivax* or some similarly harmful pathogen in prehistoric Africa. Indeed, sequencing studies confirmed that the genetic signature at the *FY* locus is consistent with African-specific positive selection (Hamblin and Di Rienzo 2000; Hamblin et al. 2002).

One of the key demographic forces influencing genome-wide patterns of genetic variation is population structure, i.e., population subdivision, migration, and subsequent admixture. Ancient population structure is a neutral process that can mimic patterns of genetic variation expected under balancing selection, because genetic drift affects allele frequencies in subdivided populations independently and not in the population as a whole. Balancing selection is loosely defined as any locus-specific process that maintains variation within a population (Harris and Meyer 2006); this is in contrast to positive selection, which reduces levels of variation at sites linked to a selected variant. Examples of balancing selection include overdominant selection, where heterozygous genotypes have higher fitness than any of the corresponding homozygous genotypes, and frequency-dependent selection, where the fitness of a particular genotype fluctuates as a function of its frequency in a population. Because loci under balancing selection tend to exhibit an excess of variants at intermediate frequencies, coalescence times for such loci are expected to be significantly longer than those for neutral loci (Slatkin 2000; Navarro and Barton 2002). This property forms the main principle underlying many statistical tests designed to detect balancing selection, such as Tajima's D statistic (Tajima 1989).

In a structured population, neutral polymorphisms can randomly drift to fixation in some subpopulations but be lost from others, so that the overall population maintains variation longer than expected by chance (Schierup et al. 2000; Muirhead 2001). Therefore, it is possible to reject a model of neutral evolution in favor of balancing selection when in fact the study populations are actually subdivided (Simonsen et al. 1995). Whereas demographic processes such as population structure impact the entire genome, natural selection affects patterns of genetic variation only at localized regions. Thus, it is theoretically possible to disentangle the effects of population structure and balancing selection by examining genome-wide patterns of variation. If a number of loci show similar patterns such as unusually long coalescence times, balancing selection need not be invoked and underlying substructure may, in fact, be a better model.

In this chapter, we synthesize recent work documenting evidence of substructure in African populations, splitting the discussion into studies of single loci and those of genome-wide patterns of variation. Several studies of genetic variation suggest that ancestral populations were geographically structured before the migration of modern humans out of Africa. A model of ancient subdivision within Africa is consistent with observations of divergent LD patterns among African populations (Tishkoff et al. 1996; Tarazona-Santos and Tishkoff 2005), because the stochastic effects of genetic drift can theoretically result in a set of alleles being positively associated in one population but negatively associated in another. Substructure in Africa is likely due to ethnicity, language, and geography as well as technological, ecological, and climatic factors. Such factors may have contributed to population expansions, contractions, fragmentations, and dispersals during recent human evolution in Africa (Mellars 2006; Hassan et al. 2008).

Particularly important to the discussion of African population structure are the timescales on which substructure has occurred. Figure 1 depicts two extreme models of population structure within Africa that are analogs of global human-origin models. The first model is analogous to the multiregional model of human origins (Wolpoff 1996; Wolpoff et al. 2000). Under this model, hominid populations existed in relative isolation across Africa for most of their histories and evolved independently into anatomically modern humans (AMH). Low levels of gene flow across structured populations may have permitted AMH to evolve independently in more than one African region, a scenario that is more feasible on the African continent than it is on a global scale. This model predicts that, in African populations, a significant number of loci have ancient times to the most recent common ancestor (tMRCAs), i.e., tMRCAs on the order of millions of years, which is much older than the expected tMRCA for a neutrally evolving autosomal locus of ~800,000 years (Harding 1999). The second model is analogous to the recent-origin model of modern humans (Stringer and Andrews 1988; Stringer 2002). Under this model, archaic human populations across Africa were com-

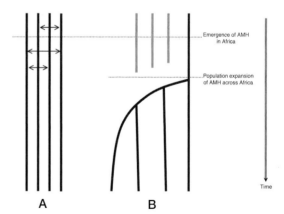

Figure 1. Extreme models of population structure within Africa. (Vertical lines) African lineages, (black lines) lineages that have survived to the present day, (gray lines) lineages that have not survived. (*A*) The first model is analogous to the multiregional model of global human origins. Under this model, hominid populations existed in relative isolation across Africa for most of their histories and evolved independently into anatomically modern humans (AMH). Horizontal arrows indicate low levels of migration between the long-structured populations. (*B*) The second model is analogous to the recent-origin model of global human origins. Under this model, archaic hominids across Africa were completely replaced with AMH, which then became structured only recently.

pletely replaced with AMH from a single geographic region sometime in the past 100,000–200,000 years. This model assumes that modern Africans became isolated and therefore structured only recently, after the replacement of archaic hominids.

Clearly, the two models in Figure 1 are extreme. Models that are intermediate between the two, stemming from migration and gene flow between archaic and modern humans, are also possible. For example, a population expansion of AMH in the past 100,000–200,000 years and subsequent gene flow with archaic populations would preclude the need to assume a complete replacement of archaic hominids. In addition, natural selection with small amounts of migration could have facilitated the emergence of AMH in geographically diverse regions of Africa. For example, a recent study of a locus adjacent to the X-chromosome centromere identified a cluster of shared derived alleles that are nearly fixed in ethnically diverse African populations but exist only at low frequencies outside of Africa (Lambert et al. 2010). Their results suggest that a single African population may have remained a relatively coherent and local entity long enough for natural selection to sweep the cluster of derived alleles to near fixation, a scenario consistent with the recent-origin model of modern humans in Africa (Fig. 1B). Alternatively, the pattern could have been caused by both natural selection and gene flow between ancient structured African populations, by which the cluster of derived alleles arose locally and then spread continentally under sustained selection across Africa. Determining the most likely scenario will require additional genome-wide data across diverse African populations.

The models in Figure 1 nevertheless illustrate a central question in human demographic history: Did all modern humans, including present-day Africans, emerge predominantly from a single ancestral population or from a set of ancient, structured populations within Africa? We begin this chapter by briefly describing common approaches for inferring population structure from genetic data, and we then summarize recent work in modeling demographic scenarios that can account for population structure as well as migration and admixture in African populations.

BAYESIAN CLUSTERING

Recently, Tishkoff et al. (2009) published a genome-wide analysis of African population structure based on DNA from 2432 individuals representing 121 geographically diverse populations. These authors examined patterns of variation at 1327 unlinked nuclear microsatellite and insertion/deletion markers and identified 14 ancestral population clusters. In another recent study of African population structure, Patin et al. (2009) focused specifically on the history of Pygmy hunter-gatherers. The authors of that study sequenced 33 kb across 24 unlinked loci in 236 individuals, representing a total of seven Pygmy populations and five African agricultural populations, and found evidence for four ancestral clusters among their 12 populations. Both of these studies used the STRUCTURE software package (Pritchard et al. 2000; Falush et al. 2003), a Bayesian clustering algorithm that identifies groups of individuals with similar allele frequency profiles while avoiding a priori population classifications. The STRUCTURE algorithm estimates the shared population ancestry of individuals based solely on their genotypes, assuming both Hardy–Weinberg equilibrium and linkage equilibrium in ancestral populations. It is based on a simple admixture model in which K theoretical ancestral populations gave rise to the individuals under analysis (Weiss and Long 2009). The algorithm places individuals into K clusters, where K is chosen in advance but can be varied across independent runs. Each sample in the data set under analysis is subsequently assigned an ancestry proportion from each of the K clusters (Friedlaender et al. 2008).

The output from STRUCTURE includes statistical support for ancestry assignments, but it is important to note that the results are fundamentally statistical in nature. The more markers and samples in a particular data set, the more fine-scale variation that STRUCTURE can infer, but this does not necessarily correspond to more complex patterns of admixture from real parental populations. The original publication describing the STRUCTURE software (Pritchard et al. 2000) provides the following example to illustrate this point: Suppose a single population has allele frequencies that vary continuously across a large geographic area but is sampled at only K distinct locations. Provided the allele frequencies at the sampling locations are different enough, STRUCTURE may infer the presence of K populations in the data set, even though there is a single biological population underlying all K samples.

STRUCTURE excels at illuminating patterns of variation within and between samples, but it was not designed to make inferences about evolutionary processes that generated those patterns of variation. Such inferences require demographic modeling and hypothesis testing, as described below.

PHYLOGENETIC ANALYSES

Before the availability of genome-wide data sets such as those analyzed by Tishkoff et al. (2009) and Patin et al. (2009), population structure in Africa was inferred by examining patterns of genetic variation at single loci. Such studies tended to be phylogenetic in nature: The authors generated trees inferred from the genetic variation and haplotypes that they observed and then estimated tMRCAs for their samples. Unusually long tMRCAs, caused by the presence of one or more highly divergent lineages, suggested that the sampled populations may have been structured throughout their histories. Table 1 lists single-locus studies in which the authors inferred tMRCAs older than 1 million years. The authors of most studies in the table ruled out the effects of balancing selection and thus attributed the long tMRCAs to ancient substructure within Africa.

Although interesting and informative, phylogenetic analyses of single loci are at best only suggestive of ancient population structure. This is because ancient tMRCAs may simply represent the tail of a genome-wide distribution that is expected due to the stochastic nature of evolution acting

Table 1. Phylogenetic Studies that Infer Ancient tMRCAs for African Populations

Gene/locus	Reference	African sample size	African populations	Length	Estimated tMRCA (millions of years)
Intergenic (22q11.2)	Zhao et al. (2000)	20	6	10 kb	1.29
CD209	Barreiro et al. (2005)	41	9	5.5 kb	2.8
CMAH (intronic)	Hayakawa et al. (2006)	13	3	7.3 kb	2.9
ASAH	Kim and Satta (2008)	15	3	11 kb	2.4
PDHA1	Harris and Hey (1999)	16 males	4	4.2 kb	1.86
Intergenic (Xp11.22)	Shimada et al. (2007)	95 males	6	10.1 kb	1.44
Intergenic (Xp21.1)	Garrigan et al. (2005a)	10 males	6	7.5 kb spanning 17.5 kb	1.3–1.9

at individual loci and thus may not be indicative of admixture between AMH and ancient hominid species. A study by Fagundes et al. (2007) illustrates this idea most effectively. The authors sequenced 500 bp at each of 50 unlinked, noncoding, autosomal loci in samples from 10 Africans, eight Asians, and 12 Native American individuals. With the data, the authors were able to test different models of modern human origins and found that a recent African replacement model best explained the patterns of genetic variation. Under their model, AMH arose ~141 kya in Africa from a founder population of ~12,800 effective individuals and then went on to replace all other hominid species. The authors also computed empirical distributions for tMRCAs, and they found that coalescence times exceeding several million years are expected under the recent African replacement model due to the stochasticity of lineages that might pass through an AMH bottleneck. Fagundes et al. (2007) concluded that a complete replacement of archaic humans during the evolution of AMH could have resulted in ancient tMRCAs for some loci, making it unnecessary to invoke admixture with archaic humans to explain such observations. The model relies on a large population size ancestral to AMH within Africa, an assumption that is consistent with findings from prior studies (Tishkoff and Williams 2002). Although it is true that the sequence data from Fagundes et al. (2007) represent short genomic regions and a sparse sampling of human variation (Garrigan and Hammer 2008), the results nevertheless show that it is possible to obtain extremely long tMRCAs under a model of recent African replacement.

A simulation study published by Wall (2000) formalizes the need for genome-wide data sets in this context. The study expands a model of population subdivision originally developed by Nordborg (2000) to show that ~50–100 unlinked, neutrally evolving, fully resequenced loci are necessary to have sufficient power for detecting ancient contributions in the human genome from either Neanderthal or *Homo erectus*. The author further found that power fluctuates as a function of both the separation time between AMH and archaic hominids and the time of admixture. For example, it is assumed that Neanderthal and AMH were separated for at least 250,000 years before living side by side in Europe 25–45 kya (McBrearty and Brooks 2000; Plagnol and Wall 2006). Wall (2000) found that, for a set number of loci, power to detect Neanderthal contributions in the human genome increases as the assumed separation time increases and decreases as the assumed time to admixture increases. The results should apply to detecting ancient substructure in African populations as well, implying that numerous loci are needed to make definitive statements about ancient population structure within Africa.

MODELING DEMOGRAPHIC HISTORY

Although the STRUCTURE program analyzes patterns of variation to detect the presence of population structure, statistical hypothesis testing is necessary to model the demographic and evolutionary processes that may have led to structured human populations. This kind of statistical modeling has typically been done using coalescent theory (Hudson 1991) as implemented in the GENETREE software package (Griffiths and Tavaré 1994). The GENETREE algorithm estimates the tMRCA for a sample of sequences under assumptions of random mating and either constant population size or a recent population expansion. This approach has been used in a variety of studies during the past decade (Harris and Hey 1999; Jaruzelska et al. 1999; Yu et al. 2002; Barreiro et al. 2005; Hayakawa et al. 2006; Yotova et al. 2007). To provide statistical evidence of population structure, some authors have also tested assumptions of panmixia directly using custom-designed coalescent simulations (Garrigan et al. 2005a; Kim and Satta 2008). In particular, Kim and Satta (2008) tested 22 sets of demographic parameters to examine models of panmixia, bottlenecks, expansions, and population structure (both ancient and recent) for an 11-kb region of the *ASAH* gene and found that a model of ancient structure in Africa best explained their data.

It is sometimes difficult to formally model demographic scenarios because the number of parameters that must be estimated becomes prohibitively large as the models become more complex. Approximate Bayesian computation, or ABC, is one approach that allows for flexible yet statistically sound comparisons of different demographic models (Beaumont et al. 2002). ABC is Bayesian in the sense that it estimates posterior probability distributions for the parameters of interest, given a demographic model and a set of prior distributions. It is approximate in two possible ways, depending on the implementation. First, because exact posterior distributions are often too complicated to calculate explicitly, ABC constructs approximate distributions numerically using stochastic simulation methods such as rejection algorithms, importance sampling, and Markov chain Monte Carlo (summarized in Marjoram and Tavaré 2006). A stochastic method simulates a data set based on the given demographic model and a parameter value sampled from the prior distribution. If the simulated data are sufficiently close to the actual data, the parameter value is stored. After simulating many data sets in this way, a posterior distribution can be built from frequencies of the stored parameter values. Second, ABC is approximate in the sense that comparisons between simulated and real data sets are often quantified using summary statistics such as the number of segregating sites. Many researchers have developed custom ABC implementations to analyze their data, although user-friendly software packages are becoming more common (see, e.g., Cornuet et al. 2008; Lopes et al. 2009).

Cox et al. (2008) used ABC to show that the pattern of variation at the X-linked pseudogene *RRM2P4* is best explained by population structure dating to 2.33 million years and rooted in East Asia. Building on previous work (Garrigan et al. 2005b), the authors sequenced 2.4 kb of the pseudogene in 131 Africans and 122 non-Africans and extended the region by sequencing two additional nearby loci, for a total 5.6 kb of resequencing data that spans 16.5 kb of the X chromosome. With the data, the authors explicitly tested whether an RAO model could explain both the ancient tMRCA that they observed and the basal lineage rooted in East Asia. They found that although an RAO model may conceivably explain the extremely long tMRCA, a model of ancient admixture best explains the East Asian root for the genealogy. The authors thus concluded that *RRM2P4* may be a remnant of admixture between AMH and an archaic hominid population in Asia such as *H. erectus*. ABC has also been recently used to compare several demographic scenarios that may have produced the patterns of genetic variation seen in contemporary African Pygmy populations (Patin et al. 2009; Verdu et al. 2009). In particular, Patin et al. (2009) used an ABC approach to estimate separation times and levels of gene flow between Western and Eastern Pygmy populations. Notably, this study used ABC in conjunction with a genome-wide data set consisting of 24 unlinked noncoding nucleotide sequences across the autosomes, sex chromosomes, and mitochondrial genome. The most likely model explaining the data involves a split ~60 kya between an ancestral Pygmy population and a population ancestral to modern-day African farmers, followed by a split between the Western and Eastern Pygmy populations ~20 kya.

A statistical framework that infers demographic parameters specifically related to population structure is the isolation with migration model, implemented in the software package IM (Nielsen and Wakeley 2001). The IM algorithm computes marginal Bayesian posterior probabilities for a suite of parameters including population sizes, migration rates, and divergence times, allowing inference of population structure both with and without the effects of migration. This approach was used by Shimada et al. (2007) to analyze 10.1 kb of resequencing data from a noncoding region on chromosome Xp11.22. The authors sequenced the locus in a panel of 672 males from 52 worldwide populations and found a surprisingly divergent haplotype distributed at low frequencies throughout Africa, Europe, and Asia. Using IM, they estimated a tMRCA of 5230 years for the haplotype, but a tMRCA of more than 1.4 million years between the haplotype and all other sequences in their sample, suggestive of archaic population structure in Africa. Although the IM approach is particularly sophisticated in its ability to test models of population structure using multilocus data sets (Hey and Nielsen 2004), it does not currently account for recombination within loci.

CONCLUSIONS

Traditionally, inferences about population structure in Africa have relied on analyses of single loci. Because population structure can affect genome-wide patterns of variation, an unusually old tMRCA at a single locus is only suggestive of ancient substructure. Extremely old tMRCAs are expected in recent replacement models that do not involve ancient population structure and may therefore represent the tails of the distribution for genome-wide tMRCAs. For this reason, large data sets consisting of hundreds of resequenced loci are needed to make definitive inferences about ancient population structure in Africa. Such data sets will allow researchers to model demographic scenarios leading to structured human populations in a statistically sound way. In particular, ABC and/or IM can be used with more genome-wide data sets to formally test whether current data are consistent with a model of ancient African substructure (Fig. 1A), a recent replacement model (Figure 1B), or demographic models intermediate between the two.

Genome-wide data sets of single-nucleotide polymorphism (SNP) genotypes, copy-number variants, and other forms of structural variation should also prove informative for elucidating patterns of genetic variation in African populations. For example, Bryc et al. (2010) genotyped 500,000 SNPs in 203 individuals from 12 West African populations and found that population structure in this region of Africa reflected both linguistic and geographic variation. It is important to note, however, that currently available SNP platforms such as the Illumina Human1M-Duo or Affymetrix Genome-Wide Human SNP Array 6.0 are derived from SNPs identified predominantly in non-African populations. This introduces an ascertainment bias when studying genetic variation in African populations. Resequencing studies of ethnically diverse African populations will identify African-specific SNPs that can subsequently be used to create SNP platforms that are more informative for analyses of African populations. Recent advances in sequencing technologies—in particular, next-generation sequencing, exome sequencing, and whole-genome sequencing—will also be extremely valuable for understanding fine-scale patterns of variation in African populations. These technologies will provide complete sequence information at multiple loci, allowing genome-wide inferences to be made about substructure in Africa. To date, the whole genomes of three African men and one

African woman have been sequenced: one Yoruban male (Bentley et al. 2008) and one Yoruban female (Drmanac et al. 2010), an indigenous hunter-gatherer from the Kalahari Desert, and a Bantu individual from southern Africa (Schuster et al. 2010). As the cost of sequencing technologies continues to decrease, it will become possible to conduct population-level analyses of ethnically diverse groups of African populations.

Finally, a thorough understanding of African evolutionary history will require sampling across a broad range of African populations. Much of what is currently known about African genetic diversity is inferred from a limited number of the ~2000 linguistically distinct ethnic groups in Africa. Extensive sampling of ethnically diverse African populations will be critical for testing models of the origin and dispersal of modern humans both within and outside of Africa. As the data sets become more diverse in terms of both population samples and genotyped loci, fine-scale inferences about African demographic history will become feasible.

ACKNOWLEDGMENTS

C.A.L. is supported by an National Institutes of Health (NIH) IRACDA postdoctoral fellowship (PAR-06-470). S.A.T. is supported by National Science Foundation grants BCS-0196183 and BCS-0827436, NIH grants R01GM076637 and 1R01GM083606-01, and an NIH Pioneer 1-DP1-OD-006445-01 award.

REFERENCES

Barnwell JW, Nichols ME, Rubinstein P. 1989. In vitro evaluation of the role of the Duffy blood group in erythrocyte invasion by *Plasmodium vivax*. *J Exp Med* **169**: 1795–1802.

Barreiro LB, Patin E, Neyrolles O, Cann HM, Gicquel B, Quintana-Murci L. 2005. The heritage of pathogen pressures and ancient demography in the human innate-immunity CD209/CD209L region. *Am J Hum Genet* **77**: 869–886.

Beaumont MA, Zhang W, Balding DJ. 2002. Approximate Bayesian computation in population genetics. *Genetics* **162**: 2025–2035.

Behar DM, Villems R, Soodyall H, Blue-Smith J, Pereira L, Metspalu E, Scozzari R, Makkan H, Tzur S, Comas D, et al. 2008. The dawn of human matrilineal diversity. *Am J Hum Genet* **82**: 1130–1140.

Bentley DR, Balasubramanian S, Swerdlow HP, Smith GP, Milton J, Brown CG, Hall KP, Evers DJ, Barnes CL, Bignell HR, et al. 2008. Accurate whole human genome sequencing using reversible terminator chemistry. *Nature* **456**: 53–59.

Bryc K, Auton A, Nelson MR, Oksenberg JR, Hauser SL, Williams S, Froment A, Bodo JM, Wambebe C, Tishkoff SA, Bustamante CD. 2010. Genome-wide patterns of population structure and admixture in West Africans and African Americans. *Proc Natl Acad Sci* **107**: 786–791.

Castri L, Tofanelli S, Garagnani P, Bini C, Fosella X, Pelotti S, Paoli G, Pettener D, Luiselli D. 2009. mtDNA variability in two Bantu-speaking populations (Shona and Hutu) from Eastern Africa: Implications for peopling and migration patterns in sub-Saharan Africa. *Am J Phys Anthropol* **140**: 302–311.

Coelho M, Sequeira F, Luiselli D, Beleza S, Rocha J. 2009. On the edge of Bantu expansions: mtDNA, Y chromosome and lactase persistence genetic variation in southwestern Angola. *BMC Evol Biol* **9**: 80.

Cornuet JM, Santos F, Beaumont MA, Robert CP, Marin JM, Balding DJ, Guillemaud T, Estoup A. 2008. Inferring population history with DIY ABC: A user-friendly approach to approximate Bayesian computation. *Bioinformatics* **24**: 2713–2719.

Cox MP, Mendez FL, Karafet TM, Pilkington MM, Kingan SB, Destro-Bisol G, Strassmann BI, Hammer MF. 2008. Testing for archaic hominin admixture on the X chromosome: Model likelihoods for the modern human RRM2P4 region from summaries of genealogical topology under the structured coalescent. *Genetics* **178**: 427–437.

de Filippo C, Heyn P, Barham L, Stoneking M, Pakendorf B. 2009. Genetic perspectives on forager-farmer interaction in the Luangwa Valley of Zambia. *Am J Phys Anthropol* **141**: 382–394.

Drmanac R, Sparks AB, Callow MJ, Halpern AL, Burns NL, Kermani BG, Carnevali P, Nazarenko I, Nilsen GB, Yeung G, et al. 2010. Human genome sequencing using unchained base reads on self-assembling DNA nanoarrays. *Science* **327**: 78–81.

Ehret C. 1998. *An African classical age: Eastern and Southern Africa in world history, 1000 B.C. to A.D. 400*. University of Virginia Press, Charlottesville.

Fagundes NJ, Ray N, Beaumont M, Neuenschwander S, Salzano FM, Bonatto SL, Excoffier L. 2007. Statistical evaluation of alternative models of human evolution. *Proc Natl Acad Sci* **104**: 17614–17619.

Falush D, Stephens M, Pritchard JK. 2003. Inference of population structure using multilocus genotype data: Linked loci and correlated allele frequencies. *Genetics* **164**: 1567–1587.

Friedlaender JS, Friedlaender FR, Reed FA, Kidd KK, Kidd JR, Chambers GK, Lea RA, Loo JH, Koki G, Hodgson JA, et al. 2008. The genetic structure of Pacific Islanders. *PLoS Genet* **4**: e19.

Garrigan D, Hammer MF. 2008. Ancient lineages in the genome: A response to Fagundes et al. *Proc Natl Acad Sci* **105**: E3; author reply, E4.

Garrigan D, Mobasher Z, Kingan SD, Wilder JA, Hammer MF. 2005a. Deep haplotype divergence and long-range linkage disequilibrium at Xp21.1 provide evidence that humans descend from a structured ancestral population. *Genetics* **170**: 1849–1856.

Garrigan D, Mobasher Z, Severson T, Wilder JA, Hammer MF. 2005b. Evidence for archaic Asian ancestry on the human X chromosome. *Mol Biol Evol* **22**: 189–192.

Griffiths RC, Tavaré S. 1994. Sampling theory for neutral alleles in a varying environment. *Philos Trans R Soc Lond B Biol Sci* **344**: 403–410.

Hamblin MT, Di Rienzo A. 2000. Detection of the signature of natural selection in humans: Evidence from the Duffy blood group locus. *Am J Hum Genet* **66**: 1669–1679.

Hamblin MT, Thompson EE, Di Rienzo A. 2002. Complex signatures of natural selection at the Duffy blood group locus. *Am J Hum Genet* **70**: 369–383.

Hammer MF, Karafet TM, Redd AJ, Jarjanazi H, Santachiara-Benerecetti S, Soodyall H, Zegura SL. 2001. Hierarchical patterns of global human Y-chromosome diversity. *Mol Biol Evol* **18**: 1189–1203.

Harding RM. 1999. More on the X files. *Proc Natl Acad Sci* **96**: 2582–2584.

Harris EE, Hey J. 1999. X chromosome evidence for ancient human histories. *Proc Natl Acad Sci* **96**: 3320–3324.

Harris EE, Meyer D. 2006. The molecular signature of selection underlying human adaptations. *Am J Phys Anthropol* (suppl.) **43**: 89–130.

Hassan HY, Underhill PA, Cavalli-Sforza LL, Ibrahim ME. 2008. Y-chromosome variation among Sudanese: Restricted gene flow, concordance with language, geography, and history. *Am J Phys Anthropol* **137**: 316–323.

Hayakawa T, Aki I, Varki A, Satta Y, Takahata N. 2006. Fixation of the human-specific CMP-*N*-acetylneuraminic acid hydroxylase pseudogene and implications of haplotype diversity for human evolution. *Genetics* **172**: 1139–1146.

Hey J, Nielsen R. 2004. Multilocus methods for estimating population sizes, migration rates and divergence time, with applications to the divergence of *Drosophila pseudoobscura* and *D. persimilis*. *Genetics* **167**: 747–760.

Hudson RR. 1991. Gene genealogies and the coalescent process.

Oxford surveys in evolutionary biology (ed. DF Antonovics), pp. 1–44. Oxford University Press, Oxford.

Jaruzelska J, Zietkiewicz E, Batzer M, Cole DE, Moisan JP, Scozzari R, Tavaré S, Labuda D. 1999. Spatial and temporal distribution of the neutral polymorphisms in the last ZFX intron: Analysis of the haplotype structure and genealogy. *Genetics* **152**: 1091–1101.

Kim HL, Satta Y. 2008. Population genetic analysis of the N-acylsphingosine amidohydrolase gene associated with mental activity in humans. *Genetics* **178**: 1505–1515.

Lambert CA, Connelly CF, Madeoy J, Qiu R, Olson MV, Akey JM. 2010. Highly punctuated patterns of population structure on the X chromosome and implications for African evolutionary history. *Am J Hum Genet* **86**: 34–44.

Lopes JS, Balding D, Beaumont MA. 2009. PopABC: A program to infer historical demographic parameters. *Bioinformatics* **25**: 2747–2749.

Macaulay V, Hill C, Achilli A, Rengo C, Clarke D, Meehan W, Blackburn J, Semino O, Scozzari R, Cruciani F, et al. 2005. Single, rapid coastal settlement of Asia revealed by analysis of complete mitochondrial genomes. *Science* **308**: 1034–1036.

Marjoram P, Tavaré S. 2006. Modern computational approaches for analysing molecular genetic variation data. *Nat Rev Genet* **7**: 759–770.

McBrearty S, Brooks AS. 2000. The revolution that wasn't: A new interpretation of the origin of modern human behavior. *J Hum Evol* **39**: 453–563.

McDougall I., Brown FH, Fleagle JG. 2005. Stratigraphic placement and age of modern humans from Kibish, Ethiopia. *Nature* **433**: 733–736.

Mellars P. 2006. Why did modern human populations disperse from Africa ca. 60,000 years ago? A new model. *Proc Natl Acad Sci* **103**: 9381–9386.

Miller LH, McGinniss MH, Holland PV, Sigmon P. 1978. The Duffy blood group phenotype in American blacks infected with *Plasmodium vivax* in Vietnam. *Am J Trop Med Hyg* **27**: 1069–1072.

Mourant AE. 1976. *The distribution of human blood groups, and other polymorphisms*. Oxford University Press, London.

Muirhead CA. 2001. Consequences of population structure on genes under balancing selection. *Evolution* **55**: 1532–1541.

Navarro A, Barton NH. 2002. The effects of multilocus balancing selection on neutral variability. *Genetics* **161**: 849–863.

Nielsen R, Wakeley J. 2001. Distinguishing migration from isolation: A Markov chain Monte Carlo approach. *Genetics* **158**: 885–896.

Nordborg M. 2000. On detecting ancient admixture. In *Proceedings of the NATO ASI Workshop* (Genes, fossils, and behaviour: An integrated approach to human evolution), pp. 123–136. IOS, Amsterdam.

Passarino G, Semino O, Quintana-Murci L, Excoffier L, Hammer M, Santachiara-Benerecetti AS. 1998. Different genetic components in the Ethiopian population, identified by mtDNA and Y-chromosome polymorphisms. *Am J Hum Genet* **62**: 420–434.

Patin E, Laval G, Barreiro LB, Salas A, Semino O, Santachiara-Benerecetti S, Kidd KK, Kidd JR, Van der Veen L, Hombert JM, et al. 2009. Inferring the demographic history of African farmers and Pygmy hunter-gatherers using a multilocus resequencing data set. *PLoS Genet* **5**: e1000448.

Pilkington MM, Wilder JA, Mendez FL, Cox MP, Woerner A, Angui T, Kingan S, Mobasher Z, Batini C, Destro-Bisol G, et al. 2008. Contrasting signatures of population growth for mitochondrial DNA and Y chromosomes among human populations in Africa. *Mol Biol Evol* **25**: 517–525.

Poloni ES, Semino O, Passarino G, Santachiara-Benerecetti AS, Dupanloup I, Langaney A, and Excoffier L. 1997. Human genetic affinities for Y chromosome P49a,f/TaqI haplotypes show strong correspondence with linguistics. *Am J Hum Genet* **61**: 1015–1035.

Plagnol V, Wall JD. 2006. Possible ancestral structure in human populations. *PLoS Genet* **2**: e105.

Pritchard JK, Stephens M, Donnelly P. 2000. Inference of population structure using multilocus genotype data. *Genetics* **155**: 945–959.

Quintana-Murci L, Quach H, Harmant C, Luca F, Massonnet B, Patin E, Sica L, Mouguiama-Daouda P, Comas D, Tzur S, et al. 2008. Maternal traces of deep common ancestry and asymmetric gene flow between Pygmy hunter-gatherers and Bantu-speaking farmers. *Proc Natl Acad Sci* **105**: 1596–1601.

Reed FA, Tishkoff SA. 2006. African human diversity, origins and migrations. *Curr Opin Genet Dev* **16**: 597–605.

Schierup MH, Vekemans X, Charlesworth D. 2000. The effect of subdivision on variation at multi-allelic loci under balancing selection. *Genet Res* **76**: 51–62.

Schuster SC, Miller W, Ratan A, Tomsho LP, Giardine B, Kasson LR, Harris RS, Peterson DC, Zhao F, Qi J, et al. 2010. Complete Khoisan and Bantu genomes from southern Africa. *Nature* **463**: 943–947.

Shimada MK, Panchapakesan K, Tishkoff SA, Nato AQ Jr, Hey J. 2007. Divergent haplotypes and human history as revealed in a worldwide survey of X-linked DNA sequence variation. *Mol Biol Evol* **24**: 687–698.

Simonsen KL, Churchill GA, Aquadro CF. 1995. Properties of statistical tests of neutrality for DNA polymorphism data. *Genetics* **141**: 413–429.

Slatkin M. 2000. Balancing selection at closely linked, overdominant loci in a finite population. *Genetics* **154**: 1367–1378.

Soodyall H, Vigilant L, Hill AV, Stoneking M, Jenkins T. 1996. mtDNA control-region sequence variation suggests multiple independent origins of an "Asian-specific" 9-bp deletion in sub-Saharan Africans. *Am J Hum Genet* **58**: 595–608.

Stringer C. 2002. Modern human origins: Progress and prospects. *Philos Trans R Soc Lond B Biol Sci* **357**: 563–579.

Stringer CB, Andrews P. 1988. Genetic and fossil evidence for the origin of modern humans. *Science* **239**: 1263–1268.

Tajima F. 1989. Statistical method for testing the neutral mutation hypothesis by DNA polymorphism. *Genetics* **123**: 585–595.

Tarazona-Santos E, Tishkoff SA. 2005. Divergent patterns of linkage disequilibrium and haplotype structure across global populations at the interleukin-13 (*IL13*) locus. *Genes Immun* **6**: 53–65.

Tishkoff SA, Verrelli BC. 2003. Patterns of human genetic diversity: Implications for human evolutionary history and disease. *Annu Rev Genomics Hum Genet* **4**: 293–340.

Tishkoff SA, Williams SM. 2002. Genetic analysis of African populations: Human evolution and complex disease. *Nat Rev Genet* **3**: 611–621.

Tishkoff SA, Dietzsch E, Speed W, Pakstis AJ, Kidd JR, Cheung K, Bonne-Tamir B, Santachiara-Benerecetti AS, Moral P, Krings M. 1996. Global patterns of linkage disequilibrium at the *CD4* locus and modern human origins. *Science* **271**: 1380–1387.

Tishkoff SA, Gonder MK, Henn BM, Mortensen H, Knight A, Gignoux C, Fernandopulle N, Lema G, Nyambo TB, Ramakrishnan U, et al. 2007. History of click-speaking populations of Africa inferred from mtDNA and Y chromosome genetic variation. *Mol Biol Evol* **24**: 2180–2195.

Tishkoff SA, Reed FA, Friedlaender FR, Ehret C, Ranciaro A, Froment A, Hirbo JB, Awomoyi AA, Bodo JM, Doumbo O, et al. 2009. The genetic structure and history of Africans and African Americans. *Science* **324**: 1035–1044.

Verdu P, Austerlitz F, Estoup A, Vitalis R, Georges M, Thery S, Froment A, Le Bomin S, Gessain A, Hombert JM, et al. 2009. Origins and genetic diversity of Pygmy hunter-gatherers from Western Central Africa. *Curr Biol* **19**: 312–318.

Wall JD. 2000. Detecting ancient admixture in humans using sequence polymorphism data. *Genetics* **154**: 1271–1279.

Weiss KM, Long JC. 2009. Non-Darwinian estimation: My ancestors, my genes' ancestors. *Genome Res* **19**: 703–710.

White TD, Asfaw B, DeGusta D, Gilbert H, Richards GD, Suwa G, Howell FC. 2003. Pleistocene *Homo sapiens* from Middle Awash, Ethiopia. *Nature* **423**: 742–747.

Wolpoff MH. 1996. Interpretations of multiregional evolution. *Science* **274**: 704–707.

Wolpoff MH, Hawks J, Caspari R. 2000. Multiregional, not multiple origins. *Am J Phys Anthropol* **112**: 129–136.

Wood ET, Stover DA, Ehret C, Destro-Bisol G, Spedini G, McLeod H, Louie L, Bamshad M, Strassmann BI, Soodyall H, Hammer MF. 2005. Contrasting patterns of Y chromosome and mtDNA variation in Africa: Evidence for sex-biased demographic processes. *Eur J Hum Genet* **13:** 867–876.

Yotova V, Lefebvre JF, Kohany O, Jurka J, Michalski R, Modiano D, Utermann G, Williams SM, Labuda D. 2007. Tracing genetic history of modern humans using X-chromosome lineages. *Hum Genet* **122:** 431–443.

Yu N, Fu YX, Li WH. 2002. DNA polymorphism in a worldwide sample of human X chromosomes. *Mol Biol Evol* **19:** 2131–2141.

Zhao Z, Jin L, Fu YX, Ramsay M, Jenkins T, Leskinen E, Pamilo P, Trexler M, Patthy L, Jorde LB, et al. 2000. Worldwide DNA sequence variation in a 10-kilobase noncoding region on human chromosome 22. *Proc Natl Acad Sci* **97:** 11354–11358.

A Defense of Sociobiology

K.R. Foster

Center for Systems Biology, Harvard University, Cambridge, Massachusetts 02138
Correspondence: kfoster@cgr.harvard.edu

To counter recent claims that sociobiology is in disarray or requires reformulation, I discuss the semantics, theory, and data that underlie the field. A historical perspective is used to identify the cause of current debates. I argue that semantic precision is required in discussing terms such as kin selection, group selection, and altruism, but once care is taken, the objections to the unity of theoretical sociobiology largely evaporate. More work is required, however, to understand group *adaptation*, which might be taken to be the process of optimizing phenotypes that is driven by group, rather than individual, context. From the empirical perspective, the eusocial insects with their fixed division between work and reproduction are often a sounding board in discussions. Here, one finds clear evidence for the role of kin selection and relatedness in both the origin of eusociality and its maintenance. Data from other systems including the social vertebrates, microorganisms, and even plants also support the role of relatedness and particularly family life in the evolution of cooperation and altruism. These data, however, in no way invalidate the claim that group selection is also a central process in social evolution and I discuss the empirical evidence for group selection. The foundations of sociobiology are solid and the future should build on these foundations. Exciting new areas include the importance of community and species-level selection in evolution and elucidating the molecular mechanisms that underlie social traits.

> Kinship plays a minor role [in social evolution], and this has been why kinship theory has produced so little over four decades in important predictions. In fact, it has made virtually none.
>
> E.O. Wilson (pers. comm.)

The first Quantitative Biology Symposium to celebrate Darwin is perhaps best known for Mayr's (1960) strong attack on theoretical population genetics, which Mayr colorfully criticized for reducing the evolutionary process to the "adding of certain beans to a beanbag and the withdrawing of others." He continued: "[Fisher, Wright and Haldane] have worked out an impressive mathematical theory of genetical variation and evolutionary change. But what, precisely, has been the contribution of this mathematical school to the evolutionary theory, if I may be permitted to ask such a provocative question?" It was this essay, along with one of Mayr's books, that led Haldane (1964) to later write the aptly named "*A Defense of Beanbag Genetics*" (Haldane did not attend the 1960 meeting because he was refused a U.S. visa).

I was privileged to attend, 50 years later, the second Quantitative Biology Symposium to celebrate Darwin. One might ask whether anything similarly controversial to Mayr's attack on population genetics has occurred. At least for me, there was an uncanny resemblance in the paper of E.O. Wilson, which was a resounding assault on the theoretical foundations of sociobiology. The above opening quotation was followed by argument for an alternative view of social evolution that instead of kinship appeals to a mix of preadaptation, emergent properties, and group selection. For students of social evolution, this assault will come as no surprise but part of a continuing theme in the recent writing of Wilson and others (Wilson 2005, 2008b; Wilson and Hollдobler 2005; Wilson and Wilson 2007). But for biologists from other disciplines, it may be shocking to hear one of the founders of sociobiology (Wilson 1975b) attempt to so strongly undermine its basis.

The goal of this chapter is to offer a defense of modern sociobiology. My discussion follows the emphasis of Wilson's original book *Sociobiology*: "This brings us to the central theoretical problem of sociobiology: How can altruism, which by definition reduces personal fitness, possibly evolve by natural selection?" (Wilson 1975b, p. 3). However, I should note up front that sociobiology now encompasses additional topics that I will largely neglect, including parent–offspring conflict (Trivers 1974), intragenomic conflicts (Hurst et al. 1996; Burt and Trivers 2006), classical game theory (Trivers 1971; Maynard Smith and Price 1973; Axelrod and Hamilton 1981; Doebeli and Hauert 2005), and perhaps even sexual selection (Darwin 1859). I seek to put the recent comments of Wilson and others into a historical context and, hopefully, offer some reassurance that the appearance of ongoing controversy is mostly illusory. Accordingly, I argue that the theoretical foundations of sociobiology are solid and that there are extensive data that support this position. Finally, I attempt to identify a few future questions that might build on this foundation. This is not the first time that I (Foster 2006; Foster et al. 2006a,b) or others (Queller 1992; Dugatkin and Reeve 1994; Reeve and Keller 1999; Lehmann and Keller 2006; Lehmann et al. 2007; West et al. 2007) have attempted to do this. Nevertheless, misunderstandings in social evolution continue to abound and our considerable progress is continually troubled by a confusion of terms and debates. This, I hope, offers some justification for what will follow.

THE HISTORY

> There will also, no doubt, be indirect effects in cases in which an animal favours or impedes the survival or reproduction of its relatives.... Nevertheless such indirect effects will in very many cases be unimportant...
>
> Fisher (1930)

Sociobiology, in name, barely existed at the time of the first Cold Spring Harbor Symposium on Darwinism, but discussions of social behavior and its evolution by natural selection began much earlier. In Darwin's writing, it is easy to find the traces of the two key concepts upon which modern sociobiology was founded—group selection and kinship—the same concepts that continue to fuel the fires of controversy. Indeed, in explaining one of the most striking examples of social behavior, the effectively sterile workers of social insects (Fig. 1), Darwin appealed to both family life and natural selection acting at the level of the insect colony (Gardner and Foster 2008). This blend of kinship and grouping in explanations of social behavior continued into the early 20th century with a near-modern perspective seen relatively early on: "The instincts of the workers can be kept up to the mark by natural selection. Those fertile females whose genes under worker diet do not develop into workers with proper instincts will produce inefficient hives; such communities will go under in the struggle for existence, and so the defective genes will be eliminated from the bee germ-plasm." (Wells et al. 1929)

It is also possible to find contemporaneous examples that play down the importance of kinship. Although recognizing its theoretical potential to affect evolution, for example, Fisher (1930) considered kinship to be an unimportant detail in the process of natural selection. Concordantly, two features are notable of the era that preceded the rise of sociobiology: a lack of controversy surrounding social evolution and, more importantly, a lack of interest. These features extend to 1959 and the Cold Spring Harbor Symposium, where despite a session entitled "Ecological systems and social organizations," there is little evidence of the modern interest in the social evolution, with its emphasis on cooperation and competition (Fig. 2). Soon after, however, the intellectual landscape began to change with two key publications—one now famous, the other infamous—which set the stage for the coming debates.

Infamy was to result for the book of Wynne-Edwards (1962) entitled *Animal Dispersal in Relation to Social Behaviour*. Replete with examples, this book made frequent use of arguments from population-level advantage to explain animal behaviors. Territoriality and dominance hierarchies, it was suggested, evolve to limit animal populations and to prevent the overuse of resources that might threaten the population with extinction. The problem with this particular brand of reasoning was rapidly

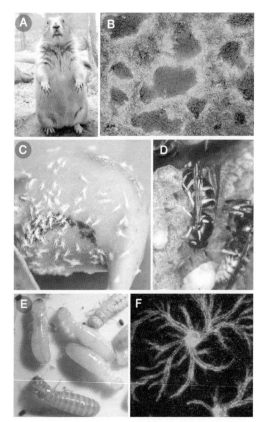

Figure 1. A selection of the social species discussed: a prairie dog (*A*), a biofilm of the bacterium *Pseudomonas aeruginosa* (*B*), and the gall-dwelling aphid *Pemphigus obesinymphae* (*C*). When disturbed, soldier aphids emerge and attack intruders. (*D*) A worker laying a male-destined egg in the social wasp *Dolichovespula saxonica*. (*E*) The termite *Cryptotermes secundus*, in which a gene required for reproductive suppression of the workers has been described (Korb et al. 2009). (*F*) Aggregation of fluorescently labeled cells in the slime mold or social amoebae *D. discoideum*. (Images created by the author, except *C* by Patrick Abbot, and *E*, by Judith Korb.)

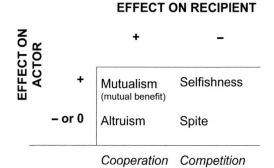

Figure 2. A social behavior is defined here as one that has a fitness effect on another individual. The four types of social actions are classified based on their average effect on the direct fitness (lifetime personal reproduction) of the actor and recipient. Altruism and spite can have either no fitness effect or a negative fitness effect on the actor. Behaviors that have no fitness effect on the recipient are not considered (based on Hamilton 1964). The +/+ interaction is typically called mutualism. However, mutualism is also used to mean a case where the actor helps the recipient and the recipient helps the actor. What the table defines, however, is when the actor helps the recipient and the actor helps himself. For this reason, West et al. (2007) renamed this case "mutual benefit."

picked up upon. Notably, Williams (1966) devoted a chapter to such "group selection" logic and argued strongly that it will typically have a weak effect on the evolutionary process. The logic is as follows. Consider a population of red grouse in which all individuals are territorial *because* this keeps population density low and heather abundant. A mutant nonterritorial grouse would do very well here because his offspring will pack themselves in and rapidly outnumber the resident birds. Over time, therefore, territoriality should be lost as the nonterritorial birds spread. Of course, territoriality does not really evolve to keep population density low but to allow males to attract females, which makes it evolutionarily stable. Nevertheless, the hypothetical example is illustrative: Evolutionary explanations that only consider the group or population level are incomplete. One must also always consider the possibility of natural selection at the level of the individual and sometimes the organelle or gene. This lesson may seem trivial to the modern evolutionary biologist. With the rise of evolutionary thinking in new disciplines, however, so rises the specter of uncritical group-level thinking. It is, for example, still seen in my own field of microbiology (West et al. 2006; Foster 2010).

Relative fame would result for Hamilton following the publication of back-to-back papers in the *Journal of Theoretical Biology* (Hamilton 1964); a briefer version of the same ideas was published a year earlier (Hamilton 1963). The result of Hamilton's Ph.D. thesis, these papers set out an argument for the evolution of altruistic behaviors "which are on average to the disadvantage of the individuals possessing them" (Fig. 2) (Hamilton 1964). Unlike William's book, Hamilton was not particularly motivated by Wynne-Edwards' book (1962), which he cited but only for a few examples. Hamilton, it seems, worked in relative isolation. Indeed, his autobiographical writings suggest that he received little support in his chosen topic from senior colleagues (Hamilton 1996). Nevertheless, he determinedly ploughed on and the papers that resulted are among the most cited in evolutionary biology. At their heart was family life, so much so that Maynard-Smith christened the resulting theory "kin selection" in a *Nature* paper (Maynard Smith 1964) somewhat controversially published while Maynard Smith was reviewing Hamilton's *Journal of Theoretical Biology* papers. Hamilton had termed his theory "inclusive fitness."

Hamilton's key insight was that evolution by natural selection is driven not only by effects on a focal individual's own reproduction (direct fitness), but also by any effects on the reproduction of individuals that are genetically similar (indirect fitness). More precisely, indirect fitness effects occur when the actor and recipient of a social action have an above average (or below average) probability of sharing alleles at variable loci. Helping or harming relatives will affect an actor's fitness in proportion to the genetic similarity between actor and recipient, where "similarity" is measured relative to the population average. This is genetic relatedness. Only by accounting for such indirect effects, as well as direct effects, can the selective effects of a behavior be understood. It is in this sense that Hamilton's fitness accounting scheme is "inclusive." A complementary fitness accounting scheme, also first envisioned by Hamilton, that again separates direct and indirect effects is that of "neighbor-modulated" fitness. The only difference to inclusive fitness is that instead of calculating the effects *from* a focal individual to its reproduction and that of neighbors, fitness accounting is done all in terms of effects *to* the focal individual's reproduction from itself and its neighbors. The difference is largely just a choice in accounting, however, and the two approaches should result in identical predictions (Frank 1998; Wenseleers et al. 2010).

The publication of Wynne Edward's book and Hamilton's papers was the spark for the controversy that continues today. Reaction to Hamilton's papers was relatively slow, but by the 1970s, there was an ever-polarizing opinion against arguments based on group- and population-level selection. This view was crystallized in the success of Dawkin's "The Selfish Gene" (1976), which to a large extent is based on Hamilton's insights. Ironically, this crystallization occurred several years after Hamilton himself had been convinced of the validity and importance of group selection in evolution. Sometime before 1970, in a letter from one solitary eccentric to another, one George Price wrote to Hamilton to explain that he had discovered a new way to view the evolutionary process (Hamilton 1996; Schwartz 2000). What would later become known as the Price equation naturally led to a way to partition the evolutionary process into effects at the level of the individual and the level of the group. Seeing this, Hamilton realized that natural selection above the level of the individual must occur and contribute to the evolutionary process. Price would soon after make a second major contribution to evolutionary biology by introducing ideas from economic game theory, although it was down to Maynard Smith and others to actually publish and popularize the idea (Maynard Smith and Price 1973; Axelrod and Hamilton 1981; Maynard Smith 1982).

By 1974, Price would be dead. It appears that Price was so affected by his discoveries to believe them miraculous. He turned to a religious selflessness that culminated in suicide. Filled with regret, Hamilton (1996) later recalled of Price: "I am pleased to say that, amidst all else that I ought to have done and did not do, some months before he died I was on the phone telling him enthusiastically that through a "group-level" extension of his formula I now had a far better understanding of group selection acting at one level or at many than I had ever had before." Hamilton published his insights on group selection in 1975 and, in the same year, D.S. Wilson published the first of his many papers that argue for the validity of group selection as a process in evolution (Hamilton 1975; Wilson 1975a). What had changed since the resounding attack on group selection by Williams and others in the 1960s? Opinions differ on this and indeed on whether anything substantial had changed (West et al. 2007, 2008; Wilson 2008a). What is clear, however, was that group selection now had a theoretical basis that showed it to be quite compatible with inclusive fitness or kin selection thinking. With this came the key observation that group selection can work as long as there is some degree of genetic differentiation among groups. That is, members of groups must be more genetically similar to their group average than to other groups, otherwise natural selection has no variation on which to act. Put another way, at the level of the group, there must be genetic relatedness (more on what relatedness exactly means here below). By 1975 then, kin and group selection had been combined, at least in some minds and the controversy was over (see next page). Or at least it should have been. In fact, we still face high-profile claims from one side that group selection as a process cannot work (Alcock 2005) and from the other that we do not need kinship-based thinking but rather a group selection framework (Wilson 2005, 2008b; Wilson and Holldobler 2005; Wilson and Wilson 2007), as again argued at Cold Spring Harbor (E.O. Wilson, pers. comm.). How can this be? There are of course many reasons both sociological and scientific, but, as is so often the case, at the heart of the confusion lies semantics. A necessary evil, therefore, in any defense of sociobiology is to discuss the semantics—as well as the theory and data—that speak to the debates.

THE SEMANTICS AND THE THEORY

Kin Selection

Either you are some kind of amnesiac capable of unconsciously fabricating an anecdote harmful to the reputation of a fellow scientist or else you are a person capable of fabricating such an anecdote conciously [sic] as part of an attempt to avoid the discomfort of admitting intellectual indebtedness to a younger man. The first supposition is the best that I can think of you.

Letter from Hamilton to Maynard Smith
(19 October 1977)

Perhaps the single most inflammatory term in modern sociobiology is the "kin" of kin selection. Hamilton was never happy with Maynard Smith's term, and they crossed swords in print and in private on several issues (Schwartz 2000). For many years, Hamilton felt Maynard Smith was trying to diminish the impact of his work. This includes the belief that Maynard Smith fabricated an anecdote about Haldane (Maynard Smith 1975) in order to create a false precedent on kin selection. Haldane had in fact published his brief thoughts on the subject (Haldane 1955) and, strangely, in an article that Hamilton cited (Hamilton 1964). Hamilton would later apologize, and, anyway, his opinion on the term "kin selection" was no match for its catchiness. To this day, it is typically used in place of inclusive fitness in discussions of Hamilton's work. The result has been an emphasis on kinship in discussions of Hamilton's work rather than the more general concept of genetic relatedness. This seemingly minor distinction proves to be critical in discussions of whether kin and group selection are equivalent. Why is this? The answer lies in the fact that kinship—being a member of the same family—is only one way to generate an above-average genetic similarity among individuals (relatedness). There are other ways, the most celebrated being the concept of a green beard gene. So named by Dawkins (1976) but dreamt up by Hamilton (1964), this is the statement that alleles that promote the energetically costly helping of others can spread if they can both identify themselves in other individuals and preferentially direct help to those individuals. Greenbeards are less likely to drive stable cooperation than kinship because of the potential for "falsebeard" individuals who do not help others but retain the signal to receive help (Gardner and West 2010). Importantly for the theory, however, greenbeards illustrate that there can be indirect fitness effects even if interacting individuals are not kin in the sense of being family relatives. This thought experiment, and the subsequent discovery of green beards in multiple species (Keller and Ross 1998; Queller et al. 2003; Smukalla et al. 2008), illustrates that Hamilton's (1964) kin selection is possible whether genetic associations occur through kinship or through other means (Foster et al. 2006a). Some proponents of group selection, however, like to take the narrow definition of kin selection that only includes family groups (Wilson and Holldobler 2005). From there, one can argue that because relatedness can occur without strict kinship, evolution can occur by group selection without kin selection, e.g., via green beard genes. If one does take a narrow definition of kin selection, this claim does follow, but hopefully, it is also clear that this is not a deep objection to the unity of sociobiology.

A more sophisticated objection to the equivalency of group and kin selection comes from the observation that group selection can occur in randomly formed groups of unrelated individuals (Wilson 1990). Consider, for example, a hypothetical Prairie dog group (known as a "town") containing unrelated individuals in which a particular Prairie dog's survival depends on the digging of tunnels that allow the group to forage. It may well pay the Prairie dog to expend her energy to dig tunnels that benefit all members of the town—what can be called a "group trait"—because this has a feedback benefit on her survival. This constitutes a form of group selection for digging behavior but does not require genetic relatedness of any sort among the members of the town. Does this run contrary to kin selection thinking? To answer this, we must first separate two possible goals that are often conflated. The first is the goal of identifying differences between kin selection and group selection as *processes,* and the other is identifying differences between kin and group selection as *theoretical frameworks*. In answer to the first goal, because the individuals in the group are unrelated by any definition, one can argue that there is indeed no kin selection, meaning the process, in this scenario. However, for the theory, the critical point is that our focal Prairie dog is a recipient of the benefits of digging (as well as is the rest of the town). She should be, hence, included in the calculation of relatedness toward recipients, which means that positive relatedness from actor to recipients does have a role simply because the Prairie dog is related to herself. As such, the example is understood equally well from a kin selection or group selection perspective. One just has to be careful about how relatedness is calculated. Again, whether or not one wants to call this kin selection a process is a matter of semantic preference, but, again, it should be clear that there is little reason for this to be a point for continued debate. What can

THE DIFFERENT WAYS OF ANALYZING SOCIAL EVOLUTION

To more formally illustrate the different methods for analyzing social evolution, I reproduce here Box 1 (with minor modifications) from Wenseleers et al. (2010), which is a more careful discussion of social evolution theory than found here. In this box, we analyze Frank's (1994, 1995) "tragedy of the commons" model, which has been successfully applied to a variety of biological problems (see, e.g., Frank 1994; Foster 2004; Wenseleers et al. 2004a,b). The tragedy of the commons states that each individual would gain by claiming a greater share of the local resources, but that the group would perish if all local resources were exhausted (Hardin 1968). Frank's model captures this tension between group and individual interests by writing individual fitness as

$$w_{ij} = (1 - g_i) \cdot (g_{ij}/g_i), \quad (B1)$$

where g_{ij} and g_i are the individual and group mean breeding values for a behavior that causes individuals to selfishly grab local resources (normalized to go from 0 to 1). In this simple model, $(1 - g_i)$ is the group's productivity, which declines as the average level of selfishness g_i increases (we assume linearly, but this can easily be relaxed) (Foster 2004), and g_{ij}/g_i is the relative success of an individual within its group. Similarly, we can write the fitness of other members in the group as

$$w' = (1 - g_i) \cdot (g'/g_i), \quad (B2)$$

where g' is the average level of selfishness of these other individuals. Note that with a group size of n, $g_i = (1/n)g_{ij} + ((n-1)/n)g')$, which we can substitute into Equations B1 and B2.

From a neighbor-modulated fitness perspective, a rare mutant that is slightly more selfish than the wild type is favored when

$$\partial w_{ij}/\partial g_{ij} + \partial w_{ij}/\partial g' \cdot r > 0 \quad (B3)$$

because an individual carrying the mutation would experience a direct cost $\partial w_{ij}/\partial g_{ij}$ but, with probability r, be paired with group mates that also carry the mutation, hence resulting in a return benefit of $\partial w_{ij}/\partial g'$.

Similarly, from an inclusive fitness perspective, a rare, slightly more selfish mutant would be favored when

$$\partial w_{ij}/\partial g_{ij} + (n-1) \cdot \partial w'/\partial g_{ij} \cdot r > 0 \quad (B4)$$

because an individual actor that expressed the mutant behavior would experience a direct cost $\partial w_{ij}/\partial g_{ij}$ but impose a cost of $\partial w'/\partial g$ to each of its $n - 1$ group mates, which are related by r to itself. It is easily checked that because $\partial w'/\partial g = (\partial w/\partial g')(g'/g_{ij})/(n-1)$ and because mutations have small effect so that $g' \cong g_{ij}$, $(n-1) \cdot \partial w'/\partial g_{ij} = \partial w_{ij}/\partial g'$, inequalities B3 and B4 are therefore equivalent.

From a levels of selection perspective, selection would be partitioned into components that are due to the differential fitness of groups with different mean levels of selfishness and the differential success of more versus less selfish individuals within groups. Specifically, if we call G group productivity and I individual fitness relative to other group members, we have $G = w_i = (1 - y_i)$, $I = w_{ij}/w_i = y_{ij}/y_i$ and individual fitness $w_{ij} = G \cdot I$. A more selfish mutant will be selected for when positive within-group selection balances with negative among-group selection:

$$\partial w_{ij}/\partial g_{ij} \cdot (1 - R) > -\partial w_i/\partial g_i \cdot R, \quad (B5)$$

where R and $1 - R$ are proportional to the between-group and within-group genetic variances and $R = (1/n) + ((n-1)/n) \cdot r$. Note that the among-group and within-group selection components are also sometimes calculated in an equivalent way as $\partial w_{ij}/\partial G \cdot dG/dg_{ij} = I \cdot \partial G/\partial g_i \cdot dg_i/dg_{ij} = I \cdot \partial G/\partial g_i \cdot R$ and $\partial w_{ij}/\partial I \cdot dI/dg_{ij} = G \cdot (\partial I/\partial g_{ij} \cdot dg_{ij}/dg_{ij} + \partial I/\partial g_i \cdot dg_i/dg_{ij}) = G \cdot (\partial I/\partial g_{ij} + \partial I/\partial g_i \cdot R)$ (cf. Ratnieks and Reeve 1992), which has the advantage that these only require the calculation of derivatives and do not involve variances.

Differently still, using contextual analysis, we can see that a more selfish mutant can invade when

$$\beta_{w_{ij}g_{ij} \cdot g_i} + \beta_{w_{ij}g_i \cdot g_{ij}} \cdot R > 0. \quad (B6)$$

Where the β terms are the partial regression of individual genotype on fitness holding group genotype constant, and the partial regression of group genotype on individual fitness holding individual fitness constant. Reassuringly, the evaluation of the partial derivatives in Equations B3–B6 for the case where $g_{ij} \cong g' \cong g_i \cong \bar{g}$ shows that no matter how we partition social evolution, the net selective effect is the same and that an equilibrium is reached when $g^* = 1 - R$, i.e., the equilibrium level of selfishness decreases as relatedness increases. At this equilibrium, no mutant that behaves slightly differently can invade in the population (Maynard Smith 1982). In addition, it can be checked that the equilibrium is evolutionarily stable, i.e., a fitness maximum, because the derivatives of the above fitness gradients D (Equations B3–B6) with respect to g_{ij} are negative. Finally, an additional stability criterion, convergence stability, specifies whether the equilibrium is an attractor or not, and is therefore attainable, and requires that the fitness gradient is positive when evaluated for g slightly below g^* and negative when g is slightly higher than g^*. Formally, this occurs when

$$\left. \frac{\partial D}{\partial g} * \right|_{g_{ij} = g' = g_i = g^*}$$

(Eshel and Motro 1981; Taylor 1996).

A strategy that is simultaneously evolutionarily and convergence stable is termed a continuously stable strategy (CSS) (Eshel 1983; Christiansen 1991), and it can be checked that the equilibrium in our example is indeed a CSS. Strategies that are convergence stable but not evolutionarily stable, however, are also possible, and can lead to disruptive selection and evolutionary branching (Metz et al. 1992; Geritz et al. 1998). Evolutionary branching points are interesting, because they provide us with the conditions under which continuous or mixed strategy ESSs (evolutionarily stable strategies) would be expected to evolve toward discrete strategy ESSs (see, e.g., Doebeli et al. 2004).

be, and is, still debated is whether kin selection or group selection theory is better for ease of calculation and making testable predictions (West et al. 2007, 2008; Wilson 2008a). Here, publication volume and impact seem to side with kin selection theory, but their utility will ultimately depend on the question at hand (Foster 2006).

The issue of distinguishing between theoretical framework and process is also raised by Nowak's *Five Rules for the Evolution of Cooperation*, which therein are kin selection, group selection, network reciprocity, reciprocal altruism, and indirect reciprocity (Nowak 2006). These five "rules" represent a mixture of both framework and process. For most biological problems, including the evolution of Hamilton's altruism (see Fig. 2), kin selection, group selection, and network reciprocity are simply different ways of conceptualizing the same processes (Lehmann et al. 2007a,b). In contrast, reciprocal altruism and indirect reciprocity are processes that are distinct from each other, as they are from the other three rules (although reciprocity can be thought of in terms of networks). A way to avoid conflating equivalent descriptions is to take a real biological problem, for example, the evolution of worker behavior in a wasp nest, and ask which of the frameworks can be applied to explain it. For the wasps, one can use kin selection, group selection, or an interacting network of wasps, but the two forms of reciprocity would not apply.

Altruism

A related bone of contention is whether the evolution of altruism strictly requires kin selection or whether it can occur with group selection alone. Answers in favor of the former (Hamilton 1964; Foster et al. 2006a) and the latter (Wilson 1990; Fletcher and Doebeli 2009) exist in the literature, with the answer again depending on how one defines altruism and indeed kin selection. To proceed, I must assume that the reader accepts that "altruism" can be defined in terms of evolutionary fitness at all. It is common to hear the lay person, humanities scholar, and even some scientists object to a fitness-based definition of altruism, arguing instead that altruism refers to an actor's conscious intention to do good. To such understandable objections, I note only that the founding discussions of the definition of altruism—shortly after its invention by Auguste Comte—include both intention-based and fitness-based usage (Dixon 2008; Foster 2008). The modern preference for the intentional definition is just that, a preference only.

Accepting a fitness-based definition of altruism, however, only leads to further choices. Here, I defer to the primacy of Hamilton's role in sociobiology. Hamilton defined altruism as an action that benefits others but will, on average, decrease the lifetime personal reproductive fitness of the actor (Fig. 2). This definition sees altruism in the social insect worker that could reproduce but instead raises her mother's offspring. However, it denies altruism in cases where there are feedback benefits to an action that eventually increase personal reproduction. Our tunnel-digging Prairie dog, for example, is not being altruistic by Hamilton's definition because she will ultimately benefit from the tunnels she builds via positive effects on her town's foraging. Hamilton's definition has been called strong altruism in order to emphasize its stringency and differentiate it from a definition that is often favored by the group selectionist (Wilson 1990). The group selection definition of altruism includes any behavior that lowers the actor's competitive ability within the group but benefits the group as a whole. Hamilton's strong altruism is included here as a subset but so are examples such as the Prairie dog where within-group fitness decreases, but overall reproduction increases (weak altruism). Returning then to the question of whether altruism can be favored without kin selection—where kin selection is natural selection effected by relatedness among individuals in a group—the answer for strong altruism is no (Foster et al. 2006a,b). Strong altruism relies on relatedness among individuals because this is the only way that an individual can gain an inclusive fitness benefit that compensates for a guaranteed reduction in lifetime reproduction.

There has been a recent claim to the contrary (Fletcher and Doebeli 2009). This appeals to an imagined scenario where different loci encode strong altruism in different individuals. Even though individuals are genetically different, altruism *can* invade if altruists are for some reason kept together and away from defectors. The problem with this argument is that this outcome will be short lived. Any modifier that causes carriers of one of the altruistic loci to no longer be altruistic would outcompete other altruist loci within any group. One might counter this with the added assumption that that new defector mutants are also somehow excluded from altruist groups. However, this then implies a causal link between "altruism" and personal reproductive benefit, via group membership; i.e., it is not altruism but mutualism (Fig. 2). The same argument can be made for cases of greenbeard cooperation, i.e., it is either unstable or a case of mutualism (see Smukalla et al. 2008). In contrast, weak altruism *can* evolve stably among unrelated individuals when helping others also provides personal reproductive benefits (Wilson 1990). Finally, "reciprocal altruism" (Trivers 1971) is something different again, where there is a delayed direct fitness benefit that comes back to the actor from the recipient of a social action. The average fitness effect of reciprocal altruism is to always increase the actor's personal fitness. These different meanings of altruism are further discussed in Foster (2008).

Group Selection

It will come as no surprise that there is also confusion and debate over the meaning of group selection. Leaving aside how to actually define a group, multiple meanings of group selection exist (for review, see Okasha 2006; Gardner and Foster 2008) and this might be taken as evidence for the insufficiency of group selection theory (West et al. 2008). The most common definition of group selection comes from the multilevel selection form of the Price equation (Price 1970, 1972), which partitions the evolutionary response to natural selection into change within groups and among groups. From this, the definition of group selection is simply the differential productivity and survival of groups. This definition, however, ascribes group selection to completely asocial individuals that just happen to be near one another. Consider some hypothetical snails that live on a patchy resource but do not affect one another. If large snails survive better and some patches contain more large snails, those groups will have a higher productivity and group selection is detected in the Price equation. For those interested in the effects of group selection because of *social* evolution, this can seem unsatisfactory (Okasha 2006).

An alternative way of assessing group selection, as developed by Heisler and Damuth, makes use of a statistical technique from the social sciences called contextual analysis (Heisler and Damuth 1987; Goodnight et al. 1992; Okasha 2006). Again, this partitions natural selection into individual and group effects, but, importantly, it only pulls out the group effects on fitness that cannot be explained by individual effects. This means that group selection is only detected when there are effects of group identity on fitness that remain once one has controlled for individual effects on fitness (one calculates the partial regression between group phenotype or genotype and individual fitness).

Further testament to the equivalence of kin and group selection comes from the fact that contextual analysis is nearly identical to the modern kin selection models that are based on neighbor-modulated fitness (see introduction, boxed text), although the two were largely independently derived within evolutionary biology. Personally, I find contextual analysis an intuitive way to define group selection, although the Price partition is more commonly used (Gardner et al. 2006; Gardner and Grafen 2009). An objection raised to contextual analysis is that it detects group selection when all groups have identical productivity but the ranking of individuals within groups matters for fitness (soft selection). There is group selection without groups actually differing in their fitness (Heisler and Damuth 1987; Goodnight et al. 1992; Okasha 2006). In the end, group selectionists must choose whether they are most interested in the differential productivities of a subdivided population (Price) or the evolution of social traits (contextual) and then go from there (Okasha 2006). But again, the existence of this choice does not undermine the enterprise.

Group Adaptation

By now, I hope that the reader is (or remains) convinced that the evolution of a strongly altruistic behavior—like a self-sacrificing insect worker—can be phrased in terms of kin or group selection. This is true even though kin selection is the more natural fitness partition for the case of strong altruism (Foster et al. 2006a). The kin selectionist can view the dying worker in terms of a beneficial trade-off between personally passing on the allele for conditional self-sacrifice versus helping her more fecund relatives to pass on the allele. The group selectionist can see that the sacrifice will have a negative fitness effect on the individuals that express it but a positive fitness effect on the groups that carry it. The two perspectives can be also combined as was done by H.G. Wells (Wells et al. 1929) above. Should self-sacrifice also be viewed as a group *adaptation*? To attempt an answer, one must first distinguish between selection and adaptation (Gardner and Grafen 2009). Natural selection is the change in allele frequencies that results from differences in survival and reproduction, where critically, these differences are caused by an individual, or group, carrying one allele (or set of alleles) rather than another. Adaptation is more difficult to define but is typically viewed as the process of optimizing a phenotype for an organism's environment (Fisher 1930), for example, the production of a more efficient wing for flight. What then distinguishes group adaptation from individual adaptation? This remains an open point of discussion in sociobiology, and legitimately so (Gardner 2009; Gardner and Grafen 2009).

If adaptation is the process of optimizing a phenotype for its environment and natural selection is the process that drives this optimization, we can reach a definition of group *adaptation* from the contextual analysis definition of group *selection* (see previous section). Group adaptation is the process of optimizing cooperative phenotypes that depends on group context. More specifically, it is the optimizing process that is dependent on the component of natural selection that contextual analysis ascribes to the group. Put this way, the products of group adaptation are those shaped by the partial effect of group phenotype on individual fitness (controlling for the effects of individual phenotype). In modern systems, this should translate roughly into the cooperative components of phenotypes that only provide a fitness benefit in the group context. These very general conditions will find evidence for group adaptation in many situations, including workers that sacrifice themselves to defend their colony, our tunnel-digging Prairie dog, and any degree of self-restraint in resource use that promotes group survival; all phenotypes that can only be explained by group-level function.

Gardner and Grafen (2009) offer a more stringent definition of group adaptation in their discussion of the superorganism: the associated idea that individuals in highly organized societies are comparable to cells in an organism (Wheeler 1911). For Gardner and Grafen, true group adaptation implies that the group is acting as the agent of natural selection, and they are only able to find that the latter is the case when societies have negligible reproductive conflict among their members. The argument is that expressed conflict means that individual-level selection is compromising group function and, accordingly, adaptation is not operating to generate function at the level of the group. Instead, Gardner and Grafen suggest that most phenotypes are best viewed as the product of individual adaptation, which, in turn, is the result of inclusive fitness maximization. Although this argument carries merit, and rescues the idea of the individual as an evolutionary agent (Gardner and Foster 2008), it excludes many derived social phenotypes from group adaptation. Stingless bees, for example, suffer horrific conflicts over queen production whereby hundreds of larvae will try to develop into new queens only to be immediately executed by the workers (Wenseleers et al. 2003; Wenseleers and Ratnieks 2004). However, these same workers will aggressively (albeit by biting rather than stinging) defend the nest in a manner intuitively consistent with group adaptation. Another potential objection to

Gardner and Grafen's (2009) argument is that true multicellular organisms are not free from conflict either (Wilson and Wilson 2008) but suffer from cancers and selfish genetic elements that may have a nonnegligible effect on organismal adaptation (Hurst et al. 1996; Burt and Trivers 2006; Merlo et al. 2006).

Rather than taking conflict to indicate a complete lack of group adaptation, therefore, one can instead ask how well adapted is a particular social trait or group? Put another way, the effectiveness of group adaptation can be assessed by asking how similar an extant group is to one that completely lacks conflict, such as a clonal group (Foster 2004; Ratnieks et al. 2006). Such thought experiments do not suppose that perfectly cooperative groups are truly optimal, just that their adaptations are the product of the best optimization *process* that is available, given realistic constraints of both biology and the evolutionary process. Even so, such assessments are typically only approximate as the exact form of a conflict-free group is unknown in most species. Moreover, even clonal groups might not be always well adapted because natural selection can disfavor clonality itself to combat disease (Hamilton 1987; Sherman et al. 1988) or improve social organization (Brown and Schmid-Hempel 2003). However, this objection may not be watertight given that natural selection on clonal groups is known to produce genetic diversity while maintaining clonality where it matters, as found in one's own immune system.

A related approach to group adaptedness is that of Queller and Strassmann (2009) who offer a two-dimensional index of "organismality," upon which social groups—of one or more species—can be placed. Their first axis "conflict" defines the group by the degree of observed conflict, which is roughly analogous to my comparing a real group to a perfectly cooperative one. Their second axis "cooperation" adds to this by asking how much do individuals appear to help one other? An illustrative example here are nongalling aphids that might exist in clonal groups—with little conflict—but do not really help one another, perhaps because there is little to be gained (Queller and Strassmann 2009). By these measures, organismality is awarded to groups with low conflict and cooperation that pays great fitness dividends. The argument that a multiple species group can be viewed as an organism (Queller and Strassmann 2009) also invites the possibility of community-level adaptation. From the perspective of contextual analysis, community-level adaptation would be when positive fitness effects on other species provide feedback benefits to a focal species that shape its phenotype, where again direct fitness effects of phenotype on the focal individual (or group) have been removed. To conclude, one can argue that the process of group adaptation is as common in social groups as is contextual group selection, but it remains an open question how adapted are different social species relative to the ideal of a conflict-free group.

THE DATA

[the problem of the superorganism] must be solved by field and laboratory research, not by mathematical modeling.
E.O. Wilson (pers. comm.)

In their recent book, Holldobler and Wilson also revisit the idea that social adaptations can be so derived that a group of animals can be considered a superorganism (Wheeler 1911). As discussed in the last section, this question is closely linked to the problem of group adaptation and, like Hamilton's papers (Hamilton 1964), recent theory has emphasized the importance of genetic relatedness in the emergence of the superorganism (Reeve and Holldobler 2007; Gardner and Grafen 2009). In criticizing theoretical sociobiology, Wilson argues not only that kinship theory has made no useful predictions but also that understanding the superorganism—and, by inference, social evolution in general—is an empirical rather than theoretical question (Wilson and Holldobler 2005; E.O. Wilson, pers. comm.). Although I clearly do not endorse these views, it *is* vital to consider the data. Does empiricism, like theory, support the mutual importance of kin and group selection? The answer again is a resounding yes.

Kin Selection

Hamilton's kin selection centers on the idea that fitness costs and benefits, weighted by genetic relatedness, are critical for social evolution. As such, proving the importance of relatedness for cooperation and altruism (see Fig. 2) does not invalidate the role of other factors that affect costs and benefits, including ecology and preadaptations. These different explanations are not in opposition, as Wilson attests, but rather all parts of one big puzzle. This point of logic is mirrored in the easily neglected distinction between *necessary* and *sufficient* that is used in genetics. An important gene can be strictly necessary for a phenotype, such as obesity, without being sufficient to produce the whole phenotype on its own. Similarly, it can be that relatedness is strictly necessary for strong altruism to evolve but that it is not sufficient without other preadaptations and ecological factors (Korb and Heinze 2008). Although I argue below that relatedness is extremely important for social evolution, therefore, I am not arguing that it is the only factor of importance. And at the risk of repetition, nor am I arguing that kin selection *theory* is only about relatedness. It is not: Fitness costs and benefits are critically important in any calculation of fitness effects (Fig. 2). The theory remains valid when one moves from nonzero relatedness among actor and recipient to zero relatedness (although the calculations do get rather easier).

The Social Insects. The evolutionary origin and maintenance of the spectacular worker castes in social insects have long been a centerpiece for discussion of social behavior and altruism. More specifically, these discussions center on the "eusocial" insects, which are those that display a fixed division between work and reproduction among individuals (Boomsma 2007, 2009). The key eusocial groups are found in the hymenoptera (bees, ants, and wasps) and the distantly related termites, but several other arthropods get close, including thrips, beetles, aphids, and even shrimp. As emphasized above, explaining the patchy distribution of eusociality will likely require consideration of multiple preadaptations, which include the potential to build a well-defended group (nest building and stings; Wilson 1971) and the possibility to guarantee that the individuals one helps will survive if one does not (Queller 1989; Gadagkar 1990). However, the potential role of relatedness in eusociality is also immediately apparent in the finding that the societies of these species are nearly always formed by immediate family. There are exceptions, including the unicolonial ants that form massive networks of interconnected nests with so many queens that the workers can end up barely related to one another. However, these are thought to be evolutionary dead ends, and, anyway, it is clear that when the worker castes originally evolved, they were living as families (Helanterä et al. 2009).

More pause for thought might come from "primitively" eusocial hymenoptera: Species that lack physically distinct queen and worker castes and whose nests can be founded by multiple unrelated foundresses (Bernasconi and Strassmann 1999). As these foundresses begin to produce daughter workers, a family group is created, but there are still many unrelated individuals in the nest. This suggests that worker behavior can evolve in the presence of relatively low relatedness, but importantly the "workers" in this case are individuals—like the queen—who have the potential for independent reproduction. That is, worker behavior in these species likely evolves through a mixture of kin-selected benefits and delayed reproductive benefits (Boomsma 2009). Nevertheless, if this were to be the route by which the truly eusocial species—such as the honeybee—were to evolve, one could argue that only low relatedness was necessary for the origin of the derived worker castes. This would not invalidate kin selection thinking but would emphasize the importance of ecological factors relative to family life. However, arguing from anecdote is not sufficient. What is needed are phylogenetic analyses and here the data are clear: The multiple ancestors of the eusocial hymenoptera were very likely all species with a single queen that mated to a single male (Boomsma and Ratnieks 1996; Foster and Ratnieks 2001; Boomsma 2007, 2009; Hughes et al. 2008). What is striking about this is that single mating appears to be an unusual state of affairs for most animals, where promiscuity is common. This led Boomsma (2007, 2009) to argue that monogamy—and the high family relatedness that results—may be a strict prerequisite for the origin of the highly eusocial insects. More data are needed to know whether strict monogamy is always absolutely required, but the evidence that monogamy favors eusociality in hymenoptera is already compelling. Eusocial species arose from family groups, and, for the cases analyzed, these ancestral societies were founded by a single mother mated to a single male.

One can also ask of the role of relatedness in the *maintenance* of eusociality. Many derived eusocial species such as honeybees, yellow jacket wasps, and leaf-cutter ants have secondarily evolved multiple mated queens, which causes low relatedness among the workers. Why then do these species not suffer from severe internal conflicts among their workers over reproduction? For the long answer to this question, one can turn to dedicated reviews (Bourke and Franks 1995; Ratnieks et al. 2006), but, briefly here, multiple mating actually helps to resolve conflicts once derived eusociality is in place. Moreover, the way that this works provides further evidence for the importance of relatedness in social evolution. In the derived eusocial hymenoptera, workers cannot mate but can typically produce eggs. These eggs develop into males because of the unusual haplodiploid genetics of the hymenoptera, in which females are diploid and males are haploid. Workers can therefore compete with the queen over male production, and what stops this from becoming a reproductive melee is that workers tend to eat each others' eggs. The workers *police* one other.

Worker policing is predicted to be prevalent in species where the queen mates many times because in haplodiploids, this reduces relatedness among the workers but not between the workers and their brothers (the queen's sons). With queen multiple mating, therefore, the workers are expected to favor the queen's sons over each others' (Ratnieks 1988). The outcome is that once distinct queen and worker castes have evolved, reduced worker relatedness is predicted to promote cooperation by promoting policing. In support of this, there is a negative correlation between worker relatedness and the effectiveness of policing and a positive correlation between relatedness and the proportion of workers that attempt to lay (Fig. 3a) (Wenseleers and Ratnieks 2006a,b). Even better, in colonies where the queen has died, the correlation between relatedness and worker reproduction reverses (Fig. 3b) (Wenseleers and Ratnieks 2006b), exactly as predicted by the basic theory (Hamilton 1964; Wenseleers et al. 2004a). Without the queen, the workers now have no reason to police for the queen's benefit, and relatedness only functions to limit the severity of competition among the workers. The outcome in the highly multiply mated honeybee is that colonies with a queen see almost no conflict among the workers, but queenless colonies are highly chaotic. In contrast, the queen's presence has little effect on worker rebellion in wasp species where the queen mates only once (Fig. 3).

Evidence for the role of genetic relatedness is also seen in other social insects, including termites, where a single king and a single queen head the colonies in a monogamous relationship, and aphids, which are clonal. Like monogamy, clonality makes raising parents' offspring just as genetically beneficial as raising one's own offspring (Boomsma 2007). Furthermore, in both termites and aphids, there is evidence of kin effects when members of different colonies meet. Aphids are less likely to altruistically defend the colony when they are in the nest of a foreign clone (Abbot et al. 2001), and, when short on food, termites will preferentially direct aid to their relatives within colonies where two different families have fused (Korb 2006). A recent paper communicated by E.O. Wilson to *Proceedings of the National Academy of Sciences* claimed that eusocial evolution in termites can occur "even when indirect fitness benefits are low or nonexistent" (Johns et al. 2009). The evidence for this was primarily the observation that colony fusion can temporarily reduce relatedness within colonies and lead to a system of serial monogamy, rather than lifetime monogamy. However, this example concerns one of the lower

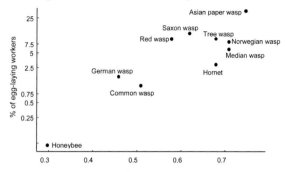

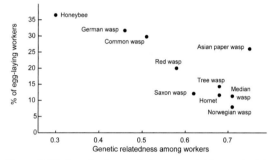

Figure 3. Worker altruism in derived eusocial insects is driven by a combination of kin selection and worker policing. (*A*) Altruistic self-restraint due to policing. In colonies where the mother queen is alive, the workers can raise either the queen's or other workers' eggs. In species where relatedness among workers is high, they tend to raise the workers' eggs because they are highly related to them, but in species where relatedness among workers is low, such as the honeybee, workers remove one anothers' eggs. This reduces the benefits to worker reproduction which, alongside indirect fitness benefits, promotes reproductive self-restraint. (*B*) Altruistic self-restraint due to inclusive fitness effects. If the queen dies, the workers compete to lay eggs. However, when relatedness is high, many show altruistic self-restraint and do not attempt to reproduce. (Reprinted, with permission, from Wenseleer and Ratnieks 2006b [©Nature Publishing Group].)

termites, which are now generally considered not to have passed the irreversible threshold toward permanent eusociality as helpers retain ample options for direct reproduction later in life (Korb 2007; Boomsma 2009). So, worker relatedness may occasionally be low in lower termites—which are essentially cooperative breeders (Korb 2007)—and also in isolated branches of the higher termites, as rare secondary developments after eusociality had become irreversibly established. However, nothing in the Johns et al. (2009) study would indicate that either cooperative breeding in the lower termites or eusocial breeding in the higher termites is not associated with full-sibling relatedness as a norm.

Other societies. Although I have focused on the eusocial insects, considerable evidence also exists for the importance of relatedness in other social species, including impressive meta-analyses on social vertebrates. The social vertebrates are bird and mammal species, including the charismatic meerkat, that live together in tight groups and help one another to raise offspring. Here, species with a higher fitness benefit to helping behaviors tend to discriminate more among relatives and nonrelatives as predicted by kin selection theory (Fig. 4a) (Griffin and West 2003). Furthermore, in those species where kin discrimination is weak, there tends to be higher average relatedness in social groups, suggesting that kin-selected benefits to helping can still occur in species that do not show kin discrimination (Fig. 4b) (Cornwallis et al. 2009).

The importance of genetic relatedness is also supported by a recent surge of studies on microorganisms (West et al. 2006; Foster 2010; Nadell et al. 2009). For example, mixing unrelated strains promotes the evolution of rapid wasteful growth in bacterial viruses (Kerr et al. 2006). Reducing relatedness among cells also promotes the success of cheater mutants that do not contribute to the common good in a host of systems, including enzyme secretion in yeast (Greig and Travisano 2004), iron scavenging and quorum sensing in the bacteria *Pseudomonas aeruginosa* (Griffin et al. 2004; Diggle et al. 2007), and in the development of the bacterium *Myxococcus xanthus* (Velicer et al. 2000) and the slime mold *Dictyostelium discoideum*, where a myriad of cheater mutants have now been found (Ennis et al. 2000; Gilbert et al. 2007; Santorelli et al. 2008). Considerable evidence also exists for kin discrimination systems, including bacteria that spitefully secrete a toxin to kill unrelated strains (Gardner and West 2004; Gardner et al. 2004), self/non-self-recognition in slime molds (Mehdiabadi et al. 2006) and single-gene green beard recognition in both slime molds (Queller et al. 2003) and yeast (Smukalla et al. 2008). Finally, there is evidence that microbes can change their social strategy in the presence of relatives versus nonrelatives. For example, it was recently shown that the allocation to spores (germ) versus stalk (soma) in slime molds will increase in the presence of unrelated strains as predicted by kin selection theory (Buttery et al. 2009).

To close the case on kin selection data, "kinship theory" has made exquisitely supported predictions in many other systems. An example is the annual plant that increases allocation to roots when its roots meet a nonrelative (Dudley and File 2007). Another comes from the wasp *Nasonia vitripennis*. Female wasps lay a very female-biased sex ratio when they parasitize fly larvae, which fits well with the predicted effects of relatedness on sex ratios. This is because sons and daughters mate in the host, and it is most efficient to have a few sons and limit competition among them (Hamilton 1967). What is even more impressive is that when a second female lays in the same host, she is able to detect the first brood and shift her sex ratio to produce many more males to compete with the unrelated males already in the host (Werren 1980). The result is a tight association between relatedness among wasps within a host and the sex ratio (Fig. 5). One can even apply kin selection thinking to the evolution of sperm, where a host of surprising social behaviors occur including sperm that hook onto one other and swim as a cooperative group (Pizzari and Foster 2008; Fisher and Hoekstra 2010).

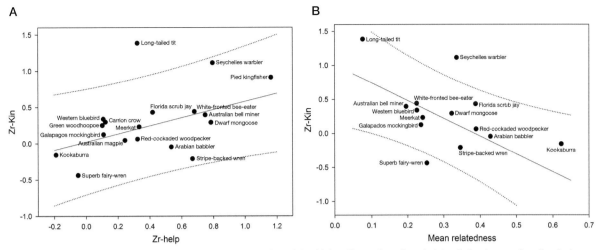

Figure 4. Helping behaviors in social vertebrates. (*A*) Species with a higher fitness benefit to helping behaviors tend to discriminate more among relatives and nonrelatives as predicted by kin selection theory (Griffin and West 2003). "Zr-Kin" and "Zr-help" are statistical measures of the degree of kin discrimination and the fitness benefits of helping behavior, respectively. (*B*) In species where kin discrimination is weak, there tends to be higher average relatedness in social groups, suggesting that kin-selected benefits to helping can still occur in species that do not show kin discrimination. (Reprinted, with permission, from Cornwallis et al. 2009 [©Wiley-Blackwell].)

Group Selection

Group selection has received less empirical attention than kin selection (West et al. 2007, 2008; Wilson 2008a), but due to the equivalency of the theories, support for one can typically be taken as support for the other. Certainly, the wealth of data showing the role of relatedness, costs, and benefits in the evolution of altruism is completely consistent with the idea that individuals are sometimes selected to favor the welfare of their group over personal reproduction.

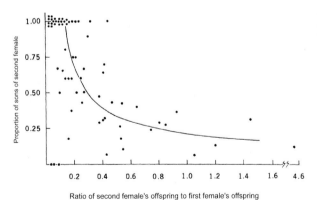

Figure 5. The sex ratio laid by the second female to parasitize a host in the wasp *Nasonia vitripennis*. The line shows the predicted optimal ratio of a second female as a function of the relative numbers of the two females' offspring, which determines average relatedness among the wasps in the host (Hamilton 1967). The plotted points show the actual empirical sex ratios. The wasps also respond as qualitatively predicted to factors such as synchrony in egg-laying and the degree of parasitism in a patch at the time of laying (Shuker et al. 2005, 2006). (Reprinted, with permission, from Werren 1980 [©AAAS].)

There are also explicit tests of group selection logic (Goodnight and Stevens 1997; Wilson and Wilson 2007). These take two key forms: experimental evolution and contextual analysis. Like the classic work on the fruit fly, experimental evolution studies on group selection take organisms through multiple generations and select on certain traits. The key difference to typical studies is that one selects on the group mean of phenotypes, rather than the individuals that best display a trait. Such studies on flour beetles have shown responses to group selection on several traits including body size, group size, and dispersal (Wade 1976, 1977; Craig 1982). A key conclusion is that group selection can work, even when group and individual selection are pitted against each other. A similar study has been performed on leaf area in the plant *Arabidopsis thaliana* (Goodnight 1985), and group selection logic found a commercial application in chickens, where selecting on groups for survival and egg-laying improved both the productivity and welfare of the birds (Craig and Muir 1996; Muir 1996). A nice aspect of the chicken work is that it makes clear that both group and family selection are central to the results: "The hypothesis was tested that selection on the basis of family means for increased survival and hen-housed egg production, when sisters with intact beaks were kept together in multiple-bird cages, would cause adaptive changes in behavior... . The evidence supported the hypothesis." (Craig and Muir 1996). Similarly, the above-mentioned study of Kerr et al. (2006) on the evolution of virulence in bacteriophage has subsequently been taken as evidence for both group (Wilson and Wilson 2007) and kin (Foster 2009) selection thinking.

Experimental group selection has also been applied to communities containing multiple species. The first such study selected on flour beetle population size and dispersal but this time with two species present. Group selection

ization of segments or appendages. Many morphological features can be dissected into distinct components, each repeated in series. These components can be added, deleted, or shuffled to create novel morphologies based on subtle changes in the timing and location of gene expression (Emlen et al. 2005; Brakefield 2006; Monteiro and Podlaha 2009). For example, variation in the expression of genes in the ectodysplasin pathway, part of a deeply conserved network of homeotic (Hox) genes, are associated with a wide array of dentition patterns in vertebrates (Fraser et al. 2009), and changes in this network have resulted in numerous adaptive gains, losses, and changes in the arrangement of teeth (Stock 2001).

Similar to the tooth example above, complex forms of behavior have been productively dissected into simpler "behavioral modules," and experimental approaches that begin with a deconstruction of behavior into component phenotypes have been successful. Ben-Shahar et al. (2004) discovered that *malvolio*, a gene related to sucrose responsiveness in *Drosophila melanogaster,* also affects the age at which honey bees (*Apis mellifera*) switch from working in the hive to foraging. *malvolio* was considered to be a candidate gene for this process because sucrose responsiveness was known to be a module of honey bee foraging behavior; an increase in sucrose responsiveness is associated with the onset of foraging (Rueppell et al. 2006). Sucrose responsiveness is just one of several modules of honey bee foraging, a complex multicomponent behavior that involves learning, communication, and changes in other sensory systems (Seeley 1995; Ben-Shahar et al. 2003).

The third insight from evo-devo is the notion of the genetic "tool kit" for development, i.e., a set of genes that perform highly conserved roles across diverse taxa, e.g., the Hox genes and their role in regulating body form (Gellon and McGinnis 1998) and the paired box (Pax) genes and their role in eye development (Pichaud and Desplan 2002). The genetic tool kit for development is thought to consist of a set of genes with highly specialized functions, especially transcription factors (Carroll et al. 2005). Similarly, there are now several cases of specific genes, pathways, or networks with conserved roles across species that are important in behavior. For example, serotonin influences aggressive behavior across a wide range of animals, from crustaceans to humans (Kravitz 2000; Lesch and Merschdorf 2000).

The fourth insight from evo-devo is the realization that many novel traits arise from evolutionary changes in gene expression (Wray 2007). It has long been known that changes in coding sequence also can be important for the evolution of novel traits (Hoekstra and Coyne 2008); evo-devo has in addition highlighted the lability of gene regulation as a major contributor to evolutionary change. Likewise, changes in gene expression—that affect timing, location, or overall levels of mRNA abundance—are receiving increasing attention for behavioral evolution (Khaitovich et al. 2006). For example, differences in brain distribution of the vasopressin receptor *V1aR* in male voles cause large species differences in affiliative pair-bonding behavior (Lim et al. 2004).

The fifth insight from evo-devo is that certain genes can exert influences on morphological development over different timescales, e.g., both developmental and evolutionary timescales. For example, *bone morphogenic factor 4* and *calmodulin* genes are involved in "real-time" beak development in birds, and they also have been selected upon during evolution to generate the wide variation in beak sizes found in Darwin's finches (Abzhanov et al. 2004, 2006). Likewise, *FOXP2* is involved in the real-time development of speech capabilities in humans and apparently also has been selected upon during evolution to generate diverse forms of animal communication (Shu et al. 2005).

Genetic Architectures for Developmental and Behavioral Traits

Molecular and genetic analyses have revealed additional similarities between developmental and behavioral traits, further arguing for the relevance of evo-devo perspectives to behavior. Both behavioral and morphological phenotypes can have a wide range of heritability estimates, from almost zero to extremely high. Furthermore, the range of heritability estimates for both behavior and morphology overlap broadly (Hoffmann 2000). It is clear that some forms of behavior, as with morphological phenotypes, are highly heritable. This has long been seen in laboratory studies of behavior (Hoffmann 2000) and is now becoming apparent for naturally occurring behavior in the field as well (see, e.g., Dingemanse et al. 2002), including a variety of human behavioral traits (Bouchard and McGue 2003).

It also is well established in both development and behavior that certain single genes can have huge effects on phenotype. In development, a mutation in the *antennapedia* gene in *Drosophila* transforms the antennae into legs (Postlethwait and Schneiderman 1971). In behavior, a mutation in the *fruitless* gene causes male *Drosophila* to court other males (Baker et al. 2001). In addition to these induced mutations, there also are examples of similar large effects for naturally occurring developmental and behavioral variation. In species of stickleback fish (genus *Gasterosteus*), for example, allelic differences in regulatory regions of the *Pitx1* gene determine whether pelvic spines are present or absent (Shapiro et al. 2004). In *Drosophila*, natural allelic variation in the *foraging* gene leads to big differences in foraging behavior, resulting in either "sitter" or "rover" phenotypes (Osborne et al. 1997).

In addition to traits affected by single genes with major effects, both developmental and behavioral traits can have more complex genetic architectures. For example, Mackay and colleagues (for review, see Mackay 2009), using transcriptomic profiling of a large set of strains of *Drosophila*, reported that 20%–40% of the genes in the *Drosophila* genome are involved in the regulation of strain-specific differences in both morphological and behavioral traits.

CHALLENGES TO THE EVO-DEVO APPROACH FOR BEHAVIOR

Despite the above similarities, some notable differences between development and behavior pose challenges for the application of the evo-devo approach to behavior. Studies on the evolution of morphological novelty were initiated only after extensive knowledge was obtained on the mechanistic—genetic and cellular—bases of development, drawing on molecular analyses of model genetic species (Carroll et al. 2005). In contrast, molecular analyses of behavioral evolution are proceeding simultaneously with mechanistic analyses (Robinson et al. 2005) and thus do not benefit from a wealth of prior information, as did evo-devo studies.

Much is known about the neural substrates of behavior, but most of this knowledge is confined to laboratory-based behavioral systems (Carew 2000; Kandel et al. 2000). For naturally occurring behavior in the field, underlying mechanisms have not been as well studied; i.e., less is known about the ways in which circuits of neurons generate a behavior than the ways in which cells and tissues generate a morphological trait. This is no doubt because behavior is generated by the brain, an exceedingly complex organ with highly specialized subregions and patterns of cell connectivity. Because of the complexity of the brain, the mapping of genes to behavior via the brain is more of a mystery than the mapping of genes to morphological trait via cells and tissues.

Another difference between development and behavior relates to timescales. Behavioral change can occur much faster than developmental change, on a timescale that is faster than the rate of gene transcription. Orchestrating changes at this timescale involves the many posttranscriptional and posttranslational changes that occur in the brain, including release of neurotransmitters and the opening and closing of various ion channels. A complete explanation of the molecular basis of behavior and its evolution will require an understanding of how both genomic and neural mechanisms interact across different timescales, from the acute physiological timescale all the way to the evolutionary timescale.

Despite these challenges, perspectives from evo-devo can be applied productively to studies of the evolution of behavior at the molecular level. Moreover, we believe that molecular analyses of behavioral evolution can also help to fill the gap of knowledge on the neural basis of naturally occurring behavior. For example, analyses of the distribution of the vasopressin receptor *V1aR* gene in the brains of male monogamous and nonmonogamous vole species (Insel and Young 2001) led to discoveries linking vasopressin signaling to dopaminergic circuits related to reward processing in the mesolimbic "reward system" of the brain (Young and Wang 2004). Similar payoffs have occurred in studies of social insects, as can be seen in the following section.

EVO-DEVO AND BEHAVIOR IN INSECT SOCIETIES

Eusociality

Previous sections included brief examples from studies of different forms of behaviors in a variety of organisms that indicate that insights from evo-devo are readily applicable to studies of the evolution of behavior at the molecular level. In the next sections, we illustrate this point in greater detail by concentrating on one particular form of social behavior: eusociality in insects. We discuss the evolutionary insights that have been gained from studies of eusocial behavior that use a key element of the evo-devo approach—a focus on gene expression analysis.

Eusociality in insects is emerging as a powerful system to study the mechanisms and evolution of social life in molecular terms (Robinson et al. 2005; Smith et al. 2008). Eusociality is a highly derived system of social organization. Eusociality in the Hymenoptera (ants, bees, and wasps) has evolved independently approximately 10 times (Brady et al. 2006; Hines et al. 2007), each time resulting in three convergent and defining traits: (1) a reproductive division of labor between queens and workers, (2) overlapping generations of individuals sharing a nest, and (3) individuals that care cooperatively for offspring that are not their own ("workers") (Wilson 1971). The behavior of many species of social insects is well understood at the ecological and evolutionary levels. Social insect behavior also is very amenable to considerable experimental manipulation, especially in terms of the ability to precisely manipulate the social environment and study its effects on individual colony members (Smith et al. 2008).

The examples we provide illustrate the evo-devo concepts described above, as follows.

1. *Co-option:* Genes involved in solitary behavior appear to have been co-opted for new functions in a social system.
2. *Modularity:* Molecular analyses have identified modules of solitary behavior as components of social behavior.
3. *Genetic tool kit:* Similar patterns of brain gene expression are associated with division of labor in independently evolved social insect lineages, supporting the idea of a tool kit for behavior in the social insects.
4. *Timescales:* Some of the same genes that are associated with behavioral differences on an evolutionary timescale (e.g., between strains of the same species) are also associated with behavioral differences on shorter timescales as well, within an individual's lifetime.

The remaining concept from above, a focus on gene expression, is implicit in our treatment of the other four concepts; i.e., the examples we provide have used gene expression analysis to derive insights into the evolution of eusocial behavior at the molecular level. A focus on gene expression might be particularly important for analyses of behavior in social insects (Robinson and Ben-Shahar 2002; Linksvayer and Wade 2005). Just as cells within an organism develop into distinct cell types on the basis of differential gene expression, individuals within a social insect colony develop into "castes" (queens and workers) that differ in both behavior and (sometimes) morphology (Wilson 1971); this also typically occurs on the basis of differential gene expression (Robinson et al. 2005). Several lines of research with social insects suggest that changes in gene regulation across several different timescales are important in the evolution of social behavior.

Co-option: **Nutrition and division of labor for foraging.** Genes involved in solitary forms of foraging behavior appear to have been co-opted for social foraging in honey bees. Evidence comes from studies of *foraging* and *malvolio* and genes in the insulin/insulin-like growth factor signaling (IIS) pathway.

Many basic forms of behavior, such as food gathering or foraging, seem to be similar between solitary and social bees—both collect pollen and nectar from flowers. However, there is a fundamental difference: A social bee's behavior is adapted to increase the fitness of its colony rather than its own personal fitness. This gives rise to important differences in when and how foraging is performed. Despite these differences, recent results suggest that genes associated with food gathering and eating in solitary insects have been co-opted to regulate social foraging in honey bees and other species of social insects.

In *D. melanogaster*, the *foraging* gene (*for*) encodes a cGMP-dependent protein kinase (protein kinase G [PKG]), and naturally occurring allelic variation in this gene results in two phenotypes: "sitters" and "rovers" (Osborne et al. 1997). Although *D. melanogaster* lives most of its life in solitary fashion, the behavioral variation associated with these allelic differences suggested an interesting parallel with the food-gathering behavior of the honey bee. Honey bees exhibit age-related changes in behavior; e.g., young bees work inside the hive (analogous to fly "sitters"), but as they age, they "rove" outside the hive in search of food.

An ortholog of *Drosophila for* was found to regulate this age-related change in behavior in honey bees. Levels of *for* mRNA in the bee brain are greater in foragers than in bees working in the hive, and experimentally activating PKG causes precocious foraging (Ben-Shahar et al. 2002). *for* expression is socially regulated in honey bees: Levels of *for* mRNA are also elevated in the brains of young bees induced to forage early in response to a lack of older bees (Ben-Shahar et al. 2002). cGMP signaling also affects feeding arousal in *Caenorhabditis elegans* (Fujiwara et al. 2002), and differences in the expression of *for* between foraging and nonforaging individuals have been reported in two ant species (Ingram et al. 2005; Lucas and Sokolowski 2009) and two wasp species (Tobback et al. 2008; AL Toth et al., in prep.), demonstrating the evolutionary lability of pathways involving PKG signaling.

mvl encodes a manganese transporter, and a mutation at this locus in *Drosophila* causes a loss of responsiveness to sucrose; this deficit is eliminated by treatment with manganese (Rodrigues et al. 1995). As stated above, sucrose responsiveness is a behavioral module of foraging in honey bees (Pankiw and Page 1999); an increase in sucrose responsiveness is associated with the onset of foraging. Quantitative trait locus (QTL) analysis has revealed that the genetic architecture of sucrose responsiveness and onset age of foraging are related (Rueppell et al. 2006). These results suggested that *mvl* also was a good candidate gene for the regulation of division of labor in honey bees. *mvl* brain expression is higher in foragers than in bees working in the hive, and manganese treatment not only increased sucrose responsiveness, but also caused an earlier onset of foraging (Ben-Shahar et al. 2004). Food-deprived bees also show increased sucrose responsiveness (Ben-Shahar and Robinson 2001), suggesting important associations among foraging behavior, nutritional state, sucrose responsiveness, and brain expression of *mlv*, as well as for other genes (Whitfield et al. 2006).

IIS genes are well known to affect feeding behavior in a wide variety of animals (Konturek et al. 2004). Ament et al. (2008) reported up-regulation of IIS genes in foragers (in brain and abdomen) relative to hive bees, and a delay in the shift from working in the hive to foraging induced by pharmacological inhibition of the TOR (target of rapamycin) pathway, which is related to IIS. Foragers have about half the lipid stores of hive bees, and a reduction in nutritional status can cause an early onset of foraging (Toth et al. 2005), indicating that high IIS is associated with low nutrient status in adult worker bees. This is opposite to what is found in larval honey bees (Wheeler et al. 2006) and opposite to the norm in other species (Ikeya et al. 2002). These results suggest evolutionary changes in the way IIS pathways are regulated across life stages and species. Measurements of IIS brain gene expression from *Polistes metricus* wasps also suggest an association between several IIS genes and lipid stores (Toth et al. 2007, 2009). In this species, there also is a causal association between reduced nutrient stores and foraging (TD Daugherty et al., unpubl.), suggesting that connections among nutrition, IIS signaling, and foraging in social insects might be widespread.

It thus appears that some well-conserved molecular pathways that influence food-gathering behavior in solitary insects such as *Drosophila* also regulate the division of labor in social insect societies. An important issue for the future is to address why certain feeding-related pathways, or components of these pathways, are more evolutionarily labile than others.

Modularity: Maternal behavior as a precursor to altruistic worker behavior. Comparative ethological analyses identified modules of solitary behavior as components of eusocial behavior more than 30 years ago (Evans and West-Eberhard 1973). Recent gene expression analyses have provided support for these ideas.

It is well established that the solitary ancestors to all social Hymenoptera were nesting wasps with well-developed maternal care (Evans and West-Eberhard 1973). The behavior of such a solitary maternal insect can be broken down into two distinct behavioral modules: (1) egg-laying and (2) maternal provisioning of brood with food collected during foraging. West-Eberhard proposed that an ancestral ovarian cycle consisting of these two basic modules of egg-laying and foraging/provisioning could be uncoupled during evolution (for review, see West-Eberhard 1996). Instead of being separated in time during the life of a solitary maternal wasp, the two behaviors could become so separated as to occur only in different individuals—"queens" that focus on egg-laying and "workers" that specialize in foraging/provisioning. In other words, worker behavior, which involves caring for siblings, may have evolved from maternal foraging/provisioning.

A molecular dimension was added to this idea with the prediction that sibling care and maternal care behaviors should be regulated by similar patterns of gene expression (Linksvayer and Wade 2005). This was tested in *P. metricus* wasps, a species that is well suited for evolutionary analyses because it is "primitively" eusocial and retains maternal behavior during the "foundress" part of its life cycle. Working with a set of 32 genes selected from microarray studies of honey bee division of labor (Whitfield et al. 2003, 2006), Toth et al. (2007) found the predicted relationship: Brain expression in worker wasps was more similar to the maternal foundresses than to queens, which do not show maternal care.

Putative ancestral modules of solitary foraging and reproduction may also have been involved in structuring various forms of division of labor among workers, which are thought to have evolved after the evolution of queens and workers (West-Eberhard 1996). As described above for honey bees, colonies of social insects often show a division of labor among young nest workers and old foragers, and although worker honey bees are mostly sterile, nest workers retain a higher reproductive capacity than foragers. When a colony of honey bees loses its queen, it is the younger workers that are more likely to begin laying (unfertilized) eggs (Winston 1987). Microarray results suggest that the brain gene expression profiles of hive bees are indeed more queen-like than those for foragers (Grozinger et al. 2007). In addition, bees that specialize on foraging for pollen have more well-developed ovaries and higher levels of expression of the egg yolk protein vitellogenin than do bees that specialize on foraging for nectar (Nelson et al. 2007). Thus, these two ancestral modules of foraging and reproduction may have been uncoupled multiple times during social insect evolution to produce the kinds of highly specialized individuals that characterize life in an insect society.

Tool kits: **Shared patterns of gene expression across social insect lineages.** The results discussed above (see section Co-option: Nutrition and division of labor for foraging), based on studies of small numbers of genes, suggest that genes related to feeding behavior may form the basis for one such tool kit for division of labor in social insects. This provides initial support for the idea of a tool kit for eusociality, i.e., core sets of genes used repeatedly during social evolution to generate novel forms of behavior (Toth and Robinson 2007). Additional evidence for this idea comes from more global studies of gene expression that explored whether the same pathways regulate different forms of division of labor in independently evolved social insect lineages.

Toth et al. (2007) provided initial evidence for this conclusion in the study of 32 genes in *Polistes* paper wasps described above. A large percentage (63%) of these genes that showed differential brain expression in the context of honey bee division of labor behavior also showed differential expression in the context of wasp division of labor. This line of study has recently been expanded with the creation of a *P. metricus* microarray (AL Toth et al., in prep.), allowing comparison of large-scale paper wasp and honey bee transcriptomic profiles. In this study, patterns of brain gene expression were compared in four different female groups that comprise a paper wasp society: foundresses (nest initiators), queens, workers, and gynes (future queens in reproductive diapause). These four groups differ in terms of foraging/provisioning and reproductive activity; queens and foundresses reproduce while workers and foundresses forage and provision the brood. Nearly 200 genes were differentially expressed as a function of foraging/provisioning. A comparison of this list of genes with microarray-derived lists of genes associated with honey bee worker foraging behavior (Whitfield et al. 2006; Alaux et al. 2009) showed a significant overlap in global brain gene expression patterns for the two species despite the vastly different forms of division of labor (AL Toth et al., in prep.). Genes associated with division of labor in honey bees are likely associated with division of labor in wasps, thus providing initial support for the existence of a "genetic tool kit" for division of labor that is shared across independent eusocial lineages.

A comparison of the functional categories of genes that affect social phenotypes across a wide range of social insects suggests broad conservation. For example, AL Toth et al. (in prep.) also found that genes for lipid metabolism, locomotory behavior, and response to heat stress were overrepresented on the list of genes regulated in association with paper wasp foraging/ provisioning, and several genes in these same categories are also differentially regulated in other social insect species. Heat shock protein and metabolic genes are differentially expressed in several honey bee brain transcriptomic studies related to foraging (Whitfield et al. 2003, 2006; Ament et al. 2008; Alaux et al. 2009). Moreover, there are similarities with whole-body gene expression analyses of other social insect species. *Polistes canadensis* wasp queens and workers were found to differ in expression of *vitellogenin,* numerous metabolic enzymes, and a heat shock gene (Sumner et al. 2006). Graff et al. (2007) discovered queen–worker differences in *vitellogenin* expression in the ant *Lasius niger.* Caste-specific differences in *hexamerin* storage proteins during early development have been reported in a wide variety of species including yellow jackets, paper wasps, and termites (Scharf et al. 2005; Hoffman and Goodisman 2007; Hunt et al. 2007). These results also provide support for the idea that certain genes and pathways have been used repeatedly during the evolution of eusociality in insects.

Gene regulation across different timescales. Some of the same genes that are associated with behavioral differences on an evolutionary timescale are also associated with behavioral differences on shorter timescales as well. This is illustrated with a recent study of aggression in honey bees.

Honey bees use their notorious stinging behavior to defend their hives against predators, both large and small. All honey bee colonies respond aggressively when their colony is attacked, but there is striking variation in the intensity of their response. In docile colonies, only a few bees may respond, whereas in more aggressive colonies, the response may involve hundreds or even thousands of stinging individuals. Colony defense begins when "guard" bees detect a disturbance at the hive entrance and release alarm pheromone, which alerts the entire colony. Older bees (who mostly forage for nectar and pollen) are more likely to respond aggressively than younger bees, but a subset of the colony's older bees, "soldiers," are the first to seek out and attack an intruder. There also are inherited differences in honey bee aggression. Africanized honey bees (AHB) have spread throughout most of the New World after the introduction in 1957 of the African subspecies *A.m. scutellata*, causing deaths of humans and animals in some parts of their newly inhabited range due to massive stinging responses. AHB are much more aggressive than European subspecies (EHB).

Certain genes can exert influences on morphological development over different timescales as discussed above, e.g., both developmental and evolutionary timescales. Alaux et al. (2009) used this perspective to explore whether environmental influences on a behavioral phenotype could have evolved into inherited differences via changes in gene regulation. One indication of this would be similar patterns of brain gene expression associated with aggression across the three timescales: acute (alarm pheromone exposure); life span (old vs. young), and evolutionary (AHB vs. EHB).

Alaux et al. (2009) found expression differences in the brain for hundreds of genes in the highly aggressive AHB compared to EHB. Similar results were obtained for EHB in response to exposure to alarm pheromone and when comparing old and young bees. Moreover, there was significant overlap of the gene lists generated from these three microarray experiments, thus supporting Waddington's (1959) genetic assimilation concept (phenotypic responses to environmental conditions can become encoded genetically) and suggesting that one element in the evolution of different degrees of aggressive behavior in honey bees involved changes in regulation of genes that mediate the response to alarm pheromone.

CONCLUSIONS

Studies of genes and social behavior, aided by new genomic resources, are coming of age, and social insects are good models for studying the evolution of social behavior at the molecular level. The relevance of some of the major insights from evo-devo suggests that using these insights to frame future studies will become increasingly fruitful. As genomic resources—including whole-genome sequencing—become more widely available, a broader array of species can be studied in this way (Hudson 2008), thus increasing the power of future comparative analyses. Well-resolved phylogenies are needed in order to carefully choose species that are both compelling in terms of social behavior and informative in a phylogenetic context, such as species in basal lineages or species that appear to have evolved similar behaviors independently.

The studies reviewed here relied extensively on transcriptomics, but other methods of molecular and genetic analysis are being used to study social behavior, in social insects and other species (Robinson et al. 2005, 2008; Page and Amdam 2007; Smith et al. 2008). Systems biology should prove to be increasingly important, allowing researchers to integrate different types of molecular information such as transcriptomics, proteomics, metabolomics, and epigenomics in order to develop models of regulatory networks (Robinson et al. 2008). Such integration will be necessary to better understand in molecular terms how various forms of behaviors evolve. Early indications suggest that, as for evo-devo, evolutionary molecular analyses of behavior have a bright future.

ACKNOWLEDGMENTS

We thank Sean B. Carroll and Trudy F.C. Mackay for comments that improved this manuscript. Research performed by the authors was supported by National Science Foundation Frontiers in Biological Research grant EF04-25852 (B.R. Schatz, PI), the Illinois Sociogenomics Initiative, NIH-DC 006395, USDA-NRI 2004-35604-14277, NSF IOS 06-41431 (G.E.R), and a USDA-NRI postdoctoral fellowship (A.L.T.).

REFERENCES

Abzhanov A, Protas M, Grant BR, Grant PR, Tabin CJ. 2004. *Bmp4* and morphological variation of beaks in Darwin's finches. *Science* **305:** 1462–1465.

Abzhanov A, Kuo WP, Hartmann C, Grant BR, Grant PR, Tabin CJ. 2006. The calmodulin pathway and evolution of elongated beak morphology in Darwin's finches. *Nature* **442:** 563–567.

Alaux C, Sinha S, Hasadsri L, Hunt GJ, Guzman-Novoa E, DeGrandi-Hoffman G, Uribe-Rubio JL, Rodriguez-Zas SL, Robinson GE. 2009. Honey bee brain gene expression supports a link between *cis* regulation and behavioral evolution. *Proc Natl Acad Sci* **106:** 15400–15405.

Ament SA, Corona M, Pollock HS, Robinson GE. 2008. Insulin signaling is involved in the regulation of worker division of labour in honey bee colonies. *Proc Natl Acad Sci* **105:** 4226–4231.

Baker BS, Taylor BJ, Hall JC. 2001. Are complex behaviors specified by dedicated regulatory genes? Reasoning from *Drosophila*. *Cell* **105:** 13–24.

Ben-Shahar Y, Robinson GE. 2001. Satiation differentially affects performance in a learning assay by nurse and forager honey bees. *J Comp Physiol A* **187:** 891–899.

Ben-Shahar Y, Robichon A, Sokolowski MB, Robinson GE. 2002. Influence of gene action across different time scales on behavior. *Science* **296:** 741–744.

Ben-Shahar Y, Leung H-T, Pak WL, Sokolowski MB, Robinson GE. 2003. cGMP-dependent changes in phototaxis: A possible role for the foraging gene in honey bee division of labor. *J Exp Biol* **206:** 2507–2515.

Ben-Shahar Y, Dudek NL, Robinson GE. 2004. Phenotypic deconstruction reveals involvement of manganese transporter *malvolio* in honey bee division of labor. *J Exp Biol* **207:** 3281–3288.

Bouchard TJ, McGue M. 2003. Genetic and environmental influences on human psychological differences. *J Neurobiol* **54:** 4–45.

Brady SG, Sipes S, Pearson A, Danforth BN. 2006. Recent and simultaneous origins of eusociality in halictid bees. *Proc R Soc Lond B Biol Sci* **273:** 1643–1649.

Brakefield PM. 2006. Evo-devo and constraints on selection. *Trends Ecol Evol* **21:** 362–368.

Carew TJ. 2000. *Behavioral neurobiology: The cellular organization of natural behavior.* Sinauer Associates, Sunderland, MA.

Carroll SB, Grenier J, Weatherbee S. 2005. *From DNA to diversity: Molecular genetics and the evolution of animal design.* Wiley-Blackwell, Hoboken, NJ.

Dingemanse NJ, Both C, Drent PJ, Van Oers K, Van Noordwijk AJ. 2002. Repeatability and heritability of exploratory behaviour in great tits from the wild. *Anim Behav* **64:** 929–938.

Emlen DJ, Hunt J, Simmons LW. 2005. Evolution of sexual dimorphism and male dimorphism in the expression of beetle horns: Phylogenetic evidence for modularity, evolutionary

lability, and constraint. *Am Nat* **166:** S42–S68.
Evans HE, West-Eberhard MJ. 1973. *The wasps*. David and Charles, Newton Abbott, United Kingdom.
Fraser GJ, Hulsey CD, Bloomquist RF, Uyesugi K, Manley NR, Streelman JT. 2009. An ancient gene network is co-opted for teeth on old and new jaws. *PLoS Biol* **7:** e1000031.
Fujiwara M, Sengupta P, McIntire SL. 2002. Regulation of body size and behavioral state of *C. elegans* by sensory perception and the *EGL-4* cGMP-dependent protein kinase. *Neuron* **36:** 1091–1102.
Gellon G, McGinnis W. 1998. Shaping animal body plans in development and evolution by modulation of *Hox* expression patterns. *BioEssays* **20:** 116–125.
Graff J, Jemielity S, Parker JD, Parker KM, Keller L. 2007. Differential gene expression between adult queens and workers in the ant *Lasius niger*. *Mol Ecol* **16:** 675–683.
Grozinger CM, Fan Y, Hoover SE, Winston ML. 2007. Genome-wide analysis reveals differences in brain gene expression patterns associated with caste and reproductive status in honey bees (*Apis mellifera*). *Mol Ecol* **16:** 4837–4848.
Hines HM, Hunt JH, O'Connor TK, Gillespie JJ, Cameron SA. 2007. Multigene phylogeny reveals eusociality evolved twice in vespid wasps. *Proc Natl Acad Sci* **104:** 3295–3299.
Hoekstra HE, Coyne JA. 2008. The locus of evolution: Evo devo and the genetics of adaptation. *Evolution* **61:** 995–1016.
Hoffman EA, Goodisman MAD. 2007. Gene expression and the evolution of phenotypic diversity in social wasps. *BMC Biol* **5:** 23.
Hoffmann AA. 2000. Laboratory and field heritabilities: Some lessons from *Drosophila*. In *Adaptive genetic variation in the wild* (ed. TA Mousseau et al.), pp. 200–218. Oxford University Press, New York.
Hudson ME. 2008. Sequencing breakthroughs for genomic ecology and evolutionary biology. *Mol Ecol Resour* **8:** 3–17.
Hunt JH, Kensinger BJ, Kossuth JA, Henshaw MT, Norberg K, Wolschin F, Amdam GV. 2007. A diapause pathway underlies the gyne phenotype in *Polistes* wasps, revealing an evolutionary route to caste-containing insect societies. *Proc Natl Acad Sci* **104:** 14020–14025.
Ikeya T, Galic M, Belawat P, Nairz K, Hafen E. 2002. Nutrient-dependent expression of insulin-like peptides from neuroendocrine cells in the CNS contributes to growth regulation in *Drosophila*. *Curr Biol* **12:** 1293–1300.
Ingram KK, Oefner P, Gordon DM. 2005. Task-specific expression of the foraging gene in harvester ants. *Mol Ecol* **14:** 813–818.
Insel TR, Young LJ. 2001. The neurobiology of attachment. *Nat Rev Neurosci* **2:** 129–136.
Kandel ER. 2007. *In search of memory: The emergence of a new science of mind*. W.W. Norton, New York.
Kandel ER, Harris Schwartz J, Jessell TM. 2000. *Principles of neural science*. Appleton and Lange, Norwalk, CT.
Khaitovich P, Tang K, Franz H, Kelso J, Hellmann I, Enard W, Lachmann M, Paabo S. 2006. Positive selection on gene expression in the human brain. *Curr Biol* **16:** R356–R358.
Konturek SJ, Konturek JW, Pawlik T, Brzozowki T. 2004. Brain-gut axis and its role in the control of food intake. *J Physiol Pharmacol* **55:** 137–154.
Kravitz EA. 2000. Serotonin and aggression: Insights gained from a lobster model system and speculations on the role of amine neurons in a complex behavior. *J Comp Physiol A* **186:** 221–238.
Lesch KP, Merschdorf U. 2000. Impulsivity, aggression, and serotonin: A molecular psychobiological perspective. *Behav Sci Law* **18:** 581–604.
Lim MM, Wang Z, Olazábal DE, Ren X, Terwilliger EF, Young LJ. 2004. Enhanced partner preference in a promiscuous species by manipulating the expression of a single gene. *Nature* **429:** 754–757.
Linksvayer TA, Wade MJ. 2005. The evolutionary origin and elaboration of sociality in the aculeate Hymenoptera: Maternal effects, sib-social effects, and heterochrony. *Q Rev Biol* **80:** 317–336.

Lucas C, Sokolowski MB. 2009. Molecular basis for changes in behavioural state in ant social behaviours. *Proc Natl Acad Sci* **106:** 6351–6356.
Mackay TFC. 2009. The genetic architecture of complex behaviors: Lessons from *Drosophila*. *Genetica* **136:** 295–302.
Monteiro A, Podlaha O. 2009. Wings, horns, and butterfly eyespots: How do complex traits evolve? *PLoS Biol* **7:** e100003.
Nelson M, Ihle K, Fondrk MK, Page RE, Amdam GV. 2007. The gene *vitellogenin* has multiple coordinating effects on social organization. *PLoS Biol* **5:** e62.
Osborne KA. Robichon A, Burgess E, Butland S, Shaw RA, Coulthard A, Pereira HS, Greenspan RJ, Sokolowski MB. 1997. Natural behavior polymorphism due to a cGMP-dependent protein kinase of *Drosophila*. *Science* **277:** 834–836.
Page RE, Amdam GV. 2007. The making of a social insect: Developmental architectures of social design. *BioEssays* **29:** 334–343.
Pankiw T, Page RE. 1999. The effect of genotype, age, sex, and caste on response thresholds to sucrose and foraging behavior of honey bees (*Apis mellifera* L.). *J Comp Physiol A* **185:** 207–213.
Pichaud F, Desplan C. 2002. *Pax* genes and eye organogenesis. *Curr Opin Genet Dev* **12:** 430–434.
Postlethwait JH, Schneiderman HA. 1971. Pattern formation and determination in the antenna of the homoeotic mutant *Antennapedia* of *Drosophila melanogaster*. *Dev Biol* **25:** 606–640.
Robinson GE, Ben-Shahar Y. 2002. Social behavior and comparative genomics: New genes or new gene regulation? *Genes Brain Behav* **1:** 197–203.
Robinson GE, Grozinger CM, Whitfield CW. 2005. Sociogenomics: Social life in molecular terms. *Nat Rev Genet* **6:** 257–270.
Robinson GE, Fernald RD, Clayton DF. 2008. Genes and social behavior. *Science* **322:** 896–900.
Rodrigues V, Cheah PY, Ray K, Chia W. 1995. *malvolio*, the *Drosophila* homolog of house *Nramp-1* (*Bcg*), is expressed in macrophages and in the nervous system and is required for normal taste behavior. *EMBO J* **14:** 3007–3020.
Rueppell O, Chandra SB, Pankiw T, Fondrk MK, Beye M, Hunt G, Page RE. 2006. The genetic architecture of sucrose responsiveness in the honeybee (*Apis mellifera* L.). *Genetics* **172:** 243–251.
Sanetra M, Begemann G, Becker M-B, Meyer A. 2005. Conservation and co-option in developmental programmes: The importance of homology relationships. *Front Zool* **2:** 15.
Scharf ME, Wu-Scharf D, Zhou X, Pittendrigh BR, Bennett GW. 2005. Gene expression profiles among immature and adult reproductive castes of the termite *Reticulitermes flavipes*. *Insect Mol Biol* **14:** 31–34.
Seeley TD. 1995. *The wisdom of the hive*. Harvard University Press, Cambridge, MA.
Shapiro MD, Marks ME, Peichel CL, Blackman BK, Nereng KS, Jónsson B, Schluter D, Kingsley DM. 2004. Genetic and developmental basis of evolutionary pelvic reduction in three-spine sticklebacks. *Nature* **428:** 717–723.
Shu W, Cho JY, Jiang Y, Zhang M, Weisz D, Elderd GA, Schmeidlerd J, De Gasperi R, Gama Sosa MA, Rabidou D, et al. 2005. Altered ultrasonic vocalization in mice with a disruption in the *Foxp2* gene. *Proc Natl Acad Sci* **102:** 9643–9648.
Smith CR, Toth AL, Suarez AV, Robinson GE. 2008. Genetic and genomic analyses of the division of labour in insect societies. *Nat Rev Genet* **9:** 735–748.
Stock DW. 2001. The genetic basis of modularity in the development and evolution of the vertebrate dentition. *Philos Trans R Soc Lond B Biol Sci* **356:** 1633–1656.
Sumner S, Pereboom JJ, Jordan WC. 2006. Differential gene expression and phenotypic plasticity in behavioural castes of the primitively eusocial wasp, *Polistes canadensis*. *Proc R Soc Lond B Biol Sci* **273:** 19–26.
Tobback J, Heylen K, Gobin B, Wenseleers T, Billen J, Arckens L, Huybrechts R. 2008. Cloning and expression of *PKG*, a candidate foraging regulating gene in *Vespula vulgaris*. *Anim Biol* **58:** 341–351.

Toth AL, Robinson GE. 2007. Evo-devo and the evolution of social behavior. *Trends Genet* **23:** 334–341.

Toth AL, Kantarovich S, Meisel AF, Robinson GE. 2005. Nutritional status influences socially regulated foraging ontogeny in honey bees. *J Exp Biol* **208:** 4641–4649.

Toth AL, Varala K, Newman TC, Miguez FE, Hutchison SK, Willoughby DA, Simons JF, Egholm M, Hunt JH, Hudson ME, et al. 2007. Wasp gene expression supports an evolutionary link between maternal behaviour and eusociality. *Science* **318:** 441–444.

Toth AL, Bilof KJ, Henshaw MT, Hunt JH, Robinson GE. 2009. Lipid stores, ovary development, and brain gene expression in *Polistes metricus* females. *Insect Soc* **56:** 77–84.

True JR, Carroll SB. 2002. Gene co-option in physiological and morphological evolution. *Annu Rev Cell Dev Biol* **18:** 53–80.

Waddington CH. 1959. Canalization of development and genetic assimilation of acquired characters. *Nature* **183:** 1654–1655.

West-Eberhard MJ. 1996. Wasp societies as microcosms for the study of development and evolution. In *Natural history and evolution of paper wasps* (ed. S Turillazzi and MJ West-Eberhard), pp. 290–317. Oxford University Press, New York.

Wheeler DE, Buck N, Evans JD. 2006. Expression of insulin pathway genes during the period of caste determination in the honey bee, *Apis mellifera. Insect Mol Biol* **15:** 597–602.

Whitfield CW, Cziko A-M, Robinson GE. 2003. Gene expression profiles in the brain predict behaviour in individual honey bees. *Science* **302:** 296–299.

Whitfield CW, Ben-Shahar Y, Brillet C, Leoncini I, Crauser D, LeConte Y, Rodriguez-Zas S, Robinson GE. 2006. Genomic dissection of behavioral maturation in the honey bee. *Proc Natl Acad Sci* **103:** 16068–16075.

Wilson EO. 1971. *The insect societies.* Belknap/Harvard University Press, Cambridge, MA.

Winston ML. 1987. *The biology of the honey bee*. Harvard University Press, Cambridge, MA.

Wray GA. 2007. The evolutionary significance of *cis*-regulatory mutations. *Nat Rev Genet* **8:** 206–216.

Young MW, Kay SA. 2001. Time zones: A comparative genetics of circadian clocks. *Nat Rev Genet* **2:** 702–715.

Young LJ, Wang ZX. 2004. The neurobiology of pair bonding. *Nat Neurosci* **7:** 1048–1054.

Cooking and the Human Commitment to a High-quality Diet

R.N. CARMODY AND R.W. WRANGHAM

Department of Human Evolutionary Biology, Harvard University, Cambridge, Massachusetts 02138
Correspondence: wrangham@fas.harvard.edu

For our body size, humans exhibit higher energy use yet reduced structures for mastication and digestion of food compared to chimpanzees, our closest living relatives. This suite of features suggests that humans are adapted to a high-quality diet. Although increased consumption of meat during human evolution certainly contributed to dietary quality, meat-eating alone appears to be insufficient to support the evolution of these traits, because modern humans fare poorly on raw diets that include meat. Here, we suggest that cooking confers physical and chemical benefits to food that are consistent with observed human dietary adaptations. We review evidence showing that cooking facilitates mastication, increases digestibility, and otherwise improves the net energy value of plant and animal foods regularly consumed by humans. We also address the likelihood that cooking was adopted more than 250,000 years ago (kya), a period that we believe is sufficient in length for the proposed adaptations to have occurred. Additional experimental work is needed to help discriminate the relative contributions of cooking, meat eating, and other innovations such as nonthermal food processing in supporting the human transition toward dietary quality.

The allocation of energy to different tissues and the total use of energy are critical variables influencing a wide range of biological functions. Here, we review evidence suggesting that humans have adapted evolutionarily to having high-energy budgets as a result of unique aspects of the diet.

Using standard metabolic equations and detailed activity data, Leonard and Robertson (1997) determined mass-specific energy expenditure for 17 nonhuman primate species and two groups of human foragers (!Kung and Ache). Using their data, we calculate that, on average, male foragers expend 44% more energy than do male chimpanzees (*Pan troglodytes*) per unit body weight (i.e., $TEE/BW^{0.792}$: male forager = 118 $kcal/d/kg^{0.792}$; male chimpanzee = 82 $kcal/d/kg^{0.792}$, where TEE = total daily energy expenditure, and BW = body weight). This is partly due to male foragers traveling 10–20 km/d and expending ~40% of their total energy budget on activity, compared to 3–5 km/d and 29% for chimpanzees (Leonard and Robertson 1997). Likewise, female foragers are estimated to expend 17% more energy than do female chimpanzees ($TEE/BW^{0.792}$: female forager = 91 $kcal/d/kg^{0.792}$; chimpanzee female = 78 $kcal/d/kg^{0.792}$). The figures for females do not include the lifetime costs of gestation and lactation, which should be higher in humans thanks to our shorter interbirth intervals (Coelho 1986; Aiello and Key 2002).

To achieve their high TEE, humans must consume more food and/or foods of higher energy density compared to chimpanzees. Because humans eat less and invest less in digestive structures than do chimpanzees, energy density is the principal solution (Wrangham 2009). Compared to nonhuman primates, humans have smaller oral cavities, reduced molar dentition, more gracile mandibles, and smaller chewing muscles relative to body size, all of which combine to confer less mastication ability (Wrangham 2009). These differences appear to be functional; whereas chimpanzees in the wild spend 4–6 h/d chewing their food, humans typically spend <1 h (Wrangham and Conklin-Brittain 2003). In addition, humans have relatively shorter, less massive, and less sacculated intestinal tracts (Milton and Demment 1988; Aiello and Wheeler 1995), and comparing the relative sizes of intestinal segments, humans have virtually no caecum and substantially reduced colons, suggesting less reliance on microbial fermentation of fiber (Milton 1987). Our physiological investment in digestion also appears to be lower than expected: Humans expend an average of 6%–7% of meal energy in digestion, compared to the mammalian average of 13%–16% (Boback et al. 2007).

Traditionally, such dietary adaptations have been attributed in large part to increased consumption of animal foods (Shipman and Walker 1989; Milton 1999; Bunn 2006). Meat eating has clearly been important in human evolution. First, it is known that stone tools were used to process meat beginning at least 2.5 million years ago (mya) (Toth and Schick 2006). Second, bone strontium/calcium (Sr/Ca) ratios, which tend to correlate inversely with trophic level, place early hominid diets between contemporary carnivores and herbivores, suggesting substantial consumption of animal foods (Sillen et al. 1995). Third, taeniid tapeworms are host-specific parasites for whom carnivores are definitive hosts and herbivores are intermediate hosts. Three species, *Taenia saginata*, *T. asiatica*, and *T. solium*, use humans exclusively as their primary host, indicating the occurrence of human meat consumption more than ~1 mya (Henneberg et al. 1998; Hoberg et al. 2001). Fourth, humans lack the ability to efficiently synthesize from plant-based raw materials the long-chain polyunsaturated fatty acids (PUFA) required for cell membrane growth, structure and function, and fetal and postnatal brain development (Clandinin 1999;

Broadhurst et al. 2002). Preformed docosahexaenoic acid (DHA) and arachidonic acid (AA), which together comprise almost all PUFA in the central nervous system, are uniquely available from animal foods (Broadhurst et al. 2002). Fifth, similar to obligate carnivores (Hedberg et al. 2007), humans exhibit diminished ability to synthesize taurine from precursor amino acids (Hayes and Sturman 1981; Chesney et al. 1998). Taurine is found only in animal tissue and is thought to have been consumed sufficiently to reduce selective pressure for in vivo synthesis (Mann 2007). Finally, meat constitutes a large percentage of the diet for modern humans. Whereas chimpanzees typically obtain <5% of their diet from animal foods, meat (hunted + fished) was found to provide ≥50% of dietary energy for 73% of 229 hunter-gatherer societies (Cordain et al. 2000).

Despite the importance of meat in human evolution, recent evidence also indicates that human diets, whether or not they include meat, are energetically inadequate if eaten raw. All human populations, regardless of environment or available materials, cook their food (Harris 1992; Wrangham and Conklin-Brittain 2003). The only groups that live for months or more on raw food are "raw foodists" living in industrial societies that forego cooked food for a variety of philosophical and perceived health reasons (Wrangham 2009). To date, all studies of raw foodists have concluded that a raw food diet provides insufficient energy for the maintenance of body weight (Carmody and Wrangham 2009). In the most extensive study—a cross-sectional survey of 572 long-term raw foodists living in Germany—Koebnick et al. (1999) found that body mass index (BMI) was inversely correlated with both the proportion of raw food in the diet and the length of time since adoption of raw foodism (Fig. 1). Odds of becoming underweight were three times greater for subjects following a 100% raw food diet compared to subjects following a <80% raw food diet. Energy deficiency in subjects following a 100% raw food diet was apparently responsible for 50% of female subjects of child-bearing age reporting amenorrhea, and an additional 10% reported suffering from menstrual irregularities. The incidence of menstrual disruption scaled with the proportion of raw food in the diet and the duration of raw foodism, conforming to the expected dependence of female reproductive capacity on energy status (Ellison et al. 1993; Ellison 2003). Critically, Koebnick et al. (1999) found no difference in the odds of being underweight or amenorrhea occurring among meat-eating, vegetarian, or vegan subjects, suggesting that the consumption of meat alone did not improve energy status.

Unlike a raw diet, lack of meat does not hinder energy status or reproductive function. In the United States, median BMIs of vegetarians eating cooked diets were 23.7 (women) and 24.3 (men), close to the median BMIs of those eating typical American mixed diets (24.8 [women] and 25.3 [men]) (Carmody and Wrangham 2009). In contrast, median BMIs for vegetarians eating predominantly raw diets were 20.1 (women) and 20.7 (men). In addition, women consuming cooked vegetarian diets exhibit no suppression of ovarian function compared to women consuming cooked diets that include meat (Barr 1999), nor are there differences in the age of menarche between vegetarian and omnivorous women eating cooked diets (Rosell et al. 2005).

Evidence of poor energy status among contemporary raw foodists is surprising. First, raw foodists tend to live in urban areas with abundant, year-round access to a wide variety of high-quality foods, including meats, bone marrow, domesticated plants, and oils (Hobbs 2005; Wrangham 2009). Second, most raw foodists process their foods heavily, using methods such as sprouting, freezing, blending, juicing, mechanically tenderizing, pickling, cold-pressing, cold-smoking, and even drying at up to ~45°C (Koebnick et al. 1999; Hobbs 2005; Wrangham 2009). These methods alter the foods in ways likely to increase energy value (see below). Third, chimpanzees would unquestionably fare well on a human raw food diet, with its superabundant and stable access to foods with lower fiber concentrations than those available in the wild (Wrangham and Conklin-Brittain 2003).

Such data suggest that in the wild, humans eating raw diets would not normally obtain sufficient energy to thrive and therefore the diet to which humans are evolutionarily adapted is obliged to include cooked food (Wrangham and Conklin-Brittain 2003). According to this hypothesis, cooking of plant and animal foods provides energetic benefits beyond those conferred by meat eating alone. This stable source of supplemental energy would have supported the adaptations for high dietary quality that we observe today, including reduced structures for mastication and digestion, coupled with a large total energy budget.

This hypothesis makes three predictions that we examine in this paper. First, for cooking to have allowed for reduced chewing ability, it must make foods easier to masticate. Second, for cooking to have supported the reduction in gut size, it must make foods more digestible. Third, cooking must lead to higher net energy gains.

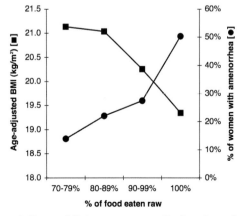

Figure 1. Energy deficiency among raw foodists. Age-adjusted BMI (*left axis*, closed box) and percentage of nonpregnant female subjects <45 years old reporting amenorrhea (*right axis*, closed circle) as a function of the percentage of food eaten raw. (Adapted from Koebnick et al. 1999.)

EFFECTS OF COOKING ON EASE OF CHEWING

Cooking reduces the structural integrity of most foods, leading to a reduction of chewing time and masticatory strain (Lieberman et al. 2004; Lucas 2004). With respect to plant materials, heat degrades the pectic polysaccharides that cause plant cell walls to adhere and hence causes a collapse of tissue structure and a loss of firmness (Jarvis et al. 2003; Waldron et al. 2003). This process applies to many plant foods regularly consumed by humans, including roots (see, e.g., Dominy et al. 2008), fruits (see, e.g., Qi et al. 2000), and stems (see, e.g., Lin and Chang 2005). As a result of this process, cooked plant foods are more easily masticated, requiring fewer chewing cycles and a shorter time in the mouth before they form a coherent bolus and are swallowed (Engelen et al. 2005). For example, Dominy et al. (2008) found that the wild tubers regularly eaten by Hadza foragers were generally too fracture resistant for human consumption until they were cooked. Light roasting over a traditional fire reduced the work of fracture by 40%–59%, depending on the species and tissue sampled, improving edibility and reducing the cost of mastication (Dominy et al. 2008).

Cooking has also been shown to tenderize several animal foods (e.g., pig, beef, goat, rabbit, and octopus; Wrangham 2009). However, the effects of heat on the mechanical properties of muscle tissue are complex. Muscle tissue consists of protein-rich muscle fibers embedded in a lattice of fat and a tough collagen-rich connective tissue that derives its strength from a triple-helix arrangement of protein strands. When heated above 40°C, muscle proteins begin to denature and coil, producing contraction of the muscle along the grain. This contraction increases with longer cooking time and higher cooking temperature, leading to progressive toughening of the muscle fibers. At the same time, the collagen surrounding each muscle fiber remains difficult to chew until heated to 60°C–70°C, when collagen begins to be hydrolyzed into gelatin. As gelatinization of collagen proceeds, muscle fibers separate and meat becomes more easily fractionated along the grain (Barham 2001; McGee 2004). Therefore, the effects of cooking on meat tenderness reflect a trade-off between tender muscle fibers and tough collagen at low temperatures and tough muscle fibers and tender collagen at high temperatures. The amount of collagen present in the meat is consequently an important factor, with collagen-rich meats (i.e., active or weight-bearing muscles, older animals) requiring longer cooking and higher temperatures to achieve maximum tenderness, whereas meats with lower collagen (i.e., infrequently used muscles, more fat marbling, younger animals) are best cooked quickly (Purslow 2005; Lepetit 2008). Wild meats available to ancestral humans would undoubtedly have been collagen rich and therefore would have required long cooking and/or high heat to achieve tenderization. Roasting meat over an open fire, a technique commonly used by traditional foragers, would likely achieve such a result.

EFFECTS OF COOKING ON DIGESTIBILITY

Starchy plant foods are important staples in almost all human societies (Atkins and Bowler 2001). Raw starch granules are semicrystalline mixtures of amylopectin and amylose plus trace amounts of protein and lipid. They are normally resistant to hydrolysis by salivary and pancreatic amylases. The degree of resistance to digestion of raw starch granules varies, with raw cereals tending to be less resistant to digestion than raw tubers, legumes, or green bananas (Carmody and Wrangham 2009). Heating acts to gelatinize starches, transforming their resistant semicrystalline structures into amorphous compounds that are more readily hydrolyzed to sugars and dextrins (Tester et al. 2006). Cooking thus functions to predigest starches, improving their nutritional availability to the human consumer, with the extent of predigestion being dependent on the temperature of processing and the amount of water present (Tester and Sommerville 2000).

To quantify the impact of cooking on starch digestibility, it is important to account for the fact that a significant and variable proportion of starch consumed by a human is digested not by the human but instead by gut microbes (Livesey 2002). Starches that escape digestion in the small intestine, called resistant starches, are fermented by microbes in the caecum and colon, generating short-chain fatty acids. This process returns only ~50% of the metabolizable energy of the starch to the human consumer, owing to a combination of the less-efficient capture of energy as ATP during mammalian oxidation of short-chain fatty acids compared to glucose, the consumption of short-chain fatty acids for fuel by gut microbes, and losses during conversion due to the production of hydrogen and methane (Livesey 1995; Silvester et al. 1995; Carmody and Wrangham 2009). These diminished returns imply that traditional measures of digestibility based on nutrients passing through into the feces (i.e., fecal digestibility) are biologically irrelevant. The biologically relevant measure of starch digestibility can instead be estimated by summing the proportion of starch digested as of the terminal ileum (i.e., ileal digestibility) together with 50% of resistant starches. Applying this index to published reports of the human ileal digestibility of plant starches, cooking increased the digestibility of all varieties, with increases ranging from 12% for oats to 35% for green bananas (Carmody and Wrangham 2009). This suggests that cooking has a profound impact on the digestibility of human plant food staples and hence on the total net energy gain from eating these foods.

The disparity between ileal and fecal measures is even more pronounced for proteinaceous foods than for starchy foods because the products of microbial fermentation of protein appear to return zero energy to the consumer (Mason 1984; McNeil 1988). Unfortunately, there has been little investigation of the effects of cooking on the ileal digestibility of protein by humans, in part because of the complexities of distinguishing food proteins from endogenous (i.e., human) protein excretions present in the ileal effluent. The only study to resolve this problem examined the effect of cooking on the ileal digestibility of isotopically labeled egg protein (Evenepoel et al. 1998, 1999). In this study, ileostomates—people who, for health reasons, have undergone surgery to connect the terminal ileum to the surface of the skin, thus allowing ileal effluent to collect in an external pouch—consumed a homogeneous mixture of egg white and yolk, served either raw or microwaved. Evenepoel et al. (1998) determined that the ileal digestibility of egg protein was 51% for raw eggs and 91% for cooked eggs, meaning that cooking increased digestibility in these patients by 78%. In addition, Evenepoel et al. (1998, 1999) controlled for the possibility that ileostomates exhibit atypical digestion by first demonstrating that recovery of isotopes in breath was positively correlated with ileal digestibility (Evenepoel et al. 1998) and then establishing that the breath recovery profiles of ileostomy patients and intact subjects were comparable (Evenepoel et al. 1999). This research revealed that, contrary to widespread belief in the high biological availability of raw egg protein, much of the raw egg protein benefited the intestinal microbes and not the human consumer.

Heat-induced denaturation of protein, the mechanism attributed by Evenepoel et al. (1998, 1999) to explain the increased digestibility of cooked eggs, is common to all forms of animal protein. When heated, proteins unwind from their tightly bound forms, losing their tertiary structures (i.e., covalent bonds, dipole-dipole interactions, and van der Waals interactions between amino acid side chains) and secondary structures (i.e., repeating patterns such as α-helices and β-pleated sheets). Denatured proteins adopt a random coil configuration that makes proteins more digestible by increasing their susceptibility to proteolytic enzymes (Davies et al. 1987). It has therefore been suggested that cooking may make all animal proteins more digestible (Lawrie 1991; Gaman and Sherrington 1996).

Heat also has contrary effects on proteins. When heated in the presence of reducing sugars, proteins participate in the Maillard reaction, a nonenzymatic condensation of amino acids and sugars that is catalyzed by heat (Maillard 1916). The Maillard reaction produces a complex variety of melanoidins and aromatic compounds responsible for food browning and the development of characteristic smells and flavors. This process lowers protein digestibility (Seiquer et al. 2006). Mechanisms include the destruction of essential amino acids (Rerat et al. 2002), structural changes in protein that prevent normal enzymatic cleavage (Oste and Sjodin 1984; Kato et al. 1986), the impedance of epithelial transport (Shorrock and Ford 1978), and the inhibition of digestive enzymes (Oste et al. 1987). Maillard reactions are largely restricted to surfaces of meat. Their effects on ileal digestibility have not been studied, and their net contribution to reducing protein digestibility remains uncertain.

EFFECTS OF COOKING ON NET ENERGY GAIN

The net energy value of a food is a function of several factors, including its gross caloric value, digestibility, diet-induced thermogenesis, and impact on basal metabolic rate. We have already presented evidence that cooking improves the digestibility of starchy plant foods and may also contribute to protein digestibility. We discuss here evidence showing that cooking also lowers the costs of food consumption, including diet-induced thermogenesis and the metabolic costs of immune defense.

Lowering Diet-induced Thermogenesis

Diet-induced thermogenesis (DIT) refers to increased metabolic expenditure as a result of ingesting and assimilating a meal. DIT is also known as specific dynamic action, postprandial thermogenesis, or the thermic effect of feeding. In humans on a Western balanced diet, DIT accounts for ~10% of the total energy budget, a proportion similar to that of locomotion (Westerterp 2004). Cooking appears to reduce DIT partly by softening and reducing the structural integrity of food, thus facilitating the mechanical work of the stomach and increasing access to foods by gastric acids and proteolytic enzymes (Secor 2003). These act to accelerate the gastric emptying rate, which corresponds closely to the overall rate and cost of digestion (Horowitz et al. 2002; Pera et al. 2002; Secor 2003).

Studies in vivo find that softer foods generate reduced DIT responses. Among rats, Oka et al. (2003) found that softening food pellets by the addition of air pockets led to reduced postprandial rises in body temperature, ultimately leading to greater weight gain and adiposity in soft-fed versus hard-fed animals despite similar levels of intake and activity. Among various species of reptiles and amphibians, meals of soft-bodied prey items led to lower DIT than equivalent meals of hard-bodied prey items (Secor and Faulkner 2002; Secor 2003; Secor and Boehm 2006).

Only one study has directly compared the effects of cooking and mechanical softening on DIT. Boback et al. (2007) fed Burmese pythons (*Python molurus*) meals of beef weighing 25% of the snake's body mass, served in one of four forms: raw and whole, raw and ground, cooked and whole, and cooked and ground. We found that cooking reduced DIT by 13% of meal energy, grinding reduced DIT by 12%, and the effects of cooking and grinding were nearly additive, with the cooked and ground meal associated with a 23% reduction in DIT compared to the raw and whole treatment. That the combined (cooked and ground) treatment produced the lowest DIT revealed that the suppressive effects conferred by cooking were above and beyond those conferred by mechanical grinding. The difference is attributable to the gelatinization of collagen-rich connective tissues, a process that can be accomplished only via heat, as discussed above. How closely the results for pythons apply to human DIT is unknown. Metabolic demands differ extensively between poikilothermic and homeothermic animals, and pythons, unlike humans, exhibit substantial gastrointestinal remodeling between meals (Secor and Diamond 1995). For these reasons, it would be useful to directly measure the effects of cooking and grinding on DIT in humans. Although experimentally more challenging due to the confounding variables of activity expenditure and body temperature regulation, human metabolic chambers have been used successfully to estimate 24-hour DIT (Westerterp et al. 1999).

Lowering Metabolic Costs of Immune Defense

Basal metabolic rate (BMR) is defined as the energy expended in a conscious, fasted body at rest in a thermally neutral environment (Kleiber 1961). BMR reflects the minimum energy sufficient to sustain vital processes, including cell function and replacement, synthesis and secretion of enzymes and hormones, maintenance of body temperature, uninterrupted work of cardiac and respiratory muscles, brain function, and immune defense (Durnin 1981). BMR accounts for 45%–70% of the total energy budget in humans, representing the largest component of energy expenditure (Black et al. 1996; FAO/WHO/UNU 2004). Cooking is proposed to reduce BMR by lowering investment in immune defense. For example, cooking kills meat-borne pathogens such as *Escherichia coli, Salmonella, Campylobacter, Staphylococcus*, and *Listeria*, which, if ingested live, typically result in multiple symptoms, including fever. Fever alone has been shown to increase resting metabolic rate (RMR) (which is conceptually equivalent to BMR but is more practical to measure because it does not require the subject to sleep overnight in situ before metabolic measurement) by ~7%–15% for each 1°C above standard temperature (DuBois 1937; Roe and Kinney 1965; Elia 1992). Using the often cited 13% relationship measured by DuBois (1937), we estimated the annual cost of fever due to consumption of cooked agricultural meat products (beef, pork, mutton, and poultry) versus the implied annual cost of fever if those same meat products were consumed raw (Carmody and Wrangham 2009). We found that, with habitual cooking of these meat products, the estimated cost of immune up-regulation due to fever was <0.01% of annual RMR. In contrast, if these products were consumed at similar rates without cooking, the estimated cost was 8.4% of annual RMR. Raw wild meat is expected to be less pathogen-bearing on average than raw meat raised and processed for mass-market consumption, so the impact of cooking may be somewhat overestimated. Nevertheless, our results suggest that meat consumption at contemporary Western levels would be energetically inefficient without cooking. Similar suppressive effects on immune regulation are likewise expected due to the role of heat in killing plant-borne pathogens, such as *Giardia lamblia*, as well as in degrading plant-borne antinutrients, such as tannins and phytates.

The role of cooking in reducing the costs of food consumption supports the hypothesis that cooking increases the net energy value of foods. Further in vivo studies focusing on the impact of cooked versus raw diets on whole-organism factors such as growth and energy balance are needed to address the possibility that the factors contributing to net energy are not independent and thus distort the assessment of net energy effects. For example, it is reasonable to expect that digestibility (positive contributor to net energy) and DIT (negative contributor) are related, because higher levels of absorbable nutrients should theoretically increase the costs of absorption and postabsorptive processes such as membrane transport, ketogenesis, amino acid deamination and oxidation, glycogenesis, urea production, and renal excretion.

OPPORTUNITY FOR EVOLUTIONARY ADAPTATION TO COOKED FOOD

The hypothesis that humans are evolutionarily adapted to a diet of cooked food depends on sufficient time having elapsed since the control of fire and subsequent adoption of cooking for such evolution to have occurred.

No consensus exists as to when fire was first controlled, in part because traces of fire usually disappear quickly and thus leave no evidence for archaeologists to find (Sergant et al. 2006). Nevertheless, fire was certainly controlled by 250 kya. Several sites dated to 250 kya or older contain evidence of fire use by hominids, including burned deposits, fire-cracked rocks, reddened areas, baked clay, ash, charcoal, fire-hardened wood, burned lithics, burned bone, and even some indication of hearths (James 1989). Older dates for the control of fire are also widely acknowledged at sites such as Beeches Pit (Preece et al. 2006), Schöningen (Thieme 2005), and Ménez-Drégan (Monnier et al. 1994), all dated to ~400 kya.

Given that wild chimpanzees have been observed to pick cooked seeds from the ashes of bush fires (Brewer 1978), it is hard to imagine that humans would have controlled fire for more than a few generations without discovering the benefits of cooking. Wobber et al. (2008) showed that captive great apes, when presented with a choice of five different foods served raw and cooked, significantly preferred the cooked item (meat, carrots, sweet potato) or were indifferent to the options (white potato, apple). In no case was the raw food preferred. At minimum, these results suggest that preference for cooked foods is not a recent evolutionary phenomenon (Wobber et al. 2008). It therefore seems reasonable to assume that hominids began cooking no later than 250 kya (Brace 1995; Ragir 2000). Given that human populations with a history of dairying developed the capacity for lactose digestion into adulthood at least twice within the last ~7000 years (Bersaglieri et al. 2004; Tishkoff et al. 2007), there has clearly been ample time for adaptation to the expected effects of cooking.

Gene expression studies on cooked versus raw diets may ultimately enable discrimination of a molecular signal of adaptation to cooking. The relationship between genomic and phenotypic features is especially difficult to study in humans and our close genetic relatives because most experimental approaches are inappropriate. However, recent work by Somel et al. (2008) has demonstrated that mice can be used to study certain aspects of human–chimpanzee dietary divergence. Somel et al. (2008) reared groups of mice for 2 weeks on one of four experimental diets: mouse chow, chimpanzee food (raw, vegetarian), human cafeteria food (cooked, mixed), and human fast food (cooked, mixed). This short period of dietary divergence led to substantial differences in liver gene expression between groups of mice. Importantly, the specific genes involved and the directional differences observed between the mice fed chimpanzee food and the mice fed either type of human food mirrored 10% of the differences in liver gene expression observed between chimpanzees and humans. Thus, Somel et al. (2008) quickly replicated in mice a sizable fraction of the expression differences observed between humans and our closest living relatives. The relevant genes were found to have evolved more quickly in their promoter and amino acid sequences than by random chance, implying that changes in diet led directly to some genetic divergence between humans and chimpanzees. Because no differences were observed between the groups of mice eating the two human diets, despite large differences in caloric density and macronutrient composition, Somel et al. (2008) reasoned that the expression differences were caused by features that unite the two human diets while simultaneously distinguishing them from both the mouse diet and the chimpanzee diet. Although Somel et al. (2008) do not elaborate, they speculate that meat eating and/or cooking may have been responsible.

CONCLUSIONS

Strategies used by humans and our closest relatives to meet daily energy requirements have been the focus of much research during the last century. However, the significance of cooking—a unique and universal human behavior—has only recently begun to be considered (Aiello and Wheeler 1995; Wrangham et al. 1999; Wrangham and Conklin-Brittain 2003; Carmody and Wrangham 2009). Few experimental studies exist that enable us to directly evaluate the effects of cooking on human mastication (see, e.g., Dominy et al. 2008), digestibility (see, e.g., Evenepoel et al. 1998), or energy gain (see, e.g., Koebnick et al. 1999). Other predictions can currently only be evaluated by reference to animal models (see, e.g., Boback et al. 2007) or with indirect logic, such as the effect of cooking on food texture and its resulting implications for DIT (see, e.g., Oka et al. 2003).

The available evidence suggests that cooking induces physical and chemical changes in food that are consistent with human adaptations toward dietary quality. Our review shows that cooking facilitates mastication of plant and animal foods, compatible with our reduced molar dentition, more gracile mandibles, and smaller chewing muscles compared to chimpanzees. Cooking also increases the digestibility of starchy plant foods and animal proteins, compatible with our shorter less-massive and less-sacculated intestinal tracts. Cooking might also improve the net energy value of human foods through other pathways, including lowering the metabolic costs of digestion and immune defense.

Meat eating is clearly also important as a source of energy. Given that textural changes are at least partially responsible for the proposed positive effects of cooking, nonthermal processing methods that manipulate food texture, such as pounding or grinding, may likewise have contributed to raising net energy gains from the diet during human evolution. Yet, current data hint that neither meat eating nor nonthermal processing is sufficient to explain all of our digestive and energetic adaptations, given that modern raw foodists who consume meat products as well as heavily process their foods using sophisticated nonthermal techniques exhibit low energy status (Koebnick et al. 1999). Further studies are required that focus on the relative impacts of cooking, meat eating, and nonthermal processing on the material properties, digestibility, and net energy value of human foods. In addition, comparative studies of diet-induced gene expression may ultimately provide a molecular signal of dietary divergence between humans and our closest relatives. Such research will discriminate more finely the pathways by which humans transitioned toward obligate dietary quality.

ACKNOWLEDGMENTS

We thank the organizers for the invitation to participate in this *Symposium* volume. We are grateful to Stephen Secor, Adrian Briggs, Corinna Koebnick, Nancy Lou Conklin-Brittain, Charles Nunn, and Katherine Zink for valuable comments and discussion.

REFERENCES

Aiello LC, Key C. 2002. Energetic consequences of being a *Homo erectus* female. *Am J Hum Biol* **14:** 551–565.

Aiello L, Wheeler P. 1995. The expensive-tissue hypothesis: The brain and the digestive system in human and primate evolution. *Curr Anthropol* **36:** 199–221.

Atkins P, Bowler I. 2001. *Food in society: Economy, culture, geography*. Arnold, London.

Barham P. 2001. *The science of cooking*. Springer, Berlin.

Barr SI. 1999. Vegetarianism and menstrual cycle disturbances: Is there an association? *Am J Clin Nutr* **70:** 549S–554S.

Bersaglieri T, Sabeti PC, Patterson N, Vanderploeg T, Schaffner SF, Drake JA, Rhodes M, Reich DE, Hirschhorn JN. 2004. Genetic signatures of strong recent positive selection at the lactase gene. *Am J Hum Genet* **74:** 1111–1120.

Black AE, Coward WA, Cole TJ, Prentice AM. 1996. Human energy expenditure in affluent societies: An analysis of 574 doubly-labelled water measurements. *Eur J Clin Nutr* **50:** 72–92.

Boback SM, Cox CL, Ott BD, Carmody R, Wrangham RW, Secor SM. 2007. Cooking and grinding reduces the cost of meat digestion. *Comp Biochem Physiol A* **148:** 651–656.

Brace C.L. 1995. *The stages of human evolution*. Prentice-Hall, Englewood Cliffs, NJ.

Brewer S. 1978. *The forest dwellers*. Collins, London.

Broadhurst CL, Wang YQ, Crawford MA, Cunnane SC, Parkington JE, Schmidt WF. 2002. Brain-specific lipids from marine, lacustrine, or terrestrial food resources: Potential impact on early African *Homo sapiens*. *Comp Biochem Physiol B* **131:** 653–673.

Bunn HT. 2006. Meat made us human. In *Evolution of the human diet: The known, the unknown, and the unknowable* (ed. PS Ungar), pp. 191–211. Oxford University Press, Oxford.

Carmody RN, Wrangham RW. 2009. The energetic significance of cooking. *J Hum Evol* (in press).

Chesney RW, Helms RA, Christensen M, Budreau AM, Han XB, Sturman JA. 1998. The role of taurine in infant nutrition. *Adv Exp Med Biol* **442:** 463–476.

Clandinin MT. 1999. Brain development and assessing the supply of polyunsaturated fatty acids. *Lipids* **34:** 131–137.

Coelho AM. 1986. Time and energy budgets. In *Comparative primate biology*, Vol 2 (Part A): *Behavior, conservation, and ecology* (ed. G Mitchell and J Erwin), pp. 141–166. A.R. Liss, New York.

Cordain L, Miller JB, Eaton SB, Mann N, Holt SHA, Speth JD. 2000. Plant-animal subsistence ratios and macronutrient energy estimations in worldwide hunter-gatherer diets. *Am J Clin Nutr* **71:** 682–692.

Davies KJA, Lin SW, Pacifici RE. 1987. Protein damage and

degradation by oxygen radicals. IV. Degradation of denatured protein. *J Biol Chem* **262:** 9914–9920.
Dominy NJ, Vogel ER, Yeakel JD, Constantino P, Lucas PW. 2008. Mechanical properties of plant underground storage organs and implications for dietary models of early hominins. *Evol Biol* **35:** 159–175.
DuBois EF. 1937. *The mechanisms of heat loss and temperature regulation.* Stanford University Press, Palo Alto, CA.
Durnin JVGA. 1981. *Basal metabolic rate in man* (Joint FAO/WHO/UNU Expert Consultation on Energy and Protein Requirements, Rome). http://www.fao.org/docrep/meeting/004/M2845E/m2845e00.htm.
Elia M. 1992. Energy expenditure to metabolic rate. In *Energy metabolism: Tissue determinants and cellular corollaries* (ed. JM McKinney and HN Tucker), pp. 19–49. Raven, New York.
Ellison PT. 2003. Energetics and reproductive effort. *Am J Hum Biol* **15:** 342–351.
Ellison PT, Panter-Brick C, Lipson SF, O'Rourke MT. 1993. The ecological context of human ovarian function. *Hum Reprod* **8:** 2248–2258.
Engelen L, Fontijn-Tekamp A, van der Bilt A. 2005. The influence of product and oral characteristics on swallowing. *Arch Oral Biol* **50:** 739–746.
Evenepoel P, Geypens B, Luypaerts A, Hiele M, Rutgeerts P. 1998. Digestibility of cooked and raw egg protein in humans as assessed by stable isotope techniques. *J Nutr* **128:** 1716–1722.
Evenepoel P, Claus D, Geypens B, Hiele M, Geboes K, Rutgeerts P, Ghoos Y. 1999. Amount and fate of egg protein escaping assimilation in the small intestine of humans. *Am J Physiol* **277:** G935–G943.
FAO/WHO/UNU. 2004. Human energy requirements (Joint FAO/WHO/UNU Expert Consultation, Rome. Food and Nutrition Technical Report Series No. 1). http://www.fao.org/docrep/007/Y5686E/y5686e00.htm.
Gaman PM, Sherrington KB. 1996. *The science of food: An introduction to food science, nutrition and microbiology.* Pergamon, Oxford.
Harris D. 1992. Human diet and subsistence. In *The Cambridge encyclopedia of human evolution* (ed. S Jones et al.), pp. 69–74. Cambridge University Press, Cambridge.
Hayes KC, Sturman JA. 1981. Taurine in metabolism. *Annu Rev Nutr* **1:** 401–425.
Hedberg GE, Dierenfeld ES, Rogers QR. 2007. Taurine and zoo felids: Considerations of dietary and biological tissue concentrations. *Zoo Biol* **26:** 517–531.
Henneberg M, Sarafis V, Mathers K. 1998. Human adaptations to meat eating. *Hum Evol* **13:** 229–234.
Hobbs SH. 2005. Attitudes, practices, and beliefs of individuals consuming a raw foods diet. *Explore* **1:** 272–277.
Hoberg EP, Alkire NL, de Queiroz A, Jones A. 2001. Out of Africa: Origins of the *Taenia* tapeworms in humans. *Proc R Soc Lond B* **268:** 781–787.
Horowitz M, O'Donovan D, Jones KL, Feinle C, Rayner CK, Samsom M. 2002. Gastric emptying in diabetes: Clinical significance and treatment. *Diabet Med* **19:** 177–194.
James SR. 1989. Hominid use of fire in the Lower and Middle Pleistocene: A review of the evidence. *Curr Anthropol* **30:** 1–26.
Jarvis MC, Briggs SPH, Knox JP. 2003. Intercellular adhesion and cell separation in plants. *Plant Cell Environ* **26:** 977–989.
Kato Y, Matsuda T, Kato N, Watanabe K, Nakamura R. 1986. Browning and insolubilization of ovalbumin by the Maillard reaction with some aldohexoses. *J Agric Food Chem* **34:** 351–355.
Kleiber M. 1961. *The fire of life: An introduction to animal energetics.* Wiley, New York.
Koebnick C, Strassner C, Hoffmann I, Leitzmann C. 1999. Consequences of a long-term raw food diet on body weight and menstruation: Results of a questionnaire survey. *Ann Nutr Metab* **43:** 69–79.
Lawrie RA. 1991. *Meat science.* Pergamon, Oxford.
Leonard WR, Robertson ML. 1997. Comparative primate energetics and hominid evolution. *Am J Phys Anthropol* **102:** 265–281.
Lepetit J. 2008. Collagen contribution to meat toughness: Theoretical aspects. *Meat Sci* **80:** 960–967.
Lieberman DE, Krovitz GE, Yates FW, Devlin M, St. Claire M. 2004. Effects of food processing on masticatory strain and craniofacial growth in a retrognathic face. *J Hum Evol* **46:** 655–677.
Lin CH, Chang CY. 2005. Textural change and antioxidant properties of broccoli under different cooking treatments. *Food Chem* **90:** 9–15.
Livesey G. 1995. The impact of complex carbohydrates on energy balance. *Eur J Clin Nutr* **49:** S89–S96.
Livesey G. 2002. Thermogenesis associated with fermentable carbohydrate in humans, validity of indirect calorimetry, and implications of dietary thermogenesis for energy requirements, food energy and body weight. *Int J Obesity* **26:** 1553–1569.
Lucas P. 2004. *Dental functional morphology: How teeth work.* Cambridge University Press, Cambridge.
Maillard LC. 1916. A synthesis of humic matter by effect of amine acids on sugar reducing agents. *Ann Chim France* **5:** 258–316.
Mann N. 2007. Meat in the human diet: An anthropological perspective. *Nutr Diet* **64:** S102–S107.
Mason VC. 1984. Metabolism of nitrogenous compounds in the large gut. *Proc Nutr Soc* **43:** 45–53.
McGee H. 2004. *On food and cooking: The science and lore of the kitchen.* Scribner, New York.
McNeil NI. 1988. Nutritional implications of human and mammalian large intestinal function. *World Rev Nutr Diet* **56:** 1–42.
Milton K. 1987. Primate diets and gut morphology: Implications for hominid evolution. In *Food and evolution: Towards a theory of human food habits* (ed. M Harris and EB Ross), pp. 93–115. Temple University Press, Philadelphia.
Milton K. 1999. A hypothesis to explain the role of meat-eating in human evolution. *Evol Anthropol* **8:** 11–21.
Milton K, Demment MW. 1988. Chimpanzees fed high and low fiber diets and comparison with human data. *J Nutr* **118:** 1082–1088.
Monnier JL, Hallegouet B, Hinguant S, Laurent M, Auguste P, Bahain JJ, Falgueres C, Gebhardt A, Marguerie D, Molines N, et al. 1994. A new regional group of the Lower Paleolithic in Brittany (France), recently dated by electron spin resonance. *C R Acad Sci Ser II* **319:** 155–160.
Oka K, Sakuarae A, Fujise T, Yoshimatsu H, Sakata T, Nakata M. 2003. Food texture differences affect energy metabolism in rats. *J Dent Res* **82:** 491–494.
Oste R, Sjodin P. 1984. Effect of Maillard reaction products on protein digestion: In vivo studies on rats. *J Nutr* **114:** 2228–2234.
Oste RE, Miller R, Sjostrom H, Noren O. 1987. Effect of Maillard reaction products on protein digestion: Studies on pure compounds. *J Agric Food Chem* **35:** 938–942.
Pera P, Bucca C, Borro P, Bernocco C, De Lillo A, Carossa S. 2002. Influence of mastication on gastric emptying. *J Dent Res* **81:** 179–181.
Preece RC, Gowlett JAJ, Parfitt SA, Bridgland DR, Lewis SG. 2006. Humans in the Hoxnian: Habitat, context and fire use at Beeches Pit, West Stow, Suffolk, UK. *J Quat Sci* **21:** 485–496.
Purslow PP. 2005. Intramuscular connective tissue and its role in meat quality. *Meat Sci* **70:** 435–477.
Qi BX, Moore KG, Orchard J. 2000. Effect of cooking on banana and plantain texture. *J Agric Food Chem* **48:** 4221–4226.
Ragir S. 2000. Diet and food preparation: Rethinking early hominid behavior. *Evol Anthropol* **9:** 153–155.
Rerat A, Calmes R, Vaissade P, Finot PA. 2002. Nutritional and metabolic consequences of the early Maillard reaction of heat treated milk in the pig: Significance for man. *Eur J Nutr* **41:** 1–11.
Roe CF, Kinney JM. 1965. The caloric equivalent of fever. II. Influence of major trauma. *Ann Surg* **161:** 140–147.
Rosell M, Appleby P, Key T. 2005. Height, age at menarche, body weight and body mass index in life-long vegetarians. *Public Health Nutr* **8:** 870–875.
Secor SM. 2003. Gastric function and its contribution to the

postprandial metabolic response of the Burmese python, *Python molurus*. *J Exp Biol* **206:** 1621–1630.

Secor SM, Boehm M. 2006. Specific dynamic action of ambystomatid salamanders and the effects of meal size, meal type, and body temperature. *Physiol Biochem Zool* **79:** 720–735.

Secor SM, Diamond J. 1995. Adaptive responses to feeding in Burmese pythons: Pay before pumping. *J Exp Biol* **198:** 1313–1325.

Secor SM, Faulkner AC. 2002. Effects of meal size, meal type, body temperature, and body size on the specific dynamic action of the marine toad, *Bufo marinus*. *Physiol Biochem Zool* **75:** 557–571.

Seiquer I, Diaz-Alguacil J, Delgado-Andrade C, Lopez-Frias M, Hoyos AM, Galdo G, Navarro MP. 2006. Diets rich in Maillard reaction products affect protein digestibility in adolescent males aged 11–14 y. *Am J Clin Nutr* **83:** 1082–1088.

Sergant J, Crombé P, Perdaen Y. 2006. The 'invisible' hearths: A contribution to the discernment of Mesolithic non-structured surface hearths. *J Archaeol Sci* **33:** 999–1007.

Shipman P, Walker A. 1989. The costs of becoming a predator. *J Hum Evol* **18:** 373–392.

Shorrock C, Ford JE. 1978. Metabolism of heat-damaged proteins in the rat: Inhibition of amino acid uptake by 'unavailable peptides' isolated from enzymic digests of heat-damaged cod fillet. *Br J Nutr* **40:** 185–191.

Sillen A, Hall G, Armstrong R. 1995. Strontium calcium ratios (Sr/Ca) and strontium isotopic ratios ($^{87}Sr/^{86}Sr$) of *Australopithecus robustus* and *Homo* species. *J Hum Evol* **28:** 277–285.

Silvester KR, Englyst HN, Cummings JH. 1995. Ileal recovery of starch from whole diets containing resistant starch measured in vitro and fermentation of ileal effluent. *Am J Clin Nutr* **62:** 403–411.

Somel M, Creely H, Franz H, Mueller U, Lachmann M, Khaitovich P, Pääbo S. 2008. Human and chimpanzee gene expression differences replicated in mice fed different diets. *PLoS ONE* **3:** e1504.

Tester RF, Sommerville MD. 2000. Swelling and enzymatic hydrolysis of starch in low water systems. *J Cereal Sci* **33:** 193–203.

Tester RF, Qi X, Karkalas J. 2006. Hydrolysis of native starches with amylases. *Anim Feed Sci Technol* **130:** 39–54.

Thieme H. 2005. The Lower Paleolithic art of hunting. In *The hominid individual in context: Archaeological investigations of Lower and Middle Paleolithic landscapes, locales and artefacts* (ed. CS Gamble and M Parr), pp. 115–132. Routledge, London.

Tishkoff SA, Reed FA, Ranciaro A, Voight BF, Babbitt CC, Silverman JS, Powell K, Mortensen HM, Hirbo JB, Osman M, et al. 2007. Convergent adaptation of human lactase persistence in Africa and Europe. *Nat Genet* **39:** 31–40.

Toth N, Schick K, Eds. 2006. *The Oldowan: Case studies into the earliest Stone Age*. Stone Age Institute Press, Gosport, IN.

Waldron KW, Parker ML, Smith AC. 2003. Plant cells walls and food quality. *Compr Rev Food Sci Food Safe* **2:** 101–119.

Westerterp KR. 2004. Diet induced thermogenesis. *Nutr Metab* **1:** 1–5.

Westerterp KR, Wilson SAJ, Rolland V. 1999. Diet induced thermogenesis measured over 24h in a respiration chamber: Effect of diet composition. *Int J Obesity* **23:** 287–292.

Wobber V, Hare B, Wrangham R. 2008. Great apes prefer cooked food. *J Hum Evol* **55:** 340–348.

Wrangham R. 2009. *Catching fire: How cooking made us human*. Basic, New York.

Wrangham RW, Conklin-Brittain NL. 2003. Cooking as a biological trait. *Comp Biochem Physiol A* **136:** 35–46.

Wrangham RW, Jones JH, Laden G, Pilbeam D, Conklin-Brittain NL. 1999. The raw and the stolen: Cooking and the ecology of human origins. *Curr Anthropol* **40:** 567–594.

The Cultural Evolution of Words and Other Thinking Tools

D.C. DENNETT

Center for Cognitive Studies, Tufts University, Medford, Massachusetts 02155
Correspondence: daniel.dennett@tufts.edu

The emergence of language and culture is one of the major transitions in evolution (Maynard Smith and Szathmary 1995), and the key to the cumulative nature of cultural transmission in *Homo sapiens* as contrasted with other species is the digital nature of language, which permits semi-understood designed entities to be preserved and transmitted. Phonemes are not the only systems of self-correcting (digitized) norms; other "alphabets" of practices also contribute to high-fidelity preservation of cultural products.

Anthropocentrism often distorts our vision of evolution, encouraging us to see our own case as special, but there are objective grounds for maintaining that the emergence of language and culture is one of the major transitions in evolution (Maynard Smith and Szathmary 1995). The key to the cumulative nature of cultural transmission in *Homo sapiens*, as contrasted with cultural transmission in other species, is the digital nature of language, which permits semi-understood designed entities to be preserved and transmitted. Phonemes and written letters are not the only systems of self-correcting (digitized) norms; other "alphabets" of practices also contribute to high-fidelity preservation of cultural products despite variable comprehension. The role of comprehension in human culture is often overestimated, and the best way to see the spectrum of possibilities is to adopt the perspective of memes (Dawkins 1976), cultural items that replicate with varying amounts of input from intelligent vectors. Words can be seen to be the foundational memes that permit the accumulation and transmission of ever more elaborate artifacts and practices.

CULTURE AS A MAJOR TRANSITION IN EVOLUTION

According to calculations by Paul MacCready (1999), at the dawn of human agriculture 10,000 years ago, the worldwide human population plus their livestock and pets was ~0.1% of the terrestrial vertebrate biomass. Today, he calculates, it is 98%! (Most of that is cattle.) His reflections on this amazing development are worth quoting:

> Over billions of years, on a unique sphere, chance has painted a thin covering of life—complex, improbable, wonderful and fragile. Suddenly we humans . . . have grown in population, technology, and intelligence to a position of terrible power: we now wield the paintbrush. (MacCready 1999, p. 19).

Some biologists are convinced that we are now living in the early days of a sixth great mass extinction event (the "Holocene"), to rival the Permian–Triassic extinction ~250 million years ago and the Cretaceous–Tertiary extinction ~65 million years ago. And because, as MacCready puts it so vividly, we wield the paintbrush, this mass extinction, if it occurs, would go down in evolutionary history as the first to be triggered by the innovations in a single species. Compared to the biologically "sudden" Cambrian explosion, which occurred over several million years ~530 million years ago, what we may call the MacCready explosion has occurred in ~10,000 years, or ~500 human generations (of course, thousands of prior generations were required to set up many of the conditions that made this possible). There is really no doubt, then, that it has been the rapidly accumulating products of *cultural* evolution—technology and intelligence, as MacCready says—that account for these unprecedented transformations of the biosphere. So Maynard Smith and Szathmary (1995) are right to put language and culture as the most recent of the "major transitions of evolution":

replicating molecules → populations of molecules in compartments
independent replicators → chromosomes
RNA → DNA and protein
prokaryotes → eukaryotes
asexual clones → sexual populations
protists → animals, plants, fungi
solitary individuals → colonies
primate societies → human societies [with language and culture] (Maynard Smith and Szathmary 1995, p. 6)

Behavioral–perceptual transmission (as contrasted with genetic transmission) occurs in many species of animals (Avital and Jablonka 2000), and behaviors earlier deemed genetically inherited "instincts" have been shown primarily by cross-fostering experiments to be mediated by behavioral and perceptual interactions between parent and offspring, not by genes. As Richerson and Boyd (2006) show, just as the standard information highway, the vertical transmission of genes, was optimized during billions of years, the second information highway from parents to offspring had to evolve under rather demanding conditions; however, once this path of vertical *cultural* transmission had been established and optimized, it could be invaded by "rogue cultural variants," horizontally or obliquely transmitted cultural items that do not have the same probability of being benign. (The comparison to spam on the internet is hard to avoid.) These rogue cultural variants are what Richard Dawkins (1976) calls *memes*, and although some of them are bound to be pernicious—parasites, not mutualists—others are profound enhancers of the native competences of the hosts they infect. One can acquire huge amounts of valuable information of which one's parents had no inkling, along with the junk and the scams.

Only in one species, *Homo sapiens*, has transmission by replication of nongenetic information taken off. In us, culture accumulates recursively, explosively, leaping thousands of miles and dozens of centuries in single steps. This hyperpotent variety of cultural evolution depends, I will argue, on language and more specifically on features of *words*, a category of cultural replicant found only in human beings (and, marginally, degenerately, in some of their domesticated animals and pets, such as parrots). It is words, I will argue, that make possible a novel system of design control never before instantiated on the planet, the difference dramatized by the comparison between the termite castle and Antonio Gaudí's La Sagrada Famiglia church in Barcelona (Fig. 1). These two animal artifacts, so outwardly similar in shape, are produced by fundamentally different processes. In the case of the termite castle, "local rules generate global order," as the slogan has it: Individual termites follow rigid rules for moving and depositing building material by detecting local pheromone signals, and no organism has, or needs, a vision or blueprint of the whole structure. In the case of La Sagrada Famiglia, there was an "intelligent designer," an individual, Antonio Gaudí, who did have a guiding vision and did draw up plans; the control of the building flowed from the top down, through *verbal representations* to subordinates and thence to their subordinates. The design and construction could not have proceeded without elaborate systems of symbolic communication.

THE NATURE OF WORDS

What then are words? Do they even exist? This might seem to be a fatuous philosophical question, composed as it is of the very items it asks about, but it is, in fact, exactly as serious and contentious as the claim that genes do or do not *really* exist. Yes, of course, there are sequences of nucleotides on DNA molecules, but does the concept of a gene actually succeed (in any of its rival formulations) in finding a perspicuous rendering of the important patterns amidst all that molecular complexity? If so, there are genes; if not, then genes will in due course get thrown on the trash heap of science along with phlogiston and the ether, no matter how robust and obviously existing they seem to us today. We live in the Age of the Gene (or the Atomic Age), but atoms have turned out not to be atomic, and the genes of Mendel are being shouldered aside, in some regards, by other ways of cutting nature at its joints (Haig 2006). Similarly, there are vocalizations that do an admirable job of purveying information from one person to another; their "moving parts" seem best described as words (or *lexical items* in the somewhat more technical idiom of Jackendoff [2002]). Genes, according to George Williams (1966, p. 25) are best seen as *the information carried* by the nucleotide sequences, not the nucleotide sequences themselves, a point that is nicely echoed by such observations as

Figure 1. (*Left*) Termite castle, (*right*) Antonio Gaudí's La Sagrada Famiglia church in Barcelona (courtesy of Diario di Viaggio).

these: A promise or a libel or a poem is identified by the *words* that compose it, not by the trails of ink or bursts of sound that secure the occurrence of those words. Words themselves have physical "tokens" (composed of uttered or heard phonemes, seen in trails of ink or glass tubes of excited neon or grooves carved in marble), and so do genes, but these tokens are a relatively superficial part or aspect of these remarkable information structures, capable of being replicated, combined into elaborate semantic complexes known as sentences, and capable in turn of provoking cognitive, emotional, and behavioral responses of tremendous power and subtlety.

Words are such familiar parts of our experience that we seldom notice how much less "concrete" they are than the rest of the furniture of the everyday world. A chair or teacup or rainbow may be made of wood, porcelain, or water droplets, but a word is an abstraction, like a method or style, a mathematical technique, or a move in a game. What is castling (in chess) "made of"? What is long division made of, or Bayesian statistics? Or songs or jokes or crossword puzzles or patents or laws or taboos? What is software made of? The human world contains a bounty of manipulable abstractions that are taken for granted and that have no counterpart in the behavioral worlds of other animals. What are words? They are not just sounds, or marks, or even symbols. They are memes (Dawkins 1976; Dennett 1991, 1995, 2006). Words are that subset of memes that can be pronounced.

The best way to see how the concept of memes clarifies and extends our understanding of the role of culture in human evolution is to compare the meme's eye perspective to the traditional wisdom—"common sense"—according to which culture is composed of various valuable practices and artifacts, inherited treasures, in effect, that are recognized as such (for the most part) and transmitted deliberately (and for good reasons) from generation to generation. Cultural innovations that are intelligently designed are esteemed, protected, tinkered with, and passed on to the next generation, whereas accidental or inadvertent combinations of either action or material are discarded or ignored as junk. This is basically an economic model, where possessions, both individual and communal, are preserved, repaired, and handed down. This familiar perspective on culture is for the most part uncritically adopted by cultural historians, anthropologists, and other theorists, and it meshes nicely, it seems, with evolutionary biology. Cultural innovations, like genetic innovations, have to "pay for themselves" to survive, by providing a fitness boost to their possessors. A new way of catching fish, whether genetically transmitted as an innate instinct or cultural transmitted as a learned practice, will go to fixation only if it is better than the old ways of catching fish.

Many celebrated elements of human culture fit this model precisely. We can even identify the intelligent designers responsible for the innovations and carve their names on the friezes of our libraries: Euclid, Pythagoras, Descartes, Newton, Einstein, Curie, Homer, Shakespeare, Austen. But there is a problem: Many of our most valuable cultural treasures have no identifiable author and almost certainly were cobbled together by many largely unwitting minds over long periods of time. Nobody invented words or arithmetic or music or maps or money. These apparent exceptions to the traditional model are typically not seen as a serious problem. The requirement of intelligent authorship can be maintained by distributing it over indefinitely many not-so-intelligent designers whose identities are lost to us only because of gaps in the "fossil record" of culture. And we can acknowledge that many of the improvements accumulated over time were "dumb luck" accidents that nevertheless got appreciated and preserved. With these concessions, the traditionalist can avoid acknowledging what ought to seem obvious: These excellent things acquired their effective designs the same way plants and animals and viruses acquired theirs—they evolved by natural selection, but not *genetic* natural selection.

The dual inheritance model, in which adaptations—fitness-enhancing innovations—can be transmitted vertically either genetically or culturally from parent to offspring, and rogue cultural variants or memes can also be transmitted obliquely or horizontally, takes us almost all the way to the memetic perspective. Rogue cultural variants need not be fitness-enhancing to their hosts in order to flourish, although of course they may be. Like other endosymbionts, they may be parasites that actually reduce fitness, or they may be neutral commensals or benign mutualists. The key improvements, then, of the memetic perspective are its recognition that

1. Excellently designed cultural entities may, like highly efficient viruses, have no intelligent design at all in their ancestry.

2. Memes, like viruses and other symbionts, have their *own* fitness. Those that flourish will be those that better secure their own reproduction, whether or not they do this by enhancing the reproductive success of their hosts by mutualist means.

Our paradigmatic memes, words, would seem to be mutualists *par excellence*, because language is so obviously useful, but we can bear in mind the possibility that some words may, for one reason or another, flourish despite their deleterious effects on this utility. (Pressed once by a student for a good example, I replied "Well, like, there might be, like, a catchphrase or, like, a verbal tic that was, like, a bad but infectious habit that could, like, spread through a subpopulation and, like, even go to fixation without, like, providing any communicative benefit at all." The student replied that he understood the point but, like, could I please give him an example?) The "syntactocentric" (Jackendoff 2002) perspective on language that has dominated theoretical linguistics since the pioneering efforts of Chomsky (see, e.g., Chomsky 1957, 1980) tends to obscure the fact that words have an identity that is to a considerable extent language-independent. Like lateral or horizontal gene transfer, lateral word transfer is a ubiquitous feature, and it complicates the efforts of those who try to identify languages and place them unequivocally in glossogenetic trees. English and French, for instance, share no ancestor later than proto-Indo-European (see Fig. 2) but have many words in common that have migrated

Figure 2. The tree of Proto-Indo-European languages (reprinted, with permission, from Fitch 2007 [© Nature Publishing Group].)

back and forth since their divergence (cul-de-sac and baton, *le rosbif* and *le football*, among thousands of others). Just as gene lineages prove to be more susceptible to analysis than organism lineages, especially when we try to extend the tree of life image back before the origin of eukaryotes (WF Doolittle, this volume), so word lineages are more tractable and nonarbitrary than language lineages (for recent work on the evolution of words, see Fitch 2007; Lieberman et al. 2007; Pagel et al. 2007).

Although words now have plenty of self-appointed guardians, usage mavens, and lexicographers who can seldom resist the temptation to attempt to legislate on questions of meaning and pronunciation, most words—almost all words aside from coinages tied tightly to particular technical contexts—are better seen as *synanthropic*, like rats, mice, pigeons, cockroaches, and bedbugs, rather than *domesticated*. They have evolved to thrive in human company, but nobody owns them, and nobody is responsible for their welfare. The exceptions, such as "oxygen" and "nucleotide," are anchored by systematic definitions fixed by convention and reproduced in the young by deliberate instruction, rehearsal, and memorization. Some domesticated species of animals—notoriously, laying hens—would become extinct without human assistance with their reproduction, and domesticated words can become extinct when the technical contexts for which they were conventionally defined fall into disuse. Who today knows what a *martingale* or a *brigantine* is without looking in a dictionary? Other terms survive with related meanings, wearing their ancestry on their sleeves: *Carriage return* and *carbon copy* are recent examples. Historical inertia permits clearly suboptimal designs to persist. English and German have appropriately terse monosyllabic words—*now* and *jetzt*—where the French and the Italians have to make do with trisyllabic indicators of the moment: *maintenant* and *adesso*. (We English often ignore this economy of course, favoring such long-winded oxymorons as "at this point in time"; our purposes are many, and brevity is seldom the determining value.)

WORDS AND OTHER DIGITIZED CULTURAL ELEMENTS

Words have one feature that has a key role in the accumulation of human culture: They are *digitized*. That is, norms for their pronunciation permit automatic—indeed involuntary—proofreading, preventing transmission errors from accumulating in much the way the molecular machines that accomplish gene replication do. A famous written example is due to Oliver Selfridge (Fig. 3). An English speaker will read this as "THE CAT" even though the second and fifth symbol are exactly the same intermediate shape. Spoken words are also automatically shoe-

Figure 3. Demonstration of involuntary use of context in correcting to norms. (Adapted from Oliver Selfridge.)

horned into phonemic sequences, depending on the language of the hearer. English-speaking audiences have no difficulty reproducing with perfect accuracy "mundify the epigastrium" on a single hearing, even when they have no inkling of what it could mean (soothe the lining of the stomach—a slang term for "have a drink" in some quarters), but they are unable to reproduce accurately the sounds that might be transliterated as "fnurglzhnyum djyukh psajj." No matter how loudly and clearly articulated, this sequence of vocal sounds has no automatic decomposition into phonemes of English. Even nonsense ("the slithy toves did gyre and gimble in the wabe") can be readily perceived and accurately transmitted, thanks to this system of norms. A similar phenomenon can also be seen to occur at higher, semantic levels of analysis, where norms can be relied upon, in the absence of understanding, to stabilize information for replication. For instance, information about kayaks is stored in Inuit brains *and in kayaks* but only on the tacit assumption that any kayak observed is or approximates the norm of a good kayak (Richerson and Boyd 2006). Speaking of Polynesian canoes, the French philosopher Alain observed

> Every boat is copied from another boat... . Let's reason as follows in the manner of Darwin. It is clear that a very badly made boat will end up at the bottom after one or two voyages, and thus never be copied... . One could then say, with complete rigor, that it is the sea herself who fashions the boats, choosing those which function and destroying the others (Alain 1908, quoted in Rogers and Ehrlich 2008).

It is this transmission of competence without comprehension that lies at the heart of all evolutionary processes of natural selection (Dennett 2009), and it is what permits human culture to accumulate geometrically while other animal traditions are barely additive. It is no mere coincidence that digitization also lies at the heart of computer engineering.

Words are not just *like* software viruses; they *are* software viruses, a fact that emerges quite uncontroversially once we adjust our understanding of computation and software. This is made easier for our imaginations by the recent development of Java, the software language that can "run on any platform" and hence has moved to something like fixation in the ecology of the Internet. The intelligent composer of Java applets (small programs that are downloaded and run on individual computers attached to the Internet) does not need to know the hardware or operating system (Mac, PC, Linux, etc.) of the host computer because each computer downloads a Java Virtual Machine (JVM), designed to translate automatically between Java and the hardware, whatever it is. The JVM is "transparent" (users seldom if ever encounter it or even suspect its existence), automatically revised as needed, and (relatively) safe; it will not permit rogue software variants to commandeer your computer. Similarly, when you acquire a language, you install, without realizing it, a Virtual Machine that enables others to send you not just data, but other virtual machines, without their needing to know anything about how your brain works. It is the English Virtual Machine or EVM (in English speakers) that automatically "fixes" the "pinched" H and the "open-ended" A in Selfridge's display (Fig. 3) and that betrays its presence in the Stroop task (Stroop 1935) by interfering with the task of identifying the colors in which the words in Figure 4 are printed. These are but a few of the myriads of microhabits that become imposed on a brain that installs the EVM.

This fact is particularly striking when we compare the practices of psychologists dealing with human subjects and their practices when dealing with other animals. As Jackendoff (2002) has pointed out, it can take hundreds or even thousands of training trials to get animals to perform a new experimental task that can be taught to human subjects in a few sentences of initial briefing. Give your subjects a handful of practice trials and they will perform flawlessly, having constructed a temporary habit that can be abandoned or adjusted as readily as it was adopted. This spectacular combination of plasticity and reliability is unparalleled in the nonhuman world, and it depends on the reliable and projectible effects of dozens of individual informational packets—words—being downloaded—heard—in a few seconds and thereupon implemented. We can readily share novel competences whether we have laboriously acquired them by trial and error or have been given them by others: How to make a cherry pie, tie a bowline, solve Sudoku puzzles, spell "epigastrium." This is the stable base of reliably reproducible elements from which we construct our knowhow and our comprehension.

Words are not just sounds or shapes. As Jackendoff (2002) demonstrates, they are autonomous, semi-independent informational structures, with multiple roles in cognition. They are, in other words, software structures, like Java applets. Unlike Java applets, they are designed by

Figure 4. Stroop task: Identify the colors of the printed words as fast as possible (Stroop 1935).

blind evolution, not intelligent designers, and they get installed by repetition, either by deliberate rehearsal or via several chance encounters. The first time a child hears a new word, it may scarcely register at all, attracting no attention and provoking no rehearsal; the second time the child hears the word, it may be consciously recognized as somewhat familiar or it may not, and in either case, its perception will begin laying down information about context, about pronunciation, and even about meaning. An average six year old has a vocabulary in the range of 5000 words, acquired in ~2000 days. Some of these words are explicitly taught to them (Johnny, this is a truck—say truck!), but most are not even deliberately adjusted in use (two *men*, Johnny, not two *mans*) and yet children reliably acquire remarkably uniform understanding and usage. Controversy rages over how to characterize the native endowment that makes this possible (is Chomsky's Language Acquisition Device a "module" and what information does it contain at the outset?). However, until these issues are sorted out, we can safely note with Deacon (1997) that the *genetic* evolution of human brains in response to the innovative behavior of language must be a response to the *cultural* evolution of words during many thousands (but not millions) of years to hold their own in the competition for rehearsal time and storage space in human brains. Cultural evolution is in principle, but not always in fact, much swifter than genetic evolution, because generation times in the former are measured in seconds, not decades. It is a truism in computer engineering that software development leads hardware development, and this principle has its parallel here: It is much easier for language applets—words and their associated informational structures—to accommodate themselves to the constraints of brain architecture than for brains to accommodate themselves to the demands of language.

THE MEMETIC PERSPECTIVE

With this brief survey of words as memes in hand, let us compare the traditional perspective on cultural evolution with the memetic perspective. Recall that the traditional view is an economic model, of possessions treasured and passed on. The memetic model does not deny the truth of the traditional model; it simply restricts it to one end of the spectrum of possible relations between cultural items and human intelligence and comprehension, including it as a limiting case, in which intelligent, appreciative minds have a strong role.

Traditional model	Memetic model
good things	good, bad, and so-so things
invented with insight	insight from 0 to genius
valued	value from −100 to +100
passed on with improvements	passed on with mutations
(an economic model)	(an economic model as a limiting case)

Darwin himself provides valuable insights into the variable role of intelligence in his introductory discussion in *On the Origin of Species* (see Darwin 1859). He begins with "methodical selection" in which plant and animal breeders with (relatively) clear intentions and expectations set out to improve the breed. He then notes that before methodical selection, there was "unconscious selection" in which human domesticators inadvertently created selection pressures, breeding their favorites without intending thereby to create any long-term adjustments to the variety. The role of intelligence is indirect in such cases and can even be orthogonal to the results obtained: Undesirable traits are often unconsciously selected. The role of intelligence is reduced to nil in Darwin's third step, *natural* selection, in which the vicissitudes of nature do all the culling.[1] We should expect to see all these phenomena in cultural evolution as well. And even if, as many believe, the vast majority of cultural practices and artifacts that persist are beneficial or at any rate not harmful to those who are their vectors, we should adopt a neutral framework from which to measure this claim: One that considers the fitness of the cultural replicators independently of the fitness of their hosts.

This neutral framework is also a valuable antidote to the ubiquitous error of overattributing understanding and intent to those who benefit from the behaviors they regularly engage in, whether these are "instincts" supplied by genes or practices supplied by memes. As the examples of the Inuit kayak or the Polynesian canoe illustrate, the brute presence—survival—of the craft is a more reliable measure of its good design than any insights its current owner may have, and this is as true of "sophisticated" designers as it is of artisans. Amory Lovins (2002) has spoken of the "infectious repetitis" that afflicts car designers; what has worked in the past is routinely assumed by automotive engineers to be optimal, often without any serious consideration of the prospect that improvements might be available.

CONCLUSION

The Gospel of John opens with "In the beginning was the word..." and if construed as a claim about *human* intelligent design, it is surely correct: Language, and especially the use of language to communicate and critique and explain design options, is a precondition for the creation and spread of elaborate artifacts. Not surprisingly, for thousands of years, our model of *all* design processes has been anthropocentric: The word-enabled activity of human designers. Darwin overthrew all that and established the priority of design processes that are not intelligent and that proceed without representation—in words or other symbols—of the reasons they uncover (Dennett 2009). Words are, in fact, a very recent innovation, themselves almost entirely the products of natural selection, not intelligent design. They made possible processes of design and construction unprecedented in

---

[1]Darwin's use of methodical and unconscious selection as bridging cases to gently introduce natural selection was a brilliant bit of pedagogy, but these passages can lead to a serious misconstrual: Both methodical and artificial selection should be seen as special cases of natural selection, not alternatives to it. They are instances of natural selection in which the psychological states of one species, *H. sapiens*, has a particularly focused role in the selection pressure on another species.

the tree of life, illustrated here by the hugely different examples of a termite castle and Gaudí's La Sagrada Famiglia (Fig. 1), and thus their emergence is truly one of the major transitions of evolution.

REFERENCES

Avital E, Jablonka E. 2000. *Animal traditions: Behavioural inheritance in evolution.* Cambridge University Press, Cambridge.

Chomsky N. 1957. *Syntactic structures.* Mouton, The Hague.

Chomsky N. 1980. Rules and representations. *Behav Brain Sci* **3:** 1–15.

Darwin C. 1859. *On the origin of species by means of natural selection,* 1st ed. Murray, London.

Dawkins R. 1976. *The selfish gene.* Oxford University Press, Oxford.

Deacon T. 1997. *The symbolic species: The coevolution of language and the brain.* Norton, New York.

Dennett D. 1991. *Consciousness explained.* Little, Brown, New York.

Dennett D. 1995. *Darwin's dangerous idea: Evolution and the meanings of life.* Simon and Schuster, New York.

Dennett, D. 2006. *Breaking the spell: Religion as a natural phenomenon.* Viking Penguin, New York.

Dennett D. 2009. Darwin's "strange inversion of reasoning." *Proc Natl Acad Sci* **106:** 10061–10065.

Fitch WT. 2007. Language: An invisible hand. *Nature* **449:** 665–667.

Haig D. 2006. The gene meme. In *Richard Dawkins. How a scientist changed the way we think* (ed. A Grafen and M Ridley), pp. 50–65. Oxford University Press, Oxford.

Jackendoff R. 2002. *Foundations of language: Brain, meaning, grammar, evolution.* Oxford University Press, New York.

Lieberman E, Michel JB, Jackson J, Tang T, Nowak MA. 2007. Quantifying the evolutionary dynamics of language. *Nature* **449:** 713–716.

Lovins A. 2002. *TED (Technology, Entertainment, Design) Talk: Ideas Worth Spreading*, February, 2002, Monterey, CA.

MacCready P. 1999. An ambivalent Luddite at a technological feast. *Designfax*, August 1999 (http://www.designfax.net/archives/0899/899trl_2.asp).

Maynard Smith J, Szathmary E. 1995. *The major transitions in evolution.* Freeman, Oxford.

Pagel M, Atkinson QD, Meade A. 2007. Frequency of word-use predicts rates of lexical evolution throughout Indo-European history. *Nature* **449:** 717–720.

Richerson P, Boyd R. 2006. *Not by genes alone: How culture transformed human evolution.* University of Chicago Press, Chicago, Illinois.

Rogers DS, Ehrlich PR. 2008. Natural selection and cultural rates of change. *Proc Natl Acad Sci* **105:** 3416–3420.

Stroop JR. 1935. Studies of interference in serial verbal reactions. *J Exp Psychol* **18:** 643–662.

Williams GC. 1966. *Adaptation and natural selection.* Princeton University Press, Princeton, New Jersey.

When Ideas Have Sex: The Role of Exchange in Cultural Evolution

M.W. RIDLEY

Blagdon Hall, Newcastle, NE13 6DD, United Kingdom
Correspondence: mwridley@gmail.com

Human economic and technological progress has been dominated for the last 100,000 years by natural selection among variants of cultures, rather than among variants of genes. Evidence suggests that cultural evolution depends on exchange and trade to bring together ideas in much the same way that genetic evolution depends on sex to spread genetic mutations, or in the case of bacteria, on horizontal gene transfer. When starved of access to a large "collective brain" by isolation from trade and exchange, people may experience not just less innovation, but even regress. The capacity for ideas to have sex on the Internet is likely to accelerate cultural evolution still further.

According to Richerson and Boyd (2005), cultural evolution can be a Darwinian process but not a genetic one, proceeding largely by the selective survival and transmission of habits, tools, and ideas. Of course, culture can select genes to coevolve, as in the selection on two separate occasions of mutations for the adult digestion of lactose following the domestication of cattle (Tishkoff et al. 2007). But the chief reason that modern society is different from Paleolithic society is down to extensive and cumulative *cultural* change, rather than extensive and cumulative *genetic* change. The entities that compete for survival in cultural evolution are variants of culture. Selective survival happens because human beings copy one another in learning such things, and occasional fortuitous improvements in relatively faithful copying can be sufficient to allow Darwinian natural selection to occur. There is less intentionality in history than human beings like to assume.

This is not a new idea. The Austrian economist Friedrich Hayek (1960) wrote that in social evolution, the decisive factor is "selection by imitation of successful institutions and habits." The evolutionary biologist Richard Dawkins (1976) in the 1970s coined the term "meme" for a unit of cultural imitation. The economist Richard Nelson in the 1980s proposed that whole economies evolve by natural selection (Nelson and Winter 1982). But until recently, Darwinian cultural evolution had not been formally modeled. Now it has been, and it is clear that the parallels with genetic evolution are closer than might have been expected (Henrich et al. 2009).

Copying and innovation in culture are thus equivalent to replication and mutation in biology. Tribalism and language divergence are equivalent to speciation (Pagel and Mace 2004), whereas functional adaptation—fitting form to function—is rampant in cultural evolution, as are Red Queen arms races in which advantages prove to be transient. Differences exist between cultural and genetic evolution, but modeling shows that they do not prevent Darwinian consequences. For example, the fact that mutation is not purely random in culture, but is sometimes intended as deliberate improvement, need not alter the outcome as long as there is a degree of random experiment (Henrich et al. 2009). In any case, the history of technology shows that innovation is a far more incremental, bottom-up, and trial-and-error process than the "heroic inventor" stories we tell one another—or to the Patent Office (see, e.g., Basalla 1988).

The steam engine thus progressed through the designs of Denis Papin to Thomas Savery to Thomas Newcomen to James Watt to Richard Trevithick to George Stephenson by relatively small, incremental steps, by "descent with modification." Yet, at each stage, the heroic inventor exaggerated the revolutionary and unfathered nature of his breakthrough in order to claim either intellectual kudos or a lucrative legal monopoly. In two cases—Thomas Savery and James Watt—aggressive prosecution of a vaguely worded patent on the use of steam enabled the inventor to slow down and set back the further improvement of the entire technology, in Watt's case to a dramatic degree (Boldrin and Levine 2007).

Another difference between cultural and biological evolution is that much cultural evolution occurs by group selection—in Darwin's words, "a tribe...always ready to aid one another, and to sacrifice themselves for the common good, would be victorious over most other tribes; and this would be natural selection" (Darwin 1859). But this makes no great difference either. Natural selection can work well among groups, provided groups have a high enough death rate, as history shows they clearly do: Witness the frequent extinction of languages. It so happens that the conditions for group selection in genetic evolution are rare, except in the special case of groups of close kin.

Here I want to explore the parallel between sex (genetic exchange) and trade (cultural exchange), in order to provide a preliminary test of the hypothesis that the invention of exchange caused a sudden acceleration of cultural evolution in the Middle Stone Age and therefore largely explains why one African hominin subspecies exploded into ecological dominance in a way that other hominins did not. This notion is not wholly original. Like all good ideas, it has evolved by descent with modification, its immediate ancestors being Ofek (2001), Henrich (2004), Richerson and Boyd (2005), and Seabright (2005), not to mention Smith (1776).

HORIZONTAL EXCHANGE

Sexual reproduction accelerates the fixation of beneficial mutations and the elimination of harmful ones through recombination, the phenomenon known as Muller's ratchet (Ridley 1993). It therefore helps to make the evolution of traits cumulative: A less-fit lineage that nonetheless has a single beneficial trait can bestow it on a fitter lineage through sex. The acquisition of new traits is accelerated by the fact that through sex, each individual in effect draws upon the entire gene pool of the species and therefore has access to a wider source of genetic innovations than are available in its vertical lineage. Asexual exceptions turn out to prove the rule; i.e., they find other ways of indulging in genetic exchange. Bdelloid rotifers, for example, have experienced massive horizontal gene transfer between species and kingdoms, which explains how they have achieved diversity and fitness without having sex for 80 million years (Gladyshev et al. 2008).

Such exceptions aside, however, animals and plants are generally limited to genetic exchange within the species, which balkanizes evolution of multicellular creatures: Much as some mammals might benefit from having ultraviolet vision and some insects might benefit from having mammary glands, the requisite genes are inaccessible to them. No such constraint applies to the microbial world, where it is now apparent that adaptation to new environments (such as the human deployment of antibiotics) occurs through the promiscuous swapping of genes between "species" of bacteria; indeed, that bacterial "species" are almost more like transient "firms" of genes than closed species.

Exchange has the same role in cultural evolution. Plenty of animals show vertically transmitted culture: Killer whales, chimpanzees, crows, and dolphins all show the ability to pass on local traditions to their young. But in none of these species is culture cumulative. The reason for this is vertical transmission equivalent to heredity in an asexual species. A chimpanzee can acquire the culture of only its own group, not that of other groups. Not only do human beings tend to learn skills from prestigious individuals other than their parents (Henrich 2004), but human culture is transmitted horizontally as well as vertically between groups through exchange and trade. The effect is that a human being draws upon a vast "gene pool" of cultural inventiveness encompassing almost the entire species. I suggest that the invention of exchange was the trigger for the explosion of human numbers, range, and ecological impact that happened in the past 100,000 years or slightly more.

THE COLLECTIVE BRAIN

Human beings are seemingly more intelligent than other animals, but this alone cannot explain the "human revolution" of the past 100,000 years when a scarce African ape at the mercy of its environment became a global species that commandeers 22% of the planet's natural primary productivity (Haberl et al. 2007). The problem is not just that big hominin brains and the capacity for language appear to be much more than 100,000 years old, with no evidence of a threshold reached or a switch thrown at a particular point, but more particularly that human ecological success is not an individual phenomenon. It is a collective phenomenon; i.e., modern technology, science, and culture are the aggregate product of millions of brains, none of which knows how to achieve them. As the economist Leonard Read put it in his essay "I, Pencil," even an ordinary pencil is made by millions of people, from loggers in Oregon and graphite miners in Sri Lanka to coffee bean growers in Brazil (who supplied the coffee drunk by the loggers). "There isn't a single person in all these millions," Read concludes, "including the president of the pencil company, who contributes more than a tiny, infinitesimal bit of know-how" (Read 1958). Knowledge, said Friedrich Hayek, "never exists in concentrated or integrated form but solely as the dispersed bits of incomplete and frequently contradictory knowledge which all the separate individuals possess" (Hayek 1945). There was a point in human history when this became true for the first time.

Cultural and technological complexity depends critically on population density (Simon 1996). This is evident at every scale and in every time period—from the complexity and inventiveness of modern cities contrasted with the simplicity of modern pastoralists; from the richness of early agricultural societies to the simplicity of hunter-gatherers; from the sophistication of dense hunter-gatherer societies that lived on rich marine resources in the Pacific Northwest to the simple technologies of sparse Australian aboriginal societies. The denser a human population becomes, the greater the chance that it will show rapid cultural innovation. Pet-grooming salons and sushi restaurants are found in large cities, not remote rural settlements. This population-density effect is not true of other species.

There is increasing interest among archaeologists in explaining the sporadic and ephemeral nature of Middle Stone Age technological advances by reference to demographic factors (Powell et al. 2009; Richerson et al. 2009). Two sudden and temporary efflorescences of new technologies in Southern Africa ~70,000 years ago both seem to coincide with population expansions, as does the explosion of innovation that happened in western Eurasia 40,000 years ago (Jacobs and Roberts 2009).

Conversely, where population falls or is fragmented, cultural evolution may actually regress. A telling example comes from Tasmania, where people who had been making bone tools, clothing, and fishing equipment for 25,000 years gradually gave these up after being isolated by rising sea levels 10,000 years ago (Henrich 2004). Henrich argues that the population of 4000 Tasmanians on the island constituted too small a collective brain to sustain, let alone improve, the existing technology. Tierra del Fuego, in a similar climatic and demographic position, experienced no such technological regress because its people remained in trading contact with the mainland of South America across a much narrower strait throughout the prehistoric period. In effect, they had access to a continental collective brain.

DID NEANDERTHALS EXCHANGE?

I therefore argue that the reason the technology used by Eurasian Neanderthals, with their big brains and probable linguistic skills (Krause et al. 2007), did not show significant evolution or local traditions was that Neanderthals did not engage in exchange. Almost all Neanderthal tools have been found—so far—within an hour's walk of their likely site of origin (Stringer 2006), whereas it is characteristic of African-origin hominins from at least 120,000 years ago that artifacts travel long distances, probably by trade. For example, shell beads in Algeria and obsidian tools in Ethiopia dating to more than 80,000 years ago have been found more than 100 miles from their likely maritime and volcanic source, respectively (Negash et al. 2006; Barton et al. 2009). In more recent times, stone axes manufactured by the Kalkadoon at a site called Mount Isa in northern Australia found their way across half the continent before western contact. Such trade networks greatly enlarge the collective brain at the disposal of each individual by giving him access to material and ideas from a wide area and a large population (Sharp 1974).

In contrast, even 30,000 years ago, European Neanderthals apparently lacked the habit of trading between groups, and simulation shows that this alone would have rendered them vulnerable to extinction in the face of competition from trading Cro-Magnons, because they were unable to survive temporary or local shortages (Horan et al. 2005). In the words of Adler et al. (2006), "we hypothesize that it is the development and maintenance of larger social networks, rather than technological innovations or increased hunting prowess, that distinguish modern humans from Neanderthals in the southern Caucasus."

There is, of course, a plethora of other candidates to explain human success: upright stance (which freed the hands), opposable thumbs, tool use, grandparental care, fire and cooking, large brains, foresight and planning, and, of course, language. The problem with each of these explanations, however, is that it occurs too early in the story or applies to other species as well, especially the Neanderthals. Cooperative rearing, for instance, is shared by many primates and is probably an ancient feature (Hrdy 2009). Cooking undoubtedly made the evolution of large brains possible (Wrangham 2009) but may have begun with *Homo erectus*. The Neanderthals stand as a powerful reminder that it is possible for a hominin to be upright, dextrous, tool-using, cooperative, fire-controlling, huge-brained, imaginative (Neanderthals buried their dead), and probably well spoken, yet still not experience rapid and cumulative cultural evolution leading to global ecological dominance. Exchange is the only feature of humanity so far described that arrives in the African lineage after the split with the ancestors of Neanderthals and coincides with takeoff.

RICARDO'S MAGIC TRICK

The cultural evolutionary process by which exchange both reduces risk and increases productivity or energy efficiency has long been known to economists. According to David Ricardo's Law of Comparative Advantage, building on the work of Adam Smith and others, exchange leads to specialization, which leads to both efficiency and innovation, which leads to further specialization, which encourages further exchange. If individual A has even a slight specialized knowledge of how to make or acquire object X, and B has a slight skill in acquiring object Y, then it pays them both for A to make or get 2X and swap one for one of B's 2Y. Remarkably, this is true even if A is better at acquiring both X and Y than is B, as long as A finds X easier to get than Y, and B finds Y easier to get than X. So, for instance, an imaginary coastal fisherman can more easily catch two fish than a fish and a banana; his best way of getting a banana is to swap his extra fish for a spare banana picked by a trading partner living inland. As Ricardo famously put it, "England may be so circumstanced, that to produce the cloth may require the labour of 100 men for one year; and if she attempted to make the wine, it might require the labour of 120 men for the same time. England would therefore find it in her interest to import wine, and to purchase it by the exportation of cloth. To produce the wine in Portugal might require only the labour of 80 men for one year, and to produce the cloth in the same country might require the labour of 90 men for the same time. It would therefore be advantageous for her to export wine in exchange for cloth. This exchange might even take place, notwithstanding that the commodity imported by Portugal could be produced there with less labour than in England" (Ricardo 1817).

Once each individual is specializing in this way, each is bound to get better still at his own speciality. The more time the fisherman spends fishing, the better he gets at it, and the more time the inland trader spends banana picking, the better he gets at that. Charles Darwin, with scant evidence, presumed that this habit of specialization and exchange is long established in human beings: "Primeval man practised a division of labour; each man did not manufacture his own tools or rude pottery, but certain individuals appear to have devoted themselves to such work, no doubt receiving in exchange the produce of the chase" (Darwin 1871). Everything that has been discovered since then implies that he was right to the extent that exchange and trade long precedes other human inventions such as farming, money, or government. But it does not precede cumulative cultural evolution. The entire story of humanity since 100,000 BC has been one of sporadic and cumulative increases in specialization and interdependence through this Ricardian process, with many setbacks along the way. It is the very definition of modern prosperity to consume diverse goods and services in exchange for singular and specialized production. The opposite—self-sufficient subsistence, in which consumption is no more diverse than production—is nowadays known as poverty.

No other animal does this. In those relatively few species where individuals do specialize—notably, social insects and other colonial creatures—the specialization is preordained by genes and environmental triggers; i.e., the number of castes in an ant colony does not increase indefinitely or change unpredictably with the life of the colony. Crucially, too, in social insects, the exchange and specialize bargain is struck between close kin, rather than between unrelated individuals. Indeed, the first and most extreme division of labor in all such species is the reproductive one, whereas reproduction is the only trait that human beings do not delegate to specialized individuals. (Not even in England do we leave breeding to a queen.) This balkanizes specialization within colonies. Human beings routinely practice exchange and specialization between strangers.

EXCHANGE IS NOT THE SAME AS RECIPROCITY

Note here that the kind of exchange required to make the Ricardian specialization engine work is emphatically not "reciprocity" of the kind that has long interested evolutionary psychologists and zoologists. Reciprocity is defined by Leda Cosmides and John Tooby (1992) as follows: "One party helps another at one point in time, in order to increase the probability that when their situations are reversed at some (usually) unspecified time in the future, the act will be reciprocated." In other words, it is the swapping of the same favor at different times. Such reciprocity is undoubtedly an important human social glue, as it is in other primates, cetaceans, some birds, and even vampire bats (Ridley 1996). It may indeed be the key to much cooperation. But it is not the same as exchange: the swapping of different objects at the same time. When an Australian gave a stone axe to his trading partner in return for a stingray barb, the bargain was simultaneous and heterogeneous, not delayed and homogeneous.

Cosmides and Tooby argue that reciprocity is harder to evolve than exchange because it requires trust during the delay between the favor and its return. Maybe, but it remains a fact that whereas homogeneous reciprocity is widespread if occasional in the animal kingdom, no other species has hit upon the trick of heterogeneous exchange between unrelated individuals (with the single exception of food-for-sex swaps between mates). "No man ever saw a dog make fair and deliberate exchange of a bone with another dog," said Adam Smith (1776). Sarah Brosnan's experiments with chimpanzees and capuchins have shown them to be quite capable of learning to exchange worthless tokens for food but incapable of learning to exchange valued foods for even more valued foods (Brosnan et al. 2008). Perhaps exchange is especially hard to evolve because it requires individuals to overcome the endowment effect, whereby individuals overvalue items in their possession relative to items that they do not yet possess (Kahneman et al. 1990; Brosnan et al. 2007).

Perhaps language is a prerequisite for exchange but not for reciprocity. Undoubtedly, language makes exchange easier, and it may be the case that human exchange could not begin until language had become sophisticated and subtle. This is, however, different from arguing that language itself causes cultural evolution. It is the power of language to assist exchange that drives cultural evolutionary acceleration, rather than the capacity for speech itself. Besides, if Neanderthals spoke but did not exchange—and it appears that they had genetic mutations possibly selected for by the use of language (Krause et al. 2007)—then widespread exchange is not a necessary consequence of language. Conversely, exchange can occur without shared language. Wrote Darwin of an encounter in Tierra del Fuego in 1834: "Some of the Fuegians plainly showed that they had a fair notion of barter. I gave one man a large nail (a most valuable present) without making any signs for a return; but he immediately picked out two fish, and handed them up on the point of his spear" (Darwin 1839).

WHEN DID HUMAN EXCHANGE AND SPECIALIZATION BEGIN?

The most universal, and probably oldest, form of human exchange and specialization is that between male hunter and female gatherer. The pattern of this sexual division of labor is highly variable but common to all humankind: In all hunter-gatherer societies yet studied, the two sexes use different foraging strategies but subsequently share the proceeds. This produces obvious Ricardian "gains from trade"; each sex can have access to both staple and reliable carbohydrates and unreliable but valuable protein. Each sex also gets access to the results of the other's specialized knowledge and skill. It is true that male foraging often looks as much like conspicuous display or the provision of a public good as individualized one-for-one exchange with individual females (Hawkes 1996), but this does not alter the consequence that males get access to (mostly) gathered food if they produce (mostly) hunted food, and vice versa.

If the sexual division of labor is the oldest form of heterogeneous exchange, perhaps it was the custom that "got us into the habit" of exchange (although it is also possible that the reverse is true and the sexual division of labor is a late application to family life of something learned through more general trade). But how old is the sexual division of labor? Glyn Isaac argued that it began with central-place foraging early in the story of the genus *Homo*, at least 2 million years ago (Isaac and Isaac 1989). Richard Wrangham (2009) argues that it follows naturally from the invention of cooking, which is a strongly gendered task that generates for females a valuable but visible and stationary product that is vulnerable to theft by males without protection, and that it dates back to the dawn of *H. erectus* more than a million years ago. On the other hand, Kuhn and Stiner (2006) argue that given the lack of evidence for "gathering" strategies in Neanderthal sites, the burden of proof is on those who would argue that Neanderthals had a sexual division of labor at all. This is not to deny the importance of cooperative foraging and food sharing in Neanderthal society but to suggest that Neanderthals of both sexes may have cooperated as hunters of megafauna, rather than foraged separately by sex for different foods.

Incidentally, dependent children do not explain the sexual division of labor well: They prove not to cramp the style of hunting females in many hunter-gatherers such as the Alyawarre (Bird and Bird 2008); even if female foraging is energetic, crèche-rearing and alloparenting—e.g., by grandparents—can free most mothers to join in foraging activities (Hrdy 2009). If Kuhn and Stiner are right, the sexual division of labor could be a specifically African invention of less than 300,000 years ago, rather than an ancient invention of early hominin species. It may have triggered, or at least prepared the psychological way for, the continuous cultural evolution through natural selection of more and more specialization and exchange.

If so, exchange between genders would have been followed by exchange between other individuals within the band, which would have been followed by exchange between bands. This would have had to overcome the instant and violent hostility between bands that characterizes most nonhuman social primates (Hrdy 2009) and that still plagues the human species. Mechanisms for overcoming such hostility, such as smiling, abound in the human behavioral repertoire.

It is striking that both anthropological anecdotes and laboratory experiments point to the conclusion that simple barter begins between individuals in different groups, rather than anonymously between groups as a whole. Thus, each Yir Yoront stingray barb trader had a single stone axe trading partner whom he met at an annual trade festival (Sharp 1974), and undergraduates at George Mason University invent mutually beneficial trade between virtual villages (in "red" or "blue" items) on a one-to-one basis (Crockett et al. 2009). Collective trade begins with individual exchange.

INTERACTION, NOT NECESSITY, IS THE MOTHER OF INVENTION

The expansion of the human collective brain continues today. Humans devote their productive hours to ever more specialized "jobs," while accessing an ever more diverse range of consumption. In both activities, the individual draws upon a huge range of innovations made across a vast geographical and historical hinterland. A typical day in a modern office uses inventions made in India (zero), Palestine (iron), Asia Minor (money), Phoenicia (alphabet), Greece (science), Mesopotamia (writing), China (printing), Italy (credit), Mexico (maize), Peru (potatoes), Germany (aspirin), Britain (electricity), Massachusetts (telephone), California (semiconductors), Scandinavia (mobile networks), and Japan (Nintendo), to name just a few. Not one of these inventions is accessible to the inhabitants of North Sentinel island in the Andaman archipelago—the only people who continue to resist all contact with globalization; all are accessible to almost everybody else, at least in theory. This is not to pass judgment on what has happened, but to describe it. The collective nature of the human enterprise is now global, so that a cultural mutation in any part of the world can spread, if popular, to all other parts of the world, just as a genetic mutation in a flu virus can. The Internet is effectively a single planetary "agora" or marketplace in which ideas are experiencing sex, or at least horizontal transfer, on an unprecedented scale, allowing the spread, replacement, and fixation of innovations. In the 1980s, computers were limited by their individual processing power. Today, they depend on shared and collective processing power: The same thing happened to human minds in the late Paleolithic.

CONCLUSION

The lineage of species leading to modern human beings experienced more than 2 million years of "normal" evolutionary change following the appearance of the first stone tools, i.e., technological change occurred no more rapidly than anatomical change. The basic design of the Acheulean hand ax typical of *H. erectus* lasted for more than a million years with very little change. Then suddenly, beginning probably between 300,000 and 100,000 years ago, there occurred a dramatic acceleration of technological change in Africa but not in other continents, accompanied by little or no change in anatomy. The timing and accelerating shape of this change does not fit the usual explanations based on some threshold or switch in human intelligence or language: it starts slowly, gathers pace, but peters out often. Instead, it is compatible with the explanation that intelligence had progressively begun to become a collective property shared by many specialized brains—a network effect. Cumulative and extensive cultural evolution was now possible because of the widespread pattern of exchanging objects, ideas, and services between individuals and the consequent specialization among them.

REFERENCES

Adler DS, Bar-Oz G, Belfer-Cohen A, Bar-Yosef O. 2006. Ahead of the game: Middle and upper palaeolithic hunting behaviors in the Southern Caucasus. *Curr Anthropol* **47:** 89–118.

Barton RNE, Bouzouggar A, Collcutt SN, Schwenninger J-L, Clark-Balzan L. 2009. OSL dating of the Aterian levels at Dar es-Soltan I (Rabat, Morocco) and implications for the dispersal of modern *Homo sapiens*. *Quat Sci Rev* **19-20:** 1914–1931.

Basalla G. 1988. *The evolution of technology*. Cambridge University Press, Cambridge.

Bird RB, Bird DW. 2008. Why women hunt: Risk and contemporary foraging in a western Desert Aboriginal community. *Curr Anthropol* **49:** 655–693.

Boldrin M, Levine DK. 2007. *Against intellectual monopoly*. (Published online at http://www.dklevine.com/general/intellectual/against.htm.)

Brosnan SF, Jones OD, Lambeth SP, Mareno MC, Richardson AS, Schapiro S. 2007. Endowment effects in chimpanzees. *Curr Biol* **17:** 1704–1707.

Brosnan SF, Grady MF, Lambeth SP, Schapiro SJ, Beran MJ. 2008. Chimpanzee autarky. *PLoS ONE* **3:** e1518.

Cosmides L, Tooby J. 1992. Cognitive adaptations for social exchange. In *The adapted mind: Evolutionary psychology and the generation of culture* (ed. JH Barkow et al.), pp. 163–228. Oxford University Press, Oxford.

Crockett S, Wilson B, Smith V. 2009. Exchange and specialization as a discovery process. *Econ J* **119:** 1162–1188.

Darwin CR. 1839. *The voyage of the beagle*. Murray, London.

Darwin CR. 1859. *On the origin of species by means of natural selection*, 1st ed. Murray, London.

Darwin CR. 1871. *The descent of man and selection in relation to sex*. Murray, London.

Dawkins R. 1976. *The selfish gene*. Oxford University Press, Oxford.

Gladyshev EA, Meselson M, Arkhipova IR. 2008. Massive horizontal gene transfer in bdelloid rotifers. *Science* **320:** 1210–1213.

Haberl H, Erb KH, Krausmann F, Gaube V, Bondeau H, Plutzar C, Gingrich S, Lucht W, Fischer-Kowalski A. 2007. Quantifying and mapping the human appropriation of net primary production in earth's terrestrial ecosystems. *Proc Natl Acad Sci* **104:** 12942–12947.

Hawkes K. 1996. Foraging differences between men and women. In *The archaeology of human ancestry: Power, sex, and tradition* (ed. J Steele and S Shennan), pp. 256–275. Routledge, London.

Hayek FA. 1945. The use of knowledge in society. *Am Econ Rev* **35:** 519–530.

Hayek FA. 1960. *The constitution of liberty*. University of Chicago Press, Chicago.

Henrich J. 2004. Demography and cultural evolution: How adaptive cultural processes can produce maladaptive losses—The Tasmanian case. *Am Antiquity* **69:** 197–214.

Henrich J, Boyd R, Richerson P. 2009. Five misunderstandings about cultural evolution. *Hum Nat* **19:** 119–137.

Horan RD, Bulte EH, Shogren JF. 2005. How trade saved humanity from biological exclusion: The Neanderthal enigma revisited and revised. *J Econ Behav Org* **58:** 1–29.

Hrdy SB. 2009. *Mothers and others: The evolutionary origins of mutual understanding*. Belknap/Harvard University Press, Cambridge, MA.

Isaac GL, Isaac B. 1989. *The archaeology of human origins: Papers by Glyn Isaac*. Cambridge University Press, Cambridge.

Jacobs Z, Roberts RG. 2009. Human history written in stone and blood. *Am Sci* **97:** 302–307.

Kahneman D, Knetsch J, Thaler R. 1990. Experimental tests of the endowment effect and the Coase theorem. *J Political Econ* **98:** 1325–1348.

Krause J, Lalueza-Fox C, Orlando L, Enard W, Green RE, Burbano HA, Hublin JJ, Hänni C, Fortea J, de la Rasilla M, Bertranpetit J, Rosas A, Pääbo S. 2007. The derived *FOXP2* variant of modern humans was shared with Neandertals. *Curr Biol* **17:** 1908–1912.

Kuhn SL, Stiner MC. 2006. What's a mother to do? A hypothesis about the division of labor and modern human origins. *Curr Anthropol* **47:** 953–980.

Negash A, Shackley MS, Alene M. 2006. Source provenance of obsidian artifacts from the Early Stone Age (ESA) site of Melka Konture, Ethiopia. *J Archaeol Sci* **33:** 1647–1650.

Nelson RR, Winter SG. 1982. *An evolutionary theory of economic change*. Harvard University Press, Cambridge, MA.

Ofek H. 2001. *Second nature: Economic origins of human evolution*. Cambridge University Press, Cambridge.

Pagel M, Mace R. 2004. The cultural wealth of nations. *Nature* **428:** 275–278.

Powell A, Shennan S, Thomas MG. 2009. Late Pleistocene demography and the appearance of modern human behavior. *Science* **324:** 1298–1301.

Read LE. 1958. I, pencil. *The Freeman*, December 1958.

Ricardo D. 1817. *On the principles of political economy and taxation*. Murray, London.

Richerson PJ, Boyd R. 2005. *Not by genes alone: How culture transformed human evolution*. University of Chicago Press, Chicago.

Richerson PJ, Boyd R, Bettinger RL. 2009. Cultural innovations and demographic change. *Hum Biol* **81:** 211–235.

Ridley M. 1993. *The red queen: Sex and the evolution of human nature*. Penguin, London.

Ridley M. 1996. *The origins of virtue: Human instincts and the evolution of cooperation*. Penguin, London.

Seabright P. 2005. *The company of strangers: A natural history of economic life*. Princeton University Press, Princeton, NJ.

Sharp L. 1974. Steel axes for Stone Age Australians. In *Man in adaptation: The cultural present* (ed. Y Cohen), pp. 116–127. Aldine, Chicago.

Simon J. 1996. *The ultimate resource 2*. Princeton University Press, Princeton, NJ.

Smith A. 1776. *An inquiry into the nature and causes of the wealth of nations* (facsimile edition). University of Chicago Press, Chicago.

Stringer C. 2006. *Homo britannicus*. Penguin, London.

Tishkoff SA, Reed FA, Ranciaro A, Voight BF, Babbitt CC, Silverman JS, Powell K, Mortensen HM, Hirbo JB, Osman M, et al. 2007. Convergent adaptation of human lactase persistence in Africa and Europe. *Nat Genet* **39:** 31–40.

Wrangham R. 2009. *Catching fire: How cooking made us human*. Perseus, Philadelphia.

An Evolutionary Approach to Financial History

N. Ferguson

Laurence A. Tisch Professor of History, Harvard University, and William Ziegler Professor of Business Administration, Harvard Business School, Cambridge, Massachusetts 02138
Correspondence: nferguson@hbs.edu

Financial history is not conventionally thought of in evolutionary terms, but it should be. Traditional ways of thinking about finance, dating back to Hilferding, emphasize the importance of concentration and economies of scale. But these approaches overlook the rich "biodiversity" that characterizes the financial world. They also overlook the role of natural selection. To be sure, natural selection in the financial world is not exactly analogous to the processes first described by Darwin and elaborated on by modern biologists. There is conscious adaptation as well as random mutation. Moreover, there is something resembling "intelligent design" in finance, whereby regulators and legislators act in a quasidivine capacity, putting dinosaurs on life support. The danger is that such interventions in the natural processes of the market may ultimately distort the evolutionary process, by getting in the way of Schumpeter's "creative destruction."

It is commonly said that finance has a Darwinian quality. "The survival of the fittest" is a phrase that aggressive market participants like to use; in the days before they themselves were rendered extinct, investment banks used to hold conferences with titles such as "The Evolution of Excellence." The financial crisis that began in August 2007 has increased the frequency of such language. U.S. Assistant Secretary of the Treasury Anthony W. Ryan was not the only person to talk in terms of a wave of financial extinctions in the second half of 2007. Andrew Lo, director of the Massachusetts Institute of Technology's Laboratory for Financial Engineering, is in the vanguard of an effort to reconceptualize markets as complex, adaptive systems similar to those we encounter in the natural world. A recent long-run historical analysis of the development of financial services also emphasizes the role of evolutionary forces in financial history (Ferguson and Wyman 2007).

The notion that Darwinian processes may be at work in the economy is not new, of course. In *On the Origin of Species*, Darwin himself explicitly acknowledged his debt to Malthus's *Essay on the Principle of Population* (1798). As Darwin put it, the observation of Victorian capitalism had left him "well prepared to appreciate the struggle for existence...it at once struck me that under these circumstances favourable variations would tend to be preserved, and unfavourable ones to be destroyed. Here, then, I had at last got a theory by which to work" (Darwin 1859). Unscientific applications of Darwinian terminology to social problems led down many blind alleys, notably biological racism and eugenics. But evolutionary economics has managed to establish itself as a respectable if minor subdiscipline, which has had its own dedicated journal for the past 17 years, the *Journal of Evolutionary Economics*.

Thorstein Veblen first posed the question "Why Is Economics Not an Evolutionary Science?" (implying that it really should be) more than a century ago (Veblen 1898). In a famous passage in his *Capitalism and Democracy*, which could equally well apply to finance, Joseph Schumpeter characterized industrial capitalism as "an evolutionary process":

> This evolutionary character...is not merely due to the fact that economic life goes on in a social and natural environment which changes and by its change alters the data of economic action; this fact is important and these changes (wars, revolutions and so on) often condition industrial change, but they are not its prime movers. Nor is this evolutionary character due to quasi-autonomic increase in population and capital or to the vagaries of monetary systems of which exactly the same thing holds true. The fundamental impulse that sets and keeps the capitalist engine in motion comes from the new consumers' goods, the new methods of production or transportation, the new markets, the new forms of industrial organization that capitalist enterprise creates. ...The opening up of new markets, foreign or domestic, and the organizational development from the craft shop and factory to such concerns as U.S. Steel illustrate the same process of industrial mutation—if I may use the biological term—that incessantly revolutionizes the economic structure *from within*, incessantly destroying the old one, incessantly creating a new on. This process of Creative Destruction is the essential fact about capitalism (Schumpeter 1987 [1943], pp. 80–84).

Seminal works in the nascent field of evolutionary economics include Alchian (1950) and Nelson and Winter (1982). Lo's recent research on hedge funds illustrates the potential benefits of importing evolutionary concepts to the realm of finance. In Lo's words, "Hedge funds are the Galapagos Islands of finance....The rate of innovation, evolution, competition, adaptation, births and deaths...occurs at an extraordinarily rapid clip" (Economist 2008).

CREATIVE DESTRUCTION

A key point that emerges from recent research is just how much creative destruction goes on in a modern economy. Approximately one in 10 U.S. companies disappears each year. Between 1989 and 1997, to be precise, 611,000 businesses vanished annually, on average, out of a total of 5.73 million firms. Ten percent is the average extinction rate, it should be noted; in some sectors of the economy, it can rise to as high as 20% in a bad year, as in the District of Columbia's financial sector in 1989, at the height of the savings and loans crisis (Ormerod 2005, pp. 180–182). According to the U.K. Department of Trade and Industry, 30% of tax-registered businesses in Britain disappear after just 3 years (Guthrie 2007).

Even if they survive the first few years of existence and go onto enjoy great success, most firms eventually fail. Of the world's 100 largest companies in 1912, 29 were bankrupt by 1995, 48 had disappeared, and only 19 were still in the top 100 (Hannah 2009). Given that a good deal of what banks and stock markets do is to provide financial services to nonfinancial companies, we should not be surprised to find a similar pattern of creative destruction in the financial world. Between one-third and one-half of all hedge funds extant in early 2007 had vanished by the end of 2009. The only reason that more banks do not fail is that they are explicitly or implicitly protected from collapse by governments.

FINANCIAL EVOLUTION AND DARWINIAN EVOLUTION: COMMON FACTORS

What are the common features shared by the financial world and a true evolutionary system? As previously proposed in Ferguson (2008), there are at least six:

- "Genes," in the sense that certain business practices perform the same role as genes in biology, allowing information to be stored in the "organizational memory" and passed on from individual to individual or from firm to firm when a new firm is created.
- The potential for spontaneous "mutation," usually referred to in the economic world as innovation and primarily, although by no means always, technological.
- Competition among individuals within a species for resources, with the outcomes in terms of longevity and proliferation determining which business practices persist.
- A mechanism for natural selection through the market allocation of capital and human resources and the possibility of death in cases of underperformance, i.e., "differential survival."
- Scope for speciation, sustaining biodiversity through the creation of wholly new "species" of financial institutions.
- Scope for extinction, with species dying out altogether.

Financial history is essentially the result of institutional mutation and natural selection. Random "drift" (innovations/mutations that are not promoted by natural selection, but just happen) and "flow" (innovations/mutations that are caused when, say, American practices are adopted by Chinese banks) play a part. There can also be "coevolution," when different financial species work and adapt together (such as hedge funds and their prime brokers). But market selection is the main driver. Financial organisms are in competition with one another for finite resources. At certain times and in certain places, certain species may become dominant. But innovations by competitor species, or the emergence of altogether new species, prevent any permanent hierarchy or monoculture from emerging. Broadly speaking, the law of the "survival of the fittest" applies. Institutions with a "selfish gene" that is good at self-replication (and self-perpetuation) will tend to endure and proliferate (Dawkins 1989).

Note that this may not result in the evolution of the perfect organism. A "good enough" mutation will achieve dominance if it happens in the right place at the right time, because of the sensitivity of the evolutionary process to initial conditions; i.e., an initial slim advantage may translate into a prolonged period of dominance, without necessarily being optimal. It is also worth bearing in mind that in the natural world, evolution is not progressive, as used to be thought (notably by the followers of Herbert Spencer). Primitive financial life-forms are not condemned to oblivion, any more than are the microscopic prokaryotes that still account for the majority of earth's species. Evolved complexity protects neither an organism nor a firm against extinction—the fate of most animal and plant species.

The evolutionary analogy is, admittedly, imperfect. When one organism ingests another in the natural world, it is just eating, whereas in the world of financial services, mergers and acquisitions can lead directly to mutation. Among financial organisms, there is no counterpart to the role of sexual reproduction in the animal world (although demotic sexual language is often used to describe certain kinds of financial transactions). Most financial mutation is deliberate, conscious innovation, rather than random change. Indeed, because a firm can adapt within its own lifetime to change going on around it, financial evolution (like cultural evolution) may be more Lamarckian than Darwinian in character. Two other key differences are discussed below. Nevertheless, evolution certainly offers a better model for understanding financial change than any other we currently have. The vogue for "behavioral finance," which draws on insights from the field of psychology, represents a welcome corrective to the unrealistic assumptions made by neoclassical models, with their utility-maximizing rational actors. But the behavioralists have little to say about the way economic institutions evolve. The tendency (see, e.g., Thaler and Sunstein 2008) is to regard institutions as exogenous frameworks that an enlightened government can modify to take better account of the nonrational biases of individual psychology.

CONCENTRATION VERSUS SPECIATION

How do financial institutions evolve? Ninety years ago, the German socialist Rudolf Hilferding predicted an inexorable movement toward more concentration of ownership in "finance capital" (Hilferding 2006 [1919]). The conventional view of financial development does indeed see the process from the vantage point of the big, successful survivor firm. In Citigroup's official "family tree," numerous small firms—dating back to the City Bank of New York, founded in 1812—are seen to converge over time on a common "trunk," the present-day conglomerate.

However, this is precisely the wrong way to think about financial evolution over the long run, which *begins* at a common trunk. Periodically, the trunk branches outward as new kinds of banks and other financial institutions evolve. The fact that a particular firm successfully devours smaller firms along the way is more or less irrelevant. In the evolutionary process, animals eat one another, but that is not the driving force behind evolutionary mutation and the emergence of new species and subspecies. The point is that economies of scale and scope are not always the driving force in financial history. More often, the real drivers are the process of speciation—whereby entirely new types of firms are created—and the equally recurrent process of creative destruction, whereby weaker firms die out.

Take the case of retail and commercial banking. Although giants like Citigroup and Bank of America exist (just), the United States and some European countries still have relatively fragmented retail banking sectors. The cooperative banking sector has seen the most change in recent years, with high levels of consolidation (especially following the savings and loans crisis of the 1980s), and most institutions moving to shareholder ownership. For a time, it seemed that the state-owned bank might become extinct in the developed world, as privatization spread from Britain to most major financial systems. But the financial crisis of 2007–2010 revived the species, as a wide range of excessively leveraged banks passed into government hands, beginning with Northern Rock in England. In other respects, the story is one of speciation, in other words, the proliferation of new types of financial institution, which is just what we would expect in a truly evolutionary system. Many new "mono-line" financial services firms have emerged, especially in consumer finance. A number of new "boutiques" now exist to cater to the private banking market. Direct banking (telephone and Internet) is another relatively recent and growing phenomenon. Likewise, even as giants have formed in the realm of investment banking, new and nimbler species—such as hedge funds and private equity partnerships—have evolved and proliferated. Finally, the rapidly accruing hard currency reserves of exporters of manufactured goods and energy are producing a new generation of sovereign wealth funds.

Not only are new forms of financial firm proliferating, so too are new forms of financial asset and service. In recent years, investors' appetite has grown dramatically for mortgage-backed and other asset-backed securities. In January 2008, there were 12 triple A-rated companies in the world but no fewer than 64,000 structured finance instruments, such as collateralized debt obligations, with the same maximum rating. Such instruments were virtually unknown 20 years ago. The use of derivatives has also increased enormously, with the majority being bought and sold "over the counter," on an ad hoc one-to-one basis, rather than through public exchanges, a tendency which, although profitable for the sellers of derivatives, proved to have unpleasant as well as unintended consequences when a crisis struck that caused a large number of options to be exercised all at once. Seven years ago, there were less than $3 trillion of credit default swap contracts outstanding; today, according to the International Swaps and Derivatives Association, there are more than $25 trillion.

In evolutionary terms, then, the financial services sector appears to have passed through a 20-year "Cambrian explosion," with existing species flourishing and new species increasing in number. As in the natural world, the existence of giants has not precluded the evolution and continued existence of smaller species. Size is not everything, in finance as in nature. Indeed, the very difficulties that arise as publicly owned firms become larger and more complex—the diseconomies of scale associated with bureaucracy, the pressures associated with quarterly reporting—provide opportunities to new forms of private firms. What matters in evolution is not your size or (beyond a certain level) your complexity. All that matters is that you are good at surviving and reproducing your genes. The financial equivalent is being good at generating returns on equity and generating imitators using a similar business model.

In the financial world, mutation and speciation have usually been evolved responses to the environment and competition, with natural selection determining which new traits become widely disseminated. Sometimes, as in the natural world, the evolutionary process has been subject to big disruptions in the form of geopolitical shocks and financial crises. The difference is, of course, that whereas giant asteroids (such as the one that eliminated 85% of species at the end of the Cretaceous period) are exogenous shocks, financial crises are endogenous to the financial system. The Great Depression of the 1930s and the Great Inflation of the 1970s stand out as times of major discontinuity, with "mass extinctions" such as the bank panics of the 1930s and the savings and loans failures of the 1980s. A comparably large disruption has clearly occurred in our time. The sharp deterioration in credit conditions in the summer of 2007 created acute problems for many hedge funds, leaving them vulnerable to redemptions by investors. But a much more important feature of the current crisis was the pressure on banks and insurance companies in the major developed economies. Estimated losses on asset-backed securities and other forms of risky debt are in excess of $1.5 trillion for American and European banks. In May 2009, the U.S. government's "stress tests" estimated American bank losses at approximately $600 billion. The equivalent figures for the United Kingdom and continental Europe were comparable in magnitude. These losses were concentrated in a relatively small number of very large banks that had become heavily involved in the market for securities and derivatives, whose value was closely linked to that of U.S. house prices. By September 2009, U.S. banks had raised capital equivalent to approximately $431 billion, leaving a shortfall of more than $150 billion. Because banks typically target a constant capital/assets ratio of between 5% and 10%, this implies that balance sheets may need to be shrunk by between $750 billion and $1.5 trillion.

The financial crisis has dashed the hopes of those who believed that the separation of risk origination and balance sheet management would distribute risk optimally throughout the financial system. At the time of writing, the situation of the U.S. banking system remains parlous. In 2009, total lending by U.S. banks fell by 7.4%, the steepest drop since 1942, a far more serous contraction than in 2007, when the phrase "credit crunch" came into common parlance. The credit withdrawn from the economy since the downfall of Lehman Brothers in September 2008 amounts to approximately $700 billion, more than double the amount so far distributed under President Barack Obama's $787 billion fiscal stimulus program. Nor is there any sign of an end to the banking crisis, as rising defaults in the residential and commercial real estate sectors impact on the balance sheets of smaller regional and local banks. According to the Federal Deposit Insurance Corporation, the number of U.S. banks at risk of failing was 702 in the spring of 2010, its highest level in 16 years. More than 5% of all bank loans are at least 3 months past due, the worst figure in 26 years.

It seems inconceivable that this crisis will end without further mergers and acquisitions, as the relatively strong devour the relatively weak. There may also be more outright extinctions. The investment banks no longer exist: Of the five biggest, Lehman failed, Bearn Stearns and Merrill Lynch were bought, and Morgan Stanley and Goldman Sachs had to become bank holding companies. Little now remains of the conduits and structured investment vehicles created by banks to hold assets off their own balance sheets. Bond insurance companies, too, seem likely to disappear. Many hedge funds have failed in the past 3 years; funds of funds are also struggling. But some hedge funds have been thriving on the return of volatility.[1] It therefore seems likely that a few large winners will emerge from the crisis in dominant positions within the "alternative investment" sector. The same may be said of private equity and venture capital. It also seems probable that new forms of financial institution will spring up in the aftermath of the crisis. As Andrew Lo has suggested, "As with past forest fires in the markets, we're likely to see incredible flora and fauna springing up in its wake" (Economist 2008).

INTELLIGENT DESIGN

There is another big difference between nature and finance. Whereas evolution in biology takes place in the natural environment, where change is essentially random (hence Dawkins's image of the blind watchmaker), evolution in financial services occurs within a regulatory framework where—to borrow a phrase from anti-Darwinian creationists—"intelligent design" plays a part. Sudden changes in the regulatory environment are rather different from sudden changes in the macroeconomic environment, which are analogous to environmental changes in the natural world. The difference is once again that there is an element of endogeneity in regulatory changes, because those responsible are often "poachers turned gamekeepers," with a good insight into the way that the private sector works. The net effect, however, is similar to climate change on biological evolution. New rules and regulations can make previously "good" traits suddenly disadvantageous. The rise and fall of savings and loans, for example, was due in large measure to changes in the regulatory environment in the United States. Regulatory changes in the wake of the 2007 crisis may have comparably unforeseeable consequences.

The primary focus of most regulators is to maintain stability within the financial services sector, thereby protecting the consumers whom banks serve and the "real" nonfinancial economy that the industry supports. Companies in manufacturing are generally less systemically important to the economy as a whole because it is banks and other financial companies that create both money and credit. The collapse of a major commercial bank, in which retail customers lose their deposits, is therefore an event that any regulator (and politician) wishes to avoid at all costs. An old question that has raised its head since August 2007 is how far guarantees to bail out banks create a problem of "moral hazard," encouraging excessive risk-taking on the assumption that the state will intervene to avert illiquidity and even insolvency if an institution is considered too big—meaning too politically sensitive or too likely to bring a lot of other firms down with it—to fail. In the words of Federal Reserve Chairman Ben Bernanke, "As the crisis has shown, one of the greatest threats to the diversity and efficiency of our financial system is the pernicious problem of financial institutions that are deemed 'too big to fail.' It is unconscionable that the fate of the world economy should be so closely tied to the fortunes of a relatively small number of giant financial firms" (http://www.reuters.com/article/idUSTRE62J0SM20100320).

This is certainly correct from an evolutionary perspective. It is in fact highly undesirable to have any institutions in the category of "too big to fail," because without occasional bouts of creative destruction, the evolutionary process will be thwarted. The experience of Japan in the 1990s stands as a warning to legislators and regulators that an entire banking sector can become a kind of economic dead hand if insolvent institutions are propped up despite economic failure and bad debts are not written off (Ferguson 2009).

CONCLUSION

Every shock to the financial system must result in casualties. Left to itself, "natural selection" should work fast to eliminate the weakest institutions in the market, which are typically devoured by the successful. But most crises also usher in new rules and regulations, as legislators and regulators rush to stabilize the financial system and to protect the consumer/voter. The critical point is that the possibility of extinction cannot and should not be removed by excessively precautionary rules. As Joseph Schumpeter wrote more than 70 years ago, "This economic system cannot do without the *ultima ratio* of the complete destruction of those existences which are irretrievably associated with the hopelessly unadapted." This meant, in his view, nothing less than the disappearance of "those firms which are unfit to live" (Schumpeter 1934). It would be highly beneficial to legislators, regulators, and financiers themselves if, in the wake of this most recent crisis of our financial system, they came to terms with the evolutionary character of financial history and ceased their efforts to suppress the process.

This is not to say that it would have been better to repeat the experience of the Great Depression, when the number

---

[1] In Andrew Lo's words, "Hedge funds are the Galapagos Islands of finance... . The rate of innovation, evolution, competition, adaptation, births and deaths, the whole range of evolutionary phenomena, occurs are an extraordinarily rapid clip."

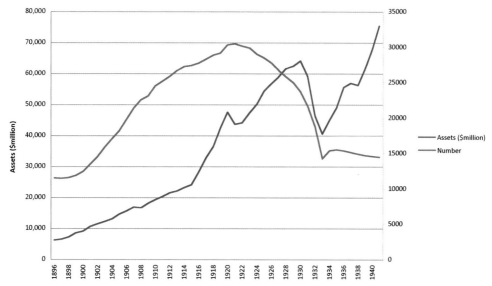

Figure 1. The number and assets of U.S. state and national banks, 1896–1941 (see Bodenhorn 2006).

of banks in the United States fell by more than half (see Fig. 1). Timely and large-scale action by the Federal Reserve and the U.S. Treasury succeeded in preventing a repeat of the early 1930s, so that the number of banks that failed in the more recent crisis was far smaller, as was the associated reduction in bank assets, i.e., credit (see Fig. 2). The problem is that such interventions have unintended consequences. Although the number of "problem institutions" in the FDIC database rose sharply in 2008 and especially in 2009, the number of mergers declined and the number of new charters fell to just 31 in 2009, compared with an average of 150 a year between 1990 and 2006 (Fig. 3). Any regulatory reforms that discourage the creation of new financial institutions will be deleterious to the long-term health of the U.S. economy.

A final pitfall that regulators must avoid is the notion that there is an ideal financial system toward which all countries are converging and which it is the duty of regulators to promote. A large literature on the relationship between law and finance encouraged the notion that there was a superior common law model that offered creditors better protection and therefore increased financial intermediation (see, e.g., La Porta et al. 1998). There is in fact some reason to doubt that this hypothesis is supported by the data that we have on creditor protection in different legal systems in the 19th century (Musacchio 2009). The historical record points to considerable diversity in the way that the financial systems of different national economies evolved (see Cameron 1972). In some countries, bank finance predominated over bond finance and stock

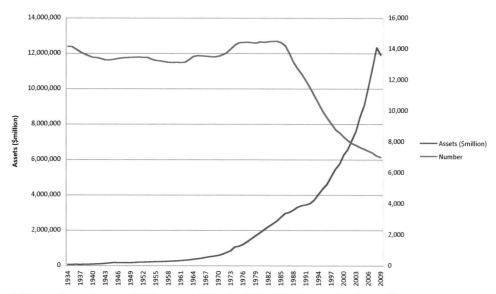

Figure 2. The number and assets of U.S. commercial banks, 1979–2009. Source: http://www.fdic.gov/bank/statistical/stats/.

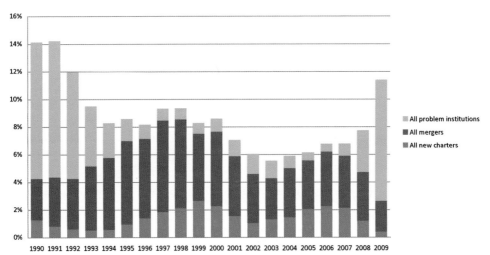

Figure 3. Changes in FDIC-insured financial institutions as percentage of total number of institutions. Source: http://www.fdic.gov/bank/statistical/stats/.

markets scarcely had any role; in others, the opposite was true. In some countries, the banking sector became highly concentrated relatively early; in others, it remains highly decentralized to this day.

If evolutionary theory has one obvious lesson for financial regulators, it may well be that diversity is desirable in finance as well as in nature. Were the whole world to converge on a template—whether it was devised in Washington, Basel, or Beijing—the international system would be considerably more vulnerable to a mass extinction than it presently is.

REFERENCES

Alchian AA. 1950. Uncertainty, evolution and economic theory. *J Political Econ* **58**: 211–222.

Bodenhorn H. 2006. State banks—Number and assets: 1896–1955. In *Historical statistics of the United States, earliest times to the present:* Millennial Edition (ed. SB Carter et al.), Table Cj, pp. 158–168. Cambridge University Press, New York.

Cameron R, Ed. 1972. Banking and economic development: Some lessons of history. Oxford University Press, London.

Darwin C. 1859. *On the origin of species by means of natural selection,* 1st ed. Murray, London.

Dawkins R. 1989. *The selfish gene,* 2nd ed. Oxford University Press, Oxford.

Economist. 2008. Fear and loathing, and a hint of hope. *The Economist*, February 16.

Ferguson N. 2008. *The ascent of money: A financial history of the world.* Allen Lane/Penguin, London.

Ferguson N. 2009. *Too big to live? Why we must stamp out state monopoly capitalism.* Center for Policy Studies, London.

Ferguson N, Wyman O. 2007. *The evolution of financial services: Making sense of the past, preparing for the future.* Oliver Wyman, London.

Guthrie J. 2007. How the old corporate tortoise wins the race. *Financial Times*, February 15.

Hannah L. 1999. Marshall's "trees" and the global "forest": Were "giant redwoods" different? In *Learning by doing in markets, firms and countries* (ed. NR Lamoreaux et al.), pp. 253–294. National Bureau of Economic Research, Cambridge, MA.

Hilferding, R. 2006 (1919). *Finance capital: A study of the latest phase of capitalist development.* Routledge and Kegan Paul, London.

La Porta R, Lopez de Silanes F, Shleifer A, Vishny RW. 1998. Law and finance. *J Political Econ* **106**: 1113–1155.

Musacchio A. 2009. *Experiments in financial democracy: Corporate governance and financial development in Brazil, 1882–1950.* Cambridge University Press, Cambridge.

Nelson RR, Winter SG. 1982. *An evolutionary theory of economic change.* Harvard University Press, Cambridge, MA.

Ormerod P. 2005. *Why most things fail: Evolution, extinction and economics.* Faber and Faber, London.

Schumpeter JA. 1934. *The theory of economic development.* Harvard University Press, Cambridge, MA.

Schumpeter JA. 1987 (1943). *Capitalism, socialism and democracy.* Harper and Row, New York.

Thaler R, Sunstein C. 2008. *Nudge: Improving decisions about health, wealth, and happiness.* Yale University Press, New Haven, CT.

Veblen TB. 1898. Why is economics not an evolutionary science? *Q J Econ,* vol. 12.

The Religious Essence of Intelligent Design

B. Forrest

*Department of History and Political Science, Southeastern Louisiana
University, Hammond, Louisiana 70402*
Correspondence: bforrest@selu.edu

Despite the protestations of its proponents, "intelligent design theory" (ID) is not science but creationism, making it in its essence a supernaturalist religious belief. This fact has been established conclusively for the legal record in *Kitzmiller et al. v. Dover Area School District* (2005) and for the public by a decade of scholarship, much of which helped to secure the *Kitzmiller* plaintiffs' victory in this first legal case involving ID. Moreover, ID is not merely a religious belief but, more specifically, a narrow form of sectarian Christianity, as specified by its own proponents. The nature of ID as a creationist, therefore religious, movement has been revealed not only by its critics, but also, most importantly, by its proponents. The explication of ID by its critics as creationism, and therefore religion, reflects the way the movement views itself.

THE RELIGIOUS ESSENCE OF INTELLIGENT DESIGN

Although its proponents issue strenuous public objections to the contrary, ID is not science but creationism, making it, in its very essence, a supernaturalist religious belief. In *Kitzmiller et al. v. Dover Area School District* (2005), the first (and so far only) legal case involving ID, Judge John E. Jones III clearly recognized the religious nature of ID:

> Not a single expert witness over the course of the six week trial identified one major scientific association, society or organization that endorsed ID as science.... We have concluded that it is not [science], and moreover that ID cannot uncouple itself from its creationist, and thus religious, antecedents (Jones 2005).

In 2004, 11 parents in Dover, Pennsylvania, filed suit to prevent their children from being taught in biology class that ID is a scientific alternative to evolution. A decade of critical scholarship, much of which helped to secure the plaintiffs' victory through the testimony of expert witnesses, has established conclusively, both for the legal record and for the public at large, the religious essence of ID (Forrest and Gross 2007; National Center for Science Education 2008c). Moreover, ID is not merely a religious belief, but, more specifically, a narrow form of sectarian Christianity, as specified by its own proponents. The nature of ID as a creationist, therefore religious, movement has been revealed not only by its critics, but also, most importantly, by its proponents. The religious essence of ID creationism as explained by its critics merely reflects the way the movement views itself.

Intelligent Design as Creationism

American creationism has developed in phases during the last 40 years, with its particular variations taking shape in response to its proponents' consistent defeats in federal courts; ID is the most recent variant (National Center for Science Education 2008a). The ID movement is headquartered at the Discovery Institute (DI), a conservative Seattle, Washington, think tank that DI president Bruce Chapman established in 1990. In 1996, the creationist Center for the Renewal of Science and Culture (CRSC) was formally established as a wing of DI. Now called the Center for Science and Culture (CSC), its small group of leading operatives have fomented virtually every major creationist outbreak in the United States since 1999 (Forrest 2007; Forrest and Gross 2007). Most of them have no scientific credentials, although a few are trained scientists. Michael Behe is a biochemist at Lehigh University. Jonathan Wells has a Ph.D. in biology but has never been a practicing scientist, choosing instead to promote ID full time. CSC director Stephen C. Meyer is a philosopher, and William Dembski, ID's chief intellectual, has degrees in philosophy, mathematics, and divinity (Forrest and Gross 2007). CSC associate director John West, a political scientist, works full time for DI (Center for Science and Culture 2009).

ID is best characterized as "progressive creationism," a form of old-earth creationism in which God periodically intervenes in natural processes to produce progressively greater complexity in living things; ID proponents contend that natural processes cannot produce complexity (Scott 2005; Forrest and Gross 2007). However, these distinguishing features merely mark ID as the most recent form of creationism; it is the direct descendant of "creation science," a biblically literalist—thus, young-earth—creationism (Forrest 2005a).

ID's Development from Creation Science

Young-earth creationism was the standard form of American creationism from the Scopes trial in 1925 until the early 1980s, when the much-publicized ruling in *McLean v. Arkansas Board of Education* (1982) proved disastrous for "creation science." Creation science developed in response to a U.S. Supreme Court ruling, *Epperson v. Arkansas* (1968), that prohibited public schools from banning the teaching of evolution. After *Epperson*, attempting to gain entry for creationism alongside evolution in public schools, its proponents shifted to calling their project "creation science," a nomenclature that they hoped would project a more scientific appearance, thus increasing their chances of surviving future judicial scrutiny. The strategy failed.

> The creation science movement reached its peak in the early 1980s. Equal time for evolution and creation science bills were proposed in at least 27 states in 1980 and 1981. Arkansas and Louisiana passed laws mandating "equal time" for the "two models" of evolution and creation science. . . . *McLean v. Arkansas* pitted a team of plaintiffs' witnesses that included eminent scientists such as Francisco Ayala, [and] Stephen Jay Gould . . . against a team of creationist defense witnesses . . . who had the impossible task of defending the scientific merits of a young earth and global flood.
>
> *McLean* put creation science on trial, and creation science lost badly. In the January 1982 decision, the judge wrote that creation science was biblical literalist Christianity in disguise, and that to teach it would be to promote a sectarian religious view. . . .
>
> Even conservative Christians recognized that creation science had been a legal disaster. . . (Scott and Matzke 2007).

Creationists now pinned their hopes on the survival of the Louisiana law. In fall 1981, already anticipating a defeat in *McLean* (Scott and Matzke 2007), the Foundation for Thought and Ethics (FTE), a small Christian think tank in Texas, had begun two publishing projects: a book entitled *The Mystery of Life's Origin: Reassessing Current Theories* (Thaxton et al. 1984) and a creationist high school textbook that was eventually published as *Of Pandas and People* (Kenyon and Davis 1993) after undergoing several name changes as FTE followed the phases of the Louisiana case (see below) (Forrest 2005b). FTE described *Mystery* as "a rigorous scientific critique of the theory of prebiotic evolution" and *Pandas* as "a two-model high school biology textbook that will fairly and impartially view the scientific evidences for creation side by side with evolution" (Scott and Matzke 2007). According to William Dembski, *Mystery* is one of the two seminal books (the other being Michael Denton's *Evolution: A Theory in Crisis*) of the ID movement (Denton 1986; Dembski 1998b). Years later, two of the *Mystery* coauthors, Charles Thaxton and Walter Bradley, along with Dembski, became founding fellows of the CRSC (see below) (Forrest and Gross 2007).

In June 1987, the U.S. Supreme Court ruled in *Edwards v. Aguillard* that since "the preeminent purpose of the Louisiana Legislature was clearly to advance the religious viewpoint that a supernatural being created humankind," the Louisiana law violated the First Amendment's separation of church and state and was therefore unconstitutional (*Edwards v. Aguillard* 1987). Creationism was thus in need of yet another public makeover. In December 1988, Thaxton decided that the ostensibly new movement would be called "intelligent design" (Witham 2005). Although his proffered rationale does not mention the *Edwards* decision, this nomenclature was clearly motivated by the post-*Edwards* necessity of referring to creationism in a way that creationists hoped would be unrecognizable, or at least legally untouchable, by federal judges. Moreover, after *Edwards*, the young earth and global flood of Genesis had to be discarded if creationism was to have any chance of surviving future litigation. However, despite such semantic tactics, ID is the recognizable descendant of its predecessor, "creation science" (Scott and Matzke 2007).

Intelligent Design Proponents Reveal ID as Creationism

The ID proponents who in 1996 became CRSC fellows have identified their movement as creationism and have referred to themselves as creationists. Before the CRSC made its first high-profile public appearance in the Kansas Board of Education's effort to remove evolution from its state science standards in 1999, its fellows had spent several years cultivating supporters by candidly stating their creationist, thus religious, agenda. The CRSC outlined this agenda to potential donors in a 1998 document, *The Wedge*, informally known as the "Wedge Document" (Forrest and Gross 2007; National Center for Science Education 2008d).

In this document, the CRSC outlines its 20-year "Wedge Strategy" for getting ID into public schools and the American cultural mainstream. The organization's understanding of ID as creationism is revealed in the opening statement: "The proposition that human beings are created in the image of God is one of the bedrock principles on which Western civilization was built" (Center for the Renewal of Science and Culture 1998). The CRSC blames "Charles Darwin, Karl Marx, and Sigmund Freud" for the ills of modern society and vows that "Discovery Institute's Center for the Renewal of Science and Culture seeks nothing less than the overthrow of materialism and its cultural legacies" (Center for the Renewal of Science and Culture 1998). Moreover, the CRSC aligns itself specifically with Christianity: "Design theory promises to reverse the stifling dominance of the materialist worldview, and to replace it with a science consonant with Christian and theistic convictions" (Center for the Renewal of Science and Culture 1998). The strategy explicitly calls for the use of "apologetics seminars" to recruit and train Christian supporters:

> Alongside a focus on influential opinion-makers, we also seek to build up a popular base of support among our natural constituency, namely, Christians. We will do this primarily through apologetics seminars. We intend these to encourage and equip believers with new scientific evidences that support the faith, as well as to "popularize" our ideas in the broader culture (Center for the Renewal of Science and Culture 1998).

"Apologetics" is an evangelical term referring to the use of argument to defend Christianity against perceived attacks (Forrest 2005a). Such seminars are used to cultivate supporters who promote the CRSC agenda at the state and local levels (Forrest and Gross 2007).

In the early 1980s, as the *Edwards* case made its way through the federal courts, Thaxton and other creationists, more or less independently of one another, were seeking to develop a new version of creationism. After the 1987 creationist defeat in *Edwards*, some of the creationists who would later comprise the ID movement became acquainted at Thaxton's 1988 conference, "Sources of Information Content in DNA" (Access Research Network 2001; Nelson 2002). Shortly afterward, they coalesced around University of California, Berkeley, law professor Phillip Johnson, who became and remains (although retired) the CRSC program advisor (Center for Science and Culture 2009). Founding CRSC fellow and young-earth creationist Paul Nelson explains Johnson's revival of creationism as a cultural force following the *Edwards* defeat:

> In that year [1987], a long cultural battle that had begun more than a quarter century earlier . . . appeared . . . to have come decisively to an end when the . . . U.S. Supreme Court declared "creation-science" to be a religious belief. . . .
>
> . . . *Edwards v. Aguillard* . . . seemed to shut the door permanently on creationism (at least as an admissible dissent in public school science teaching). . . .
>
> . . . A revolution from an unexpected quarter, however, was about to occur. . . (Nelson 2002).

That revolution was Johnson's becoming the catalyst for the formal organization of what is now the ID creationist movement in the United States (Forrest and Gross 2007).

In 1992, prefiguring what became a formal part of the CRSC's strategy in the Wedge Document, Johnson and several of his new ID associates—Dembski, Behe, and Meyer—invited mainstream, proevolution scientists and philosophers to participate in the movement's first conference, "Darwinism: Scientific Inference or Philosophical Preference?," at Southern Methodist University (Forrest and Gross 2007). FTE, a conference cosponsor, described the event as "a remarkable exchange of views in a symposium between Darwinists and Intelligent Design proponents" (Foundation for Thought and Ethics [undated]). The Wedge Strategy calls for "direct confrontation with the advocates of materialist science through challenge conferences in significant academic settings" (Center for the Renewal of Science and Culture 1998).

Conferencing has therefore been central to the ID movement's modus operandi, especially in its early years, when it was used as a way of enticing legitimate—and unsuspecting—scientists and scholars to share public platforms with ID creationists, allowing the CRSC to construct a façade of academic legitimacy for public relations purposes. However, in a 1995 article about the SMU conference by CRSC fellow Mark Hartwig for the religious *Moody Magazine*, aimed at the Wedge Strategy's "natural constituency" of Christians, Hartwig described Dembski and his ID associates as "evangelical scholars"; he wrote that "creationists and evolutionists met as equals to discuss serious intellectual questions" (Hartwig 1995). The word "creationists" referred to his fellow ID proponents. In this initial stage of the Wedge Strategy, candor about the movement's true nature and goals was essential to securing political and financial backing for its advancement (Forrest and Gross 2007).

In no single person is the transfiguration of creation science into ID more visible than in biologist Dean H. Kenyon, who created a stir by teaching creationism in his biology classes in the early 1980s (Associated Press 1980). In 2007, Kenyon explained that his transition "from evolution to a non-evolutionary view of cosmic and biologic origins . . . began with exposure to creationist literature for the first time in my eighth year of teaching evolution in the Biology Department of San Francisco State University after joining the faculty as a convinced Darwinist and chemical evolutionist" (Kenyon 2007). In the *Edwards* case, he assisted the defense by submitting a sworn affidavit in 1984, attesting to the scientific integrity of "creation-science":

> It is my professional opinion, based on my original research, study, and teaching, that creation-science is as scientific as evolution. . . . Moreover, I believe that a scientifically sound creationist view of origins is not only possible, but is to be preferred over the evolutionary view (Kenyon 1984).

At the same time, he was working as a coauthor on *Of Pandas and People* (Forrest 2005a). After the Supreme Court declared the Louisiana law unconstitutional in its *Edwards* decision, Kenyon became one of the earliest CRSC fellows, transitioning seamlessly into an ID proponent (Forrest and Gross 2007). However, as recently as 2000, he admitted that "scientific creationism, which in its modern phase began in the early 1960s, is actually one of the intellectual antecedents of the Intelligent Design movement" (Wiker 2000). Kenyon is a young-earth creationist, of which there are a few others, such as Paul Nelson, in the ID movement. Still a CSC fellow, he also serves on the Advisory Council of the Kolbe Center, a Catholic "lay apostolate," which—putting itself at odds with the Catholic Church on both biblical interpretation and evolution—promotes the literal interpretation of Genesis, including the belief that Earth is only a few thousand years old (Kenyon 2007; Kolbe Center for the Study of Creation [undated]).

Kenyon's transition from "creation science" to ID in the wake of the creationists' defeat in *Edwards* is also illustrated in his role as *Pandas* coauthor (Kenyon and Davis 1993). As a central piece of evidence in the *Kitzmiller* plaintiffs' case, *Pandas* provided direct proof that ID is understood by its proponents as creationism. Subpoenaed documents that plaintiffs' attorneys obtained from the FTE yielded no fewer than five early drafts of this book—the first dated 1983, a second dated 1986, with the third, fourth, and fifth dated 1987—all of which Kenyon coauthored. Except for the fifth draft, all were written using explicit creationist terminology such as "creationist" and "creation biologist." In the fifth, which was produced after the Supreme Court's June 1987 *Edwards* ruling, the creationist language had been expunged and replaced with ID terminology (Forrest 2005b; National Center for Science Education 2008b). However, except for this selective alteration, the fifth draft was the same as the pre-*Edwards* drafts in which creationist terminology had been used; thus, this draft constituted a virtual "smoking gun" for the plaintiffs.

The post-*Edwards* alteration of *Pandas* points clearly to the ID movement's anticipation of the need to evade future court challenges. Hence, it shows that they planned to continue their effort to insert creationism into public schools—but as "intelligent design."

Intelligent Design as Supernaturalist Religion

ID proponents at the Discovery Institute deny to the media and to mainstream audiences that they regard the intelligent designer as a supernatural being; indeed, they assert that ID cannot and need not determine the designer's identity. A Discovery Institute "Truth Sheet" written by DI staffer Casey Luskin during the 2005 *Kitzmiller* trial exemplifies such denials:

Truth Sheet # 09-05

Does intelligent design postulate a "supernatural creator?"

> **Overview: No.** The ACLU, and many of its expert witnesses, have alleged that teaching the scientific theory of intelligent design (ID) is unconstitutional in all circumstances because it posits a "supernatural creator." Yet actual statements from intelligent design theorists have made it clear that the scientific theory of **intelligent design does not address metaphysical and religious questions such as the nature or identity of the designer** (Luskin 2005).

However, as early as the 1992 SMU conference, Dembski had committed himself in writing to specifying the designer's supernatural identity. Remarkably, not only did he divulge this openly, but in his conference presentation, "The Incompleteness of Scientific Naturalism," he also stressed that the designer's supernatural nature is essential:

> I want here to examine scientific naturalism. I am going to argue that this view has a serious defect—it is *incomplete*. As a consequence of this defect I shall argue that it is legitimate within scientific discourse to entertain questions about supernatural design....
>
> We are asking a transcendental question in the Kantian sense: What are the conditions for the possibility of discovering design (i.e., supernatural intervention, nonmaterial interference, divine meddling, call it what you will) in the actual world? This question must be answered at the outset....
>
> ... By a super-intelligence I mean a supernatural intelligence, i.e., an intelligence surpassing anything that physical processes are capable of offering. This intelligence exceeds anything that humans or finite rational agents in the universe are capable of even in principle (Dembski 1994).

When the Discovery Institute formally established the CRSC in 1996, Phillip Johnson explicitly defined ID in religious terms, making God as creator the core of his definition:

> My colleagues and I speak of "theistic realism"—or sometimes, "mere creation"—as the defining concept of our movement. This means that we affirm that God is objectively real as Creator, and that the reality of God is tangibly recorded in evidence accessible to science, particularly in biology (Johnson 1996).

Johnson's defining ID as theistic realism confirms its dependence on the existence of a supernatural deity. His alternate definition of ID as "mere creation" is a thinly veiled reference to its biblical basis (see below). The concept of "mere creation" is so important to the ID movement that another of its early conferences was devoted to it. Entitled "Mere Creation: Reclaiming the Book of Nature," this 1996 event was held at Biola University, formerly the "Bible Institute of Los Angeles" (hence its current name) (Mere Creation 1996). Biola employs several CRSC fellows, holds pro-ID events, and offers a Master of Arts in Science and Religion that incorporates ID (Biola University 2009).

Dembski's and Johnson's supernaturalist stipulations concerning ID explain why its proponents reject utterly the naturalistic methodology of modern science. Although creationists have always rejected the naturalistic methodology that has made modern science so successful at explaining the natural world, Johnson's attack on naturalism, along with his crafting of the Wedge Strategy, is his signature contribution to the ID movement. His earliest article, in 1990, 3 years after the *Edwards* decision, set the tone for the message that he has consistently broadcast as the movement's organizer, advisor, and now doyen. According to Johnson, scientists have stacked the deck against creationism by arbitrarily defining science as naturalistic; consequently, having made an a priori philosophical commitment to naturalism the condition for admission to the inner circle of mainstream science, they have unfairly shut creationists out of the (putative) debate about evolution. In 1990, before the ID movement coalesced around him and he became the CRSC advisor, Johnson was not yet so cautious as to avoid the term "creationism":

> Creationists are disqualified from making a positive case, because science by definition is based upon naturalism. The rules of science also disqualify any purely negative argumentation designed to dilute the persuasiveness of the theory of evolution. Creationism is thus out of court—and out of the classroom—before any consideration of evidence. Put yourself in the place of a creationist who has been silenced by that logic, and you may feel like a criminal defendant who has just been told that the law does not recognize so absurd a concept as "innocence."
>
> With creationist explanations disqualified at the outset, it follows that the evidence will always support the naturalistic alternative (Johnson 1990).

ID proponents also consistently—and therefore, one must conclude, deliberately—conflate the *methodology* of science with the *metaphysical* commitment to naturalism (hence, atheism) that takes one beyond methodology. They further conflate naturalism with a crude materialism, as Robert Pennock points out:

> ID creationists typically use the term *naturalism* interchangeably with *materialism*, even though metaphysical naturalism is a richer concept that says that nature and its

laws are all that exist, but it allows that nature may not be limited to matter *per se*. More important, they regularly conflate these *metaphysical* concepts with the related *methodological* norms that are actually employed by science (Pennock 2007).

Yet, ID proponents go much further than simply rejecting the methodological naturalism of science. Dembski goes so far as to assert that using the naturalistic methodology of science without a "transcendent" designer—i.e., without invoking a supernatural agent as an explanatory principle—actually "stifles" scientific inquiry.

> [D]esign pervades cosmology and biology. Moreover, it is a transcendent design, not reducible to the physical world. Indeed, no intelligent agent who is strictly physical could have presided over the origin of the universe or the origin of life.
>
> Unlike design arguments of the past, the claim that transcendent design pervades the universe is no longer a strictly philosophical or theological claim. It is also a fully scientific claim. . . .
>
> Demonstrating transcendent design in the universe is a scientific inference, not a philosophical speculation. . . . (1) Intelligent agency is logically prior to natural causation and cannot be reduced to it. (2) Intelligent agency is fully capable of making itself known against the backdrop of natural causes. (3) Any science that systematically ignores design is incomplete and defective. (4) Methodological naturalism, the view that science must confine itself solely to natural causes, far from assisting scientific inquiry actually stifles it. . . (Dembski 1998a).

Although Dembski wrote this in 1998, his views have undergone no development since then.

Dembski's contention that science cannot advance if its methodology is limited to the search for only natural processes raises an unavoidable question: What, according to Dembski, is essential to a truly scientific understanding of nature, which is to say, an understanding of nature as intelligently designed? His answer is succinct: miracles. In fact, he says, without miracles, ID is "incoherent" (Dembski 1999a), i.e., it makes no sense. In *Intelligent Design: The Bridge between Science and Theology*, in which Dembski explains ID to his popular Christian audience, he rejects the "naturalistic critique of miracles" by 17th century philosopher Benedict Spinoza and 19th century theologian Friedrich Schleiermacher because the outcome of their respective critiques was to "render miracles incoherent" (Dembski 1999a). Therefore, the value of miracles as "the most direct evidence for divine activity in the world" was destroyed (Dembski 1999a). Having destroyed the evidential value of miracles, Spinoza and Schleiermacher thereby destroyed their value as evidence for the Christian faith. Finally, says Dembski, "By rendering miracles incoherent Spinoza and Schleiermacher undercut all nonnaturalistic [i.e., supernatural] modes of divine activity and thereby rendered design incoherent as well" (Dembski 1999a).

Dembski's message is clear: ID cannot be properly understood without invoking miracles, and science cannot be properly conducted without invoking ID. This is the presiding consensus of the ID movement. Even biochemist Behe, who says that he became an ID proponent purely through his dispassionate analysis of the (insufficient) evidence for evolution, views the supernatural as an acceptable scientific explanation:

> The philosophical argument (made by some theists) that science should avoid theories which smack of the supernatural is an artificial restriction on science. Their fear that supernatural explanations would overwhelm science is unfounded. . . . The philosophical commitment of some people to the principle that nothing beyond nature exists [i.e., atheism] should not be allowed to interfere with a theory that flows naturally from scientific data [i.e., intelligent design]. The rights of those people to avoid a supernatural conclusion should be scrupulously respected, but their aversion should not be determinative (Behe 1998).

ID as Sectarian Christianity

The DI visually revealed the designer as the Christian God on its first CRSC Web site in 1996. The first website banner featured the well-known section of Michelangelo's Sistine Chapel painting in which God creates Adam with a touch of his finger (Center for the Renewal of Science and Culture 1996). The second banner, posted in ~1999, featured Michelangelo's God (the Sistine Chapel version) creating DNA. As the ID movement moved more prominently into the public spotlight, DI progressively cleansed the overtly religious references from both the organization's website and its nomenclature. In 2002, Michelangelo's God was replaced by the Hubble Space Telescope photograph of the MyCn 18 Hourglass Nebula, and the CRSC's name was shortened to "Center for Science and Culture" (CSC) (National Center for Science Education 2002). The word "renewal" had all too obvious religious connotations as spelled out in the Wedge Document, in which "spiritual and cultural renewal" is one of the CRSC's Five Year Objectives (to be met by 2003), along with a specification of this objective's relevance to Christianity:

Spiritual and cultural renewal:

- Mainline renewal movements begin to appropriate insights from design theory and to repudiate theologies influenced by materialism.
- Major Christian denomination(s) defend(s) traditional doctrine of creation and repudiate(s) Darwinism.
- Seminaries increasingly recognize and repudiate naturalistic presuppositions . . . (Center for the Renewal of Science and Culture 1998).

Today, the CSC website has the visual façade of a secular think tank, but ID's substance cannot be sufficiently camouflaged. Not only is ID religious, but, even more specifically, it is also fundamentally Christian as defined by Dembski, the ID movement's principal intellectual. Just as Dembski's 1998 article (see above) makes clear the religious essence of ID, he also makes Christianity integral to ID in the same article, as one example suffices to show:

> Now, within Christian theology there is one and only one way to make sense of transcendent design, and that is as a divine act of creation. . . . My aim is to use divine cre-

ation as a lens for understanding intelligent agency generally. God's act of creating the world is the prototype for all intelligent agency (creative or not). Indeed, all intelligent agency takes its cue from the creation of the world. How so? God's act of creating the world makes possible all of God's subsequent interactions with the world, as well as all subsequent actions by creatures within the world. God's act of creating the world is thus the prime instance of intelligent agency (Dembski 1998a).

This article was absorbed into Dembski's 1999 book *Intelligent Design: The Bridge between Science and Theology*, in which Jesus Christ has an essential role vis-à-vis both ID and mainstream science. With respect to ID, Jesus's incarnation and bodily resurrection are part of what Dembski calls the premodern "logic of signs," i.e., signs of God's (the designer's) divine agency—the virgin birth being the sign of the incarnation and the resurrection a sign of both Jesus's and humankind's mastery over death (Dembski 1999a). Rejecting both modernity, "with its commitment to rationality and science," and post-modernity, which "offers a plurality of separate discourses of which none is privileged," Dembski prefers a premodern worldview, which is "rich enough to accommodate divine agency" (Dembski 1999a). His goal is to resituate both science and ID on the foundation of this premodern logic of signs:

> My aim in this book then is to take this premodern logic of signs and make it rigorous. In doing so, I intend to preserve the valid insights of modern science as well as the core commitments of the Christian faith.
>
> The rigorous reformulation of the premodern logic of signs is precisely what intelligent design is all about. The premodern logic of signs used signs to identify intelligent causes. Intelligent design is the systematic study of intelligent causes and specifically of the effects they leave behind... (Dembski 1999a).

Having located ID's premodern foundation in the person of Jesus, Dembski subsequently explains Jesus's essentiality to modern science:

> My thesis is that all disciplines find their completion in Christ and cannot be properly understood apart from Christ.
>
> If we... view Christ as the *telos* toward which God is drawing the whole of creation, then any view of the sciences that leaves Christ out of the picture must be seen as fundamentally deficient....
>
> Christ is indispensable to any scientific theory, even if its practitioners don't have a clue about him.... [T]he conceptual soundness of the theory can in the end only be located in Christ... (Dembski 1999a).

In another publication, also aimed at a religious audience, Dembski explicitly reveals the Christian foundation of ID in his definition of it: "Intelligent design . . . embraces the sacramental nature of physical reality. Indeed, intelligent design is just the Logos theology of John's Gospel restated in the idiom of information theory" (Dembski 1999b). In a 2007 interview with Focus on the Family, a Christian Right organization, he stated forthrightly that "the Designer of intelligent design is, ultimately, the Christian God" (Williams 2007). He followed with the immediate but wholly implausible assurance that "the focus of my writings is not to try to understand the Christian doctrine of creation; it's to try to develop intelligent design as a scientific program" and that his research is "going to change the national conversation" (Williams 2007). (As seen below, his own CSC colleague Paul Nelson had undermined that assurance several years earlier.) Most recently, Dembski says that his 2008 book *Understanding Intelligent Design: Everything You Need to Know in Plain Language* (Dembski and McDowell 2008) "is geared at Christian young people (junior high and high schoolers) as well as for Church groups (e.g., Sunday Schools) to help get out the word about ID" (Dembski 2008).

Despite Dembski's belated assurance in 2007 of his scientific intentions for ID, his work heretofore shows undeniably that, whereas "creation science" is based on Genesis, the biblical basis of ID is the New Testament Gospel of John. (To avoid disputes with their young-earth creationist allies regarding the age of the earth, the CRSC supplanted Genesis with the New Testament Gospel of John as the biblical basis of ID.) We thus have Dembski's testimony, written in his own hand, that ID is not only a religious belief, but also a sectarian Christian belief.

The Truth about ID: From Its Own Proponents

In the same publication in which Dembski had defined ID in terms of John's Gospel in 1999, Nelson gave a candid answer in 2004—in an interview that included Dembski—when asked, "Where is the ID movement going in the next 10 years? What new issues will it be exploring, and what new challenges will it be offering Darwinism?"

> **Nelson:** Easily the biggest challenge facing the ID community is to develop a full-fledged theory of biological design. We don't have such a theory right now, and that's a real problem. Without a theory, it's very hard to know where to direct your research focus. Right now, we've got a bag of powerful intuitions, and a handful of notions such as "irreducible complexity" and "specified complexity"—but, as yet, no general theory of biological design (Macosko 2004).

Nelson had been equally candid in 2003, when he told a Dartmouth College audience that he opposed teaching ID in schools: "It isn't a fully-fledged theory—there isn't yet enough there to actually teach" (Barry 2003). (In the 2004 interview, responding to the same questions as Nelson, Dembski made a dire prediction about the future of evolutionary theory: "In the next five years, molecular Darwinism—the idea that Darwinian processes can produce complex molecular structures at the subcellular level—will be dead.... I therefore foresee a Taliban-style collapse of Darwinism in the next ten years. ID will of course profit greatly from this" [Macosko 2004]).

In 2006, only a few months after the *Kitzmiller* ruling handed ID proponents a sound defeat, Phillip Johnson, who by that time had devoted 16 years to promoting ID, was equally candid. Revealing that he had viewed the *Kitzmiller* case as "a loser from the start," he added,

I also don't think that there is really a theory of intelligent design at the present time to propose as a comparable alternative to the Darwinian theory, which is, whatever errors it might contain, a fully worked out scheme. There is no intelligent design theory that's comparable. Working out a positive theory is the job of the scientific people that we have affiliated with the movement. Some of them are quite convinced that it's doable, but that's for them to prove. . . . No product is ready for competition in the educational world (D'Agostino 2006).

Just as ID proponents themselves have confirmed the religious essence of intelligent design, so have they confirmed its scientific bankruptcy.

ACKNOWLEDGMENTS

Partial funding for my participation in the 74th Cold Spring Harbor Symposium was provided by Southeastern Louisiana University.

REFERENCES

Access Research Network. 2001. *Origins & Design* issue 39 now available, featuring 'In Pursuit of Intelligent Causes' by Charles B. Thaxton. *ARN Announce* **16**. Accessed June 26, 2009, at http://www.arn.org/announce/announce0601no16.htm.

Associated Press. 1980. Teacher of Bible creation theory arouses storm. *New York Times*, December 25: A21.

Barry C. 2003. "Intelligent design" may underlie life. *The Dartmouth.com*. February 21. Accessed July 7, 2009, at http://thedartmouth.com/2003/02/21/news/intelligent/.

Behe MJ. 1998. *Darwin's black box: The biochemical challenge to evolution*. Touchstone, New York.

Biola University. 2009. Intelligent design theory and Biola. Accessed July 8, 2009, at http://www.biola.edu/id/.

Center for the Renewal of Science and Culture. 1996. Welcome to the Discovery Institute's Center for the Renewal of Science and Culture. Accessed July 5, 2009, at http://web.archive.org/web/1103063546/www.discovery.org/crsc.html.

Center for the Renewal of Science and Culture. 1998. The wedge. Accessed July 5, 2009, at http://ncseweb.org/creationism/general/wedge-document.

Center for Science and Culture. 2009. Fellows. Accessed June 26, 2009, at http://www.discovery.org/csc/fellows.php.

D'Agostino M. 2006. In the matter of Berkeley v. Berkeley. *Berkeley Sci. Rev.* **Spring:** 31–35. Accessed July 7, 2009, at http://sciencereview.berkeley.edu/articles/issue10/evolution.pdf.

Dembski WA. 1994. The incompleteness of scientific naturalism. In *Darwinism: Science or philosophy?* (ed. J Buell and V Hearn). Accessed July 5, 2009, at http://www.leaderu.com/orgs/fte/darwinism/chapter7.html.

Dembski WA. 1998a. The act of creation: Bridging transcendence and immanence. Access Research Network. Accessed July 6, 2009, at http://www.arn.org/docs/dembski/wd_actofcreation.htm.

Dembski WA. 1998b. The intelligent design movement. *Cosmic Pursuit* **1 (2):** 22–26. Accessed June 30, 2009, at http://www.arn.org/docs/dembski/wd_idmovement.htm.

Dembksi WA. 1999a. *Intelligent design: The bridge between science and theology*. InterVarsity Press, Downers Grove, IL.

Dembski WA. 1999b. Signs of intelligence: A primer on the discernment of intelligent design. *Touchstone* **12 (4):** 76–84.

Dembski WA. 2008. *Understanding intelligent design*—now available at Amazon.com! June 29. Uncommon descent weblog. Accessed July 7, 2009, http://www.uncommondescent.com/intelligent-design/understanding-intelligent-design-now-available-at-amazoncom/.

Dembski WA, McDowell S. 2008. *Understanding intelligent design: Everything you need to know in plain language*. Harvest House, Eugene, OR.

Denton M. 1986. *Evolution: A theory in crisis*. Alder and Alder, Chevy Chase, MD.

Edwards v. Aguillard. 1987. U.S. Supreme Court. Accessed July 5, 2009, at http://www.law.cornell.edu/supct/html/historics/USSC_CR_0482_0578_ZO.html.

Epperson v. Arkansas. 1968. U.S. Supreme Court. Accessed June 30, 2009, at http://www.law.cornell.edu/supct/html/historics/USSC_CR_0393_0097_ZO.html.

Forrest B. 2005a. Expert witness report: Kitzmiller et al. v. Dover Area School District. Accessed June 30 2009, at http://ncseweb.org/files/pub/legal/kitzmiller/expert_reports/2005_04_01_Forrest_expert_report_P.pdf.

Forrest B. 2005b. Supplement to expert witness report: Kitzmiller v. Dover Area School District. Accessed June 30, 2009, at http://ncseweb.org/files/pub/legal/kitzmiller/expert_reports/2005-07-29_Forrest_supplemental_report_P.pdf.

Forrest B. 2007. Understanding the intelligent design creationist movement: Its true nature and goals. Center for Inquiry Office of Public Policy, Washington, DC. Accessed July 7, 2009, at http://www.centerforinquiry.net/uploads/attachments/intelligent-design.pdf.

Forrest B, Gross PR. 2007. *Creationism's Trojan horse: The wedge of intelligent design*, 2nd ed. Oxford University Press, New York.

Foundation for Thought and Ethics. [Undated]. Groundbreaking symposium. Accessed July 5, 2009, at http://www.fteonline.com/symposium.html.

Hartwig M. 1995. Challenging Darwin's myths. *Moody Magazine*. **May**. Accessed June 30, 2009, at http://www.leaderu.com/orgs/arn/dardoc1.htm.

Johnson PJ. 1990. Evolution as dogma: The establishment of naturalism. *First Things* **6:** 15–22. Accessed July 5, 2009, at http://www.firstthings.com/article/2007/09/002-evolution-as-dogma-the-establishment-of-naturalism-43.

Johnson PJ. 1996. Starting a conversation about evolution. Access Research Network. Accessed July 5, 2009, at http://www.arn.org/docs/johnson/ratzsch.htm.

Jones JE. 2005. Memorandum opinion. *Kitzmiller et al. v. Dover Area School District* (2005). Accessed June 30, 2009, at http://www.pamd.uscourts.gov/kitzmiller/kitzmiller_342.pdf.

Kenyon DH. 1984. Affidavit of Dr. Dean H. Kenyon in biology and bio-chemistry. *Aguillard et al. v. Edwards et al.* Accessed June 30, 2009, at http://talkorigins.org/faqs/edwards-v-aguillard/kenyon.html.

Kenyon DH. 2007. Reflections on macroevolution. Kolbe Center for the Study of Creation. Accessed June 26, 2009, at http://www.kolbecenter.org/kenyon_reflections.htm.

Kenyon DH, Davis PW. 1993. *Of pandas and people*. Foundation for Thought and Ethics, Richardson, TX.

Kolbe Center for the Study of Creation. [Undated]. Kolbe Center mission statement. Accessed June 30, 2009, at http://www.kolbecenter.org/missionstatement.htm.

Luskin C. 2005. Truth sheet # 09–05. Does intelligent design posit a "supernatural creator?" Accessed July 5, 2009, at http://www.discovery.org/scripts/viewDB/filesDB-download.php?command=download&id=565.

Macosko J. 2004. The measure of design. *Touchstone* **17 (6):** 60–65. Accessed July 7, 2009, at http://www.touchstonemag.com/archives/article.php?id=17-06-060-i.

McLean v. Arkansas Board of Education. 1982. Accessed June 30, 2009, at http://www.talkorigins.org/faqs/mclean-v-arkansas.html.

Mere creation: Reclaiming the book of nature. 1996. Conference overview. Accessed July 6, 2009, at http://www.origins.org/mc/menus/overview.html.

National Center for Science Education. 2002. Evolving banners at the Discovery Institute. Accessed July 6, 2009, at http://ncse.com/creationism/general/evolving-banners-at-discovery-institute.

National Center for Science Education. 2008a. Creationism and

the law. Accessed June 21, 2009, at http://ncse.com/creationism/legal/creationism-law.
National Center for Science Education. 2008b. Forrest's testimony: "Creationism and ID." Accessed July 4, 2009, at http://ncse.com/creationism/legal/forrests-testimony-creationism-id.
National Center for Science Education. 2008c. Intelligent design on trial: Kitzmiller v. Dover. Accessed June 30, 2009, at http://ncse.com/creationism/legal/intelligent-design-trial-kitzmiller-v-dover.
National Center for Science Education. 2008d. The wedge document. Accessed June 24, 2009, at http://ncse.com/creationism/general/wedge-document.
Nelson P. 2002. Life in the big tent: Traditional creationism and the intelligent design community. *Christian Research J.* **24 (2):** 1–7. Accessed June 30, 2009, at http://www.equip.org/articles/life-in-the-big-tent.
Pennock RT. 2007. God of the gaps: The argument from ignorance and the limits of methodological naturalism. In *Scientists confront creationism: Intelligent design and beyond* (ed. A Petto and L Godfrey), pp. 309–338. W.W. Norton, New York.
Scott EC. 2005. *Evolution v. creationism: An introduction.* University of California Press, Berkeley.
Scott EC, Matzke NJ. 2007. Biological design in science classrooms. *Proc Natl Acad Sci* **104:** 8669–8676.
Thaxton CB, Bradley WL, Olsen RL. 1984. *The mystery of life's origin: Reassessing current theories.* Philosophical Library, New York.
Wiker B. 2000. A new scientific revolution. *Catholic World Report* **10 (7)**. Accessed June 30, 2009, at http://web.archive.org/web/20070930205501/www.catholic.net/rcc/Periodicals/Igpress/2000-07/intrview.html.
Williams D. 2007. Friday five: William A. Dembski. *Citizen link*. **December 14.** Accessed July 7, 2009, at http://www.citizenlink.org/content/A000006139.cfm.
Witham L. 2005. *Where Darwin meets the Bible: Creationists and evolutionists in America.* Oxford University Press, New York.

Deconstructing Design: A Strategy for Defending Science

K.R. MILLER
Department of Molecular Biology, Cell Biology, and Biochemistry, Brown University, Providence, Rhode Island 02912
Correspondence: kenneth_miller@brown.edu

Despite its legal and scientific failings, the "intelligent design" (ID) movement has been a public relations success story in the United States. By first creating doubts about the adequacy of evolution to account for the complexity of life, the ID movement has invoked the values of "fairness" and "openness" to argue for inclusion in the classroom and curriculum. In this way, it has attempted to lay claim to the very principles of critical analysis and open discussion at the heart of the scientific enterprise, leaving many researchers in doubt as to how to respond to these challenges.

Specific case studies, including the blood-clotting cascade and data from the human genome, show how scientists can have a leading role in deconstructing the arguments advanced in favor of ID. The key to this strategy is remarkably simple and was at the heart of the landmark 2005 Kitzmiller v. Dover trial on ID. It is for researchers to take the claims made by ID proponents seriously, and then to follow them to their logical scientific conclusions. When this is done effectively, the hypothesis of "design" can be publicly falsified in ways that are understandable to laypeople and decision makers in education.

We live and work in the midst of a remarkable dualism. Today, 150 years after the publication of *On the Origin of Species*, the scientific foundations of evolutionary biology have never been stronger. Indeed, as Theodosius Dobzhansky (1973) famously wrote, "Nothing in biology makes sense except in the light of evolution," and the papers presented at this conference are eloquent testament to the validity of that assessment. Nonetheless, in the public mind, evolution remains a "controversial" idea, a mere "theory" rejected by as many as half of all Americans. Widespread opposition to evolution has led some states to weaken their science education standards, forced teachers to deemphasize evolutionary principles in biology, and placed pressure on authors and publishers to include "alternate" theories in their textbook offerings. Although this is primarily an American phenomenon, it is worth noting that antievolution movements have made major gains in Europe as well (Graebsch and Schiermeier 2006), suggesting that in the near future, this may become a truly international issue for scientists and educators.

Although I am a cell biologist, my own work as a textbook coauthor with my colleague Joseph S. Levine (Miller and Levine 2008) has forced us to confront these issues and to develop effective responses to a number of antievolution arguments and movements. In 2002, for example, one of our textbooks was chosen for use in the high schools of Cobb County, Georgia. Community reaction against the treatment of evolution in several textbooks, including ours, led to a petition drive to include creationism in the county's curriculum. The Cobb County School Board attempted to deal with the popular pressure by fashioning what they viewed as a compromise. The Board required that a sticker be affixed to each book warning students that "evolution is a theory, not a fact, regarding the origin of living things" (Holden 2002). Several parents, contending that these stickers represented a government attempt to advance a particular religious point of view, sued the School Board in Federal court, and a week-long trial resulted. The court ruled (Holden 2005) in favor of those plaintiffs, and the stickers have now been removed. Late in 2005, a more highly publicized trial, known legally as Kitzmiller v. Dover Area School District, took place in another Federal court, and I discuss some of the details of that trial below.

In many states, the struggle over evolution has also found its way into the political arena. In fact, evolution was *the* pressing election issue in two American states in 2006—Ohio and Kansas. Each elects their state board of education in highly politicized contests, and in 2006, the candidates' positions on evolution seemed to be the only issue that mattered to many voters. According to a newspaper report (Stephens 2006), a radio talk show host in Cleveland described one of these contests involving Deborah Owens Fink, the leader of antievolution forces on the Ohio Board, like this: "If you believe in God, creation, and true science, vote for Debbie. If you believe in evolution, abortion, and sin—vote for her opponent."

With rhetoric like that, one might have expected Ms. Fink to cruise to an easy win. In reality, proevolution candidates, including Ms. Fink's opponent, swept to victory in Ohio, and proevolution candidates also took control of the Kansas Board of Education. Evolution supporters in Kansas further strengthened their hold on the Board in the recent 2008 elections. The reasons, in each state, were effective proscience campaigns mounted by coalitions of scientists, educators, health professionals, and others interested in quality science education. Given a choice, the American people will choose science every time, but only if we in the scientific and educational communities put the issue on the table clearly and forcefully. Being an optimist by nature, I hope we can continue to do just that.

THE DEVIL IN DOVER

Of all the recent battles over evolution, by far the most spectacular took place in the small community of Dover, Pennsylvania in 2004 and 2005. Late in 2004, Dover's Board of Education voted to instruct the teachers in the Dover Area High School to prepare a biology curriculum that included an antievolution concept known as "intelligent design" (ID). Although the Dover science faculty courageously refused to go along, the Board persisted. They purchased classroom sets of an ID textbook known as *Of Pandas and People* (Davis and Kenyon 1993) and wrote a four-paragraph statement on ID to be read to students. When it was clear that the Board would go ahead with this policy, 11 parents filed a lawsuit asking that the ID policy be rescinded. The case moved rapidly to trial, gaining media attention all the while, and convened in the Federal Courtroom of Judge John E. Jones III in Harrisburg on September 26, 2005.

What is the concept called "intelligent design" that was at the heart of this battle? The best way to begin might be by defining what ID does not mean. Most theists, those individuals who believe in a God of any sort, would argue that there is a plan and pattern to existence. As such, they might well agree, in a certain sense, that there is indeed an "intelligent" order to existence. Valid or not, this is a philosophical argument that lies outside the purview of the natural sciences. It does not so much challenge the theory of evolution as define a view of how evolutionary science may be viewed in a philosophy of nature.

That is not, however, how ID was presented to the citizens and schoolchildren of Dover. In the context of public discourse in the United States, ID is a claim that "design," meaning outside intelligent intervention, is required to account for the origins of living organisms. As such, it clearly is a doctrine of *special creation*. The reason for this assessment is that when one states that the bacterial flagellum, or the blood-clotting cascade, or even the animals of the Cambrian period were "designed," what one really means is that they were *created*. One cannot speak of the "design" of a biochemical system without also claiming that the genes to specify that system were, in the most direct sense, created by an intelligence outside of nature. Pointing out that ID is a form of creationism does not, of course, mean that it is wrong; rather, it is only to call it by a proper and accurate name.

For advocates of ID, the looming court case was their chance to crush those whom they scorned as "Darwinists" in front of a conservative Republican judge. John E. Jones III, who would preside over the case, had been named to the bench in 2002 by President George W. Bush. William Dembski, a leading advocate of ID who at first agreed to appear as an expert witness in the trial, even proposed a "strategy for interrogating the Darwinists to, as it were, squeeze the truth out of them." Dr. Dembski did not appear in court, citing disagreements with attorneys representing the Dover Board, and the case certainly did not go as he expected. The actual result of the 7-week trial was a crushing defeat for ID, as described by the Judge himself in a recent interview (Gitschier 2008). So completely did the case for ID as science collapse that the citizens of Dover did not feel the need to wait for the judge's decision. Only a few days after arguments in the trial concluded, voters turned out the pro-ID school board, replacing them with a reform slate that had strongly opposed the ID policy. Six weeks later, the judge filed his own opinion. As reported in *The New York Times*, "In the nation's first case to test the legal merits of intelligent design, the judge, John E. Jones III, issued a broad, stinging rebuke to its advocates and provided strong support for scientists who have fought to bar intelligent design from the science curriculum" (Goodstein 2005).

There were many elements to the success of the plaintiffs in the Dover trial, some of which have been discussed by other speakers at this meeting (Kevin Padian, Barbara Forrest, and I served as expert witnesses in the trial, and Eugenie Scott, Director of the National Center for Science Education, had a key role in coordinating the case). In particular, the religious origins of the ID movement were laid bare, clearly demonstrating that the intentions of the Dover Board were in clear violation of the First Amendment of the U.S. Constitution. For legal reasons, this may have been the single most important element of the trial. Many Americans, however, might not be bothered by such connections. After all, if one has genuine scientific evidence of the work of a "designer," it only follows that people of faith would seek to use the public schools to spread the word. And, they might ask, if the science were legitimate, what would be the harm of that?

For that reason, the Dover Board argued that ID was in fact sound science and that its presence in the classroom would serve a legitimate secular purpose. Aside from pointing out that ID is not generally accepted by the scientific community—an important point to be sure—how might one counter that assertion? As I suggest below, the answer is remarkably simple. We should take the suggestion of "design" in biological systems as seriously as we do any scientific proposal, follow it up, and see where it leads.

AN ENDURING APPEAL

It is abundantly clear to the members of the scientific community that the advocates of ID have not made a convincing scientific case. Indeed, just a few months after the conclusion of the trial, even law professor Phillip Johnson, one of the founders of the ID movement, admitted frankly that the scientific people in ID had let him down:

> I also don't think that there is really a theory of intelligent design at the present time to propose as a comparable alternative to the Darwinian theory, which is, whatever errors it might contain, a fully worked out scheme. There is no intelligent design theory that's comparable. Working out a positive theory is the job of the scientific people that we have affiliated with the movement. Some of them are quite convinced that it's doable, but that's for them to prove… No product is ready for competition in the educational world.
>
> *As quoted in D'Agostino 2006.*

Nonetheless, despite these scientific failings, ID has been a public relations success story. A recent study (Miller et al. 2006) placed the United States second to last in the extent to which the citizens of 34 different nations accepted the theory of evolution. Among the countries studied, only Turkey ranked lower in terms of support for evolution. The reasons for this should be obvious. Not only do individuals in the United States show a much higher degree of religious belief than those of most other industrialized countries, but these individuals also seem to be at the very center of the antievolution movement. Organizations such as *Answers in Genesis* and *The Discovery Institute* turn out steady streams of antievolution material, much of it freely available on the web. Adding to this, one might include the recently opened Creation Museum in northern Kentucky, and *Expelled*, a popular 2008 documentary purporting to show links between evolutionary theory and the Nazi Holocaust. Given such steady and skillful promotion, it seems clear that the appeal of ID creationism will endure.

What is the source of this appeal? I contend that ID succeeds in the public imagination because it seems to fill a vacuum in our understanding of biology. Any biological structure or process that is not yet fully understood contributes to this vacuum, and ID fills it at a stroke. Indeed, the critics of evolution find it easy to point to complex molecular machines such as the ribosome and then challenge the scientific community to provide detailed, step-by-step evolutionary explanations for their origins. When such explanations are not forthcoming, they announce that "design" must be the answer. Using a strategy such as this, the vast reservoir of unsolved and unexplored scientific problems becomes grist for the creationist mill. In the minds of many members of the general public, it actually becomes "evidence" for the hypothesis of ID.

The appeal of the "design" argument, therefore, is the closure that it seems to provide to such questions. Where evolution seems to offer open-ended inquiry and unresolved questions, ID brings things to a neat and tidy conclusion. Its appeal is that it seems to provide answers where science supplies only questions and certainty where science calls for doubt.

DECONSTRUCTING "DESIGN"

One of the keys to the public success of the ID movement has been the tacit agreement the scientific community has given to the creationist argument that "design requires a designer." Because, to most laypeople, the form and function of everything from the human body to a muscle cell amount to "design," the scientist seems forced to argue that there is no design in nature and that the exquisite architecture of life is some sort of illusion. This approach fails as common-sense argument, but more importantly, it fails as science. There is indeed a "design" to living systems—but it is not the top-down design that would be produced by an architect or craftsman; it is a bottom-up design that is the result of evolution.

We should begin our deconstruction of the design argument by pointing out the obvious—that living systems do show a correlation between structure and function that a reasonable person might indeed call "design." The structural biologist David DeRosier (1998) acknowledged this point exactly when he reviewed the organization of the bacterial flagellar motor, stating that "… more so than other motors, the flagellum resembles a machine designed by a human." The question, of course, is whether this resemblance implies the sort of intelligent agent that the advocates of ID would suggest. It does not, as Bruce Alberts has pointed out (Alberts 1998). Among the questions he would have scientists ask about "protein machines," Alberts wrote, was "to what extent has the design of present-day protein machines been constrained by the long evolutionary pathway through which the function evolved, rather than being optimally engineered for the function at hand?" I believe that Alberts was on to something. As his words suggest, biological complexity can indeed show "design," but a design revealed and constrained by the process of evolution itself.

In his book *Your Inner Fish*, paleontologist Neil Shubin addressed the issue of biological design at the physiological level by bringing evidence together from fossils, developmental biology, and molecular genetics (Shubin 2008). There is indeed a design to the body, as Shubin demonstrated, a design reflecting the evolutionary history of our species. Our skeletal structure results from a modification of the fish body plan; our muscles are laid out in segments that reflect the blocks of tissue associated with each segment of the vertebrate body, and even the complex and confusing pathways of cranial nerves can be explained by comparison with our evolutionary relatives.

Even proteins can fairly be said to possess a design, yet once again, that design makes sense only in evolutionary terms. This fact was brilliantly exploited in a study that used evolutionary relationships between present-day organisms to reconstruct the actual gene for an ancestral corticoid receptor existing some 450 million years ago (Ortlund et al. 2007). Using the comparative structures of two different receptor proteins, one that binds glucocorticoid and another specific for mineralocorticoid, they reconstructed the ancestral receptor from which both are derived. The comparative study not only proved the value of exploiting the evolutionary design of protein structures, but also provided new insights into the mutational pathways by which gene duplication generates new biochemical systems with novel functions.

In short, the scientific community can make the case for evolution by accepting the concept of design and then demonstrating that the design of living things is an evolutionary one. One element of the power of this approach is that it clearly rises above the appeal to ignorance inherent in ID. Another equally important aspect is that it does not require the scientific specialist to solve the evolutionary origins of every conceivable structure, pathway, or organ. By demonstrating that well-understood cases display clear evidence of their evolutionary ancestry, the point is made and it is made convincingly.

DARWIN IN THE BLOOD: A CASE STUDY

An example of this approach can be fashioned from one of the arguments used by the ID movement itself: the supposed "irreducible complexity" of the vertebrate blood-clotting cascade (Fig. 1). In humans, more than a

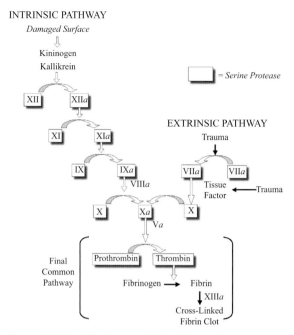

Figure 1. A simplified representation of the major components of the blood-clotting pathway. Each of the components of the pathway represents a "clotting factor," a portion of the pathway that triggers the next step. The horizontal arrows represent conversion of a factor from its inactive form to the active form, and the vertical arrows indicate the factors that trigger such conversions. The final result of the pathway, shown at the bottom, is the formation of a clot of cross-linked fibrin proteins that stops bleeding. Intelligent design (ID) contends that the pathway cannot work until all of these factors are in place, so it could not have been produced by a gradual step-by-step evolutionary process. This contention is refuted by recent research on the evolution of clotting factors.

dozen proteins and cofactors are involved in the clotting process, and serious disorders result when any of these components is missing or damaged.

Michael Behe, a leading advocate of ID, discussed the clotting system in a chapter he helped to write for the widely distributed ID textbook *Of Pandas and People*:

> However, biochemical investigation has shown that blood clotting is a very complex, intricately-woven system containing a score of interdependent protein parts. The absence or defective operation of any of several of these components will cause the system to fail, and blood will not clot at the proper time or at the proper place (Davis and Kenyon 1993, p. 141).

Consider the nature of this argument: If each and every part of the system must be present simultaneously for blood to clot, the clotting system could never have been produced by gradual step-by-step evolution. It is indeed "irreducibly complex" and therefore unevolvable. If Darwinian evolution could not have produced it, what could have? The answer, according to ID enthusiasts, must be intelligent design. Behe (1996) made this point even more directly in a popular book on intelligent design, *Darwin's Black Box*. As he wrote, "…in the absence of any of the components, blood does not clot, and the system fails" (p. 86) and "Since each step necessarily requires several parts, not only is the entire blood-clotting system irreducibly complex, but so is each step in the pathway" (p. 87).

The blood-clotting system provides a perfect example of how to make a case for ID. We find a system that is not only complex, but *irreducibly* complex, a system in which the absence or loss of a single component would destroy function. Such a system would not just be difficult to evolve, it would be impossible. Darwinian evolution, in the words of Darwin himself, requires "numerous graduations" on the way to a complex system, and every one of those graduations must be advantageous—they all have to work. The irreducible complexity of blood clotting, however, shows that absence of even a single part of the pathway would be fatal. Find as many fossils as you like, one might say, but it does not matter if evolution cannot clot the blood.

As with other claims made against evolution, the most effective way to deconstruct "design" is to take the ID argument seriously. In this case, it would involve investigating ID's bold prediction that *all* of the clotting components must be present for the system to function. Unfortunately for the ID argument, this prediction is now known to be wrong.

A report from the 1960s suggested that whales and dolphins lacked one of the clotting factors (Robinson et al. 1969), but ID advocates could easily have explained that away as the unreliable product of research in the premolecular age. However, the loss of factor XII was confirmed (Semba et al. 1998) in a study demonstrating that pseudogene conversion accounts, in molecular terms, for the factor's absence from the cetacean bloodstream. In 2003, the case against irreducible complexity was further strengthened when Russell Doolittle's laboratory demonstrated that the genome of *Fugu*, the puffer fish, lacks three of the clotting factors but nonetheless has a functional clotting system (Jiang and Doolittle 2003). More recently, the same lab has studied the lamprey genome and discovered that lampreys lack even more of the components of the supposedly "irreducible complex" clotting system (Doolittle et al. 2008). These investigators wrote, "In summary, the genomic picture presented here suggests that lampreys have a simpler clotting scheme than later diverging vertebrates. In particular, they appear to lack the equivalents of factors VIII (or V) and IX, suggesting that the gene duplication leading to these factors, synchronous or not, occurred after their divergence from other vertebrates."

The existence of a partial pathway that not only has a useful function, but also performs what we might call the final function (blood clotting) demonstrates beyond any doubt that complex pathways can be built up a few steps at a time from simpler ones. Furthermore, Doolittle's lab has also shown that the genome of the sea squirt *Ciona intestinalis*, which does not have functional clotting factors, nonetheless contains copies of nearly all of the protein domains from which those factors are built (Jiang and Doolittle 2003). In effect, we find the raw materials for clotting exactly where evolution tells us they should be, in the last group of organisms to split off from the vertebrates before blood clotting appeared. By taking the claim of "design" seriously, we discover that even one of the ID movement's favorite examples was clearly the product of evolution. The clotting system is but one of many cases to which this approach can be applied.

DARWIN'S GENOME

In the popular imagination, the principal evidence for human evolution is thought to come from the fossil record of prehuman primates. Although the evidence is indeed compelling, an even more powerful case can be made from the record of human ancestry in our own genome. Just 2 weeks before the Dover ID trial was called to order, researchers added to this evidence the DNA sequence of the chimpanzee. The utility of this new information in establishing the validity of evolutionary theory could hardly be understated. As the authors of the lead article on this breakthrough observed,

> More than a century ago Darwin and Huxley posited that humans share recent common ancestors with the African great apes. Modern molecular studies have spectacularly confirmed this prediction and have refined the relationships, showing that the common chimpanzee (*Pan troglodytes*) and bonobo (*Pan paniscus* or pygmy chimpanzee) are our closest living evolutionary relatives (Mikkelsen et al. 2005).

To bring the weight of this evidence into the courtroom, we chose a simple example that provides a direct test of the hypothesis of common ancestry for our species. As any biology student knows, we humans normally have 46 chromosomes. If we do indeed share common ancestry with organisms such as the gorilla, orangutan, and chimpanzee, an interesting question must be answered. All of the great apes have 48 chromosomes. If we really do share a common ancestor with these species, then what happened to that extra pair of chromosomes?

One might suggest that in the lineage leading to our species, a pair of chromosomes simply was lost or discarded. Unfortunately, in genetic terms, this is not a realistic suggestion. There are so many important genes on every primate chromosome that the loss of both members of a chromosome pair would be fatal. The only realistic possibility is that two different primate chromosomes were accidentally fused into one at some point in human evolution. Chromosome fusions of this sort are not at all uncommon and would indeed have reduced the chromosome number from 48 to 46. But if this sort of fusion did take place in the recent past, it should have left unmistakable evidence behind. Somewhere in the human genome there should be a chromosome still bearing the marks of that fusion, and therein lies an opportunity to put the hypothesis to a scientific test.

What would a fused chromosome look like? Telomeres, the tips of chromosomes, contain unique, repeating DNA sequences that are especially easy to recognize. If two chromosomes fused into one, the fusion site would contain telomere DNA sequences where they simply do not belong, on either side of the fusion site. In addition, each chromosome also contains a region known as the centromere where chromosomes attach to the machinery that separates them during cell division. Centromeres likewise have distinctive DNA sequences that enable them to be easily identified. If one of our chromosomes had indeed been produced by the fusion of two others in the recent past, that chromosome should contain telomere sequences near the middle of the chromosome and should also contain two centromere sequences.

Now the task gets interesting. We can scan the human genome and see if any of our chromosomes fit this very precise description. If we do not find such a chromosome, the hypothesis of common ancestry for our species might be cast into serious doubt. But if we do find a fused chromosome, a specific evolutionary prediction is fulfilled. So, which is it?

The answer—provided in dramatic detail by the human genome project—is that evolution got it exactly right (Hillier et al. 2005). The solution is found in human chromosome 2, which does indeed contain telomere DNA sequences at the fusion point and carries the remnants of two centromere sequences, as illustrated in Figure 2. One of these is still active in humans and corresponds to the centromere for chimp chromosome 12. The other has been inactivated, which makes the fused chromosome more stable during cell division, but it is still recognizable as corresponding to the centromere from chimp chromosome 13 (Hillier et al. 2005). The conclusion from these data is unavoidable: We do indeed share a common ancestor with these species, a common ancestor that possessed, in the recent past, 48 chromosomes. No fingerprint left at the scene of a crime was ever more decisive than this genetic evidence. We evolved.

CONCLUSIONS

For science, I believe that the collapse of "intelligent design," so evident in the Dover trial, carries a clear

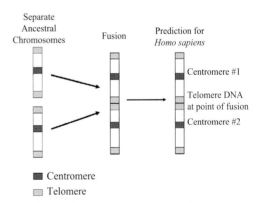

Figure 2. An accidental fusion between two chromosomes could explain why humans possess 46 chromosomes, rather than 48 as do the great apes. However, such a fusion event would leave distinct marks in the new chromosome. Chromosomes contain recognizable regions at their tips known as telomeres and regions near their midpoints called centromeres. If two complete chromosomes fused together, telomere sequences would be expected to remain near the fusion site. In addition, the fused chromosome would be expected to carry two centromeres. The second human chromosome displays each of these predicted elements, providing strong support for the evolutionary hypothesis of common ancestry.

meaning, i.e., that the process of science should be respected. Challenges to evolution—or any other scientific theory—are very much within the scope and tradition of science. If the practitioners of ID actually seek to displace evolution scientifically, they need only to produce the data to support their case, to carry the fight to the scientific community in a way that would win the battle of evidence in the free marketplace of scientific ideas. Instead, they have consistently rejected that route in favor of public relations activity and the generation of political support. Scientifically, it should be obvious that no idea deserves a place in the classroom that it cannot win for itself on the basis of the evidence. The lesson for science is that organized attempts to skirt the scientific process of debate and peer review can and must be resisted. Not just because they happen to be wrong, as is the case with ID, but because they subvert the very process of science itself.

One of the most effective scientific responses, as I have suggested, is to defend science by deconstructing the design argument. The structure, composition, and organization of living systems do indeed reveal a kind of design, but it is a living architecture produced by the evolutionary process itself. Science is necessarily incomplete, and the opponents of evolution will always be able to point to unsolved problems as evidence that the evolutionary narrative is incomplete as well. But these challenges should be seen as opportunities. We make the best case for science when we show how evolution accounts for the realities of living systems in a way that pretenders such as ID simply cannot. If we do this effectively, in the final analysis, the vast majority of Americans may come to realize, as Charles Darwin did, that there is indeed beauty, wonder, and grandeur in the evolutionary view of life.

REFERENCES

Alberts B. 1998. The cell as a collection of protein machines: Preparing the next generation of molecular biologists. *Cell* **92:** 291–294.

Behe M. 1996. *Darwin's black box: The biochemical challenge to evolution*. The Free Press, New York.

D'Agostino M. 2006. In the matter of Berkeley vs. Berkeley. *Berkeley Science Review* (spring issue), pp. 31–35. University of California, Berkeley.

Davis P, Kenyon DH. 1993. *Of pandas and people: The central question of biological origins*. The Foundation for Thought and Ethics, Richardson, Texas.

DeRosier D. 1998. The turn of the screw: The bacterial flagellar motor. *Cell* **93:** 17–20.

Dobzhansky T. 1973. Nothing in biology makes sense except in the light of evolution. *Am Biol Teacher* **35:** 125–129.

Doolittle RF, Jiang Y, Nand J. 2008. Genomic evidence for a simpler clotting scheme in jawless vertebrates. *J Mol Evol* **66:** 185–196.

Gitschier J. 2008. Taken to school: An interview with the Honorable Judge John E. Jones, III. *PLoS Genet* **4:** e1000297.

Goodstein L. 2005. Judge rejects teaching intelligent design. *The New York Times* (December 21, 2005), p. 1.

Graebsch A, Schiermeier Q. 2006. Anti-evolutionists raise their profile in Europe. *Nature* **444:** 406–407.

Hillier LW, Graves TA, Fulton RS, Fulton LA, Pepin KH, Minx P, Wagner-McPherson C, Layman D, Wylie K, Sekhon M, et al. 2005. Generation and annotation of the DNA sequences of human chromosomes 2 and 4. *Nature* **434:** 724–731.

Holden C. 2002. Georgia county opens door to creationism. *Science* **298:** 35–36.

Holden C. 2005. Teaching evolution. Judge orders stickers removed from Georgia textbooks. *Science* **307:** 334.

Jiang Y, Doolittle RF. 2003. The evolution of vertebrate blood coagulation as viewed from a comparison of puffer fish and sea squirt domains. *Proc Natl Acad Sci* **100:** 7527–7532.

Mikkelsen TS, et al. (The Chimpanzee Sequencing and Analysis Consortium). 2005. Initial sequence of the chimpanzee genome and comparison with the human genome. *Nature* **437:** 69–87.

Miller JD, Scott EC, Okamoto S. 2006. Public acceptance of evolution. *Science* **313:** 765–766.

Miller KR, Levine JS. 2008. *Biology*. Prentice Hall, Boston.

Ortlund EA, Bridgham JT, Redinbo MR, Thornton JW. 2007. Crystal structure of an ancient protein: Evolution by conformational epistasis. *Science* **317:** 1544–1548.

Robinson AJ, Kropatkin M, Aggeler PM. 1969. Hageman factor (factor XII) deficiency in marine mammals. *Science* **166:** 1420–1422.

Semba U, Shibuya Y, Okabe H, Yamamoto T. 1998. Whale Hageman factor (factor XII): Prevented production due to pseudogene conversion. *Thromb Res* **90:** 31–37.

Shubin N. 2008. *Your inner fish: A journey into the 3.5-billion-year history of the human body*. Pantheon, New York.

Stephens S. 2006. Normally low profile context in spotlight. *Cleveland Plain Dealer* (October 22, 2006), p. 1.

Summary

B. CHARLESWORTH

*Institute of Evolutionary Biology, School of Biological Sciences,
University of Edinburgh, Edinburgh, EH9 3JT, United Kingdom*
Correspondence: brian.charlesworth@ed.ac.uk

Advances in molecular biology have revolutionized the study of evolution. Detailed comparative studies of genomes are facilitating the analysis of phylogenies and raising new questions such as the extent of lateral gene transfer. Evolutionary analyses of development show that innovations frequently involve the reuse of existing gene products and gene networks in new ways and that changes in gene expression are important in morphological evolution. Population genetic studies are shedding increasing light on the genetic basis of traits subject to both artificial and natural selection. Laboratory models of evolution are being applied to both molecular and whole-organism systems, yielding insights into the evolution of adaptations, which complement those arising from reconstructions of evolutionary paths using molecular sequence or paleontological data. Overall, the Symposium portrayed evolution as a field that, while retaining its Darwinian roots, is exploring ever-wider areas of biology as new techniques and ideas emerge.

The 100th anniversary of the publication of *On The Origin of Species* was celebrated by the 24th Cold Spring Harbor Symposium on Quantitative Biology "Genetics and Twentieth Century Darwinism." This was attended by many of the leading lights of the Modern Evolutionary Synthesis, including Theodosius Dobzhansky, Ernst Mayr, Bernhard Rensch, George Gaylord Simpson, G. Ledyard Stebbins, and Sewall Wright, all of whom have now departed from the scene. The Symposium was as much a celebration of the Modern Synthesis as of Charles Darwin. In his meeting summary, Stebbins (1960) remarked that the participants "have reached substantial agreement on concepts remarkably similar to those which Darwin himself held, not only about the existence of evolution and the course which it followed, but also about the basic processes responsible for it. The only major qualitative difference between our knowledge and that possessed by Darwin lies in our recognition of particulate Mendelian inheritance, determined by chromosomal genes, as the basis of nearly all of the hereditary variability upon which selection acts."

It is interesting to look back at the 24th Symposium to see how much the intellectual and methodological framework for studying evolution has changed during 50 years. There was very little mention of DNA, with the exception of Dobzhansky's impressively forward-looking paper (Dobzhansky 1960). He gave an estimate of the size of the human genome in terms of numbers of base pairs that was reasonably close to the current value. He recognized that this gives an enormous number of possible combinations of the four possible nucleotides at each site: "The four-letter 'genetic alphabet' is, to use Leonardo's words, a beautiful, economical and direct means to create an ample supply of genetic raw materials from which evolutionary changes can be constructed." Dobzhansky also compiled data on the genome sizes of different species and pointed out what is now known as the C-value paradox; there is little relationship between the perceived complexity of multicellular organisms and the sizes of their genomes. This has received essentially no recognition in the later literature on genome size evolution (see, e.g., Cavalier-Smith 1985). Dobzhansky went on to speculate about gene homology and sequence divergence in evolution, foreshadowing the subsequent rise of studies of molecular sequence evolution.

The contents of other papers presented at the 24th Symposium reveal astounding differences in the problems under consideration, and the methods used to solve them, from the papers in the 74th Symposium. Although few present at the latter event would probably dissent from the basic principles of the Modern Synthesis, the enormous advances in molecular biology since 1959 have revolutionized both the techniques used by evolutionary biologists and the questions that they ask. In addition, field and experimental studies at the whole-organism level have greatly advanced our understanding of phenotypic evolution. There have been substantial advances in the theoretical understanding of evolutionary processes, from models of molecular evolution and variation at one extreme to social behavior at the other. Modern methods of genome sequence comparisons and genome scans of variation within populations are using evolutionary methods to identify functional changes in the genome, so that there is now a two-way interplay between evolutionary and functional biology. All of these themes were represented at the 74th Symposium; as its title suggests, molecular aspects of evolution were especially emphasized.

The Symposium spanned an enormous range of topics, from the RNA world to human social behavior, making it an exceptionally challenging task to summarize coherently. I will begin by asking, What did it show us about the impact on evolutionary biology of the spectacular advances in

---

All authors cited here without dates refer to chapters in this volume.

molecular biology since 1959, culminating in whole-genome sequencing? One aspect of this revolution has been our ability to conduct ever-more detailed comparative studies of genomes and to relate these to changes in biological organization. This has greatly facilitated the analysis of phylogenies. For example, it now seems clear that the unicellular choanoflagellates are the sister group to metazoans (N King, pers. comm.). But as well as helping to settle old questions about evolutionary relationships, some challenging problems have emerged from sequence comparisons, particularly in prokaryotes. The evidence for lateral transfer of genetic information among archaebacteria and eubacteria, as well as among less distant prokaryote groups, is now overwhelming. One response to this is to abandon any concept of a classic Darwinian "Tree of Life" (WF Doolittle); another is to search for sets of genes with consistent phylogenies that shed light on the evolutionary histories of extant taxa (Koonin et al.). Although lateral gene transfer seems far less rampant in eukaryotes and poses few problems for phylogeny reconstruction, some surprising examples have nevertheless come to light, especially in the mitochondrial genomes of fungi and flowering plants (J Palmer, pers. comm.). A spectacular example is provided by the mitochondrial genome of *Amborella*, which has expanded in size to 4 Mb by incorporating pieces of DNA up to the size of whole mitochondrial genomes from other flowering plants, algae, and mosses.

Comparative sequencing of whole genomes and parts of genomes is also shedding light on the evolution of the organization of the genome itself. One outcome of this work has been the discovery that transposable elements (TEs), and relics of TEs, comprise a large fraction of the genomes of many vertebrates (D Haussler, pers. comm.) and flowering plants (S Wessler, pers. comm.). There is increasing evidence that "domesticated" TEs have a significant functional role in gene regulation and contribute to conserved noncoding DNA. It is also clear from these studies, as well as from laboratory analyses of the rates of movement of TEs, that the numbers of copies of elements can expand very rapidly in evolutionary time when checks to their multiplication within the genome are removed, consistent with the idea that they are maintained in the genome primarily by their ability to self-replicate. Rapid progress has been made in understanding how TE activity is regulated by small interfering RNAs (Malone and Hannon; Cibrián-Jaramillo et al.), showing both how the host genome has evolved mechanisms to restrain TE activity and how TEs can evade these mechanisms.

Sex chromosome systems are another aspect of genome evolution whose understanding has recently advanced, with light being shed on their astonishing diversity within vertebrates alone. Mammals and birds are known to have diverged more than 300 million years ago; both groups have chromosomal sex determination, with male and female heterogamety, respectively. Sequencing of the chicken Z chromosome shows that it has homology with three different autosomes in humans, but not to the X chromosome (Bellott and Page). Despite the independent origin of the X and Z, they show interesting similarities in their organization, including an unusually low gene density and the acquisition of testis-expressed genes from elsewhere in the genome. These observations raise important questions concerning the evolutionary forces involved.

Genome evolution has long been known to involve a large amount of chromosomal rearrangements and changes in copy numbers. Modern technology allows much more detailed studies of these processes. Large segmental duplications of relatively recent origin form a significant portion of the human genome, and comparative studies of humans and other higher primates suggest an accelerated rate of occurrence of these in the common ancestor of humans, chimpanzees, and gorillas (Marques-Bonet and Eichler). This has resulted in the formation of new gene families, in some cases associated with signatures of positive selection. The presence of these duplications creates the opportunity for the generation of chromosome rearrangements by unequal exchanges, contributing to copy-number variation within the human population. Some regions of the human genome that contain unusual concentrations of duplications contain many different copy-number variants associated with genetic diseases, suggesting that the establishment of selectively advantageous duplications can result in an increased load of harmful variants (Dumas and Sikela). This illustrates the fact that Darwinian evolution is a process that lacks foresight, making use of whatever variants are currently selectively favorable or have a sufficiently weak disadvantage that they can be fixed by drift, without reference to long-term costs.

An objection to Darwinian evolution that is frequently made by the creationists is that it explains biological complexity by a random process. This of course ignores the fact that natural selection is decidedly nonrandom. Nevertheless, selection is impotent without the genetic variability that is thrown up by mutation, a much more random process. We now have a very detailed understanding of mutational mechanisms at the molecular level and of the elaborate cellular machinery that ensures the high accuracy of DNA replication and the correspondingly low rate of mutation (Kunkel). This accuracy reflects the fact that most mutations with any phenotypic effects are deleterious, because selection has had a long time to perfect most features of the biological machinery. But evolutionary change in response to a changing environment requires genetic variation on which selection can act, so that a zero mutation rate would bring evolution to a halt. Most evolutionists believe that the supply of mutational variability is rarely a rate-limiting factor in adaptive evolution (see, e.g., Muller 1949). Lindquist challenges this view, presenting evidence for special mechanisms that facilitate the production of novel phenotypes.

The opportunistic nature of evolutionary change was brought out by several contributions, at a variety of levels of biological organization. One major insight that has emerged from our advancing understanding of the functions of genes and the genetic control of development, is that evolutionary innovations frequently involve the reuse of existing gene products and gene networks, rather than the emergence of completely new genes and pathways. This was strikingly brought out by analyses of the genomes of choanoflagellates mentioned above and sponges. The choanoflagellates are unicellular, but some species form colonies by cell division. They possess genes that code for proteins involved in both cell adhesion and signaling between cells. The latter appear to be used in sensing and responding to the environment in the choanoflagellates. The genome sequence of the demosponge *Amphimedon queenslandica* (Richards and Degnan) reveals that, despite lacking nerve cells, this organism possesses many of the components needed for synapses. In addition, most of genes that code are present for the major components of the signaling systems used in the major developmental pathways of more complex animals. At a lower taxonomic level, the unusual mode of embryonic development of sea urchins is associated with a remodeling of a regulatory circuit present in other echinoderms (Davidson and Erwin). Evidence from bacteria for the evolution of the control of different regulatory circuits by the same upstream regulator is described by Goley et al. Yet another example of a shift in the role of a developmental regulator is provided by the regulation of apoptosis in horned beetles (where is it is involved in the remodeling of tissues used for horn development) by the *Scr* homeobox gene (Moczek). In insects, the product of the *Dorsal* gene binds to numerous other genes, whose pattern of activity depends on the identity of activation proteins that bind close to Dorsal protein-binding sites in their enhancers (Perry et al.). Comparisons of different insect species suggest that changes in enhancer sequences, not the evolution of new enhancers, are responsible for changes in expression of genes regulated by *Dorsal*.

Nevertheless, there are cases where novel genes with important functional roles have emerged during evolution, other than by the mechanism of gene duplication and divergence mentioned earlier. For example, the snaR family of small noncoding RNA genes is restricted to humans and the African great apes and has evolved from TEs in a complex series of steps, which have been reconstructed in detail (Parrott and Mathews). In other cases, the details remain obscure, as in the case of *nodal* (Grande and Patel). This gene is involved in the control of left/right asymmetry in chordates; it is now known to be present in molluscs but absent from ecdysozoa, which include insects and nematodes.

There is increasing evidence concerning the role of changes in gene expression in morphological evolution, a topic that was well represented at the Symposium. S Carroll (pers. comm.) argued that changes in *cis*-regulatory elements allow tissue-specific changes in gene expression to occur, without the harmful pleiotropic effects that may be caused by mutational changes in widely expressed, *trans*-acting regulatory proteins. He presented evidence for this thesis from several studies of the evolution of *Drosophila* pigmentation and pattern differences. In one example, the intraspecific evolution of darker abdominal pigmentation in *D. melanogaster* populations is associated with the spread of a haplotype surrounding an enhancer of *ebony*, with a total of five mutations that confer the new phenotype. Examples of this kind are relevant to the old controversies concerning micromutations versus macromutations in morphological evolution (see, e.g., Sheppard 1960). There are now several cases in which multiple changes in a small regulatory region are required for the full development of a novel phenotype, yet classical genetic analyses would have suggested that the phenotypic difference is under major gene control. Stepwise Darwinian evolution at the sequence level is thus likely to be pervasive at the level of changes in gene expression.

Linnen and Hoekstra described a somewhat contrasting genetic scenario for the adaptation of the deermouse *Peromyscus polionotus* to the beach environment of the Florida Gulf and Atlantic coast, a classic example of natural selection for cryptic coloration that protects against predation. Here, the white coat coloration associated with the beach environment involves mutations affecting three different genes, one of which (in *Mc1r*) is a single-amino-acid mutation; the two others involve regulators of gene expression. Examination of the genotype–phenotype relationships enabled reconstruction of the most likely route to the evolution of the complete phenotype, by a stepwise path in which each change increases fitness, in accordance with Darwin's views on the evolution of adaptations.

Another example of the successful genetic dissection of the evolution of novel phenotypes in a new environment with respect to predator pressure is provided by sticklebacks, which have repeatedly invaded freshwater habitats (D Kingsley, pers. comm.). Here, the adaptive phenotypes involve loss of function, an evolutionary phenomenon also discussed by Cronk. Loss of the hindfin is a repeatable characteristic of these populations; a major (but not sole) contributor to this phenotype is a reduction in the expression of the *Pitx1* gene in the hindlimb, due to deletions of portions of the enhancer that evolved independently in different populations. In contrast to this use of new mutations, the loss of the lateral armour plate, which has also evolved in freshwater populations, appears to have involved the repeated spread of the same mutation at a locus controlling expression of the ectodysplasin signaling molecule. The evidence for this comes from the fact that the same haplotype with respect to sequence variants surrounding this locus is found in different populations and is also found at low frequencies in ocean populations.

This population-genetic approach of searching for signatures of selection from the effects of the spread of a selectively favorable variant on patterns of variation at linked sites is an increasingly used tool for identifying both targets of selection at sites with unknown phenotypic effects and the sites associated with changes in a specific phenotype. This provides a paradigm for integrating molecular and developmental approaches with evolutionary genetics methods, and it is likely that we will see an explosive growth in studies of this kind with the advent of new technologies. The theoretical basis for this approach was worked out long ago by Maynard Smith and Haigh (1974), well before the advent of modern methods for screening populations for large numbers of polymorphic markers. This method is being widely used in human evolutionary genetic studies (C Bustamante, pers. comm.) and (complemented by more traditional genetic mapping) in studies of the genetic basis of traits that have evolved during animal and plant domestication, as exemplified by dogs (Quignon et al.; C Bustamante, pers. comm.) chickens (Andersson), maize (J Doebley, pers. comm.), and forest trees (Sederoff et al.).

The joint effects of linkage and selection have long been studied theoretically by evolutionary geneticists interested in the evolutionary origin and maintenance of genetic recombination and sexual reproduction. Barton reviewed the population-genetic processes that are thought to be involved, especially the "Hill–Robertson effect," a probable major player in creating an evolutionary advantage to recombination. This involves the fact that two mutations at different loci which increase fitness are likely to be present in different haplotypes, given the finite sizes of natural populations. In the absence of recombination, the spread of a favorable mutation to a high frequency may be hindered by the presence of one or more harmful mutations in the same genetic background. Similarly, an absence of recombination makes it easier for deleterious mutations scattered across the genome to increase in frequency as a result of genetic drift. A new model of this process makes predictions about levels of variability in nonrecombining genomic regions that appear to be consistent with observations on DNA sequence variation in *Drosophila* (Charlesworth et al.).

The evolutionary interactions between parasites and their hosts are an important source of ongoing selection pressures that are thought by many to provide an advantage to genetic recombination. The importance for medicine and agriculture of understanding the biology of pathogens is another reason for the increasing interest in studies of their evolutionary biology. This topic was covered by several contributors. Plant bacterial pathogens such as *Pseudomonas syringae* use the same type III effectors as many animal pathogens for attacking host cells, and their distribution among different strains provides an excellent system for the study of host–pathogen evolution (J Dangl, pers. comm.). The water flea *Daphnia*, with its recently sequenced genome and facility for clonal reproduction, is an unusually favorable system for the dissection of the genetics and evolution of host–parasite interaction, providing strong evidence for interactions between host and parasite genotypes (Allen and Little). The fact that pathogens are often exceedingly specialized with respect to the hosts that they can infect raises important practical and evolutionary questions. These are being tackled using the evolution of RNA viruses on alternative hosts as a model system (Ogbunugafor et al.).

Experimental models of evolution in the laboratory have long been used to test hypotheses in evolutionary biology, dating back to experiments on artificial selection experiments in mice and *Drosophila* (Greenspan). A spectacular example of this is provided by the 48,000 generation experiment on 12 replicate lines of *Escherichia coli*, started from a single isogenic strain (Barrick and Lenski). This experiment has shown how the fitness of a population in a new environment steadily increases, eventually approaching an asymptotic value as the supply of new mutations that cause adaptations to the environment runs out. Because ancestral populations can be kept as frozen samples, these populations provide a unique window on the time course of evolutionary change that is now being explored with next-generation sequencing technology. This allows the tracking of individual mutations that spread through the population and reveals others that rise to high frequencies and then disappear, probably because of Hill–Robertson interference from the other, more successful mutations. Theoretical studies of "adaptive walks" of this kind, which extend Fisher's classic model of adaptive evolution in a multidimensional phenotype space (Fisher 1930), predict that mutations of relatively large effect are likely to be the main players in the first few hundred generations from the start, and this is being confirmed by genetic analyses of several different evolutionary experiments (Bell).

Experimental evolution also is being extensively applied to the evolution of protein molecules. This is especially relevant to the problem of "irreducible complexity" raised by critics of neo-Darwinism such as Michael Behe (pers. comm.), a participant in the Symposium. How can a polypeptide that performs a function requiring a specific combination of amino acids evolve from a sequence that lacks all of them? This question was beautifully answered by experiments challenging cytochrome P450 from *Bacillus megaterium* to perform new functions (Arnold). By gradually increasing the severity of the challenge, and selecting for induced mutations that meet it, it was possible to evolve a fatty acid hydroxylase into a protein that acted as an efficient propane hydroxylase, involving a stepwise accumulation of 23 mutations. Darwin and Fisher would have been delighted!

Another approach to this problem is to use phylogenetic comparisons to reconstruct an "ancestral sequence" of a protein, synthesize it, and express the protein in an experimental system that permits tests of its functional efficiency (J Thornton, pers. comm.). This has allowed reconstructions of the pathways of functional evolution of steroid hormone receptors. More traditional methods for studying molecular evolution have long been applied to the very complex pathway involved in vertebrate blood coagulation, showing that there has been a progressive increase in complexity from early vertebrates to mammals, often involving gene duplications (RF Doolittle).

Again, this work shows that biological complexity can be built up in a series of graded steps, which was also brought out by the paleontological studies described by Kevin Padian (unpubl.) in his Dorcas Cummings Lecture to the public. The human fossil record has been uncovered by discoveries made since Darwin's time, and it provides further evidence of graded evolution, especially for cranial capacity (White). Genetic studies of human populations are yielding increasing insights into population histories, as well as signatures of selection in relation to local environments, from both mitochondrial (Wallace) and nuclear gene (Lambert and Tishkoff) markers. The genetic and paleontological evidence for an African origin of modern humans is now overwhelming, confirming Darwin's insights of 150 years ago. The sequencing of the genome of Neanderthal humans is in progress and will undoubtedly shed further light on many questions in human evolution at the genetic level (S Pääbo, pers. comm.).

The advent of large-scale genome resequencing projects on populations of humans and other species will increase the amount of data on within-species variability and species differences by several orders of magnitude over the next few years. This presents the new generation of evolutionary biologists with a set of huge challenges: How to organize these data in an accessible fashion, how to interpret them by means of sensible models of the evolutionary processes involved, and how to integrate the results of inferences from model-fitting exercises into programs of experimental work designed to test them. We have in our grasp the potential to answer a large number of questions that have long concerned evolutionary geneticists, especially the interplay among mutation, genetic drift, and natural selection in shaping natural variability and evolutionary divergence at both the genomic and phenotypic levels. The impressive set of posters presented at the meeting, and the talks by the Symposium Fellows, give one great confidence that this challenge will be met.

Evolution is probably the area of biology that probably attracts the most attention outside of the scientific community, and the question of the public understanding of evolution was also well represented at the Symposium. The idea that we are ourselves simply exceedingly complex machines, produced by impersonal evolutionary forces acting over billions of years, is still repugnant to many people, especially in the United States and the Islamic world, where fundamentalist religious beliefs are widely held. It is one of the paradoxes of our time that the country with the greatest concentration of scientific talent in the world is the only developed nation where a large fraction of the population espouses creationism, a doctrine that is tantamount to rejecting the scientific exploration of nature. A series of contributions to the Symposium presented the background to the "Intelligent Design" movement, and the Dover trial that represented a major setback to this attempt to introduce creationism into the U.S. public school curriculum by the back door (Forrest; Miller; E Scott, pers. comm.).

Many questions about the evolution of humans concern mental qualities that seem to divide us sharply even from our nearest primate relatives, especially the use of language and symbolic reasoning. These problems were discussed by several participants (Dennett; M Hauser; S Pinker; both pers. comm.), as well as some provocative extensions of Darwinism to social questions (Ridley; Ferguson). There is an inherent difficulty in studying evolutionary processes when there is no replication of the events involved across independent lineages, as is the case for human language and cognition. There is also no possibility of major experimental manipulations of humans to investigate the genetic control of behavior, in contrast with what can now be done with model organisms (Grant). Ingenious and (sometimes) convincing hypotheses can be proposed, but rigorous tests are problematic. This should not be taken as a reason for questioning the proposition that these aspects of humans have evolved in the same way as other traits, but merely that the nature of the case makes scientific investigation especially difficult. There is, however, hope that the use of genome-wide scans to detect genetic variants associated with individual differences in human behavior and cognition will provide insights into their genetics and the mechanisms involved.

There is a similar lack of evolutionary replication in the origin and early evolution of life itself, but here, laboratory experiments can illuminate the probable nature of the processes involved, with increasing success in providing models of the origins of self-replicating informational molecules, cell membranes, and the translational machinery (Cech; Joyce; Ramakrishnan; Mansy and Szostak).

One aspect of humans that has commonalities with other organisms is our social behavior, in particular "altruism," whereby individual fitness is sacrificed with benefits to others in the social group (this is perhaps less prevalent in human society than we would like to think). The most spectacular examples are found in social insects, where sterile workers comprise most of the members of a colony. The molecular developmental basis of differences among workers in bee colonies is starting to be understood (Toth and Robinson), which should greatly aid in understanding the evolution of these traits. Darwin himself proposed that the adaptations of different types of workers to their roles in the colony, which often reach bizarre extremes (Wilson), are the result of what animal breeders call "family selection": The worker phenotypes enhance the reproductive success of the breeding individuals in the colony, to whom they are usually closely related. This interpretation represents a special case of "kin selection," a process whose importance in social evolution was vigorously defended by Foster. In contrast, Wilson dismissed kin selection and argued that social insects represent superorganisms that have evolved by a form of group selection based on colony-level characteristics. This disagreement probably left the audience somewhat bemused and showed that modern advances have not completely eliminated the ability of evolutionary questions to generate controversy.

Darwin's work went well beyond the evolutionary ideas presented in *Origin*, including a large body of experimental work on plants (Browne). Nevertheless, *The Origin of Species* is his book that captured most public attention. It has, however, frequently been pointed out that Darwin's concept of what constitutes a species was very different from current views. He never arrived at a satisfactory theory of the origin of reproductive isolation, regarded by most modern evolutionary biologists as the primary criterion for speciation in sexual organisms. The species problem was covered by several contributors to the Symposium, although there was regrettably no mention of the spectacular work that has recently been done in *Drosophila* in establishing the molecular genetic basis of reproductive isolation among closely related species and the important role of natural selection in this process (see, e.g., Tang and Presgraves 2009). We are likely to see much more work of this kind, in a wider range of taxa, as more genome sequences become available.

Work toward establishing a systematic database of species identified by "bar coding" using short mitochondrial DNA sequences was described by P Hebert (pers. comm.). Soltis discussed the important role of hybridization followed by polyploidy in plant speciation and documented examples of rapid genomic alterations following very recent events of this kind. Ehrenreich et al. described the remarkable genetic system in *Caenorhabditis elegans*, where two different haplotypes at a pair of tightly linked genes are present in natural populations. Crosses between strains with different haplotypes result in lethality of one-quarter of the F_2 embryos, as a result of a toxic interaction between the two loci. This represents a type of "Dobzhansky–Muller" incompatibility that is normally associated with divergence between two geographically isolated populations that are on the way toward speciation.

I hope that this summary brings out some of the amazing advances in evolutionary biology that have been made since 1959, which are nevertheless consistent with much of what Darwin proposed in 1859. These advances owe much to molecular biology. But what Ernst Mayr said in introducing the 24th Symposium remains true today (Mayr 1960): "We live in an age that places great value on molecular biology. Let me emphasize the equal importance of evolutionary biology. The very survival of man on this globe depends on a correct understanding of the evolutionary forces and their application to man."

REFERENCES

Cavalier-Smith T. 1985. *The evolution of genome size*. Wiley, Chichester, United Kingdom.
Dobzhansky T. 1960. Evolution of genes and genes in evolution. *Cold Spring Harbor Symp Quant Biol* **24:** 15–30.
Fisher RA. 1930. *The genetical theory of natural selection*. Oxford University Press, Oxford.
Maynard Smith J, Haigh J. 1974. The hitch-hiking effect of a favourable gene. *Genet Res* **23:** 23–35.
Mayr E. 1960. Where are we? *Cold Spring Harbor Symp Quant Biol* **24:** 1–14.
Muller HJ. 1949. Redintegration of the symposium on genetics, paleontology, and evolution. In *Genetics, paleontology and evolution* (ed. GL Jepsen et al.), pp. 421–445. Princeton University Press, Princeton, NJ.
Sheppard PM 1960. The evolution of mimicry: A problem in ecology and genetics. *Cold Spring Harbor Symp Quant Biol* **24:** 131–140.
Stebbins GL. 1960. The synthetic approach to organic evolution. *Cold Spring Harbor Symp Quant Biol* **24:** 305–311.
Tang SW, Presgraves DC. 2009. Evolution of the *Drosophila* nuclear pore complex results in multiple hybrid incompatibilities. *Science* **323:** 779–792.

Author Index

A

Allen D.E., 169
Andersson L., 319
Arnold F.H., 41

B

Barbazuk W.B., 215
Barrick J.E., 119
Barton N.H., 187
Behringer R.R., 297
Bell G., 139
Bellott D.W., 345
Betancourt A.J., 177
Boettiger A.N., 275
Browne J., 1
Buggs R.J.A., 215

C

Cande J.D., 275
Cardona A., 235
Carmody R.N., 427
Cech T.R., 11
Charlesworth B., 177, 469
Chase K., 327
Chen C.-H., 297
Cibrián-Jaramillo A., 267
Cretekos C.J., 297
Cronk Q.C.B., 259

D

Davidson E.H., 65
Degnan B.M., 81
Dennett D.C., 435
Doolittle R.F., 35
Doolittle W.F., 197
Dumas L., 375

E

Ehrenreich I.M., 145
Eichler E.E., 355
Erwin D.H., 65

F

Ferguson N., 449
Forrest B., 455
Foster K.R., 403

G

Gerke J.P., 145
Goley E.D., 55
Gordo I., 177
Grande C., 281
Grant S.G.N., 249
Greenspan R.J., 131

H

Hannon G.J., 225
Hartenstein V., 235
Hoekstra H.E., 155

J

Johnson G.S., 327
Joyce G.F., 17

K

Kaiser V.B., 177
Kirst M., 303
Koonin E.V., 205
Kruglyak L., 145
Kunkel T.A., 91

L

Lambert C.A., 395
Lark K.G., 327
Lenski R.E., 119
Levine M., 275
Lindquist S., 103
Linnen C.R., 155
Little T.J., 169

M

Malone C.D., 225
Mansy S.S., 47
Marques-Bonet T., 355
Martienssen R.A., 267
Mathews M.B., 363
McAdams H.H., 55
McBride R.C., 109
Miller K.R., 463
Moczek A.P., 289
Mosher D.S., 327
Myburg A., 303

O

Ogbunugafor C.B., 109
Ostrander E.A., 327

P

Page D.C., 345
Parker H.G., 327
Parrott A.M., 363
Patel N.H., 281
Perry M.W., 275
Puigbò P., 205

Q

Quignon P., 327

R

Ramakrishnan V., 25
Rasweiler IV J.J., 297
Richards G.S., 81
Ridley M.W., 443
Robinson G.E., 419

S

Saalfeld S., 235
Schnable P.S., 215
Schoenebeck J.J., 327
Sederoff R., 303
Shapiro L., 55
Sikela J.M., 375
Soltis D.E., 215
Soltis P.S., 215
Szostak J.W., 47

T

Tishkoff S.A., 395
Tomancak P., 235
Toro E., 55
Toth A.L., 419
Turner P.E., 109

W

Wallace D.C., 383
White T.D., 335
Wilson E.O., 9
Wolf Y.I., 205
Wrangham R.W., 427

Subject Index

1q21.1 genome location, 375–376, 378, 381

A

A Defense of Beanbag Genetics (Haldane), 403
Aardvark, 85
ABC (approximate Bayesian computation), 398
Acanthaceae, 260
Acetyl-CoA, 388, 389
Acoel flatworms, 66, 70
Adaptation
  basis of quantitative variation, 140
  conclusions, 143
  effect of beneficial mutations, 140–141
  gradualist view of evolution, 139
  identifying genes responsible for, 141–142
  in natural populations, 142
  theory of polygenic variation, 139
Adaptive evolution
  background to current research, 41–42
  conclusions, 46
  cytochrome P450 BM3 model, 42–44
  described, 42
  functional promiscuity of proteins and, 45
  novel protein generation by recombination, 45–46
  role of neutral mutations in, 45
Adenomatous polypotis coli (APC), 83–85
Afar Depression, Ethiopia, 335, 340
Africanized honey bees (AHB), 424
African populations. *See* Genetic structure in African populations
African violet (*Saintpaulia*), 264
AGO3, 227–229
Agouti, 165
Agrobacterium, 307
AHB (Africanized honey bees), 424
Alberts, Bruce, 465
Allele swaps, 151
Altruism and sociobiology, 408
Alu (sc*Alu*), 366
Alu RNAs, 369–370
Alu SINE, 363, 366–367
Alu/snaR-related (ASR), 363, 366
Amborella, 270
Amphimedon
  Hedgehog signaling and, 86–87
  Notch pathway and, 82–83
  TGF-β signaling pathway and, 86, 88
  WNT pathway and, 83, 85
Amplicons, 345, 348, 349–350, 351
Angelman syndrome, 390
Ant (*Lasius niger*), 423
Antirrhinum (snapdragon), 264
APC (adenomatous polypotis coli), 83–85
Ape evolution and copy-number variation, 375–377

Apolipoprotein(a), 38
Approximate Bayesian computation (ABC), 398
Arabidopsis sp., 262, 263, 309, 310
Arabidopsis thaliana, 104–105, 304, 413–414
Ardipitheus ramidus, 340
Argonaute, 226–227
Argonaute, 272
Artificial selection. *See* Adaptive evolution
Aspergillus, 106
ASR (*Alu/snaR*-related), 363, 366
Association mapping, 150
ATP, 388, 389
Au. africanus and hominid diversity, 339–340
Aubergine, 227–229
Australopithecus, 336, 340
Autism, 378
Aveling, Edward, 1
Axin, 83

B

Bacterial cell cycle of *Caulobacter crescentus*
  conclusions, 62
  entry into S phase
    master regulators' functions, 55–56
    modes of DnaA and CtrA regulation, 56–58
    prevention of improper replication initiation in other bacteria, 58
    replication licensing in eukaryotes, 58
  final cell division stage
    MipZ communication of information, 60–61
    systems used by other bacteria, 61–62
  life cycle, 55
  post-S phase
    chromosome segregation process, 59
    discovery that the chromosome is structured, 59–60
    DNA mobility within the nucleus, 60
    timing of replication and segregation, 58–59
Bantu expansion, 395
Bar locus, 355
Basal chordates, 36–37
Basal metabolic rate (BMR), 431
Bat limb development. *See Carollia perspillata* (short-tailed fruit bat)
Bauhinia, 261
Bayesian clustering, 397
β-catenin, 84, 85
Beckwith-Wiedemann syndrome, 390
Beetle horns. *See* Horned beetles

Behavior and evolution. *See* Cultural evolution of words; Evo-devo and the evolution of social behavior; Sociobiology
Behe, Michael, 455, 466
Benzer, Seymour, 131
Bioenergetics and epigenomics
  bioenergetic hypothesis for common diseases, 390
  calorie intake and cellular growth and reproduction, 387–389
  cellular evolution and, 385–386
  conclusions, 392
  epigenome origins and, 387–389
  epigenomic disease and, 389–391
  in human origins and disease, 386–387
  intraspecific energy resource adaptation, 383–385
  mtDNA mutations accumulations, 387
  paradigm in animal evolution and human disease, 391–392
Biola University, 458
Biological catalysis
  "disappearing RNA" hypothesis, 14–15
  essential activities of genetic material, 11
  evidence of new catalytic RNPs, 15
  questions regarding evolution of chirality establishment, 12
  encapsulation of self-replicators, 12–13
  environmental conditions, 12
  first self-replicating molecules, 12
  RNP enzyme defined, 11–12
  scenarios for evolution of an RNP World
    early binding of small molecules, 13
    evolution of an RNP enzyme, 14
    evolution of telomerase RNP, 14
    middle step of RNA synthesis, 13
    pathways, 13
Biological Species Concept (BSC), 198
Biston betularia (peppered moth), 159, 164
BLADE ON PETIOLE (*BOP*), 262
Blood coagulation evolution in vertebrates
  arrangements of subsidiary domains, 35
  basal chordates, 36–37
  challenges to "irreducible complexity" argument, 465–467
  complexity of system, 39
  conclusions, 39–40
  fibrinogen γ chain, 38
  jawless fish, 37–38
  origins of clotting, 38–39
  pathway in humans, 35, 36f
  search for relevant genes in genome database, 35–36, 37t
BMR (basal metabolic rate), 431
Boyd, E., 443
Bradley, Walter, 456
Brassicaceae, 262
Brinker (*brk*), 277

477

Brosnan, Sarah, 446
BSC (Biological Species Concept), 198
Bulk segregant analysis, 151
Butterfly pea (*Clitoria ternatea*), 264
BYxRM cross in yeast
    gene expression traits showing complex inheritance, 146–147
    genetic dissection of traits, 147–149
    molecular mechanisms of gene expression variation, 147
    QTL hot spots importance, 149
    summary, 149

C

Cadia, 264, 265
Caenorhabditis elegans, 238, 422
Cameron, Julia Margaret, 2
Candida albicans, 106
CAN family, 86
Capitalism and Democracy (Schumpeter), 449
Capitella, 88
Carbonaria, 164
Carollia perspillata (short-tailed fruit bat)
    conclusions, 301
    enhancer switch and deletion by gene targeting, 300–301
    gene expression during limb development, 299–300
    limb diversity and development between mice and bats, 297–299
    Prx1, 300–301
    use as a model organism, 297
Catarrhine ancestor of *snaR* (CAS), 363, 365–366
Caulobacter crescentus cell cycle
    conclusions, 62
    entry into S phase
        master regulators' functions, 55–56
        modes of DnaA and CtrA regulation, 56–58
        prevention of improper replication initiation in other bacteria, 58
        replication licensing in eukaryotes, 58
    final cell division stage
        MipZ communication of information, 60–61
        systems used by other bacteria, 61–62
    life cycle, 55
    post-S phase
        chromosome segregation process, 59
        discovery that the chromosome is structured, 59–60
        DNA mobility within the nucleus, 60
        timing of replication and segregation, 58–59
Center for Science and Culture (CSC), 455
Center for the Renewal of Science and Culture (CRSC), 456
Chaetodipus intermedius (pocket mice), 160
Chaperone proteins. *See* Protein folding and evolutionary change
Chapman, Bruce, 455
Charcot-Marie-Tooth, 390
Choanoflagellates, 66
Chondrodysplastic breeds, 330–331
Chordin, 86
Chorionic gonadotropin (CG) genes, 370–371

Ciona intestinalis (sea squirt), 37
Clitoria ternatea (butterfly pea), 264
Cnidarians, 66
Coat color in mammals, 320–322. *See also* Phenotypic evolution in domestic animals
Cognition
    behavioral adaptation to the environment and, 249–250
    behavioral roles for synapse evolution, 255
    conclusions, 256
    defined, 249
    DUF1220 domains link to cognitive brain disease, 378–380
    evolutionary basis of synapse composition, 254
    implications of postsynaptic complexity as a source of synapse diversity, 255–256
    molecular evolution and development of the postsynaptic proteome, 252–254
    molecular organization of the postsynaptic proteome, 250–252
    neurophysiological consequences of expansions in postsynaptic complexity, 254–255
    patterns of neural activity and, 250
    in unicellular organisms, 249–250
Cold Spring Harbor Laboratory, 335, 336–338, 403, 469
Complementation and robustness, 111–113
Concerted evolution, 217
Continuous in vitro evolution, 18–19
Cooking and a high-quality diet
    conclusions, 432
    digestibility of food and, 429–430
    ease of chewing and, 429
    energy expenditure for humans versus chimpanzees, 427
    evolutionary adaptation to cooked food, 431–432
    heat effects on protein, 430
    human need to consume high-energy-density foods, 427
    meat eating's importance in human evolution, 427–428
    meatless diet's impact on energy status and reproductive function, 428
    net energy gain and
        lowering of diet-induced thermogenesis, 430
        lowering of metabolic costs of immune disease, 431
    raw diet's impact on energy status and reproductive function, 428
Copy-number gain and loss, 359, 375–377
Cos2, 87
Cosmides, Leda, 446
Creationism. *See* Intelligent Design
Creation science, 455–456. *See also* Intelligent Design
Crenicichla alta (pike predator), 159
CRSC (Center for the Renewal of Science and Culture), 456
Cryogenian Period, 67
CSC (Center for Science and Culture), 455
Ctenophores, 66
CtrA, 56–58
Cultural evolution of words
    conclusions, 440–441
    culture as a major transition in evolution, 435–436
    digitized cultural elements and, 438–440
    memetic perspective, 440
    nature of words, 436–438
CYCLOIDEA (*CYC*), 263
Cystoviridae, 113
Cytochrome P450 BM3 model
    creation of $P450_{PMO}$, 43
    $P450_{PMO}$ evolutionary trajectory, 43–44
    usefulness in experiments, 42–43

D

Dachshund (*dac*), 291–292. *See also* Horned beetles
Daphnia magna–Pasteuria ramosa host–parasite model. *See* Host–parasite coevolution
Dart, Raymond, 336
Darwin, Charles
    artificial selection arguments, 41
    celebrity due to his theory of evolution, 2
    documentation through letter writing, 3–4
    love of experimentation, 2–3
    recruitment of family, 5
    on regressive evolution, 259
    reliance on commonplace features of Victorian life, 4–5
    research on natural selection, 1
    research on sexual selection and human emotions, 5–6
    view on hominids, 335–336
    work environment, 3, 6
Darwin, Emma, 5
Darwin, Francis, 5, 6
Darwin, Henrietta, 5
Darwin's Black Box, 466
Dawkins, Richard, 405, 436, 443
Descent of Man (Darwin), 3, 5
Deconstructing Intelligent Design
    absence of a scientific ID theory, 464
    antievolution movements, 463
    appeal of the "design" argument, 465
    challenges to "irreducible complexity" argument, 465–467
    conclusions, 467–468
    court case testing legal merits of ID, 464
    design described in evolutionary terms, 465
    fused chromosomes evidence supporting evolution, 467
    ID's public relations successes, 464–465
    ID's relation to creationism, 464
Deer mouse (*Peromyscus maniculatus*), 161
Delta/Serrate/Lag domain (DSL), 82–83
Dembski, William, 455, 456, 458, 459, 460, 464
Denton, Michael, 456
De Queiroz, K., 198
DeRosier, David, 465
Destruction complex, 83–85
Detarieae, 263
Deuterostomes, 66
Deuterostomia, 281. *See also* Left/right asymmetry in bilateria
De Vries, H., 139
dGRN kernels, 72–73. *See also* Eumetazoan evolution
DI (Discovery Institute), 455
Diabetes, 390
Diastereomers, 12

SUBJECT INDEX

Dictyostelium, 85
Dictyostelium discoideum, 412
Diet-induced thermogenesis (DIT), 430
Directed (adapted) evolution
 background to current research, 41–42
 conclusions, 46
 cytochrome P450 BM3 model, 42–44
 described, 42
 functional promiscuity of proteins and, 45
 novel protein generation by recombination, 45–46
 role of neutral mutations in, 45
Discovery Institute (DI), 455
Dishevelled (Dsh), 83
Dispatched (Disp), 86
Distalless (*Dll*), 262, 291–292. *See also* Horned beetles
Dlg proteins, 255
DnaA, 56–58
DNA replication fidelity
 conclusions, 98–99
 fidelity of DNA synthesis
  extrinsic proofreading, 96
  insertion–deletion errors, 94
  major replicative polymerases, 93
  nucleotide selectivity, 93–94
  proofreading by replicative polymerases, 94–95
 fidelity of repair polymerases
  mismatch repair, 98
  repair pathways, 96–97
  translesion synthesis, 97–98
 polymerases' function, 91–93
 replication asymmetry and fidelity, 96
Dobzhansky, Theodosius, 336, 463, 469
Dogs. *See* Fine mapping of leg morphology in dogs
Dollo's Law, 261–262
Dominant white (*I*), 320. *See also* Phenotypic evolution in domestic animals
Doolittle, Russell, 466
Dorsoventral (DV) patterning in *Drosophila*
 control by Dorsal, 275
 shadow enhancers, 277
 sim enhancer evolution, 275–276
 sim enhancer location constancy, 276–277
 stalled pol II and, 278
 transcriptional synchrony, 277–278
Doushanto formation biota, 68–70
Drosophila
 appendage formation, 291
 brain architecture (*see Drosophila* brain development)
 co-option and the foraging gene, 422
 dorsoventral patterning in (*see* Dorsoventral [DV] patterning in *Drosophila*)
 gene network studies (*see* Gene networks)
 Hsp90 buffering of polymorphisms in, 104–105
 regressive evolution and, 262
 signaling pathway and, 83, 87
 transposon control pathways in (*see* Transposon control pathways in *Drosophila*)
Drosophila brain development
 conclusions, 247
 connectivity and network motifs in the VNC microvolume, 242, 244

lineage-based analysis of, 236–238
microcircuitry
 analysis techniques, 238
 data analysis and reconstruction, 238–239
 density, branching, and directionality of neurites, 242, 243f
 neurite profiles and synapses distribution, 240–242
 quantitative analysis of connectivity in microvolumes, 240
 nervous system structure, 235–236
 neuropile architecture in mammalian cortex and fly brain, 244–246
 phylogeny of neuropile architecture, 246–247
Dsh (Dishevelled), 83
DSL (Delta/Serrate/Lag domain), 82–83
Dubois, Eugene, 336
DUF1220 domains
 conclusions, 381
 core duplicons in the human genome and, 380
 finishing of the 1q21.1 genome sequence, 381
 gene-copy-number gain and loss, 375–376
 human lineage-specific amplification of, 377–378
 implications of instability in 1q21.1, 381
 lineage-specific gene-copy-number variation, 376–377
 link to brain size, 378
 link to cognitive brain disease, 378–380
 model linking sequences to diseases and brain evolution, 380–381
Duffy-O, 396
Duplicons, 356–358. *See also* Human segmental duplications

E

E (Extension), 320. *See also* Phenotypic evolution in domestic animals
Ecdysozoa, 281. *See also* Left/right asymmetry in bilateria
Ectdysoplasin (*Eda*), 142, 156, 164
Edicaran biota, 70–71
Edicaran Period, 65, 67
Edwards v. Aguillard, 456, 457
EF-G, 27
EF-Tu, 27, 28
EGF (epidermal growth factor), 35
Eldredge, Niles, 339
Eliminative pluralism, 203
Enantiomers, 12
Endler, J.A., 155
Engrailed (*En*), 262
Environmental genomics, 313
Ep, 320
Ephedra, 270
Epidermal growth factor (EGF), 35
Epigenomics. *See* Bioenergetics and epigenomics
Epperson v. Arkansas, 456
Escherichia coli population diversity study
 conclusions, 128
 emerging technologies used in evolution experiments, 119
 materials and methods, 120–121
 results
  changes in genetic diversity over time, 127–128

 distinguishing SNPs from sequencing errors, 123–124
 examination of genetic diversity through data sets, 122–123
 limits to mutations in evolving populations, 121–122
 SNP predictions, 124–127
E site, 31
Eucalyptus, 306
Eucalyptus sp., 304, 307, 311
Eumetazoan evolution
 bilateria fossil record
  Doushanto formation biota, 68–70
  earliest evidence of Metazoans, 68
  Edicaran biota, 70
  Edicaran trace fossils, 70–71
  evidence of environmental effects, 67–68
  summary, 71
 bilaterian developmental evolutionary events, 65–66
 fossil record information, 65
 integration of fossil and developmental data
  Bilaterian evolution, 71–72
  developmental characteristics of dGRNs, 71
  Eumetazoan lineages and dGRN kernels, 72–73
 mechanistic aspects of dGRN evolution
  building of dGRNs, 75–76
  DNA level change, 75
  rates of morphological change, 76–77
 Metazoan phylogeny and molecular clocks, 66–67
 stages of evolution of dGRNs, 73–75
 summary, 77–78
European wild boar, 319. *See also* Phenotypic evolution in domestic animals
Evo-devo and the evolution of social behavior
 challenges to, 420–421
 conclusions, 424
 eusociality in insects
  co-option, 421–422
  gene regulation, 423–424
  genetic tool kits, 423
  modularity, 422–423
  overview of concepts, 421
  study of social life in molecular terms, 421
 genetic architectures for developmental and behavioral traits, 420
 molecular analyses of behavior and, 419–420
Evolutionary biology
 24th Symposium at CSHL, 469
 biological catalysis (*see* Biological catalysis)
 evolutionary process in biology, 9–10
 molecular biology's impact on
  duplications and copy-number variants, 470
  experimental models of evolution, 472
  genome resequencing projects, 472–473
  genome sequencing, 470
  joint effects of linkage and selection, 472
  mutational mechanisms, 470

Evolutionary biology (*continued*)
    opportunistic nature of evolutionary change, 470–471
    phylogeny analysis, 470
    public understanding of evolution, 473
    role of changes in gene expression in morphological evolution, 471
    sex chromosome systems, 470
    social behavior study, 473
    species problem, 473–474
  natural selection and, 9
  problem solvers versus naturalists, 10
Evolutionary economics. *See* Financial history from an evolutionary perspective
Evolution: A Theory in Crisis (Denton), 456
Evolution of development. *See* Evo-devo and the evolution of social behavior
Exchange and cultural evolution
  collective accumulation of knowledge, 444–445
  conclusions, 447
  cultural versus biological evolution, 443
  decisive factors in social evolution, 443
  entities that compete for survival in cultural evolution, 443
  exchange versus reciprocity, 446
  horizontal exchange, 444
  impact of absence of exchange on Neanderthals, 445
  interaction as leading to innovation, 447
  sexual division of labor, 446–447
  specialization and exchange, 445–446
Expression of Emotions in Man and Animals (Darwin), 3, 6
Extension (E), 320, 321. *See also* Phenotypic evolution in domestic animals

F

Factors XI and XII, 38
Fatty acids
  duplex melting inside fatty-acid-based vesicles, 51
  protocells and, 48, 49–50
  thermal stability of vesicles, 51
Fgl8, 299–300
Fibrinogen γ chain, 38
Fibrinogen-related domains (FREDs), 37
Financial history from an evolutionary perspective
  concentration versus speciation, 450–452
  conclusions and future prospects, 452–454
  extinction rate of businesses, 449–450
  features shared by financial and evolutionary systems, 450
  industrial capitalism, 449
  regulation's impact on financial services, 452
Finch (*Geospiza fortis*), 142, 157–158
Fine mapping of leg morphology in dogs
  canine model system, 327
  conclusions, 332
  information from pedigree studies, 327–328
  materials and methods, 328
  study results
    identification of associated haplotypes, 330–331
    LD reduction techniques, 330
    sequencing to discover new SNPs, 331–332
    SNPs identification, 328–330
Fisher, R.A., 134, 139
Flicker, 261, 264
Follistatin, 86
foraging, 422
Forest of Life (FOL). *See* Phylogenetic Forest of Life
Forest trees evolution
  application of genomics in forest trees, 305–306
  areas of genome-related investigation in, 303
  conclusions and future prospects, 312–313
  diversity and domestication of, 303–304
  forward genetics approaches to study, 306
  genomes being studied, 307–308
  reverse genetics approaches to study, 307
  somatic sector analysis, 307
  transcriptomics
    adapative stress response, 308–309
    association studies in tree populations, 310–311
    genetical genomics, 311–312
    integration of quantitative and molecular genomics, 310
    transcript changes in development, 309
    wood formation, 309–310
  wood development and evolution, 304
Forrest, Barbara, 464
Foundation for Thought and Ethics (FTE), 456
FOXOs, 389
FREDs (fibrinogen-related domains), 37
Fringe genes, 82
Frizzled (Fzd), 83
FtsZ, 60
Fundamental unit of evolution (FUE), 210–212
FY gene, 395–396

G

Gardner, A., 409
Gasterosteus aculeatus (stickleback), 142, 156
Gaudi, Antonio, 436
GcrA, 56
Gene networks
  classical models of selection, 134
  concept of a normal system state, 136
  conclusions, 136–137
  expression differences used to study discrepant genes, 131–133
  flexibility, robustness, and degeneracy in, 135–136
  large-effect mutations, 134
  reconciling views on modes of selection, 135
  relationship between selected variants and induced mutants, 135
  technology used to study available genes, 133
Genetic recombination and molecular evolution
  discussion and future prospects, 183–184
  Hill–Robertson effects explained, 177–179
  reformulation of background selection model regarding variability, 179–182
  relation of recombination to genetic variation, 179
  selection in low recombination genomic regions, 182
Genetic structure in African populations
  Bayesian clustering, 397
  conclusions, 399–400
  demographic history modeling, 398–399
  genetic and phenotypic history, 395
  phylogenetic analyses, 397–398
  population structure variation and, 396
  shaping of patterns of genetic variation, 395
  timescales related to population structure, 396–397
GENETREE software, 398
Geospiza fortis (finch), 142, 157–158
Gesneriaceae, 264
Gli/Ci transcription factors, 87
Gloxinia (*Sinningia*), 264
Glycerol nucleic acid (GNA), 48–49
Glycogen synthase kinase (GSK), 83–85
Glycosylation in Notch pathway, 82
Gnetophyta, 270
Gnetum, 270
Goldschmidt, R., 139
Gould, Stephen J., 336, 339, 342
Grafen, A., 409
Green beard gene, 406, 408
Groucho, 84
Group adaptation, 409–410
Group selection and sociobiology, 408
GSK (glycogen synthase kinase), 83–85
Guppies (*Poecilia reticulata*), 159
Gypsy, 228–229

H

Hagfish, 37–38
Haldane, J.B.S., 134, 403, 406
Hamilton, W.D., 405, 406, 408, 410
Hayek, Friedrich, 443, 444
hChorionic gonadotropin (hCG) genes, 370–371
HDACs, 84
Heat effects on proteins, 430
Heat-shock proteins. *See* Protein folding and evolutionary change
Hedgehog interference protein (Hhip), 87
Hedgehog signaling, 86–87
Hedgling (Hling), 86
Helianthus (sunflower), 164
Helminthorhaphe, 70
Heterochrony, 262
hexamerin, 423
Hhip (Hedgehog interference protein), 87
High-molecular-weight kininogen (HMWK), 38
Hildferding, Rudolf, 450
Hill–Robertson effects (HR), 177–179, 188–189
Hirsch, Jerry, 131, 134
Hling (Hedgling), 86
H.M.S. Beagle, 1, 5
HMWK (high-molecular-weight kininogen), 38

Hog domain, 88
Holldobler, B., 410
Homo erectus, 336
Homo floresiensis, 343
Homoplasy, 260
Homoscleromorph sponges, 66
Homothorax (*hth*), 291–292. *See also* Horned beetles
Honey bees, 423
Hooker, Joseph, 5
Horned beetles
    conclusions, 294–295
    exaptation and the origin of adult thoracic horns, 293–294
    natural history of, 289, 290f
    ontogeny of horns, 289, 291
    regulation of horn growth, 291–292
    regulation of pupal remodeling, 292–293
    trade-offs during development and evolution of horned beetles, 294
Host–parasite coevolution
    Daphnia studies
        characterization of the immunome, 172–173
        coevolution potential, 172
        purpose of studying, 171–172
        experimental genomics applied, 173–174
    lack of a thorough understanding of the coevolutionary dynamic, 169–170
    methods of gene discovery
        gene expression use, 171
        phylogenetic candidate gene approaches, 170
        quantitative trait loci use, 170–171
    molecular evolution analysis applied to, 173
Hoxd13, 299–300
HR (Hill–Robertson) effects, 177–179, 188–189
Hsp90. *See* Protein folding and evolutionary change
Human evolution and DUF1220 domains. *See* DUF1220 domains
Human origins and evolution
    Afar Depression discoveries, 340–341
    Au. africanus and hominid diversity systematics, 339–340
    characterization of hominid paleontology, 336
    conclusions and future prospects, 343
    Darwin's view on hominids, 335–336
    evolutionary species concept, 338
    focus of 1950 CSH symposium, 336–338
    hominid fossil discoveries, 336
    hominids in the modern synthesis, 336
    hominid species debate, 341–343
    On the Origin of Species, 335
Human segmental duplications
    Bar locus, 355
    compared to mouse properties, 355–356
    conclusions, 359–360
    disease consequences and copy-number variation, 359
    duplicons and duplication organization, 356–358
    primate comparisons, 356
    primate sequencing of LCR16a, 358
Hutchinson-Guildford progeria, 390

Huxley, Julian, 336
Huxley, Thomas Henry, 2, 3, 139
Hybrid linkage/association approaches, 150–151
Hydra vulgaris, 82–83
Hymenoptera, 410–412. *See also* Sociobiology

I

I (Dominant white), 320. *See also* Phenotypic evolution in domestic animals
Insects and social behavior
    co-option, 421–422
    gene regulation, 423–424
    genetic tool kits, 423
    modularity, 422–423
    overview of concepts, 421
    study of social life in molecular terms, 421
Intelligent Design (ID)
    basis in creationism, 455
    development from creation science, 455–456
    legal case challenging the teachings of, 455
    proponents' acknowledgment of lack of scientific basis, 460–461
    rejection of the methodological naturalism of science, 459
    revealed as creationism, 456–458
    science-based analysis of
        absence of a scientific ID theory, 464
        antievolution movements, 463
        appeal of the "design" argument, 465
        challenges to "irreducible complexity" argument, 465–467
        conclusions, 467–468
        court case testing legal merits of ID, 464
        design described in evolutionary terms, 465
        fused chromosomes evidence supporting evolution, 467
        ID's public relations successes, 464–465
        ID's relation to creationism, 464
    as sectarian Christianity, 459–460
    as supernaturalist religion, 458–459
Intelligent Design: The Bridge between Science and Theology (Dembski), 459, 460
Internet, 447
Introgression and targeted recombination, 151
Ipomoea, 262
Isaac, Glyn, 446

J

JAGGED, 262
Jawless fish, 37–38
Johnson, Phillip, 457, 458, 460, 464
Jones, John E. III, 464
Journal of Researches (Darwin), 1

K

Kansas Board of Education, 463
Kenyon, Dean H., 457

Kimberella, 65, 70
KIT, 320–322. *See also* Phenotypic evolution in domestic animals
Kitzmiller et al. v. Dover Area School District, 455, 460, 463
Kringles, 35, 38, 39
Kuhn, S.L., 446, 447

L

Lahn, Bruce, 347
Laminopathies, 390
Lamprey, 37–38
Language and culture. *See* Cultural evolution of words
Large White pig, 319. *See also* Phenotypic evolution in domestic animals
La Sagrada Famiglia, 436
Lasius niger (ant), 423
Last Human, 339
Law of Comparative Advantage, 445
LCR16a, 358
LD (linkage disequilibrium), 327. *See also* Fine mapping of leg morphology in dogs
LEAFY, 262
Left/right asymmetry in bilateria
    components of the LR pathway, 281–282
    conclusions, 286
    DV inversion hypothesis and, 285–286
    Nodal pathway in bilateria, 283–284
    Nodal-Pitx gene cascade conservation, 284–285
    in snails, 282–283
Lefty, 281. *See also* Left/right asymmetry in bilateria
Leg morphology in domestic dogs. *See* Fine mapping of leg morphology in dogs
Levine, Joseph S., 463
Linanthus sect. *Leptosiphon*, 261
Linaria (toadflax), 264
Linkage disequilibrium (LD), 327. *See also* Fine mapping of leg morphology in dogs
Linkage mapping, 150
Lo, Andrew, 449, 452
Long interspersed nucleotide element (LINE), 229
Lophotrochozoa, 281. *See also* Left/right asymmetry in bilateria
Loss/reversal/regain (LRR), 260. *See also* Regressive evolution
Lottia, 88
LRP5/6, 83
Luskin, Casey, 458
Lymnaea stagnalis (snail), 282–283

M

MacCready, Paul, 435
MAGUK-associated signaling complexes (MASCs), 250–252. *See also* Cognition
MAGUKs (membrane-associated guanylate kinases), 250–252. *See also* Cognition
Maillard reaction, 430
malvolio, 420, 422
Mammalian diversity studied through bats. *See Carollia perspillata*
Marinoan glaciation, 67, 68

MASCs (MAGUK-associated signaling complexes), 250–252. *See also* Cognition
Mastermind, 83
Maynard Smith, J., 405, 406, 435
Mayr, Ernst, 198, 336, 340, 403, 469, 474
McLean v. Arkansas Board of Education, 455–456
Meat tenderness and cooking, 429
Medicago, 261
Melamspora medusae, 308
Melanocortin 1 receptor (MC1R), 321
Membrane-associated guanylate kinases (MAGUKs), 250–252. *See also* Cognition
Memes, 436, 437, 440
Metabolic syndrome, 390
Metapopulation lineage concept, 198
Metazoan phylogeny and molecular clocks, 66–67
Metazoan signaling pathways evolution
 conclusions, 88–89
 Hedgehog signaling, 86–87
 Notch pathway, 82–83
 overview of signaling analysis, 81–82
 receptor and ligand evolution, 87–88
 TGF-β signaling pathway, 85–86
 WNT pathway, 83–85
Meyer, Stephen C., 455
Mice and bats
 human segmental duplications compared to mouse properties, 355–356
 limb diversity and development between, 297–299
Microarchitecture of the *Drosophila* brain. *See Drosophila* brain development
Middle Awash study, 340–341
MipZ, 60–61
Mismatch repair (MMR), 98–99
Mitochondrial DNA (mtDNA), 383. *See also* Bioenergetics and epigenomics
Mitochondrial permeability transition pore (mtPTP), 386
Mollusca, 70
Monosiga, 83, 85, 87
Morgan, Thomas Hunt, 345
Muller, Hermann, 345, 346
Muller's ratchet, 444
Mycosphaerella cryptica, 308
Myristoleic acid model, 50, 51–52
Mystery of Life's Origins: Reassessing Current Theories (Thaxton, Bradley, and Olson), 456
Myths of Human Evolution (Eldredge and Tattersall), 339
Myxococcus xanthus, 119, 412

N

NAHR (nonallelic homologous recombination), 380
Nasonia vitripennis (wasp), 412
Natural History (Gould), 339
Naturalists, 10
Natural selection and wild genotypes and phenotypes
 accounting for adaptation, 155
 changes in allele frequencies
  genotypes, 159
  phenotypes, 159–160
  conclusions, 165
 fitness differences
  genotypes, 156
  phenotypes, 156–158
 identifying the agent of selection, 162–163
 integration across levels and timescales of selection, 163–165
 selection estimates based on DNA sequence data, 161–162
 spatial patterns of allele frequencies and phenotype means, 160
Natural selection's role in biology, 1, 9
Nature, 339
Nearly universal trees (NUTs), 205, 206–207
Nelson, Paul, 457, 460
Nelson, Richard, 443
Nematostella
 Hedgehog signaling and, 87
 TGF-β signaling pathway and, 88
 WNT pathway and, 85
Neomorphisms, 263
Neoproterozoic Period, 65, 67
NF90 and NF110, 363–364
NICD (Notch intercellular domain), 82–83
NMDA receptor complex (NRC), 250–252. *See also* Cognition
N-methyl-D-aspartate (NMDA), 250–252. *See also* Cognition
Nodal pathway, 281, 283–284. *See also* Left/right asymmetry in bilateria
Noggin, 86
Nonallelic homologous recombination (NAHR), 380
Notch (*N*), 262
Notch intercellular domain (NICD), 82–83
Notch pathway, 82–83
Nowak, M.A., 408
NRC (NMDA receptor complex), 250–252. *See also* Cognition
NRC/MASC, 255
Nuclear DNA (nDNA), 383. *See also* Bioenergetics and epigenomics
Numb, 83
NUTs (nearly universal trees), 205, 206–207

O

Of Pandas and People (Kenyon and Davis), 456, 457–458, 464, 466
Ohio Board of Education, 463
Ohno, Susumu, 346, 349
Oligogenic view of adaptation
 adaptation in natural populations, 142
 basis of quantitative variation, 140
 conclusions, 143
 effect of beneficial mutations, 140–141
 gradualist view of evolution, 139
 identifying genes responsible for adaptation, 141–142
 theory of polygenic variation, 139
Onthophagus, 292. *See also* Horned beetles
On the Origin of Species (Darwin), 1, 5, 319, 335
OXPHOS, 384, 386, 387, 388

P

P450 BM3 model
 creation of $P450_{PMO}$, 43
 $P450_{PMO}$ evolutionary trajectory, 43–44
 usefulness in experiments, 42–43
Padian, Kevin, 464
PAN (plasminogen-apple-nematode), 35
Parallel evolution, 260
Parallel innovation, 260
Paranthropus robustus, 336
Pastrana, Julia, 2
Patched (Ptch), 86
PCD (programmed cell death), 292–293
Pennock, Robert, 458
Peppered moth (*Biston betularia*), 159, 164
Peptidyl transferase center (PTC), 25, 27, 29–30
Peromyscus (wood mouse), 142, 160, 163
Peromyscus maniculatus (deer mouse), 161
PGC-1α, 389
Phenotypic evolution in domestic animals
 genetic study of muscle and fat using QTL mapping, 322–324
 selection of coat color between wild and domesticated pigs, 320–322
 study description, 319
Phospholipid membranes, 49–50
Phylogenetic Forest of Life (FOL)
 arguments for fundamental units of evolution, 211–212
 concept of the TOL, 205
 conclusions, 212
 fundamental unit of evolution defined, 210–211
 phylogenetic depth and inconsistency between trees, 208
 phylogenetic trees studied with FOL and NUTs, 206–207
 TOL concept and comparative genomics, 205–206
 a tree as an isomorphous representation of replication history, 209–210
 trends in the FOL, 208–209
Pike predator (*Crenicichla alta*), 159
Piltdown forgery, 336
Pines, 306
Pinus sp., 307
Pithecanthropus erectus, 336
Pitx2, 281, 284–285. *See also* Left/right asymmetry in bilateria
Piwi, 227–229
Piwi-interacting RNAs (piRNAs), 227–229. *See also* Transposon control pathways in *Drosophila*
Plasminogen-apple-nematode (PAN), 35
Plasmodium vivax, 396
Pocket mice (*Chaetodipus intermedius*), 160
Poecilia reticulata (guppies), 159
Polistes canadensis (wasp), 423
Polistes metricus (wasp), 422–423
Polyploid populations and evolution
 conclusions and future prospects, 221
 observed recurrent formation of the same polyploid species, 215
 Tragopogon system
  compared to well-studied systems, 220–221
  genomic approaches to studying, 219
  history of the polyploids, 216
  homeolog loss and gene silencing, 217–218
  mechanism of gene loss, 219
  multiple origins of the polyploids, 216–217

SUBJECT INDEX

rDNA loci and concerted evolution, 217
synthetic lines creation, 217
tissue-specific silencing, 219
Poplars, 306
Populus sp., 307, 309, 310, 311
Populus tremuloides (quaking aspen), 303
Portuguese water dog (PWD), 327. *See also* Fine mapping of leg morphology in dogs
Power of Movement in Plants (Darwin), 6
Prader-Willi syndrome, 390
Prekallikrein, 38
Price, George, 405
Prions, 107–108
Programmed cell death (PCD), 292–293
Prokaryotic systematics and evolution
adopting a species-free approach, 203
approaches to studying the existence of bacteria species, 199–200
arguments against applying species category realism to prokaryotes, 199–200
arguments against the Tree of Life, 201–203
contested history of the prokaryote/eukaryote dichotomy, 200–201
eliminative pluralism, 203
flaws in defining species disjunctively, 198
ribo-essentialism and, 198, 201
species taxon realism versus species category antirealism, 197–198
tree thinking and metapopulation lineage concept, 198
typological versus population thinking, 198
understanding that pattern is comes before process, 203
Protein folding and evolutionary change
heat-shock proteins' function, 103–104
Hsp90 buffering of polymorphisms, 104–105
Hsp90's chaperoning of new mutants, 105
Hsp90's chaperoning of proteins that respond to mutations, 106
identification of genetic polymorphisms affected by Hsp90, 106–107
prions, 107–108
Protocells
compartment division process, 52–53
conclusions, 53
nonenzymatic template copying, 48–49
nucleic acid replication, 48
nutrient acquisition by heterogeneous membrane composition, 50
heterotrophy, 50
permeability of membranes, 49–50
spatial compartmentalization and, 47, 48f
strand separation
duplex melting inside fatty-acid-based vesicles, 51
means of achieving, 50–51
thermal stability of fatty acid vesicles, 51
vesicle permeability at high temperatures, 51–52
synthesis of, 48
template copying inside vesicles, 50
Protochordates, 36–37
Protostomes, 66
Prx1, 300–301
[*PsI*⁺] prion, 107–108
Pseudomonas, 113
Pseudomonas aeruginosa, 412
PTC (peptidyl transferase center), 25, 27, 29–30
Ptch (Patched), 86
PWD (Portuguese water dog), 327. *See also* Fine mapping of leg morphology in dogs

Q

Q81X162, 378
Quaking aspen (*Populus tremuloides*), 303
Quantitative Biology Symposium, 403
Quantitative trait loci (QTL), 140, 170–171
Queller, C., 410

R

R3C ligase, 19
Read, Leonard, 444
Recent African Origin (RAO) model, 395
Reciprocal hemizygosity, 151
Recollections of the Development of My Mind and Character (Darwin), 1
Recombination. *See* Genetic recombination and molecular evolution; Sex and recombination
Redwoods, 304
Regressive evolution
conclusions, 265
cryptic innovation, 265
ecological and developmental factors, 261
gain-/loss-of-function axis, 262–263
heterochrony and, 262
importance of, 259
irreversibility of loss, 261–262
loss, reversal, and regain defined, 259–260
loss-with-vestige, 263
phylogenetic versus developmental approaches to loss and reversal, 260–261
reassessment at the morphological and molecular developmental level, 265
reversals from monosymmetry to poly-symmetry in flowers, 264–265
soft versus hard wiring, 263–264
Reilander, Oscar, 6
Rensch, Bernhard, 469
Resting metabolic rate (RMR), 431
Rett syndrome, 390
Reverse transcriptase, 18
Ribo-essentialism, 198, 201
Ribonucleoprotein (RNP) complexes, 11–14
Ribosomes
binding sites for tRNA, 25
in biological catalysis, 11–12
decoding by tRNA
distortions in tRNA during decoding, 28, 29f
minor-groove recognition by RNA, 27–28
minor-groove recognition's role in evolution, 28
evolution of, 31–32
peptidyl release, 29–30
peptidyl transfer, 28
ratio of RNA to protein in, 25
relative roles of protein and RNA in, 27
RNA World hypothesis, 27
translation stages, 25, 26f
translocation, 31
Ricardo, David, 445
Richerson, P.J., 443
RM-11. *See* Yeast traits analysis
RMR (resting metabolic rate), 431
RNA and evolutionary processes in biology. *See* Biological catalysis
RNA-dependent RNA polymerase 6 (RDR6), 272
RNA enzymes evolution
advances in studies, 17–18
artificial genetic system test, 20–21
continuous in vitro evolution, 18–19
limitations of the artificial genetic system, 22
making replication contingent on other functions, 21, 22f
self-sustained replication of RNA, 19–20
RNA interference (RNAi) pathways role in evolution of plants
conclusions, 272
discussion, 270–272
methods used in the phylogenetic study, 268–269
phylogenetic framework usefulness, 267–268
results of phylogenetic study, 269–270
transposon regulation and (*see* Transposon control pathways in *Drosophila*)
RNase P, 14–15
RNA virus φ6
conclusions, 117
investigation of robustness mechanism, 115–116
link between robustness and evolvability in, 113–115
used to demonstrate evolution of robustness, 111–115
RNA World
evolution in (*see* RNA enzymes evolution)
hypothesis, 27
questions regarding (*see* Biological catalysis)
RNP (ribonucleoprotein) complexes, 11–14
RNP World, 13–14

S

Saccharomyces cerevisiae, 94, 106, 117
Sachs, Julius, 6
S-adenosylmethionine (SAM), 388, 389
Saintpaulia (African violet), 264
Schizophrenia, 378
Schleiermacher, Friedrich, 459
Schultz, Adolph, 336–337
Schumpeter, Joseph, 449, 452
Schwartz, J., 340
Scopes trail, 455

Scott, Eugenie, 464
Scrophulariaceae, 264
Sea squirt. See *Ciona intestinalis*
Secreted frizzled receptor proteins (SFRPs), 84
Segmental duplications (SD). See Human segmental duplications
Selection and flexible networks. See Gene networks
Serine threonine kinases (STKRs), 85
Sex and recombination
  conclusions and future prospects, 193–194
  evolutionary purpose of, 187
  fixation probability and the linear genetic map, 192–193
  Hill–Robertson effects, 188–189
  infinitesimal model, 190–191
  maximum rate of adaptation determination, 189–190
  recombination's influence on variance, 188
  sexual versus asexual reproduction, 191–192
Sex chromosomes in vertebrates
  amplicons, 345, 348, 349–350, 351
  conclusions and future prospects, 350–351
  degeneration of sex-specific chromosomes, 347
  theoretical models of sex chromosome evolution, 345–347
  X chromosome innovation, 349
  Y chromosome conservation and recombination, 347–349
SFRPs (secreted frizzled receptor proteins), 84
Short-tailed fruit bat. See *Carollia perspillata*
Shubin, Neil, 465
Sim enhancer, 275
Simpson, George, 336, 469
Single-nucleotide polymorphisms (SNPs) in *E. coli*. See *Escherichia coli* population diversity study
Sinningia (Gloxinia), 264
Smad ubiquitin regulatory factor (Smurf), 86
Small NF90-associated RNAs (snaRs). See *snaR* genes
Smith, Adam, 446
Smoothened (Smo), 86, 87
Smurf (Smad ubiquitin regulatory factor), 86
Snails and left/right asymmetry, 282–283
Snapdragon (*Antirrhinum*), 264
snaR genes
  Alu SINE, 366–367
  biology and possible functions, 369–370
  chorionic gonadotropin genes, 370–371
  conclusions, 371–372
  function and regulation of hCG, 371
  function of noncoding RNAs, 363
  NF90 and NF110, 363–364
  NF90 binding to cellular RNA family, 364
  as paralogs, 369
  predicted structures, 368–369
  primate speciation and stepwise evolution, 365–366
  segmental duplication and, 367–368

snaR family of noncoding RNAs, 364, 365t
Sociobiology. See also Evo-devo and the evolution of social behavior
  altruism, 408
  analyzing social evolution, 407
  conclusions, 414–415
  green beard gene, 406, 408
  group adaptation, 409–410
  group selection, 408–409
  history of arguments against, 403
  influence on evolution by natural selection, 405–406
  key concepts, 404
  kin selection, 406, 408
  kin selection and fitness costs and benefits
    group selection, 413–414
    relatedness in other social species, 412, 413f
    social insects and eusociality, 410–412
  kinship as an element in natural selection, 404
  molecular mechanisms study and, 414–415
  Nowak's rules for the evolution of cooperation, 408
  objection to the equivalency of group and kin selection, 406
  reciprocal altruism and indirect reciprocity, 408
Sociobiology (Wilson), 403
Sog, 277
Somatic sector analysis, 307
Spalt (*Sal*), 262
Spinoza, Benedict, 459
Spliceosome, 11–12
Sponges, 66
Spruce, 306
Stalled pol II, 278
Starch digestibility and cooking, 429–430
Stebbins, G., 469
Stern, Curt, 346
Stickleback (*Gasterosteus aculeatus*), 142, 156
Stiner, M.C., 446, 447
STKRs (serine threonine kinases), 85
Strassmann, J.E., 410
Structure of Evolutionary Theory (Gould), 342
STRUCTURE software, 397
Sturtevant, Alfred, 345
Sturtian glaciation, 67, 68
SuH, 275. See also Dorsoventral (DV) patterning in *Drosophila*
Sunflower (*Helianthus*), 164
Szathmary, E., 435

T

T7 RNA polymerase, 18
Tattersall, Ian, 339, 340
Tcf/Lef, 84
TEE (total daily energy expenditure), 427
TERT protein subunit, 14
TGF-β signaling pathway, 85–86, 88
Thaxton, Charles, 456
Theory of evolution and Darwin, 1–2
Thrifty gene hypothesis, 322
Thrombin, 38–39
Tianzhushania, 68
Toadflax (*Linaria*), 264

Tooby, John, 446
Total daily energy expenditure (TEE), 427
Tragopogon system
  compared to well-studied systems, 220–221
  genomic approaches to studying, 219
  history of the polyploids, 216
  homeolog loss and gene silencing, 217–218
  mechanism of gene loss, 219
  multiple origins of the polyploids, 216–217
  rDNA loci and concerted evolution, 217
  synthetic lines creation, 217
  tissue-specific silencing, 219
Translesion synthesis polymerases, 97–98
Transposon control pathways in *Drosophila*
  coevolution of transposons and their control pathways, 227–229
  conclusions, 232
  hybrid dysgenesis and the piRNA pathway, 231–232
  long- and short-term evolution of transposon silencing, 229–231
  positive effects of transposon activity, 225
  potential detrimental effects of unregulated transposon activity, 225
  small RNA-based pathways and, 226–227
  speciation and the piRNA pathway, 231–232
  threat posed by transposons, 228
  transposable element diversity and, 226
  transposon definition and mechanism of function, 225
  uses of soma-derived RNAs, 231
Tree of Life (TOL)
  arguments against, 201–203
  concept of, 205
Trichoplax
  Hedgehog signaling and, 87
  Notch pathway and, 82
  TGF-β signaling pathway and, 86, 88
  WNT pathway and, 85
tRNA
  binding sites for, 25
  decoding by
    distortions in tRNA during decoding, 28, 29f
    minor-groove recognition by RNA, 27–28
    minor-groove recognition's role in evolution, 28
Typological thinking in prokaryotic evolution, eradicating. See Prokaryotic systematics and evolution

U

Understanding Intelligent Design: Everything You Need to Know in Plain Language (Dembski and McDowell), 460

V

Variation of Animals and Plants Under Domestication (Darwin), 319, 322

SUBJECT INDEX

VA RNA$_{II}$, 363–364
Veblen, Thorstein, 449
Ventral nerve cord (VNC), 242, 244
Vernanimalcula, 68
Vertebrate blood coagulation. *See* Blood coagulation evolution in vertebrates
Vestigial structures, 263
Virus evolution prediction
   conclusions, 117
   demonstration of evolution of robustness, 111–113
   experimental evolution and evolutionary biology, 110
   investigation of robustness mechanism, 115–116
   link between robustness and evolvability in RNA virus φ6, 113–115
   mathematical theory applied to evolutionary biology, 109–110
   relationship between robustness and evolvability, 113
   robustness and the genotype-to-phenotype translation, 110–111
vitellogenin, 423
VNC (ventral nerve cord), 242, 244
Vnd, 277

v-Src, 105

W

Walker A cytoskeletal ATPase (WACA), 60
Wallace, Alfred Russel, 1, 5, 139
Washburn, S., 337, 338
Wasp *(Nasonia vitripennis)*, 412
Wedge Strategy, 456, 457
Wells, Jonathan, 455
Welwitschia mirabilis, 270
Werner syndrome, 390
West, John, 455
Whole-genome sequence (WGS) databases, 35, 37t
Wilberforce, Samuel, 2
Williams, G.C., 414
Wilm's tumor, 390
Wilson, D.S., 405
Wilson, E.O., 403, 410, 411
WNT pathway, 83–85
Wood mouse *(Peromyscus)*, 142, 160, 163
Wrangham, Richard, 446
Wright, Sewall, 134, 469
Wynne-Edwards, V.C., 404, 405

X

X chromosome. *See* Sex chromosomes in vertebrates

Y

Yates, Edmund, 6
Y chromosome. *See* Sex chromosomes in vertebrates
Yeast traits analysis
   approaches to dissecting complex traits
      mapping complex traits to genomic loci, 150–151
      refining QTLs to genes or causal polymorphisms, 151
   challenges associated with genetic analysis, 145
   conclusions and future prospects, 149–150, 151–152
   insights from BYxRM cross
      gene expression traits showing complex inheritance, 146–147
      genetic dissection of traits, 147–149
      molecular mechanisms of gene expression variation, 147
      QTL hot spots importance, 149
      summary, 149
Your Inner Fish (Shubin), 465

WITHDRAWN